CW00959405

Plant Engineer's Reference Book

Plant Engineer's Reference Book

Edited by

Eur Ing **Dennis A. Snow**

CEng, MIMechE, HonFIPlantE, HonFIIPE

Teaching Fellow, Loughborough University of Technology, UK; Senior Lecturer, Leicester Polytechnic, UK

Butterworth-Heinemann Ltd
Linacre House, Jordan Hill, Oxford OX2 8DP

 PART OF REED INTERNATIONAL BOOKS

OXFORD LONDON BOSTON
MUNICH NEW DELHI SINGAPORE SYDNEY
TOKYO TORONTO WELLINGTON

First published 1991

British Library Cataloguing in Publication Data
Plant engineer's reference book.
1. Engineering plants
I. Snow, Dennis A.
620

ISBN 0 7506 1015 8

Library of Congress Cataloging-in-Publication Data
Plant engineer's reference book / edited by Dennis A. Snow.
p. cm.
Includes bibliographical references and index.
ISBN 0 7506 1015 8
1. Plant engineering—Handbooks, manuals, etc. I. Snow, Dennis A.
TS184.P58 1991
696—dc20 90-2105
CIP

Filmset by Mid-County Press, Salisbury
Printed and bound in Great Britain by Hartnolls Ltd,
Bodmin, Cornwall

Foreword

The opportunity to write the foreword to a new book falls to very few people and I consider myself very fortunate to be able to introduce you to the *Plant Engineer's Reference Book*.

It has been many years since a book was written specifically for the engineer who is involved with so many facets of the engineering profession, and I am of the firm opinion that this publication is long overdue.

The many chapters covering the multidisciplinary role of the plant engineer have been written by practising experts and specialists, and the whole has been edited by Dennis Snow, a plant engineer of international reputation.

This reference book is so written as to take the reader through the design, planning, installing, commissioning and operating stages of both major and minor engineering projects. It covers all aspects of maintenance, management, health and safety, and finance and gives a guide to where further information on these subjects may be obtained.

It is a book which should be in the hands of all practising plant engineers and senior management who are involved in the efficient running and control of their respective establishments.

As a plant engineer of many years' experience, I recommend this book to you as the definitive Reference Book for all members of our profession.

Richard J Wyatt
President, The Institution of Plant Engineers

Preface

The preparation of a reference book such as this could not possibly be acheived without the total cooperation of so many individual authors and the backing of their various employers, especially where company contributions have been made, bringing together a wealth of professional knowledge and experience.

An acknowledgement such as this can only scratch the surface and cannot really portray the grateful thanks I wish to acknowledge to all these people and concerns who have devoted so much time and effort to place their ideas and contributions on paper.

Plant Engineering is such a broad subject incorporating a multitude of disciplines and a wide variety of solutions to virtually every problem or situation, unlike some subjects that have clear cut methods.

In compiling the initial suggested guidelines for each of the contributions, I posed the questions to myself for what information and assistance had I found difficult to locate during over 30 years in industry as a Plant Engineer responsible for plant in the UK and abroad and how could it be best presented to assist others in their profession.

As an active member of the Institution of Plant Engineers for many years I have been able to seek the knowledge of my colleagues when problems have occurred and I hope that I have been able to help them on occasions.

I would therefore like to take this opportunity to thank those members of the Institution for their patience and assistance in helping me to structure this publication, and in particular, members of the Publication Committee, together with the following companies:

ABB Power Ltd
W S Atkins Consultants Ltd
J B Auger (Midlands) Ltd
APV Baker Ltd
The Boots Company plc
BP Oil (UK) Ltd
British Coal
British Gas
British Compressed Air Society
Colt International Ltd
Cortest Laboratories Ltd
Davy McKee (Stockton) Ltd
Film Cooling Towers Limited
Heaton Energy Services
The Institution of Plant Engineers
I T I Anti-Corrosion Ltd
Liquefied Petroleum Gas Industry Technical Association
Loughborough University of Technology
National Vulcan Engineering Insurance Group Ltd
Ove Arup & Partners, Industrial Division
Pilkington Insulation Ltd
Royal Insurance (UK) Ltd
SBD Construction Products Limited
Saacke Ltd
Senior Green Limited
Spirax-Sarco Limited
Support Services
Taylor Associates
Thorn Lighting

I would also like to include my wife Betty, who has been extremely tolerant whilst I have spent many hours word processing.

Dennis A Snow

List of Contributors

A Armer
Spirax Sarco Ltd

B Augur, IEng, FIPlantE, MBES
J B Augur (Midlands) Ltd

H Barber, BSc
Loughborough University of Technology

D A Baylis, FICorrST, FTSC

J Bevan, IEng, MIPlantE

R J Blaen
Senior Green Limited

British Compressed Air Society

Eur Ing **G Burbage-Atter,** BSc, CEng, FInstE, HonFIPlantE, FCIBSE
Heaton Energy Services

P D Compton, BSc, CEng, MCIBSE
Colt International Ltd

I G Crow, BEng, PhD, CEng, FIMechE, FIMarE, MemASME
Davy McKee (Stockton) Ltd

P Fleming, BSc(Eng), ARSM, CEng, MInstE
British Gas plc

C Foster
British Coal

C French, CEng, FInstE, FBIM
Saacke Ltd

F T Gallyer
Pilkington Insulation Ltd

Eur Ing **R R Gibson,** BTech, MSc, CEng, FIMechE, FIMarE, FRSA
W S Atkins Consultants Ltd

B Holmes, BSc(Tech), PhD, CEng, FIChemE, FInstE
W S Atkins Consultants Ltd

A P Hyde
National Vulcan Engineering Insurance Group Ltd

H King
Thorn Lighting

B R Lamb, CEng, MIChemE
APV Baker Ltd

S McGrory
BP Oil UK Ltd

R J Neller
Film Cooling Towers Ltd

Ove Arup & Partners, Industrial Division

G Pitblado, IEng, MIPlantE, DipSM
Support Services

R S Pratt, ALU, CEng, MIMfgE, MBIM, MSAE
Secretary-General, The Institution of Plant Engineers

G E Pritchard, CEng, FCIBSE, FInstE, FIPlantE, MASHRAE

Risk Control Unit
Royal Insurance (UK) Ltd

R Robinson, BSc, CEng, FIEE
The Boots Co. plc

M J Schofield, BSc, MSc, PhD, MICorrT
Cortest Laboratories Ltd

J D N Shaw, MA
SBD Construction Products Ltd

R H Shipman, MIMechE, MIGasE, MInstE
Liquefied Petroleum Gas Industry Association

K Shippen, BSc, CEng, MIMechE
ABB Power Ltd

G Solt, FIChemE, FRSC
Cranfield Institute of Technology

K Taylor, CEng, FCIBSE, FIPlantE, FIHospE, FSE, FIOP, MASHRAE, FBIM, ACIArb
Taylor Associates Ltd

L W Turrell, FCA

K Turton, BSc(Eng), CEng, MIMechE
Loughborough University of Technology

E Walker, BSc, CEng, MIMechE
Senior Green Limited

R C Webster, BSc, MIEH
Environmental Consultant

D Whittleton, MA, CEng, MIMechE, MHKIE
Ove Arup & Partners, Industrial Division

Contents

34 Insurance: Plant and Equipment

35 Insurance: Buildings and Risks

36 Health and Safety

37 Education and Training

38 Lubrication

INDEX

1A Physical Considerations in Site Selection

Ove Arup & Partners,
Industrial Division

Contents

1A.1 Environmental considerations of valley or hillside sites

1A1.1 The effect of topography on prevailing winds and strengths

Apart from the obvious influence of topography in producing shelter or the enhanced exposure to wind (discussed later), the influence of large topographic features can be sufficient to generate small-scale weather systems which are capable of producing significant winds. Three types of wind are associated with topography:

Diurnal winds
Gravity winds
Lee waves

Diurnal winds

Under clear skies in daytime the slopes of hills and mountains facing the sun will receive greater solar heating than the flat ground in valley bottoms. Convection then causes an upslope flow, called anabatic wind, which is generally light and variable but which can often initiate thunderstorms. At night, the upper slopes lose heat by radiation faster than the lower slopes and the reverse effect happens, producing downslope katabatic winds. However, the denser cold air falling into the warmer valley can produce strong winds in a layer near the ground. The higher the mountains, the stronger is the effect. As the mountains in the UK are not very high, it is not surprising that the speeds of katabatic winds do not approach those of large depressions.

Gravity winds

The effect of katabatic winds can be much enhanced if greater differences in air temperature can be obtained from external sources. A continuous range of mountains can act as a barrier to the passage of a dense mass of cold air as it attempts to displace a warmer air mass. Cold air accumulates behind the mountain range until it is able to pour over the top, accelerating under gravity to give strong winds down the lee slope. The mountains of the UK are not high enough to produce gravity winds of speeds sufficient to damage buildings.

Lee waves

Under certain conditions of atmospheric stability, standing waves may form in the lee of mountains. This wave motion is an oscillating exchange of kinetic and potential energy, excited by normal winds flowing over the mountain range, which produces alternately accelerated and retarded flow near the ground. Sustained lee waves at the maximum amplitude are obtained when the shape of the mountain matches their wavelength, or when a second range occurs at one wavelength downstream. The existence of lee waves is often indicated by clouds which are unusual in that they remain stationary with respect to the ground instead of moving with the wind. These clouds are continuously forming at their upwind edge as the air rises above the condensation level in the wave and dissipating at the downward edge as the air falls again.

Conditions are frequently suitable for the formation of lee waves over the UK, an effect that is routinely exploited by glider pilots to obtain exceptionally high altitudes. The combination of lee waves with strong wind, sufficient to produce damage to structures in the bands of accelerated flow, is fortunately rare, but an analysis of the Sheffield gale of 16 February 1962 by the UK Meteorological Office showed that this was a case where strong winds were further enhanced by lee waves.

Other factors

Other factors to account for topography with regard to valley or hillside sites should include possible inversion and failure to disperse pollutants. Temperature inversion occurs when the temperature at a certain layer of the atmosphere stays constant, or even increases with height, as opposed to decreasing with height, which is the norm for the lower atmosphere. Inversions may occur on still, clear nights when the earth and adjacent air cool more rapidly than the free atmosphere, or when throughout a layer high turbulence causes rapid vertical convection so that the top of the turbulent layer may be cooler than the next layer above it at the interface.

The running of a cool air flow under a warm wind is another cause of temperature inversion. As a rule, the presence of an inversion implies a highly stable atmosphere: one in which vertical air movements are rapidly damped out. In such a situation, fog and airborne pollutants collect, being unable to move freely or be dissipated by convection.

Additional dispersal problems may occur when the prevailing wind occurs perpendicular (or nearly so) to the valley or hill ridge line. This may lead to speed up and turbulence over the valley or it may simply reduce the effect of airflow carrying away airborne pollutants.

It is possible to obtain wind data for almost any location in the world, although these frequently require modification and interpretation before they can be used. Addresses of advisory offices for the UK are listed below.

Advisory offices of the Meteorological Office

For England and Wales:
Meteorological Office
Met 0 3
London Road
Bracknell
Berkshire RG12 2SZ
Tel. 0344 420242 (Extn 2299)

For Scotland:
Meteorological Office
231 Corstorphine Road
Edinburgh EH12 7BB
Tel. 031 344 0721 (Extn 524)

For Northern Ireland:
Meteorological Office
Progressive House
1 College Square East
Belfast BT1 6BQ
Tel. 0232 28457

Advisory service of the Building Research Establishment

The Advisory Service
Building Research Station
Bucknalls Lane
Garston
Watford WD2 7JR
Tel. 0923 67 6612

1A.1.2 Design for wind

A structure may be designed to comply with any of the following information:

1. No specific details available.
2. Specified basic wind speed and relevant site data.
3. Specified design wind speed, with or without FOS.
4. Specified survival wind speed, with or without FOS.

When details are given they should be checked, if only by comparison with equivalent wind speeds derived from first principles, to ensure that they are reasonable.

Depending on the specified requirements the wind speeds may or may not utilize gust wind speeds as in CP3 (3) or mean hourly wind speeds, v, with applied gust factors.

1A.1.3 The factored basic wind speed approach

The current British Standard Code of Practice (CP3: Chapter V: Part 2: 1972) uses a basic gust wind speed, V, multiplied by a series of S factors which adjust the basic values to design values for the particular situation. CP3 uses up to four S factors:

S_1: Topography factors
S_2: Ground roughness, building size and height above ground factors
S_3: Statistical factor
S_4: Directional factor

S_1 – Topography factors

The effect of local topography is to accelerate the wind near summits or crests of hills, escarpments or ridges and decelerate it in valleys or near the foot of steep escarpments or ridges. The extent of this effect on gust wind speeds is generally confined to the region 1.0–1.36. Local topography is considered significant when the gradient of the upwind slope is greater than 5%.

The shape of the upwind slope affects the degree of shelter expected near the foot of the slope when the slope is shallow and the flow remains attached. When the changes in slope are sudden, so that upwind slope can be approximated by a single straight line for more than two-thirds of its length, then the shape is 'sharp'. Otherwise the changes of slope are gradual and the shape is 'smooth'. This distribution is relevant for sites close to the foot of the upwind slope, where 'sharp' topography offers a greater degree of shelter.

S_2 – Ground roughness, building size and height above ground factors

The factor S_2 takes account of the combined effect of ground roughness, the variation of wind speed with height above ground and the size of the building or component part under consideration. In conditions of strong wind the wind speed usually increases with height above ground. The rate of increase depends on ground roughness and also on whether short gusts or mean wind speeds are being considered. This is related to building size to take account of the fact that small buildings and elements of a building are more affected by short gusts than are larger buildings, for which a longer and averaging period is more appropriate.

S_3 – Statistical factor

Factor S_3 is based on statistical concepts and can be varied from 1.0 to account for structures whose probable lives are shorter (or longer) than is reasonable for the application of a 50-year return-period wind.

S_4 – Directional factor

In the latitudes occupied by the UK (50–60°N) the climate is dominated by westerly winds and a band of low pressure between Scotland and Iceland. The basic wind speed may be adjusted to ensure that the risk of it being exceeded is the same for all directions. This is achieved by the wind speed factor S_4.

When applying S_4, topography factor S_1 and the terrain roughness, building size and height above ground factor S_2 should be appropriately assessed for that direction.

1A.2 Road, rail, sea and air access to industrial sites

1A.2.1 Introduction

Many industrial processes and factories require specific accessibility for one particular form of transport. Examples of the above include distribution warehousing, transport operations (particularly intermodal such as through seaports) and those industries dealing with bulk commodities (e.g. oil refineries). For other industries access to strategic modal networks is important in order to be competitive where cost of transport and time savings are a significant factor. Examples of these operations include air freighting and fresh-food deliveries. A third category would include those establishments which would require high-visibility sites to enhance their reputation in the marketplace.

1A.2.2 Design considerations

It is difficult to give specific advice on this subject as there is a very large range of industrial undertakings. The awareness for, and acceptability of, access is dependent on the types of goods to be moved and the frequency and method of movement. In some undertakings there is a major movement between different transport modes which is concentrated either at ports or at major road/rail interchanges.

In addition to the amount of commercial traffic it is vital to consider the movement associated with employees and visitors, which themselves can generate large numbers of vehicular and pedestrian movements. For very large manufacturing sites there will also be the need for accessibility for public transport, which, for a large workforce, may need to be supplemented by investment in subsidized travel.

Site access will reflect the nature of the existing local transport system and will need to be designed to cater for the anticipated future traffic flows associated with on-site development. At the extreme of the range this could include a significant on-site infrastructure, potentially involving small bus stations for staff or private rail sidings for goods heavily committed to using the rail network. Special consideration might also need to be given to customs facilities, where operations include cross-border movements with or without bonding operations.

1A.2.3 Forms of site access

Access to the road network can range from a simple factory gate or location on a business park to a major industrial complex requiring its own major grade separated interchange due to the high traffic volumes on the strategic road network. New site developments will need to cater for future traffic growth and must be adequate to deal with a design life over the foreseeable future.

Access to the rail system must be negotiated with British Rail and may involve the use of a Section 8 Grant, where it can be demonstrated that the use of the rail network minimizes lorry traffic on the road system. It will be necessary to incorporate security arrangements to prevent trespass on the rail line and some form of signalling arrangement will be necessary for the junction and sidings. Major complexes may run their own railway network, including a private shunting engine.

Access to a seaport will be limited by the ability of total traffic generated by the docks and the incorporation of these traffic movements into the local road system.

Air traffic access may be constrained by the operational aspects of the airport. Otherwise, the road-related traffic will be dealt with in a manner similar to that of seaports, except that the vehicles are likely to be smaller in size and of lower traffic volumes, reflecting the higher-value goods being transported by air.

1A.2.4 Providing access to the road system

Before access is obtained to any road it is necessary to obtain the consent of the relevant highway authority. In England the Department of Transport is the highway authority for all motorways and trunk roads. Direct access to motorways is generally prohibited and the policy regarding access to trunk roads is to minimize the number of accesses and to encourage the free flow of traffic on the major road. Therefore careful consideration needs to be given to the ability of the proposed access to cover traffic capacity and road safety adequately.

The local county council is the highway authority, in non-metropolitan areas, for all other roads, although, in many instances the local authority (generally, city or district) may have agency powers for the roads within its area.

It will be necessary to forecast the amount of traffic to be generated by the development within the site and to propose a form of junction which not only deals with the site's traffic but also adequately caters for the existing traffic on the road. Tests for capacity are required and attention should also be given to the safety of operation of the proposed access.

Various types of junctions are available and include simple priority T-junctions, traffic signals and roundabouts. In proposing any junction improvements or new junction it is necessary to be aware of land ownership in order to ensure that all improvement works can be carried out within highway land or land within the proposed development site.

As part of the planning approvals it is increasingly common to provide road-improvement schemes which are sometimes off-site and are necessary to deal with site-generated traffic, which has detrimental effects on the local road network. These off-site improvement schemes can be obtained within legal agreements referred to as Section 106 Agreements. These are undertaken with the local planning authority or with Section 278 Agreements, undertaken directly with the highway authority. Generally, these Agreements require the applicant of the proposed site to carry out specified highway improvement schemes to an agreed timetable relevant to the planning application.

In the above examples the location of the site adjacent to a strategic route network is an important consideration.

1A.2.5 Selection of sites

Suitable sites are normally limited to those areas designed in Development Plans as being for industrial or commercial uses. Such land should be capable of being accessed directly from the primary or secondary distributor roads in the area. Segregation of lorries and lorry access from residential areas should be achieved where possible.

The utilization of existing or the provision of new rail heads will also be a determining factor for some operators, and frequently the rail sidings do not have good road access. In these cases extensive improvement measures may be necessary to provide adequate space and geometrical requirements.

1A.2.6 Checklist

The following list, while not exhaustive, identifies many of the issues which will need careful consideration. In many instances it might be necessary to seek the advice of a specialist traffic consultant, either in the design of a scheme or in access, or to negotiate with the highway authority the impact of a proposed development and any attendant road-improvement schemes.

1. *Types of operation to be carried out*
- Number of lorries
- Staff cars
- Visitors' cars
- Rail/water/air access
- Public transport provision
- Cyclists
- Pedestrians

2. *Types of site*
- Large single site
- Industrial estate
- Segregation of access for lorries and cars
- Capacity of access and need for improvement
- Ensure no queueing back onto highway
- Ensure sufficient on-site space for all vehicles to enter highway in forward gear
- Ensure off-highway loading/unloading
- Access for emergency vehicles

3. *Access arrangements*
- Junction types – priority
  – roundabout
  – signals
- Access width should be a minimum of 6.1 m to allow lorries to pass each other (7.3 m ideal)
- Single access could cope with up to 250 lorry movements per day
- Any gate or security barrier to be set in at least 20 m from public highway to avoid blockage or interference to pedestrians
- Minimum centre line radius to be 12 m
- Minimum entire live radius to access road to be 60 m. Widening on bends may be required

4. *Manoeuvring space*
- Turning circle for articulated vehicles to be 26 m diameter minimum
- For draw-bar vehicles this can be reduced to 21 m
- Turning head for rigid lorries only needs to be 35 m long
- Turning head for articulated vehicles should be 53 m long. Kerb radii need to be 9 m
- Loading bays at 90° to road should be 31 m deep including the road width. Bay should be 3.5 m wide
- Strong site management is required to ensure manoeuvring space is kept clear of storage/goods/debris at all times
- Headroom clearance should be a minimum of 4.65 m with careful consideration to ensure all pipework, etc. is above that level. Approach gradients to flat areas will reduce the effective height.

It is emphasized that the above checklist is not exhaustive. Any reductions in the standards identified above will lead to difficulty of operation, tyre scrub, potential damage to vehicles and buildings, and general inefficiency. Cost effectiveness could also be hindered due to loss of time caused by blocked-in vehicles. Safety is also a highly important factor which should be prominent in any decision making.

1A.3 Discharge of effluent and general site drainage

1A.3.1 Effluent

Introduction

The control of drainage and sewerage systems and of sewage disposal is governed entirely by Parliamentary Acts and statutory regulation. The Building Regulations 1985 and Public Health Acts 1936 and 1961 cover sewerage, sewage disposal, drainage and sanitation for buildings and other public health matters. The Public Health Act 1937 (Drainage of Trade Premises) and the Control of Pollution Act 1974 deals, among other things, with the disposal of trade and industrial effluent.

Methods of treatment

Two methods of treatment can be considered:

1. On-site treatment and disposal; and
2. Off-site treatment and disposal.

Where on-site treatment is to be undertaken consideration should be given to the following:

1. Where large volumes of effluents are produced and/or different types of contaminants, large equipment areas may be required. Sufficient space must also be allowed for maintenance and inspection of such equipment.
2. Settlement/storage areas for effluent need to be sized not just for average flow but also for peak periods. Where production is based on a shift system, peak flows created during holiday periods (shutdown, major maintenance, etc.) should be considered.
3. Where effluents require primary, secondary and possibly additional tertiary treatment, attention should be paid to the various treatment processes with regard to personnel safety and public sensitivity to on-site treatment.
4. Where concentrated alkali and/or acids are stored and used on-site as part of the treatment process, care should be exercised to prevent misuse, fire, and security and health hazards. The provision of emergency showers, eye-wash stations, etc. needs careful consideration.
5. If equipment malfunctions during the treatment process, adequate precautions should be taken to prevent the discharge of untreated effluent. Such precautions should be the provision of emergency collection tasks or the use of approved, licensed, effluent-disposal traders.
6. Where accidental discharge of untreated effluent does occur the water authority and/or environmental

health officer should be advised immediately. All steps should be taken to limit the extent and intensity of any potential contamination.

7. Where small and/or single contaminant effluents are encountered, packaged treatment plants may be acceptable. Consideration should, however, be given to capital cost, payback period, reliability of equipment, maintenance, plant-life expectancy and contaminant-removal efficiencies.
8. Pipework material for conveying effluent to treatment plants should exhibit resilience to corrosive attack by the effluent as well as scouring and erosion created by the material content of the effluent.
9. Consideration should be given to plant operation in a shift system and any requirements for an analyst to be present during operational/non-operational periods.
10. Precautions must be taken against freezing for external pipework, tanks, meters, gauges, and monitoring equipment.
11. Assessments should be made for electrically operated process equipment which may require an essential power supply in the event of a mains failure.
12. The quality of the effluent discharge must be regularly checked. Depending on the quantity and type of discharge, this may require an in-house laboratory and analysis room.
13. The quantity of final treated effluent may be limited by the water authority, and monitoring of the final outfall may have to be considered in conjunction with a holding tank.
14. Large or small on-site treatment plants will create sludge concentrates which require disposal. Where large quantities of sludge occur, on-site de-watering filters may be considered with dry sludge cakes properly removed from site by licensed contractors. Alternatively, small quantities of wet sludge concentrates may be removed and disposed of by similar contractors.

Where off-site treatment is undertaken the following should be considered:

1. Cost comparison with on-site treatment.
2. Availability of approved, licensed contractors to handle the type of effluents being considered.
3. Reliability of licensed contractors during emergency, weekend and holiday periods.
4. Space requirements for holding untreated effluent prior to removal from site.
5. Accessibility, safety and security associated with the holding vessels by the vehicles of the licensed contractors.
6. Suitable pumps may be required to pump from holding tanks into licensed contractor vehicles.

1A.3.2 Site drainage

The discharge of surface water from a site may originate from three potential sources:

Rainwater from building(s)
Surface-water runoff from paved/hard standing areas
Subsoil drainage (groundwater)

1. The rainwater runoff from buildings is dependent on the geographical location and storm-return period specified. (Reference should be made to BS 6367.) Rainwater runoff from a roof is relatively clean and can discharge directly to a watercourse, lake, etc. without passing through an interceptor.
2. The surface water runoff from paved/hard standing areas is also dependent on rainfall intensity calculated from the geographical locations of the site and storm-return period. However, the return period for a site will be far higher than for a building in order to ensure prevention of persistent flooding of the site. In many instances the local authority may specify the storm-return period as the design criterion. The water authority and/or National Rivers Authority will determine the maximum quantity of surface water that may discharge to a watercourse or river. Where development of a greenfield site or an extension to an existing building takes place, the rate of stormwater runoff may necessitate the provision of a balancing pond or reservoir.

   While the drainage design may be able to cater for minimal surcharging, any substantial rise in floodwater can be contained by the balancing pond and minimize flooding to the site and damage to plant ecology. Lining of a balance pond must be considered to prevent water seepage. Suitable linings are clay or butyl rubber sheeting.

   Where potential flooding to the site is minimal an alternative to the balance pond could be an open-trench system which could provide on-site storage and added security to watercourses.

   The open-trench system provides easy maintenance and may obviate the need for highway kerbs and gulleys, but requires adequate security to prevent vehicle collision.

   Surface water from hard standing areas subject to petrol and oil contamination (e.g. car parks) must pass through an interceptor prior to discharge to a watercourse, surface-water drain, etc. The sizing of petrol/oil interceptors varies between local authorities/District Council areas. Many agree that the first 5 minutes of surface-water runoff is the most contaminated, and accept a reduced interceptor size, while allowing a higher flow rate, caused by increased storm intensity, to bypass the interceptor chamber (see Figures 1A.1 and 1A.2).
3. Subsoil drainage may be required for the following reasons:
   (a) Seasonal fluctuations in water-table level may cause isolated flooding. Subsoil drainage may be considered to keep the water-table level relatively constant.
   (b) Where underground springs occur, laying of subsoil drainage may be considered to maintain water-level equilibrium within the subsoil to permit building construction or similar activities.
   (c) Draining of permanently flood land such as marsh or bogland.

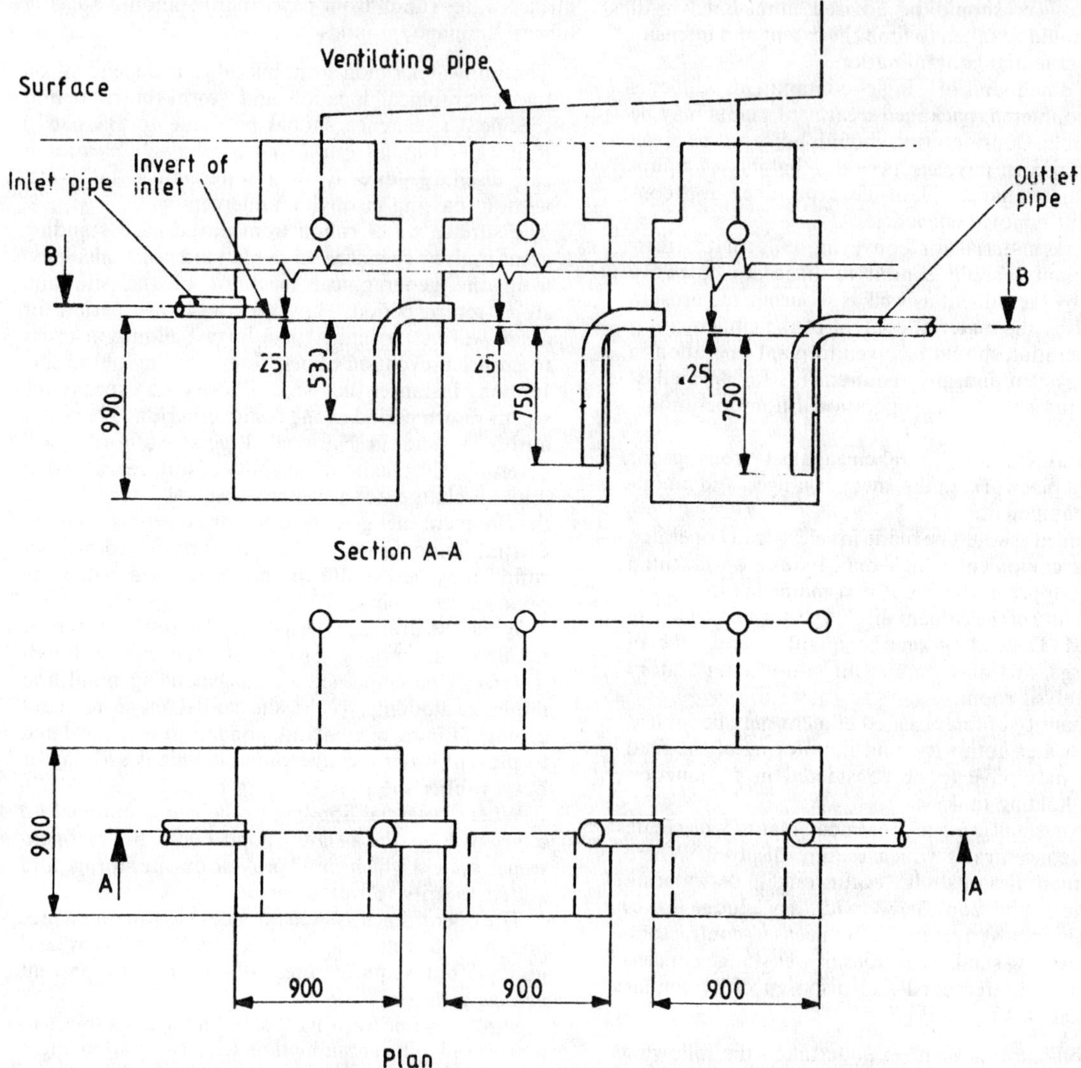

Figure 1A.1 A traditional three-chamber petrol interceptor. (Dimensions are in millimetres.) All pipes within the chambers through which liquid passes should be of iron or another equally robust petrol-resistant material

Gravity or pumped discharge

Wherever practical, surface-water runoff should be designed for gravity discharge, preferably located at the lowest part of the site. However, where a gravity system is impractical or impossible a pumped discharge must be considered.

When sizing a pumped discharge the following should be evaluated:

1. Pumps, valve controls, etc. to be in duplicate;
2. Maximum discharge permitted into watercourse by water authority;
3. Discharge duty to be twice dry weather flow subject to (2) above;
4. Will essential power supply be required to both pumps?
5. Space requirements for pumps, including installation, maintenance and inspection.

Figures 1A.3–1A.7 show edge and bank details for storm-retention reservoirs and Figure 1A.8 and Table 1A.1 relate rainfall intensity, duration and return period.

1A.4 Natural water supplies, water authority supplies and the appropriate negotiating methods and contracts

There are three major sources of water supply:

Borehole
Rivers
Service reservoir

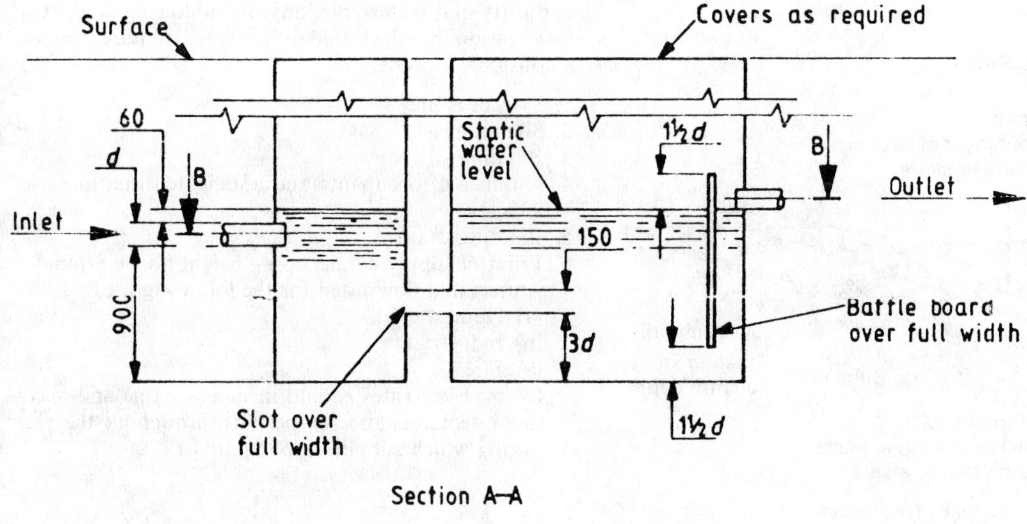

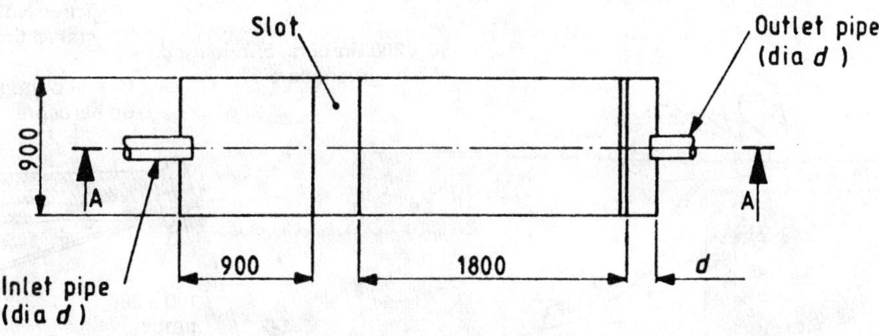

Figure 1A.2 An alternative two-chamber petrol interceptor. (Ventilation not shown: dimensions are in millimetres)

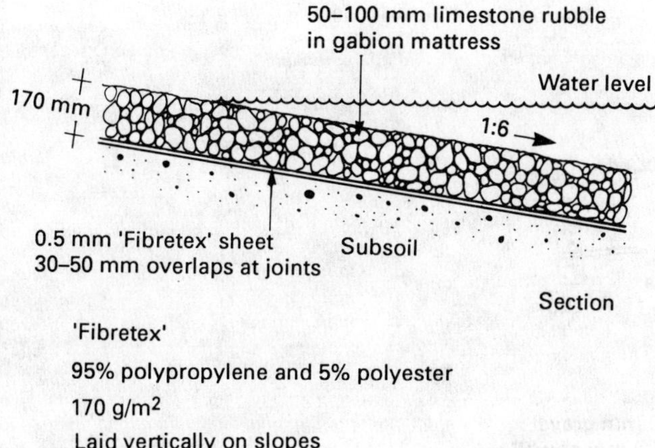

Figure 1A.3 Rip-rap bank ('Fibretex'/stone)

The majority of water supplies are provided from a combination of all three.

A wide variety of Acts of Parliament govern the supply of water to domestic and industrial premises. Within these Acts the statutory undertakings for preventing waste, undue consumption, misuse and contamination of water supplied by the various authorities.

Having established the respective local regional water

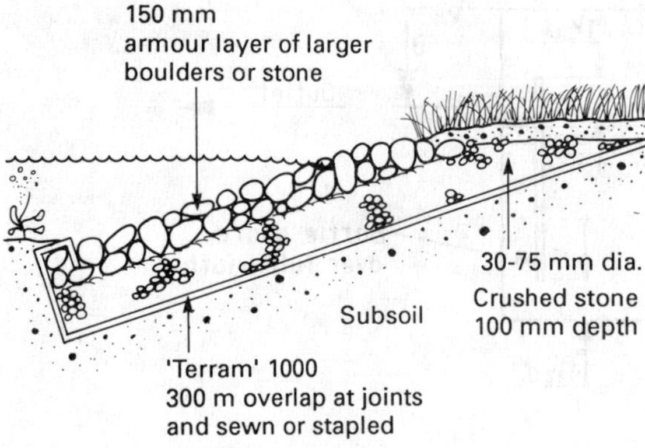

Figure 1A.4 Rip-rap bank ('Terram'/stone)

authority and water company, the following is a list of information that the water authority will require to assess a consumer's needs:

1. Project name
2. Site address
3. Building size (m^2)
4. Number of occupants and anticipated multiple use/ tenancy
5. Building usage
6. Point of supply connection – height above ground
7. Anticipated flow rated for the following uses:
   (a) Domestic
   (b) Industrial
   (c) Fire
   (*Note*: Flow rates should include average and maximum demands and the periods throughout the year during which supplies are required.)

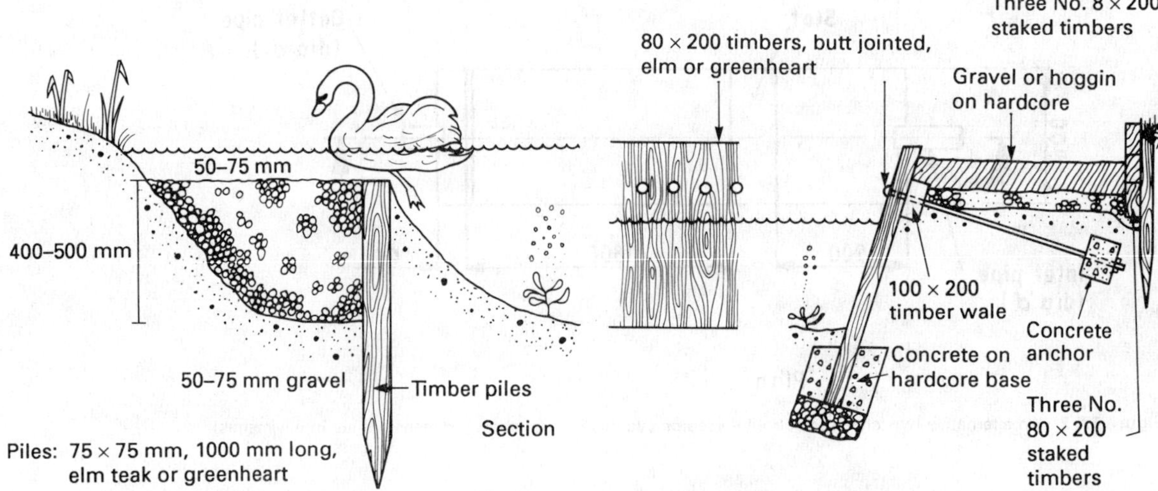

Figure 1A.5 Piling margin

Figure 1A.7 Board wall/path

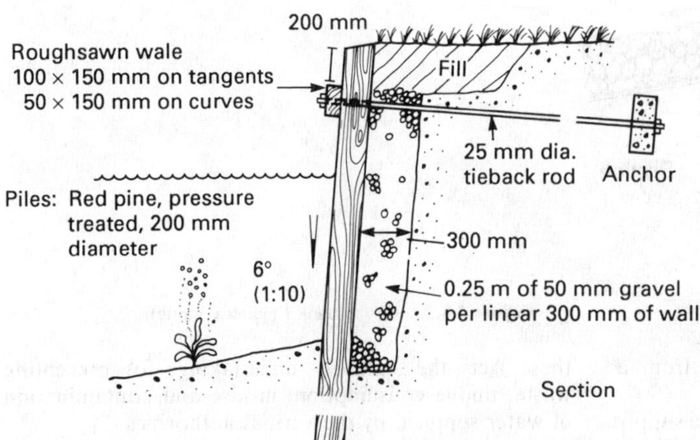

Figure 1A.6 Piling edge

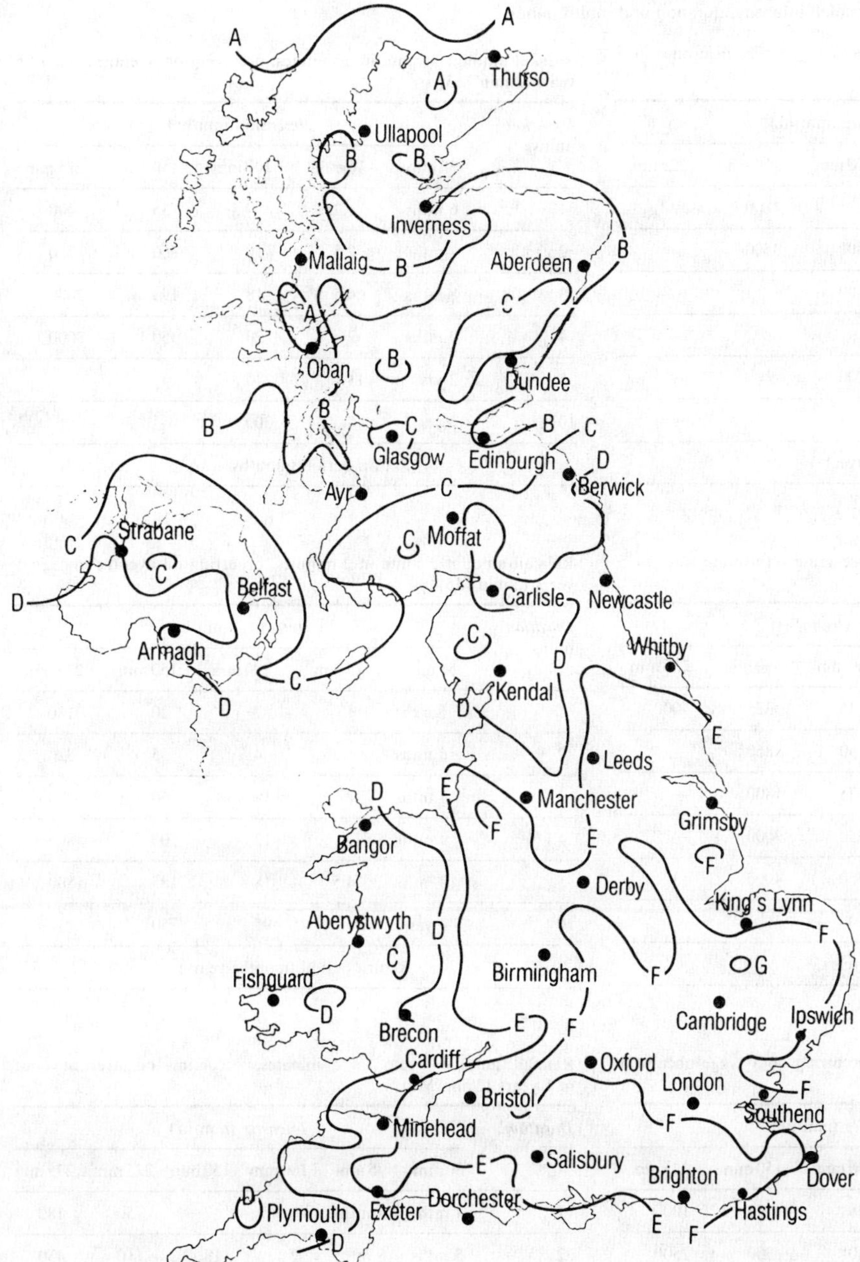

Figure 1A.8 Key to rainfall tables (to be read with Table 1A.1)

8. Cold water storage volume
   (a) Domestic
   (b) Industrial
   (c) Fire

Once the water authority has established that a water supply is available, the following information should be obtained by the consumer:

1. Copy of water authority bye-laws.
2. Marked-up layout drawing indicating location, size, depth and maximum and minimum pressure of the main in the vicinity of the site.
3. An indication of which mains may be considered for:
   (a) Domestic use
   (b) Industrial use
   (c) Fire use
4. Confirmation as to whether the water main passes through any adjacent property which may involve easements to gain access for maintenance, etc.
5. Confirmation as to whether the water flow and

Table 1A.1 Series of tables relating rainfall intensity, duration and return period

Rainfall amount 1.5 mm in 2 minutes, occurring on average once in 5 years (2 min M5)

Duration (mins)	*Intensity* (mm/hr)				
	50 mm	75 mm	100 mm	150 mm	225 mm
1	4.5	35	185	1800	—
2	5.5	55	300	3500	—
3	10	175	800	—	—
4	20	210	—	—	—
5	40	400	2000	—	—
10	300	300	—	—	—
	Return periods (years)				

Rainfall amount 3.0 mm in 2 minutes, occurring on average once in 5 years (2 min M5)

Duration (mins)	*Intensity* (mm/hr)				
	50 mm	75 mm	100 mm	150 mm	225 mm
1	6 mths	1.75	4.5	35	400
2	7 mths	2.5	10	60	750
3	8 mths	4.5	18	190	2000
4	9 mths	7	20	350	3000
5	2 yrs	11	40	—	—
10	7 yrs	40	300	—	—
	Return period (months/years)				

Rainfall amount 2.0 mm in 2 minutes, occurring on average once in 5 years (2 min M5)

Duration (mins)	*Intensity* (mm/hr)				
	50 mm	75 mm	100 mm	150 mm	225 mm
1	1.75	7	35	325	4000
2	2	18	60	850	—
3	4.5	35	175	1800	—
4	7	45	350	3000	—
5	8	60	750	4000	—
10	45	450	2500	—	—
	Return period (years)				

Rainfall amount 3.5 mm in 2 minutes, occurring on average once in 5 years (2 min M5)

Duration (mins)	*Intensity* (mm/hr)				
	50 mm	75 mm	100 mm	150 mm	225 mm
1	5 mths	1	2	20	180
2	6 mths	1.75	4.5	35	320
3	7 mths	2	19	60	700
4	9 mths	4	12	100	800
5	10 mths	4.5	18	185	1800
10	4 yrs	20	95	750	—
	Return period (months/years)				

Rainfall amount 2.5 mm in 2 minutes, occurring on average once in 5 years (2 min M5)

Duration (mins)	*Intensity* (mm/hr)				
	50 mm	75 mm	100 mm	150 mm	225 mm
1	7 mths	3	8	90	1000
2	1.5	5	20	300	2500
3	2	7.5	40	500	—
4	3	15	60	800	—
5	4	20	100	1000	—
10	15	150	700	—	—
	Return period (months/years)				

Rainfall amount 4 mm in 2 minutes, occurring on average once in 5 years (2 min M5)

Duration (mins)	*Intensity* (mm/hr)					
	50 mm	75 mm	100 mm	150 mm	225 mm	275 mm
1	4 mths	7 mths	1.5	7	50	180
2	5 mths	8 mths	2	18	110	450
3	5 mths	1.75	4	35	250	700
4	6 mths	1.75	6	40	350	1000
5	7 mths	2	7.5	50	650	—
10	2 yrs	7.5	40	300	3500	—
	Return period (months/years)					

Table 1A.1 continued

Rainfall amount 4.5 mm in 2 minutes, occurring on average once in 5 years (2 min M5)

Duration (mins)	*Intensity* (mm/hr)						
	50 mm	75 mm	100 mm	150 mm	225 mm	275 mm	300 mm
1	3 mths	5 mths	9 mths	4.5	40	75	350
2	3 mths	6 mths	2	7	55	190	750
3	4 mths	1	3	18	120	350	1100
4	6 mths	1.5	4.5	20	200	450	1600
5	6 mths	2	5	38	250	650	—
10	1 yr	6	50	200	1800	1900	—
	Return period (months/years)						

Notes: For intensities of 225 mm per hour and above, alternative means of protecting the building other than increasing the size of rainwater pipes etc. should be considered, such as the provision of overflows or additional freeboard.

The average roof should have a duration/runoff time not exceeding two minutes. Periods of greater duration shown for information and completeness only.

pressure will be affected by usage from adjacent buildings.

6. Confirmation as to whether a guarantee of security of supply can be provided throughout the year, including any anticipated periods of drought.
7. Confirmation of the source(s) of water supply and provision of a current water analysis.
8. Confirmation that the authority will supply the water meter and housing.
9. Confirmation of the authority's preferred pipe material for incoming mains and any materials not recommended due to aggressive soil.
10. Confirmation that direct mains boosting is allowed.
11. Provision of rates of charges (depending on annual consumption), including any proposed EC levy.
12. Confirmation of costs of maintenance and metering charges.
13. Provision of details relating to any future authority plans which may affect supplies to the site.
14. Confirmation of total capital cost of supplying a water supply.
15. Provision of a fixed time-frame installation programme.

While the above list is not exhaustive, it provides an

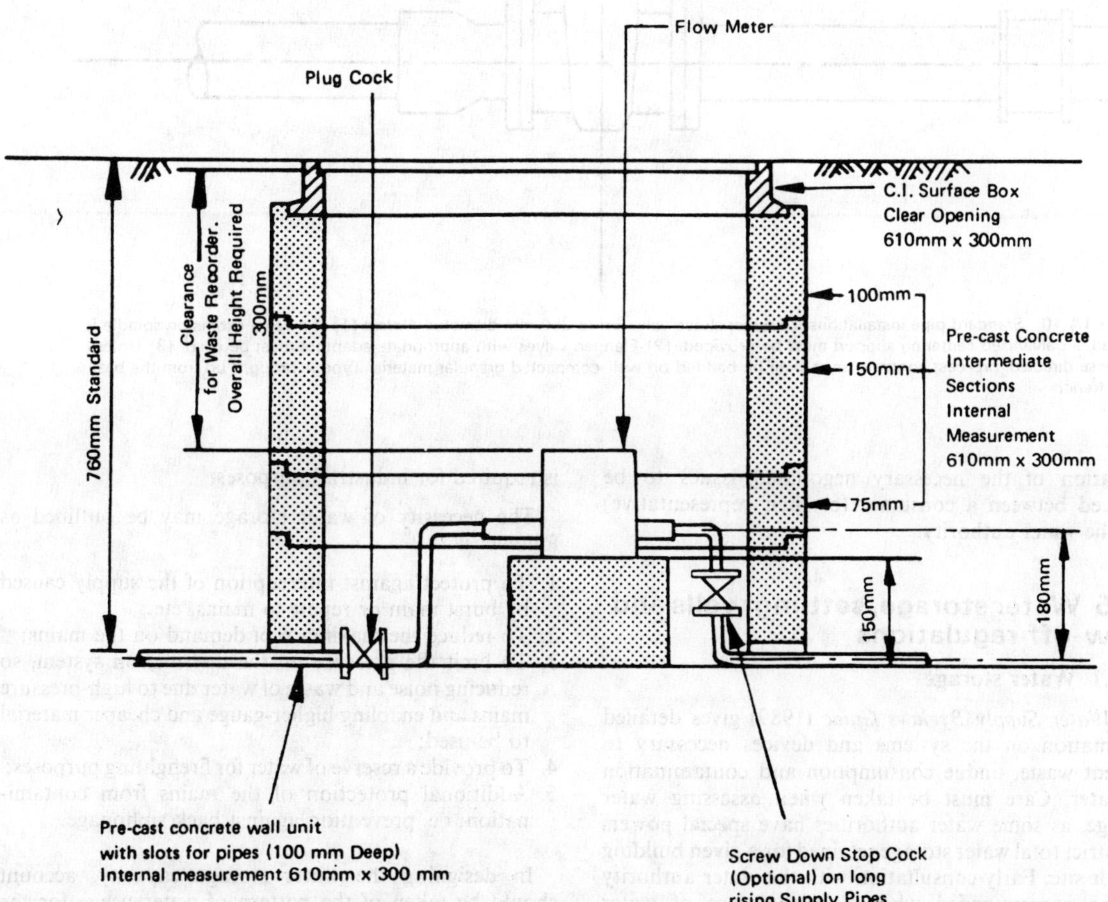

Figure 1A.9 Standard pipe installations for water flow meter and frost cock

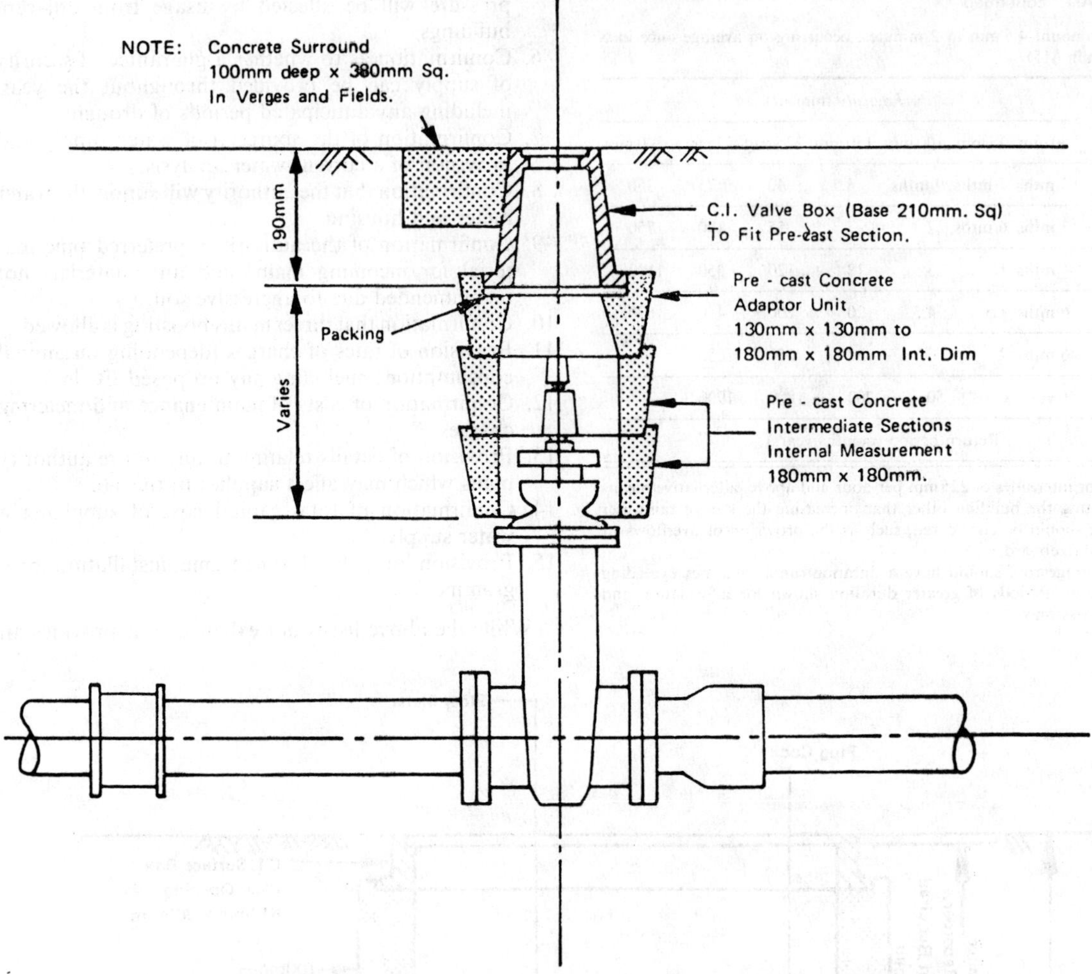

Figure 1A.10 Standard pipe installations for water sluice valve up to 250 mm diameter. *Notes:* (1) If a valve extension spindle is required, a galvanized centering support must be provided. (2) Flanged valves with appropriate adaptors must be used. (3) Unless otherwise directed, pre-cast concrete units shall be bedded on well-compacted granular material Type A brought up from the base of the trench

indication of the necessary negotiated issues to be resolved between a consumer (or their representative) and the water authority.

1A.5 Water storage, settling wells and draw-off regulations

1A.5.1 Water storage

The Water Supply Byelaws Guide (1989) gives detailed information on the systems and devices necessary to prevent waste, undue consumption and contamination of water. Care must be taken when assessing water storage, as some water authorities have special powers to restrict total water storage retained for a given building and/or site. Early consultation with the water authority is also recommended when large volumes of water storage/usage are anticipated, especially if heavy demand is required for industrial purposes.

The necessity of water storage may be outlined as follows:

1. To protect against interruption of the supply caused by burst main or repair to mains, etc.;
2. To reduce the maximum of demand on the mains;
3. To limit the pressure on the distribution system, so reducing noise and waste of water due to high-pressure mains and enabling higher-gauge and cheaper material to be used;
4. To provide a reserve of water for firefighting purposes;
5. Additional protection of the mains from contamination, i.e. prevention against back siphonage.

In designing the water storage capacity, account should be taken of the pattern of water usage for the premises and, where possible, to assess the likely

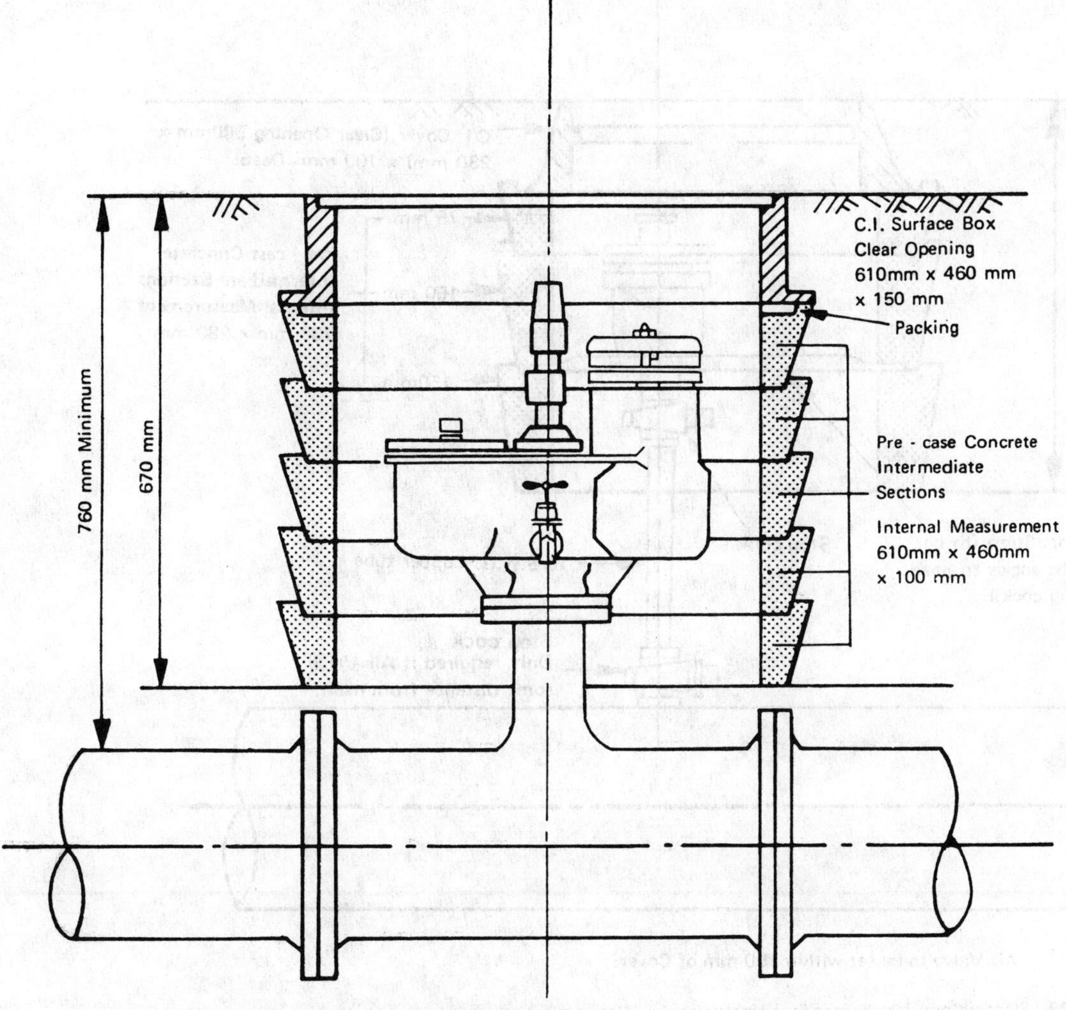

Figure 1A.11 Standard pipe installations for a double air valve. *Note:* Unless otherwise directed, pre-cast concrete units shall be bedded on well-compacted granular material Type A brought up from the base of the trench

frequency and duration of breakdown from the water authority mains. When dealing with domestic water storage this is usually provided to meet a 24-hour demand and, when considering health care premises, Hospital Technical Memorandum 27 outlines minimum recommended water-storage capacities for differing types of hospital wards.

However, to apply the same philosophy to industrial buildings would be incorrect, and due consideration must be given to the effect of loss of water supply to the process/manufacturing production.

When assessing water storage the following should be considered:

1. Where there is a requirement for large volumes of water, storage may be sized on quantity of usage per shift;
2. Frequency of interruption to mains supply;
3. Space requirements;
4. Cost of water storage tank and associated supports;
5. Protection against frost;
6. Type of industrial usage and effect of loss of water supply to production;
7. Minimizing the risk of Legionnaires' Disease.

1A.5.2 Draw-offs

In the UK, maximum draw-off from water mains differs throughout the various water authorities. What may be acceptable in one area may not be acceptable in another. Water mains may generally be able to supply direct to domestic buildings but not to industrial premises, where constant flow and negligible fluctuation in pressure is required. Contamination of water supply by back siphonage is of major concern to water authorities.

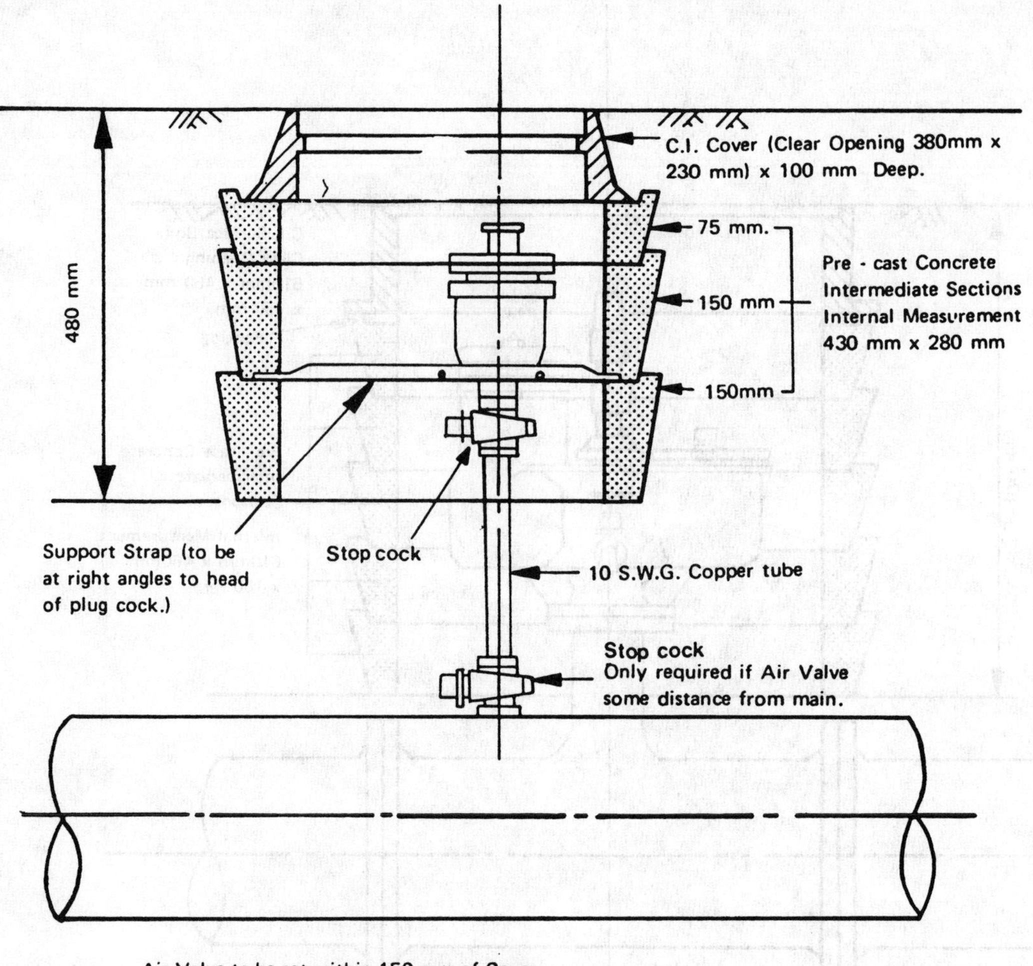

Figure 1A.12 Standard pipe installations for a single air valve. *Note:* Unless otherwise directed, pre-cast concrete units shall be bedded on well-compacted granular material Type A brought up from the base of the trench

Water consumption for commercial and industrial premises are metered and costed on a per cubic metre of usage basis, with additional costs for reading and maintenance of the meter station(s). It is important that assessment of water usage is as accurate as possible, as water charges are based on a sliding scale, i.e. as volume of water usage increases, cost per unit decreases.

Where production in new premises is to be phased with peak usage only being attained after a period of years it is important to consider this during initial discussions with the relevant water authority when negotiating unit rate costs.

See Figures 1A.9–1A.17.

1A.6 Problem areas associated with on-site sewage treatment for isolated areas

The outfall of a foul-water drainage system should discharge into a foul-water or combined drainage system (foul and surface water). Where such a drainage system is not conveniently available and cannot economically be extended to a site, other methods of foul-water disposal will be necessary, either a cesspool or a septic tank.

1A.6.1 Cesspools

A cesspool is a tank normally located underground which is designed to store the entire foul drainage discharge from premises between disposals by tank vehicles(s). The following points should be noted when considering the use of a cesspool:

1. Minimum capacity of 18 m^3 measured below the level of the inlet (or as equivalent to approximately 45 days' capacity);
2. It must be covered and impervious to rainwater, groundwater and leakage;
3. It must be ventilated;

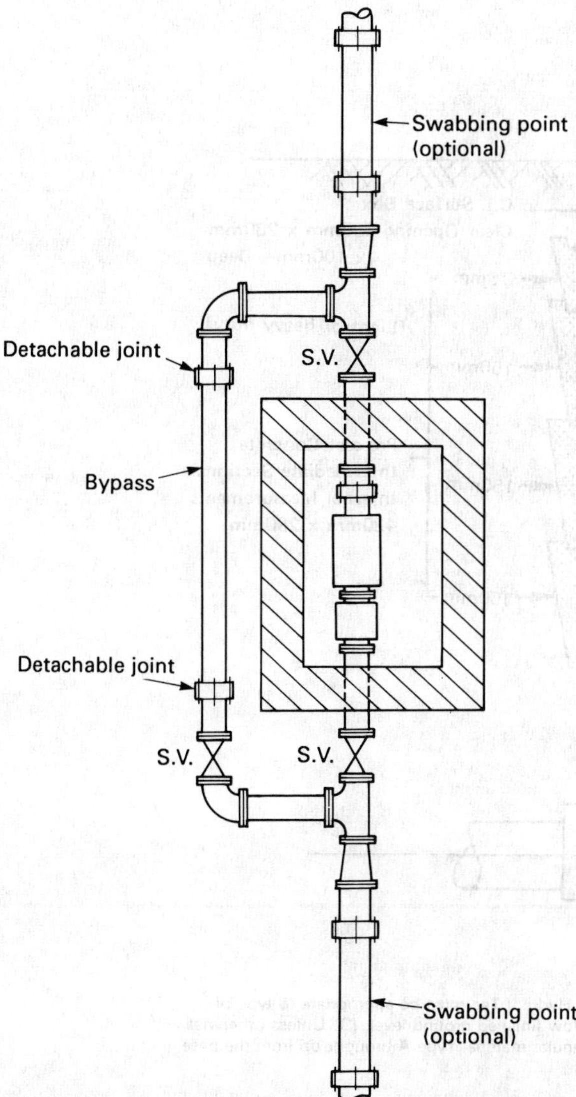

Figure 1A.13 Standard bulk meter layout for a brick-built chamber (230 mm brickwork on a concrete base). *Notes*: (1) The meter should be either on straight or bypass to suit site (i.e. to keep meter off road). (2) Chamber cover should be approximately 685 mm × 510 mm

4. It must be positioned such as not to pollute any water sources or cause a public nuisance;
5. It must be sited so that the contents may be removed other than through a building, with reasonable access for tanker vehicles where required;
6. Its position should also be considered relative to future sewerage systems.

Cesspools should only be utilized in extremely remote areas and after consideration of all other forms of effluent disposal has been undertaken.

1A.6.2 Septic tanks

A septic tank is a purification installation designed to accept the whole sewage/trade discharge from premises. Its construction is such that it allows the settling out of the sludge content of the incoming effluent and renders the final effluent acceptable, by prior agreement with the water authority, for discharge to a watercourse or soakaway. The sludge within the septic tank decreases in volume by the action of microorganisms changing the sludge from aerobic to anaerobic.

The following points should be noted when considering the use of a septic tank:

1. It should be impervious to rainwater, groundwater and leakage;
2. It must be positioned such that the contents may be periodically removed by a tanker vehicle;
3. A percolation filter and/or sub-surface irrigation may be required to provide aeration and final purification;
4. The change of bacterial action from aerobic to anaerobic produces methane gas which must be ventilated to atmosphere;
5. Its position should also be considered relative to future sewerage systems;
6. Minimum capacity of 2700 l to be provided. Subject to local authority approval, the effluent outfall from a septic tank may be connected to an adopted sewer subject to quality of outfall and size and location of drainage.

1A.7 Landscaping on industrial and reclaimed land

1A.7.1 General

In the context of overall landscaping, here we will concentrate on what is normally recognized as soft landscaping, i.e. that area which includes waterbodies and growing plants. The primary problems which have to be solved in any scheme for creating a new growing, living landscape out of a reclaimed land are:

1. Contamination in the ground. (There may be contamination in the air, but this is beyond the scope of this chapter.)
2. Soil structure and land drainage within the reclaimed ground.

1A.7.2 Contaminated land

It is not unusual for industrial land or land formed from tipped waste to be contaminated to some degree. Heavy contamination will have to be dealt with as a particular engineering problem: i.e. to seal, bury or remove any highly contaminated material. Moderate contamination can normally be dealt with *in situ* by dilution of the concentrated contaminant. Whatever treatment is utilized, steps must be taken to prevent the leaching of any contaminants through the soil.

As far as the effects of contamination are concerned, the main problems associated with harm to plant growth

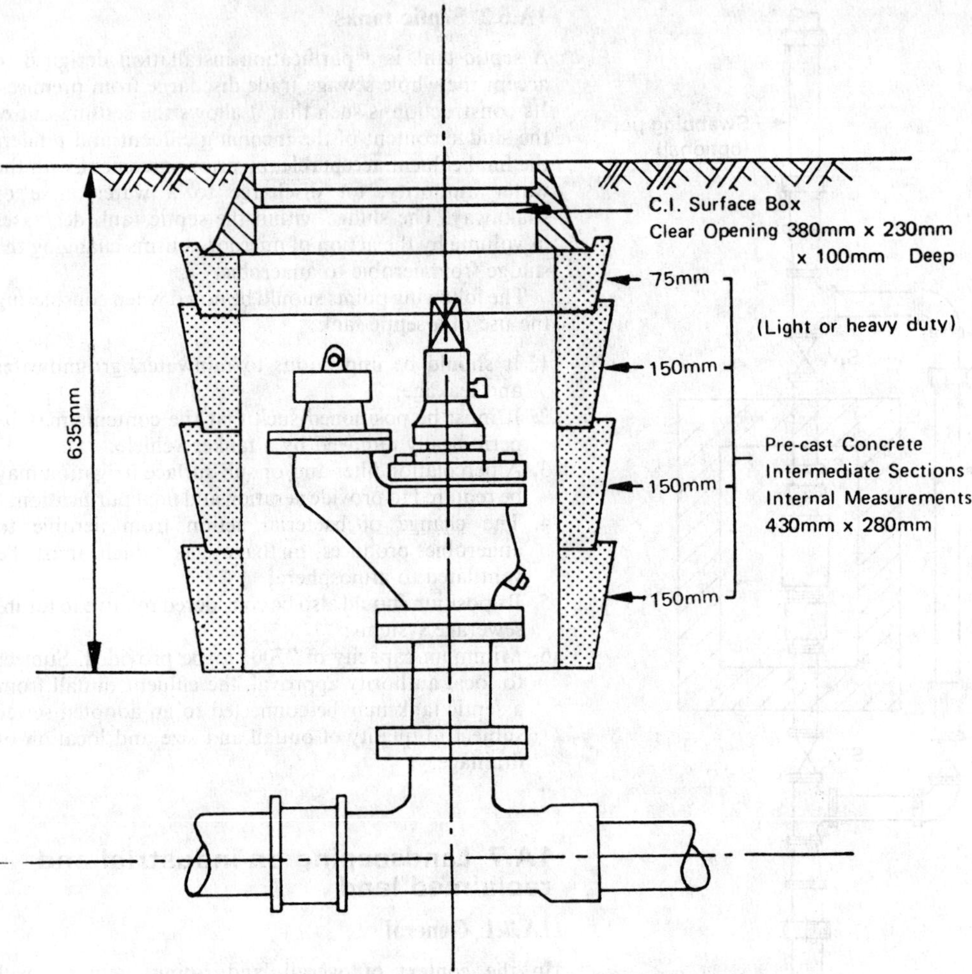

Figure 1A.14 Standard pipe installations for a fire hydrant. *Notes:* (1) Hydrant Tee must be appropriate to type of main installed. (2) Depth of hydrant outlet must not exceed 300 mm below finished ground level. (3) Unless otherwise directed, pre-cast concrete units shall be bedded on well-compacted granular material Type A brought up from the base of the trench

are the presence of particular heavy metals which are poisonous to plant life and/or lack of oxygen in the soil growing medium due to the presence of gases produced by the contaminated ground (for example, methane). Lack of oxygen can also be the result of a poor soil structure, over-compaction (no air voids), etc.

The Department of the Environment sets guidelines for safe limits to heavy metals content in the soil for reasons of both public and plant health. Before planting regimes are implemented in any reclaimed ground these heavy metal levels must be assessed and, if high, reduced. One acidic or alkaline soil also will need neutralizing and the type of proposed planting chosen for the soil type.

As indicated earlier, heavy contamination can be buried, sealed or removed. Burying of the material should be well below the root growth zone, and this is normally taken as 3.0 m below the final ground-surface level. Sealing for heavy contamination to prevent vertical or lateral leaching through groundwater flow can be with compacted clay or proprietary plastic membranes. Removal from site of the contaminants is normally only contemplated in a landscaped scheme where the material, even at depth, could be a hazard to public health directly or phytotoxic to plant life.

Where there is a landfill gas-generation problem, active (i.e. pumped) or passive venting, by way of stone-filled trenches at least 1.0 m deep, may be needed. Where active venting is appropriate, the commercial use of the gas extracted can be considered.

Contaminated groundwater (leachate) should be kept below the root growth zone. Only rainwater or clean irrigation water should meet the needs of plants.

Any waterbodies constructed on the contaminated land which are to support aquatic life will need to be completely sealed against the underlying ground and inlet and outlet water provisions designed so that they are sealed against any flow of contaminated water into the waterbody.

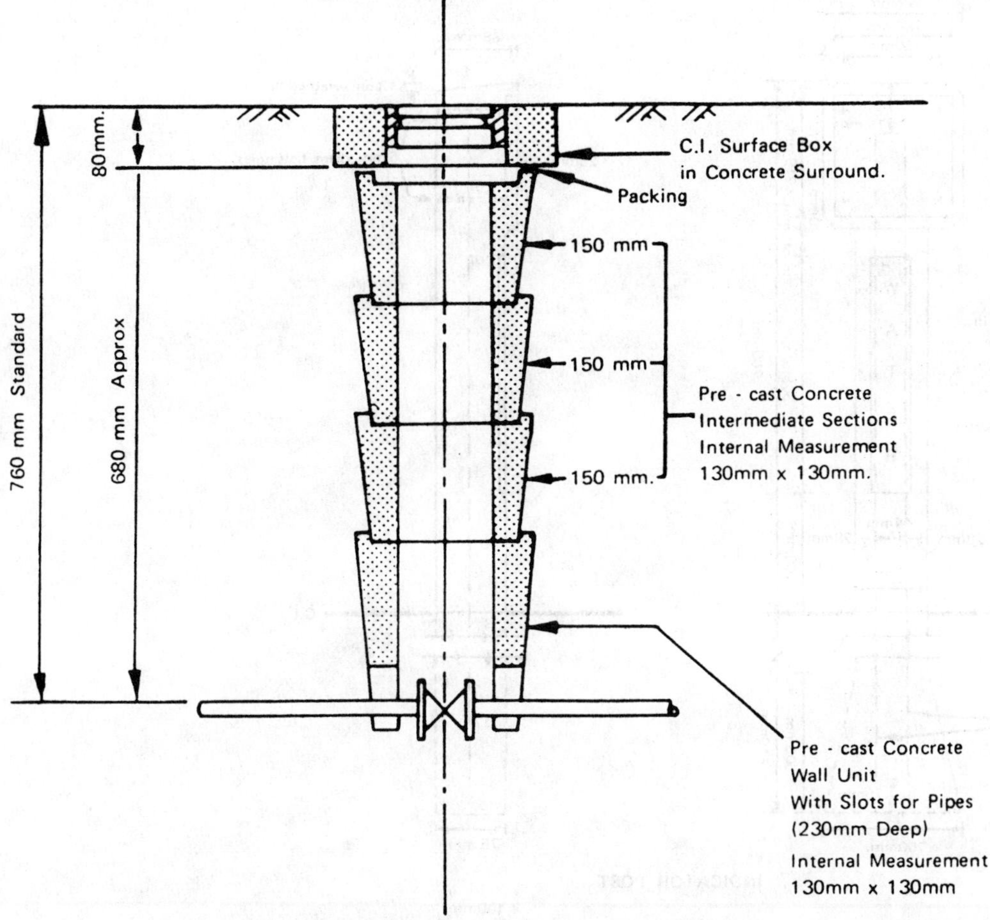

Figure 1A.15 Standard pipe installations for a communication pipe stopcock. (It may be necessary to provide a base unit where polythene is being used)

As a general rule, young, immature plants should be chosen for any landscape planting scheme on reclaimed, previously contaminated, land. This allows the plants to adapt gradually to such an environment. It is not normally appropriate to plant mature shrubs and trees to create an instant 'mature' landscape in such an environment.

1A.7.3 Non-contaminated land

For land that is reclaimed with 'inert'/non-contaminated materials the main problem in creating a good growing medium tends to be in producing a soil structure profile within a limited period of time. It its normal natural state such a profile would take many many years to develop. A good profile will provide an oxygen- and nutrient-rich soil, not too acid or alkaline, which is well drained. Any reclamation project should aim to supply this ideal environment as soon as possible. To do this, a clear and firm specification for the reclamation material and how it is to be placed will be needed at the beginning. Extensive drainage with both the land form and the subsoil will also be required.

Imported topsoil or topsoil manufactured from imported nutrients mixed into existing soil will normally be required. Topsoil depths will vary in accordance with the planting, from 400 mm for grassed areas to 1.5 m for major trees. Under these conditions instant, mature landscapes can be constructed immediately.

In any scheme for creating a landscape on reclaimed land the ultimate aim will be not only to create a green landscape but also to provide an environment in which the ecosystem will develop quickly. Some important factors which must be taken into account here are:

1. Early structural planting to provide windbreaks and shelter for younger plants and wildlife to develop.
2. A balanced landscape needs waterbodies. These should be designed to have shallow margins to provide the appropriate conditions for wetland planting which, in turn, provide the balanced habitats for a rich wetland ecosystem.

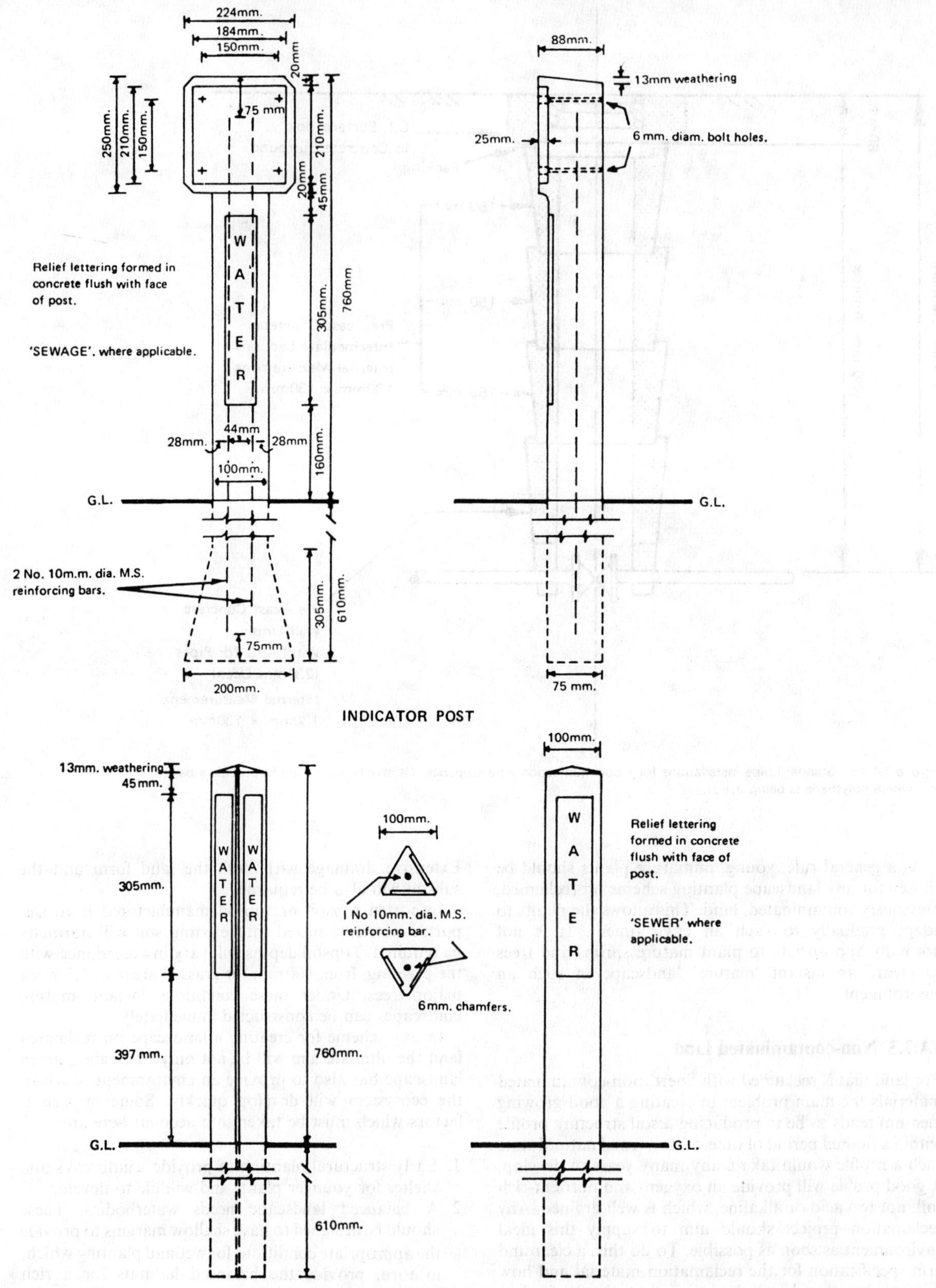

Figure 1A.16 Pre-cast concrete indicator and marker posts

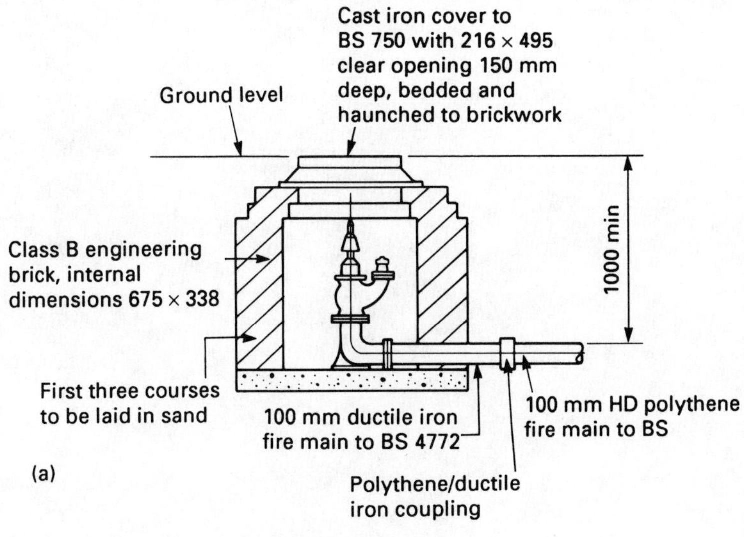

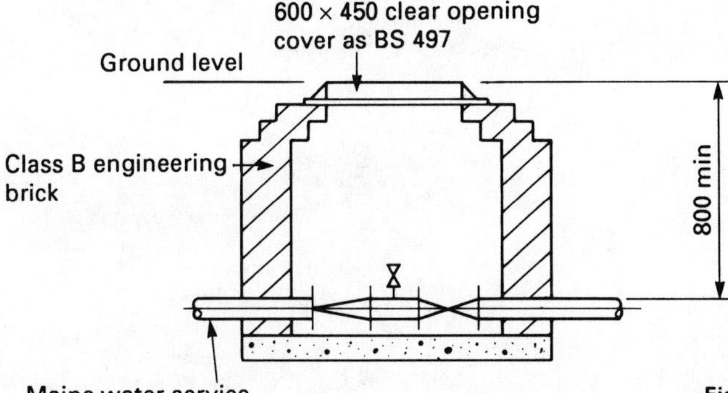

Figure 1A.17 (a) Fire hydrant pit; (b) stopvalve pit. *Note*: To each fire hydrant an indicator plate with post to BS 3251 must be provided and installed

3. A properly designed planting regime away from the water bodies will provide cover, space and the natural habitats for a wider wildlife system.
4. It is not normally necessary to seed waterbodies with fish. Fish will be introduced by natural means (e.g. by aquatic bird life).

Finally, in any projected landscape due consideration should be given to the future maintenance of the completed scheme. Unlike a normal building development, a landscape progressively develops with time. In order to maintain the original design concept a clear maintenance regime and organization should be established early in the project development process to start as soon as the landscape has been initially established.

Further reading

BS 6297, Code of Practice for the design and installation of small sewage treatment works and cesspools
BS 6700, Design of water services
The Building Regulations 1985
Chartered Institute of Building Services Engineers (CIBSE) Guides
Code of Practice on Legionnaires' Disease for health care premises
DHSS and Welsh Office Health Technical Memorandum 27
Institute of Plumbing Data Book
Institution of Water and Environmental Management
Minimising the risk of Legionnaires' Disease, Technical Memorandum CIBSE 13

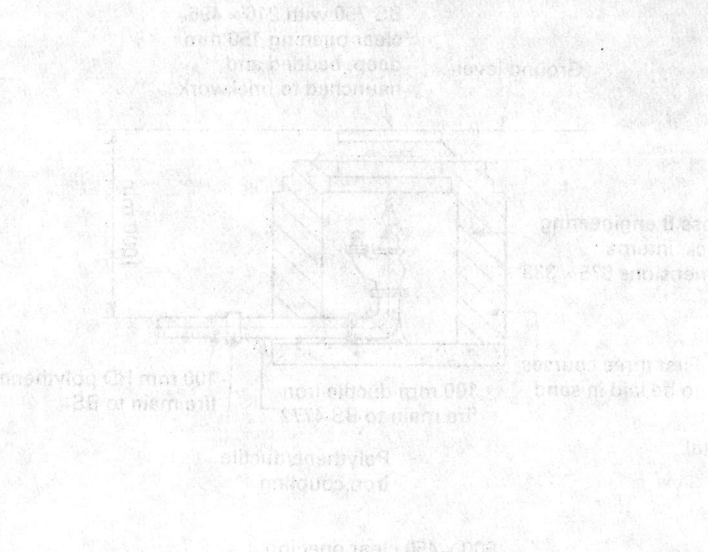

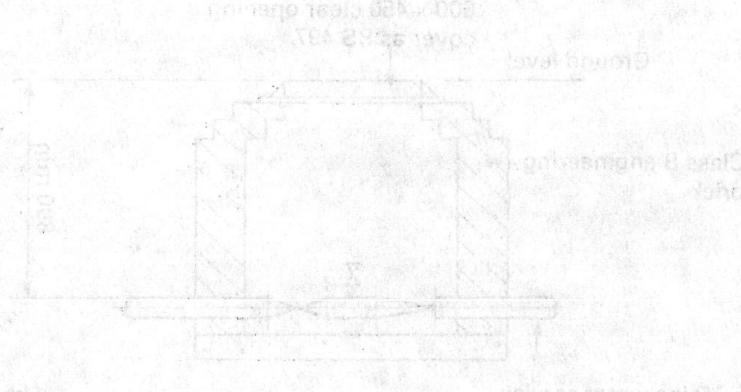

1B

Plant Location

Barry Holmes BSc(Tech), PhD, CEng, FIChemE, FInstE
W.S. Atkins Consultants Ltd

Contents

1B.1 Selecting the location

Location is a strategic issue, and the decision where to locate cannot be taken lightly. It is the first decision in the implementation of a project (that is, once it has been agreed to go ahead). Upon taking this decision, an investment is made which is irreversible. That investment in 'bricks and mortar' cannot physically be transferred to another location if the decision turns out to be wrong. In time, the assets may be realized but probably at much lower values than paid.

Traditionally, companies have sought to acquire competitive advantage over their rivals through their choice of location. In a historical context, firms tended to establish their factories for reasons of economic geography, e.g.:

- Proximity to raw material source
- Proximity to relatively cheap and abundant energy
- Availability of relatively inexpensive manpower or specialist skills
- Good transport links with materials suppliers and markets.

In more recent times the location decision has been influenced by government intervention, e.g.:

- High tariff duties imposed on imports which encourage exporters to that country to consider setting up local operations
- Investment incentives in the form of tax relief and grants
- The provision of infrastructure, especially improved transport communication

In today's world the decision is more complex. Markets are more sophisticated, skills can be in short supply, technological change can soon outdate newly installed processes, and there has been a phenomenal revolution in communications, both in terms of the physical movement of goods and people and of information around the world. As a consequence, companies often have to consider a wide number of options, and the eventual decision is based on optimizing the perceived net benefits.

The following are the key factors in selecting a location, and these are discussed in more detail below:

- Factor costs
- Protected markets and economies of scale
- Government influences
- Corporate matters
- People matters

1B.1.1 Factor costs

All other things being equal, the company will site its plant where it is cheapest to manufacture. This is determined by the nature of the product and of the manufacturing process. For instance:

- Where the process discards significant quantities of raw material (especially if it is bulky) the economics favour locations close to sources of the raw material (for example, sugar milling, cotton ginning, the beneficiation of non-precious minerals).
- Where considerable quantities of energy are used in the process, relatively cheap and abundant sources of energy need to be available (for example, alumina and aluminium smelting, steel making).
- Where the process is labour intensive, low-cost workers are required in numbers (for example, textiles and clothing). The cost of capital has little effect on the choice of location as sourcing of capital can be from anywhere. The ability, however, to repatriate profits and the proceeds from the sales of assets and exposure to foreign exchange risk are important if the location options are abroad. This factor becomes more relevant, the greater the capital intensity of the project.

Associated with factor costs are linkages with material suppliers, subcontractors and support services (e.g. temporary staffing agencies, travel agents, reference libraries, office-maintenance services). The availability and cost of these need to be considered.

1B.1.2 Protected markets and economies of scale

All markets are protected to some extent. Domestic suppliers have, at times, the natural protection of lower transportation costs, a greater understanding of the local marketplace through its experience and knowledge of social customs and culture, and an easier ability to respond to domestic customer requirements. Added to this, tariff duties may have to be paid on imports.

Where products are demanded in volume, process technologies have been developed to minimize unit costs. The concept of the 'minimum economic scale' of plant (MES: sometimes referred to as 'minimum efficient size') is the point beyond which increases in scale do not significantly reduce unit costs.

Where domestic markets are large enough, companies can establish MES plants or larger. When they are not, several alternatives need to be assessed, namely:

- Setting up a plant of sub-optimal size to serve the local market based on the premise that the higher unit costs can be offset against factor cost advantages and the protection that the market affords the local producer.
- Establishing a MES plant and seeking export opportunities.
- Supplying, or continuing to supply, the local market from a MES plant established abroad.

1B.1.3 Government influences

As already indicated, governments (regional and national) try to attract inward manufacturing investment in various ways. For example, in the 1950s, 1960s and 1970s, this was largely developed behind high tariff duties on imports. Infant-industry arguments are used by countries to adopt protectionist policies, but companies must expect eventual pressures 'to grow up' and face open market competition. With trends for countries to join economic unions such as entrance of Spain, Portugal and

Greece into the EC, the advantages of tariff protection cannot always be guaranteed.

Local and regional governments are frequently involved in the promotion of their areas. In some places, particularly in Western developed countries, they are able (and often keen) to provide useful information on their locality, covering available sites, rents, land prices and details on transport communications. These bodies can be very effective in helping to finalize choice. For example, Derbyshire County Council played a significant role in persuading Toyota to establish its first European car plant in Derby.

Governments often offer monetary incentives to prospective manufacturing investors. The types of inducements offered are:

- Tax-free holidays; periods of possibly 3–15 years of no tax liability on profits (this also can apply to local, property or any other tax normally imposed on a firm);
- Deferred tax allowances; where capital and operating costs in early loss-making years can be offset against profits later;
- Liberation from payment of import duties on capital plant and equipment, and on materials;
- Capital grants and/or low-cost loans for the purchase of plant and equipment;
- Liberal depreciation allowances;
- Training grants and/or facilities; to assist towards the cost of acquiring a suitably skilled workforce;
- The provision of land and services at zero or nominal prices (these sometimes extend to factory units and accommodation for senior management staff);
- The provision of 'free zones', usually in port locations, where goods can be imported duty free, processed and the bulk (if not all) exported.

Although many of these incentives are generally limited to the poorest regions of the European Community, where natural economic advantages to meet present-day opportunities are few, they are also widespread in developing countries.

Investment is more easily attracted to areas with adequate infrastructure. Governments encourage inward investment through the provision of industrial estates, sometimes with standard factory units ready for occupation; energy and water supply; roads suitable to withstand heavy vehicle traffic; other means of transport, airports, seaports, railways; and waste-disposal facilities. Progressive government departments concerned with industrial development make available a wide range of information about their areas, including local services, amenities, housing and education.

When locating in a foreign country the national government may insist on a certain level of 'local content' being achieved in a given period of time. This requirement, which is often subject to negotiation, can involve the company in a significant amount of administrative work in order to demonstrate that the conditions are being fulfilled. In addition, local content schemes invariably mean higher costs of production.

1B.1.4 Corporate matters

The opportunity cost of senior management time is high, and new projects usually take up a disproportionate amount of senior management time. This may be the critical factor in choosing between the best candidates in the 'last round' of selection.

Another critical factor could be 'image'. High-technology companies like to reinforce their image by having addresses that are synonymous with education and science (e.g. Cambridge). Others may want to associate themselves with an area traditional for high-quality manufacturing (e.g. Sheffield) or research and development such as in the pharmaceutical industry, which is concentrated in Switzerland.

1B.1.5 People matters

A prospective location may appear to be right in all other respects but fails because of the manpower resource. This refers to not only the economic aspects such as the availability of labour and skills but also to the qualitative ones of being able to understand the local culture and customs, appreciation of people's attitudes and values and confidence in the ability to blend these with the culture and aspirations of the company to form a cohesive production unit that will work. A successful processing concept in one cultural environment may not be a success in another.

Japanese investors in overseas manufacturing operations have been careful in their approach as well as in their choice of location. It is notable that the UK has received more than its share of Japanese investment in Europe in terms of population. An important factor here is language, as English is the most commonly spoken foreign language in Japan.

Stability of labour relations is an important criterion for location selection. The Japanese have negotiated single-union plants and 'no-strike' deals with trade unions to try to ensure such stability.

Probably the ultimate factor behind the choice of location is the influence key personnel can bring to bear on the decision. Project success largely depends on people, especially those at the top. With limited options on the person and/or team to lead the project's implementation, their location preference could be final.

1B.2 Services

1B.2.1 Availability of water

A water supply is an essential service for all manufacturing and process industries for domestic, cooling or process use. In 1988 over 300 million m^3 of water was supplied to the industries of England and Wales with a further 2000 million m^3 to the CEGB. About half the total for industry was provided as metered mains supply with the remainder obtained as privately licensed abstractions from surface or groundwaters. Almost all of the waters used by the CEGB for cooling purposes was taken from surface waters.

The significance of a water resource on the location of an industrial plant is essentially cost and security of supply. Certain industries (e.g. beer and mineral water production) may consider a water supply with particular chemical characteristics as essential to their location. Water supply costs are related directly to quantity but can rise almost exponentially in relation to quality. Therefore, subject to other commercial considerations, there is an advantage to industries who use large volumes of relatively low-quality water for process or cooling (e.g. paper and pulp, textiles, chemicals and steel) to locate in areas where surface or groundwaters may be exploited. Their need for higher-quality water is met from the public metered supply or from on-site treatment facilities. For the majority of industries, however, water supply is not a prime consideration in plant location. Traditionally, their demand for water is met from the public metered supply supplemented, where appropriate to availability and quality for use, by surface or groundwaters. Nevertheless, the cost of water provided by the ten water companies in England and Wales varies significantly. The current average metered cost for the ten companies is 31.85 p/m^3 in the range of 25.58–33.5 p/m^3, i.e. a 31% cost difference. Recent legislation and regulations will undoubtedly increase substantially the cost of public water supplies and also the cost differences between the water companies.

In addition to varying in costs, the chemical composition of the water provided from the mains supply also varies between the water companies, as may that between independent supplies within each company's area. The current criterion on potable water quality requires it to be 'wholesome', i.e. it should not create a health hazard, with relatively wide limits on particular constituents. The cost of removing these constituents (e.g. calcium, magnesium, chlorides, iron and silica) increases with concentration and variability. This imposes a cost burden on, for example, the semiconductor and electronic component industries and on the operation of high-pressure boilers. Therefore both the potential cost of metered water supply and the chemical composition of the supply waters may influence future decisions on the water company's area in which an industry may wish to locate.

Security of supply in relation to quantity is governed for surface and groundwaters by the granting of an abstraction licence by the National Rivers Authority (NRA). The licence limits the total daily quantity and the rate at which the waters may be abstracted, taking account of the natural resource and the needs of other abstractors. Metered water supply is subject to contract with the water companies, who may impose quantity and draw-off rate limitations. The limitations imposed by the licence or contract may influence plant location.

With regard to surface-water abstraction the following additional points are worth noting:

- The character of surface waters can change significantly and rapidly throughout the year, and any process or cooling-water system should be designed accordingly.
- The higher the quality of the surface waters providing the supply, the more stringent will be the treated wastewater discharge consent standards imposed by the NRA on effluents returned to the surface waters.

1B.2.2 Trade effluent disposal

It will be seen from Section 1B.2.1 above that industry is an intensive user of water. Almost all of this water produces a wastewater, requiring to be disposed as either domestic waste or trade effluent. Domestic wastewater is invariably discharged to the municipal sewers for treatment at the municipal sewage-treatment facilities by the water companies. The cost of this service is currently generally incorporated into the rating system, with the water companies using the rating authorities as an agent. Introduction of the unified business rating system may require metering and direct payment to the water companies.

Trade effluents may be discharged to the municipal sewers, to surface waters or on or into the land. Discharges to surface waters, including estuaries and coastal waters, or into the land are controlled by means of Consents to Discharge with the NRA and Agreements with the water companies. Discharges to controlled landfill sites are by agreement with the local waste disposal authority. In all cases the Consents and Agreements will impose conditions on the quantity, rate of discharge and chemical composition of the trade effluents acceptable for discharge.

The quantity, rate of discharge and chemical constituent limits incorporated into the Consents and Agreements for surface-water discharges are set by the NRA to prevent pollution and to meet the river quality objectives (RQOs) of the receiving waters. Limits on land discharges are set to protect groundwater aquifers. Currently, limits are imposed on organics, suspended solids, ammonia, toxic metals, pH and temperature in the trade effluent discharges. The actual limits on each of these constituents take account of available dilution, the surface or groundwater use and the impact of other dischargers. The limits can therefore vary significantly between discharge locations. Nevertheless, the constraints imposed by the NRA on discharges to surface waters and to land are generally considerably more stringent than those imposed by the water companies on sewer discharges. Impending and future EC directives on environmental protection and drinking-water standards may not only reduce the constituent acceptance limits but also extend the list of constituents constrained. At present the NRA does not levy a charge on the discharge of trade effluents to surface waters or to land.

Disposal of industrial effluents to controlled landfill sites is generally confined to slurries and sludges. The quantity and composition of the wastes acceptable for disposal is controlled by licences issued by the waste disposal authority.

The quantity, rate of discharge and chemical constituent limits incorporated into the Consents and Agreements for discharges to the sewerage systems are set by the water companies to protect the health of sewer and treatment plant personnel, the fabric of the sewers and the operation of sewage-treatment processes. The constraints take account of the hydraulic capacity of the sewers and the treatment works, the organic and solids-handling capacity of the treatment works, the ultimate disposal route for sewage sludge and the needs of other trade effluent dischargers. In addition to limiting the concentration of specific constituents, including those mentioned for surface-water discharges, materials which could be dangerous to the health of sewer workers are prohibited from discharges. Also, the water companies are obligated to inform Her Majesty's Inspectorate of Pollution (HMIP) of the presence of any chemical included in a 'Red List' of some 30 chemical constituents in a trade effluent before the effluent can be granted a Consent to Discharge. HMIP have the powers to prohibit or limit the concentration of any of these chemicals in effluents discharged to municipal sewers.

Municipal sewage-treatment effluents discharge to surface waters and are subject to the same NRA control on quality and quantity as independent industrial surface-water discharges. Any tightening of NRA Consent standards may therefore result in more stringent controls on industrial effluents discharged to sewers. All the water companies levy charges on industry for the reception, conveyance and treatment of the industrial effluents.

The significance of these industrial effluent disposal options on the location of an industrial plant is essentially cost. As previously stated, the NRA do not, as yet, impose a cost on effluents complying with the Consent standards discharged to surface waters or to land. However, the cost of installing and operating treatment facilities, including sludge removal, to achieve and maintain the Consent standards can be considerable. The cost of treatment escalates significantly with increasingly stringent treated effluent standards. The actual cost of this disposal route will, of course, depend on the quantity and character of wastewaters generated by a particular industry in addition to the treated effluent standards to be met. The conditions imposed in the Consents to Discharge could therefore influence the location of an industrial plant adopting a surface-water or land-disposal route for its effluents.

In addition to the treatment costs, it should be remembered that provision of treatment plant occupies site space which may be more profitably used and that the penalties for infringements of the consent standards are increasing.

The cost of depositing waste on controlled landfill sites is relatively cheap. However, the expense of road transport to suitable sites generally limits this disposal route to relatively low-volume applications.

The cost of industrial effluent disposal to the municipal sewers is based on a 'Polluter Pays' policy, which takes account of the quantity and pollution loads in the discharge. All the water companies calculate their trade waste charges in accordance with:

$$C = R + V + \frac{(O_t)B}{O_s} + \frac{(S_t)S}{S_s}$$

where

C = total charge for trade effluent treatment (p/m^3),
R = reception and conveyance charge (p/m^3),
V = volumetric and primary treatment cost (p/m^3),
O_t = the chemical oxidation demand (COD) of trade effluent after one hour settlement at pH 7 (mg/l),
O_s = COD of crude sewage after one hour settlement (mg/l),
B = biological oxidation cost (p/m^3) of settled sewage,
S_t = total suspended solids of trade effluent at pH 7 (mg/l),
S_s = the total suspended solids of crude sewage (mg/l),
S = treatment and disposal of primary sludges generated by the trade waste (p/m^3).

However, all of the functions in the formula, excluding the trade effluent functions O_t and O_s, are different for each of the water companies and are reviewed annually. Therefore the unit costs of charges vary between the companies. In the period 1988–1989 the unit costs for treating a trade waste equivalent to domestic-strength sewage ranged from 14.47p to 59p per cubic metre. Subject to other commercial considerations, the cost of trade waste charges could influence the water company area in which to locate an industrial plant. The conditions imposed in the Consent to Discharge could influence the location within an area most beneficial to plant location.

1B.2.3 Electricity

Previous considerations concerning the purpose engineering of the facility will have established the size and characteristics of the electrical load in terms of consumption and maximum demand. These will be essential data with which to enquire of the electricity authority as to the availability of an electrical supply.

Generally, there will be no insurmountable difficulty in obtaining an electrical power supply, the principal factors being the cost of providing it and the time delay in making it available. The scale and nature of the project will dictate the amount of power required, the load characteristics and the most suitable voltage for its supply. The state of the public supply network in the area of the proposed development will then constrain the supply authority in its ability to quote for a suitable supply in the short, medium or long term.

A relatively small power demand at low or medium voltage in a built-up area might be provided in the short term by teeing from an existing main feeder in the locality, or providing a radial feed from an existing substation. If the load is large or remote from the existing supply network, or if the local network is fully loaded, then a new incoming supply brought from a distance might be a medium- to long-term project, with time and cost dictated by legal wayleave considerations as well as the technical aspects involved. It is therefore necessary to make the earliest possible approach to the relevant electricity supply authority to establish the availability and cost-environment situation, as this may influence the initial engineering and planning factors to be accounted.

1B.3 Ecology and pollution

1B.3.1 Introduction

Urban, rural and industrial developments may have profound effects on the surrounding environment. Such effects can defeat the object of development, in that the disbenefits may outweigh the benefits. In the case of natural resources, inappropriate development may even destroy the resource base. However, if environmental matters are accorded adequate consideration during the planning and management of development programmes and projects it is possible for pollutants to be assimilated, and the whole development to be accommodated by the environment in such a way that adverse effects are minimized and the economic and social benefits of development are maximized.

1B.3.2 Baseline studies and modelling

The basic requirement of any precommissioning environmental study is information on the existing state of the environment prior to any new development. Where such data are not already available, baseline studies must be carried out. Essentially, such studies are designed to provide baseline data from which the physico-chemical and ecological effects of development may be assessed and against which changes due to development may be measured after commissioning of the project.

Computer programs are available to develop these studies to provide accurate predictions of future pollutant loadings in the environment. For example, using the program AIRPOLL[1] ground-level concentrations of stack gases are calculated at various distances from proposed plants under a variety of meteorological conditions. The program CAFE[2] is similarly used to predict the dispersion of thermal and other effluents in the aquatic environment, and OXBAL[3] is employed to determine the ecological consequences of effluent discharges and engineering works in estuaries. Emission-control equipment and waste-treatment plant can then be designed to meet any local standards for ambient air and water quality.

1B.3.3 Environmental standards

In some parts of the world the need to exploit natural resources is so urgent that it has preceded the formulation of adequate environmental controls. The pace of natural resource exploitation and the growth of associated industries has overtaken the evolution of institutions which would have the authority to exert such controls.

The problem has been recognized by many of the developers concerned, who have consequently themselves adopted the environmental standards of other industrialized nations. In the absence of national controls this is a responsible and laudable approach. However, the piecemeal adoption of standards taken from elsewhere does not take account of local conditions. These conditions may either enhance or limit the ability of the environment to disperse and attenuate/assimilate pollutants (e.g. the occurrence of temperature inversions will limit the dispersion of air pollutants). Similarly, the use to which local resources are put may demand particularly high standards of environmental quality (e.g., the use of sea water or river water as the basis of potable water supply). The choice of standards must also take into account local practices and existing local administration.

At the plant level, in-plant monitoring of unique compounds and the modelling of plant conditions to develop appropriate working practices and internal environmental quality standards may be needed.

1B.3.4 Environmental assessment

In the case of a large-scale development it may be desirable to combine several environmental services for a full environmental impact assessment (EIA). This is the process of examining, in a comprehensive, detailed and systematic manner, the existing environment (natural, built and social) and the development which it is proposed to place within it. By integrating the two, an objective estimate can then be made of the likely effects of the development upon the environment, including benefits and disbenefits. Special techniques may be employed to help identify or quantify these impacts (e.g. the use of interaction matrices, overlays, screening tests, checklists, etc.).

An EIA can be particularly useful in distinguishing the relative environmental impact of alternative sites, processes and strategies for industrial, rural or urban development. This essential environmental information, together with financial and political considerations, can then be used by decision makers to choose the alternative which will provide maximum economic and social benefits with the minimum of environmental disturbance. Guidelines for assessing industrial environmental impact and environmental criteria for the siting of industry have been prepared for the United Nations Environment Programme.

1B.3.5 Resource planning

Ecology studies employ information from all the other environmental sciences to draw conclusions about environmental change which will result from development. However, when the exploitation of natural resources is being planned, the nature of the ecosystem and its response to change are themselves the most important elements to be considered. Industrial developments need to be assessed in the context of their ecological setting to ensure that the proposed development, first, will be feasible within local constraints and, second, will not bring about irreversible and unacceptable changes to other essential parts of the ecosystem.

The activity of ecologists in resource planning has three forms: preservation, integration and conservation.

Preservation

In this case, individual new developments are designed in such a way as to preserve discrete ecological systems which have been identified as of importance as 'life-support systems', as regionally/internationally important wildlife habitats or as sources of rare natural materials, etc.

Integration

In the case of development projects which are on a large scale (e.g. mining or dam construction) or which cover a considerable area, ecological planning attempts to achieve integration between the project and the ecosystem. The object is to ensure not only the continued functioning of the ecosystem as a whole but also those elements of the ecosystem which provide the development project with 'goods and services' (e.g. fuel, raw materials and the assimilation of waste).

Ecological planning in this context will help to avoid ecological disasters (e.g. excessive weed growth in stored water) and maximize the lifespan of the benefits derived from the development. It will be clear that this holistic approach to development has economic as well as ecological advantages.

Conservation

Where regional development is to be undertaken it is reasonable to adopt a positive ecological approach. This can begin with the formulation of a conservation policy with specific goals. These may include the conservation of individual animal/plant species or habitat types which are threatened by development. The goals will be met by a programme of conservation projects (e.g. the setting up of national parks, country parks or natural reserves). This positive approach to the ecosystem will not only benefit wildlife but will also create the opportunities for tourism and leisure, which are a vital adjunct to most development projects.

1B.3.6 Environmental management

Once commissioned, even the best-planned industrial development requires monitoring and management to ensure that its operation continues to be environmentally acceptable. This applies equally to established industries. When unexpected environmental problems arise a rapid response is required to assess the cause and magnitude of the problem and to devise remedial measures.

Air pollution

Dusts produced by quarrying and fluorides emanating from oil refineries are typical pollutants which need regular monitoring. A range of portable equipment for the identification and quantification of toxic and other gases can be used on an *ad hoc* basis.

When unpleasant odours resulting from manufacturing processes or waste-disposal operations give rise to public complaints they should be identified and quantified prior to deriving methods of abatement. Such work is often innovative, requiring the design and fabrication of new equipment for the sampling and analysis of pollutants.

Water pollution

Consultants are equipped to monitor the quality of freshwater, estuarine and marine environments and can make field measurements of a variety of water-quality parameters in response to pollution incidents. For example, reasons for the mortality of marine shellfish and farmed freshwater fish have been determined using portable water-analysis equipment. Various items of field equipment are, of course, also employed in baseline studies and monitoring, respectively, before and after the introduction of new effluent-disposal schemes.

Where extreme accuracy is required in the identification of pollutants or in the quantification of compounds which are highly toxic or conservative, laboratory analysis of samples is conducted. Highly sophisticated techniques have, for example, been employed in the isolation of taints in drinking-water supplies.

Land pollution

As development proceeds, land is coming under increasing pressure as a resource, not only for the production of food and the construction of new buildings but also for disposal of the growing volume of industrial and domestic waste. The design and management of sanitary landfill and other waste-disposal operations requires an input from most of the environmental sciences, including geologists and geotechnicians, chemists and physicists, biologists and ecologists. Such a team can deal with the control and treatment of leachate, the quantification and control of gas generation, and the placement of toxic and hazardous wastes. This may be needed in designs for the treatment of industrially contaminated land prior to its redevelopment.

Ecological studies

The acceptability of some industrial and ephemeral development projects such as landfill or mineral extraction may depend upon an ability to restore the landscape after exploitation has been completed. As more rural development projects come to fruition, ecologists will become increasingly involved in resource management to ensure that yields are sustained and to avert the undesirable consequences of development. Some industrial developments and rearranged plant layout schemes will not be complicated, but when ecology studies are needed the employment of specialist consultants is recommended.

References

1 AIRPOLL, W. S. Atkins Consultants Ltd, Woodcote Grove, Ashley Road, Epsom, Surrey
2 CAFE, As above
3 OXBAL, As above

2 Industrial Buildings

Dave Whittleton MA, CEng, MIMechE, MHKIE
Ove Arup & Partners, Industrial Division

Contents

2.1 Introduction

Industrial activity encompasses an enormous variety of operations, and industrial buildings provide the required protection from the external environment. Of necessity, therefore, they take many shapes and structural forms and may be composed of a variety of materials.

2.1.1 Procurement

Industrial buildings can be procured in one of three basic forms:

1. Specific design
2. 'Off the shelf'
3. Speculative.

Specifically designed industrial buildings

Many industrial operations will require buildings with a large number of specific attributes. Such requirements will include many of the following:

Location
Plan dimensions
Height to eaves
Column layout
Services provision
Drainage provision
Provision of cranes
Superstructure-imposed load capacity
Floor loading capacity (point load, uniformly distributed load, line load)
Floor flatness
Floor abrasion/impact/chemical/slip resistance
Access facilities
Loading/unloading facilities
Floor pits
Intermediate platform arrangements
Suitability for automatic guided vehicles
Suitability of environmental controls
Ease of cleaning
Corrosion resistance
Machine base availability
Sound insulation
Provision for future expansion.

Any one of the above may require a specific building to be designed and constructed due to the unsuitability of available facilities falling into the option (1) and (2) categories.

Apart from performance criteria, an industrial building might also be required to project a 'corporate identity' by a striking appearance. This generally requires a specifically designed facility.

'Off the shelf' industrial buildings

'Off the shelf' buildings are generally of set modular form designed to a standard set of criteria. The available degree of variation may be limited, but they are perfectly suitable for several industrial uses.

Speculative industrial buildings

These are often developed on new industrial estates with no particular tenant in mind. They may be quite adequate for general light engineering/warehousing/retail units. However, prospective industrial occupiers with any but the most basic general requirements are likely to have to spend considerable sums adapting the building to their needs, if indeed this is physically and economically viable.

2.1.2 Structural materials

The majority of industrial building superstructures are framed in structural steel, although a small percentage are in precast concrete. Steel is used primarily for its large strength-to-weight ratio, enabling it to span large distances economically. Steelwork is easily modified, which provides for a degree of adaptability not always available from concrete structures.

Ground slab and foundations are invariably reinforced concrete, though some ground bearing slabs are constructed with no reinforcement. Industrial buildings for containment of toxic of other processes may require construction primarily from reinforced concrete.

A dwarf wall of concrete, blockwork or brickwork is often constructed around the building perimeter to minimize cladding damage from forklift trucks, etc.

2.1.3 Structural form

Stability

Building stability can be provided by:

1. Framing action from rigid connections between columns and roof members;
2. Vertical cross-bracing or shear walls;
3. Columns cantilevering from foundations; or
4. A combination of these techniques.

Roof bracing may be required for rafter and purlin stability or load transfer to vertical bracing.

A roof diaphragm can also be provided by a composite concrete slab or a stressed-skin system, but the latter may severely restrict the provision of subsequent roof penetrations. Concrete roof slabs are unusual, due to the greatly increased mass over the more normal metal decking/insulation/waterproof membrane or insulated metal decking options.

Column layout

It is self-evident that fewer permanent obstructions within the perimeter provide a greater potential flexibility of operations. However, larger spans are bought only at the expense of structure.

Columns may also perform other functions, apart from roof support:

1. Support of intermediate-level platforms;
2. Restraint to partition walls;
3. Support for cranes/jibs;
4. Services (pipework, ductwork, etc.) support.

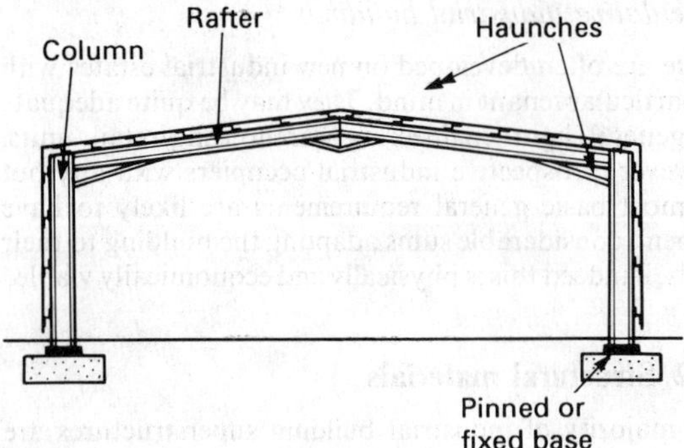

Figure 2.1 A typical portal frame

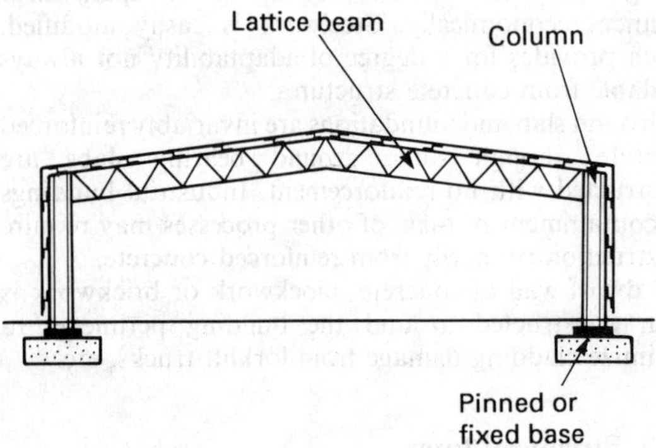

Figure 2.2 A typical lattice girder frame

Roof shape

Flat roofs. Flat roofs are popular with architects due to their reduced visual impact. However, they generally require a much higher standard of waterproofing, and may be susceptible to ponding, moss growth and dirt/debris entrapment. Inevitably they also require internal downpipes for rainwater disposal.

Falls to gully positions may be provided by the natural deflections of the roof structure or small slopes to the roof steelwork (e.g. 1:40–1:60).

Pitched roofs. Pitched roofs are typically sloped at a minimum of 6° to ensure the weather resistance of lapped sheeting without sophisticated seals or a waterproof membrane. Portal frames are also more liable to 'snap through buckling' at very shallow pitches. A pitched roof means a greater 'dead' volume to heat, although there is additional space for high-level services distribution.

Roof construction

1. *Flat roofs*

Typically, framing is achieved with:

(a) Universal beams (span < 15 m);
(b) Castellated beams (20 m > span > 15 m);
(c) Lattice girders (span > 20 m).

A recent development has been the practice of using site butt-welded, closely spaced, continuous secondary beams spanning over continuous primary beams and replacing the traditional purlins. Secondary spans of 20 m have been achieved with this method.

Other considerations include:

(a) Castellated beams may require reinforcement at high shear regions;
(b) Castellated and lattice beams provide a services zone within the structural depth;
(c) Lattice girders may be formed into V beams of triangular cross section.

2. *Pitched roofs*

Typically, framing is achieved with:

(a) Portal frames (Figure 2.1);
(b) Pitched lattice girders (Figure 2.2);
(c) Trusses (Figure 2.3);

Portal frames may be single-bay, multi-bay (Figure 2.4), propped, tied, or mansard with pinned or fixed foot. They may be composed of universal beams or tapered plate girders.

Trusses may be the best solution for very high imposed loads. Frame action with columns is not possible with trusses. Although trusses are generally the lightest form

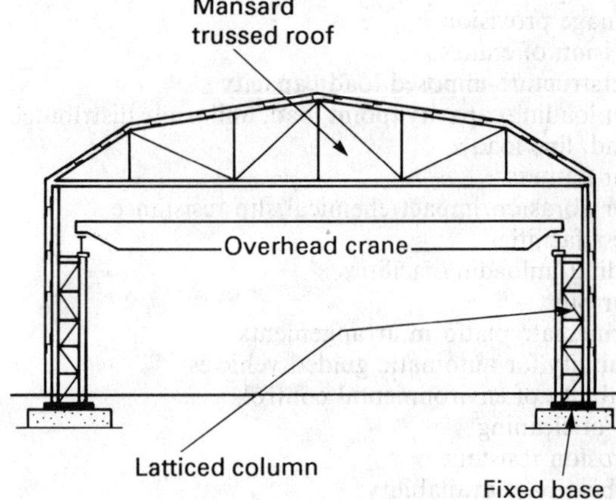

Figure 2.3 A typical trussed roof

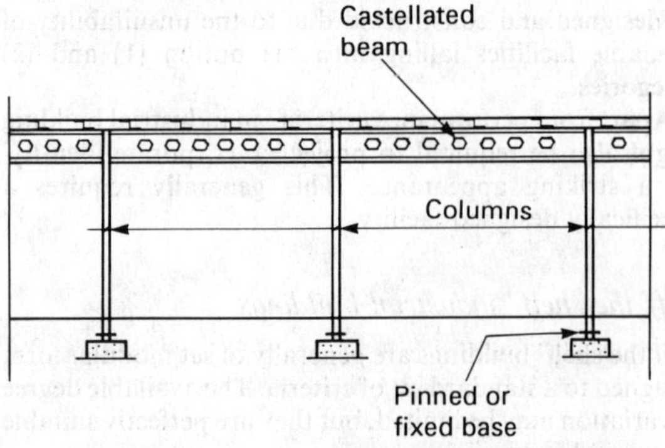

Figure 2.4 A typical multi-bay beam and post construction

of roof construction, they may be the most expensive due to high fabrication cost. A combination of lattices or lattice and truss may form a sawtooth roof profile for incorporation of northlights.

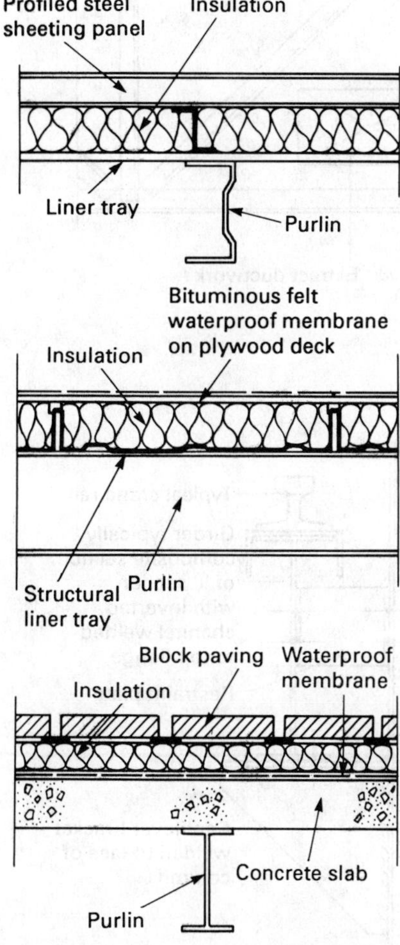

Figure 2.5 Typical roof constructions

3. *Space trusses*
Substantial two-directional spans can be achieved by three-dimensional space trusses utilizing proprietary nodal joints. Typically, they are too expensive for normal industrial buildings.

2.2 Specifying an industrial building

Many industrial buildings, and industrial projects in the broader sense, are completed entirely to the owner's satisfaction in terms of the key criteria of time, cost and quality. It is also true that many are not. There can be various reasons for this, some related to the nature of the construction industry in general, some to the performance of professional teams and others related to the manner in which the project was initially set up. The most common cause of problems with the finished product is the way in which the owner's requirements were specified, and this extends to include the selection of the project team.

These problems cannot always be attributed to the owner. An industrial client is not usually a regular developer of buildings, which are, in most cases, secondary to whatever process or production facility he wishes to construct. He is, after all, not engaged in the construction industry on a day-to-day basis. He should be entitled to expect good advice from the team he appoints, and to expect them to 'ask the right questions'. But whatever the project, and however limited the owner's resources, it is vital that he is able to provide the necessary input to ensure that he gets what he wants.

We are not concerned here with contractual issues, nor with questions of responsibility. What follows is a set of practical guidelines which seek to address the main areas where problems can arise.

2.2.1 Procurement

From the time when the owner decides that a project is desirable there are many approaches to the procurement strategy, ranging from a conventional set-up with

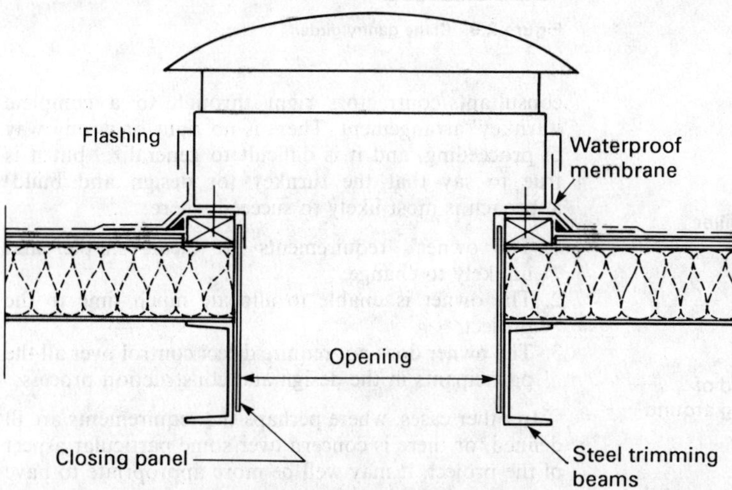

Figure 2.6 Typical roof penetration

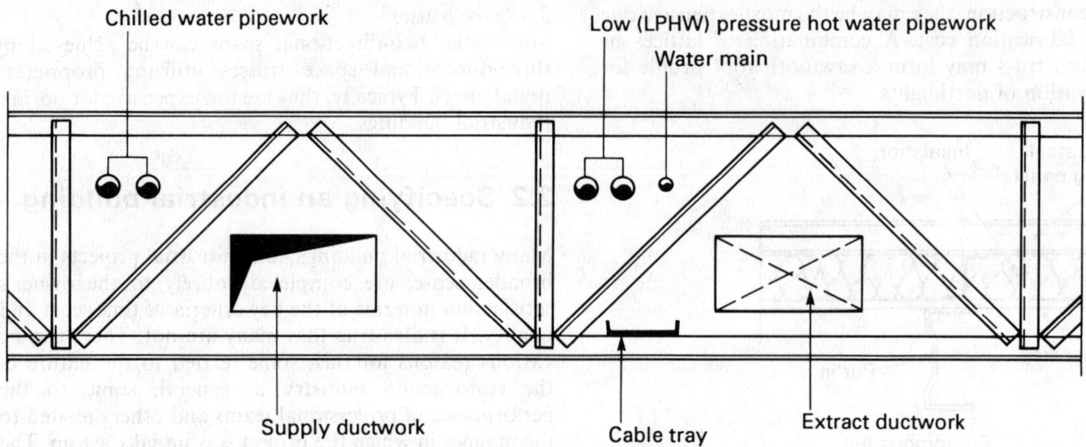

Figure 2.7 Typical services coordination through a lattice girder

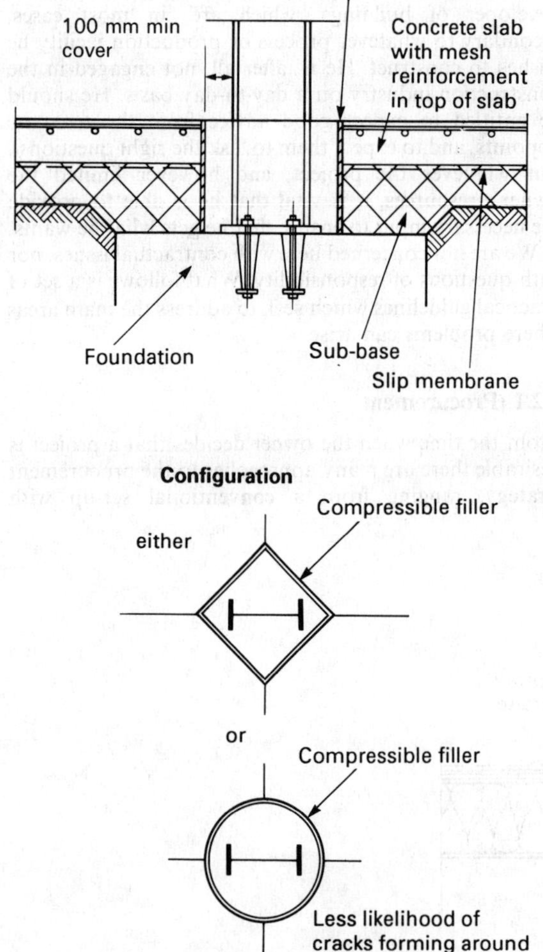

Figure 2.8 Isolation joint at foundation

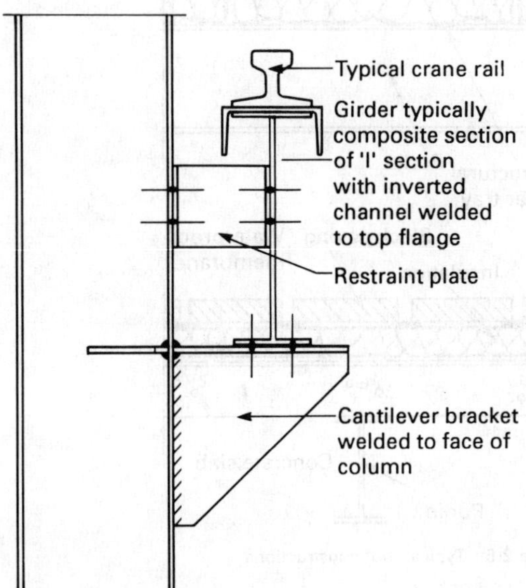

Figure 2.9 Crane gantry girder

consultants/contractors right through to a complete 'turnkey' arrangement. There is no right or wrong way of proceeding, and it is difficult to generalize, but it is true to say that the turnkey (or design and build) approach is most likely to succeed where:

1. The owner's requirements are clear, simple, and unlikely to change;
2. The owner is unable to allocate much time to the project;
3. The owner does not require direct control over all the participants in the design and construction process.

In other cases, where perhaps the requirements are ill defined, or there is concern over some particular aspect of the project, it may well be more appropriate to have the project designed and/or managed by independent professionals. For example, most industrial building projects are best suited to an engineer-led approach. If, however, the owner is particularly concerned about the architectural aspects of the scheme, he may prefer to make his own selection and appointment of an architect.

Whichever approach is adopted, the owner must be sure that he has considered those issues most important to him and has set up the project in a way that gives him the level of control he wishes to exercise. He should also take care to ensure that the team involved has adequate experience, not only of the type of project under consideration but also of the approach being adopted.

2.2.2 Budgets

The construction industry has a poor reputation in terms of cost control, and tales of budget overspends are legion. This is often attributed to deficiencies in the industry and among the professionals who work within it.

Whoever is to blame at the end of the day, the problem can often be traced to the fact that the initial budget was unrealistic at the outset. To set an inadequate budget, watch it overrun, and then look for a scapegoat may have become common practice but it achieves nothing.

It is therefore absolutely vital that a sensible cost plan be prepared and agreed at the earliest possible stage. Obviously, an owner will prepare rough budgets when considering basic project viability, but as soon as he is able he should obtain advice from his selected team.

Good professionals will not simply agree with the rough budget already prepared. They will give objective advice, including the level of confidence in the estimate and an assessment of the likely effects of changes to the specification, so that a realistic view can be taken by the owner. It is better to recognize the problem immediately, and modify the scope or specification to suit, than to be forced into late changes or even omissions which reduce the effectiveness of the finished project.

2.2.3 Specification

The level of briefing which an owner can give is variable, depending on his resources and his detailed knowledge of building or facilities requirements. The most important point here, however, is clarity. The brief may have to be simple, but it must be clear – and it must also be clear who has the responsibility for developing it.

The best way to ensure controlled development of the brief, and of the cost plans which necessarily accompany it, is to set up a system of data sheets which define, in increasing detail as the project develops, the spatial, functional and servicing requirements of each component part of the project. A well-resourced technical client may wish to do this himself. More commonly, it is a task which should be entrusted to the design team. Whoever does it, regular joint review of this information will help to ensure satisfaction with the finished product.

2.2.4 Schedules

Much of what has been said about budgets applies equally to the project programme. It should be realistic from the outset – again independent professional advice should be sought.

Modern 'fast-track' techniques of design and procurement have taken all the slack out of the traditional methods, and this is generally to the benefit of all concerned. There is a price to pay, however, in terms of flexibility. The owner, and his design and construction team, have to work within a discipline in respect of changes; programmes and budgets are very vulnerable to changes in this environment. This relates very closely to the earlier comments on clarity of briefing and information flow.

2.2.5 Procedures

Having decided on the appropriate way to proceed with the project, the owner needs to satisfy himself that adequate systems and procedures have been put in place to give him the necessary control and confidence in the project. These will, of course, vary considerably from owner to owner and project to project, but certain key points are worth remembering.

There is a clear need for the owner to appoint a single project officer. He may well be part-time as far as this activity is concerned, and will often need to draw on the expertise of others in his own organization. However, a single point of contact between the owner and his project team has been shown to be a vital ingredient in successful projects. Of equal importance is the need for regular reviews to ensure that things are going to plan. It is also necessary to have an agreed mechanism for change control, since once a project is under way, particularly if, as is usual, it is being run to a tight programme, any changes are likely to have an impact on both programme and budget.

The theme running through the above is that the key ingredient is clarity, both in specification and organizational terms. This does not need to mean a major effort or input on the part of the owner, but rather careful consideration, at an early stage, of what is actually required.

If what is required is unclear at the outset, then it is wise to say so, and allow for that fact in the development of budgets and programmes, rather than to pretend that the project is fully specified and suffer the consequences later. An industrial client, like any other, will get what he pays for; but he also needs to ensure that he gets what he needs.

2.3 Security

2.3.1 Objectives

The object of all security systems is to guard the company against loss. The principal risks that are faced include:

1. *Criminal action:* This may include robbery, vandalism, fraud and industrial espionage.
2. *Direct damage to assets:* This may be caused by explosion, rainwater ingress and fire. The level of security is dependent upon good building details and the fire-suppression system used.

2.3.2 Criminal action

This section will consider the options available to deal with criminal action. There are several levels of security. The higher the level of security, the greater the cost. This expenditure will relate not only to the capital cost of the security system but also the restriction that is placed upon the efficiency of the employees. In planning a security system it should be recognized that no system can guarantee detection of, or immunity from, intruders. The purpose of a system is to confer on the property or persons at risk a level of security that, through careful planning, is in balance with the degree of determination and expertise of the intruder. In making this evaluation, all physical circumstances of the risk need to be assessed. However, the degree of security decided upon has to be consistent with a system which is, in every sense, practical and reliable and which cannot become an unacceptable burden to the general public or police.

Furthermore, the cost of the security system must bear a relationship to the risk, particularly since an increase in the cost of a system does not necessarily cause an increase in the level of security. Cost effectiveness should be of prime consideration.

In planning a security system its requirements should be ascertained as accurately as possible by consultation between appropriate interested parties, which may include one or more of the following:

1. The insurers
2. The local police authority
3. The Health and Safety Executive
4. The local public authority
5. The customer
6. Any statutory regulations
7. Consultants
8. Security companies

In assessing how to apply the requirements of a security system the following items should be considered both individually and collectively:

1. Physical barriers and deterrents
2. Electronic detection systems
3. Security patrol and guarding
4. Controlling entry and egress
5. Employee awareness, procedures and training
6. Local, remote and delayed alarms.

All parts of the security system should interact with each other to form an integrated whole.

2.3.3 Layers of protection

The security system can be considered as layers of protection. Physical barriers and deterrents constitute not only the building structure itself, made up of walls, doors, windows, floor and a roof, but the yard around the building and probably a perimeter fence or wall.

Electronic detection systems may range from simple intruder-detection devices monitored by basic control units to a variety of complex systems monitored by sophisticated computer-operated controls linked to 24-hour manned stations. Intruder-detection devices can be arranged into the following groups:

1. Static detection devices
2. Movement-detection devices
3. Trap and object devices

Static detection devices would include the following:

1. Tube and wire frames are used to protect windows, skylights and similar glass areas and constitute a high level of security but may be aesthetically unacceptable.
2. Aluminium foil is used to protect windows and glass doors. Although usually acceptable, this is subject to tampering if accessible to the public.
3. Break-glass detectors detect the high frequencies produced when glass is broken or cut with a glass cutter. These are the most cost-effective methods of glass protection in most cases and are especially suited to areas accessible to the public.
4. Vibration sensors are similar to break-glass detectors but are fixed to walls and doors against violent attack. The structure to which they are fitted must not be subject to vibration from external sources, otherwise false alarms may result.
5. Closed-circuit wiring comprises plastic-covered single-strand wire strung at regular intervals across doors, walls and ceilings. This is prone to failure in damp conditions and is mainly used in high-security buildings.
6. A pressure pad, also used as a form of trap protection, is a flat rectangular, plastic-covered contact placed at strategic points underneath a carpet.
7. Magnetic reed contacts are glass-encapsulated magnetically operated reed switches used on doors and windows. The contacts can be overcome by cutting the door, leaving the magnet undisturbed.

Movement-detection devices are designed to detect the movement of an intruder in the interior of the building:

1. Beams of invisible infrared light shone across an opening will detect movement that breaks the beam. Although relatively cheap, beam devices are easily recognized and avoided, and therefore constitute low security.
2. Microwave movement detectors utilize the principle of the Doppler effect on high-frequency low-power radio waves. These units are moderate in cost and suitable for large-volume coverage. Microwaves, however, penetrate certain materials easily, such as plasterboard, and careful siting is required to avoid false alarms.
3. Ultrasonic movement detectors utilize the principle of the Doppler effect on high-frequency sound waves. Ultrasonic movement detectors do not penetrate solid objects, but have smaller volume of coverage than microwave movement detectors. These units may also be affected by moving hot or cold air pockets in a room.
4. Passive movement detection senses radiated heat, such as that from a human body. These units are also sensitive to heat emitted by radiator, convection

heaters and direct sunlight, so that careful siting is required.
5. Dual-technology detectors combine two of the techniques described above, except for beams, and provide good rejection of unwanted alarms.

Trap- and object-protection devices provide protection to an area and object within a building and should never be used on their own:

1. A contact on a safe will recognize attack by drilling, sawing, filing, explosions and oxyacetylene or thermic lance.
2. CCTV cameras are very expensive and require a manned station, but with careful siting and the latest technology, they can see in moonlight and can set off alarms when picking up motion.

Security patrol and guarding is only satisfactory if properly managed and controlled with trained guards and complemented by electronic detection and monitoring equipment. The guards would generally be available to react to the unexpected, and should be well briefed as to responsibility and how to obtain help as well as how to deal with any type of unexpected situation.

Controlling entry and egress can be very useful for ensuring that only authorized people have entered the secured area as well as identifying who is actually in the building in an emergency situation such as a fire. There are several methods of controlling entry and egress, some of which are listed below:

1. *Guards:* These are probably the least efficient, primarily because of the human element. However, there are obvious psychological strengths to human presence.
2. *Mechanical key system:* This provides a basic level of security for heavily used areas. Although complicated mechanisms can be designed, their operation is slow, and hence are suitable only for limited access.
3. *Magnetic card systems:* These are very good for busy areas and can be made very difficult to forge. The level of security can be increased, at the expense of efficiency in busy areas, by incorporating the necessity for a PIN number.
4. *Automatic systems:* These require no direct contact between the user and the device reading the code. They provide a high degree of convenience in use.

For high-security buildings it is imperative that employees are made aware not only of the need for security but also what procedures have to be followed and instructed on basic security awareness.

The type of alarm system used is dependent on the expected type of security breach and the method employed in responding to one. In unguarded premises, requiring only a low level of security, an alarm which operates immediately a device detects a security breach may be sufficient to ward off vandals, burglars and crimes of opportunity. On the other hand, where breaches of security may involve more determined criminals, such as fraud or industrial espionage, delayed alarms on the premises may give time for security personnel and/or police to apprehend the criminal in the process of committing the crime.

2.3.4 Reliability

No alarm system, however well planned and installed, can be completely reliable or tamperproof. The successful operation of a security system requires the active cooperation of those involved in carrying out the necessary procedures carefully and thoroughly. The usefulness of the whole system can be jeopardized by lack of care or inadequate attention to routine procedures in maintenance and servicing. This care has to be extended to the security of keys and of information regarding the system, its installation and method of operation.

2.4 Leases

2.4.1 Relationships

The relationship of landlord and tenant is created by a lease, whereby the landlord grants a portion of his interest in the land and allows the tenant to occupy and enjoy that land for a period or term usually in return for a monetary payment called rent. The relationship is governed by the law of landlord and tenant, which is covered by common law dating from feudal times and beyond, and, more importantly, statute law. The statute law is updated and revised by successive governments and legal advice should be sought when drafting or entering into leases.

2.4.2 The preparation of a lease

A lease or contract of tenancy can be verbal but is usually in writing, as there is the possibility of misunderstandings and thus disputes. In some cases this is overcome by a deed which sets out the conditions of the letting. This is a contractual document which is made by indenture, i.e. in two parts. One part, called 'the lease', is signed and sealed and delivered by the landlord to the tenant while the other, called 'the counterpart', is signed, sealed and delivered by the tenant and handed to the landlord. The two are identical and, being legal documents, use formal words such as 'lessor', 'lessee', etc. and generally commence with the words 'This lease' or 'This indenture'.

The lease by deed sets out the conditions of the letting in a formal manner and should include:

1. Generally, the description of the property;
2. The names of the parties;
3. When it is to commence and how long it will last;
4. The covenants which consist of promises and agreements by the landlord and tenant to do or not to do certain things;
5. The rent that is to be paid;
6. Any exceptions and reservations which the landlord wishes to retain;
7. The conditions showing how the term may be ended, extended or created (which may include implied conditions);
8. The provisos, which are any express conditions.

The deed is dated, signed, sealed, witnessed and delivered and operates by the devise of the landlord's interest in the property to the tenant.

The understanding of these constituent parts of the lease is very important, and they should be read very carefully. Expert advice should be sought for a thorough understanding of them.

2.4.3 Covenants

Covenants may be express or implied, i.e. those expressly written into the lease in a deed and those not expressed but implied in a deed by law in order to give effect to the intention of the parties or to remedy an obvious omission. Express covenants could be:

1. To pay rent;
2. To pay rates and taxes;
3. To repair and keep in repair;
4. To paint/decorate within certain times;
5. To insure and produce receipt of insurance;
6. To permit the lessor and his agents to enter and view;
7. Not to alter the structure;
8. Not to assign or sub-let;
9. To yield up at the end of the term.

Implied covenants could be implied into a lease by common law or by statute, and in the case of the latter cannot easily be varied or excluded by agreement between the parties. Some examples of covenants by the landlord are:

1. That the landlord has good title and therefore to give possession of the date fixed for commencement of the term;
2. To give the tenant quiet enjoyment during the tenancy;
3. In the case of residential tenancies to give at least 4 weeks' notice to quit;
4. In the case of furnished premises to ensure at the commencement of the tenancy that the premises are fit for human habitation.

2.4.4 Rent

Land is granted by a landlord to a tenant for a variety of purposes, and several names are given to the 'rent' due from the tenant to the landlord. Premises for occupation are generally let at a 'rack rent' which is assessed upon the full annual value of the property, including land and buildings, the 'best rent' being the highest rack rent that can reasonably be obtained for the whole lease.

Sometimes land and buildings are let on a long lease to a tenant who proposes to sub-let on short terms such as weekly or monthly tenancies. In this case, the rent paid on lease is less than the rack rent, and is called a head rent.

When land is let for construction it is usual for a 'peppercorn rent' of no actual value to be agreed while the houses, etc. are built. Although cases do occur where a peppercorn rent is received throughout a long lease it is more usual for a 'ground rent' to be charged. This is less than the rack rent, usually because a premium has been paid at the start of the lease. Ground rents are much higher than peppercorn rents and may include increases during the lease.

Rent is due on the days appointed for its payment in the lease, and unless the lease expressly provides for rent to be paid in advance it will be payable in arrears.

2.4.5 Reviews

Many leases, especially business leases, contain rent-review clauses which provide for regular reviews at, for example, three-, five- or (more usually) seven-yearly intervals. A clause will define rent-review periods and set out the formula and machinery for assessing the rent in each period.

2.4.6 Assigning and sub-letting

Any tenant who has not contracted to the contrary may sub-let his land (or part of his land) for any term less than that which he holds. A sub-letting is therefore distinguished from an assignment in that a tenant who assigns, disposes of the whole of his interest under the lease (and thereby puts the assignee into his place), whereas a tenant who sub-lets does not.

2.5 Obtaining approval to build

When contemplating the construction of a new industrial building or alterations to an existing one the statutory requirements and the powers available to the authority responsible for issuing approval should be taken into account. The exact procedures will, necessarily, be dependent upon the location (i.e. district and/or country) and the type of industrial building proposed. The approval process in the UK can, however, be divided into two stages. These are to obtain:

Planning approval
Building control approval.

We shall consider each stage in the context of general requirements. The standard and detail needed to obtain approval is contained within statutory documents that can be obtained from the relevant authority.

2.5.1 Planning approval

Obtaining planning approval provides permission to build and, more importantly, justification to pursue the preparation of detailed plans and documentation necessary for building control approval. Planning permission prior to detailed design of the building is not a requirement, but rejection of planning submission would probably render the design useless.

The authority granting permission to build will depend upon the type and size of building. Major projects, such as nuclear power stations or large industrial complexes, may involve a public enquiry and the need to produce a study into the environmental impact. Generally speaking, though, the planning submission will be considered by a committee nominated by the local authority.

The following is a list of items that could (and would usually) form the basis for a planning submission:

1. Site location
2. Building footprint
3. Height
4. Use of building
5. Aesthetic appearance
6. Utility requirements (e.g. gas, electricity, etc.)
7. Employment requirements.

The list is not exhaustive, but it will be apparent that the submissions content implies that any decision will be subjective. A knowledge of the current planning policy of the local or national government is therefore advisable.

2.5.2 Building control approval

Having obtained planning permission, building control approval involves producing a set of detailed documents to satisfy a list of statutory requirements. These lists, although amended and updated according to current practice, if complied with, should gain approval. The decision to grant approval is therefore much less subjective than that needed for planning permission, being more closely defined by statute.

The need for building control arose primarily to protect the health and safety of the population and, latterly, to conserve fuel and power and prevent waste. The Building Regulations (as they are known in the UK) give standards of performance necessary in the building and, as such, make reference to national and international Standards and Codes of Practice. The regulations cover two areas of health and safety and can be subdivided as follows:

Safety	*Health*
Fire	Toxic substances
Structure	Ventilation
Stairs, ramps and guards	Hygiene
	Waste disposal and drainage

The specific technical requirements within each category will depend upon the type of building and the use to which it will be put. For example, the requirements for fire escape will be different for an office from those for a factory.

Finally, the regulations are not exhaustive and there will be circumstances, especially relating to fire safety, where approval can only be obtained by negotiation.

2.6 Extending existing buildings

2.6.1 Viability

The extension of existing facilities can often present an attractive alternative to relocation, with all its attendant disruption and potential impacts on production, industrial relations and morale. However, the viability of extension is predicated on the consideration of various strategic and tactical issues:

1. Land availability and location in relation to existing buildings;
2. Nature of activities proposed in extension (e.g. linear, cellular, storage, bulk process, administrative, etc.);
3. Topographical and geotechnical conditions;
4. Existing structures (their condition, type of construction, etc.);
5. Infrastructure and services such as roads, telecommunications, water, foul and storm drainage, gas, power, etc., location, routing, capacity (used and spare), state of repair, etc.;
6. Local and statutory authority issues (e.g. planning, building control, service tariffs);
7. Environmental concerns (effluent treatment, noise, etc.).

2.6.2 Facility brief

Once the primary considerations of size, height, conceptual layout, structural loads, servicing (mechanical, electrical, communications, public health, statutory services) requirements, access, material and personnel traffic, etc. have been addressed, a facility brief can be produced to allow collation of the basic project planning information:

1. Scheme, site plan, layouts and elevations including infrastructure modifications;
2. Basic plant services;
3. Budget costs;
4. Tentative planning assessment;
5. Preliminary project programme.

These items should be formulated against optional approaches, each option being tested against one another (and rated accordingly) on the basis of:

1. Capital expenditure, cash flow and operating cost analyses;
2. Impacts on existing operations (productivity, quality, personnel, etc.);
3. Project-completion times (including the options for phased completion);
4. Comparison of operational, spatial, servicing simplicity, internal environmental performance (daylighting, thermal transmittance, acoustic break-in/out, etc.), parameters between options.

The adoption of the best option should be based on all these parameters.

2.6.3 Other considerations

Many extraneous and site-specific considerations may have to be taken into account. Certainly, the construction and layout of existing facilities will play a major role in deciding the basic choice of location and configuration of extensions.

Particular attention should be paid to existing site arrangements, bay widths, material handling, foundation design and construction methods. For instance, it may prove impractical to extend the bays of a traditional steel-framed lightweight structure if the impact on road rerouting, statutory services extensions, substation requirements or other existing facilities seriously affect the overall cost or programme time.

Equally, the most logical (and least externally disruptive) extension method may conflict with desired material handling or work flow requirements. Generally, economic bay widths should be adopted (typically *c.* 15 m), if possible. Roof design should be adapted to accommodate roof-supported plant and services while minimizing steel tonnage. The integration of daylighting, service routes, cranage, etc. should all be early design issues, solved during the evolution of the basic structural design.

Attempting to modify structure or services to accommodate structural idiosyncrasies later is always problematic. Protection and isolation of existing facilities from the disruption of construction should be resolved early in the design. The reconciliation of building and production activities is never easy, but early planning for construction traffic, personnel, site screening, security, site access and site communications can minimize these adverse effects.

Where existing buildings are of an age which would indicate that their design and construction were performed in accordance with statutory requirements, planning, building control, standards or codes of practice subsequently superseded, care must be taken to ensure that extensions to such facilities are designed and constructed in accordance with *current* requirements.

2.6.4 Re-use of materials

Re-use of cladding or other construction materials, smoke ventilation, drainage, fire detection/suppression systems or techniques to reflect the existing installations may not be in compliance with current requirements, particularly with respect to:

1. Fire protection of structural steel
2. Means of escape
3. Proscribed materials
4. Thermal performance
5. Fire detection and suppression
6. Fire compartmentation
7. Structural design.

The requirements of the property insurers must also be adhered to if potentially adverse impact to premiums is to be avoided.

2.7 Fire detection and suppression

2.7.1 Fire regulations

As a general statement, it is unwise to assume particular requirements for specific cases in relation to either statutory requirements or preferential desires of bodies having jurisdiction in fire-related matters. The fire regulations pertain to building use and location. Hence requirements for urban offices will vary considerably from those for rural industrial sites.

The Fire Precautions Act coordinates the statutory requirements. Under its provisions any building may be designated a special case. The auspices of the Act cover the fire-prevention services, the Building Regulations and Factory Inspectorate insofar as these areas impinge on the scope of the Act.

It is generally true that the statutory instruments and authorities are concerned with the preservation of life as their primary objective. Consequently, the requirements of insurance companies with regard to the preservation of the building or its material contents may be more involved. Adherence to such requirements (or otherwise) will be reflected in the insurance premiums quoted.

2.7.2 The influence of design

The design of the structure (and its operation) will greatly influence the effects of a fire. For instance:

1. Heat sources should be kept remote from flammable materials;
2. Compartmentation of the structure (and maintenance of such compartmentation) may contain or minimize fire spread;
3. Fire protection of the structure (discussed elsewhere);
4. Installation of detection, alarm and sprinkler (or other suppression) systems;
5. Separation of buildings one from another.

2.7.3 Types of system

There are many types of detection and suppression systems. The one selected should be compatible with the likely fire source and be consistent with the likely locations and fire size within the building.

It is essential to discuss the requirements for structural protection, compartmentation, emergency lighting, detection, alarms, call points, suppression, means of escape and signage with the applicable local authority, fire brigade or insurance company personnel before finalizing designs.

2.7.4 British Standards

The following British Standards are applicable to the design of fire-related systems:

BS 1635: 1970: Graphical symbols and abbreviations for fire protection drawings

BS 5306, Parts 0 to 5, inclusive: Fire extinguishing installations and equipment on premises

Loss Prevention Council and Fire Protection Association Publications

BS 5839: Fire detection and alarm systems in buildings

BS 5266: Emergency lighting.

Design of fire-suppression systems in the UK is generally required to be consistent with the Rules of the Fire Officers Committee (FOC). However, with increasing investment by companies of foreign origin, compliance with external design guides such as the National Fire Protection Association (NFPA) of the USA is increasingly required for specific projects.

Such documents require the definition of potential fire risks in buildings under specific 'hazard' classifications. Against these classifications, types of suppression systems, flow rates, water storage, distribution systems and other specific design data are provided. Typically, the design of suppression systems will require the prior approval of insurance companies (and possibly other bodies having jurisdiction) before the commencement of work.

2.7.5 Fire protection of structures

The principal construction materials used in industrial buildings are

1. Steel
2. Concrete

Two different types of fire protection systems can be considered

1. *Active:* This is a fire-*suppression* system (e.g. sprinklers) and is described elsewhere.
2. *Passive:* In this system the aim is to provide protection to the structural material for a specified period of time, and will be considered in this section.

Fire protection of concrete

Although not used extensively in superstructure work within industrial buildings, the fire-resistance characteristics of concrete are excellent. Failure of a concrete member is caused by a loss of strength in the steel reinforcement associated with its rise in temperature. The aim in providing fire resistance is, therefore, to reduce the temperature rise of the steel. This may be achieved by:

1. Judicious selection of the shape, size and distribution of reinforcement within the element;
2. Providing adequate cover to the reinforcement;
3. Adopting a lightweight or limestone aggregate which is less susceptible to spalling than a siliceous aggregate such as flint.

Fire protection of steel

Apart from concrete or masonry encasement of steelwork, the following options are available:

1. Sprays (up to 4 h)
2. Boards (up to 4 h)
3. Intumescents (up to 2 h: this can be the most expensive, depending upon the application)
4. Preformed casings (up to 4 h)

The spray is based upon either a natural plate-like material, such as vermiculite bound together with cement, or mineral fibres. Application is fast but not precise or clean, and is generally only suitable for areas where the steel will be hidden (by a false ceiling, for example). Sprays for external applications are available. However, the steel must first be provided with a compatible corrosion protection system.

The boards are based upon the same constituent materials used in sprays. They are suitable for situations in which only 'dry trades' are allowed. The boards are cut to suit on site and mechanically fixed to the steel (e.g. by screws and straps). The system produces a smooth surface which is suitable for decoration.

Intumescents are thin films or mastics which swell under heat to many times their original thickness. Their major use is in circumstances where the architectural statement of steel is to be preserved. As such, their costs vary considerably. Their application is fast and can be either by spray or brush. They are generally used internally, but external intumescents are available.

Preformed casings are very similar to fire boards but are tailored to the particular needs of the member being protected and, as such, permit fast application. They are, however, expensive.

2.8 Cost comparisons and contract procedure

Building costs per square metre for the construction of new buildings can readily be obtained from reference books such as *Spon's Architects' and Builders' Price Book* and *Laxton's Building Price Book*. Both are updated and issued on an annual basis, with the costs quoted based on the previous year's statistics.

The costs given are for a range of prices for average building work obtained from past records, and can vary by as much as 50% between minimum and maximum costs. They serve therefore as an initial rough guide to the probable cost of a new building. *Spon's* also gives details of fitting-out prices within similar ranges, but these are generally limited to office fit-outs. The introduction to the sections dealing with prices per square metre in both books should be carefully studied before applying the figures to new structures or costing alterations/extensions to existing buildings on a pro rata basis.

As stated elsewhere in this chapter, it is of vital importance to assess the cost effectiveness of any proposed design (whether it be for a new building and/or alterations/extensions to an existing one) at an early stage. This can best be accomplished by seeking professional advice at the earliest opportunity. Such advice is normally provided by a quantity surveyor, who can work alongside the building owner and his engineers in assessing the cost of alternative designs/use of various materials/construction methods/construction times, etc. and evaluate the various proposals as the design develops. This will ensure that the most cost-effective solution is arrived at in the minimum time.

The correct selection of the form of contract to be used when inviting tenders is of paramount importance, as it constitutes the signed agreement between the employer and the contractor and forms part of the conditions of contract. The JCT (Joint Contract Tribunal) issue several different forms of building contracts while the Joint IMechE/IEE Committee on Model Forms of General Conditions of Contract (Institution of Mechanical Engineers/Institution of Electrical Engineers) have recently issued Model Form MF/1, replacing both the original Model Forms A and B3. Broadly speaking, the JCT forms are for General Building Works (from large contracts to minor contract) while Form MF/1 should only be used for works which are substantially of a mechanical or electrical nature.

2.9 Structural and services supports

When considering the introduction of supplementary structures for *any* purpose there are a number of general items of design data which must be examined and formulated:

1. *Support function:* What are the functional usage loads likely to be applied (Figure 2.10)?
2. *Existing conditions:* What are the existing spatial, civil, structural and access conditions within the facility? Typically, the following data should either be accessed from existing records and design requirements defined by the support function assessment or acquired by new site survey/investigations:
   (a) Ground conditions (soil-bearing capacity, water table, etc.) for ground-based support systems;
   (b) Existing construction elements; state of repair, materials, composition, bearing capacity, etc., allowable attachment methods, existing loading conditions;
   (c) Current access and people/materials flow in the area. Assessment of disruption, modifications to existing operations required both *during* installation process and *after* completion of installation.
3. *Alternative optional approaches:* Based on the constraints of space, cost, time, disruption, etc., what are the viable options? The following questions are pertinent when considering installation options:
   (a) Effects on facility downtime;
   (b) Use of proprietary, modularized systems;
   (c) Phased installation capabilities;
   (d) Cost comparisons;
   (e) Capabilities for later expansion;
   (f) Consolidation of support functions (e.g. multiple service gang hangers, access/personnel safety features (Figure 2.12)).
4. Where possible the 'best option' solution should provide the optimum mix of:
   (a) Cost effectiveness;
   (b) Personnel safety;
   (c) Minimum impact (downtime, disruption, structural alterations, ease of installation, etc.);
   (d) Expansion capability;
   (e) Multiple use;
   (f) Fitness for purpose;
   (g) Off-the-shelf materials, devices, etc.

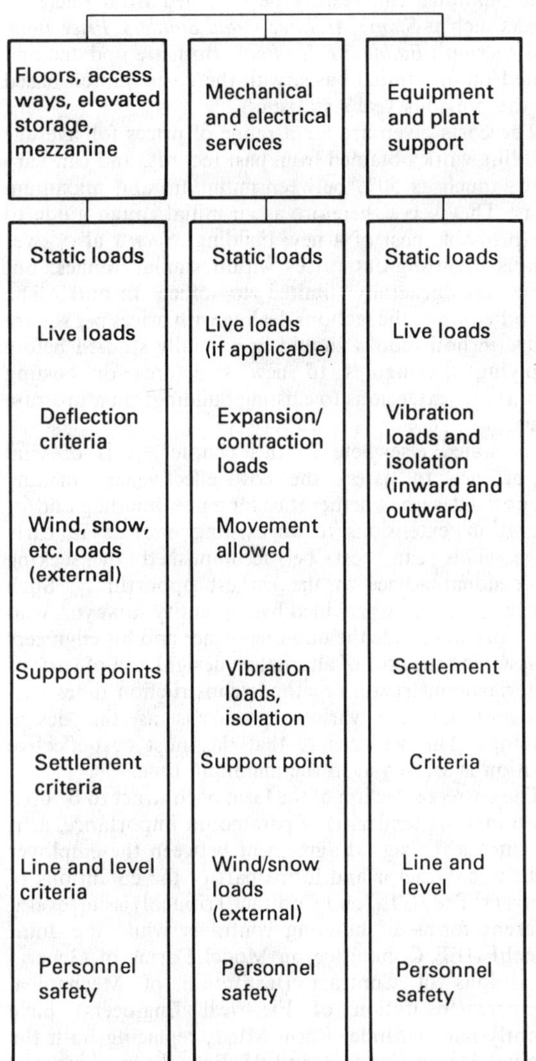

Figure 2.10

While the statements made above are necessarily general, it is essential that the conceptual approach delineated be adhered to. The necessity for compiling adequate design data for the new installation, followed by examination of existing conditions against the design requirements, is paramount.

In many instances, adequate existing civil/structural or services installation information may not be available. Guessing the likely bearing capacity of floor slabs, foundations, structural steel or sub-slab ground, etc. can prove disastrous. Equally, assuming likely operational conditions for new installations can lead to embarrassing (and quite possibly dangerous) underperformance. The cheapest or quickest installation is not always the best or most cost effective when effects on current operations or potential future expansion plans are examined.

Care must also be taken where large live loadings are concerned such as gantry cranes, hoists, materials handling, etc. to ensure that structural and safety implications are properly addressed during both design and construction stages.

Where multiple service support systems are concerned, apart from operational movement, sufficient separation must also be provided to preclude electrical interference, water damage, inadequate clearance for insulation, cladding, cable de-rating (due to inadequate ventilation), mechanical damage during maintenance, etc. (Figures 2.12 and 2.13). It is essential that competent professional engineering personnel with applicable design experience be utilized for design of the installations.

Figure 2.11 A typical modular mezzanine floor (courtesy of The Welconstruct Co. Ltd)

Figure 2.12 Multiple piping system anchor of guide installation (courtesy of Industrial Hangers Ltd)

2.10 Natural ventilation

2.10.1 Usage

Natural ventilation can be described as a process for providing fresh air movement within an enclosure by virtue of air pressure differentials caused primarily by the effects of wind and temperature variations in and around the enclosure. The primary usages of natural ventilation are to provide:

1. Fresh air introduction for odour and carbon dioxide dissipation;
2. Fire control/smoke clearance;
3. Heat dissipation/cooling;
4. Provision of air for fuel-burning devices;
5. Control of internal humidity (primarily condensation reduction/exclusion).

2.10.2 Requirements

For engineering purposes the composition of fresh air is generally taken at the following 'standard' conditions:

20.94% Oxygen (O_2)
0.03% Carbon dioxide (CO_2)
79.03% Nitrogen and inert gases (N_2 + xenon, zeon and others)

These fractions are, of course, somewhat variable, dependent on geographic location (i.e. urban versus rural).

Figure 2.13 A typical multi-service gang hanger (courtesy of Industrial Hangers Ltd)

Table 2.1 Recommended outdoor air supply rates for air-conditioned spaces (extracted from BS 5925: 1980)

Type of space	*Smoking*	*Outdoor air supply (l/s)*		
		Recommended	*Minimum (the greater of the two should be taken)*	
		Per person	*Per person*	*Per square metre floor area*
Factories[a,b]	None	8	5	0.8
Open-plan offices	Some	8	5	1.3
Private offices	Heavy	12	8	1.3
Boardroom, executive offices and conference room	Very heavy	25	18	6
Toilets[c]	—	—	—	—
Corridors[c]	—	—	—	10
Cafeterias[a,d]	Some	12	8	—
Kitchens[a,c]	—	—	—	20

[a] Rate of extraction may be overriding factor.
[b] See statutory requirements and local byelaws.
[c] A per capita basis is not appropriate in these cases.
[d] Where queuing occurs in the space, the seating capacity may not be the appropriate total occupancy.

Generally, average humidity for Britain is as follows:

75–95% sat Winter
55–75% sat Summer

General guides to fresh air requirements for odour control and general ventilation are given in Table 2.1 and Figure 2.14. Note that requirements for removal of lavatory odour or for other highly concentrated areas are dealt with elsewhere. Figure 2.14 pertains to generally ventilated spaces only.

2.10.3 Threshold limit values

With regard to limiting the concentration of particular contaminants, these are generally referred to in terms of

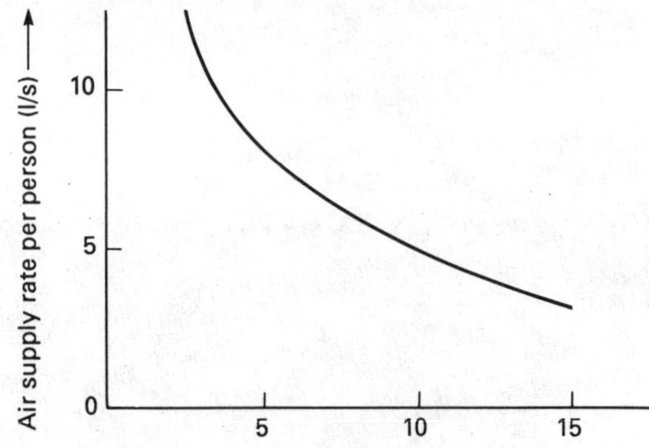

Figure 2.14 Approximate air supply rate for (human) odour removal (extracted from BS 5925: 1980)

'threshold limit values' (TLV), expressed either in terms of a time-weighted average (TLV – TWA) which represents the average concentration for a normal working day over a 40-hour week to which nearly all workers may be exposed on a repeated basis or as a short-term exposure limit (TLV – STEL), which represents the maximum concentration to which workers can be exposed for up to 15 minutes. Clearly, these value vary, depending on the particular contaminant.

Reference should be made to *Threshold Limit Values, Guidance Note EH17/78*, issued by the Health and Safety Executive (HSE), or *Industrial Ventilation* (American Conference of Governmental Industrial Hygienists). In all cases, proposals should be reviewed by and submitted to the relevant local authority agencies.

Space does not permit a detailed investigation of all the requirements concerning ventilation for the remaining usages listed. However, a brief description of the rationale for provision is given below, and further details can be obtained from the references listed at the end of this chapter.

2.10.4 Fire control/smoke clearance

Depending on the requirements of local fire officers,

statutory regulations and insurance bodies (or any other documents or bodies having jurisdiction), there may well be a need to address the clearance of smoke from escape routes, the control of smoke spread generally and the removal of smoke during and after firefighting activities.

2.10.5 Heat dissipation/cooling

Human occupants, electrical/electronic equipment and process plant all emit varying quantities of sensible and latent heat. Equally, these various elements require (or can tolerate) differing environmental conditions. Depending on these operational constraints, the need may well exist to provide natural (or powered) ventilation to maintain environmental conditions (temperature and/or humidity) consistent with the occupational/process requirements.

2.10.6 Provision of air for fuel-burning devices

All combustion equipment (oil, gas, solid fuel) requires primary air to support combustion and secondary air to permit adequate velocities in flue ways, etc. These requirements are governed by the minimum air/fuel ratio and operating flue-way parameters. There are also published recommended minimum requirements (British Standards, British Gas, etc.) which are generally in excess of these.

2.10.7 Control of internal humidity

Where human occupancy or 'wet' process plant is present, the emission of water vapour will occur. Depending on external conditions and building fabric construction, the attendant potential for excessive ambient humidity or surface condensation may exist.

Consequently the introduction of external air ventilation may be required in order to reduce ambient moisture contents. However, this will need to be balanced against energy costs and the use of other design solutions such as building fabric moistureproof membranes, local air exhausts, etc.

2.10.8 General design considerations

The use of openings within the building fabric to provide natural ventilation is predicated on various primary factors:

1. Location and orientation of building;
2. Building height and configuration;
3. Shape, size and location of openings;
4. Wind and temperature effects.

Identifying specific quantified building ventilation rates can be difficult due to the number of non-linear parallel paths that may have to be simultaneously considered.

However, simple-case type solutions (which are often good enough to establish whether mechanical ventilation is required or excessive air infiltration problems may exist) can be identified. These simple solutions are based on a knowledge of:

1. Wind direction and velocity;
2. Building opening locations and size;
3. Temperature differentials (internal and external).

These solutions are discussed in more detail in many publications, including Building Research Establishment (BRE) *Digest No. 210*, the American Society of Heating, Refrigerating and Air Conditioning Engineers (ASHRAE) *Fundamentals Handbook* and Chartered Institute of Building Services *Engineers Guide* (Books A and B).

2.11 Building durability

The durability of a building (i.e. its life expectancy and its resistance to deterioration) is determined by deliberate design decisions relating to structure and choice of materials as much as to the natural or precipitate process of ageing. Within certain limits, the design of an industrial building can and should take into account the predicted use or lifespan of the process or method of operation which it is to accommodate. To aim at durability beyond that has ascertainable cost implications. These may be acceptable if the building is to serve future known or even unpredictable purposes.

2.11.1 Durability and plan form

If the plan of the building (and the same applies to its cross-sectional features) is precisely and inflexibly related to the initial plant layout and the production processes known at design stage, non-adaptability to later desirable changes in use is clearly a durability factor.

2.11.2 Structure and durability

The structural design of an industrial building will, first, reflect the requirements of plant layout and manufacturing procedures. Apart from holding up the building envelope, structural members will be designed and placed to support particular static and dynamic loads. As with the plan form, the shape and disposition of the structural features may either curtail or allow extension of the life of the building. Load-bearing perimeter walls exemplify the former and maximum freedom from internal columns the latter. The cost implications of reducing structural constraints and thereby increasing durability need to be considered.

2.11.3 Cost of construction materials

There is a clear correlation between quality or cost of materials and the durability/life expectancy of buildings. Greater resistance of better materials to wear and tear can be assumed, with obvious implications on future maintenance. Striking the best balance in each case between initial capital outlay and maintenance cost calls for complex calculations which take into account such intangibles as future interest rates and taxation of building operations.

There are many studies of lifecycle costs of buildings. One was published by the College of Estate Management of Reading University in 1985.

2.11.4 Durability of materials

Apart from selecting better, more durable and expensive (rather than cheaper) materials, their life can be prolonged by appropriate treatment, coating or other protection, e.g.:

Mild steel: Galvanizing/plastic coating
Extruded aluminium: Anodizing/polyester powder coating
Concrete: Integral or applied hardeners
Brickwork: Joint formation/masonry paint
Cement screeds: Mineral or metallic aggregate hardeners
Timber: Impregnation/staining.

Obviously, durability is greatly enhanced by specifying stainless rather than mild steel, engineering rather than common bricks, etc. Comparative life expectancies of materials are tabulated in reference books such as that published by NBA Construction Consultants Limited in 1985.

2.11.5 Damage and protection

Industrial buildings, more than most, are prone to damage. Causes range from mechanical impact to chemical attack resulting from the production process – quite apart from heavy wear in normal use. Examples of causes and available protective measures are as follows.

Mechanical impact

Vehicles (external): Buffer rails, ramps
Transporters (internal): Bollards, route demarcation, column base plinths, guide rails, flexible/automatic doors
Production processes: Wall/floor shields pipe cages, resilient rails
Craneage: Electromechanical limiters
People: Wall rails, door kick plates, footmats (shopfloor offices)

Chemical/atmospheric attack

Plant/process emissions: Anti-corrosion coatings heat shields, extractors
Humidity (condensation): Dehumidifiers, insulation, ventilation, warmed surfaces
Atmosphere (external): Corrosion-resistant coatings, fungicidal treatment, overcladding
Fire: Smoke detectors, intumescent coatings/seals

2.11.6 Details and specification of construction

Design for durability, as defined above, will be negated if it does not take into account the essential need for ease of maintenance or ignores sound specification of workmanship and constructional details. Deterioration factors notorious for shortening the life of a building include the ingress of moisture and structural or thermal movement. Necessary safeguards against these and others are covered by Building Regulations and Codes of Practice, all fully referred to in the National Building Specification.

2.12 Building maintenance

Maintenance is estimated to represent more than one third of all building expenditure, which underlines its significance in the economic lifecycle of any building, the vital importance of designing for it and skilfully managing its execution. There is a wealth of published literature on every aspect of maintenance, and more recent studies are listed at the end of the chapter.

2.12.1 Building records

The prerequisite for proper planning and performance of building maintenance is the comprehensive set of drawings, specifications and servicing manuals compiled by architects, engineers, contractors and suppliers, respectively. Those responsible for maintenance must ensure that all records are kept up to date; computers are increasingly used for this purpose.

2.12.2 Maintenance work

Depending on the scope and complexity of maintenance as well as economic considerations, inspections and performance will be (both or either) the responsibility of in-house permanently employed staff or outside specialists or contractors. In the latter case particular expertise is required, and this should be reflected in the selection/tendering procedure.

2.12.3 Maintenance planning

Whoever carries out the work, basic strategy, programming and procedures will be established by the plant engineer and his maintenance manager. Inspection checklists and record sheets (of which those designed by the Department of the Environment/Property Services Agency are very good examples) are necessary for systematic location of critical elements/potential defects and setting out frequencies of inspection. Computers and work processors are an obvious aid to effective maintenance management. Inspection-based preventive methods are demonstrably cost effective.

2.12.4 Monitoring equipment

Most, if not all, of the equipment listed below is required, even where outside specialists are generally retained to carry out maintenance (for either access or measurement), including:

1. Tubular towers, ladders, long-arms, cleaning cradles;
2. Levels, tapes, thermometers, hydrometers, light meters, smoke detectors, audio-sensors.

Features such as cradle rails, safety eyes and climbing irons should be incorporated into the design of the building.

2.13 Building repairs

The need for building repair arises from the inevitable process of ageing, and such deterioration is brought to light by systematic maintenance. There is also repair work necessitated by damage from observable causes such as physical impact (discussed elsewhere in this chapter).

2.13.1 Repair work

Depending on its complexity and urgency, repair work may best be carried out by in-house staff employed and trained for this purpose (but not necessarily to the exclusion of other work) or by directly employed labour skilled in particular trades or by specialist contractors. In many cases immediate (but adequate) steps can be taken by permanent staff, sufficient to deal with an acute problem while deferring more comprehensive or permanent repairs to be carried out by outside specialists (possibly as part of a regular maintenance programme with resultant economies). Useful reference books include *Building Failures: Diagnosis and Avoidance*, by W. H. Ransom, Spon, London (1987) and *Defects in Building*, published by DOE/PSA (1989).

2.13.2 Repair: works specification

Leaving aside stop-gap emergency repairs, the work should be done to a standard set by the original material and workmanship specification (which must be referred to by whoever carries out the work), unless this has been found to be inadequate. Any deviation from original materials or methods must be assessed by qualified professionals with regard to compatibility, cost and compliance with building regulations. The specification must also take into account limitations of the structure as originally designed.

If faulty specification or unsuitable materials are found to be the cause of defects requiring repairs, replacement work should be considered in anticipation of problems arising on a wider scale. A less disruptive repair programme may thereby be adopted compatible with routine maintenance operations.

2.13.3 Advisory bodies

The assessment of underlying causes of defects and/or the optimum methods of rectifying them are sometimes not within the competence of the original designers of the building or specialist contractors. In such cases reference is best made to organizations which have detailed or broader expertise, such as the Building Research Establishment at Garston.

Examples are given below of two kinds of body which can be approached for independent advice. More can be found in reference books published by the RIBA and other professional institutions.

Testing laboratories

British Ceramic Research Limited, Stoke-on-Trent: Tel (0782) 45431
BSI Test House, Hemel Hempstead: Tel (0442) 3111
Chatfield Applied Research Laboratories, Croydon: Tel (081) 688 5689
Laing Technology Group Limited: Tel (081) 959 3636
Paint Research Association, Teddington: Tel (081) 977 4427
Harry Stanger Limited, Elstree: Tel (081) 207 3193

Advisory associations

Aluminium Coatings Association, Birmingham: Tel (021) 4550311
Asbestos Removal Contractors Association, Chelmsford: Tel (0245) 259744
British Flat Roofing Council, Haywards Heath: Tel (0444) 41668
Brick Development Association, Winkfield Row: Tel (0344) 885651
British Cement Association, Fulmer: Tel (02816) 2727
Construction Fixings Association, Sheffield: Tel (0742) 663084
External Wall Insulation Association, Haslemere: Tel (0428) 54011
Fire Protection Association: Tel (071) 606 3757
Galvanizers Association: Tel (071) 499 6636
Institute of Concrete Technology, Beaconsfield: Tel (04946) 4572
Metal Roof Deck Association, Haywards Heath: Tel (0444) 440027
National Association of Lift Makers: Tel (071) 437 0678
Partitioning Industry Association, Solihull: Tel (021) 7059270
Patent Glazing Contractors' Association, Epsom: Tel (03727) 29191
Steel Construction Institute, Ascot: Tel (0990) 22944
Steel Window Association: Tel (071) 637 3572
Suspended Ceilings Association, Hemel Hempstead: Tel (0442) 40313

2.13.4 Flat roof defects

Unlike other defective building elements, roofs generally require prompt repairs. Latent defects in roofs can go undetected with correspondingly more serious consequences when they do manifest themselves. Regular inspection of roofs is therefore doubly necessary, as well as a recognition of potential causes of damage, e.g.:

1. Inadequate falls and unimpeded water discharge;
2. Ultraviolet ray attack of unprotected roofing;
3. Cracking caused by alternating expansion and shrinkage;
4. Distortion of roofing due to heat rejection (insulation);
5. Traffic and impact damage of unprotected roofing;
6. Perimeter constraint/upstand and flashing details.

2.13.5 Roof repairs

Diagnosis of defects by inspection of visible causes or more sophisticated means (tracer-dye inundation to find sources and routes of leaks, infrared photography, etc.) is a prerequisite for correct repair specification. Greater skill in detailing and specifying suitable materials and methods are required in the design of flat roofs and the same applies to repairs. Here, too, durability and effectiveness are, to some extent, cost related.

Specialists should be employed to carry out roof repair work who are able and willing to guarantee its quality and life.

The Technical Guide to Flat Roofing, published by the Department of the Environment/PSA in 1987, is one of the many useful reference books.

2.14 Domestic facilities

2.14.1 Toilet accommodation

Toilet accommodation encompasses the provision of water closets, urinals, wash basins and/or washing troughs and, depending on a particular trade or occupation, the provision of baths and/or showers. Provision of toilet accommodation is specified under Acts of Parliament, British Standards, regulations and HM Inspector of Factories.

The provisions given in the various standards are essentially minimum requirements only, and when assessing quantities and type, the following should be considered:

1. Number of personnel and male/female ratio;
2. Is a shift system operating within the building?
3. Type of process in operation – dirty or clean, wet or dry;
4. Specialized usage – sterile conditions, radioactive contamination;
5. Provision for remote-actuated appliances (i.e. foot operated, photoelectric cell, etc.);
6. Ethnic considerations relating to siting of toilet accommodation;
7. Provision for handicapped personnel;
8. Provision of facilities for female hygiene;
9. Changing-room facilities for male/female personnel.

2.14.2 Canteens

Canteen facilities for staff personnel can generally be divided into two categories:

1. Provision of cold meals (sandwiches, cold buffets, etc.);
2. Provision of both hot and cold meals prepared by resident staff or delivered under contract by outside caterers.

Many industrial buildings offer canteen facilities free of charge or at subsidized rates. However, before embarking on the provision of canteen facilities the following points should be considered:

1. Proximity of building relative to adjacent village, town facilities;
2. Hot and/or cold food and selective menu;
3. Number of meals to be catered for and operating hours;
4. Ethnic requirements (i.e. vegetarian, etc.);
5. Secondary catering facilities for senior staff and/or visitors;
6. Provision of separate hot, cold and drinking-water services for canteen facilities.

2.14.3 Rest rooms

Statutory requirements dictate that a space be set aside within an industrial building for use as a rest room. The usage and facilities available are often extremely varied, and can be sub-divided as follows:

1. Use of room for morning/afternoon breaks;
2. Consumption of food at lunch time in the absence of canteen facilities;
3. Combined usage as a recreation room.

In providing rest room facilities the following points should be considered:

1. Prime use of room as defined above;
2. Number of persons to be accommodated;
3. Provision of drinking and snack facilities (i.e. tea-preparation sink, vending machine, automated snack bar, etc.);
4. Self-service or resident catering staff;
5. Male/female segregation;
6. Provision of recreation activities (i.e. darts, table tennis, television, etc.);
7. Provision of room furniture.

2.14.4 First-aid facilities

It is a statutory requirement that first-aid facilities be available at all times for staff personnel. This facility is very diverse and can extend from a first-aid cabinet to the provision of a first-aid room and attendant semi-resident nurse/doctor. The extent of first-aid facilities are dependent on the following:

1. Type of building and nature of business (i.e. dirty production line, computer components);
2. Number of staff personnel to be serviced and male/female ratio;
3. Proximity of building to easily accessible external medical facilities;
4. Provision of first-aid trained staff from within the building personnel;
5. Provision of emergency equipment for specialized building activities (i.e. chemical, radio-active contamination);
6. Provision of bed furniture, stretchers, respirators, etc.;
7. Provision of separate toilet facilities for resident nurse/doctor;
8. Ethnic considerations.

2.14.5 Detail design

Having established the design parameters for the domestic facilities as outlined above, the following references should be used in the development of detail design:

1. Factories Act of 1961 and revised 1971;
2. Sanitary Accommodation Regulations SR and O 1938, No. 611;
3. Sanitary Accommodation (Amendment) Regulations 1974, No. 426;
4. HM Inspector of Factories recommendations;
5. BS 6465: Part 1: 1984: Sanitary installation.

Quantity of sanitary fittings shall be based on total numbers of personnel per shift and not total number of employees per day.

Where dirty or wet processing occurs continuously within a building, consideration may be given to the provision of baths and showers for use by personnel at the end of the day or shift. Discussion with the operator and HM Inspector of Factories would be an advantage in assessing quantities required.

Working areas within certain industrial buildings may have restricted access (i.e. sterile laboratories, radioactive areas). Separate toilet accommodation may be required in these areas, an assessment of which may be obtained from the operator and reference to such publications as Atomic Energy Code of Practice, Laboratory Practice. It may also be necessary to operate such appliances remotely by photo-electric cell, sonic control or foot control.

Where building employees are from mainly ethnic communities consideration should be given to the positioning of toilet facilities so that religious beliefs are not compromised.

In addition to the general employees of a building, the following specialized personnel should also be considered:

1. Disabled
2. Elderly
3. Administration
4. Catering
5. Visitors.

It is a requirement of the Building Regulations 1985, Schedule 2, to provide facilities for disabled people. Reference should also be made to BS 5619: 1978 and BS 5810: 1979, which are codes of practice for housing of disabled and access requirements to buildings for disabled persons.

Where provision within industrial buildings for elderly persons are required, reference should be made to the DSS. Consideration should also be given to providing water at lower temperatures to appliances exclusively for use by elderly and/or disabled persons.

Sanitary-towel disposal must be provided in each female toilet. The type of disposal can be by macerator, incinerator, bag or chemical.

Separate changing-room facilities for male and female personnel would include the following:

1. Personal locker with lock for change of clothes at the end of a shift or day;
2. Bench/chair for seating.

The location of an industrial building relative to adjacent shops, village and town may well influence the provision of canteen facilities as well as the type of service to be provided.

Where full canteen facilities are provided an accurate assessment of the hours of operation must be obtained so that plant and services are correctly sized. The operating hours will also greatly influence the need to provide separate plant from the main building, especially if meals are to be provided at the end of each shift.

2.15 Lifts

2.15.1 Introduction

The planning of lift systems should be one of the first to be considered, usually alongside the initial building structure, bearing in mind the long lead in times and structural requirements. It is important that the correct configuration is established early, as changes at a later date can be difficult to incorporate.

The first essential step in the system design is goods and pedestrian planning, i.e.

1. How many people require transportation?
2. What will be the peak demand and how and when will it occur?
3. How many floors will the lift system serve?
4. What are the loads/sizes of goods/equipment requiring transportation?

The responses to these questions can be used to determine the number of lifts required, the size of the care, the door arrangement and the speed of car travel. Other major considerations should be the location of lifts within the building and accessibility for disabled users.

2.15.2 Lift system design – drive systems

There are various types of drive systems employed throughout the lift industry, and the following are the most commonly used:

1. Electric a.c. motor with gearbox;
2. Gearless electric d.c. motor;
3. Hydraulic

Electric a.c. motor with gearbox

Lifts of this type are normally rope driven and are the most common drive used in lift installations. The gearbox is fitted with a grooved traction sheave on which a series of steel ropes are wound. The ropes are attached to the top of the lift car, wrapped around the traction sheave (and occasionally secondary sheaves) and attached to a counterweight of equal mass to the car and a percentage of its rated load (normally 40%).

This drive can be used for a wide range of car sizes but has a limit to the speed of travel of approximately 2.5 m/s. The lift machine room is usually located above the lift shaft. However bottom-of-shaft machine rooms are not unknown, although they tend to be more costly.

Gearless electric d.c. motor

Lifts of this type are of a configuration similar to the geared drive systems mentioned above. The difference lies in the increase in the physical size of the drive as the required operating torque of a gearless d.c. motor is produced at a fraction of the rpm of a geared a.c. motor.

The cost of these drives is significantly greater than that of a comparable geared drive, but the quality of ride and maximum speed of car travel are much improved. Gearless drives usually have applications in prestige offices and hotels where the lift travel is in excess of 35 m.

Hydraulic lifts

Lifts of this type are generally used for low-rise applications and also if good levelling accuracy at each landing is required. Hydraulic lifts are either the direct-acting type, where a hydraulic ram acts on the underside of the lift car, or the suspended type, where the ram acts on the top of the car via a series of ropes or chains and pulleys. Hydraulic lifts can be used for very large goods lifts when two or more hydraulic rams are used, but are limited to a slow car speed, which improves to a maximum of 1.5 m/s for small lift applications. Hydraulic lifts have an advantage over rope-traction lifts in that the lift machine room can be situated remotely from the lift shaft with a connection to the lift ram via flexible hydraulic pipework. Hydraulic lifts have a limited maximum travel which is approximately 12 m for (up to) 1000 kg car size and 8 m for car sizes over 1500 kg.

2.15.3 Lift machine rooms

Each lift or group of lifts requires a lift machine room which should be designed complete with:

1. Drive system machinery, pulley mechanisms and control panels;
2. Lifting beams tested for safe working loading;
3. Adequate emergency lighting, small power and 110 V sockets;
4. Adequate ventilation/cooling/heating/frost protection as required by BS 5655;
5. Fire detection and alarms;
6. An electrical switchgear panel which provides an electrical supply to each lift machine.

The lift motor room should be identified with safety signs as required in BS 5655 and should be secured to allow access by authorized persons only.

2.15.4 Lift control in the event of fire

In the event of fire within a building the controller for each lift should isolate all manually operated inputs and return automatically to the evacuation level, usually the ground floor. An output from the building fire alarm panel is 'hard wired' to the lift controller, giving the signal for a fire condition. The lift remains disabled at the evacuation level and the car doors open. If control of a lift is required by a fire officer a key switch or break-glass unit should be used to re-activate the lift.

2.15.5 Firefighting lifts

These should not be confused with lifts fitted with a fire officer's override control. A firefighting lift is required for any building exceeding 18 m in height or 9 m in depth below the evacuation level, and is to be used to allow fire officers to manoeuvre personnel and equipment during firefighting operations.

A firefighting lift must:

1. Not be a goods lift under normal operating conditions;
2. Have a minimum size of 8 persons/630 kg;
3. Be installed in a firefighting shaft with 2-hour fire-rated walls;
4. Have suitable fire-retarding finishes for the lift car;
5. Have a separate emergency electrical supply and feeder cable;
6. Have a machine room located above the lift shaft;
7. Be able to complete a journey in less than 60 seconds.

2.16 Site health and safety

2.16.1 The Health and Safety at Work Act 1974

Apart from the obvious benefits deriving from a well-organized and safely run site, it is in the interests of both employer and employee to work towards a safe working environment. In addition, there are statutory obligations. Predominant among these is the Health and Safety at Work Act 1974.

A detailed examination of the Act cannot be attempted here, but some of the major provisions are as follows:

1. Generally speaking, the Act pertains to the protection of *employees* and discusses the responsibilities of *employers* in this regard. However, the Act also addresses the need to ensure that site activities pose no hazard to the public.
2. The employer must ensure the health and safety of employees is protected as well as possible. The employer must also provide sufficient information and training to employees in order to promote such protection.
3. On site, operatives must have sufficient training to perform their work safely, be subject to competent supervision and know their duties.
4. Lifting operations must be performed by competent individuals trained in slinging and signalling and be properly briefed regarding the operation. Proper communications between lifting operatives must be provided and provision made for adequate load-bearing bases, tie-ups, etc. Potential overload warnings must be provided and vigilance maintained to monitor unforeseen occurrences.
5. Site conditions must be monitored/rectified to ensure that the ground is capable of sustaining loads applied during all activities, including backfilling, etc. This may require the provision of supplementing support, fill materials, spreaders, etc. Ground levels must be

adjusted to accommodate safe working of plant (e.g. cranes, forklifts, etc.).
6. Sub-ground services must be investigated and adequate provision made (where possible) to preclude impacts on services (cabling, gas mains, etc.) or other obstructions.
7. For electrical equipment notice must be taken (and compliance, where required) of manufacturer's requirements with particular reference to:
   (a) Equipment ratings;
   (b) Loads and fault conditions;
   (c) Fault level at supply;
   (d) Environmental conditions.

   All electrical equipment must be suitably isolated prior to maintenance, etc.
8. All temporary structures such as scaffolding must be properly designed and installed with all necessary protective devices (handrails, toe boards, etc.).

Compliance with the requirements of the Act are monitored by Health and Safety Inspectors who possess powers conferred by statute.

2.16.2 The Factories Act 1961

The Factories Act also contains pertinent requirements for construction activities. These primarily relate to:

1. Environmental conditions (lighting, etc.);
2. Machinery safety (where applicable);
3. Fumes;
4. Welfare facilities;
5. Accident notification.

However, due to the age of the Act, in many instances particular sections have been overtaken by more recent specific legislation. Reference should be made to the listings given below.

2.16.3 Other sources of information

Useful information can be obtained from various codes of practice, some of which are listed at the end of the chapter. In addition, the National Federation of Building Trades Employers (NFBTE) publish a handbook on construction safety. Also useful is the book, *Managing Health and Safety in Construction*, published by the Construction Industry Advisory Committee of the Health and Safety Executive. These documents provide fairly comprehensive guidelines to construction industry employers regarding safety policy formulation and implementation.

From a practical viewpoint it is essential to ensure that all contractors working on a site are well organized and have a clearly stated safety policy. Where possible, a safety officer should be appointed to coordinate all safety-related issues, monitor compliance, hold safety meetings, etc. Where the employer has existing corporate safety guidelines these should be made part of all construction contracts and compliance monitored.

In addition, it is a good practice for each contractor to submit 'Method Statements' for the work they are to perform. These should include definitive statements regarding safety policy and implementation on the site, including named personnel, individual responsibilities, organizational arrangements, safety measures to be adopted and training to be performed. A safety officer should be identified.

2.17 Sub-ground pits and basements

Sub-ground pits may be required for access, service distribution or process requirements. Depending on pit dimensions and water table levels, problems that may have to be considered at the design stage include 'heave' due to relief of overburden, buoyancy from water pressure and waterproofing.

2.17.1 Waterproofing

There are three standard solutions to waterproofing pits and basements:

1. Watertight concrete construction;
2. Drained cavity construction;
3. External continuous impervious membrane (tanking).

Watertight concrete construction

This relies on the integrity of the concrete construction itself with reference to design (crack control, joint spacing and detailing, concrete mix) and skilled workmanship and supervision.

Drained cavity construction

This is also based on watertight concrete construction but accepts that some seepage may occur through the external concrete wall. The water is kept away from a separate inner-skin wall by an intervening drained cavity. An internal vapour barrier may be applied to the inner-skin wall to limit humidity, but the cavity may then require ventilation to prevent condensation.

External waterproof membrane

Two possible forms of membrane are hot applied mastic asphalt or bitumen/butyl rubber sheeting with welded or glued joints. The membrane under the floor slabs has to be lapped with that around the walls. It is essential that the membrane is protected during construction, and a typical arrangement is as shown in Figure 2.15.

2.18 Internal and external decoration

2.18.1 Internal

Ceiling, wall and floor finishes can be separated into 'wet' and 'dry' finishes. Below is a listing of the common types of wet and dry finishes currently in use.

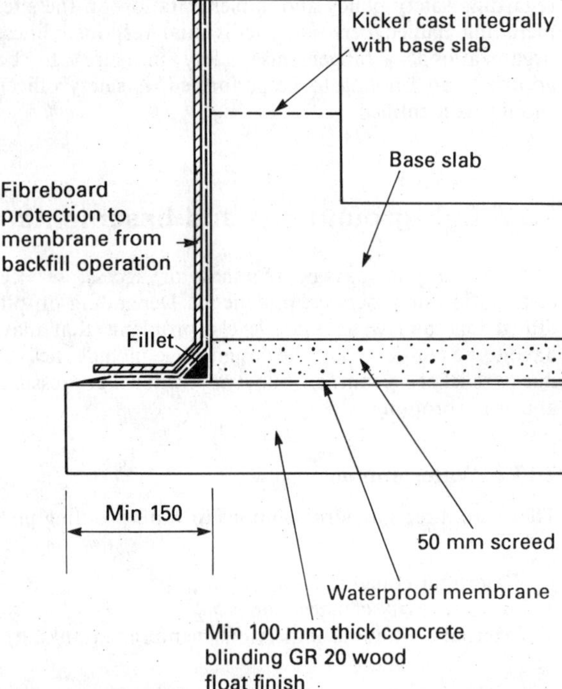

Figure 2.15 A typical external waterproof membrane detail

Wet finishes

1. Plasterwork
2. Screeds
3. Tiling
4. Paint finishes
5. Wet structural fire-protection systems

Dry finishes

1. Dry lining (plasterboard)
2. False (suspended) ceilings
3. Wall claddings
4. Dry floor finishes (thermoplastic sheeting, carpeting, raised deck flooring, etc.)
5. Dry structural fire-protection systems

The selection of finishes is generally made against the following parameters or constraints:

1. Acoustic performance
2. Environmental conditions
3. Life expectancy/maintenance requirements
4. Safety requirements (both installation and use)
5. Fire rating/flame spread
6. Aesthetic considerations
7. Flexibility of use
8. Load-bearing requirements
9. Costs
10. Installation time/disruption.

Environmental conditions

All finishes should be selected to be resilient to expected average and worst-case environmental exposure such as high/low humidity or temperature, airborne contaminants, vibration, possible aggressive liquid spillages, cleanliness/hygiene requirements, etc.

Life expectancy/maintenance requirements

Many finishes exhibit low maintenance requirements (e.g. plasticized metallic sheetings, epoxy coatings, continuous tiling systems, etc.). Others may be more maintenance intensive and may provide lower durability. However, selection must also consider the other operating parameters such as acoustic performance (which may mandate heavier mass or more porous-surfaced materials) or load-bearing capabilities, etc.

Safety requirements

Materials containing toxic or irritant materials should be avoided. Flooring finishes should exhibit non-slip properties but be cleanable and not provide excessive crenellations, etc., for contaminants. Materials will need to be certificated in terms of fire resistance (to the passage of fire) and flame spread. They are classified when tested against BS 476 and the National Building Regulations. Some materials (such as PVC) emit toxic fumes when burning. Structural integrity must also be assessed where secondary cladding or suspended finishes are being utilized, particularly where load-bearing abilities are required (such as suspended ceilings or wall claddings).

Aesthetic considerations

While this may appear to be self-explanatory, the need to change colour schemes or textures over the lifetime of the installation will need to be reviewed. Self-coloured materials do not typically take secondary coatings such as paint or stain.

Flexibility of use

Consideration should be given to likely layout or usage changes. Generally, wet materials will provide least flexibility in terms of subsequent removal and reinstallation. However, they may be more conducive to changes in usage of the space.

Costs and installation time

It is generally accepted that the use of dry materials results in shorter and less disruptive installations, particularly if modular systems are used. However, these materials may well not offer the least expensive solution.

2.18.2 External

In many ways, the same criteria apply to the selection of external finish materials as to internal finishes, and can be identified as 'wet' or 'dry' types. However, increasingly, dry cladding (roof and wall) and bonded roof membrane systems are being used in preference to more traditional rendering and liquid applied roofing membranes.

If extensions are being made to existing facilities or buildings adjacent to 'traditional' structures then this may require the use of brickwork cladding or block and render in order to satisfactorily 'marry' with the existing structures. Care must be taken to ensure such extensions/additions meet current building regulations, bye-laws, standards and codes of practice.

Of major importance when selecting suitable exterior finish materials are their weather resistance and 'weathering' characteristics (with particular reference to ultraviolet resistance and moisture staining), including resistance to biological/vegetable growth. Cladding systems, particularly, should be examined against their wind and moisture penetration resistance. External building elements, either individually or when viewed as a collective system, must comply with the thermal transmittance requirements of the building regulations.

Corrosion resistance should also be considered with particular reference to the ambient environment. Painting schemes, coating materials and other protective skins (along with the substrate bonding method) should all be reviewed against the prevailing conditions.

2.19 Industrial ground floors

The ground floor of an industrial building is arguably the most important structural element. Failure of a floor slab or even a topping can lead to enormous expense and disruption due to repair or replacement. Considerations in the design of a ground slab may include:

1. Loading (uniformly distributed loads and point loads)
2. Abrasion/impact resistance
3. Chemical resistance
4. Flatness
5. Ease of cleaning.

It is important that these are accurately assessed since underspecification may lead to severe maintenance problems while overspecification may result in substantial cost increases.

2.19.1 Structural form

Dependent on ground conditions, loading requirements and settlement limits, industrial ground floors may be ground bearing, raft construction or piled. The piled option should, if possible, be avoided, as the cost is comparatively very high.

Rafts

Rafts fall into a category between fully stiff (behaving as a rigid body) and fully flexible. The raft may also support the superstructure. A soil/structure interaction study must be carried out to estimate the degree to which loads are distributed by the raft. Hence the resulting slab forces and required reinforcement are calculated.

Ground-bearing slabs

Ground-bearing slabs are typically the cheapest and most common form of industrial ground floor, and rely on the tensile capacity of concrete. Any reinforcement provided is to reduce the potential for cracking from shrinkage and thermal effects. The degree of reinforcement is therefore primarily dependent on slab thickness and joint spacing (Figure 2.16).

The detailing arrangement of joints is of prime importance in the slab performance. The purpose of these joints is to concentrate cracks at specific locations, isolate the slab from other structural elements and sometimes to allow for expansion or articulation.

The main elements of a ground-bearing slab are:

1. *Sub-grade:* the material at formation level after excavating or infilling;
2. *Sub-base:* selected imported granular material to form a stable level surface on which to construct the slab. Where lean-mix concrete is not used the surface is normally blinded with sharp sand;
3. *Slip membrane:* serves to reduce friction between the slab and sub-base to minimize subsequent slab cracking. This is normally a polyethylene sheet. It may also serve as a dampproof membrane;
4. *Structural slab*, usually with single layer of reinforcement in the top;
5. *Wearing surface:* may be the finished slab surface or the surface of an applied topping.

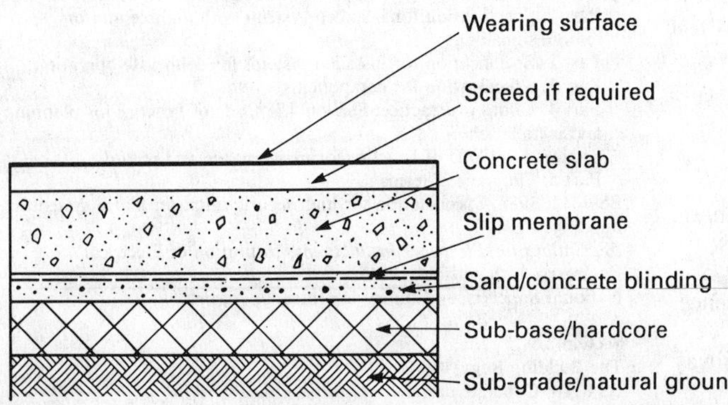

Figure 2.16 A typical section through a bearing slab

2.19.2 Abrasion resistance

Abrasion resistance can be influenced by:

1. Concrete mix
2. Method of finishing the concrete
3. Method of curing
4. The addition of surface hardeners or sprinkle finishes (metallic or non-metallic)
5. The addition of toppings.

2.19.3 Chemical resistance

Chemical resistance is normally achieved by synthetic resin toppings, polymer or resin-modified cementitious toppings or modular (tiled or paviour) toppings.

2.19.4 Flatness

Flatness is of critical importance to some sections of industrial operations such as VNA (very narrow aisle) high rack warehouses or automatic guided vehicles. The degree of flatness achieved is dependent on construction techniques, workmanship and supervision.

Traditionally, global tolerances have been specified for any point on the floor relative to datum and local tolerances specified relative to a 3 m straight edge. For VNA type requirements tolerances distinguish between areas of defined movement (i.e. predetermined vehicle routes) and areas of free movement (the rest of the slab).

2.19.5 Dusting

High-quality concrete slabs that have been well finished should not give rise to dusting. Complete avoidance of dust, however, will only be achieved by the addition of a resin or chemical sealer. Chemical sealers (such as silicofluorides and silicates) give little increased abrasion resistance, and can themselves be prone to dusting on overapplication.

2.20 Ground considerations

2.20.1 Existing site conditions

The existing site conditions which can influence the design, construction and performance of industrial buildings and their operations include:

1. Topography – surface topology, watercourses, surrounding property, vegetation, access;
2. Geology;
3. Groundwater conditions;
4. Site history – industrial contamination, mining, domestic refuse, old foundations, cellars, tunnels;
5. Existing buried or overhead services;
6. Liability to flooding, wind or sea erosion, subsidence, slope instability;
7. Liability to vibration from earthquakes, railway tunnels.

2.20.2 Contaminated sites

Extensive available level sites have sometimes previously accommodated gas works, chemical works, munitions factories, or industrial or domestic refuse tips. These sites require special consideration as regards protection of the structure to be built, the construction workforce and the building occupants.

Refuse tips invariably result in the production of methane from the decay of organic matter. Measures are necessary to avoid trapping the methane in or beneath the building. This can be achieved by a system of methane vent pipes in the fill together with an impermeable membrane at sub-ground-floor level. Alternatively, a vented cavity can be constructed sub-ground floor.

2.20.3 Foundations

The foundation system will be based on such information as can be determined together with physical and chemical testing carried out on samples procured from a site investigation. In regions of very poor ground conditions there is sometimes an option of utilizing a ground-improvement technique such as dynamic consolidation.

2.20.4 Existing services

Existing services are important in two respects:

1. They may have to be rerouted or the structural scheme amended to avoid clashes;
2. The services required for the new industrial operation must usually tie into the existing infrastructure.

2.20.5 Car parking

Car parking area per car (space + road), is approximately 25 m^2. Standard car parking dimensions are 4.8 m × 2.3 m (3.3 m for disabled persons) with access roads 6 m wide. Petrol interceptors for surface water from car parking areas are required where 50 spaces or more are provided.

References

Section 2.3

BS 4166: Automatic intruder alarm terminating equipment in police stations

BS 4747: Intruder alarms systems in buildings

Part 1: Specification for installed systems with audible and/or remote signalling

Part 2: Specification for installed systems for deliberate operation

Part 3: Specification for components

Part 4: Codes of Practice: Section 4.1: Code of Practice for planning and installation

Section 4.2: Code of Practice for maintenance and records

Part 5: Glossary of terms

BS 7042: 1988: Specification for high security intruder alarm system in buildings

Regulations for Electrical Installations, Institution of Electrical Engineers, 15th edition (1981)

National Supervisory Council for Intruder Alarms

Section 2.6

The Building Regulations

BS 8110: Concrete design

BS 5950: Steelwork design
CP 3, Chapter 5, Part 2: Wind loading for building
CP 102: Protection of buildings from water from the ground

Section 2.7
Fire protection: *BSC S754 5M2.84*, BSC Sections and Commercial Steels
Fire and steel construction: Elliott, D. A., *An Introduction to the Protection of Structural Steelwork*, 2nd edition, Constrado (1981); Law, M. and O'Brien, T., *Fire Safety of Bare External Structural Steel*, Constrado (1981)

Section 2.9
IEE Regulations, 15th edition
ANSI Code for Pressure Piping B31.1
Grinnell Piping Design and Engineering Manual
CIBSE Guides (B and C)
HSE Memorandum of Guidance on the Electricity at Work Regulations (1989)
BS 6399: Design loading for buildings
BS 5950: Steelwork design
BS 6180: Handrailing
BS 5395: Access stairs and ladders
BS 466: Gantry crane construction
BS 5744: Gantry crane installation and use

Section 2.10
BRE Digest No. 210
ASHRAE *Fundamentals Handbook*
BS 5925: 1980: Code of practice for design of buildings: ventilation principles and designing for natural ventilation
BS 5250: Code of practice for design of buildings: control of condensation in dwellings
BS 5588: Part 9: 1989: Fire precautions in the design and construction of buildings: code of practice for ventilation and air conditioning ductwork
Industrial Ventilation Manual of Recommended Practice, American Conference Governmental Industrial Hygienists
CIBSE Guide, Books A and B
US Naval Civil Engineering Laboratory (NCEL) Technical Report: *Natural Ventilation Cooling of Buildings*

Section 2.12
Mills, E. D. (ed.), *Building Maintenance and Preservation Guide to Design and Management*, Butterworths (1980)
Direct Labour Organizations Maintenance, Department of the Environment (Audit Inspectorate) HMSO, London (1983)
Estimating Handbook for Building Maintenance, Department of the Environment/PSA (1988)
Managing Building Maintenance, Chartered Institute of Buildings (1985)
Maintenance Cycles and Life Expectancies of Building Components and Materials, NBA Construction Consultants Limited (1985)
Collins, R. L., *Building Maintenance Management* (1987)
Seeley, I. H., *Building Maintenance*, Macmillan, London (1987)
Spedding, A. H. (ed.), *Building Maintenance Economics and Management*, Spon, London (1987)
Armstrong, J. A., *Maintaining Building Services: a Guide for Managers*, Mitchell, London (1987)
Miles, D. and Syagga, P., *Building Maintenance: a Management Manual*, Intermediate Technology Publications, London (1987)

Section 2.14
Water Supply Byelaws Guide, 2nd edition (1989)
BS 6700: 1987
BS 5572: 1978 (1983)
BS 4118: 1981
BS 6367: 1983 (1984)
BS 6455
BS 6465: Part 1: 1984
BS 8005: Part 1: (1987)
BS 8233: 1987
BS 8301: 1987
CP 312: Part 1: 1973
CP 413: 1973 (1980)
Health and Safety Executive Publication Booklet HS(G)34
Health and Safety Executive Guidance Notes C54, C56 and C58
Liquefield Petroleum Gas Industry Code of Practice
The Gas Safety (Installation and Use) Regulations 1984
BS 1123: Part 1: 1987
BS 4250
BS 6129: Part 1: 1981
CP 342
BS 5810: 1979
Plumbing Services for the Disabled Community
Code of Practice: Legionnaires' disease

Section 2.15
BS 5655: Part 1: Safety rules for the construction and installation of electric lifts (implementing EN 81-1), together with PD6500 'Explanatory supplement to BS 5655: Part 1'
Part 2: Safety rules for the construction and installation of hydraulic lifts (implementing EN 81-2)
Part 3: Specification for electric service lifts
Part 5: Specification for dimensions of standard electric lift arrangements (implementing ISO 4190/1 and ISO 4190/2)
Part 6: Code of practice for selection and installation
Part 7: Specification for manual control devices, indicators and additional fittings (implementing ISO 4190/5)
Part 8: Specification for eyebolts for lift suspension
Part 9: Specification for guide rails (implementing ISO 7465)
Part 10: Specification for testing and inspection of electric and hydraulic lifts
Part 11: Recommendations for the installation of new and the modernization of electric lifts in existing buildings
Part 12: Recommendations for the installation of new and the modernization of hydraulic lifts in existing buildings
Part 13: Code of practice for vandal resistant lifts
BS 5588: Part 5: Fire Precautions in the design and construction of buildings. Code of practice for firefighting stairways and lifts

Section 2.16
BS 6100: Glossary of building and civil engineering terms
BS 5228: Noise control on construction and open sites
BS 5531: Safety in erecting structural frames
BS 5405: Electrical code of practice for maintenance
HMSO: The Health and Safety at Work Act 1974
HMSO: The Factories Act 1961
NFBTE: *Construction Safety Manual* (amended 1983)
Building Employers' Confederation: *Guidance Note on Safety Responsibilities for Sub Contractors on Site*
Health and Safety Executive (HSE): *Avoiding Danger from Underground Services*
Royal Society for the Prevention of Accidents: *Supervisors' Guide to the Construction Regulations*

Section 2.17
Tomlinson, M. J., *Foundation Design and Construction*
CP 102: Protection of buildings against water from the ground
CIRIA Guide 5: *Guide to the design of waterproof basements*
BS 8110: Structural use of concrete

Section 2.18
British Standard Codes of Practice
CP 202: Tile flooring and slab flooring
CP 203: Sheet and tile flooring
CP 204: In situ floor materials
CP 209: Care and maintenance of floor surfaces
CP 211: Internal plastering
CP 212: Wall tiling
CP 221: External rendered finishes
CP 231: Painting of buildings
2525–27: Ready mixed undercoating and paints (white lead based)
2528–32: Ready mixed undercoating and paints (oil based)
CP 290: Suspended ceilings
CP 143: Sheet wall and roof coverings

British Standards
BS 1053: Water paints and distemper (interior)
BS 381C: Colours for paints for specific purposes
BS 2929: Safety colours for industry
BS 1248: Wallpapers
BS 1763: Thin PVC sheeting
BS 476: Fire tests on building materials and structures

BS 5442: Ceiling coverings
BS 4841: Ceiling linings
BS 5247: Corrugated asbestos cement wall and roof
(Pt 14) coverings
BS 5427: Profiled sheeting in building
BS 4868: Profiled aluminium sheet for building

Section 2.19

Concrete Society Technical Report 34: Concrete Industrial Ground Floors
Deacon, R. C., *Concrete Ground Floors: their design, construction and finish*, 3rd edition, Cement and Concrete Association (1986)
Chandler, J. W. E., *Design of Floors on Ground*, Cement and Concrete Association (1982)

3 Industrial Flooring

J D N Shaw MA
SBD Construction Products Limited

Contents

3.1 Introduction

There is a bewildering number of special proprietary floor treatments available for the architect and engineer to consider, and much of the brief technical literature describing them suggests that many products would appear to offer the same improved service at greatly differing costs. It is therefore not surprising that many specifiers are totally confused and tend to stick to the products they know rather than consider some of the novel treatments based on new technology, which often offer distinct improvements over the materials traditionally used as flooring materials.[1]

Before attempting to classify the various types of special floorings available it is important to consider the concrete substrate itself. By proper use of good mix designs and admixtures, with careful control of the water/cement ratio and careful attention to laying, finishing and curing techniques, concrete itself can serve as a highly durable flooring material under many industrial service conditions without the need for special separately applied finishes.[2,3] Properly laid concrete provides an abrasion-resistant floor surface which has good resistance to attack by alkalis and reasonable resistance to mineral and vegetable oils, although oils do cause some staining and impair appearance. However, irrespective of how well it has been laid, concrete has poor resistance to acids and many other chemicals far too numerous to mention here. Where spillage of such materials is envisaged, the concrete must be protected by the application of a special flooring.

3.1.1 Selection of flooring required

Before selecting a flooring material it is imperative to consider carefully the precise service conditions to which the floor will be subjected. Conditions that must be considered include:

1. Service temperature;
2. Rate of change of temperature, as rapid temperature changes can cause some heavy-duty finishes to crack up due to the high stresses developed by thermal shock;
3. Nature and concentrations of any materials likely to come into contact with the floor;
4. Accuracy of laying the concrete sub-floor to levels to allow spillages to run away reliably to drains;
5. Grade of concrete laid for the sub-floor;
6. Nature of traffic (maximum loads and types of wheels using floor) and traffic concentration;
7. Degree/ease of cleaning required;
8. Non-slip characteristics required.

Without such precise information and, on occasions, even with it, inappropriate floorings are all too often used, resulting in rapid breakdown of the floor in service. In some instances, some of the performance requirements are contradictory (especially 7 and 8 above) and where this happens, a compromise may have to be accepted.

3.1.2 Requirements of concrete substrate

If, at the specification stage, it is decided that the service conditions for the concrete floor do require special floorings to be applied subsequently, then it is imperative that the contractor laying the concrete is fully aware of this factor and does not use a conventional spray-applied resin solution curing membrane, as this could seriously affect the adhesion of any special flooring to be applied subsequently, being difficult to remove uniformly and reliably.[4,5] In these circumstances, and also for industrial buildings where the eventual use of the floor is not known but may therefore require a special finish, overlapping polythene sheets or another efficient curing method which does not affect the adhesion of any subsequently applied finishes should be used.

It is important to examine carefully the surface of the concrete substrate prior to applying the special flooring. Although the concrete laid by the contractor may indeed have cube strengths well in excess of that specified, it is still possible for the concrete slab to have a very weak surface due to overtrowelling, for example. It is the top few microns of the concrete to which the special flooring will be applied, and it is essential that any weakness in the surface is removed by a technique which is appropriate to the type of flooring to be applied.[7] Most specialist flooring contractors have sufficient experience to assess the quality of concrete surface without site testing. However, if there is any doubt, the surface strength of the concrete should be tested using a simple pull-off tester or other appropriate means. In general, a concrete substrate should have a tensile strength (by pull-off) of at least 0.75 N/mm^2.[8,9]

As a general rule, any concrete base which will be subsequently treated with a special flooring must not be subject to rising damp, and thus any ground-supported slabs must incorporate an efficient damp-proof course. If there is any doubt, the concrete should be tested using a direct-reading concrete moisture meter (maximum 6% moisture) or an Edney Hygrometer (reading not exceeding 75% relative humidity after 4 hours). It should be stressed that these figures are based on the practical experience of a number of specialist flooring contractors and serve only as a guide. Other factors such as the depth of the slab, time elapsed since placing and degree of weather protection all have an influence on the moisture content within the concrete substrate.[19]

Finally, before considering flooring materials, mention should be made of the application of the right joint filler in all movement joints.[8] Far too often, with a carefully laid concrete floor for industrial service, no detailed attention to joint filling is given, and this results in unfilled or wrongly filled joints rapidly spalling at the edges under heavy loads of rigid or semi-rigid wheel traffic, leading to expensive repairs. Apart from preventing spalling at the floor edges, the right joint filler will also improve cleanliness, help smooth running and prolong wheel life of forklift trucks, for example, and contribute to safety. The selection of the right joint filler is, however, a difficult problem, and, in general, it is true to say that there is no one single ideal material for floor joints, since it is impossible to combine all the performance characteristics ideally required in one product.

There is now an active trade federation, the Federation of Resin Formulators and Applicators (FERFA) to which many of the specialist formulators and industrial flooring contractors belong. If specifiers/users carefully list all the service conditions which the flooring must meet, FERFA members should be able to recommend the most cost-effective flooring materials to give good service under the conditions indicated. FERFA have recently produced two Flooring Guides, Application Guide No. 4, *Synthetic Resin Floor Screeds*, and Application Guide No. 6, *Polymer Floor Screeds*. These guides cover in considerable detail all aspects of laying high-performance resin and polymer floorings.[5]

3.1.3 Special finishes

In this chapter the author has attempted to classify the various types of finishes available for concrete floors in terms of increasing applied costs.

3.2 Thin applied hardener/sealers

3.2.1 Sodium silicate and silico fluoride solutions as concrete surface hardeners

Both sodium silicate and silico fluoride solutions are applied to clean, dry, sound concrete floors as dilute aqueous solutions (10–15% solids) in two to three applications, taking care to ensure that all material penetrates and is absorbed into the concrete surface. The silicate or silico fluoride reacts with the small amount of free lime in the cement to form glassy inert materials in the surface, and the successful application of both materials depends upon filling the micropores in the surface of good-quality concrete, leaving its surface-appearance and non-skid characteristics virtually unchanged. The main difference between the two types are that the reaction products of the silico fluoride types are less soluble in water and are also harder, which may give better in-service performance but at a slightly higher material cost. However, with recent developments in floor-laying techniques, the concrete substrates for industrial floors are laid with much more dense low-porosity surfaces, so that neither silicate nor silico fluoride treatments are as effective as they used to be, say, ten years ago, when the concrete used had slightly more open finish and hence was more receptive to these treatments. With modern concrete floors it is imperative to wash any material not absorbed into the surface within a short period. Otherwise, unpleasant white alkaline deposits, which are difficult to remove, may occur.

It is important to stress that neither sodium silicate nor silico fluoride will improve the performance of a poor, low-strength, dusty concrete floor and if the surface is too porous, there is no way that all the material applied can react with the relatively small quantity of free lime in the concrete surface. All that will happen is that the pores will be filled with unreacted powder, producing a most unpleasant alkaline dust, which can be very irritating to the skin and eyes when the floor is put into service.

Finally, it is important to note that sodium silicate or silico fluoride treatments properly applied to clean and sound concrete floors can improve their performance, wear resistance and resistance to mild aqueous chemicals and oils, at a relatively low cost. However, they are not the answer to all industrial flooring problems, as many specifiers appear to believe.[17]

3.2.2 Low-viscosity resin-based penetrating in-surface finishes

Liquid resin-based systems which, like the chemical surface hardeners, penetrate into the surface of a concrete topping or directly finished slab and protect the acid-susceptible cement matrix from attack and, at the same time, strengthen the surface of the concrete are now being increasingly used. These in-surface seals leave the slip resistance of the concrete floor virtually unchanged but the treated floors are easier to clean and are more durable.

3.2.3 Non-reactive and semi-reactive resin solutions

Resin solution penetrating sealers are now available which, for very large warehouse floors, are comparable in applied costs with the concrete surface hardeners and are now being increasingly specified. Experience indicates that certain acrylic resin solutions are proving more durable and offer better protection to chemical and oil spillages than concrete surface hardeners. Acrylic resin solution sealers can markedly improve the abrasion resistance of concrete floors and have 'rescued' a number of poor-quality floors.

Other resin solutions, in white spirit or stronger solvent blends, used as penetrating floor sealers include:

1. Air-drying alkyds (similar to the resins in conventional gloss paints);
2. Styrene butadiene resins;
3. Urethane oils;
4. Styrene acrylates.

All such resin solutions are based on flammable solvents and are becoming increasingly less acceptable on health and safety grounds. There is therefore increased interest in water-based polymer dispersion floor sealers, but, to date, none offer the same improvement to flooring performance that some of the resin solution can provide.

3.2.4 Polymer dispersions

Polyvinyl acetate (PVA), acrylic and other polymer dispersions have been widely used as anti-dust treatments for concrete floors for many years. In general, the polymer dispersions have been similar to those used in the manufacture of emulsion paints, and until recently have tended to be based on dispersions of relatively large polymer particles (particle size $0.15–0.25 \times 10^{-6}$ m). Dispersions are now becoming available which offer superior performance as floor sealers. The chemical and water resistance of the various polymer dispersions which have been used in the past various considerably from the PVA types, which are rapidly softened and eventually washed out by water, to acrylic and SBR types which exhibit excellent resistance to a wide range of chemicals. Water-based sealers are gaining wider acceptance because of the increased handling problems associated with polymer solutions based on hydrocarbon solvents.

3.2.5 Epoxy resin dispersions

Two-component epoxy resin water thinned dispersions are now being used as floor sealers. They have good adhesion to concrete and good chemical resistance but the particle size of the dispersion is comparatively large (approximately 1–1.5 microns) and consequently penetration into good-quality concrete is minimal and an 'on-surface seal' is obtained. However, with porous low-quality concrete substances, considerable binding/strengthening, etc. of the surface can be achieved with water-dispersible epoxy resin-based floor sealer.

3.2.6 Reactive resin solutions

The two-pack low molecular weight epoxy resin systems in volatile solvents have proved very effective for improving the wear and chemical resistance of both good- and poor-quality concrete floors. The epoxy resin solutions (approximately 20% solids) are high-strength systems, very similar to those used in heavy chemically resistant trowelled epoxy floors and, depending on the concrete, can penetrate a significant depth into the surface of the concrete, where the solvent evaporates and the resin cures to form a tough, chemically resistant seal, with a compressive strength of up to 70 N/mm^2, thus reinforcing the concrete surface. One-pack low-viscosity resin solutions based on moisture-curing polyurethane systems are available which perform in a manner similar to epoxy resin solutions. Some of these polyurethane resin solutions demonstrate a greater ability to penetrate and bind the surface of suspect concrete floors measurably better than other penetrating sealers.[17,18]

3.3 Floor paints

Floor paints, in a wide range of colours and based on a number of different binder systems, are used extensively for concrete floors in light industrial applications.

3.2.1 Chlorinated rubber paints

Chlorinated rubber floor paints are probably the most common of the lower-cost floor paints on the market. They produce tough and chemically resistant coatings, but their adhesion to concrete is not always good. They tend to wear off in patches and cannot be considered as a durable floor treatment except under light traffic conditions. However, recoating is a simple job and floors can easily be repainted over weekend shutdowns, for example. Similar paints based on other resins such as acrylics, vinyls and styrene butadiene are also used.

3.2.2 Polyurethane floor paints and multi-coat treatments

Solvents containing moisture-cured or two-pack polyurethane resin paints are also used extensively. They combine excellent abrasion resistance with good chemical resistance, and are normally applied in two coats to give a coating thickness of 0.10–0.15 mm. In addition, moisture-cured polyurethane resin solutions are used for quite thick durable decorative floorings.

Several coats of resin are applied to the prepared substrate at approximately 4- to 6-hour intervals, with one or more coats being dressed with coloured paint flakes which are sealed in by the next coat and then lightly sanded. This type of flooring was widely marketed in the UK about ten years ago but, in the main, they were considered unsatisfactory due to rapid discoloration of the floor because of the lack of ultraviolet stability of the urethane resins used, which rapidly turned yellow-brown and looked dirty. However, ultraviolet-stable urethane resins which do not suffer this discoloration are now available, and this type of durable decorative flooring is gaining re-acceptance (for example, for kitchens, toilets and reception areas).

3.2.3 Epoxy resin high-build floor paints

Solvent-free high-build floor paints are available which can readily be applied by brush, roller or spray to a prepared concrete substrate to give a thickness of 0.10–0.20 mm per coat. Normally, two coats are applied and the first is often lightly dressed with fine sand or carborundum dust to give a non-slip, chemically resistant and durable coloured floor, ideal for light industrial traffic conditions (for example, rubber-shod wheels).[6]

3.4 Self-levelling epoxy, polyester or reactive acrylic resin systems

Like the high-build epoxy paints, these are solvent-free low-viscosity systems which are readily applied onto a prepared level substrate to provide a jointless thin (thickness approximately 1.5 mm) chemically resistant flooring in a single application.[6] The term 'self-levelling' by which they are commonly described is somewhat of a misnomer, as they require spreading out to a near-level finish with a squeegee or the edge of a steel trowel, and by themselves they flow out to give a smooth finish. Perhaps a better description is 'self-smoothing'. Before the system is cured, the surface is normally lightly dressed with fine abrasion-resistant grit. Without a non-slip dressing there is a tendency to produce a slippery, very glossy surface, which shows every scratch mark. This can be overcome to some extent by careful formulating and also by the application of a slip-inhibiting industrial floor polish on a regular basis when the floor is in service.

This type of flooring is widely used in laboratories,

pharmaceutical factories and food-processing areas where easily cleaned, chemically resistant durable floors are required.

In recent years more heavily filled flowing epoxy resin mortar flooring systems laid at 3–5 mm thickness are increasingly being used instead of the more traditional trowelled epoxy resin mortar flooring systems described below. The laying costs of the flowing mortars are signficantly lower and do not require the same degree of skill to achieve a satisfactory floor finish.

Polyester resin systems and, more recently, acrylic resins are also used. Polyester resin-based systems have a tendency to shrinkage during and after application and application is very critical.

Acrylic resin systems developed in Germany are similar to polyester resins but, by careful formulation, the problems due to shrinkage have been largely overcome. The acrylic resin-based systems are currently based on highly flammable materials (flash point 10°C) which can present hazards during laying. However, there are systems available which can take foot traffic 2–3 hours after application and full service conditions within 24 hours, even at very low temperatures.

3.5 Heavy-duty flooring

A considerable range of different toppings are available for heavy-duty service. The correct selection of the most appropriate topping on a cost/performance basis can only be made if service conditions are very clearly defined. In general, heavy-duty toppings require a sound (preferably 35 N/mm^2 strength) concrete substrate.

3.5.1 Granolithic toppings

In effect, granolithic toppings are just a method of producing a high cement content concrete wearing surface on a concrete substrate. The application of separately laid granolithic toppings is always fraught with the danger of debonding and curling and, therefore, monolithic grano-toppings are generally essential. However, for many industrial floors, where good resistance to abrasion under heavy traffic is specified, a suitable floor could be achieved more economically by direct finishing of a high cement content, high-strength (40–60+ N/mm^2) concrete. This was borne out by recent work carried out by Chaplin,[11] who found the abrasion resistance measured by a number of different methods to be directly related to the compressive strength of the concrete. This work also showed that the abrasion resistance and compressive strength could also be related to Schmidt hammer test results, which could be of considerable interest for the future non-destructive site testing of concrete floors.

Where it is considered essential to apply a granolithic topping onto an existing concrete substrate the danger of debonding can be much reduced by the use of polymer-based bonding aid. Two types of bonding aid are commonly used:

1. Epoxy resin adhesive specially formulated for bonding freshly mixed cementitious materials to well-prepared cured concrete substrates. With the right epoxy resin bonding aid the strength of the bond achieved is greater than the shear strength of both the topping and the concrete substrate.
2. A bond coat of a polymer latex (also called polymer emulsions or dispersions) such as styrene butadiene (SBR), polyvinyl acetate (PVA) acrylics or modified acrylics. These are applied to the prepared concrete as neat coats of emulsion or, more commonly, as slurries with cement.

Polymer latex bonding aids are cheaper and more simple to use than epoxy resins and give a good, tough bond which is 'less structural' than that achieved by the right epoxy bonding aid. The so-called 'universal' PVA bonding aids are not recommended for external or wet service conditions, as there is a danger of the polymer breaking down.

3.5.2 Bitumen emulsion-modified cementitious floors

The use of specially formulated bitumen emulsions as the gauging liquid for graded aggregate/sand/cement screeds can produce a dustless, self-healing, jointless surface for industrial areas subject to heavy wheeled traffic under normally dry conditions. This type of topping is normally laid approximately 12 mm thick and has been used very successfully for more than 30 years, particularly in warehouses. The bitumen-modified cementitious floor topping is less hard underfoot than concrete, and has proved a very popular improvement with warehouse staff. However, with recent trends towards high-rise tracking, heavier forklift trucks and narrow aisles between racks, the topping tends to indent or shove, and the truck forks become misaligned with the pallets stacked on the higher shelves of the racks, so that it is not possible to get the goods down.

Loading levels above about 8 N/mm^2 for short term and 4 N/mm^2 for an indefinite period are, therefore, not recommended for bitumen emulsion-modified cementitious floors.

3.5.3 Mastic asphalt floors

Hot applied mastic asphalt floors have been used for many years in industrial environments, where a good degree of chemical resistance under normally wet conditions is required. Properly laid mastic floors are totally impervious to a wide range of chemicals but not solvents. In terms of mechanical performance, mastic asphalt floors are similar to the bitumen-modified cementitious floors, but they are generally laid at a minimum of 25 mm thickness and tend to shove and corrugate in service under heavy loads. Mastic floors are not very commonly used now, except where the floor is essentially tanked, such as car park decks over shopping precincts.

3.5.4 Polymer-modified cementitious floor toppings

Polymer-modified cementitious floor toppings are now widely used instead of separately laid granolithic toppings. The polymers used are normally supplied as milky white dispersions in water and are used to gauge a carefully selected sand/aggregate/cement mix as a whole or partial replacement of the gauging mortar. They must always be mixed in a forced-action mixer.

The polymer latex acts in several ways:

1. It functions as a water-reducing plasticizer, producing a flooring composition with good workability at low water/cement ratios.
2. It ensures a good bond between the topping and the concrete properly prepared.
3. It produces a topping with good tensile strength and toughness.
4. It produces (based on the right polymer latex) a topping with good water and chemical resistance.
5. It acts to a significant degree as an integral curing aid, much reducing the need for efficient curing. (Curing is, however, essential in dry, draughty conditions.)

Polymer-modified cementitious floor toppings are normally laid 6–12 mm thick. Two polymer latex types are most commonly used – styrene butadiene and acrylics – which have been specifically developed for incorporation into cementitious compositions. The principal difference between the two types is that acrylic lattices are available, which gave higher early strengths than can be achieved with the current SBR lattices. Toppings based on acrylic lattices are used in food-processing industries, particularly meat processing. Toppings can be laid in a Friday–Monday weekend shutdown and are reported to be capable of withstanding full service conditions 48 hours after laying, although longer cure periods are desirable.

When adequately cured, polymer-modified cementitious toppings based on acrylic/SBR lattices can be cleaned with steam-cleaning techniques without problems of thermal shock breakdown, which has been observed with other heavy-duty polymer toppings. They are resistant to many chemicals encountered in the food and printing industries but, being based on an acid-sensitive cement matric, their resistance to organic or mineral acids is limited. There are now becoming available special polymer powders derived from the polymer lattices which can be preblended with sand, aggregates, cement and other additives and then gauged on site with water to produce factory-quality controlled flooring compositions with performance similar to the materials based on the addition of the milky lattices on site.

3.5.5 Epoxy resin mortar floorings

Trowelled epoxy resin flooring approximately 6 mm thick is used extensively where a combination of excellent chemical resistance and good mechanical properties are required, particularly abrasion and impact resistance and resistance to very heavy rolling loads. Epoxy toppings are available with compressive strengths up to 100 N/mm^2 and tensile strengths up to 30 N/mm^2. This is achieved by careful formulation of the binder and the incorporation of high-strength blended fillers.[6] When formulating a system for optimum abrasion resistance, both the epoxy/resin hardener binder system and the filler blends used appear to have an influence. The simulation of abrasive service loads on industrial floor toppings in a laboratory is not simple, and numerous wear test machines have been devised. Correlation between different wear test machines is not always good, although most laboratory tests on abrasion resistance give an indication of the floor's likely performance in service in a qualitative rather than quantitative manner.[10,11]

In one series of laboratory tests carried out to find the optimum wear resistance of heavy-duty epoxy resin flooring compositions, a number of different abrasion resistant materials were evaluated using BS 416, employing three different epoxy resin binders which themselves had significantly differing chemical compositions and mechanical properties. The results of this work, which was carried out under dry conditions, are given in Table 3.1. As can be seen from the table, the selection of the abrasion-resistant material and the resin matrix both influence the abrasion resistance of the system, although the abrasive material incorporated appears to play a more crucial role.

In wet abrasive conditions, which often occur with heavy-duty industrial flooring, a small quantity of abrasion-resistant material tends to be carried on the wheels of trucks and produces a grinding paste between the heavy-duty wheel and the surface. Since the abrasion-resistant material in the surface is generally harder than any sand or grit carried into the factory on wheels, the grinding paste tends to become more abrasive as the binder is worn away. Abrasion resistance tests under wet grinding paste conditions, however, do indicate a similar order of resistance, although the binder appears to play a more significant part. In applications where the flooring is flooded with water for long periods, the resin binder plays a more important part, since the

Table 3.1 Abrasion values of trowelled epoxy resin flooring compositions, using BS 416

Aggregate	*Epoxy binder compositions*	*Mass loss after abrasion (g)*
Graded sand, Grade C	A	4.10
Graded sand, Grade M	B	2.75
Graded sand, Grade C	C	2.85
Graded sand, Zone 2	A	5.5
Graded sand, Zone 3	A	5.7
Gritstone	A	9.95
Granite	A	1.45
Calcined bauxite	A	0.95
Basalt	A	1.5
Cast iron grit	A	0.45
Copper slag	A	2.25
Sand (Zone 1) gritstone (50/50 by mass)	A	1.35

strength of the adhesive bond between the particles of abrasion-resistant materials can, if the wrong resin binder system is used, drop markedly under prolonged wet conditions. In formulating resins for heavy-duty floors it would appear that the adhesive properties of the resin binder used to bond the resistant particles firmly together is the more important factor when selecting a resin system. In the selection of systems for highly abrasive service conditions, costs must also be considered and, on this basis, bauxite, calcined under defined temperature conditions, has often been used as the abrasion-resistant aggregate.

Another aspect of epoxy resin mortar floorings which needs careful attention is that their coefficients of thermal expansion are approximately three times that of concrete. This, coupled with the relative low thermal conductivity of epoxy mortar, can cause stresses to be induced at the resin mortar/concrete interface under conditions of thermal shock (e.g. thermal cleaning), resulting in break-up of the flooring due to initial failure in the concrete. Two approaches have been tried to overcome this problem:

1. Using a lower modulus epoxy resin mortar and applying the topping at a thickness of 3–4 mm;
2. Applying a stress distributing flexible epoxy layer 1–2 mm thick between the rigid epoxy topping and the concrete.

Both approaches have been used with some success but in (1), the lower modulus topping also tends to have lower chemical resistance, which can be a problem, while the technique (2) is significantly more costly in terms of both material and labour.[10]

3.5.6 Polyester resin mortars

Polyester resin floor toppings, similar in performance to the epoxy toppings, have been used but, as indicated earlier, polyester systems tend to shrink and, without careful formulation and laying, shrinkage stresses with polyester resin systems can develop at the interface between the topping and the concrete substrate. Coupled with the additional stresses due to the differences in their coefficients to thermal expansion, this can cause failure at the surface of the concrete substrate.[11,12] Several years ago one UK company had considerable success with a carefully formulated polyester mortar topping specifically designed to minimize these stresses, but found that, unless it was laid with meticulous care, failures could occur.

Polyester resin mortars, however, cure within 2 hours of placing to give greater strength than concrete and are widely used for the rapid repair of small areas of damaged concrete floors and, with the use of 'igloos', even in cold stores in service. Another polyester resin-based heavy-duty topping which has proved very satisfactory in service is based on a unique approach. It comprises a blend of treated Portland cement and a dispersion of a special water-soluble catalyst system in an unsaturated polyester resin. This blend is mixed on site with graded aggregates and a measured quantity of water. The water addition dissolves the catalyst and, in the presence of free alkali from the cement, releases free radicals which trigger the curing of the unsaturated polyester resin. The cured product gives a tough floor topping which, over the past ten years, has been widely used in abattoirs, dairies and food-processing plants. Recently, another system based on a similar approach has been introduced.

Polyurethane mortar flooring systems based on somewhat similar technology to this special polyester system have also been used in chemical plants and have given excellent service. The basic urethane polymer is more elastomeric than either epoxy and polyester resins and, as such, is reported to have excellent thermal properties up to at least 140°C and good resistance to thermal shock. The adhesion of the urethane systems to damp concrete is, to some extent, suspect, and dry substrates are therefore normally essential, although systems with improved adhesion to damp substrates are becoming available.[14]

3.5.7 Reactive acrylic resins

Reactive acrylic resins similar to polyester resins are also being used increasingly in heavy-duty floors. Acrylic resins primarily based on methyl methacrylate monomer are low in viscosity and wet out fillers very efficiently, enabling the production of heavily filled flooring compositions, which are easily laid. The high filler loadings much reduce the danger of problems due to shrinkage. Most acrylic resin systems in the uncured state are highly flammable (flash point below 10°C) and special precautions need to be taken.

3.5.8 Industrial tile floorings

There are industrial flooring situations where the service requirement or the time allowed for laying do not permit the use of jointless floor toppings. For such applications a wide range of industrial tiles are available which will meet most requirements, in terms of either mechanical properties or chemical resistance. When tiles are used in very aggressive chemical environments the main problem is grouting between the tiles with a grout having adequate resistance. Grout systems based on specially formulated furane resins (which, in particular, resist very strong acids) and epoxy resins are available for this purpose, and tiles laid in very fast-setting mortar bedding and properly grouted can be installed and returned to service in under 48 hours. A typical application for tiles is in dairies, many of which operate 364 days a year. By using a fast-setting fondue-based mortar bed bonded to the underlying substrate with an epoxy adhesive one day and then laying quarry tiles bonded and grouted with an epoxy resin system the following day, it is possible to repair a completely broken down and impossible to clean floor to a good standard with no interruptions to production.

3.6 Comparative applied costs

It is difficult to give precise costs of floor treatments at

Table 3.2

Floor treatment/flooring	*Comparative cost index*
Concrete surface hardeners, two to three coats	1–1.8
In-surface seals:	
Resin solution non-reactive, two coats	1.2–2.5
Reactive resin solutions, two coats	2–3
Paints:	
Chlorinated rubber, one to two coats	1.7–3.2
Polyurethane, two coats	1.8–4
High-building epoxy, one to two coats	3.5–7
Multicoat polyurethane flake, four-plus	5–12
Epoxy self-levelling, 1–2 mm	10–16
Polyester, 2–3 mm	9–12
Bitumen-modified cementitious, 12–16 mm	7–10
Mastic asphalt, 25 mm	9–13
Polymer-modified cementitious, 12 mm	8–15
Epoxy trowelled, 6 mm	18–24
Polyester, 6–9 mm	15–20
Industrial tiles, various	15–30

size of total area, areas to be coated at one time, degree of surface preparation required and other factors all influence costs. Table 3.2 is a rough guide to comparative applied costs in the UK.

3.7 Conclusion

The range of special flooring materials available is very wide and, to many, extremely confusing. Specifiers, however, will find that if they carefully list all the service requirements for the flooring, FERFA members should be able to recommend and apply suitable products for most service conditions.

References

1 This chapter is based on 'Special finishes for concrete floors', presented by J. D. N. Shaw at the International Conference on Advances in Concrete Slab Technology, Dundee University, April 1979. (Proceedings published by Pergamon Press, Oxford, 1980, pp. 505–515)

2 Barnbrook, G., 'Durable non slip concrete floors for low maintenance', *Building Trades Journal*, 12 January (1979)

3 Anon., 'Good warehouse floors don't just happen', *Concrete Construction*, 493–494, September (1977)

4 Barnbrook, G., 'The concrete slab – a base for resin flooring', paper presented at the Symposium on Resin for Industrial Problem Areas. Flooring and Anti-Corrosion Application, organized by the Federation of Resin Formulators and Applicators Ltd, London, November (1980)

5 FERFA, 241 High Street, Aldershot, Hants GU11 1TJ, Guide No. 4, *Synthetic Resin Floor Screeds*, Guide No. 6, *Polymer Floor Screeds*

6 Phillips, G., 'Caring for concrete floors protects profits', *Building, Maintenance and Services*, 37–38, April (1978)

7 Berger, D. M., 'Preparing concrete surfaces', *Concrete Construction*, 481–494, September (1978)

8 Shaw, J. D. N., 'Epoxy resin based flooring compositions', *Flooring and Finishing News*, 7–11, January (1967)

9 Simmonds, L. B., 'Concrete sub-floor testing', *Architect and Building News*, July (1964)

10 Metzer, S. N., 'A better joint filler', *Modern Concrete*, 45–47, May (1978)

11 Chaplin, R. G., 'Abrasion resistant concrete floors', paper presented at the International Conference on Concrete Slabs, University of Dundee, April (1979)

12 Shaw, J. D. N., 'Epoxy resins for construction: recent developments', paper presented to the Japanese Society of Materials, November (1971)

13 Hewlett, P. C. and Shaw, J. D. N., *Developments in Adhesives*, Chapter 2, pp. 25–72, Applied Science, London (1977)

14 McCurrich, L. H. and Kay, W. M., 'Polyester and epoxy resin concrete', paper presented at the Symposium on Resins and Concrete, University of Newcastle-upon-Tyne, 17–18 April (1973)

15 Nutt, W. O., 'Inorganic polymer structures', British Patent No. 1,065,053, April (1967)

16 Benson, L. H., 'Urethane concrete in the chemical industry', Acema Conference, Frankfurt, August (1973)

17 Shaw, J. D. N., 'Polymer concretes – UK experiences', paper presented at the Conference on Polymers in Concrete, Fourth International Congress, September (1984)

18 Sadegzadeh, M. and Kettle, R. J., 'Abrasion resistance of polymer impregnated concrete', *Concrete*, 32–34, May (1987)

19 FERFA Technical Report, *Osmosis in Flooring*, August (1989)

4 Planning and Plant Layout

Eur Ing Roland R. Gibson
BTech, MSc, CEng, FIMechE, FIMarE, FRSA
Engineering Consultant,
W. S. Atkins Consultants Ltd

Contents

4.1 Introduction

Plant layout can affect the total operation of a company, including the production processes, equipment, storage, dispatch and administration. It has a direct effect upon production efficiency and economics of the operation, the morale of employees and can affect the physical health of operatives.

A production facility will be considered as a facility for processing pharmaceuticals or food products or manufacturing engineering products or consumer goods. The facility must utilize real estate, equipment, materials and labour to generate profit for investors and, philosophically, to enrich the life of all associated with it.

Layout planning involves knowledge of a wide range of technologies which will extend beyond those of individual planners and the full range of expertise may not exist in a production facility. Consultants can provide the expertise but guidance can be found in the published works listed in the References.

The design methods presented here allow a layout plan to be quickly formulated. The methods rely upon a thorough understanding of factory operations gained from experience and a good understanding of the relationship between people and equipment. When such an understanding is not present a more rigorous approach is recommended. Muther[1] published a formalized procedure in 1973 and Tompkins and White[2] a more academic method in 1984, both valuable contributions to problems of layout planning.

The first step in any design is to identity the real need, and this is often the most difficult task. Without it, designs can be produced which do not satisfy the requirements and the end result is often unsatisfactory. It is essential to clearly define the objectives of the task and to re-confirm the objectives as time progresses. A useful aid is a value analysis at the end of the concept design stage. This assesses the design for value for money while meeting the defined project objectives. A good source document is *A Study of Value Management and Quantity Surveying Practice*, published by The Royal Institute of Chartered Surveyors.

A criterion of effective plant operation used to be efficient utilization of capital equipment. The main requirement is now recognized as short door-to-door times, not short floor-to-floor times. The prime need is therefore to achieve a plant layout which facilitates reception of raw materials and dispatch of finished goods in the shortest possible time with minimum capital tied up in work in progress (WIP). This involves access to the site, reception of goods vehicles, raw material goods storage and issue to production, procurement of component parts and sub-assemblies from sub-contractors, process technology and process routes, integrating the sub-contractors' supplies, finished goods storage and dispatch to the customer.

4.2 Technological development and its effect upon plant layout

There have been important world developments in the philosophy of factory operation and they affect plant layouts. These developments, employed by companies now called World Class Manufacturers (WCM),[3] must be examined by the company management who are contemplating establishing a new factory or expanding or reorganizing an existing plant.

Much of the new thinking on strategy has centred on Advanced Technology and the exploitation of the computer, but the following recommendations, requiring minimum investment in new plant, should be considered first:

1. Establish the operational priorities:
   (a) Production to be sales driven;
   (b) Make only what can be immediately sold;
   (c) Make every part right first time.
2. Reorganize production equipment:
   (a) Consider process orientation but employ product-oriented flow when possible;
   (b) Organize products into groups which require the same manufacturing equipment;
   (c) Arrange equipment into product cells.
3. Re-form the structure of production teams:
   (a) Form accountability teams for each product;
   (b) Bring staff functions onto the shopfloor and into the teams, form quality circles and establish total quality control (TQC).

These ideas, interestingly described by Goldratt and Cox,[4] provide flexible production and cater for customer requirements. They have become known as Just In Time (JIT) manufacture, but are often overlooked by many who think that investment in computer-controlled equipment is the only way to go. The concept, simplify before you automate, can provide significant improvements in production and can point the way to later Advanced Manufacturing where:

1. Equipment cells can be replaced by computer-coordinated machines – direct numerical control (DNC), flexible manufacturing systems (FMS), flexible manufacturing cells (FMC);
2. Warehousing automated to include computer-controlled storing and retrieval (AS/AR) and movement by automatic guided vehicles (AGV);
3. Whole manufacturing processes computer linked (CAM);
4. Computer-aided design (CAD);
5. Sales and management computerized, leading to computer-integrated manufacture (CIM).

The development of these technologies can be followed in the proceedings of the professional institutions and the many magazines, some available free.

You should define the state of technological awareness in the company and consider how this affects the plant layout.

4.3 Layout planning concepts

Designs are often executed by a process of analysis. The methods vary from the traditional and well tried to experimental techniques. These include:

1. A conventional plan drawing;
2. Cardboard cut-outs depicting blocks of buildings pinned to an outline drawing of the site;
3. Three-dimensional scale models;
4. Software calculation and simulation packages.

Proprietary systems are available[10,11] employing Lego type blocks which can be constructed and placed on a base board to represent a three-dimensional visualization of the layout. Another system employs aluminium castings of the actual items of equipment and operatives. Cardboard cut-outs have been the most commonly used method, particularly with those who rarely plan layouts.

The above methods allow proposed layouts to be visualized, analysed and altered. These methods are often a try-it-and-see process, and the associated analysis is rarely structured. The old saying that 'If it looks right, it is right' is often used to justify results, but it may be more true to say that if it looks right it is conventional, and the analysis method will rarely produce something original. This applies, of course, to any design activity, whether it is a plant layout, a piece of machinery or a consumer product.

Computer-aided methods are increasingly being used to assist design and a process of synthesis, described later, employs techniques to produce a design which will satisfy predetermined criteria. This produces a layout which can be subjected to critical review, and suggested changes to the layout can be examined for violation of the criteria.

The introduction of computers to many companies allows proprietary software to be used for layout design.[5] Spreadsheet, mathematical modelling and computer-aided design (CAD) techniques are available and greatly assist the design process, and have added to the resources available to planners. However, the traditional scale models described above will still be useful to present the final result to management and shopfloor personnel.

4.4 Plant data

4.4.1 Collection

The objective of the production process is to produce the right goods of the right quality and at the right price in order to generate a financial profit from the capital investment, but the plant layout fundamentally affects this objective.

The essential requirement for any design exercise is a thorough understanding of the working medium. For layouts this means collecting sufficient data to describe those characteristics of the company which affect the layout. Data collection is time consuming and difficult, and it requires the contribution of factory personnel from many departments.

The data required includes:

1. *The company organization structure*
   All departments, including maintenance
   This should be readily available from the personnel department but may need to be updated.
2. *Number of employees in each department and shift*
   Administration staff
   Technical staff
   Factory workers direct
   Factory workers indirect
   This and the next item will be obtained from the personnel department, but the actual working situation will probably need to be confirmed with the individual managers of the various process departments.
3. *Hours and the arrangements for shift working*
   Administration and technical staff on day shift
   Difficulty is often experienced when workers are shared between departments. Some information will not be documented but since employees are generally paid for attendance hours these are meticulously recorded.
   Factory workers on each of the worked shifts
   Consider the arrangements for holidays
   Quantify work contracted to outside organizations
   Work contracted out is usually difficult to quantify. The accounts department will probably have records of the cost of such work and this can be employed to provide an indication of the significance of the work in relation to the operations within the company.
4. *Process flow diagram*
   Description of the products
   Description of the processes
   Flow diagram showing process material flow, scrap and recycled material. Brief descriptions of the products are usually sufficient for the early stages of analysis of the company operation. The information will be amplified during the subsequent stages of the study.
5. *Layout plans for all areas of the site*
   Calculate the covered areas for all process areas. Calculate the areas taken up by services such as the boiler house, fuel-storage farm, etc. Layout plans are often available but the areas devoted to the various activities in the plant will need to be discussed with the individual area managers. The actual area in square metres will need to be estimated by scaling from the drawings and measurement on site. List the items of equipment on the site. Identify the location of the items on the layout plan. This information may exist. It will be needed for designing the layouts of the internals of the factory buildings and it is useful to initiate the quest for the information during the early stages of the project.
6. *Vehicle movements*
   The *types and numbers of vehicles* moving around the site and vehicles arriving from outside the plant. Identify interdepartmental vehicle movements. It is useful to start with the records from the gatehouse, which will indicate the number of vehicles arriving at the plant each day. This will need to be confirmed with the plant manager who will know if abnormal conditions existed during the time for which the records are valid. Aim to establish average and maximum numbers.

7. *Consumables*
The *quantities* of main consumables used. Electricity, water and gas will be included in this item but of more importance to the plant layout are those items which occupy site space. Fuel oil is such an item, and it is important to establish whether the existing facilities, e.g. a fuel-storage farm, are sufficient for the planned operations in the factory.
8. *Effluent*
It should be established whether the area allocated to effluent treatment is sufficient, whether the technology is up to date and determine what influence the findings will have upon the effluent treatment area and the whole plant layout.
9. *Site features*
Site features can be obtained by observation and ordnance maps may be available from the drawing office. The features which need to be recorded are listed in Section 4.7.2.
10. *Future plans*
It might be assumed that if an examination of the plant layout is being carried out, the future plans, as they affect the site layout, will be known. This is not always the case, therefore management should be questioned at the beginning of the project. Particular attention should be paid to hazardous substances which may be used on the site, and the health and safety document[6] should be consulted.

Pareto, the Italian economist, observed that 85% of wealth is owned by 15% of the population, and it is sometimes stated that 15% of a company's products generate 85% of the turnover. Some workers have interpreted this to mean that 15% of the data will suffice, but Tompkins and White[2] demonstrated that this is not always true, so data collectors should use their judgement regarding what is sufficient.

The data needed to plan the site layout are shown in the form of a questionnaire in Figure 4.1.

After the site layout has been determined the layout of equipment inside the buildings will need to be considered. Details of the individual items of equipment will be needed, and Figure 4.2 lists the required information in the form of a questionnaire.

The areas for administration and amenities will need to be identified and in some countries space will be required for accommodation and a mosque.

4.4.2 Analysis

The collection and tabulation of data will provide an opportunity to become familiar with the company and will generate a good understanding of the operation of the plant. Much of the data will be needed to establish the models needed to size the various areas of the plant. For instance, hours worked should be analysed to identify shift work, direct and indirect effort, and the normal and overtime activity. Outside contractors may be employed, and this contribution will need to be determined, sometimes by visiting the contractor's facilities. When expansions are being planned it may be necessary to

Description: Manufacturer:

Sketches or Manufacturing literature attached Y/N: ....

Dimensions: Equipment Size (m)

Working envelope (m) ... L x W x H Equipment weight (kg) ....

LOCATION: Building No: Zone:

ELECTRICITY:

Single phase, isolated	Flameproof (Zone 1) ....
Three phase, isolated	Flameproof (Zone 2) ....
Three phase & neutral, isolated	Emergency supply ....
Direct current	Controls required ....
Waterproof	Emergency stop ....
Clean supply	Voltage ....
Special earthing	Power requirement (kW) ....
Uninterruptible supply	Other voltage ....
	Other frequency ....

WATER: FLOW: lit/sec PRESSURE: bar CONNECTION: Details

Cold/hot ....

Soft ....

Demineralised .(Silica free Y/N ...) ....

Distilled ....

Drinking ....

Circulating Cooling .... (Softened Y/N ....)

Flow temp (°C) .. Return temp (°C) ....

DRAINAGE:

Gravity	Discharge flow rate (litres/min) ....
Pumped	Discharge temperature (°C) ....
Solids/Sediments	Outlet connection size (mm) ....
Oil/Grease/Wax	Number of outlet connections ....
Collection Tank	

Hazardous (eg. radioactive/solvent/biological/chemical) give details ....

....

....

STEAM:

CONDENSATE RECOVERY: Y/N	Kg/hour	psi
COMPRESSED AIR:	litres/sec	bar
NATURAL GAS:	litres/sec	bar

Model No. .... Equipment Identification No ....

SPECIAL GASES: .... Y/N If YES complete separate SPECIAL GASES DATA SHEET

FUME EXTRACTION:

Located in fume cupboard ....Y/N If YES complete FUME CUPBOARD DATA SHEET

Fume hood ....Y/N If YES complete FUME/DUST HOOD DATA SHEET

Local extract ductwork required	Scrubbing required ....
Discharge to central system	Filtration required ....
Discharge through wall (inbuilt fan)	Extraction rate (m^3/s) ... .
Connection of ductwork to equipt	
Connection size and type	Type of fumes ....

MOUNTING:

Free standing	Wall mounting ....
Static/Mobile	Antivirbation mounts (type) ....
Floor fixing	Concrete plinth (m) .....L x W x H...
Bench mounting	Depth of embedment of bolts ....

Specify specific spacing or access requirements ....

SAFETY AND OPERATIONAL CONSIDERATIONS:

Heat (kW)	Radiation ....
pn Noise (decibels)	Dust/Fumes/Vapours ....
Chemical	Fire/Explosion ....
Moving parts	Biological ....
Vibration (frequency)	Others (specify) ....
(amplitude)	(acceleration) ....

ADDITIONAL INFORMATION/COMMENTS:

Frequency of use of equipment ....

Duration of usage ....

Please list other users of this equipment ....

....

Will this equipment require:	Specialist disconnection? ....
	Specialist reconnection? ....
	Specialist commissioning? ....
	Specialist transport? ....

Figure 4.1 Equipment data sheet

The organisation is segregated into divisions, departments, sections, rooms etc. A separate Basic User Requirement (BUR) form should be filled in for each section so that those with special needs can be identified by the planner.

DEPARTMENT/SECTION - NAME: ....

STAFF LEVELS: Indicate split function/responsibilities of staff.

Grade	Number
....	....
....	....

FLOOR AREA: (square metres) excluding corridors and common areas etc.

	Current/Projected		Current/Projected
Office	....	Workshop	....
Machine Room	....	Lab Wash-Up	....
Clean Room	....	Computer Room	....
Tank Farm	....	Storage - General	....
- Lockers	....	- Disks	....
- Fuel	....	- Acid	....
- Solvent	....	- Paint	....
- Temp. Controlled	....		
- Process areas - List	....		....
Car park	....		....
Other -	....		....

FLOOR TYPE: No Special Requirement ....
Vehicular traffic (detail heavy, long, awkward loads) ....
Fork Lift .... Drain ....
Acid Resistant .... Solvent resistant ....
Static Conductive .... Other ....

REQUIRED LOCATION:
Ground Floor .... First Floor ....
Basement..................... Not Critical ....
Proximity to other facilities ....
Other eg. proximity to other facilities ....

CEILING: No special requirement ....
Special requirement - act as plenum ....
Lighting - No special requirement ....
Integral..................... tear drop ....
Yellow .... flameproof ....
Non-Fluorescent .... other ....
Lighting levels see Room Environment

INTERNAL WALLS AND WINDOWS:
Internal Walls - Standard partition walls unless otherwise stated.
Acoustic .... Soundproof ....
Stainless Steel ....Electro Static Discharge...............
Other - Non Reflective? ....

EXTERNAL WINDOWS: No Special Requirement ....
Special Requirements ....
Fixed ....Opening ....
Sealed ....Grilles ....
Blinds ....Curtains ....

INTERNAL DOORS: No Special Requirements ....
Special Requirements ....

ROOM ENVIRONMENT: - Not critical ....
Noise Level 40dBA ... 45dBA ... 50dBA ... 55dBA ... 60dBA ... Other ...
Temperature - Non critical ....
Special requirement .... Temp °C .... tolerances ....
Relative Humidity
Not Critical .... Special Requirement ....%
Lighting Levels (LUX) ....
Cleanliness Based on British Standard 5295 measured at working level:
Class 1 .... 2 .... 3 .... 4 .... 5 ....
Air Lock ....Air shower ....
Storage - Lockers .... Overalls .... Other ....
Security - Secure Area ....Contraband ....
No Smoking ....Controlled ....
Recorded Access .... Locks .... Grilles ....
Secure Communications System .... Other ....

GASES:
Oxygen .... Carbon Dioxide ....
Nitrogen - Ordinary .... High purity ....Ultra pure ......
Hydrogen .... Helium ....
Argon .... Air: .... Press .... (Bars)............
Other ....

Figure 4.2 Basic user requirements schedule

discuss the contributions which outside contractors make to the factory and assess the increase which they can accommodate.

4.5 Process/site layout modelling

Site and factory layout designs are needed when new installations are being planned or changes to existing facilities are investigated. New layouts are examined when increased efficiency is being sought and when plant managers are planning expansions to the existing installations. There is also a need for plant managers to anticipate the changing requirements of both site and plant over future years.

When a new installation is being planned the site requirements will be predetermined, but when future expansions are being investigated a model for sizing is invaluable.

4.5.1 Computer modelling

The use of computer models to assist in plant layout decisions can often be helpful, but a clear understanding of what they can and cannot do is needed. Used blindly, they can lead to solutions which are, in some sense, optimal but which have little practical merit.

In discussing the benefits and limitations of modelling we distinguish between three types:

1. Optimizing calculations for plant layout;
2. Flowsheet models for sizing the individual items of equipment;
3. Simulation models for understanding the interaction between different manufacturing units.

Optimizing plant layouts

Historically, modelling techniques have had little impact on the problem of designing plant layouts. This has been for two reasons. First, the calculations are difficult, and second, practical constraints and considerations essential to the decision are often ignored.

To understand the mathematics, consider a large empty space into which a number of production units are to be placed, and assume that the major variable to be optimized is the cost of transporting materials between them. If the manufacturing process is essentially a

SERVICES SUPPORT:
Vacuum pressure ....
Water - Town mains .... Specific pressure ....
- Softened .... De-mineralised ....
- Chilled (open circuit) . Distilled ....
- Heating ....
- Process cooling water (closed circuit) ....
Other ....
Fume Cupboards Ordinary ...... Ordinary..............................
Acid .... Solvent ....
Water Drench .... Radioactive ....
Other ....

EXTRACTS:
to atmosphere .... scrubbed..............................
filtered .... dedicated ....

CRANES: (Tonnes)
Overhead .... Gantry ....
Manual .... Electric ....

ELECTRICS:
Total power KW ....
Process/machine load KW ...... No of machines ....
Other load KW ....
Bus bar system/Trunking system...................................
3 Phase .... 3 phs + Neutral ....
Clean Supply .... Uninterruptable ....
Stand-by Generator .... Emergency Lighting ....
13A Sockets (do not specify number) ...... with RCCB ....
Earthing requirements ....
Faraday Cage ....
Special Lightening requirements......Other ....

CONTAMINATION: Do you need to be separate from:
Noise .... Vibration ....
Electrical Interference ..... Fumes ....
Gases .... Others ....
Vibration Requirements ....
Equipment Specific .... Room in total ....
Detune Structure ....
Electro Magnetic Interference Counter-Measures ....
Electro Magnetic Interference from Satellites aircraft etc. ....
Screened enclosures .... GHz ....

WASTE AND EFFLUENT: (Large quantities - please specify)
Paper .... Cardboard ....
Glass .... Metal ....
Solvents .... Acids ....
Cutting Fluids .... Other ....
Powders .... Radioactive material ....
Special wastes ....
Do you generate contamination? Please specify levels where known:
Noise .... Vibration ....
Electrical Interference...... Fumes ....
Gases .... Others.................................

FIRE:
Local Fire regulations will apply throughout but special requirements which need to be drawn to the Fire Adviser's attention should be stated below:
No Special Requirement ....
Special Requirement ....

SAFETY HAZARDS - not brought out in answers above:
....

FLEXIBILITY: How often do you anticipate changing your layout ... yrs

Figure 4.2 *(continued)*

flow-line operation, then the order in which units should be placed is clear (from the point of view of transport costs), and the problem is simply to fit them into the space available. In a job-shop, where materials are flowing between many or all the production units, the decision is more difficult. All the potential combinations of units and locations could be enumerated on a computer and the resultant cost of each option obtained. The number of combinations quickly becomes large, however, and more sophisticated methods, such as branch-and-bound techniques, are required for complex problems. Alternatively, the location of subsets of plant, within which the layout is assumed to be fixed, could be optimized.

Solutions of this nature ignore practical considerations of noise, hazardous areas, etc. unless these are specifically entered as constraints. Furthermore, if the site is geometrically complex, then additional detail will need to be included in the problem formulation.

Flowsheet models for plant sizing

The consistent sizing (i.e. balancing the capacities) of the equipment which is to make up a plant is obviously of importance in the overall design. Models used for this type of decision are usually an extension of the manual calculations that a designer would normally make.

They are based on a network of activities (or machines) with flows of materials between them. Relationships between the activities are expressed in terms of yields or unit consumptions and, for a given output of finished products, the required output from each of the other activities can be calculated.

The complexity of the plant and the type and number of sensitivity tests to be carried out will determine whether a formal model or a simple calculation is needed.

Flowsheet models assume that activities operate at a constant rate. This is an important limitation, and may be an oversimplification. To overcome this, simulation techniques can be employed.

These models, however, can be useful if estimates are to be made of operating costs. By assigning fixed and variable costs to each activity, average and marginal unit costs at each stage of the process can be easily calculated, which will assist in decisions regarding pricing policies or whether to buy in components and materials or make them on site. ATPLAN[12] is an example of a network-based model of this type.

Simulation models

Simulation models aim to replicate the workings and logic of a real system by using statistical descriptions of the activities involved. For example, a line may run at an average rate of 1000 units per hour. If we assume that this is always the case, we lose the understanding of what happens when, say, there is a breakdown or a halt for routine maintenance. The effect of such a delay may be amplified (or absorbed) when we consider the effect on downstream units.

A simulation model has 'entities' (e.g. machines, materials, people, etc.) and 'activities' (e.g. processing, transporting, etc.). It also has a description of the logic governing each activity. For example, a processing activity can only start when a certain quantity of working material is available, a person to run the machine and an empty conveyor to take away the product. Once an activity has started, a time to completion is calculated, often using a sample from a statistical distribution.

The model is started and continues to run over time, obeying the logical rules that have been set up. Results are then extracted concerning throughputs, delays, etc.

It is clear that simulation models can replicate a complex production system. They can be used to indicate the level of shared resources needed by the operation (e.g. forklift trucks or operators), the speed of lines, sizes of vessels or storage tanks, etc.

A number of packages are available for quickly building simulation models. HOCUS[13] and SEE-WHY[14] are two major UK systems. They allow graphical displays to be used to show, for example, the movement of men and materials between machines. This can be useful for understanding the way in which the operation is reacting to particular adverse circumstances and can assist in designing methods of avoiding them (for example, by building in redundancy).

Simulation models can be expensive to build and the results obtained need to be analysed with care because they are statistical in nature. For example, two runs of the model may give different results – just as the performance on two real days in a factory can vary. Sufficiently large samples need to be taken therefore for a proper understanding of the performance of the plant.

4.5.2 Model construction

A simple model can be produced using the areas required for the various site activities for different scenarios. These might include redistribution of a company's manufacturing facilities, changes in the market demand, or simply increased factory output.

Constructing the model is an area where innovative effort is needed to maximize the validity of the model. The model will vary from industry to industry, but a simple one can start with a relationship between the number of employee hours and the output produced in each discrete activity area on the site, e.g.

Activity area	*Output*	*Manning hours*
Raw material store	Tonnage/volume handled	No. and shift no.
Process areas	Output tonnage or volume	No. and shift no.
Finished goods store	Tonnage/volume handled	No. and shift no.
Maintenance	Specific area output	No. and shift. no.

Service facilities, electrical sub-station, water treatment, fuel storage, etc. will depend upon process parameters, mainly tonnage or volume handled, and may need to be assessed from the summation of the individual activity areas, but it will be affected by technology when improved techniques are being introduced.

Office accommodation will depend upon the factory manning and the car parking facilities upon the total manning during the day shift.

4.5.3 Model for change

The recent introduction of inexpensive desk-top computers has allowed their extensive use throughout many companies. The standard spreadsheet packages which accompany these machines enables the above data to be laid out in an interactive way, so that 'what if' situations can be explored at the planning stage and the implications of, for example, market trends in the food industry, to be examined over the long term for its effect on the plant layout. The model may include a factor to take into account improvements in technology and working practices in both the office and factory.

In determining the area required for increased production the relationship between the output and the area may not be linear, and needs to be examined in some detail. When the output required exceeds the maximum output of a process line then increased shift working may be considered. Alternatively, a second process line may be needed.

The model for the maintenance requirements of a large metallurgical plant was recently constructed by relating the man hours to the production output, but the relationship was not linear, and a more sophisticated model was needed to examine the site requirements for different output tonnages.

Mechanical wear is a feature which is significant among the causes of machinery failure and the James Clayton Lecture delivered to the Institution of Mechanical Engineers in 1981 discussed friction and wear and identified tribological losses as 45–50% of all maintenance costs in metallurgical plants. This suggested that mechanical transmission of energy could be investigated to obtain the relationship between production output and the maintenance requirements.

The model equation produced was:

$$H_2 = H_1 \times K \times \sqrt{(P_2/P_1)}$$

where

H_2 = Future man hours,

H_1 = Current man hours,
K = Technology or practices factor,
P_2 = Future production output,
P_1 = Current production output.

This produces a global requirement chart and a chart showing the discrete, departmental requirements in Figure 4.3.

New plants and expanding plants rarely produce maximum output on start-up. In large capital plants it may need many months for staff training and equipment commissioning before output is sufficient to need maximum manning. The technology factor K in the above equation can be used to reflect this. In Figure 4.3(a) a technology factor of 0.9 has been used.

It is important to test models, and in the above case, figures were obtained for other plants and the validity of the model confirmed.

4.5.4 Determination of factory areas

When existing factories are being examined the floor areas occupied by the various items of equipment can be measured on site during the data-collection period. It is important to consider and include the equipment outline (maximum travels), space for maintenance activity, operator movement and the handling of material and workpieces. In many cases the space needed for handling exceeds that required to accommodate the operative part of the equipment. Aisle space is also needed for through-shop movement.

When planning new installations manufacturers' catalogues may be the only information source of equipment data, but they rarely indicate the space needed outside the equipment outline. This requires careful consideration from an experienced facility planner to establish realistic space requirements.

In the absence of reliable information and during the preliminary planning stage (feasibility studies, etc.) the following occupancy values can be employed:

Activity	*Occupancy* (m^2/*person*)
Service industries	15–30
Manufacturing	20–35
Machine tool shop	20
Electrical/plastics	25
Mechanical fitting	27
Pottery/glass	35
Transport	30–65
Cars/mobiles	33
Locos	37
Wagon shop	55
Lorries	65
Distributive trades (average)	–80

These values will change with the future move towards unmanned factories.

Dividing the areas established in Section 4.5.3 by these occupancy values will give building areas. Bay widths can be established for each type of building and dividing the areas by this width will establish a building length which can be arranged into a convenient building block shape.

4.6 Design synthesis

4.6.1 Plant activities and intercommunications

In 1973 Richard Muther[1] published a method of analysing the interrelationships of activities within industrial plants, and the method allows a high degree of detail to be examined. The method proposed here is similar in that a relationship grid is constructed, but this technique employs the power of the modern desk-top computer to rapidly examine alternative layouts to obtain best solutions.

The design synthesis starts by listing the activities or areas in the plant and indicating the access which each activity needs to the other activities. The access may be required for internal process, traffic, people movement or plant services. Rather than go into the detail of examining all interrelationships for all of the above requirements, the experienced layout planner can decide which requirement predominates and employ it to produce quick results. The method allows inexperienced planners to iteratively examine the implications of changing requirements as the project proceeds and plant knowledge increases.

Table 4.1 shows the construction of a typical activity chart using one of many computer spreadsheet packages. The cross (+) identifies an activity in the plant and stars (*) in the vertical lines denote where access to other activities is needed or an interrelationship exists. A wide scattering of the stars away from the diagonal line of crosses indicates large distances between the activities and thus large communicating distances.

A feature of PC spreadsheets is that the position of rows or columns can be changed at the touch of a key, so the sequence of the rows of activities can be rapidly altered, and the position of the columns varied to preserve the diagonal pattern of crosses. This exercise can be executed a number of times to reduce the scatter of the stars away from the diagonal line and reduce the communicating distances between the activities. This has been carried out, and the result is shown in Table 4.2, where it can be seen that the stars are clustered more closely around the diagonal line. The list now shows the activities arranged in a preferred order of sequence.

4.6.2 Location criteria and boundary groups

There may be reasons why individual activities cannot be located near to those which are adjacent in the determined sequential list, so design criteria are needed to determine which activities are compatible and can be near, and which are incompatible and cannot.

Particular plants will have characteristics which will determine the importance of particular criteria, but

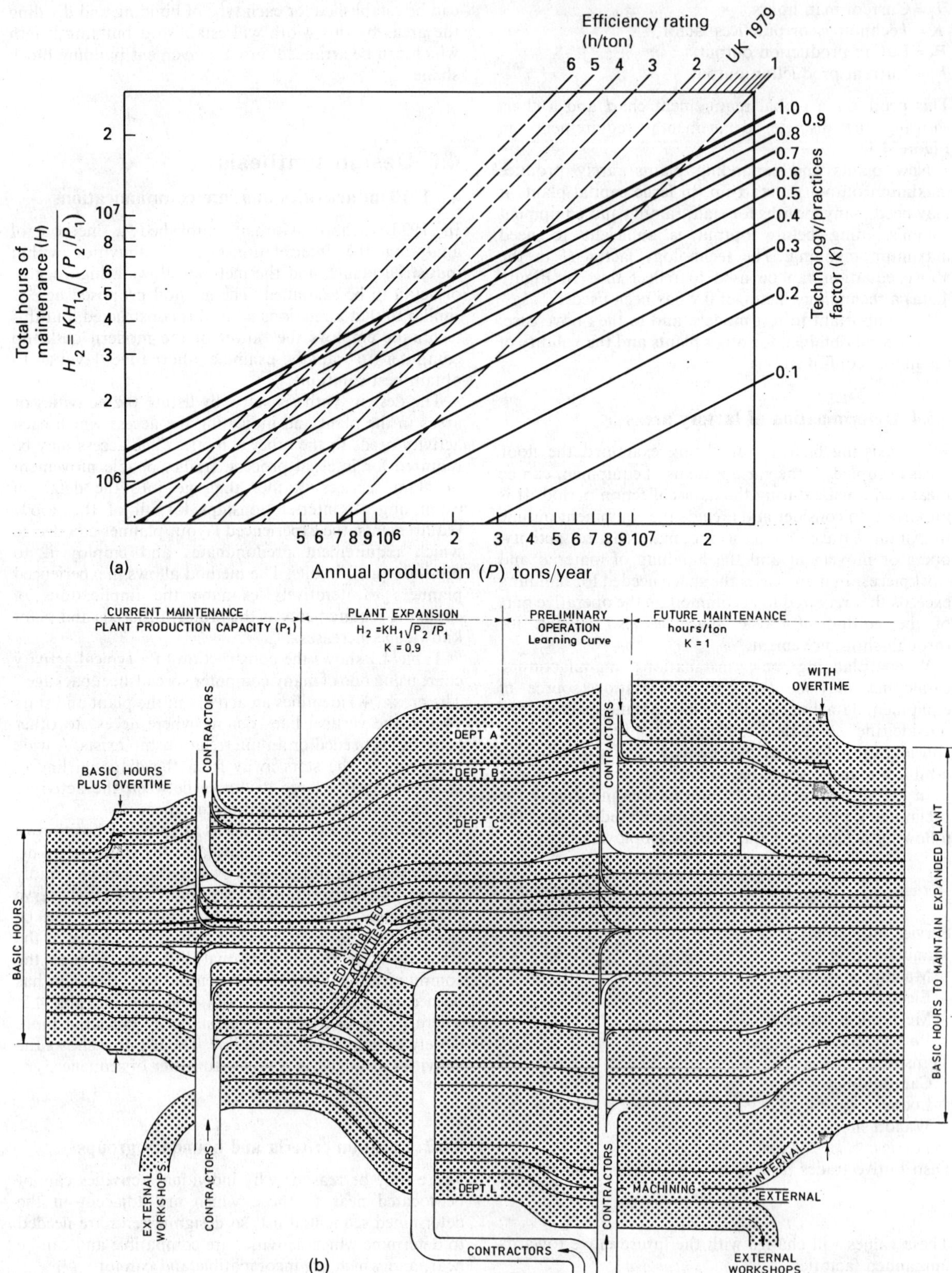

Figure 4.3 (a) Model of manning requirements related to plant output; (b) Sankey diagram of the changing manning requirement in an expanding plant

Table 4.1 Plant activity intercommunications: first design criteria

Activity	*Access to each other*																									
Heavy plant repairs	+	*	*		*		*				*	*		*		*							*	*		
Vehicle hot washing	*	+		*		*			*							*								*		
Vehicle cold washing	*		+	*		*			*															*		
Public health repairs		*	*	+	*							*	*	*		*							*	*		
Store for large tyres	*			*	+						*									*			*	*		
Service station		*	*			+										*							*	*		
Electrical repairs	*						+			*						*							*	*		
Bus washer								+								*										*
Vehicle test area		*	*						+														*	*		
Mechanical repairs							*			+		*		*		*							*	*		
Tyre shop	*				*						+					*				*			*	*		
Welding shop	*			*						*		+	*		*	*							*	*		
Panel beater				*								*	+		*	*							*	*		
Radiator shop	*			*						*				+		*							*	*		
Paint area												*	*		+	*							*	*		
Compressor house	*	*		*		*	*	*		*	*	*	*	*	*	+										
Tailor																	+					*	*	*		
Carpenter																		+	*				*	*		
Store – carpenter																		*	+				*	*		
Store – tyres					*						*									+			*	*		
Store – tailor																	*				+		*	*		
Store – paint															*							+	*	*		
Store – general	*			*	*	*	*		*	*	*	*	*	*	*		*	*	*	*	*	*	+	*		
Administration block	*	*	*	*	*	*	*		*	*	*	*	*	*	*		*	*	*	*	*	*	*	+		
Filling station																									+	*
Vehicle park								*																	*	+

Table 4.2 Rearranged activities to improve plant operation

Activity	*Access to each other*																									
Filling station	+	*																								
Vehicle park	*	+	*																							
Bus washer		*	+						*																	
Vehicle hot washing				+		*			*	*	*		*	*												
Vehicle cold washing					+	*				*	*		*	*												
Service station				*	*	+			*		*	*														
Tyre shop							+	*	*		*	*	*													
Store – tyres							*	+			*	*			*											
Vehicle test area				*	*				+	*	*	*														
Compressor house			*	*		*	*		*	+			*	*		*	*	*	*	*	*					
Administration block				*	*	*	*	*		*	+	*	*	*	*	*	*	*	*	*	*	*	*	*	*	*
Store – general						*	*	*		*	*	+	*	*	*	*	*	*	*	*	*	*	*	*	*	*
Heavy plant repairs				*	*		*		*		*	*	+		*	*		*	*							
Public health repairs				*	*				*		*	*		+	*			*	*	*						
Store for large tyres								*			*	*	*	*	+											
Electrical repairs									*		*	*	*			+	*									
Mechanical repairs									*		*	*				*	+	*	*							
Welding shop									*		*	*	*	*			*	+		*	*					
Radiator shop									*		*	*	*	*			*		+							
Panel beater									*		*	*		*				*		+	*					
Paint area									*		*	*						*		*	+	*				
Store – paint											*	*									*	+				
Carpenter											*	*											+	*		
Store – carpenter											*	*											*	+		
Tailor											*	*													+	*
Store – tailor											*	*													*	+

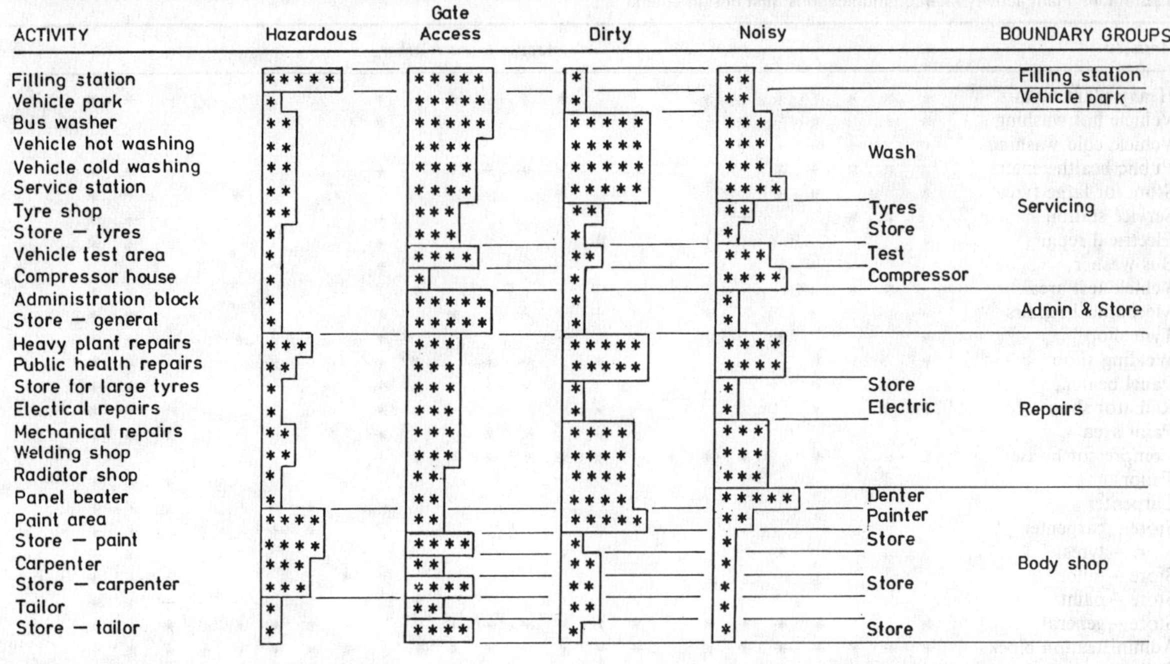

Figure 4.4 Compatible activities and boundary groups: secondary design criteria

typical criteria will include the following:

1. The hazardous nature of the activity;
2. The ease of access to the entry gate which the activity needs;
3. The amount of dirt and potential contamination which the activity would generate;
4. The amount of noise and potential disturbance which the activity will normally generate.

Each of the listed plant activities can be assessed for each of the chosen criteria and a star rating determined for each assessment. An experienced engineer will find that a qualitative assessment is sufficient, but quantitative measurements can be used if necessary. The results of this exercise are shown in Figure 4.4.

For each of the criteria a two-star rating difference between an activity and an adjacent activity is taken to indicate incompatibility. Activities which are hazardous, dirty, noisy or needing good access to the entry/exit gate are consequently rated with many stars. In the site layout they should not be located near other activities with fewer stars because they could adversely affect that activity. However, an activity which is not critical and has a low star rating may be located next to an activity with many stars.

Boundary maps can be drawn around groups of stars which are compatible and horizontal lines drawn through the incompatibility points to show groups of compatible activities on the right-hand side of the chart Figure 4.4.

Groups of compatible activities together with their major characteristic can be listed, e.g. for a transportation maintenance facility, as follows:

Boundary group	*Characteristic*
Filling station	Hazardous
Vehicle park	High traffic density
Servicing facilities	Dirty
Administration and stores	Quiet and clean
Repair facilities	Medium traffic, dirt and noise
Body shop	Noisy
Accommodation	Quiet

The synthesized layout can be described in terms of the characteristics of the various zones within a hypothetical site boundary as shown in Figures 4.5 and 4.6.

4.7 Site layout realization

4.7.1 Process and traffic flows

Traffic and material movement is a major consideration, and should be arranged so that cross flows are minimized and the potential for congestion and accidents reduced. Much useful information on both the analysis and practice of movements is contained in the References.

Using the example of a transportation maintenance facility, Figure 4.7 shows that a counter-clockwise rotation of the traffic flow could achieve minimum crossing and give a graduation of decreasing traffic density in the counter-clockwise direction as vehicles are diverted off the main stream (Figure 4.8).

The circulation diagram indicates that, in a left-hand drive environment, activities with a high traffic density

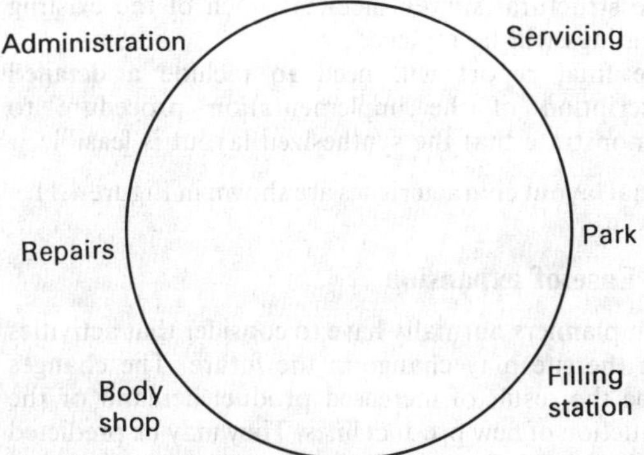

Figure 4.5 Synthesized location of plant activities

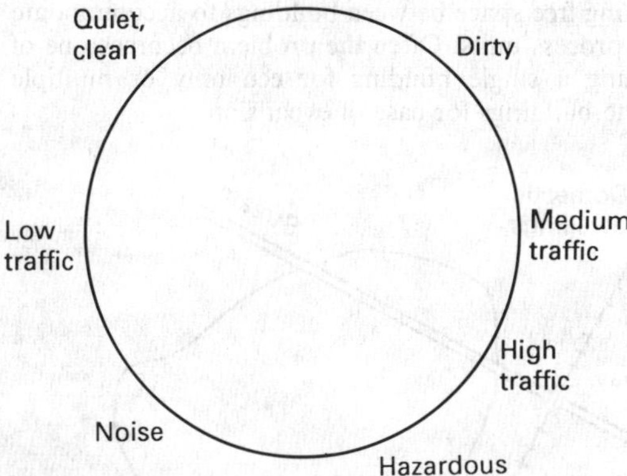

Figure 4.6 Characteristics of the synthesized layout

should be located on the right-hand side of the traffic circulation diagram and the activities with low traffic on the left-hand side.

With this arrangement, the vehicle needing no servicing or maintenance could be parked on the right-hand side of the site in a park designated Overnight Park, and the filling station which handles all the vehicles could be located towards the centre of the site in the flow line of the vehicles leaving the area.

The servicing area could be located on the right-hand side of the site but remote from the gate, and part of the park near to the servicing area could be allocated to vehicles which arrive for their scheduled routine service.

The left-hand side of the site has a reducing traffic density and could thus accommodate the vehicle repair shops. The body shop is a repair function, and thus would be located on the left-hand side but remote from the existing administration office.

The technical administration office needs to be quiet and clean and central to the facilities which it administers. Hence it should be located in the central area between the service and repair shops. This is remote from the dirty, washing facility and the noisy body shop while providing good access to the gate.

The remaining corner on the left-hand side of the site can be arranged to be sufficient distance from the body shop so that it can be quiet and thus suitable for staff accommodation and the plant administration office. Thus the synthesized layout becomes as shown in Figure 4.9.

4.7.2 Site constraints

The real design must consider design criteria derived from the physical constraints of the chosen site and the current

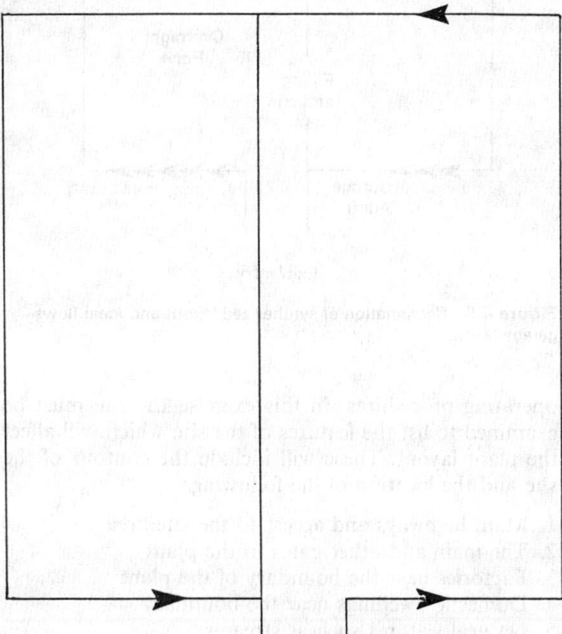

Figure 4.7 Ideal traffic or material flow

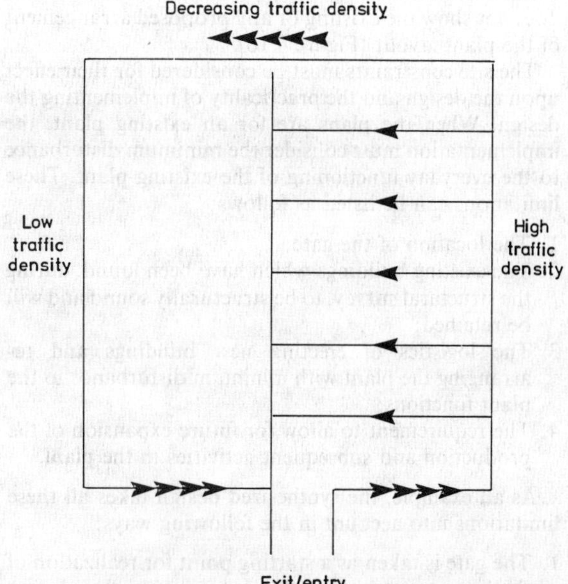

Figure 4.8 Traffic density zones

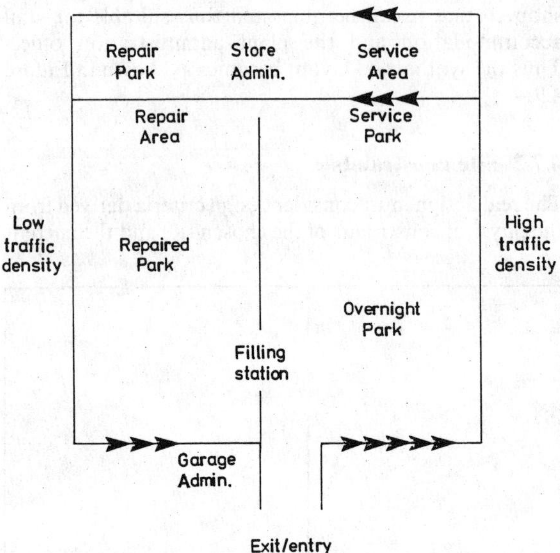

Figure 4.9 Combination of synthesized layout and ideal flows – garage layout

operating procedures. In this exercise the site must be examined to list the features of the site which will affect the plant layout. These will include the contour of the site and the location of the following:

1. Main highways and access to the site area
2. The main and other gates to the plant
3. Factories near the boundary of the plant
4. Domestic dwellings near the boundary
5. Natural features such as streams
6. Effluent routes
7. Prevailing wind directions.

These can be drawn onto a plan of the site area, which does not show the existing or any proposed arrangement of the plant layout (Figure 4.10).

The site constraints must be considered for their effect upon the design and the practicality of implementing the design. When the plans are for an existing plant, the implementation must consider the minimum disturbance to the everyday functioning of the existing plant. These limitations can be listed as follows:

1. The location of the gate;
2. The existing buildings which have been found, during the structural survey, to be structurally sound and will be retained;
3. The logistics of erecting new buildings and rearranging the plant with minimum disturbance to the plant functions;
4. The requirement to allow for future expansion of the production and subsequent activities in the plant.

As an example, the synthesized design takes all these limitations into account in the following ways:

1. The gate is taken as a starting point for realization of the layout and high traffic densities are placed near to it;
2. The structural survey dictates which of the existing buildings will be replaced;
3. The final report will need to include a detailed description of the implementation procedure to demonstrate that the synthesized layout is feasible.

The final layout characteristics are shown in Figure 4.11.

4.7.3 Ease of expansion

Layout planners normally have to consider that activities within the site may change in the future. The changes may be the result of increased product demand or the introduction of new product lines. They may be predicted from the model designed in Section 4.5.2 or allowance for unpredictable changes may be needed.

The dilemma facing the planner is between designing the layout with minimum distance between buildings and allocating free space between buildings to accommodate future process units. Often the problem becomes one of providing a single building for economy or multiple separate buildings for ease of expansion.

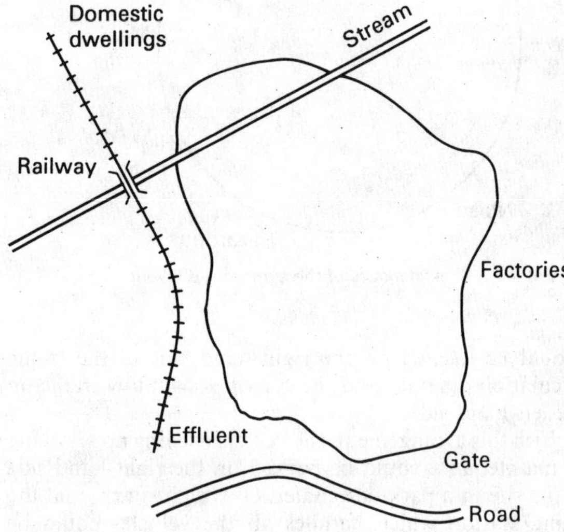

Figure 4.10 Site constraints

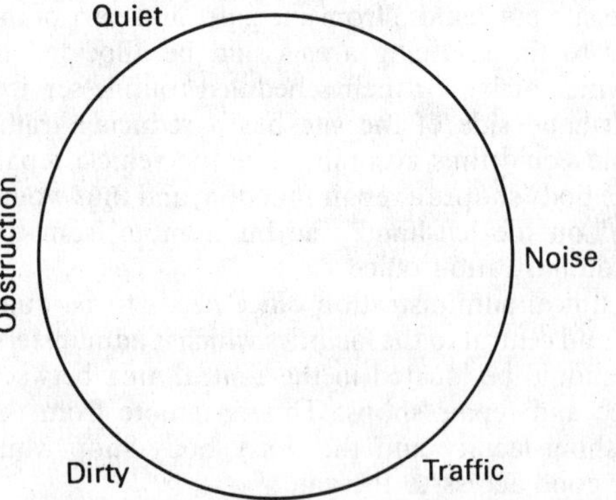

Figure 4.11 Conceptual layout design on the real site

The solution may be a policy decision, but this decision will be assisted by the results of the internal layout configurations, which may indicate how simply or otherwise expansion can be accommodated.

4.7.4 Options

Management usually wish to see alternatives to a particular design in order to say that the alternatives have been considered. Unique design solutions are rare, consequently the layout planner will normally be able to interpret the results of the finding of the study to show alternative possibilities. Each alternative option should be accompanied by a critique listing the advantages and disadvantages.

4.8 Internal layouts of buildings

Material, traffic and people are involved, and the constraints on internal layouts will be different to those for site layouts but will not affect the design process. The constraints will include organization of the factory work tasks, which may involve trade practices and the type of organization adopted for the control of the process equipment. This is discussed in Section 4.2.

4.8.1 Process flows and performance indicators

The same method as described above can be employed to plan the layout of factory building interiors. Establishing the interior activities and characteristics of the activities must, however, derive from a thorough understanding of the products and processes.

The management of production is well described by Hill,[7] but the prime consideration is product flow, and all features of the layout must assist flow. There are some indicators which can be used to measure the quality of the production facilities and these can be employed to demonstrate the viability of the proposed new layout.

Performance indicators

The relative importance of each indicator will vary according to the type, quantity, end quality, variety and value of the product and the capital cost, flexibility and required utilization level of the plant. There are three variables which will serve most industrial processes:

1. *The stock value*. The value of the stores stock represents a significant investment for many manufacturers and financing the work in progress can also represent a significant cost. It is a useful indicator of the financial health of a company and is a prime target for cuts in the drive for increased economic efficiency. Section 4.2 highlighted this.
2. *The cycle time*. This represents the maximum time interval between the start of a single operation on a product on a particular flow line and the start of the same operation on the next product on the same flow line. The cycle time is calculated by dividing the target production volume (annual or monthly) by the working time (hours, minutes or seconds) and then adjusting for the reject rate and the percentage of lost time (down-time). If every operation is conducted within the cycle time then the steady-state functioning of the flow line will produce the required production rate. The number of sequential process operations multiplied by the cycle time is the actual process time.
3. *The door-to-door time*. This represents the total elapsed time from the delivery of the raw materials or component parts to the despatch of the finished product. This figure can be adjusted to control the value of the stock of materials, parts, work-in-progress (WIP) and finished goods in the process. The gross excess stock carried can be found by dividing the door-to-door time by the actual process time and multiplying the result by the stock value. The difference between the gross and the net excess stock values will depend on individual circumstances. Reducing these figures to a minimum requires strategies for purchasing, stock-level monitoring and process control.

4.8.2 Process equipment

When the idealized flow routing has been determined, the equipment must be located and the practicalities of the interrelationships between, and integration of, equipment, services and people must be considered.

Equipment information would have been collected during the data-gathering exercise described in Section 4.4, but a thorough understanding of the process is vital during this phase of the work.

Adequate space for the equipment must include:

1. The machine itself with the maximum movement of all machine elements;
2. Space for maintenance activities, withdrawing shafts, etc.;
3. Process material movement;
4. Operator mobility.

4.8.3 Material handling

Knowledge of material-handling techniques is vital to the layout planner, and detailed consideration will need to be given to the various techniques and equipment which are available. References 2, 8 and 9 contain valuable information and trade journals report the current state of the market.

In the absence of reliable information and during the preliminary planning stage (feasibility studies, etc.) the following space requirements for aisles can be employed:

Traffic type	*Aisle width (m)*
Tow tractor	3.7
3-ton forklift truck	3.4
2-ton forklift truck	3.0
1-ton forklift truck	2.75
Narrow-aisle truck	1.85
Manual truck	1.6
Personnel	1.0
Access to equipment	0.85

Allow extra for door openings

4.8.4 Storage

Storage in plants range from simple facilities to fully automated flow through warehouses. Simple facilities are installed to contain the day-to-day requirements of, for instance, maintenance workers, and the just-in-time supply of components on the production floor. Fully automated flow-through warehouses are employed in the food industry where the shelf life of many products is short and a high throughput is involved. Much has been written about storage and References 8 and 9 are excellent.

4.8.5 Electrical distribution

The plant layout design and whether the processes involve dry, wet or contaminated environments with possibly elevated temperatures will affect the principles adopted for the support and containment of electric cables and bus bars. Cables from the incoming supply sub-station to load centres would normally be run in the ground, either directly buried with suitable route marking and protection, or in ducts or concrete trenches. Depending on the number of cables, the latter may incorporate support trays or cleats fixed to the trench walls.

Inside the factory buildings, below-floor trenches are frequently the preferred method for main routes, but where there is danger of flooding, due to storms or process fluids, above-ground routing must be considered. Single or few cable runs could be clipped to walls and building structures but multiple cable runs need traywork, involving space equivalent to major pipes and air ducting, and routing must be coordinated with other services.

Busbar trunking systems are frequently employed in machine shops to facilitate relocation of the equipment. The busbars are located overhead, suspended from the roof structure, and arranged, in conjunction with the lighting system, in a suitable grid pattern. Fused plugs and sockets provide outlets to individual items of equipment.

In some installations (e.g. clean rooms) the electrical distribution system can be located in the sealed ceilings, allowing maintenance access without affecting the clean facilities (Figure 4.12).

4.8.6 Building utilities

These include the canteen, toilets, workers' rest rooms, access routes and the car parks. Hygiene, traffic density, traffic flow and sizing need to be addressed and have been discussed in the section on the sizing model.

4.9 Selling the concept

The site layout needs to be presented to management, union, staff and possibly others to obtain approval, and the design procedure described here, together with a visualization, will assist the presentation.

Photographs of the proposed site are useful and a site layout drawing is needed, but a perspective artistic impression showing buildings with architectural facades, vehicles and other site activities improves the presentation. Employing three-dimensional models[10,11] discussed in Section 4.3 helps communication and allows layout options to be easily demonstrated and discussed.

4.10 Implementation

One of the principal tasks facing the manager responsible for implementing a new plant layout is the planning and scheduling of the many activities involved in the project. Modern computerized systems have evolved into easy-to-use tools that avoid the need for a detailed technical knowledge of planning techniques.

Project plans for the installation and commissioning of new plant should be prepared in order to:

1. Provide a readily understood and concise description of the scope of work for all involved in the project in order to facilitate easy communication between all parties;
2. Provide an informative management tool for the monitoring and control of progress, including the effect of changing circumstances that so often occur in the form of extended delivery timescales, revised requirements and new external constraints on the project completion date.

The project plan should be started as soon as possible so that all aspects of the early layout planning can be incorporated into the plan.

Most detailed programmes for plant-related projects are prepared using simple Critical Path Analysis (CPA) techniques. CPA is a well-known and familiar technique to most engineers and is incorporated into most modern PC-based planning and scheduling systems. Such a system typically has the following features that a plant project manager will find useful:

1. Easy screen-based data entry of related task details;
2. Ability to allocate material and labour resources to each activity to allow a picture to be built of the total resource requirements over time;
3. Simple production of barcharts or Gantt charts detailing the task timescales;
4.. The ability to schedule the tasks logically within the constraints of available resources.

The definition of what constitutes a resource from a scheduling point of view may vary, but would normally include materials, labour, special tools, temporary services and also access to space on the plant floor if the layout is particularly tight. This last item is of great importance, as it is easy to forget that a given amount of space will limit the number of people working in it, and the scheduling system can take it into account when re-scheduling start dates.

For the more sophisticated requirements of complex one-off projects, techniques such as PERT, resource, levelling and precedence diagrams should be investigated.[15]

The most important aspect of planning the work is that the plan should be constantly monitored for progress

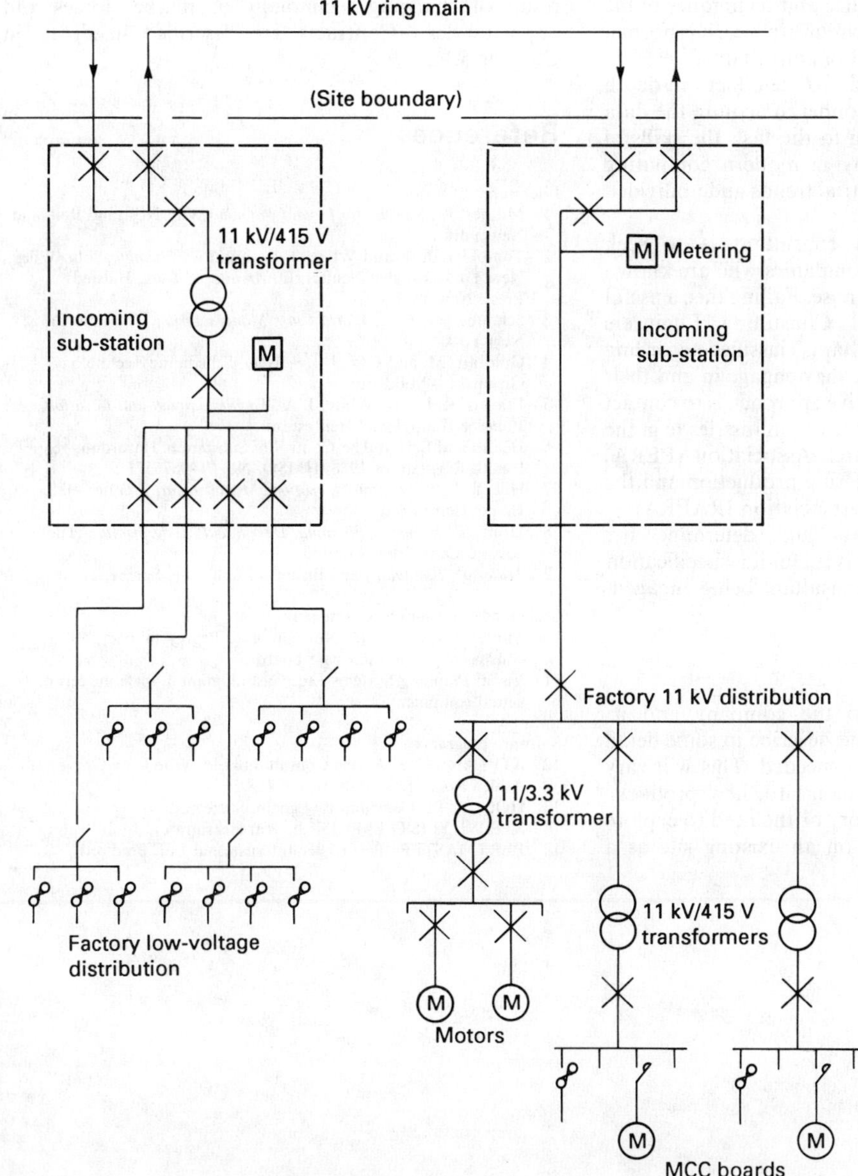

Figure 4.12 Typical power distribution system

and the individuals responsible for particular areas of work should be kept well informed regarding delays or advances in the programme.

Frequent updates and reports will ensure that surprises are kept to a minimum and allow smooth progress of the project towards completion.

Should problems occur in the project at any time, the modern planning systems allow rapid changes to the logic or task details and the consequent update to the project programme.

4.11 Consultants

4.11.1 The case for

Planning the layout of a manufacturing facility should be an infrequent activity for an operational manufacturing company. The staff of such a company will normally be fully committed to the day-to-day activity of assisting factory output, and they will rarely have the time to consider the design problems associated with plant layout. They do have intimate knowledge of the products and processes but not the practice and techniques of the design of plant layout. In these circumstances, it is normal to engage the services of a firm of consultants.

The consultants will need to conduct in-depth discussions with company personnel to acquire the data described above, but will bring to the task the skills of years of practice, often employing modern computing techniques, knowledge of industrial trends and undivided attention to the task.

The best way to select a consultant is to seek recommendations from other companies who are known to have carried out a similar exercise. Failing this, a useful guide is the Association of Consulting Engineers yearbook, which lists the consultants classified according to the type of engineering work they engage in and their specialist discipline. An alternative approach is to contact associations set up to serve particular industries (e.g. the Production Engineering Research Association (PERA) in Melton Mowbray for engineering production and the Rubber and Plastics Research Association (RAPRA)).

The consultant will discuss and determine the objectives of the activity, but it is useful if a specification is prepared in advance of the consultant being engaged.

4.11.2 Specification

After a short introduction to the company and its products the specification should describe in some detail the reasons why a new layout is needed. This will vary from the introduction of, or change to, new products, the establishment of a new factory or the need to explore the changing future demands on an existing site as a result of changing technology or market forces and opportunities. (Contracts are described in detail in Chapter 5.)

References

Literature

1 Muther, R., *Systematic Layout Planning*, Van Nostrand Reinhold, New York
2 Tompkins, J. A. and White, J. A., *Facilities Planning*, John Wiley, New York; see also Neufert, E., *Architects' Data*, Halsted Press, New York
3 Schonberger, R. J., *World Class Manufacturing*, The Free Press, New York
4 Goldratt, M. and Cox, J., *The Goal*, Scheduling Technology Group Ltd, Middlesex
5 Francis, R. L. and White, J. A., *Facility Layout and Location*, Prentice-Hall, Hemel Hempstead
6 Health and Safety; The Control of Substances Hazardous to Health Regulations 1988, HMSO, No. 0110876571
7 Hill, T. J., *Production/Operations Management*, Prentice-Hall, Hemel Hempstead
8 Drury, J., *Factories Planning, Design and Modernisation*, The Architectural Press, London
9 *Material Handling*, Reed Business Publishing, Surrey

Scale models (proprietary systems)

10 Modulux Systems Ltd, Northampton; Lego symbolic constructions on a flat base board
11 Visual Planning Systems Ltd; Cast aluminium scale models of actual equipment

Software programs

12 ATPLAN; W. S. Atkins Consultants Ltd, Woodcote Grove, Ashley Road, Epsom, Surrey KT18 5BW
13 HOCUS; PE Consultants, Egham, Surrey
14 SEE-WHY; ISTEL, Redditch, Warwickshire
15 PERTMASTER; Pertmaster International Ltd, Bradford

5

Contracts and Specifications

Ken Taylor CEng, FCIBSE, FIPlantE, FIHospE, FSE, FIOP, MASHRAE, FBIM, ACIArb
Taylor Associates Ltd

Contents

5.1 Contracts

In the course of his career the plant engineer will be a party to a contract. His employment is a contract between himself and his employer unless he is self-employed. What is a contract? A contract is an agreement between at least two parties to a matter. The important aspect is that the agreement is to the same thing. This is known as *consensus ad idem*, and follows an offer which is accepted. An invitation to tender is not an offer, the offer being made by the tenderer in submitting his tender. English law says that certain essential elements need to be met before a contract can be enforceable, and these are:

1. *Agreement* – as mentioned above;
2. *Intention* – The parties must intend to create legal relations;
3. *Consideration* – This is essential in simple contracts, which most are. An example is that *A* makes a contract to purchase a car from *B* at a cost of £3000. In paying the £3000 *A* is fulfilling the essentials of consideration. A contract would not exist if *B* gave his car to *A* because there would be no consideration.

The law of contract is exceedingly complicated, and the pitfalls experienced by engineers keep a substantial number of solicitors, barristers and judges in the style to which they have become accustomed over the past few hundred years or so.

Contract law is a combination of common law and statute law. Common law is judges' law, which is continually being revised as important cases come up in the High Court and above. Statute law is made by Parliament and Acts of Parliament which are current Acts going back to the thirteenth century.

This chapter is concerned with contracts and specifications in engineering and the way in which they should be approached by the plant engineer.

5.2 Approaching the contract

A contract is made when the essential elements have been fulfilled and the offer and acceptance communicated, and, apart from contracts involving land, can be verbal. A common misconception is that a party signs a contract. In reality, a party signs a document which contains the terms of the contract being entered into. Almost all contract documents omit many matters which are important, and the reason for this is that the law says that certain matters are implied to be contained within a contract. An example is that if *x* buys an item of plant from *y* it is implied that *x* will receive it within a reasonable time, and that it will be of merchantable quality when he receives it.

While express terms in a contract are advisable, some are of no value at all, and in fact can work against the author. An example is an unfair contract term which may be included, but if it is not sufficiently brought to the attention of the other party before the contract is made, it is not enforceable.

The object of contractual documents is to have a clear and unambiguous agreement to supply and install plant which will be in accordance with the specification and drawings, and installed within the time agreed, or within a reasonable period if none is stated.

What documents will be drawn together to form a contract? These will include:

1. Drawings of the installation sufficient for the tenderer to be able to provide a tender;
2. A detailed technical specification of the works, which will contain other matters such a defining the parties, standards of work, programme and basic contractual positions;
3. Tender forms, which will include provision for the tenderers to submit a lump sum tender, or a rate for labour and materials, or both;
4. Letter of invitation to tender.

5.3 Types and forms of contract

In English law there are two main kinds of contract – Simple and under seal. Simple contracts require consideration, whereas contracts under seal do not.

There are many standard forms of contract which are used in the engineering industry, most of which are published by professional bodies or groups of professional bodies. It is usual for large companies and government organizations to produce their own, to suit their needs and protect their interests. In the case of air conditioning, because the works are likely to be carried out alongside other building works, it is likely that either the JCT Domestic Sub-Contract or Model Form A will be used.

A contract document cannot go against a common law rule or Act of Parliament, by the introduction of an express term.

The Unfair Contract Terms Act 1977 will ensure that any onerous term must be sufficiently brought to the attention of the other party, otherwise it will have no effect.

There are several ways to employ the contractor, and some of these are outlined below:

1. Lump sum fixed price. This is the most commonly used method, and ensures that the client will know exactly how much he will be paying;
2. Lump sum fixed price with a schedule of rates to cater for variations to the works;
3. Measured Bill. A bill of quantities based on a measurement of the amount of work, with remeasuring at the end of the job;
4. Daywork or Quantum Meruit. Based on the agreed cost of labour per hour and on cost of plant and materials. This is sometimes the only fair way, because the amount of work is uncertain, or cannot be sufficiently estimated by the tenderer.

5.4 The specification and drawings

The smooth running of the contract and the satisfactory operation of the system will depend to a large extent on the way the plant engineer designs the plant, and expresses it in his specification and drawings. There is a tendency for designers to adopt standard specification formats, and while this is to be applauded, a 50-page specification for a small job can mean that only a small percentage is applicable to that particular project. It should be remembered that the foreman on the job will have the task of interpreting what the designer had in mind.

The General Specification must coincide with the drawings to be credible. It is all very well inserting a clause to the effect that 'The Engineer shall determine the interpretation of any clause' which implies that he has the power of an arbitrator, but in reality, if he has negligently specified anything by being ambiguous, then his client or the contractor may have a claim against him if either of them suffers loss as a result.

The General Specification itself can be divided into the following headings:

Conditions of Contract
Technical Clauses
Specification of the Works.

The following is a checklist and brief description of the various clauses:

(a) *Conditions of Contract*

1. *General* – This is a description of the type of contract and the relationship with other parties. For example, whether the contract is to be nominated, main or domestic, fixed price or fluctuating.
2. *Interpretation of Terms* – Name of all known parties and their positions. Some terms may be outlined, i.e. 'month' means calendar month and 'shall' means action by the contractor.
3. *Form of Tender* – A more detailed description is given regarding the form of tender and of prime cost and provisional sums, handling and profit on cost.
4. *Bills of Quantities* – The client can require a successful tenderer to supply a priced-up bill of all items tendered for within a certain time after being requested to do so. This is useful in a lump sum contract where variations to the contract occur. The bill in this case is provided by the tenderer.
5. *Tender make-up* – Invitation to the tenderers to clarify points prior to tender. Alternative makes of equipment may be offered instead of the ones specified, but the tenderer must tender for what was specified, and include the alternatives as a separate list.
6. *Attendance* – In the case of a sub-contract it should be made clear to the tenderer any attendance which will be provided to him by the main contractor, and any conditions attached.
7. *Programme* – The programme of works should be outlined, and any special provisions, such as working in phases to suit either the client or the main contractor.
8. *Site Meetings* – The frequency of site meetings should be outlined and a commitment to attend any *ad hoc* meetings which may be called by the client or main contractor.
9. *Related Documents* – A list of all the documents which make up the contract. Interpretation of terms.
10. *Payments* – Details of when valuations and payments are to be made.
11. *Completion* – Practical completion and defects liability period details.
12. *Arbitration Clause* – This is likely to be mentioned in most standard forms of contract, and if repeated here, should be verbatim.

(b) *Technical Clauses*

1. *Scope* – general description of the works.
2. *Regulations* – a general list of regulations which apply to the particular type of works (e.g. Health and Safety at Work Act 1974), etc.
3. *British Standards and Codes of Practice* – A general note that the installation shall comply with all relevant British Standards and Codes of Practice.
4. *Drawings* – Details of drawings and manuals which the contractor shall supply to other parties (e.g. building work, drawings and working drawings, and a description of details to be supplied).
5. *Site visit* – Invitation to contractor to visit site to see for himself any possible site problems. A disclaimer against extras in the event that the contractor does not appreciate obvious site problems.
6. *Tests/commissioning* – Details of any specific tests required, and commissioning format, and who is to witness and reports to be submitted.

5.5 Estimates and bills of quantities and estimates

On some contracts, mainly connected with general building projects, rather than specifically plant and machinery, a quantity surveyor is appointed to draw up a schedule of items which require pricing, and which the total is the tender figure.

A qualified quantity surveyor has a first-class knowledge of building construction, and is an expert in drawing up a bill of quantities, and very often has a good knowledge of contract law, which makes him invaluable where there are complicated issues. While the plant engineer may not often be presented with bills of quantities, it may be useful to have a grasp of the general approach. There are three stages to a bill: taking off, abstracting and billing. Taking off is a detailed measurement of drawings or on site of the subject matter. This is carried out in conjunction with a manual used by all quantity surveyors called the *Standard Method of Measurement*. The items contained are complete in that they include for all labour and materials. The abstract is a sheet where all similar items are grouped together, and the bill is where the items are listed for the tenderer to add his prices. The tenderer is also given the opportunity in the bill to add his profit and on-cost for other contractors under his control.

5.6 Specific sums stated in tender documents

In most engineering contracts for works to be carried out on site, there are unknown costs at the time of tender and costs which are known, where the work is to be carried out by others, but to be under the control of the successful tenderer. These sums are shown below:

5.6.1 Prime cost (PC) sum

This is a sum which has been determined prior to tenders being obtained for the main works, and which can be expressed accurately as a sum to be included in the tender build-up. The tenderer has the opportunity to add a sum within his tender for profit or on-cost in administering the works.

An example would be where the consultant had already invited competitive quotations for supplying air-handling units so that he could design a particular make into the system and ensure that his client received the best value for money. The tenderer would include in his tender, either separately shown or in the body of his figures, a sum to cover profit and on-cost in handling the equipment. The contractor appointed would be responsible for organizing delivery and would give a warranty on the equipment, in accordance with the contract. This aspect is amplified in Section 5.20.

5.6.2 Provisional sum

This is a specific sum included in the tender documents to indicate to the tenderer that it is contemplated that the particular work referred to will be carried out, but that, at present, no firm plans have been made as to cost, and usually the work has not been detailed either. The contractor is not entitled to profit or on-cost, unless the works go ahead. An example of this would be a general intention by the client to provide special ventilation to a laboratory, which is part of a larger scheme, but that no firm plans as to the full requirements have been drawn up at the time of inviting tenders.

Let us assume that £5000 has been included as a provisional sum. At a later date a scheme is produced and priced up by the successful tenderer, and his price is £4875, including profit and on-cost. Then at the end of the contract, when the final account is submitted, the contractor will show that of the £5000 provisional sum, £4875 has been spent, and the client will be £125 better off than expected.

5.6.3 Contingency sum

In most contracts of supply and install, a sum of money is included in the tender figure for which there is no allocated use. This is known as a contingency sum. The amount of the sum varies according to the size of the contract and the degree of complexity of the works to be carried out. For example, the installation of a piece of plant which will be delivered to site fully tested and ready to run on its own, once services are installed to it, will not need a high sum. On the other hand, a system which is a one-off, designed and built specially for the occasion on site, depending on items such as distribution ductwork or pipework or unknown structural and services matters will need a substantial sum of contingencies. There is no hard and fast rule, but the sums range from 4% to 10% of the contract sum. Generally, where there are known problems which cannot be accurately assessed (such as diverting an underground water main) a provisional sum should be put in rather than a contingency sum.

The danger is that often many issues are left to the contingency sum, which means that it is spent before other known matters come to light. It is better to report a higher tender figure which has the possibility of being reduced at the end of the job, rather than a lower one which is almost certain to increase.

5.7 Tender documents

These are documents to be completed by the tenderer after he has studied the specification and drawings, where appropriate, and which contain the special commitment of him with respect to the particular job. His chances of being successful will depend largely on how he fills in the forms contained in the documents. Very special care should be taken in preparing these documents, and equally important is analysing them after tender, as will be discussed in Section 5.12.

Standard forms of tender such as the JCT 1980 contain information and questions which are carefully put together and understandable in the industry. Contents of tender documents should include:

1. Clear indication of what the document is (i.e. 'Form of Tender');
2. Who the employer (client) is with full address. He will be party to the contract;
3. What the tender is for, briefly but specifically, and reference to the documents which will form part of any contract;
4. Date and time by when tenders are to be received;
5. Where the tenders are to be sent to and the manner in which they are to be submitted (for example, in a plain or specially marked envelope);
6. A note that the employer does not bind himself to accepting the lowest or any tender, and that the tenderer will not receive any renumeration or expenses for providing a tender.

The tenderer should be invited to submit the following:

7. His full name and trading address;
8. His lump sum tender for the works, where appropriate;
9. His rate for labour, materials and on-cost, separately or as a composite rate in the case of a bill of quantities;
10. A statement of how long he needs to do the work following receipt of an order or the contract being signed;
11. Date and signature, with position of signatory (see Figure 5.1).

(1) Job title: ....

(2) Name and address of employer: ....
....

(3) Location of project: ....

(4) Tender for: ....

(5) (a) This is a fixed/fluctuating price tender (delete as appropriate)

(b) Schedules of rates are/are not attached (delete as appropriate)

(c) Labour and material rates and on-costs are shown in a separate document attached

(6) I/we agree to keep this tender open for acceptance within .... weeks of the date of this tender.

(7) I/we agree to commence the works within .... weeks of receiving an official order, and to complete the works within .... weeks of commencement

(8) I/we intend to employ the following subcontractors to carry out part of the works, if this tender is accepted:

Name Address Description of works

(9) I/we agree to carry out the works referred to in the drawings and specification attached to the invitation letter dated .... for the sum of £ .... Amount in words ....
....excluding VAT and which sum includes the following:

Contingency sum £ ....
Provisional sum £ ....
Prime cost sum £ ....

(10) Name of tenderer ....
Address ....
Signed by .... position .... Date ....

(11) Tender to be returned to .... in the endorsed envelope provided, not later than 12 noon, on the .... day of ....

(12) The employer does not bind himself to accepting the lowest or any tender, and the tenderer will not receive any renumeration or expenses for providing a tender.

Figure 5.1 Typical tender document for a principal contractor

5.8 Direct and bulk purchasing contracts

To bypass the contractor or installer by purchasing plant directly from the manufacturer usually has an advantage of a saving in cost. The drawback is that the contractor does not have control of delivery or of quality, and this can lead to disputes or a defence against a claim for faulty workmanship or materials. Any saving can quickly become meaningless if the works are delayed, and a loss and expense claim is submitted. In a competitive situation a tenderer will keep his 'mark-up' to the absolute minimum, and a saving by approaching the manufacturer direct will be small. It is always possible to obtain a price with a manufacturer following competitive quotations, and then include a PC sum in the tender documents.

This gives the tenderer the opportunity to add a modest mark-up, and also place him in a position where he is responsible for ordering, obtaining delivery and then paying for the equipment, in addition to installing it. There is still a responsibility on the client or his agent for nominating the equipment.

Bulk purchase of equipment is carried out in order to bring the price down per item. Having established that a large number are required, a client can then negotiate or invite competitive tenders to supply the number involved over a certain period. Care needs to be taken that the number agreed are actually required within the period, otherwise the supplier can sue for damages if he is frustrated from supplying the number agreed.

5.9 Programme of works

Prior to writing tenders, the client should decide when he wishes the work to be carried out. Factors affecting his decision will be availability of funds either from loans or grants, or in the case of local authorities or central government departments it may be that a capital or revenue allocation needs to be spent within a particular financial year. Other factors will be other trades such as a building to house the plant, whose timing is obviously important.

Where an architect is involved, he will draw up a general programme incorporating design time as well as installation, and this will be submitted to the plant engineer for his views before design commences. Having agreed the programme, the tenderers will be informed in the tender documents how it is intended that the works will proceed. However, where a main contractor is involved it is vital that he be given the opportunity to submit his own programme of works, incorporating all the various trades, keeping within the programme completion date as laid down in the tender documents.

Difficulties arise when delays in obtaining tender approval occur, because time moves very quickly, and the work may not actually commence on the date stated in the tender documents. This causes problems when the contractors have carefully arranged their own work schedules. It is better not to be too optimistic regarding start and completion dates, unless various stages to a contract document being signed are clear.

5.10 Selection of tenderers

There are four ways in which a tender list can be drawn up for a particular project:

1. Advertise in the local press or trade journal for firms to apply and then draw up a short list.
2. Seek out firms of good reputation.
3. Draw up a list from firms who have shown an interest in working for you.
4. Use firms who are known to you and who have given service in the past.

Vetting prospective tenderers has become a science, and for those who work in local government the rules are laid down by statute in the Local Government Act 1988. The procedure in inviting tenders under (1) would be to invite forms on an annual, bi-annual or job-by-job basis to apply for a questionnaire in order to consider whether they should be placed on the list of tenderers. Any questionnaire sent out by or on behalf of a local authority must not contain questions of a commercial nature. For example, the question 'Are you a Member of the Heating and Ventilating Contractors' Association?' would be wrong in law. However, the question 'Are you registered with the National Inspection Council for Electrical Installation Contractors?' is permissible because it is connected with quality of work.

All other clients are able to ask their own questions. A consulting engineer who did not properly assess a firm prior to recommending that they be placed on the tender list would be liable in negligence to his clients for not discharging his duties with diligence. Questions which the plant engineer should ask when compiling a questionnaire are:

1. Address from which firm would carry out contracts.
2. How long in business.
3. Number of office and site personnel and their trades.
4. Expertise professed by the firm and size of projects which they can undertake.
5. Sample of six jobs recently undertaken.
6. Names of three referees who can be approached.
7. Any pending litigation or recent court judgment against the firm.
8. Name and address of bankers.

Having received completed questionnaires a shortlist can now be drawn up. After discarding the non-starters, the hopefuls should be vetted.

Remember that firms will put their best face forward. Where possible, it is advisable to visit at least two jobs recently carried out, and to contact at least two referees. A banker's reference can be misleading, although perfectly accurate. Since *Hedley Byrne & Heller* (1966) banks will usually give references 'Without Responsibility'.

A banker's reference will say that the firm is in a trading position, and sometimes that the firm is in a position to carry out the sizes of contract envisaged, but it will mention nothing about the reputation of the firm, its quality of work or administrative ability. A bank will not deal directly with a client, but only through his bank.

A different questionnaire needs to be sent to the referees, remembering to enclose a stamped addressed envelope. The referee is doing an unpaid favour in completing the questionnaire, and for his sake it should enable him to return it without having to write a thesis with an attached letter. A box to be ticked for most questions will suffice, with a space at the end for general comments. Questions which should be asked of the referee are:

9. How long have you known the firm?
10. What size of contracts have they carried out for you in the past 3 to 5 years?
11. How do you rate their workmanship, administrative ability to keep to programme, after-sales service?

12. Would you use them again?
13. General comments.

Having passed the test so far, the way is now open to visit the premises occupied by the firm and meet the management of the company. Once a firm is accepted for the list it should be placed in the size of contract category and within the range of its financial capabilities.

5.11 Inviting tenders

Most large organizations, particularly in central and local government, have set procedures in the form of standing orders, and this area is usually of most interest to internal and external auditors. The object of inviting competitive tenders is to obtain the lowest price for the job, based on a fair method which enables every tenderer to tender for the same thing.

A design and build tender can cause problems in that different design engineers have their own views on approaching a design. To help overcome this, the client or his agent should issue a design brief which limits the design parameters. For example, in a scheme to relight a warehouse, the brief could say that the lighting levels should be in accordance with the *CIBSE Guide* or could quote actual lighting levels to be achieved.

When drawing up the tender list, note should be made of the points referred to below before deciding to place a firm on the list:

1. Financial capability;
2. Location with respect to project;
3. Workforce and administrative strength to carry out the project.
4. Check that the firm has carried out similar projects in the past;
5. It should be ensured that a firm is not offered too many contracts at once. An overstretched firm can cause severe problems of performance.

Prior to inviting tenders each firm should be contacted, preferably in writing, to seek their agreement to tender. Information to be given to them will include:

1. Approximate value of works;
2. Type of contract;
3. When tenders will be sent out and closing date;
4. Programme of works, start and finish dates.

Having agreed to tender, if a firm does not submit a tender without good reason consideration should be given to removing it from the tender list for a short period, because when a firm does not tender, it reduces the competitive element, possibly increasing the cost of the job. It also denies another firm the opportunity of tendering.

The number of tenderers invited to tender should be determined by the size of the works. Good contractors should be encouraged to tender and not be disillusioned by too many tenderers. For most contracts, a tender list of six firms should be sufficient to keep the tendering competitive and allow for the odd firm which does not return its tender.

On important contracts it may be advisable to invite tenderers to pick up the tender documents, or they could be delivered to them. Documents do go astray in the postal system.

Tenders should be returned in endorsed envelopes so that it can be seen that a tender for a named project is enclosed, but should not disclose whose tender it is. In large organizations it is normal for tenders to be returned by a certain date and time, to a certain place, which is usually either a legal or an administrative department. Tenders should be opened by someone other than the person involved in the job itself and then handed to the person involved, after details of tenders have been noted. Tenderers cannot expect to be paid for the time or other costs in ordinary circumstances for preparing their tenders, as in *William Lacey* v. *Davis* (1957).

5.12 Analysing tenders

When the plant engineer receives all tenders returned, he will need to analyse them to ensure that they are arithmetically correct, and that each tenderer has understood what is being tendered for. A predesigned analysis sheet can be helpful and a typical sheet is shown in Table 5.1. All tenderers are shown on one sheet with their prices compared together. The first task is to check the arithmetic, which seems to cause problems even in these days of computers. Where the arithmetic is incorrect it should be brought to the attention of the firm quickly to enable it to either withdraw or stand by the figure. If the correct addition would have meant lower costs, the firm should be given the opportunity of revising the tender, but only after consultation with the legal or administrative department. Action taken now without consultation can be hard to explain later.

The addition may be correct, but a tenderer may have underestimated a particular item. With an itemized tender it is possible to spot an error of this kind. For example, referring to Table 5.1 and looking at item 3, tenderer 2 has less than half in his tender for item 3 than any other tenderer. On the other hand, item 2 appears to have made up for it. Where an apparent mistake has

Table 5.1 Analysis of tenders for replacing a drying machine

Item	*Detail*	*1*	*2*	*3*	*4*
1	Strip out	£2 000	£1 600	£1 900	£2 200
2	Steelwork	£1 500	£2 500	£1 300	£1 700
3	Rein. conc.	£2 000	£950	£2 100	£1 900
4	Machine	£99 000	£98 000	£97 000	£101 000
5	Elect.	£3 000	£2 900	£2 950	£3 100
6	Piping	£2 500	£2 300	£2 600	£2 700
7	Ductwork	£2 700	£4 700	£5 200	£5 100
8	Insulation	£2 000	£1 950	£1 900	£2 000
9	Painting	£1 000	£970	£1 100	£1 050
10	Test/commission	£1 000	£990	£995	£1 100
11	Contingency	£5 000	£5 000	£5 000	£5 000
12	Provisional	£500	£500	£500	£500
13	Prime cost	£1 000	£1 000	£1 000	£1 000
Total tender		£123 200	£123 360	£123 545	£128 350

occurred, the firm concerned should be contacted, and should be asked to look closely at a certain item which appears low. For example, tenderer 1 has a much lower amount for item 7 than any other, and if he had tendered similarly to the other tenderers for that item, he would not be the lowest tenderer. It is better to lose a low price than have the firm lose money on a job which can cause difficulties with performance.

Where the scheme was designed before being sent to tender, the tenders should be grouped fairly closely together, because design decisions have already been made. Design-and-build tenders will need much more attention to find out which is the best value for money. For example, in lighting the warehouse mentioned earlier tenderers may have been advised of the lighting level, but if they have not been told of the type and manufacture of fittings, there can be a substantial difference in cost.

Tenders received can vary by up to 100% over a range of firms. The higher tenders may be from firms who do not want the job but wish to keep contact with the client. Where there is a difference of more than 10% between the lowest and second-lowest tenders, this can cause problems. For example, the Electrical Contractors Association offer a guarantee of completing through their members, at no extra cost, in the event that a member is unable to do so, but only if the difference between the contractor appointed and the next tender is not more than 10%. In some instances this could be enough to seriously consider not accepting the lowest tender.

5.13 Selection of the contractor

Having established that the lowest tenderer is in fact the lowest, after taking all matters into consideration there are other aspects to consider before agreeing to a contract.

A qualified tender can make the offer void, and most organizations would disqualify a tender which was qualified. An example of a qualified tender would be where the specification says that electrical attendance is part of the contract, whereas the tenderer says in his tender documents or in an accompanying letter that he has not included for electrical work, and is not prepared to carry it out. A common qualification is where a contractor puts forward a manufacturer of plant different to that specified. The plant may be just as good, but the tender is nonetheless qualified, and would give an unfair advantage to the offeror.

A *bona fide* qualification is where the specification is unclear.

An area of increasing importance financially is the programme. Large sums are being claimed by contractors with respect to delays, and it is important to take note before the contractor is appointed of any qualifications he may make to the tender. For example, if a tenderer says in his tender that he can carry out the works in 40 weeks and it is known that the period envisaged is 52 weeks to coincide with other trades, then if he is frustrated from doing the work he can sue for loss and/or expense for the 15 weeks when he will have, in theory, to lay off men. The solution is to challenge the period inserted, and if the tenderer will not amend his period, then do not appoint him.

Tenders are normally kept open for a specific period of time, which is usually 4 weeks. In any case, without a specific period being mentioned in law, a tender must be accepted within a reasonable period, otherwise it will lapse. In the event that a decision cannot be reached, or an appointment made for some time, the tenderers should be approached in order to ensure that their tenders are kept open. The problem for the contractors is that wages and salaries increase regularly, as do materials and other on-costs.

In contract law the acceptance of an offer must be unequivocal. Where a tender is accepted subject to an amendment on, say, programme, the client is making a counter-offer, and acceptance is then made by the tenderer.

The successful tenderer should be informed as soon as possible that his tender is to be accepted, to give him the opportunity to plan his operation. Equally, the unsuccessful tenderers should be sent a letter expressing appreciation of the time they have spent in preparing their tenders, but informing them that they have been unsuccessful.

5.14 Making a contract

A simple contract for any value need not be in writing, but can be verbal and is known as a parole contract. Many contracts are made which are verbal. When a person goes into a supermarket, takes an item off the shelves, carries it to the checkout and makes an offer to purchase it at the price shown, this constitutes an offer in contract law. The checkout cashier takes the money and is, at the same time, accepting the offer. Verbal contracts are fraught with problems and should be avoided where possible. For example, to prove that a verbal contract has been entered into without the work having been carried out will need parole evidence in court. Where arbitration is entered into later, a verbal contract does not come under the Arbitration Acts 1950 and 1979.

Letters of intent can be the cause of problems, and there are varying views as to their validity. These are sent in advance of contractual documents, being signed because either the documents are not fully prepared or a decision has not yet been confirmed. For example, in local government a tender may have been accepted by a committee but needs ratification by the full council, which can be up to six weeks later. It is rare for a council to overturn a decision made by a committee, but it does happen. In an attempt to expedite matters it is common policy to send out letters of intent directly following the committee. These will be marked either 'Subject to Contract' or 'Subject to Council Approval', and in these cases they are useless as a contract if the client does not go ahead. Where the contractor has reasonably relied upon the letter and the client has encouraged him to purchase the equipment, then a court may order that he be paid for the work done under *quantum meruit*. A letter of intent without a qualification will infer a contract and will commit the client to the whole works.

5.15 Relationships between contractor and other parties

Where only one contractor is involved, then his relationship with the client is direct and contractual. Most contracts have more than one contractor, and the client must decide how he intends this to be arranged. A few of the possible compilations are:

1. *Direct contract* (sometimes known as a principal contract), where only one contractor is involved;
2. *Main and nominated sub-contract.* This is where one contractor is appointed and known as the main contractor and the client nominates or instructs this contractor to appoint a named firm to carry out part of the works. Sometimes tenders will have been sought separately prior to the involvement of the main contractor;
3. *Main and nominated supplier.* The client may wish a certain firm to supply materials for the contract which may be for installation by the main or a sub-contractor.
4. *Principal contractor with domestic sub-contractor and supplier.* The expression 'domestic' does not necessarily infer housing, although housing contracts can be carried out in this manner. A domestic sub-contractor is a contractor who is appointed by the principal contractor. The client has no relationship with a domestic sub-contractor, unless he interferes with his selection such that the appointment becomes nominated. This type of appointment has become very common.

The contractors have no contractual relationships with any independent consulting engineers who are appointed by the client. Where the client decides to appoint a firm to 'design-and-build' there is a different duty on the designer. When a consulting engineer is appointed he must exercise reasonable skill and care in carrying out his duties, as an average firm of this type would do. However, where the client entrusts his design and installation to a contractor, the contractor owes a strict duty to his client to ensure that the installation works, and that it is fit for its purpose, as in *Independent Broadcasting Authority* v. *EMI* (1981).

A nominated sub-contractor has no contractual relationship with the client, but the case of *Junior Books* v. *Veitchi* (1982) placed an 'almost' contractual duty upon the nominated sub-contractor. Where a separate agreement has been signed between the client and nominated sub-contractor then a direct contractual relationship applies. Problems can arise where a main contractor says that he is being delayed by his nominated sub-contractor or supplier. Because the client has instructed him to use a firm which was not of his choice, the main contractor could claim damages in the form of loss and/or expenses from his client. Where the principal contractor has brought in his own domestic sub-contractor, then he is responsible for his actions. That is the main reason domestic sub-contracts have become popular. The client must decide upon which trade will become the main contractor. In building matters it is usual for the main contractor to be a firm who can carry out the most substantial works, and this will be a general building firm. Some general building firms carry their own mechanical and electrical departments, which reduces the need for sub-contractors.

There are occasions when it is more appropriate to appoint a mechanical or electrical firm to be the main contractor. An example would be the laying of a district heating mains, where the largest value and complexity is in mechanical engineering. However, experience shows that not all specialists are geared to this role, and this can cause problems, especially from a main contractor who is more familiar with the role of sub-contractor. On the whole, it is recommended that the main contractor be a building or civil engineering firm, unless those works are very small.

5.16 Site meetings

Meetings are a necessity to keep a contract moving. The first meeting is normally at the offices of either the client or consulting engineers, and is a pre-contract meeting. A pre-contract meeting is a gathering of all interested parties to discuss the format of the contract and is a chance for views to be expressed, prior to the actual formation of a contract, which very often takes place after the meeting.

Matters which are likely to be on the agenda of a pre-contract meeting are:

1. Names and addresses of all parties and introductions;
2. Names of specific persons from each firm who will be handling matters, such as site engineers and clerks of works;
3. Format of site meetings to be agreed (who will take minutes and who will chair meetings);
4. General lines of communication agreed;
5. How often site meetings will occur;
6. Accommodation for sub-contractors, site engineer, clerks of works and meetings;
7. General programme discussed. At this stage the main contractor may not have drawn up a bar chart or critical path programme;
8. Procedures for variations and interim valuations may need to be cleared up, depending upon the form of contract adopted;
9. Where nominated suppliers or sub-contractors are involved, the main contractor has the opportunity to voice his opinion of those selected by the client.

There are generally two kinds of site meetings. One is where all parties are present and the other where the professional team meets with the main contractor and then the main contractor organizes his own contractors' meeting with his sub-contractors. The job itself will determine the kind of site meeting to be held. *Ad hoc* meetings are held throughout the job, and these meetings should always be with the knowledge of the main contractor, if not chaired by him.

Full site meetings may be held monthly or weekly, depending on the pace of the job. A typical site meeting will be chaired by the main contractor or the project leader, who could be the architect or engineer, and matters for the agenda will include:

1. Apologies for absence;
2. Previous minutes read;
3. Matters arising from previous meeting;
4. Main contractor's report;
5. Sub-contractors' reports;
6. Progress (various clerks of works' reports);
7. Consulting engineers' reports;
8. Any other business;
9. Date and time of next meeting.

Minutes are then published through the chairman and circulated to all parties, even to those not present at that meeting. When receiving the minutes it is worth studying them to ensure that they are accurate and to note any action needed. It is usual to have a column drawn on the right-hand side of the minutes for the names of the parties who need to take action. The smooth running of the project can depend on all parties complying with action noted in the minutes. Inaccurate minutes should be challenged at the next meeting. Site meetings and publication of minutes are vital to a large or complicated project to avoid misunderstandings, which can cost money.

There will be *ad hoc* meetings, particularly inspections by visiting clerks of works. Points worth noting at these inspections are:

1. The site is under the control of the main contractor, and he must be informed of all visitors, who are advised to comply with requests (e.g. wearing hard hats).
2. An apparently simple visit can result in disputes later. Therefore it is wise to meticulously record all salient points of discussion. The time and date of arrival and time of leaving may seem pedantic, but can avoid difficulties later. Clerks of works' diaries are often admitted as contemporaneous evidence in litigation and arbitration. This is discussed further in Section 5.17.

5.17 Progress and control

The client will have decided at the very beginning when he wants the job to be finished on site. Circumstances such as approvals, preparation of contract documents or availability of labour and materials will determine when the works start.

At the beginning of the contract, before work starts on site, the main contractor will produce a programme of how he intends to carry out the works. In some tenders the contractor is invited to state what his programme will be if he is successful. This method has merit where the programme is very tight. The most common type of programme is a bar chart, and will show at what periods the various sub-contractors will be on site. A typical bar chart is shown in Figure 5.2.

The value of having one contractor to control the progress of the work can be seen where there are several contractors on a site at different periods. Progress can suddenly come to a halt if a key contractor does not comply with an agreement. For example, if the electrical sub-contractor does not provide a temporary electrical supply, there may be no light and power, causing other contractors to be delayed.

In the matter of progress the main contractor is in charge of the works. However, the client or his consulting engineer will need to monitor the works to satisfy themselves that programmes are being adhered to. There is a temptation for a client or his consulting engineer to assist the contractor by instructing him how to carry out the works. This can amount to interference, and if the

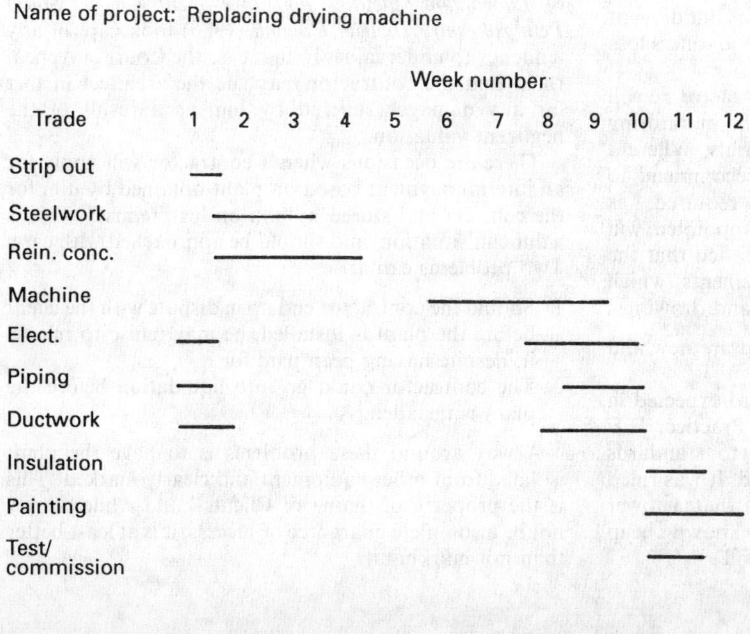

Name of project: Replacing drying machine

Week number

Trade 1 2 3 4 5 6 7 8 9 10 11 12

Strip out

Steelwork

Rein. conc.

Machine

Elect.

Piping

Ductwork

Insulation

Painting

Test/ commission

Figure 5.2 Typical bar chart programme

contractor suffers loss as a result he can sue for damages, as in *Oldschool* v. *Gleeson* (1976).

There are varying degrees of monitoring which a consulting engineer can recommend to his client:

1. Occasional inspection visit by the design engineer, as necessary, on a small job;
2. Regular clerk of works visits;
3. On-site engineer or clerk or works.

It is the duty of the consulting engineer to make a recommendation to the client as to the type of supervision required, and the client must not refuse a reasonable request by him for on-site supervision. Where a client refuses a reasonable request for on-site supervision, the consulting engineer should warn him of the possible consequences, otherwise he may face a claim for negligence later.

On most contracts there are requests by the contractor for clarification of the specification or for instructions with regards to variations. These requests need to be answered promptly to avoid a claim for damages by the contractor which may result from delay of the works.

All progress should be carefully noted in the diary of the supervisor or clerk of works. Points worth noting are:

1. Number of men on site;
2. Weather (where appropriate);
3. Name and title of any person spoken to. Note that instructions should generally be given through the main contractor's site agent;
4. Progress of job;
5. Any items discussed.

5.18 Quality control

The contractor is under a duty in contract to provide and install plant which is of merchantable quality. There is no duty on the client to ensure how the contractor carries out those duties, although the consulting engineer can be sued in negligence by his client, if he is not diligent in detecting faulty workmanship, such that he suffers loss as a result.

Where an industrial firm employs a contractor he will have a chief engineer who will organize the monitoring of the quality of workmanship and materials. Where a consulting engineer is involved, he will recommend to his client the degree of supervision required, as mentioned in Section 5.17. The person appointed will need to consider the following and be satisfied that the works being installed meet the requirements which should be laid down in the specification and drawings:

1. Materials and plant which are specified are new and unused;
2. That workmanship is up to the standard expected in this type of contract. The Codes of Practice, IEE regulations and other bodies related to standards should be used as a measure of standard. It was ruled in the case of *Cotton* v. *Wallis* (1955) that a lower standard of work can be accepted in a known cheap job, as long as the work is not 'rank bad'.

A problem which seems to occur frequently is that contractors do not give sufficient thought to allowance for maintenance in the future. To install plant without means of maintaining it is bad practice and should be avoided. The plant engineer should make sure that the design allows for adequate maintenance. This item comes under the scope of the clerk of works to interpret the design.

5.19 Interim payments

In a lump sum contract the contractor does not have a common law right to interim payments. Most larger contracts have an express term within them giving the right to interim payments, or payments on account, notwithstanding that they have not fulfilled their contractual duties in completing the works. The matter of interim payments and their frequency should be resolved before the contract is signed, and not left until a dispute arises.

Interim payments on the contract follow a valuation of the work done. Valuations are normally carried out by the plant engineer, the quantity surveyor, the architect or the consulting engineer and, in the light of recent case law, should be assessed very carefully.

Until 1974 an architect was considered in law to be a quasi-arbitrator, and his valuation was final. However, the case of *Sutcliffe* v. *Thackrah* (1974) saw an end to this role. In that case the architect overvalued the work which had been carried out, and because the client ordered the contractor off the job he suffered loss, as a result of which he successfully sued the architects for damages. Architects and engineers suddenly had a greater interest in professional indemnity insurance.

An obvious solution would be to undervalue the work done, in order to avoid a claim from the client. The case of *Lubenham Fidelities and Investment Co.* v. *South Pembrokeshire District Council* (1986) took care of any tendency to undervalue. In that case the Court of Appeal ruled that the contractor may sue the architect in tort for any damages suffered by him as a result of the negligent valuation.

There are occasions when a contractor will apply for an interim payment based on plant obtained by him for the contract and stored by him on his premises. This is a difficult situation, and should be approached with care. Two problems can arise:

1. Should the contractor end up in dispute with the client before the plant is installed, he may refuse to release it, despite having been paid for it.
2. The contractor could go into liquidation before the plant is installed.

A way around these problems is to have the plant isolated from other equipment and clearly marked 'This is the property of (name of Clients)' and while it may not be a complete guarantee of success, it is at least better than not marking it.

5.20 Budget control and variations

At the beginning of the contract it will be known to the client and his consulting engineer how much money is budgeted for the project. The contractor will be aware of the contract value, of course, and he will also know of any contingency sums which have not yet been allocated.

Contingency sums are included in a contract in case there are any additional works which need to be carried out and have not been covered in the contract. This may be because the client wishes to have more items of plant than he originally envisaged, or there may be unknown features, or it could be for items which have been forgotten in the original design. (See Section 5.6 for a description of contingency sums.) Where a consulting engineer forgets to include an item which is necessary for the proper completion of the works he can be sued for negligence by his client, or advised to pay for this work himself. Where a contractor forgets an item which is necessary he will have to install it himself at no extra cost.

There is a temptation for the contractor to see contingencies as extra work, or there is always a danger that it will be spent by the end of the contract because the client will see it as available monies.

Provisional sums are already allocated when the contract is made, and while the true value is not known at tender, the estimate should be reasonable. Where a schedule of rates or a bill of quantities is available, there is no problem in arriving at the figure after the contract has been let. However, where the contract is simply lump sum, there is no basis to determine any variation, and this will sometimes mean a difficult negotiation. When a variation to the contract becomes apparent it should always be agreed in writing, preferably with a firm agreement on price or at least an agreed rate. A major part of a dispute when the final account is being agreed can be variations. Where the variation is too difficult to assess accurately, such as an underground gas or water main being found which was not known previously, the contractor will have to be instructed to carry it out on day work (*quantum meruit*). In the tender there will be a rate for labour and an on-cost figure for labour, materials, plant and other expenses, and the purpose of this is for items of daywork which may occur. Certain standard forms of contract such as the JCT form contain specific documentation when dealing with variations.

5.21 Safety on site

Unfortunately, deaths occur every year in industry due to accidents which are preventable. Accidents are more likely to occur where firms are under pressure to meet targets, and safety measures tend to be ignored. The Health and Safety Executive (HSE) was set up to administer the Health and Safety at Work, etc. Act 1974 and incorporates the Factories Inspectorate. General duties of employers described in section 2 of the Act are that 'It shall be the duty of every employer to ensure so far as is reasonably practicable the health, safety and welfare at work of all his employees'. The emphasis is on the expression 'reasonably practicable', and courts interpret the meaning of it, under the particular circumstances.

The HSE prosecutes firms or their employees in the Magistrates or Crown Courts when safety measures are not taken by them and the courts impose fines of up to £2000 with 2 years' imprisonment for convicted offenders. Where gross negligence occurs such that someone is killed, then the Department of Public Prosecutions can bring a charge of manslaughter against a person, or more recently against a corporate body. The maximum penalty for manslaughter is life imprisonment. Prosecutions of this type are also being pursued.

While Codes of Practice and the IEE regulations are not statutory, in a situation where injury was caused through neglect of them the HSE can prosecute under section 16 of the Act.

A plant engineer may become involved with an inspector from the HSE when he arrives on site to carry out an inspection of the works under powers given to the Executive by section 20 of the Act. The HSE will visit at any reasonable time, or at any time where the situation may be dangerous in the inspector's opinion. He may take with him a police officer if he has reasonable cause to apprehend any serious obstruction in the execution of his duty.

Having carried out an inspection of the works, the inspector has powers under section 21 to serve an improvement notice if, in his opinion, a person is contravening one or more of the relevant statutory provisions, or has contravened one or more of those provisions in circumstances that make it likely that the contravention will continue or be repeated. The person concerned then has a duty to remedy the contravention or as the case may be, the matters occasioning it within such a period as may be specified in the notice.

An inspector can order under section 22 that activities be prohibited which have not yet commenced, and may be about to be carried on or are already being carried on, by serving a prohibition notice.

To ignore either of the notices above is a criminal offence, and a prosecution is likely to follow such contravention. There is an appeal to an industrial tribunal against an order.

An HSE inspector has the power under section 25 to enter any premises where he has reasonable cause to believe that, in the circumstances, any article or substance is a cause of imminent danger of serious personal injury. He may then seize the article or substance and cause it to be rendered harmless, whether by destruction or otherwise.

5.22 Supply of Goods and Services Act 1982

The general theme of all the legislation contained within the current Acts with respect to supply or sale of goods and services is connected with quality of goods or services, supplied or sold to the private and business consumer. Current legislation related to this subject includes:

1. Trading Stamps Act 1964
2. Supply of Goods (Implied Terms) Act 1973 (replacing Sale of Goods Act 1893)
3. Consumer Credit Act 1974
4. Health and Safety at Work, etc. Act 1974
5. Unfair Contract Terms Act 1977
6. Sale of Goods Act 1979
7. Supply of Goods and Services Act 1982

The Acts mentioned in (1) and (3) above are unlikely to affect the plant engineer in his business environment. Any Act of Parliament will supersede a common law precedent made in the High Court or above, unless the Act specifically states otherwise.

The most significant Act to affect the plant engineer is the Supply of Goods and Services Act 1982, and this will be discussed in detail, being described simply as 'the Act'.

The Act is in two parts, the first part being connected with the supply of goods and the second regarding supply of services. Building contracts and those for the supply and installation of plant are included within the Act.

Under the Act, any goods transferred by the transferor to the transferee must be of 'merchantable quality' (section 4(2) Part 1). They must also be fit for the purpose or purposes for which goods of that kind are commonly supplied as it is reasonable to expect, having regard to any description applied to them, the prices (if relevant) and all the other relevant circumstances (section 4(9)). In addition, goods must be suitable for their particular purpose (section 4(5)).

The transferee would need to make known to the transferor the particular purpose of the goods. Under privity of contract, an action in contract will lie against the transferor, who may not be the manufacturer. Where a manufacturer offers a guarantee to the transferee, the transferee has a choice to sue either parties. An important part in section 4(2) is that it carries a strict liability. Where a consulting engineer is appointed to carry out a design, that design must be up to the standard of a reasonable or average consulting engineer, and must not fall below that standard. Where a firm supplies a design, and the goods as well, then the standard applied is strict, and in the event of a faulty design, the client can sue, regardless of how reasonable the design was.

An example would be as follows. *B* appoints a consulting engineer *C* to design a particular plant and briefs him fully on the performance he wants from the plant. A specification is produced by *C* and he invites tenders on behalf of *B* from contractors. *D* is the contractor employed to supply and install the plant. When installed by *D* the plant operates but does not perform as the brief from *B*, because *D* ordered the incorrect size, although properly specified by *C*. In this case the plant is of merchantable quality as section 4(2) of the Act, but is not fit for its purpose as in section 4(5). *C* is in the clear on design because he can rely on the specification provided, although he can be sued by *B* for negligent supervision, and *D* can be sued in contract by *B* for breach of contract. In the event that the specification was correct, but that the machinery was defective, *B* can rely on section 4(2) of the Act for merchantable quality. *B* may have a case against *C* for negligence in not supervising the works properly if the works were installed badly by *D*.

5.23 Delays and determination

Contractors are all too frequently delayed in their completion for a variety of reasons, some of which are:

1. Client adds extra work or changes his mind;
2. Client or consulting engineer takes his time in giving instructions or clarifying the specification;
3. The contractor starts on site later than agreed;
4. Sub-contractors do not coordinate properly;
5. Site conditions are not as expected;
6. Materials are not delivered as promised;
7. Exceptionally bad weather;
8. Slow progress generally by the contractor.

Delays to contracts can be costly to both client and contractor, and, in some cases, suppliers. There has been a tendency in the past few years for an injured party to sue for damages under the contract, and where there is no contract, in tort. Since the case of *Marden* v. *Esso Petroleum* (1978) it is possible to sue both in contract and tort at the same time.

The various forms of contract make a provision for registering a delay or possible delay, and this aspect of a contract should be taken very seriously, because it can be the greatest source of financial claim, particularly by the contractor. Some contracting firms examine contracts very carefully to see if a claim of delays can be made, and this starts on the day that the contract has been signed.

Close control of the contract should be kept at all times. This is where formal site meetings, with reports of progress, can be most useful and can help to steer a late contract back on the right course. When a contract begins to fall behind the agreed programme it is important to find out why and to rectify the problem. It is in everyone's interest to see the plant commissioned on the agreed date, and a casual attitude by the project leader should never be taken. He should be strict with anyone who causes a contract to slip behind schedule.

There are legitimate reasons why a contract can fall behind, and some of these are referred to in (1) to (8) above. However, item (8) must not be tolerated, and the contractor should be taken to task early in the contract when this occurs. Close monitoring will show how the contractor is progressing, and where the contractor has fallen down so badly that there seems to be no end to the contract, then the client should seriously consider determining the contract.

Most standard forms of contract contain specific procedures for determination, and these should be rigidly adhered to. For example, the contract document may lay down that the contractor be given 14 days' notice in writing to rectify his progress before determination is carried out. Any letters should be sent by recorded delivery or served by hand. While liquidated or unliquidated damages can be claimed with respect to late or incomplete work, expense claims usually follow the events shown below, which are also mentioned in Section 5.24:

1. Client adds extra work or changes his mind. It may seem reasonable that where the client authorizes extra work and pays for that extra work, he should not then be required to pay, because the contractor needs to stay on site longer. This is, however, the case, and the claim because of the delay very often exceeds the actual extra costs of the works. Where a client simply changes his mind and cancels some of the work, then the contractor can insist on his profit, and overhead costs for the time he will be without work as a result.
2. Delay in instructions being issued to the contractor can be a loss to the contractor if he is waiting for the instructions, and he can sue for damages for his loss.

5.24 Liquidated damages and loss and/or expense

This is an aspect of a contract which should not be contemplated by either party at the outset of a contract. However, it is common for the client to set a figure for liquidated damages which he considers that he will genuinely lose if the contract is not completed on time, and should not be confused with a penalty clause, which is for punitive damages.

The expression 'unliquidated damages' means actual losses which are incurred by a client, and could be claimed in the absence of an agreement to a specific amount in the contract. Loss and/or expense can be suffered by a contractor due to other parties who caused him to be delayed. These sums per week are invariably far in excess of any liquidated damages amount set in the contract.

In practice, what happens is that where a contract is falling behind programme the contractor will apply to the client for an extension to his period for completing the works. Should the contractor not make an application then he will be in breach of his contract to complete the works by the date agreed, and the client can claim liquidated or unliquidated damages as a result. Where no specific date for completion has been set, in a dispute situation, the court will set a date which, under the circumstances, was or ought to have been in the contemplation of the parties at the outset of the contract. For example, a firm agrees to supply and install a machine, and while no particular date is set for installation, the firms knows that production would be affected badly if it was not installed within three months. Without extenuating circumstances prevailing, the firm could be in breach of its contract if, say, the machine was not installed within three months.

The contract could include a penalty clause for non-completion by the contractor, and paid by him, as well as a bonus clause paid by the client to the contractor in the event that he completes his contract on or before the agreed completion date. This bonus clause can be on a reducing scale (say, from 75% of the time onwards).

Liquidation of the contractor has nothing to do with liquidated or unliquidated damages, although these can still be claimed to a company in liquidation. Some forms of contract, such as the JCT form, make provision for automatic determination in the event of a firm going into liquidation but that the contractor may be reinstated under certain circumstances.

5.25 Practical and final certificates

The essence of a plant engineering contract will be that the plant actually performs to the criteria laid down in the design, and this must be established before a Practical Completion Certificate is even considered. Tests will need to be made to the plant, and then it will be commissioned by the contractor and left in a condition whereby it can be operated as specified. When the works within the contract are completed, a certificate acknowledging that fact should be issued by the client (or employer) to the contractor. This duty is normally performed by the architect or consulting engineer, acting as agent. Standard forms of contract have particular formats, but the points which should be noted are:

1. Names and addresses of parties
2. Name and reference of job
3. Date
4. Brief outline of contract and note that the works are now practically complete
5. Period of guarantee from contractual documents
6. Signature.

The works should be fully complete when the Practical Completion Certificate is issued, without any defect or omission at that point. This is clear from the High Court decision in *H. W. Nevill* (*Sunblest*) *Ltd* v. *William Press & Son Ltd* (1981). Other cases which went to the House of Lords substantiate this, as in *Jarvis & Sons* v. *Westminster Corporation* (1970) and *P. M. Kaye Ltd* v. *Hosier & Dickinson Ltd* (1972).

There is a tendency in the building industry, for expediency, to issue a Practical Completion Certificate subject to a list of defects. Very often this is done in order to see the end of the main works.

From the date of the Practical Completion Certificate the contractor is no longer liable for liquidated damages, and the defects liability begins to run. Latent defects which appear for the next six or twelve months, depending upon the agreed period, have to be made good by the contractor, free of charge to the client.

The final certificate is issued when the end of the defects period has been reached, and when all defects which have appeared within that period have been attended to.

The client is not prevented from making a claim for latent defects after the twelve-month period ends, but may be prevented from making such claims by the Limitation Act 1980 and the Latent Damages Act 1986, after a period of six years for a simple contract and twelve years for a contract made under seal.

5.26 Disputes and arbitration

It is unfortunate when a dispute arises between any of the parties. The two most common areas are a claim by the client that the work has been done badly or taken too long, and from the contractor that his contract has been interfered with by the client or the nominated sub-contractors.

Most disputes take the form of strongly worded letters between the parties, with eventual threats of litigation or arbitration, very often in an attempt to cause a party to capitulate. Where there is a genuine attempt to settle a dispute the letters between the parties can be marked 'Without Prejudice' and letters so worded cannot be used in evidence without the consent of both parties. If a binding contract has been effected as a result of 'without prejuduce' interviews or letters, this may be proved by means of the 'without prejudice' statement, as in *Tomlin* v. *STC Ltd* (1969). The standard forms of contract invariably contain agreements to arbitrate in the event of a dispute and some even name the arbitrator or a professional body who will appoint an arbitrator, upon the application of one of the parties. Where no agreement to arbitrate exists, the injured party may sue in either a County Court or High Court, depending upon the value of the claim, and he must then prove his case before a judge.

A more speedy method, although not necessarily less expensive, can be for the parties to make an agreement to arbitrate, and to appoint their own arbitrator, who should not have been previously connected in any way with them. Where a written arbitration agreement is made, the arbitration will be conducted under the Arbitration Acts 1950 and 1979 and, unless otherwise agreed, before a single arbitrator, whose powers are almost those of a High Court judge.

An arbitration hearing can be held anywhere convenient to the parties, and at times agreed. While the arbitrator will not meet any one party separately or take instructions from him, he will take instructions within the law from all of the parties. For example, if the parties wish the hearing to be held in a certain place, as long as the facilities are adequate for the hearing for the parties to meet their legal advisors, the arbitrator is likely to agree. On matters of the conduct of the hearing, the arbitrator is in complete control, subject to a limited appeal by the parties to the High Court, who can, on matters of law and conduct, direct the arbitrator to take certain action.

The hearing itself will be conducted in a less formal manner than a court trial. The room itself is set out like a conference room, with a seat at the head for the arbitrator and places (called 'the box') at the opposite end for the witnesses. To the left of the arbitrator is usually the claimant and his advisors and to the right the respondent and his advisors. It is normal for all present to stand when the arbitrator enters or leaves the room and address him as 'Sir', referring to him during the hearing as the 'Learned Arbitrator'. A judge needs to be treated with more respect and a procedure of bowing upon entering and leaving is normal. A County Court judge is referred to as 'Your Honour' and a High Court judge as 'My Lord'.

The plant engineer may be asked to give expert evidence, which means giving his opinion to the court or tribunal on the subject matter. Prior to giving evidence a proof of evidence will be prepared by the expert and will contain simply the following:

1. Identification, qualifications and experience of witness;
2. Details of his appointment;
3. General description of subject matter in dispute;
4. Any tests, photographs, samples or other documents relied upon;
5. Conclusions and opinions.

There is a distinct difference in the way that expert evidence is received by a tribunal to that of witnesses of fact. The plant engineer is strongly advised not to agree to accept an appointment and subsequently give expert evidence on matters for which he is not fully experienced. In accepting an appointment to write a report for a client, his legal advisors should be made fully aware of the plant engineer's background and expertise. Having become committed to the case, the client could, if necessary, apply for a *subpoena ad testificandum* to ensure that the expert gives evidence. There may be occasions when the plant engineer will be called upon to give evidence of fact, and under these circumstances he is not an expert witness, although he may be able to give opinion on matters for which he is considered an expert. An example of this is where the plant engineer is an employee of the claimant, and he needs to give evidence that a machine delivered to his works would not perform satisfactorily. It would be a matter of fact that the machine would not perform, and the evidence would be strengthened because the plant engineer is an expert in this type of machine.

Upon entering the witness box the plant engineer will be invited to take the oath, and can either swear on the Bible, affirm or take the kind of oath suitable to his religion. The advocate acting for him will then take him through his proof of evidence, but will be careful not to ask him leading questions, except his name, qualifications and experience. The opposing advocate will then cross examine the witness to test the evidence given. He will ask leading questions and may attempt to shake the evidence given. The advice to the witness is to stay calm, courteous and truthful, remembering that the advocate is only doing his job.

Following cross examination the advocate may ask a few more questions to clear up points already raised. The judge or arbitrator may then ask a question in clarification.

5.27 Common problems and solutions

There is no sure way of avoiding claims with respect to contracts, but the following summary may be of assistance to the plant engineer:

1. Vet prospective tenderers carefully prior to invitations to tender to ensure as much as possible that they are capable of carrying out the work.

2. Select the type of contract most appropriate for the work.
3. Where possible, avoid nominating a contractor or supplier.
4. Make sure that the drawings, specifications and tender documents are clear and unambiguous. Do not rely on exclusion clauses in the specification or tender document. Have designs checked and use methods and equipment which comply with British Standards and Codes of Practice.
5. Unless there is a clear programme of works which has been mentioned in the tender documents, before accepting tenders make sure that the contractor or supplier has specifically stated his commencement and completion dates. In the event that the completion date is too early or too late, resolve the matter before entering into the contract.
6. Ensure that site meetings are as short as possible and organized to assist the project, and that they serve the purpose for which they were intended (i.e. sort out problems and ensure satisfactory reports).
7. When a contractor indicates that he is or may be delayed, take him seriously and solve the problem without delay. Do not interfere with the way in which the contractor carries out his work, unless he is patently incompetent.
8. Make sure that all site supervisors, clerks of works and site engineers keep good records, and that they are kept contemporaneously, so that, if necessary, they will be admissible in evidence in the unfortunate event that a dispute arises.
9. Be careful on the matter of interim valuations for payments to contractors. To undervalue can be as serious as overvaluing.
10. Keep up to date with variations to the contract. Do not wait until the end of the contract before extras are added up.
11. Safety should be completely in the hands of the contractor controlling the site, unless the client is in occupation. Keep a high profile in dealings on safety matters, and ensure that staff and any other parties are conscious of the need for safety.
12. The plant engineer should be careful of giving casual advice to parties with whom he has no contract. Under the *Hedley Byrne* v. *Heller* rule the plant engineer could be sued for a negligent mis-statement if his advice was wrong, and the receiver of the advice could reasonably rely upon it, and he suffered loss as a result.
13. When attempting to settle a dispute the heading 'Without Prejudice' can be used in correspondence which, subject to the rules of privilege, cannot be used in court.

6 Finance for the Plant Engineer

Leon Turrell FCA
Financial Consultant

Contents

6.1 Accounting

Ultimately, all activities of an organization will be expressed in money, the common denominator within the accounts of the proprietor, be it a company, a partnership or a statutory organization. The accounting for these activities will be under the control of the finance department, which, in turn, will be controlled, through the financial director, by the board of directors.

A typical accounting organization is shown in Figure 6.1. These divisions, depending on the size of the organization, may be broken down further to provide accounting services at each plant or operating facility. There will normally be an accountant at each plant, or, in the case of very large organizations, at each major profit or cost centre at plant level. By this means, close liaison can be maintained with and management services information provided for the locally based operating management, in addition to providing the flow of information and appropriate documentation to enable the organization to prepare timely and informative accounts.

Many of the financial functions will tend to be centralized in order to enjoy economies of scale available through the use of specially qualified staff and of computers.

Prompt and accurate accounting is vital for the well-being of the organization and for the early detection of problem areas needing corrective treatment by management. The achievement of this promptitude and accuracy depends to a very large extent on the cooperation and attention to detail by staff in non-accounting departments responsible for the recording of information (e.g. fuel and water consumptions, meter readings, the taking of stocks, etc.). Such staff may not recognize their involvement in these areas as one of their prime concerns. Each of these departmental chains will normally be staffed and headed by appropriately qualified people.

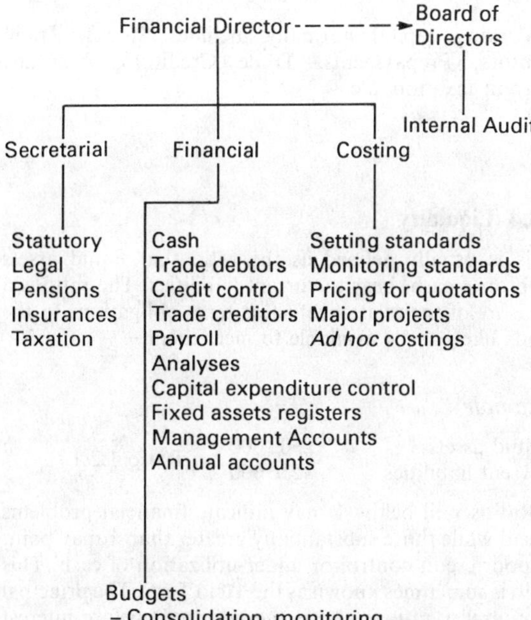

Figure 6.1 A typical accounting organization chart for a large company

6.2 Types of organization

The most common of the organizations operating in today's industrial scene is the joint stock company with limited liability. In different countries there exist variants of this format, but there is a wide spread of organizations based upon broadly similar principles.

The partnership and sole proprietor types of organization are suitable for businesses while they remain small and can be managed and financed by the owners, but as expansion takes place and the business demands more capital, then the ability of relatively small numbers of proprietors to provide additional capital from their own resources becomes increasingly difficult. In the search for additional funds recourse may be had to a variety of institutions such as banks and the specialist companies which have been set up, often by the banks themselves in conjunction with insurance companies and merchant banks, to cater for this need.

These specialist lenders will often wish to take a share in the capital or 'equity' of the business as part of the arrangement by which they will advance capital. In order to achieve this the business will normally be required to incorporate.

In the UK the share capital of a company is expressed as a sum of money, the 'Authorized Share Capital', made up of a stated number of shares, each of which has a 'face' or nominal value.

On incorporation, a small, family-owned firm may start with an initial capital of, say, £10 000 divided into 10 000 shares of £1 each. A large public company, on the other hand, may have a share capital of many millions of pounds, sub-divided into many more millions of shares or stock units, of perhaps as little as 25p.

Not all the shares in a company may, or need to be, issued. By the issue of more shares, at an appropriate stage in the company's growth, capital may be raised from the institutions mentioned above or, through a flotation on the Stock Exchange, from the investing public at large. When this happens the price paid per share by the purchasers may be many times greater than the nominal value.

Example

Company A has an authorized and issued share capital of 300 000 ordinary shares of £1 each held by five shareholders. After many years of successful trading it is decided to seek a listing on the London Stock Exchange. The capital is increased to £3 million and is then divided into 12 million shares of 25p. Of these, 8 million are offered for sale to the public at £1 each. Subject to the costs of the issue, this will raise £8 million of fresh capital, some of which will be used in the business and some will go to the original owners of the business.

Before the issue:

	Shares of £1	
	No.	*Value*
Original shareholders	300 000	£300 000

After the issue:

	Shares of 25p	
	No.	*Value*
Original shareholders	4 million	£1 million
Outside shareholders	8 million	£2 million

The original shareholders' ownership of the business has declined from 100% to 33.3%, and it may be necessary for the owners to accept the appointment of further directors from outside the business, possibly in a non-executive capacity.

A price for the shares will be set by the market, based upon its assessment of the prospective performance of the company and the perceived quality of its management, and trading in the shares will begin. This price does not affect the money value received by the company for the sale of the shares at flotation.

6.3 Definitions

6.3.1 Capital

Capital is variously described, but is generally taken to be the amount of finance provided by the proprietors or outside lenders to enable the business to operate and earn a profit, from which the lenders will expect to receive an income. Share capital is that part which is regarded as fixed in the sense that it changes infrequently, and is increased only when the business requires additional permanent finance. In a limited company the amount of accumulated profits left in the business by the proprietors (i.e. not drawn in dividends) is grouped with the share capital and termed 'Shareholders' Funds'. Under present-day legislation it is also possible for a company to reduce its share capital by buying back its own shares or, by applying to the court, to have its capital reduced to enable a reorganization of the company's structure to take place. In this context, proprietors may be shareholders, stockholders, partners or sole traders. Capital is often termed the 'equity'.

6.3.2 Capital and revenue expenditure

The difference between capital and revenue expenditure can be compared broadly with that between expenditure on fixed and current assets. Capital expenditure is incurred in the acquisition of fixed assets and positioning the organization in order to start or continue its operations. Revenue expenditure relates to the purchase of goods and materials for manufacture or resale and in bringing those goods or materials to a condition suitable for sale, providing the facilities to bring about the sales and in the general administration and running of the business.

6.3.3 Cash flow

This is the amount of cash passing through the hands of an organization in an accounting period. The cash flow statement analyses the sources and the disposition of cash during a given period. It is akin to the Sources and Application of Funds Statement found in the published accounts of companies.

In addition to their historical use, Cash Flow Statements are prepared as part of the budgeting process in order to identify the effects upon the cash facilities of the proposed activities for the period under review. A typical, simplified, statement would give the following information.

Cash flow statement

Operating profit	×
Depreciation	×
Cash flow from operations	×
Fixed assets bought	(×)
Fixed assets sold	×
Loans received	×
Loans repaid	(×)
Corporate taxes paid	(×)
Interest paid	(×)
Interest received	×
Increases in working capital*	(×)
Decreases in working capital*	×
Dividends paid	(×)
Net cash inflow (outflow)	×
Opening cash balances	×
Closing cash balances	×

* Working capital normally includes Stocks, Trade Debtors, Prepayments, Trade Creditors, Accruals, Current taxation, etc.

6.3.4 Liquidity

This is usually defined as the ratio that liquid assets (debtors + cash) bear to current liabilities. The ratio is a measure of the relation of short-term obligations to the funds likely to be available to meet them.

Example

Liquid assets	£300 000	Ratio 1.2:1
Current liabilities	250 000	

Ratios well below 1 may indicate financial problems ahead while those substantially greater than 1 may point to poor credit control or under-utilization of cash. This ratio is sometimes known as the 'Acid Test'. The principal profitability ratio in use is the net profit before interest and tax (NPBIT) to net assets or return on capital employed.

6.3.5 Financial and operating ratios

Ratios are widely used to compare the performance of a business with predetermined objectives set by the business itself, with standards used by banks and other lenders, with other businesses in the same segment of industry (Inter-firm Comparison) and by the Stock Exchange and its attendant analysts. The principal criteria are profitability, return on capital, liquidity and growth. To undertake growth by acquisition a high standing in the market is required. This is often a reflection of the performance against these criteria.

The ratios which follow are illustrated by reference to the following balance sheet and profit and loss account.

Balance sheet		£000
Current assets		
Cash		1 175
Stocks		4 700
Debtors and prepayments		4 935
		10 810
Current liabilities		
Creditors		3 170
Short-term loans		600
Current taxation		1 056
Dividends		354
		5 180
Net current assets		5 630
Fixed assets		6 250
		11 880
Financed by:		
Share capital – Preference	825	
Ordinary	3 050*	3 875
Reserves		4 700
Shareholders' funds (Net worth)		8 575
Long-term loans		2 500
Deferred taxation		805
		11 880

* 12 200 000 shares of 25p each.

Profit and Loss Account		£000
Sales		22 000
Cost of sales (72%)		15 800
Net profit before interest and taxation (NPBIT)		2 555
Loan interest		200
Taxation		980
Net profit after taxation (NPAT)		1 375
Dividends – Preference	47	
– Ordinary	330	377
Retained profits		998

1. Gearing – the ratio of fixed rate capital and borrowings to other capital, that is:

$$\frac{\text{Preference capital} + \text{long-term loans}}{\text{Ordinary capital} + \text{reserves}}$$

$825 + 2500 : 3050 + 4700 = 43\%$

If the gearing is high the capacity to borrow may be affected, since the risk to creditors is high and the company may already be burdened with heavy interest charges.

2. Outside liabilities to shareholders' funds

	£000
Long-term loans	2500
Taxation – deferred	805
Current liabilities	5180
	8485
Shareholders' funds	8575
Ratio	0.99:1

The total of capital and reserves is the amount by which the assets can fall below the balance sheet value without depleting the amount available for creditors. A high ratio will reduce borrowing capacity.

3. Current assets to current liabilities

$10\,810 : 5180 = 2.1 : 1$

Shows the margin of safety available to short-term creditors. Significantly higher ratios than 2:1 may indicate excess stocks, poor credit control or inadequate control of cash resources.

4. Stocks, debtors, creditors to sales

The ratio of the minimum net assets required to support the sales volume of the business. From these ratios can be calculated the additional working capital needed if sales are to be increased:

$4700 + 4935 - 3170 = 0.294 : 1$

or

Stocks plus debtors minus creditors = 29.4% of sales

The average for manufacturing industries is usually around 25% of sales.

Changes in sales levels will give rise to movements in the working capital required to support them. The amounts can be calculated:

Increase in sales 15% of 22 000 = 3300
Increase in working capital
15% of (4700 + 4935 − 3170) = 970
or 29.4% of 3300 = 970

5. Profitability

The most widely used gauge of profitability is the ratio of profit to net assets (or return on capital employed):

$$\frac{\text{NPBIT} \times 100}{\text{Shareholders' funds} + \text{Long-term loans} + \text{Deferred taxation}}$$

$2555 \times 100 / 11\,880 = 21.5\%$

Two ratios amalgamate to provide the profitability ratio.

	£000
Sales	22 000
Net assets	11 880
NPBIT	2 555

Return on capital	255 500:11 880 = 21.5%
Profit/Sales	255 500:22 000 = 11.6%
Sales/Net assets	22 000:11 880 = 1.85
	1.85 × 11.6% = 21.5%

These two ratios indicate funds usage efficiency and provide a basis for deciding the order in which improvements can be made.

6. Stock ratios
   Expressed as a 'Stock turn', representing how many times stocks are turned round in a given period and/or as the number of days' stock of finished goods available to meet sales demand.

		£000
Stocks	– Raw materials	1530
	– Work in progress	1300
	– Finished goods	1740
	– Other	130
		4700

$$\text{Stock turn} = \frac{\text{Cost of sales}}{\text{Finished goods}}$$

$= 15\,800/1740 = 9.1$ times

Days' stock $= 1740 \times 365/15\,800 = 40$ days

7. Debtors' ratio
   Expressed as days of sales outstanding or 'Debtors' Days'. Used in conjunction with prior figures for measurement of credit control efficiency and trends:

	£000
Debtors	4 935
Sales	22 000

$$\text{Ratio} \quad \frac{4935 \times 365}{22\,000} = 82 \text{ days}$$

Sometimes refined by deducting the sales for the current period as being not yet payable, and therefore leaving figure to represent *overdue* debt only. The ratio is also known as 'Collection Period'.

8. Liquidity
   Liquidity can be expressed as the ratio of liquid assets (cash plus debtors) to current liabilities. Such assets are also known as 'Quick Assets', i.e. capable of swift realization.

$$\frac{(1175 + 4935)}{5180} = 1.2{:}1$$

The ratio shows the ability to settle short-term liabilities and should not normally be lower than 1:1, though it very often is! The lower the ratio, the greater indication of possible financial strain. A high ratio could mean poor credit control or under-utilized cash.

9. Earnings per Share (Net Basis)

$$\frac{\text{NPAT} - \text{Preference dividend}}{\text{Number of issued shares}}$$

$$\frac{1375 - 47}{12\,200} = £0.109$$

Growth in earnings per share is used widely by analysts as a measure of a company's success. Adjustments must be made where issues of shares have taken place to satisfy acquisition purchase considerations.

10. Yield
    In this example Advance Corporation Tax has been calculated at 25:75.

$$\frac{\text{Earnings per share} \times 100/75 \times 100}{\text{Market Price}}$$

$$\frac{0.109 \times 100/75 \times 100}{1.40} = 10.4\%$$

This demonstrates the current return on the stock market price.

11. Dividend Yield

$$\frac{\text{Dividend} \times 100}{\text{Market Price}} = \frac{0.036 \times 100}{1.40} = 2.58\%$$

The ratio gives the return which the dividend provides on the market price.

12. Dividend Cover

$$\frac{\text{Gross Earnings}}{\text{Gross Dividend}} = \frac{0.145}{0.036} = 4.0 \text{ times}$$

The ratio indicates by how much earnings can fall before the dividend must be reduced and also shows the company policy towards the payment of dividends and profit retention.

13. Price/Earnings Ratio (P/E Ratio)

$$\frac{\text{Market Price}}{\text{Earnings (Net)}} = \frac{1.40}{0.109} = 12.8$$

This shows the number of years' earnings represented by the current market price. It evidences the Stock Market's assessment of the company's ability to maintain or increase earnings. A low P/E Ratio will normally indicate a high-risk business, a high ratio a company with potential for growth.

6.4 Budgetary control

6.4.1 Preparation

Financial data on their own provide only an historical record of the transactions of a business. The accounts of a company published annually are mainly of historical value but must, by law, include comparisons with the figures for a prior corresponding period.

Until information can be compared with similarly classified data its use must be limited. In order to plan ahead, a business will prepare a strategy or budget for the next trading year and probably several years thereafter, with that for the next year broken down into the business's scheme of accounting periods.

The final form of the budget will include profit and loss account, balance sheet and cash flow statement for the planning periods together with such supporting statements as are deemed necessary (for example, proposed capital expenditure). These will provide management with advance warning of points of stress upon resources, and enable steps to be taken to ease their effects or to avoid them altogether. The financial controller or chief accountant will have the responsibility for bringing together the relevant information in the final budgets to be presented to senior management.

Budgeting practices vary from one business to another. Imposed budgets will be drawn up by senior management and, as the name implies, be imposed on those lower down the hierarchy. Participation budgets require the input of data and information at the formulatory stages by the people who will be responsible for bringing about the achievement of the results envisaged in the budgets.

The process of compiling the budgets will start with the preparation of estimates of the physical requirements of the plan, be they manpower, materials, tonnes of fuel, cubic metres or therms of gas, units of electricity or cubic metres of water. These estimates will be prepared in relation to the estimates of output by manufacturing or process departments which, in turn, will have been based on the quantities or other measures forecast in the sales budgets.

It is therefore vital that detailed and accurate records are kept by the operating and engineering departments of the usage and consumption of fuel, water and other services in order that performance against budget can be properly measured and so that data are available for use in compiling future estimates.

The budgets will also require estimates of expenditure upon equipment of a capital nature in the plan periods. This expenditure will normally fall into the following categories:

1. Replacement of existing machinery and equipment at the end of its useful life. (Straight replacement is rarely possible, since improvements in technology will usually have taken place.)
2. Items requiring replacement because of changes in the law (for example, those relating to fire prevention, safety measures for the protection of employees and public).
3. Items which will improve profitability by saving costs or by carrying out processes faster or by using less manpower (see Section 6.5).
4. Expenditure in connection with new projects or necessary to provide increased throughput.

Motor vehicles can also fall within the above categories but businesses will usually have specific policies for their replacement.

As said elsewhere, the importance of realistic estimates of costs cannot be over-stressed, if, in turn, realistic comparisons can be made as the budget periods progress.

The compilation of figures for the budget should almost always be done from 'the bottom up', that is, by calculating the costs for the individual accounting periods by taking estimated quantities and prices for each period and summing for the whole of the budget. Thus, for example, holiday periods and known peaks and troughs can be recognized and catered for within the figures. Estimates made in the first instance on an annual basis can rarely be analysed satisfactorily to individual periods except where the costs themselves are expressed in annual terms (e.g. rents and rates).

The sources of information from which the data are obtained are infinitely variable, but will include manufacturers' and suppliers' price lists, direct contact with suppliers and long-term contracts with sources of fuel and power.

At all times the interrelationship between budgets must be maintained – sales with finished products – products with consumption of materials – consumption of materials with labour using them, machine time and power. Machine time will influence maintenance, which will have its own content of labour, materials and work by outside specialist contractors.

When all the budgets have been prepared they will be consolidated by the finance department for submission to the approving authority. In large businesses there will be several levels at which authorization will be made before the final total plan for the business and its component divisions is agreed.

In group organizations, company budgets will themselves be consolidated into group form for final approval by the group board. A typical budget programme might be based on profit centres taking the form shown in Figure 6.2. (In multinational organizations this consolidation will continue worldwide.)

6.4.2 Control

During the progress of the financial year budgets will be compared at intervals with the actual performance to enable adjustments to be made to ensure that the planned outcome is achieved as nearly as possible. Ideally, because of changing circumstances after the start of the

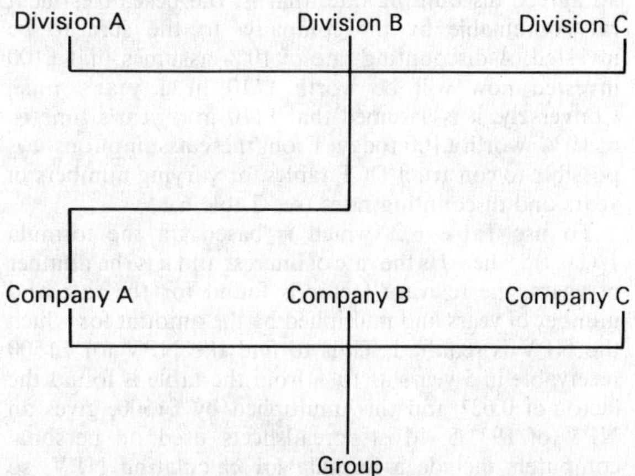

Figure 6.2 A typical group's profit centres

budget year, the budgets should be revised, possibly at three-monthly intervals. In practice, particularly in large and complex organizations, this is rarely practicable, given limitations of staff and time. An alternative is to require to be given a forecast of the immediately following period(s) to be included with each report to identify areas diverging from planned performance and to alert management to the need for correction.

6.5 Capital expenditure – appraisal methods

There are a number of accepted methods available for the comparison and appraisal of the virtues of proposed capital expenditure projects. Those considered here are as follows:

1. *Pay-back period.* This consists of calculating how long it will take for the profits generated by the capital outlay to equal the outlay itself, such profits usually being calculated after taking into account tax and any grants receivable. The defects of this method are that it takes no account of the profitability of the schemes after the break-even point is reached and the same value is placed upon each pound of profit, whether it is earned in year 1 or year 10. This latter shortcoming is avoided by the use of Discounted Cash Flow assessment.
2. *Rate of return.* A rate of return is calculated on the profits remaining after the initial outlay has been written off. This method suffers from the same defect mentioned in (1) above and also from the use of arbitrary periods for the writing down of the initial expenditure and of arbitrary rates of interest for the calculation of the rate of return.
3. *Discounted cash flow (DCF).* This method recognizes that £1000 income in five years' time is worth less than £1000 receivable this year. The use of DCF in appraising two or more competing projects offers two methods of assessment: the Net Present Value (NPV) or DCF Rate of Return.

The principle of NPV is best understood by applying an agreed discounting rate, that is, the best investment rate obtainable by the company, to the sum to be invested. A discounting rate of 10% assumes that £100 invested now will be worth £110 in a year's time. Conversely, it is assumed that £110 in a year's time is, at 10%, worth £100 today. From these assumptions it is possible to construct DCF tables for varying numbers of years and discounting rates (see Table 6.1).

To use Table 6.2, which is based on the formula $1/(1+i)^n$, where i is the rate of interest and n is the number of years, the relevant factor is found for the rate and number of years and multiplied by the amount for which the NPV is required. Thus to find the NPV for £1500 receivable in 5 years at 10% from the table is found the factor of 0.621 and this, multiplied by £1500, gives an NPV of £931.5. Most spreadsheets used on personal computers include a formula for calculating NPV, so avoiding the need to construct tables.

Table 6.1 Application of NPV. To illustrate the uses of the two DCF methods the following example assumes the following data relating to two competing projects

Year	*Net cash flows*	
	Project A £	*Project B* £
0	−100 000	−200 000
1	−15 000	−50 000
2	20 000	5 000
3	25 000	10 000
4	35 000	25 000
5	40 000	40 000
6	50 000	75 000
7	−5 000	75 000
8	65 000	90 000
9	85 000	110 000
10	100 000	130 000
Totals	300 000	310 000
Year	*Net present values at 10%*	
	Project A £	*Project B* £
0	−100 000	−200 000
1	−13 367	−45 455
2	16 528	4 132
3	18 783	7 513
4	23 905	17 075
5	24 836	24 836
6	28 225	42 338
7	−2 566	38 490
8	30 323	41 985
9	36 049	46 651
10	38 550	50 115
Totals	101 266	27 680

Although Project B shows a greater total of net cash inflows over the whole period, at net present values Project A indicates a more satisfactory return, all other factors being ignored.

The use of NPV (or DCF) leaves unresolved an important problem, that of determining the rate of interest or return to be used. Different rates of return could alter the ranking of the projects by changing the point at which the returns shown by the projects are in balance. If the company's own rate of return on capital is higher than that revealed by the NPV calculation then the apparently more viable scheme may not prove to be the more acceptable.

An alternative is therefore to use another method, using the same principles, by calculating a DCF rate of return. This has the advantage of not involving any assumptions as to interest rates, but calculates the effective rate of return on each project. The DCF rate of return is defined as the rate which reduces the NPV to zero. This method is more difficult to calculate in that it necessitates taking several trial values until two are found giving values on either side of zero. A weighted average can then be applied to 'fine tune' the result.

Where there are constraints upon the provision of

Table 6.2

n	*5%*	*10%*	*15%*	*20%*	*25%*	*30%*
1	0.9524	0.9091	0.8696	0.8333	0.8000	0.7692
2	0.9070	0.8264	0.7561	0.6944	0.6400	0.5917
3	0.8638	0.7513	0.6575	0.5787	0.5120	0.4552
4	0.8227	0.6830	0.5718	0.4823	0.4096	0.3501
5	0.7835	0.6209	0.4972	0.4019	0.3277	0.2693
6	0.7462	0.5645	0.4323	0.3349	0.2621	0.2072
7	0.7107	0.5132	0.3759	0.2791	0.2097	0.1594
8	0.6768	0.4665	0.3269	0.2326	0.1678	0.1226
09	0.6446	0.4241	0.2843	0.1938	0.1342	0.0943
10	0.6139	0.3855	0.2472	0.1615	0.1074	0.0725
11	0.5847	0.3505	0.2149	0.1346	0.0859	0.0558
12	0.5568	0.3186	0.1869	0.1122	0.0687	0.0429
13	0.5303	0.2897	0.1625	0.0935	0.0550	0.0330
14	0.5051	0.2633	0.1413	0.0779	0.0440	0.0253
15	0.4810	0.2394	0.1229	0.0649	0.0352	0.0195
16	0.4581	0.2176	0.1069	0.0541	0.0281	0.0150
17	0.4363	0.1978	0.0929	0.0451	0.0225	0.0116
18	0.4155	0.1799	0.0808	0.0376	0.0180	0.0089
19	0.3957	0.1635	0.0703	0.0313	0.0144	0.0068
20	0.3769	0.1486	0.0611	0.0261	0.0115	0.0053
21	0.3589	0.1351	0.0531	0.0217	0.0092	0.0040

The cash flow expected for each period is discounted by the factor for the rate of interest chosen and the number of periods in which the cash flows will occur. The number of periods is calculated from the commencement of the capital expenditure. The factors are arrived at from the formula $1/(1+i)^n$, where i is the rate of interest expressed as a decimal and n is the number of periods. In reality, the factors assume that the cash flow passes on the last day of each period but can be adopted where the flow is roughly even throughout the period.

funds then the DCF rate of return method will be the more appropriate. Where the organization has ready access to finance then the NPV method, using the known long-term borrowing rate, should be used.

6.6 Control of capital expenditure

Significant capital expenditure usually represents a substantial commitment of the resources of a business, both financially and in terms of man-hours. It is therefore incumbent upon management to ensure that proposals for such outlays receive proper and full consideration of all the relevant implications before implementation. Once policies as to levels of authorization and commitment are laid down, there should follow the formal appraisal of the financial effects of the proposal. These can be formulated only after detailed discussion with the appropriate departments as to all the physical, technical and environmental factors involved in making the final decision. There will also be brought into consideration, where pertinent, the marketing and sales effects.

Most organizations will have sets of forms for use in the authorization and control of capital expenditure and these will vary in design and content. A basic guide for such documents would include the following:

1. *Capital variation proposal* (Figure 6.3). This should embrace the following:
   (a) Description;
   (b) Amount for which authority is requested;
   (c) Reasons for application (e.g. to increase production, to maintain production, to reduce costs, for the introduction of a new product, etc.);
   (d) Expenditure or losses not included in the proposal, such as staff amenities, transport requirements, self-competition, etc.;
   (e) Summary of the cost;
   (f) Disposal or modification of existing assets;
   (g) Details and timing of the outlays;
   (h) Extra working capital demands which will arise from the implementation of the project;
   (i) Index of the documents supporting the application.
2. *DCF calculations* (Figure 6.4). These will normally be in the form of working sheets and/or graphs, the latter being used for interpolation where the calculations are too numerous or too detailed for tabulation. Calculations will normally be carried out using computers but where this is not possible, DCF tables should be utilized.
3. *Calculation document* (Figure 6.5). This document provides a means of calculating the DCF rate of return or a net present value where these are required, usually where profitability of the project is of major concern.
4. *DCF rate of return graph* (Figure 6.6). This may not be necessary in all cases but would be used for the interpolation of rates of return where a precise answer is not obtained from the detailed DCF workings described above.
5. *Payback graph* (Figure 6.7). This provides a view of the profile of the cash flows emanating from the project and forms a useful adjunct to the DCF information.
6. *Summary of cash flows* (Figure 6.8). This document is an essential part of every capital expenditure project evaluation. The forecast thus provided will be needed in the preparation of the overall cash budget.

6.6.1 Current and post-event monitoring

Large and complex projects will be monitored as the expenditure proceeds while smaller outlays will be looked at after completion. The purpose of such monitoring is to provide a comparison of actual expenditure with that estimated when the project was sanctioned. This will normally be carried out by the accounting or internal audit departments. However, for a sensible comparison to be achieved it is essential that the planned expenditure demonstrated in the project be classified in the same way that the actual expenditure will be analysed. While somewhat obvious, this is an area where very great difficulty is often met in practice, especially where large projects are involved.

Considerable thought should be given to this aspect at the planning stage, with consultation taking place between the engineering and accounting functions. Apart from its obvious use to prevent serious over-runs of expenditure, 'post-event monitoring' provides useful lessons for the future preparation of capital projects.

Description	Amount	£
	New	
	Other	
	Total	
	Budgeted	

Reason
New product
Cost reduction
Legal requirement

RELATED EXPENDITURE not included here
Staff amenities

Transport and distribution costs

SUMMARY OF EXPENDITURE	£	ASSETS TO BE DISPOSED OF	Value £
Buildings			
Other			
Sub-total			
Working capital			
Grand total			

Prepared by	Recommended by
Approved by Board	Date

Details of expenditure for approval

	£	Comments	Timing

Details of additional working capital required

	£	Comments	Timing
Stocks increase			
Debtors increase			
Minus creditors increase			
Total			

Figure 6.3 A capital variation proposal

6.7 Standards and standard costing

Standard costing is usually thought of in connection with manufacturing production but can be used with advantage in the measurement of the efficiency of supporting plant and equipment. Most readers of this book will already be familiar with the measurement of efficiencies against, for example, manufacturers' standards for a specific item of equipment. The standards related to such plant will themselves play a part in setting the production standards mentioned above.

It is not proposed here to provide instruction in the techniques of standard costing but merely to illustrate the uses to which certain of the methods can be put. This can best be shown by the use of a worked example.

The following is the statement for a service department

Cash outflow

Year	0% Cash	Factor	5% NPV	Factor	10% NPV	Factor	15% NPV	Factor	20% NPV	Factor	NPV
0		1.0		1.0		1.0		1.0		1.0	
1		0.952		0.909		0.870		0.833			
2		0.907		0.826		0.756		0.694			
Total £		£		£		£		£		£	

Cash inflow

Year	Cash	Factor	NPV	Factor	NPV	Factor	NPV	Factor	NPV	Factor	NPV
1		0.952		0.909		0.870		0.833			
2		0.907		0.826		0.756		0.694			
3		0.986		0.751		0.658		0.579			
4		0.823		0.683		0.572		0.482			
5		0.784		0.621		0.497		0.402			
6		0.746		0.564		0.432		0.335			
7		0.711		0.513		0.376		0.279			
8		0.677		0.467		0.327		0.233			
9		0.645		0.424		0.284		0.194			
10		0.614		0.386		0.247		0.162			
Total £		£		£		£		£		£	

Figure 6.4 DCF calculations

Project____________________________________

Year ending__________	1	2	3	4	5	6	7	8	9	10	
Revenue from project											
Other income arising											
Total income											
COSTS											
Operating profit (loss)											
Capital allowances											
Taxable profits											
	1	2	3	4	5	6	7	8	9	10	11
Tax											

Figure 6.5 Profit calculation

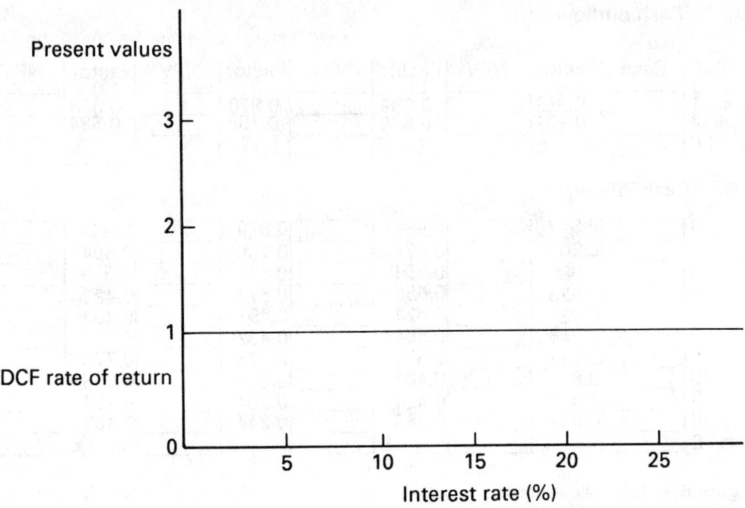

Figure 6.6 Interpolation of DCF rate of return

supplying a factory which uses both fuel and materials.

	Budget	*Standard*	*Actual*
Output units	32 000	33 600	29 400
Operating days	20	21	21
Cost per unit	83.5p	83.5p	90.0p
Fuel – Consumption	200	210	213
– Unit price	£50	£50	£49
– Cost	£10 000	£10 500	£10 437
Materials – Consumption	800	840	700
– Unit price	£2	£2	£2.10
– Cost	£1 600	£1 680	£1 470
Labour – Hours	80	84	105
– Hourly rate	£5	£5	£7
– Cost	£400	£420	£735
Repairs			
– Rate per output unit	10p	10p	11p
– Cost	£3 200	£3 360	£3 234
Variable overheads			
– Rate per output unit	20p	20p	22p
– Cost	£6 400	£6 720	£6 468
Fixed overheads			
– Rate per output unit	16p	16p	14p
– Cost	£5 120	£4 704*	£4 116
Total cost	£26 720	£27 384	£26 460

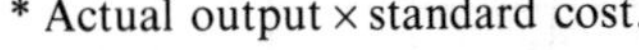

* Actual output × standard cost.

Variances from standards are then calculated as follows:

			£	
Fuel				
Usage variance	SP (SQ – AQ)	= 50 (210 – 213)	= 150A	
Price variance	AQ (SP – AP)	= 213 (50 – 49)	= 213F	63F
Materials				
Usage variance	SP (SQ – AQ)	= 2 (840 – 700)	= 280F	
Price variance	AQ (SP – AP)	= 700 (2.00 – 2.10)	= 70A	210F
Labour				
Rate variance	AH (SR – AR)	= 105 (5 – 7)	= 210A	
Efficiency	SR (SH – AH)	= 5 (84 – 105)	= 105A	315A
Repairs				
Volume variance	SC (SV – AV)	= 0.10 (33 600 – 29 400)	= 420F	
Price variance	AV (SC – AC)	= 29 400 (0.10 – 0.11)	= 294A	126F

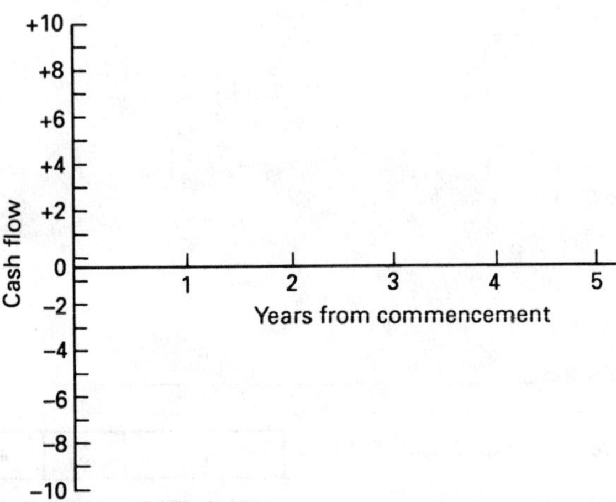

Figure 6.7 Cumulative cash flow

Variable overheads				
Volume variance	SC (SV – AV)	= 0.20 (33 600 – 29 400)	= 840F	
Price variance	AV (SC – AC)	= 29 400 (0.20 – 0.22)	= 588A	252F
Fixed overheads				
Expenditure variance	AFO – BFO	= 4 116 – 5 120	= 1004F	
Volume efficiency	SC (AQ – SQ)	= 0.16 (32 000 – 33 600)	= 256A	
Yield	SC (SY – AY)	= 0.16 (32 000 – 29 400)	= 416A	
Capacity	SC (RBQ – SQ)	= 0.16 (33 600 – 33 600)	= NIL	
Calendar variance	SC (RBQ – BQ)	= 0.16 (33 600 – 32 000)	= 256F	588F
				924F

SC	= Standard cost	BQ	= Budgeted quantity
SQ	= Standard quantity	AQ	= Actual quantity
SP	= Standard unit price	AP	= Actual unit price
AH	= Actual hours		
SH	= Standard hours	AH	= Actual hours
SV	= Standard volume	AV	= Actual volume
AFO	= Actual fixed overheads	BFO	= Budgeted fixed overheads
SR	= Standard rate per hour	AR	= Actual rate per hour
SY	= Standard yield	AY	= Actual yield
BQ	= Budgeted quantity	RBQ	= Revised budgeted quantity
A	= Adverse	F	= Favourable

Fixed overheads require some amplification. Budgeted output for the year is 400 000 units, budgeted fixed overheads for the year are £40 000. There are 50 working weeks in the year with 4 weeks in the period being reviewed, with one extra day being worked. The actual fixed overheads for the period amounted to £4116.

Actual output for the period	29 400 units
Budgeted output for the period	32 000 units
Standard output for the period	33 600 units

(Budgeted output is 1600 (400 000/250) units per day, so 21 days' standard output is 33 600 units.)

It is necessary to present this information in an understandable way, and this may take the following form:

		£
Budgeted cost of output		26 720
Output variances		
Volume	2 171F	
Unit cost	1 911A	260F
Actual cost of output		26 460
Less: Standard cost of output		
Fuel	10 500	
Materials	1 680	
Labour	420	
Repairs	3 360	
Variable overheads	6 720	
Fixed overheads	4 704	27 384
Favourable variance		924

Analysis of variances:

Fuel	– Usage	150A	
	– Price	213F	63F
Materials	– Usage	280F	
	– Price	70A	210F
Labour	– Rate	210A	
	– Efficiency	105A	315A
Repairs	– Expenditure		126F
Variable overhead			252F
Fixed overhead			
	– Expenditure	1 004F	
	– Volume efficiency	256A	
	– Yield	416A	
	– Calendar	256F	588F
			924F

Comments

Fuel suffered from excessive usage, which was partly offset by a lower unit price.
Materials showed lower usage but a higher than expected price was paid.
Labour both cost more per hour and was utilized longer.
Repairs and variable overheads both benefited from lower than expected expenditure.
Fixed overheads were less than forecast but this saving was partially offset by lower volume and yields. The extra day's working increased the net advantage gained from these factors.

6.8 Capital

Businesses require funds for day-to-day operations ('working capital') and for expansion by acquisition and for the provision of plant and machinery, buildings, etc. Most working capital needs are normally (and should be) met from the company's own cash generated from its own operations. Indeed, the need to meet this criterion serves as a discipline upon the company's standard of cash management in relation to credit control, payment of suppliers, etc.

Short-term needs will be met mainly from the company's own cash flow and from overdraft facilities provided by the company's bankers. Medium-term facilities can be provided by entering into arrangements for the hire purchase or leasing of specific items of equipment. The facilities available are many and varied. Long-term resources will normally be met by the raising of further capital by way of Stock Exchange placings or flotation or by the securing of loans through debentures secured upon the assets of the company, bearing fixed rates of interest and redeemable at a date in the future.

6.8.1 Short term

The main sources include:

1. *Overdrafts.* Usually the least expensive; interest is charged on the daily balance usually at a premium (often substantial) over the bank's base lending rate.
2. *Factoring.* Usually applied to sales invoices (effectively, the debtors). Specialist finance houses provide a service whereby the company is paid promptly a percentage of its sales value for a period and the finance house then collects the full amount from the debtor. In some cases the finance house will undertake the invoicing of the customers direct, providing a complete sales ledger service. Confidentiality as to the arrangement is lost in this case. There are many schemes with considerable variations available.
3. *Acceptance credits.* Provided mainly for exports by the major banks and specialist accepting houses.

6.8.2 Medium term

This may be provided by:

1. *Hire-purchase.* Normally for periods of up to three or five years. The ownership of the goods remains with the finance company until the instalment payments are complete and a final option to buy is exercised. Payments remain fixed throughout the term of the agreement with a predetermined unchanging rate of interest included.
2. *Leasing.* The asset is leased for a term of years at a fixed rental. The term can then be extended for secondary or tertiary periods at nominal rentals. Recently-introduced accounting standards in the UK require that assets acquired under contracts of this sort must appear in the balance sheet at cost and be depreciated even though not legally owned by the company.
3. *Lease purchase.* There are on offer many variants of the two previously illustrated sources, giving choices of both methods in combination with each other. The underlying principles are the same.

Year	Fixed assets £	Working capital £	Other expenses £	Operating profit or loss £	Taxation £	Net profit or loss £	Cumulative totals £
0							
1							
2							
3							
4							
5							
6							
7							
8							
9							
10							
Totals							

Figure 6.8 Summary of cash flows

4. *Contract hire.* Although not strictly a source of outside finance, this is a method of avoiding capital outlay, especially favoured for vehicles. Again, there are many varying systems to be found.
5. *Sale and lease-back.* Once considered only for land and buildings, schemes are now available for other assets, notably sizeable fleets of vehicles.

6.8.3 Long term

The principal sources include:

1. *Flotation on stock markets.* This involves transferring in whole or in part the ultimate ownership of the business to persons outside the company and, dependent upon the proportion sold, may leave the business vulnerable to takeover by competitors or unwelcome bidders.
2. *Share placings.* Shares are issued for a consideration to known and specific parties at an agreed price, usually through the medium of a merchant bank. This reduces the element of vulnerability to takeover bids but does not eliminate the possibility, should any of the 'placees' be prepared to sell their holdings at a future date.
3. *Issue of debentures.* These will normally carry a fixed rate of interest and have a predetermined date of redemption, possibly at a premium. The holders of debentures will usually require security perhaps by means of a fixed charge over specific assets (or all the assets), and will have a right of prior payment in the event of a liquidation. Debenture holders can also sometimes exercise their rights on the occurrence of certain events. Widespread security given to one class of lender can militate against the provision of short-term finance from other lenders who require collateral.
4. *Loans from specialist companies.* There is a number of companies and institutions who specialize in providing long-term finance to industry. The forms of loans offered vary from one deal to another and, in some cases, will be accompanied by a request for some degree of participation in the equity of the borrower.

6.9 Value Added Tax

The calculation of VAT is straightforward when the rate at present in force is applied to a value to which it is to be added. The formula is:

Value + (Value × Rate/100)

Where the amount of Value Added Tax included in a value is required the calculation is as follows:

Value × Rate/(100 + Rate)

Where the amount exclusive of Value Added Tax included in a value is required the calculation is:

Value/[(100 + Rate)/100]

6.10 Break-even charts

It is often useful to prepare break-even charts in order to illustrate and give a clear picture of the position of a business. They can be adapted to help in showing the viability and profiles of individual projects.

The two most common designs of break-even charts are shown in Figures 6.9 and 6.10. In both cases the *y*

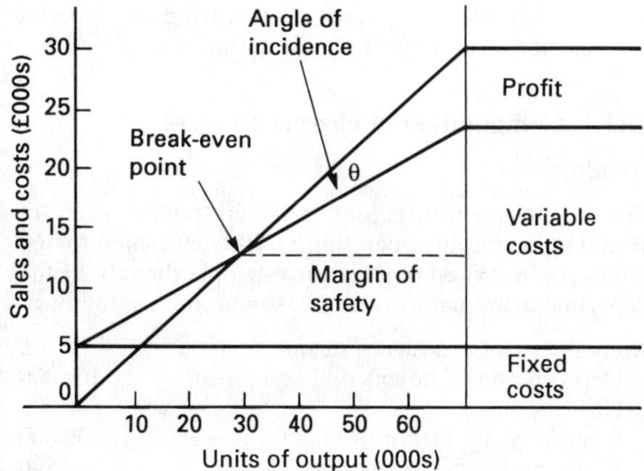

Figure 6.9 Break-even chart (1)

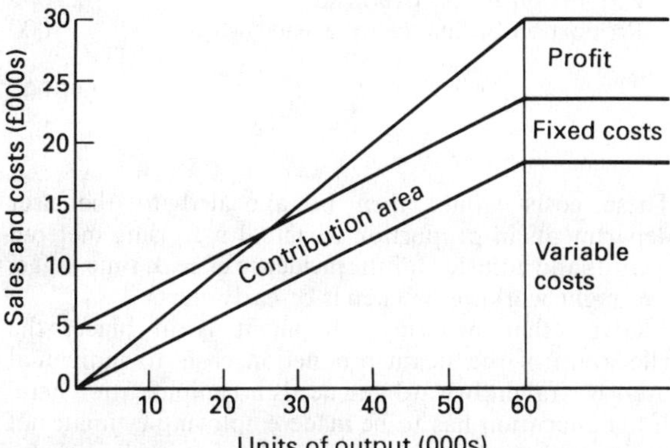

Figure 6.10 Break-even chart (2)

(vertical) axis is used for sales (output) and costs while the *x* (horizontal) axis is used for volumes, capacity or time.

Illustration

Output (000s)	10	20	30	40	50	60
Fixed costs	5 000	5 000	5 000	5 000	5 000	5 000
Variable @ 30p	3 000	6 000	9 000	12 000	15 000	18 000
Sales @ 50p	5 000	10 000	15 000	20 000	25 000	30 000

The angle of incidence, if wide, will indicate a high rate of profitability. If this is coupled with a high margin of safety an extremely advantageous investment is demonstrated.

The position of the break-even point serves as an indicator. If it is well to the right of the chart then the margin of safety will be poor. Conversely, a position to the right of the chart will hold out opportunities for expansion.

The guidance obtainable from break-even charts is limited and should be used only with other criteria in formulating decisions. The effects of policy decisions may alter the data used and vitiate or amend the results disclosed.

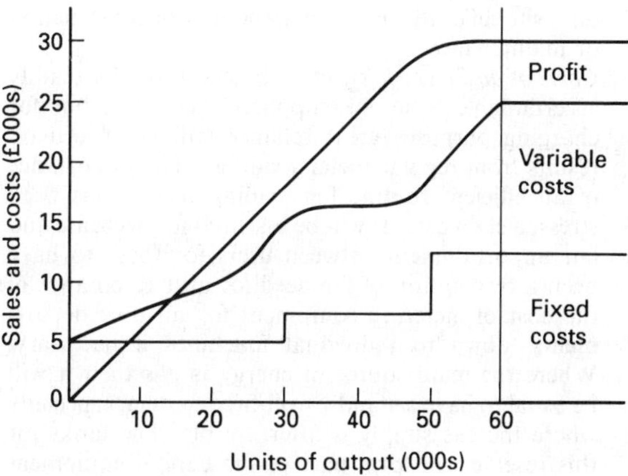

Figure 6.11 Break-even chart (3)

In Figure 6.10 fixed costs are shown above the variable costs and the non-recovery of the fixed costs below the break-even point is more clearly demonstrated. The 'contribution' to fixed costs is of significance in the consideration of marginal costing.

In practice, costs do not increase smoothly or remain absolutely constant. Fixed costs frequently move in a series of steps and variable costs also change unevenly. There may be several break-even points at different levels of sales and outputs (Figure 6.11).

6.11 Supply of steam, power, water, etc. to other departments

The plant engineer is frequently called upon to assist and advise on the bases for charging to other departments within the plant or to outside purchasers where there is a central generation plant for a variety of users, as may occur in an industrial estate or where the plant providing the service has a surplus. The elements to be considered and brought together in formulating a basis for charge can be summarized as follows, always bearing in mind individual needs and circumstances:

1. *Expenses associated with the capital cost of equipment.* Depreciation will be provided over the useful life of the equipment, varying according to the engineer's estimate. Present-day custom is to write assets down on a straight-line basis, but there are several accepted ways of providing the necessary depletion of the asset value. For tax purposes, a reducing-value method is employed in the UK.
2. *Costs relating to the space occupied by the equipment.* These will usually include rent and rates or similar taxes, together with the upkeep of the buildings housing the machinery. Where the buildings are owned by the operator a notional rent should be included to reflect the full cost of occupancy. Such costs will normally need to be allocated to the department by reference to the area occupied, but may require more complex calculations where the buildings

on a site differ in age, or in standards of construction or maintenance.

3. *Costs of gas, electricity, etc.* These are usually readily ascertainable from the suppliers' accounts, but for charging purposes great reliance will be placed on results from regular meter readings. The importance of an efficient routine for reading meters has been stressed elsewhere. It will be essential for accurate and fair apportionment between users for there to have been a recognition of the need for and acceptance of the cost of metering equipment for all user departments, down to individual machines if necessary. Where the main source of energy is gas then it will be usual to have a stand-by oil-fired system, especially where the gas supply is interruptible. The tanks for this reserve will form part of the capital equipment and the oil itself will need to be included when considering the capital employed. Where electricity is produced on site as a by-product in periods of low demand for steam, more cheaply than that obtained from the utility company, adjustments to the charges on a seasonal basis will be required.
4. *Water.* Costs are available from the supplier's invoices but where the operator has reservoirs and/or licences for extraction from canals or rivers the annual fees and penalties which are sometimes leviable should not be forgotten. Projections of future demands will need to be carefully considered where this type of arrangement is made, since such contracts are often available only on long-term bases. Water-treatment plant will produce its own range of costs across the whole field of depreciation, materials, electricity, labour, etc. and these will need to be apportioned to the steam cost departments before final allocation to the user departments.
5. *Labour costs.* In arriving at the total labour costs there must be included social security costs, pension costs and any provisions for holiday pay. It may be desirable, where users take steam overnight, to apportion costs in such a way as to reflect the higher cost of labour employed on night work.
6. *Maintenance expenses.* The nature of these is generally self-evident, but long-term maintenance contracts, insurance inspection fees and any subsequent requirements arising therefrom should not be overlooked. Inspections are commonly undertaken at weekends, at premium labour rates.
7. *Recovery of appropriate proportion of overheads.* The overhead charge to the supplying department will usually be provided by central finance in accordance with the company's policy for allocation of overheads.
8. *Recovery of proportion of exceptional financing and interest charges.* This will normally depend upon the policy of individual companies and will probably have been included, if at all, in the general overheads referred to in (7). It should not be left out of consideration where provision has been made for a particular user's requirements involving, say, special extension or modification of existing plant.

6.11.1 Computation of charges to users

Assumptions

For a site operating four gas-fired boilers each of 10 000 kg/h output, operating at 70% efficiency for 60 hours per week, employing three people directly in the department, annual costs may be summarized as follows:

Annual cost of producing steam	£
Depreciation of boilers and equipment	100 000
Gas	850 000
Electricity (for steam production)	30 000
Water and water treatment	12 500
Labour for steam production	36 000
Repairs, maintenance and insurance	35 000
Stand-by oil	1 500
Proportion of site overhead	25 000
Proportion of finance overhead	6 000
	1 096 000

These costs would then be allocated to the user departments in proportion to actual use, using metered records and adjusted for the incidence of peak-time usage, overnight working, weekends or early starts.

Given that metering equipment is in place, the allocation of the steam production costs to individual users is straightforward and needs no amplification here. If the allocation has to be made employing estimates of usage, perhaps by using the ratings of individual pieces of equipment, then everything will turn upon the general and continuing acceptance of those estimates by all the users concerned. Should these conditions not be met or maintained then the life of the plant engineer will be an uneasy one!

6.12 Charges for effluent and environmental services

These services are now very often a significant part of the costs of a factory and are worth a note here. Projecting accurate costs can be difficult, since water authorities commonly render accounts quarterly with charges fluctuating from month to month in relation to both volumes and chemical content ('toxicity') of the effluent, all of which can be highly variable. Monthly liaison with the authority's representative and inspection of sampling reports will often enable more accurate information to be available, both for the assessment of costs where charges are to be made to individual departments and for the accruing of costs by the management accountant.

7 Industrial Boilers

E Walker CEng, BSc, MIMechE
and
R J Blaen
Senior Green Limited

Contents

7.1 Terminology

The following explain some of the more fundamental terms encountered when considering boilers.

7.1.1 Shell boiler

This is a boiler in which the products of combustion or hot gases pass through a series of tubes surrounded by water. All are contained in an outer shell.

7.1.2 Watertube boiler

In a watertube boiler water circulates through small-bore tubes constructed in banks and connected to drums or headers. The external surfaces of the tubes are exposed to the products of combustion or hot gases.

7.1.3 Dryback boiler

This is a horizontal shell boiler where the gas-reversal chamber from the combustion tube to the first pass of tubes is external to the rear tube plate and is formed by a refractory lined steel chamber.

7.1.4 Wetback boiler

A wetback boiler is a horizontal shell boiler where the gas-reversal chamber from the combustion tube to the first pass of tubes is integral within the boiler shell and surrounded by water.

7.1.5 Economic boiler

This is a term applied to the early free-standing shell boilers of two- and three-pass construction. Originally, they were dry back and later wet back. These boilers superseded the brickset Cornish and Lancashire boilers. The earliest economic boilers were also brickset. The gases from the front smokebox returning across the lower external part of the shell were contained within the brick setting to form a third pass.

7.1.6 Packaged boiler

A packaged boiler is a concept of a factory-built and assembled shell boiler complete with its combustion appliance, feedwater pump and controls, valves, base frame and insulation. Before this, the economic boiler was delivered to site as a bare shell and assembled *in situ*.

Originally, in the early 1960s package boilers were designed to make them as compact as possible, resulting in some inherent faults. Since then, design criteria have greatly improved and the present packaged boiler is constructed to acceptable commercial standards.

7.1.7 Evaporation

This is the quantity of steam produced by the boiler at temperature and pressure. It may be quoted as 'actual evaporation' or 'evaporation from and at 100°C'. Actual evaporation is the quantity of steam passing through the crown valve of the boiler.

Evaporation from and at 100°C is a figure taken for design purposes, and is based on the actual evaporation per pound of fuel multiplied by the factor of evaporation.

7.1.8 Factor of evaporation

This is the figure obtained by dividing the total heat of steam at working condition by the latent heat of steam at atmospheric condition (i.e. 2256 kJ/kg). Then

$$\text{factor of evaporation} = \frac{H-T}{2256}$$

where H = Total heat in 1 kg of steam at working pressure above 0°C taken from steam tables in kJ/kg.
T = Heat in feedwater (kJ/kg).
2256 = The latent heat of steam at atmospheric conditions.

7.1.9 Availability

This is the period of time that a boiler may be expected or required to operate before being shut down for cleaning or maintenance. This will vary with the type of boiler, the fuel being used and the operating load on the boiler.

7.1.10 Priming

This is when the water surface in the boiler shell becomes unstable. Vigorous surging will occur and this may cause the boiler to go to low water and cut out or possibly lock out. This, in turn, will exacerbate the condition.

There are two possible causes. The first could be incorrect control of water treatment and blowdown. This can result in excessive levels of suspended solids in the boiler water, organic matter in the boiler water or high alkalinity. The second can be mechanical. If the boiler is operated below its designed working pressure it will increase the efflux velocity of the steam leaving the water surface area to a point where it may lift the water surface and drop the water level. It is important therefore to give due consideration to the steam load required from the boiler.

7.1.11 Thermal storage

Thermal storage is a method of supplying a steam load in excess of the maximum continuous rating of the boiler for short periods. The boiler shell diameter is increased to provide a greater height of water than normal above the top line of heating surface to the normal working water level. When an excess load is then imposed on the boiler it allows this extra water to flash to steam while lowering the working pressure in the boiler. Safety-level controls protect the boiler against an excessive draw-off of steam. As the load decreases, the boiler maintains its maximum firing rate and allows the extra water to return to the normal (higher) working water level and the higher working pressure regained.

7.1.12 Accumulator

This may be likened to an unfired boiler. Steam is generated in the boilers at a pressure higher than that required for the process and fed to the process through pressure-reducing valves. This higher-pressure steam is also supplied to the accumulator, where it heats and pressurizes the water in the accumulator. When a steam load in excess of the boiler maximum firing rate occurs, steam may be supplied from the accumulator until such time as the load may be met from the boilers. As the steam load falls away the accumulator will recharge from the boilers. This system enables boilers to be installed with ratings to match an average load while the peaks will be met from the accumulator.

7.1.13 Cavitation

This is a condition which occurs when the feedwater pump is unable to deliver feedwater to the boiler although the feed tank has water available. The temperature of the feedwater coupled with the possible suction effect from the feedwater pump in the line between the feed tank and the pump effectively drops the pressure, causing the feedwater to flash to steam. The pump then loses its water supply.

In most cases this condition may be avoided by arranging a sufficient heat of water and by correct sizing of the feedwater pipework. Clean filters must also be maintained.

7.2 Heat transfer in industrial boilers

Heat is transferred from the hot products of combustion to the boiler heating surfaces through the plate and tube walls and to the water by various mechanisms which involve conduction, radiation and convection.

7.2.1 Conduction

The rate at which heat is transferred by conduction through a substance without mass transfer is given by the Fourier Law. This states that the heat flow rate per unit area, or heat flux, is proportional to the temperature gradient in the direction of heat flow. The relationship between heat flux and temperature gradient is characterized by the thermal conductivity which is a property of the substance. It is temperature dependent and is determined experimentally.

For a plate of area A (m^2), thickness e (m) and with hot and cold face temperatures of T_1 and T_2 (°C), respectively, the normal heat flux ϕ and heat transfer rate Q are given by:

$$\phi=\frac{Q}{A}=\frac{k(T_1-T_2)}{e}\ (\mathrm{W/m^2})$$

where K = thermal conductivity (W/mK).

7.2.2 Radiation

Thermal radiation takes place by the emission of electromagnetic waves, at the velocity of light, from all bodies at temperatures above absolute zero. The heat flux from an ideal or 'black body' radiating surface is proportional to the fourth power of the absolute temperature of the surface. The constant of proportionality is the Stefan–Boltzmann constant, which has a value of 5.6696×10^{8} (W/m^2 K^4).

The heat flux radiated from a real surface is less than that from an ideal 'black body' surface at the same temperature. The ratio of real to 'black body' flux is the normal total emissivity. Emissivity, like thermal conductivity, is a property which must be determined experimentally.

Although the rate of emission from a surface is independent of the condition of the surroundings, the net overall exchange of radiant heat between surfaces at different temperatures depends on a number of factors. The continuous interchange of energy is a result of the reciprocal processes of radiation and absorption, and these are dependent on geometrical relationships, emissivity differences and the presence of any absorbing and emitting gases in the intervening space.

7.2.3 Convection

Convective heat transmission occurs within a fluid, and between a fluid and a surface, by virtue of relative movement of the fluid particles (that is, by mass transfer). Heat exchange between fluid particles in mixing and between fluid particles and a surface is by conduction. The overall rate of heat transfer in convection is, however, also dependent on the capacity of the fluid for energy storage and on its resistance to flow in mixing. The fluid properties which characterize convective heat transfer are thus thermal conductivity, specific heat capacity and dynamic viscosity.

Convection is classified according to the motivating flow. When the flow takes place as a result of density variations caused by temperature gradients, the motion is called natural convection. When it is caused by an external agency such as a pump or a fan the process is called forced convection.

At a convection heat transfer surface the heat flux (heat transfer rate per unit area) is related to the temperature difference between fluid and surface by a heat transfer coefficient. This is defined by Newton's law of cooling:

$$\phi=\frac{Q}{A}=h_c\Delta T_m$$

where ϕ = heat flux (W/m^2),
Q = heat transfer rate (W),
A = surface area (m^2),
ΔT_m = mean temperature difference between fluid and surface (K),
h_c = convective heat transfer coefficient (W/m^2 K).

The heat transfer coefficient is correlated experimentally with the fluid transport properties (specific heat, viscosity, thermal conductivity and density), fluid velocity and the geometrical relationship between surface and fluid flow.

7.2.4 Furnace heat transfer

Heat transfer in the furnace is mainly by radiation, from the incandescent particles in the flame and from hot radiating gases such as carbon dioxide and water vapour. The detailed theoretical prediction of overall radiation exchange is complicated by a number of factors such as carbon particle and dust distributions, and temperature variations in three-dimensional mixing. This is overcome by the use of simplified mathematical models or empirical relationships in various fields of application.

For industrial boilers the mean gas temperature at the furnace exit, or at the entrance to the convection section of the boiler, may be calculated using the relationship:

$$T=k(H/A)^{0.25}$$

where T = gas temperature (°C),
H = heat input rate (W) based on the net calorific value of the fuel,
A = effective (projected) water-cooled absorption surface area (m^2),
K = a constant which depends on the fuel and the excess air in the combustion products.

The value of k is determined experimentally by gas temperature measurement. The measurement error of a simple pyrometer can be 250–300 K, due to re-radiation to water-cooled surroundings, and the values given below are based on measurement by a 'Land' multi-shielded high-velocity suction pyrometer. Typical values for normal excess air at or near full boiler load are:

Natural gas	$k=52.4$
Gas oil	$k=49.1$
Heavy fuel oil	$k=48.3$
Coal	$k=40.3$

In calculating the smoke tube inlet gas temperature of a shell boiler, A includes the effective water-cooled surface in the reversal chamber. In coal-fired boilers any water-cooled surface below the grate is excluded from A.

The total furnace heat absorption may be estimated by using the calculated furnace exit gas temperature and analysis to determine the enthalpy (excluding the latent heat of water vapour) and thus deducting the heat rejection rate from the net heat input rate.

7.2.5 Boiler tube convection heat transfer

The radiant section of an industrial boiler may typically contain only 10% of the total heating surface, yet, because of the large temperature difference, it can absorb 30–50% of the total heat exchange. The mean temperature difference available for heat transfer in the convective section is much smaller. To achieve a thermally efficient yet commercially viable design it is necessary to make full use of forced convection within the constraint of acceptable pressure drop.

Forced convection heat transfer has been measured under widely differing conditions, and correlation of the experimental results is made by using the dimensionless groups:

Nusselt number $$Nu=\frac{h_c D}{k}$$

Reynolds number $$Re=\frac{GD}{\mu}$$

Prandtl number $$Pr=\frac{C_p \mu}{k}$$

where h_c = heat transfer coefficient (W/m^2 K),
D = characteristic dimension (m),
K = thermal conductivity (W/m K),
G = gas mass velocity (kg/m^2 s),
μ = dynamic viscosity (kg/m s),
C_p = specific heat at constant pressure (J/kg K).

In applying the correlations use is made of the concept of logarithmic mean temperature difference across the boundary layer. For a boiler section, or pass, this is given by:

$$\Delta T_m=\frac{(T_1-T_w)-(T_2-T_w)}{\log_n((T_1-T_w)/(T_2-T_w))}\ (\mathrm{K})$$

where T_1 = inlet gas temperature (°C),
T_2 = outlet gas temperature (°C),
T_w = tube wall temperature (°C).

The difference in temperature between the tube wall and the water is small, typically less than 10 K in the convective section. Therefore little error is introduced by using the water temperature as T_w in the evaluation of the gas transport properties.

The representative gas temperatures used in the correlations are the bulk temperature and the film temperature. These are defined as:

Bulk temperature $T_b=T_w+\Delta T_m$
Film temperature $T_f=(T_b+T_w)/2$

For longitudinal flow in the tubes of shell boilers the mean heat transfer coefficient may be determined from:

$$Nu=0.023Re^{0.8}Pr^{0.4}(1+(D/L)^{0.7})$$

where D/L is the tube inside diameter-to-length ratio and the characteristic dimension in Nu and Re is the tube inside diameter. Gas properties are evaluated at the film temperature.

Correlations for forced convection over tubes in crossflow are complicated by the effect of the tube bank arrangement. For the range of Reynolds numbers likely to be encountered in industrial boilers the following equations may be used:

In-line arrays $Nu=0.211Re^{0.651}Pr^{0.34}F_1F_2$
Staggered arrays $Nu=0.273Re^{0.635}Pr^{0.34}F_1F_2$

In these cases gas properties are evaluated at the bulk temperature, the characteristic dimension in Nu and Re is the tube outside diameter, and the Reynolds number is based on the mass velocity through the minimum area for flow between tubes. F_1 is a correction factor for wall

Table 7.1 Transport properties: natural gas products of combustion

Temp. (°C)	*Spec. heat* (J/kg K)	*Viscosity* (kg/m s × 10^6)	*Conductivity* (W/m K × 10^3)	*Sp. vol.* (m^3/kg)
100	1098	20.01	27.27	1.1
200	1133	23.97	34.45	1.395
300	1166	27.55	41.34	1.69
400	1198	30.83	47.94	1.985
500	1227	33.89	54.25	2.28
600	1255	36.74	60.29	2.575
700	1281	39.44	66.09	2.87
800	1305	41.99	71.61	3.164
900	1328	44.43	76.86	3.459
1000	1348	46.75	81.86	3.754
1100	1367	48.98	86.6	4.049
1200	1384	51.13	91.08	4.344
1300	1400	53.2	95.31	4.639
1400	1413	55.2	99.25	4.934

Table 7.2 Transport properties: gas oil products of combustion

Temp. (°C)	*Spec. heat* (J/kg K)	*Viscosity* (kg/m s × 10^6)	*Conductivity* (W/m K × 10^3)	*Sp. vol.* (m^3/kg)
100	1061	20.32	27.24	1.058
200	1096	24.29	34.4	1.342
300	1128	27.88	41.22	1.625
400	1159	31.16	47.73	1.909
500	1188	34.2	53.92	2.192
600	1215	37.05	59.81	2.476
700	1240	39.72	65.42	2.76
800	1263	42.26	70.71	3.043
900	1284	44.67	75.73	3.327
1000	1303	46.98	80.46	3.61
1100	1320	49.19	84.89	3.894
1200	1336	51.32	89.02	4.177
1300	1349	53.37	92.88	4.461
1400	1361	55.35	96.43	4.745

to bulk property variation which can be calculated from the relationship:

$$F_1 = \frac{(Pr_b)^{0.26}}{(Pr_w)}$$

where Pr_b and Pr_f are Prandtl numbers at the bulk and wall temperatures, respectively. F_2 is a correction factor for the depth of the tube bank in the direction of flow. For bank depths of 10 rows or more, $F_2 = 1$. For smaller bank depths the following values of F_2 may be used:

No. of rows	1	2	3	4	5
In-line tubes	0.64	0.8	0.87	0.9	0.92
Staggered tubes	0.68	0.75	0.83	0.89	0.92

No. of rows	6	7	8	9
In-line tubes	0.94	0.96	0.98	0.99
Staggered tubes	0.95	0.97	0.98	0.99

Gas transport properties for the products of combustion of the common fuels, fired at normal excess air at or near-full boiler load, may be obtained from Tables 7.1–7.4. Non-luminous gas radiation has a small overall effect in the convective section, typically 2–5% of total convection. It may therefore be neglected for a conservative calculation.

7.2.6 Water-side conditions

In the radiant section of a boiler the fourth power of the wall temperature is typically less than 2% of the fourth power of the mean flame and gas temperature. The effect of water-side conditions and wall thickness on the heat transfer rate are therefore negligible.

Even the presence of a dangerous layer of water-side scale reduces the heat flux only by a few per cent. Although this means that scale has little effect on radiant section performance, it also indicates that the metal temperature escalation due to the presence of scale is not self-limiting but is almost proportional to scale thickness.

The thermal conductivity of an average boiler scale is 2.2 (W/m K) and that of complex silicate scales is 0.2–0.7

Table 7.3 Transport properties: heavy fuel oil products of combustion

Temp. (°C)	*Spec. heat* (J/kg K)	*Viscosity* (kg/m s × 10^6)	*Conductivity* (W/m K × 10^3)	*Sp. vol.* (m^3/kg)
100	1054	20.37	27.22	1.05
200	1088	24.34	34.37	1.332
300	1121	27.93	41.17	1.613
400	1152	31.21	47.66	1.895
500	1181	34.25	53.82	2.176
600	1207	37.09	59.69	2.458
700	1232	39.44	66.09	2.87
800	1255	42.3	70.51	3.02
900	1276	44.71	75.47	3.302
1000	1294	47.01	80.15	3.583
1100	1311	49.22	84.51	3.865
1200	1326	51.35	88.59	4.146
1300	1339	53.40	92.38	4.428
1400	1351	55.38	95.86	4.709

Table 7.4 Transport properties: bit coal products of combustion

Temp. (°C)	*Spec. heat* (J/kg K)	*Viscosity* (kg/m s × 10^6)	*Conductivity* (W/m K × 10^3)	*Sp. vol.* (m^3/kg)
100	1031	20.82	27.43	1.034
200	1065	24.83	34.63	1.312
300	1096	28.44	41.39	1.589
400	1125	31.73	47.78	1.866
500	1152	34.78	53.8	2.143
600	1177	37.63	59.5	2.421
700	1201	40.3	64.88	2.698
800	1222	42.83	69.93	2.975
900	1242	45.24	74.68	3.252
1000	1259	47.55	79.11	3.53
1100	1275	49.75	83.23	3.807
1200	1289	51.87	87.05	4.084
1300	1301	53.92	90.56	4.361
1400	1311	55.89	93.77	4.638

(W/m K). Since the furnace peak wall flux can be over 300 000 (W/m^2) it may readily be seen that a small thickness of scale can raise the metal temperature into the creep region, resulting in very expensive repairs.

In the convective section the gas-side heat transfer coefficient controls the heat flux distribution since the water-side coefficient and the thermal conductance of the tube walls are very large in comparison. For this reason, it is usually satisfactory to make an allowance by adding 10 K to the water temperature in steam boilers. In hot water generators the allowance should be about 20 K, because sub-cooled nucleate boiling generally takes place only on the radiant walls and in shell boilers on the reversal chamber tubeplate. Water-side heat transfer on the major part of the convective heating surface in these units is by convection without boiling.

7.2.7 Further reading

A good introduction to the extensive literature on the science and technology of heat transfer, with 87 further references, is given in Rose, J. W. and Cooper, J. R., *Technical Data on Fuel*, 7th edition, British National Committee, World Energy Conference, London, p. 48 (1977).

7.3 Types of boiler

As this covers industrial boilers, only units of 500 kg/h of steam, or equivalent hot water, and above will be considered. There are nine categories of boiler available. In order of evaporation these are:

Cast iron sectional boilers
Steel boilers
Electrode boilers
Steam generators
Vertical shell boilers
Horizontal shell boilers
Watertube boilers
Waste heat boilers
Fluid bed boilers

7.3.1 Cast iron sectional boilers

These are used for hot water services with a maximum operating pressure of 5 bar and a maximum output in the order of 1500 kW. Site assembly of the unit is necessary and will consist of a bank of cast iron sections. Each section has internal waterways.

The sections are assembled with screwed or taper nipples at top and bottom for water circulation and sealing between the sections to contain the products of combustion. Tie rods compress the sections together. A standard section may be used to give a range of outputs dependent upon the number of sections used. After assembly of the sections, the mountings, insulation and combustion appliance are fitted. This system makes them suitable for locations where it is impractical to deliver a package unit, e.g. basements where inadequate access is available or rooftop plant rooms where sections may be taken up using the elevator shafts. Models available use liquid, gaseous and solid fuel.

7.3.2 Steel boilers

These are similar in rated outputs to the cast iron sectional boiler. Construction is of rolled steel annular drums for the pressure vessel. They may be of either vertical or horizontal configuration, depending upon the manufacturer. In their vertical pattern they may be supplied for steam raising.

7.3.3 Electrode boilers

These are available for steam raising up to 3600 kg/h and manufacture is to two designs. The smaller units are element boilers with evaporation less than 500 kg/h. In these, an immersed electric element heats the water and a set of water-level probes positioned above the element controls the water level being interconnected to the feedwater pump and the element electrical supply.

Larger units are electrode boilers. Normal working pressure would be 10 bar but higher pressures are available. Construction is a vertical pattern pressure shell containing the electrodes (Figure 7.1). The lengths of the electrodes control the maximum and minimum water level. The electrical resistance of the water allows a current to flow through the water which, in turn, boils and releases steam. Since water has to be present within the electrode system, lack of water cannot burn out the boiler. The main advantage with these units is that they may be located at the point where steam is required and, as no combustion fumes are produced, no chimney is required. Steam may also be raised relatively quickly, as there is little thermal stressing to consider.

7.3.4 Steam generators

While the term 'steam generator' may apply to any vessel raising steam, this section is intended to cover coil type boilers in the evaporative range up to 3600 kg/h of steam. Because of the steam pressure being contained within the tubular coil, pressures of 35 bar and above are available, although the majority are supplied to operate at up to 10 bar. They are suitable for firing with liquid and gaseous fuels, although the use of heavy fuel oil is unusual.

The coiled tube is contained within a pressurized combustion chamber and receives both radiant and convected heat. A control system matches the burner firing rate proportional to the steam demand. Feedwater is pumped through the coil and partially flashed to steam in a separator. The remaining water is recirculated to a feedwater heat exchanger before being run to waste. Because there is no stored water in this type of unit they are lighter in weight and therefore suitable for siting on mezzanine or upper floors adjacent to the plant requiring steam. Also, as the water content is minimal, steam raising can be achieved very quickly and can respond to fluctuating demand within the capacity of the generator. It must be noted that close control of suitable water

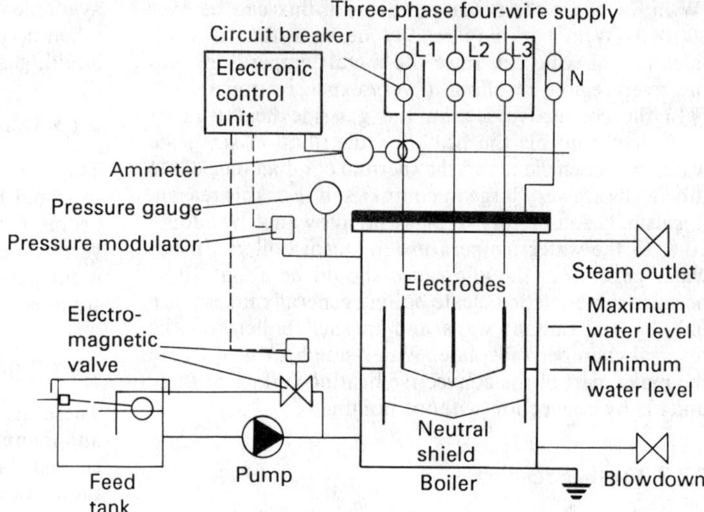

Figure 7.1 Diagrammatic layout of electrode boiler

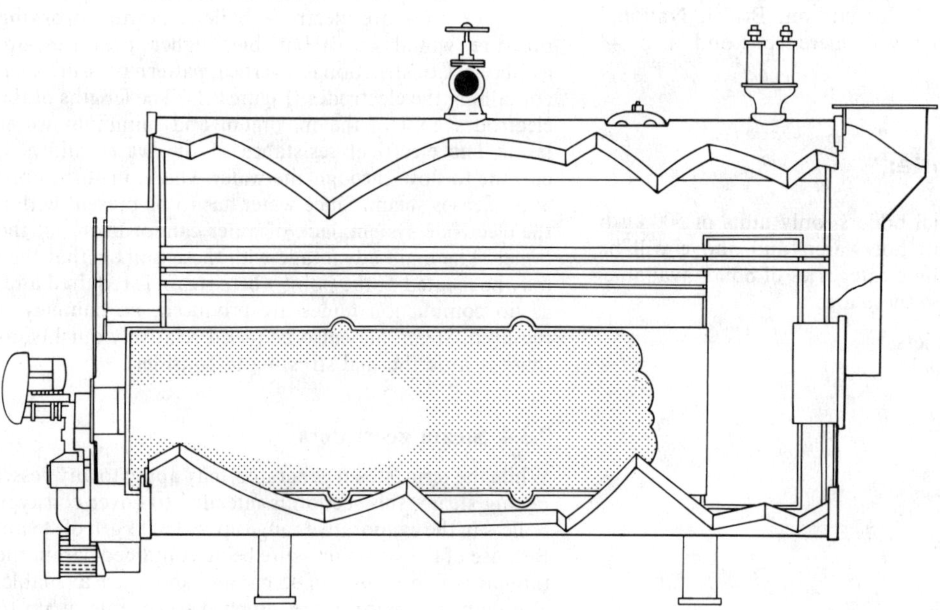

Figure 7.2 A three-pass wetback shell

treatment is essential to protect the coil against any build-up of deposits.

7.3.5 Vertical shell boilers

This is a cylindrical boiler where the shell axis is vertical to the firing floor. Originally it comprised a chamber at the lower end of the shell which contained the combustion appliance. The gases rose vertically through a flue surrounded by water. Large-diameter (100 mm) cross tubes were fitted across this flue to help extract heat from the gases which then proceeded to the chimney. Later versions had the vertical flue replaced by one or two banks of small-bore tubes running horizontally before the gases discharged to the chimney. The steam was contained in a hemispherical chamber forming the top of the shell.

The present vertical boiler is generally used for heat recovery from exhaust gases from power generation or marine applications. The gases pass through small-bore vertical tube banks. The same shell may also contain an independently fired section to produce steam at such times as there is insufficient or no exhaust gas available.

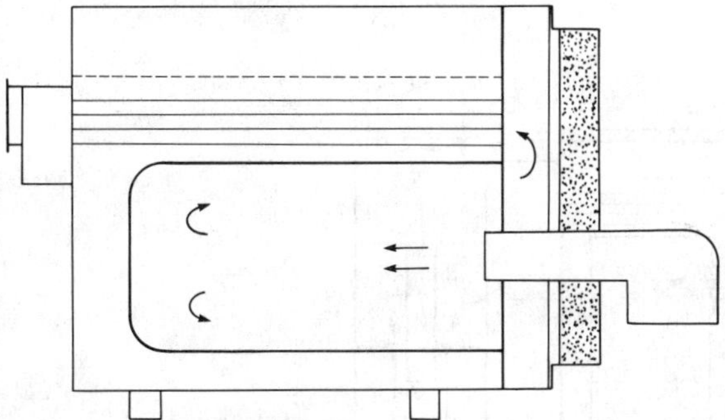

Figure 7.3 A reverse flame shell

7.3.6 Horizontal shell boilers

This is the most widely used type of boiler in industry. The construction of a single-flue three-pass wetback shell is illustrated in Figure 7.2. As a single-flue design boiler evaporation rates of up to 16 300 kg/h *F* and *A* 100°C are normal on oil and gas and 9000 kg/h *F* and *A* 100°C on coal.

In twin-flue design these figures are approximately double. Normal working pressures of 10–17 bar are available with a maximum working pressure for a shell boiler at 27 bar. The outputs of larger boilers will be limited if high pressures are required.

The boilers are normally despatched to site as a packaged unit with the shell and smokeboxes fully insulated and painted and mounted on a base frame. The combustion appliance and control panel will be fitted together with the feedwater pump, water-level controls and gauges and a full complement of boiler valves. Additional equipment may be specified and incorporated during construction. Larger boilers may have to have certain items fitted at site due to site restriction or weight.

Some variations of the three-pass wetback design exist. The most common is the reverse flame boiler, and Figure 7.3 illustrates this shell. In this design the combustion appliance fires into a thimble-shaped chamber in which the gases reverse back to the front of the boiler around the flame core. The gases are then turned in a front smokebox to travel along a single pass of smoketubes to the rear of the boiler and then to the chimney. In order to extract heat from these gases, turbulators or retarders are fitted into these tubes to agitate the gases and help produce the required flue gas outlet temperature. Evaporative outputs up to 4500 kg/h *F* and *A* 100°C on liquid and gaseous fuels are available.

Other variations of the three-pass wetback design are the two-pass, where only one pass of smoketubes follows the combustion tube, and the four-pass, where three passes of smoketubes follow the combustion tube. Neither of these are as widely used as the three-pass design.

Dryback boilers are still occasionally used when a high degree of superheat is required, necessitating a rear chamber to house the superheater too large for a semi-wet-back chamber. A water-cooled membrane wall chamber would be an alternative to this.

With twin-flue design boilers it is usual to have completely separate gas passes through the boiler with twin wetback chambers. It is then possible to operate the boiler on one flue only, which effectively doubles its turndown ratio. For example, a boiler rated at 20 000 kg/h *F* and *A* 100°C may reasonably be expected to operate down to 2500 kg/h *F* and *A* 100°C on oil or gas providing suitable combustion equipment and control is incorporated. It would be good practice to alternate on a planned time scale which flue takes the single-flue load if prolonged periods of single-flue operation occur.

Shell boilers are supplied with controls making them suitable for unattended operation, although certain operations such as blowdown of controls are called for by the insurance companies to comply with safety recommendations.

Oil-, gas- and dual-fired boilers are available with a range of combustion appliances. The smaller units have pressure jet-type burners with a turndown of about 2:1 while larger boilers may have rotary cup, medium pressure air (MPA) or steam-atomizing burners producing a turndown ratio of between 3:1 and 5:1, depending upon size and fuel. The majority have rotary cup-type burners, while steam- or air-atomizing burners are used where it is essential that the burner firing is not interrupted even for the shortest period.

For coal-fired boilers, chain grate stokers, coking stokers and underfeed stokers are supplied. An alternative to these is the fixed-grate and tipping-grate boiler with coal being fed through a drop tube in the crown of the boiler (Figure 7.4). With the fixed grate, de-ashing is manual while with the tipping grate a micro-sequence controller signals sections of the grate to tip, depositing the ash below the grate, where it is removed to the front by a drag-link chain conveyor and then to a suitable ash-disposal system.

It is possible to have boilers supplied to operate on liquid, gaseous and solid fuels, although there may be a time penalty of two or three days when converting from solid to liquid and gaseous fuels and vice versa.

Access to both waterside and fireside surfaces of the boiler is important. All boilers will have an inspection opening or manway on the top of the shell with inspection openings in the lower part. Some larger boilers will have a manway in the lower part of the shell or end plate. With a three-pass wetback boiler all tube cleaning and maintenance is carried out from the front. The front smokebox doors will be hinged or fitted with davits. On most sizes of boilers bolted-on access panels are sufficient on the real smokebox. As the majority of shell boilers operate under forced-draught pressurized combustion, steam raising is relatively quick. While good practice could require a cold boiler to come up to pressure over a period of several hours once it is hot, it may be brought up to pressure in minutes, not hours.

For hot water shell boilers the above still applies. The shells would be slightly smaller for equivalent duties due to the absence of steam space. There are three accepted operated bands for hot water boilers. Low-temperature hot water (LTHW) refers to boilers having a mean water temperature (between flow and return) of below 95°C;

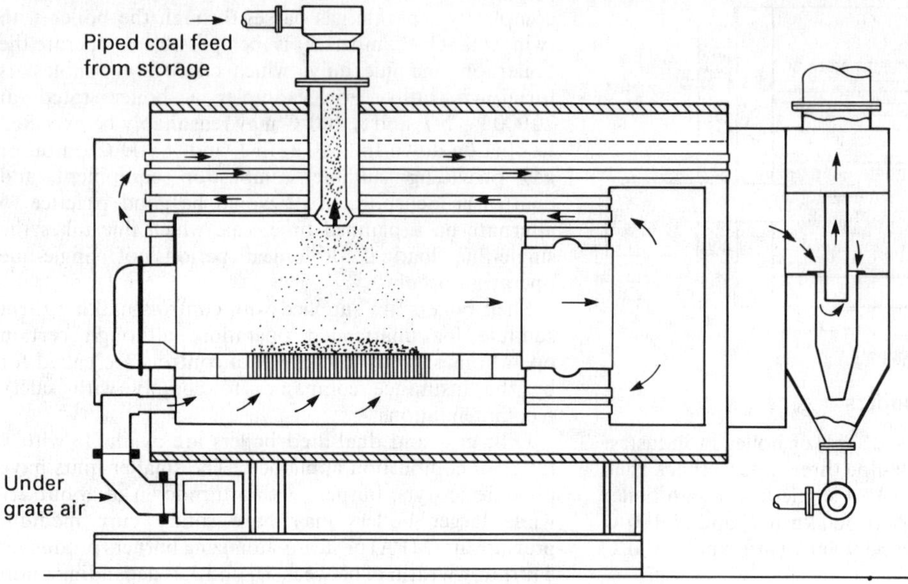

Figure 7.4 A fixed-grate coal-fired boiler

medium-temperature hot water (MTHW) would cover the range 95–150°C; high-temperature hot water (HTHW) covers applications above 150°C.

The flow and return connections will be designed to suit the flow rates and temperature differentials required. The water-return connection will be fitted with either an internal diffuser or a venturi nozzle to assist mixing of the water circulating within the shell and prevent water stratification. The flow connection will incorporate the temperature control stat to signal control of the firing rate for the burner.

Hot water boilers are potentially more susceptible to gas-side corrosion than steam boilers due to the lower temperatures and pressures encountered on low- and medium-temperature hot water boilers. With low-temperature hot water especially, the water-return temperature may drop below the water dewpoint of 50°C, causing vapour in the products of combustion to condense. This, in turn, leads to corrosion if it persists for long periods of time. The remedy is to ensure that adequate mixing of the return water maintains the water in the shell above 65°C at all times. Also, if medium or heavy fuel oil is to be used for low- or medium-temperature applications it is desirable to keep the heat transfer surfaces above 130°C, this being the approximate acid dewpoint temperature of the combustion gases. It may be seen, therefore, how important it is to match the unit or range of unit sizes to the expected load.

7.3.7 Watertube boilers

Originally, watertube boilers would have been installed for evaporation of 10 000 kg/h of steam with pressures as low as 10 bar. At that time this would have been the maximum evaporation expected from a shell-type boiler. Now shell boilers are available at much greater duties

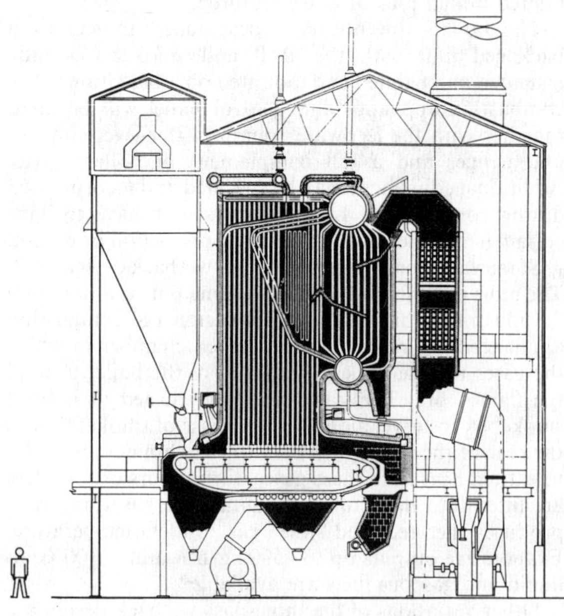

Figure 7.5 A stoker-fired watertube boiler of 36 300 kg/h steam capacity at 28 bar and 385°C (from British Coal Publication, *Boiler House Design for Solid Fuel*, 1980, and kind permission from College of Fuel Technology)

and pressures as described in Section 7.3.6. It may be appreciated that there will be an overlap of types of boiler in this area, with watertube covering first for high-pressure applications and ultimately the larger duties. Figure 7.5 illustrates a stoker-fired unit.

Generally, outputs up to 60 MW from a single unit may be considered for industrial installations. Higher duties are available if required. Watertube boilers supplied for national power generation will have outputs up to 900 MW, pressures of 140 bar and final steam temperatures of 500°C. For the smaller industrial unit we are considering, the maximum pressure would be 65 bar with final steam temperature up to 500°C. This is the maximum temperature and pressure likely to be required for small turbine-driven generating units, although turbines are available to operate at much lower pressures of, say, 17 bar.

Construction is a water-cooled wall combustion chamber connected to a steam drum at high level. The bottoms of the walls are connected to headers. Sometimes a bottom or mud drum is incorporated, but improved water treatment now available does not always necessitate this.

The chamber is externally insulated and clad. Combustion equipment for solid fuel may be spreader or travelling-grate stokers or by pulverized fuel or fluid bed. Oil and gas burners may be fitted either as main or auxiliary firing equipment. The boilers will incorporate superheaters, economizers and, where necessary, air preheaters and grit arresters and gas-cleaning equipment to meet clean air legislation.

Where watertube boilers are used to recover waste heat (for example, exhaust gases from reciprocating engines) lower gas temperatures may be involved, and this, in turn, could obviate the need for water-cooled walls. In this case, tube banks may be contained within a gas-tight insulated chamber.

There are two basic types of watertube boilers: assisted and natural circulation. Assisted circulation might apply where heat is from a convection rather than a radiation source such as a waste heat application. Natural circulation is more suited where radiant heat and high gas temperatures are present.

Depending upon the required duty and the site, units may be shop assembled or of modular construction. Site-erected units may be designed to have their main components arranged to fit in with the space available.

7.3.8 Waste-heat boilers

These may be horizontal or vertical shell boilers or watertube boilers. They would be designed to suit individual applications ranging through gases from furnaces, incinerators, gas turbines and diesel exhausts. The prime requirement is that the waste gases must contain sufficient usable heat to produce steam or hot water at the condition required.

Supplementary firing equipment may also be included if a standby heat load is to be met and the waste-gas source is intermittent. Waste-heat boilers may be designed to use either radiant or convected heat sources. In some cases, problems may arise due to the source of waste heat, and due consideration must be taken of this, with examples being plastic content in waste being burned in incinerators, carry-over from some type of furnaces causing strongly bonded deposits and carbon from heavy oil fired engines. Some may be dealt with by maintaining gas-exit temperatures at a predetermined level to prevent dewpoint being reached and others by sootblowing. Currently, there is a strong interest in small combined heat and power (CHP) stations, and these will normally incorporate a waste-heat boiler.

7.3.9 Fluid-bed boilers

The name derives from the firebed produced by containing a mixture of silica sand and ash through which air is blown to maintain the particles in suspension. The beds are in three categories, shallow bed, deep bed and recirculating bed. Shallow beds are mostly used and are about 150–250 mm in depth in their slumped condition and around twice that when fluidized. Heat is applied to this bed to raise its temperature to around 600°C by auxiliary oil or gas burners. At this temperature coal and/or waste is fed into the bed, which is controlled to operate at 800–900°C. Water-cooling surfaces are incorporated into this bed connected to the water system of the boiler.

The deep bed, as its name implies, is similar to the shallow bed but in this case may be up to 3 m deep in its fluidized state, making it suitable only for large boilers. Similarly, the recirculating fluid bed is only applicable to large watertube boilers.

Several applications of the shallow-bed system are available for industrial boilers, the two most used being the open-bottom shell boiler and the composite boiler. With the open-bottom shell the combustor is sited below the shell and the gases then pass through two banks of horizontal tubes.

In the composite boiler the combustion space housing the fluid bed is formed by a watertube chamber directly connected to a single-pass shell boiler. In order to fluidize the bed the fan power required will be greater than that with other forms of firing equipment.

To its advantage, the fluid bed may utilize fuels with high ash contents which affect the availability of other systems. It is also possible to control the acid emissions by additions to the bed during combustion. They are also less selective in fuels and can cope with a wide range of solid-fuel characteristics.

7.4 Application and selection

Figure 7.6 illustrates the selective bands for various types of boilers. The operating pressure will govern the steam temperature except where superheaters are used. For hot water units the required flow temperature will dictate the operating pressure. It is important that when arriving at the operating pressure for hot water units due allowance is made for the head of the system, an anti-flash steam margin of 17°C and a safety-valve margin of 1.5 bar.

In arriving at a decision to install one or more boilers the following should be considered. The first choice (providing the load is within the duty range of the boiler) will be a single unit. This is economically the most attractive in capital cost, providing account is taken of

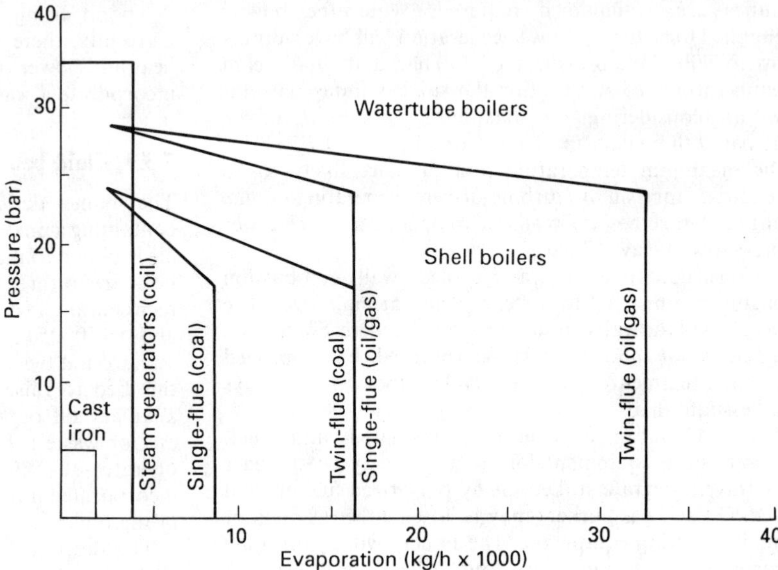

Figure 7.6 Guide to boiler capacities

the following:

1. If there is a breakdown on the boiler will production be seriously affected immediately?
2. Will adequate spare parts for the boiler be held in stock or available within an acceptable time, and will labour be available to carry out the repair work?
3. Will time be available to service the boiler?
4. The duty will preferably fall within the modulating firing rate of the burner.
5. Prolonged periods of intermittent operation are avoided.
6. Is there an existing standby unit?

If any or all of these points are not accepted, then the next consideration for a shell boiler could be a twin-flue unit suitable for single-flue operation. This has the advantage of using less space than two smaller boilers and having only one set of services.

Moving now to two boilers, the heat load may comprise two elements. One may be a production process whose interruption would cause problems and the other, say, a heating load where any interruption would not be noticed immediately. Assuming the two elements were of equal duty, it would be reasonable to install two boilers each 50–50% of the total load. One boiler would then be able to cover the process load.

An extension to this is to install two boilers, each capable of handling the total combined load. Depending upon the boiler size, there may be only a relatively small difference in total capital cost between the above two schemes.

Further options involving three or more boilers must take into account minimum and maximum loads in order to run the plant efficiently. When considering hot water it may be advantageous to consider units in a range of outputs. This will help in operation, so that a unit may be brought into duty to match the load and thus avoid low-load conditions and consequent danger of dewpoints. Also if the plant is fired on solid fuel it will help in maintaining a more even firing rate and a clean stack.

7.5 Superheaters

Steam produced from a boiler is referred to as dry saturated, and its temperature will correspond with the working pressure of the boiler. In some instances, particularly with shell boilers, this is perfectly acceptable. There are occasions, however, where it is desirable to increase the temperature of the steam without increasing the pressure. This function is performed by a superheater. Steam from the drum or shell of the boiler is passed through a bank of tubes whose external surfaces are exposed to the combustion gases, thus heating the steam while not increasing the pressure.

Where a superheater is fitted the boiler working pressure must be increased to allow for the pressure drop through the elements. This will be between 0.3 and 1.0 bar.

In a watertube boiler the superheater is a separate bank of tubes or elements installed in the area at the rear or outlet of the combustion chamber. Saturated steam temperature may be increased by 200C° with a final steam temperature of up to 540°C.

For shell boilers, superheaters may be one of three types, depending upon the degree of superheat required. The first and simplest is the pendant superheater installed in the front smokebox (Figure 7.7). The maximum degree of superheat available from this would be around 45C°. The second pattern is again installed in the front smokebox but with this the elements are horizontal 'U' tubes which extend into the boiler smoketubes. The degree of superheat from this pattern is around 80C°. Third, a superheater may be installed in the reversal chamber of the boiler. A wetback chamber presents

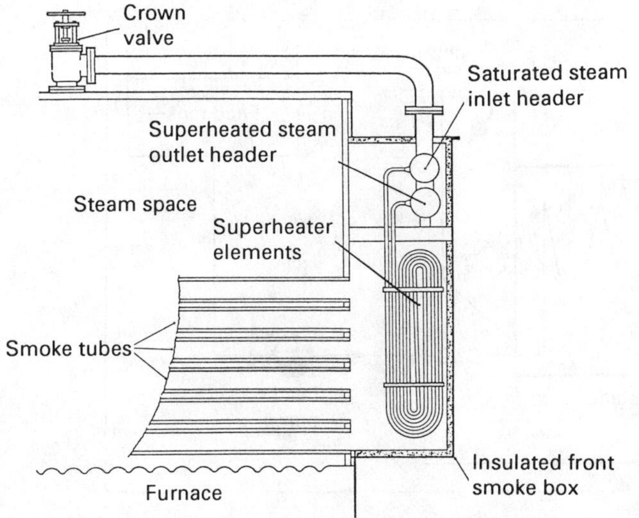

Figure 7.7 Front smokebox pendant superheater (with permission of Senior Thermal Engineering Ltd)

problems with lack of space, and therefore either a semi-wetback, dryback or water-cooled wall chamber may be considered. Maximum degree of superheat would be around 100C°.

Superheater elements are connected to inlet and outlet headers. The inlet header receives dry saturated steam from the steam drum of a watertube boiler or the shell of a horizontal boiler. This steam passes through the elements where its temperature is raised and to the outlet header which is connected to the services. A thermometer or temperature recorder is fitted to the outlet header.

It should be appreciated that a steam flow must be maintained through the elements at all times to prevent them burning away. If a single boiler is used then provision to flood the superheater during start-up periods may be required.

Superheated steam may be needed where steam distribution pipework in a plant is over extended distances, resulting in a loss of heat and increase in wetness of the steam. Another case may be where a process requires a temperature above the working pressure of the plant. The third case is where steam is used for turbines. Here it improves the performance of the turbine, where for every 6C° increase in steam temperature it can produce a saving of about 1% reduction in steam consumption. Superheaters may also be supplied as independently fired units. These may be used when either the amount of superheated steam required is much less than the boiler evaporation or is only needed on an intermittent basis.

7.6 Economizers

Economizers are installed in the exhaust gas flow from the boiler. They take heat from the flue gases which they transfer via extended surface elements to the feedwater immediately prior to the water entering the boiler. They therefore increase the efficiency of the boiler and have the added advantage of reducing thermal shock. In watertube boilers they may be incorporated within the structure of the boiler or supplied as a free-standing unit. With shell boilers they will be a separate unit fitted between the boiler flue gas outlet and the chimney.

Figure 7.8 is a schematic illustration of such a unit. It is desirable for each boiler to have its own economizer. Where one economizer is installed to take the exhaust gases from more than one boiler special considerations must be taken into account. These will include gas-tight isolation dampers. Consideration must be made of flue-gas pressures at varying loads and maximum and minimum combined heat load to match economizer and a pumped feedwater ringmain. Economizers may be used for both forced-draught and induced-draught boilers, and in both cases the pressure drop through the economizer must be taken into account when sizing the fans.

Economizers are fitted to most watertube boilers. An exception is on a waste-heat application, where it may be desirable, due to the nature of the products being burned, to maintain a relatively high gas outlet temperature to prevent corrosive damage to the boiler outlet, ductwork and chimney.

With watertube boilers economizers may be used when burning coal, oil or gas. The material for the economizer will depend on the fuel, and they may be all steel, all cast iron or cast iron protected steel. All steel would be used for non-corrosive flue gases from burning natural gas, light oil and coal. Cast iron may be used where the feedwater condition is uncertain and may attack the tube bore. Fuels may be heavy fuel oil or coal, and there is a likelihood of metal temperatures falling below acid dewpoint. Cast iron protected steel is used when heavy fuel oil or solid fuel firing is required and feedwater conditions are suitably controlled. As cast iron can withstand a degree of acid attack, these units have the advantage of being able to operate without a gas bypass, where interruptible natural gas supplies are used with oil as standby.

With shell boilers, economizers will generally only be fitted to boilers using natural gas as the main fuel and

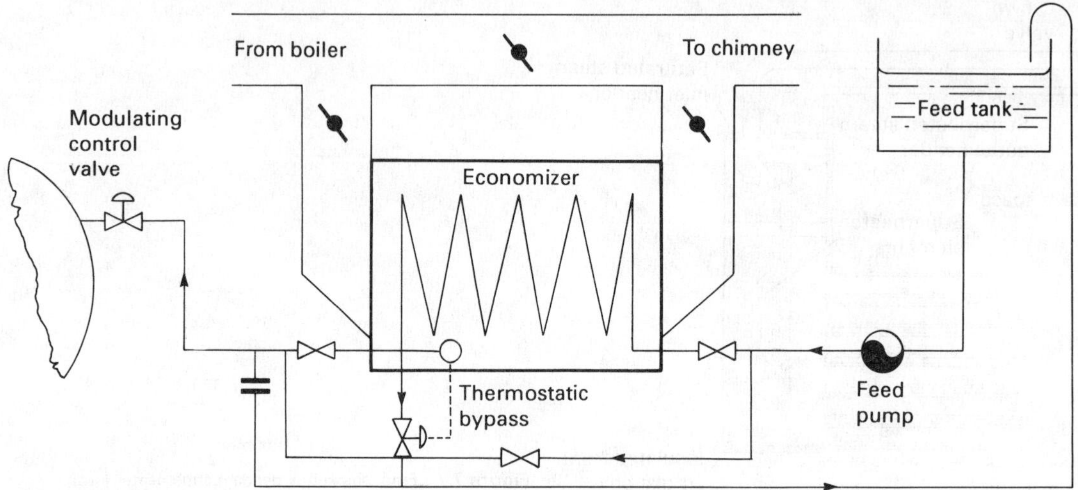

Figure 7.8 Schematic illustration of an installed economizer

then only on larger units. It would be unlikely that a reasonable economic case could be made for boilers of less than 4000 kg/h *F* and *A* 100°C evaporative capacity. The economizer will incorporate a flue gas bypass with isolating dampers to cover for periods when oil is used and for maintenance. The dampers require electric interlocks to the selected fuel.

As the majority of shell boilers operate in the pressure range 7–10 bar, the flue gas outlet temperature will be in the range of 190–250°C. It may be appreciated from this that the boiler needs to operate at 50–100% of its maximum continuous rating for most of the working day to produce an economic return.

Where an economizer is installed it is essential to have water passing through the unit at all times when the burners are firing to prevent boiling. Therefore boilers fitted with economizers will have modulating feedwater control. Even then, it is possible that the water flow requirement can become out of phase with the burner firing rate. To prevent damage, a temperature-controlled valve allows a spillage of water back to the feedwater tank, thus maintaining a flow of water through the unit. Each economizer will be fitted with a pressure-relief safety valve.

Due to the amount of water vapour produced when natural gas is burned, it is important not to allow the exhaust gas temperature to fall below 80°C, otherwise the water dewpoint will be reached. Not only the economizer but also the ductwork and chimney must be considered and provision incorporated for drainage.

In the event of a separate use for low-grade hot water being available, it is sometimes practical to install a secondary condensing economizer. With this, the material of which the economizer is constructed allows for condensate to form and drain away without excessive attack from corrosion.

A recent development in heat recovery has been the heat tube. This is a sealed metal tube which has been evacuated of air and contains a small quantity of liquid which, for boiler applications, could be water. When heat from the flue gases is applied to one end of the heat pipes the water in the tube boils, turning to steam and absorbing the latent heat of evaporation. The steam travels to the opposite end of the tube which is surrounded by water, where it gives up its latent heat, condenses and returns to the heated end of the tube. Batteries of these tubes can be arranged to form units, usually as a water jacket around a section of a flue.

7.7 Water-level control

Water-level controls continuously monitor the level of water in a steam boiler in order to control the flow of feedwater into the boiler and to protect against a low water condition which may expose the heating surfaces with consequent damage. The controls may be either float operated or conductivity probes.

With watertube boilers the control of the water level needs to be precise and sensitive to fluctuating loads due to the high evaporative rates and relatively small steam drums and small water content. Control will be within ± 10 mm on the working water level and will be two- or three-element control. Two-element control will comprise modulating feedwater control with first low water alarm and high–low control with low-water cut-out and alarm. The second element will be monitoring of the steam flow to give early indication of any increase in steam demand. This signal may then be linked to the firing rate of the burners and the feedwater modulating valve. The third element senses a drop in feedwater demand, which would signal the firing rate of the burners to modulate down.

Shell boilers will have two external level controls each independently attached to the shell. Boilers up to about 9000 kg/h *F* and *A* 100°C will have a dual control and either a single or high–low control. The dual control instigates the feedwater pump, which operates on an on–off cycle over a water level band of ± 15 mm and also operates the first low-water alarm. The single or high–low control will incorporate a second low-water alarm with burner lockout and with the high–low control also an indication of high water, which may be linked to shut down the feedwater pump with automatic restart when the water level drops to normal.

Boilers of larger evaporations will have modulating and high–low control. The modulating level control monitors the working water level in the shell and operates a control valve in the feedwater line, allowing water to enter the boiler from a continuously running feed pump. It will also incorporate the first low-water alarm. The high–low control operates as before.

The advantage of modulating control is that it maintains a constant working water level and therefore the boiler is always in its best condition to supply steam for peak loading. These controls may also be fitted to boilers below 9000 kg/h *F* and *A* 100°C particularly if severe loads are present or when the working pressure is above 10 bar.

With water-level controls it is important to check they are functioning correctly, and they will be operated daily to simulate low-water condition. On shell boilers this is invariably a manual operation but may be motorized on watertube boilers.

Shell boilers may be fitted with internal level controls. Here controls are mounted on the crown of the boiler with the floats or probes extending to the water surface through the steam space. To check the operating function of these, it is necessary to drop the water level in the boiler, or, alternatively, a separate electronic testing device can be fitted. With fully flooded hot water boilers a single level control or switch is fitted to protect against low-water condition.

7.8 Efficiency

This is the ratio of heat input to the useful heat output of the boiler, taking account of heat losses in the flue gases, blowdown and radiation. It is covered in BS 845: Part 1, which fully details the concise or losses method of determining the efficiency of a boiler.

7.9 Boiler installation

Figure 7.9 illustrates the services required for a boiler. The individual items are covered separately in other sections and only the general concept is being considered here.

If boilers are being replaced it is reasonable to expect that existing services will be available, apart from possible upgrading in some areas. For new installations a new building is likely to be required, and it will first be necessary to determine the overall size and type of construction.

Starting with the boiler(s), these will be set out giving due consideration to space between boilers and other items of plant in order to give adequate access to all equipment, valves and controls for operation and maintenance. For small- to medium-sized boilers 1 m may be considered a reasonable space between items of plant where access is required. With large boilers (including watertube) this may be increased up to 3 m. The width of firing aisle will be dependent upon the size and type of boiler.

Having now established the space requirement for the boilers, we must consider the other items of plant to be housed in the same structure. These may include water-treatment plant, heat-recovery system, tanks and pump, instrument and controls. With oil- and gas-fired installations there is little additional plant to consider within the boiler house. With solid fuel then the handling, distribution and perhaps storage could also form part of the building with provision for ash handling and removal.

Access for air into the boiler house must be considered. Air for combustion will be drawn in through purpose-made louvres and care should be taken that this air does not create a hostile environment to personnel in the building. Also, passage of air across cold water pipes should be avoided, as in cold weather this may aggravate freezing in the pipes.

Ventilation will also be required. With gas-fired installations this is mandatory, and specified free areas must be provided.

The fabric of the building may range from a basic frame and clad structure offering minimum weather

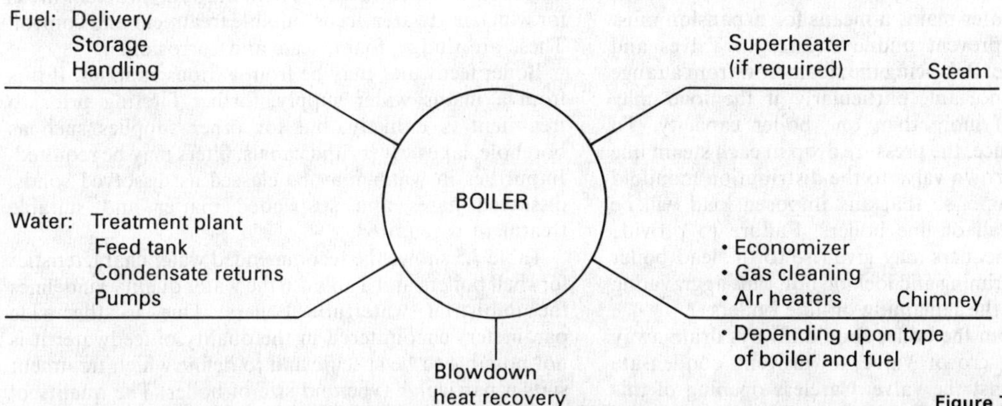

Figure 7.9 Boiler services

protection to a brick and concrete construction. While it is possible to weatherproof boilers for outside installation, it should be remembered that the occasion will arise when someone is going to carry out maintenance or service work, and if weather conditions are inclement, this may prove impractical. Two instances where outdoor installation may be considered are where the climatic conditions of the country permit and in the case of waste-heat boilers where controls are minimal.

Foundations and trenches will be formed prior to boiler delivery. The boiler house structure may be erected before or after delivery of the boilers, dependent upon site conditions and the type of boiler. Due consideration will be given to access openings and doorways and provision for lifting any items of plant once the boiler house is complete.

Noise may be considered in two areas, one being an unacceptable level of noise to adjacent areas and the other an excessive noise level for personnel inside the boiler house. With the first, the type of construction of the building together with position of openings and the fitting of acoustic louvres for air ingress will all help to reduce external noise. As an extention to this and to also reduce noise for personnel within the building, all items of plant generating noise may be considered. These will generally be forced-draught fans and feedwater pumps. Various degrees of silencing are available for combustion appliances and feed pumps may be housed in a separate enclosure within the building.

External to the boiler house, provision for the storage and handling of solid and liquid fuel is required with access for delivery vehicles. With some small boilers such as electrode or steam-coil generators, where the boiler only serves a single item of plant or process, it is practical to install them immediately adjacent to that process. The electrode boiler is eminently suitable here as no combustion gases are produced.

7.10 Boiler house pipework

Pipework layout should be designed so that it does not restrict access to the plant or building. Walkways and gantries should be considered when deciding on pipe service routes.

Where pipework is subject to temperature such as steam and hot water mains a means for expansion must be included to prevent undue stresses to valves and building structure. Balancing the steam flow from a range of boilers is important, particularly if the load may include peaks of more than one boiler capacity. To achieve this balance, the pressure drop in each steam line from the boiler crown valve to the distribution manifold should be the same, so that any imposed load will be shared between all on-line boilers. Failure to provide balanced steam headers may give rise to the lead boiler losing pressure, priming and locking out, thus aggravating the situation for the remaining on-line boilers.

Steam lines from the boiler should always drain away from the boiler crown value to prevent condensate building up against the valve. Careless opening of this valve to allow steam to pass when the line is flooded can result in splitting the valve and pipework. If the steam line rises from the crown valve a tee trap must be installed and fitted with a steam trap and drain valve. The drain valve must be operated to check that the steam line is clear of water before the crown valve is opened.

Safety-valve vent pipework should be run on the shortest possible route. Where bends or long runs occur, the pipe size may have to be increased to prevent back pressure on the safety valve during operation.

Feedwater pipework will normally be gravity head suction from the hotwell or feed tanks to the pumps and at a pressure in excess of the boiler working pressure from the pumps to the boiler. Few problems occur on the pipework between pumps and boilers. However, inadequately sized suction lines can give rise to cavitation at the feed pump with subsequent boiler shutdown. The feed tank should always be positioned to suit the temperature of the feedwater and the pipework sized to give free flow at that head, taking account of bends, valves and filters.

Fuel lines should be run where they will not be subject to high temperatures and must include protection with automatic and manual shut-off fire valves at the perimeter of the boiler house. Individual boilers should also be capable of isolation from this supply. Local requirements may vary on fuel supply systems, and these should be ascertained before installation.

Blowdown and drains may be taken to either a sump or blowdown tank prior to discharging into the drains. The purpose of this is to reduce the temperature by dilution and dissipate the pressure to prevent damage to the drains.

All pipework must be installed to the appropriate standard and codes of practice. The degree of insulation will be suitable for the temperature of the pipework, although the finish may vary, depending upon site preferences. Valve boxes are recommended, although on some low-pressure installations these are not always included.

7.11 Feedwater requirements

Poor or unsuitable water can be a major factor where failure in a boiler occurs. There are four problem areas for which feedwater needs suitable treatment and control. These are sludge, foam, scale and corrosion.

Boiler feedwater may be from various supplies. If it is from a mains water supply, further filtering prior to treatment is unlikely, but for other supplies such as borehole, lakes, rivers and canals, filters may be required. Impurities in water may be classed as dissolved solids, dissolved gases and suspended matter and suitable treatment is required.

Table 7.5 shows the recommended water characteristics for shell boilers and Table 7.6 the water quality guidelines for industrial watertube boilers. Due to the wide parameters encountered in the quality of feedwater it is not possible to be specific and to define which treatment suits a particular type and size of boiler. The quality of

Table 7.5 Recommended water characteristics for shell boilers

For pressures up to 25 bar[a]			
Total hardness in feedwater, mg/l in terms of $CaCO_3$ max.	2	20	40
Feedwater			
pH value	7.5–9.5	7.5–9.5	7.5–9.5
Oxygen	[b]	[b]	[b]
Total solids, alkalinity, silica	[b]	[b]	[b]
Organic matter	[b]	[b]	[b]
Boiler water			
Total hardness, mg/l in terms of $CaCO_3$ max.	ND	ND	ND
Sodium phosphate, mg/l as Na_3PO_4[c]	50–100	50–100	50–100
Caustic alkalinity, mg/l in terms of $CaCO_3$ min.	350	300	200
Total alkalinity, mg/l in terms of $CaCO_3$ max.	1200	700	700
Silica, mg/l as SiO_2 max.	Less than 0.4 of the caustic alkalinity		
Sodium sulphite, mg/l as Na_2SO_3	30–70	30–70	30–70
or			
Hydrazine, mg/l as N_2H_4	0.1–1.0	0.1–1.0	0.1–1.0
Suspended solids, mg/l max.	50	200	300
Dissolved solids, mg/l max.	3500	3000	2000

—: Not applicable.
ND: Not detectable.
[a] 1 bar = 105 N/m^2 = 100 kPa = 14.5 lb/in^2.
[b] Numerical values depend upon circumstances but the comments are relevant.
[c] Phosphate is usually added as sodium phosphate but determined as phosphate (PO_4^{3}); $Na_3PO_4 = 1.73 \times PO_4^{3}$.
Based on Table 2 of BS 2486: 1978, by permission of BSI.

make-up and percentage of condense returns in a system will both have to be taken into consideration.

For some small boilers it may be possible to supply internal dosing subject to a suitable water supply and other conditions being favourable. However, for anything other than very small installations, external treatment is recommended.

For shell boiler installations a simplex or duplex base exchange system with suitable dosing is usual, although on larger installations or if the water is excessively hard and there is little condense return then a de-alkalization plant may be used.

For watertube boilers base exchange or de-alkalization may be used providing the water quality is suitable and the boilers are not operating at pressures in excess of 30–35 bar. With modern watertube boilers demineralized water is recommended. Where boilers are operating at high pressures or are used for power generation it is essential to use demineralized water in order to prevent build-up of deposits, particularly silica, on turbine blades.

With hot water installations it is equally important that water suitably treated for hardness and corrosion should be used. Even when cleaning or flushing a new or modified system, care must be taken to prevent premature corrosion occurring by the addition of a suitable treatment. Few (if any) hot water systems are completely sealed, and provision should be designed into the system to treat all make-up water. Draw off of hot water directly from the system should never be done and a calorifier must always be used. Analysis of the water in the boiler and system should be carried out at least monthly and more frequently during the commissioning period of a new installation or where an existing system has been refilled.

Where steam or hot water boilers are not required to operate for a period of time it is important that suitable measures are taken to prevent water-side corrosion. For periods of a few days the water may be left at its normal level but daily testing should be carried out as if the boiler were in use and corrective treatment added as necessary. If the period is for several months then the boiler should be fully flooded to exclude all air and the water treated. Regular testing of this water should be carried out and corrective treatment used. For longer periods, boilers should be drained completely and thoroughly dried out. The boiler may then either be left vented with the addition of a small electric heater inside or sealed and trays of moisture-absorbing chemicals such

Table 7.6 Water-quality guidelines recommended for reliable, continuous operation of modern industrial watertube boilers

Boiler feedwater

Drum pressure (lb/in^2 g)	*Iron* (ppm Fe)	*Copper* (ppm Cu)	*Total hardness* (ppm $CaCO_3$)
0–300	0.100	0.050	0.300
301–450	0.050	0.025	0.300
451–800	0.030	0.020	0.200
601–750	0.025	0.020	0.200
751–900	0.020	0.015	0.100
901–1000	0.020	0.015	0.050
1001–1500	0.010	0.010	ND[d]
1501–2000	0.010	0.010	ND[d]

Boiler water

Drum pressure (lb/in^2 g)	*Silica* (ppm SiO_2)	*Total alkalinity*[a] (ppm $CaCO_3$)	*Specific conductance* (μmho/cm)
0–300	150	350[b]	3500
301–450	90	300[b]	3000
451–600	40	250[b]	2500
601–750	30	200[b]	2000
751–900	20	150[b]	1500
901–1000	8	100[b]	1000
1001–1500	2	NS[c]	150
1501–2000	1	NS[c]	100

[a] Minimum level of hydroxide alkalinity in boilers below 1000 lb/in^2 must be individually specified with regard to silica solubility and other components of internal treatment.
[b] Maximum total alkalinity consistent with acceptable steam purity. If necessary, the limitation on total alkalinity should override conductance as the control parameter. If make-up is demineralized water at 600–1000 lb/in^2 g, boiler water alkalinity and conductance should be shown in the table for the 1001–1500 lb/in^2 g range.
[c] NS (not specified) in these cases refers to free sodium- or potassium-hydroxide alkalinity. Some small variable amount of total alkalinity will be present and measurable with the assumed congruent control or volatile treatment employed at these high-pressure ranges.
[d] None detectable.

as hydrated lime or silica gel laid inside. In potentially damp atmospheres such as near coasts the dry method is preferred, as keeping the boiler full of cold water will cause condensation to be continuously present on the fire-side, giving rise to surface corrosion.

7.12 Feedwater supply and tanks

The design, size and siting of the boiler feed tank or hotwell must be compatible with the boiler duty capacity and system temperatures. They should be installed giving sufficient space for access to controls, valves and manways.

The tank will normally be of fully welded construction from mild steel with the internal surfaces shot blasted and a protective plastic coating applied. It is important that all connections together with any additional future connections be included before the coating is applied to prevent breaking the surface, thus giving a source of future corrosion. The tank should have a sealed top with adequate bolt-on access covers. This will keep out foreign bodies and also protect the water surface from the atmosphere, where it would absorb oxygen. Adequate venting must be included to prevent any build-up of pressure from either heating or condense returns.

Alternative materials may be used, including stainless steel and plastic. Sectional tanks may be installed for convenience in restricted places, but great care must be taken if they are subject to temperature variations which can give rise to joint problems.

The following connections will be required in addition to the vent already referred to:

1. Feedwater take-off. This will be from the bottom of the tank and preferably from the base with a weir pipe extending 50–100 mm up into the tank. If the take-off is from the side of the tank the bottom of the pipe bore must be a similar dimension up from the base. This allows sludge to remain in the tank which should be inspected and cleaned as required.
2. Drain pipework will be directly from the bottom of the tank.
3. Overflow pipework should be sized to meet the supply quantity to the tank and run so that visible evidence is obvious of an overflow condition.
4. Condense returns. Where returns arrive at the feed tank from several sources it is desirable to bring them together with a collection manifold at the top of the tank. The manifold then has a dip pipe extending down into the tank terminating in a sparge pipe. This method reduces the risk of entraining air in the returns and also helps to promote mixing of the water in the tank.
5. Make-up water. Water from the treatment plant will be fed to the main tank either through a separate make-up tank or by a semi-sealed section within the main tank. In either case, the water level will be the same in both sections, excluding the ullage left for condense returns. Control of the make-up water level may be by float valve, float switches or conductivity probes. These methods allow water to flow through the treatment plant, although conductivity probes or switches permit a positive flow and avoid the risk of slippage.
6. Chemical dosing. This is a connection into the tank with an internal diffuser for any corrective treatment required for the water.
7. Level indicator. This may be either visual or electrical or both.

Additional connections may also be required for the following:

8. Flash steam. This will result from the blowdown heat recovery referred to in Section 7.13. Flash steam is introduced into the tank through a dip pipe terminating in a distribution manifold near the bottom of the tank.
9. Tank heating. This is referred to in detail later in this section.
10. Thermometer.

Connections should be arranged to avoid short circuiting between the tank supply and take-off to the boiler feed. Also, adequate mixing within the tank is required to prevent stratification.

With the capacity of the feed tank, assuming that there is an adequate supply of water either from mains or local storage, a 1 h supply should be considered a minimum. This would be based on a continuous supply of make-up water being available such as from a duplex treatment plant. If the supply of make-up is not continuous then the tank capacity must be increased. For example, a base exchange plant will take about 75 min to regenerate, and in that case the tank capacity should be able to sustain a supply of water to the boiler for 90 min in order to give a safety margin. The treatment plant then has to supply a quantity of water in excess of the boiler demand rate in order to replenish the tank. On top of this capacity a volume has to be allowed for condense returns when the system starts in order to prevent wastage of heat and water. As steam capacities increase, the feed tank becomes correspondingly larger, and it may be considered that on steam outputs in excess of 20 000 kg/h a duplex treatment plant should be installed in order to keep the feed tank to a reasonable size.

In areas where the water demand of the plant may exceed the supply capacity of the local mains it will be necessary to install additional storage tanks. It is likely that if the local supply is subject to low flow rates additional storage is already available for services such as fire protection.

Feedwater to the boiler should not be cold, as this can cause harmful thermal stresses to the boiler. A minimum feedwater temperature of 70–80°C should be designed into the system. This increased temperature has the added advantage that it accelerates some water-treatment reactions and also helps to remove oxygen and other gases from the feedwater. Once the system is working, this higher temperature may be achieved from the condense returns, but this condition is not effective until the plant has been running for some time. A tank heating system should therefore be installed. This is best achieved by direct steam injection into the feed tank. Steam is

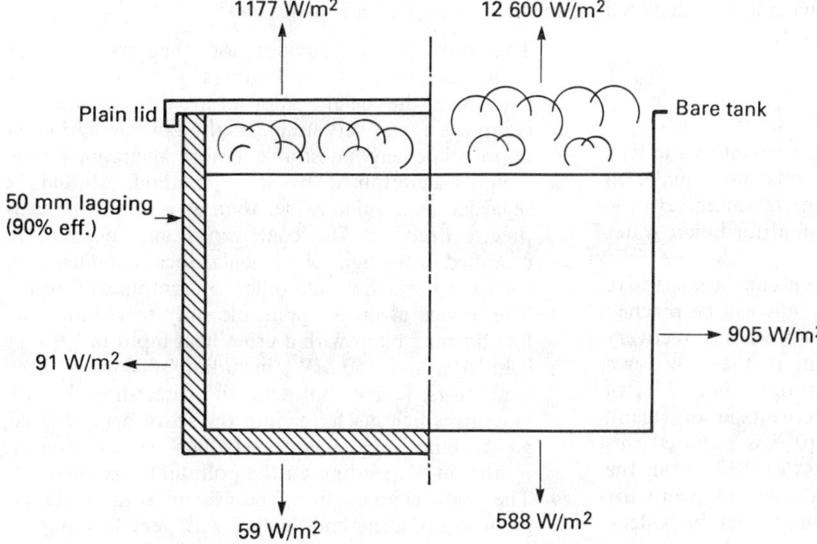

Figure 7.10 Heat losses from a feedwater tank operating at 93°C with ambient 21°C

taken from the boiler and reduced in pressure to 1–2 bar to reduce noise in the tank. Passing through a thermostatic control valve, the steam mixes with the water in the tank through a sparge pipe or steam nozzle.

With watertube boilers economizers are normally used, therefore the situation of cold feed to the boiler would not apply. However, the feed to the economizer should be treated similarly to that to a boiler to reduce thermal shock.

With the feedwater being raised in temperature it is necessary to avoid cavitation at the inlet to the boiler feed pump. One method of reducing this problem is to elevate the feed tank to give an adequate positive head at the pump. The following are recommended heights to the minimum water level in the tank at various temperatures:

88°C – 1.6 m
93°C – 3.1 m
99°C – 4.6 m
100°C – 5.2 m

The feed tank should always be insulated to reduce heat loss to a minimum. Figure 7.10 illustrates typical heat losses from a feed tank operating at 93°C in an ambient temperature of 21°C with and without 50 mm of insulation.

7.13 Blowdown requirements, control and tanks

In order to maintain the level of dissolved and suspended solids within the boiler as recommended in Section 7.11 it is necessary for the boiler to be blown down. This is an operation where a quantity of water is drained from the boiler while the boiler is operating at pressure and may be achieved by various methods.

The simplest and that applied to small boilers is for the main bottom blowdown valve to be opened for a set period of time at regular intervals (e.g. 20 s every 8 h). This method may also extend to larger boilers where conditions are such that there is little build-up of solids. Such conditions could be high condense returns and good-quality make-up feedwater.

The second method could be automatic intermittent blowdown. With this, a timer-controlled valve is installed at the bottom of the boiler prior to the main blowdown valve. A programme is then designed to operate this valve in short bursts which disperses any sludge and controls the levels of solids. This method is preferred for boilers having internal treatment.

The third method would be continuous blowdown through a regulating or micrometer valve. The take-off position for this should preferably be about 250 mm below the working water level and may either be on the side of the shell or on the crown with a dip pipe down to the correct level. If a connection is not available it is possible to instal the valve on the bottom connection prior to the main blowdown valve.

All these methods will require careful monitoring initially to set up and determine the correct rate of blowdown once the plant is operating. In order to take the necessary sample from the boiler the boiler(s) should be fitted with a sample cooler. To automate the continuous blowdown a conductivity-controlled system may be installed. Here a controller continuously compares the boiler water electrical conductivity with a value set in the controller. Depending on whether this is above or below the set rate, it will automatically adjust the blowdown flowrate.

While the above methods control the level of dissolved and suspended solids in the boiler, it will still be an insurance requirement to operate the main blowdown valve periodically.

The minimum amount of blowdown may be calculated as a percentage of the evaporation rate by:

$$\text{Blowdown rate} = \frac{F}{B-F} \times 100\%$$

where F = the total dissolved solids content of the feed in parts per million allowing for the mixture of make-up and condensate plus any chemical treatment and B = the maximum recommended solids content for boiler water in parts per million.

While vitally necessary, blowdown can be expensive in terms of lost heat. Therefore a point will be reached when it is economical to instal a blowdown heat recovery system. Generally, the heat content in the blowdown water for a shell boiler will represent only about 25% of the heat content in the same percentage of steam. Therefore if a blowdown rate of 10% is required this represents an approximate heat loss of 2.5% from the boiler capacity. This differential reduces and eventually becomes insignificant on high-pressure watertube boilers.

The blowdown from the boiler(s) will be run to a flash-steam vessel mounted adjacent to the feed tank. Flash steam will be introduced into the feed tank through a dip pipe terminating in a distribution manifold. The drain from the flash vessel may then be taken to a residual blowdown heat exchanger. Any remaining heat is then transferred to the make-up water to the tank before the blowdown runs to drain.

Blowdown from the boiler(s) should always be taken to either a blowdown sump or blowdown vessel before discharging into drains. Both should be adequately sized to give cooling by dilution and be fitted with vent pipes to dissipate pressure safely. The boiler(s) should have independent drain lines for the main manually operated blowdown valve and the drains from a continuous blowdown system. Where more than one boiler is connected to either system the line should be fitted with a check or secondary valve capable of being locked.

Blowdown sumps will be constructed from brick and/or concrete and the blowdown lines will drain under gravity. Where the blowdown lines enter the sump they will turn to discharge downwards and the bottom of the sump will be protected below this area with a cast iron tray to prevent erosion. The drain or overflow from the pit will be at such a level to produce a weir effect, thus holding water for dilution.

With a blowdown vessel these may be installed at ground level, and thus the water in them can be above the boiler blowdown valve. In this case, a drain valve for maintenance purposes must be installed at the lowest point in the line between the boiler and the vessel.

7.14 Clean Air Act requirement for chimneys and flue designs

7.14.1 Introduction

The function of a chimney is to discharge in a manner to give adequate dispersal to the products of combustion in accordance with the third edition of the 1956 Clean Air Act Memorandum on Chimney Heights. The scope of the memorandum is as follows:

1. The publication provides for the use of local authorities, industry and others who may need to determine the height appropriate for certain new chimneys a relatively simple method of calculating the appropriate height desirable in normal circumstances.
2. Heights determined by these methods should be regarded as a guide rather than as a mathematically precise decision. The conclusions may need to be modified in the light of particular local circumstances such as valleys, hills and other topographical features.
3. The advice given is applicable only to chimneys of fuel burning plant with a gross heat input of between 0.15 MW and 150 MW, including stationary diesel generators. It does not deal with direct-fired heating systems which discharge into the space being heated, gas turbines or incinerators (which require separate treatment, depending on the pollutants emitted).
4. The main changes from the second edition are the inclusion of a method dealing with very low sulphur fuel and extensions of the methods for taking into account the height of nearby buildings and of the range of the size of furnace included.

7.14.2 Gas velocity

In order to maximize the chimney height, the efflux velocity of the gases leaving the chimney should be designed on 12 m/s at Maximum Continuous Rating (MCR) of the boiler. On some very small boilers this may be impractical to achieve, but a target velocity of not less than 6 m/s at MCR should be aimed for. With boilers at the top end of the range a velocity of 15 m/s at MCR is required. Some inner-city authorities may stipulate higher efflux velocities and some plants have been installed with gas velocities of 22 m/s.

7.14.3 Chimney height

Originally, the height of the chimney was designed to produce a draught sufficient to produce induced- draught air for combustion. With modern boiler plant forced-draught and/or induced-draught fans are used. This allows for the greater degree of control of the air to be designed into the combustion appliance. The chimney is therefore only required to disperse the gases.

When using gaseous fuel it is normally sufficient to terminate the chimney 3 m above the boiler house roof level, subject to there being no higher buildings adjacent to the boiler house. In such cases these buildings may need to be considered.

On medium-sized boiler plant where gas is to be the main fuel it may have oil as a secondary standby fuel. In this case, the chimney height must be based on the grade of fuel oil capable of being burned.

The methods of calculating proposed chimney height are clearly laid out in the Clean Air Act Memorandum, and will be based on:

1. Quantity of fuel burned;
2. Sulphur content of fuel burned;

3. District category;
4. Adjacent buildings;
5. Any adjacent existing emissions.

Application for approval of the proposed chimney height should be made to the appropriate authority at an early stage of a project in order to ascertain their approval or other height they may require. Failure to do this can result in an embarrassing situation where insufficient finance has been allocated due to their requiring a larger chimney than was included in the planned costings.

Where waste products are being incinerated special consideration may have to be given to the resulting flue gases. This may involve having to arrive at a chimney height in conjunction with HM Inspectorate of Factories for Pollution.

7.14.4 Type and combustion

Where a multi-boiler installation is being considered a multi-flue chimney is preferred. This is where the required number and size of flues are enclosed in a single windshield. It is preferred by planning offices and has the advantage of a greater plume rise than from separate stacks.

Several boilers discharging into a single-core chimney are to be avoided. At times of low load the efflux velocity will be very low, which, in turn, will allow the chimney to cool. This may then drop to dewpoint temperatures and, where sulphur is present in the fuel, acid will form. If the chimney is unprotected steel it will suffer rapid corrosion. Even worse, as the boilers increase their load the efflux velocity will increase and start to discharge the acid droplets, which quickly fall out and cause damage to surrounding property.

Construction of chimneys for industrial boilers will mainly be of steel, being either single or multi-flue. A single-core chimney should be suitably insulated in order to maintain a maximum temperature at the chimney outlet to prevent corrosion and will be finish clad with aluminium sheet. A multi-flue chimney will have each flue suitably insulated and enclosed in an outer windshield. Provision for drainage should be incorporated at the base of each flue. Large watertube boilers may have concrete windshields containing internal flues.

7.14.5 Connecting ducts

The duct between the boiler outlet and the chimney connection should be designed to have a gas velocity no greater than the chimney flue and be complete with insulation and cladding, access openings and expansion joints. It should have the least number of bends and changes of section possible and should preferably not fall between the boiler outlet and the chimney connection.

With the use of multi-flue chimneys, dampers are not required in each flue as they were for boiler isolation when several boilers discharge into a common duct or chimney. This should not be confused with the fitting of dampers for heat conservation in a boiler during off-load periods.

7.14.6 Grit and dust emissions

Solids emissions from solid and liquid fuel fired plant are covered in the HMSO publication *Grit and dust – The measurement of emissions from boiler and furnace chimneys*. This states levels of emissions which should be achieved in existing plant and specified for new plant. Suitable sampling connections should be incorporated into the flue ducting for the use of test equipment if permanent monitoring is not included.

7.15 Steam storage

Most boilers built now, together with their combustion equipment, are quick to respond to local fluctuations. Occasionally, where very rapid load changes occur, the firing rate of a gas or oil burner can be virtually instantaneous by the use of special control equipment. This control will have to work in conjunction with the boiler, and therefore the boiler should have adequate steam space and water surface area to help accommodate the rapid changes in steam demand. Good water treatment is especially important here in order to reduce the risk of priming during peak draw-off periods. A boiler with a large shell will have an advantage over one with a smaller shell, assuming equal heating surfaces, but it will give no more than a slight buffer against severe loads.

Most boiler plants can be installed using one or more boilers which can accommodate minimum to maximum loads. Occasionally, heavy peak loads occur for only relatively short periods, and here there may be an advantage on economic running grounds to install boilers whose firing rate will not meet these peaks. In these cases, there are two methods which may be used: one is thermal storage and the other is with an accumulator.

7.15.1 Thermal storage

This is briefly described in Section 7.1.11. Its principle is based on a special feedwater control system which allows the water level in the shell to fall during periods of steam demand in excess of the maximum firing rate. Conversely, during periods of low steam demand the control system allows the water level to re-establish itself. This is achieved using a constant burner firing rate which should match the average steam demand, thus allowing maximum efficiency. It is claimed it is possible with this system to control the limits of boiler working pressure to within ± 0.07 bar.

7.15.2 The accumulator

This was briefly described in Section 7.1.12. Unlike thermal storage, it depends upon differential pressures, and it is suited to a situation where both high- and low-pressure steam systems are required (for example, 17 bar and 7 bar). Alternatively, if no high-pressure steam is required then the boilers must be designed to operate at a higher pressure with all steam supplies going to process through a pressure-reducing station. Any high-pressure surplus then goes to the accumulator to

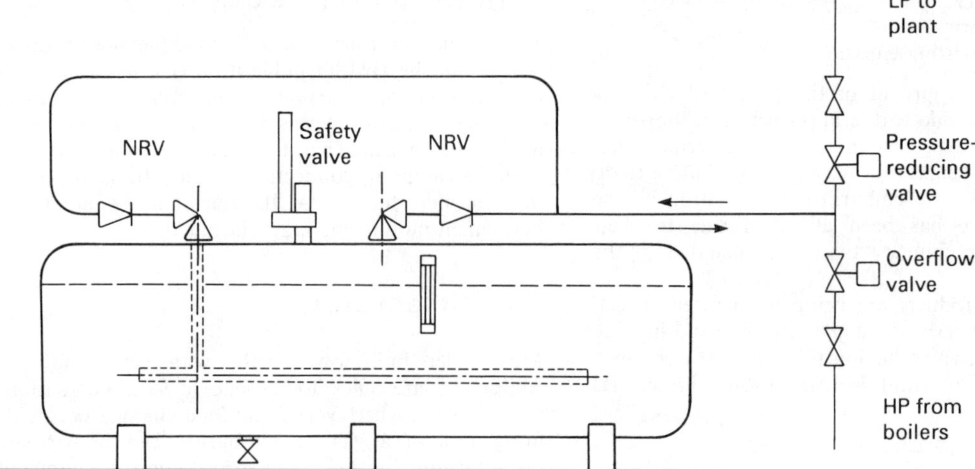

Figure 7.11 Diagrammatic layout of a steam accumulator

help meet peak loads. Figure 7.11 shows the layout of an accumulator.

The storage vessel is filled to around 90% of its volume with water and the overflow valve is controlled by the pressure of the boilers. On rising steam pressure (indicating that the boilers are producing more steam than the process requires) a signal to the overflow valve allows all surplus high-pressure steam to flow into the accumulator via a non-return valve and internal distribution header. Here it is condensed and its thermal energy stored. If a peak load develops on the high-pressure system, controls will close the overflow valve and allow steam to discharge from the accumulator through the pressure-reducing valve set to meet the low-pressure steam requirement. Similarly, if the peak develops on the low-pressure system, then high-pressure steam may pass directly to the pressure-reducing set to supplement steam from the accumulator.

Every accumulator will be designed to meet its specified duty. It will be appreciated that the greater the differential pressure, the smaller the vessel will need to be.

7.16 Automatic controls on boilers

Whether the boiler is fired on oil, gas or solid fuel, it may be expected that it will operate automatically. When boiler plant is not run continuously initial start-up may be manual, time clock or through an energy-management system. Manual attendance may be limited to maintenance functions dictated by the type and size of plant. Automatic controls will cover three areas:

1. Combustion appliance
2. Water level
3. Blowdown

7.16.1 Combustion appliance

When using oil or gas this may operate in one of three modes. On/off, high/low/off or modulating. Only small units will have on/off operation and usually all units of 4000 kg/h evaporation and above will have modulating control. The burners will have safety systems incorporated to prove satisfactory fuel supply before ignition takes place, flame proving at point of ignition and continuous flame monitoring thereafter.

High/low burners will have between 2:1 and 3:1 turndown on the maximum firing rate while a modulating burner will give between 3:1 and 5:1, depending upon the unit size and the fuel used. A higher turndown ratio is available from some burners, but if the above-mentioned ratios are exceeded care must be taken to consider the effect on the final gas temperature leaving the boiler. Dewpoint temperatures are to be avoided. Coal-fired units by nature of the fuel will have modulating control whether chain or travelling grate, underfeed or fixed/tipping grate.

7.16.2 Water level

The control of feedwater into the boiler is automatic. Feedwater will be delivered from a pump with on–off operation or from a continuously running pump delivering water through a modulating feed control valve. In both cases the water requirement is continuously monitored by level controls directly connected to the shell or drum of the boiler.

In addition to controlling the feedwater, the level controls will also monitor low- and high-water conditions. It is essential that all heated surfaces are always fully immersed, therefore two stages of low-water alarms are fitted on two independently mounted controls. The first low water will give audible and visual indication and shut down the combustion appliance. If normal water level is automatically restored it will allow the boiler to go back to firing mode. If the second low water is reached again, audible and visual alarms are instigated, but now the boiler will go to lock-out condition, requiring manual restart after the problem has been resolved. Watertube boilers may have additional monitoring controls to give early indication of load changes as referred to in Section 7.7.

7.16.3 Blowdown

Blowdown on a boiler is mandatory. On small boilers the required operation of the main blowdown valve may be sufficient to control the quality of water within the boiler. On medium and large plants additional systems are employed.

The simplest is a preset continuous blowdown valve to maintain a suitable water quality in the boiler. It is necessary for water quality to be checked frequently and the rate of continuous blowdown adjusted as may be found necessary.

A second method is a time-controlled valve allowing regular intermittent blowdown of the boiler. Again, regular checks need to be carried out to monitor the quality of water in the boiler.

The third and most automatic system is the conductivity-controlled blowdown. This constantly measures the level of solids in the water and instigates an automatic variable blowdown on a continuous or intermittent basis.

7.17 Automatic boiler start

In order to control the operating times of a boiler it is a simple matter to fit each with time clock control. Alternatively, they may be controlled through a central energy management system. Either way, a boiler or boilers may be shut down at the end of each day and programmed to restart the following day or when required. Special considerations need to be made if standing periods are extended, allowing a boiler to go cold.

With hot water units time clock control can operate satisfactorily as automatic bypass valves built into the distribution system will help the heater to achieve its working temperature quickly. With steam boilers it is important that the boiler reaches a reasonable working pressure before steam is allowed into the distribution system. For example, if boilers are left open to a system for an extended length of time while not firing they will quickly lose their pressure. This is not only wasteful of energy but eventually creates a problem on start-up. To start a boiler on a zero-pressure system with all valves open will undoubtedly cause the boiler to prime and go to lock-out condition, but not before condensate has at least partly flooded the system.

Therefore where a time clock is incorporated it is recommended that the crown valve(s) be closed at the end of each working day and only opened after the boiler has reached working pressure the next time it is required. This operation can be automated by the use of motorized or similar valves. These valves may be fitted to each steam supply line from a manifold adjacent to the boilers and, provided adequate safeguards are incorporated to protect the boilers, the on-line boiler(s) may be left open to the manifold. Alternatively, each boiler may have its own automatic motorized start-up valve. Figure 7.12 illustrates a valve suitable for either system.

Each valve would have a control panel incorporating a timer. This may initially be set to a 'crack' position timed to open after the boiler has started to fire and is already building up pressure. This will allow gentle warm-up of the system while enabling the boiler to achieve working pressure. After this, the valve may be set to open in timed adjustable steps to its fully open position. At the end of a timed period, coinciding with the time clock fitted to the boiler, the valve automatically closes at the end of the working cycle. Where multiple valves are used their control may be incorporated into a single panel or, alternatively, become part of an energy-management system.

7.18 The automatic boiler house

As described in Section 7.16, the boiler will operate automatically and we saw in Section 7.17 that it can be programmed to operate to suit various cycles. There are, however, other areas within the boiler house which still require a degree of attendance.

The first of these is blowdown control. Assuming that control of the water quality within the boiler is being carried out by a separate system, we are left with the main blowdown and blowdown of the water-level controls and gauge glasses. It is possible to automate these valves but it is unusual. Before proceeding, the advice of the insurance company must be sought and agreement reached on the operation of the valves. When blowing down the level controls it should take the boiler to lock-out. If this, in turn, is overridden, a further proving system will need to be incorporated to prove correct functioning of the level control. In addition to the above, it will still be necessary to carry out an evaporation test on the boiler to prove correct operation of the level controls at defined periods.

The second area will be feedwater pumps. It is normal to have a duplicate standby feedwater pump. Sometimes this may be two for each boiler or one duplicate pump to serve any selected boiler. These will usually require manual changeover in the event of failure of the duty pump. It is practical to automate this changeover by using pressure sensors and motorized valves. The same can apply to oil-circulating pumps, gas boosters, water-treatment plant and any other valves and motors. It is possible to do most things, but in the end there is the cost to be considered. An energy-management system coupled to a suitable maintenance programme can still prove the best option for the great majority of installations.

7.19 Safe operation of automatic boiler plant

The most frequent cause of damage and even explosion in boilers is a low-water condition. This will expose the

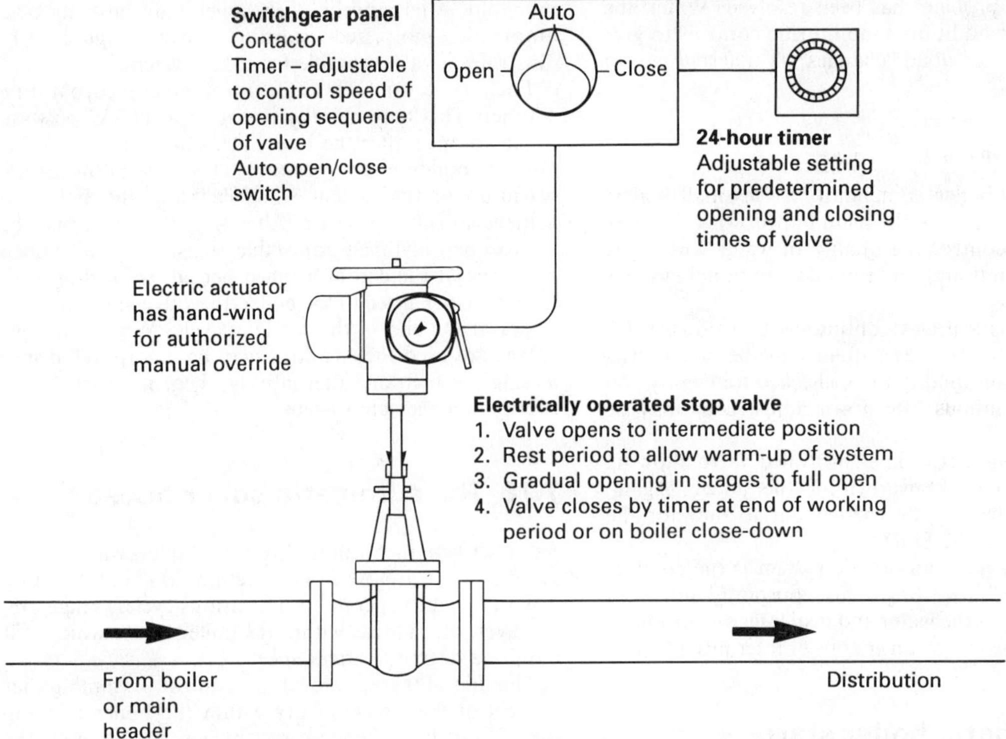

Figure 7.12 Diagram of electrically operated automatic steam stop valve (with permission of Hopkinsons Valves Ltd, Huddersfield)

heating surfaces which ultimately overheat and rupture under the operating pressure. Experience has shown that since the introduction of controls for unattended automatic operation of boilers the accident rate has increased. Investigation invariably shows that lack of maintenance has been the main contributing factor. It is therefore imperative that personnel responsible for the running of the boiler plant be fully trained and conversant with its safe operation.

The minimum recommended requirements for automatic controls for boilers not under continuous supervision are:

1. Automatic water-level controls arranged to positively control the flow of water into the boiler and maintain a set level between predetermined limits;
2. Automatic firing controls arranged to control the supply of fuel and air to the combustion appliance. A shutdown will occur in the event of one or more of the following occurring:
   (a) Flame/pilot flame failure on oil- or gas-fired boilers. The control should be of the lock-out type requiring manual reset;
   (b) Failure to ignite fuel on oil- or gas-fired boilers within a set time. Again, lock-out type with manual reset;
   (c) At a preset high pressure (or temperature in the case of hot water) at or below the safety valve setting;
   (d) When the water level falls to a preset point below the normal working water level. This control will operate on audible alarm but will automatically go back to firing mode once the water level is re-established;
   (e) Failure of combustion air fans or automatic flue dampers;
3. An independent overriding second low-water control. This will be set below the level control in (2). It will give an audible alarm on shutting off the fuel and air supply and require manual reset before the boiler may be brought back into operation;
4. All electrical equipment for the water-level and firing controls should be designed so that any fault will cause the boiler to shut down and require manual resetting before the boiler may be restarted.

For hot water boilers the burner controls will be similar but controlled by a combination of pressure and temperature signals. A single overriding level control will be fitted to the flooded boiler to protect against any accidental low-water condition.

In order to monitor the safe operation of the boiler a daily and weekly programme of tests should be drawn up and log sheets completed as verification of the tests being carried out. Items checked should include:

Boiler pressure;
Water level gauges – visual and blowdown;
Sequence valves opened and level controls operated;
Feedwater pump operating;
Operate main blowdown valve;

First low-water alarm and burner off;
Second low-water alarm and burner lock-out.

When a boiler may not be shut down for maintenance of the level control chambers isolating valves can be fitted between the water-level control and the steam space. In this instance, the valves must be capable of being locked in the open position and the key retained by a responsible person. When these valves are closed during maintenance periods the boiler must be under manual attendance. Fitting of these valves should only be with the agreement of the insurance company responsible for the boiler. Drains from the water-level controls and level gauges should be collected at a manifold or sealed tundish before running to the blowdown vessel.

As the boilers are designed for unattended operation, it follows that if a fault occurs and an alarm activated this alarm may need to be duplicated at a secondary location where it will be intercepted and acted upon.

Safety valves should be trouble-free. No attempt should be made to alter the set relief pressure without reference to the insurance company and the boiler manufacturer.

With waste-heat boilers fitted to incinerators care must be taken not to overfire. It is possible to introduce additional heat either by increasing the quantity of waste or by a change in the composition of the waste. The resultant increase in gas volume and/or temperature is then capable of imparting more heat to the waste-heat boiler. As the boiler will have been designed for a specified duty it could be possible to raise an amount of steam in excess of that which may be safely controlled.

Sometimes it is not practical to blowdown the level controls and shut down the incinerator. In this case, the situation should be discussed with the insurance company and the boiler supplier. It is possible to include for an extra high working water level giving a safety margin above the heating surfaces. The controls may then be blown down and checked for satisfactory operation with a predetermined time delay before it shuts down the incinerator or operates a bypass in the event of a fault.

7.20 Energy conservation

Energy conservation in the boiler house can be considered in two areas. One is the selection and installation of suitable equipment and the other is good operation and management.

7.20.1 Plant installation

The boiler, flues and chimney, pipework and hotwell, where installed, should all be insulated to adequate standards and finish. Valves should be enclosed in insulated boxes, although on small installations this can prove disproportionately expensive. The boilers may be fitted with either inlet or outlet air-sealing dampers. These will prevent the flow of ambient air through the boiler during off-load and standby periods, thus helping to maintain the heat already in the boiler.

Economizers may be installed, particularly if gas is the main fuel. It is unlikely that an economic case can be made for a single boiler if there is less than 4000 kg/h evaporation. An economizer can produce fuel savings of 4–5%, but it must be remembered that this will be at MCR, and if the load factor of the installation is lower, the savings will also be proportionately lower.

Combustion controls such as oxygen trim help to maintain optimum operating conditions, especially on gaseous fuels. Instrumentation can give continuous visual and recorded information of selected boiler and plant functions. To be effective, it must be maintained and the data assessed and any required action taken before the information is stored.

Energy-management systems will form an important part of a multi-boiler installation, whether on steam or hot water. Boiler(s) for base load will be selected and further boilers brought on- or taken off-line as required. The important feature of these systems is that the selection of boilers coming either on- or off-line will be ahead of the load and programmed to anticipate rising or falling demands.

Computer monitoring and control systems have recently been introduced. These are designed to operate in place of conventional instrumentation. Using intelligent interface outstations connected to a desktop computer, many plant functions may be programmed into the computer and controlled centrally.

7.20.2 Operation and maintenance

As most boiler plants installed today are designed for unattended operation it is even more important that early action is taken in the event of the boiler requiring adjustment of combustion or other maintenance. If full instrumentation is not installed then a portable test kit should be used and the plant checked and logged daily or weekly. Perhaps the most obvious waste to look for after steam leaks is a rise in the flue-gas outlet temperature. The boiler will progressively have deposits adhering to its heating surfaces, but at an increase in temperature of no more than 16C° above its design outlet temperature it should be cleaned. The time period between cleaning will vary according to the type of fuel and operational load.

Comparatively small air leaks into the combustion spaces of a boiler can produce localized problems. All access opening seals and sight glasses and seals must be monitored to prevent this.

7.21 Noise and the boiler house

Noise can constitute a danger to health, and therefore adequate precautions must be taken to protect personnel who are required to be in such an environment. The Health and Safety at Work, etc. Act 1974 has the power to control noise emissions, but the subject is complex. If it is anticipated that noise will exceed acceptable levels then specialist knowledge should be employed at an early stage in the design and layout of the plant.

As a guide, Figure 7.13 shows a noise-rating curve for

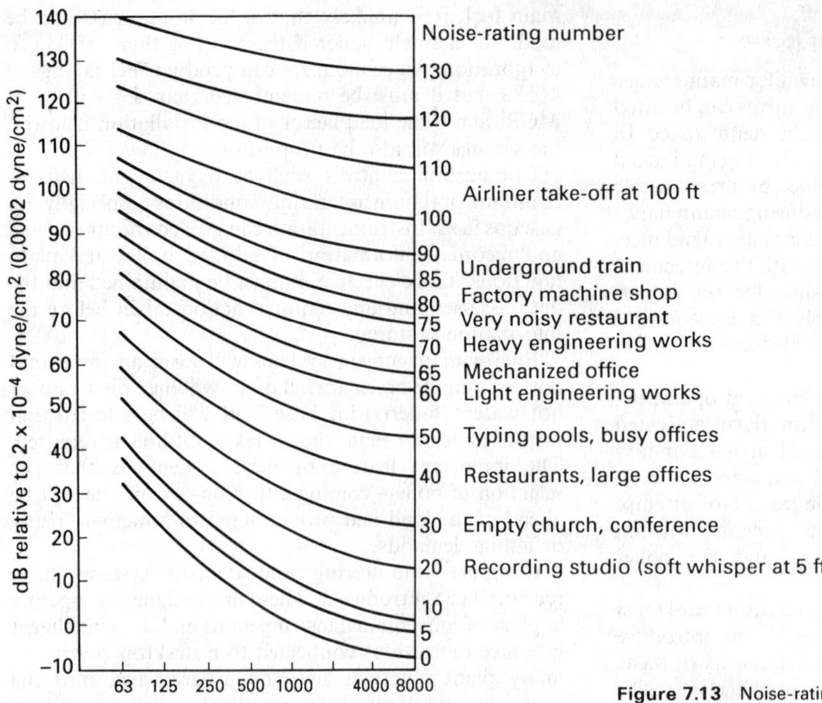

Figure 7.13 Noise-rating curves for some everyday sounds (with permission of Saacke Ltd, Portsmouth)

everyday noises. At the lower sound levels these may be considered acceptable, and it will be found that to produce overall sound levels below this will prove costly. Alternatively, sound levels exceeding the upper figures need careful consideration if personnel are required to be in that environment. For some machinery, such as reciprocating engines, turbines and compressors, it is not practical to contain the noise at source and then ear protection has to be utilized. Generally, however, industrial boilers and their associated plant can be designed to meet acceptable noise levels.

It is generally the case that small boiler plants will create less noise than large ones. Nevertheless, due to the practice of siting small boilers close to working areas, the matter of noise must not be overlooked. Noise sources within the boiler house will be from the following:

Combustion appliance, including forced- and induced-draught fans;
Feedwater pumps;
Oil pumps;
Gas boosters;
Water-treatment plant;
Steam lines and PRV stations;
Safety valve vent pipes;
Gas lines;
Flue ducting.

Small oil and gas burners are generally supplied without any form of silencing. Some may be enclosed within an acoustic hood and air inlet silencers are normally available if specified. Oil and gas burners fitted to boilers of outputs above 3000 kg/h evaporation can be supplied with progressive degrees of silencing equipment. Figure 7.14 illustrates sound-pressure levels of one burner manufacturer when fitting three basic types of silencing. The use of variable-speed motor drive for the combustion air fan can also reduce the overall noise level, specifically when the unit is not operating at maximum continuous rating. Coal-fired plant will normally be quieter than oil- or gas-fired plant, although, where fixed grate boilers are employed, the same criteria will apply to the forced-draught fans as for oil and gas.

Feedwater pumps will not normally constitute a noise problem unless the area is particularly sensitive. Two alternatives are then available. One is to install reduced-speed pumps and the other is to site the pumps in a separate acoustic enclosure within the boiler house. Oil-circulating pumps are usually low speed and, as such, do not cause a noise problem.

Gas boosters may be fitted with motor enclosures or designed to operate at lower rotating speeds. Alternatively, they may be housed in an acoustic enclosure within the boiler house. The booster and drive unit can be supplied with anti-vibration mountings to isolate it from the floor or steelwork and flexible bellows fitted to the gas inlet and outlet connections.

Water-treatment plant may include one or more electric motors driving feedwater pumps or dosing pumps. Dosing pumps are usually very small and will not cause noise problems. Water pumps may be considered as for the feedwater pumps already described.

Steam transmission will produce noise. Low steam

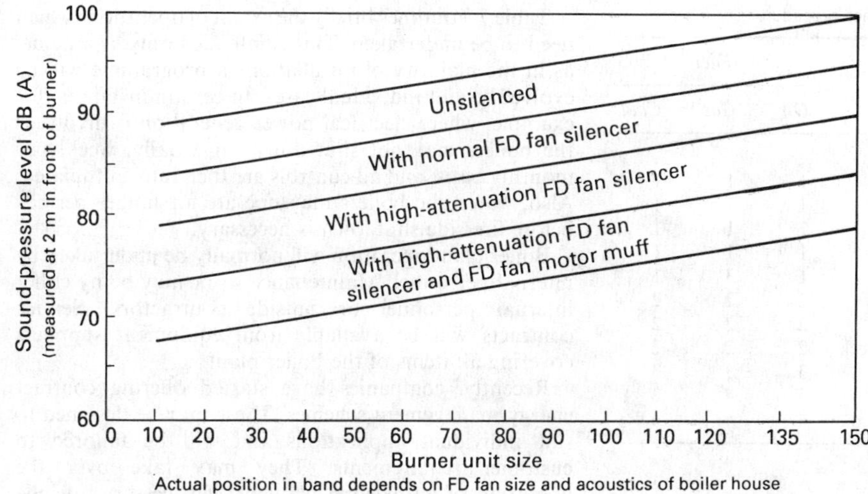

Figure 7.14 Sound-pressure levels for Saacke rotary cup burners (with permission of Saacke Ltd, Portsmouth)

velocity will help to reduce this. One specific noise area will be if a pressure-reducing valve set is employed. Therefore it is desirable to have this within the boiler house or other building and not outside. Some types of pressure-reducing valves operate at lower sound levels than others, and therefore if environmental noise could be a problem then this should be investigated. Insulated valve boxes will also help to contain any noise.

Safety valve venting is noisy. It should only happen infrequently and therefore may be acceptable. If the location is near to offices or housing it may be advisable to fit silencer heads. Noise levels from gas lines and meters should not normally prove a problem on industrial premises.

Flue gas ductwork will be insulated, which helps to reduce any transmitted combustion noise. Where large rectangular flue ductwork is used it must be adequately stiffened and insulated to reduce any risk of reverberation being set up from the combustion system.

Finally, if the boiler house is to be within an industrial complex and there is no adjacent residential development, the building need only be suitable for adequate weatherproofing of the plant and for maintenance. Alternatively, if it is sited adjacent or near to housing or offices then the building structure should take account of this.

Ideally, the walls should be of brick or concrete with an insulated roof. Large areas of glass should be avoided. It is desirable that all routine maintenance and operating functions can be carried out without having to open doorways. Openings and access to the boilerhouse should be away from sensitive areas. Free areas for air for both combustion and ventilation should be through acoustic louvre panels and preferably again away from sensitive areas. Discharge points for vehicles delivering fuel should be carefully considered so as to cause the minimum nuisance.

7.22 Running costs

With the wide range of boiler types and outputs being considered together with the ever-changing costs for fuel it is difficult to arrive at a specific figure which can relate running costs to the installed plant. In order to arrive at a guide in this area several case studies were carried out. It quickly became apparent that few operators maintain separate costs relating to the boiler plant and only keep overall factory maintenance costs. Fortunately, sufficient information was obtainable to arrive at an overall pattern of costs.

On an annual basis, with running costs covering fuel, operation, electricity, water treatment and maintenance, fuel can account for 90% on a plant operating 168 h per week and 80% on a plant running 40–80 h per week, the remaining 10–20% covering the other items. These proportions were consistent for modern plant burning coal, oil or natural gas. With the recent increase in natural gas tariffs (1989) for large consumers it will now mean that natural gas will account for a larger percentage of the annual figure.

An alternative cost also became apparent from the case studies. If the mean hourly steaming rate in kilograms is multiplied by 2.25 it produces a figure roughly equivalent to the annual operating costs in pounds, excluding the cost of fuel.

7.23 Management and operation

The boiler plant, regardless of its size, is an essential part of the plant otherwise it would not be there. Having established this, it should be covered by a planned maintenance schedule to produce maximum availability. Most boiler plant is supplied for unattended operation,

Table 7.7 Recommended maintenance procedure

Operation	*Fuel*		
	Oil	*Gas*	*Coal*
Blowdown			
Main	1	1	1
Water-level controls	1	1	1
Water-level gauges	1	1	1
Visual check	1	1	1
Check/clean burner	1	3	1
Water-treatment check	2	2	2
Combustion check	2	2	2
Water-level control operational check	2	2	2
Inspect refractories	3	3	3
Open fire side and clean	3	4	3
Open water side for inspection	4	4	4

1: Daily.
2: Weekly.
3: 6–12 weeks.
4: Annually.

but certain procedures still need to be carried out. The extent of this programme will depend upon the size of the plant and the fuel used.

Table 7.7 outlines briefly the extent of operations which need to be undertaken. This is intended only as a guide, as in the majority of installations a programme will be evolved to suit individual cases. In certain instances (for example, where electrical power generation is involved) the boilers may not shut down on a daily, weekly or monthly basis, and all controls are then fully automated. Also, watertube boilers may operate for longer periods before fire-side shutdown is necessary.

Boiler plant operation will normally be undertaken by internal personnel. Maintenance work may be by either internal personnel or outside contractors. Service contracts will be available from equipment suppliers covering all items of the boiler plant.

Recently, companies have started offering contract energy-management schemes. These may be designed to suit individual applications and will be tailored to customer requirements. They may take over the operation of an existing plant or, if necessary, include for a new replacement plant. They will usually operate over a 3–10-year contract period. Dependent upon the terms of contract, all fuels, electricity, repairs and replacements may be covered.

8 Combustion Equipment

Colin French CEng, FInstE, FBIM
Saacke Ltd

Contents

8.1 Introduction

A burner is a device for liberating heat by the combustion of fuel. Fuels are predominantly hydrocarbons which release their heat exothermically when oxidized in a controlled manner. The most freely available oxidant is air, which contains only 21% oxygen, the remaining 79% being nitrogen which does not contribute to the process. The nitrogen, because it is heated at the same time, reduces the maximum flame temperature that would have been possible with a pure oxidant. Similarly, the combustion products, when discharged from the process, contain nitrogen which increases the volume of the gases and hence the sensible heat loss.

A burner, then, comprises a means to inject the fuel, a fan to provide the air for the combustion reaction, a register or stabilizer assembly which provides for the mixing of air and fuel and the stability of the flame, and a means for controlling the air–fuel ratio and fuel input. In addition, on automatic burners, a management system is necessary to ensure programmed start-up and shut-down together with supervision while firing via a flame-detection system and interlocks to prove that certain parameters are maintained such as air and fuel pressure. An almost infinite number of types of burner have developed over the years, but broad categories exist characterized by the type of fuel being burnt, the principle of the fuel injection and mixing system, and often the application for which they were designed.

Although some integration has taken place where the applicance or boiler maker has assumed responsibility for the combustion system, on the whole, the market is served by specialist manufacturers of combustion equipment who have developed products for each application such as boilers, furnaces, kilns and dryers, etc. The burner makers have manufactured products which provide a packaged solution to the combustion requirement, looking after not just the burners and controls but also the fuel supply system, which may involve pumping, heating of the fuel, filtration and other peripheral equipment and functions.

8.2 Oil burners

With the exception of the vaporizing burner, these are normally characterized by the method of atomizing, which itself is dependent on the grade of fuel being combusted.

8.2.1 Vaporizing burners

This principle is confined to domestic applications where kerosene or premium gas oil is concerned. The simplest type uses a number of concentrically arranged wicks which promote vaporization of kerosene into an air/vapour mixing zone enclosed within a perforated drum arrangement. Normally, these burners obtain their air by natural draught.

Another type utilizes a pot and may have natural draught or a fan. The pot burner is essentially an open-topped drum into which fuel is fed at constant head. The vapour rising from the surface is mixed with air being discharged from a perforated drum or pot (Figure 8.1). A further development is the wall flame burner in which the fuel is thrown against the wall of the pot by a central rotating distributor which is driven from the forced-draught fan motor.

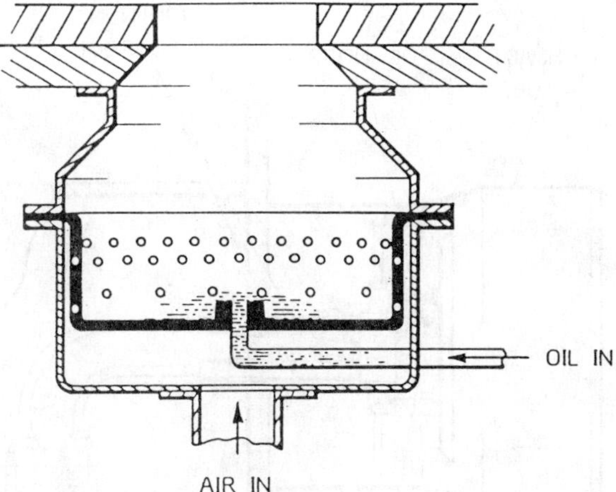

Figure 8.1 Vaporizing pot burner

8.2.2 Pressure-jet atomizers

Oil is fed at high pressure to a nozzle in which the oil passage is positioned to feed oil radially inward via a number of slots which are arranged at a tangent to the swirl chamber. The high rotational velocity given to the oil as it exits at high pressure through the central discharge hole provides the means for droplet disintegration via a conical sheet formed at discharge. Limitations of this principle include restriction to gas oil for small sizes and poor turndown caused by a limited range of pressures over which the atomization is satisfactory. There is a choice of spray pattern, notably solid, semi-hollow and hollow cone, and a reasonable range of spray angles is available (Figure 8.2), often used in a

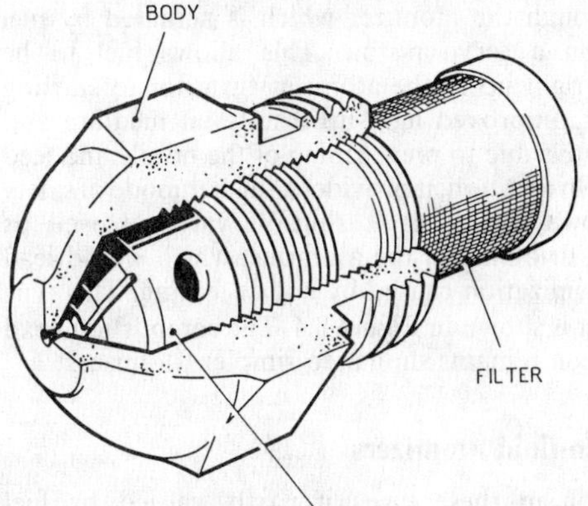

Figure 8.2 Pressure-jet nozzle

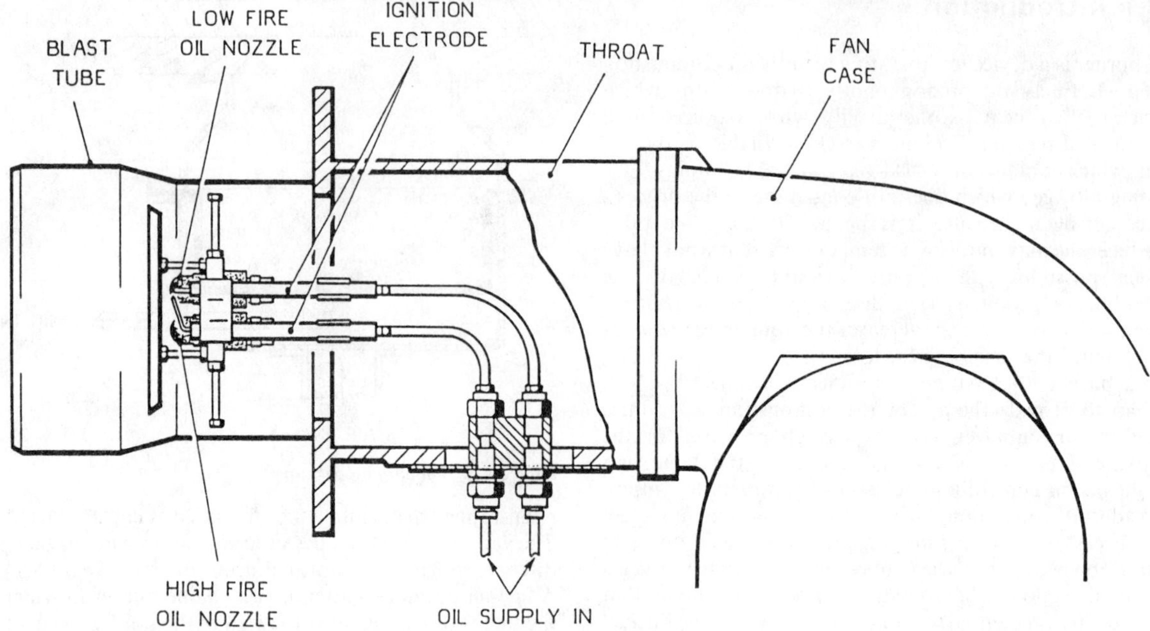

Figure 8.3 Combustion head pressure-jet burner

two-nozzle head configuration to improve the turndown ratio from 1.4:1 for a single nozzle to 2:1 using two nozzles at constant pressure. The orthodox arrangement for the combustion head is shown in Figure 8.3. The simplex pressure-jet atomizer is also used in power station boilers firing heavy fuel oils in arrays of up to 60 burners.

8.2.3 Spill return atomizers

These partially overcome the weakness of the simplex pressure jet regarding turndown ratio by spilling back the unconsumed fuel at part load. In this way, the swirl velocity in the exit chamber is maintained constant but the diameter of the exit hole remains the same. A further advantage is that it is possible to add a central shut-off needle through the atomizer which is actuated by fuel pressure on a servo piston. This allows fuel to be circulated right up to the atomizer tip prior to starting the burner. Improved light-up results on medium and heavy fuel oils due to prewarming of the nozzle and feed pipework. In addition, it provides a further mode of safety on shutdown acting as a shut-off valve as well as preventing dribbling of the atomizer, which would lead to poor atomization caused by nozzle fouling. This type of atomizer is shown in Figure 8.4. The combustion head configuration remains similar to simplex atomizers.

8.2.4 Twin-fluid atomizers

Atomization in these types if partly caused by fuel pressure, but this is enhanced by the kinetic energy provided by another fluid which is normally air or steam. At present, this secondary fluid is at a medium or high pressure, the low-pressure method being largely superseded. Pressures are around 1–2 bar for those categorized as medium pressure and 6–10 bar for high-pressure types. Oil pressures are also typically 6–12 bar.

Steam is the preferred atomizing medium, since it is more economic than compressed air. Steam consumption is typically less than 0.5% of the fuel burnt on a mass basis, although this rises in direct proportion to turndown ratio. On very large burners the steam flow is modulated in proportion to fuel burnt. Turndown ratios range from about 5:1 for small shell boilers to 12:1 in watertube applications, making this one of the most versatile burners. The steam condition is important in that it must be dry saturated or slightly superheated at the nozzle to avoid condensate formation. On small or non-continuously running plant where no steam is available for start-up a compressed air supply must be provided until steam becomes available from the boiler.

Possibly the best-known version of this principle is the Y-jet atomizer developed by Babcock Energy, in which between four and ten exit holes are arranged circumferentially, each consisting of two converging passages arranged in a Y formation. Possible limitations of the principle in spray angle and pattern have restricted its use to larger boilers and watertube types. A cross section of this nozzle is shown in Figure 8.5 and a typical register arrangement is given in Figure 8.6 for a dual-fuel burner.

8.2.5 Rotary cup atomizers

A shaft rotating at 4000–6000 rev/min carries a primary air fan and an atomizing cup. The cup, typically of about 70–120 mm diameter, is tapered by a few degrees to

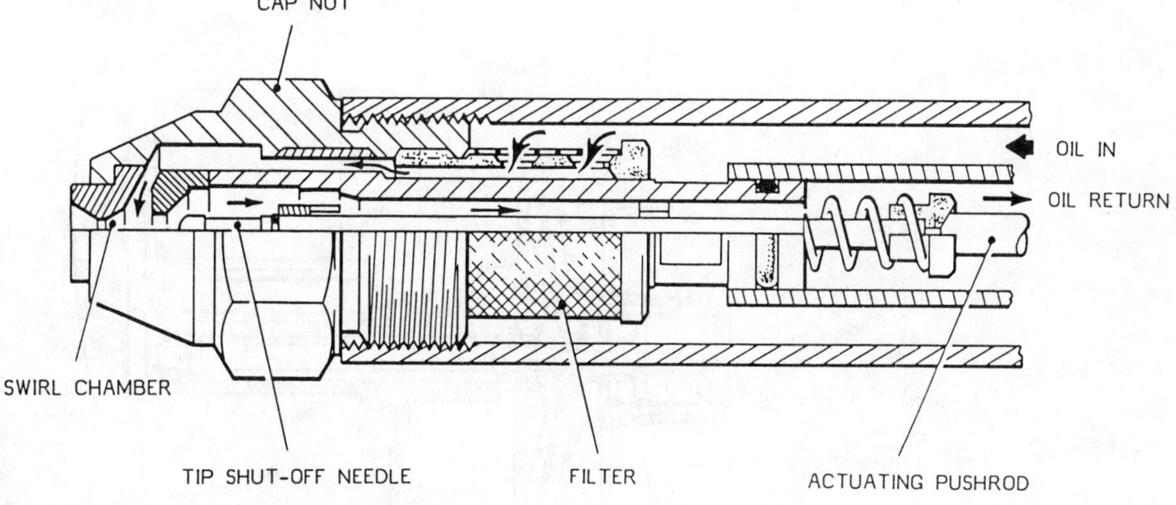

Figure 8.4 Spill return nozzle with tip shut-off needle

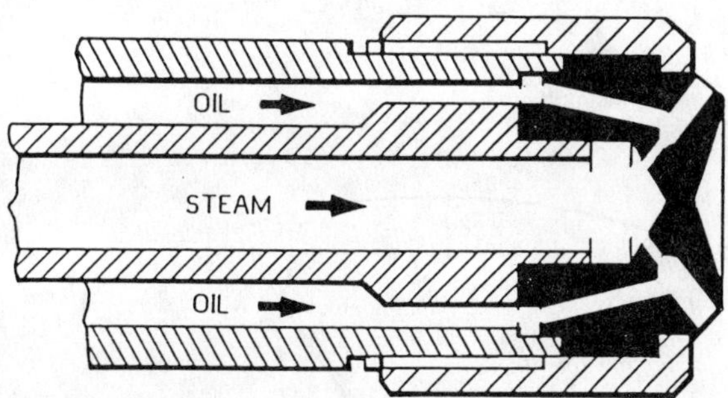

Figure 8.5 Steam-atomizing nozzle

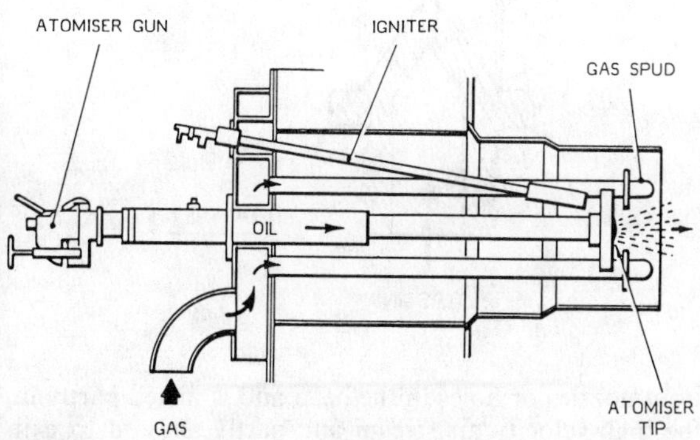

Figure 8.6 Steam-atomizing burner register

increase in diameter at the exit. Oil is fed to the inner surface by a stationary distributor which projects oil onto the smaller-diameter end of the cup. The oil, influenced by centrifugal force, forms a thin film which passes towards the cup lip. Atomization occurs as the oil leaves this lip. In addition, a primary air supply, normally in the range of 5–12% of stoichiometric (chemically correct) air, is arranged to exit over the cup outer surface, at a velocity of about 50–90 m/s. The primary air is swirled to oppose the rotation of the cup. Droplets shattered by the combined centrifugal action of the cup and the primary air blast are propelled axially into the furnace.

Advantages of this type include an ability to burn all fuels including those containing solid particles, good turndown ratio (4 to 10:1 typically) and an insensitivity to oil conditions such as pressure and temperature. It is widely used in shell boilers, and the only real limitation is that the cup surface has to be cleaned daily. The most common atomizer layout is shown in Figure 8.6. Variants include direct driven cup and separate mounting of the primary air fan.

8.3 Gas burners

Industrial gas burners are mainly of the nozzle mix configuration. Beneath industrial burners, which are used for raising steam and hot water in the power and process industries, lies a large array of types and principles. The most common types are normally characterized as to whether they are aerated or non-aerated.

8.3.1 Non-aerated types

Otherwise known as diffusion flame or post-aerated, these normally comprise a simple nozzle supplying gas at a controlled pressure into a chamber where air is made available via entrainment into the flame by natural draught. Common types are the Bray jet, Aeromatic and Drew jet (Figure 8.8).

8.3.2 Aerated types

These are otherwise known as atmospheric or premix burners. Primary air is entrained into the gas stream prior to exit from the nozzle. The most well known of

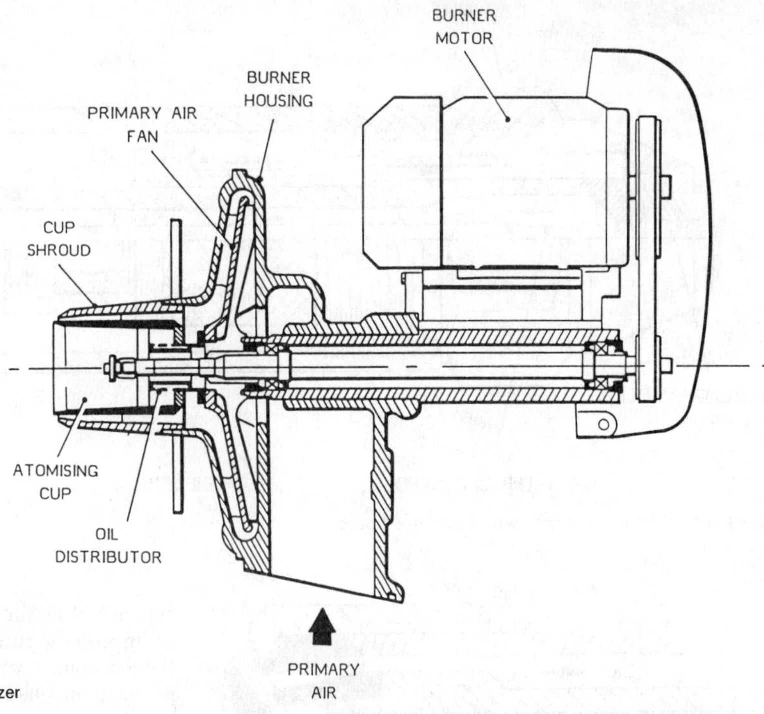

Figure 8.7 Section through rotary cup atomizer

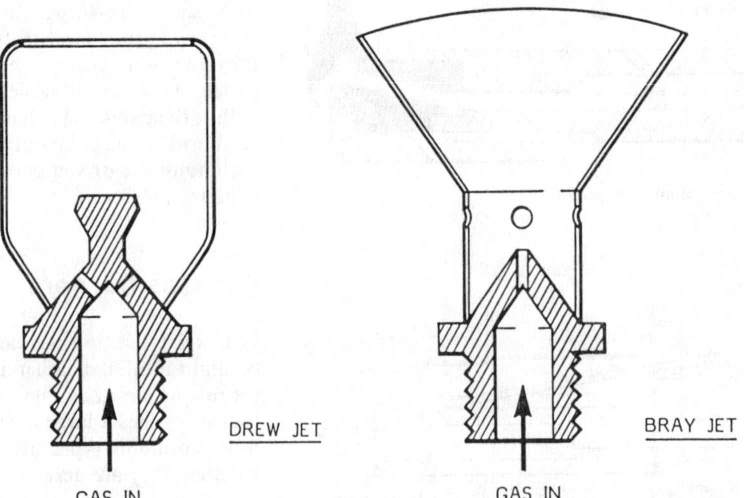

Figure 8.8 Non-aerated burners

these types is the Bunsen burner, and the most common is the ring-type domestic cooker hob arrangement. Both aerated and non-aerated types are often found in a bar configuration. Typical applications are heating of tanks and process uses involving direct heating of the product. Figure 8.9 shows a typical aerated bar burner.

8.3.3 Nozzle-mix types

Utilizing a forced-draught fan, the burner has a gas head arranged to mix the fuel and air in a blast tube which controls the stability and shape of the flame. Gas exits from nozzles or holes in the head and is mixed partly in the high-velocity air stream and partly allowed to exit into an area downstream of a bluff body. Behind the bluff body a relatively quiescent zone forms which provides a means for flame stability. Many configurations exist, but the most frequent are those which are designed around the most common types of oil burner. This allows for the burner to easily be converted to oil or dual-fuel (gas and oil) firing. As gas is a relatively easy fuel to burn, the design is strongly influenced by the optimum oil burner configuration. Two typical types are shown in Figures 8.10 and 8.11 based, respectively, around pressure-jet and rotary cup atomizers.

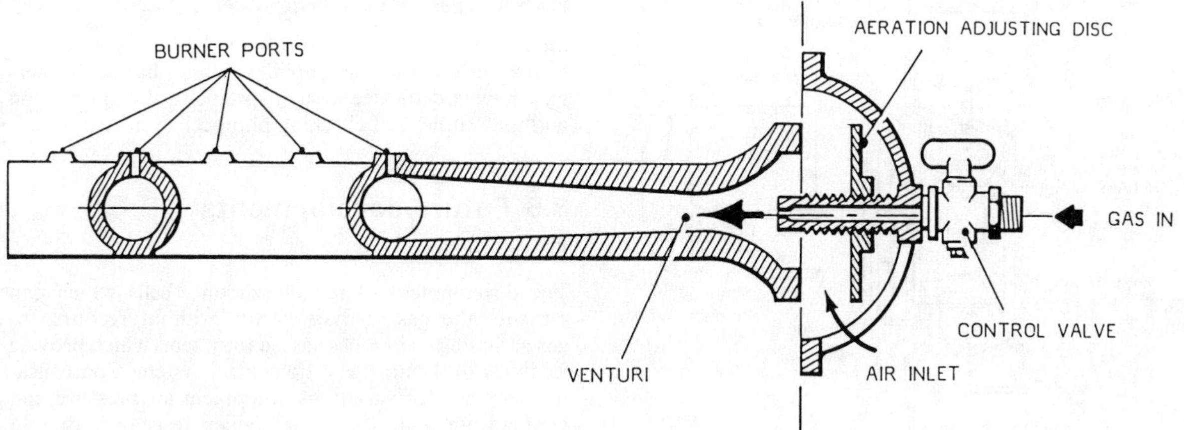

Figure 8.9 Aerated bar burner

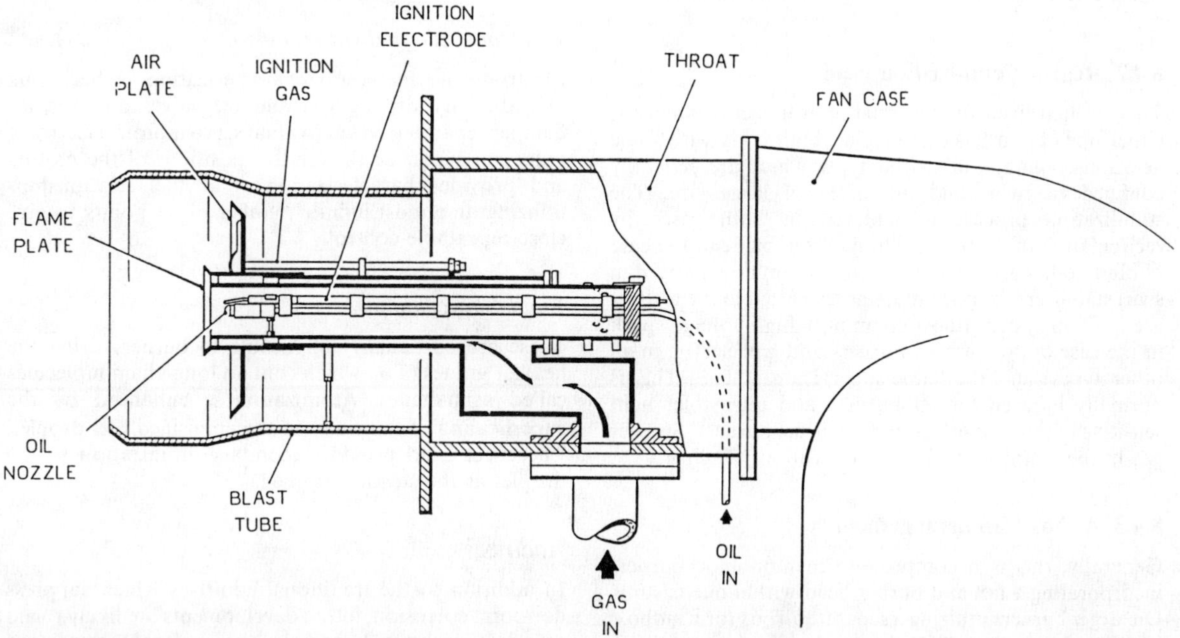

Figure 8.10 Dual-fuel burner based on pressure-jet configuration

Performance on gas is normally limited in dual-fuel applications to that of the oil burner. In gas-only applications the performance is better, notably in lower excess air factors and better turndown ratios (3:1 small burners, 15:1 very large burners).

Problems can occur with highly rated boilers converted to gas firing where tube end and tube plate cracking can occur due to overheating. This can normally be overcome by modifying the tube attachment arrangement. Another problem that is quite common is resonance or pulsation, where the burner acoustically couples with the natural resonant frequency of the combustion chamber. This is normally easily overcome by modifying the burner head to change the rate of mixing.

8.4 Burner design considerations

8.4.1 Atomizer

Characteristics of various atomizers are given in Table 8.1. Primary considerations are selecting the best principle for the type of fuel, the size of the burner/boiler and the type of application. Other important characteristics are ability to operate with the minimum of excess air, turndown ratio and questions of durability and maintenance.

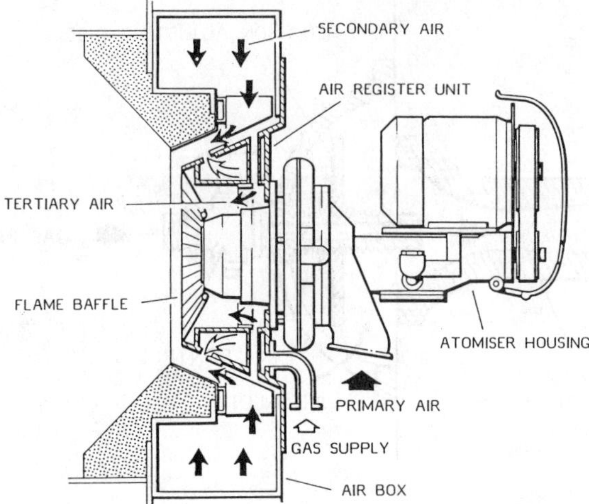

Figure 8.11 Dual-fuel burner based on rotary cup configuration

8.4.2 Register/combustion head

This is often fixed for the type of atomizer. Stability is either one of two basic principles: bluff body with some aeration/cooling, and swirler types which are generally confined to twin fluid atomizers of large size. The stabilization process is achieved in both cases by recirculation of vortices spilling off the baffle in the case of bluff bodies and by a full recirculation flow pattern in swirl stabilizers. Important items also fitted to the register are ignition system (most commonly high-voltage spark in the case of pressure-jet burners and gas/electric in all other types) and the flame-supervision system. This is normally infrared for oil burners and ultraviolet light sensitive in the case of gas and dual-fuel burners. Smaller gas burners utilize the flame-rectification principle.

8.4.3 Airbox/fan arrangement

Generally, this is a compromise in monobloc burners incorporating a fan and burner head within one casting. On larger burners utilizing a separate airbox (or windbox, as it is sometimes called) the air-distribution quality is important to maintain low excess air operation and good flame shape throughout the turndown range. Two approaches are common: designs utilizing a constant velocity approach, the alternative being a large plenum chamber in which low velocities provide an equalization effect. The shape of the flame can deteriorate at part load due to the register pressure reducing on a square law with turndown.

8.4.4 Air/fuel ratio control

This is often the Achilles' heel of burners, since poor design can lead to hysteresis. Correct sizing of the control valves and fan size is essential to maximize damper travel/backlash ratios and give good linearity throughout the firing range. Robust characterization cams and linkages are essential.

8.4.5 Burner management system

Small monobloc burners usually have a proprietary control box/photo cell amplifier system. Larger burners may have a dedicated system specific to the application and may utilize self-checking photocell systems.

8.5 Future developments

Oxygen trim

The development of reliable zirconia cells which can measure the gas analysis *in situ* without recourse to gas-sampling techniques has led to systems which provide feedback to the air/fuel ratio control system. Compensation is made for variations in ambient air pressure and temperature, calorific value, boiler resistance due to fouling and burner performance drift by trimming the air damper with a separate servo motor.

Electronic air/fuel ratio control

Electronic air/fuel ratio characterization is becoming available. By driving gas and oil valves and the air damper separately via individual servo motors, electronic units can supervise the relative positions of the motors and provide characterization of air/fuel relationships utilizing an almost infinite number of set points to give close repeatable control.

Emulsion techniques

These are particularly applicable to burners firing the heavier grades of oil which contain long-chain molecules called asphaltenes. Atomization is enhanced by the superheating of the water in the emulsified fuel droplet. The effect is to provide secondary atomization to the droplet as the steam is formed.

Additives

In addition to the traditional additives which suppress dewpoint corrosion, future developments are likely to aid atomization by reducing droplet surface tension and stimulating combustion catalytically.

Air and exhaust gas sealing dampers

These positively shut off the draught which naturally occurs due to chimney buoyancy when the burner is in its off cycle, thereby reducing standby losses. Burners incorporating shut-off dampers are becoming increasingly common.

Emission control

Future legislation will stimulate burner development in the areas of carbon monoxide, No_x and particulate generation. Techniques will include flue-gas recirculation, staged combustion, additives to reduce the No_x and more sophisticated controls. Controls over the sulphur

Table 8.1 Summary of burner characteristics

Oil burners

Type	*Size range*	*Dual-fuel capability*	*Fuel type*	*Atomizing pressure* (bar)	*Atomizing viscosity* (cS)	*Turndown ratio*	*Flame characteristics*	*Main applications*
Vaporizing	10–40 kW	No	Kerosene, gas oil	—	<5	On/off	Normally blue flame	Central heating
Pressure jet commercial market	30 kW–3 MW	Larger sizes	Gas oil (class D)	6–12	<5	On/off or <1.5:1 by pressure control	Soft yellow and radiant fairly wide-angle spray	Domestic and commercial hot water boilers
Pressure jet utility boilers	10–50 MW	Possible but gas not burnt on utility boilers	Heavy oil (classes G and H)	20–40	~20	On/off or <1.5:1 by pressure control	Highly radiant low excess air operation	Power station and large petro-chemical water-tube boilers
Twin-fluid medium pressure	1–10 MW	Yes	All types (mainly class G)	1–4	~20	<5:1	Wide range of shapes mainly used for process applications	Kilns, furnaces, processes requiring special flame characteristics
High pressure	2–50 MW	Yes	All types (mainly classes G and H)	6–20	~20	3:1 to 12:1	Mainly wide-angle sprays in WT boilers, low excess operation	Large shell boiler and watertube boiler process applications
Rotary cup	1–40 MW	Yes	All types (mainly class G)	2–3	~60	3:1 to 15:1	Medium intensity, shape varied by register design	Very popular for all sizes of shell boiler. Some use on process and WT boilers

Gas burners

Type	*Size*	*Dual-fuel*	*Fuel type*	*Gas pressure*	*Turndown ratio*	*Flame characteristics*	*Main applications*
Non-aerated	0.01 kW–50 kW	No	N gas, LP gas	5 mbar–75 mbar	<2:1	Semi-luminous lambent, low intensity	Pilot flame and domestic applications
Aerated	0.5 kW–150 kW	Larger sizes (yes)	N gas, LP gas	20 mbar–1 bar	<20:1	Non-luminous shape depends on application	Wide use in heating and direct-contact process applications
Nozzle mix	30 kW–40 MW	Yes	All gases	15 mbar–1.5 mbar	<20:1 depends on oil turndown when dual-fuel	Normally non-luminous shape depends on burner and register configuration	Wide use in packaged burners of all sizes. Common in hot water and steam boilers of all sizes

Coal burners

Type	*Size range* (MW)	*Fuels*	*Grate thermal loading* (mW/m^2)	*Bed thickness*	*Turndown ratio*	*Ashing system*	*Main applications*
Underfeed	0.3–2	600–900 Rank singles and doubles	1.0	Depends on retort depth	On/off	Manual or side screw automatic	Small hot water and steam boilers
Fixed grate	0.6–7	600–900 Rank singles	2.7	25 mm live on up to 200 mm ash	2:1 to 5:1	Manual or automatic tipping grate or robot rake	Hot water and steam shell boilers
Chain grate	1–30	500–900 Rank normally washed smalls, but other types possible including high ash	1.7	80–150 mm	3:1 to 8:1	Manual drop tube boiler or drag link	Shell and water boilers
Coking stoker	1–4.5	300–900 Rank washed or untreated smalls and singles	1.4	250–350 mm	5:1	Manual drop tube in boiler or drag link	Shell boilers
Pulverized fuel burners	1–50	All types including anthracite lignite and bituminous	—	—	<15:1	None	Limited use in shell boilers mainly kilns and power stations
Fluidized beds	1–50	All types of coal and waste solid fuels	3.0	0.3–1.0 m	2:1 and then by bed sectoring	Removal and clean-up from bed	Special vertical shell boilers and watertubes

generated do not affect burner design greatly since the sulphur dioxide is a natural product of combustion and can only be reduced by lower fuel sulphur contents or sulphur removal from the exhaust gases.

8.6 Coal burners

Coal burners demand design consideration in all the aspects mentioned under gas and oil burners, but, in addition, need attention to the aspect of ash removal. The extent to which ash removal plays a part in the combustion system design often determines the ability of the burner to burn specific coals, particularly those with a high ash content. Principal types are as follows.

8.6.1 Underfeed stoker

Mainly applied to sectional and small shell boilers, this principle is limited to doubles or singles coal. Simple in construction, it consists of a hopper normally mounted close to the appliance which feeds via an Archimedean screw to a retort where the combustion takes place. Coal is fed up through the retort where the coal is burned progressively until the residue is mainly clinker and ash. The tapering nature of the retort allows sufficient residence time in the combustion zone before the ash overflows from the top of the retort into a surrounding area of deadplates from which the ash is removed periodically. The double-walled construction of the retort allows combustion air to be fed into the combustion zone through holes in the internal walls. These holes are known as tuyeres, and are sometimes supplemented by secondary air jets which transfer air from the retort via vertical tubes onto which are mounted nozzles to direct this overfire air onto the top of the bed. Recent improvements have allowed auto de-ashing by arranging for the ash and clinker to fall down to one side of the retort only, so allowing extraction by a screw. A typical underfeed stoker is shown in Figure 8.12.

8.6.2 Fixed-grate burner

Very popular in the main industrial shell boiler market, the fixed-grate stoker utilizes a bed of firebars which allow primary air to be fed through from underneath while not allowing coal or ash to fall through. This is achieved via a special ash-retaining pocket. Part of the grate at the front and rear comprises dead bars which prevent air passing through. Two types of coal feed predominate, both fed with coal which is transported by air after metering by a variable-speed screw. The coal is propelled along a pipe normally of 80–100 mm dia. by air travelling at about 20 m/s. This air is discharged over the grate to form part of the secondary air supply. In one principle the coal is spread on the grate via a deflecting nozzle which is characterized electronically to provide an even bed. The other type uses a drop tube which has a fixed deflector cone to spread the coal. Nozzles are arranged to provide secondary air over the bed.

De-ashing is achieved by stopping the burner and manually raking the ash off the bed after first moving the live coal onto the deadplate. This live coal is respread onto the live grate after de-ashing. Recent developments have included a tipping version of the grate to discharge the ash and an automatic system using a robot to rake ash off the grate.

Using the front-feed system it is possible to retrofit this principle onto boilers designed for oil and gas with some derating. Performance in terms of turndown ration, response rate, etc. can be as good as gas and oil. Best performance is achieved using 600–900 Rank coals containing 3–8% ash. Figures 8.13 and 8.14 illustrate the two feed arrangements common with fixed-grate stokers.

8.6.3 Chain grate

This is a very versatile stoker common in shell boilers and is also used with modification in watertube boilers as the travelling-grate stoker. In shell boilers the air supply is via forced- and induced-draught fans in combination with a balanced or slightly negative condition in the furnace.

Coal is fed from a front-feed hopper onto the moving grate through a guillotine door which controls bed thickness. Nowadays a fire-break system is arranged in the hopper to prevent burnback, which was relatively common. This is achieved via a rotating vane feeder. The grate comprises hundreds of links arranged into a chain over the whole width of the stoker, so forming a continuous, flexible mat. Driven via sprockets arranged across the width of the bed, the bed is supported on an airbox which supplies the air to the grate via slots in the individual links. The air in the airbox is staged via dampers and/or baffles to be released as required down the length of the grate.

An important principle is that the coal becomes ignited from the fuel which is on top of it and further down the grate. This reduces carry-over of small unburnt particles,

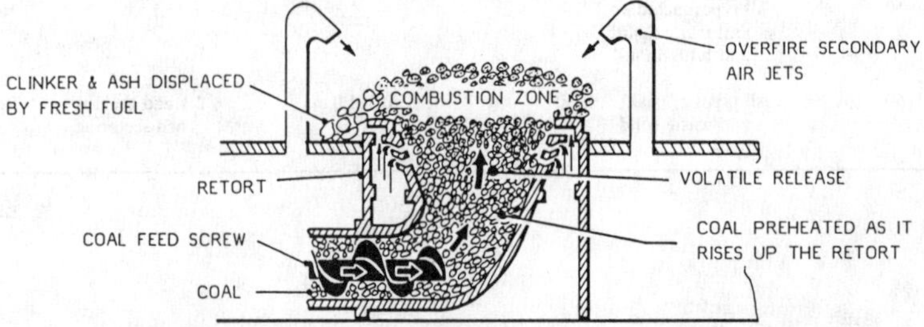

Figure 8.12 Underfeed stoker

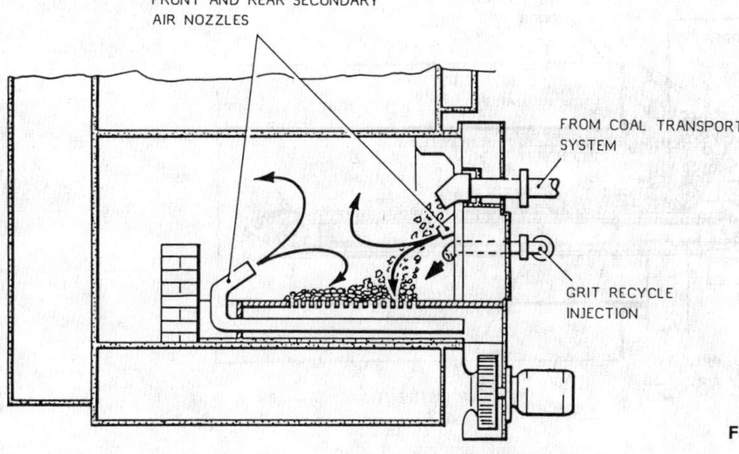

Figure 8.13 Fixed-grate coal burner with front feed

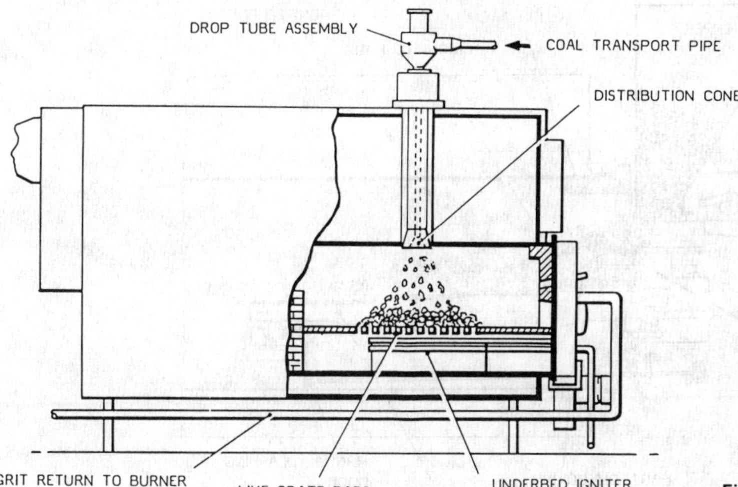

Figure 8.14 Fixed-grate coal burner with top feed

as they are filtered out by the burning coal on top. Volatiles released from the fresh coal are also ignited and consumed in the burning layer. This minimizes smoke formation which is caused by incomplete combustion of the volatiles.

Ash removal can be accomplished via a drop tube through the boiler shell. Suitable fuels include most singles and washed or unwashed smalls up to a top size of about 25 mm, but not anthracite or strongly caking coals. Ash content must be sufficiently high to protect the grate bars (>6%) and the coal should contain free moisture. An important specification criterion is the material of the chain links, particularly with low-ash coals and with variable loads. Figure 8.15 illustrates the main features of a chain-grate stoker.

8.6.4 Coking stoker

Less popular nowadays, the coking stoker resembles the chain-grate stoker in many respects. Coal is fed from a hopper onto the coking plate via a reciprocating ram. Transformation into the coke phase on these plates takes place by the release of volatiles.

The ignited coke then passes further down the grate via the reciprocating action of the grate bars, which extend longitudinally for the full length of the grate. The grate bars are driven sequentially by cams formed in a shaft which is driven via an electric motor and gearbox. This stoker is capable of burning a wide range of coals, and ash can be removed without stopping the combustion process. Air supply is via an induced-draught fan. The coking stoker is shown in Figure 8.16.

8.6.5 PF burners

These types are most common in power station-sized boilers and process industries such as cement and gypsum, where the residual ash is absorbed in the process. As their application is at the larger end of the combustion field, on-site milling and preparation is technically feasible and economic, using ball, roller or impact mills. Small applications can utilize pre-milled coal delivered to site, but the choice of suppliers is very limited.

Initial preheating of the combustion chamber by gas or oil is normally required in order to provide the necessary temperature environment to release the volatiles which provide the stabilization in the base of

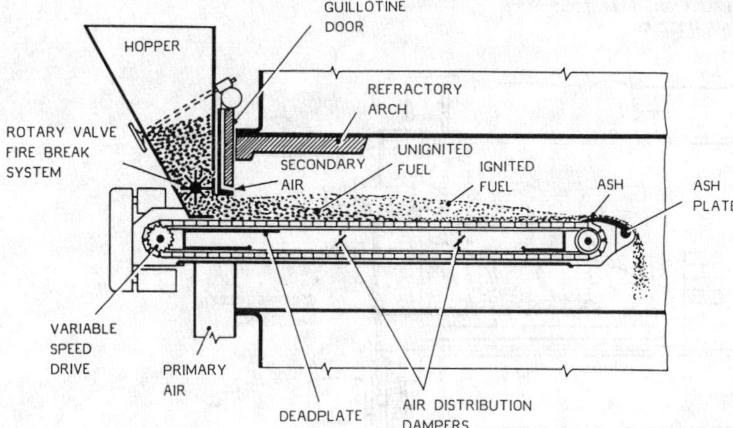

Figure 8.15 Chain-grate stoker

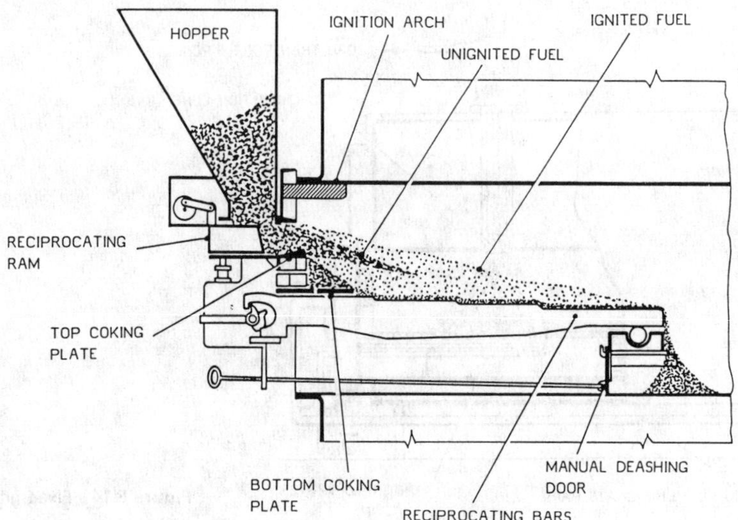

Figure 8.16 Coking stoker

the flame. Some small PF systems have used another fuel for flame support, but this compromises the economics. A typical pulverized fuel burner is shown in Figure 8.17.

8.6.6 Fluidized-bed firing

This is normally a very integrated system of combustion which has received considerable development in the last few years. The combustion chamber is arranged to allow combustion of the coal within a fluidized bed of inert mineral matter, normally sand. The sand is maintained in a fluidized state by an air supply from beneath the bed via nozzles or distributors. Rapid combustion is ensured via the vigorous bubbling of the bed, which is an ideal combustion environment. To avoid sintering of the bed material, it is necessary to prevent the bed from exceeding the ash-fusion temperature of the coal. This requires the bed temperature to be maintained in the range of 900–1000°C. As the temperature of the products of combustion would be about 1400°C naturally, it is fundamental that heat be removed with in-bed cooling tubes to suppress the temperature to less than 1000°C. The turbulent nature of the bed provides for high convective heat transfer rates, so minimizing the surface area. A number of developments have taken place with fluid bed systems which have included deep-bed technology, recirculating beds and reduction of emissions by in-bed additives such as limestone and dolomite to retain sulphur. Problem areas are ash removal, which takes place by recirculating some material from the bed for clean-up and erosion. Positive attributes include the ability to burn virtually any solid combustible material. Figure 8.18 illustrates the principle.

8.7 Dual- and triple-fuel firing

This is very common nowadays to allow bargaining on fuel price or to arrange an interruptible gas tariff which is backed up at times of peak demand with a stored oil supply. Most types of oil and gas burner are available in dual-fuel form, normally with gas burner design

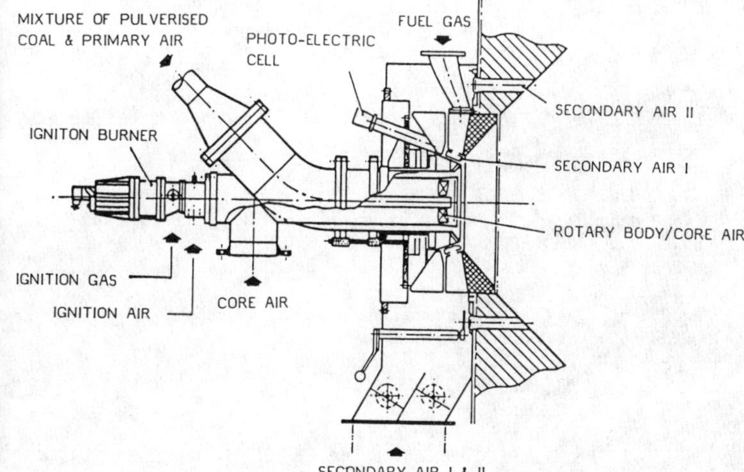

Figure 8.17 Pulverized fuel burner

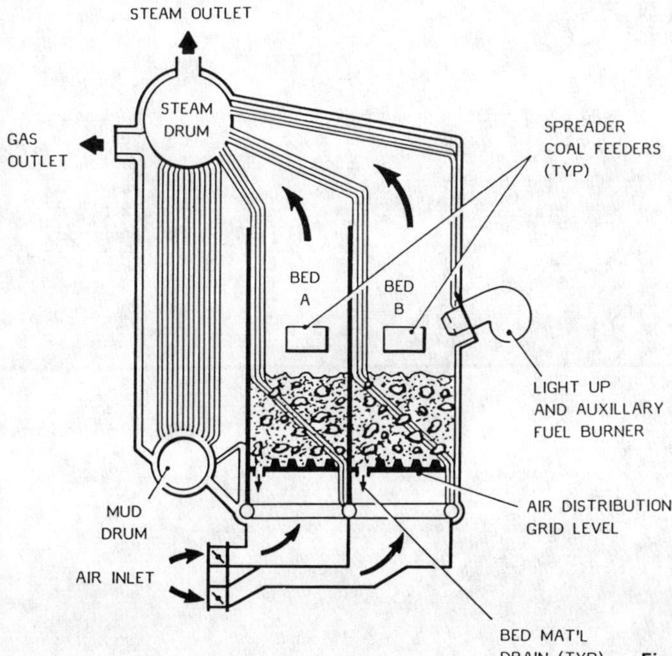

Figure 8.18 Fluid-bed combustion system with sectioned bed

'wrapped around' the arrangement for oil firing. This is usually the more difficult fuel to burn, particularly in the case of residual heavy oils. Fuel selection is normally by a switch on the burner control panel after isolation has taken place of the non-fired fuel. To avoid the cost and complexity of the fuel preheating on oil firing, smaller systems use gas oil as the standby fuel.

Multi-fuel operation with coal-firing equipment is more difficult to achieve, partly because the grate obscures some of the boiler heating surfaces and partly due to the volume that it also occupies. Systems do exist for this requirement, but changeover is not as instant as oil and gas, as it is normally necessary to remove part of the coal-firing equipment.

Requirements for dual- and triple-fuel firing including solid fuel as one option are best met by PF burners and fluid beds. PF burners are particularly suitable, as no static grate exists to compromise the design. They also have a combustion geometry which is similar to gas and oil, and therefore the flame can be arranged to allow full development of flame shape and maximum radiant heat transfer surface utilization.

Fluid beds can be fired with gas and oil across the top of the slumped bed since sufficient freeboard exists with coal firing to prevent particle elutriation. Oil, gas or dual-fuel burners so arranged can also provide the means for bed preheating, especially if the flame is redirected down to the fluidization zone.

9 Oil

Eur Ing Roland Gibson BTech, MSc, CEng, FIMechE, FIMarE, FRSA
W. S. Atkins Consultants Ltd

Contents

9.1 Distribution and delivery

Crude oil is processed at oil refineries, generally located around coastal areas, and then transferred to a nationwide network of oil fuel terminals, from where it is distributed to customers. Fuel supplies to the customer can be made by road delivery vehicle, rail tank wagon or coastal tanker, depending on the location of the customer's storage installation.

9.1.1 Road delivery vehicles

Deliveries of oil fuels by road are made by vehicles with capacities ranging from 11 800 to 28 700 l (2600 to 6300 gal). The vehicles discharge fuel into customers' storage by pump or compressed air at rates up to 1050 l (230 gal) per minute, and can deliver to a height of 10.7 m (35 ft) above the vehicle.

There are occasions where the noise of vehicle discharge needs to be minimized, examples being night discharge at hospitals, hotels and adjacent to residential areas where there are parking restrictions on daytime deliveries or where a customer agrees to a 24-hour delivery service. In such cases, an offtake suction pump can be installed by the customer as part of his storage facilities, thus eliminating the use of the vehicle engine for vehicle pump operation.

There must, in all cases, be safe road access to sites where tanks are situated and suitable hard standing provided for the vehicle during delivery.

9.1.2 Rail tank cars

The capacities of rail tank cars are generally 45 t gross laden weight (approx. 30 t payload) or 100 t gross laden weight (approx. 70 t payload). Cars for carrying heated fuel oil are insulated with fibre-glass lagging covered with galvanized sheeting, and are equipped with finned tubes for steam heating.

A customer's steam supply must be capable of providing an adequate amount of heat to discharge a full train-load of oil. Considering individual rail tank cars, a 100 t gross laden weight (glw) car requires 385 kg/h (850 lb/h) and a 45 t glw car requires 154 kg/h (350 lb/h) of steam. Because of the risk of contamination by oil, the condensate should be drained from the outlet connection and run to waste.

Rail tank cars have bottom outlets terminating in a quick-acting coupling, and are designed for pump off-loading only. A flow indicator should be inserted between the terminal outlet and the fill line. Flexible connections for steam and oil lines should be provided for the maximum number of cars that will be discharged simultaneously.

A block train consists of a series of rail tank cars of either 45 t glw or 100 t glw (or a mixture of both) coupled together for movement from source to destination and back as a unit. The size of the trains will be determined by contractual agreement, but will never be less than 300 t payload, and will normally be 500 t.

9.1.3 River and coastal tankers

Where a customer has suitable berthing facilities and satisfactory arrangements can be made, deliveries can be by either river or coastal tanker. River tankers currently in use are of 960 t capacity but may be larger if conditions permit. The customer is responsible for supplying oil hose and/or discharge arms capable of discharging the full contents of the barge under low-water conditions.

Coastal tankers are available up to 2000 t capacity, equipped with pumps for discharging into customer's storage. The size of the vessels to be used will be dependent upon local conditions and the depth of water in the approaches.

9.2 Storage tanks

9.2.1 Type of tanks

There are four main types of storage tank available for industrial and commercial fuel oils, namely:

Mild steel welded
Mild steel sectional
Cast iron sectional
Reinforced concrete

Of these four, the mild steel welded tank is the most popular and is widely used for every type of application.

The majority of storage tanks are of the horizontal cylindrical type, as shown in Figure 9.1. Where ground space is a limiting factor, vertical cylindrical tanks may be used.

9.2.2 Construction

The principal British Standard used for the design and construction of oil storage tanks and associated fittings is BS 799, Part 5. Other relevant standards are:

BS 1564, Pressed steel sectional rectangular tanks
BS 2594, CS welded horizontal cylindrical tanks
BS 2654, Vertical steel welded tanks for the petroleum industry

Limitations on the tanks covered by the above-mentioned codes are as follows:

1. *BS 799, Part 5*
   Integral tanks, which form part of a complete oil-fired unit, having a capacity not exceeding 25 l;
   Service tanks, which isolate the main storage tanks from the burner installation, having a capacity not exceeding 1000 l;
   Storage tanks of unlimited capacity but with a maximum height of 10 m;
   Design pressure not exceeding the equivalent head of water acting at the bottom of the tank, i.e. height of tank, plus full height of vent pipe.
2. *BS 1564*
   Constructed of pressed steel plates 1220 mm square;
   Design pressure not exceeding the static head corresponding to the depth of the tank;
   Maximum depth of 4880 mm.

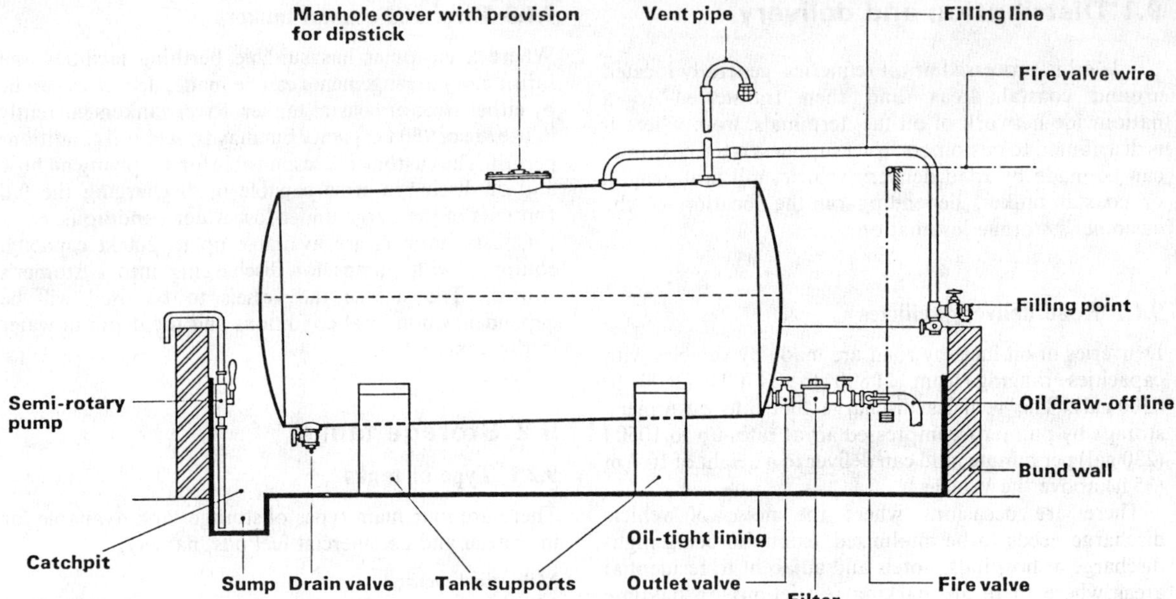

Figure 9.1 Typical storage tank for a distillate fuel oil

3. *BS 2594*
   Maximum internal working pressure not exceeding 0.4 bar at top of tank and maximum internal vacuum not exceeding 10 mbar.
4. *BS 2654*
   Design pressure and vacuum pressure not exceeding 56 mbar and 6 mbar, respectively;
   Design metal temperature not lower than −10°C.

When using mild steel rectangular welded tanks, internal bracing is required. Great care is necessary in the design and installation of supports for sectional steel tanks to prevent distortion under load.

Design criteria

Basic design criteria, which need to be taken into consideration when designing a storage tank, will include:

1. Geographical location of tank (e.g. indoors/outdoors, height above ground level, available space, accessibility, etc.);
2. Principal dimensions or capacity;
3. Type of roof (e.g. fixed or floating, cone, dome, membrane, pontoon, etc.);
4. Properties of fluid;
5. Minimum and maximum design temperatures;
6. Requirement for heating elements;
7. Design vapour pressure and/or vacuum conditions;
8. Minimum depth of fluid;
9. Maximum filling/emptying rates;
10. Venting arrangements;
11. Connections/manhole requirements;
12. Access ladders;
13. Corrosion allowance;
14. Number and type of openings;
15. Type of support or foundation;
16. Internal/external coating or lagging.

9.2.3 Capacities

The capacity of storage tanks for oil-fired installations is an important consideration. BS 779: Part 5 recommends that the minimum net storage capacity should be calculated either by taking

1. Three weeks' supply of oil at the maximum rate of consumption; or
2. Two weeks' supply at the maximum rate of consumption, plus the usual capacity ordered for one delivery, whichever is the larger.

Where the maximum weekly offtake is less than 200 gal (910 l) the capacity should still not be less than 650 gal (2950 l) in order to accept a standard 500-gal (2270-l) tanker delivery. In some circumstances it may be desirable to provide more than one tank, each of sufficient capacity to accept at least a full delivery.

9.2.4 Tank supports

Horizontal cylindrical tanks should be installed on brick or reinforced concrete cradles with a downward slope of 1 in 50 from the draw-off end towards the drain valve, as shown in Figure 9.1. Cradles should be constructed on foundations adequate for the load being supported and the type of soil. A reinforced concrete raft equal to the plan area of the tank, and of adequate thickness to bear the load, is normally suitable for all but the weakest soils. Cradles should not be placed under joints or seams of the tank plates and a layer of bituminized felt should be interposed between the cradle and tank. The height of the tank supports should provide at least 450 mm space between the drain valve and ground level to allow access for painting or draining the tank.

9.2.5 Tank fittings

Oil-level indicators

A brass dipstick is recommended as a cheap and reliable means of determining the contents of a storage tank. A dipstick, when required, is usually provided ready calibrated by the tank manufacturer before installation of the tank.

In many cases it is inconvenient to use a dipstick, due to the position or location of the tank, and there are a variety of direct and remote contents gauges available, including gauge glasses, float and weight, float and swing arm, float and indicator, hydrostatic, electrical capacitance, etc.

Filling connection

Filling pipes should be as short as possible and free from sharp bends. The terminal should be in a convenient position to allow easy coupling of the vehicle hose connection, wherever possible within 5 m (15 ft) of the hard standing for vehicle delivery. The most suitable height for a filling pipe is about 1 m (3 ft) above ground level and clear of all obstructions.

A non-ferrous dust cap with chain should be provided to close the end of the filling pipe and protect the thread when not in use. When the filling pipe is not self-draining into the tank a gate valve should be fitted as close as possible to the hose coupling. Where there is any possibility of damage or misuse of the terminal equipment a lockable fill cap should be fitted and, if considered necessary, the terminal enclosed in a lockable protective compartment.

All filling lines should be self-draining. Where this is not possible with residual fuel installations, trace heating and lagging should be applied, and this is particularly important in exposed positions. This will ensure that any oil fuel remaining in the filling pipe will be at pumping viscosity when delivery is made.

The filling pipe should enter at the top of most horizontal storage tanks. For vertical tanks up to 3 m (10 ft) diameter the filling pipe should be fitted at the top and then bend through 90°. This directs the incoming oil fuel down one side of the tank and minimizes air entrainment. In larger vertical storage tanks containing residual fuels, bottom inlet filling may sometimes be used. A non-return valve must be fitted in the filling line as close to the tank as possible. The entry position of the filling pipe into the tank in relation to the position of the draw-off connection must be carefully selected to avoid air entrainment into the oil-handling system.

Where delivery vehicles do not have access adjacent to the storage area a permanent extended filling pipe should be provided from the tank to a position where the vehicle can stand in safety. In cases where the length of this pipe will exceed 30 m (100 ft) careful attention should be paid to draining requirements. Where lines are not self-draining, a drain valve should be situated at the lowest point in that section. Provision should be made for lagging and tracing to be applied to extended fill lines that will carry residual grades of industrial oil fuel. Careful attention should also be paid to the use of correctly sized pipe diameters. Where the storage tank is not visible from the filling point, an overfill alarm should be fitted.

Ullage

The air space between the oil surface and the top of the tank is known as ullage; there should always be a small ullage remaining when the contents gauge registers full. This prevents the discharge of oil from the vent pipe due to any frothing and surging of the liquid during delivery. The ullage should provide not less than 100 mm (4 in) between the oil surface and the top of the tank, or be equivalent to 5% of the total contents, whichever is the greater.

Vent pipes

A vent pipe must be fitted at the highest point of every storage tank. Wherever possible, it should be visible from the filling point and terminate in the open air, in a position where any oil vapour will not be objectionable and, in the event of an overflow, there will be no damage to property, fire risk or contamination of drains.

The vent pipe bore must be equal to or greater than the bore of the filling pipe, and never less than 50 mm (2 in) diameter. It should be as short as possible and free from sharp bends. It should terminate in a return bend or 'goose neck' fitted with a wire cage for protective purposes (fine gauze must never be used for this purpose).

Where, of necessity, the vent pipe rises to a considerable height, excessive internal pressure on the tank may result, due to the pressure head of oil should an overflow occur. To prevent any possible tank failure due to such an occurrence a vent pipe pressure-relief device must be provided. It should be self-draining to reduce the risk of blockage, particularly when using heavy oil fuel. BS 799 requires that these devices should not place any restriction on oil flow and must discharge within the bund area.

Draw-off connection

The draw-off connection to the oil-burning plant should be at the raised end of horizontal tanks. Where heating facilities are not provided, the lowest point of the draw-off connection should never be less than 75 mm (3 in) above the bottom of the tank. For tanks fitted with heating elements it is essential that these and their associated thermostatic control probes should always remain below the oil surface.

To enable the contents of the tank to be isolated a screw-down gate valve should be fitted adjacent to the draw-off connection. Since stresses may be applied to valves during any tank settlement or movement of pipework, cast-iron valves should not be used for this purpose.

Drain valve

A screw-down gate valve with a bore similar to that of the draw-off connection should be fitted to every storage tank at its lowest point to permit complete draining. The valve should be readily accessible with a clear space below to facilitate its use. Extension pipes to or from drain valves should be avoided if possible, but where these are necessary, the pipe should be lagged and, if necessary, traced to ensure that residual grades of oil will flow during adverse weather conditions. Valves and extension pipes should be fitted with a plug or blank flange to prevent inadvertent discharge of the tank contents.

Tanks containing heated grades of oil fuel require regular draining. This is due to the small amount of moisture which accumulates over a period of time, by condensation formed on the sides of the tank. The quantity of moisture formed will be dependent upon the relative humidity conditions, the amount of 'breathing' which takes place, and the time allowed for settling. It is recommended that the tank is checked for accumulated water prior to a fuel delivery.

With storage tanks which are filled from the bottom such as large vertical tanks, the stirring action of incoming oil will carry any water (which would normally settle out) into the oil-handling system. In these circumstances any daily service tank used in the system should receive regular inspection to ensure that water has not accumulated. Any oil/water mixtures should be drained into suitable containers and subsequently removed for disposal into a separator or interceptor.

Manhole

Every storage tank must have a manhole in an accessible position, preferably on top. It may be circular, oval or rectangular, and not less than 460 mm (18 in) diameter if circular or 460 mm (18 in) long and 410 mm (16 in) wide if oval or rectangular. The manlid must be securely fixed by bolts, studs or set-screws, and have a liquid and vapour-tight joint. Close-woven proofed asbestos, graphited on both sides, is a suitable jointing compound for this purpose.

Vertical tanks over 3.65 m (12 ft) high should have a further manhole fitted near the bottom to provide access for cleaning and maintenance of any storage-heating facilities.

9.2.6 Heating requirements

Distillate grades of oil fuel may be stored, handled and atomized at ambient temperatures, and do not require heating facilities to be provided in storage tanks and handling systems. However, exposure to extreme cold for long periods should be avoided, since oil flow from the tank may become slightly restricted. This is particularly important for storage serving stand-by diesel generating plant where summer cloudpoint specification fuel may be in both storage and oil feed lines during the most severe winter weather. In very exceptional conditions, even winter cloudpoint specification fuel may present occasional problems in unlagged storage and handling systems. Such systems should be lagged and, if necessary, traced to ensure faultless starting.

Table 9.1 Recommended storage and handling temperatures for residual fuel oils (based on BS 2869: 1970)

Grade of oil	*BS classification*	*Minimum storage temperature*	*Minimum temperature at outflow from storage and for handling*
Light Fuel Oil	Class E	10°C (50°F)	10°C (50°F)
Medium Fuel Oil	Class F	25°C (77°F)	30°C (86°F)
Heavy Fuel Oil	Class G	35°C (95°F)	45°C (113°F)

Heating facilities are required for all residual grades of oil fuels, such as Light Fuel Oil, Medium Fuel Oil, and Heavy Fuel Oil. Table 9.1 gives the recommended minimum storage and handling temperatures for residual oil fuels (from BS 2869: 1970).

Where oil is to be maintained at minimum storage temperature, an outflow heater will be necessary to raise the temperature of the oil leaving the tank to that required for handling. It is not good practice to store oil fuel at unnecessarily high temperatures, and the temperature given under the column 'Minimum temperature at outflow from storage and for handling' should not be exceeded by more than 16.7°C (30°F). This is particularly important in relation to Light Fuel Oil. Maximum heat losses from storage tanks can be determined from Figure 9.2. These losses can be translated into maximum steam consumption rates of the heating coils from knowledge of the latent heat of steam appropriate to the steam pressure used in a given installation. The tank heating arrangements must be capable of maintaining the oil storage temperature with the appropriate rate of heat loss.

Another important point is that concerning the occasional requirement for heating the oil from ambient temperature. A heat input several times greater than that required to maintain handling temperatures will be necessary, depending on the minimum time stipulated for heat-up. Tank heating arrangements should therefore be capable of providing this heat input if the contingency is a possibility.

Heating methods

Storage tanks can be heated with thermostatically controlled steam coils, hot-water coils, electric immersion heaters or a combination of these. The elements and their thermostats should be positioned below the level of the oil draw-off line, so that they never become uncovered during normal operation. The temperature-sensitive element of the thermostat should always be situated

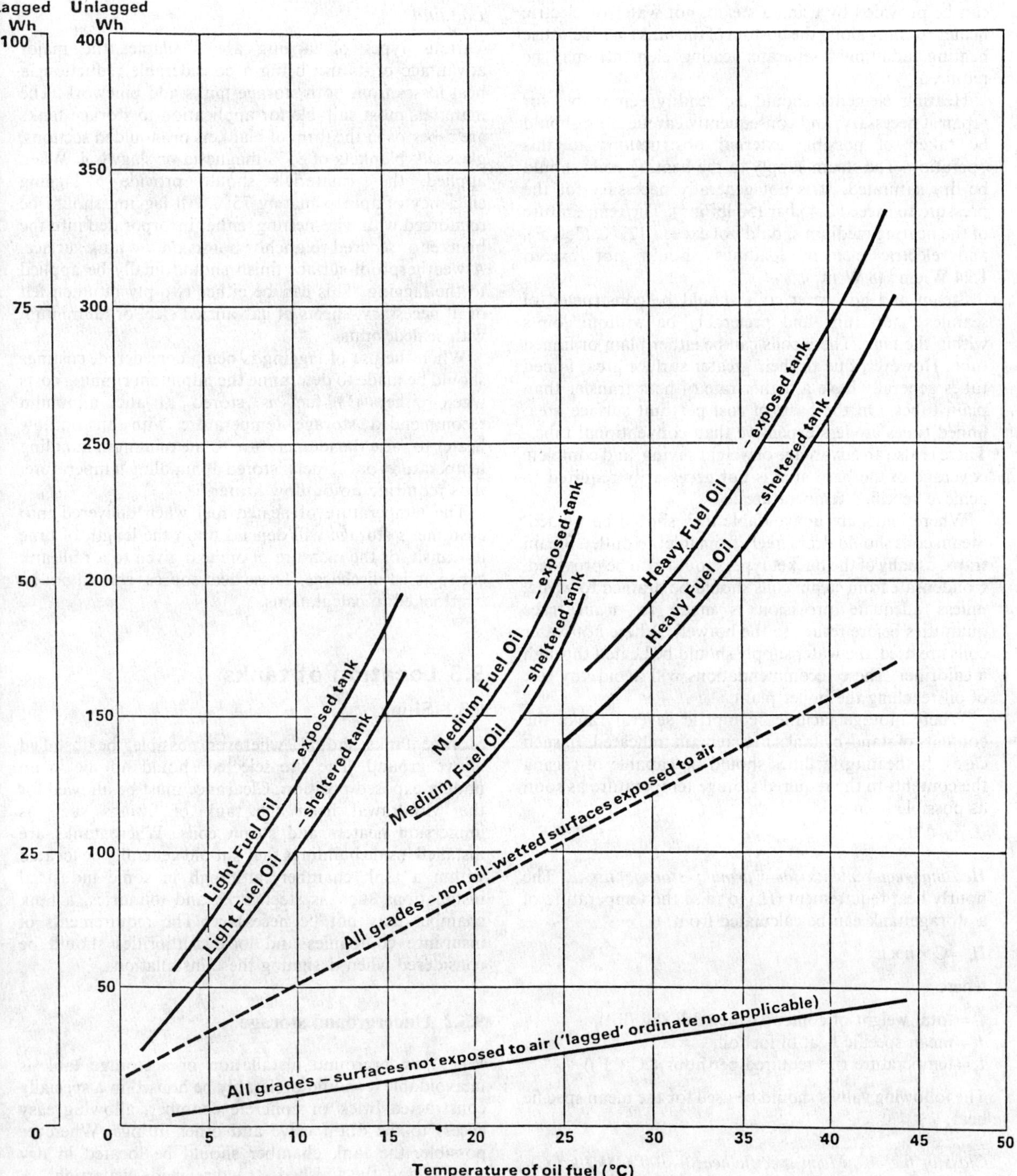

Figure 9.2 Heat losses from storage tanks plotted against oil fuel temperature

above and to one side of the heating element. The heating elements should be spaced evenly over the bottom of the tank or concentrated towards the draw-off end. A combination of steam and electric heating can be used for installations where periods may occur during which steam is not available.

Where a residual oil fuel is stored in tanks with outflow heaters provision should also be made to maintain the oil at or above the minimum storage temperature shown in Table 9.1. Excessive heating is not recommended, particularly for Light Fuel Oil, and will increase running costs unnecessarily. Additional heating to outflow heaters can be provided by using a steam, hot water or electric heater running along the bottom of the tank. With electric heating, additional separate heating elements may be required.

Heating elements should be readily removable for repair if necessary, and consequently careful note should be taken of possible external obstructions to this operation. The steam supply to the heating coils should be dry saturated. It is not generally necessary for the pressure to exceed 3.45 bar (50 lbf/in^2). The temperature of the heating medium should not exceed 177°C (350°F), and electric element loadings should not exceed 1.24 W/cm^2 (8 W/in^2).

Steam and hot-water coils should be constructed of seamless steel tube and preferably be without joints within the tank. These coils can be either plain or finned tube. However, due to their greater surface area, finned tubes generally have a higher rate of heat transfer than plain tubes. On the basis of cost per unit surface area, finned tubes are less expensive than conventional tubes. There is also an advantage of weight saving, and complete coverage of the base area is not necessarily required to achieve specified temperatures.

Where joints are unavoidable they should be welded. Steam coils should drain freely from inlet to outlet. Steam traps, usually of the bucket type, should also be provided. Condensate from steam coils should be drained to waste, unless adequate provision is made to drain trace quantities before return to the hotwell. Where hot-water coils are used, the water supply should be heated through a calorifier. These recommendations will avoid any risk of oil reaching the boiler plant.

Where storage facilities comprise several tanks the contents of stand-by tanks may remain unheated. In such cases the heating facilities should be capable of raising the contents to the required storage temperature as soon as possible from cold.

Heating requirements for warming storage tanks. The hourly heat requirement (H_r) to raise the temperature of a storage tank can be calculated from:

$$H_r = C \times h \times t_r$$

where

C = total weight of contents when full (kg (lb)),
h = mean specific heat of fuel oil,
t_r = temperature rise required per hour (°C (°F)).

The following values should be used for the mean specific heat:

Class of fuel	*Mean specific heat 0–100°C (kJ/kg°C)*
D	1.99
E	1.99
F	1.97
G	1.95

The rate of temperature rise (t_r) will depend on the particular circumstances of the installation and on how quickly the contents of the tank must be brought up to temperature.

Lagging

Various types of lagging are available; the major advantage of its use being a considerable reduction in heat losses from both storage tanks and pipework. The materials most suitable for application to storage tanks are asbestos in the form of blankets or moulded sections, glass silk blankets of 85% magnesia or slagwool. When applied, these materials should provide a lagging efficiency of approximately 75%. All lagging should be reinforced with wire-netting, either incorporated into the blanket or secured to anchor points on the tank surface. A weatherproof surface finish should finally be applied to the lagging. This can be either two-ply bitumen felt or, if necessary, sheets of galvanized steel or aluminium with sealed joints.

When the use of lagging is being considered, costings should be made to determine the minimum running costs when either (1) fuel is stored at the minimum recommended storage temperature with an outflow heater to raise the temperature to the minimum handling temperature, or (2) fuel is stored at handling temperature, thus requiring no outflow heater.

The temperature of heated fuel when delivered into customer's storage will depend upon the length of time in transit, or the measure of preheat given to a rail tank car to assist discharge. These heat sources should not be overlooked in calculations.

9.3 Location of tanks

9.3.1 Siting

Storage tanks should, wherever possible, be installed above ground. The site selected should not be in an unduly exposed position. Clearance must be allowed for the withdrawal from the tank of fittings such as immersion heaters and steam coils. Where tanks are installed inside buildings they should generally be located within a tank chamber, although in some industrial installations such as steelworks and foundries, a tank chamber may not be necessary. The requirements of insurance companies and local authorities should be considered when designing these installations.

9.3.2 Underground storage

Where underground installation of a storage tank is unavoidable it should preferably be housed in a specially constructed brick or concrete chamber, allowing easy access to the drain valve and other fittings. Wherever possible, the tank chamber should be located in dry ground and the finished structure made watertight. A sump must be provided in the floor of the chamber at one end to collect any water which may enter the chamber in exceptional circumstances, and the floor should slope slightly downwards towards the sump. Water collected in the sump can be removed by using a semi-rotary pump. The lower part of the chamber should form an oil-resistant catchpit, as with the more usual above-ground storage tank.

9.3.3 Buried tanks

It is recommended that storage tanks should not be buried directly in contact with soil, since it is almost impossible to avoid corrosive attack. Where, for some reason, it is not possible to provide a tank chamber as already described, adequate corrosion protection must be applied to all exterior surfaces of the tank, fittings and pipework.

9.3.4 Bund area

Where overfilling or leakage from the tank would contribute to a fire hazard, cause damage to property or contaminate drains or sewers, a bund wall should be constructed around the tank. This should be of brick or concrete with an oil-tight lining, and sealed to the concrete base under the tank supports. The capacity of the bunded area should be at least 10% greater than that of the storage tanks contained within it.

The bund walls must be oilproof and be capable of withstanding considerable liquid pressure in the event of an overflow or other emergency. No permanent drain must be incorporated into a bund area, but suitable facilities should be provided to remove rainwater which may accumulate. A sump and semi-rotary hand pump are suitable for this purpose.

9.4 Pipework systems

9.4.1 General

Oil fuel pipeline systems transfer oil from storage to the oil burner at specified conditions of pressure, viscosity, temperature and rate of flow. There can be considerable variety in the choice of system, but its design (particularly correct pipe sizing and temperature control) is most important if it is to function satisfactorily.

9.4.2 Handling temperatures

Distillate grades are usually handled at ambient temperatures, provided these are not below the cloudpoint when using gas oil. Residual grades, on the other hand, are handled at temperatures above ambient. The recommendations of Table 9.1 should always be followed regarding minimum recommended handling temperatures. Residual grades can also be handled at temperatures above those recommended as minimum handling levels in order to reduce oil viscosity, improve regulation and control of oil flow, reduce friction losses in pipelines and, when necessary, provide oil at the correct atomizing temperature for the oil-burning equipment.

9.4.3 Handling equipment

Two items of equipment should always be inserted in the handling system as close to the storage tank as possible. These are a filter and a fire valve.

9.4.4 Filters

To prevent foreign matter from damaging components and choking valves or atomizer orifices, filters must be incorporated into the handling system. There are usually two stages of filtration. The first provides protection for pumps and fire valves which handle oil at temperatures below those required at the oil-burning equipment. Second-stage filtration protects the atomizer orifice and burning equipment, and is sometimes incorporated as part of the burner assembly.

For distillate grades of fuel oil the first-stage filter should protect pumps and valves and can be placed in the draw-off line from the storage tank. Paper filters can be used where appropriate, providing filter apertures are not less than 0.1 mm (0.004 in). The paper element should be regularly replaced when necessary. For residual grades of oil, first-stage filters should be placed in the draw-off line as near to the storage tank as possible, and incorporate a filtering medium equivalent to a circular hole 2.5–0.75 mm (0.1–0.03 in) diameter. Fine filtration must not be employed at the first stage, since pressure drop across the filter may be excessive due to the higher viscosity of these grades at low temperatures. Second-stage filtration is usually incorporated into the oil burner assembly, often in the hot oil line after the preheater so that it filters the oil at atomizing temperature.

Reliable enclosed filters of ample effective filtering area should be used for both stages. The filter design should not allow the pressure drop to exceed 7 kN/m^2 (1 lbf/in^2) across second-stage fine filters in clean conditions. Filters should preferably be of the duplex or self-cleaning type, and installed so that oil spillage will not occur during operation.

They must be readily accessible for cleaning, which should be carried out as frequently as necessary to ensure that pressure drop across the filter does not affect normal oil flow rates. The filtering medium should be of corrosion-resistant material such as monel metal, phosphor-bronze or stainless steel. All first-stage filters should be provided with isolating valves.

9.4.5 Fire valves

A valve which closes in case of fire should be inserted in the oil fuel line to the oil-burning equipment and fitted as close to the tank as possible. It may be held open mechanically, pneumatically or electrically. Temperature-sensitive elements should be arranged to close the valve at a fixed maximum temperature, and sited close to the oil-fired plant and well above floor level. The operating temperature of the heat-sensitive elements should not be greater than 68°C (155°F) except where ambient temperatures in the vicinity of the plant may exceed 49°C (120°F), in which case the operating temperatures may be 93°C (200°F).

Where an electric circuit is employed the valve should close on breaking, not making, the circuit and be reset manually. A warning device should be included to indicate that resetting is necessary, to cover the eventuality of temporary electrical failure.

Where a valve is closed by the action of a falling weight, a short free fall should be allowed before the weight begins to close the valve. This assists in overcoming any tendency for the valve to stick in the open position.

Fire valves should preferably be of glandless construction. If a gland is incorporated into the system it should be of a type that cannot be tightened to an extent which would prevent the correct functioning of the valve. A manual release should be installed on all fire valves so that they may be tested regularly.

9.4.6 Types of system

There are two main types of oil-handling system in common use. These are gravity and ring main.

Gravity systems

Gravity systems are of three basic types: gravity, pump-assisted gravity and sub-gravity. A gravity system is one in which the oil flows directly from the storage or service tank through a gravity feed pipeline. The static head on the feed line will vary with the depth of oil in the tank, and the system should therefore only be used for burners which will operate satisfactorily between such limits.

A pump-assisted gravity system is one in which oil flows by gravity from the storage or service tank to a pump. The pump supplies oil to the combustion equipment through a pipeline passing only the quantity required to feed the oil-firing equipment. The inclusion of a pump in this system will not reduce the static head due to the fuel supply in the tank. The working pressure required at the oil-burning equipment should therefore be greater than the maximum static head available when the storage or service tank is full.

A sub-gravity system is one in which a pump associated with the oil-burning equipment is used to suck the oil from a tank in which the level of the oil can be below the level of the pump.

Gravity systems in general will handle distillate grades at ambient temperatures, and residual oils at pumping or atomizing temperatures. Lagging and tracing will be required with residual grades to prevent cooling of the oil to below handling temperatures.

Ring main systems

A ring main system draws oil from storage and circulates it to each consuming point in turn, the balance of the oil being returned either to the suction side of the circulating pump or to storage. A diagram of a typical hot oil ring main system is shown in Figure 9.3. Each offtake point is connected to the burner it supplies by a branch line. Pressure conditions are maintained approximately constant at each offtake point by a pressure-regulating valve situated after the last offtake point, and circulating a quantity of oil one and a half to three times the maximum offtake from the circuit. By this means, stopping offtake at one consuming point will not have a marked effect on the pressure at other consuming points. Pressure conditions should be calculated at each consuming point for all conditions of operation and offtake. The bore of the ring main should be such that pressure variations are not excessive for the equipment served. If these variations at each offtake point are likely to be critical they can be accommodated by the use of individual pressure-regulating valves on each branch.

Ring main systems are of three types: hot oil, warm oil and cold oil. Hot oil ring mains circulate oil at atomizing temperature, warm oil ring mains at a temperature between minimum pumping and atomizing temperature, and cold oil ring mains at ambient temperature.

Hot oil ring mains

This is the most important of the three types of ring main system, since it offers economies of installation and running costs. The smallest pipe sizes and fittings can be used with low-viscosity high-temperature oil, and the number of line heaters is minimized. The oil is circulated at atomizing temperatures and in consequence the system is less liable to be affected by pressure fluctuations due to variations in viscosity, or small changes in the viscosity of the oil as delivered. The system should usually be considered as the first possibility and only discarded in favour of warm or cold oil systems where these will give some real advantage.

Where there is a large temperature difference between a hot oil line and ambient temperature conditions,

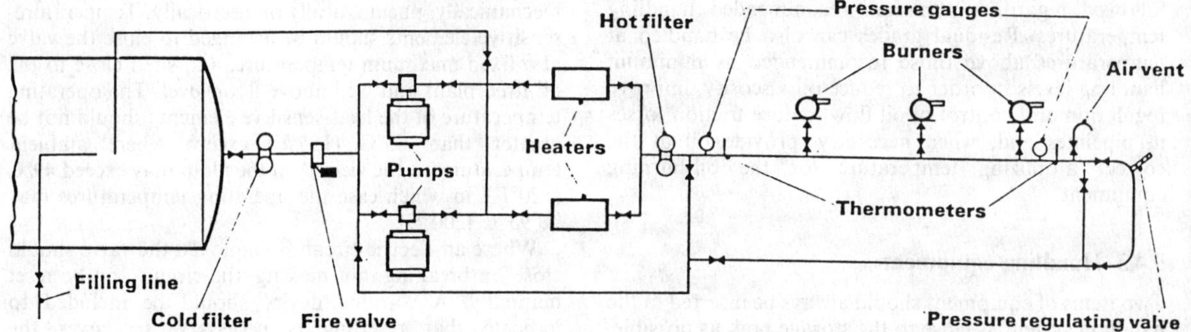

Figure 9.3 Typical hot oil ring main layout

particularly in the long ring main system, lagging will prevent excessive heat loss. Electric or steam tracing, in addition, will balance the heat loss from the pipeline. Residual oils should not be allowed to drop below their minimum handling temperature at any time. Where it is necessary to shut down any part of the plant, such as when servicing pumps and other equipment, a drain valve and air vent must be provided at the lowest point in the system to prevent cold oil becoming static.

With some installations it is possible to maintain circulation during periods when the oil-firing plant is not required to operate. The oil is circulated at reduced temperature to minimize the heat loss from the ring main and, consequently, the amount of heat required to compensate for this loss. The reduced temperature should be such that the increased oil viscosity does not result in an excessive increase in pressure.

Warm oil ring mains

This system is similar to a cold oil ring main but includes provision for heaters in the circuit to maintain oil temperature between minimum handling and atomizing levels. This provides a reduction in oil viscosity and reduces pipe friction. The circulation temperature of the oil should be chosen to give the minimum pressure drop consistent with the system design when circulating one and a half to three times the maximum offtake and using a suitable pipe diameter. This circulation temperature should allow a reasonable margin below the specified atomizing temperature to facilitate the selection of the necessary line heaters for branch lines between the ring main and the oil-burning equipment.

The volumetric capacity is adjustable to provide uniform heat transfer to the oil passing through. Heaters are flexible in operation down to about 50% of their designed temperature rise, when the quantity of oil passing through is approximately double. Where the temperature level required with a reduced throughput is outside this range, the heater manufacturer should be consulted.

Cold oil ring mains

This system is used mainly where different atomizing temperatures are required at various consuming points or where a branch line would be unacceptably long. The system should only be used where the length of pipeline involved and the quantity of oil circulated will not cause an excessive pressure drop due to friction. When designing a cold oil ring main system, care must be taken to ensure that the pressure variation between offtake points, due to changes in the oil consumption rate, do not affect burner performance. This is achieved by circulating one and a half to three times the maximum offtake required. The system is widely used with distillate grades but rarely with residual grades.

Sub-circulating loops

Where it is not possible or convenient to arrange for the ring main to be carried near the burners an alternative to using a warm oil ring main with branch heaters is a hot oil ring main with sub-circulating loops. The branch line from the high-pressure leg of the ring main is extended past the burners and returned into the low-pressure or return side after the pressure-regulating valve. This ensures that circulation of hot oil is maintained past the consuming point. The amount circulated through the sub-circulating loop must be carefully regulated to avoid the pressure-regulating valve becoming inoperative by short-circuiting too high a percentage of the oil in the ring main.

The regulation of oil flow in a sub-circulating loop is controlled by a secondary pressure-regulating valve situated after the last oil burner served by the loop. Alternatively, a fixed orifice or regulating valve, either hand-controlled or of the lock shield type, may be used. With the former method it is sometimes difficult to obtain balance between the primary and secondary regulating valves. The latter method is generally satisfactory providing the offtake rate of the burners does not fluctuate appreciably, and ample time can be allowed to establish stable pressure/temperature conditions in the sub-circulating loop after a shutdown period.

Branch lines

Branch lines transfer oil from a ring main circuit to the oil-burning equipment. Where a residual oil fuel is to be used, there will be some cooling of the oil immediately adjacent to the pipe surfaces and this will show as a small increase in viscosity. To keep this variation to a minimum and so prevent any difficulties in atomization at the oil burner, care should be taken over the length and diameter of branch lines. Provision should always be made to isolate and drain branch lines.

9.4.7 Pipe sizing

When a fluid is flowing through a pipe, resistance to flow is caused by friction. The pipebore selected for each section must be such that under any operating conditions, the initial head, either static head of oil in the supply tank or the pump delivery pressure, will be adequate to ensure the required flow rate. Additionally, any change of flow rate and consequent variation in loss of head must not adversely affect the operation of the associated oil-burning equipment.

The following factors must be taken into account when assessing pressure drop.

Viscosity

Pressure drop is directly proportional to viscosity. The effect of heat loss from pipelines and consequent increase in viscosity should also be considered.

Flow conditions

The handling system should be designed to provide streamline flow at all times when steady and predictable pressure conditions are essential.

Table 9.2 Equivalent length of pipeline fittings

Nominal pipe bore (mm (in))	*Equivalent length of fitting*				
	Elbow (m (ft))	*Tee* (m (ft))	*Easy bend* (m (ft))	*Gate valve* (m (ft))	*Non-return valve* (m (ft))
25 (1)	0.9 (3)	1.8 (6)	0.5 (1.5)	0.3 (0.8)	0.6 (2)
32 (1.25)	1.1 (3.5)	2.1 (7)	0.6 (1.75)	0.3 (1)	0.8 (2.5)
40 (1.5)	1.2 (4)	2.4 (8)	0.6 (2)	0.4 (1.2)	0.9 (3)
50 (2)	1.5 (5)	3.0 (10)	0.9 (3)	0.5 (1.7)	1.2 (4)
65 (2.5)	1.8 (6)	3.7 (12)	1.2 (4)	0.6 (2.1)	1.5 (5)
80 (3)	2.4 (8)	4.9 (16)	1.5 (5)	0.8 (2.5)	1.8 (6)
100 (4)	3.0 (10)	6.1 (20)	1.6 (6)	1.0 (3.3)	2.4 (8)
150 (6)	4.6 (15)	9.1 (30)	2.4 (8)	1.5 (5)	3.0 (10)
200 (8)	5.5 (18)	11.0 (36)	3.0 (10)	2.1 (7)	3.7 (12)
250 (10)	6.7 (22)	13.4 (44)	3.7 (12)	2.7 (9)	4.3 (14)

Flow rate

Pressure drop under streamline flow conditions is directly proportional to the quantity of oil flowing. The effect of reduced flow rate after offtake points, as compared with full flow rate throughout the full length of the pipeline when there is no offtake, should be taken into account to ensure that variation in pressure is within the specified pump output. Special consideration is necessary with gravity and ring main systems serving several offtake points.

Length of pipeline

Pressure drop is directly proportional to the length of the pipeline. All fittings used in the system should be included in the determination of 'effective pipeline length'. Table 9.2 should be used to determine the equivalent length to be added to the actual length of the pipeline for various types of fitting. The resulting figure is the total 'effective length' of the system.

The following empirical formula for estimating pressure drop under streamline flow conditions may be used:

$$H_f = \frac{1.2 \times 10^5 VT}{d^4} = \frac{0.0765VT}{d^4}$$

Metric (SI) Imperial

where

H_f = loss of head in metres per 100 m (feet per 100 ft) effective length of pipeline ((m)(ft)),
V = oil fuel viscosity at handling temperature (Redwood I scale),
d = internal diameter of pipe (mm (in)),
T = oil flow rate (tonne/h (ton/h)).

Note: The flow is normally streamline when

$$\frac{4Q}{dV} < 2000 \qquad (30T < dV)$$

Metric (SI) Imperial

where

Q = oil flow rate (m^3/s),
d = internal diameter of pipe (m),
V = kinematic viscosity at handling temperature.

When it is necessary to estimate the pressure drop in a pipeline where turbulent flow conditions exist, the following formula will give an approximation:

$$H_f = \frac{7.98 \times 10^4 G^2}{d^5} = \frac{G^2}{6.4d^5}$$

Metric (SI) Imperial

where

H_f = loss of head in metres per 100 m (feet per 100 ft) of effective pipe run,
G = oil flow rate (l/min (gal/min)),
d = internal pipe diameter (mm (in)).

Branch lines

Branch lines transfer oil from a ring main circuit to the oil-burning equipment. Where a residual oil fuel is to be used, there will be some cooling of the oil immediately adjacent to the pipe surfaces and this will show as a small increase in viscosity. To keep this variation to a minimum and so prevent any difficulties in atomization at the oil burner, care should be taken over the length and diameter of branch lines. In general, the following empirical formula should be used when designing branch lines for hot residual oils:

$$\text{Length} = 7L < \sqrt{M} \qquad (L < \sqrt{M})$$

Metric (SI) Imperial

where

L = 'equivalent' length of branch line in m (ft) (Table 9.2),
M = minimum oil consumption (l/h (gal/h)).

Internal diameter $D < 10.4L$ ($D < L/8$) where D is in mm (in). Provision should always be made to isolate and drain branch lines.

The author is grateful to BP for much of the information given in this chapter.

10

Gas

Peter F Fleming BSc(Eng), ARSM, CEng, MInstE
British Gas plc

Contents

10.1 Selection and use of gas as a fuel

10.1.1 Advantages of gas

The advantages of gas can be summarized as:

1. Reliable fuel of constant composition;
2. Used as required (does not have to be ordered in advance or stored on the users' premises);
3. Clean fuel-emissions are low compared with combustion of most other fossil fuels.

10.1.2 Availability

Mains gas is widely available throughout the mainland of the United Kingdom and in the Isle of Wight. Thinly populated rural areas, particularly in Scotland, Wales, and south-west England, do not have access to mains gas although liquid petroleum gas from a central supply may be available. All major industrial areas are within the gas supply area, and less than 15% of domestic dwellings are outside this area.

10.1.3 Metering

Payment for gas consumed is made based on the consumption as registered on a meter. Meters used for billing purposes are checked for accuracy and badged by the Department of Energy. These meters are known as primary meters. Those used by the user for monitoring the consumption of plant, etc. are known as check meters (often erroneously referred to as secondary meters).

It is current policy to meter and bill for gas in Imperial units. The meter registers in cubic feet and gas consumption is calculated in therms; 1 therm (105.506 MJ) is 100 000 Btu.

The meter location should be located as close as is sensibly practicable to the site boundary adjacent to the gas main. It is generally preferable for all but small low-pressure installations to be located in a separate purpose-built structure or compound and, wherever possible, away from the main buildings. Installations should be protected from the possibility of accidental damage, hazardous substances and extremes of temperature or vibration.

With the exception of certain very large loads which may use such metering devices as orifice plates, metering will be by means of diaphragm, rotary displacement or turbine meters.

Diaphragm meters

These are the traditional gas meter which incorporates diaphragms contained within a steel case. The diaphragms are alternately inflated and then deflated by the presence of the gas. The movement of the diaphragms is linked to valves which control the passage of gas into and out of the four measuring compartments.

These meters are suitable for low-pressure applications (<75 mbar) and low flow rates. Meters rated at up to 16 m^3/h (5650 ft^3/h) are available but it is not usual to use them above about 85 m^3/h (3000 ft^3/h). An important consideration is that the meters are physically large for their rating.

Rotary positive displacement meters

These meters incorporate two 'figure-of-eight' impellers rotating in opposite directions inside a casing. The impellers are made of either cast iron or aluminium. RPD meters are available with ratings from about 22 cm^3/h (800 ft^3/h) upwards but are not normally used as primary meters below about 85 m^3/h (3000 ft^3/h).

Turbine meters

These meters are only used for large loads which also have a nearly constant gas consumption. The gas flow impinges on a specially shaped turbine and is streamlined by the contour of the casing on either side of the turbine. Turbine meters are very small for their rating. Disadvantages are that long lengths of straight pipework upstream are necessary to ensure the correct flow profile, and a fluctuating load will cause inaccuracies in registration.

Correction

Meters are accurate within close limits as legislation demands. However, gas is metered on a volume basis rather than a mass basis and is thus subject to variation with temperature and pressure. The Imperial Standard Conditions are 60°F, 30 inHg, saturated (15.56°C, 1013.7405 mbar, saturated). Gas Tariff sales are not normally corrected, but sales on a contract basis are. Correction may be for pressure only on a 'fixed factor' basis based on Boyle's Law or, for larger loads, over 100 000 therms per annum for both temperature and pressure using electronic (formerly mechanical) correctors. For high pressures the compressibility factor Z may also be relevant. The current generation of correctors correct for pressure on an absolute basis taking into account barometric pressure.

10.1.4 Contracts

For annual loads of up to 25 000 therms per annum gas is sold subject to published tariffs. Above this, the price of gas is subject to contract of which there are two types:

1. *Firm contract.* The gas sale is on a firm basis, that is, there is a firm commitment to supply gas under all except the most exceptional circumstances.
2. *Interruptible contract.* There is an obligation for the user to provide an alternative standby fuel to be used at the request of the gas supplier. The maximum period of 'interruption' is one of the conditions agreed in the contract. The advantage to the user is that he or she can obtain a more favourable gas price. The advantage to the supplier is that at times of peak demand, such as a cold winter, he can balance supply and demand without investing in excessive storage capacity.

10.1.5 Range of uses of gas

Gas is used for a diverse range of applications in the domestic, commercial and industrial sectors. Commercially, the main uses are space heating, water heating and catering, together with some chilling. In larger premises such as hospitals, etc. there may be large central boiler plant. Industrially, there will be applications similar to those in commerce. In addition, there are many process applications. These may be high-temperature kilns and furnaces (ceramics, metal reheating, metal melting, glass, etc.) or low-temperature (drying, baking, paint stoving, etc.).

The theoretical flame temperature is 1930°C, but in practice this is not obtainable, and a maximum process temperature of perhaps 1300°C is realistic without recuperation. Gas is used for 'working flames' in such processes as flame hardening and glass bulb manufacture. In industrial process heating either direct or indirect heating (i.e. using a heat exchanger) may be used according to the process requirements. Gas can be used as a chemical feedstock. On a small scale this can be to produce protective atmospheres by controlled partial combustion, on a large scale as a base for ammonia production, etc. Gas is, however, sold as a source of heat and not as a chemical of constant quality. It is suitable for the production of shaft power from either reciprocating engines or turbines.

The cleanliness of the products of combustion is such that the use of heat-recovery equipment is possible without the risk of corrosion. This has led to the development of combined heat and power packages where the overall efficiency is high.

Small reciprocating engines from 15 kW upwards can be used, the heat being recovered from the engine coolant and exhaust. Large reciprocating engines normally exhaust into a waste-heat boiler.

Gas turbines are available with power outputs of 1 MW upwards, and the exhaust is used to fire waste-heat boilers. The high oxygen content of the exhaust enables supplementary firing to be used to increase the heat/power ratio as desired.

10.2 Theoretical and practical burning and heat transfer

10.2.1 Types of burner

Efficient combustion of gas under varying conditions demands the use of a wide variety of burners. However, these can all be categorized as natural draught or forced draught.

The majority of domestic burners, together with a large number of commercial and many industrial burners, are of the natural-draught type. In these the gas passes through a jet situated in a venturi such that primary air is mixed with the gas. The resulting mixture passes through the burner nozzle where mixing with secondary (and, in some designs, tertiary) air takes place together with ignition and combustion. Such burners have the advantage of simplicity but have a limited turndown ratio and their poor mixing characteristics lead to a rather low efficiency. Indeed, to ensure that complete combustion takes place without the formation of soot or carbon monoxide it is necessary to allow a 'margin' of excess air well above that strictly needed for combustion.

The low gas pressure available at the injector (typically, 17.5 mbar) allows a primary aeration of only about 40%. The resulting flame envelope is rather large and the intensity of combustion low. It is possible to increase the degree of primary aeration, producing a more intense flame, if a higher gas pressure is used. To produce complete primary aeration a gas pressure of the order of 1 bar will be needed.

For certain processes a flame may be produced by a burner in which there is no primary aeration. These 'laminar' flames have a very low intensity of combustion and a luminous appearance.

The majority of larger industrial burners, including furnace and boiler applications, are of the forced-draught type. These employ a combustion air fan to provide all the air needed for complete combustion. The burners are usually sealed into the combustion chamber so that there is no access to secondary air from the atmosphere as with natural-draught burners. Forced-draught burners may be of the premix type, where air and gas are mixed prior to the burner, or, more commonly, of the nozzle mix type, where the mixing takes place within the burner.

A refractory quarl is usually an integral part of forced-draught burners. Suitable design of burner and quarl can determine the flame characteristics. Long, short, pencil or even flat flames are possible.

For general purposes, including firing hot water boilers and warm air space heaters, 'package' burners are commonly used. In these the burner is ready assembled together with all its controls and air fan. To install such a burner it is only necessary to connect up to gas and electric supplies and controls (thermostats, etc.).

10.2.2 Turndown ratio

The turndown ratio is an indication of the ability of the burner to maintain a stable flame at lower firing rates, and is a ratio of the maximum and minimum firing rates. Turndown can be fairly low for average burners of both natural- and forced-draught burners, 3:1 being a typical figure with 5:1 a maximum although up to 40:1 is possible with special burners.

It is important to remember that although a burner can be fired at low rates it is probable that the efficiency at low fire will be reduced because the excess air is invariably higher at turndown. This will affect the selection of controls (e.g. on/off or modulating).

10.2.3 Heat transfer

The carbon/hydrogen ratio of gas is considerably lower than oil or coal, which results in a flame of very low luminosity. Radiation from the flame is therefore low and furnace design must allow for heat transfer to be primarily by convection and conduction, together with re-radiation from hot surfaces.

Burners can be designed to produce a luminous flame by means of laminar mixing and partial cracking of the gas, but the radiation is still low. A typical forced-draught burner used for boiler firing will be essentially transparent.

10.2.4 Water vapour in products of combustion

The high hydrogen/carbon ratio of gas means that the quantity of water vapour in the products of combustion is greater than most other fossil fuels. The latent heat of this cannot be released in conventional appliances leading to a fairly low net/gross ratio of calorific value of 90%. (It is normal practice to quote *gross* CV; in Europe *net* CV is often used. If net CV is quoted, efficiencies of over 100% are possible.)

The cleanliness of products of combustion from gas enable recovery of latent heat by means of condensing appliances in which the products are cooled below the dewpoint of 55°C. The condensate is only weakly acidic and a suitable choice of materials of manufacture permit it to be dealt with. Most other fuels produce a condensate which is too acidic to allow condensing appliances to be used.

10.3 Pressure available to user

10.3.1 Low-pressure supply

The majority of users are supplied with gas at low pressure, and the categories are defined as:

Low pressure	(<75 mbar)
Medium pressure	(75 mbar to 7 bar)
High pressure	(>7 bar)

Normal supply pressure to industrial and commercial users is 21 mbar. Allowing for pressure losses in the system, at least 17.5 mbar should be available at the point of use. Higher pressures can often be supplied by agreement where available. There may be process advantages in having a higher pressure.

It is not normal practice to supply direct from a high-pressure transmission main rather than a local distribution system (known in the USA as 'farm taps').

10.3.2 Boosters and compressors

If a pressure higher than that available is wanted then a booster or compressor will be required. Boosters are normally considered as adaptions of centrifugal fans and raise pressure by typically 75 mbar for a single-stage machine. Higher pressures require a compressor which will be a positive displacement or screw type. The use of a booster should not be regarded as a substitute for the correct design and engineering of pipework within a site.

Requirements for the installation of boosters and compressors are detailed in the British Gas publication *Guidance Notes on the Installation of Gas Pipework Boosters and Compressors in Customers' Premises* (IM/16). The most important requirement is that a low-pressure cut-off device must be installed at the inlet to the booster or compressor to comply with the requirements of the Gas Act 1986.

10.4 Energy conservation

Energy use can be minimized by a combination of various measures. These can be categorized as those reducing energy used and those recovering heat from a process.

10.4.1 Reduction of energy used

Air/gas ratio

Any process using a fossil fuel will involve the rejection of the products of combustion following heat transfer. These flue products will contain sensible heat which is lost and represents inefficiency in the process. Unless some form of recuperation is practised, the flue products must be at a higher temperature than the process, and this cannot be reduced. The amount of excess air can, however, be controlled.

Figure 10.1 indicates the flue losses to be expected for different temperatures and excess air. It is seen that considerable savings can be made, particularly at higher temperatures, by reducing excess air levels to a practical minimum. It is also evident that a reduction in air/gas ratio to below stoichiometric will cause a rapid deterioration in efficiency caused by the energy remaining in the incomplete combustion of fuel.

The ideal air/gas ratio is that which is marginally higher than stoichiometric. It is not possible to run a burner with no excess air for various reasons (e.g. changing ambient temperature, a slight change in calorific value, variation in barometric pressure, wear of control equipment, etc.). All of these and other factors dictate that the burner is operated with sufficient excess air to avoid the production of carbon monoxide in any quantity.

Figure 10.2 shows how the production of CO can vary with excess air for two typical burners. It is seen that to limit CO to, say, 50 ppm with burner B, 3% oxygen in the flue is needed, and with burner A, which exhibits

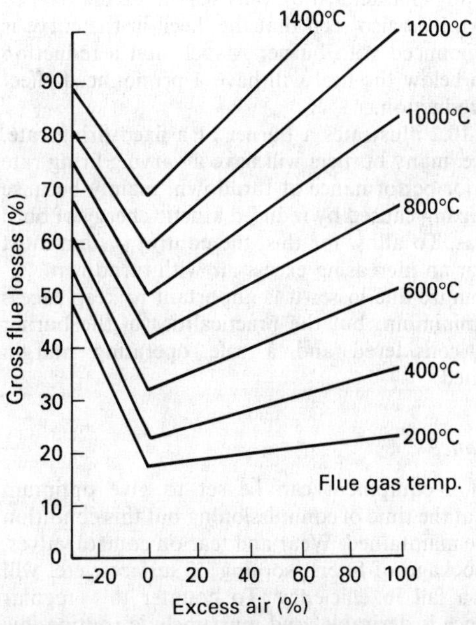

Figure 10.1 Gross flue losses versus excess air

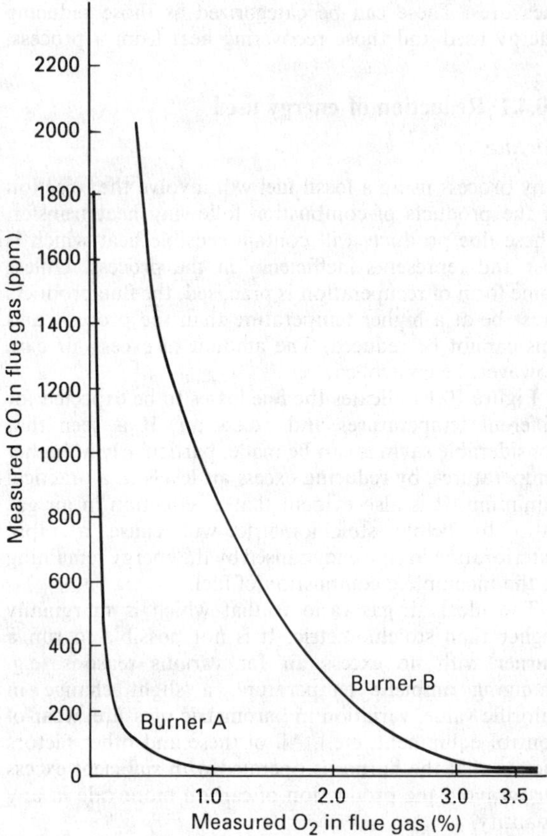

Figure 10.2 Measured percentage of O_2 in flue gas. Range of CO versus O_2 for a variety of burners

better mixing characteristics, only 0.75% excess oxygen is required. It is also seen that the 'heel' in the curve is more pronounced with burner A such that a reduction in oxygen below the heel will have a pronounced effect on CO production.

Figure 10.2 illustrates a burner at a fixed firing rate. In practice, many burners will have a varying firing rate with inferior performance at turndown, mainly because of poor mixing caused by reduced kinetic energy of both air and gas. To allow for this, the control system must provide for an increasing excess air with turndown.

To minimize flue losses it is important to keep excess air to a minimum, but the practicalities of the burner must be considered and a safe operating margin incorporated.

Maintenance

Combustion equipment can be set to give optimum efficiency at the time of commissioning but this condition will not be maintained. Wear and tear on control valves, partial blockage of filters, sooting of surfaces, etc. will all cause a fall in efficiency. To counter this, regular maintenance is desirable, and must include routine flue analysis and burner adjustment.

Insulation

Sensible heat losses from thermal plant should be kept to a realistic minimum by the use of correctly specified insulation. There will be a point beyond which further insulation is not economically viable. Careful analysis of the properties of insulation materials is necessary to prevent, for example, the adding of more insulation to the cold face of a furnace wall, causing the maximum service temperature of intermediate insulation to be exceeded.

Modern low-density insulation such as those based on ceramic fibres can be used to save energy in plant operating on a batch basis. The low thermal mass permits a rapid heating and cooling period which can save a substantial amount of energy. With continuously operating plant the advantages are not so pronounced.

Procedures

Energy can be conserved by operation of plant in such a way as to minimize part loading. Various practices can be adopted that can be described as 'good housekeeping'. In addition to maintenance, this will include such factors as the avoidance of plant operating in the standby mode for long periods, operation at correct temperature, ensuring doors are closed where applicable, etc.

For production plant the energy used per unit produced is lowest when operated at design capacity. At low throughputs the energy used increases markedly because the standing losses are constant, irrespective of throughput.

For plant such as boilers operating in parallel controls should ensure that one boiler acts as a 'lead' to minimize part-load operation. As far as is possible, plant should be sized to meet the load and oversizing should be avoided.

10.4.2 Heat recovery

General points

Technology exists to recover heat from processes operating at all temperatures, from regenerators on high-temperature plant to heat pumps using low-temperature effluent as a heat source. The problem in many cases is to find a use for the heat recovered. The best solution is to recycle the heat within the same plant, as the supply will always be matched to the demand. An alternative is to use the heat recovered in associated plant (for example, the heat recovered from a melting furnace can be used to dry feedstock for the furnace).

In general, 'high grade' recovered heat is more valuable than 'low grade'. The latter can often take the form of large quantities of warm water for which there is a finite need.

Recuperation

Figure 10.1 shows the heat carried away in flue gases, particularly for high-temperature processes. For example, for a furnace operating at 800°C with no excess, air losses are 40%. Recovery of a proportion of these losses is possible by means of recuperation.

The simplest form of recuperation is load recuperation, but this is not suited to retrofit and is incorporated at the design stage. Flue products from the highest temperature zone are used to preheat incoming stock on the counterflow heat exchanger principle. Such techniques are well established in pottery tunnel kilns, etc.

In many processes load recuperation is not practicable, and combustion air is preheated in a heat exchanger by means of the outgoing flue products. Figure 10.3 gives an indication of the savings to be made for different

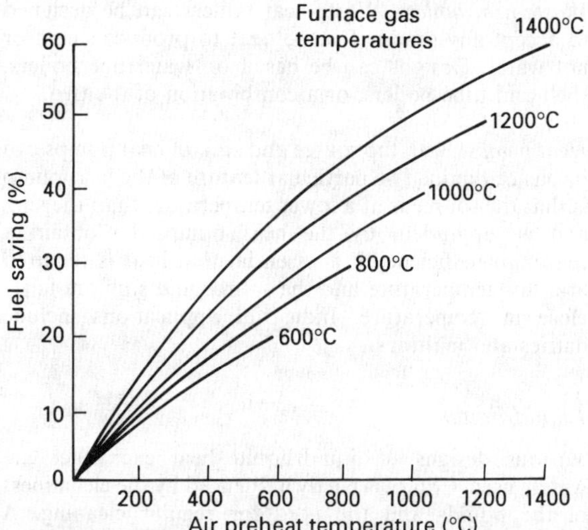

Figure 10.3 Fuel saving versus air preheat temperature (stoichiometric conditions)

operating temperatures. It is not normally considered economic to operate a recuperator at flue temperatures below about 750°C.

The recuperator can be positioned in the flue of the furnace or be integral with the burner (i.e. a recuperative burner). The separate recuperator is usually less costly, particularly on a multi-burner furnace. The recuperative burners avoid the need for lagging pipework, and reduce the risks of leakage upsetting the air/gas ratio. The flow pattern of hot gases in the furnace may, however, be disturbed as the flue gases must exit through the burner.

Regenerators

Recuperators are limited in their performance, partly by problems with materials operating for long periods at elevated temperatures and also by the efficiency of fairly simple gas to gas heat exchangers. For high-temperature applications a regenerator has advantages.

Regenerators have long been established in such processes as steel melting in open-hearth furnaces and glass-melting tanks. They consist of checkerwork brickwork which act as a heat sink for the high-temperature flue gases. On reversing the cycle this brickwork acts as a heat source for the incoming cold air. The regenerator is thermally efficient but only suited to very large, intensively used plant.

Regenerative burners have been developed to widen the range of application of the regenerator principle. They consist of a pair of gas burners each with its own regenerator consisting of a bed packed with refractory balls. Regular cycling between the two burners, only one of which fires at a time, gives high efficiency with up to 90% of the heat in the flue products recoverable. Figure 10.4 shows typical performance figures for these burners.

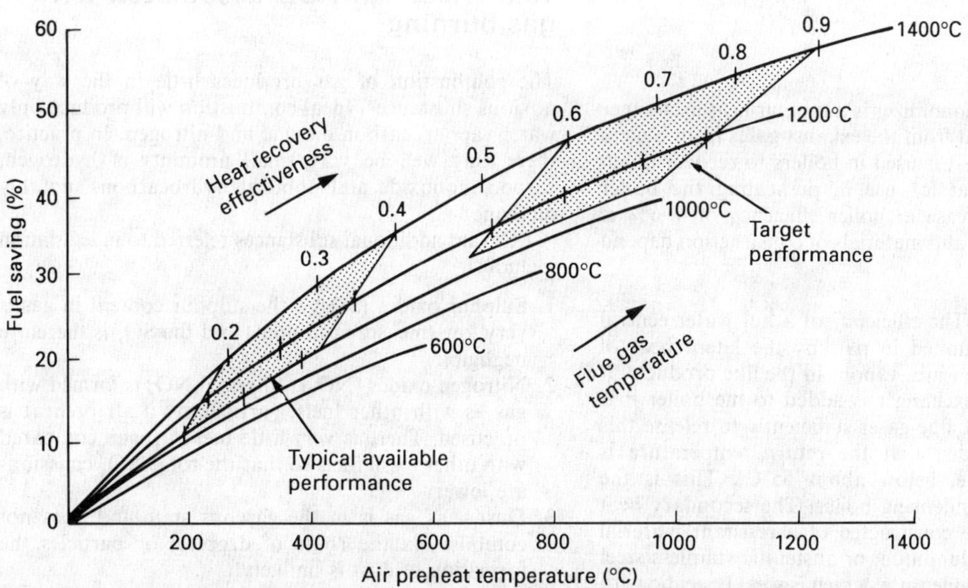

Figure 10.4 Fuel saving versus air preheat temperature

Heat exchangers

There is a wide range of heat exchangers available to cater for most temperature ranges and for gas/gas, gas/liquid or liquid/liquid, as appropriate. High exhaust temperatures normally involve the use of a recuperator or regenerator and lower temperatures use other heat exchangers.

Gas/gas

Simple heat exchangers. These can be of the parallel flow, crossflow or counterflow pattern and constructed of materials to suit the temperature.

Rotary heat exchangers. These consist of a slowly revolving wheel of diameter 0.6–4 m driven by an electric motor. The wheel consists of a metallic matrix which absorbs sensible heat (and, in some designs, latent heat) from the outgoing stream and transfers it to the ingoing stream. With these exchangers 75–90% efficiency is possible.

Heat pipes. The use of heat pipes involves the incoming cold air stream and the outgoing warm air stream being immediately adjacent and parallel, and between the two is a battery of heat pipes. These contain a liquid and operate on the thermosyphon principle. The liquid takes in latent heat and evaporates and the vapour travels to the cold end of the tube where condensation releases the latent heat. Generally, heat pipes are restricted to 400°C, and effectiveness can be up to 70%.

Run-around coils. Where the incoming and outgoing airstreams are remote it is necessary to use a run-around coil to couple them. A heat exchange coil in the warm exhaust is connected by a pumped liquid to a heat exchanger in the cold stream. The effectiveness can be up to 60%.

Gas/liquid

Economizer. The economizer is a tubular heat exchanger used to recover heat from the exhaust gases from boilers or some processes. It is used in boilers to recover much of the sensible heat for use in preheating the boiler feedwater. An increase in boiler efficiency of 4–6% is typical. The design and materials of construction depend on the application.

Condensing boiler. The efficiency of a hot water central heating boiler is limited in part by the latent heat of vaporization in the water vapour in the flue products. If a secondary heat exchanger is added to the boiler it is possible to cool the flue gases sufficiently to release this latent heat, provided that the return temperature is sufficiently low (i.e. below about 55°C). This is the principle of the condensing boiler. The secondary heat exchanger must be constructed of a resistant material such as anodized aluminium or austenitic stainless steel to withstand the condensate, which is weakly acidic with a pH 3.3–3.8. The condensing boiler concept is not applicable to oil or coal, which normally have a more acidic condensate. Fuel savings of from 14% to 18% can be achieved.

Spray recuperator. The principle of the spray recuperator is similar to that of the condensing boiler, and again is restricted to gas firing. In this case the condensation takes place by direct contact of a water spray with the flue gases. The resultant water is at temperatures up to 50°C and is recycled through a heat exchanger. If a suitable use can be found for the low-grade heat (e.g. boiler feedwater preheating) then fuel savings of up to 17% can be made.

Waste-heat boilers. Waste-heat boilers can be designed to accept any grade of waste heat to produce steam or hot water. Designs can be based on watertube boilers, shell and tube boilers, or a combination of the two.

Heat pumps. Both the source and sink of heat pumps can be gas or liquid. The particular feature of the heat pump is that the source is at a lower temperature than the sink and is 'upgraded' by the heat pump. To obtain a reasonable efficiency it is essential that heat is required at a low temperature and the source and sink are fairly close in temperature. Industrial applications include dairies and maltings.

Liquid/liquid

Various designs of liquid/liquid heat exchanger are widely used. Choice is partly influenced by the cleanliness of the liquids and the need for regular cleaning. A compact and commonly used type is the plate heat exchanger.

10.5 Clean Air Acts associated with gas burning

The combustion of gas produces little in the way of noxious substances. Ideal combustion will produce only water vapour, carbon dioxide and nitrogen. In practice, there may well be very small amounts of hydrogen, carbon monoxide and unburnt hydrocarbons, notably methane.

Relevant additional substances referred to in legislation include:

1. Sulphur oxides (SO_x) – the sulphur content in gas is very low (measured in ppm) and the SO_x is therefore negligible.
2. Nitrogen oxides (NO_x) – thermal NO_x is formed with gas as with other fuels, particularly if air preheat is practised. There is very little fuel nitrogen compared with other fossil fuels so that the total NO_x emissions are lower.
3. Dust – as gas is in the gaseous state and does not combust via the route of droplets or particles the formation of dust is unlikely.

Nevertheless, gas is subject to legislation.

10.5.1 EC Directive on emissions from large combustion plant

This legislation was formally adopted on 26 November 1988. It applies to plant for the production of energy (effectively, boiler plant and similar) and specifically excludes heating furnaces, drying plant, etc. The Directive applies to plant of thermal input 50 MW and over. If two or more separate plants are in close proximity such that they could share a common flue then if their total input is 50 MW or over then they are also subject to the Directive.

The Directive applies to new plant (defined as that planned after 1 July 1987) with provision for retrofitting existing plant. Limits for new plant are:

Sulphur dioxide: 35 mg/N m^3
NO_x: 350 mg/N m^3
Dust: 5 mg/N m^3

10.5.2 UK legislation

UK legislation referring to emissions from gas-fired plant is currently rather limited. The most important is The Health and Safety (Emissions into the Atmosphere) Regulations 1983 (SI No. 943, 1983). In Schedule 2 is listed substances deemed to be noxious, which include combustion products, dust, etc.

Schedule 1 lists those works covered by the regulations. The only categories which concern the combustion products of gas are:

1. Electricity works (excluding compression-ignition engines burning distillate fuel with a sulphur content of <1%);
2. Boilers with an aggregate not less than 200 tonnes/h of steam being used wholly or in part for electricity generation.

It is clear that very few gas-fired installations will fall within these categories. *Notes on the Best Practicable Means* for meeting the requirements were published by HM Inspectorate of Pollution in 1988.

For gas-fired plant large boilers and furnaces 50–700 MW thermal:

NO_x limit of 100 ppm v/v (measured at 6% oxygen 15°C and 1 bar without correction for water vapour), (approx. 200 mg/m^3);
Chimney efflux velocity not less than 15 m/s at MCR;
Temperature of gases entering chimney not less than 120°C.

For large boilers and furnaces >700 MW thermal:

Chimney efflux velocity not less than 18 m/s at MCR;
Temperature of gases entering chimney not less than 80°C.

(*NB:* There is no limit for NO_x for gas-fired plant. In practice, little or no such plant currently exists. The conditions for measurement of NO_x are clearly defined.)

10.6 Chimney requirements: codes of practice and environmental considerations

10.6.1 Functions of a flue

The overriding function of the flue is to remove the products of combustion. This involves the creation of sufficient draught, either mechanical or thermal, to move the flue products from the combustion chamber to the terminal of the flue. A secondary function is, by means of suitably locating the flue terminal, to ensure adequare dispersion of the products of combustion.

10.6.2 Operating principles

To operate effectively, the flue has to apply a pressure differential sufficient to overcome the system resistance and enable the products of combustion to flow from the combustion chamber to the terminal. This pressure differential can be mechanical (by forced or induced draught or a combination of the two) or thermal, possibly combined with mechanical.

The 'natural-draught' flue operates on the thermal principle. The pressure differential is caused by the difference in density between the column of hot gases within the flue and a column of air of the same height. Within limits, the taller the flue, the greater the draught, but the upwards movement is opposed by the resistance to flow inherent in the geometry and friction of the flue.

Successful flue design involves the balancing of the draught against the resistance, possibly for a range of thermal inputs, the avoidance of condensation, and a location of the terminal which ensures unrestricted dispersion of the flue products.

Factors affecting flue performance

Flue height. Raising the height of a flue increases the flue draught but also adds to the pressure loss due to friction. The net effect is that increasing height is very beneficial for short flues but has progressively less effect as the height is increased. With very tall flues there can be a reduction in flow rate due to excessive heat loss. The point where an increase in height produces no additional flow rate depends on many factors. In the case of small commercial-size appliances an increase beyond 6–9 m will normally produce no benefit.

Cross-sectional area. There is a direct relationship between flue flow and the cross-sectional area. The draught is unaffected by the cross-sectional area but the frictional losses decrease as the area is increased, resulting in greater flow. Too great an area, however, will lead to a low velocity with its attendant problems of downwash and possible condensation.

Thermal input. The flue draught rises with temperature but the pressure losses increase because of the greater volume of the flue gases. The net result is that increased temperature leads to a greater flow rate, but only up to about 260°C, beyond which the flow is reduced.

If the heat input from a given appliance is increased both the temperature and flow rate of the gases entering the flue will rise. The net result is that an increased heat input leads to a greater flow rate, but this is more noticeable at low heat inputs. There is a limit to the rate that can be vented into a given flue before the pressure losses overcome the flue draught and 'spillage' (failure of the flue products to be satisfactorily cleared) occurs.

Flue route. The only parts of the flue route that contribute to flue draught are the vertical sections. All sections, however, contribute to the pressure losses. It follows therefore that to increase the effectiveness of the flue it is important to adhere as closely as possible to a vertical route and to have a minimum of bends, changes of section, and horizontal runs.

Heat losses. Heat losses through the fabric of the flue will lower the mean temperature of the flue gases, causing both the flue draught and the pressure losses to be reduced. The net effect is quite small (perhaps a 10% difference between a well-insulated and an unlagged flue).

It is, however, important to avoid condensation. Although the condensate from gas combustion is generally considered to be non-acidic, if the flue has not been designed for conditions of continual wetting condensation should be avoided. Intermittent condensation under start-up conditions can usually be tolerated.

The dewpoint for the flue products from gas is a maximum of 60°C and will, in most cases, be lower because of excess air. There is no acid dewpoint because the sulphur content is negligible. It would be thought that condensation would be most unlikely, and this is so for the bulk flue gases. However, the temperature gradient across the flue wall can be such that the 'skin temperature' at the inside wall can be considerably less than the bulk temperature, and condensation will take place.

To minimize the possibility of condensation it is desirable to:

1. Keep the flue as short as possible;
2. Avoid the use of bends and horizontal runs;
3. Insulate the flue;
4. Keep the surface area as small as possible.

These features must be consistent with the other design parameters as far as possible.

Termination. A suitable design and location of the flue terminal is essential for optimum performance. This is primarily to counter adverse wind effects.

The wind effect may assist or oppose the natural-draught flue, causing increased draught up the flue, or down-draught. For this reason, it is important to minimize their effects by careful attention to the terminal position. In general, this means a position at least 0.25 m above any nearby obstacle.

For smaller appliances flue terminals are necessary. These have the main functions of:

1. Keeping the resistance as low as possible;
2. Preventing blockage;
3. Being unaffected by wind of any speed or direction.

Proprietary terminals are available up to 200 mm, but above this are not used, a 'Chinese hat' terminal being considered adequate to keep out rain and larger foreign bodies. For large flues, particularly forced draught, terminals are not used.

Down-draught diverters. Any flue should be designed as far as possible to produce a constant draught. If the draught is insufficient then spillage of combustion products will take place, and if too great, then excess air will be induced, leading to an unnecessary loss of efficiency.

A typical down-draught diverter used with natural-draught flues is shown in Figure 10.5. This consists of a baffle plate mounted at the top of the primary flue. There are three modes of operation:

1. Normal conditions – the products of combustion mix with diluent air at the diverter and diluted products pass up the secondary flue.
2. Adverse conditions with down-draught – the down-draught mixes with the products of combustion and diluted products enter the room for the brief period

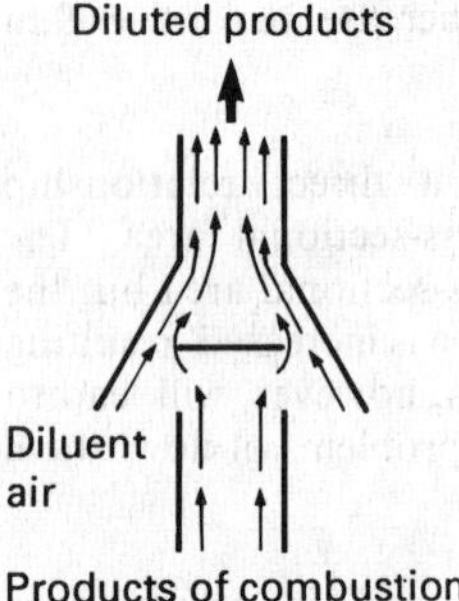

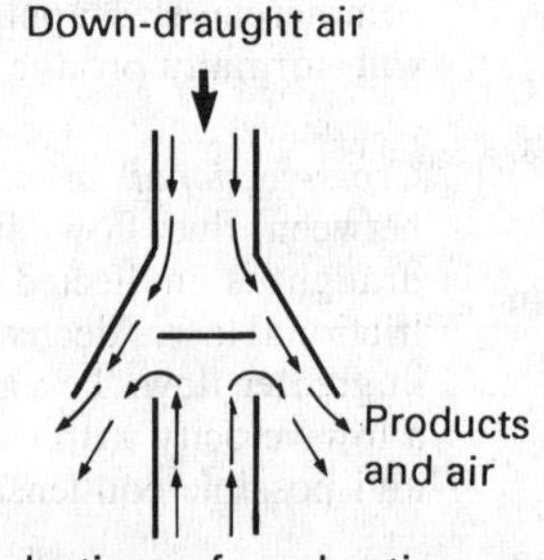

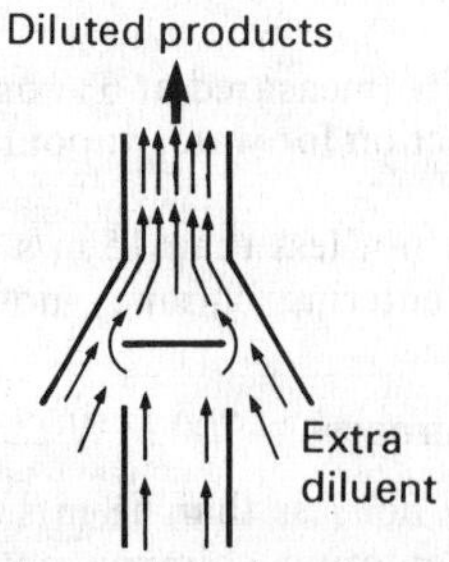

Figure 10.5 The function of the draught diverter

that the down-draught persists. The draught in the primary flue (and hence the performance of the appliance) is unaffected.

3. Adverse conditions with up-draught – the additional draught in the secondary flue causes excess air to be pulled through the diverter from the room. The baffle plate prevents the draught in the primary flue from being affected.

For small appliances the down-draught diverter is often incorporated into them; for larger appliances it is external.

In larger appliances the use of a down-draught diverter can be impractical if only because of size, and a draught stabilizer may then be used. This will protect against up-draught by opening to allow air to be pulled in directly from the room. If it is of the double-swing type it will protect against down-draught by opening to relieve the down-draught into the room.

Forced-draught flues. The above design parameters are relevant to natural-draught flues. With forced-draught flues it is possible by choice of a fan, either forced or induced draught, to overcome system resistance so that the flue will still clear the products. A crude rule-of-thumb is to allow 1 mm^2 of flue area for each 2.2–3.7 kW for natural draught and for 4.5–13.6 kW for each forced draught.

10.6.3 Design procedure

The design procedure will vary with the type of flue, but the basic procedure will not change. It is necessary to consider the relevant parameters discussed above and decide on flue dimensions and materials of construction to satisfy them. The parameters to consider are:

1. Type of flue to be used;
2. Pressure needed at inlet to flue for correct operation of appliance;
3. Diameter of flue (or dimensions if rectangular);
4. Length of flue and route taken;
5. Materials of construction (this will influence both heat transfer and flow characteristics);
6. Heat input and efficiency of the appliance, together with the flue temperature and excess air. A consideration of all or some of these allows the flue gas volume to be assessed;
7. Termination position.

This design procedure is detailed in the British Gas publication *Flues for Commercial and Industrial Gas Fired Boilers and Air Heaters* (IM/11). This publication addresses itself to:

1. Chimney heights;
2. Natural-draught flues;
3. Induced-draught flues;
4. Modular boiler systems;
5. Fan-diluted flues.

Other relevant aids to design include:

BS 5440: 1, Code of practice for flues and air supply for gas appliances of input not exceeding 60 kW: flues

This is only applicable to small appliances.

BS 5854, Code of practice for flues and flue structures in buildings

This applies to larger appliances of 45 kW output and above but does not cover flues not integral to a building.

10.6.4 Chimney Heights Memorandum

The relevant legislation to consider when determining the termination of a flue from a larger item of plant is *Chimney Heights: Third edition of the 1956 Clean Air Act Memorandum*. This deals mainly with the adequate dispersion of gases containing sulphur, of which there is very little in gas. However, gas is included and defined as a 'very low sulphur (<0.04%) fuel' (VLS) together with premium kerosene and LPG.

The Memorandum gives methods of calculating the uncorrected chimney height (*U*) based on heat input and sulphur content. The height is then corrected on the basis of the type of area in which the plant is located and the influence of nearby buildings. Important points to note are:

1. The scope of the document is for heat inputs of 150 kW to 150 MW.
2. An efflux velocity sufficient to prevent downwash is necessary. This is defined as:
   (a) For boilers up to 2.2 MW input, not less than 6 m/s at MCR;
   (b) For boilers and similar plant with induced-draught fans:
   Rating up to 9 MW, not less than 7.5 m/s at MCR.
   Rating above 135 MW, not less than 15 m/s at MCR.
   Rating 9–135 MW, pro rata.
3. Where there are several adjacent plants in the same works, the waste gases should normally be discharged from a common chimney, preferably multi-flued.
4. A chimney should terminate at least 3 m above the level of any adjacent area to which there is general access.
5. A chimney height should never be less than *U*.
6. A chimney should never be less than the height of any part of an attached building within a distance of 5*U*.
7. Fan-diluted systems are covered.

10.6.5 New flue systems

Various flue systems have been developed to overcome the need for a conventional flue, the provision for which can be both difficult and expensive in some locations.

Balanced flue

In the balanced flue, which may be natural draught or fan-assisted, the flue outlet and combustion air inlet are integral in a housing to be mounted on an outside vertical wall. As the flue and the inlet are subject to the same outside pressure, they are unaffected by it and are in balance. No provision for ventilation is needed and the appliance is room-sealed.

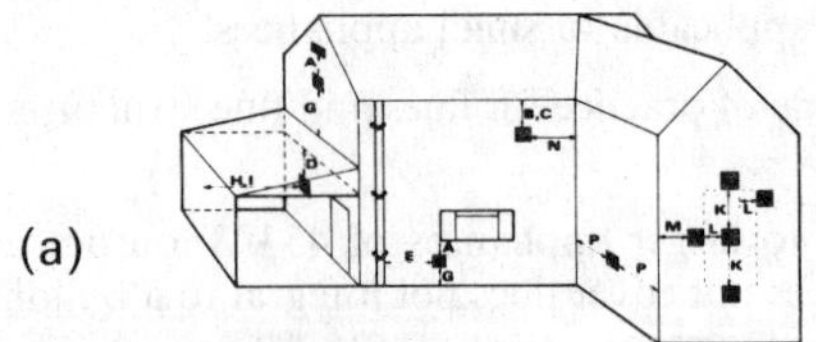

If appliance manufacturer's installation instructions do not give specific siting dimensions then the dimensions shown on this chart should be used.

TERMINAL POSITION	MINIMUM DISTANCE
A — Directly below an openable window or other opening e.g. air brick	300 mm
B — Below gutters, soil pipes or drain pipes.	200 mm
C — Below eaves.	200 mm
D — Below balconies or car port roof.	400 mm
E — From vertical drain pipes and soil pipes.	75 mm
G — Above ground, roof or balcony level.	300 mm
H — From a surface facing a terminal.	600 mm
I — From a terminal discharging towards another terminal.	600 mm
K — Vertically from terminal on the same wall.	1,500 mm
L — Horizontally from a terminal on the same wall.	300 mm
M* — From a single external corner.	100 mm
N* — From a single internal corner.	300 mm
P* — From double corners (both sides of the terminal).	500 mm

*Plan view of corners:-

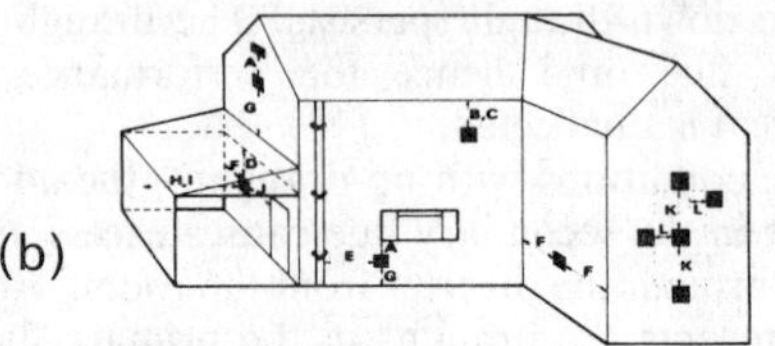

If appliance manufacturer's installation instructions do not give specific siting dimensions then the dimensions shown on this chart should be used.

TERMINAL POSITION	MINIMUM DISTANCE NATURAL DRAUGHT	MINIMUM DISTANCE FANNED DRAUGHT
A — Directly below an openable window or other opening e.g. air brick	300 mm	300 mm
B — Below gutters, soil pipes or drain pipes.	300 mm	75 mm
C — Below eaves.	300 mm	200 mm
D — Below balconies or car port roof.	600 mm	200 mm
E — From vertical drain pipes and soil pipes.	75 mm	75 mm
F — From internal or external corners.	600 mm	300 mm
G — Above ground, roof or balcony level.	300 mm	300 mm
H — From a surface facing a terminal.	600 mm	600 mm
I — From a terminal facing a terminal.	600 mm	1,200 mm
J — From an opening in the car port (e.g. door, window) into dwelling.	1,200 mm	1,200 mm
K — Vertically from a terminal on the same wall.	1,500 mm	1,500 mm
L — Horizontally from a terminal on the same wall.	300 mm	300 mm

Figure 10.6 Balanced flue terminal positions for appliances with a maximum heat input of (a) 3 kW and (b) 60 kW

Balanced flues are only available for low ratings (up to 60 kW input). Possible locations for the terminal are restricted as shown in Figure 10.6.

Balanced compartment

This can be regarded as an extension of the balanced flue principle to larger plant. A conventional chimney is used for clearing flue products, and the air inlet to the boilerhouse is integral with this so that the inlet and outlet are balanced. It is important for correct operation that there is no additional ventilation and that the room is sealed with all doors closed.

Fan-diluted flues

If the products of combustion can be diluted so that the carbon dioxide content is not greater than 1% it is permissible to discharge them at ground level. This is the principle of the system shown in Figure 10.7, in which fresh air is drawn in to dilute the flue products which are discharged, preferably on the same wall as the inlet to balance against wind effects. It is essential to interlock the air flow switch with the burner controls.

The volume flow rate of the fan (Q) is given by:

$Q = 9.7 \times (\text{Rated input in kW of boiler})/3600\ \text{m}^2/\text{s}$

Additional points to note are:

1. The rating of the plant must not exceed 6 MW.
2. The emission velocity must be at least $75/F$ m/s. (F is the fan dilution factor, defined as $F = V/V_0$, where V is the actual flue gas volume and V_0 is the stoichiometric combustion volume, $0.26Q$ m^3/s for gas.)
3. The outlet must not be within $50U/F$ of a fan-assisted intake (except for intakes for combustion air or fan dilution air).
4. The outlet must not be within $20U/F$ of an operable window.
5. The distance to the nearest building must be at least $60U/F$.

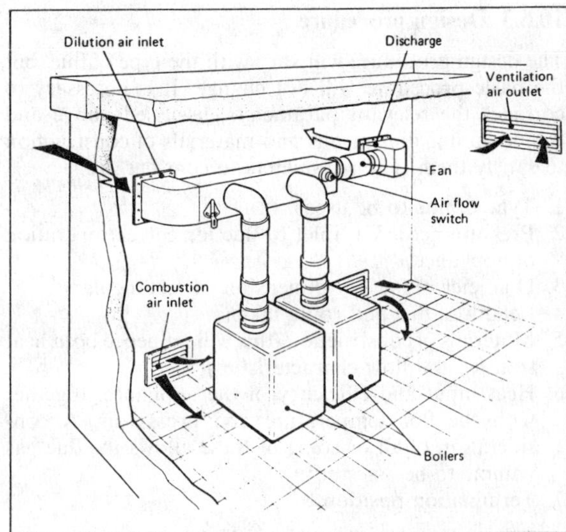

Figure 10.7 A typical arrangement of a fan-diluted flue system

6. The outlet must be at least 3 m above the ground, except if the input is less than 1 MW, where 2 m is permissible.
7. The outlet must be directed at an angle above the horizontal (preferably at 30°) and not under a canopy.
8. The outlet should not discharge into an enclosed (or almost totally enclosed) 'well' or courtyard.

Flues for condensing systems

Boilers and air heaters can be designed to operate in the condensing mode, that is, the outlet temperature of the products entering the flue can be at or close to the dewpoint. The usual design criteria for flues, which attempt to avoid condensation, are no longer applicable and it is necessary to incorporate additional features. These will include:

1. Provision must be made to drain off and dispose of the condensate, which is produced at a rate of up to 0.15 l/h for every kilowatt of input. This is slightly acidic, and plastic is the most suitable for the pipework.
2. The flue will be wetted by the condensate, and should be constructed of a suitable material such as stainless steel, aluminium or suitable plastic.
3. All sockets in the flue should face upwards and the joints sealed.
4. The low temperature of the flue products will contribute little flue draught.
5. The flue gases will readily form a plume at the flue terminal.

Additional guidance in the design of flues for condensing systems is given in the British Gas publication *Installation Guide for High Efficiency (Condensing) Boilers – Industrial and Commercial Applications* (IM/22).

10.6.6 Dual-fuel installations

In a dual-fuel installation the alternative fuel will usually have a higher sulphur content than gas, and it is the alternative fuel which will usually dictate the chimney height, materials of construction, etc. In a changeover situation, that is, where plant previously fired by another fuel is changed over to gas, chimneys or flues that have been satisfactory on oil or solid fuel will also be satisfactory on gas firing, unless condensing or direct contact boilers are to be fitted. The chimney should be checked to ensure that there is no restriction, that no leakage is occurring, and that on the previous fuel draught was adequate. In some cases it may be necessary to reduce an excessive draught with double-swing draught stabilizers, etc.

It is permissible to fire with other fuels into the same chimney provided that:

1. The other fuel is not solid fuel;
2. The common flue is of adequate size and construction;
3. The burner system of the gas- and oil-burning boilers are operating under similar draught conditions and preferably in the same room;
4. The burner-control equipment for both oil and gas is to the current standard for this type of equipment.

10.6.7 Flue dampers

The use of flue dampers may be considered as a means of energy conservation by preventing ventilation of the boiler in the non-firing state. There is a possible safety hazard if the damper remains closed when the burner ignites and fires. Safe operation with flue dampers is covered in the British Gas publication *Automatic Flue Dampers for Use with Gas Fired Space Heating and Water Heating* (IM/19). An assessment of the fuel savings to be expected from the use of flue dampers has been made by Dann, R. G., Lovelace, D. E. and Page, M. W., 'The effect of flue dampers on natural ventilation heat losses from boilers', *The Heating and Air Conditioning Journal* (May 1984).

10.7 Health and safety in the use of gas

10.7.1 Legislation

Health and Safety at Work etc. Act 1974

This has relevance to the supply of gas and plant burning gas, in particular section 6, as amended by the Consumer Protection Act 1987.

Gas Act 1986

This is the primary legislation concerning the gas industry, together with certain sections of the Gas Act 1972 and the Gas Act 1965 which have not been repealed. The Gas Act does not have great relevance to safety aspects of gas utilization but it is an enabling Act under which specific regulations can be published.

An exception is Schedule 5, Public Gas Supply Code, which deals with meter installation, use of anti-fluctuators and valves, reconnection of gas supplies, escapes of gas, etc.

Gas Safety (Installation and Use) Regulations 1984 (SI 1984, No. 1358)

These Regulations cover the safe use of gas in users' premises. They apply to domestic and commercial premises but not to industrial, although it is practice to apply the intent of the Regulations to such premises. The Regulations cover most aspects of installation and safety, including gas fittings, meters, installation pipes, gas appliances, ventilation and flueing. They use the term 'competent person' but do not define it. It is probable that in the near future the term will be defined as one who has undergone a training course that complies with the HSE Approved Code of Practice on Standards of Training in Safe Gas Installation.

Gas Safety Regulations 1972 (SI 1972, No. 1178)

These Regulations were largely repealed by the later 1984 Regulations. Certain sections were retained, however, notably Part II, which deals with service pipes (i.e. the pipes to carry gas from the main to the user's property).

Gas Quality Regulations 1983 (SI 1983, No. 363)

These Regulations concern the purity and distinctive smell of gas, the uniformity of calorific value, and the minimum pressure which must be made available. For service pipes of internal diameter 50 mm or more this is 12.5 mbar.

Gas (Meters) Regulations 1983 (SI 1983, No. 684)

These deal with the accuracy of meters together with fees for testing, etc.

Gas (Declaration of Calorific Value) Regulations 1972 (SI 1972, No. 1878)

These cover the declared calorific value and the methods of making known changes in its value.

Gas Safety (Rights of Entry) Regulations 1983 (SI 1983, No. 1575)

These concern the rights of an 'officer authorized by the relevant authority' to enter premises for purposes connected with safety of fittings and appliances.

The Building Regulations 1985

Approved Document J, Section 2, deals with gas appliances of rated input up to 60 kW. It should therefore be noted that most commercial and industrial plant is not covered by these Regulations.

10.7.2 Potential hazards in the use of gas

Properties

Natural gas is a colourless gas to which has been added a distinctive odour. It is composed principally of methane (88–95%) together with small proportions of higher hydrocarbons (ethane, propane, butane), and nitrogen and/or carbon dioxide. It has a calorific value generally in the range 38–39 MJ/m^3 and is lighter than air with a specific gravity (with regard to air) of 0.59–0.64.

Health hazard data

Natural gas does not in itself constitute a hazard to health. However, exposure to concentrations in excess of approximately 10% in air may cause dizziness and nausea. Higher concentrations may lead to asphyxiation.

Exposure to carbon monoxide resulting from inadequate ventilation and/or leakage of combustion products may cause headaches, chronic tiredness or muscular weakness. High concentrations or long-term exposure may be fatal. Personnel suffering from these effects should be treated by normal resuscitation methods and medical advice sought.

Safety hazard data

Natural gas forms flammable mixtures with air in the concentration range 5–15% in air. In the event of a natural gas fire, steps should be taken to shut off the gas supply, and the local fire brigade and the gas supplier notified.

Handling

Natural gas must always be contained in appropriate pipes or vessels and precautions taken to ensure that leakage cannot occur. If a gas leak does occur, the main gas supply should be shut off, the area ventilated and the gas supplier informed. Electrical switches should not be turned on or off, portable electrical appliances including hand-held torches should not be operated and all other possible sources of ignition removed or rendered inoperable and the affected area ventilated. In circumstances of excessive leakage the building should be evacuated.

Application

Appliances and other equipment intended to be used with natural gas should be purpose designed and built to recognize safety standards, installed in properly ventilated locations and supplied with the necessary flueing systems or other means of disposal of combustion products. Lack of adequate ventilation and/or leakage of combustion products into the work space may give rise to carbon monoxide poisoning. In this event, affected personnel should be evacuated from the area, the gas supply shut off, the supplier informed and medical advice sought. Attention is drawn to the Gas Safety (Installation and Use) Regulations, relevant British Standards and appropriate British Gas Standards, codes of practice and guidance notes, all of which deal with the design, application and safe use of gas and gas-fired equipment.

10.7.3 Safety procedures

These have been partly covered in Section 10.7.2. All gas-fired equipment should be designed to be ignited, operated and shut down in a safe manner. Instructions to this effect should be clearly displayed. This is a requirement of the Health and Safety at Work etc. Act 1974. In addition, such plant should comply with all relevant standards.

In the event of an incident occurring it is essential in the first instance to shut off the gas supply. This is recognized in the Gas Safety (Installation and Use) Regulations by the need for an 'emergency control valve' sited close to the entry to a building. In smaller premises this may be the meter control valve. In larger premises it is a requirement of the Regulations that for every building with a gas supply of 50 mm or more and more than one self-contained area then a valve must be installed at the position where the gas enters the building. In such premises a line diagram of the pipe layout should also be prominently displayed, at least close to the primary meter. Such a diagram, indicating all emergency control valves and all pipework of internal diameter

25 mm or more, can be used to locate areas to be isolated in an emergency.

It is additionally recommended in the British Gas publication *Guidance Notes on the Installation of Gas Pipework, Boosters and Compressors in Customers' Premises* (IM/16) that for buildings containing plant over 2 MW total heat input and being supplied with gas at pressures above 1 bar a remotely operable valve shall be fitted in the gas supply to the building. In the case of large boilerhouses provision for remote operation of the valves shall be provided both inside and outside the building.

10.7.4 Maintenance

Safe operation of all plant is improved by means of regular maintenance by competent personnel in conjunction with the manufacturer's maintenance instructions. This should always be followed by recommissioning in a safe manner. It is particularly important that valves such as safety shut-off valves, non-return valves, etc. are regularly checked for soundness.

10.8 Pressure control

10.8.1 Governors

Constant-pressure governors are required at various stages of the gas supply within the user's premises from the first pressure reduction from distribution pressure to supply pressure at the meter installation to the appliance governor. The complexity and design of the governor installation depends on pressure, throughput, duty, etc. In addition to governors, pressure-relief, slam-shut and non-return valves may be relevant in some installations.

Low-pressure governors used for smaller appliances and smaller pressure-regulation installations should comply with BS 6448: 1, Specification for pressure governors with nominal connection size up to 50 mm for gas appliances with inlet pressures up to and including 200 mbar. For larger installations, but still low pressures, governors should comply with BS 3554: 2, Specification for gas governors: independent governors for inlet pressures up to 350 mbar.

Pressure-regulating installations

Detailed requirements for pressure-regulating stations are contained in two Institution of Gas Engineering publications:

TD/9: Offtakes and Pressure-regulating Installations for Inlet Pressures between 7 and 100 bar
TD/10: Pressure-regulating Installations for Inlet Pressures between 75 mbar and 7 bar

These references give details of the requirements for safely maintaining a constant pressure outlet over a wide range of flow rates. It should be noted that 'governor' and 'regulator' are largely interchangeable terms.

Requirements become more demanding as the pressure at the inlet rises. A governor alone has its characteristics which include a pressure 'droop' at high throughputs and an increase above set point at very low throughputs. In many cases there must additionally be protection against the failure of a governor. This can be provided by:

1. 'Active and monitor' governors (defined as two governors in series whose settings are stepped so as to allow the active governor to control the outlet pressure and the monitor governor to assume control in the event of failure of the active governor to the open position);
2. Pressure-relief valve (this will vent to atmosphere when a predetermined maximum pressure is attained);
3. Slam-shut valve, this will close quickly in the event of an abnormal pressure (usually excess) being detected.

The essential requirements for lower-pressure installations can be summarized as below. In most cases the governor installation is incorporated with the metering installation.

1. *Inlet pressure not exceeding 75 mbar*. The only requirement is for a single governor and a single stream, although twin streams will be advantageous for larger loads.
2. *Inlet pressures 75 mbar to 2 bar*. A single or multi-stream installation including:
   (a) Single-stage governor with internal valves open at rest;
   (b) Relief valve with capacity not more than 1% of the stream fault capacity;
   (c) Slam-shut valve.
3. *Inlet pressures 2–7 bar, governor of 50 mm bore or less and stream fault flow rates less than 300 m³/h*. As in (2) above *or* duplicate streams, each having two-stage governing and a slam-shut valve;
   (a) A monitor override pilot governor to be fitted to the first-stage governor impulsed from the outlet of the second-stage governor;
   (b) A relief valve of capacity not more than 1%, the fault stream capacity to be included in each stream.
4. *Inlet pressures 2–7 bar, larger flow rates*. Duplicate streams with two-stage governing as in (3) above *or* duplicate streams each having monitor and active governors and a slam-shut valve;
   (a) The monitor governor to be upstream of the active governor;
   (b) A relief valve of capacity not more than 1%, the fault stream capacity to be included in each stream.

Legal requirements

The Gas Safety (Installation and Use) Regulations require that the gas supply to the meter be governed in all premises other than mines, quarries and factories (Regulation 13). It is practice to govern all premises, but exceptions may be made in categories where it is permissible, and there is advantage in so doing.

10.8.2 Pressure-relief valve

The pressure-relief valve has the physical appearance of a pressure governor. It operates in the reverse manner to a governor, that is, on exceeding the set pressure the valve lifts off its seat, rather than being forced onto its seat, as in a governor. Gas is therefore allowed to flow through the valve.

The main function of the relief valve is to act as a back-up to a pressure governor to prevent the pre-set downstream pressure being exceeded. In particular, it will relieve the small amount of 'creep' that will occur by slippage past the valve seating in a governor.

Safe venting of the gas relieved is of great importance. The British Gas publication *Guidance Notes on the Installation of Gas Pipework, Boosters and Compressors in Customers' Premises* (IM/16) gives guidance on this. The main requirements are:

1. Vents must be terminated in a safe place, preferably in the open air above roof level, remote from possible sources of ignition.
2. Vents must not be manifolded together.
3. Terminals must be designed to minimize the risk of blockage and ingress of water.
4. The vent pipe, particularly if over 20 m in length, must be designed to minimize back-pressure.

Certain pressure governors have integral vent valves.

10.8.3 Slam-shut valves

The slam-shut valve cuts off the gas supply in the event of predetermined pressure criteria being exceeded. It must be manually reset, having made safe the downstream pipework and the cause of the abnormal pressure removed.

It consists of a spring-loaded cut-off valve which is held open in the open position under normal conditions. Should the pressure deviate sufficiently, the diaphragm will move and disengage the trigger mechanism, releasing the latch on the valve, which will close under the force of the spring.

Slam-shut valves are usually set to operate under conditions of high downstream pressure. They can be used for low pressures, and also combined high and low. Although they normally protect the supply pressure to premises using gas, slam-shut valves are sometimes used on process plant.

10.8.4 Non-return valves

It is a requirement of the Gas Act 1986, Schedule 5, that where 'air at high pressure' or 'any gaseous substance not supplied by the supplier' is used, it is necessary to install an 'appliance' to prevent the admission of that gas into the supplier's gas main. This requirement is usually met by means of a non-return valve installed at the meter outlet, where it will protect the supplier's mains. There are advantages to the user in having the non-return valve installed close to the point of use so that, in the event of a gas at higher pressure entering the gas supply, it will be checked by the non-return valve at an early stage rather than diffusing throughout the user's installation pipework. A non-return valve should meet the requirements of British Gas *Standard for Non-Return Valves* (IM/14).

It is important to note that the requirements of this standard are specific to gas, and include a low-pressure differential under forward-flow conditions. It is unlikely that a non-return valve suitable for compressed air, etc. will be acceptable for gas.

Reverse-pressure requirements specified in IM/14 are:

Up to 25 mm:	7 bar
25–150 mm:	2 bar
Over 150 mm:	1 bar

In the particular case of oxygen, valves shall be resistant to exposure for up to 12 h at all pressures up to 2 bar at 20°C. Requirements for non-return valves for oxygen are also discussed in the British Gas publication *Guidance Notes on the Use of Oxygen in Industrial Gas Fired Plant and Working Flame Burners* (IM/1).

It is essential for the safe operation of a non-return valve that it be regularly checked and be shown to be resistant to reverse pressures. To enable this check to be carried out, the non-return valve should be installed with a manual isolation valve at both inlet and outlet.

10.9 Gas specification and analysis

Natural gas as distributed in the UK is obtained from various sources. These comprise primarily the southern North Sea basin, northern North Sea fields (both British and Norwegian), Morecambe Bay from the Irish Sea (used primarily for winter peaks) and gas from world sources imported in small quantities as liquefied natural gas (LNG). Gas from the different fields is of very consistent quality, and further blending, conditioning, etc. allows a gas of very consistent quality and specification to be distributed.

10.9.1 Calorific Value

The Calorific Value is constantly monitored, and it is a condition of the Gas (Declaration of Calorific Value) Regulations 1972 that alterations in the declared Calorific Value (CV) are publicly made known. It is customary to quote the Gross (Upper) CV rather than the Net (Lower) CV. The difference between the two represents the heat contained in the latent heat of vaporization of the water vapour in the products of combustion which can only be recovered in condensing appliances.

The ratio of net/gross CV is about 90%, reflecting the high hydrocarbon/carbon ratio for gas. The range of declared CV (gross) is 38–39 MJ/m^3 with a typical value of 38.63 MJ/m^3. This latter figure represents a net CV of 34.88 MJ/m^3.

10.9.2 Wobbe Number

The Wobbe Number of gas is defined as 'The heat release when a gas is burned at a constant gas supply pressure'. It is represented by:

$$\text{Wobbe No.} = \frac{\text{Gross CV}}{\text{Square root of relative density of gas}}$$

The relative density (with respect to air) has no units, so that the units of the Wobbe Number are the same as Calorific Value. The value for gas is 49.79 MJ/m^3.

The Wobbe Number is of interest because it represents the heat released at a jet. A gas of varying CV but constant Wobbe Number would give a constant heat release rate.

10.9.3 Analysis of natural gas

The analyses of various gases from the North Sea are shown in Table 10.1. It is seen that the analysis is very constant, with methane being the dominant constituent at 91–96%.

10.9.4 Properties of natural gas

The main properties of gas and its combustion products are shown in Table 10.2. For comparison, the properties of propane and butane are included. It can be seen that the calorific values of these three fuels on a mass basis are very similar.

The properties of natural gas are dominated by those of methane, notably a low maximum flame speed of 0.33 m/s. This strongly influences burner design, which must ensure that the mixture velocity is sufficiently low to prevent 'blow-off'. 'Light-back', on the contrary, is very unlikely with such a low flame speed.

The range of satisfactory operation for a gas burner, defined by light-back, blow-off and incomplete combustion are limited. The variation in gas analyses, particularly higher hydrocarbons and inerts, can influence the range of operation. This has led to the definition of different groups of natural gas. A practical effect is that burners designed for the continent may not be suitable for the UK without adjustment. This does not apply to forced-draught burners.

10.9.5 Standard reference conditions

If properties are being accurately quoted it is important that the reference conditions are defined, as these can vary. This also has implications for metering. The main reference conditions are:

1. *Metric standard reference conditions*
   15.0°C, 1013.25 mbar, dry
   Symbols: MSC, m^3 (st)
2. *Imperial standard conditions*
   60.0°F, 30 inHg, saturated with water vapour
   Symbols: ISC, standard cubic foot, sft^3 (src in old references)
   SI equivalent: 15.56°C, 1013.7405 mbar, saturated
3. *Normal temperature and pressure*
   0°C, 760 mmHg (at 0°C and g=9.80665 m/s^2)
   Symbols: nm^3, N m^3, m^3 (n), STP, NTP
   SI equivalent: 0°C, 1013.25 mbar

   Unless otherwise specified, the gas may be assumed to be dry.
4. *US gas industry 'standard cubic foot'*
   60°F, 30 inHg (at 32°F, g=32.174 ft/s^2), saturated
   SI equivalent: 15.56°C, 1015.92 mbar, saturated
   Symbols: scf, sft^3

Table 10.1 Analysis of natural gas

	Mean North Sea	*Bacton*	*Easington*	*Theddlethorpe*	*St Fergus*
Composition (vol %)					
Nitrogen (N_2)	2.72	1.78	1.56	2.53	0.47
Helium (He)		0.05	—	—	—
Carbon dioxide (CO_2)	0.15	0.13	0.54	0.52	0.32
Methane (CH_4)	92.21	93.63	93.53	91.43	95.28
Ethane (C_2H_6)	3.6	3.25	3.36	4.10	3.71
Propane (C_3H_8)	0.9	0.69	0.70	0.99	0.18
Butanes (C_4H_{10})	0.25	0.27	0.24	0.33	0.03
Pentanes (C_5H_{12})	0.07	0.09	0.07	0.10	0.01
Higher hydrocarbons	0.1	0.11	—	—	—
Composition (wt %)					
Nitrogen (N_2)	4.38	2.90	2.55	4.05	0.79
Helium (He)		0.003	—	—	—
Carbon dioxide (CO_2)	0.38	0.33	1.39	1.31	0.84
Methane (CH_4)	84.89	87.37	87.28	83.62	91.10
Ethane (C_2H_6)	6.21	5.69	5.88	7.03	6.65
Propane (C_3H_8)	2.28	1.59	1.80	2.49	0.47
Butanes (C_4H_{10})	0.83	0.91	0.81	1.09	0.11
Pentanes (C_5H_{12})	0.29	0.38	0.29	0.41	0.04
Higher hydrocarbons	0.74	0.82	—	—	—
Ultimate composition (wt %)					
Carbon	72.16	73.23	72.93	71.98	74.38
Hydrogen	23.18	23.62	23.51	23.02	24.21
Nitrogen	4.38	2.91	2.55	4.05	0.79
Oxygen	0.28	0.24	1.01	0.95	0.61

Table 10.2 Properties of gaseous fuels and combustion products

Properties at 1013.25 mbar (15°C)	*Mean North Sea*	*Typical commercial propane*	*Typical commercial butane*
Specific gravity (air = 1)	0.602	1.523	1.941
Gross calorific value			
MJ m^{-3} (st)	38.63	93.87	117.75
Btu ft^{-3} (ISC) dry	1036	2500	3270
MJ kg^{-1}	52.41	50.22	49.41
Btu lb^{-1}	22530	21500	21200
Therms ton^{-1}	505	480	477
Btu $gall^{-1}$	—	110040	124352
Net calorific value			
MJ m^{-3} (st)	34.88	86.43	108.69
Btu ft^{-3} (ISC) dry	935	2310	3030
MJ kg^{-1}	47.32	46.24	45.61
Btu lb^{-1}	20340	19800	19650
Liquid SG (water = 1) at 60°F	—	0.51	0.575
Volume gas/volume liquid (at 0°C)	—	274	233
Wobbe No.			
MJ m^{-3} (st)	49.79	76.06	84.52
Btu ft^{-3} (ISC) dry	1335	2026	2347
Theoretical air required for combustion			
(v/v)	9.76	23.76	29.92
(w/w)	16.5	15.6	15.3
m^3 (st) kg^{-1} of fuel	13.24	12.73	12.50
f^3 (st) lb^{-1} of fuel	212.0	203.92	200.23
Theoretical stoichiometric dry flue gas CO_2%	11.86	13.8	14.1
Volumetric composition of wet theoretical flue gas			
% CO_2	9.66	11.7	12.0
% H_2O	18.63	15.4	14.9
% N_2	71.71	72.9	73.1
Dewpoint of flue gas (°C)	59	55	54
Theoretical flame temperature (°C)	1930	2000	2000
Limits of flammability (% by volume to form combustible mixture)	5–15	2–10	1.8–9

NB: Calorific Value (saturated) = 0.9826 × Calorific Value (dm).

10.10 Control of efficiency

10.10.1 Monitoring of combustion

Reference to Figure 10.1 shows how efficiency can be adversely affected by deviation from the optimum air/gas ratio. By maintaining combustion close to stoichiometric, efficiency will be improved, but the practical limitations of burners discussed above must be noted.

The quality of combustion can be measured with suitable instrumentation, on either a periodic or a continual basis. If continuous analysis is practised then there may be feedback to continuously adjust the air/gas ratio and/or ratio and/or record the data derived.

Flue gases analysed

The flue gases analysed will be one or more of carbon dioxide, carbon monoxide and oxygen. If carbon dioxide alone is measured it is possible to draw erroneous conclusions, as the level will peak at stoichiometric and reduce in both the excess air and air deficiency regions. It is essential to measure another flue gas to obtain a reliable assessment of burner performance.

Oxygen does not have this disadvantage, being, in theory, zero at stoichiometry and increasing in the excess air region. Sampling of carbon monoxide additionally is advocated during initial commissioning, if not always on a routine basis.

Carbon monoxide is usually sampled as the second parameter in conjunction with carbon dioxide or oxygen. In theory, as the optimum is usually to have near-stoichiometric combustion without 'CO breakthrough' it is the most reliable gas to sample. A problem is that although small quantities of CO usually indicate the need for additional air, they can also be caused by flame chilling and careful interpretation of results is needed.

Wet and dry analysis

The relatively high hydrogen content of gas, contained in the methane, leads to a water vapour content of approximately 18% by volume in the flue products. The analysis of the other constituent gases is affected by whether or not the water vapour is included.

The 'dry' analysis is on the basis of the water being

removed from the sample prior to analysis, and the maximum theoretical carbon dioxide content is 11.87%. Most flue analyses are carried out on the dry basis.

The 'wet' analysis assumes that the water vapour is present. The maximum theoretical carbon dioxide content is 9.66%. The zirconia cell method of measuring oxygen is on the wet basis.

Instruments used

A wide range of analysis instruments can be used, either of the portable or permanently installed type. The latter will frequently be recording instruments and may have control capabilities.

Various principles are employed in analysis equipment, including:

1. Wet chemical analysis;
2. Electrolytic cell;
3. Chromatography;
4. Paramagnetism – for oxygen;
5. Non-dispersive infrared absorption;
6. Thermal conductivity.

10.10.2 Control of combustion

In the great majority of installations flue gas analysis is carried out for monitoring reasons only. In some larger installations, primarily boilers, a deviation in the reading of the gas measured is used for control purposes. The safety aspects of such system must be satisfactory, in particular that the controller cannot erroneously drive the burner to fire rich under any circumstances. The controller should be capable of being set to follow the optimum curve for excess air with firing load which recognizes that more excess air is needed at lower firing rates, as shown in Figure 10.8.

Much responsibility rests with the commissioning engineer who must have an intimate knowledge of the burner, the boiler, and the combination of the two.

A decision to use an accurate method of combustion control will, in general, be based on economics. The fuel saving is unlikely to be more than about 2% and this must be balanced against the total costs of operating the control system.

In the case of larger boilers such an installation can often be justified.

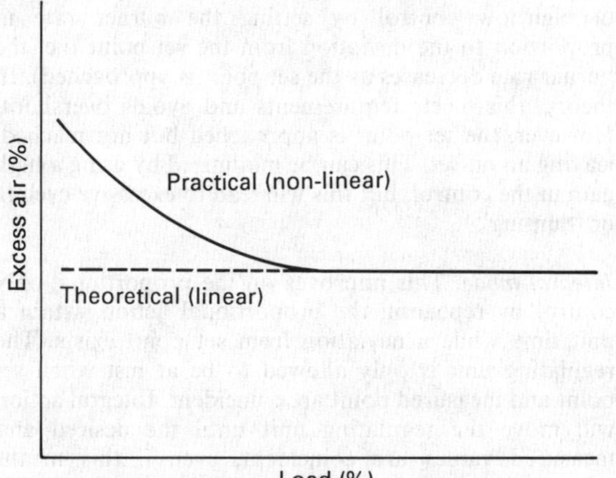

Figure 10.8 Variation of excess air with load

Oxygen trim systems

Most oxygen trim systems interpose an additional link in the air/gas ratio controller. Others use an additional valve. Most types are based on the zirconia cell installed in the flue, while others use paramagnetic or electrolytic cell methods. The zirconia type has the advantage that there is no time lag in sampling, nor is there a risk of contamination of the sample.

If the flue is operating under negative pressure, which is often the case, care should be taken that no 'tramp' air is allowed to enter the flue upstream of the sampling point, as this will give erroneous measurements. The trim system should be set to follow the practical firing curve and, in the event of a malfunction, be 'disabled' so that it ceases to have any influence on the air/gas ratio, which reverts to the normal load control only.

Carbon monoxide trim systems

There is limited application for CO trim systems which are widely used on utility and other large watertube boilers. The principle of operation is for an infrared beam to traverse the flue from emitter to sensor. The absorption of the infrared radiation is proportional to the CO content.

For gas-fired shell boilers it is difficult to justify these trim controllers on an economic basis. Equally important, the position to control based on CO in such a boiler is ideally in the reversal chamber and not the flue. However, the temperature and stratification of flue products here make it impracticable.

Control without trim systems

Most combustion equipment is not controlled by means of a feedback from flue gas analysis but is preset at the time of commissioning and preferably checked and reset at intervals as part of a planned maintenance schedule. It is difficult to set the burner for optimum efficiency at all firing rates and some compromise is necessary, depending on the control valves used and the control mode (e.g. on/off, fully modulating, etc.).

In the absence of a trim system with feedback it is likely to prove cost-effective to use a simple portable 'efficiency monitor' on a regular basis, perhaps weekly for small boilers. The change of reading is as important as the actual value of the reading. A deviation from what is known to be a good post-commissioning setting will indicate a drift from ratio and the need for remedial action.

Types of air/gas ratio control. There are various types of air/gas ratio device commonly used, including:

1. *Linked butterfly valves.* This simple technique is only suited to high/low controls.
2. *Linked square port valve.* This system has a more linear relationship between flow rate and rotation than butterfly valves. There is no backlash, as the air and gas valves are on a common spindle and fine tuning is possible.
3. *Linked characterized valves.* The relationship between angular rotation of the valve spindle and open area of the valve can be adjusted over different portions of the flow range using a series of screws. In this way, the air/gas ratio can be characterized to any desired profile over the whole firing range. These valves are suited to fully modulating systems and are commonly used on steam boilers.
4. *Balanced pressure-control systems.* Various control systems are based on the balanced pressure principle in which the air pressure (or a portion of it) is applied to the diaphragm of a zero governor in the gas supply. As the air pressure is varied so is the gas pressure in proportion, as shown in Figure 10.9. The air/gas ratio is set by means of an adjustable orifice valve in the gas supply. The pressure divider system will maintain air/gas ratio over much of the turndown ratio. At very low firing rates there is a need for excess air to maintain satisfactory combustion, and this is achieved by means of tensioning the spring within the governor to give an offset (Figure 10.9). The effect of this offset is considerable at very low rates but insignificant at higher ones. The pressure divider principle is exploited in more complex multi-diaphragm controllers. In these, the pressures balanced are differential pressures across orifices, enabling the ratio to be maintained under conditions of varying back-pressure.
5. *Electronic ratio controller.* In this type of controller a proportion of both gas and air is diverted through a by-pass in which a thermistor sensor measures the flow. The air and gas flows can be compared and the ratio calculated and displayed. A deviation from the pre-set ratio will be automatically restored by a ratio control valve in the air or gas supply, depending on whether the mode of operation is gas- or air-led. The electronic controller maintains ratio over a 10:1 turndown. The principle of operation is based on mass flow, so that it can be used with preheated air in recuperative systems.

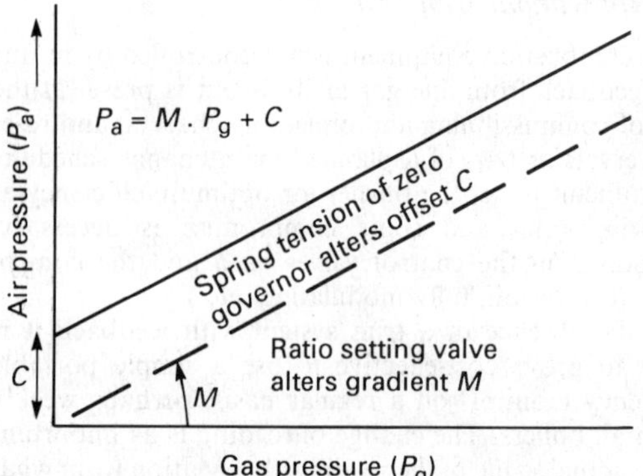

Figure 10.9 Relationship between air and gas pressures for cross-loaded governor system

10.10.3 Process controls

The choice of the mode of control of heating plant can have a considerable effect on the efficiency of the process. The process itself may dictate the controls. For example, a simple high/low temperature controller is unlikely to satisfy a process tolerance requirement of within a few degrees Centigrade. Similarly, a more complex process controller cannot be coupled with a burner only capable of firing high/low.

Types of process control

High/low control. This may be on/off in certain applications. Control is fairly crude, with the burner moving to high fire at a point somewhere below the nominal set point and returning to low fire somewhere above it. The result is a sinewave control pattern with a band rather than a point of control. This is, however, still an adequate control mode for many purposes.

Most burners are efficient at high fire but less so at intermediate rates and particularly at low fire. An on/off burner is therefore apparently efficient from an energy-utilization viewpoint. However, when the burner is called on to fire, in the case of forced-draught burners a purge is usually necessary which will both cool down the process and cause a delay in response, and in the case of natural draught there will be heat losses due to ventilation in the 'off' period.

In general, a fully modulating burner will be thermally more efficient, particularly if it maintains air/gas ratio accurately at intermediate rates. For natural-draught burners this may not be the case, as, in general, the air is not modulated, leading to progressively more excess air at lower rates.

Proportional-only control. This avoids the cycling effect of high/low control by setting the burner rate in proportion to the deviation from the set point (i.e. the burner rate decreases as the set point is approached). In theory, this meets requirements and avoids overshoot. However, the set point is approached but not reached, leaving an off-set. This can be minimized by using a high gain in the control, but this will lead to excessive cycling or 'hunting'.

Integral mode. This improves on the proportional-only control by repeating the proportional action within a unit time while a deviation from set point exists. The regulating unit is only allowed to be at rest when set point and measured point are coincident. Integral action will move the regulating unit until the desired and measured values are coincident, even if this means moving the regulating unit to the fully open or closed position.

Derivative mode. This improves on the proportional-only control by responding solely to the rate of change of the deviation but not in any way to the actual value of the deviation. Derivative action is always used with proportional control.

Proportional plus integral plus derivative action. Proportional action provides a controller output proportional to the error signal. Integral action supplies a controller output which changes in the direction to reduce a constant error. Derivative action provides a controller output determined by the direction and rate of change of the deviation. When all these are combined into one controller (three-term or PID) there is an automatic control facility to correct any process changes.

Controls for space heating in buildings

The selection of controls for space heating a building shows how the overall efficiency of the process can be improved. Whether or not it is cost effective, to do this must be examined in detail. At its most basic, such a system might include one or more LPHW boilers, a single thermostat in the heated space controlling the circulating pump on/off, and domestic hot water from a calorifier.

Boiler selection. Although not directly concerned with controls, the sizing of boilers is very important. There is a tendency for oversizing with consequent unnecessary cycling. Controls should ensure that if more than one boiler is installed then one boiler should act as 'lead'. Consideration should be given to a higher-efficiency boiler, possibly a condensing type.

In most cases, splitting the heating and domestic hot water loads will be advantageous. This avoids the inefficient use of a large boiler solely to heat a calorifier in the summer months.

Controls selection. The selection of controls will be influenced to an extent by the characteristics of the building, particularly the insulation material and the number and disposition of windows. Control options to be considered include:

1. *Zone control valves*. In most buildings different areas will be subject to varying levels of solar gain, occupancy, etc. Dividing the building into different zones each with its own thermostat and control valve will be beneficial.
2. *Compensating controller*. The heat requirements for a building vary according to the outside conditions, particularly the outside air temperature. The compensating controller measures the outside air temperature and, together with other parameters, including the thermal response of the building, adjusts the circulating water temperature to meet the demand.
3. *Three-port mixing valves*. The valve takes hot water from the boiler and mixes it with the cooler return water to provide a circulation of mixed water of sufficiently high temperature to meet the heating demand. The system operates with a constant flow rate and variable temperature. In the case of condensing boilers the return water should pass through the condensing heat exchange of the boiler prior to flowing to the mixing valve. An alternative for condensing boilers that does not use a mixing valve is to set the boiler flow temperature, by means of an outside air temperature sensor and compensating control box, in inverse proportion to the outside air temperature. Both these alternatives cause the boiler to operate in its efficient condensing mode for the maximum period.
4. *Optimum start controller*. Any building will heat up more quickly on a warm day: to start the heating system unnecessarily early is uneconomic. The optimum start controller monitors the outside and inside temperatures and computes the start time that will just bring the temperature up to design by the start of occupancy. The controller is only used during the preheat period during which other modulating controls are not operative. Some versions are available with a self-learning capability, which corrects the initial programme until it matches the building responses, and an optimum off control, which takes account of the allowable drop in space temperature at the end of occupancy.
5. *Building management systems*. These can monitor and control most of the building services by means of sensors in the various building service plants for collecting data and carrying out control functions.

10.11 Automation

10.11.1 Requirements for automatic burners

Process or heating plant may have controls ranging from manual operation with some processes supervised by interlocks to semi-automatic and fully automatic operation. Deciding factors will include temperature of operation, frequency of ignition, degree of operator supervision, and rating of the plant. For example, boiler plant, both steam and water, will invariably have automatic control whereas tunnel kilns operated continuously at high temperature are unlikely to require it.

The definition of an automatic burner is: 'A burner where, when starting from the completely shut-down condition, the start-gas flame is established and the main gas safety shut-off valves are activated without manual intervention.' This means that a burner is only automatic if it is ignited by means of a remote interlock (e.g. thermostat, timeswitch, etc.) closing. A burner is not automatic if it has a pilot burner that remains ignited in the 'off' condition. Nor is a burner strictly automatic if a start button needs to be pushed, even though the controls may comply with all requirements for automatic burners.

10.11.2 Automatic burner sequence

The sequence of operation for automatic burners is based on that which a knowledgeable, conscientious and alert operator would perform. This involves checks at stages to ensure that no hazard has developed. This sequence can be summarized as:

1. Check that the flame detector (ultraviolet or flame rectification) is not giving a spurious signal. This is continuous throughout the purge period.
2. Check that the safety shut-off valves are in the closed position (where applicable).
3. Check that the air flow detector is not giving a false indication.
4. Start the combustion air fan.
5. Prove the combustion air flow (at high rate, if variable).
6. Purge for a predetermined period (sufficient to give five volume changes of the combustion chamber and associated flueways).
7. Modulate the combustion air flow to the low fire rate (where applicable).
8. Prove the burner throughput control valves to be in the low fire position (where applicable).
9. Energize the ignition source.
10. Energize the start gas safety shut-off valve(s).
11. De-energize the ignition source.
12. Check that the flame detector is registering the presence of a flame.
13. Energize the main gas safety shut-off valves.
14. In the case of an interrupted pilot, de-energize the safety shut-off valves.
15. Check that the flame detector is registering the presence of a flame. This is continuous until the burner is shut down.
16. Release the burner controls to modulation so that the burner can be driven to high fire, according to demand.

In the event of any of the checks not being satisfactory the burner is shut down or locked out, as appropriate.

10.11.3 Burner Standards and Codes of Practice

Codes of Practice

Gas burners should comply with the relevant Codes of Practice, depending on whether the plant is low or high temperature. Low-temperature plant is defined as that having a normal working temperature insufficient to ignite the fuel, that is, below 750°C at the working temperature walls. The British Gas Codes of Practice are:

Code of Practice for the Use of Gas in High Temperature Plant (IM/12)

Code of Practice for the Use of Gas in Low Temperature Plant (IM/18)

Automatic Burner Standards

British Standards exist for automatic burners, i.e. those requiring no manual intervention to be ignited. The Standard is:

BS 5885, Automatic gas burners
Part 1. Specification for burners with input rating 60 kW and above (1988)
Part 2. Specification for packaged burners with input rating 7.5 kW up to but excluding 60 kW (1987)

Part 2 deals only with packaged burners while Part 1 covers all types.

The number and quality of the safety shut-off valves required is specified in the relevant Standards and Codes of Practice. These are usually electrically operated valves conforming to class 1 or class 2 as defined in BS 5963, Specification for electrically-operated automatic gas shut-off valves (1981). This specifies forward pressures and reverse pressures that the valve must withstand in the closed position together with closing times.

For start gas supplies two safety shut-off valves are needed on all but the smallest burners. For automatic burners the main gas safety shut-off valve requirements are:

From 60 kW up to 600 kW:	One class 1 and one class 2
Above 600 kW up to 1 MW:	Two class 1
Above 1 MW up to 3 MW:	Two class 1 with a system check
Above 3 MW:	Two class 1 with proving system

A 'system check' is usually satisfied by closed-position indicators on the valves. These show that the valves are nominally in the closed position but do not actually prove that there is no leakage. A 'proving system' can be met by various systems which prove the soundness of the upstream and downstream safety shut-off valves by sequential pressurization, evacuation, mechanical over-travel with limit switches and a normally open vent valve, or other means.

Valves other than electrically operated (e.g. pneumatically operated ball valves) are equally acceptable if the pressure and closing time requirements of BS 5963 are met.

10.11.4 Standards to be applied

Automatic or semi-automatic

The amount of process plant that can be defined accurately as automatic is relatively small, and manual intervention is often involved at some stage. The relevant design criteria are therefore often IM/12 or IM/18. In practice, fully automatic burner controllers tested and certified by British Gas are available that comply with the requirements of BS 5885. Although these have features which may not be applicable to non-automatic plant it may be more appropriate to use such a controller, particularly as its safety is well proven. It may also be less expensive than buying and installing separate timers, relays, etc. For some processes (for example, those that do not need and cannot tolerate a long purge) such controllers may not be appropriate.

Appliance Standards

In most cases Codes of Practice or British Standards should be applied as appropriate. A few industrial and commercial gas appliances have relevant British

Standards and these should take precedence over more general codes and standards. These include:

BS 5978, Safety and performance of gas-fired hot water boilers (60 kW to 2 MW input) (1983)
BS 5990, Specification for direct gas fired forced convection air heaters for space heating (60 kW to 2 MW input) (1981)
BS 5991, Specification for indirect gas fired forced convection air heaters for space heating (60 kW to 2 MW input) (1983)

10.12 Fire and explosion hazards

10.12.1 Limits of flammability

Gas will form a flammable and explosive mixture with air or oxygen and the band within which this will happen will vary with pressure and temperature. At atmospheric pressure and 15°C the properties are:

	Lower explosive limit (LEL) (% gas)	*Upper explosive limit* (HEL) (% gas)
In air	5	15
In oxygen	5	61

In the combustion chamber it is essential to purge down to below the LEL before initiating the ignition source. The usual criterion is 25% LEL.

Outside the combustion chamber there should be no gas under any circumstances. Such a presence will indicate an uncontrolled leakage into the building. A gas level of 20% LEL is normally sufficient grounds for evacuating the building.

The human nose is often considered to be the best gas sensor, as the odorant contained in the gas permits detection at levels as low as 100 ppm or 0.01% in air. Gas detectors are, however, sometimes used to give additional warnings of any leaks. The following points should be borne in mind when considering such an application:

1. Location is of the greatest importance. It is not known from where a gas leak will emanate and such a leak will disperse rapidly. In general, a high-level location should be chosen, as gas is lighter than air. More than one detector will usually be advisable.
2. Detectors should comply with BS 6020, Instruments for the detection of combustible gas, specifically Parts 4 and 5.
3. Detectors may respond to other gases and vapours giving a false reading.
4. Detectors will require frequent maintenance and calibration to perform accurately.
5. The use of gas detectors should not be regarded as a substitute for proper ventilation.

10.12.2 Flame speed and flame traps

The flame speed of gas in air is 0.338 m/s. This is lower than most common flammable gases and is an indication of the low propensity for light-back of natural gas.

In certain applications a flame trap may be required, and further guidance is given in HSE publication HS(G)11, *Flame Arrestors and Explosion Reliefs*. The flame trap is constructed of strips of corrugated and flat steel ribbon wound in a spiral. It breaks up a flame in the very small passages between the corrugations where the flame is quenched and extinguished.

A flame trap is employed where premixed air and gas is used in combustion equipment and prevents the flame passing upstream into the pipe system. Flame traps should be situated as near as possible to the gas burner. This is so that the flame does not have a long pipe run in which it might accelerate to such a speed as to form a detonation wave and make the trap useless.

Thermocouple elements can be incorporated into the flame trap so that a flame that has lit back to the flame trap element and continued to burn there without being quenched can be detected. This sensor can be used to close an upstream safety shut-off valve. Flame traps will be the cause of a significant pressure drop for which allowances must be made in low-pressure systems.

Typical applications for flame traps include:

1. The outlet pipework from premix machines, but not fan-type mixers as they may adversely affect the operation of such machines;
2. Vents from pressure-relief systems and governor vents;
3. Purge outlet pipes.

10.12.2 Fire valves

Unlike oil burners, there is no requirement to install a fire valve. With oil burners this is usually met by having a fusible link sited over the burner and connected in a tensioned wire which holds open a weighted valve in the oil supply so that the valve closes if the link melts.

The use of automatically closing valves operated from gas-, fire- or smoke-detection systems is not normally required for gas installations. The reasons for this are:

1. There are very few recorded instances of fire caused by gas in installations such as boilerhouses;
2. A leak in the oil supply will cause oil to collect on the floor fuelling a fire in a predictable location, most probably at the burner itself. With gas a leak is possible at any joint, but such a leak will rapidly disperse. A predictable location is not possible, and a fire is not likely.
3. Gas can feed a fire caused by other sources. The integrity of pipework should withstand an external fire for sufficient time for an external emergency control valve to be operated.
4. Gas does not have the lubricant qualities of fuel oil. Consequently whereas a dropweight valve in the oil supply can be expected to close when called upon, this is not the case with a relatively large-bore gas ball valve or plug valve used for gas.

However, if it is desired to protect a self-contained area such as a boilerhouse against fire the best method is to use suitably located fusible links as interlocks in the controls of the burners, designed to BS 5885. The burner valves should be to BS 5963 and mounted in a non-vulnerable position.

If an automatic isolation valve is specified, the selection of the valve and its operating system must be carefully considered, particularly with respect to the design and methods of restoring the gas supply in those cases where appliances do not incorporate automatic flame safeguards. Where possible, valves should be to BS 5963 and systems in compliance with the British Gas publication *Weep By-Pass Pressure Proving Systems* (IM/20).

10.12.4 Explosion reliefs

HSE publication HS(G) 16, *Evaporating and Other Ovens*, gives guidance on the use of explosion reliefs in plant where flammable solvent is present as a potential source of explosion. Consideration should also be given to utilizing explosion reliefs in other plant, particularly low-temperature ones and where recirculation is practised. Correctly designed controls will minimize the risk of an explosion but explosion reliefs will reduce to a minimum adverse effects from such an incident. Further guidance on the use of explosion reliefs is given in the HSE publication HS(G) 11, *Flame Arrestors and Explosion Reliefs*.

The main requirements of a relief are:

1. If obstructions exist in the plant the relief should be located in a plane which allows an unobstructed flow of gases to the relief vent (e.g. if there are horizontal shelves the relief should be in a side, not the roof).
2. The relief size should be as large as possible, ideally occupying the whole of one side, and minimum vent area = 60 × area of side/P_{max} (m^2) (where P_{max} is the pressure that the plant can withstand without damage – if unknown, assume 140 mbar).
3. The relief should be as light as possible (the weight per unit area should not exceed 24 kg/m^3).
4. The relief should be held in position with the minimum of force.
5. There must be sufficient space around the plant to allow gases to vent freely and not impair the effectiveness of the relief.
6. Reliefs must not form dangerous missiles when called upon to operate.
7. The thermal resistance of the structure should not be significantly altered by the relief.

10.13 Maintenance

10.13.1 Need for maintenance

It is desirable that plant is regularly maintained to ensure that:

1. Thermal efficiency remains at an optimum.
2. Availability and reliability is at a maximum, with no avoidable unplanned downtime.
3. Safety devices are reliable.

The time intervals for maintenance will obviously vary widely. In all cases the work should be part of a clearly determined schedule.

10.13.2 Thermal efficiency

Combustion equipment, when first commissioned, can be set to operate at its optimum efficiency. With time, however, there will be a deterioration due to blockage of air filters and breather holes, wear in valve linkages, etc. Such changes may have safety implications if gas-rich firing is a consequence.

Regular checking of measured parameters and comparison with commissioning data will enable adjustments to be made to optimize efficiency. Replacement of items subject to wear, filters, etc. and items subject to thermal distortion such as some burner components will be necessary.

10.13.3 Reliability

To ensure that plant is not subject to breakdown it is important that there are no unnecessary 'failures to safety', that is, that correctly operating safety equipment only operates because of exceptional circumstances, not avoidable faults. Safety shut-down will occur for various reasons, i.e.:

1. No flame or shrinkage of flame during operation;
2. Failure to ignite flame initially;
3. Failure of any interlock, leading to (1) or (2).

Preventative maintenance that can be carried out that will minimize the chance of nuisance shutdowns includes:

1. Attention to air filters;
2. Cleaning of all burner ports liable to blockage;
3. Replacement of mechanical and electrical components known to have a finite life. For example, the average service life of an ultraviolet flame detector may be 10 000 h, and much less if used above its recommended temperature.

After any maintenance work involving replacement or adjustment of components it is essential that the plant be recommissioned in a safe way, including 'dry runs' with the gas turned off to ensure that flame failure devices operate correctly and pilot turndown tests to ensure that if the pilot can energize the flame detector then it can also smoothly ignite the main burner.

10.13.4 Safety

When initially commissioned all safety devices should be proved to be operating correctly. At intervals these should be checked to ensure that no undetected failure has taken place. Checks that are necessary include:

1. Visual examination of ignition of pilot burner and main burner;
2. All interlocks checked for correct setting and operation. These will include thermostats, pressurestats, limit switches, pressure switches, process interlocks, etc.;
3. Condition of flame sensor and its location with relation to pilot burner and main burner;
4. Soundness of non-return valve with regard to reverse flow;

5. Soundness of safety shut-off valves in the closed position for both forward and reverse flow. Note that this should be done with the valve *in situ* and without dismantling it, by means of a bubble leak detector, etc. The valve should not be dismantled to check for swarf on the seat, etc.

After modifying the controls in any way during maintenance it is essential to carry out a recommissioning of the plant as outlined in Section 10.13.3 above.

10.13.5 Training of maintenance staff

All staff involved with maintaining gas-fired equipment should be capable of doing so in a safe and responsible manner. The term 'competent' has not yet been defined in this context, but personnel should be qualified by both training and experience to carry out work on any plant which they are to maintain.

A code of practice outlining the training programme that should be undergone by persons carrying out work on gas-fired plant is Approved Code of Practice on Standards of Training in Safe Gas Installation. The Code offers guidance on the standards of training to produce competence in gas installation.

10.13.6 Manufacturer' instructions

For much plant the manufacturers will provide detailed maintenence schedules which should be followed at all times. This may necessitate the use of spares and tools specific to that appliance. Attention is drawn to Regulation 25(7) of the Gas Safety (Installation and Use) Regulations, which states that 'No person shall carry out any work in relation to a gas appliance which bears an indication that it conforms to a type approved by any person as complying with safety standards in such a manner that the appliance ceases to comply with those standards'.

10.14 Statutory requirements

The most important statutory requirements concerning gas distribution and utilization are:

Health and Safety at Work etc. Act 1974
Gas Act 1986
Gas Act 1972
Gas Act 1965
Gas Safety (Installation and Use) Regulations 1984 (SI 1984, No. 1358)
Gas Safety Regulations 1972 (SI 1972, No. 1178)
The Building Regulations 1985
Gas Safety (Rights of Entry) Regulations 1983 (SI 1983, No. 1575)
Gas Quality Regulations 1983 (SI 1983, No. 363)
Gas (Meters) Regulations 1983 (SI 1983, No. 684)
Gas (Declaration of Calorific Value) Regulations 1972 (SI 1972, No. 1878)
Oil and Gas (Enterprise) Act 1982
Control of Pollution Act 1974
Health and Safety (Emissions into the Atmosphere) Regulations 1983 (SI 943 of 1983)
Consumer Protection Act 1987
Reporting of Injuries, Diseases and Dangerous Occurrences Regulations 1985 (SI 2023 of 1985)

10.15 Testing

10.15.1 Soundness testing

It is essential that gas is not admitted to an installation pipe which is newly installed or has had work carried out on it unless the pipe is proved to be sound. This is a requirement of the Gas Safety (Installation and Use) Regulations, specifically Regulation 21. A suitable procedure is set out in the British Gas publication *Soundness Testing Procedures for Industrial and Commercial Gas Installations* (IM/5). This recognizes that there is a difference between new installations, which must be sound within practical limitations of measurement, and existing installations which may have inherent small leakages at, for example, valve glands.

It is important that any paint, which may provide temporary seals to leaks, is applied after the testing procedure.

An allowance has to be made for temperature, as an increase in temperature will cause a rise in pressure of an enclosed volume of pipework. In practice, this will mean avoiding the effects of direct sunlight and allowing an adequate time for stabilization.

Soundness testing procedure

For both new and existing installation the soundness testing procedure would normally consist of:

1. Estimation of the system volume;
2. Establishment of the test procedure;
3. Selection of a pressure gauge. The sensitivity of the pressure gauge will determine the minimum leak rate detectable;
4. Determination of the permitted leak;
5. Determination of test period;
6. Carrying out the test.

New installations and new extensions

1. There should be no perceptible drop in pressure as shown by the test gauge, although a small leakage rate is defined which allows for testing technique limitations.
2. Air or inert gas should be used and not fuel gas.
3. The test pressure should be 50 mbar or 1.5 times the working pressure or the maximum pressure likely to occur under fault conditions, whichever is the greater. It is assumed that the pipework has been designed and installed to withstand the test pressure to which it is to be subjected.
4. Decide on the permitted leak rate. This is taken to be 0.0014 dm^3/h.
5. Calculate the test period. On large installations it may be necessary to section the system.

Existing installations

1. Existing installations may be tested with fuel gas or inert gas. Air may also be used subject to correct purging procedures.
2. The test pressure is as with new installations.
3. Existing installations may not meet the requirements laid down for new installations because of minor leaks that have developed. It is necessary to relate such leak rates to the pipe environment.
4. Permitted leak rates are determined by installation location:
   (a) Potentially hazardous areas: no leakage is permitted; all joints in the area must be tested.
   (b) Occupied areas: permitted leakage rate is 0.75 dm^3/h per cubic metre of space. But in no case shall the permitted leak rate be taken as greater than 30 dm^3/h.
   (c) Large open work areas and exposed pipes outside: leak rates in excess of those in (b) may be acceptable from the safety point of view.
   (d) Buried pipes: The measured leak rate must be assessed in the light of ground conditions. In particular, special consideration shall be given to the proximity of cellars, ducts, cable runs, sewers, etc.

10.15.2 Purging procedures

After testing for soundness it will be necessary to safely introduce gas into the pipework displacing the air or inert gas that is in it. Similarly, if pipework is decommissioned for any reason fuel gas must be displaced by air or inert gas. This is a requirement of the Gas Safety (Installation and Use) Regulations, Regulation 21. Guidance on recommended procedures is given in the British Gas publication *Purging Procedures for Non-Domestic Gas Installations* (IM/2).

Purging can be defined as:

1. The displacement of air or inert gas by fuel gas;
2. The displacement of fuel gas by air or inert gas;
3. The displacement of one fuel gas by another fuel gas.

Important requirements for purging include:

1. A full knowledge of the pipework layout;
2. Adequate provision of vent points and their termination in a safe location;
3. Preferably a written procedure, appropriate to the installation, should be prepared and followed;
4. Ignition sources should be prohibited from within 3 m of vents and test points.

Purge volumes

For pipework the purge volume may be taken as 1.5 times the pipework volume. For diaphragm meters the purge volume may be taken as 5 revolutions of the meter. For other meters the purge volume may be taken as 1.5 times the volume of a length of pipe equal to a flange-to-flange dimension of the meter.

Purging methods

Purging may be by either the direct or slug method, and may involve the use of an intermediate inert gas purge. At all stages the vent point is sampled and the purge is deemed to be satisfactory when certain criteria are met.

Direct purge. This refers to the displacement of an air/gas mixture with fuel gas. The purge is complete with:

Oxygen less than 4%
Combustibles levels greater than 90%

Inert purging. This refers to the displacement of fuel gas or air with inert gas (complete displacement) or the formation of a barrier of inert gas (less than 4% oxygen) between fuel gas and air during purging (slug purging):
Gas to inert, purge complete with: combustibles levels less than 7%.
Inert to air, purge complete with: oxygen level greater than 16%.
Air to inert, purge complete with: oxygen level less than 4%.
Inert to gas, purge complete with: combustible levels greater than 90%.

Slug purging. Slug purging is appropriate only for long pipe runs. It should not be used to purge installations with branches unless each one can be (and is) valved off and purged separately.

Air purging. This refers to the displacement of gas or gas/air mixture with air. The purge is complete with:

Oxygen level greater than 20.5%
Combustibles levels less than 40% LEL.

Fuel gas to fuel gas purging. The purge may be carried out by burning the gas under supervision on an appliance or appliances. If the two fuel gases have different burning characteristics and Wobbe Numbers, correct burner selection is important.

10.15.3 Commissioning

After soundness testing and purging, the item of plant must be commissioned by a competent person. It is a requirement of the Gas Safety (Installation and Use) Regulations, Regulation 33, that an appliance is fully commissioned at the time that gas is made available to it or that it is isolated in such a manner that it cannot be used.

Commissioning sequence

Commissioning procedures will be specific to particular plant, but the basic procedure is general to all. There are four consecutive stages:

1. *Inspection period*. All relevant documents, the installation and its components are visually and physically examined, with all energy sources isolated. The commissioning programme is finalized.

2. *Activation period.* This is in two parts, the dry run and the live run, the change between the two being when fuel is made available to the combustion space. The activation period commences with all energy sources isolated from the plant. These are gradually made available as the dry run progresses. Fuel must be isolated from the combustion space until the live run commences.
3. *Operation period.* Final adjustments are made and checks carried out to ensure satisfactory operation of the plant, including acceptance trials.
4. *Completion period.* Operating staff are instructed. A report including details of operating levels is prepared and the user takes direct control of the plant.

10.5.4 *In-situ* testing

Regular *in-situ* testing of various parameters can be an aid to maintaining plant in an optimum condition regarding both efficiency and safety. Commissioning data, which should be on record, must be compared with measurements made on a regular basis for such parameters as gas and air pressures, pressure switch settings, limit switch settings, timer settings, flue gas analyses, etc. Adjustments should be made as necessary.

The period between such tests will depend on the maintenance schedule for the plant, which should also include the *in-situ* soundness testing of non-return valves and safety shut-off valves by bubble leak detectors, etc.

10.16 The gas grid system and distribution networks

10.16.1 The National Transmission System

The majority of gas consumed in the United Kingdom originates in the North Sea, both northern and southern basins, and is brought ashore at four terminals, St Fergus, Easington, Theddlethorpe and Bacton. At these terminals the gas is metered, odorized and conditioned (i.e. condensates are removed, it is blended to give a constant calorific value, etc.). Gas can also be brought ashore in the form of liquefied natural gas (LNG) to the Canvey Island terminal, where it can be stored, vaporized and conditioned.

The gas of constant quality is passed into the National Transmission System (NTS) which conveys the gas to all areas of the country where it is passed to regional transmission systems operating at somewhat lower pressures. The length of the NTS is approximately 5000 km with diameters up to 900 mm and operating pressures up to 70 bar. To maintain pressure and to allow for frictional losses, compressor stations, powered by gas turbines, are positioned approximately every 65 km. The National Transmission System is shown in Figure 10.10.

10.16.2 Storage of gas

The rate of use of gas varies considerably both on a seasonal and on a diurnal basis. The gas supplier must ensure that the supply meets the demand under all foreseeable conditions. To an extent, this is achieved by the rates at which gas is taken into the NTS which is determined by contracts specifying daily consumptions throughout the year. Additional fluctuations must be taken up by storage.

Methods of storage

Low-pressure holders. These are the conventional cylindrical multi-stage gas holders. They operate at low pressure and are usually water-sealed. The storage capacity is low for the large physical size and fewer of these holders are used now as their storage capacity can be met by other means.

High-pressure holders. A relatively small number of high-pressure holders are in use. These 'bullets' operate at pressures of up to 30 bar and consequently have a higher storage capacity for their size.

Liquefied natural gas. LNG is stored at certain locations as a means of 'peak-shaving'. Advantages are that the liquid gas occupies a volume equal to 600 volumes of gas under standard conditions. The main disadvantage is that it is necessary to lower the temperature to −162°C to liquefy the gas. It is also necessary to maintain the gas at this low temperature, usually in insulated aluminium vessels.

Salt cavities. Where the geological formation is favourable it is possible to leach out salt to create underground impermeable cavities in which gas can be stored at high pressure. These can be likened to large underground high-pressure holders.

Rough storage. The Rough gas field is a largely depleted field in the southern North Sea basin which is now used as a storage facility. Gas from other fields is re-injected into the Rough field during the summer period of low demand and drawn on during periods of peak demand.

Morecambe Bay. The Morecambe Bay field is a recently developed gas field in the Irish Sea connected to the NTS by means of a gas pipeline to Barrow terminal. While it is possible to operate this field throughout the year it is currently being drawn on only in the winter months as a form of peak-shaving.

Interruptibles. As outlined in Section 10.1.4, some users are offered interruptible contracts. A condition of these is that the user must provide a standby fuel and that, when requested, he or she must switch to this fuel. Under peak demand conditions the large quantity of gas that would be burnt (typically, for steam raising) is available for use by firm contract and tariff users.

Line packing. The NTS, together with the regional transmission systems, constitute a considerable length of pipework operating at pressures of up to 70 bar. Maximum pressures are set by design and minimum pressures by operational constraints. Between these two the pressure can be permitted to fluctuate.

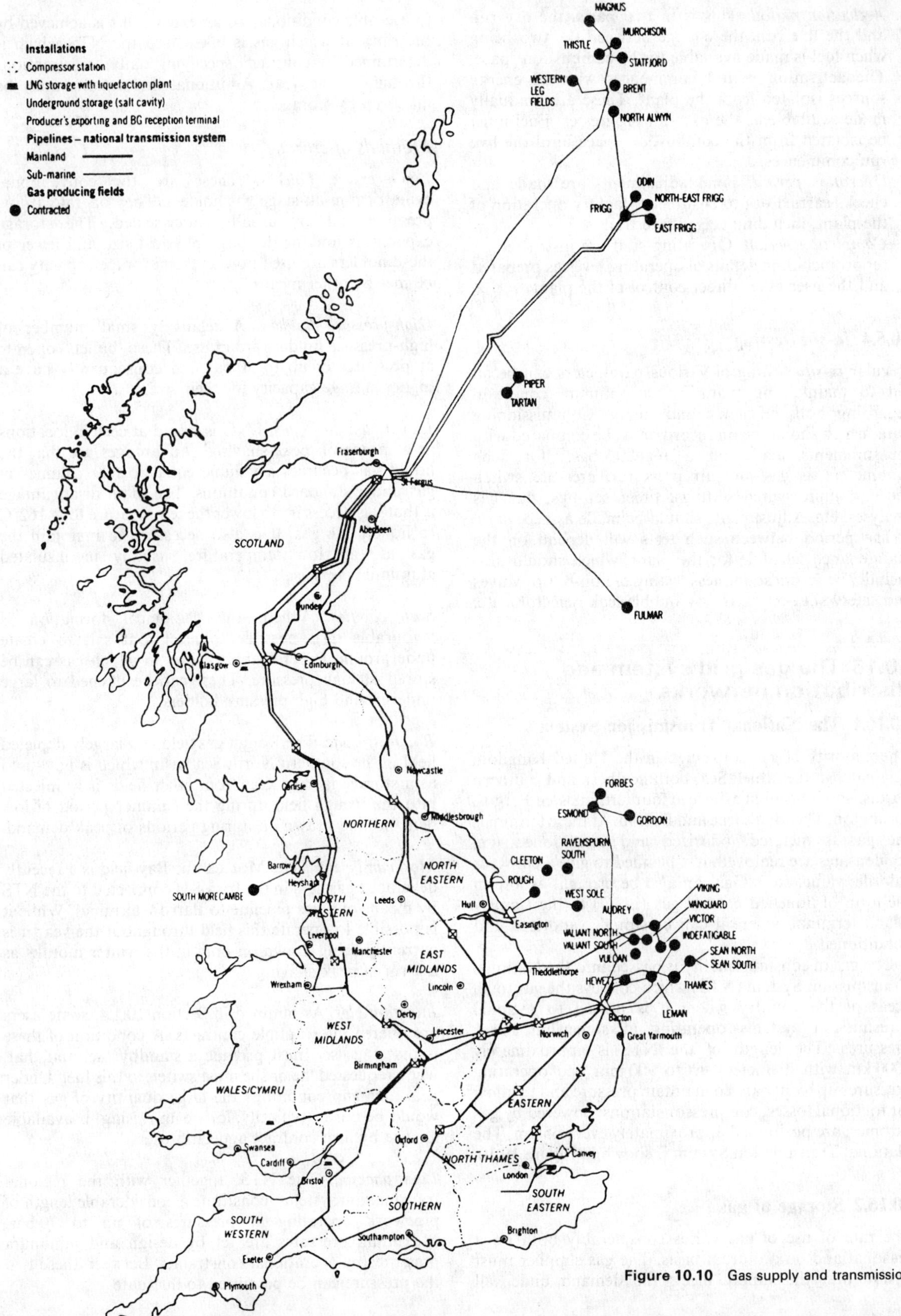

Figure 10.10 Gas supply and transmission

When it is considered that the volume of 1 km of 900 mm pipe is 636 dm^3, and if the pressure is allowed to fall from 42 bar to 19 bar, the equivalent volume of gas at standard conditions is 14 628 dm^3, then this should be compared to the capacity of a low-pressure gas holder of typically 28 320 dm^3 (1 million ft^3). Two kilometres of pipeline operated under these conditions is therefore the equivalent of one gas holder.

Line packing is generally practised by building up the pressure during the night and reducing it during the day.

10.16.3 Distribution networks

The NTS is used for the conveyance of gas from the terminals to regions sited in different parts of the country. Regional transmission systems are used to convey gas across each region. For local conveyance of gas to the final points of use, the distribution system is used.

The primary difference between transmission and distribution systems is one of pressure. Transmission systems operate at high pressures (>7 bar) while distribution systems operate at low (<75 mbar) and medium (75 mbar to 7 bar) pressures. The functions are also different in that the transmission system is used to convey gas over distances and store it, while the distribution system is used to convey gas to the user over a local network. The pipe which conveys gas from the main of the distribution system to the meter control valve of the user is the service.

Operating pressures

Distribution systems can be operated at pressures up to 7 bar but it is practice to operate at as low a pressure as is consistent with supplying users with the agreed supply pressure, which is 21 mbar at the meter inlet in many cases. Pressure reduction takes place at offtakes from the transmission system, at 'district' governor installations serving a large number of users within a geographical area and, for many larger users, at the users' premises.

Pressure-regulating installations will incorporate regulators, pressure-relief valves, slam-shut valves, etc. as necessary and as outlined above. They should comply with the requirements of the Institution of Gas Engineers recommendations:

TD/9: *Offtakes and Pressure-Regulating Installations for Inlet Pressures between 7 and 100 bar*
TD/10: *Pressure-Regulating Installations for Inlet Pressures between 75 mbar and 7 bar*

Pipework design criteria

1. Transmission pipelines operating at high pressures are always constructed of welded steel. Recommendations are contained in the Institution of Gas Engineers publication TD/1, *Steel Pipelines for High Pressure Gas Transmission.*
2. Distribution mains may be constructed of steel, ductile iron or polyethylene depending on operating pressure, etc. Recommendations are contained in the Institution of Gas Engineers publication TD/3, *Distribution Mains.*
3. Services may be constructed of steel, ductile iron or polyethylene depending on operating pressure, etc. Recommendations are given in the Institution of Gas Engineers publication TD/4, *Gas Services.*

Requirements for gas services are also covered by the Gas Safety Regulations 1972, Part 2.

10.17 Emergency procedures

Emergency procedures will be called upon following any abnormal occurrence such as an explosion, fire, release of unburnt gas or production and release of toxic products of combustion. It is essential to carry out an emergency shutdown procedure in a safe way that prevents the hazard continuing.

10.17.1 Emergency control valve

It is a requirement of the Gas Safety (Installation and Use) Regulations that an emergency control valve is provided to which there is adequate access (Regulation 8). In addition, it is stipulated (Regulation 23) that for large users such a valve must be employed for every separate self-contained unit where the gas service is of 50 mm or more and that a line diagram be provided to show their location. By means of such emergency control valves it is possible to safely isolate the gas supply to any building in the event of an emergency.

In addition to the use of such manual valves, automatically operated valves may be considered as an additional safeguard in certain installations. These can be connected to (among other interlocks) panic buttons sited within the building which they are protecting. This is a recommended requirement for buildings containing plant of over 2 MW total heat input and supplied with gas at above 1 bar.

10.17.2 Normal shutdown procedures

All gas-fired plant should be provided with operating instructions for shutting down as well as starting up. Such instructions will ensure that the correct sequence of operations is carried out to both avoid a hazard during the shutdown and to leave the plant in a safe condition. These procedures should also contain instructions for actions to be carried out in emergencies. Such actions may differ from the normal shutdown.

10.17.3 Instruction and training

It is essential that relevant staff receive adequate training for the start-up and shutdown of plant for which they are responsible. This will include emergency procedures.

10.18 Pipework

10.18.1 Design criteria

Gas pipework in a user's premises serves the function of transporting the gas from the meter to the point of use in a safe way and without incurring an avoidable pressure loss. For low-pressure installations the permitted pressure loss is only 1 mbar from the meter to the plant manual isolating valve at maximum flow rate. The pipework must be sized adequately to allow for this. Boosters are sometimes used to overcome pressure losses, but the use of a booster should never be considered a satisfactory substitute for correct design of pipe sizes. Where gas is available at higher pressures it may be permissible to tolerate pressure losses of more than 1 mbar.

Guidance on pipework in users' premises is given in the British Gas publication *Guidance Notes on the Installation of Gas Pipework, Boosters and Compressors in Customers' Premises* (IM/16). Certain Institution of Gas Engineers publications are also of relevance, particularly:

TD/3, *Distribution Mains*
TD/4, *Gas Services*

10.18.2 Materials

Materials used for gas pipework will depend on application, location, environment and operating pressure. General guidelines are as below.

Polyethylene

This is to be used only for buried pipework or in ducts, as it is adversely affected by ultraviolet radiation and its use inside buildings is prohibited by the Gas Safety (Installation and Use) Regulations. For pressures up to 4 bar.

Copper

Copper may be used above ground and for buried pipework at pressures up to 75 mbar and outside diameters of 67 mm. If utilized for buried pipework it must be factory sheathed and should not be attached to buried steel pipe and fittings.

Ductile iron

This may be used for buried pipework, in ducts and above ground at pressures up to 2 bar. Not to be used for plant pipework. Ductile iron pipe and fittings shall be to BS 4772, Specifications for ductile iron pipes and fittings.

Steel

Steel may be used in all applications. The working pressure is related to the specification and grade used. However, pipe to BS 1387 heavy (up to 150 mm) and to BS 3601 (150 mm and above) is suitable for applications up to 7 bar.

10.18.3 Jointing

The method of jointing will depend on the material, the pipe size, the operating pressure and the location. General guidelines are given below.

Spigot and socket connections

This is to be used for ductile iron in all applications and must be of a type suitable for rubber jointing rings to BS 2494. Where the working pressure is over 75 mbar the connections shall be anchored by means of purpose-designed anchor blocks or self-anchoring joints.

Fusion welding

This method to be used for polyethylene pipes only by companies and persons specializing in this field.

Capillary joints

These can be used for copper in all applications and must be to BS 864, Part 2, or BS 2051, Part 1. In ducts all such joints must be accessible within the duct or chase. Not to be used within vertical ducts in high-rise buildings.

Compression joints

These can be used for copper in all applications where the joints are accessible and are not therefore recommended for underground use. In ducts all such joints must be accessible within the duct or chase. Not to be used within vertical ducts in high-rise buildings.

For polyethylene pipe compression couplings may be used provided they comply with British Gas Specification PS/PL3.

Screwed and welded jointing for steel pipework

The type of jointing may be restricted by the grade of pipe used. For example, some grades of steel pipe are not suitable for the application of screwed threads to BS 21 or ISO R7. The use of welded pipework is based on a combination of pipe size and pressure. Up to 25 mm screwed connections are permitted at all pressures up to 5 bar, whereas above 100 mm pipe must be welded at all pressures.

Other points to consider include:

1. Where pipework is welded, the number of flanged joints should be kept to a minimum and such flanges shall be welded to the pipe.
2. Where pipework is welded, the minimum standard shall be: up to 2 bar: BS 2971 or BS 2640; above 2 bar up to 5 bar: BS 2971 with 10% of welds subjected to non-destructive testing, or BS 2640 with inspection and 10% of welds subjected to non-destructive testing.
3. For screwed connections, jointing materials meeting the following specifications shall be used:

   BS 1560, 1832, 2815
   BS 5292 – Type A up to 0.7 bar or Type C

   with the exception of vulcanized fibre and rubber reinforced jointing, which shall not be used.

With screwed joints of 50 mm and above it is recognized that, in addition to a jointing paste, the use of hemp may be necessary to ensure a sound joint. However, hemp should never be used in conjunction with PTFE tape.

10.18.4 Pipes in ducts

1. It is not permissible to install an installation pipe in an unventilated shaft, void or duct (Regulation 18(5) of the Gas Safety (Installation and Use) Regulations).
2. Adequate ventilation is provided by means of openings at the top of high points and the bottom of low points. These openings shall have a free open area of 1/150 of the cross-section area of the duct or $0.05\ m^2$, whichever is the greater.
3. For ducts smaller than $0.05\ m^2$ the openings shall be not less than $3000\ mm^2$.
4. In exceptional circumstances if it is not possible to ventilate the duct then:
   (a) The pipe shall be continuously sleeved through the duct with the sleeve ventilated at one or both ends into a ventilated area; or
   (b) The unventilated duct or void shall be filled with a crushed inert infill to reduce to a minimum the volume of any gas which may accumulate. Suitable material is crushed slate chippings or dry washed sand.

10.18.5 Pipe supports

Pipe in whatever plane shall be adequately supported. Reference should be made to the relavant parts of BS 3974, Part 1. In some instances short vertical lengths in otherwise horizontal pipe runs may be self-supporting.

Pipe supports shall be provided throughout the length of the pipe and in such a way as to allow thermal movement without causing damage to any corrosion protection that may be on the pipe.

10.18.6 Corrosion protection and identification

Buried pipework must be protected against accidental and physical damage from sharp material, etc. and chemical action from corrosive soils, etc. It must be protected against corrosion by means of wrapping, cathodic protection, etc. for metal pipes. Above-ground pipework should be protected with suitable paint after preparation.

Pipework should be easily identifiable in accordance with BS 1710. Where the normal pressure in the pipe exceeds 75 mbar consideration should be given to labelling the pipe with the normal operating pressure. In buildings in which there are no other piped flammable gas supplies it is sufficient to paint the pipe yellow ochre or to band it with appropriately coloured adhesive tape. In large complex installations it is desirable to identify pipe contents more precisely, and in those instances the base colour should be supplemented with a secondary code band of yellow and/or its name or chemical symbol.

10.18.7 Flexible connections and tubes

In all instances where it is known or expected that pipework will be subjected to vibration, expansion or strain the use of flexible connections should be considered. However, flexible connections should not be used where there is a practical alternative.

Attention is drawn to Regulation 4(2) of the Gas Safety (Installation and Use) Regulations, which prohibits the use of non-metallic materials for gas pipework other than for small portable appliances such as bunsen burners and lighting torches.

Types of flexible connections

Various types of flexible connections are available for different purposes. Careful consideration is needed for suitability for purpose. General comments on connections available can be made. All connections should:

1. Be readily available for inspection;
2. Be kept to the minimum length;
3. Be protected from mechanical damage and the effects of the environment;
4. Not pass through walls, etc.

Semi-rigid couplings and flange adaptors. Restraints are needed to prevent separation of the pipes when installed above ground and to prevent angularity.

Bellows. The pipework either side of a bellows joint shall be supported so that the bellows itself is not supporting any of the weight of the attached pipework.

Swivel joints. The axis of rotation of the swivel must be accurately aligned on both sides of the joint and the joint should have lateral freedom and be free from side bending moments.

Quick-release couplings. The coupling shall be of the type having self-sealing valves in both the plug and the body with a flexible tube fitted to the downstream connection which shall be the male plug section. A manual valve shall be fitted immediately upstream of the coupling.

Flexible tubes. Metallic tubes complying with BS 669, Part 2 or BS 6501 shall be used other than with small portable appliances or with domestic-type appliances, in which case they shall comply with BS 669 Part 1. A manual valve shall be fitted on the inlet side in close proximity to the tube. The pipework shall be adequately supported such that the tube does not support the weight of the attached pipework Tubes shall be installed so that they are neither twisted nor subjected to torsional strain, have flexing in one plane only, and are not subject to sharp bends near end fittings.

10.18.8 Purge points

Purge points shall be fitted at section isolation valves, at the end of pipe runs and at other suitable positions to facilitate the correct purging of pipes. Pressure test points should be fitted at the outlet of each section-isolating valve.

10.18.9 Syphons and condensate traps

These are not normally required. Where wet gas is supplied it will be necessary to fit a vessel at low points in the installation to collect any condensate or fluid. This vessel shall be in a readily accessible position and a valve, suitably plugged or capped, shall be fitted to its drain connection. In the exceptional case where hydraulic pressure testing is to be carried out, similar provisions must be made.

10.18.10 Commissioning

After pipework has been installed it is essential that it be tested for soundness and purged, as required by Regulation 21 of the Gas Safety (Installation and Use) Regulations. Detailed procedures for these two operations are given in the British Gas publications *Purging Procedures for Non-Domestic Gas Installations* (IM/2) and *Soundness Testing Procedures for Non-Domestic Gas Installations* (IM/5).

10.19 Flow charts for use with gas

10.19.1 Need for flow charts

It is essential when designing the pipe layout for gas distribution that unavoidable pressure losses are not incurred. For low-pressure gas the pressure available at the meter inlet will be only 21 mbar, and the allowable pressure loss to the point of use only 1 mbar, although higher pressures may be available in some circumstances. If such a low-pressure loss is not to be exceeded it is essential that the pipework be sized correctly. It is preferable to oversize pipework rather than undersize, particularly as this allows scope for future expansion. For routine purposes, the use of flow charts is much to be preferred to formulae based on first principles.

10.19.2 Theory of pressure loss

The theory of pressure losses can be established by developing Bernoulli's theorem for the case of a pipe in which the work done in overcoming frictional losses is derived from the pressure available. For a fluid flowing in a pipe the pressure loss will depend on various parameters. If

L = length of pipe,
D = diameter of pipe,
S = specific gravity of gas,
Q = flow rate of gas,
f = friction factor,
P = pressure loss

then P is proportional to L
$1/D^5$
S
Q^2
f

For a particular material of pipeline construction the friction factor can be assumed. This allows a fairly simple formula to be derived:

$$P = kQ^2SL/D^5$$

where k is a constant taking into account the friction factor, units of length, units of pressure, etc. This is the Pole equation, and is frequently expressed in the form:

$$Q = k\sqrt{(PD^5/SL)}$$

Forms of this equation have been drawn up for use with different units (SI or Imperial). Typical assumptions are:

1. Flow is fully turbulent with $Re > 3500$ (the transition from laminar flow starts at about $Re = 2000$);
2. Viscosity is known and allowed for in k;
3. Pressure is < 25 mbar;
4. If SG is known this can be incorporated into k;
5. Pipe is assumed to be to BS 1387. This determines f, which can be incorporated into k.

For uniform circular pipe the Fanning equation can be used:

$$P = (4fL/D)U^2/2g$$

where U = linear velocity,
g = gravitational constant.

The Fanning Friction factor (f in the above equation) varies with Reynolds number and relative roughness of the internal surface of the pipe. This variation is plotted in the Moody diagram.

The relative roughness is expressed as E/D, where E = the surface roughness and D = the internal pipe diameter. Typical values of E/D are 0.0015 for drawn tubing, 0.046 for commercial steel and 0.12 for asphalted cast iron.

10.19.3 Practical methods of sizing pipework

The use of formulae derived from first principles is time consuming and cannot normally be justified in comparison with approximate methods, which have a sufficient degree of accuracy. It is necessary to know:

1. Flow rate;
2. Pressure loss;
3. Pipe diameter;
4. Pipe length;
5. Pipe material (if this is not known, an assumption that it is steel will normally be sufficiently accurate).

Using these parameters, there are three main ways of expressing the relationships between them:

1. A tabulated form as in Table 10.3. This shows flows versus pipe length and diameter for both copper and steel. Such tables are included in British Gas IM/16 and British Standard BS 6891. Note that Table 10.3 includes allowances for elbows, tees and bends. Allowances, equivalent to numbers of pipe diameters, must be made for all pipe fittings which cause an additional pressure loss. Further details for allowances to be made for pipe fittings, including valves and non-return valves, are given in Table 10.4.
2. A graphical representation as shown in Figure 10.11. This is drawn in the logarithm–logarithm format and allows a rapid estimate of pressure loss to be expected. Note that this particular chart is in imperial units and

Table 10.3 Gas flow through pipes and fittings

Flow in a straight horizontal steel pipe (to BS 1387, Table 2, Medium) with 1.0 mbar differential pressure between the ends, for gas of relative density 0.6 (air = 1)

Nominal size mm (in)	*Length of pipe* (m) 3	6	9	12	15	20	25	30	40	50
	Discharge (m^3/h)									
6 (1/8)	0.29	0.14	0.09	0.07	0.05	—	—	—	—	—
8 (1/4)	0.80	0.53	0.49	0.36	0.29	0.22	0.17	0.14	0.11	0.08
10 (3/8)	2.1	1.4	1.1	0.93	0.81	0.70	0.69	0.57	0.43	0.34
15 (1/2)	4.3	2.9	2.3	2.0	1.7	1.5	1.4	1.3	1.2	1.0
20 (3/4)	9.7	6.6	5.3	4.5	3.9	3.3	2.9	2.6	2.2	1.9
25 (1)	18	12	10	8.5	7.5	6.3	5.6	5.0	4.3	3.7
32 (1¼)	39	27	21	18	16	14	12	11	9.2	8.1
40 (1½)	59	40	32	27	24	21	18	16	14	12
50 (2)	110	76	61	52	46	39	34	31	26	23
65 (2½)	220	150	120	100	92	79	70	63	54	47
80 (3)	340	230	190	160	140	120	110	97	83	73
100 (4)	690	470	380	330	290	250	220	200	170	150

Flow in a straight horizontal copper tube (to BS 2871, Part 1, Table X) with 1.0 mbar differential pressure between the ends, for gas of relative density 0.6 (air = 1). Tube sizes given below are outside diameters

Nominal size (mm)	*Length of pipe* (m) 3	6	9	12	15	20	25	30	40	50
	Discharge (m^3/h)									
6	0.12	0.06	—	—	—	—	—	—	—	—
8	0.52	0.26	0.17	0.13	0.10	0.07	0.06	0.05	—	—
12	1.5	1.0	0.85	0.82	0.69	0.52	0.41	0.34	0.26	0.20
15	2.9	1.9	1.5	1.3	1.1	0.95	0.92	0.88	0.66	0.52
22	8.7	5.8	4.6	3.9	3.4	2.9	2.5	2.3	1.9	1.7
28	18	12	9.4	8.0	7.0	5.9	5.2	4.7	3.9	3.5
35	32	22	17	15	13	11	9.5	8.5	7.2	6.3
42	54	37	29	25	22	18	16	15	12	11
54	110	75	60	51	45	38	33	30	26	23
76	280	190	150	130	120	98	86	78	66	58
108	750	510	410	350	310	260	230	210	180	160

Note: $1\ m^3/h = 1000\ dm^3/h$. For smaller rates of flow it may be more convenient to use dm^3/h.

The effects of elbows, tees or bends inserted in a run of pipe (expressed as the approximate additional lengths to be allowed)

Nominal size				
Cast iron or mild steel	Up to 25 mm (1 in)	32 mm to 40 mm (1¼ in to 1½ in)	50 mm (2 in)	80 mm (3 in)
Stainless steel or copper	Up to 28 mm (1 in)	35 mm to 42 mm (1¼ in to 1½ in)	54 mm (2 in)	76 mm (3 in)
Elbows	0.5 m (2 ft)	1.0 m (3 ft)	1.5 m (5 ft)	2.5 m (8 ft)
Tees	0.5 m (2 ft)	1.0 m (3 ft)	1.5 m (5 ft)	2.5 m (8 ft)
90° bends	0.3 m (1 ft)	0.3 m (1 ft)	0.5 m (2 ft)	1.0 m (3 ft)

is drawn for use with town gas. A correction for specific gravity would be needed for natural gas.

3. Flow calculators. These are available in the form of plastic concentric discs which can be rotated against each other.

The parameters of pipe bore, flow, length of pipe, and specific gravity can be used to determine pressure loss. The calculators are versatile in that they can be used for pressure losses of under 1 mbar and also for up to 50 bar. Scales for both cast iron and steel pipe are available for greater accuracy.

10.19.4 Discharge through orifices

The Bernoulli theorem can be used as the basis for a means of determining the flow through an orifice. The equation will be of the form:

$Q = kC_dD^2\sqrt{(P/S)}$

Table 10.4 Allowances for fittings and entrance and exit losses in pipes

Fitting		*S* *Additional velocity heads*	*N* *Additional length: pipe diameters $f=0.025$*
Elbows of various angles	$\alpha=90°$	1.0	40
	60	0.36	14
	45	0.18	7
	30	0.07	3
	15	0.02	1
90° bends	$r/D=0.5$	1.0	40
	1.0	0.6	24
	2.0	0.4	16
	4.1	0.3	12
Standard tee 90°, equal areas:			
From barrel to branch		1.2	48
From branch to barrel		1.5	60
Through barrel		0.25	10
(Third leg stagnant in each case)			
Close return bend, 180°	$r/D=1.0$	2.0	80
Globe valve, full open		10	400
Angle valve, 45° open		5	200
Flap non-return valve		2	80
Gate valve, full open		0.15	6
$\frac{3}{4}$ open by area		0.8	32
$\frac{1}{2}$ open by area		4.5	180
$\frac{1}{4}$ open by area		28	1 120
Square entrance from a tank		0.5	20
Bell mouth entrance from a tank		0.05	2
Re-entrant (Borda) mouthpiece		0.8	32
Pipe discharge into a tank		1.0	40

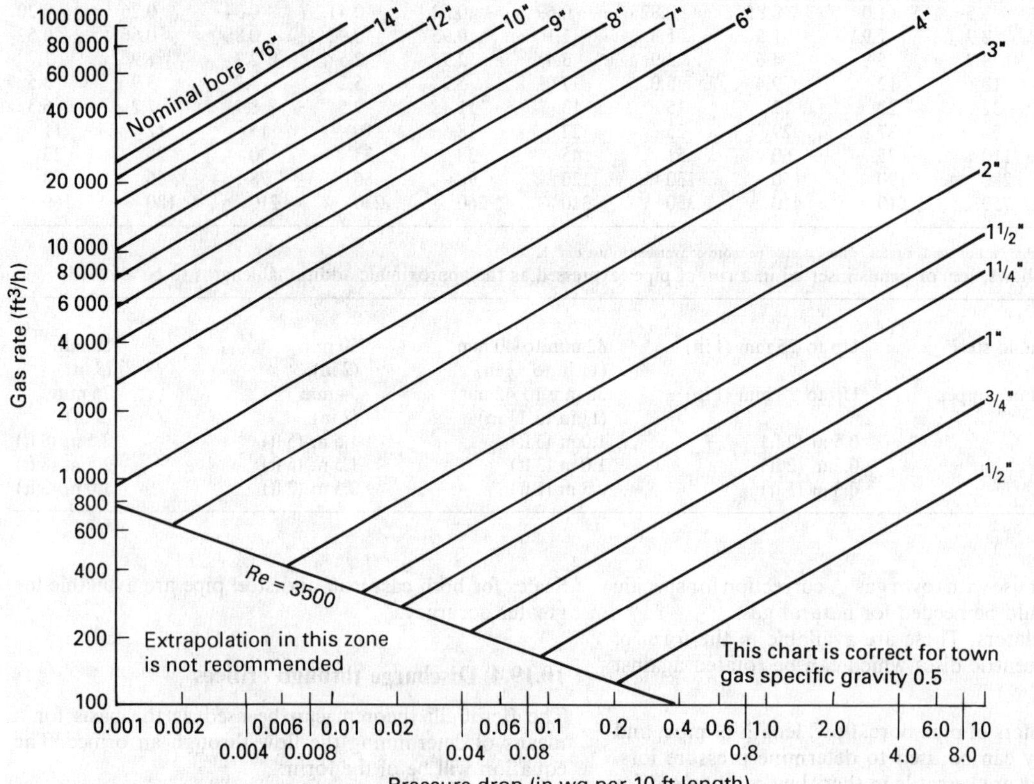

Figure 10.11 Chart for flow of town gas in pipes

where C_d = coefficient of discharge,
D = diameter of the orifice

This equation can be used for an orifice plate introduced into pipework as a measuring, throttling or balancing device, and for a jet discharging gas into the injector of a burner at atmospheric or sub-atmospheric pressure.

The major unknown is the value of the coefficient of discharge. In the case of the orifice plate this will be of the order of 0.6. With the jet this may be as low as 0.6 but can be up to 0.9 for jets with profiled entries.

Tables are available for discharges from jets for different diameters and pressures, both on a general basis and for proprietary jets.

10.20 Conversion factors

SI units are widely used in the gas industry. Imperial units are also employed, particularly for measuring gas and for its calorific value. In some areas SI and Imperial units can co-exist, particularly thermal ratings, which can equally be expressed in kW and MW or in Btu/h and therm/h.

American combustion equipment is frequently used. This tends to use Imperial units as well as degrees Fahrenheit and ounces per square inch pressure.

European combustion equipment utilizes metric units but these are frequently non-SI. For example, heat and power ratings often use calories and kilocalories/h instead of joules and watts.

It is not intended to include a comprehensive set of conversion factors, but rather those which are frequently to be found in the energy industries.

Pressure

Non-SI unit	*Conversion factor to N/m²*
bar	100 000
inch of Hg	3386.39
inch of water	249.089
Pascal	1
millibar	100
pound-force/in² (psi)	6894.76
ton-force/in²	15 444.3
torr	133.322
ounce-force/in²	430.922 5

Linear

Non-SI unit	*Conversion factor*
foot	0.304 8 m
inch	25.4 mm
yard	0.914 4 m
mile	1.609 34 km

Volume and capacity

Non-SI unit	*Conversion factor*
cubic foot	0.028 316 8 m³
gallon	4.546 09 dm³
gallon (US)	3.785 41 dm³

Heat and energy

Non-SI unit	*Conversion factor*
British thermal unit	1.055 06 kJ
calorie	4.186 8 J
therm	105.506 MJ
thermie	4.185 5 MJ

Power

Non-SI unit	*Conversion factor*
Btu/h	0.293 071 W
therm/h	29.307 1 kW
kilocalorie/h	1.163 W
horse power	745.700 W
ton of refrigeration	3516.85 W
frigorie	1.1626 W
cal/s	4.186 8 W

Calorific value

Non-SI unit	*Conversion factor*
Btu/ft³	37.258 9 kJ/m³
Btu/lb	2326 J/kg
therm/gal	23.208 GJ/m³

Temperature

Non-SI unit	*Conversion factor*
degree Fahrenheit	$(T-32)5/9$°C
degree Rankine	5/9 K
degree Celsius	$(T+273.15)$ K

Thermal conductivity

Non-SI unit	*Conversion factor*
Btu in/ft² h °F	0.144 228 W/mK
Btu in/ft² s °F	519.220 W/mK
Btu/ft² h	3.154 59 W/m²
Btu/ft² h °F	5.678 26 W/m²K
Btu ft/ft² h °F	1.730 73 W/mK

Thermal resistivity

Non-SI unit	*Conversion factor*
ft² h °F/Btu in	6.933 47 m K/W
ft² h °F/Btu ft	0.577 89 m K/W

11 Liquefied Petroleum Gas

R H Shipman MIMechE, MIGasE, MInstE
Liquefied Petroleum Gas Industry Technical Association

Contents

11.1 Introduction

Liquefied petroleum gases (often referred to as LPG or LPGas) are a constituent of crude oil or the condensate of natural gas fields (NGL). They are the C_3 and C_4 hydrocarbons, propane and butane, respectively, which have the property of being gases at normal ambient temperature but can be liquefied and kept in the liquid state by quite moderate pressure. Gases released from crude oil are called 'associated gases' while those found without heavier hydrocarbons are known as 'unassociated gases'. The North Sea gas wells are good examples of the latter.

Methane is the major constituent of both associated and unassociated gas at source. There will be higher (heavier) hydrocarbons in varying amounts present in the gas, associated gas having more than unassociated gas.

When crude oil is refined some of the processes yield additional gaseous products. The C_3 and C_4 constituents differ from those released from crude oil or from NGLs, which are 'saturated' hydrocarbons. Refinery gases are high in 'unsaturates', e.g. propene (propylene) and butene (butylene). These unsaturated hydrocarbons are a valuable source of chemical process intermediates and enjoy a large market alongside naphtha.

The major constituents of LPG can be shown by the following molecular diagrams:

```
  H  H  H
  |  |  |
H—C—C—C—H
  |  |  |
  H  H  H
```

C_3H_8 Propane Saturated

```
  H  H  H  H
  |  |  |  |
H—C—C—C—C—H
  |  |  |  |
  H  H  H  H
```

C_4H_{10} Butane Saturated

```
     H  H  H
      \ | /
H       C       H
  \           /
 H—C—C—C—H
  /   |   \
H     H     H
```

C_4H_{10} ISO Butane Saturated

```
  H  H  H
  |  |  |
H—C—C=C
  |     |
  H     H
```

C_3H_6 Propene (Propylene) Unsaturated

```
H  H  H  H
|  |  |  |
C=C—C—C—H
|     |  |
H     H  H
```

C_4H_8 Butene (Butylene) Unsaturated

No doubt the most successful exploitation of the peculiar properties of LPG has been its use as a fuel gas. Originally used predominantly in refineries for process heating, its value as a fuel was first realized before the Second World War, when it was sold in small portable containers. From primarily domestic use this spread after the war to commercial and industrial utilization and the introduction of bulk storage on-site. In order to develop such a market it was necessary to refine the product to an agreed standard. Today, LPG sold into the UK commercial fuel gas market meets the British Standard BS 4250: Part 1: 1987 or Part 2: 1987 for automotive LPG. The Standard sets the limits of the constituents for commercial propane and commercial butane since they are always mixtures of C_3 and C_4 with one or the other predominating. The following are derived from BS 4250: Part 1.

11.2 Composition

Commercial butane. This product shall consist of a hydrocarbon mixture containing predominantly butanes and/or butenes.

Commercial propane. This product shall consist of a hydrocarbon mixture containing predominantly propane and/or propene.

11.3 Requirements

11.3.1 General

When tested in accordance with the methods given in Table 11.1 the properties of the commercial butane and commercial propane shall be in accordance with the limiting requirements given in that table. For gauge vapour pressure either the direct measurements method described in BS 3324 or the calculation procedure described in Appendix C of this standard shall be used. The method described in BS 3324 is the referee method and shall be used in cases of dispute.

11.3.2 Water content

Commercial butane and commercial propane shall not contain free or suspended water on visual inspection. Additionally, for commercial propane, the content of dissolved water shall not be such as to cause failure when tested in accordance with the valve freeze method.

Note: The addition of up to 0.125% (v/v) of methanol to commercial propane will normally ensure that it complies with the specified limit for water content.

11.3.3 Odour

When tested in accordance with the procedure described in Appendix B, the odour of the gas shall be distinctive and unpleasant, and the odour in a gas/air mixture shall be such that it is detectable down to a concentration of 20% of the concentration corresponding to the lower limit of flammability for the hydrocarbon mixture concerned.

Note: Odorants such as ethanediol, tetrahydrothiophene or dimethyl sulphide may be added so that the gas complies with the specified requirements for odour.

Table 11.11 General requirements for commercial butane and commercial propane

Property	*Limit*		*Test method*	
	Commercial butane	*Commercial propane*	*British Standard*	*Technically equivalent to IP method*
Gauge vapour pressure, at 40°C (measured or calculated) (in kPa), max. (see E.1)	505	1560	BS 3324 or appendix C	–
Total sulphur content (in mg/kg), max.	200	200	BS 5379	–
Mercaptan sulphur content (in mg/kg), max.	50	50	BS 2000: Part 272	IP 272
Hydrogen sulphide content (in cm^3/m^3), max. (see E.2)	0.5	0.5	Appendix A	–
Copper corrosion, 1 h at 40°C	Class I	Class I	BS 6924[a]	–
Tendency to freeze in valves	–	Pass	Appendix D	–
Dienes content, mole %, max.	10.0	–	BS 3156: Part 4	–
Ethylene content, mole %, max.	–	1.0	BS 3156: Part 4	–
Alkynes content, mole %, max.	0.5	0.5	BS 3156: Part 4	–
C_4 and higher hydrocarbons content, mole %, max.	–	10.0	BS 3156: Subsection 11.1.1	–
C_5 and higher hydrocarbons content, mole %, max.	2.0	2.0	BS 3156: Subsection 11.1.1	–

[a] BS 6924 is in preparation. Pending its publication, the identical International Standard, ISO 6251, should be used.

11.3.4 Filling ratios and developed pressures

The maximum filling ratios and developed pressures for the containers in which the commercial butane or propane is supplied shall comply with BS 5355.

11.4 Typical properties of LPG

Typical properties are shown in Table 11.2. These are important in setting the requirements for the storage, handling and use of LPG and require further elaboration.

11.4.1 Vapour pressure

This is one of the most important properties of LPG since it determines the pressure that will be exerted by the gas at ambient temperature, and therefore affects the requirements for handling and the design working pressures of storage vessels. It constitutes the main difference in physical characteristics between commercial propane and butane. The vapour pressure is the pressure at which a liquid and its vapour are in equilibrium at any given temperature. The boiling point of a liquid is, in fact, the temperature at which the vapour pressure is equal to the external ambient pressure.

Commercial propane and butane often contain substantial proportions of the corresponding unsaturated analogues and smaller amounts of near-related hydrocarbons, as well as these hydrocarbons themselves. Figure 11.1 shows vapour pressure/temperature curves for commercial propane and commercial butane. Due to its lower boiling point, higher rates of vaporization for substantial periods are obtainable from propane than from butane, and at the same time appreciable pressures are maintained even at low ambient temperatures.

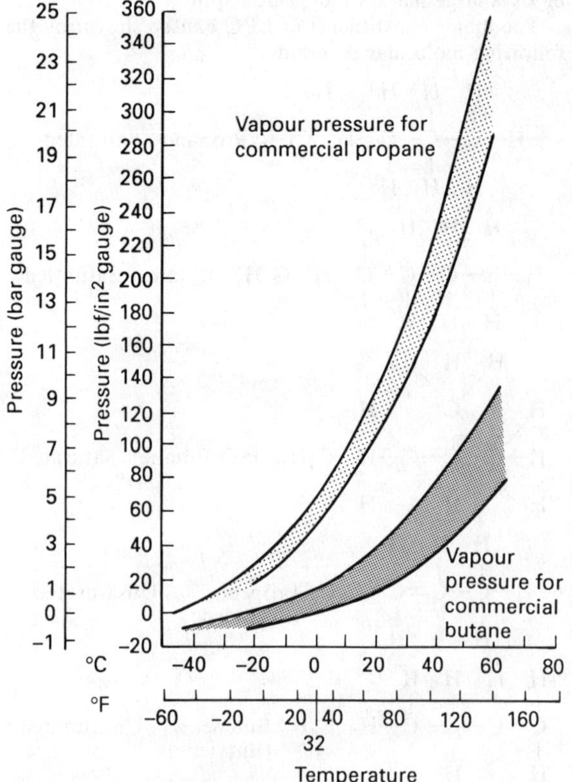

Figure 11.1 Vapour pressure/temperature curves for commercial propane and commercial butane

11.4.2 Gross calorific value

This is defined as the amount of heat liberated when unit volume (or unit mass) of the gas is burned at a standard

Table 11.2 Properties of commercial propane and butane

	Propane	*Butane*
Relative density of liquid at 15°C (60°F)	0.50–0.51	0.57–0.58
Imperial gallons/ton at 60°F	439–448	385–393
Litre/tonne at 15°C	1965–2019	1723–1760
Relative density of gas compared with air at 15°C (60°F) and 1016 mbar (30 inHg)	1.40–1.55	1.90–2.10
Volume of gas (ft^3) per lb of liquid at 60°C and 30 inHg	8.6–8.7	6.5–6.9
Volume of gas (litres) per kg of liquid at 15°C and 1016 mbar	537–543	406–431
Ratio of gas volume to liquid volume at 15°C (60°F) and 1016 mbar (30 inHg)	274	233
Boiling point at atmospheric pressure		
°C (approx.)	−45	−2
°F (approx.)	−49	28
Latent heat of vaporization (Btu/lb) at 60°F	154	160
Latent heat of vaporization (kJ/kg) at 15°C	358	372
Specific heat of liquid at 60°F (Btu/lb °F)	0.60	0.57
Specific heat of liquid at 15°C (kJ/kg °C)	2.512	2.386
Sulphur content, % weight	Negligible to 0.02	Negligible to 0.02
Limits of flammability (percentage of volume of gas in a gas–air mixture to form a combustible mixture)	Upper 10.0 Lower 2.2	Upper 9.0 Lower 1.8
Calorific values		
Gross		
Btu/s ft^3 dry	2500	3270
MJ/s m^3 dry	93.1	121.8
Btu/lb	21 500	21 200
MJ/kg	50.0	49.3
Net		
Btu/s ft^3 dry	2310	3030
MJ/s m^3 dry	86.1	112.9
Btu/lb	19 900	19 700
MJ/kg	46.3	45.8
Therm/ton (gross CV)	482	475
GJ/tonne	50.0	49.3
Air required for combustion		
ft^3 to burn 1 ft^3 of gas at STP	24	30
m^3 to burn 1 m^3 of gas at STP	24	30

temperature and pressure. It is usually expressed in terms of megajoules per cubic metre at 15°C and 1016 mbar, i.e. MJ/s m^3 dry or megajoules per kilogram. Typical calorific values of LPG and other gaseous fuels are shown in Table 11.3 and allow a comparison of their heating values to be made.

The great advantage of LPGs is that they are stored as a liquid. However, they are almost always used as a gas. One volume of liquid propane when released at STP gives 274 volumes (233 for butane) of high calorific value fuel gas.

Table 11.3 Typical gross calorific values of fuel gases

Gas	*MJ/s m^3 dry*	*Btu/s ft^3 dry*	*MJ/kg*	*Btu/lb*
Commercial propane	93.1	2500	50.0	21 500
Commercial butane	121.8	3270	49.3	21 200
Producer gas (cold)	4.7–6.1	125– 165	–	–
Town gas	14–20.5	375– 550	–	–
Natural gas	31.6–46.6	850–1250	–	–
Acetylene	56	1500	49.9	21 460

11.4.3 Sulphur content

Compared to most other widely available fuels, with the exception of natural gas, LPG has a very low sulphur content which is strictly controlled within tight specification limits. This makes LPG a particularly useful fuel where the products of combustion are intended to be released directly into living accommodation.

11.4.4 Relative density – liquid

The density of liquid butane and propane is about half that of water, and as such is much lower than other liquid fuels.

11.4.5 Relative density – vapour

This is the one major property where there is an important difference between LPG vapour and natural gas. Natural gas is lighter than air; propane and butane vapour are, respectively, one and a half times and twice as heavy as air. This is of importance in two areas. First, when converting equipment running on natural gas to run on propane or butane the amount of gas which issues from a fixed orifice at a fixed pressure is inversely proportional to the square root of its density. Second, if there is a leak of LPG vapour it will collect at ground level or in depressions, drains or cellars if appropriate precautions are not taken. This is a major safety consideration when designing or installing LPG systems.

11.4.6 Limits of flammability

When mixed with air, LPG can form a flammable mixture. The flammable range at ambient temperature and pressure extends between approximately 2% of the vapour in air at its lower limit and approximately 10% of the vapour in air at its upper limit. Outside this range any mixture is either too weak or too rich to propagate flame. However, over-rich mixtures resulting from accidental releases can become hazardous when diluted with air. At pressures greater than atmospheric, the upper limit of flammability is increased but the increase with pressure is not linear.

The limits of flammability for propane and butane are much narrower than most other gaseous fuels, making LPG safer in this respect.

11.4.7 Coefficient of expansion

This is defined as the increase in volume of unit volume of a substance when its temperature is raised by one degree. It is important in that the coefficient of expansion of LPG in its liquid form is relatively high, so that when filling a storage vessel adequate space must always be provided to allow for possible thermal expansion of the liquid.

11.4.8 Other physical characteristics

LPG is both colourless and odourless. However, a distinctive odour is added to aid detection in the case of leakage. BS 4250 requires sufficient stench to be added so that the odour of the gas can be detected in air at concentrations down to one-fifth of the lower limit of flammability, i.e. about 0.4% gas in air.

Neither propane nor butane is toxic, but they do possess anaesthetic properties. There is a threshold limiting value for LPG at 1000 ppm given as an occupational exposure standard in the Health and Safety Executive Guidance Note EH40/89, for an 8-hour time-weighted average. This calls for consideration in particular by LPG cylinder-filling plant operators of the requirements of the COSHH Regulations (Control of Substances Hazardous to Health).

Liquid propane and butane will vaporize rapidly if released into the open air, and if they come into contact with bare skin will cause painful freeze burns. Therefore gloves and goggles should always be worn if there is a danger of liquid LPG being released or spilt.

11.5 Transport and storage

The great advantage of LPG is that for use it can be regarded as a gas but for conveyance it can be treated as a liquid. It is therefore transported by road or rail in a manner similar to liquid fuels. Thus the most usual method is by road tanker (up to 17 tonnes of product) or rail tank wagon (from 20 tonnes to 100 tonnes per wagon). Figure 11.2 shows a typical 15-tonne capacity propane road tanker. In order to maintain the product in its liquid state, however, it has to be kept under pressure and the tanks are therefore pressure vessels. This pressure must be maintained during all activities, from storage at the refinery, to transfer into the road tanker or rail tank wagon, and then in the storage vessel at the users' premises.

In addition to the modes of distribution described above there is also the transportable container which is now a familiar feature in all kinds of activity. These vary from tiny disposable cartridges of less than 100 g to the larger refillable steel or aluminium alloy cylinders ranging

Figure 11.2 LPG road tanker. (Courtesy of BP Oil Ltd)

Figure 11.3 Refrigerated LPG storage at 'Flotta'. (Courtesy of the Motherwell Bridge Group)

up to around 50 kg of product. These are filled at purpose-designed filling plants and distributed by road or rail transport to the retail outlet.

At the other end of the scale there is also the transport and storage of LPG in a refrigerated state. Large international sea transport cargoes are usually refrigerated, while refrigerated land storage is only economically justified for storage measured in thousands of tonnes and for import/export terminals, as illustrated by Figure 11.3. For design, construction and installation of refrigerated storage, references are given to the Institute of Petroleum, and the Engineering Equipment and Materials Users Association (EEUMA) at the end of this chapter.

Thus for most inland transport and storage the liquid is held at ambient temperature under pressure. The pressure is simply created by the natural vapour pressure of the product which varies with its temperature. Butane has a lower vapour pressure than propane. The carrying vessels, whether road tankers, static storage, cylinders or cartridges, must be designed to accept the highest pressure likely to occur in service. A knowledge of the highest ambient temperature and the grade of product is thus essential.

This illustrates the need for an agreed national standard for product specification which provides assurance of the highest pressure likely to be attained in service from a knowledge of ambient conditions. Against this knowledge, Codes of Practice can safely recommend the pressures which should be used for the design of storage vessels both for static use and for road/rail tankers. Guidance on the design parameters is given in the LPGITA (UK) Code of Practice No. 1, Installation and Maintenance of Bulk LPG Storage at Consumer's Premises and Code of Practice No. 2, Safe Handling and Transport of LPG in Bulk by Road.

Storage vessels for LPG range from those for major industrial users and refineries of 1000 tonnes capacity or more down to small industrial or commercial users of 1 tonne. Even smaller vessels down to 380 litres designed for on-site refilling are used for domestic purposes, i.e. heating and cooking. Similar vessels are used for permanent caravan sites and holiday homes. Figures 11.4 and 11.5 illustrate a storage installation for industrial and domestic use, respectively.

Deliveries to these installations are usually by road tanker. The tanker carries is own off-loading pump driven by the vehicle engine and the transfer is by a flexible hose which is connected to the receiving vessel by a gas-tight coupling.

LPG storage vessels have to be designed as pressure vessels to a recognized standard. In the UK the currently used standards are BS 5500 for static storage and BS 7122 for transport storage (i.e. for road tankers). The design limits for static storage are based on a maximum product temperature in service of 38°C, which gives a design pressure for propane of not less than 14.5 bar gauge. Those for transport storage are defined in BS 5355 which, for barrels greater than 5 m^3, sets a temperature of 42.5°C, giving a design pressure for propane of 16.75 bar gauge.

Figure 11.4 Typical industrial LPG storage installation. (Courtesy of Esso Petroleum Co. Ltd)

Figure 11.5 LPG storage vessel for a domestic installation. (Courtesy of the Calor Group Plc)

It is apparent from the vapour pressure curves that to obtain a gas supply from a storage vessel, or from a cylinder, it is only necessary to release it from the top vapour space. Propane can usually be supplied from storage to the plant by this natural vaporization, without any pumps or compressors, throughout the year in the UK.

However, in the case of bulk storage of butane the vapour pressure is generally too low for this simple method, and use is made of a liquid feed from the bottom of the storage vessel via a pump to a vaporizer. The vaporizer is simply a heat exchanger either using hot water, steam, electricity or even direct flame as the source of heat. Figure 11.6 shows a typical vaporizer.

A glance at the vapour pressure curve for butane will, however, reveal that in winter there is a possibility of butane vapour liquefying after the vaporizer if the temperature is allowed to fall in the pipeline, even at moderate pressure. For this reason, such pipework is usually heated, either by electrical tapes or, if available, by steam or hot-water lines.

There is an alternative to using neat butane vapour, however, which overcomes the need for pipework heating. This is to use a gas–air mixture. A special gas–air mixer is used which ensures that a preset ratio is maintained at all demand rates. The ratio chosen must be well outside the flammable limits. A typical LPG–air mixing plant is illustrated by Figure 11.7.

The effect of the air is to depress the vapour dewpoint temperature. A further advantage is that the physical properties of the gas can be made to 'simulate' another gas, e.g. natural gas or manufactured town gas. Such a simulated gas will produce the same heat release through a burner if the supply pressure is the same. This is characterized in the UK by a term known as the Wobbe Number (W):

$$W = \frac{\text{Calorific value}}{\sqrt{\text{Specific gravity of the gas (air} = 1)}}$$

The gas–air ratio is therefore chosen to achieve the desired Wobbe Number. However, the use of a simulated gas for plant conversion may not entirely avoid burner adjustment or modification if other properties differ markedly (e.g. flame speed).

The supply of gas from storage or vaporizer to plant is usually governed down to a medium pressure of around 1.5 bar, for final reduction to 28 mbar for butane or

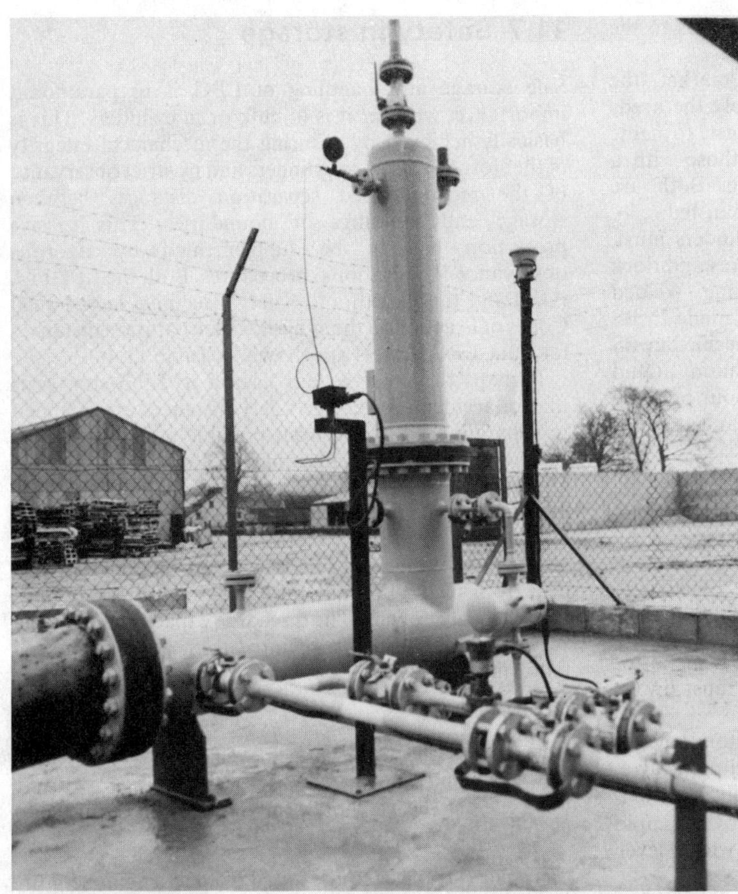

Figure 11.6 Hot-water heated LPG vaporizers. (Courtesy of Esso Petroleum Co. Ltd)

Figure 11.7 An LPG–air mixing unit. (Courtesy of John Wigfull & Co. Ltd)

37 mbar for propane. However, the versatility of LPG is such that plant and combustion equipment designers can, for specialized purposes, select virtually whatever pressures are most suitable for their requirements.

An unusual but successful alternative system has been applied for some very large high-energy consuming plants. This involves using liquid-phase LPG right up to the burner.

11.6 Cylinder storage

The earliest market for LPG was developed by the use of small transportable cylinders filled at purpose-built filling plants and returned there for refilling. This market has grown steadily over its 50-year or more lifetime. Cylinders are available for either commercial propane or commercial butane, and in a variety of sizes from easily portable ones with a capacity of around 11 litres to the largest with a capacity of around 108 litres.

In parallel with the refillable cylinder market, the non-refillable cylinder has emerged to supply the needs of the leisure market and the DIY enthusiast, to name but two. These are broadly divided into those with a valve outlet and the pierceable cartridge. Both are designed to be disposed of when finally exhausted.

Both the refillable and non-refillable cylinders must, of course, be designed for the internal pressure conditions likely to be experienced during their lifetime. Welded steel cylinders in the UK are designed and made to BS 5045: Part 2 and for developed pressures specified in BS 5355. They are based upon a maximum liquid temperature of 55°C and the associated vapour pressure of 22.13 bar gauge for propane and 7.52 bar gauge for butane.

It is now common practice in the UK to design all refillable cylinders to propane standards. Furthermore, the use of safety relief valves has been introduced for all cylinders carrying over 3 kg of LPG.

The outlet valves of refillable butane cylinders are different from those of refillable propane cylinders. The valve for propane is a handwheel type, with a female 5/8 in BSP LEFT-HAND thread for a POL metal-to-metal connector. A compatible half-coupling must always be used to ensure a leak-free joint.

Butane cylinders, which are so widely used for domestic service, are now provided with a self-sealing clip-on valve. The pressure regulator, which is normally attached directly onto the cylinder outlet, is fitted by a simple push-on or snap-on action and is provided with a lever which will open or shut the cylinder valve.

There are a number of different clip-on valve designs used by LPG distributors and care is necessary to ensure that only the correct mating regulator is employed.

11.7 Safety in storage

Safe storage and handling of LPG is of paramount importance, whether it is in bulk or in cylinders. This is basically achieved by ensuring the mechanical integrity of the storage vessels or cylinders and by strict observance of the recommended separation distances between storage and buildings or boundaries. This passive protection has to be supplemented by rigorous observance of operational procedures. Both the LPGITA (UK) and the Health and Safety Executive have issued Codes of Practice on the subject. The separation distances for bulk LPG vessels are shown in Table 11.4.

Perhaps the most serious hazard to LPG storage is that of accidental fire. The safety distances are intended to separate the storage from possible adjacent fires so that the risk of a fire affecting the storage is very low. However, this residual risk has to be catered for. The mechanical integrity needs to be assured under severe fire attack. For this reason, the vessels are provided with relief valves designed to cope with fire engulfment. The heat from such a fire may raise the stored pressure until the relief valve opens. The discharge capacity of the relief valve when fully open is required to meet or exceed the following:

$$Q = 10.6552A^{0.82}$$

where A is the total surface area of the vessel (m^3), and
Q is the discharge (m^3/min) of air reduced to 15°C and 1 atm.

The heat transfer to the liquid from an engulfment fire has been estimated at around 100 kW/m^2, and the above formula equates this to the vapour produced from this input as latent heat. The exponential is an area exposure

Table 11.4 Location and spacing for vessels

Maximum water capacity					*Minimum separation distances*						
Of any single vessel in a group			*Of all vessels in a group*		*Above-ground vessels*				*Buried or mounded vessels*		
									From buildings, etc. to		*Between vessels*
Litres	*Gallons*	*Nominal LPG capacity (tonnes)*	*Litres*	*Gallons*	*From buildings boundary, property line or fixed source of ignition (a)*		*With firewall (b)*	*Between vessels (c)*	*Valve assembly (d)*	*Vessel (e)*	*(f)*
					m	(ft)	m	m	m	m	m
150–500	28–100	0.05–0.25	1 500	330	2.5	(8)	0.3	1	2.5	0.3	0.3
>500–2500	100–500	0.25–1.1	7 500	1 650	3	(10)	1.5	1	3	1	1.5
>2500–9000	500–2000	1.1 –4	27 000	6 000	7.5	(25)	4	1	7.5	3	1.5
>9000–135 000	2000–30 000	4 –60	450 000	100 000	15	(50)	7.5	1.5	7.5	3	1.5
>135 000–337 500	30 000–75 000	60–150	1 012 500	225 000	22.5	(75)	11	$\frac{1}{4}$ of sum of the dia. of two adjacent vessels	11	3	–
>337 500	>75 000	>150	2 250 000	500 000	30	(100)	15	–	15	3	–

factor which recognizes that large vessels are less likely to be completely exposed to flames.

The safety relief valve will protect the liquid-wetted areas of the storage vessel. The metal temperature will not significantly exceed the liquid temperature which will be absorbing the latent heat of vaporization. However, above the liquid line no such cooling will take place. The metal temperature at the top of the vessel could therefore exceed safe limits.

The usual protection for large installations is to provide a water-spray system. For small bulk storage, fire hoses or monitors are often adequate. However, for installations over 50 tonnes of storage (and all major cylinder-filling plants) it is accepted that a fixed water-spray system needs to be provided which is automatically initiated by a system capable of detecting a fire threatening the vessels and/or the adjacent tanker loading or unloading area. The deluge rate to provide protection against fire engulfment is 9.8 $l/min/m^2$ of vessel surface area, and this should be capable of being sustained for at least 60 min. The spray pattern adopted for a fixed installation normally includes four longitudinal spray bars, two at the upper quadrants and two at the lower quadrants of horizontal cylindrical vessels, with nozzles spaced to give uniform coverage.

An alternative means of avoiding the hazard from fire is to bury the vessels or to employ the increasingly popular method of mounding. In either case acknowledgment of the reduced hazard is indicated by the reduced separation distances (see Table 11.4). Since both burial or mounding preclude the possibility to monitor continuously the external condition of the vessels, very high-quality corrosion protection needs to be applied, often supplemented by cathodic protection, depending on soil conditions.

The use of burial or mounding is sometimes employed to overcome visual environmental objections since the mounding, for instance, can be grassed over. Indeed, this is the method often adopted to prevent erosion of the mounding material. A typical mounded storage is seen in Figure 11.8.

Codes of Practice also provide guidance on the storage of LPG in cylinders. Again, it is based on the adequate separation of stacks of cylinders from buildings and boundaries. These distances depend on the total tonnage stored or the tonnage in the largest stack in the storage area. The Codes also provide guidance on width of gangways, height of stacks and storage within buildings.

11.8 Uses of gaseous fuels

LPG has many advantages over the alternative liquid fuels, and is regarded as a 'premium fuel'. Some of the premium characteristics are:

Very low sulphur
Easy ignition
Simplicity of burner design
Versatility (e.g. very small flames, localized heating)
Wide turndown of burners
Clean combustion gases
Ease of control

These properties apply equally to natural gas as for LPG. However, since LPG can be used anywhere in the country which is accessible by road or rail it is not subject to the geographic limitations of a piped supply. The versatility of gaseous fuels has resulted in much development work in the design of burners for a wide range of applications. Some are highly specialized. Much of this development has been primarily for natural gas since its introduction to the UK in the late 1960s. The properties of LPG and natural gas are different, which means that LPG cannot be used in an appliance or burner designed for natural gas. However, a key property, flame speed, is of the same order, unlike the very high flame speed of coal gas. This similarity means that most natural gas applicances or burners can be adapted for use with LPG with minor

Figure 11.8 Mounded LPG storage. (Courtesy of Southern Counties Gas)

modifications to allow for the much higher calorific value and higher gas density. Aeration needs attention, but for a given heat input rate the majority of gaseous fuels require approximately the same amount of air for combustion. Of course, in the domestic and leisure use of LPG particular appliances have been developed quite specifically for use with LPG. Thus LPG equipment and appliances are available for all the gas-fired applications in commercial, industrial and domestic use.

When drawing off LPG vapour from storage, whether from bulk storage, cylinders or even non-refillable cartridges, the liquid will begin to vaporize but will need latent heat to do so. It will take this heat from itself, and the container, and both will cool. As the LPG liquid cools its pressure must fall. Equilibrium will be obtained when the heat flow from the ambient air equals the demand for latent heat.

This, together with the natural variations of vapour pressure with the climatic conditions, results in a source of gas with quite wide variations in supply pressure. Except for some small heating appliances, which are designed to attach directly to a small cylinder or cartridge, this pressure variation needs to be eliminated as far as the appliance is concerned. Therefore a pressure regulator is used. For permanent installations two-stage pressure regulation is recommended. The first is usually located immediately at the outlet from the storage vessel or cylinder.

LPG cylinders are filled only at purpose-designed filling plants and (if refillable) are returned there when exhausted. To provide a continuous gas supply, two, four or more cylinders are connected to a changeover valve so that empty cylinders can be exchanged without interrupting the supply. Automatic changeover devices are available which switch to the reserve supply when necessary, and indicate that they have done so. A typical wall-mounted ACD is illustrated in Figure 11.9.

Codes of Practice restrict the use of LPG cylinders indoors for commercial and domestic use to commercial butane. Propane cylinders should always be located outside the premises except for special industrial purposes. A typical four-cylinder pack is shown in Figure 11.10.

The use of propane as a motor vehicle fuel has been highly developed in some overseas countries, particularly in the USA, Holland and Italy. It is, of course, an entirely lead-free fuel. Very high efficiencies can be obtained using a gaseous fuel in spark-ignition engines since intimate mixing of the fuel and air is much more easily achieved than with a liquid fuel. This results in a much cleaner exhaust, with considerable reductions in CO and hydrocarbons.

The low level of harmful exhaust emissions has, for instance, been one of the spurs to the wide use of LPG as a forklift truck engine fuel in all countries because of their wide use inside buildings such as warehouses, railway station forecourts, etc. Another advantage is the high power continuously available compared with electric battery-driven equivalents. Similar arguments favour its use for other engine-driven workhorses such as mechanical sweepers, in large warehouses, railway stations, etc.

Figure 11.9 Wall-mounted automatic changeover device. (Courtesy of Sperryn & Co. Ltd)

Figure 11.10 A 2 × 2 LPG cylinder pack supply domestic premises. (Courtesy of the Calor Group Plc)

Vehicle fleet owners often have their cars and/or vans converted to run on LPG and have an on-site bulk storage vessel installed complete with a dispensing pump. This is an attractive economic proposition for taxis, ambulances, local authority fleets, etc. Virtually any spark-ignition engine can be converted to run on LPG. It can therefore be considered for any engine-driven machinery.

11.9 Safety and legislation

Safety in the supply and use of LPG, as with all forms of energy supply, is of paramount importance. The guidance given by the LPGITA (UK) and the Health and Safety Executive needs to be fully appreciated and observed. Safety for industrial plants is maximized by following the requirements on storage and plant layout, plant design, system design, operational procedures, maintenance and periodic inspection procedures, and management control. The LPGITA (UK) has produced a series of Codes of Practice which provides the basis for safe storage, handling and use of LPG in the UK. A list of current issues is given at the end of this chapter.

There are many statutory requirements which impose on the design, installation and operation of the LPG plants. There are the general duties under the Health and Safety at Work Act, of course, but specifically, the storage of LPG at 'factories' as defined in the Factories Act 1961 are required to comply with the Highly Flammable Liquids and Liquefied Petroleum Gas Regulations. This requires suitable storage vessels, and their marking with the words 'LPG Highly Flammable'. The design of LPG cylinders is also covered in these Regulations.

The Pressure Systems and Transportable Gas Containers Regulations 1989 will be implemented over the next few years and will place duties on the designers and suppliers of pressure systems for use at work and on the owners or users of such systems to ensure they do not present any danger. It will require a written scheme of periodic examination drawn up or endorsed by a competent person.

For new sites with 50 tonnes or more of LPG the Control of Major Accident Hazards Regulations 1984 apply. For sites which will have 300 tonnes or more (shortly to be reduced to 200 tonnes or more) these Regulations impose additional duties on site operators which include the prior submission of a 'safety report'. This report has to set out the potential hazards of the plant and the means by which the risks are reduced to an acceptable level. The LPGITA has produced a *Guide to the Writing of LPG Safety Reports* which supplements the general guidance in the HSE booklet HS(R)21.

Consumer installations which supply LPG to more than one user (e.g. metered estates, holiday home parks, caravan sites) require the gas supplier (i.e. the site owner/operator, not the LPG supplier) to obtain prior consent from the Office of Gas Supply (Ofgas) and, for the necessary pipework installation, from the Department of Energy, Pipelines Inspectorate.

Shortly to be introduced is the need to obtain Hazardous Substances Consent under the Housing and Planning Act 1986 from the local authority for any site intended to hold 25 tonnes or more of LPG. This will be in addition to any requirement for planning consent and to the need to notify the HSE under the Notification of Installations Handling Hazardous Substances Regulations 1982.

The conveyance of LPG by road is also subject to control via a number of Regulations aimed at potentially hazardous substances. There are the Dangerous Substances (Conveyance by Road in Road Tankers and Tank Containers) Regulations 1981 and the Road Traffic (Carriage of Dangerous Substances in Packages, etc.) Regulations 1986.

11.10 British Standards

BS 4250 Liquefied petroleum gas
Part 1: Specification for commercial butane and propane
Part 2: Specification for automative LPG
BS 5500 Unfired fusion welded pressure vessels
BS 7122 Welded steel tanks for the road transport of liquefiable gases
BS 5355 Filling ratios and developed pressures for liquefiable and permanent gases
BS 5045 Transportable gas containers
Part 2: Steel containers up to 450 litres water capacity with welded seams
BS 4329 Non-refillable metallic containers up to 1.4 litres capacity for liquefied petroleum gases

11.11 Codes of Practice and Guidance Notes

HSE

Guidance Note HS(G)34, *The storage of LPG at fixed installations*
Guidance Note CS4, *The keeping of LPG in cylinders and similar containers*
Guidance Note HS(R)21, *A guide to the Control of Industrial Major Accident Hazards Regulations 1984*

LPGITA (UK)

An Introduction to Liquefied Petroleum Gas
Code of Practice No. 1, Installation and Maintenance of Fixed Bulk LPG Storage at Consumer's Premises
Part 1: Design and Installation
Part 2: Small Bulk Installations for Domestic Purposes
Part 3: Periodic Inspection and Testing.
Code of Practice No. 2, Safe Handling and Transport of LPG in Bulk by Road
Code of Practice No. 3, Prevention and Control of Fire Involving LPG
Code of Practice No. 4, Safe and Satisfactory Operation of Propane – Fired Bitumen Boilers, Mastic Asphalt Cauldrons/Mixers, and Hand Tools
Code of Practice No. 7, Storage of Full and Empty LPG Cylinders and Cartridges
Code of Practice No. 9, LPG–Air Plants
Code of Practice No. 10, Recommendations for Safe Handling of LPG in Small Trailer Mounted Vessels, Skid Tanks and Small Bulk Storage Vessels Permanently Attached to Mobile Gas-fired Equipment
Code of Practice No. 10, Part 1 (first stage of revision), Containers Attached to Mobile Gas-fired Equipment
Code of Practice No. 11, Safe handling of LPG used as an Internal Combustion Engine Fuel for Motor Vehicles
Code of Practice No. 12, Safe filling of LPG cylinders at depots
Code of Practice No. 14, Hoses for the transfer of LPG in bulk. Installation, inspection, testing and maintenance.
Code of Practice No. 15, Valve for LPG cylinders. Part 1. Safety valves for LPG cylinders. Part 2. Outlet valves for butane cylinders – quick coupling types

Code of Practice No. 17, Purging LPG vessels and systems
Code of Practice No. 18, Safe use of LPG as propulsion fuel for boats, yachts and other craft
Code of Practice No. 19, Liquid measuring systems for LPG
Code of Practice No. 20, Automotive LPG refuelling facilities
Code of Practice No. 21, Guidelines for caravan ventilation and flueing checks
Code of Practice No. 22, LPG piping system design and installation

Institute of Petroleum
Model Code Safe Practice Part 9, *Large bulk pressure storage and refrigerated LPG*

Engineering Equipment and Materials Users Association (EEMUA)
Recommendations for the design and construction of refrigerated liquefied gas storage tanks

12

Coal and Ash

C Foster
British Coal

Contents

12.1 Introduction

One of the most important considerations to be taken into account when using coal as a fuel for industrial boilerhouses is the provision of coal- and ash-handling equipment. Although such equipment was traditionally manual, with only larger installations making use of mechanized handling systems for the coal, the higher expectations of modern boilerhouse operators have led to the increasing use of automatic coal- and ash-handling systems for all coal-fired boilerplant. The cost of handling equipment is not insignificant, and may well represent a considerable proportion of the overall cost of a boilerhouse. It is therefore important that the appropriate type of equipment is selected and that it is properly installed. It is also important that the use of coal- and ash-handling equipment is environmentally acceptable in terms of noise, dust and visual impact. The relatively low manning levels now encountered in industrial boilerhouses are not easily able to cope with breakdowns in equipment which may be handling coal at more than 100 tonnes per hour. Similarly, modern automatic firing equipment is less able to tolerate coal which is significantly out of specification due to degradation caused by the handling system. Reliability and performance, both a function of correct design and installation, are therefore of major importance for coal-handling systems.

Names and addresses of manufacturers of proprietary equipment discussed in this chapter are available from British Coal.

12.2 Characteristics of coal

Several different grades of coal are currently supplied to the UK industrial market. These include washed singles coal (38–12 mm, or 25–12 mm) and various washed, blended and untreated smalls coals with size distributions ranging from 50–0 mm to 12–0 mm. Because of its size and consistency, washed singles coal presents fewer handling problems than smalls, although care has to be taken to avoid excessive degradation. Smalls coal can contain a high proportion of fine particles (more than 10% below 0.5 mm), and typical moisture contents of between 10% and 14%. The cohesion between the fine particles can be very high, particularly at moisture contents of about 12% which are exhibited by most UK smalls. Untreated smalls coal may also contain clay and shale which reduces the handleability of the fuel. Smalls coals therefore require more sophisticated handling techniques than singles, and this can become an important consideration during the planning of a boilerhouse.

12.3 Delivery

Most of the coal for industrial use in the UK is delivered by road, although a few large installations use rail, ship or barge. The use of these latter methods will depend upon the availability of suitable track or waterway close to the boilerhouse, and are only generally applicable to installations burning in excess of 50 000 tonnes per year. There are several types of road-delivery vehicle, of which the tipper lorry is the most economical. The preferred size of tipper is 26 tonnes capacity, although smaller units are used for sites with limited access or stocking facilities. Other types of vehicle include conveyor lorries and pneumatic delivery lorries, but because of their lower versatility, a small surcharge (typically, $2\frac{1}{2}$%) is usually levied on delivery by these vehicles, and this restricts their use to smaller installations.

Conveyor lorries incorporate a hydraulically driven belt conveyor located at the back of the vehicle. This conveyor can be swung round through 180°, and can deliver up to a height of 2.8 m, thus providing greater flexibility of discharge than the tipper lorry. The conveyor can be fed either by tipping in the normal way or by a belt conveyor running from front to rear of the lorry bed. Pneumatic delivery lorries incorporate a positive displacement blower to provide the air supply for pneumatic conveying.

Coal is fed into the conveying airstream of a rotary valve from the bed of the lorry either by tipping or by use of a belt conveyor. The coal (which is normally singles for this type of delivery) is initially conveyed through 125 mm diameter flexible pipe. The flexible pipe is then connected to a fixed pipe at the customer's site which conveys the fuel to storage hoppers up to 40 m distant from the vehicle. Pneumatic delivery is particularly useful to the smaller consumer because the coal can be delivered directly to the service hopper, from where it can flow by gravity to the boiler, and thus obviate the requirement for any on-site coal-handling equipment.

Inevitably, with this form of delivery, times are extended in comparison with simple tipping vehicles and, depending on the type and condition of the vehicle, may take more than one hour to complete. Because of the extra equipment required for pneumatic delivery these vehicles have a slightly smaller capacity than tipper vehicles, typically ranging between 12 and 20 tonnes. Noise is an important factor to be taken into account with pneumatic delivery. In particular, worn rotary valves allow significant air leakage and this requires higher blower (and hence engine) speeds to be maintained. The air leakage also reduces the rate at which coal can be delivered, thereby extending delivery time. In order to reduce the noise of delivery it is possible for a site-installed electrically driven blower to be used instead of the vehicle-mounted unit. Some additional noise is created by the coal being conveyed within the fixed site pipework, and this can be significantly reduced by the use of ordinary boilerhouse pipework lagging. British Coal have published a report on pneumatic delivery[1] which provides detailed guidance for existing and potential users of this delivery method.

In addition to the above types of delivery, containerized delivery using 20-tonne capacity ISO containers can be employed. The advantage of using containers is that the coal need not be handled directly (thereby causing degradation) at intermediate transfer points, and is

Figure 12.1 Container unloading system

therefore available for use in as good a condition as when the container was first loaded.

Several systems are available for container handling and unloading at user sites, an example of which is shown in Figure 12.1. Two reception bays are provided at this site so that the delivery vehicle can deliver a full container and immediately load on the adjacent empty unit. Together with a small-capacity site storage system, the container itself can thus provide a useful degree of additional storage. In operation, the vehicle backs into the unloading zone, and chains are secured to the container by means of twist-locks. Hydraulic rams lift the container clear of the vehicle, which can then leave. Further operation of the hydraulic rams tips the container so that coal is discharged into the boilerhouse conveying system.

12.4 Coal reception

12.4.1 Open stocking

Where ground space is available, large industrial users have traditionally stored coal in open stockpiles. In this case, the coal is tipped directly onto the ground and then picked up and stocked using a tractor and bucket. For quantities of coal of less than about 500 tonnes which are used as working stock, a tractor and bucket is probably the most effective means for stock reclamation, although automatic reclamation can be achieved by using grab-skip-hoists which operate from an overhead monorail. A British Coal document[2] is available which provides guidance on the management of open stockpiles, particularly with respect to the prevention of spontaneous heating. Where quantities of coal greater than about 1000 tonnes are required as strategic stock, then it is recommended that specialist advice be sought from British Coal.

12.4.2 Underground bunkers

Underground reception bunkers are the traditional method of accepting coal from tipper lorries. They are usually sized to hold about 30 tonnes (one and a half lorry loads) so that a continuous supply of coal can be maintained to the coal-conveying system. A grid screen is usually fitted to the top of the bunker in order to prevent oversize material from entering the conveying system. If the vehicle is to run onto the screen (which should have a mesh size of 100 mm), then due consideration should be given to the axle weights likely to be encountered. In order to maximize the capacity of the bunker, retaining walls are normally used to surround the grid screen, and these can also be used to support a roof structure which will prevent rain from entering the bunker, as well as reducing dust in the near vicinity. More specific detail concerning the design of the hopper and its outlet is provided in Section 12.5.

A variant of the underground bunker is the boot bunker, where the vehicle backs up a ramp until the rear wheels are about 2 m above ground level. In this way the depth below ground level, and hence the cost of excavation, is reduced. The disadvantage of the boot type of bunker is that the ramp takes up a considerable space on the boilerhouse site.

12.4.3 Tipping hoppers

The disadvantage of underground bunkers is that expensive civil works are usually necessary to provide drainage and access to the bunker base. In order to overcome this problem several UK manufacturers have developed end-tipping hoppers which provide a controlled feed from coal which has been tipped at ground level. In operation, the hopper is lowered so that one side becomes a reception platform onto which coal is tipped from the lorry. After the load has been tipped, the entire hopper is elevated by a hydraulic jacking system into a vertical position, and can then discharge its contents by gravity into the boilerhouse conveying system as shown in Figure 12.2.

Commercial tipping hoppers of this type often incorporate a screening system on the outlet to remove any oversize material which might otherwise block the conveying or combustion system. An alternative design of side-tipping hopper is available for sites where a limited headroom prevents the use of an end-tipping unit. In this case, coal is tipped onto a platform at ground level as previously described, and is then removed from one side of the hopper using an open-screw conveyor to feed the boilerhouse conveying system. When the depth of coal above the screw conveyor has fallen so that no more coal is being conveyed, the platform is tilted by hydraulic jacks. The remainder of the coal is thus tipped onto the conveyor which can then empty the hopper.

These types of reception system can provide significant savings for larger installations which use singles coal, by using tipper lorry delivery instead of pneumatic delivery. The actual saving will depend on the levy imposed by the coal distributor for the use of the pneumatic delivery vehicle, but they are generally viable where the annual coal consumption is greater than about 3000 tonnes.

12.4.4 Walking Floor

The advantage of this system, which is shown

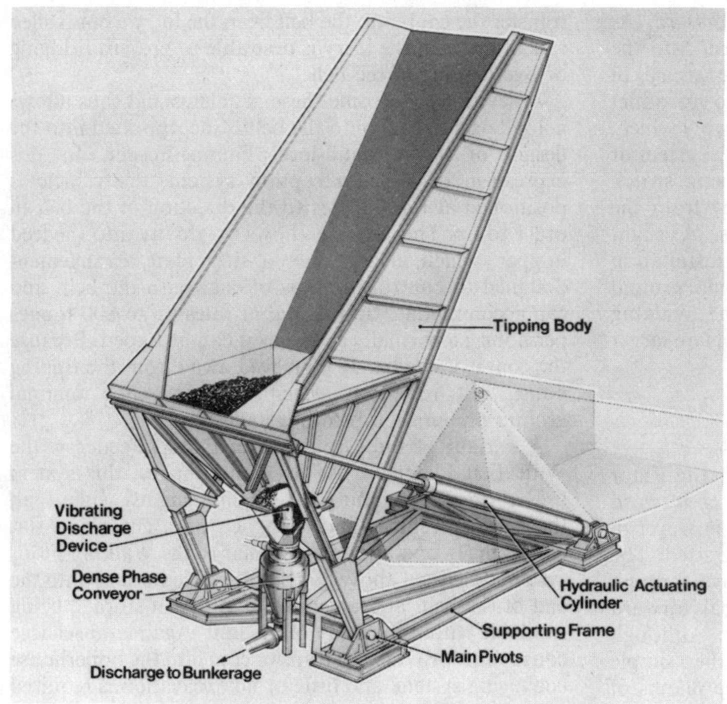

Figure 12.2 Tipping hopper

Figure 12.3 Walking Floor coal reception system

schematically in Figure 12.3, is that tipped coal can be stored under cover within the body of the unit. In operation, the delivery vehicle back up to the Walking Floor unit and is then tipped. The floor is constructed of a number of sliding longitudinal planks which reciprocate in sequence to provide a forward motion to bulk materials. Each third plank moves backwards in turn, sliding beneath the coal, and then all planks move forward together, thus 'walking' the coal forward. As the coal is tipped onto the floor it is 'walked' into the body of the unit, where it can be stored if necessary. In order to discharge, the floor feeds coal onto an outlet conveyor (such as a screw or *en-masse* conveyor) which, in turn, either feeds the boilerhouse conveying system or coal directly to the boilers. A simple reversing switch allows coal to be conveyed backwards away from the outlet conveyor for maintenance purposes. A slight incline (up to about 6°) on the floor enables installation to be carried out with no excavation and minimal ground preparation. Typical unloading rates for the Walking Floor are 250 tonnes per hour, reducing slightly on larger units holding up to 60 tonnes of coal.

12.4.5 Push-Floor

The Push-Floor system is similar to the Walking Floor in that a reciprocating movement provides a forward motion to coal as it is tipped. In this case, however, a number of shaped bars carried in frames are used. The bars are wedge shaped so that they can move through the coal in one direction but push the coal forward (against their flat faces) in the other direction. Although the flow of coal cannot be reversed, the simple wedge-shaped bars are more robust than the planks of the Walking Floor, and it is possible to drive a vehicle over them without causing damage.

12.4.6 Belt reception systems

Where it is necessary to convey coal rapidly away from the reception point, coal can be tipped directly onto the moving belt of a belt elevator which then conveys it directly to storage. When using a normal feed hopper to transfer the coal onto the belt from the lorry a controlled rate of tip from the lorry is desirable to prevent bridging or overloading of the belt.

A device to overcome these problems and thus allows a flood feed of coal onto the belt is incorporated into the design of the Ground-level Dump Feeder. In this expression of a belt reception system, the vehicle is positioned at right angles to the direction of the belt in order to tip. The coal then flows by gravity into the feed hopper, which incorporates a steel plate arrangement designed to control the flow of coal onto the belt, and can accommodate tipped coal at rates up to 800 tonnes per hour, i.e. virtually as fast as it can be tipped. Because the coal is immediately conveyed away from the tipping point, belt reception systems require only minimal ground preparation before installation.

A variant of the Ground-level Dump Feeder is the Wide Belt Unloader. As its name implies, this system uses a wide belt (about 3 m), which is slightly wider than the delivery vehicle, and is carried on a chain and slat conveyor. In operation, it is similar to the Walking Floor system described above, with the vehicle tipping onto the end of the belt, and a certain quantity of storage being available within the body of the unit. Again, a discharge conveyor is provided to transfer coal into the boilerhouse conveying system, and little or no excavation is required for installation. Unloading rates are also similar to the Walking Floor, although the maximum capacity of the body is unlikely to exceed one vehicle load. A schematic diagram of the system is shown in Figure 12.4.

12.4.7 Auger reception systems

Screw conveyors have been used for a number of years to extract coal from bunkers and coal heaps, usually for

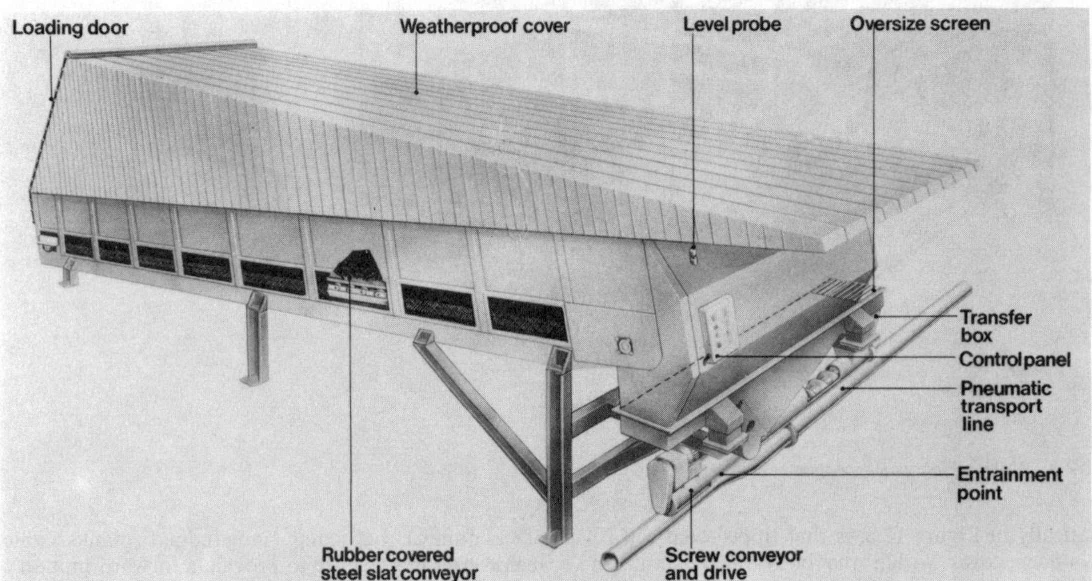

Figure 12.4 Wide Belt Unloader

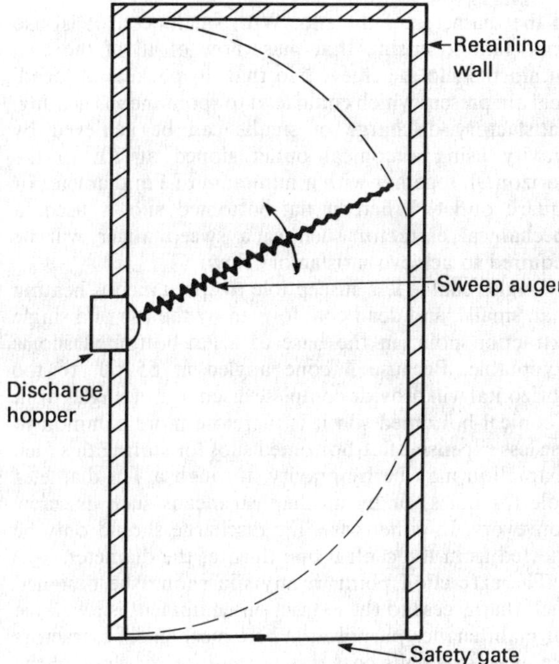

Figure 12.5 Ground sweep reclaimer

feeding directly to boilers. Examples of these systems are pneumatic drop feed or sprinkler stokers, underfeed stokers or as conveyors to serve the boiler feed hopper. Fixed screws, however (the design of which are discussed in Section 12.6), will only recover a proportion of the coal within the tipped area, and this has led to the development of sweeping screw conveyors which traverse through the coal.

The example shown in Figure 12.5 is based upon the use of such a sweeping screw, and uses a special drive mechanism to limit the force applied to achieve a semi-circular sweep. This prevents damage caused by the sweeping motion from forcing the screw to move out of the plane of the sweep. Coal is tipped directly onto a gridscreen from where it flows into a shallow (1 m deep) bunker contained within a retaining wall enclosure. The rotary sweep auger works through 180° across the floor of the enclosure, drawing coal off into the boilerhouse conveying system.

12.5 Coal storage

The quantity of coal stored at any particular site is dictated by the likely risk of cuts in supply to any installation, and may vary from a few tonnes held in the boiler hopper to stockpiles of many thousands of tonnes.

12.5.1 Spontaneous heating

When storing any quantity of coal greater than about 50 tonnes and for longer than a few weeks, care has to be taken to avoid contamination, breakage of singles coal, and spontaneous heating. Spontaneous heating arises from partial oxidation of the coal, and occurs when ventilation within the stock is sufficient to provide the oxygen necessary for oxidation but is insufficient to disperse the heat generated in the process. Signs of spontaneous heating are generally exhibited within a 10- to 12-week period after the coal is laid down. If there is no sign of heating after six months, it can be assumed that the coal is safe.

In order to prevent spontaneous heating it is necessary that either adequate ventilation should be provided or, alternatively, air should be almost completely excluded. For singles coal, ventilation is normally encouraged so that the air carries away any heat. In storing smalls coal, however, open stockpiles should be compacted and silos sealed so that air is virtually excluded from the stock.

Generally, the younger soft, bright lignitic coals are more susceptible to spontaneous heating than older coals such as anthracite. If there is a possibility of spontaneous heating in a silo then the condition of the coal should be monitored by measuring the carbon monoxide content of the air at the top of the silo. (The use of thermocouples to indicate a rise in temperature due to spontaneous heating is ineffective.) The occurrence of spontaneous heating is indicated by a relatively substantial increase in the level of carbon monoxide, and in this event the silo should be emptied as quickly as possible and the coal spread thinly on the ground to cool. On no account should water be pumped into the stored coal in an attempt to extinguish the fire as water gas may be formed, although if spontaneous heating is detected sufficiently early, nitrogen purging of the silo can reduce the risk of the heating spreading. Further information on spontaneous heating of stored coal is available from British Coal.[2–4]

12.5.2 Coal-storage silos

Limited ground space together with the requirement to store coal in any considerable quantity has led to the increased use of storage silos up to sizes in excess of 1000 tonnes capacity. Silos above 500 tonnes capacity should be constructed of reinforced concrete and may be circular or rectangular in cross section. Capacities between 300 and 500 tonnes are usually most economically obtained by using cylindrical silos constructed of concrete staves bound together with steel hoops. Below this capacity, vitreous enamel-coated steel silos are generally less expensive than concrete, and for smaller storage units of up to 80 tonnes, low-cost polyester silos may be used.

The information provided below is intended as a general guide to users of silos, but further more detailed information is available from British Coal.[4]

Filling of silos

The method selected for filling will be governed largely by the height of the silo. Where single coal is delivered into the silo by means of a vehicle-mounted pneumatic delivery system, the maximum height for delivery should not exceed 18 m. Taller silos can be considered when filled using a mechanical conveyor or elevator, or pneumatically using a purpose-designed on-site system as described in Section 12.3. Where smalls coal is to be stored in silos, lean-phase pneumatic delivery is impractical and other mechanical or dense-phase pneumatic systems must be used.

Except when constructed of reinforced concrete, all silos should be filled from within the centre third of the diameter to reduce the risk of eccentric loadings. The action of filling a silo or bunker results in breakage of the coal particles. Simple gravitational fall will cause some degradation but most breakage occurs when some form of pneumatic system is used to fill the silo. The presence of fine coal particles can then result in segregation, the finer particles concentrating in a heap directly below the fill position while the larger ones roll to the outside of the pile. Such a concentration of fine particles results in a pillar of coal within the silo which (particularly if the coal is moist) will not flow. Segregation must therefore be avoided, and for lean-phase pneumatic delivery it is recommended that the coal be spread by allowing it to fall onto the apex of a steel cone mounted below a velocity-reducing cyclone as shown in Figure 12.6. An impact and wear resistant rubber can be used to line the cyclone which will reduce degradation as well as contributing to a significant reduction in noise caused by coal entering the cyclone.

When filling an empty silo (or tall bunker), breakage can be reduced by filling as far as possible from a side access door, thus reducing the distance the coal has to fall from the permanent fill system. Good management in maintaining high coal levels with the silo will also minimize breakage caused by free fall.

Extraction from silos

In order to prevent asymmetric loadings on the silo walls coal should only be discharged from the central one third of the diameter of the silo. With smalls coal it is also important to ensure that mass flow of all of the coal within the silo is achieved so that no pockets of 'dead' coal are present which could lead to spontaneous heating. Satisfactory discharge of smalls can be achieved by gravity using a conical outlet sloped at 70° to the horizontal, together with a minimum of 1 m diameter or square outlet. Where a flat-bottomed silo is used, a mechanical extractor such as a sweep auger will be required to achieve satisfactory flow.

Singles coal is less susceptible to spontaneous heating than smalls, and dead coal formed by the use of a single extraction point in the base of a flat-bottomed silo is acceptable. Because a cone angled at 55–60° to the horizontal will provide complete discharge of singles from a conical-bottomed silo it is therefore more common to use less expensive flat-bottomed silos for storing this fuel. Extraction may be by gravity, through a 1 m diameter hole (or holes) or by mechanical means such as screw conveyors. In either case, the discharge should only be effected from the central one third of the diameter.

The extraction point in any silo should be designed such that access to the extraction equipment is available for maintenance purposes. Where mechanical extractors are used for smalls coal it is normal for the floor of the silo to be constructed at a height of 2.5–3 m above ground level, thereby providing space for conveyors to be mounted beneath.

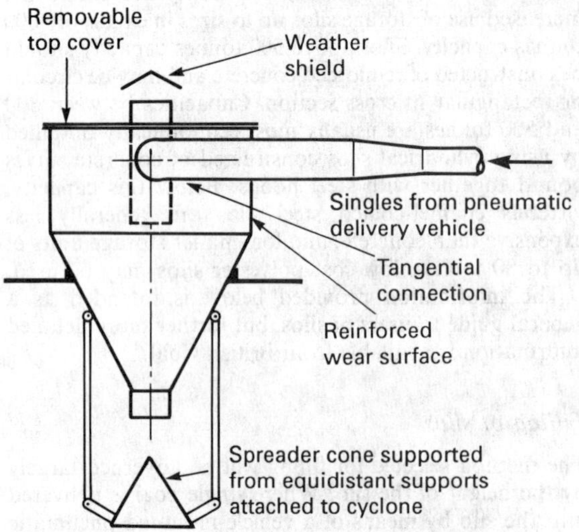

Figure 12.6 Velocity-reducing cyclone

Methane venting

Some coals, particularly when freshly mined, are liable to give off methane gas. In order to prevent an explosive mixture forming within the silo headspace a 150 mm diameter vent should be provided at the highest point within the roof of the silo.

Safety

As with all coal-storage systems, silos represent a 'confined space' and the Health and Safety Executive Note GS5 should be fully complied with whenever a silo is entered. Particular care should be taken to avoid personnel having to walk over coal which may have a void below the surface. Any portable lighting used should be approved battery-operated lamps.

12.5.3 Hopper design

All silos and bunkers will eventually terminate in a hopper to feed the stored coal into a conveyor or boiler. In the case of singles coal the hopper should have a minimum cross-sectional dimension of 600 mm and a minimum cone angle of 55° from the horizontal. In the case of smalls coal these figures should be increased to 1 m and 70°, respectively. It is possible to reduce the cone angles slightly (by about 5°) by using a low-friction lining material such as ultra-high molecular weight (UHMW) polyethylene. This material, which should only be used in 100% virgin state, is widely used for improving the flow characteristics of hoppers and chutes, and can be attached to virtually any surface, including concrete and steel.

It should be borne in mind, however, that UHMW polyethylene is combustible, and that both spontaneous heating and burn-back (from a stoker hopper) may give rise to poisonous fumes. Detailed information regarding its installation may be obtained from British Coal.[5]

Storage systems designed to promote mass flow of the stored coal using the hopper outlet sizes recommended above typically have to feed into the relatively small-sized inlets of conveying systems. For singles coal, this transition from mass flow to controlled flow can be achieved by gravity, using a suitably steep-sided inlet hopper. Smalls coal, however, will normally require the use of a mechanical discharge system.

A number of such devices are available to promote flow from hopper outlets, all working on the principle of agitating the coal close to where it is likely to bridge. The most common system consists of a vibrator attached in various ways to the side of the hopper and designed to create slip at the hopper wall. These can be very effective, although care has to be taken to avoid compacting the coal. A different system which is currently becoming more popular is the use of a bin activator, as shown in Figure 12.7. Again, a vibrator is used, but in this case the vibration is applied to an inverted cone close to the bunker outlet, and this generally promotes better mass flow of the coal than the side vibrator systems.

An alternative method, more commonly used where flow problems exist due to poor hopper design or a change in fuel, is the use of one or more air blasters. These devices direct jets of compressed air into the coal in the vicinity of expected blockages. The sudden release of relatively small quantities of compressed air can be very effective in dislodging coal which has bridged across hopper outlets.

With any contained coal-storage system it will occasionally be necessary to carry out maintenance on equipment installed beneath the outlet. In order to isolate this equipment from the flow of coal it is recommended that a rod gate valve be installed as shown in Figure 12.8. This valve is more robust than a slide valve, and is easier to free if it becomes corroded.

12.6 Coal conveying

Industrial boilerhouse coal-conveying systems are either mechanical or pneumatic, and the selection of one of these for a particular boilerhouse will depend on the following factors:

1. *Required coal-conveying rate.* Most boilerhouses will require a minimum coal-conveying rate of three times the maximum coal-burning rate in order to provide adequate time for maintenance or repairs. Mechanical systems are able to convey coal at rates ranging from zero up to several thousand tonnes per hour. Pneumatic systems such as conveyors have capacities ranging from about 5 tonnes per hour up to a maximum of about 60 tonnes per hour. (When pneumatic conveyors are used for feeding direct to boilers, the flow rate may reduce to zero, but such low conveying rates are liable to cause increased coal breakage.)
2. *Type of coal.* Mechanical systems are generally able to convey both singles and smalls coal, although careful attention to design is necessary to minimize degradation of singles coal, and to provide self-cleaning systems for smalls coals, particularly where the fines (<0.5 mm) content is high Pneumatic systems are selected according to the type of coal being conveyed. Lean-phase systems are generally only suitable for singles coal, whereas dense-phase systems are most suitable for smalls. These features are discussed more fully in Section 12.6.2.
3. *Allowable power consumption.* Although, by itself, this factor is unlikely to determine the choice of system, it should be borne in mind that mechanical conveyors typically use only 20% of the power required for equivalent pneumatic conveyors.

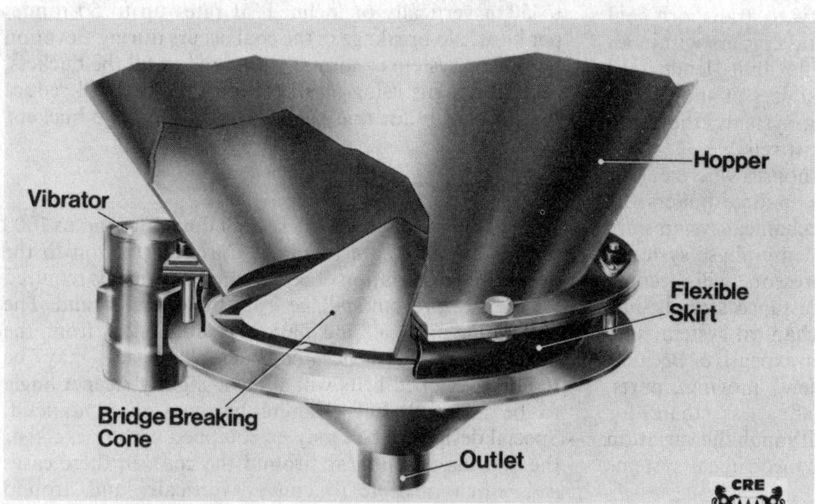

Figure 12.7 Bin activator

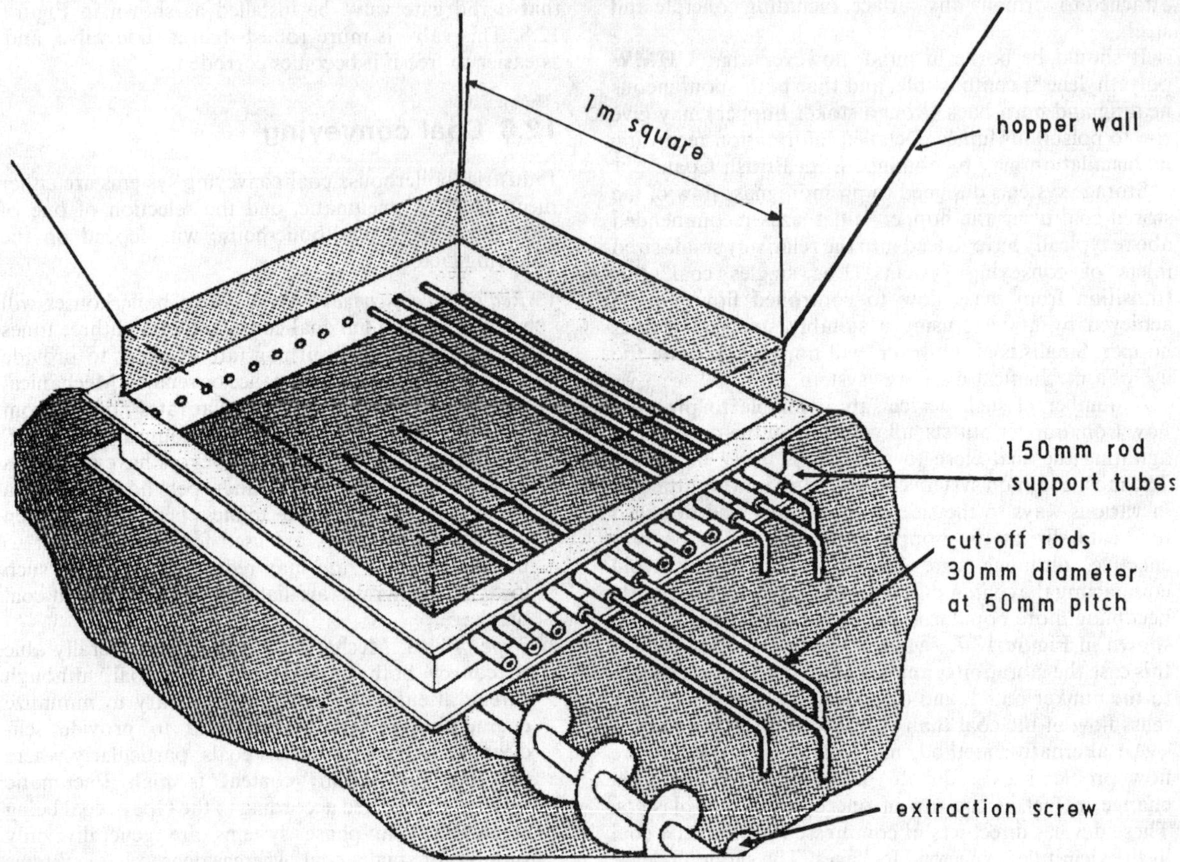

Figure 12.8 Rod gate valve

4. *System complexity and layout*. Mechanical systems can generally only convey coal in straight lines, although some designs can bend within a vertical plane. The more recent popularity of pneumatic conveyors is due in no small part to their ability to transport coal horizontally, vertically and around corners within an enclosed pipeline. Although this flexibility lends itself to the aesthetic design of boilerhouses, it should not be abused. As with all handling systems, the more simple designs tend to be the most reliable.
5. *Costs*. For most industrial boilerhouses where coal is required to be conveyed to two or three boilers the capital cost of pneumatic and mechanical systems are broadly similar. For very long dense-phase systems, however, the cost of air compressors will become significant, and for very short-distance straight-line conveying it is likely that a mechanical system such as a screw conveyor will be least expensive. Because pneumatic systems contain few moving parts, maintenance costs are normally less than for equivalent mechanical systems, although the variation in maintenance requirements for mechanical systems is considerable.

12.6.1 Mechanical systems

Bucket elevator

These systems are commonly used for elevating coal up to 50 m vertically or inclined, at rates up to 50 tonnes per hour. No breakage of the coal occurs during elevation but a feed system is normally required to fill the buckets. Special designs using plastic or steel baskets will reduce the propensity for fine particles to stick to the buckets.

Belt conveyors

Belts are normally used where high flow rates (up to 1000 tonnes per hour) are required. Careful attention to the design and installation of belt-cleaning devices is required in order to prevent spillage from the return strand. The maximum angle of inclination is about 20° from the horizontal, and hence considerable space may be required. Ribbed belts will allow a slightly steeper angle to be attained, but cleaning becomes more difficult. Special designs of belt may be equipped with pockets, or the belt may even wrap around the coal. In these cases it is often possible to convey vertically and around

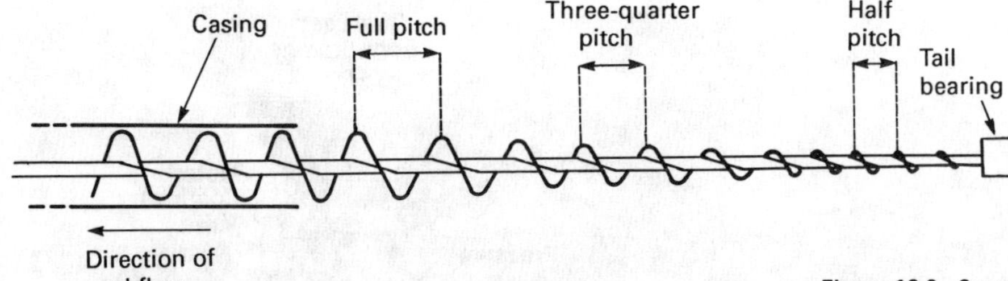

Figure 12.9 Screw conveyor pick-up for singles coal

corners, but the capital cost is higher than the standard belt. No coal breakage will occur while the coal is being conveyed, although a feed system is usually required.

En-masse conveyors

These systems can convey coal at up to several hundred tonnes per hour, although they are more commonly used at 15–20 tonnes per hour. They can elevate vertically or inclined, or they will operate as horizontal transfer conveyors. A steel casing renders *en-mass* conveyors dust-tight, although coal rubbing against the casing during conveying causes a small degree of breakage. Because no feed system is required, they can be mounted directly beneath bunkers or silos.

Screw conveyors

Although only able to transport coal in straight lines over relatively short distances (up to about 8 m maximum), screw conveyors provide an inexpensive means of conveying or feeding coal. The use of intermediate bearings for increasing the length of screw is not recommended for coal because they cause breakage of singles and can cause blockages with smalls coal (due to the build-up of fine particles). Difficulties of feedings smalls coal into the relatively narrow inlet of screws makes them more commonly (but not exclusively) used for singles coal. In this case, careful attention to design is required in order to minimize degradation.

It is recommended that the pick-up length be restricted to less than ten diameters, and that the screw is tapered along this length from shaft diameter at the tail end to full diameter at the casing inlet. The pitch of the flights should also be increased by a factor of about two at the tail end, reducing towards the casing inlet as shown in Figure 12.9. Screws designed in this way will reduce the tendency for coal to be picked up preferentially from the tail end, which causes recirculation and considerable breakage at the casing inlet.

Other mechanical systems

The majority of mechanical handling systems currently in use in industrial boilerhouses are represented above. There are, however, a number of other systems occasionally used, which including vibratory conveyors, grab-skip hoists, drag-link conveyors and chain and bucket conveyors.

12.6.2 Pneumatic conveying systems

Pneumatic conveying systems for coal fall into two categories: dense phase and lean phase. By definition, lean-phase conveying occurs when coal particles are conveyed in suspension by virtue of the conveying air velocity, whereas dense-phase conveying only occurs when coal is conveyed using velocities lower than those required for suspension flow. Because relatively high-pressure air is generally required for dense-phase conveying when compared to lean-phase conveying, different types of equipment have evolved for each category.

One of the most critical applications for pneumatic conveyors is the transport of single coal. A combination of relatively high conveying velocities together with a considerable degree of inter-particulate movement and rubbing against pipe walls inevitably causes a certain amount of breakage. The actual breakage depends upon a number of factors, including conveying velocity and friability of the coal, but, using a properly designed system, it is usually well within acceptable limits. Occasionally, however, pneumatic handling systems are installed where their performance is adversely affected by poor overall design, particularly where manufacturers and installers are obliged to follow the wishes of boilerhouse designers who are unaware of the limitations of such systems.

Both dense- and lean-phase systems can be 'tuned' to a certain extent to achieve minimum coal breakage. In principle, because of its lower conveying velocities a dense-phase system should cause less coal breakage than lean phase. However, the variation in control of both coal and air flow rates available from lean-phase systems generally leads to a greater tolerance of poor design than dense-phase ones. For the latter it is therefore even more important that the overall system design be correct.

Dense-phase systems

Both singles and smalls coal can be conveyed using dense-phase conveyors. In operation, coal enters a pressure vessel (shown schematically in Figure 12.10) by gravity. The size of the vessel depends upon the required conveying rate and distance, but would typically be 0.1 m^3 capacity for 10 tonnes per hour. The inlet valve at the top of the vessel is then closed and sealed, and compressed air (typically at 2–3 bar) is introduced into the vessel. The air causes the coal to be conveyed out of the vessel and along the pipeline in the form of a coherent

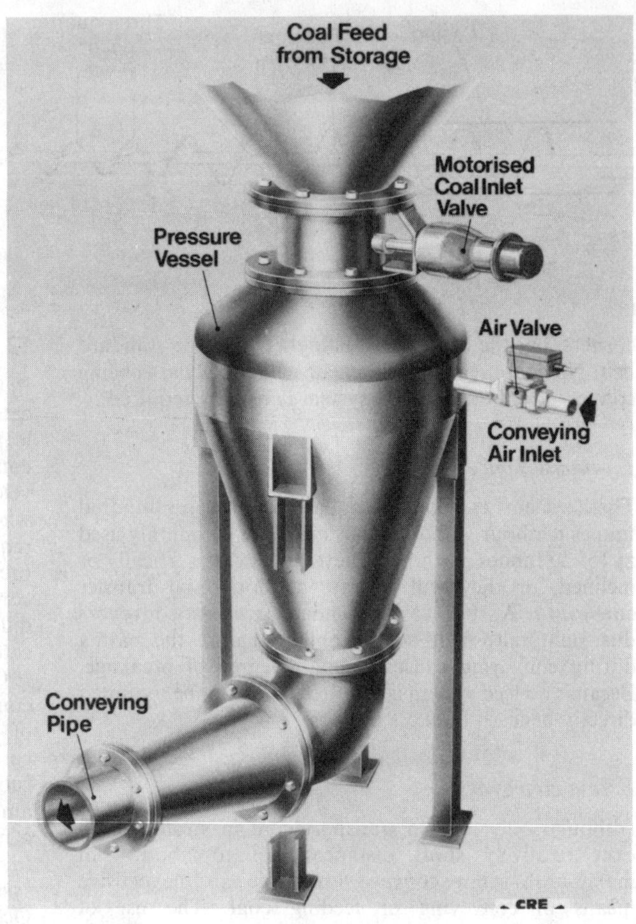

Figure 12.10 Dense-phase pneumatic conveyor

aerated slug. The supply of air is continued until the coal slug has been pushed through the pipe to the boiler hopper, and is then turned off. The sequence is then repeated until sufficient coal has been conveyed. The coherence of the slug is important to the control of the conveying process – if the slug breaks down for any reason then it will become more porous to the conveying air. Some of the air will then escape through the slug, and reach very high velocities at its leading edge where there is little pressure resistance. The high-velocity air will then entrain the leading particles, conveying them in the lean phase, and leading to further breakdown of the slug and consequent breakage of the conveyed coal. The porosity of singles coal to the conveying air is far greater than that for smalls, and conveying singles coal in the dense phase as a uniform and coherent slug is therefore more difficult to achieve.

Most conveying systems include a requirement to elevate coal to a high level. When using a dense-phase system it is desirable for the coal to enter a vertical riser immediately upon leaving the pressure vessel. At this point the coal is still being conveyed at a relatively low velocity, and the vertical riser provides the best opportunity for a coherent slug to be formed. In order to minimize the number of bends in the system it is also desirable that this vertical riser should encompass the total lift required.

Dense-phase conveyor pressure vessels are manufactured to BS 5500 and require periodic inspection. The vessel itself, however, is generally maintenance free, except where certain smalls coals are conveyed, when it is possible for fines to build up on the walls. The effect of such a build-up is to reduce the capacity of the vessel and hence to reduce the conveying rate. Methods used to prevent the fines from sticking include special finishes to the inner surface of the vessel, complete linings for the vessel walls, the use of water jets and the provision of access hatches for cleaning. All these methods are effective, but special finishes which are applied by painting may have to be renewed periodically.

Most pressure vessels are fed by means of vibrating feed chutes, and these have to be mechanically isolated from the pressure vessel. Failure of the closure valve at the top of the vessel to seal properly can result in the presence of high-pressure air in the inlet chute. This can lead to failure of the flexible joint between the chute and pressure vessel and the ejection of coal particles from the failed joint. It is therefore recommended that some form of personnel protection be provided at this point in order to prevent the ejection of coal particles. High-pressure

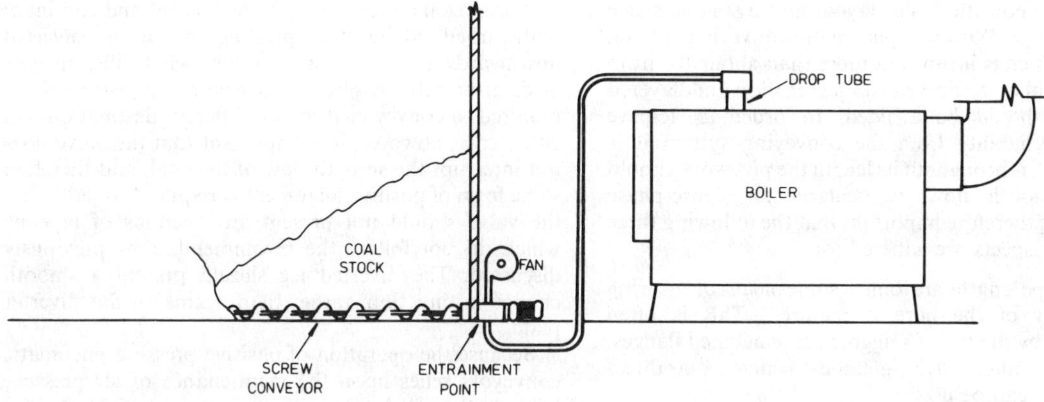

Figure 12.11 Lean-phase pneumatic conveyor

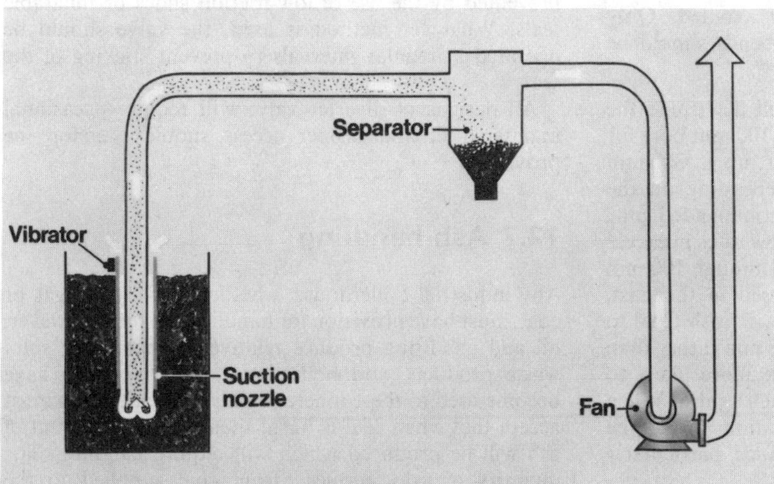

Figure 12.12 Suction nozzle pneumatic conveyor

air in the inlet chute can also lead to coal bridging and access to the chute to clear blockages may be useful.

Lean-phase conveyors

Lean-phase pneumatic conveyors are used for conveying singles coal, either by positive pressure or by suction from low-pressure centrifugal fans. They are not suitable for smalls coal – wet fine particles tend to adhere and build up in bends and discharge systems.

The most common example of a lean-phase conveyor is that fitted to pneumatic delivery vehicles as described in Section 12.3, where a rotary valve is used to feed coal into the airstream. Fixed boilerhouse systems, however, are generally fed by screw conveyor, as shown schematically in Figure 12.11. In this case the screw conveyor is controlled by the boiler combustion system, and the pipework delivers coal directly to the grate where the conveying air is subsequently used for the combustion process. It is always important to minimize coal breakage in these systems, and the screw conveyor should be designed as described in Section 12.6.1. Rotary valves and screw conveyors can be used for entraining coal into either positive (blowing) or suction systems. A more simple entrainment device is the suction nozzle developed by British Coal. This system, shown in Figure 12.12, offers the advantage of no mechanical parts at the pick-up zone, although a rotary valve or lock-hopper arrangement is required at the discharge point.

One of the most important factors influencing coal breakage is conveying velocity. In order to obtain minimum breakage consistent with reliable conveying it is recommended that the conveying air velocity be 27 m s^{-1} for lean-phase suction systems and 24 m s^{-1} for lean-phase positive-pressure systems. (The difference is due to the difference in conveying air density in each type of system.)

General design considerations for pneumatic conveying

The layout of pipework for any particular boilerhouse

will depend upon the boiler layout and available space for coal storage. With any pneumatic conveying system, pipework which is inclined at more than about 10° from the horizontal or vertical causes segregation and 'layered flow', and should be avoided. In order to achieve maximum reliability from the conveying system it is important that throughout its length the pipework should promote smooth flow, particularly for dense-phase systems. It is therefore important that the following three installation aspects are adhered to:

1. Where pipe lengths are joined, some means of ensuring continuity of the bore is required. This is often achieved by the use of spiggoted or machined flanges, but any alternative method which guarantees alignment can be used.
2. Welded butt joints of pipe should be avoided, as it is impossible to maintain a completely smooth internal surface at such a joint.
3. Kinks in the pipework should be avoided. Only straight sections or properly swept bends should be used.

The required coal conveying rate will determine the pipework diamter. This will range from 100 mm bore for smalls coal at (say) 10 tonnes per hour, up to 300 mm bore at about 80 tonnes per hour (depending on the conveying distance). The minimum recommended pipe diameter for conveying singles coal at flow rates in excess of about 5 tonnes per hour is 125 mm. Although 100 mm bore systems have been successfully used in the past, some of the singles coal now supplied by British Coal to the industrial market has a top size of 38 mm rather than the previous 25 mm, and blockages are more likely to occur with the lower pipe/particle diameter ratio. When using larger pipe diameters, consideration should be given to the cost of pipework maintenance, particularly replacement of bends.

Washed coals used in industrial boilerhouses do not generally cause significant wear of the pipeline, and depending on the length of the system and pipework diameter, ordinary mild steel bends should be capable of providing a lifetime equivalent to at least 20 000 tonnes of coal conveyed. Where unwashed coals are used, however, this figure may be somewhat reduced. Because these higher-ash fuels are more likely to be used on larger boiler plant where the coal throughput is much higher, some form of bend reinforcement is desirable. For large-diameter pipelines, consideration should be given to manufacturing bends in sections to facilitate maintenance. In this case it is only necessary to replace the failed section rather than the complete bend.

Bends for pneumatic conveying of coal should always be of the long-radius sweeping type. It is recommended that a radius/diameter ratio of 12 be used wherever possible.

It is likely that any pneumatic conveying system will be required to convey coal to more than one destination. Where the conveying pipeline passes directly over a row of boiler feed hoppers, each hopper can be fed almost directly from the conveying pipe by means of a two-position diverter valve. The valve can be switched such that coal either passes the feed point and continues undisturbed along the pipeline or it is diverted downwards into the hopper. A somewhat different type of diverter valve is often used where a pipeline split is required to convey coal to two different destinations. In either case, however, it is important that the valve does not interrupt the smooth flow of the coal, and therefore some form of positive alignment is required. In addition, the valve should not present any sections of pipeline which do not follow the recommendations previously discussed. The diverted leg should present a smooth change in direction, rather than a kink or flat diverter plate.

Because the operation of positive-pressure pneumatic conveyors relies upon the maintenance of air pressure behind the coal, it is important (particularly for dense-phase systems) that air leakage from diverter valves out of the 'closed' leg be minimal. Such leakage may be prevented by the use of low-friction slides or inflatable seals. Whatever method is used, the valve should be operated at regular intervals to prevent sticking of the seals.

All designs of diverter valve will require occasional maintenance, and proper access should therefore be provided.

12.7 Ash handling

Any industrial boilerhouse, whether fired by oil, gas or coal, must have provision for handling the fuel. However, oil and gas firing produce relatively insignificant solid waste products, and boilerhouse operators who have become used to these inherently high-amenity fuels must accept that when coal is burnt then a certain amount of ash will be produced which will require handling. The quantity of ash produced from coal supplied to the UK industrial market is generally somewhat less than 10% of the total amount of coal burnt, and its physical characteristics are determined by the firing method used for the coal. Some combustion systems are deliberately operated to form hard pads of clinker to facilitate manual removal and handling, but in other cases large clinkers may only form under faulty operation.

Because the quantity of ash requiring handling is small when compared to the coal burnt, there is often a reluctance to invest additional capital in ash-handling equipment. Furthermore, any equipment which is installed must be capable of dealing with a material which may still contain some burning carbon, and is dusty and highly abrasive, and may be corrosive when wet. For smaller boilerhouses some manual involvement using a shovel, wheelbarrow and dustbin is therefore often cost effective.

Where automatic ash handling is required, either pneumatic or submerged mechanical systems are normally used. Ordinary dry mechanical conveyors are not generally suitable for ash handling because of the likely ingress of dust into the working parts and of the possibility of high-temperature ash.

When burning coal in a fluidized bed it is likely that

Figure 12.13 Pneumatic ash-conveying system

a proportion of the ash will remain in the bed as large (>4 mm) particles. The removal of this material is a specialized technology which is beyond the scope of this chapter. Detailed information on such systems is available from both British Coal and the boiler manufacturers.

12.7.1 Pneumatic handling of ash

Before being pneumatically conveyed a mechanical clinker breaker or crusher is required to reduce the maximum particle size to less than about 25 mm. (Clinkers from some combustion systems may exceed 500 mm in size.) Care also has to be taken to ensure that any burning or hot carbon which may be present in the ash is sufficiently cooled to prevent re-ignition during the pneumatic conveying process, when oxygen is intimately available.

Both dense- and lean-phase pneumatic systems are currently widely used for handling ash in the UK, although the wear rate for the pipework is high, and even specially reinforced bends often require replacement several times each year. One of the major factors influencing bend wear rate is the conveying velocity of the ash particles, and for lean-phase systems some form of reactive control is recommended to maintain the air velocity at about 27 m s^{-1} for clinker ash and at about 12 m s^{-1} for grits and fly-ash. For the latter materials, where the particle size is small and the porosity of the material consequently low, dense-phase systems may have the advantage of conveying at lower velocities than lean phase. However, in the inevitable event of the pipework wearing through, a suction lean-phase system as shown in Figure 12.13 offers the advantage of in-leakage of clean air into the system, rather than dusty air leaking out into the boilerhouse. Reinforcement of bends can be carried out using alumina or basalt tiles, or concrete or other mixtures contained within a channel welded to the extrados of the bend. Specially designed wear-resistant bends are also available using materials such as Ni-hard.

Ash is normally pneumatically conveyed from the boilerhouse to an overhead bunker for discharge into lorries or skips for disposal, and in order to separate the fine ash from the conveying air, high-efficiency cyclones and filters are required. Dust emitted from the ash during discharge from the bunker can be reduced by conditioning with paddle-screw mixers and water sprays.

Because the conveyed ash may be hot, it is important that unburnt carbon is not allowed to enter the ash silo. If a boilerhouse vacuum-cleaning system is fitted to a suction pneumatic ash-removal system then a separate silo should be provided for this purpose. Explosions have occurred in ash silos where coal has been cleaned from boilerhouse floors and conveyed into silos containing hot ash. If this practice is likely then a suitable explosion-relief device should be incorporated into the silo.

12.7.2 Submerged mechanical systems

Where mechanical systems are used for ash handling they are normally partially submerged under water, as shown in Figure 12.14. This allows a water seal to be maintained at the ash outlet of the boiler which effectively isolates

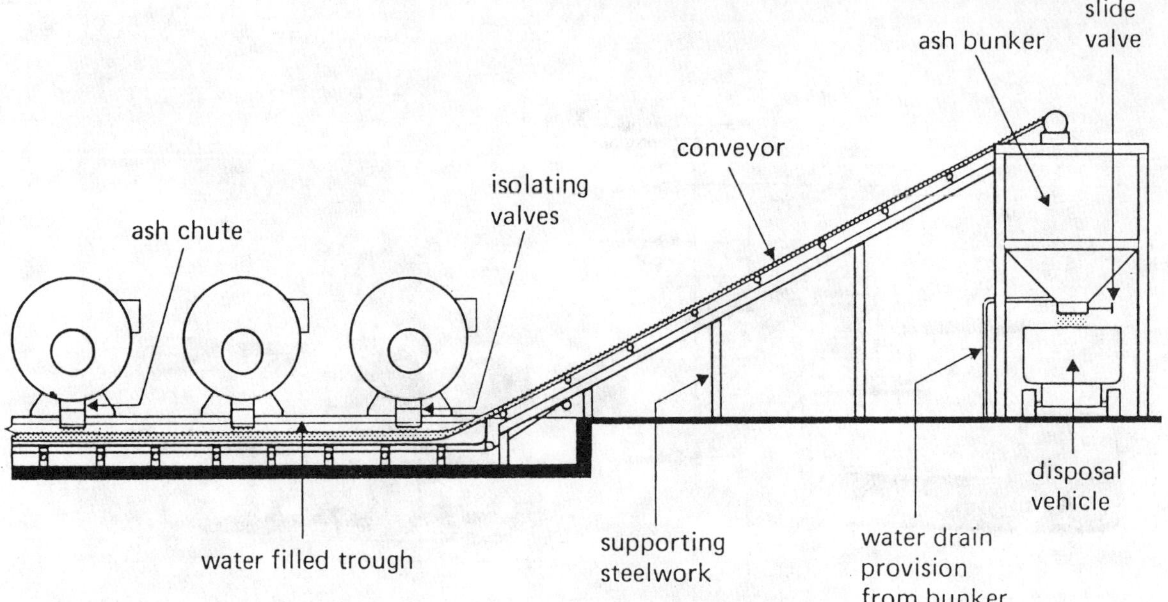

Figure 12.14 Submerged mechanical ash conveyor

the combustion chamber from atmospheric pressure. A further advantage of these submerged systems is that they are inherently dust free, and any burning carbon in the ash will be immediately extinguished. There is also no risk of an explosion occurring in the storage silo due to the presence of hot combustible material.

Both *en-masse* and belt conveyors are used for submerged ash conveying. They are operated very slowly (typically, about 0.1 m s^{-1}), and this allows water to drain from the ash as the conveyor is inclined up towards the discharge point. Although submerged conveyors are used for all types of ash, very fine ash can float on the water and some means of wetting may be required.

Acknowledgement

This chapter is published by permission of the British Coal Corporation. The views expressed, however, are those of the author and not necessarily those of British Coal.

References

1 Technical Report TD 131, *Pneumatic Delivery of Coal*, British Coal
2 Industrial Development Report ID2, *The Storage of Coal in Heaps*, CRE, British Coal (1981)
3 Guidance Note GN10, *The Use of Bunkers and Silos for Coal Storage*, British Coal (1988)
4 Technical Report TD 121, *Silos for Coal Storage*, British Coal
5 Technical Report TD 122, *The Use of Ultra-High Molecular Weight Polyethylene Linings*, British Coal

13 Steam Utilization

Albert Armer
Spirax Sarco Ltd

Contents

13.1 Introduction

The use of steam provides an ideal way to deliver just the right amount of heat energy to a point of use, to help with product manufacture or to provide an acceptable environment. The story of steam efficiency begins at the boiler (which is discussed in Chapter 7) but it does not end until the energy carried by the steam is transferred appropriately and the resulting hot condensate is returned to the boiler. Control of steam is necessary throughout the plant if the performance designed into the system is to be achieved and maintained. However, the natural laws which steam obeys are few in number and are easy to understand. Following them allows efficiencies to be achieved which would be envied with most other energy-distribution systems.

Two of the laws of thermodynamics and two of the laws of motion will cover almost all needs. In the simplest terms, it can be said that:

1. Energy is not destructible. It can always be accounted for, and if it disappears at point A then it reappears in equal amount at point B. This ensures that the world is a consistent place, that energy does not mysteriously appear from or disappear to nowhere, and that the Steam Tables can be relied on always to provide information on the properties of steam.
2. Heat flows from higher-temperature to lower-temperature objects without any help. It follows that the rate of flow will vary directly with temperature differences, and inversely with any resistances to this flow. A temperature difference is necessary for heat to flow!
3. Any matter tends to move in the direction in which it is pushed, and, in particular, because of the effects of friction, any fluid will only flow from high-pressure to lower-pressure regions. Again, the rate of flow will vary directly with pressure differences, and inversely with any resistances to this flow.
4. Gravity acts downwards! The denser constituents in a mixture often tend to move to the bottom of a space, unless other forces acting on them oppose such motion.

If these basic laws can be accepted it is easy to build up an understanding of steam systems and the way that steam behaves. The ground rules for an effective (and efficient) steam system then quickly become apparent. Specific heats and weights for various solids, liquids, gases and vapours, as well as pipeline capacities, are shown in Tables 13.1–13.4.

13.2 What is steam?

The datum point, when considering the steam/water substance, is usually taken as water at the temperature of melting ice, at normal atmospheric pressure and so at a temperature of 0°C. Adding heat energy raises its temperature; some 419.04 kJ will raise 1 kg to 100°C, when any further addition of heat evaporates some of the water. If 2257 kJ are added to each kg of water, then all the water becomes the dry gas, steam. Equally, if only part of this extra energy is added – say, 90% – then 90% of the water evaporates and the other 10% remains liquid. The specific volume of steam at atmospheric pressure is 1.673 m^3/kg, so the mixture of 90% steam and 10% water would occupy a volume of $(0.9 \times 1.673)\,m^3$ plus $(0.1 \times 0.001)\,m^3$. Clearly, almost all the volume of this 1 kg of steam and water is occupied by the steam, at 1.5057 m^3, and the mixture would be described as steam with a Dryness Fraction of 0.9.

If the water is kept at a pressure above atmospheric, its temperature can be raised above 100°C before boiling begins. At 10 bar gauge, for example, boiling point is at about 184.1°C. The extra energy needed to convert water at this pressure and temperature into steam (the enthalpy of evaporation) is now rather less at 2000.1 kJ/kg, while the volume of 1 kg of pure steam is only 0.177 m^3. These figures are all recorded in the Steam Tables at the end of this chapter.

When steam at the saturation temperature contacts a surface at a lower temperature, and heat flows to the cooler surface, some of the steam condenses to supply the energy. With a sufficient supply of steam moving into the volume which had been occupied by the steam now condensed, the pressure and temperature of the steam will remain constant. Of course, if the condensate flows to a zone where it is no longer in contact with the steam it can cool below steam temperature while supplying heat to a cooler surface.

Equally, if steam at the saturation temperature were to contact a surface at a higher temperature, as in some boilers, its temperature could be increased above the evaporating temperature and the steam would be described as superheated. Superheated steam is very desirable in turbines, where its use allows higher efficiencies to be reached, but it is much less satisfactory than saturated steam in heat exchangers. It behaves as a dry gas, giving up its heat content rather reluctantly as compared with saturated steam, which offers much higher heat transfer coefficients.

The condensate produced within the heat exchangers, as also within the steam lines, is initially at the saturation temperature and carries the same amount of energy as would boiler water at the same pressure. If it is discharged to a lower pressure, through a manual or automatic drain valve (steam trap) or even through a leak, it then contains more energy than water is able to hold at the lower pressure if it is to remain liquid. If the excess of energy amounted to, say, 5% of the enthalpy of evaporation at the lower pressure, then 5% of the water would be evaporated. The steam released by this drop in pressure experienced by high-temperature water is usually called flash steam. Recovery and use of this low-pressure steam, released by flashing, is one of the easiest ways of improving the efficiency of steam-utilization systems.

It is equally true that condensate, even if it has been released to atmospheric pressure, carries the same 419.04 kJ/kg of heat energy that any other water at the same temperature would hold. Allowing condensate to drain to waste, instead of returning it to the boiler feed tank or de-aerator, makes no economic sense. Condensate

Table 13.1 Specific heats and weights of various solids

Material	*Specific gravity*	*Btu per lb per °F*
Aluminium	2.55–2.8	0.22
Andalusite		0.17
Antimony		0.05
Apatite		0.20
Asbestos	2.1–2.8	0.20
Augite		0.19
Bakelite, wood filler		0.33
Bakelite, asbestos filler		0.38
Barite	4.5	0.11
Barium	3.5	0.07
Basalt rock	2.7–3.2	0.20
Beryl		0.20
Bismuth	9.8	0.03
Borax	1.7–1.8	0.24
Boron	2.32	0.31
Cadmium	8.65	0.06
Calcite, 32–100°F		0.19
Calcite, 32–212°F		0.20
Calcium	4.58	0.15
Carbon	1.8–2.1	0.17
Carborundum		0.16
Cassiterite		0.09
Cement, dry		0.37
Cement, powder		0.20
Charcoal		0.24
Chalcopyrite		0.13
Chromium	7.1	0.12
Clay	1.8–2.6	0.22
Coal	0.64–0.93	0.26–0.37
Cobalt	8.9	0.11
Concrete, stone		0.19
Concrete, cinder		0.18
Copper	8.8–8.95	0.09
Corundum		0.10
Diamond	3.51	0.15
Dolomite rock	2.9	0.22
Fluorite		0.22
Fluorspar		0.21
Galena		0.05
Garnet		0.18
Glass, common	2.4–2.8	0.20
Glass, crystal	2.9–3.0	0.12
Glass, plate	2.45–2.72	0.12
Glass, wool		0.16
Gold	19.25–19.35	0.03
Granite	2.4–2.7	0.19
Hematite	5.2	0.16
Hornblende	3.0	0.20
Hypersthene		0.19
Ice, −112°F		0.35
Ice, −40°F		0.43
Ice, −4°F		0.47
Ice, 32°F		0.49
Iridium	21.78–22.42	0.03
Iron, cast	7.03–7.13	0.12
Iron, wrought	7.6–7.9	0.12
Labradorite		0.19
Lava		0.20
Lead	11.34	0.03
Limestone	2.1–2.86	0.22
Magnetite	3.2	0.16
Magnesium	1.74	0.25
Malachite		0.18
Manganese	7.42	0.11
Marble	2.6–2.86	0.21
Mercury	13.6	0.03
Mica		0.21
Molybdenum	10.2	0.06
Nickel	8.9	0.11
Oligloclose		0.21
Orthoclose		0.19
Plaster of paris		1.14
Platinum	21.45	0.03
Porcelain		0.26
Potassium	0.86	0.13
Pyrexglass		0.20
Pyrolusite		0.16
Pyroxylin plastics		0.34–0.38
Quartz, 55–212°F	2.5–2.8	0.19
Quartz, 32°F		0.17
Rock salt		0.22
Rubber		0.48
Sandstone	2.0–2.6	0.22
Serpentine	2.7–2.8	0.26
Silk		0.33
Silver	10.4–10.6	0.06
Sodium	0.97	0.30
Steel	7.8	0.12
Stone		0.20
Stoneware		0.19
Talc	2.6–2.8	0.21
Tar	1.2	0.35
Tellurium	6.0–6.24	0.05
Tin	7.2–7.5	0.05
Tile, hollow		0.15
Titanium	4.5	0.14
Topaz		0.21
Tungsten	19.22	0.04
Vanadium	5.96	0.12
Vulcanite		0.33
Wood	0.35–0.99	0.32–0.48
Wool	1.32	0.33
Zinc blend	3.9–4.2	0.11
Zinc	6.9–7.2	0.09

is a form of distilled water, requiring little chemical feed treatment or softening, and it already holds energy which may amount to 15% of the energy which would have to be supplied to cold make-up feedwater, even in relatively low-pressure systems. An installation which makes good use of both flash steam and condensate will see the benefits on the bottom line! To summarize:

1. Saturated steam provides a heat source at a temperature which is readily controlled, by the control of pressure;
2. It carries very large amounts of heat as enthalpy of evaporation or latent heat in relatively small weights of steam;
3. It supplies heat, by condensing, at a constant temperature and with high heat transfer coefficients, so it maximizes the effectiveness of heat exchangers;
4. Recovery and re-use of the condensate and its heat content is usually simple and effective.

Table 13.2 Specific heats and weights of various liquids

Liquid	*Specific gravity*	*Btu per lb per °F*
Acetone	0.790	0.51
Alcohol, ethyl, 32°F	0.789	0.55
Alcohol, ethyl, 104°F	0.789	0.65
Alcohol, methyl, 40–50°F	0.796	0.60
Ammonia, 32°F	0.62	1.10
Ammonia, 104°F		1.16
Ammonia, 176°F		1.29
Ammonia, 212°F		1.48
Ammonia, 238°F		1.61
Anilin	1.02	0.52
Benzol		0.42
Calcium chloride	1.20	0.73
Castor oil		0.43
Citron oil		0.44
Diphenylamine	1.16	0.46
Ethyl ether		0.53
Ethylene glycol		0.53
Fuel oil	6.96	0.40
Fuel oil	0.91	0.44
Fuel oil	0.86	0.45
Fuel oil	0.81	0.50
Gasoline		0.53
Glycerine	1.26	0.58
Kerosene		0.48
Mercury	13.6	0.033
Naphthalene	1.14	0.41
Nitrobenzole		0.36
Olive oil	0.91–0.94	0.47
Petroleum		0.51
Potassium hydrate	1.24	0.88
Sea water	1.0235	0.94
Sesame oil		0.39
Sodium chloride	1.19	0.79
Sodium hydrate	1.27	0.94
Soybean oil		0.47
Toluol	0.866	0.36
Turpentine	0.87	0.41
Water	1	1.00
Xylene	0.861–0.881	0.41

13.2.1 The working pressure in the boiler and mains

Where steam is generated as a source of energy for turbines or engines the boiler pressure is usually high and often the steam is superheated. Steam for process use or for heating may then be supplied through pressure-reducing valves or be available at pass-out pressures from the engine or at the exhaust pressure. In other cases where steam is wanted for process or heating use only, use of a packaged boiler, or sometimes a single-pass watertube steam generator is almost universal. The most suitable operating pressure for the boiler has then to be decided. It should be said that steam of the best quality is provided by boilers operating at or close to their designed working pressure. When the steam is to be used at a lower pressure, then pressure-reducing valves or other control valves may be utilized, close to the steam-using points.

The reason for this becomes clear when the specific volume of steam at varying pressures is noted from the steam tables and Figure 13.1 is considered. This shows a boiler operating at either high or lower pressure while producing the same weight of steam from the same energy input. Energy flows through the outer surface of the tubes into the boiler water, and when this water is at saturation temperature any addition of energy means the formation of steam bubbles. These bubbles rise to the surface and break, releasing steam into the steam space.

The volume of a given weight of steam contained in these bubbles is inversely related to the pressure at which the boiler operates. If this pressure is lower than the design pressure, the volume in the bubbles is increased. This raises the apparent water level in the boiler, reducing the volume of the steam space. The greater volume of the bubbles enlarges the turbulence at the surface, and as the bubbles burst, splashing of droplets into the steam space increases. The larger volume of steam, flowing towards the boiler crown valve through the reduced steam space volume, moves at higher velocity. All these factors increase the carry-over of water droplets with the steam into the distribution system.

Table 13.3 Gases and vapours

Gas or vapour	*Specific heat, Btu per lb per °F at constant pressure*	*Specific heat, Btu per lb per °F at constant volume*
Acetone	0.35	0.315
Air, dry, 50°F	0.24	0.172
Air, dry, 32–392°F	0.24	0.173
Air, dry, 68–824°F	0.25	0.178
Air, dry, 68–1166°F	0.25	0.184
Air, dry, 68–1472°F	0.26	0.188
Alcohol, C_2H_5OH	0.45	0.398
Alcohol, CH_3OH	0.46	0.366
Ammonia	0.54	0.422
Argon	0.12	0.072
Benzene, C_6H_6	0.26	0.236
Bromine	0.06	0.047
Carbon dioxide	0.20	0.150
Carbon monoxide	0.24	0.172
Carbon disulphide	0.16	0.132
Chlorine	0.11	0.82
Chloroform	0.15	0.131
Ether	0.48	0.466
Hydrochloric acid	0.19	0.136
Hydrogen	3.41	2.410
Hydrogen sulphide	0.25	0.189
Methane	0.59	0.446
Nitrogen	0.24	0.170
Nitric oxide	0.23	0.166
Nitrogen tetroxide	1.12	1.098
Nitrous oxide	0.21	0.166
Oxygen	0.22	0.157
Steam, 1 psia, 120–600°F	0.46	0.349
Steam, 14.7 psia, 220–600°F	0.47	0.359
Steam, 150 psia, 360–600°F	0.54	0.421
Sulphur dioxide	0.15	0.119

Table 13.4 Pipeline capacities at specific velocities (metric SI units)

Pressure (bar)	*Velocity (m/s)*	kg/h										
		15 mm	*20 mm*	*25 mm*	*32 mm*	*40 mm*	*50 mm*	*65 mm*	*80 mm*	*100 mm*	*125 mm*	*150 mm*
0.4	15	7	14	24	37	52	99	145	213	394	648	917
	25	10	25	40	62	92	162	265	384	675	972	1 457
	40	17	35	64	102	142	265	403	576	1 037	1 670	2 303
0.7	15	7	16	25	40	59	109	166	250	431	680	1 006
	25	12	25	45	72	100	182	287	430	716	1 145	1 575
	40	18	37	68	106	167	298	428	630	1 108	1 712	2 417
1.0	15	8	17	29	43	65	112	182	260	470	694	1 020
	25	12	26	48	72	100	193	300	445	730	1 160	1 660
	40	19	39	71	112	172	311	465	640	1 150	1 800	2 500
2.0	15	12	25	45	70	100	182	280	410	715	1 125	1 580
	25	19	43	70	112	162	295	428	656	1 215	1 755	2 520
	40	30	64	115	178	275	475	745	1 010	1 895	2 925	4 175
3.0	15	16	37	60	93	127	245	385	535	925	1 505	2 040
	25	26	56	100	152	225	425	632	910	1 580	2 480	3 440
	40	41	87	157	250	357	595	1 025	1 460	2 540	4 050	5 940
4.0	15	19	42	70	108	156	281	432	635	1 166	1 685	2 460
	25	30	63	115	180	270	450	742	1 080	1 980	2 925	4 225
	40	49	116	197	295	456	796	1 247	1 825	3 120	4 940	7 050
5.0	15	22	49	87	128	187	352	526	770	1 295	2 105	2 835
	25	36	81	135	211	308	548	885	1 265	2 110	3 540	5 150
	40	59	131	225	338	495	855	1 350	1 890	3 510	5 400	7 870
6.0	15	26	59	105	153	225	425	632	925	1 555	2 525	3 400
	25	43	97	162	253	370	658	1 065	1 520	2 530	4 250	6 175
	40	71	157	270	405	595	1 025	1 620	2 270	4 210	6 475	9 445
7.0	15	29	63	110	165	260	445	705	952	1 815	2 765	3 990
	25	49	114	190	288	450	785	1 205	1 750	3 025	4 815	6 900
	40	76	177	303	455	690	1 210	1 865	2 520	4 585	7 560	10 880
8.0	15	32	70	126	190	285	475	800	1 125	1 990	3 025	4 540
	25	54	122	205	320	465	810	1 260	1 870	3 240	5 220	7 120
	40	84	192	327	510	730	1 370	2 065	3 120	5 135	8 395	12 470
10.0	15	41	95	155	250	372	626	1 012	1 465	2 495	3 995	5 860
	25	66	145	257	405	562	990	1 530	2 205	3 825	6 295	8 995
	40	104	216	408	615	910	1 635	2 545	3 600	6 230	9 880	14 390
14.0	15	50	121	205	310	465	810	1 270	1 870	3 220	5 215	7 390
	25	85	195	331	520	740	1 375	2 080	3 120	5 200	8 500	12 560
	40	126	305	555	825	1 210	2 195	3 425	4 735	8 510	13 050	18 630

There is much to be said in favour of carrying the steam to the points where it is to be used, at a pressure close to that of the boiler. The use of a high-distribution pressure means that the size of the steam mains is minimized. The smaller mains have lower heat losses, so that better-quality steam at the usage points is more readily achieved and the smaller pipes are often much lower in capital cost.

Pressure reduction to the levels needed by the steam-using equipment can then be achieved by the use of pressure-reducing stations located close to the steam users themselves. The individual reducing valves will tend to be smaller in size than would valves at the supply end, will give closer control of the reduced pressures and emit less noise. Problems which may arise if a whole section of a plant were dependent on a single reducing valve are avoided. The effects on the steam-using equipment of pressure drops in the pipework, varying in amount at different loads, are eliminated.

13.3 The steam load

Before selecting the size of a steam-control valve, a supply main or even a steam boiler it is necessary to know at least approximately (or better) how much steam is to be supplied. Where steam is to be used as the energy source for heat exchangers this is the same as knowing the heat load which is to be met.

Almost all heat loads fall into one of two categories. Either some material is to be heated from a lower temperature to a higher one or it is to be maintained at a high temperature while heat is supplied at a rate sufficient to balance the heat losses. In the first case the amount of heat needed to produce the change in temperature is given by

$$Q = W \times Sp \times \delta t$$

where Q = heat load (kJ),

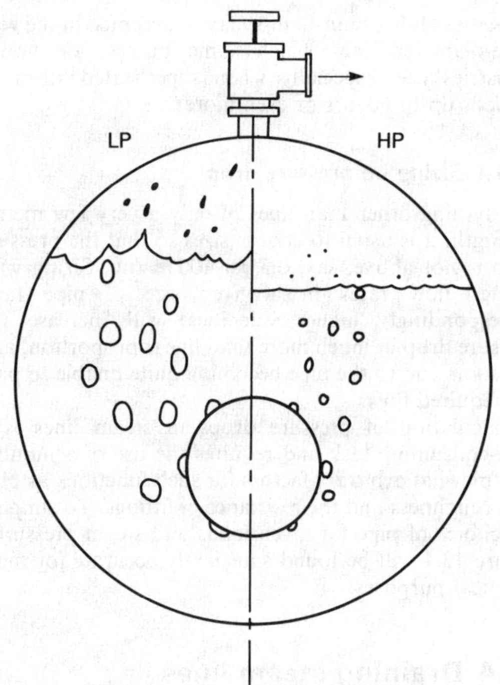

Figure 13.1 Boiler operation at low or high pressure

W = weight of material (kg),
Sp = specific heat (kJ/kg °C),
δt = temperature rise (°C).

More usefully, since the steam flow rate is sought rather than the weight of steam,

$$q=\frac{Q}{h}=\frac{W\cdot Sp\cdot \delta t}{h}$$

where q = heat flow (kJ/h),
h = time available (h).

Then if hfg = enthalpy of evaporation of the steam at the pressure involved,

$$Ws=\frac{W\cdot Sp\cdot \delta t}{hfg\cdot h} \qquad (13.1)$$

where Ws = flow rate of steam (kg/h),
hfg = enthalpy of evaporation (kJ/kg).

The second case is the supply of heat at a rate which balances the heat losses, and here the heat load is given by

$$q_1=U\cdot A\cdot \Delta t \qquad (13.2)$$

where q_1 = heat flow (kJ/h),
U = heat transfer coefficient for heat flow through the surface involved (kJ/m² °C · h),
Δt = temperature difference across surface involved (°C),
A = area of surface involved (m²).

It must be said that heat transfer coefficients are affected by so many variables that they are best regarded as approximations, unless measurements have been made on identical equipment under similar conditions. Further, it will be noticed that equations (13.1) and (13.2) each include a temperature difference term, Δt or δt, but two different quantities are meant. To avoid confusion, it is useful to construct a diagram of the temperatures around the heat exchanger, as shown in Figure 13.2. At temperature T_1 a horizontal line represents the temperature of steam within the exchanger. Points T_2 and T_3 are the inlet and outlet temperatures of the heated material (or initial and final temperatures). A line is drawn from T_2 to T_3, and the difference between T_2 and T_3 is the temperature increase of the material, δt in equation (13.1). The midpoint of the line T_2 to T_3 gives the arithmetic mean of the material temperatures, T_m, and the difference between T_m and T_1 is then the arithmetic mean temperature difference Δt in equation (13.2). Note that although theoretical considerations lead to the use of logarithmic means in heat transfer calculations, arithmetic means are usually sufficiently accurate when sizing steam supply equipment.

Some typical specific heats are listed in the tables, and a few examples of heat load calculations may be useful.

1. An air heater battery raises 2 m³/s of air from −5°C to 30°C using steam at a gauge pressure of 6 bar. What is the hourly steam load?

$$2\text{ m}^3/\text{s}=2\times 3600\text{ m}^3/\text{h}=\frac{2\times 3600}{0.76}\text{ kg/h}$$

$$=9474\text{ kg/h}$$

(0.76 m³ of air at −5°C weighs approx 1 kg)

$$Ws=\frac{9474\times 1.0\times (30+5)}{2066\times 1} \quad \text{(from equation (13.1))}$$

$$=160.5\text{ kg/h}$$

(1 kg of air is heated through 1°C by 1.0 kJ).

2. A calorifier holding 2000 l is heated from 10°C to 60°C in 2 h using steam at a gauge pressure of 3 bar. What is the hourly steam load?

$$Ws=\frac{W\cdot Sp\cdot \delta t}{hfg\cdot h}=\frac{2000\times 4.186\times (60-10)}{2133.4\times 2}=98.1\text{ kg/h}$$

3. A calorifier has a heating surface area of 2 m² and is required to heat a flow of water from 65°C to 80°C. The U value has been found to be 1250 W/m² °C, and steam can be supplied at a gauge pressure of 2 bar in the steam chest. What rate of water flow can the

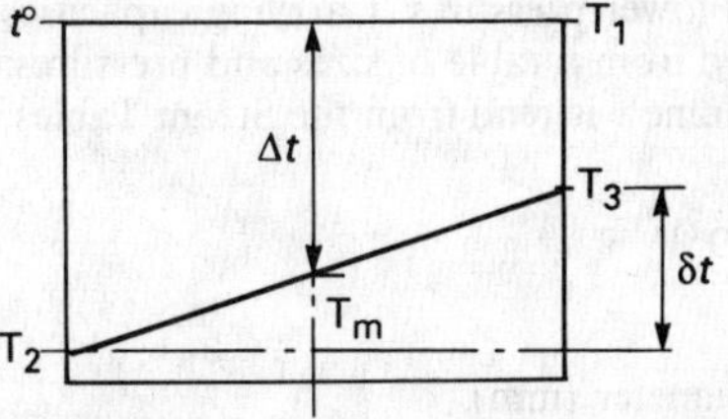

Figure 13.2

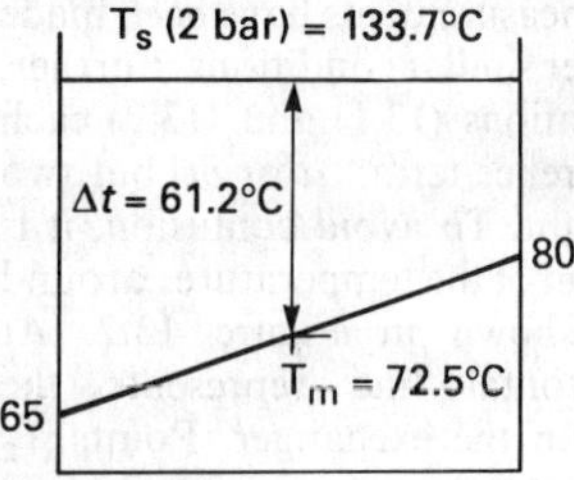

Figure 13.3

calorifier handle (see Figure 13.3). From equation (13.2):

$$q = UA\Delta t$$
$$= 1250 \times 2 \times 61.2$$
$$= 153\,000 W$$
$$= 550\,800 \text{ kg/h}$$

(This heat flow implies the condensing of some 550 000/2163.3 or 254.6 kg/h of steam.) From equation (13.1):

$$550\,000 = W \times 4.186 \times (80 - 65)$$
$$W = 8772 \text{ kg/h of water}$$

13.3.1 Sizing the steam lines

The appropriate size of pipe to carry the required weight of steam at the chosen pressure must be selected, remembering that undersized pipes mean high-pressure drops and high velocities, noise and erosion. Conversely, when pipe sizing is unduly generous, the lines become unnecessarily expensive to install and the heat lost from them will be greater than it need be.

Steam pipes may be sized so that the pressure drop along them is within an acceptable limit, or so that velocities along them are not too high. It is convenient and quick to size shorter mains and branch lines on a velocity basis, but the pressure drops along longer runs should also be checked to verify that sufficient pressure is available at the delivery points.

13.3.2 Sizing on velocity

In lines carrying saturated steam reasonable maxima for velocities are often taken at up to 40 m/s in larger pipes and at higher pressures. However, 25 m/s may be more appropriate in the middle ranges and even 15 m/s with small-size lines and lower pressures. Carrying capacities of pipes may be read from a table of sizes and pressures, or if the specific volume v is read from the Steam Tables, then

$$W \text{ kg/h} = \frac{0.002827 D^2 V}{v}$$

where D = pipe diameter (mm),
V = velocity (m/s),
v = specific volume (m^3/kg).

Velocities higher than 40 m/s may be accepted in the very large-diameter lines which some process or power industries use, especially when superheated steam is carried (up to 60 m/s or even more).

13.3.3 Sizing on pressure drop

In anything other than lines of only a very few metres in length, it is usual to choose sizes so that the pressure drop is not above, say, 0.3 bar/100 m run. Sizing with greater flow rates in a given size of pipe (and correspondingly higher velocities) will increase the pressure drop in much more than linear proportion, and can soon lead to the pipe becoming quite unable to pass the required flow.

Calculation of pressure drops in steam lines is a time-consuming task and requires the use of a number of somewhat arbitrary factors for such functions as pipe wall roughness and the resistance of fittings. To simplify the choice of pipe for given loads and steam pressures, Figure 13.4 will be found sufficiently accurate for most practical purposes.

13.4 Draining steam lines

Draining of condensate from steam mains and branch lines is probably the most common application for steam traps. It is important for reasons of safety and to help achieve greater plant efficiency that water is removed from steam lines as quickly as possible. A build-up of water can, in some cases, lead to slugs of water being picked up by the steam flow and hurled violently at pipe bends, valves or other fittings, a phenomenon described graphically as waterhammer. When carried into heat exchangers, water simply adds to the thickness of the condensate film and reduces heat transfer. Inadequate drainage leads to leaking joints, and is a prime cause of wiredrawing of control valve seating faces.

13.4.1 Waterhammer

Waterhammer may occur when water is pushed along a pipe by the steam instead of being drained away at the low points and is suddenly stopped by impact on a valve, pipe tee or bend. The velocities which are achieved by such slugs of water can be very high, especially on start-up, when a pipe is being charged with steam. When these velocities are destroyed the kinetic energy of the water is converted into pressure energy and a pressure shock is applied to the obstruction. Usually, there is a banging noise, and perhaps movement of the pipe. In severe cases the fitting may fracture with an almost explosive effect and consequent loss of live steam at the fracture.

Fortunately, waterhammer may be avoided completely if steps are taken to ensure that water is drained away before it accumulates in sufficient quantity to be picked up by the steam. Avoiding waterhammer is a better alternative than attempting to contain it by choice of materials and pressure ratings of equipment.

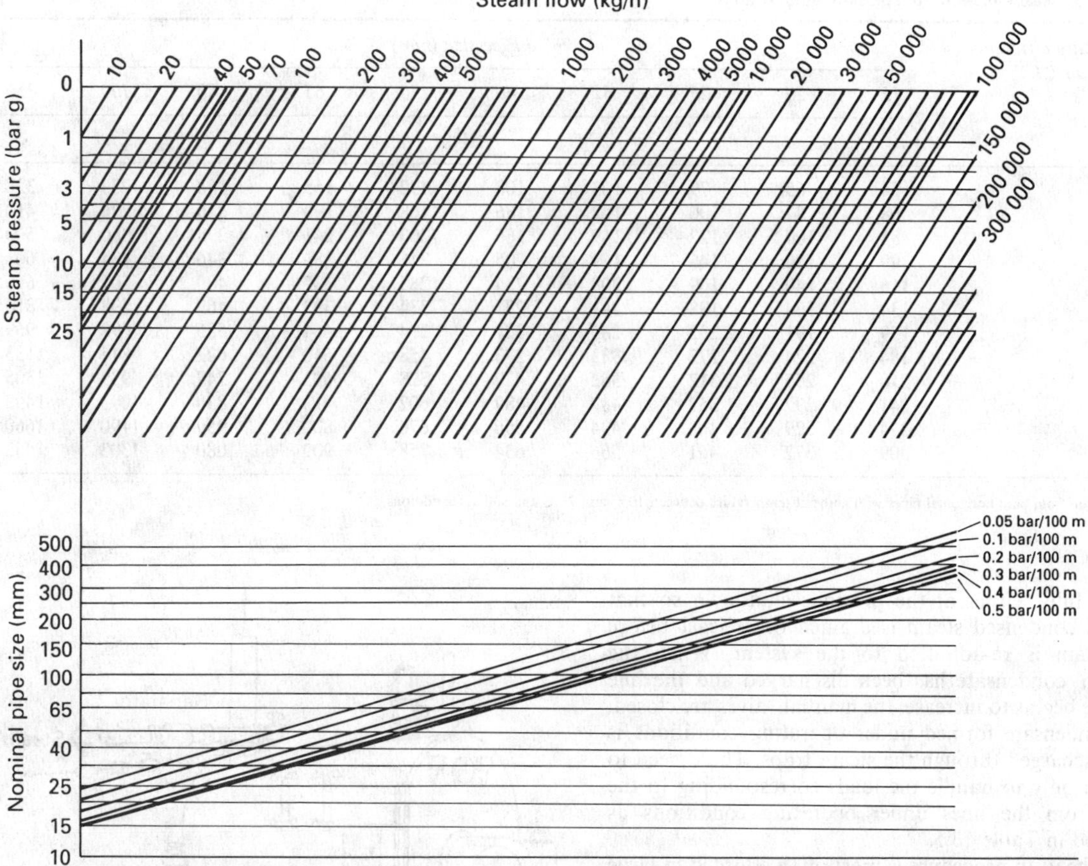

Figure 13.4 Pressure drop in steam lines

The steam traps (used to drain water from a separator where this is fitted, perhaps to deal with carry-over from a boiler, or to drain condensate from collecting legs at intervals of not more than, say, 50 m along the steam line) clearly must have adequate capacity. For the separator at the boiler offtake or at a header supplied from the boiler(s) and itself supplying the steam mains, it may be necessary to have capacity in the trap(s) of 10% or even more of the boiler rating. On the steam lines themselves it is rare for a trap larger than 1/2 in or 15 mm size to be needed. Very often, the low-capacity versions of the 1/2 in steam trap are more than adequate.

The capacity of any steam trap will depend on the difference in pressure between its inlet and outlet connections. Under system start-up conditions the steam pressure in the line will at first be only marginally above atmospheric. If the trap discharge line rises to a higher level, or delivers to a pressurized return pipe, no condensate will flow through the trap until the line pressure exceeds the back pressure. It is important that steam traps which can drain by gravity, with zero back pressure, are fitted if condensate is to be removed as the system is being heated from cold. Where this means that the start-up condensate can only be drained to waste, additional steam traps connected to the return line may also be fitted to enable recovery of the condensate to be made during normal running.

13.4.2 Steam line drainage

The use of oversized steam traps giving very generous 'safety factors' does not ensure safe and effective steam line drainage. A number of points must be considered if a satisfactory installation is to be achieved, including:

1. The heat-up method employed;
2. The provision of suitable collecting legs or reservoirs for the condensate;
3. Provision of a minimum differential pressure across the trap;
4. Choice of steam trap type and size;
5. Proper installation of the trap.

13.4.3 Heat-up method – supervised

In this case a manual drain valve is fitted at each drain point on the steam system, bypassing the trap and discharging to atmosphere. These manual drains are

Table 13.5 Heat emission from pipes (metric SI units)

Temperature difference steam to air (°C)	*Pipe size (mm)*									
	15	*20*	*25*	*32*	*40*	*50*	*65*	*80*	*100*	*150*
	W/m									
56	54	65	79	103	108	132	155	188	233	324
67	68	82	100	122	136	168	198	326	296	410
78	83	100	122	149	166	203	241	298	360	500
89	99	120	146	179	205	246	289	346	434	601
100	116	140	169	208	234	285	337	400	501	696
111	134	164	198	241	271	334	392	469	598	816
125	159	191	233	285	321	394	464	555	698	969
139	184	224	272	333	373	458	540	622	815	1133
153	210	255	312	382	429	528	623	747	939	1305
167	241	292	357	437	489	602	713	838	1093	1492
180	274	329	408	494	556	676	808	959	1190	1660
194	309	372	461	566	634	758	909	1080	1303	1852

Heat emission from bare horizontal pipes with ambient temperature between 10°C and 21°C and still air conditions.

opened fully, either at the previous shutdown so that residual condensed steam is drained or at least before any steam is re-admitted to the system. When the 'heat-up' condensate has been discharged and the line pressure begins to increase, the manual valves are closed. The condensate formed under operating conditions is then discharged through the steam traps. These need to be sized only to handle the loads corresponding to the losses from the lines under operating conditions as indicated in Table 13.5.

This heat-up procedure is most often adopted in large installations where start-up of the system is an infrequent (perhaps only an annual) occurrence. Large heating systems and chemical processing plants are typical examples.

The procedure can be made more automatic by using a temperature-sensitive liquid expansion steam trap in place of the manual valve. Such an arrangement is a compromise between supervised start-up and the automatic start-up discussed below.

13.4.4 Heat-up method – automatic

One traditional method of achieving automatic start-up is simply to allow the steam boiler to be fired and brought up to pressure with the crown valve wide open. Thus the steam main and branch lines come up to operating pressure without supervision, and the steam traps must be arranged so that they can discharge the condensate as it is formed. This method is usually confined to the smaller installations that are regularly and frequently shut down and started up again. Typically, the boilers in many laundries and dry-cleaning plants often are shut down each night and restarted each morning.

In anything but the smallest installations the flow of steam from the boiler into the cold pipes at start-up (while the boiler pressure is still very low) will lead to excessive carry-over of boiler water with the steam. Such carry-over may be enough to overload a separator at the

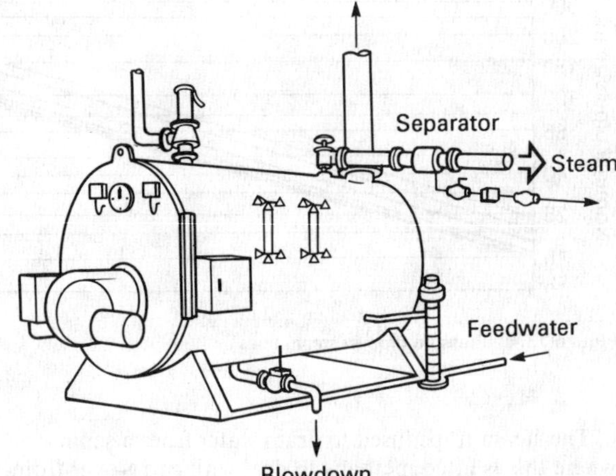

Figure 13.5 Boiler with separator at steam offtake

boiler offtake point or its steam trap (Figure 13.5). Great care (and even good fortune) are necessary if problems are to be avoided.

Modern steam practice calls for an automatic valve to be fitted in the steam supply line, arranged so that the valve remains closed until a sufficient pressure is attained in the boiler. The valve can then be made to open over a timed period, so that steam is admitted to the pipework at a controlled rate. The pressure within the boiler may be climbing quickly but the slow opening valve protects the pipework.

Where these valves are used, the time available to warm up the pipework will be known, as it is set on the valve control. In other cases, details of the start-up procedure must be known so that the time may be estimated. Thus boilers started from cold may be fired for a short time, shut off for a period while temperatures equalize, and then fired again. Boilers may be protected from undue stress by these short bursts of firing, which extend the warm-up time and reduce the rate at which the condensate in the mains must be discharged at the traps.

13.4.5 Sizing steam traps for warm-up

Whatever automatic start-up method is adopted, steam will only flow into the mains and discharge air from the air vents at a pressure at least a little above atmospheric. It then reaches the pipe and condenses on the cold metal, and at first the condensate temperature will be well below 100°C (212°F). This means that the steam traps will have a greater capacity than their normal rating when handling saturated condensate. If the traps are fitted to a collecting leg, at about 700 mm below the main, then filling of the collecting leg will provide a hydraulic head of about 0.07 bar or 1 lb/in² in addition to the line pressure.

Sizing the traps requires an estimate of the condensate produced in bringing the main up to temperature and then determining the hourly rate of flow by allowing for the time available for this warm-up. Suitable traps can then be chosen to give at least this rate with a total pressure difference of, say, 0.15 bar or 2 lb/in² across the trap. Warm-up loads for the most common pipe sizes and pressures are listed in Table 13.6. The maximum condensation rate occurs when the mains are cold and the pressure (of the air/steam mixture) within them is still, say, 0.07 bar (1 lb/in²). As the mains have been heated to some temperature below the corresponding saturated steam mixture, the amount of condensate formed will not have reached the values shown in Table 13.3. Further, although some of the condensate is flowing through the trap, the remainder is filling the collecting leg.

It seems reasonable to take the Table 13.6 value and to subtract the volume of the collecting leg, and to use steam traps capable of passing this nett amount in the available time with a pressure difference of 0.15 bar (2 lb/in²). This may be clarified by an example.

A length of 150 mm (6 in) main carries steam at 17 bar (250 lb/in²). Drain points are located at 45 m (150 ft) intervals, with collecting legs 100 mm (4 in) dia. × 700 mm (28 in) long. The main is brought up to pressure from 21°C (70°F) in 30 min.

	(Imperial)
Warm-up load to: 100°C	= approx. 21.7 kg/45 m: 212°F
	= 31.9 lb/100 ft
	= 47.8 lb/150 ft
Capacity of collecting leg	= approx. 5.7 kg
	= 12.7 lb
Nett condensate to be discharged	= 16.0 kg
	= 35.1 lb
Temperature rise to 17 bar	= 207° − 21° = 186°C
	= 406 − 70 = 336°F
Temperature rise to	= 79°C
	= 142°F

Table 13.6 Warm-up load in kg of steam per 30 m of steam main

Steam pressure (bar)	*Main size (mm)* 50	65	80	100	125	150	200	250	300	350	400	450	500	*−18°C correction factor*
0	2.8	4.4	5.8	8.3	11.2	14.5	21.8	31	41	49	64	80	94	1.5
0.33	3.1	5.0	6.4	9.2	12.4	16.2	24.1	35	45	54	71	89	104	1.44
0.67	3.4	5.4	7.0	10.0	13.6	17.6	26.4	38	49	59	77	97	114	1.41
1.00	3.7	5.9	7.6	10.8	14.8	19.1	28.7	42	53	64	83	105	124	1.36
1.5	4.0	6.4	8.1	11.8	16	20.7	30.3	43	58	69	90	114	135	1.35
2.0	4.3	6.9	9.0	12.8	17.3	22.4	32	45	64	75	98	124	146	1.34
3.0	4.7	7.5	9.8	13.9	18.9	24.4	37	52	70	82	107	135	159	1.32
4.0	5.0	8.0	10.4	14.8	20.1	26.0	39.0	55	74	87	114	144	169	1.29
5.0	5.3	8.4	11.0	15.6	21.1	27.4	41	58	78	91	120	151	178	1.28
6.0	5.5	8.8	11.5	16.4	22.2	28.8	43	61	82	96	126	159	187	1.27
7.0	5.8	9.2	12.1	17.2	23.3	30.2	45	65	85	100	132	166	196	1.26
8.0	6.1	9.6	12.7	18.0	24.1	31.6	47	68	89	105	138	173	205	1.25
10.0	6.6	10.5	13.6	19.5	26.4	34.0	51	73	96	114	149	188	221	1.24
12.0	6.9	11.0	14.4	20.5	27.8	36.0	54	77	102	120	158	199	233	1.22
14.0	7.3	11.5	15.0	21.4	30.0	38	57	80	106	126	164	223	263	1.21
16.0	8.6	13.5	17.8	25.6	36.0	47	71	102	138	165	212	276	334	1.21
18.0	10.0	15.5	20.6	29.8	42	56	85	124	170	203	261	333	404	1.20
20.0	11.3	17.4	23.3	34.0	47	65	98	146	201	241	309	388	474	1.20
25.0	12.2	18.8	25.1	37.0	51	70	106	158	217	259	333	418	511	1.19
30.0	13.1	20.1	26.9	39.0	55	75	114	169	232	278	357	448	548	1.18
40.0	14.8	22.7	30.5	44.0	62	85	129	191	263	315	405	507	620	1.16
60.0	18.0	27.6	37	54	98	131	230	338	476	569	728	923	1120	1.145
80.0	21.3	32.5	43	63	115	154	257	378	532	638	815	1034	1255	1.14
100.0	23.5	36	48	70	127	171	284	418	588	707	902	1145	1390	1.135
125.0	28.8	44	59	86	159	208	347	510	718	863	1100	1400	1700	1.127

Ambient temperature 20°C. For outdoor temperature −18°, multiply load value in table by correction factor. Loads based on ANSI Sch. 40 pipe for pressures up to 16 bar. Sch. 80 pipe for pressures above 16 bar, except Sch. 120 above 40 bar in sizes 125 mm and over.

$$\text{Time to reach } 100°\text{C} = \frac{79}{186} \times 30 \text{ min} = 12.7 \text{ min}$$

$$= \frac{142}{336} \times 30 = 12.7$$

$$\text{Trap discharge rate} = \frac{16.0 \text{ kg}}{12.7 \text{ min}} \times 60 \frac{\text{min}}{\text{h}}$$

$$= \frac{35.1}{12.7} \times 60$$

$$= 75.6 \text{ kg/h}$$

$$= 166 \text{ pph}$$

The capacity of a 1/2 in TD trap at 0.15 bar differential (2 lb/in^2) is about 115 kg/h (255 pph) and this trap would have ample capacity.

It is clear that in most cases other than very large distribution mains, 1/2 in TD traps are sufficiently large. With shorter distances between drain points, or smaller diameters, then 1/2 in low-capacity traps more than meet even start-up loads. On very large pipes it may be worth fitting 3/4 in traps, or two 1/2 in traps in parallel. Low-pressure mains often are drained through float/thermostatic steam traps, and these traps are now available for use at much higher pressures than formerly, where it is known that waterhammer will not be present.

13.4.6 Drain point layout

Condensate-collecting legs can be of the same diameter as the main, up to, say, 100 mm (4 in) size. Larger pipes can have collecting legs two or three sizes smaller than the main but not less than 100 mm (4 in) size. The length of the collecting legs used with automatic start-up is usually 700 mm (28 in) or more, to give a hydraulic head of 0.15 bar (1 lb/in^2). With supervised start-up, the length of the legs can be 1.5 pipe diameters but not less than about 200 mm (8 in).

The spacing between the drain points often is greater than is desirable. On a long horizontal run, drain points should be provided at intervals of about 45 m (150 ft) with a maximum of, say, 60 m (200 ft). Longer lengths should be split up and additional drain points fitted. Any low points in the system, such as the foot of risers and in front of shut-off or control valves, must also be drained.

In some cases the ground contours are such that a steam main can only be run uphill. This will mean that the drain points should be closer together over the uphill section (say, 15 m (50 ft) apart) and the size of the main increased so that the steam velocity is not more than about 15 m/s (50 ft/s). The lower steam velocity may then allow condensate to drain in the direction opposite to the steam flow.

13.4.7 Air venting the steam lines

Air venting of steam mains is of paramount importance and is far too often overlooked. Air is drawn into the pipes when the steam supply is shut off and the residual steam condenses. A small amount of air is dissolved in the feedwater entering the boiler, and even if this amounts to only a few parts per billion, the gas can accumulate in the steam spaces since it is non-condensable. Further, unless the feedwater is demineralized and decarbonated, carbonates can dissociate in the boiler and release carbon dioxide into the steam supply. Automatic air vents fitted above any possible condensate level at the ends of the steam mains and branches will allow discharge of these non-condensable gases, so promoting high heat transfer rates in the exchangers and helping to minimize corrosion (Figure 13.6).

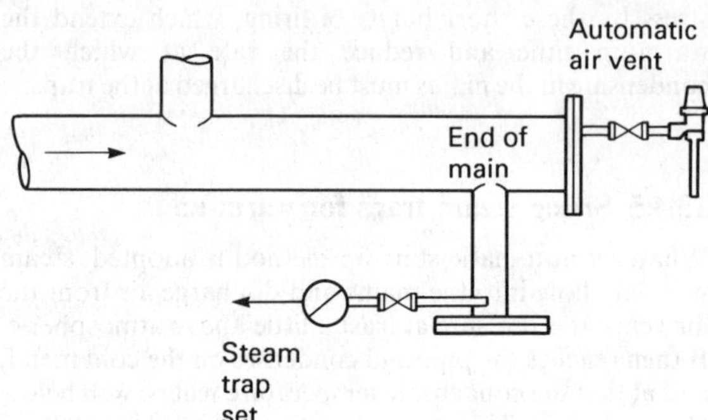

Figure 13.6 Draining and venting mains

13.5 Low-pressure systems

With steam generated at or close to the boiler design pressure it is inevitable that some of the steam-using equipment will have to be supplied at a lower pressure. In some cases the plant items themselves have only been designed to withstand a relatively low pressure. Sometimes a reaction will only proceed when the steam is at a temperature below a certain level or an unwanted reaction will occur above a certain level. For these and similar reasons, steam often is distributed at a relatively high pressure which must then be lowered, close to the point of use. Pressure-reduction stations incorporating pressure-reducing valves are fitted to perform this function.

Pressure-reducing valves may be either of the simple, direct acting pattern or be pilot operated (sometimes described as relay operated) (Figures 13.7 and 13.8). In the simple type, the force produced by the pressure downstream of the valve seat acting on a diaphragm or bellows is balanced by a control spring. When this force exceeds that of the spring at a given setting the diaphragm or bellows moves, compressing the spring further until balance is regained at a higher spring load – and at the same time moving the valve disc towards its seat. At the new balance point the steam flow through the valve is lessened and the downstream pressure is a little higher than the initial setting. Equally, at higher demand rates the downstream pressure falls by a small amount. A new balance point is reached when the lower pressure balances the spring force at a lower compression, with the valve disc now further away from the seat.

The pressure downstream of a simple reducing valve must vary, then, from a maximum at no-load with the

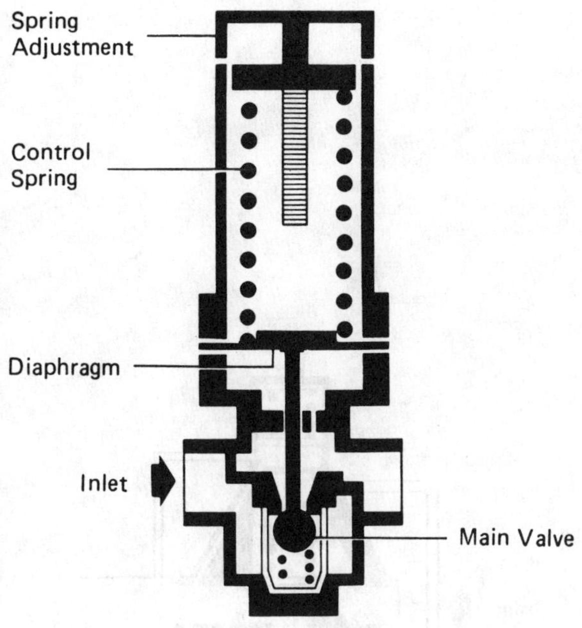

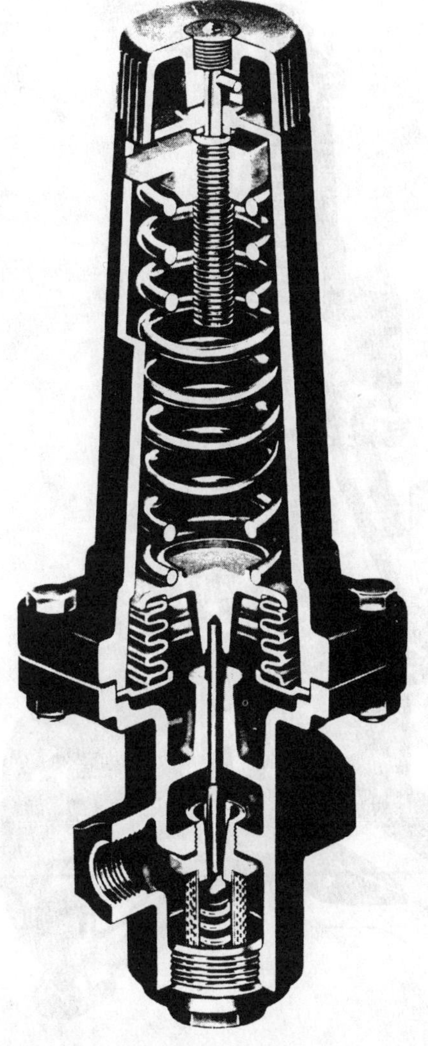

Figure 13.7 Direct-acting reducing valve

valve closed to a lower value on-load with the valve open. The change in reduced pressure may be called 'droop' or sometimes 'accuracy of regulation'. Simple valves are used to supply fairly small loads, or loads which are fairly steady at all times, to minimize the droop effect or where the reducing valve is followed by a temperature control valve which can compensate for the droop.

In other cases of larger and varying loads the droop of a large valve of simple pattern may be unacceptable. A very small valve of simple pattern can then be used as a pilot or relay valve to operate a much larger valve. The combination then has the droop characteristic of the very small pilot valve. Typically, a 1/2 in (15 mm) simple reducing valve might pass full flow with a droop of 20% of the reduced pressure, while a 3 in (80 mm) pilot-operated valve can have a droop of only 0.2 bar (3 lb/in^2).

It is a mistake to install even the best of pressure-reducing valves in a pipeline without giving some thought to how best it can be helped to give optimal performance. First, the valve chosen must have a large enough capacity that it can pass the required steam flow, but oversizing should be avoided. The weight of steam to be passed in a given time must be calculated or estimated, and a valve selected which can cope with this flow from the given upstream pressure to the required downstream pressure. The size of the valve is usually smaller than either the upstream or downstream pipe size because of the very high velocities which accompany the pressure drop within the valve.

Second, any valve which has been designed to operate on steam should not be expected to work at its best when supplied with a mixture of steam, water and dirt. A separator, drained through a steam trap, will remove virtually all the water from the steam entering the pressure-reducing set. The baffle type separators are found to be effective over a wide range of flow rates.

A stop valve is needed so that the steam supply can be shut off when necessary, and if this valve follows the separator, the steam supply will be drained even when

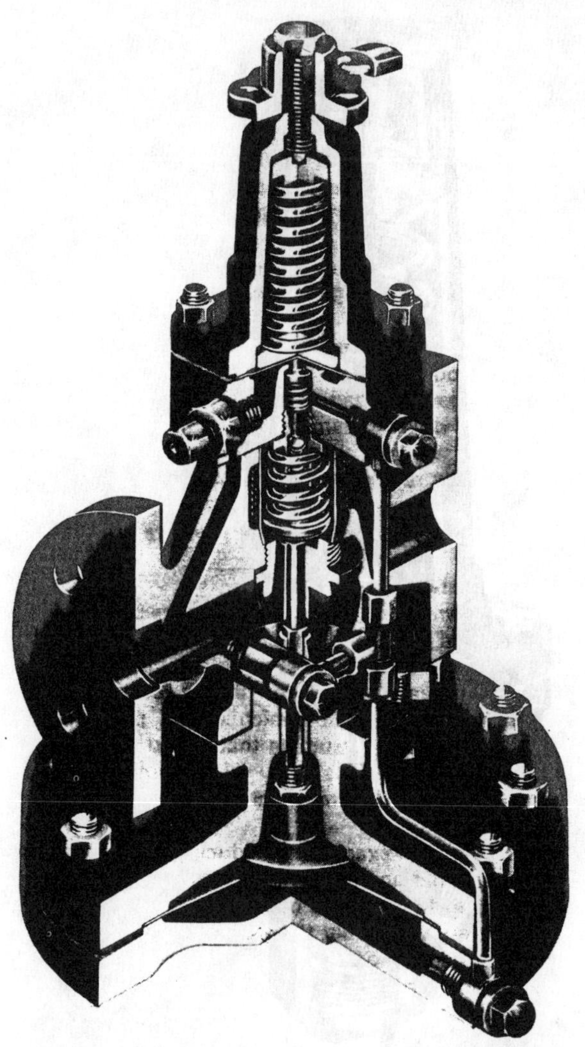

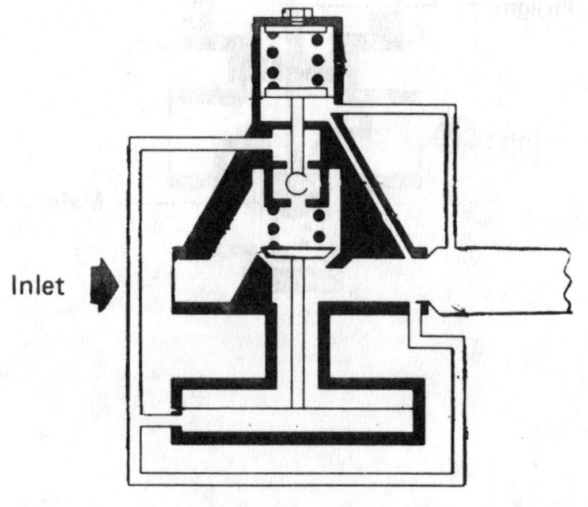

the valve is closed. After the valve a line size strainer is needed. It should have a 100-mesh stainless steel screen to catch the fine dirt particles which pass freely through standard strainers. The strainer is installed in the pipe on its side, rather than in the conventional way with the screen hanging below the pipe. This is to avoid the screen filling with condensate, which is often pumped between the valve plug and seat of a just-opening control valve as the flow (and pressure drop) through a strainer increases. This leads rapidly to wiredrawing of the control valve seat faces, and it is avoided by fitting the strainer horizontally to be self-draining. Pressure gauges at each side of a pressure-reducing valve both allow the downstream pressure to be set initially and aid in checking the functioning of the valve in service.

At the low-pressure side of the reducing valve it is usually essential to fit a relief or safety valve. If any of the steam-using equipment connected to the low-pressure range is designed to withstand a pressure below that of the upstream steam supply, then a safety valve is mandatory. Further, it may be called for when it is sought to protect material in process from over-high temperatures (Figure 13.9).

13.5.1 Safety relief valve sizing

When selecting a safety valve the pressure at which it is to be set must be decided. This may be sufficiently above the maximum normal operating pressure, under no-load conditions, to ensure the safety valve will reseat and not continue to blow steam, but it *must not* exceed the design pressure of the low-pressure equipment. The valve will give its rated capacity at an accumulated pressure of 10% above the set pressure. This capacity must equal or exceed the maximum possible capacity of the pressure-reducing valve if it should fail wide open, when passing steam from the upstream pressure to the accumulated pressure at the

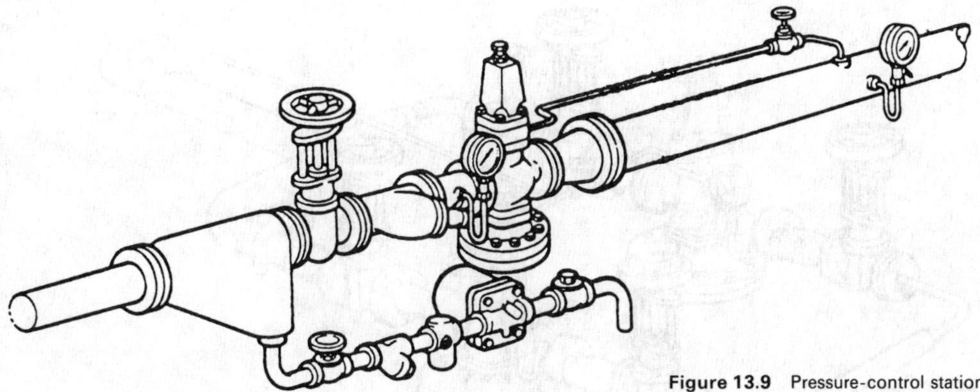

Figure 13.9 Pressure-control station

relief valve. For more information on this important subject, reference should be made to BS 6759: Part 1 and BS 5500.

13.5.2 Parallel and series operation of reducing valves

Parallel operation

In steam systems where load demands fluctuate over a wide range, parallel pressure control valves with combined capacities meeting the maximum load perform better than a single large valve. Maintenance needs, downtime and lifetime total costs can all be minimized with such an arrangement.

Any reducing valve must be capable of both meeting its maximum load and modulating down towards minimum loads when required. The load turndown with which a given valve will satisfactorily cope is limited. There are no rules which apply without exception, but when the low load is less than about 10% of the maximum load, two valves should always be preferred.

Whether a reducing valve is wide open or nearly closed, movement of the valve head through a given distance in relation to the seat will be in response to a given change in the controlled pressure. This valve movement changes the flow rate by a given amount, the pressure-reducing valve having a nearly linear characteristic. However, the change in the controlled pressure follows from a given percentage change in the flow rate. A valve movement which changes the steam flow just enough to match the demand change at high loads will change the flow by too much at low loads. It follows that instability of 'hunting' becomes more likely when a single valve is asked to cope with a high turndown in load.

A single valve then tends to keep opening and closing, on light loads. This leads to wear of both the seating and guiding surfaces and reduces the life of the diaphragms which operate the valve. Where valves make use of pistons sliding within cylinders to position the valve head the situation is worsened. Friction and sticking between the sliding surfaces mean that the valve head can only be moved in a series of discrete steps. The flow changes resulting from these movements are likely to be grossly in excess of the load changes which initiate them. Turndown ratios possible with piston-operated valves are inevitably smaller than those available with diaphragm-operated ones.

Stable control of reduced pressures is readily achieved by the use of two (or more) pressure-reducing valves is parallel (Figure 13.10). At full load and loads not too much below this level both valves are in use. As the load diminishes, the controlled pressure begins to increase and the valve which is set at the lower pressure begins to close. When the load can be supplied completely by the valve set at the higher pressure, the other valve closes. Any further load reduction causes the remaining valve to modulate through its proportional band.

Automatic changeover is achieved by the small difference between the pressure settings of the valves. For example, a maximum load of 5000 kg/h at 2 bar g might be supplied through one valve capable of passing 1200 kg/h and set at 2.1 bar g, and a parallel valve capable of passing 4000 kg/h set at 1.9 bar g. When the steam demand is at any value up to 1200 kg/h the smaller valve will supply the load at a controlled pressure just below 2.1 bar g. An increase in demand above 1200 kg/h will then lead to the controlled pressure falling to 1.9 bar or lower, and the second valve then opens to supplement the supply.

Equally, if the demand only rarely exceeded 4000 kg, or fell below about 1000 kg, it would be possible to set the larger valve at the higher pressure and supplement the supply through the smaller valve on those few occasions when this was necessary.

Sometimes the demand pattern is not known, except for the minimum and maximum values. Usually valves are then chosen with capacities of one third and two thirds of the maximum. The smaller valve is set at the slightly higher pressure.

Series operation

When the pressure reduction is through a ratio of more than about 10:1 the use of two valves in series should be considered. Much will depend on the valve being used, the pressures involved and the variations in the steam demand. Diaphragm-operated reducing valves have been used successfully with a pressure ratio as high as 20:1. They could perhaps be employed on a fairly steady load

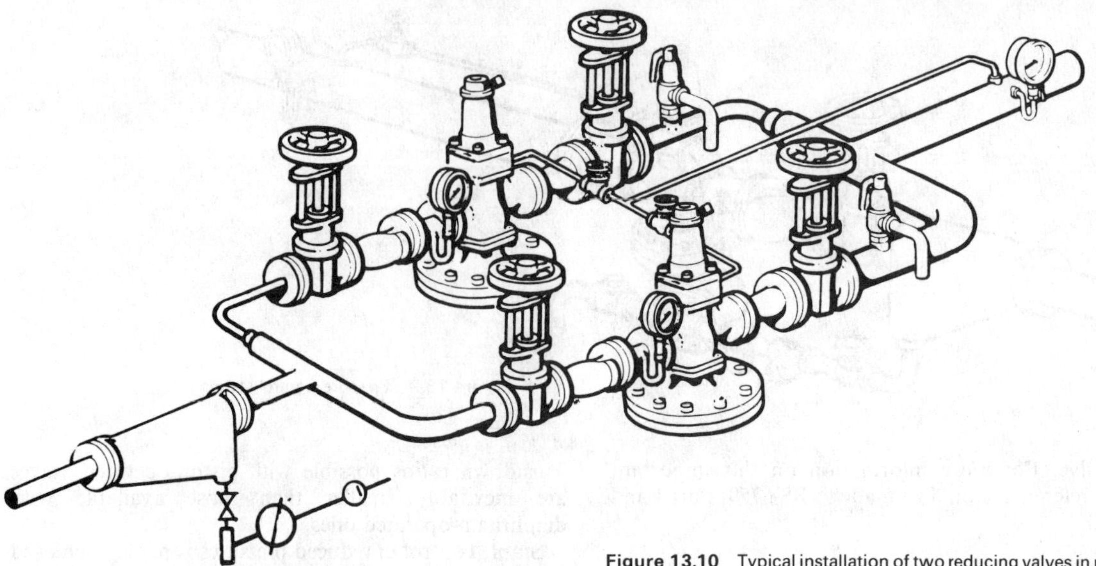

Figure 13.10 Typical installation of two reducing valves in parallel

to reduce from 6.8 bar g ($100 lb/in^2$) to 0.3 bar g ($5 lb/in^2$). However, the same valve would probably be unstable on a variable load reducing from 2.75 bar ($40 lb/in^2$) to 0.13 bar g ($2 lb/in^2$).

There is no hard-and-fast rule, but two valves in series can be expected to provide more accurate control. The second, low-pressure, valve should give the 'fine control' with a modest turndown, due consideration being given to valve sizes and capacities.

13.5.3 Series installations

For stable operation of the valves some appreciable volume between them is necessary. A length equal to 50 diameters of the appropriately sized pipe is often recommended or the same volume of larger pipe.

The downstream pressure-sensing pipe of each valve is connected to a straight section of pipe 10 diameters or 1 m downstream of the nearest tee, elbow or valve. This sensing line should be pitched down, to drain into the low-pressure line. If it cannot drain when connected to the top of this line it can often be connected instead to the side of the pipe.

The pipe between the two control valves must be drained through a steam trap, just as would the foot of any riser downstream of the pressure-reducing station.

13.5.4 Bypasses

The use of bypass lines and valves is best avoided. Bypass valves are often found to be leaking steam because of wire drawing of the seating faces of valves which have not been tightly closed. If they are used, the capacity of the bypass valve should be added to that of the pressure-reducing valve when sizing relief valves.

If it is thought essential to maintain a steam supply even if a pressure-reducing valve should be faulty, or undergoing maintenance, consideration should be given to fitting a reducing valve in the bypass line. Sometimes the use of a parallel reducing station avoids the need for bypasses.

13.5.5 Selecting control valves for steam

The choice of a suitable temperature or pressure control valve for steam application will depend on the supply side pressure, the downstream pressure, and the flow rate of steam to be passed. In the case of temperature control valves the first of these is usually known and the third can be calculated, but the appropriate pressure drop through the valve is often to be decided. Sometimes the maker's rating of a heater will specify that it transfers heat at a certain rate when supplied with steam at a certain pressure. This pressure is then the pressure downstream of the control valve, and the valve may be selected on this basis.

If the pressure drop across the valve is to be more than 42% of the inlet absolute pressure the valve selection is the same as if the pressure drop were only 42%. With this pressure ratio the steam flow through the valve reaches a critical limit, with the steam flowing at sonic velocity, and lowering the downstream pressure below 58% of the inlet absolute pressure gives no increase in flow rate. When the heater needs a higher pressure, or when the pressure required in the heater is not known, it is safer to allow a smaller pressure drop across the control valve. If the necessary heater pressure is not known, a pressure drop across the control valve of 10–25% of the absolute inlet pressure usually ensures sufficient pressure within the heater. Of course, in the case of pressure-reducing valves the downstream pressure will be specified.

Valve capacities can be compared by use of the *Kv* (or *Cv* when Imperial units are used) values. These factors are determined experimentally, and the *Kv* value is the number of cubic metres per hour of water that will flow through a valve with a pressure drop of one bar. The *Cv* value is the number of gallons per minute of water that will flow through the valve with a pressure drop of one pound f. per square inch. As the gallon is a smaller unit in the USA, the number of gallons passed is greater, and the US *Cv* is 1.2 times the UK *Cv*. The *Kv* is about 0.97 of the UK *Cv* value.

The steam flow through a valve at critical pressure drop when $P2$, the downstream pressure, is 0.58 times $P1$, the supply pressure, or $P2/P1=0.58$, is given very closely by $W=12KvP1$, where W=steam flow (kg/h), Kv=flow coefficient and $P1$=supply pressure (bar). In the USA, $W=1.5CvP1$, where W=steam flow (pph), Cv=flow coefficient and $P1$=supply pressure (psia).

The factors 1.5 (US) and 12 (English metric) are not exactly equivalent but are sufficiently close for practical valve sizing purposes.

When the pressure drop ratio $(P1-P2)/P1$ is less than 0.42, the steam flow is less than the critical flow. While BS 5793: Part 2: Section 2.2: 1981 provides an approximate and a rigorous method of calculating flows, it is found in practice that the best results are not distinguishable from the empirical formula

$$W_x=(\text{Critical flow rate})\sqrt{[1-5.67(0.42-x)^2]}$$

where x is the pressure drop ratio and W_x is the corresponding flow rate.

13.5.6 Draining condensate from heat exchangers

If heat exchangers are not to steadily fill with condensate and the flow of heat to cease, then the condensate must be drained from them, and steam traps are normally used for this purpose. However, it is clear that a positive pressure difference must always exist between the condensate within the heat exchanger and the outlet side of the steam trap. Where the steam is supplied through a pressure-reducing valve, to meet a steady load, and the condensate is drained to a return line at lower level than the outlet and running to a vented receiver, this positive differential pressure will exist. In a great number of applications the steam flow into the exchanger must be controlled in response to the temperature of the heated fluid, and if the heat load on the exchanger is less than the maximum, then the steam flow must be proportionately reduced.

With such a reduction in steam flow the steam pressure within the exchanger will fall. A balance is reached when the steam pressure is such that the steam temperature gives an appropriately lowered temperature difference between the steam and the mean temperature of the heated fluid. Under no-load conditions, if the flow of the fluid being heated were to cease or if the fluid were already at the required temperature as it entered the exchanger then this temperature difference would have to be zero and the steam at the same temperature as the fluid leaving the exchanger.

As a first approximation, the expression

$$q_1=UA\Delta t$$

is used, and since A is a constant and U does not vary very much, the temperature of the steam must fall along a straight line from full-load steam temperature to the control temperature, as the load on the exchanger varies from maximum to zero. Clearly, as the steam temperature falls, the corresponding pressure in the steam space drops, and when it equals the pressure at the outlet side of the steam trap, condensate flow will cease. If the heat exchanger is not to flood with condensate it is necessary to ensure that the steam trap draining the exchanger is sufficiently far below the condensate drain point as to provide a hydrostatic head to push the condensate through the trap. Further, if the pressure at the outlet side of the steam trap is that of the atmosphere, and the pressure within the steam space is likely to fall below atmospheric, then a vacuum breaker is needed to admit air at a point above the maximum possible level of any condensate so that gravity can then clear the condensate through the steam trap.

It follows that condensate at the outlet side of the steam trap can only be at atmospheric pressure, and if it is to be lifted to a high-level return line, or into a return line in which a positive pressure exists, then a condensate pump is needed for this purpose.

Equally, if insufficient height is available to provide the hydrostatic head in front of the steam trap, and fall from the steam trap to the receiver of a condensate pump, it may be possible to allow condensate to fall directly from the exchanger to the body of a steam-powered 'alternating receiver' condensate pump, which can then under low-load low-pressure conditions temporarily isolate collected condensate from the exchanger and use high-pressure steam to push it through the steam trap. Figure 13.11 illustrates the alternative condensate removal methods, which are used for the differing pressure conditions.

13.5.7 Air venting

The existence of air (using this word to cover atmospheric air and also any other non-condensing gases) in steam systems was mentioned in Section 13.4.7. It is even more important that the presence of air in the heat exchangers themselves is recognized and arrangements made for its discharge. Air will exert its own partial pressure within the steam space, and this pressure will be added to the partial pressure of the steam in producing the total pressure present. The actual steam pressure is lower than the total pressure shown on a gauge by virtue of the presence of any air. Since the temperature of saturated steam must always be the condensing temperature corresponding with its partial pressure, the effect of air in a steam space is always to lower the temperature below the level which would be expected for pure steam at the total pressure present.

This means that thermostatic balanced pressure steam traps which open when their element senses a temperature somewhat below that of saturated steam at the pressure existing within the steam space are very effective when used as automatic air vents. They are connected to a steam space at any location where air will collect. Usually this means at any 'remote point' from the steam entry, along the path the steam takes as it fills the steam space.

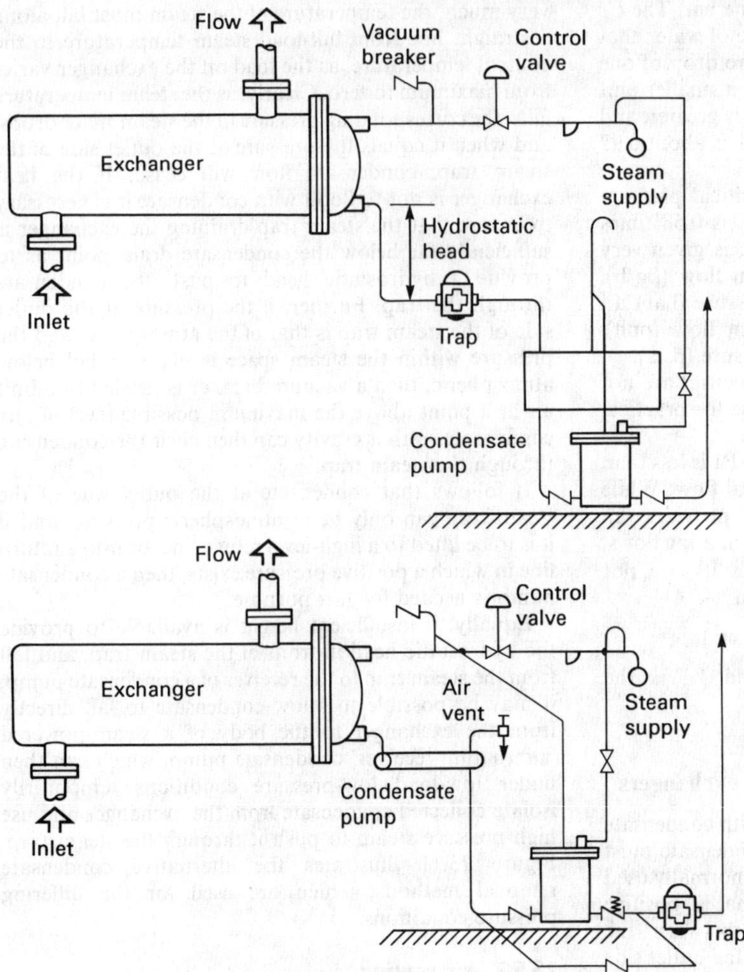

Figure 13.11 Lifting condensate drained from exchangers

Air vents are most effective when they are fitted at the end of a length of 300 mm or 450 mm of uninsulated pipe which can act as a collecting/cooling leg. Air is an excellent insulating material, having a thermal conductivity about 1300 times less than that of iron. The last place where it can be allowed to collect is in the steam space of heat exchangers. Further, as it contains oxygen or carbon dioxide, which dissolve readily in any subcooled condensate that may be present, the presence of air initiates corrosion of the plant and the condensate return system.

13.6 Flash steam

13.6.1 The release of flash steam

High-pressure condensate forms at the same temperature as the high-pressure steam from which it condenses, as the enthalpy of evaporation (latent heat) is transferred from it. When this condensate is discharged through a steam trap to a lower pressure the energy it contains is greater than it can hold while remaining as liquid water. The excess energy re-evaporates some of the water as steam at the lower pressure. Conventionally, this steam is referred to as 'flash steam', although in fact it is perfectly good steam, even if at low pressure.

13.6.2 Flash steam utilization

In an efficient and economical steam system this so-called flash steam will be utilized on any load which can make use of low-pressure steam. Sometimes it can be simply piped into a low-pressure distribution main for general use. The ideal is to have a greater demand for low-pressure steam at all times than the available supply of flash steam. Only as a last resort should flash steam be vented to atmosphere and lost.

This means that condensate from high-pressure sources should usually be collected and led to a flash vessel which operates at a lower pressure (but high enough to be useful). Remember, the flashing off does not normally take place within the flash vessel. It begins within the seat of the steam trap and continues in the condensate line. Only when the high-pressure steam traps are very close to the flash vessel does any flashing take place within it.

Instead, the flash vessel is primarily a flash steam separator. Its shape and dimensions are chosen to encourage separation of the considerable volume of low-pressure steam from the small volume of liquid.

13.6.3 Proportion of flash steam released

The amount of flash steam which a given weight of condensate will release may be readily calculated. Subtracting the sensible heat of the condensate at the lower pressure from that of the condensate passing through the traps will give the amount of heat available for use as the enthalpy of evaporation (latent heat). Dividing this amount by the actual enthalpy of evaporation at the lower pressure will give the proportion of the condensate which will flash off. Multiplying by the total quantity of condensate being considered gives the weight of low-pressure steam available.

Thus if, for example, 1000 kg of condensate from a source at 10 bar is flashed to 1 bar we can say:

Specific enthalpy of water (10 bar) = 781.6
Specific enthalpy of water (1 bar) = 505.6
Energy available for flashing = 276
Enthalpy of evaporation at 1 bar = 2201.1

Proportion evaporated $= \dfrac{276}{2201.1}$
$= 0.125$

Flash steam available $= 0.125 \times 1000$ kg
$= 125$ kg

Alternatively, tables such as Table 13.7 or charts such as that shown in Figure 13.12 will allow the proportion of flash steam available between two pressures to be read off directly, within the more usual pressure ranges.

13.6.4 Sub-cooled condensate

Note that the method described assumes that the high-pressure condensate has not been sub-cooled. If any sub-cooling has taken place, then the figure taken for the enthalpy of water at the higher pressure is reduced by the amount of sub-cooling. The chart or table can still be used if the upstream pressure is taken as that corresponding to saturated steam at the same temperature as the sub-cooled condensate.

In the previous example condensate sub-cooled by 20°C would be at 164.13°C instead of 184.13°C. Steam at this temperature would be at a pressure of some 5.83 bar, so that the chart in Figure 13.12 would be used for a drop from 5.83 bar to 1 bar instead of from 10 bar to 1 bar, to allow for the sub-cooling.

13.6.5 Making use of flash steam

The steam/process air heater with multiple coils typifies the kind of application on which recovery of flash steam from the condensate is most readily effected. Condensate from the high-pressure coils is taken to a flash steam recovery vessel. Low-pressure steam leaving the vessel is supplied to the first coil at the air-inlet side of the heater or to an extra coil acting as a pre-heater unit.

Condensate from the low-pressure coil together with that from the flash vessel will then drain to a collecting tank, or direct to a condensate pump, for return to the boiler plant. If the pressure of the flash steam is left to find its own level it will often be sub-atmospheric. As the condensate must then drain by gravity through the steam traps these also must be sufficiently below the condensate drain points to provide an appropriate hydraulic head, and a vacuum breaker fitted above the coil. The alternatives are to allow the condensate to drain directly to a condensate pump, or to supply additional low-pressure steam through a pressure-reducing valve, to maintain a positive pressure in the coil and flash vessel.

13.6.6 Space heating

Somewhat similar arrangements can be used when large areas are heated by radiant panels or unit heaters. Some

Table 13.7 Condensate proportion released as flash steam

Pressure at traps (bar G)	*Pressure in flash vessel (bar G)*												
	0	*0.2*	*0.5*	*0.75*	*1.0*	*2*	*3*	*4*	*5*	*6*	*7*	*10*	*15*
1	0.038	0.029	0.017	0.008	0								
2	0.063	0.054	0.042	0.033	0.025	0							
3	0.082	0.073	0.061	0.053	0.045	0.02	0						
4	0.098	0.089	0.077	0.069	0.061	0.036	0.017	0					
5	0.111	0.102	0.091	0.083	0.078	0.05	0.03	0.014	0				
6	0.123	0.119	0.108	0.1	0.093	0.068	0.049	0.032	0.019	0			
7	0.133	0.125	0.113	0.106	0.098	0.074	0.054	0.038	0.024	0.012	0		
10	0.16	0.152	0.141	0.133	0.125	0.101	0.083	0.067	0.053	0.04	0.029	0	
15	0.192	0.183	0.172	0.165	0.157	0.134	0.116	0.1	0.087	0.075	0.014	0.035	0
20	0.222	0.214	0.203	0.195	0.188	0.165	0.148	0.133	0.12	0.108	0.097	0.069	0.032
25	0.245	0.237	0.226	0.219	0.212	0.189	0.172	0.157	0.144	0.133	0.122	0.095	0.058
30	0.261	0.253	0.243	0.235	0.228	0.206	0.189	0.174	0.162	0.15	0.14	0.113	0.077
40	0.296	0.288	0.278	0.271	0.264	0.242	0.226	0.212	0.2	0.189	0.179	0.153	0.118

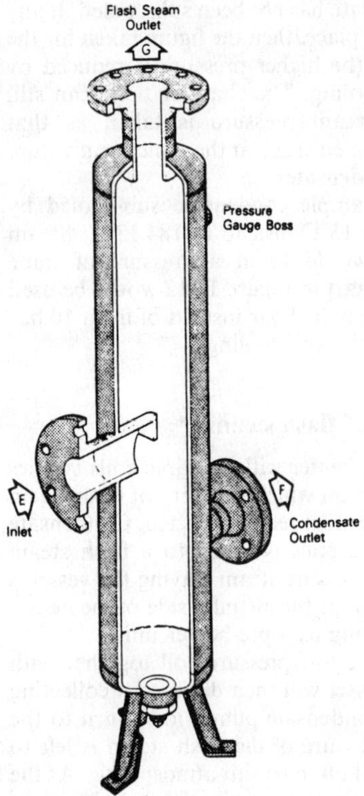

Figure 13.12 Flash vessels

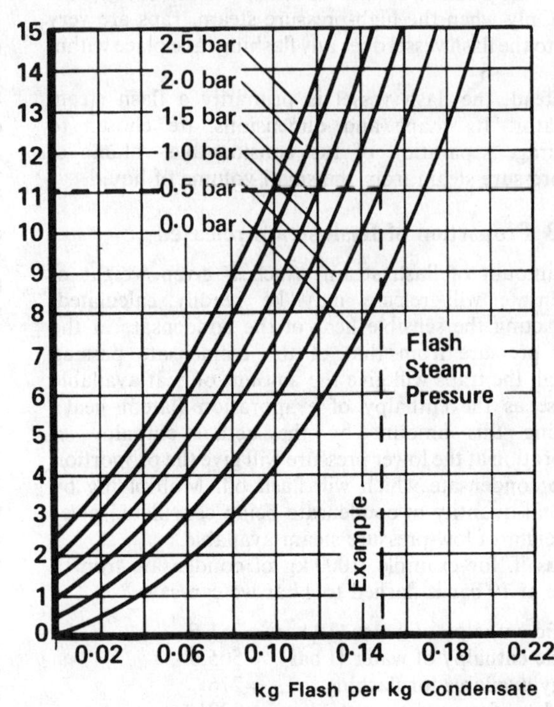

10–15% of the heaters are separated from the high-pressure steam supply and supplied instead with low-pressure steam flashed off the high-pressure condensate. The heating demands of the whole area remain in step, so supply and demand for the flash steam are balanced.

13.6.7 The general case

In other cases flash steam is utilized on equipment which is completely separated from the high-pressure source. Often the low-pressure demand does not at all times match the availability of the flash steam. A pressure-reducing station is often needed to make up any deficit and a surplussing valve is required to vent any flash steam in excess of the amount being condensed.

13.6.8 Steam traps

The wide choice of steam trap types which are available must at first seem confusing, and it is useful to first define the term 'steam trap'. In practice, a steam trap has two separate elements. The first of these is a valve and seat assembly, which can provide a variable orifice through which the condensate can be discharged at such a rate as to match the rate of condensation in the equipment being drained. The opening may be modulated continually to provide a continuous flow of condensate, or may operate in an on/off fashion so that the average rate over a period of time matches the condensation rate.

The second element is a device which will open or close the valve by measuring some parameter of the fluid reaching it and 'deciding' whether this may or may not be discharged. It would be found that the controlling elements mainly fall into one of three categories. The steam trap can decide automatically whether to open or to close to the fluid reaching it on the basis of:

1. The density of the fluid, by using a float which will float in water or sink in steam;
2. By measuring the temperature of the fluid, closing the valve at or near to steam temperature, and opening it when the fluid has cooled to a temperature sufficiently far below that of steam;
3. By measuring the kinetic effects of the fluid in motion, since at a given pressure drop, low-density steam will move at a much greater velocity than will high-density condensate, and the conversion of pressure energy into kinetic energy can be used to position a valve.

The groupings then may be described as *mechanical*, which will include both ball float and inverted bucket steam traps; *thermostatic*, which will include both balanced pressure and bimetallic elements; and *thermodynamic* or *disc* pattern traps (Figure 13.13). Each type of trap has its own characteristics, and these will make one pattern of trap more suitable for use on a given application than another. In practice, it is usual to find that the applications in any given plant fall into a small number of categories, and it often is possible to standardize on a quite small number of trap types.

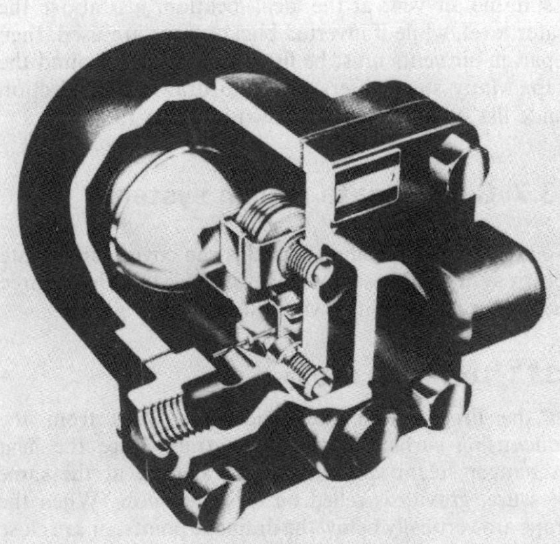

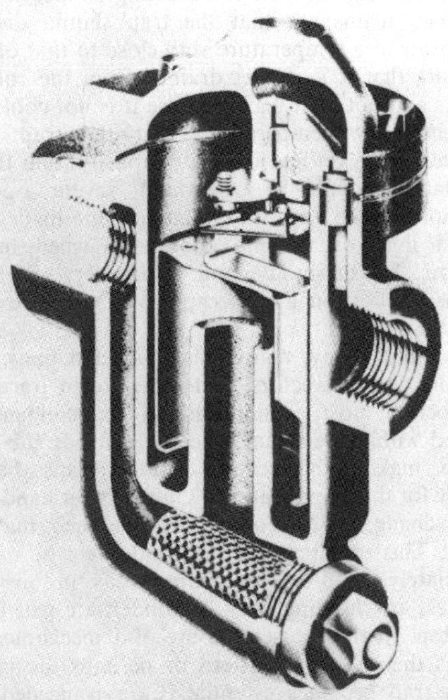

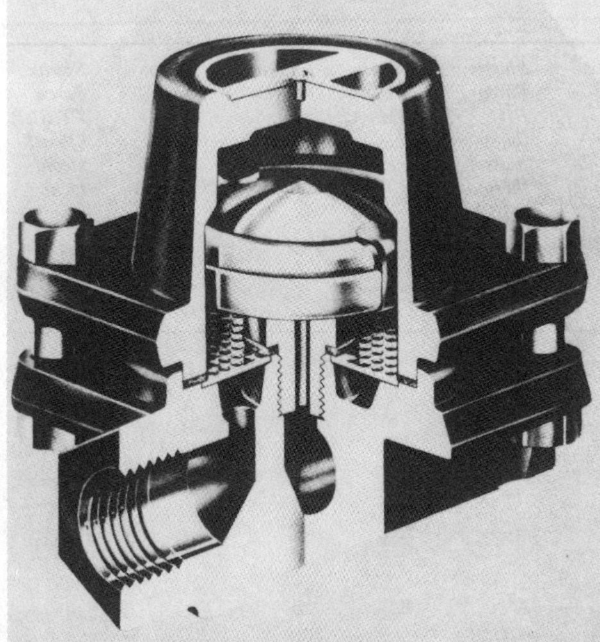

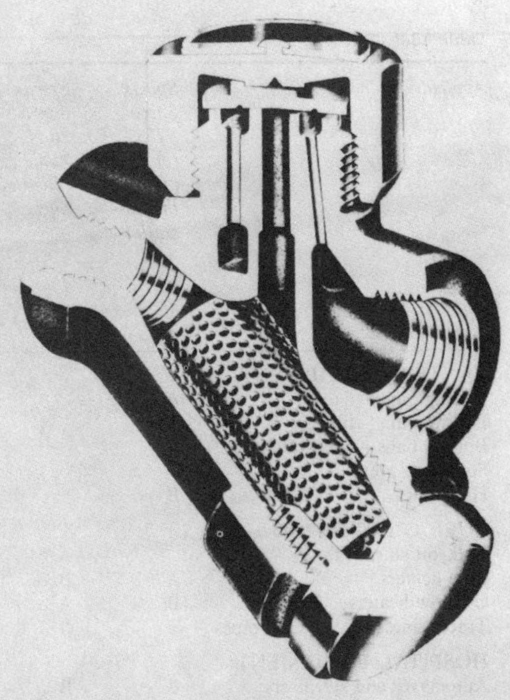

Figure 13.13 Steam traps

Thus the requirements for draining condensate from the steam mains are that the trap should discharge condensate at a temperature very close to that of steam to ensure that it is in fact drained from the collecting pockets and not held back because it is not cooled to a sufficiently low temperature; that the trap is not physically large yet has adequate capacity; that the trap is robust enough to withstand severe operating conditions, such as waterhammer in inadequately drained lines or freezing conditions when installed outdoors. The thermodynamic trap is very widely used for this application and is capable of giving excellent results.

In a similar way, many small jacketed pans, steam radiators and convectors, and some steam tracer lines can operate most economically if the condensate is retained within the steam space until it has sub-cooled a little, making thermostatic pattern traps the most suitable for these applications. On the other hand, many heat exchangers are required to give their maximum output. This requires that the condensate be removed immediately it forms, or perhaps, as in air heater batteries, any holding back of condensate will lead to corrosion. In these cases the use of a mechanical trap such as the ball float pattern or perhaps the inverted bucket trap becomes essential. Care is needed when inverted bucket traps are chosen for jobs where the steam control valve may at times close down, lowering the pressure in the steam space. This can lead to re-evaporation of the water seal in an inverted bucket trap, which would then blow steam. Further, on many heater exchangers it is important to provide air-venting capacity at the trap. Float thermostatic traps incorporate a suitable air vent at the ideal location, just above the water level, while if inverted bucket traps are used, then separate air vents must be fitted in bypasses around the trap. Many steam users prefer to draw up a selection guide list such as the one shown in Table 13.8.

13.7 Condensate return systems

No single set of recommendations can cover condensate return systems. These divide naturally into at least three sections, each with its own requirements.

13.7.1 Drain lines to traps

In the first section the condensate flows from the condensing surface to the steam trap. Since the heat exchanger steam space and the traps are at the same pressure, gravity is relied on to induce flow. When the traps are vertically below the drainage points, or are close to them, it may be satisfactory to use a line of the same size as the inlet connection of the trap. However, if the trap must be located a little further away from the drainage point, then the line between the trap point and the trap can be laid with a slight fall (say, 1 in 250), and Table 13.9 shows the water-carrying capacities of the pipes with such a gradient. It is important to allow for the passage of incondensables to the trap and for the extra water which is carried during cold-start conditions.

Table 13.8 Selecting steam traps

Application	*Spirax Sarco FT range (float/ thermo-static)*	*Spirax Sarco IB range (inverted bucket)*	*Spirax Sarco TD range (thermo-dynamic)*	*Spirax Sarco BPT (balanced pressure thermo-static)*	*Spirax Sarco SM (bi-metallic)*	*Spirax Sarco No. 8 (liquid expan-sion)*	*Spirax Sarco FT/TV/ SLR (float/ thermo-static with steam lock release)*	*Spirax Sarco FT/SLR (float/ steam lock release)*
CANTEEN EQUIPMENT								
Boiling pans – fixed	A		B[1]	B			B	B[1]
Boiling pans – tilting				B			A	B[1]
Boiling pans – pedestal	B			A[2]			B	B[1]
Steaming ovens				A[2]				
Hot plates	B			A[2]			B	B[1]
FUEL OIL HEATING								
Bulk/oil storage tanks		A[1]	B[1]					
Line heaters	A	B[1]						
Outflow heaters	B	A[1]						
Tracer lines and jacketed pipes			B	A[3]	B	B		
HOSPITAL EQUIPMENT								
Autoclaves and sterilizers	B	B		A			B	B[1]
INDUSTRIAL DRYERS								
Drying coils (continuous)	B	A		B	B			
Drying coils (grid)		B[1]		A				

Table 13.8 (*continued*)

Application	*Spirax Sarco FT range (float/ thermo-static)*	*Spirax Sarco IB range (inverted bucket)*	*Spirax Sarco TD range (thermo-dynamic)*	*Spirax Sarco BPT (balanced pressure thermo-static)*	*Spirax Sarco SM (bi-metallic)*	*Spirax Sarco No. 8 (liquid expan-sion)*	*Spirax Sarco FT/TV/ SLR (float/ thermo-static with steam lock release)*	*Spirax Sarco FT/SLR (float/ steam lock release)*
Drying cylinders	B	B[1]					A	B[1]
Multi-bank pipe dryers	A	B[1]		B				
Multi-cylinder sizing machines	B	B[1]					A	B[1]
LAUNDRY EQUIPMENT								
Garment presses	B	B	A					
Ironers and calenders	B	B[1]	B[1]	B			A	B[1]
Solvent recovery units	A	B	B					
Tumbler dryers	A	B[1]					B	B[1]
PRESSES								
Multi-platen presses (parallel connections)	B	B	A					
Multi-platen presses (series connections)		B[1]	A[1]					
Tyre presses		A	B	B				
PROCESS EQUIPMENT								
Boiling pans – fixed	A		B[1]	B			B	B[1]
Boiling pans – tilting							A	B
Brewing coppers	A	B[1]					B	B[1]
Digesters	A	B[1]	B[1]					
Evaporators	A	B[1]					B	B[1]
Hot tables			B	A				
Retorts	A	B[1]						
Bulk storage tanks	B	A[1]	B[1]					
Vulcanizers	B	A[1]						
SPACE HEATING EQUIPMENT								
Calorifiers	A	B[1]					B	B[1]
Heater batteries	A	B[1]					B	B[1]
Radiant panels and strips	A	B[1]	B[1]				B	B[1]
Radiators and convention cabinet heaters	B			A	B			
Overhead pipe coils	B	B[1]		A				
STEAM MAINS								
Horizontal runs	B	B	A	B[1]				
Separators	A	B	B	B[2]				
Terminal	B	B[1]	A[1]	B[2]				
Shutdown drain (frost protection)				B[3]		A		
TANKS AND VATS								
Process vats (rising discharge pipe)	B	B	A	B				
Process vats (discharge pipe at base)	A	B	B	B				
Small coil-heated tanks (quick boiling)	A	B		B				
Small coil-heated tanks (slow boiling)						A		

A = best choice
B = acceptable alternative
1. With air vent in parallel.
2. At end of cooling leg. Minimum length 1 m (3 ft).
3. Use special tracing traps which offer fixed temperature discharge option.

Table 13.9 Water-carrying capacity of pipes (SI units) (approximate frictional resistance in mbar per m of travel)

Steel tube (mm)	*0.3 (30 Pa)*	*0.5 (50 Pa)*	*0.6 (60 Pa)*	*0.8 (80 Pa)*	*1.0 (100 Pa)*	*1.4 (140 Pa)*
15	95	130	140	160	180	220
20	220	290	320	370	420	500
25	410	540	600	690	790	940
32	980	1 180	1 300	1 500	1 700	2 040
40	1 360	1 790	2 000	2 290	2 590	3 100
50	2 630	3 450	3 810	4 390	4 990	6 000
65	5 350	6 950	7 730	8 900	10 150	12 100
80	8 320	10 900	12 000	13 800	15 650	18 700
100	17 000	22 200	24 500	28 200	31 900	38 000

In most cases it is sufficient to size these pipes on twice the full running load.

13.7.2 Trap discharge lines

At the outlet side of the traps the 'condensate' lines have to carry both the condensate and any incondensable gases, together with the flash steam released from the condensate. Wherever possible, these pipes should drain by gravity from the traps to the condensate receiver, whether this be a flash recovery vessel or the vented receiver of a pump. With atmospheric pressure in the return line, and pressures upstream of the traps of up to about 4 bar (60 lb/in^2), the sizing method described above is more than adequate. At higher upstream pressures the volume of the flash steam released from the condensate becomes significant and must be given due consideration. If the pressure at the discharge side of the traps is above atmospheric pressure, then sizing these lines on twice the full running load may still be adequate even with pressures well above 4 bar (60 lb/in^2) at the inlet side. The chart shown in Figure 13.14 allows the lines to be sized as flash steam lines, since the volume of the condensate is so much less than that of the steam released.

Draining condensate from traps which serve loads at differing pressures to a common condensate return line is a concept which is often found difficult. Many users assume that the high-pressure condensate will prevent the low-pressure condensate from passing through the low-pressure traps and give rise to waterlogging of the low-pressure systems.

However, the terms 'high pressure' and 'low pressure' can only apply to the conditions on the upstream side of the seats in the steam traps. At the downstream or outlet side of the traps the pressure must be the common pressure in the return line. This return line pressure will be the sum of at least three components:

1. The pressure at the end of the return line, either atmospheric or of the vessel into which the line discharges;
2. The hydrostatic head needed to lift the condensate up any risers in the line;
3. The pressure drop needed to carry the condensate and any flash steam along the line.

Item 3 is the only one likely to give rise to any problems if condensate from sources at different pressures enters a common line. The return should be sufficiently large to carry all the liquid condensate and the varying amounts of flash steam associated with it, without requiring excessive line velocities and excessive pressure drop. If this is accepted, then the total return line cross-sectional area will be the same, whether a single line is used or if two or more lines are fitted with each taking the condensate from a single pressure source.

Return lines may become undersized, requiring a high pressure at the trap discharges and restricting or even preventing discharge from the low-pressure traps, if it is forgotten that the pipe must carry flash steam as well as water and that flash steam is released in appreciable quantity from high-pressure condensate.

13.7.3 Pumped return line

Finally, the condensate is often pumped from the receiver to the boiler house. Pumped condensate lines carry only water, and rather higher water velocities can often be used to minimize pipe sizes. The extra friction losses entailed must not increase back pressures to the point where pump capacity is affected. Table 13.10 can be used to help estimate the frictional resistance presented by the pipe.

Condensate pumps usually operate with an on/off action, so that the instantaneous flow rate during the 'on' period is greater than the average rate of flow of condensate to the pump receiver. This increased instantaneous flow rate must be kept in mind when sizing the delivery lines.

Where long delivery lines are used, the water flowing along the pipe as the pump discharges attains a considerable momentum. At the end of the discharge period when the pump stops, the water tends to keep moving along the pipe and may pull air or steam into the delivery pipe through the pump outlet check valve. When this bubble of steam reaches a cooler zone and condenses, the water in the pipe is pulled back towards the pump. When the reverse flow reaches and closes the check valve, waterhammer often results. This problem is greatly reduced by adding a second check valve in the delivery line some 5 or 6 m from the pump. If the line

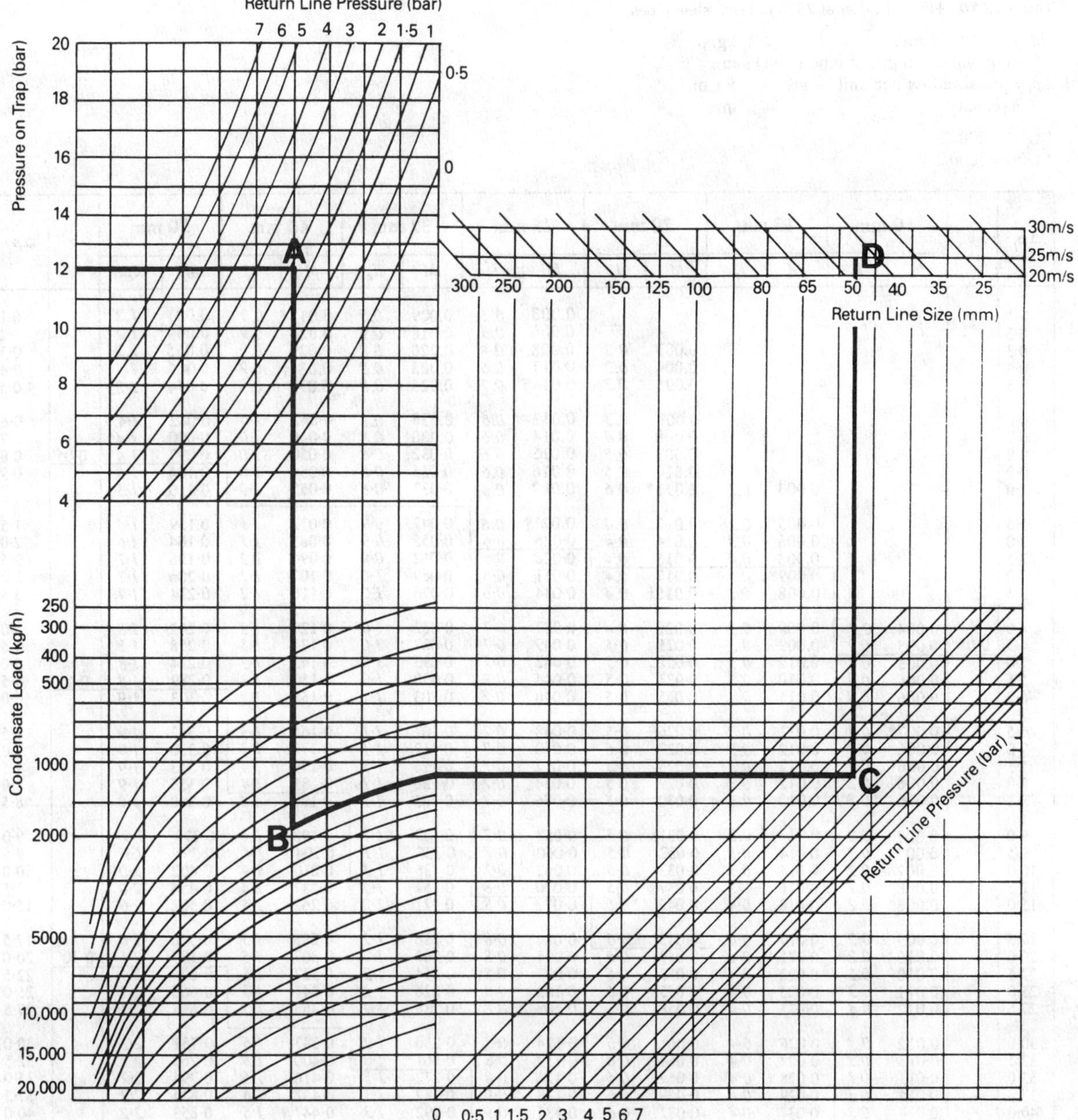

Figure 13.14 Condensate line sizing chart where pressure at traps is above 4 bar (SI units). 1: From pressure upstream of trap move horizontally to pressure in return line (A). 2: Drop vertically to condensate load in kg/h (B). 3: Follow curve to RH scale and across to same return line pressure (C). 4: Move upward to return line flash velocity – say, 25 m/s maximum (D). 5: Read return line size

lifts to a high level as soon as it leaves the pump then adding a generously sized vacuum breaker at the top of the riser is often an extra help. However, it may be necessary to provide means of venting from the pipe at appropriate points the air which enters through the vacuum breaker.

The practice of connecting additional trap discharge lines into the pumped main is usually to be avoided. The flash steam which is released from this extra condensate can lead to waterhammer. Preferably, the trap should discharge into a separate gravity line which carries the condensate to the receiver of the pump. If this is impossible one may arrange for the condensate and associated flash steam to enter the pumped main through a small 'sparge' or diffuser fitting so that the flash steam can be condensed immediately in the pumped water.

Table 13.10 Flow of water at 75°C in black steel pipes

M = mass flow rate kg/s
l_e = equivalent length of pipe ($\zeta = 1$) m
Δp_l = pressure loss per unit length Pa/m
v = velocity m/s

* $(Re) = 2000$
† $(Re) = 3000$

Δp_l	v	10 mm		15 mm		20 mm		25 mm		32 mm		40 mm		50 mm		v	Δp_l
		M	l_e	M	l_e	M	l_e	M	l_e	M	l_e	M	l_e	M	l_e		
0·1								0.003	*0.1*	0.009	*0.5*	0.017*	*0.9*	0.031	*1.2*		0·1
0·2								0.006	*0.3*	0.018*	*0.9*	0.024	*0.9*	0.044†	*1.2*		0·2
0·3						0.003	*0.2*	0.008	*0.4*	0.020	*0.8*	0.029	*0.9*	0.055	*1.2*		0·3
0·4						0.004	*0.2*	0.011	*0.6*	0.023	*0.8*	0.034†	*0.9*	0·065	*1·3*		0·4
0·5						0.005	*0.3*	0.014*	*0.7*	0.025	*0.8*	0.038	*0.9*	0·074	*1·3*		0·5
0·6						0.007	*0.3*	0.013	*0.6*	0.028	*0.8*	0·042	*1·0*	0·082	*1·4*		0·6
0·7						0.008	*0.4*	0.014	*0.6*	0.030†	*0.8*	0·046	*1·0*	0·090	*1·4*		0·7
0·8						0.009	*0.5*	0.015	*0.6*	0·032	*0·8*	0·050	*1·0*	0·097	*1·4*	0·05	0·8
0·9						0.010	*0.5*	0.016	*0.6*	0·035	*0·8*	0·054	*1·0*	0·104	*1·5*		0·9
1·0				0.003	*0.2*	0.011*	*0.6*	0.017	*0.6*	0·037	*0·8*	0·057	*1·0*	0·110	*1·5*		1·0
1·5				0.005	*0.2*	0.012	*0.4*	0.021†	*0.6*	0·047	*0·9*	0·072	*1·1*	0·139	*1·6*		1·5
2·0				0.006	*0.3*	0.014	*0.4*	0·025	*0·6*	0·055	*0·9*	0·085	*1·1*	0·164	*1·6*		2·0
2·5				0.008	*0.4*	0.015	*0.4*	0·028	*0·6*	0·062	*0·9*	0·096	*1·2*	0·186	*1·7*		2·5
3·0				0.009*	*0.5*	0.017	*0.4*	0·031	*0·6*	0·069	*1·0*	0·107	*1·2*	0·206	*1·7*		3·0
3·5				0.008	*0.3*	0.018†	*0.4*	0·034	*0·6*	0·076	*1·0*	0·116	*1·2*	0·224	*1·7*		3·5
4·0		0.004	*0.2*	0.009	*0.3*	0.020	*0.4*	0·037	*0·7*	0·082	*1·0*	0·126	*1·3*	0·242	*1·8*		4·0
4·5	0.05	0.005	*0.2*	0.009	*0.3*	0·021	*0·5*	0·039	*0·7*	0·087	*1·0*	0·134	*1·3*	0·258	*1·8*		4·5
5·0		0.005	*0.3*	0.010	*0.3*	0·022	*0·5*	0·042	*0·7*	0·093	*1·0*	0·142	*1·3*	0·274	*1·8*		5·0
5·5		0.006	*0.3*	0.010	*0.3*	0·023	*0·5*	0·044	*0·7*	0·098	*1·0*	0·150	*1·3*	0·289	*1·8*	0·15	5·5
6·0		0.006	*0.3*	0.011	*0.3*	0·025	*0·5*	0·046	*0·7*	0·103	*1·1*	0·158	*1·3*	0·303	*1·9*		6·0
6·5		0.007*	*0.4*	0.011	*0.3*	0·026	*0·5*	0·048	*0·7*	0·107	*1·1*	0·165	*1·3*	0·317	*1·9*		6·5
7·0		0.006	*0.2*	0.012	*0.3*	0·027	*0·6*	0·050	*0·7*	0·112	*1·1*	0·172	*1·3*	0·330	*1·9*		7·0
7·5		0.006	*0.2*	0.012	*0.3*	0·028	*0·5*	0·052	*0·7*	0·116	*1·1*	0·179	*1·4*	0·343	*1·9*		7·5
8·0		0.006	*0.2*	0.012	*0.3*	0·029	*0·5*	0·054	*0·7*	0·120	*1·1*	0·185	*1·4*	0·355	*1·9*		8·0
8·5		0.006	*0.2*	0.013	*0.3*	0·030	*0·5*	0·056	*0·7*	0·125	*1·1*	0·191	*1·4*	0·368	*1·9*		8·5
9·0		0.007	*0.2*	0.013†	*0.3*	0·031	*0·5*	0·058	*0·7*	0·129	*1·1*	0·198	*1·4*	0·379	*1·9*		9·0
9·5		0.007	*0.2*	0·014	*0·3*	0·032	*0·5*	0·060	*0·7*	0·133	*1·1*	0·204	*1·4*	0·391	*1·9*		9·5
10·0		0.007	*0.2*	0·014	*0·3*	0·033	*0·5*	0·062	*0·7*	0·136	*1·1*	0·210	*1·4*	0·402	*2·0*		10·0
12·5		0.008	*0.2*	0·016	*0·3*	0·037	*0·5*	0·070	*0·8*	0·154	*1·1*	0·237	*1·4*	0·454	*2·0*		12·5
15·0		0.008	*0.2*	0·018	*0·4*	0·042	*0·6*	0·077	*0·8*	0·171	*1·2*	0·262	*1·5*	0·502	*2·0*		15·0
17·5		0.009	*0.2*	0·019	*0·4*	0·045	*0·6*	0·084	*0·8*	0·186	*1·2*	0·285	*1·5*	0·546	*2·0*		17·5
20·0		0.010	*0.2*	0·021	*0·4*	0·049	*0·6*	0·091	*0·8*	0·200	*1·2*	0·307	*1·5*	0·587	*2·1*	0·30	20·0
22·5		0.010†	*0.2*	0·022	*0·4*	0·052	*0·6*	0·097	*0·8*	0·214	*1·2*	0·327	*1·5*	0·626	*2·1*		22·5
25·0		0.011	*0.3*	0·023	*0·4*	0·055	*0·6*	0·103	*0·8*	0·226	*1·2*	0·347	*1·5*	0·663	*2·1*		25·0
27·5		0.012	*0.3*	0·025	*0·4*	0·058	*0·6*	0·108	*0·8*	0·238	*1·2*	0·365	*1·5*	0·698	*2·1*		27·5
30·0		0.012	*0.3*	0·026	*0·4*	0·061	*0·6*	0·114	*0·8*	0·250	*1·2*	0·383	*1·6*	0·731	2·2		30·0
32·5		0·013	*0·3*	0·027	*0·4*	0·064	*0·6*	0·119	*0·8*	0·261	*1·3*	0·400	*1·6*	0·763	2·2		32·5
35·0		0·013	*0·3*	0·028	*0·4*	0·067	*0·6*	0·124	*0·8*	0·272	*1·3*	0·416	*1·6*	0·794	2·2		35·0
37·5		0·014	*0·3*	0·029	*0·4*	0·069	*0·6*	0·129	*0·9*	0·282	*1·3*	0·432	*1·6*	0·824	2·2		37·5
40·0		0·014	*0·3*	0·031	*0·4*	0·072	*0·6*	0·133	*0·9*	0·292	*1·3*	0·447	*1·6*	0·853	2·2		40·0
42·5	0·15	0·015	*0·3*	0·032	*0·4*	0·074	*0·6*	0·138	*0·9*	0·302	*1·3*	0·462	*1·6*	0·882	2·2		42·5
45·0		0·015	*0·3*	0·033	*0·4*	0·077	*0·6*	0·142	*0·9*	0·312	*1·3*	0·477	*1·6*	0·909	2·2		45·0
47·5		0·016	*0·3*	0·034	*0·4*	0·079	*0·6*	0·146	*0·9*	0·321	*1·3*	0·491	*1·6*	0·936	2·2		47·5
50·0		0·016	*0·3*	0·035	*0·4*	0·081	*0·6*	0·150	*0·9*	0·330	*1·3*	0·504	*1·6*	0·962	2·2		50·0
52·5		0·017	*0·3*	0·036	*0·4*	0·083	*0·6*	0·155	*0·9*	0·339	*1·3*	0·518	*1·6*	0·987	2·2	0·50	52·5
55·0		0·017	*0·3*	0·036	*0·4*	0·085	*0·6*	0·159	*0·9*	0·347	*1·3*	0·531	*1·6*	1·01	2·2		55·0
57·5		0·018	*0·3*	0·037	*0·4*	0·088	*0·6*	0·162	*0·9*	0·356	*1·3*	0·544	*1·6*	1·04	2·3		57·5
60·0		0·018	*0·3*	0·038	*0·4*	0·090	*0·6*	0·166	*0·9*	0·364	*1·3*	0·556	*1·6*	1·06	2·3		60·0
62·5		0·018	*0·3*	0·039	*0·4*	0·092	*0·7*	0·170	*0·9*	0·372	*1·3*	0·569	*1·6*	1·08	2·3		62·5
65·0		0·019	*0·3*	0·040	*0·4*	0·094	*0·7*	0·174	*0·9*	0·380	*1·3*	0·581	*1·6*	1·11	2·3		65·0
67·5		0·019	*0·3*	0·041	*0·4*	0·096	*0·7*	0·177	*0·9*	0·388	*1·3*	0·592	*1·7*	1·13	2·3		67·5
70·0		0·020	*0·3*	0·042	*0·4*	0·098	*0·7*	0·181	*0·9*	0·395	*1·3*	0·604	*1·7*	1·15	2·3		70·0
72·5		0·020	*0·3*	0·042	*0·4*	0·099	*0·7*	0·184	*0·9*	0·403	*1·3*	0·616	*1·7*	1·17	2·3		72·5
75·0		0·020	*0·3*	0·043	*0·4*	0·101	*0·7*	0·188	*0·9*	0·410	*1·3*	0·627	*1·7*	1·19	2·3		75·0
77·5		0·021	*0·3*	0·044	*0·4*	0·103	*0·7*	0·191	*0·9*	0·418	*1·3*	0·638	*1·7*	1·21	2·3		77·5

Table 13.10 (*continued*)

Δp_l	v	10 mm		15 mm		20 mm		25 mm		32 mm		40 mm		50 mm		v	Δp_l
		M	l_e	M	l_e	M	l_e	M	l_e	M	l_e	M	l_e	M	l_e		
80·0		0·021	*0·3*	0·045	*0·4*	0·105	*0·7*	0·194	*0·9*	0·425	*1·4*	0·649	*1·7*	1·24	*2·3*		80·0
82·5		0·021	*0·3*	0·046	*0·4*	0·107	*0·7*	0·197	*0·9*	0·432	*1·4*	0·659	*1·7*	1·26	*2·3*		82·5
85·0		0·022	*0·3*	0·046	*0·4*	0·108	*0·7*	0·201	*0·9*	0·439	*1·4*	0·670	*1·7*	1·28	*2·3*		85·0
87·5		0·022	*0·3*	0·047	*0·4*	0·110	*0·7*	0·204	*0·9*	0·446	*1·4*	0·680	*1·7*	1·30	*2·3*		87·5
90·0		0·023	*0·3*	0·048	*0·4*	0·112	*0·7*	0·207	*0·9*	0·452	*1·4*	0·691	*1·7*	1·31	*2·3*		90·0
92·5		0·023	*0·3*	0·049	*0·4*	0·113	*0·7*	0·210	*0·9*	0·459	*1·4*	0·701	*1·7*	1·33	*2·3*		92·5
95·0		0·023	*0·3*	0·049	*0·4*	0·115	*0·7*	0·213	*0·9*	0·466	*1·4*	0·711	*1·7*	1·35	*2·3*		95·0
97·5		0·024	*0·3*	0·050	*0·4*	0·117	*0·7*	0·216	*0·9*	0·472	*1·4*	0·721	*1·7*	1·37	*2·3*		97·5
100		0·024	*0·3*	0·051	*0·4*	0·118	*0·7*	0·219	*0·9*	0·479	*1·4*	0·731	*1·7*	1·39	*2·3*		100
120		0·026	*0·3*	0·056	*0·4*	0·131	*0·7*	0·242	*0·9*	0·527	*1·4*	0·805	*1·7*	1·53	*2·4*		120
140	0·3	0·029	*0·3*	0·061	*0·5*	0·142	*0·7*	0·262	*0·9*	0·572	*1·4*	0·873	*1·7*	1·66	*2·4*		140
160		0·031	*0·3*	0·065	*0·5*	0·152	*0·7*	0·282	*1·0*	0·614	*1·4*	0·937	*1·7*	1·78	*2·4*		160
180		0·033	*0·3*	0·070	*0·5*	0·162	*0·7*	0·300	*1·0*	0·654	*1·4*	0·997	*1·8*	1·89	*2·4*		180
200		0·035	*0·3*	0·074	*0·5*	0·172	*0·7*	0·317	*1·0*	0·691	*1·4*	1·05	*1·8*	2·00	*2·4*	1·0	200
220		0·037	*0·3*	0·078	*0·5*	0·181	*0·7*	0·334	*1·0*	0·727	*1·4*	1·11	*1·8*	2·10	*2·4*		220
240		0·039	*0·3*	0·081	*0·5*	0·189	*0·7*	0·349	*1·0*	0·761	*1·4*	1·16	*1·8*	2·20	*2·4*		240
260		0·040	*0·3*	0·085	*0·5*	0·198	*0·7*	0·364	*1·0*	0·793	*1·5*	1·21	*1·8*	2·29	*2·4*		260
280		0·042	*0·3*	0·088	*0·5*	0·206	*0·7*	0·379	*1·0*	0·825	*1·5*	1·26	*1·8*	2·38	*2·4*		280
300		0·044	*0·3*	0·092	*0·5*	0·213	*0·7*	0·393	*1·0*	0·855	*1·5*	1·30	*1·8*	2·47	*2·5*		300
320		0·045	*0·3*	0·095	*0·5*	0·221	*0·7*	0·407	*1·0*	0·884	*1·5*	1·35	*1·8*	2·55	*2·5*		320
340		0·047	*0·3*	0·098	*0·5*	0·228	*0·7*	0·420	*1·0*	0·913	*1·5*	1·39	*1·8*	2·64	*2·5*		340
360		0·048	*0·3*	0·101	*0·5*	0·235	*0·7*	0·433	*1·0*	0·941	*1·5*	1·43	*1·8*	2·71	*2·5*		360
380	0·5	0·049	*0·3*	0·104	*0·5*	0·242	*0·7*	0·445	*1·0*	0·970	*1·5*	1·47	*1·8*	2·79	*2·5*		380
400		0·051	*0·3*	0·107	*0·5*	0·248	*0·7*	0·457	*1·0*	0·994	*1·5*	1·51	*1·8*	2·87	*2·5*		400
420		0·052	*0·3*	0·110	*0·5*	0·255	*0·7*	0·469	*1·0*	1·02	*1·5*	1·55	*1·8*	2·94	*2·5*		420
440		0·054	*0·3*	0·113	*0·5*	0·261	*0·7*	0·481	*1·0*	1·04	*1·5*	1·59	*1·8*	3·01	*2·5*	1·5	440
460		0·055	*0·3*	0·115	*0·5*	0·267	*0·7*	0·492	*1·0*	1·07	*1·5*	1·63	*1·8*	3·08	*2·5*		460
480		0·056	*0·3*	0·118	*0·5*	0·273	*0·8*	0·503	*1·0*	1·09	*1·5*	1·66	*1·8*	3·15	*2·5*		480
500		0·057	*0·3*	0·120	*0·5*	0·279	*0·8*	0·514	*1·0*	1·12	*1·5*	1·69	*1·8*	3·22	*2·5*		500
520		0·059	*0·3*	0·123	*0·5*	0·285	*0·8*	0·524	*1·0*	1·14	*1·5*	1·73	*1·8*	3·28	*2·5*		520
540		0·060	*0·3*	0·125	*0·5*	0·291	*0·8*	0·535	*1·0*	1·16	*1·5*	1·77	*1·8*	3·35	*2·5*		540
560		0·061	*0·3*	0·128	*0·5*	0·296	*0·8*	0·545	*1·0*	1·17	*1·5*	1·80	*1·8*	3·41	*2·5*		560
580		0·062	*0·3*	0·130	*0·5*	0·302	*0·8*	0·555	*1·0*	1·21	*1·5*	1·83	*1·8*	3·47	*2·5*		580
600		0·063	*0·3*	0·133	*0·5*	0·307	*0·8*	0·565	*1·0*	1·23	*1·5*	1·87	*1·8*	3·53	*2·5*		600
620		0·064	*0·3*	0·135	*0·5*	0·312	*0·8*	0·575	*1·0*	1·25	*1·5*	1·90	*1·8*	3·59	*2·5*		620
640		0·065	*0·3*	0·137	*0·5*	0·318	*0·8*	0·584	*1·0*	1·27	*1·5*	1·93	*1·8*	3·65	*2·5*		640
660		0·066	*0·3*	0·139	*0·5*	0·323	*0·8*	0·594	*1·0*	1·29	*1·5*	1·96	*1·8*	3·71	*2·5*		660
680		0·067	*0·3*	0·142	*0·5*	0·328	*0·8*	0·603	*1·0*	1·31	*1·5*	1·99	*1·9*	3·77	*2·5*		680
700		0·069	*0·3*	0·144	*0·5*	0·333	*0·8*	0·612	*1·0*	1·33	*1·5*	2·02	*1·9*	3·83	*2·5*		700
720		0·070	*0·3*	0·146	*0·5*	0·338	*0·8*	0·621	*1·0*	1·35	*1·5*	2·05	*1·9*	3·88	*2·5*		720
740		0·071	*0·3*	0·148	*0·5*	0·343	*0·8*	0·630	*1·0*	1·37	*1·5*	2·08	*1·9*	3·94	*2·5*		740
760		0·072	*0·3*	0·150	*0·5*	0·347	*0·8*	0·639	*1·0*	1·39	*1·5*	2·10	*1·9*	3·99	*2·5*	2·0	760
780		0·073	*0·3*	0·152	*0·5*	0·352	*0·8*	0·648	*1·0*	1·41	*1·5*	2·14	*1·9*	4·04	*2·5*		780
800		0·074	*0·3*	0·154	*0·5*	0·357	*0·8*	0·656	*1·0*	1·42	*1·5*	2·17	*1·9*	4·10	*2·5*		800
820		0·075	*0·4*	0·156	*0·5*	0·362	*0·8*	0·665	*1·0*	1·44	*1·5*	2·19	*1·9*	4·15	*2·5*		820
840		0·075	*0·4*	0·158	*0·5*	0·366	*0·8*	0·673	*1·0*	1·46	*1·5*	2·22	*1·9*	4·20	*2·5*		840
860		0·076	*0·4*	0·160	*0·5*	0·371	*0·8*	0·681	*1·0*	1·48	*1·5*	2·25	*1·9*	4·25	*2·5*		860
880		0·077	*0·4*	0·162	*0·5*	0·375	*0·8*	0·689	*1·0*	1·50	*1·5*	2·27	*1·9*	4·30	*2·5*		880
900		0·078	*0·4*	0·164	*0·5*	0·379	*0·8*	0·698	*1·0*	1·51	*1·5*	2·30	*1·9*	4·35	*2·5*		900
920		0·079	*0·4*	0·166	*0·5*	0·384	*0·8*	0·706	*1·0*	1·53	*1·5*	2·33	*1·9*	4·40	*2·5*		920
940		0·080	*0·4*	0·168	*0·5*	0·388	*0·8*	0·713	*1·0*	1·55	*1·5*	2·35	*1·9*	4·45	*2·5*		940
960		0·081	*0·4*	0·170	*0·5*	0·392	*0·8*	0·721	*1·0*	1·56	*1·5*	2·38	*1·9*	4·50	*2·5*		960
980		0·082	*0·4*	0·172	*0·5*	0·397	*0·8*	0·729	*1·0*	1·58	*1·5*	2·40	*1·9*	4·55	*2·5*		980
1 000		0·083	*0·4*	0·173	*0·5*	0·401	*0·8*	0·737	*1·0*	1·60	*1·5*	2·43	*1·9*	4·59	*2·5*		1 000
1 100		0·087	*0·4*	0·182	*0·5*	0·421	*0·8*	0·774	*1·1*	1·68	*1·5*	2·55	*1·9*	4·82	*2·6*		1 100
1 200		0·091	*0·4*	0·191	*0·5*	0·441	*0·8*	0·809	*1·1*	1·75	*1·5*	2·67	*1·9*	5·04	*2·6*		1 200
1 300	1·0	0·095	*0·4*	0·199	*0·5*	0·459	*0·8*	0·844	*1·1*	1·83	*1·5*	2·78	*1·9*	5·25	*2·6*		1 300
1 400		0·099	*0·4*	0·207	*0·5*	0·477	*0·8*	0·876	*1·1*	1·90	*1·5*	2·89	*1·9*	5·46	*2·6*		1 400
1 500		0·102	*0·4*	0·214	*0·5*	0·495	*0·8*	0·908	*1·1*	1·98	*1·5*	2·99	*1·9*	5·65	*2·6*		1 500
1 600		0·106	*0·4*	0·222	*0·5*	0·511	*0·8*	0·939	*1·1*	2·03	*1·5*	3·09	*1·9*	5·84	*2·6*		1 600
1 700		0·109	*0·4*	0·229	*0·5*	0·528	*0·8*	0·968	*1·1*	2·10	*1·5*	3·19	*1·9*	6·02	*2·6*	3·0	1 700
1 800		0·113	*0·4*	0·236	*0·5*	0·543	*0·8*	0·997	*1·1*	2·16	*1·6*	3·28	*1·9*				1 800
1 900		0·116	*0·4*	0·242	*0·5*	0·559	*0·8*	1·03	*1·1*	2·22	*1·6*	3·37	*1·9*				1 900
2 000		0·119	*0·4*	0·249	*0·5*	0·574	*0·8*	1·05	*1·1*	2·28	*1·6*	3·46	*1·9*				2 000

Table 13.10 *(continued)*

Δp_l	ν	65 mm		80 mm		90 mm		100 mm		125 mm		150 mm		ν	Δp_l
		M	l_e	*M*	l_e	*M*	l_e	*M*	l_e	*M*	l_e	*M*	l_e		
0·1		0·061	*1·5*	0·096	*2·0*	0.144	*2·5*	0·200	*2·9*	0·362	*4·1*	0·600	*5·3*		0·1
0·2		0·091	*1·7*	0·144	*2·2*	0·215	*2·7*	0·298	*3·3*	0·544	*4·5*	0·889	*5·8*	0·05	0·2
0·3		0·115	*1·8*	0·181	*2·3*	0·271	*2·9*	0·375	*3·4*	0·685	*4·7*	1·12	*6·1*		0·3
0·4		0·136	*1·9*	0·214	*2·4*	0·319	*3·0*	0·442	*3·6*	0·805	*4·9*	1·31	*6·4*		0·4
0·5		0·154	*2·0*	0·243	*2·5*	0·362	*3·1*	0·501	*3·7*	0·913	*5·0*	1·49	*6·5*		0·5
0·6	0·05	0·171	*2·0*	0·269	*2·6*	0·401	*3·2*	0·556	*3·8*	1·01	*5·1*	1·65	*6·6*		0·6
0·7		0·187	*2·1*	0·294	*2·6*	0·438	*3·2*	0·606	*3·8*	1·10	*5·2*	1·79	*6·7*		0·7
0·8		0·202	*2·1*	0·317	*2·7*	0·472	*3·3*	0·653	*3·9*	1·19	*5·3*	1·93	*6·8*		0·8
0·9		0·216	*2·1*	0·339	*2·7*	0·504	*3·3*	0·698	*4·0*	1·27	*5·4*	2·06	*6·9*		0·9
1·0		0·229	*2·2*	0·359	*2·8*	0·535	*3·4*	0·740	*4·0*	1·34	*5·5*	2·18	*7·0*		1·0
1·5		0·288	*2·3*	0·451	*2·9*	0·671	*3·6*	0·928	*4·2*	1·68	*5·7*	2·73	*7·3*	0·15	1·5
2·0		0·338	*2·4*	0·530	*3·0*	0·787	*3·7*	1·09	*4·3*	1·97	*5·9*	3·20	*7·5*		2·0
2·5		0·383	*2·4*	0·600	*3·1*	0·891	*3·8*	1·23	*4·4*	2·23	*6·0*	3·61	*7·6*		2·5
3·0		0·424	*2·5*	0·664	*3·1*	0·985	*3·8*	1·36	*4·5*	2·46	*6·1*	3·99	*7·7*		3·0
3·5		0·462	*2·5*	0·723	*3·2*	1·07	*3·9*	1·48	*4·6*	2·68	*6·2*	4·34	*7·9*		3·5
4·0	0·15	0·498	*2·6*	0·778	*3·2*	1·15	*3·9*	1·59	*4·6*	2·88	*6·3*	4·66	*8·0*		4·0
4·5		0·531	*2·6*	0·830	*3·3*	1·23	*4·0*	1·70	*4·7*	3·07	*6·3*	4·97	*8·0*		4·5
5·0		0·563	*2·6*	0·880	*3·3*	1·30	*4·0*	1·80	*4·7*	3·25	*6·4*	5·26	*8.1*	0·30	5·0
5·5		0·594	*2·7*	0·927	*3·3*	1·37	*4·1*	1·90	*4·8*	3·42	*6·4*	5·54	*8·2*		5·5
6·0		0·623	*2·7*	0·973	*3·4*	1·44	*4·1*	1·99	*4·8*	3·59	*6·5*	5·81	*8·2*		6·0
6·5		0·651	*2·7*	1·02	*3·4*	1·51	*4·1*	2·08	*4·9*	3·75	*6·5*	6·06	*8·3*		6·5
7·0		0·678	*2·7*	1·06	*3·4*	1·57	*4·2*	2·16	*4·9*	3·90	*6·6*	6·31	*8·3*		7·0
7·5		0·704	*2·7*	1·10	*3·4*	1·63	*4·2*	2·24	*4·9*	4·05	*6·6*	6·55	*8·4*		7·5
8·0		0·729	*2·7*	1·14	*3·5*	1·69	*4·2*	2·32	*4·9*	4·19	*6·6*	6·78	*8·4*		8·0
8·5		0·754	*2·8*	1·18	*3·5*	1·74	*4·2*	2·40	*5·0*	4·33	*6·7*	7·00	*8·4*		8·5
9·0		0·778	*2·8*	1·21	*3·5*	1·80	*4·2*	2·48	*5·0*	4·46	*6·7*	7·22	*8·5*		9·0
9·5		0·801	*2·8*	1·25	*3·5*	1·85	*4·3*	2·55	*5·0*	4·60	*6·7*	7·43	*8·5*		9·5
10·0		0·824	*2·8*	1·29	*3·5*	1·90	*4·3*	2·62	*5·0*	4·72	*6·7*	7·63	*8·5*		10·0
12·5		0·930	*2·9*	1·45	*3·6*	2·14	*4·4*	2·96	*5·1*	5·32	*6·8*	8·60	*8·7*	0·50	12·5
15·0	0·30	1·03	*2·9*	1·60	*3·6*	2·37	*4·4*	3·26	*5·2*	5·87	*6·9*	9·47	*8·8*		15·0
17·5		1·12	*3·0*	1·74	*3·7*	2·57	*4·5*	3·54	*5·2*	6·37	*7·0*	10·3	*8·8*		17·5
20·0		1·20	*3·0*	1·87	*3·7*	2·76	*4·5*	3·80	*5·3*	6·84	*7·1*	11·0	*8·9*		20·0
22·5		1·28	*3·0*	1·99	*3·8*	2·94	*4·6*	4·05	*5·3*	7·28	*7·1*	11·7	*9·0*		22·5
25·0		1·35	*3·0*	2·11	*3·8*	3·11	*4·6*	4·28	*5·4*	7·69	*7·1*	12·4	*9·0*		25·0
27·5		1·42	*3·1*	2·22	*3·8*	3·27	*4·6*	4·50	*5·4*	8·09	*7·2*	13·0	*9·1*		27·5
30·0		1·49	*3·1*	2·32	*3·8*	3·43	*4·6*	4·71	*5·4*	8·47	*7·2*	13·6	*9·1*		30·0
32·5		1·56	*3·1*	2·42	*3·8*	3·58	*4·7*	4·92	*5·4*	8·84	*7·3*	14·2	*9·1*		32·5
35·0		1·62	*3·1*	2·52	*3·9*	3·72	*4·7*	5·12	*5·5*	9·19	*7·3*	14·8	*9·2*		35·0
37·5	0·50	1·68	*3·1*	2·61	*3·9*	3·86	*4·7*	5·31	*5·5*	9·53	*7·3*	15·3	*9·2*		37·5
40·0		1·74	*3·1*	2·70	*3·9*	3·99	*4·7*	5·49	*5·5*	9·86	*7·3*	15·9	*9·2*		40·0
42·5		1·80	*3·1*	2·79	*3·9*	4·12	*4·7*	5·67	*5·5*	10·2	*7·4*	16·4	*9·3*		42·5
45·0		1·85	*3·2*	2·88	*3·9*	4·25	*4·7*	5·84	*5·5*	10·5	*7·4*	16·9	*9·3*		45·0
47·5		1·91	*3·2*	2·96	*3·9*	4·37	*4·8*	6·01	*5·6*	10·8	*7·4*	17·4	*9·3*		47·5
50·0		1·96	*3·2*	3·04	*3·9*	4·49	*4·8*	6·17	*5·6*	11·1	*7·4*	17·8	*9·3*	1·0	50·0
52·5		2·01	*3·2*	3·12	*4·0*	4·61	*4·8*	6·33	*5·6*	11·4	*7·4*	18·3	*9·3*		52·5
55·0		2·06	*3·2*	3·20	*4·0*	4·72	*4·8*	6·49	*5·6*	11·6	*7·4*	18·8	*9·4*		55·0
57·5		2·11	*3·2*	3·28	*4·0*	4·83	*4·8*	6·64	*5·6*	11·9	*7·5*	19·2	*9·4*		57·5
60·0		2·16	*3·2*	3·35	*4·0*	4·94	*4·8*	6·79	*5·6*	12·2	*7·5*	19·6	*9·4*		60·0
62·5		2·20	*3·2*	3·42	*4·0*	5·05	*4·8*	6·94	*5·6*	12·5	*7·5*	20·0	*9·4*		62·5
65·0		2·25	*3·2*	3·50	*4·0*	5·16	*4·8*	7·08	*5·7*	12·7	*7·5*	20·5	*9·4*		65·0
67·5		2·30	*3·2*	3·57	*4·0*	5·26	*4·9*	7·22	*5·7*	13·0	*7·5*	20·9	*9·4*		67·5
70·0		2·34	*3·2*	3·63	*4·0*	5·36	*4·9*	7·36	*5·7*	13·2	*7·5*	21·3	*9·4*		70·0
72·5		2·38	*3·2*	3·70	*4·0*	5·46	*4·9*	7·50	*5·7*	13·5	*7·5*	21·7	*9·5*		72·5
75·0		2·43	*3·3*	3·77	*4·0*	5·56	*4·9*	7·63	*5·7*	13·7	*7·5*	22·0	*9·5*		75·0
77·5		2·47	*3·3*	3·83	*4·0*	5·65	*4·9*	7·77	*5·7*	13·9	*7·5*	22·4	*9·5*		77·5

Table 13.10 *(continued)*

Δp_l	v	65 mm		80 mm		90 mm		100 mm		125 mm		150 mm		v	Δp_l
		M	l_e	M	l_e	M	l_e	M	l_e	M	l_e	M	l_e		
80·0		2·51	*3·3*	3·90	*4·0*	5·75	*4·9*	7·90	*5·7*	14·2	*7·6*	22·8	*9·5*		80·0
82·5		2·55	*3·3*	3·96	*4·1*	5·84	*4·9*	8·02	*5·7*	14·4	*7·6*	23·2	*9·5*		82·5
85·0		2·59	*3·3*	4·02	*4·1*	5·93	*4·9*	8·15	*5·7*	14·6	*7·6*	23·5	*9·5*		85·0
87·5		2·63	*3·3*	4·09	*4·1*	6·02	*4·9*	8·27	*5·7*	14·8	*7·6*	23·9	*9·5*		87·5
90·0		2·67	*3·3*	4·15	*4·1*	6·11	*4·9*	8·40	*5·7*	15·0	*7·6*	24·2	*9·5*		90·0
92·5		2·71	*3·3*	4·21	*4·1*	6·20	*4·9*	8·52	*5·7*	15·3	*7·6*	24·6	*9·5*		92·5
95·0		2·75	*3·3*	4·27	*4·1*	6·29	*4·9*	8·64	*5·7*	15·5	*7·6*	24·9	*9·6*		95·0
97·5		2·79	*3·3*	4·32	*4·1*	6·37	*4·9*	8·75	*5·8*	15·7	*7·6*	25·2	*9·6*		97·5
100·0		2·82	*3·3*	4·38	*4·1*	6·46	*4·9*	8·87	*5·8*	15·9	*7·6*	25·6	*9·6*	1·5	100·0
120·0		3·11	*3·3*	4·82	*4·1*	7·10	*5·0*	9·75	*5·8*	17·5	*7·7*	28·1	*9·6*		120·0
140·0	1·0	3·37	*3·4*	5·22	*4·2*	7·69	*5·0*	10·6	*5·8*	18·9	*7·7*	30·4	*9·7*		140·0
160·0		3·61	*3·4*	5·60	*4·2*	8·25	*5·0*	11·3	*5·9*	20·3	*7·7*	32·6	*9·7*		160·0
180·0		3·84	*3·4*	5·95	*4·2*	8·76	*5·0*	12·0	*5·9*	21·6	*7·8*	34·6	*9·7*	2·0	180·0
200·0		4·05	*3·4*	6·29	*4·2*	9·25	*5·0*	12·7	*5·9*	22·7	*7·8*	36·5	*9·8*		200·0
220·0		4·26	*3·4*	6·60	*4·2*	9·72	*5·1*	13·3	*5·9*	23·9	*7·8*	38·4	*9·8*		220·0
240·0		4·46	*3·4*	6·91	*4·2*	10·2	*5·1*	14·0	*5·9*	25·0	*7·8*	40·1	*9·8*		240·0
260·0		4·65	*3·4*	7·20	*4·2*	10·6	*5·1*	14·5	*6·0*	26·0	*7·9*	41·8	*9·8*		260·0
280·0		4·83	*3·4*	7·48	*4·3*	11·0	*5·1*	15·1	*6·0*	27·0	*7·9*	43·4	*9·9*		280·0
300·0	1·5	5·00	*3·5*	7·75	*4·3*	11·4	*5·1*	15·6	*6·0*	28·0	*7·9*	45·0	*9·9*		300·0
320·0		5·17	*3·5*	8·01	*4·3*	11·8	*5·1*	16·2	*6·0*	29·0	*7·9*	46·5	*9·9*		320·0
340·0		5·34	*3·5*	8·27	*4·3*	12·2	*5·2*	16·7	*6·0*	29·8	*7·9*	47·9	*9·9*		340·0
360·0		5·50	*3·5*	8·51	*4·3*	12·5	*5·2*	17·2	*6·0*	30·7	*7·9*	49·4	*9·9*		360·0
380·0		5·65	*3·5*	8·75	*4·3*	12·8	*5·2*	17·7	*6·0*	31·6	*7·9*	50·7	*9·9*		380·0
400·0		5·80	*3·5*	8·99	*4·3*	13·2	*5·2*	18·1	*6·0*	32·4	*7·9*	52·1	*9·9*		400·0
420·0		5·95	*3·5*	9·22	*4·3*	13·6	*5·2*	18·6	*6·0*	33·2	*7·9*	53·4	*9·9*		420·0
440·0		6·09	*3·5*	9·44	*4·3*	13·9	*5·2*	19·0	*6·0*	34·0	*7·9*	54·7	*9·9*	3·0	440·0
460·0		6·24	*3·5*	9·66	*4·3*	14·2	*5·2*	19·5	*6·0*	34·8	*8·0*	55·9	*9·9*		460·0
480·0		6·37	*3·5*	9·87	*4·3*	14·5	*5·2*	19·9	*6·0*	35·6	*8·0*	57·2	*10*		480·0
500·0		6·51	*3·5*	10·1	*4·3*	14·8	*5·2*	20·3	*6·0*	36·3	*8·0*	58·4	*10*		500·0
520·0		6·64	*3·5*	10·3	*4·3*	15·1	*5·2*	20·7	*6·1*	37·1	*8·0*	59·5	*10*		520·0
540·0	2·0	6·77	*3·5*	10·5	*4·3*	15·4	*5·2*	21·1	*6·1*	37·8	*8·0*	60·7	*10*		540·0
560·0		6·90	*3·5*	10·7	*4·3*	15·7	*5·2*	21·5	*6·1*	38·5	*8·0*	61·8	*10*		560·0
580·0		7·02	*3·5*	10·9	*4·3*	16·0	*5·2*	21·9	*6·1*	39·2	*8·0*	62·9	*10*		580·0
600·0		7·15	*3·5*	11·1	*4·3*	16·3	*5·2*	22·3	*6·1*	39·9	*8·0*	64·0	*10*		600·0
620·0		7·27	*3·5*	11·3	*4·4*	16·6	*5·2*	22·7	*6·1*	40·5	*8·0*	65·1	*10*		620·0
640·0		7·39	*3·5*	11·4	*4·4*	16·8	*5·2*	23·1	*6·1*	41·2	*8·0*	66·2	*10*		640·0
660·0		7·50	*3·5*	11·6	*4·4*	17·1	*5·2*	23·4	*6·1*	41·9	*8·0*	67·2	*10*		660·0
680·0		7·62	*3·5*	11·8	*4·4*	17·3	*5·2*	23·8	*6·1*	42·5	*8·0*	68·2	*10*		680·0
700·0		7·73	*3·5*	12·0	*4·4*	17·6	*5·2*	24·1	*6·1*	43·1	*8·0*	69·2	*10*		700·0
720·0		7·85	*3·5*	12·2	*4·4*	17·8	*5·2*	24·5	*6·1*	43·7	*8·0*	70·2	*10*		720·0
740·0		7·96	*3·5*	12·3	*4·4*	18·1	*5·2*	24·8	*6·1*	44·4	*8·0*	71·2	*10*		740·0
760·0		8·07	*3·5*	12·4	*4·4*	18·4	*5·3*	25·1	*6·1*	45·0	*8·0*	72·2	*10*	4·0	760·0
780·0		8·17	*3·5*	12·6	*4·4*	18·6	*5·3*	25·5	*6·1*	45·6	*8·0*				780·0
800·0		8·28	*3·6*	12·8	*4·4*	18·8	*5·3*	25·8	*6·1*	46·2	*8·0*				800·0
820·0		8·39	*3·6*	12·9	*4·4*	19·1	*5·3*	26·2	*6·1*	46·7	*8·0*				820·0
840·0		8·49	*3·6*	13·1	*4·4*	19·3	*5·3*	26·5	*6·1*	47·3	*8·0*				840·0
860·0		8·59	*3·6*	13·3	*4·4*	19·6	*5·3*	26·8	*6·1*	47·9	*8·0*				860·0
880·0		8·69	*3·6*	13·5	*4·4*	19·8	*5·3*	27·1	*6·1*	48·4	*8·0*				880·0
900·0		8·80	*3·6*	13·6	*4·4*	20·0	*5·3*	27·4	*6·1*	49·0	*8·0*				900·0
920·0		8·89	*3·6*	13·8	*4·4*	20·2	*5·3*	27·7	*6·1*	49·6	*8·1*				920·0
940·0		8·99	*3·6*	13·9	*4·4*	20·5	*5·3*	28·0	*6·1*	50·1	*8·1*				940·0
960·0		9·09	*3·6*	14·1	*4·4*	20·7	*5·3*	28·3	*6·1*	50·6	*8·1*				960·0
980·0		9·19	*3·6*	14·2	*4·4*	20·9	*5·3*	28·6	*6·1*						980·0
1 000·0		9·28	*3·6*	14·4	*4·4*	21·1	*5·3*	28·9	*6·1*						1 000·0
1 100·0		9·74	*3·6*	15·1	*4·4*	22·2	*5·3*	30·4	*6·1*						1 100·0
1 200·0	3·0	10·2	*3·6*	15·8	*4·4*	23·2	*5·3*	31·7	*6·1*						1 200·0
1 300·0		10·6	*3·6*	16·4	*4·4*	24·1	*5·3*								1 300·0
1 400·0		11·0	*3·6*	17·0	*4·4*	25·0	*5·3*								1 400·0
1 500·0		11·4	*3·6*	17·6	*4·4*										1 500·0
1 600·0		11·8	*3·6*	18·2	*4·4*										1 600·0
1 700·0		12·2	*3·6*	18·8	*4·4*										1 700·0
1 800·0		12·5	*3·6*												1 800·0
1 900·0		12·9	*3·6*												1 900·0
2 000·0		13·2	*3·6*												2 000·0

13.7.4 Condensate pumping

In nearly all steam-using plants condensate must be pumped back to the boilerhouse from the location where it is formed, and even in those cases where gravity drainage to the boilerhouse is practical, often the condensate must be lifted into a boiler feed tank. Where de-aerators are used they usually operate at a pressure of about 0.3 bar above atmospheric, and again a pump is needed to lift condensate from atmospheric pressure to de-aerator pressure.

Most pump units comprise a receiver tank which conventionally is vented to atmosphere and one or more motorized pumps. It is important with these units to make sure that the maximum condensate temperature as specified by the manufacturer is not exceeded, as well as that the pump has sufficient capacity to handle the load. Condensate temperature usually presents no problem with returns from low-pressure heating systems. There, the condensate is often below 100°C even as it leaves the traps, and a little further sub-cooling in the gravity return lines and in the pump receiver itself means that there is little difficulty in meeting the maximum temperature limitations.

On higher-pressure systems the gravity return lines often contain condensate at a little above 100°C, together with some flash steam. The cooling effect of the piping is limited to condensing a little of the flash steam, and the remainder passes through the vent at the pump receiver. The water must remain in the receiver for an appreciable time if it is to cool sufficiently, or sometimes the pump discharge may be throttled down to reduce the pump's capacity if cavitation is to be avoided.

The absolute pressure at the inlet to the pump is usually the atmospheric pressure in the receiver, plus the static head from the water surface to the pump inlet and minus the friction loss through the pipes, valves and fittings joining the pump to the receiver. If this absolute pressure exceeds the vapour pressure of water at the temperature at which it enters the pump, then a nett positive suction hand (NPSH) exists. If this NPSH is above the value specified by the pump manufacturer, the water does not begin to boil as it enters the pump suction and cavitation is avoided. If the water entering the pump is at a higher temperature, its vapour pressure is increased and a greater hydrostatic head over the pump suction is needed to ensure that the necessary NPSH is obtained.

However, since in most cases pumps are supplied coupled to receivers and the static head above the pump inlet is already fixed by the manufacturer, it is only necessary to ensure that the pump set has sufficient capacity at the water temperature expected at the pump. Pump manufacturers usually have a set of capacity curves for the pump when handling water at different temperatures, and these should be consulted.

Where steam systems operate at higher pressures than those used in low-pressure space heating systems, as in process work, condensate temperatures can be 100°C, or more when positive pressures exist in the return lines. Electric pumps are then used only if their capacity is downrated by partial closure of a valve at the pump outlet; by using a receiver mounted well above the pump to ensure sufficient NPSH; or by sub-cooling the condensate through a heat exchanger of some type.

All these difficulties are avoided by the use of steam-powered alternating receiver pumps of the type illustrated in Figure 13.16. These pumps are essentially alternating receivers which can be pressurized using steam (or air or other gas). The gas pressure displaces the condensate (which can be at any temperature up to and including boiling point) through a check valve at the outlet from the pump body. At the end of the discharge stroke, the internal mechanism changes over to close the pressurizing inlet valve and open the exhaust valve. The pressurizing gas is then vented to atmosphere, or to the space from which the condensate is being drained. When the pressures are equalized, condensate can flow by gravity into the pump body to refill it and complete the cycle. As the pump fills only by gravity, there can be no cavitation and pumps of this type readily handle boiling water or other liquids, compatible with the materials of construction.

Most often, the pump uses steam as the operating medium. This steam is exhausted at the end of the discharge stroke to the same pressure as the space from which the condensate is being drained. It is often possible where the larger condensate loads are being handled to dedicate a single pump to each load. The pump exhaust line can then be directly connected to the steam space of the heat exchanger, with condensate draining freely to the pump inlet and with any steam trap at the pump outlet.

During the discharge stroke the inlet check valve is closed. Condensate draining from the steam space then fills the inlet piping. Unless the piping is sufficiently large or contains a receiver section (reservoir), condensate could back up into the steam space being drained.

A similar reservoir is desirable where multiple loads discharge through individual steam traps to a common pump. Provision should then be made for venting incondensables and flash steam which reach the reservoir. The exhaust line from the pump can then be connected to the same vent line.

13.7.5 Allowance for expansion

All pipes will be installed at ambient temperature. Pipes used to carry hot fluids, whether water, oil or steam, operate at higher temperatures. It follows that they expand (especially in length) with the increase from ambient to working temperatures. The amount of the expansion is readily calculated or read from charts, and Table 13.11 may be helpful.

The piping must be sufficiently flexible to accommodate the movements of the components as it heats up. In many cases the piping has enough natural flexibility, by virtue of having reasonable lengths and plenty of bends, that no undue stresses are set up. In other installations it is necessary to build in some means of achieving the required flexibility.

Where the condensate from a steam main drain trap is discharged into a return line running alongside the steam line the difference between the expansion of the

Figure 13.15 Condensate-recovery unit

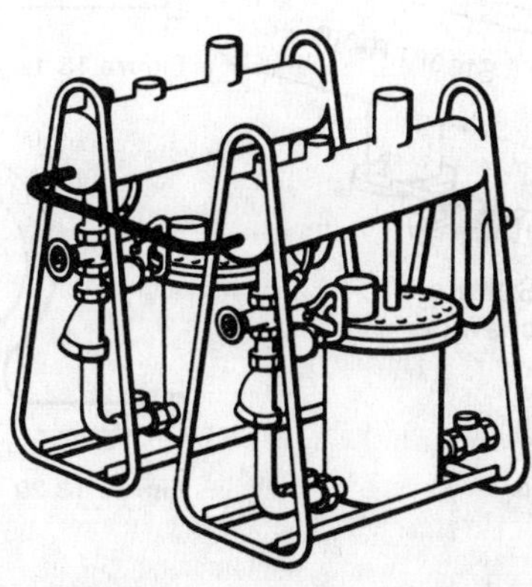

Figure 13.16 Spirax Ogden packaged unit

two lines must be remembered. The steam line may be at a temperature very much above that of the return line, and the two connection points can move in relation to each other during system warm-up. Some flexibility should be incorporated into the steam trap piping so that branch connections do not become overstressed, as in Figure 13.17.

The amount of movement to be taken up by the piping and any device incorporated into it can be reduced by the use of 'cold draw'. The total amount of expansion is first calculated for each section between fixed anchor points. The pipes are left short by half this amount, and stretched cold, as by pulling up bolts at a flanged joint, so that at ambient temperature the system is stressed in one direction. When warmed through half the total temperature rise, and having expanded by half the total amount, the piping is unstressed. At working temperature and having fully expanded, the piping is stressed in the opposite direction. The effect is that instead of being stressed from zero to $+f$ units, the piping is stressed from $-1/2f$ to $+1/2f$.

In practical terms the piping is assembled with a spacer piece, of length equal to half the expansion, between two flanges. When the piping is fully installed and anchored the spacer is removed and the joint pulled up tight (Figure 13.18). The remaining part of the expansion, if not accepted by the natural flexibility of the piping, will call for the use of an expansion fitting. These can take several forms.

13.7.6 Full loop (Figure 13.19)

This is simply one complete turn of the pipe and should preferably be fitted in a horizontal rather than a vertical position to prevent condensate building up. The downstream side passes below the upstream side and great care must be taken that it is not fitted the wrong way round. When full loops are to be fitted in a confined space, care must be taken in ordering, otherwise wrong-handed loops may be supplied.

The full loop does not produce a force in opposition to the expanding pipework as in some other types but with steam pressure inside the loop, there is a slight tendency to unwind, which puts an additional stress on the flanges.

13.7.7 The horse-shoe or lyre loop (Figure 13.20)

Where space is available this type is sometimes used. It is best fitted horizontally so that the loop and main are all in the same plane. Pressure does not tend to blow the ends of the loop apart but there is a very slight straightening out effect. This is due to the design but causes no misalignment of the flanges.

In other cases the 'loop' is fabricated from straight lengths of pipe and 90° bends. This may not be as effective and requires more space but it meets the same need.

If any of these arrangements are fitted with the loop vertically above the pipe then a drain point must be provided on the upstream side.

13.7.8 Sliding joint (Figure 13.21)

These are often used because they take up little room but it is essential that the pipeline is rigidly anchored

Table 13.11 Expansion of pipes

Final temperature (°C)	(°F)	*Expansion per 30 m* (*mm*)	(*100 ft*) (*in*)
66	150	19	0.75
93	200	29	1.14
121	250	41	1.61
149	300	50	1.97
177	350	61	2.4
204	400	74	2.91
232	450	84	3.3
260	500	97	3.8

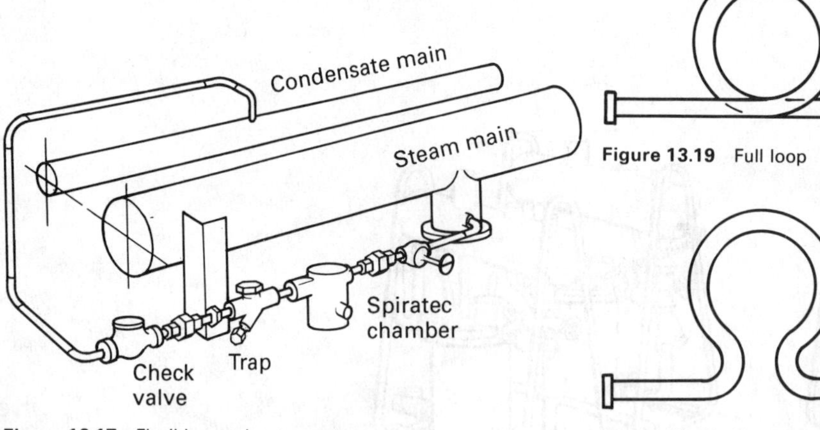

Figure 13.17 Flexible trapping arrangements

Figure 13.19 Full loop

Figure 13.20 Horse-shoe loop

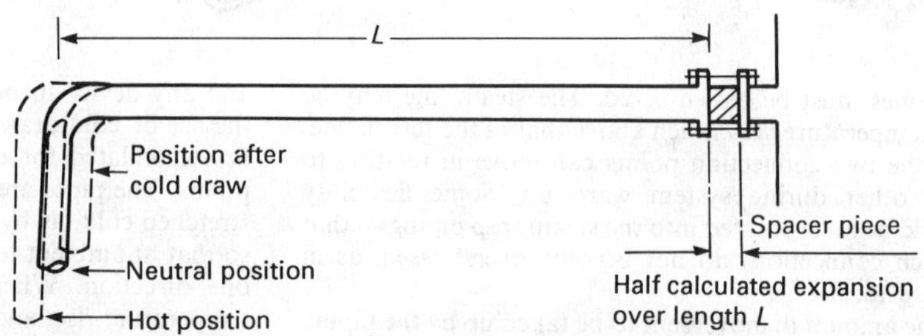

Figure 13.18 Cold draw

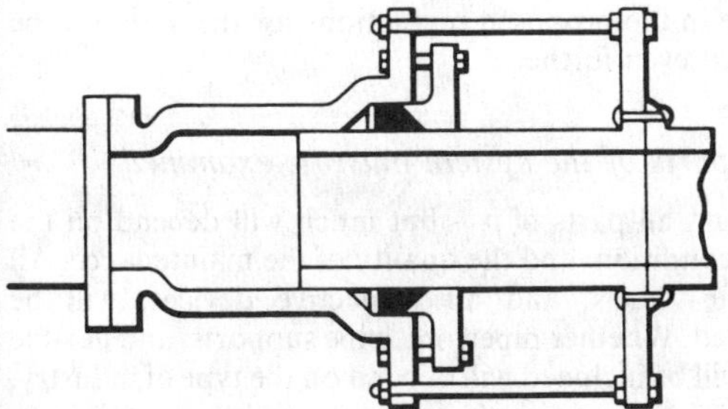

Figure 13.21 Sliding joint

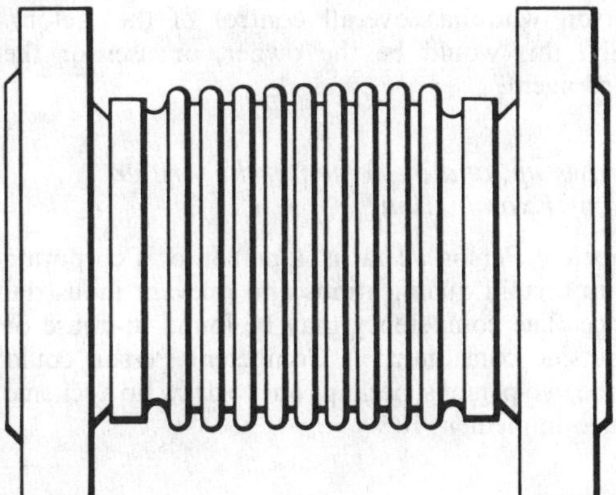

Figure 13.22 Bellows

and guided. This is because steam pressure acting on the cross-sectional area of the sleeve part of the joint tends to blow the joint apart in opposition to the forces produced by the expanding pipework. Misalignment will cause the sliding sleeve to bend, while regular maintenance of the gland packing is also needed.

13.7.9 Bellows (Figure 13.22)

A simple bellows has the advantage that it is an in-line fitting and requires no packing as does the sliding joint type. But it does have the same disadvantage as the sliding joint in that pressure inside tends to extend the fitting so that anchors and guides must be able to withstand this force.

The bellows can, however, be incorporated into a properly designed expansion fitting as shown in Figure 13.23, which is capable of absorbing not only axial movement of the pipeline but some lateral and angular displacement as well.

If expansion fittings are to work as intended, it is essential that the steam line is properly anchored at some point between the expansion fittings. Guiding is also important to ensure that any movement does not interfere with the designed fall towards the drain points.

Detailed design is clearly outside the scope of this section but Figure 13.24 shows some typical anchor points utilizing pipe flanges or lugs welded onto the pipe.

13.8 Proposed Pressure Systems and Transportable Gas Containers Regulations

What are these proposed regulations?

A set of legally enforceable rules, to ensure the integrity of pressure systems and certain containers for gases under pressure, by a written scheme of examination.

Is there a Code of Practice?

Yes, and compliance with the Code of Practice will be acceptable providing that the requirements of the Health

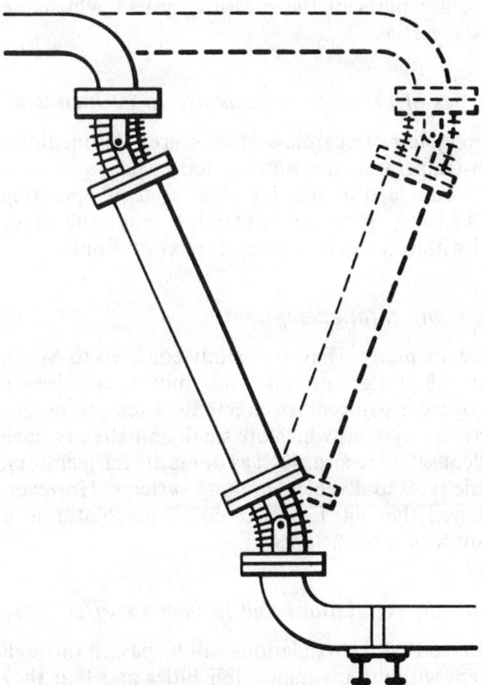

Figure 13.23 Expansion fitting

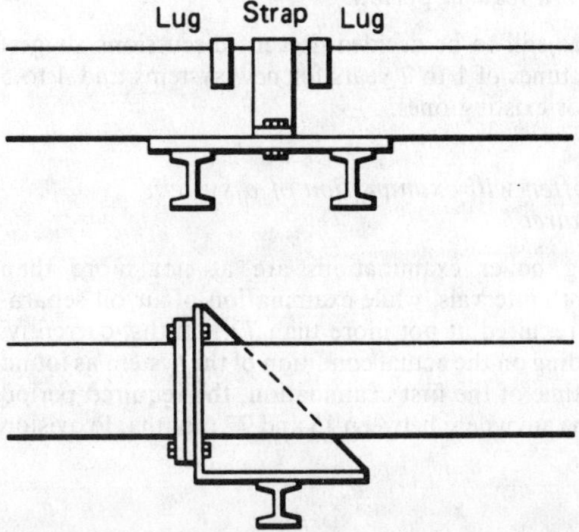

Figure 13.24 Anchor points

and Safety at Work Act 1974 and the Pressure Systems Regulations are fulfilled.

Why do we need more regulations?

The existing Factories Act has many grey areas, which can allow certain potentially dangerous systems to escape examination. This was highlighted, for example, by the enquiry which followed the Flixborough incident. It was then established that many pressure systems containing or carrying corrosive, flammable, explosive or toxic gases are not covered by the Factories Act.

Typically, the Factories Act applies to the boiler or other pressure-containing vessels. However, there may well be other parts of the pressure system which are equally dangerous.

What types of pressure systems are to be included?

All steam systems (regardless of pressure). All medium- and high-temperature hot water systems where an escape of liquid could lead to 'flashing'. Gas systems operating above 0.5 bar g pressure containing any substance included within the definition of a 'relevant fluid'.

Are there any total exemptions?

Yes, there are many. They are mainly confined to MoD, merchant shipping, aircraft and industries already covered by more stringent requirements. Exemptions also apply to some systems which are small and already have an excellent safety record, such as domestic refrigerators, pneumatic tyres and vehicle braking systems. However, it is believed that an 'Espresso' coffee percolator in a restaurant would be included.

When are the regulations due to become effective?

It is expected that the regulations will be passed through Parliament without any major difficulties and that they should become law in the first half of 1989.

Is there a lead-in period?

This has still to be decided. Recent discussions suggest lead-in times of 1 to 2 years for new systems and 4 to 5 years for existing ones.

How often will examination of a system be required?

Existing boiler examinations are at not more than 13-month intervals, while examination of air/oil separators is required at not more than 72 months, currently. Depending on the actual condition of the system as found at the time of the first examination, the required period could be anywhere between 13 and 72 months. Provision is made in the proposed regulations for the period to be extended even further.

What parts of the system must be examined?

In theory, all parts of it – but much will depend on the actual condition, and the quality of the maintenance. All pressure vessels, and all protective devices, will be examined. Whether pipework, pipe supports, and flexible hoses will be included will depend on the type of industry.

Whose responsibility is it to have a written system of examination?

The person who has overall control of the factory. Generally, this would be the owner, or user or the appointed agent.

Who draws up, and implements, the written Scheme of Examination?

A 'Competent Person'. That is, a person or a corporate body with certain qualifications and relevant industrial experience. The competency may be found in-house or be an outside consultant. A Competent Person could actually be two persons, perhaps one to draw up a scheme and one to implement it.

What are the minimum qualifications of a Competent Person?

For minor systems, a Tech. Eng.
For intermediate and major systems, a Chartered Engineer.

What is involved in implementation of a written scheme?

A critical scrutiny of a pressure system, or of part of it, whether or not in service, and including testing as necessary, to determine the actual condition and fitness for use during the period until the next examination. A written report must be made and copies of this kept by both the examiner and the user. In any case where the report indicates that the system is unsafe it will then be an offence to operate the system until the necessary repairs are carried out.

What changes in working practices will be needed in a steam-using plant?

Where preventive maintenance and good housekeeping are already the normal practice, the new regulations will have little effect on most steam users. On the other hand, where this is not the case, then upgrading of practices and maintenance of proper records will, quite rightly, become enforceable requirements.

Steam Tables (metric SI units)

Pressure		Temperature °C	Specific enthalpy			Specific volume steam
bar	*kPa*		*Water* (h_f) *kJ/kg*	*Evaporation* (h_{fg}) *kJ/kg*	*Steam* (h_g) *kJ/kg*	m^3/kg
absolute						
0.30	30.0	69.10	289.23	2336.1	2625.3	5.229
0.50	50.0	81.33	340.49	2305.4	2645.9	3.240
0.75	75.0	91.78	384.39	2278.6	2663.0	2.217
0.95	95.0	98.20	411.43	2261.8	2673.2	1.777
gauge						
0	0	100.00	419.04	2257.0	2676.0	1.673
0.10	10.0	102.66	430.2	2250.2	2680.4	1.533
0.20	20.0	105.10	440.8	2243.4	2684.2	1.414
0.30	30.0	107.39	450.4	2237.2	2687.6	1.312
0.40	40.0	109.55	459.7	2231.3	2691.0	1.225
0.50	50.0	111.61	468.3	2225.6	2693.9	1.149
0.60	60.0	113.56	476.4	2220.4	2696.8	1.083
0.70	70.0	115.40	484.1	2215.4	2699.5	1.024
0.80	80.0	117.14	491.6	2210.5	2702.1	0.971
0.90	90.0	118.80	498.9	2205.6	2704.5	0.923
1.00	100.0	120.42	505.6	2201.1	2706.7	0.881
1.10	110.0	121.96	512.2	2197.0	2709.2	0.841
1.20	120.0	123.46	518.7	2192.8	2711.5	0.806
1.30	130.0	124.90	524.6	2188.7	2713.3	0.773
1.40	140.0	126.28	530.5	2184.8	2715.3	0.743
1.50	150.0	127.62	536.1	2181.0	2717.1	0.714
1.60	160.0	128.89	541.6	2177.3	2718.9	0.689
1.70	170.0	130.13	547.1	2173.7	2720.8	0.665
1.80	180.0	131.37	552.3	2170.1	2722.4	0.643
1.90	190.0	132.54	557.3	2166.7	2724.0	0.622
2.00	200.0	133.69	562.2	2163.3	2725.5	0.603
2.20	220.0	135.88	571.7	2156.9	2728.6	0.568
2.40	240.0	138.01	580.7	2150.7	2731.4	0.536
2.60	260.0	140.00	589.2	2144.7	2733.9	0.509
2.80	280.0	141.92	597.4	2139.0	2736.4	0.483
3.00	300.0	143.75	605.3	2133.4	2738.7	0.461
3.20	320.0	145.46	612.9	2128.1	2741.0	0.440
3.40	340.0	147.20	620.0	2122.9	2742.9	0.422
3.60	360.0	148.84	627.1	2117.8	2744.9	0.406
3.80	380.0	150.44	634.0	2112.9	2746.9	0.389
4.00	400.0	151.96	640.7	2108.1	2748.8	0.374
4.50	450.0	155.55	656.3	2096.7	2753.0	0.342
5.00	500.0	158.92	670.9	2086.0	2756.9	0.315
5.50	550.0	162.08	684.6	2075.7	2760.3	0.292
6.00	600.0	165.04	697.5	2066.0	2763.5	0.272
6.50	650.0	167.83	709.7	2056.8	2766.5	0.255
7.00	700.0	170.50	721.4	2047.7	2769.1	0.240
7.50	750.0	173.02	732.5	2039.2	2771.7	0.227
8.00	800.0	175.43	743.1	2030.9	2774.0	0.215
8.50	850.0	177.75	753.3	2022.9	2776.2	0.204
9.00	900.0	179.97	763.0	2015.1	2778.1	0.194
9.50	950.0	182.10	772.5	2007.5	2780.0	0.185
10.00	1000.0	184.13	781.6	2000.1	2781.7	0.177
10.50	1050.0	186.05	790.1	1993.0	2783.3	0.171
11.00	1100.0	188.02	798.8	1986.0	2784.8	0.163
11.50	1150.0	189.82	807.1	1979.1	2786.3	0.157
12.00	1200.0	191.68	815.1	1972.5	2787.6	0.151
12.50	1250.0	193.43	822.9	1965.4	2788.8	0.148
13.00	1300.0	195.10	830.4	1959.6	2790.0	0.141
13.50	1350.0	196.62	837.9	1953.2	2791.1	0.136
14.00	1400.0	198.35	845.1	1947.1	2792.2	0.132
14.50	1450.0	199.92	852.1	1941.0	2793.1	0.128
15.00	1500.0	201.45	859.0	1935.0	2794.0	0.124
15.50	1550.0	202.92	865.7	1928.8	2794.9	0.119
16.00	1600.0	204.38	872.3	1923.4	2795.7	0.117

Steam Tables (*continued*)

Pressure		*Temperature*	*Specific enthalpy*			*Specific volume*
bar	*kPa*	*°C*	*Water* (h_f) *kJ/kg*	*Evaporation* (h_{fg}) *kJ/kg*	*Steam* (h_g) *kJ/kg*	*steam* m^3/kg
17.00	1700.0	207.17	885.0	1912.1	2797.1	0.110
18.00	1800.0	209.90	897.2	1901.3	2798.5	0.105
19.00	1900.0	212.47	909.0	1890.5	2799.5	0.100
20.00	2000.0	214.96	920.3	1880.2	2800.5	0.0949
21.00	2100.0	217.35	931.3	1870.1	2801.4	0.0906
22.00	2200.0	219.65	941.9	1860.1	2802.0	0.0868
23.00	2300.0	221.85	952.2	1850.4	2802.6	0.0832
24.00	2400.0	224.02	962.2	1840.9	2803.1	0.0797
25.00	2500.0	226.12	972.1	1831.4	2803.5	0.0768
26.00	2600.0	228.15	981.6	1822.2	2803.8	0.0740
27.00	2700.0	230.14	990.7	1813.3	2804.0	0.0714

14 Heating

G E Pritchard CEng, FCIBSE, FInstE, FIPlantE, MASHRAE
Chartered Engineer

Contents

14.1 Introduction

The past forty years have seen significant advances in every aspect of space heating, resulting from increased demands for the provision within buildings of more closely controlled environments and services of progressively increasing complexity. Although it remains impossible in a chapter of this size to cover adequately the extent of the subjects associated with these services, hopefully it will provide a useful reference.

14.2 Statutory regulations

Except for some defined types of accommodation, the use of fuel or electricity to heat premises above a temperature of 19°C is prohibited by the Fuel and Electricity (Heating) (Control) Order 1980. The current Order is an amendment to an earlier Regulation, which limited the temperature to a maximum of 20°C, and although 19°C is generally taken to refer to air temperature the Order does not specify this. The minimum temperature was laid down in the Factories Act 1961 and should be reached one hour after the commencement of occupation.

14.3 Building regulations

Unfortunately, the optimum results in cutting down space heating energy usage can often be obtained only when a building is at the design stage. Insulation, draught exclusion and the best possible heating system can then be built in at minimum cost. It is usually more expensive to add to (or modify) an existing building. Space heating is probably the largest usage of energy in buildings, so this section considers what can be done to improve insulation and other thermal properties. When energy was relatively cheap, little thought was given to conservation, and these omissions now have to be rectified.

In 1957 the Thermal Insulation (Industrial Buildings) Act laid down standards of insulation for roofs of new buildings. This first attempt to minimize heat losses did not cover walls, floors or windows. However, in 1978, Amendments to the Building Regulations rectified this by specifying standards for walls and windows. At this point it is necessary to define the terms '*U* value', or the insulation characteristic of the building material. This measures the rate at which energy flows through the material when there is a temperature difference of 1°C between the inside and outside faces, and this value is measured in watts (the unit of energy) per square metre of surface area, i.e. W/m^2 °C or W/m^2 K.

Symbol 'K' = °C temperature difference.

The amendments can briefly be summarized in Table 14.1.

The *U* values for walls, roofs and floor are intended as average figures, so it is permissible to have some areas of the structure underinsulated (i.e. with higher *U* values) providing other areas have sufficient extra insulation to bring the average of all areas down to (or below) the Regulation values.

Table 14.1

Industrial and commercial buildings
External walls of building enclosing heated spaces, internal walls exposed to unheated ventilated spaces, floors where the undersurface is exposed to outside air or an unheated ventilated space, and roofs over heated spaces (including the cases of ceilings with an unheated ventilated space above them).
Maximum average U value
For factories and storage buildings, such as warehouses, the *U* value is laid down to be 0.7. For shops, offices, institutional buildings and places of assembly, such as meeting halls, theatres, etc., the maximum average *U* value is to be 0.6.

Table 14.2

Type of building	*Maximum permitted glazed area*	
	In walls as percentage of wall area	As rooflights as percentage of roof area
Factories and storage	15	20
Offices, shops and places of assembly	35	20
Institutional, including residential	25	20

Note: Where figures for both rooflights and windows in walls are given, these really apply as a combined total. If the full wall window allowance is not used the balance can be reallocated to rooflight areas and vice versa. For example, a factory with only 10% of wall area as windows could add the other 5% of wall area as an increase to the permitted 20% of roof area that could be rooflights.

Limits are also imposed on window areas and apply to all buildings above 30 m^2 floor area. For the first group, industrial and commercial buildings, these limits apply both to rooflights and to windows in the walls. These percentages for windows or rooflights assume single glazing, and somewhat larger values can be used if double or triple glazing is to be fitted. However, calculations must be produced to show that the total heat loss from such units would be no greater than the single-glazed unit complying with the set limits (Table 14.2).

In most single- and two-storey buildings the largest proportion of heat loss from the building structure is usually through the roof. (In buildings of three storeys or more the losses through walls and windows may overtake the roof loss.)

If we first consider typical roofs, the methods of insulation break down into four groups:

1. Under-drawing, involving the fitting of insulation below the existing roof, as rigid self-supporting slabs, semi-rigid sheets supported by framing or a combination of an insulating blanket on top of sheets. This insulating blanket could be of mineral or glass fibre or a flexible 'foamed' plastic.
2. External, where insulation is added on top of the existing roof. This can be done with sheets or slabs of

insulating material finished with some waterproofing layer or very conveniently for corrugated or shaped roofing sheets by using a spray system to apply both the insulation and final waterproofing layer.

3. Where there is a ceiling below the actual roof, to place insulation above this ceiling. Where there are wooden ceiling joists, the insulation can be a flexible blanket laid between joists or loose granular fill spread between them. Alternatively, an overall quilt or blanket can be laid over the ceiling, taking care to check that the ceiling and its supports can safely carry the extra weight.
4. Possibly in a small minority of cases, it could be advantageous to install a false ceiling which could be of insulating sheets or panels. This can reduce the volume to be heated, particularly with steeply pitched roofs. Obviously, this idea could not be recommended where valuable daylight was available from roof-lights and where this daylight significantly reduced the artificial illumination necessary. Often such a false ceiling unfortunately conflicts with the existing architecture.
5. New Building Regulations for the Conservation of Fuel and Power for England and Wales came into operation on 1 April 1990. The new maximum U values of the elements (W/m^2 K) are shown in Table 14.4.

Table 14.3 gives some of the insulation properties for various building materials. The property given is for the rate at which energy would pass through a unit area of the material. In the standard units it becomes the number of watts that would be transferred through a square metre of the material of normal

Table 14.3 *U values*

Roofs	
Pitch covered with slates or tiles, roofing felt underlay, foil-backed plasterboard ceiling	1.5
Pitched covered with slates or tiles and roofing felt underlay, foil-backed plasterboard ceiling with 100 mm glass-fibre insulation between joists	0.35
Corrugated steel or asbestos cement roofing sheets	6.1–6.7
Corrugated steel or asbestos cement cladding with 75 mm fibreglass lightweight liner	0.38
Corrugated steel or asbestos cement roofing sheets with cavity and aluminium foil-backed 10 mm plasterboard lining	1.9–2.0
Corrugated double-skin asbestos cement sheeting with 25 mm glass-fibre insulation between with cavity and aluminium foil-backed 10 mm plasterboard lining; ventilated air space	0.8
Steel or asbestos cement roofing sheets, no lining with rigid insulating lining board 75 mm	0.4
Asphalt 19 mm thick or felt/bitumen layer on solid concrete 150 mm thick	3.5
Asphalt 19 mm thick or felt/bitumen layer on 150 mm autoclaved aerated concrete roof slabs	0.9
Flat roof, three layers of felt on chipboard or plasterboard	1.54
Flat roof, three layers of felt on rigid insulating board 100 mm thick	0.29
Timber roof with zinc or lead covering and 25 mm plaster ceiling	0.96

Table 14.3 *continued*

Walls		
Steel or asbestos cement cladding		5.3–5.7
Steel or asbestos cement cladding 75 mm fibre glass lightweight liner		0.37
Steel or asbestos cement cladding with plaster-board lining and 100 mm fibre insulating roll		0.4
Solid brick wall unplastered 105 mm		3.3
Solid brick wall unplastered 335 mm		1.7
Solid brick wall 220 mm thick with 16 mm lightweight plaster on inside face		1.9
Brick/cavity/brick (260 mm total thickness)		1.4
260 mm brick/mineral fibre-filled cavity/brick		0.5
260 mm brick/cavity/load-density block		1.0–1.1
Brick/expanded polystyrene board in cavity/low-density block/inside face plastered		0.5
Weather boarding on timber framing with 10 mm plasterboard lining, 50 mm glass-fibre insulation in the cavity and building paper behind the boarding		0.62
Glazing		
Single glazing	Wood frame	4.3
	Metal frame	5.6
Double glazing	Wood frame	2.5
	Metal frame	3.2
Triple glazing		2.0
Roof skylights		6.6
Floors		
20 mm intermediate wood floor on 100 mm × 50 mm joists 10 mm plasterboard ceiling allowed for 10% bridging by joists		1.5
150 mm concrete intermediate floor with 150 mm screed and 20 mm wood flooring		1.8
The heat loss through floors in contact with the earth is dependent upon the size of the floor and the amount of edge insulation. Insulating the edge of a floor to a depth of 1 m can reduce the *U* value by 35%. Following are some typical *U* values for ground floors. Effectively, most of the heat loss is around the perimeter of the floor.		
Solid floor in contact with the earth with four exposed edges:		
150 m × 50 m		0.11
60 m × 60 m		0.15
15 m × 60 m		0.32
15 m × 15 m		0.45
7.5 m × 15 m		0.62
3 m × 3 m		1.47
Suspended timber floors directly above ground. Bare or with linoleum, plastic or rubber tiles:		
150 m × 60 m		0.14
60 m × 60 m		0.16
15 m × 60 m		0.37
15 m × 15 m		0.45
7.5 m × 15 m		0.61
3 m × 3 m		1.05
Suspended timber floors directly above ground with carpet or cork tiles:		
150 m × 60 m		0.14
60 m × 60 m		0.16
15 m × 60 m		0.34
15 m × 15 m		0.44
7.5 m × 15 m		0.59
3 m × 3 m		0.99

Table 14.4

Building type	*Ground floors*	*Exposed walls and floors*	*Semi-exposed walls and floors*	*Roofs*
Industrial storage and other buildings, excluding dwellings	0.45	0.45	0.60	0.45

Note: An exposed element is exposed to the outside air; a semi-exposed element separates a heated space from a space having one or more elements which are not insulated to the levels in the table.

Maximum window areas for single glazing in buildings other than dwellings will be unchanged.

thickness in the form it would be used, if the air at either side of the material shows a temperature difference of 1°C. In SI units this becomes W/m^2 °C, which, in this case, is commonly known as the *U* value. The larger the *U* value, the more energy it will transfer, so the worse are its insulation properties.

A very poor insulating material can be detected very simply. If the material is at normal room conditions and a hand placed upon its surface feels cold, heat is being conducted away from it as the *U* value is very high. A low *U* value is shown by no cooling affect. To try this, if one places a hand on a window and a wooden table there should be a notable difference between the two showing a difference in the *U* value. Wood's *U* value is about 1 W/m^2 °C and glass has a value of over 5.5 W/m^2 °C.

The *U* values are given in W/m^2 °C for various building material under normal weather conditions. There will always be slight variations around these values, dependent on particular manufacturers of the materials. With any insulation which is being fitted, advice should be sought regarding the fire risk and condensation problems.

14.4 Estimation of heat losses from buildings

The normal procedure in estimating the heat loss from any building is as follows:

1. Decide upon the internal air temperature to be maintained at the given external air temperature.
2. Decide the heat transmission coefficient (*U* values) for the outside walls and glass, roof and bottom floor, and the inside walls, ceilings, or of heated spaces adjacent to non-heated spaces.
3. Measure up the area of each type of surface and compute the loss through each surface by multiplying the transmission coefficient by the measured area by the difference between the inside and the outside temperatures.
4. Calculate the cubic contents of each room and, using the appropriate air change rate, the amount of heat required to warm the air to the desired temperature by multiplying the volume of air by the difference between the inside and outside temperatures and the specific heat of air.

The above calculations will give the heat losses after the building has been heated. Under conditions in which the heating system will operate continuously, satisfactory results will be obtained if the heating system is designed to provide heat equivalent to the amount calculated above. Suitable allowance must be made for losses from mains.

When, however, operation is intermittent, safety margins are necessary. These are, of course, speculative, but the following suggestion has frequently proved satisfactory. When it is necessary to operate after a long period of vacancy, as may happen in certain types of substantially built buildings, it is necessary to add up to 30% to the 'steady state' heat transmissions. In buildings of light construction this margin may be reduced.

In selecting the appropriate *U* values we must pay due regard to the exposure and aspect of the room. It appears reasonable to make allowance for the height of a room, bearing in mind that warm air rises towards the ceiling. Thus in a room designed to keep a comfortable temperature in the lower 1½ or 2 m, a higher temperature must exist nearer the ceiling, which will inevitably cause greater losses through the upper parts of windows, walls and roof. This effect is greatest with a convective system, i.e. one which relies on the warming of the air in the room for the conveyance of heat. This would occur in the case of conventional radiators, convectors and warm air systems. In the case of radiant heated rooms, this does not occur, and a much more uniform temperature exists from floor to ceiling.

14.5 Allowance for height of space

In heat loss calculations a uniform temperature throughout the height of the heated space is assumed, although certain modes of heating cause vertical temperature gradients which lead to increased heat losses, particularly through the roof. These gradients need to be taken into account when sizing appliances. Additions to the calculated heat loss to allow for this are proposed in Table 14.5. However, these percentages should not be added to replacement heat to balance that in air mechanically exhausted from process plant. Attention is also drawn to the means of reducing the effect of temperature stratification, discussed in Section 14.10.13.

14.6 Characteristics of heat emitters

Designers will need to decide whether it is necessary to add a margin to the output of heat emitters. During the warm-up cycle with intermittently operated heating systems, emitter output will be higher than design because space temperatures are lower. Also, boost system temperatures may be used to provide an emission margin during warm-up. The need for heat emitter margins to meet extreme weather conditions will depend on the design parameters used in determining heat losses.

In summary, although the addition of a modest margin

Table 14.5 Allowances for height of heated space

Method of heating and type or disposition of heaters	*Percentage addition for following heights of heated space* (m)		
	5	5–10	>10
Mainly radiant			
Warm floor	Nil	Nil	Nil
Warm ceiling	Nil	0–5	[a]
Medium- and high-temperature downward radiation from high level	Nil	Nil	0–5
Mainly convective			
Natural warm air convection	Nil	0–5	[a]
Forced warm air			
Cross flow at low level	0–5	5–15	15–30
Downward from high level	0–5	5–10	10–20
Medium and high-temperature cross radiation from intermediate level	Nil	0–5	5–10

[a] Not appropriate to this application.

to heat emitter output would add little to the overall system cost and a margin on the heat generator or boiler output can only be utilized if the appropriate emitter capacity is available, the decision should be based on careful discrimination rather than using an arbitrary percentage allowance. In general, for buildings of traditional construction and for the incidence of design weather in normal winters in the UK an emitter margin in excess of, say, 5% or 10% is unlikely to be justified. However, for well-insulated buildings the heat loss reduces in significance relative to the heat stored in or needed to warm up the structure. For such applications a larger heating system margin is required, and the emitter margin provided would need to be considered accordingly.

14.7 Central plant size

In estimating the required duty of a central plant for a building it should be remembered that the total net infiltration of outdoor air is about half the sum of the rates for the separate rooms. This is because, at any one time, infiltration of outdoor air takes place only on the windward part of the building, the flow in the remainder being *outwards*.

When intermittent heating is to be practised the pre-heating periods for all rooms in a building will generally be coincident. The central plant rating is then the sum of the individual room heat demands, modified to take account of the *net* infiltration.

If heating is to be continuous some diversity between the several room heating loads can be expected. The values listed in Table 14.6 are suggested. When mechanical ventilation is combined with heating, the heating and the ventilation plant may have different hours of use, and the peak loads on the two sections of the plant will often occur at different times.

The central plant may also be required to provide a domestic hot water supply and/or heat for process purposes. These loads may have to be added to the net heating load to arrive at the necessary plant duty, but careful design may avoid the occurrence of simultaneous peaks. In large installations the construction of boiler curves may indicate whether savings in boiler rating can be made. In many cases little or no extra capacity may be needed for the hot water supply, its demands being met by 'robbing' the heating circuits for short periods.

14.8 Selective systems

In some cases the various rooms of a building do not all require heating at the same time of day and here a so-called 'selective system' may be used. The supply of heat is restricted to different parts of the building at different times of the day; the whole building cannot be heated at one time. A typical application is in dwellings where the demands for heat in living spaces and bedrooms do not normally coincide..

In a selective system the individual room appliances must be sized as indicated above, to provide the appropriate output according to heat loss, gains and intermittency. The central plant need only be capable of meeting the greatest simultaneous demands of those room units which are in use at the same time. This will generally lead to a large power being available to meet the demands of those units which form the lesser part of the load. These units may then be operated with a high degree of intermittency.

14.9 Multiple-boiler installations

Load variation throughout the season is clearly large, and consideration should be given to the number of boilers required in the system. Operation at low loads leads to corrosion and loss in efficiency and should be avoided. On the other hand, a number of smaller boilers gives an increase in capital costs.

It has been shown that when boilers are chosen which have a fairly constant and good efficiency over a working range of 30–100%, then the effects on overall costs (running + capital) of varying the number and relative sizes of boilers in the system is less than 5%. The optimum

Table 14.6 Diversity factors for central plant (continuous heating)

Space or building served by plant	*Diversity factor*
Single	1.0
Single building or zone, central control	0.9
Single building, individual room control	0.8
Group of buildings, similar type and use	0.8
Group of buildings, dissimilar uses[a]	0.7

[a] This applies to group and district heating schemes where there is substantial storage of heat in the distribution mains, whether heating is continuous or intermittent.

number depends on the frequency of occurrence of low loads.

Under these circumstances the engineer is free to choose the number of boilers in the system based on practical rather than economic considerations. The following procedure is recommended:

1. Choose a type of boiler with a fairly constant and high efficiency over its full turndown range. Obtain its efficiency curve.
2. On the basis of avoiding acid corrosion and obtaining required standby, choose the number of boilers required and their relative sizes. Equally sized boilers should be used except where the provision of domestic hot water in summer requires one smaller boiler. Table 14.7 gives suggested relative sizes based on turndown to 30%.
3. The boilers should be controlled in sequence, the switching points for bringing boilers on line occurring whenever an additional boiler makes the system more efficient (for an evaluation of this see Figure 14.1). Full boiler load is not usually the most economic switching point, but switching points too close to full turndown should also be avoided. At any given system load the boilers on-line should share the load between them in proportion.

14.10 Heating systems

14.10.1 Warm and hot water heating systems

Warm water or low-, medium- or high-temperature hot water systems are categorized in Table 14.8. Warm water systems may use heat pumps, fully condensing boilers or similar generators, or reclaimed heat. In many cases the system design may incorporate an alternative heat generator for standby purposes or for extreme weather operation. Under such circumstances the system may continue to function at warm water temperatures or could operate at more conventional LTHW ones.

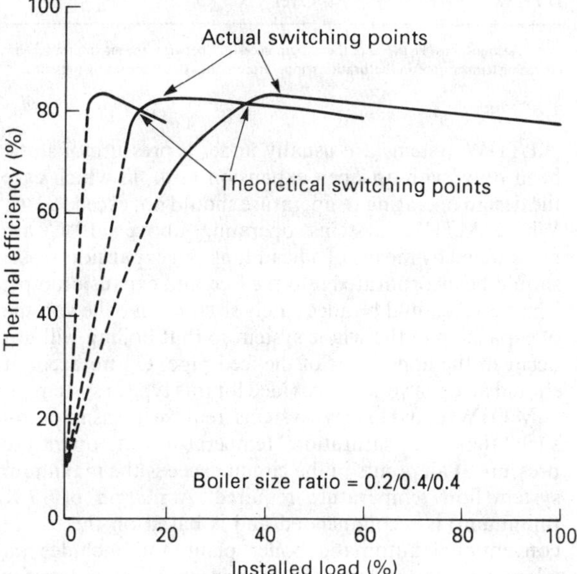

Figure 14.1 Evaluation of optimum boiler control

Table 14.7 Appropriate boiler size ratios assuming turndown to 30% of full load

Number of boilers	*Installed plant ratio*[a]	*Heating only*			*Heating plus domestic hot water*			*Remarks*
		Installed boiler size ratios	*Lowest load (%)*	*Load if largest fails (%)*	*Installed boiler size ratios*	*Lowest load (%)*	*Load if largest fails (%)*	
1	1.0	1.0	30	0	—			Use only if load is seldom < 30%. No standby
2	1.0	0.5/0.5	15	50	—	—	—	Impossible to gain reasonable standby and meet low loads
	1.0	0.33/0.67	10	33	0.30/0.70	9	30	
	1.25	0.3/0.7	11.3	37	0.25/0.75	9.4	31	
	1.5	0.25/0.75	11.3	37	—	—	—	
3	1.0	0.33/0.33/0.33	10	67	0.2/0.4/0.4	6	60	
	1.25	0.2/0.4/0.4	7.5	75	0.2/0.4/0.4	7.5	75	
	1.50	0.2/0.4/0.4	9	90	0.2/0.4/0.4	9	90	
	1.67	0.2/0.4/0.4	10	100	0.2/0.4/0.4	10	100	
4	1.0	0.25/0.25/0.25/0.25	7.5	75	0.25/0.25/0.25/0.25	7.5	75	
	1.25	0.25/0.25/0.25/0.25	9.4	94	0.1/0.3/0.3/0.3	3.7	87	
	1.50	0.1/0.3/0.3/0.3	4.3	105	0.1/0.3/0.3/0.3	4.5	105	
	1.33	0.25/0.25/0.25/0.25	10	100	0.1/0.3/0.3/0.3	4	93	
5	1.0	0.2/0.2/0.2/0.2/0.2	6	80	0.2/0.2/0.2/0.2/0.2	6	80	
	1.25	0.2/0.2/0.2/0.2/0.2	7.5	100	0.2/0.2/0.2/0.2/0.2	7.5	100	
	1.50	0.2/0.2/0.2/0.2/0.2	9.0	120	0.2/0.2/0.2/0.2/0.2	9.0	120	

[a] Installed plant ratio = $\frac{\text{Total installed boiler capacity}}{\text{Design maximum heat requirement}}$

Table 14.8 Design water temperatures for warm and hot water heating systems

Category	*System design water temperatures* (°C)
Warm	40– 70
LTHW	70–100
MTHW	100–120
HTHW	Over 120

Note: Account must be taken of the margin necessary between the maximum system operating temperature and saturation temperature at the system operating pressure.

LTHW systems are usually under a pressure of static head only, with an open expansion tank, in which case the design operating temperature should not exceed 83°C. Where MTHW systems operating above 110°C are pressurized by means of a head tank, an expansion vessel should be incorporated into the feed and expansion pipe. This vessel should be adequately sized to take the volume of expansion of the whole system so that boiling will not occur in the upper part of the feed pipe. On no account should an open vent be provided for this type of system.

MTHW and HTHW systems require pressurization such that the saturation temperature at operating pressure at all points in the circuit exceeds the maximum system flow temperature required. A margin of 17 K (minimum) is recommended and is based on the use of conventional automatic boiler plant and includes an allowance for tolerances on temperature set points for the automatic control of heat-generation output. A check must be made on actual tolerance used in the design of a control system to ensure that this allowance is adequate.

When selecting the operating pressure, allowance must be made for the effect of static head reduction at the highest point of the system and velocity head reduction at the circulating pump section, to ensure that all parts of the system are above saturation pressure within an adequate anti-flash temperature margin. Additionally, the margin on the set point of the high-temperature cut-out control should be 6 K, except for boilers fired with solid fuel automatic stokers, where it should be at least 10 K.

Medium- and high-temperature systems should be fully pressurized before the operating temperature is achieved and remain fully pressurized until the temperature has dropped to a safe level. In all systems the heat generator or boiler must be mechanically suitable to withstand the temperature differentials, and the return temperature to the boiler must be kept high enough to minimize corrosion. Automatic controls may be used to achieve this.

14.10.2 Design water flow temperature

For low-temperature heating systems using natural convective or radiant appliances the normal design water flow temperature to the system is 83°C (see also Table 14.8). Boost temperatures may be used on modulated-temperature systems because of the changes in heat output characteristics with varying temperatures. Additionally, comfort aspects must be borne in mind, as forced convective emitters operating on modulated temperature systems can deliver airstreams at unacceptably low temperatures.

For MTHW and HTHW systems heat emitters may be as for LTHW systems, except that, for safety reasons, units with accessible surfaces at water temperature would not normally be employed. Embedded panel coils may be used in conjunction with a MTHW or HTHW distribution system, with insulating sleeves around the coil piping to reduce the heat flow. Alternatively, the coils can be operated as reduced temperature secondary systems by allowing only a small, carefully controlled proportion of flow temperature water to be mixed with the water circulating in the coils. Design arrangements for reduced-temperature secondary systems (sometimes referred to as injection circuits) include fixed provisions for minimum dilution rates. Conventional system-balancing devices with three-port automatic modulating valves to regulate mixed water temperatures and, hence, heat output are used. Automatic safety controls must prevent excessive temperatures occurring in the coil circuits, as floor fabrics or finishes could be damaged very rapidly.

14.10.3 Maximum water velocity

The maximum water velocity in pipework systems is limited by noise generation and erosion/corrosion considerations. Noise is caused by the free air present in the water, sudden pressure drops (which, in turn, cause cavitation or the flashing of water into steam), turbulence or a combination of these. Noise will therefore be generated at valves and fittings where turbulence and local velocities are high, rather than in straight pipe lengths.

A particular noise problem can arise where branch circuits are close to a pump and where the regulating valve used for flow-rate balancing may give rise to considerable pressure differences. Oversizing regulating valves should be avoided, as this will result in poor regulation characteristics; the valve operating in an almost shut position and creating a very high local velocity.

High water velocities can result in erosion or corrosion due to the abrasive action of particles in the water and the breakdown of the protective film which normally forms on the inside surface of the pipe. Erosion can also result from the formation of flash steam and from cavitation caused by turbulence.

Publishing data on limiting water velocities are inconclusive. Table 14.9 summarizes the available information.

14.10.4 Minimum water velocity

Minimum water velocities should be maintained in the upper floors of high-rise buildings where air may tend to come out of solution because of reduced pressures. High velocities should be used in down-return mains feeding into air-separation units located at a low level in the system. Table 14.10 can be taken as a guide.

Water velocities shown in Tables 14.9 and 14.10 are indicative parameters only; on the one hand, to limit

Table 14.9 Limiting water velocities in pipework

Pipe diameter (mm)	*Steel pipework*		*Copper pipework* (m/s)
	Non-corrosive water (m/s)	*Corrosive water* (m/s)	
50 and below	1.5	1	1
Above 50	3	1.5	1.5
Large distribution mains with long lengths of straight pipe	4	2	—

Table 14.10 Minimum water velocities

Pipe diameter (mm)	*Minimum water velocity* (m/s)
50 and below	0.75
Above 50	1.25

noise problems and erosion and, on the other, to try to ensure air entrainment. Within these parameters the design engineer will need to discriminate on the selection of water velocities in a distribution system based on other considerations. It is particularly necessary to bear in mind the effect of low water velocities on flow-measuring components used in balancing flow rates in systems.

14.10.5 System temperature drop

British practice on LTHW systems uses a typical system temperature drop of 11 K and a maximum system temperature of 17 K. Continental practice has tended to use higher drops (up to 40 K). An advantage of a higher system temperature drop is the reduction in water flow rates. This will result in reduced pipe sizes with savings in capital cost and distribution heat losses and a reduced pump duty, with savings in running costs. A disadvantage of higher system temperature drops is the need for larger and consequently more expensive heat emitters. However, if it is possible to raise the system flow temperature so that the mean water temperature remains the same, then with certain types of emitter only a small increase in size is required. With large system temperature drops the average water temperature in a radiator tends to fall below the mean of flow and return temperature and, thus, a larger surface is needed. Furthermore, on one-pipe circuits the progressive reduction in temperature around the circuit may lead to excessively large heat emitters.

Higher system temperature drops can be used with MHTW and HTHW systems since the mean temperature of the heat emitters will be correspondingly higher. Additionally, these media are well suited to use for primary distribution systems, conveying heat over long distances.

Precautions should be taken to prevent the danger of injury from contact with hot surfaces. The safe temperature for prolonged contact is relatively low and reference should be made to BS 4086 and other sources.

14.10.6 Use of temperature-limiting valves on emitters

On some group and district heating schemes, outlet limiting valves which permit flow only when the water temperature has dropped to a specified low level are used. This procedure minimizes the water quantity to be pumped and permits indicative heat metering by water quantity alone. In such cases care must be taken to size emitters to suit the available water temperatures. The effect of low water velocities through the emitter must also be taken into consideration, since the heat output of some convective appliances is greatly reduced under such conditions.

14.10.7 Miscellaneous components

Data regarding relief valves, feed and expansion cisterns, etc. are given in Tables 14.11 and 14.12. Cistern sizes shown in Table 14.12 are based on typical system designs and are approximate only. An estimate of the water content of the particular system should always be made where there is any doubt regarding these typical data, to ensure that the cistern capacity is adequate to contain the expansion volume.

14.10.8 Distribution system design

The design of pipework distribution systems must allow for the following:

1. Future extensions, where required, by the provision of valved, plugged or capped tee connections.
2. Provision for isolation for maintenance. Where it is necessary to carry out maintenance on a 'live' system, valves must be lockable and may need to be installed in tandem.
3. Thermal expansion.
4. Provision for distribution flow rate balancing for initial commissioning or rebalancing to meet changed operational requirements. Typical provisions for

Table 14.11 Hot water heating boilers – recommended sizes of relief valves

Rated output of boiler (kW)	*Minimum clear bore of relief valve* (mm)	*Equivalent area* (mm^2)
Up to 250	20	310
250– 350	25	490
350– 450	32	800
450– 500	40	1250
500– 750	50	1960
750–1000	65	3320

Note: The above sizes apply to boilers fired with solid fuel. For oil- and gas-fired boilers the relief valve should be one size larger.

Table 14.12 Approximate sizes of feed and expansion cisterns for low-pressure hot water heating systems

Boiler or water-heating (kW)	*Cistern size* (l)	*Ball-valve size* (mm)	*Cold-feed size* (mm)	*Open-vent size* (mm)	*Overflow size* (mm)
15	18	15	20	25	25
22	18	15	20	25	32
30	36	15	20	25	32
45	36	15	20	25	32
60	55	15	20	25	32
75	68	15	25	32	32
150	114	15	25	32	32
225	159	20	32	40	40
300	191	20	32	40	40
400	227	20	40	50	40
450	264	20	40	50	50
600	318	25	40	50	50
750	455	25	50	65	50
900	636	25	50	65	50
1200	636	25	50	65	80
1500	910	25	50	65	80

Notes: (1) Cistern sizes are actual.
(2) Cistern sizes are based on radiator-heating systems and are approximate only.
(3) The ball-valve sizes apply to installations where an adequate mains water pressure is available at the ball-valve.

balancing comprise the following:

(a) A measuring station – which may be an orifice plate, a venturi, an orifice valve or other proprietary device – provided with a pair of tappings to permit the measurement of upstream and downstream system dynamic pressures.

(b) An associated regulating valve – preferably a double-regulating valve or other arrangement which permits the required setting to remain undisturbed by closure.

5. Provision for drainage, including drainage after pre-commission flushing; water circulation during flushing must be in excess of design flow rates and, in order to discharge the flushing effluent effectively, drainage connections must be full diameter.
6. Removal of air from the system by provision of:

   (a) Air separators, one form of which uses the principle of centrifugal force to separate the heavier constituent (water) from the lighter one (non-condensable gases). Best results are achieved by locating the separator at the highest temperature point of the system where air has a greater tendency to come out of solution. The velocity of the medium requires to be above the minimum stated by the manufacturer (usually about 0.25 m/s).

   (b) Automatic air vents for systems operating at temperatures below atmospheric boiling point.

   (c) Air bottles with manually operated needle valves to release accumulated air, for systems operating at temperatures in excess of atmospheric boiling point.

7. Provision of test points for sensing temperature and pressure at selected locations.

14.10.9 Sealed heating systems

Pressurization of medium- and high-temperature hot water sealed heating systems referred to above may take the following forms.

Pressurization by expansion of water

The simplest form of pressurization uses the expansion of the water content of the system to create a sufficient pressure in an expansion vessel to provide an anti-flash margin of, say, 17°C at the lowest pressure (highest point) of the system. The main disadvantage of a naturally pressurized expansion vessel is the ability of water to absorb air and the consequent risk of oxygen corrosion.

A diaphragm expansion vessel is divided into two compartments by a special membrane or diaphragm of rubber or rubber composition which prevents the water coming into contact with the air. On one side of the diaphragm the vessel is filled with air or nitrogen at the required pressure. The other section of the vessel is connected directly to the water system. A correctly positioned air separator will assist in de-aerating the water in the system.

Pressurization of elevated header tanks

Given very careful attention to design, installation and commissioning, MTHW systems may be operated with the necessary system pressure provided by an elevated feed and expansion tank. Where the system operating temperature exceeds 110°C an expansion vessel should be sized to absorb the volume of expansion for the complete system, thus preventing water at operating temperatures entering the feed and expansion tank and causing boiling. On no account should an open vent be provided for this type of system.

Gas pressurization with spill tank

This form consists of a pressure cylinder maintained

partly filled with water and partly with gas (usually nitrogen) which is topped up from pressure bottles. Water expansion is usually arranged to discharge from the system through a pressure-control valve into a spill tank open to atmosphere or to a closed cylinder lightly pressurized with nitrogen. A pump is provided to take water from the spill tank and return it under pressure to the system as cooling-down results in a pressure drop. The pump operation is regulated by a system pressure sensor.

Hydraulic pressurization with spill tank

In this form the pressure is maintained by a continuously running centrifugal pump. A second pump under the control of a pressure switch is provided to come into operation at a predetermined pressure differential and as an automatic standby to the duty pump. Surplus water is delivered to or taken from a spill tank or cylinder as described previously.

Example of pressure differential

Assume system flow temperature of 120°C
Allow 17 K anti-flash margin – 137°C

Corresponding absolute pressure	3.4 bar
Assume static absolute pressure on system	2.0 bar
Minimum absolute pressure at cylinder	5.4 bar
Allow operating differential on pressure cylinder, say –	0.5 bar
Minimum operating absolute pressure of system	5.9 bar

Example of water expansion

Assume water capacity of system 200 000 l
Assume ambient temperature of 10°C
Assume system maximum flow temperature of 120°C
Assume system minimum return temperature of 65°C
Increase in volume from 10°C to 65°C

$$200\,000\ \frac{(999.7-980.5)}{980.5} = 3916\ \text{l}$$

Increase in volume from 65°C to 120°C

$$200\,000\ \frac{(980.5-943.1)}{943.1} = 7931\ \text{l}$$

Total increase in volume = 11 847 l

14.10.10 Maintenance of water heating systems

A common practice in many hot water heating installations is to drain the complete system during summer months. This practice, involving a complete change of raw water every year, is to be deprecated. It introduces additional hardness salts and oxygen to the system, resulting in very significant increases in scaling and corrosion. Where it is necessary to drain the boiler or heat generator or other parts of the system for inspection or maintenance purposes, isolating valves or other arrangements should be used to ensure that the section drained is kept to a minimum.

14.10.11 Steam heating systems

These are designed to use the latent heat of steam at the heat emitter. Control of heat output is generally by variation of the steam saturation pressure within the emitter. For heating applications with emitters in occupied areas low absolute pressures may be necessary in order to reduce the saturation temperature to safe levels.

The presence of non-condensable gases in steam systems (e.g. air and CO_2) will reduce the partial pressure of the steam, and hence its temperature, thus affecting the output of the appliance. A further adverse effect is the presence of a non-condensable gas at the inside surface of a heat emitter. This impedes condensation and, hence, heat output. It is therefore imperative that suitable means are provided to prevent formation of CO_2 and to evacuate all gases from the system.

Superheat, which must be dissipated before condensation occurs, can be used to reduce condensation in the distribution mains.

On–off control of steam systems can result in the formation of a partial vacuum, leading to condensate locking or back feeding, and infiltration of air which subsequently reduces the heat transfer.

When using modulating valves for steam, heat emitter output must be based on the steam pressure downstream of the valve, which often has a high-pressure drop across it, even when fully open.

Steam traps must be sized to cope with the maximum rate of condensation (which may be on start-up) but must perform effectively over the whole operational range, minimizing the escape of live steam.

Partial waterlogging of heater batteries can lead to early failure due to differential thermal expansion. Steam trap selection should take account of this.

Where high temperatures are required (e.g. for process work) and lower temperatures for space heating, it is desirable to use flash steam recovery from the high-temperature condensate to feed into the low-temperature system, augmented as required by reduced pressure live steam.

Steam as a medium for heating is now seldom used. Hot water, with its flexibility to meet variable weather conditions and its simplicity, has supplanted it in new commercial buildings. Steam is, however, often used for the heating of industrial buildings where steam-raising plant occurs for process or other purposes. It is also employed as a primary conveyor of heat to calorifiers such as in hospitals, where again steam boiler plant may be required for sundry duties such as in kitchens, laundry and for sterilizing. Heating is then by hot water served from calorifiers.

14.10.12 High-temperature thermal fluid systems

Where high operating temperatures are required, high-temperature thermal fluid systems may be used instead of pressurized water or steam systems. These systems operate at atmospheric pressure using non-toxic media such as petroleum oil for temperatures up to 300°C or synthetic chemical mixtures where temperatures in excess of this are required (up to 400°C). Some advantages and disadvantages of thermal fluid or heat transfer oil systems are listed below.

Advantages

No corrosion problems.
Statutory inspections of boilers/pressure vessels not required.
No scale deposits.
No need for frost protection of system.
Cost of heat exchangers/heat emitters less, as only atmospheric pressures are involved.
Better energy efficiency than steam systems.
Operating temperature can be increased subsequent to design without increasing operating pressure.

Disadvantages

Medium more expensive than water (but no treatment costs).
Medium is flammable under certain conditions.
Heat transfer coefficient is inferior to that of water.
Care necessary in commissioning and in heat-up rates due to viscosity changes in medium.
Circulating pump necessary (not required for steam systems).
Air must be excluded from the system.
In the event of leakage the medium presents more problems than water.

14.10.13 Warm air heating systems

These may be provided with electric or indirect oil- or gas-fired heaters or with a hot water heater or steam battery supplied from a central source. Because the radiant heat output of warm-air systems is negligible, the space air temperature will generally need to be higher for equivalent comfort standards than for a system with some radiant output. This will increase energy use, and legislative standards for limiting space temperatures should be considered. Attention is drawn to the vertical temperature gradient with convective systems and, when used for cellular accommodation, the likelihood of some spaces being overheated due to the difficulty of controlling such systems on a room-by-room basis.

With the advent of natural gas, direct-fired warm air systems are used where the heat and products of combustion, diluted by fresh air introduced into the system, are distributed to the heated spaces. In designing such installations account must be taken of the requirements of the Building Regulations 1985, Part J, and of the Regional Gas Authority. Care must also be taken in design and application to ensure that the moisture in the products of combustion will not create condensation problems. Direct-fired systems are more suited to large, single-space low-occupancy applications such as warehouses and hangers and should not be used to serve sleeping accommodation.

14.10.14 Reducing the effect of temperature stratification

As with all convective systems, warm air heating installations produce large temperature gradients in the spaces they serve. This results in the inefficient use of heat and high heat losses from roofs and upper wall areas. To improve the energy efficiency of warm air systems, pendant-type punkah fans or similar devices may be installed at roof level in the heated space. During the operational hours of the heating system these fans work either continuously or under the control of a roof-level thermostat and return the stratified warm air down to occupied levels.

The energy effectiveness of these fans should be assessed, taking into account the cost of the electricity used to operate them. The following factors should also be borne in mind:

1. The necessary mounting height of fans to minimize draughts;
2. The effect of the spacing of fans and the distance of the impeller from the roof soffit;
3. Any risk to occupants from stroboscopic effects of blade movements;
4. The availability of multi- or variable-speed units.

Punkah fans may also be operated during summer months to provide air movement and offer a measure of convective cooling for occupants.

14.10.15 High-temperature high-velocity warm air heating systems

These systems, best suited to heating large, single spaces, may use indirect heating by gas or oil or direct gas heating. Relatively small volumes of air are distributed at high temperature (up to 235°C) and high velocity (30–42.5 m/s from heater unit) through a system of well-insulated conventional ductwork. Air outlets are in the form of truncated conical nozzles discharging from the primary ductwork system into purpose-designed diffuser ducts. The high-velocity discharge induces large volumes of secondary air to boost the outlet volume and reduces the outlet temperature delivered to the space, thereby reducing stratification. Most of the ductwork thermal expansion is absorbed by allowing free movement and long, drop-rod hangers are used for this purpose. Light, flexible, axial-bellows with very low thrust loads can also be employed where free expansion movement is not possible. System design and installation is generally handled as a package deal by specialist manufacturers.

14.11 Heating equipment – attributes and applications

14.11.1 Water system heating equipment

The range of heat emitters may be divided into three generic groups:

1. Radiant
2. Natural convective
3. Forced convective

Table 14.13 lists the principal types of appliance in each group, together with descriptive notes. Typical emission

Table 14.13 Characteristics of water system heating equipment

Type	*Description*	*Advantages*	*Disadvantages*	*Emission range*
Radiant				
Radiant panel	Consists of steel tube or cast-iron waterways attached to a radiating surface. Back may be insulated to reduce rear admission or may be left open to give added convective emission. Particularly useful for spot heating and for areas having high ventilation rates (e.g. loading bays), the radiant component giving a degree of comfort in relatively low ambient air temperatures.	No moving parts, hence little maintenance required; may be mounted at considerable height or, in low-temperature applications, set flush into building structure.	Slow response to control; must be mounted high enough to avoid local high intensities of radiation (e.g. onto head).	350 W/m^2 to 15 kW/m^2 of which up to 60% may be radiant.
Radiant strip	Consists of one or more pipes attached to an emissive radiant surface. Is normally assembled in long runs to maintain high water flow rates. The back may be insulated to reduce rear emission. When using steam, adequate trapping is essential together with good grading to ensure that tubes are not flooded. Multiple-tube types should be fed in parallel to avoid problems due to differential expansion. Hanger lengths should be sufficient to allow for expansion without lifting the ends of the strip. Heating media may be steam, hot water or hot oil.	No moving parts, hence little maintenance required; may be mounted at considerable height or, in low-temperature applications, set flush into building structure.	Slow response to control; must be mounted high enough to avoid local high intensities of radiation (e.g. onto head).	150 W/m to 5 kW/m of which radiant emission may be up to 65% of total.
Natural convective				
Radiators	Despite their name, 70% of the emission from these devices is convective. Three basic types are available; column, panel, and high output, the last incorporating convective attachments to increase emission. Panel radiators offer the least projection from the wall but emission is higher from column and high-output units. In application they should be set below windows to offset the major source of heat loss and minimize cold down-draughts.	Cheap to install; little maintenance required.	Fairly slow response to control. With steel panel radiators there is a risk of corrosive attack in areas having aggressive water, which may be accentuated by copper swarf left in the radiator. This leads to rapid failure unless a suitable inhibitor is used. Not suitable for high-temperature water or steam.	450–750 W/m^2.
Natural convectors	Compact units with high emissions. Often fitted with damper to reduce output when full emission not required, usually to about 30% of full output. Heat exchangers normally finned tube. Units may be built into wall of building.	May be used on high-temperature hot water or low-pressure steam without casing temperature becoming dangerously high: fairly rapid response to control.	Take up more floor space than radiators. Likelihood of fairly high-temperature gradients when using high-temperature heating media.	200 W to 20 kW.
Continuous convectors	Comprise single-or double-finned tube high-output emitters in factory-made sill-height sheet metal casings or builders' work enclosures which may be designed to fit wall to wall. Can be fitted with local output damper control, which reduces the emission to approximately 30% of the full output. They should be placed at the point of maximum heat loss, usually under	Take up relatively little space; give even distribution of heat in room. May be used with medium-temperature hot water or low-pressure steam without casing temperatures	May produce large temperature gradients on high-temperature heating media if poorly sited.	500 W/m to 4 kW/m.

Table 14.13 *continued*

Type	*Description*	*Advantages*	*Disadvantages*	*Emission range*
	windows. The wall behind the unit should be well insulated. For long-run applications distribution of flow water must be provided to modular sections of the unit to ensure that input remains reasonably consistent over full length of unit. For builders' work casings inlet and outlet apertures must have the free-area requirements stipulated by the manufacturer.	becoming dangerously high. Return pipework may be concealed within casing.		
Skirting heating	Finned-tube emitters in a single or double skirting height sheet metal casing, usually with provision for a return pipe within the casing. Applications and distribution of flow water similar to continuous convectors.	May be used on water or low-pressure steam. Give low-temperature gradients in the room. All pipework concealed.	Relatively low output per metre of wall. More work involved when installing in existing building as existing skirting has to be removed.	300 W/m to 1.3 kW/m.
Forced convective				
Far convectors	These units give a high heat output for volume of space occupied by the unit, together with the ability to distribute the heat over a considerable area using directional grilles. May be used to bring in heated fresh air for room ventilation. Leaving air temperatures should be above 35°C to avoid cold draughts. Where mixed systems of radiators and fan convectors are installed it is advisable to supply fan-assisted units on a separate constant-temperature circuit to avoid the above problems. To minimize stratification, leaving air temperatures above 50°C should be avoided. Must not be used on single-pipe systems. Care must be taken at design stage to avoid unacceptable noise levels. Control by speed variation or on/off regulation of fan.	Rapid response to control by individual thermostat. By use of variable speed motors rapid warm-up available in intermittent systems; filtered fresh air inlet facility.	Electric supply required to each individual unit.	2 to 25 kW.
Unit heaters	A unit with a large propeller or centrifugal fan to give high air volume and wide throws. Louvres direct the air flow in the direction required. May be ceiling mounted, discharging vertically or horizontally or floor mounted. Can be used with fresh air supply to ventilate buildings. Large units may be mounted at a considerable height above the floor to clear travelling cranes, etc. May be used with steam or hot water but care should be taken to restrict leaving air temperature, usually 40–55°C to avoid reduction of downward throw and large temperature gradients in the building. The air flow from the units should be directed towards the points of maximum heat loss. Control as for fan convectors.	Rapid response to control by individual thermostat; by use of multi-speed motors rapid warm-up available on intermittent systems; filtered fresh air inlet facility.	Electric supply required for each individual unit.	3 to 300 kW.

Table 14.14 Electric heating equipment

Type	*Description*	*Advantages*	*Disadvantages*	*Emission range*
Radiant				
High-intensity radiant heaters	Consist of high-temperature elements mounted in front of polished reflector. Element can be silica or metal sheathed wire (up to 900°C) or quartz lamp (up to 2200°C)	Fast response time. Little regular maintenance required. May be mounted at considerable height. Quartz lamp heaters have improved beam accuracy allowing higher mounting heights. Especially suitable for spaces with high air movement.	Must be mounted sufficiently high to avoid local high intensities of radiation (e.g. onto head).	0.5 to 6 kW per heater.
Low-temperature radiant panels	Consist of low-temperature elements (300°C and below) mounted behind a radiating surface. Thermal insulation behind the elements minimizes heat loss. Very low-temperature elements (40°C) used in ceiling heating applications.	No regular maintenance required. Set flush into building structure and unobtrusive.	Slow response time.	Up to 200 W/m^2.
Natural convective				
Storage heater	Consists of a thermal storage medium which is heated during off-peak electricity periods. A casing and insulation around the medium enables the heat to be gradually released throughout the day. Manual or automatic damper control allows 20% of heat output to be controlled.	No regular maintenance required.	Not intermittent. Limited control of heat charging and output.	Storage element sizes 1.4 to 3.4 kW.
Panel heaters, convectors or skirting heaters	Consist of a heating element within a steel casing with air grilles allowing the natural convection of air across the element. Generally controlled by an integral room thermostat.	No regular maintenance required. Cheap to install. Suitable for low heat loss applications. Fast response time for intermittent operation.	High surface temperatures. High-temperature gradients if poorly sited.	0.5 to 3 kW output.
Forced convective				
Storage fan heaters/ electricaire	These storage heaters have increased core thermal insulation and contain a fan which distributes warm air to the space to be heated. The fan is usually controlled by a room thermostat. Storage fan heaters are single room units but can heat an adjacent room with a stub duct. Electricaire units are larger and can be ducted to servce several areas. Up to 80% controllable heat.	Use off-peak electricity. Suitable for intermittent operation.	Heavyweight.	Fan storage heaters: 3 to 6 kW. Electricaire 6 to 15 kW. Industrial models up to 100 kW.
Fan convectors	These wall-mounted or free-standing units incorporate a fan which forces air over the heating elements into the space. High output rate relative to size. Can incorporate integral room thermostat control	Low maintenance. Fast response. Accurate temperature control. Suitable for highly intermittent heating applications. Low surface temperatures. Fan-only operation for summer use.		2 to 3 kW.
Downflow fan convectors	These units are forced air convectors mounted to direct the heated air downwards. High air flow and heat output. Often used to provide a hot air curtain over entrance doorways.	Suitable for localized heating.	Can be noisy.	3 to 18 kW.

Table 14.15 Direct gas-fired heating equipment

Type of heater	*Usual rating* (kW)	*Surface temperature* (°C)	*Flue system*	*Notes*
Radiant convector gas	4.4–7.3	1100 at radiant tips	Conventional	Wall mounted or at low level
Overhead radiant heaters	3.1–41	850–900	Flueless[a]	High level or ceiling mounted
Overhead tubular radiant	10–15	315 mean	Conventional or fan-assisted flue or flue-less[a]	High level or ceiling mounted
Convector heaters	2.5–16.7	—	Conventional or balanced flue or flueless[a]	Wall mounted or at low level
Fan convectors	1.4–3.7	—	Balanced flue or fan-assisted flue	Wall mounted or at low level
Make-up air heaters[b]	49–250	—	Flueless[a]	Mounted at high level and fan assisted
Unit air heaters[b]	17–350	—	Flued or flueless[a]	Mounted at high level and fan assisted

[a] The use of flueless appliances should be discouraged, since they discharge much moisture into the heated space.
[b] The installation of flueless appliances in excess of 44 kW is not permitted by the Building Regulations (1985) and application should be made to the local authority for the necessary waiver where installations of this size are contemplated.

Table 14.16 Direct oil-fired heating equipment

Type of heater	*Usual rating* (kW)	*Surface temperature* (°C)	*Flue system*	*Notes*
Radiant paraffin heaters	1.5	Less than 121[b]	Flueless[a]	Mounted at low or floor level
Flued radiant convector heaters	8.5–10.9	Less than 121[b]	Conventional	Mounted at low or floor level
Convective paraffin heaters	1–4	Less than 121[b]	Flueless[a]	Mounted at low or floor level
Air heaters	8.5–13.5	Less than 121[b]	Conventional	Mounted at low or floor level
Blown air heaters	10.7–16.7	Less than 121[b]	Conventional	Mounted at low or floor level
Warm air heaters	50–450	Maximum 60[c]	Flued	Floor mounted or overhead
Make-up air heaters	90–450	Maximum 60[c]	Flued	Can be flueless Floor or overhead mounted

[a] The use of flueless appliances should be discouraged since they discharge much moisture into the heated space.
[b] See BS 799.
[c] See BS 4256.

ranges are quoted for each type over its normal span of working temperatures. These are intended as a guide only and manufacturers' catalogues should be consulted for detailed performance values.

14.11.2 Electric heating equipment

Where electric heating equipment is installed within the space to be heated the total electrical input is converted into useful heat. There are two categories of electric heating equipment, direct acting and storage heating. The two types of electric heating can be used independently or to complement one another to meet particular heating requirements. Table 14.14 gives a brief description of the different types of electric heating.

14.11.3 Gas- and oil-fired heating equipment

Where gas or oil appliances are used for heating and installed within the heated space, between 70% and 90% of the total energy content of the fuel input will be converted into useful heat. Table 14.15 gives particulars of some gas-fired equipment types and Table 14.16 gives similar details for some oil-fired heaters. The first three types of equipment detailed in Table 14.15 and the first two in Table 14.16 are usually used for local warming of individuals rather than to provide a particular temperature throughout the space.

Acknowledgement

The author wishes to thank the Chartered Institute of Building Services Engineers for permission to reproduce data from the Heating and Plant Capacity Sections of the *CIBSE Guide* (Fig. A9.3, Tables A9.10, A9.11, B1.3, B1.12, B1.13, B1.14, B1.15, B1.16, B1.23, B1.24, B.125, B1.26) and Formecon Services Limited for their kind permission to reproduce data from *The Businessman's Energy Saver* data book (Tables 1–3).

15 Ventilation

P D Compton BSc, CEng, MCIBSE
Colt International Limited

Contents

15.1 Introduction

Any building, even with all windows and doors shut, will have a degree of ventilation (referred to as natural infiltration) by virtue of pressure differences across cracks or permeable materials in the structure causing air to flow through the structure. The degree of infiltration is governed by the quality of design and build of the structure and by the pressures generated by wind and thermal buoyancy.

This fortuitous infiltration is essential in otherwise unventilated buildings, since it allows ingress of air providing oxygen for us to breathe and for combustion equipment to burn. The infiltration is, however, effectively uncontrolled, and is often insufficient in quantity, at the wrong temperature or too contaminated to maintain satisfactory internal environmental conditions.

We must therefore be able to define the conditions we need and to design, install and operate systems to provide them. These systems cover three main areas: heating, ventilation and air conditioning. Heating and air conditioning are dealt with in Chapters 14 and 16. This chapter covers ventilation systems, defined as systems providing air movement through a space without artificially heating or cooling the air. It must be said, however, that, in practice, there is often a large degree of overlap, since office ventilation systems often provide heating in winter and complex ducted ventilation systems share much equipment and design procedures with air-conditioning systems.

This chapter is intended to provide guidance towards defining needs, assessing whether ventilation is the correct solution and selecting equipment and systems to match these requirements in as economic a manner as possible.

15.1.1 Reasons for ventilation

Ventilation is used to maintain a satisfactory environment within enclosed spaces. The environmental criteria controlled may be:

Temperature – relief from overheating
Humidity – prevention of condensation or fogging
Odour – dilution of odour from smoking, body odour, processes, etc.
Contamination – dilution or removal of dangerous or unpleasant fumes and dust

The required values for these criteria will depend upon the reason the space is being ventilated. It may be for the benefit of people, processes, equipment, materials, livestock, horticulture, building preservation or any combination of these. Guidance on selection of these values is provided by CIBSE[1] and ASHRAE.[4]

15.1.2 Definitions

Aerodynamic area – The effective theoretical open area of an opening. It is related to the measured area by the coefficient of entry or discharge (C_d).
Air-handling unit – A self-contained package incorporating all equipment needed to move and treat air, requiring only connection to ductwork and services to provide a complete ventilation system.
Coefficient (entry or discharge) – The ratio of aerodynamic (effective) area to the measured area of an opening. The value for a square-edged hole of 0.61 is used for most building openings.
Capture velocity – The air velocity needed to capture a contaminant at source, overcoming any opposing air currents.
Automatic fire ventilation – See *Smoke ventilation.*
Dilution ventilation – A ventilation strategy whereby contaminants are allowed to escape into the ventilated space and are then diluted to an acceptable level by means of the ventilation system.
Industrial ventilation – A term used to cover any ventilation system designed to remove contaminants. Its use is sometimes restricted to local extract systems.
Maximum Exposure Limit (MEL) – Maximum limits of concentration of airborne toxic contaminants, listed by the Health and Safety Executive[18] which must not be exceeded.
Occupational Exposure Standards (OES) – Limits of concentration of airborne toxic contaminants, listed by the Health and Safety Executive[18] which are regarded as safe for prolonged exposure for 8 hours per day.
Infiltration – Movement of air through a space with no specific ventilation openings by natural forces.
Local extract – A ventilation strategy whereby heat, steam or contaminants are captured at source and ducted to discharge outside the space.
Mechanical ventilation – See *Powered ventilation.*
Natural ventilation – A ventilation system in which air movement is produced through purpose-designed openings by natural forces (wind and thermal buoyancy).
Powered ventilation – A ventilation system in which air movement is induced by mechanical means – almost invariably a fan.
Smoke logging – The filling of a space with smoke in the event of fire.
Smoke ventilation – A ventilation system designed to remove smoke and heat in the event of fire to prevent or delay smoke logging allowing personnel to escape and firefighters to attack the fire.
Spot cooling – A ventilation strategy whereby the space temperature is allowed to rise and air movement is induced locally to provide comfort conditions within a limited area.
Threshold Limit Value (TLV) – Maximum values of concentrations of airborne toxic contaminants, listed by the American Conference of Governmental Industrial Hygienists[5] (ACGIH), regarded to be safe for 8 hours per day exposure.
Transport velocity – The air velocity required in a duct to transport a contaminant without it falling out of suspension.

15.2 Ventilation systems and controls

15.2.1 Natural ventilation

How natural ventilation works

Natural ventilation operates by means of airflows

generated by pressure differences across the fabric of the building. An airflow will occur wherever there is a crack, hole or porous surface and a pressure difference.

For the relatively large openings in which we are interested the flow rate can be found from the velocity of airflow generated through the aerodynamic area of the opening from the formulae:

$$V=\sqrt{\frac{2\Delta P}{\rho}} \qquad (15.1)$$

where V = velocity (m/s),
ΔP = pressure difference (Pa),
ρ = density (kg/m³).

Then flow rate:

$$\dot{V}=AC_d V \qquad (15.2)$$

where $\dot{V}$ = volumetric flow rate (m³/s),
A = measured area of opening (m²),
C_d = coefficient of opening.

For purpose-built ventilators the manufacturer will be able to provide values of C_d. For other openings it is conventional to use the value for a sharp-edged square orifice of 0.61.

The pressure can be generated by three mechanisms:

1. Powered ventilation equipment (see equation (15.3));
2. Buoyancy (temperature difference);
3. Wind.

In still air conditions the source of pressure difference to drive ventilation is buoyancy due to the decrease in density of heated air. In any occupied building there will be a higher temperature inside than outside due to heat gains from people, plant and solar radiation. The lighter heated air will try to rise, causing an increase in internal pressure at high level and a reduction at low level with a neutral plane between the two conditions. Any opening above the neutral plane will therefore exhaust air and any opening below the neutral plane will provide inlet air. Under steady heat load conditions a balance will be achieved with a throughput of air dependent upon the heat load and the size and location of the openings. Conditions at this balance point can be readily calculated using one of the following formulae:

For more than one opening (inlets all at one height, exhausts all at one height)

$$\dot{V}=A_e C_e\sqrt{\frac{2gH\Delta t}{\bar{T}}} \qquad (15.3)$$

For a single opening

$$\dot{V}=\frac{AC_d}{3}\sqrt{\frac{gh\Delta t}{\bar{T}}} \qquad (15.4)$$

where g = acceleration due to gravity (m/s²),
H = height between centre lines of inlet and outlet openings (m),
Δt = temperature difference between inside and outside (°C),
$\bar{T}$ = average of inside and outside temperatures (absolute) (K),
h = height of single opening (m),
$C_e A_e$ = overall effective opening size calculated from

$$\frac{1}{(C_e A_e)^2}=\frac{1}{(\sum C_i A_i)^2}+\frac{1}{(\sum C_v A_v)^2}$$

(subscript i denotes inlet opening, subscript v exhaust opening).

Under wind conditions a complex system of pressures is set up on the external surfaces of the building which will vary with wind speed and direction. Pressure coefficients Cp[6,7] define the relationship according to the formula:

$$\dot{V}=A_e C_e U_r\sqrt{(\Delta Cp)} \qquad (15.5)$$

where U_r = reference wind speed,
ΔCp = difference between coefficients at ventilation openings.

The coefficients Cp will vary across each surface of the building and, except for very simple shapes, can only be found by model or full-scale test. Since the coefficients will change with wind direction, complete calculation of wind-induced ventilation is very unwieldy, needing computer analysis.

When both wind and temperature difference act on ventilation openings the result is very complex, but a reasonable approximation of flow rate is made by taking the higher of the two individual flow rates. This means that we can, for ventilation design purposes, generally ignore wind effects and design on temperature difference only, since wind effects can be assumed only to increase the ventilation rate.

Advantages and disadvantages

Advantages	*Disadvantages*
Quiet	Variable flow rate and direction dependent upon wind conditions
Virtually no running cost	Filtration is generally impractical
Self-regulation (flow rate increases with heat load)	Limited ducting can be tolerated
Low maintenance cost	Effectiveness depends on height and temperature difference
Provides daylight when open (roof vent)	
Psychological appeal of clear sky (roof vent)	
Easy installation	

When to use natural ventilation

Natural ventilation is used in a number of situations:

1. Shallow-plan offices – by opening windows to remove heat and odour;
2. Large single-storey spaces (factories, warehouses, sports halls, etc.) – by roof and wall ventilators – to remove heat, contaminants, smoke, steam;
3. Plant rooms.

It is not suitable in situations where:

1. Dust, toxic or noxious contaminants must be removed at source;

2. Unfavourable external conditions exist requiring treatment to incoming air – e.g. noise, dust, pollution;
3. A steady controlled flow rate is required – e.g. hospitals, commercial kitchens;
4. Existing mechanical ventilation will affect the flow adversely;
5. Abnormal wind effects can be anticipated due to surrounding higher buildings;
6. The space is enclosed so as to have no suitable source of inlet air.

In many of these situations a system of natural inlet/powered exhaust or powered inlet/natural exhaust will be the best option.

Control

Low-level ventilation openings, whether windows, doors or ventilators, are generally manually operated for simplicity and economy, allowing personnel to control their own environment. High-level openings can also be manually controlled by means of rod or cable operation, although this has generally lost favour (except in the case of simple windows) and automatic operation is preferred.

Automatic operation may be by means of compressed air, operating a pneumatic cylinder, or electricity. Pneumatics are generally favoured for industrial applications and electricity for commercial premises. Economy of installation is normally the deciding factor, since running costs are low for either system.

Automatic control allows a number of options to be considered to provide the best form of control for the circumstances. Generally available controls offer the following features:

1. Local control by personnel;
2. Automatic thermostatic control (single or multiple stage);
3. Fire override to open ventilators automatically by means of a connection to the fire-detection system or fireman's switch. This normally overrides all other control settings;
4. Timeswitch control to shut ventilators during unoccupied periods;
5. Weather override to close ventilators during rain or snow;
6. Wind override to shut high-level exhaust ventilators on windward walls (mainly used for smoke ventilation).

15.2.2 Powered (mechanical) ventilation

How powered ventilation works[9]

By definition, a powered ventilation system includes a mechanical means of inducing an airflow using an external power source. This is invariably an electrically driven fan. When a fan blade rotates it does work on the air around it, creating both a static pressure increase (P_s) and an airflow across the fan. The airflow has a velocity pressure associated with it, defined as $P_V = \frac{1}{2}\rho V^2$, and the fan can be described as producing a total pressure $P_T = P_s + P_V$. The pressure generated is used to overcome pressure losses (resistances) within the ventilation system.

Each fan has a unique set of characteristics which are normally defined by means of a fan curve produced by the manufacturer which specifies the relationship between airflow, pressure generation, power input, efficiency and noise level (see Figure 15.1). For geometrically similar fans the performance can be predicted for other sizes, speeds, gas densities, etc. from one fan curve using the 'fan laws' set out below.

For a given size of fan and fluid density:

1. $\frac{\dot{V}_1}{\dot{V}_2} = \frac{N_1}{N_2}$ Volume flow is directly proportional to fan speed

2. $\frac{P_1}{P_2} = \left(\frac{N_1}{N_2}\right)^2$ Total pressure and static pressure are directly proportional to the square of the fan speed

3. $\frac{W_1}{W_2} = \left(\frac{N_1}{N_2}\right)^3$ Air power and impeller power are directly proportional to the cube of the fan speed

For changes in density:

4. $\frac{P_1}{P_2} = \frac{W_1}{W_2} = \frac{\rho_1}{\rho_2}$ Pressure and power are directly proportional to density and therefore for a given gas are inversely proportional to absolute temperature

For geometrically similar fans operating at constant speed and efficiency with constant fluid density:

5. $\frac{\dot{V}_1}{\dot{V}_2} = \left(\frac{D_1}{D_2}\right)^3$ Volume flow is directly proportional to the cube of fan size

6. $\frac{P_1}{P_2} = \left(\frac{D_1}{D_2}\right)^2$ Total pressure and static pressure are directly proportional to the square of fan size

7. $\frac{W_1}{W_2} = \left(\frac{D_1}{D_2}\right)^5$ Air power and impeller power are directly proportional to the fifth power of fan size

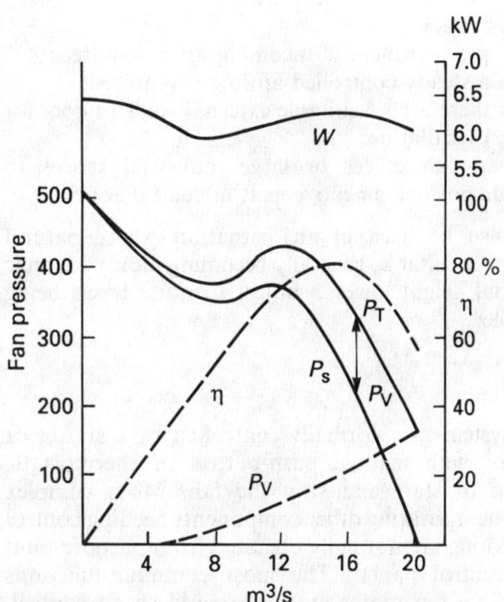

Figure 15.1 Typical fan curve for an axial fan

where $\dot{V}$ = volumetric flow (m^3/s),
P = pressure (kN/m^2),
W = power (W),
D = size parameter (diameter) (mm).

In passing through the fans gases are compressed slightly due to the increase in pressure. For absolute accuracy note should be taken of this effect using the gas compressibility factor which will affect flow rate, static and total pressure and power. However, in most fan systems the effect is very small, since the pressure increase through the fan is insignificant compared to atmospheric pressure. By convention, compressibility effects are therefore normally ignored.

Since the pressure generated by most fans is far in excess of pressure differences due to buoyancy and wind, the performance of a powered ventilation system is effectively independent of these, and flow rates and directions can be confidently predicted and will be constant regardless of conditions. The high-pressure generation also allows resistive components such as heater batteries, filters and attenuators to be used within the system.

Advantages and disadvantages

Advantages	*Disadvantages*
Weatherproof	Fixed air flow – not self-regulating
Predictable constant performance	Running costs (electrical and maintenance)
Air treatment can be incorporated	Noise
Fresh air can be delivered at optimum volume, velocity and temperature	

When to use powered ventilation

Powered ventilation is essential in some instances:

1. Local extract;
2. When pre-treatment of incoming air is required;
3. When a steady controlled airflow is required;
4. When there are no suitable external walls or roof for natural ventilation;
5. In deep-plan offices or large industrial spaces to provide positive air movement in central zones.

It can also be used in any situation where natural ventilation is suitable, generally becoming more economic as the roof height lowers, subject to noise levels being acceptable.

Control

Simple systems are normally controlled by a starter or contractor with manual push-button or thermostatic operation to start and stop the fan. More complex systems incorporating other components needing control or monitoring are normally operated from purpose-built central control panels. The most common functions provided are fan motor stop, start and speed control, damper control, filter-condition indication and heater battery control. For optimum control the system should be automatically controlled from thermostats or other sensors and a timeswitch.

15.3 Powered ventilation equipment

This falls into two basic groupings: supply air systems and extract systems. The equipment used for both is similar, comprising, as a minimum, a fan and weatherproof cowl, plus ducting, air-treatment equipment and grilles as required.

15.3.1 Fans

Five main types of fan are used in ventilation systems as described below.

Centrifugal

In this type air enters the impeller axially and is discharged radially into a volute casing. The airflow therefore changes direction through 90°, which can make this type of fan difficult to use within a ducted system. Two blade types are used, backward curved providing high-pressure at low volume flow and forward curved providing medium-pressure and volume flow. Typical static efficiencies are 70–75% and 80–85%, respectively.

Axial

Air enters and leaves the fan axially giving a straight-through configuration. Duties are usually high- to medium-volume flow rates at medium to low pressures. In its simplest form there is an impeller and its drive motor only mounted within the cylindrical casing, and the discharge flow usually contains a fairly pronounced element of rotational swirl which may, if not corrected, materially increase the resistance of the downstream part of the system. More sophisticated versions include either downstream or upstream guide vanes to correct the swirl. Typical static efficiencies are 60–65% or 70–75% with guide vanes.

Propeller

This is really a simple form of axial fan but with its impeller mounted in a ring or diaphragm which permits it to discharge air with both axial and radial components. Duties covered are high volume and low pressure. Static efficiency is normally under 40%.

Mixed flow

This is a fan in which the air path through the impeller is intermediate between the axial and centrifugal types giving the benefit of increased pressures but capable of being constructed to provide either axial or radial discharge. Static efficiency is typically 70–75%.

Cross flow

This type normally has a long cylindrical impeller having a relatively large number of shallow forward-curved blades. Due to the shape of the casing surrounding this impeller, air enters all along one side of the cylindrical surface of the impeller and leaves on another side. Static efficiency is typically 40–50%.

In general, axial fans are used for roof extract units and small ducted systems and centrifugal fans for large-ducted systems.

15.3.2 Roof extract units

These are the most commonly used powered ventilators in large open buildings such as factories, warehouses and sports halls. Mounted directly onto the roof or wall, they comprise an axial fan, a safety grille and a weatherproof casing. Two forms are normally available, the vertical-discharge type which tends to have a complex casing arrangement but which throws the exhaust clear of the building, and the low-discharge type which has a simple casing but directs the exhaust onto the roof of the building. Vertical discharge is essential when smoke or fumes are being exhausted (Figures 15.2 and 15.3).

These ventilators can normally be used with limited ducting or accessories. A variation, fitted with a centrifugal fan, is available for more extensive ducting and is often used for duties such as toilet extract in commercial buildings.

15.3.3 Roof inlet units

This specialized form of supply air system is often used in large open industrial spaces. It comprises a modular system of components which can be built up into simple systems. A typical system might have a roof inlet cowl, a recirculation damper, a heater battery, a fan, one or two outlet grilles and short sections of connecting ductwork, and would handle airflows up to 3–4 m^3/s, depending on size. A number of individual systems would be used to provide the total airflow required in the space (Figure 15.4). Systems are normally manufactured with aluminium casings to reduce the roof load.

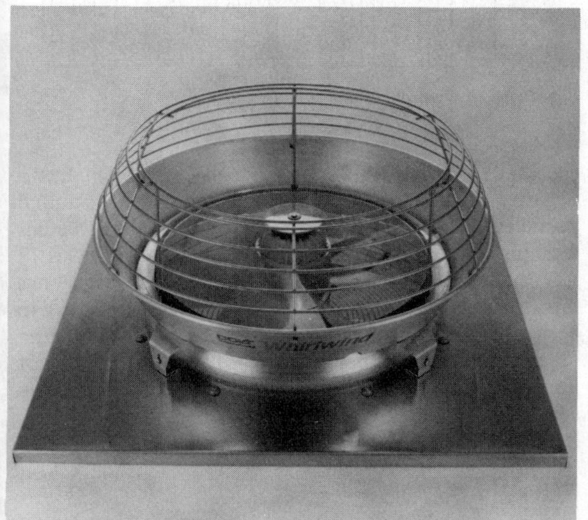

Figure 15.2 Vertical discharge powered roof ventilator

Figure 15.3 Low discharge powered roof ventilator

15.3.4 Ducted systems[10,11]

Larger ducted ventilation systems, as used in offices and commercial premises using a central air-handling unit and fabricated distribution ductwork, are akin to air-conditioning systems but with less treatment to the air at the AHU (see Chapter 16).

15.3.5 Local extract systems[12]

Local extract systems are designed specifically to remove fumes, dust, mists, heat, etc. at source from machinery and fume cupboards. The main design considerations are capture of the contaminant which will normally involve special hoods or cabins, and extract at sufficient velocity to satisfactorily transport the contaminant. Ductwork must be manufactured to resist abrasion or corrosion and sufficiently well sealed to prevent leakage. Welded ductwork is often needed. The fan may also need protection and the motor may need to be flameproof or out of airstream if the contaminant is flammable or corrosive. Treatment of the exhaust may be required to reduce pollution and nuisance and to comply with legislation.

Figure 15.4 Roof inlet system

15.3.6 Air cleaners

A wide range of types of air cleaners are available to match the number of contaminants needing removal from air. Figure 15.5 shows typical particle size ranges and the range of operation of each type of air cleaner.

Filters

Filters are a type of air cleaner in which a membrane of some kind is placed across the duct so that all airflow has to pass through it. In passing through, the particles can be separated.

There are three main types of filter: viscous, dry and HEPA. Viscous filters are normally a coarse weave of glass, plastic or metal strands coated with oil. As the air passes through the filter, particles impact against the material and are held in the oil. The panels are normally washable and can be re-used after re-oiling. Viscous filters provide good general filtration for air inlet to buildings.

Dry filters are normally a finer-weave fabric or fibrous material and will separate smaller particles than viscous filters, typically down to 0.5 μm. They may be thrown away or washed when dirty, depending upon type.

HEPA (or absolute) filters provide the greatest degree of air cleaning. These filters, which are akin to paper or felt, will capture particles down to 0.01 μm with efficiencies of up to 99.995%. They are generally used for clean room applications and are expensive.

If a high degree of filtration is required it is normal practice to have several grades of filter in series so that the coarser, cheaper filters can take most of the dust load, extending the life of the more expensive filters. Filters add a fairly high resistance to the ducting system, increasing the energy use and running cost of the system, and they require regular maintenance.

Electrostatic precipitators

In these devices the incoming air is passed through an ionizer which gives each particle a positive charge

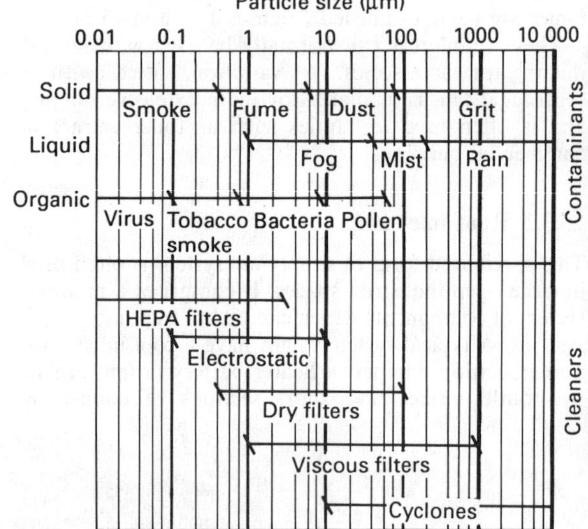

Figure 15.5 Typical particle sizes

(+6000 V) and then through a collector which has a negative charge (−6000 V). The positively charged particles are attracted to the collector walls. Regular washing is required to remove the collected particles.

Very small particles can be collected but large ones (above 5 μm) can cause arcing across the narrow gap between collector plates. These large particles should first be removed by a coarse filter. Precipitators are not normally suitable for wet environments since excessive moisture can cause electrical tracking and failure. Resistance to airflow is negligible.

Activated carbon filters

These are used to provide adsorption of gases and vapours, most commonly to remove odours from extracts from kitchens or industrial processes. An efficiency of 95% is obtainable and the carbon has a long life since it can be reactivated by heating.

Cyclones

These are dynamic setting chambers in which a vortex is generated by the high-velocity tangential inlet airflow. Centrifugal force holds particles close to the wall where they collect and drain down if wet or spiral down if dry. The contaminant is collected at the bottom and the cleaned air is exhausted through the top. They range in size from small 200 mm diameter units used to collect oil mist from machinery to the large units seen outside paper and woodworking factories.

15.4 Natural ventilation equipment

The natural ventilation equipment commercially available can be split into two basic groups, fixed and controllable. The fixed ventilation, normally a weathered louvre system, is used in applications where ventilation is constantly needed, winter and summer, with no need for control to maintain human comfort conditions. Typical examples are plant rooms and building block drying rooms. Louvre systems are also often used for inlet or exhaust to powered ventilation or air-conditioning systems and for architectural cladding. A number of years ago, before energy efficiency became important, many factories had uncontrollable ridge vents of various types, but most have now been replaced. Controllable ventilation, which may be weatherproof if required, is used in buildings which are normally occupied and where waste process heat is not sufficient to heat a ventilated building in winter (i.e. most buildings).

15.4.1 Fixed ventilation

Fixed ventilation has to be weatherproof to some degree. If it were not, then a simple hole in the wall or roof would be sufficient. The normal form of fixed ventilator is the louvre panel. At its simplest, this may be a number of slats of wood mounted at 45–60° in a door or wall opening, and at its most sophisticated an aerodynamically designed two- or three-bank 'chevron' of roll-formed or extruded aluminium (Figure 15.6).

With all louvres there is a balance between their weatherproofing qualities and their airflow, and it is important to select a louvre with the correct balance for each application. No louvre can be guaranteed to be 100% waterproof under all conditions, but the best ones approach this standard under normal flow and wind conditions.

Louvres can be supplied in various materials and finishes, the most common being anodized or painted aluminium, since this provides good corrosion resistance and light weight. Other options are galvanized steel or, for more rigorous conditions, stainless steel.

Most louvre systems have a pitch between 30 mm and 100 mm. Where noise control is required, louvres can be supplied either in the form of acoustic louvre with a pitch of 300 mm and acoustic material in the louvre blade or as standard louvre with an attenuator section behind it. A bird or insect guard (as applicable) is normally mounted behind the louvre.

Figure 15.6 Fixed louvre system

15.4.2 Controllable ventilation

Controllable ventilation is normally provided by one of three types of ventilator: louvred, opening flap and weathered. The most common is the louvred ventilator which has a number of centre hinged louvre blades controlled from a pneumatic ram or electric actuator which can be fully opened when ventilation is needed and fully closed to a weatherproof condition when it is not. These may normally be roof or wall mounted (Figure 15.7).

Opening flap ventilators have one or two hinged flaps (opaque or glazed) which are normally held closed but which may be opened to between 45° and 90°, depending upon design. Control is normally pneumatic or electric. Specific designs are available for roof or wall mounting (Figure 15.8).

Weathered ventilators need to have a complex air path in order to prevent rain entry so they therefore provide a restricted airflow. To overcome this, multifunctional ventilators are often used providing a direct airpath during dry weather and a restricted one during rainfall.

Figure 15.7 Louvred roof ventilator

Figure 15.8 Flap type roof ventilator

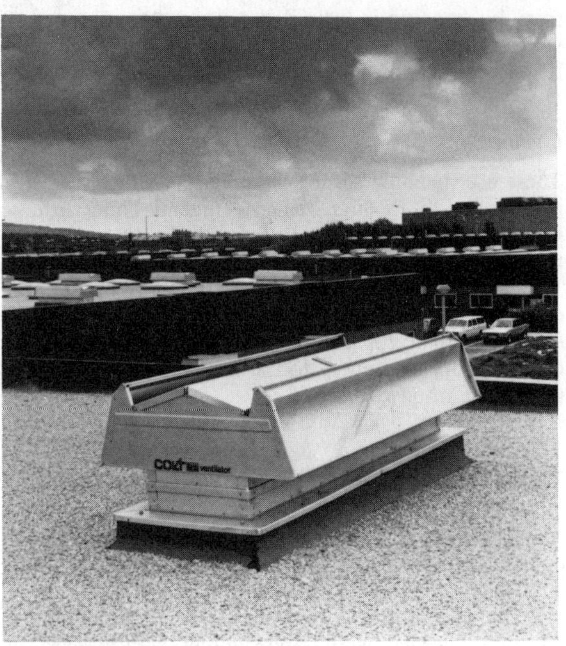

Figure 15.9 Weathered roof ventilator

These ventilators are normally only roof mounted (Figure 15.9).

Most ventilators are powered in one direction with spring return. Thus a pneumatic ventilator might be described as 'pressure to open' or 'pressure to close'. With the pressure to open type the ventilator will fail to the closed position under the influence of the return spring, ensuring the building remains weatherproof. Where ventilators are installed mainly as smoke ventilators it is important that they fail to the open position, so pressure to close ventilators should be used. A fusible link is normally fitted into the controls so that in the event of fire, affected ventilators will open automatically to release smoke and heat.

Most ventilators are made of aluminium because of its good corrosion resistance and light weight. On industrial premises they are normally left in mill finish and allowed to oxidize. On commercial premises a poly-powder paint finish is normally specified for aesthetic reasons. To reduce energy losses during the heating season many ventilators are available either as standard or optionally with polypropylene pile or rubber seals to reduce leakage, and with insulated surfaces.

A limited range of accessories may be provided, normally bird guard, insect guard or attenuators, but since airflow through the ventilators is driven by very small pressure differences (perhaps 5 Pa), great care has to be taken to ensure the ventilation remains effective.

15.5 System design

15.5.1 Overheating

Overheating can be due to a number of causes, but is usually from solar gain and machinery heat losses. If ventilating to provide thermal comfort in a space for humans or animals there are a number of considerations apart from simply air temperature:

1. Air temperature
2. Mean radiant temperature
3. Radiant temperature asymmetry
4. Air velocity
5. Humidity

Thus someone working near a window will receive hot solar radiation through the window in summer and cold radiation from the cold window surface in winter, causing uncomfortable radiant asymmetry even if the room temperature is perfect for personnel working away from the window. Ventilation will not be the correct solution for this problem – sun shading and double glazing would be more effective.

It is therefore important when considering ventilation design to consider the whole thermal environment and not to simply assume that reduction of air temperature to 20–22°C will cure all problems.

Factories, warehouses, etc.

Two methods of preventing overheating are available: either general ventilation to provide fairly even conditions over the whole space at working level or local ventilation to give spot cooling for localized hot spots.

General ventilation is designed to keep air temperatures within a few degrees Centigrade of the outside shade temperature over the majority of the floor area. The ventilation rate required, which is not directly related to the space floor area or volume, is calculated from the formula:

$$\dot{V}=\frac{Q_T}{C_p\Delta t}\ \mathrm{m^3/s} \tag{15.6}$$

where $\dot{V}$ = ventilation rate required (m^3/s),
Q_T = total heat load in the building (W),
C_p = specific heat of air
Δt = temperature difference between inlet (°C) and extract air.

The heat load (Q_T) includes all sources of heat in the building – electrical equipment, furnaces, people, solar gain, etc. The temperature difference (Δt) is selected to give the most economical design without compromising comfort and is generally found by allowing a 1.8°C temperature rise per metre height between inlet and extract ventilators. For natural ventilation the inlet and extract areas required can then be calculated from equation (15.3). Either powered or natural ventilation may be used. See Section 15.3 for a discussion of the benefits of each system.

The positioning of equipment is important to ensure that the whole space is ventilated without any significant dead spots of stagnant air. Generally, extract ventilators should be positioned as high as possible and spread evenly over the roof area with no more than 15 m between adjacent units. In smaller buildings the inlet is usually provided through existing openings such as loading-bay doors unless security precludes this. In larger buildings, inlets should be provided on at least two sides and, if any area is more than 20 m from an outside wall, consideration should be given to providing some local powered inlet to provide a positive airflow in that area.

If the heat load in the building is high enough for ventilation to be needed in autumn and spring, when air temperatures are fairly low, then the design must either ensure that airspeeds through the inlets are not high enough to be felt as a cold draught, which limits velocity to around 1 m/s, or provide for some heating of the inlet air to raise it to an acceptable temperature. (Partial recirculation can achieve this economically.)

Local ventilation is used where conditions in most areas are acceptable or unimportant but a small area or areas require cooling, generally because a workstation is close to a source of heat. In this situation there are three methods of treatment:

1. Reduce heat output or provide shielding. Radiant heat can be reduced by several methods such as insulating the equipment's hot surfaces, painting it with aluminium paint to reduce heat emission or by installing a screen (two layers of expanded metal) between the workstation and the equipment.
2. Provide local air movement. Creating local air movement, even with warm air, provides a reduction in effective temperature and is inexpensive. A 'punkah' type fan may be used, although this can have the effect of drawing down warmer air from a high level, or a portable 'fan in a box'. It is advisable to have a multi-speed fan to allow the effect to be adjusted by the operator for maximum comfort.
3. Provide fresh air. For maximum effect, a ducted fresh-air inlet system will provide cool air in summer and, if fitted with a recirculation unit, allow a flow of tempered or warm air in winter. The air outlet should be positioned to ensure that the airflow does not cause a draught on the back of the neck.

Offices and public spaces

In factories, air movement can be generated by fairly unsophisticated equipment, since there is generally reasonable scope for equipment location and noise levels are often not critical. In offices and public spaces, the space, aesthetic, noise and air movement criteria tend to be much tighter, and additional natural ventilation above that already provided by windows is normally impractical. The ventilation rate needed can still be calculated using equation (15.6), although the acceptable temperature difference will be much lower (typically, 5°C).

It is generally only practical to ventilate office spaces by means of a ducted system. If sufficient opening window area is available, either an inlet or extract system alone may be satisfactory, but if optimum control of conditions is required, then both mechanical inlet and extract should be provided.

Inlet systems should incorporate heating if winter ventilation is needed and air filtration if the ambient air is not clean enough, as is common in city centres. Extract grilles are normally at high level and inlet may be high or low level. The design and position of the grilles or diffusers is critical for successful ventilation to avoid draughts and stagnant areas, and flow patterns will vary, depending upon whether the airflow is heating or cooling. ISO 7730 recommends that for light sedentary work the mean air velocity in the occupied area should not exceed 0.25 m/s in summer and 0.15 m/s in winter, and temperature gradients should not exceed 3°C between the ankle and neck.

Two design methods are commonly used: displacement ventilation, where clean air is introduced at low velocity at low level, picking up heat and odour as it passes through the room and rising by buoyancy up to ceiling extracts; and diffusion ventilation, whereby clean incoming air at higher velocity is mixed with the air in the space, with the resulting mixture extracted at ceiling level.

15.5.2 Fume dilution

Factories, warehouses, etc.

Dilution of fumes in these areas is generally required for one of two reasons: either to reduce the level of harmful (toxic or irritant) fumes to a safe level, normally below the OES[18] (Occupational Exposure Standard) or to dilute offensive odours. Care must be taken with the latter to ensure that the problem is not merely passed on to neighbours. If it is, then a local extract with air-cleaning equipment will be preferred if it is practical. Indeed, a local extract system is always preferable, since it removes the problem at source, resulting in a cleaner environment within the building.

The starting point for design of a dilution ventilation scheme is normally a hygiene survey in which levels of pollutants and the ventilation rate are measured under worst conditions. The required ventilation rate is then calculated by ratio of level of contamination measured and required, where

$$\dot{V}_r = \frac{\dot{V}_m \times C_m}{C_r} \quad \text{m}^3/\text{s} \qquad (15.7)$$

where $\dot{V}_r$ = ventilation rate required (m^3/s),
$\dot{V}_m$ = ventilation rate measured (m^3/s),
C_r = contamination level required,
C_m = contamination level measured.

The ventilation should be designed so that airflow is directed from clean to dirty areas to keep the majority of the building as clean as possible. However, this form of ventilation does little for operatives working at the source of contamination.

Normal ventilation equipment can be used unless the fumes are corrosive or flammable. Powered equipment is normally employed to ensure that a steady airflow is provided.

Since the ventilation will be needed both summer and winter, there will be an energy penalty for the high level of airflow, and either the inlet airflow must be pre-heated or additional space heating will be required. The extra heat requirement is given by

$$Q = 1.2 \times (\dot{V}_r - \dot{V}_m) \times (t_c - -1) \quad \text{kW} \qquad (15.8)$$

where Q = extra heat input needed (kW),
t_c = winter room temperature (°C).

Offices and public spaces

In these areas fume dilution is normally required to reduce body odour and tobacco odours. CIBSE[1] gives current UK recommendations for fresh air requirements for various rooms to overcome these, varying from 5 to 25 l/s per person. These figures give a balance between energy loss and air purity, and higher levels of ventilation may be needed if a clean atmosphere is of overriding importance.

In most older offices the natural ventilation achieved by infiltration is sufficient to provide reasonable conditions, but areas of high occupancy such as conference rooms, theatres, bars and restaurants normally require mechanical ventilation. Sufficient ventilation must be provided to match the highest demand, whether for odour dilution or summer overheating.

An alternative to ventilation to clear tobacco smoke is the use of electrostatic air cleaners which clean and recirculate air within a room. These provide smoke dilution without the energy penalty of extra ventilation but require frequent cleaning to keep operating at maximum efficiency.

15.5.3 Prevention of condensation

Condensation occurs whenever moist air comes into contact with a surface which is colder than the 'dewpoint' of the air. The dewpoint is the minimum temperature at which the air can contain the amount of moisture within it, and it will vary with moisture content. Condensation can also occur in the air when warm moist air meets cold air, when it is known as 'fogging'.

Condensation will appear on the inside surface of porous or impervious materials, forming first on the worst-insulated surfaces (normally glazing or steelwork). On porous surfaces condensation can occur within the material or at an internal boundary. This is known as interstitial condensation, and it is especially dangerous, since it is often not known about until it has caused noticeable damage.

Condensation can be avoided or reduced by several methods:

1. *Insulation* – Increase the insulation of surfaces to raise their temperature above the dewpoint of the internal air. This should be treated with care, since insulation of one surface may merely shift condensation to the next worst-insulated surface, or cause interstitial condensation.
2. *Reduce moisture emission* – If practical, this is always the best method of avoiding condensation. It can be achieved by good housekeeping, reducing steam leaks, covering open tanks when not in use or by changing working practices or modifying equipment either to

contain the moisture more efficiently or remove it directly from the space.

3. *Ventilation* – Increasing the ventilation rate through a space while keeping the air temperature high will reduce the moisture content of the air without lowering surface temperatures and, by reduction of the dewpoint, will limit or avoid condensation. The hot, moist air exhausted takes a lot of energy with it, which it may be practical to recover using a heat-recovery system.
4. *Dehumidification* – Increasing the ventilation rate through a building has a high-energy penalty. Dehumidifiers do not increase the ventilation rate but recirculate air while removing the moisture from it by a refrigeration process. Their electrical cost is high, but since most of the energy is released into the room as heat, the overall running costs can be significantly lower than for a ventilation system.

Calculation for condensation problems is complex but is covered in the CIBSE Guide[1] in some depth.

15.5.4 Local extract

Local extract is used to remove contaminants directly from a process to the exterior without passing through personnel breathing zones. It thus provides a high degree of safety and because small volumes of air are extracted relative to a dilution ventilation system it is energy efficient.

Where dusts, grits, sawdust or other large particles are being extracted it is normally mandatory to include an air-cleaning device before the air is exhausted. Unless they are toxic or odorous fumes can normally be simply exhausted into the atmosphere at a suitable location well away from anywhere with normal personnel access.

The design of a system can be broken into three main areas: capture, transport and cleansing. Capture of the contaminant is of paramount importance. Depending upon the source of the contaminant, capture may be via a hood, slot, booth or enclosure, with the airflow designed to take the contaminant from the source into the duct system without passing through the operatives' breathing zone. Some typical examples of good design are shown in Figure 15.10. To capture the contaminant, a minimum air velocity, referred to as the 'capture velocity', is needed at the source. The value of the capture velocity depends upon the type of source, enclosure and local air movement. Typical values are given in Table 15.1. Full guidance in the design of capture systems is given by the ACGIH.[12]

Once captured, the contaminant has to be carried along a duct system. If the duct velocity is too low, particles will tend to drop out of suspension and collect in or fall back down the duct. It is therefore essential that a suitable minimum duct velocity, referred to as the transport velocity, is maintained. Typical values are given in Table 15.2. Nothing is gained by velocities far in excess of the recommended transport velocity, and in some cases as much is lost, since abrasion of particles on the duct can cause premature erosion and failure.

Removal of particles from the airstream is generally carried out in a cyclone, positioned outside the building and taking particles from all sources in the building. Care should be taken to position cyclones away from noise-sensitive areas since both ducts and cyclones can be noisy, especially if grits or chips are being carried.

Selection of fans and ducting for local extract must be more rigorous than for other systems. The fan must be capable of withstanding abrasion or corrosion from the contaminants, and if they are flammable must have a flameproof or out-of-airstream motor. The ducting also must be able to withstand abrasion or corrosion and must be fully sealed to prevent escape of contaminants within the building. For specialist applications such as fume cupboard extract, ducting and fans are often of plastic construction.

Table 15.1 Capture velocities

Condition of dispersion of contaminant	*Examples*	*Capture velocity (m/s)*
Released with practically no velocity into quiet air	Evaporation from tanks; degreasing, etc.	0.25–0.5
Released at low velocity into moderately still air	Spray booths; intermittent container filling; welding; plating	0.5–1.0
Active generation into zone of rapid air motion	Spray painting in shallow booths; barrel filling	1.0–2.5

Table 15.2 Transport velocities

Contaminant	*Typical transport velocity (m/s)*
Vapours, gases	5
Fumes	7–10
Light dusts	10–12
Dusts	12–17
Industrial dusts	17–20
Heavy or moist dusts	20+

15.5.5 Smoke ventilation

Smoke-ventilation systems are designed to clear smoke and heat from a building in the event of fire. In large open spaces it is impossible to 'smother' a fire (as is often recommended in domestic situations) by closing doors and windows. The aim of smoke ventilation is to minimize damage due to smoke staining and heat and to assist evacuation and firefighting by providing a layer of clear air below the smoke. Without smoke ventilation a space can become 'smoke logged' from ceiling to floor in only a few minutes.

The actual design of a smoke-ventilation system is very complex, and although much published guidance is available[14–17] it should be left to experts, and the design should be vetted by a fire-prevention officer.

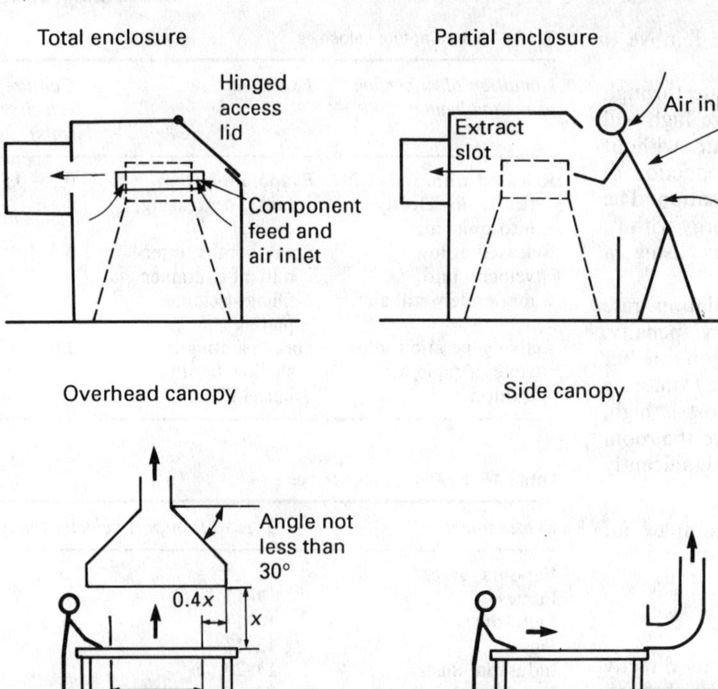

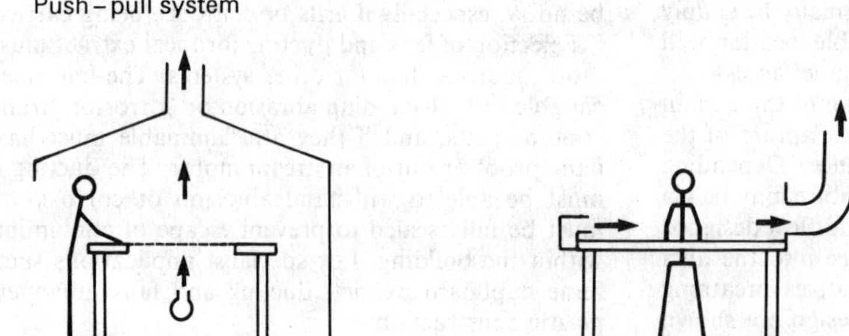

Figure 15.10 Typical local extract hoods and enclosures

In discussing design or vetting tenders, there are a number of important points to consider:

1. *Design fire size* – An accurate assessment is needed of the maximum size of fire which is likely and which can be designed for. Since the whole design is based upon this value it is essential that this be carefully considered.
2. *Interaction with sprinklers* – The likely size of fire will be smaller in sprinklered buildings and the temperature of the smoke produced will be lower since the smoke is cooled by the sprinkler flow.
3. *Fire detection* – For efficient operation, control of the ventilation must be linked to a fire-detection system or sprinkler flow switch to ensure that the ventilators are operated as early as possible. Natural ventilators must incorporate a fusible link or bulk as a back-up fail-safe device, but this should not be considered as the main form of emergency operation.
4. *Fire resistance* – Fans, motors, cabling and controls which are expected to operate under fire conditions must be suitably rated for the temperature expected.
5. *Air inlet* – A suitable low-level inlet ventilation area must be provided for the expected air flow rate. Where personnel escape routes will be used for inlet, the inlet velocity must be low enough not to impede progress.

Smoke ventilation is not intended to replace other forms of fire prevention and control but to work as an important component in an overall scheme.

15.6 Legislation and codes of practice

15.6.1 Legislation

Most legislation regarding ventilation is aimed at controlling the environment within the workplace, and until very recently there was a plethora of assorted and outdated regulations in force. Most of these have now been repealed under COSHH.

Legislation regarding the state of the air exhausted into the external environment is limited, the Environmental Protection Act being the main Act in force. The major pieces of existing legislation are discussed briefly below.

Health and Safety at Work etc. Act 1974 (HASAWA)

This does not directly relate to ventilation but places duties upon employers and employees regarding health and safety and is an enabling Act for further Regulations.

Control of Substances Hazardous to Health Regulations 1988 (COSHH)

Enabled under HASAWA, these Regulations provide a requirement for adequate control of hazardous substances and therefore directly cover use of ventilation to keep the atmosphere clean within the workplace. Ventilation equipment installed to comply with COSHH must be regularly inspected and tested at no more than 14-month intervals, and monitoring may be required to ensure its continued effectiveness.

Factories Act 1961

This Act requires effective and suitable provision of ventilation to supply fresh air and render harmless, so far as practical, fumes, dusts, etc.

Highly Flammable Liquids and Liquefied Petroleum Gases Regulations 1972

These require mechanical ventilation to be provided, preferably by local extract from cabinets or enclosures, to avoid dangerous concentrations of vapours occurring in workrooms.

Woodworking Machines Regulations 1974

These Regulations require local extract to be provided at designated machines to remove sawdust, chips, etc. and for all solid particles collected to be discharged into suitable receptacles.

Offices, Shops and Railways Premises Act 1963

This Act requires that in all workrooms there must be effective and suitable means of ventilation.

Environmental Protection Act 1990

This Act controls emission of pollutants from ventilation systems by classification as 'Statutory Nuisances'. Local council environmental health departments acting on a complaint or otherwise can classify an emission as a statutory nuisance and issue an Abatement Order requiring emission to be stopped or reduced. This strengthens and supersedes the provisions of the Public Health Act 1936.

Control of Pollution Act 1974

This Act allows local authorities to require the occupier of any premises to provide estimates or other information as specified concerning emission of pollutants.

Note: These synopses are the author's interpretation only. Neither the author nor the publishers can take any responsibility for any result of any actions taken as a result of reference to this section.

15.6.2 Codes of Practice

BS 5925: 1980: Design of buildings: ventilation principles and designing for natural ventilation

This British Standard gives recommendations on the principles which should be observed when designing natural ventilation of buildings for human occupation and provides a basis for choice between natural and powered systems.

BS 5720: 1979: Code of Practice for mechanical ventilation and air conditioning in buildings

This deals with design, planning, installation, testing and maintenance of systems.

BS 6540: Part 1: 1985: Methods of test for atmospheric dust spot efficiency and synthetic dust weight arrestance

This gives standard test methods for filters used in ventilation and air-conditioning systems.

ISO 7730: Moderate thermal environments – determination of the PMV and PPD indices and specification of the conditions for thermal comfort

This gives a method of measuring and evaluating moderate thermal environments to which people are exposed.

EH22 (Revised May 1988): Ventilation of the workplace

This Health and Safety Executive Guidance Note provides information on standards of general ventilation and fresh air requirements in the workplace.

EH40/89: Occupational Exposure Limits 1989

This gives advice on limits to which exposure to airborne substances hazardous to health should be controlled in workplaces. It is revised annually.

15.7 After installation

15.7.1 Commissioning and testing

Once any system has been installed it is important that it is properly commissioned to ensure that everything is working satisfactorily and to specification. A guide to commissioning ducted systems is available from BSRIA. Commissioning of other systems should be in accordance with the manufacturers' recommendations, but typical commissioning lists are shown below.

Natural ventilation systems

1. Check incoming electrical supplies to isolators.
2. Check pneumatic and electrical connections between panels and ventilators.
3. Check ventilators and ensure that any transport closure pieces and packing are removed and ventilators are correctly installed.
4. Commission compressor (if fitted) according to manufacturer's instructions.
5. Operate ventilators and check that all open and close correctly.
6. Check operation of all controls.
7. Check pneumatic system for leakage.

Powered ventilation systems

1. Check incoming electrical supplies to isolators.
2. Check electrical connections between panel and fans.
3. Check that fan blades rotate freely and ventilators are correctly installed.
4. Operate ventilators and ensure that all run correctly and in the correct direction.
5. Check full load current on all phases on all ventilators.
6. Check operation of all controls.

Once a system is installed, no specific regular testing of day-to-day ventilation systems is required unless they are installed to comply with COSHH Regulations. This testing would normally be carried out in conjunction with annual maintenance.

Smoke-ventilation systems must be regularly tested, since failure in an emergency cannot be allowed. The system must be fully tested in accordance with the manufacturer's instructions at intervals of not more than 1 month or as agreed with the fire authorities. Any faults found must be rectified immediately.

15.7.2 Maintenance

Most simple ventilation systems require only annual maintenance unless some form of air cleaning or filtration is incorporated, although where systems are pneumatically operated the compressor will need weekly checking. Air cleaners or filters will need regular emptying, cleaning or replacement to maintain efficiency and prevent clogging up. This can either be carried out on a regular schedule (based on the manufacturer's guidance and site experience) or when indicated by a pressure differential gauge or alarm. The cost of a gauge is easily repaid by savings in maintenance costs by maximizing intervals between cleaning or replacement. The task of maintenance can be made easier by taking care in design to provide good access.

Compressors will need a weekly oil level and receiver auto drain check and oil changes and filter cleaning at (typically) 500-hour intervals, although for compressors used only to operate ventilation this can be only an annual task due to the limited usage. Typical annual maintenance for systems are shown below.

Natural ventilation systems

1. Clean ventilators with a soft brush or cloth and remove any debris (leaves, moss, etc.) from rain channels or drainage holes.
2. Check security of fixings.
3. Check operation of ventilators and, if necessary, clean or oil linkages.
4. Examine hinges, pulleys, cables, anchorages, etc. for wear.
5. Check pneumatics for air leaks and fully service compressor.
6. Check that all controls operate correctly.

Powered ventilation systems

1. Clean ventilators with a soft brush or cloth and remove any debris from rain channels or drainage holes.
2. Check security of fixing.
3. Check that fan impeller turns freely.
4. Check electrical connections to fan.
5. Check operation of moving components (other than fan). Clean, oil or adjust as appropriate.
6. Run fan and check current against data plate.
7. Check that all controls operate correctly.
8. Check and clean filters, heater batteries, etc.

15.7.3 Running costs

Running costs of ventilation systems can be broken down into three main areas:

1. Maintenance costs;
2. Electrical costs;
3. Heat loss due to ventilation use during heating season.

Maintenance costs

Regular service checks such as filter cleaning and compressor oil level are normally carried out in-house and time can be allocated for these tasks once some experience has been gathered. Annual maintenance may be carried out in-house or by specialist service engineers employed either by manufacturers or HVAC service companies. A service contract can often include breakdown cover, which has the advantage of reducing risk of unexpected bills and ensuring that prompt repairs are effected at the cost of a higher annual premium.

Electrical costs

The only significant electrical cost involved in ventilation systems is operation of fans. Other electrical equipment such as dampers, compressors, etc. generally run for such short periods that costs are negligible.

Electrical costs for fans can be estimated from the following formula:

$$C = V \times \phi \times A \times \text{fuel cost} \times \text{hours run} \quad (15.9)$$

where C = cost per annum (£),
V = motor voltage per phase (V),
ϕ = number of phases,
A = operating current (A),
Fuel cost = cost of electricity (in £/kWh),
Hours run = total running hours per annum.

Heat loss

Where a ventilation system is required to run during periods when heating is provided then there is an energy cost associated with the heated air being exhausted from the building. This is related to the extra heat input needed from the heating system to balance the heat loss through the extra ventilation. A calculation method is available in section B18 of the CIBSE Guide.[2]

References

1 CIBSE Guide, Volume A, *Design Data* (1986)
2 CIBSE Guide, Volume B, *Installation and Equipment Data* (1986)
3 ASHRAE Handbook, *Fundamentals* (1989)
4 ASHRAE Handbook, *HVAC Systems and Applications* (1987)
5 *Threshold Limit Values and Biological Exposure Indices for 1988–1989*, American Conference of Governmental Industrial Hygienists
6 BRE Digest 119, *Assessment of Wind Loads* (July 1970)
7 BRE Digest 210, *Principles of Natural Ventilation* (B. B. Daly) (February 1978)
8 BRE Digest 346 (7 parts), *The Assessment of Wind Loads* (1989)
9 Daly, B. B., *Woods Practical Guide to Fan Engineering*, Woods of Colchester Ltd (1978)
10 *Ductwork Specification DW142*, HVCA (Heating and Ventilation Contractors Association)
11 CIBSE Technical Memorandum TM8, *Design Notes for Ductwork*
12 *Industrial Ventilation, A Manual of Recommended Practice*, 20th edition, American Conference of Governmental Industrial Hygienists (1984)
13 ISO 7730: 1984, Modern thermal environments – determination of the PMV and PPD indices and specification of the conditions for thermal comfort
14 Fire Paper 7, 'Investigations into the flow of hot gases in roof venting', HMSO (now available from Colt International Ltd) (1963)
15 Fire Paper 10, 'Design of roof venting systems for single storey buildings', HMSO (now available from Colt International Ltd) (1964)
16 BS 7346, Components for smoke and heat control systems. Part 1: Specification for natural smoke and heat exhaust ventilators
17 BS 7346, Components for smoke and heat control systems. Part 2: Specification for powered smoke and heat exhaust ventilators
18 EH 40/89 Occupational Exposure Limits, Health and Safety Commission (1989)

16 Air Conditioning

J Bevan IEng, MIPlantE

Contents

16.1 Basic principles and terms

16.1.1 Abbreviations

In addition to the abbreviations used in SI, the following are employed in air conditioning work:

db	Dry bulb temperature
wb	Wet bulb temperature
dp	Dewpoint temperature
rh	Relative humidity
kg/kg	Kilograms water vapour per kilogram dry air (absolute moisture content)
TH	Total heat
SH	Sensible heat
LH	Latent heat
SHR	Sensible heat ratio
ON OFF	The condition of air or water entering or leaving a coil or heat exchanger
TR	Tons of refrigeration capacity
TRE	Tons of refrigeration capacity extracted
TRR	Tons refrigeration rejected (at final cooler)
HP	High pressure (refrigerant)
LP	Low pressure (refrigerant)
DX	Direct expansion cooling
ΔT	Temperature difference
ach	Air changes (room volumes) per hour
ahu	Air-handling unit
swg	Static water gauge
NR	Noise rating. One of a series of curves relating noise level and frequency to speech interference
NC	Noise criteria. Similar to NR, but differing, particularly at the low-frequency end

16.1.2 Terms

Mechanical ventilation

The movement of air by fan, conveying outside air into the room or expelling air or both. Filtration, heating and control of the distribution pattern may be included. It is not cooling in the sense of temperature reduction but can be used to limit temperature rise when the outside air is below that of the space being treated.

Full air conditioning

This necessitates plant capable of control of temperature by being able to add or subtract heat from the air and control of humidity by being able to add or subtract moisture. The system also comprises fan(s), filtration, a distribution system and may include noise control. Other terms such as 'cooling' or 'comfort cooling' may be met and these can be taken to mean an ability to lower the temperature of the air by refrigeration but without full control of humidity. Moisture may be removed as an incidental characteristic of the cooling coil. The term 'air conditioning' is sometimes used where control of humidity is not included. It is essential to employ clear specifications of performance.

Air

Atmospheric air is a mixture of gases, mainly nitrogen and oxygen together with water vapour. It normally carries many millions of dust particles per cubic metre.

Temperature

A measure of the average energy of the molecules of a substance. The heat intensity.

Heat

A form of energy which, when given to a body, raises its temperature or changes its state from solid to liquid or liquid to gas.

Heat flow

Heat flows from a body at one temperature to a body at a lower temperature. Materials have the property of resistance to the rate of heat flow. It differs from material to material.

Sensible heat

The heat energy causing a change in temperature, as in raising a kettle of water from cold to boiling point.

Latent heat

The heat necessary to change the state of a substance from solid to liquid or from liquid to gas, or the heat given up during the reverse process. There is no change in temperature during these processes. For example, continuing to boil a kettle of water previously raised to 100°C to steam requires the addition of latent heat, but there is no change in temperature if the pressure remains constant.

Total heat

The sensible heat plus latent heat in such a mixture as moist air. In air-conditioning work it is referred to a base a little below 0°C, not absolute zero.

Sensible heat ratio

Sensible heat flow divided by the total heat flow.

Enthalpy

The heat content of a substance per unit mass.

Dry bulb temperature

The temperature of air as indicated by a dry sensing element such as a mercury-in-glass thermometer.

Psychrometrics

The study of moist air. The psychrometric chart shows the relationship between the various properties of moist air in graphical form and can be used for the solution of problems.

Wet bulb temperature

The temperature of air as indicated by a thermometer when its bulb is enclosed by a water-wet wick. If the surrounding air is not saturated water will evaporate, taking the latent necessary latent heat from the thermometer bulb which then gives a lower reading than a dry bulb in the same air. The depression in wet bulb temperature is proportional to the amount of moisture in the air.

Normal practice is to arrange a flow of air over the wick by using a sling (whirling) or fan-assisted instrument. If the thermometer is stationary an area of higher saturation builds up around the wick but the reading may be referred to tables for screen instead of sling readings.

Partial pressure

The contribution by each constituent gas to the total air pressure. Standard air pressure is 1013 mbar.

Vapour

A gas which is below its critical temperature and which can therefore be turned to liquid by an increase in pressure.

Saturation

There is a limit to the amount of water vapour air can hold. It is higher at higher dry bulb temperatures. At the limit, air is said to be saturated.

Relative humidity

This compares the amount of moisture in a sample of air with the amount it would contain if saturated. More accurately, relative humidity is the partial pressure of vapour present divided by saturation vapour pressure × 100%. Saturation = 100% relative humidity.

Dewpoint

The temperature to which a sample of air has to be reduced to bring it to saturation. It is fixed by the moisture content of the air sample.

Absolute humidity

This measures the quantity of water in a sample of air in kg moisture per kg air. The relative humidity then depends on the air dry bulb temperature. Air at 25°C containing 0.01 kg/kg is at 50% relative humidity (rh). If now cooled to 14°C the air would be at its dewpoint (i.e. saturated). If cooled further, moisture is condensed out, the sample remaining saturated as it cools. If now reheated back to 25°C its rh would be lower than 50%. If cooling had not been continued to condense moisture its rh would return to 50% at 25°C.

System resistance

The resistance to air flow which causes a static pressure drop. It is similar to electrical resistance and voltage drop (see Section 16.5). The term 'resistance' is often used erroneously when pressure drop is meant.

Upstream, downstream

Used to denote positions earlier or later in the system relative to the direction of air flow.

Condensing unit

A refrigeration compressor and condenser on one chassis complete with controls.

Split system

As above but with a remote condenser.

Chiller

A compressor, water-chilling evaporator and condenser on one chassis.

16.1.3 The plant

These divide broadly into two types:

1. The direct expansion plant where the air-cooling coil is fed with cold refrigerant;
2. The chilled water plant where the cold refrigerant first chills water (or other liquor) which is fed to the air-cooling coil.

A block diagram of the DX system is shown in Figure 16.1. It has two main circuits – the air circuit and the refrigerant circuit. In the chilled-water system there are additional circuits:

1. Of chilled water between the refrigerant and the air-cooling coil;
2. Of water carrying heat from the refrigerant to the heat-rejecting device.

Figure 16.2 shows a block diagram of the system.

A temperature difference must exist between each stage to cause a heat flow from one to the other. The air and refrigeration side are described in further detail below.

16.1.4 The air-handling plant

The plant may comprise one or more complete factory-made units or may be built up on-site from sub-

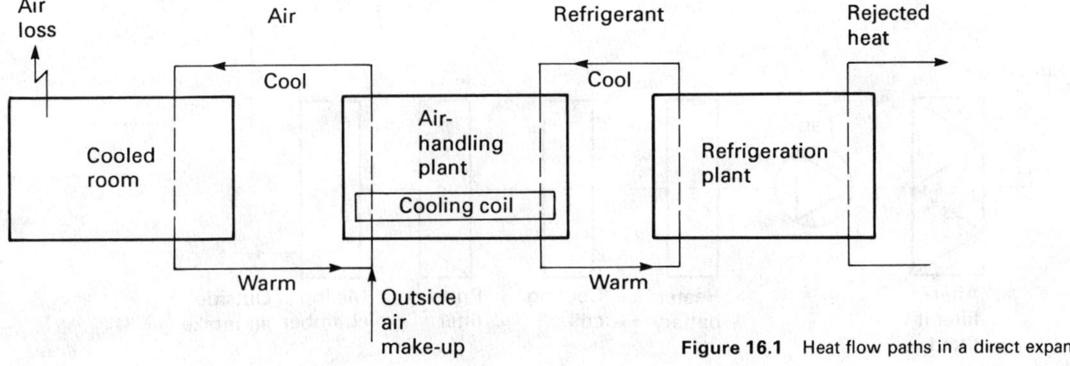

Figure 16.1 Heat flow paths in a direct expansion system

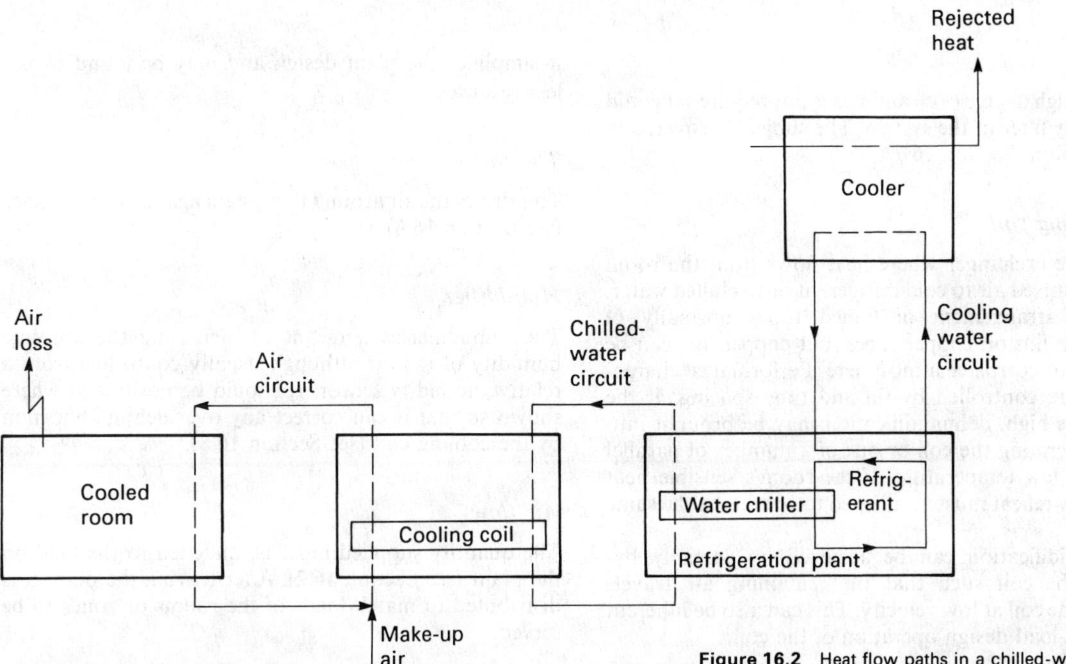

Figure 16.2 Heat flow paths in a chilled-water system

assemblies. There can be variations from the arrangements discussed below and shown in Figure 16.3. The condition or quantity of air input to the conditioned space (referred to below as the room) must be varied such that after it has gained or lost heat or moisture by the applied load its condition and therefore the room condition is as specified.

Outside air intake

The quantity is discussed in Section 16.2. Its purpose is to keep the room fresh and to pressurize it against the ingress of unconditioned air. Its psychrometric condition during most of the year will differ from that required. It can be introduced into a chamber, mixing it for treatment with return air from the room. Alternatively, it can be treated in a separate plant before being introduced to the system, but care then has to be exercised in design, since the sometimes small quantity of air has to control the full humidity load. An advantage is that corrosive wet processes are kept out of the main plant.

Mixing chamber

This is where the outside air intake and recirculated air are brought together before proceeding to the next stage of treatment. As shown in Figure 16.3, it is a low-pressure area which will induce outside air without the use of another fan if the route has low resistance.

When two samples of air are brought together the condition of the mixture may be arrived at arithmetically by adding the heat flow of each and dividing by the total mass flow; and similarly for the moisture flow. Alternatively, plot the condition of each onto a psychrometric chart. The mixed condition lies on a straight line between the two in a position proportional to the two quantities.

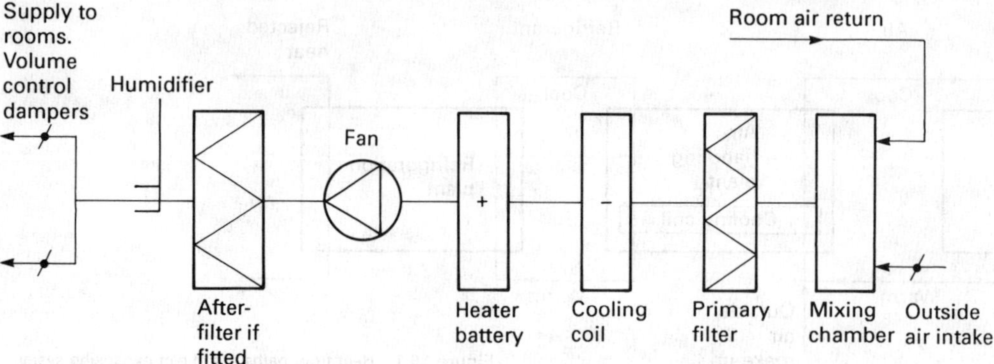

Figure 16.3 Air flow path in an air-conditioning system

Pre-filter

Where a high degree of cleanliness is not required it could be the only filter in the system. The subject is covered in more detail in Section 16.7.

The cooling coil

This is the exchanger where heat flows from the room return or mixed air to cold refrigerant or to chilled water. It is an arrangement of finned tubes normally of aluminium fins on copper tubes, but copper fins can be specified for corrosive atmosphere. Performance characteristics are controlled by fin and tube spacing. If the room rh is high, dehumidification may be brought into use by operating the coil or one of a number of parallel coils at a low temperature. If the room's sensible heat load is low reheat must be allowed to operate at the same time.

Dehumidification can be achieved by partially by-passing the coil such that the remaining air travels through the coil at low velocity. This can also be inherent in the full-load design operation of the coil.

Heater battery

This is used when (1) the room needs heating instead of cooling or (2) for reheat as described above. It is vital in close control systems that its capacity is sufficient to maintain room temperature under these conditions, otherwise the system may fall into a loop, with the controls continuing to see high rh due to temperature. Using only part of the cooling coil for dehumidification will alleviate this situation. A heater capacity of the sensible heat extracted during dehumidification plus half the peak winter fabric loss is recommended where the room load could be nil in winter such as a start-up situation.

Separate reheat batteries may be placed in branch ducts where one plant supplies both a main area calling for cooling and an auxiliary room without heat load. Correct rh in the auxiliary rooms results (only) if it is correctly controlled in the main room and they require the same dry bulb temperature. While wasteful of energy, it simplifies the plant design and may be found to use less resources.

The fan

This drives the air around the system against its resistance (see Section 16.6).

Humidifier

The humidifier is a means of increasing the absolute humidity of the air although usually controlled from a relative humidity sensor. It should be positioned where shown so that it can correct any over-dehumidification by the cooling coil (see Section 16.8).

Air flow

The quantity supplied must be matched to the load on the plant (see Section 16.2). After leaving the plant it is distributed to match loads of the rooms or zones to be served.

16.1.5 The refrigeration plant

The basic circuit is shown in Figure 16.4 and the principal items are described below.

Evaporator

This is the device where the air or water being cooled gives up its heat to provide the latent heat of evaporation to the refrigerant. Superheat is also added to the refrigerant at this point to prevent damaging liquid forming on the way to the compressor.

Compressor

A compressor circulates the refrigerant around the system, raising its pressure such that the refrigerant can be condensed by removal of latent heat. It may also be considered as raising the temperature of the refrigerant above that of the final cooling medium to which heat is

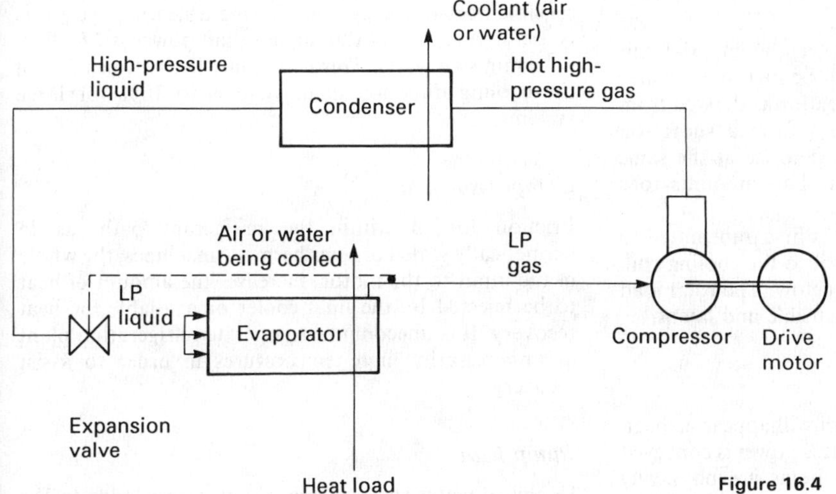

Figure 16.4 Refrigerant flow path in a cooling system

rejected. Lubricating oil is contained in the crankcase but, being miscible in the refrigerant, is carried around the system and returned.

Condenser

This is the vessel where the refrigerant rejects its heat to waste or reclaim, turning back to liquid in the process. Sub-cooling is practised by the removal of further heat. This prevents liquid flashing back to vapour on return to the evaporator.

Expansion valve

A reduction in pressure and hence in temperature takes place across this item before the refrigerant re-enters the cooling coil via distributor pipes.

16.1.6 Controls

Room condition

Dry bulb temperature This is sensed by a thermostat in the conditioned space or in the return duct. Where underfloor air return is practised it is strongly recommended that the sensor be placed close under a return grille to prevent changes of condition occurring between the room and the sensing position. In large rooms separate thermostats can be arranged to give an average signal but individual zones of control each separately treated are much better. Dry bulb sensors may be bi-metal strips, thermistors or refrigerant-filled phials or bellows responding to pressure differences caused by temperature change. These, in turn, provide an electrical or mechanical signal. The mechanical items are used to alter the value of potentiometers or make-or-break contacts. The signals are transmitted to amplifiers which respond to the degree of error. An important feature is the proportional band of temperature over which the controls call for up to full plant capacity.

Humidity Humidity sensors may be animal or plastic skins varying in length with changes in rh or lithium chloride coating changing in electrical resistance. The former are prone to lose calibration. Other comments above apply equally to rh control.

Control at the evaporator A phial senses the temperature of the outlet suction line to the compressor and controls the expansion valve opening to maintain a constant temperature in the coil.

Control at chilled water coil

A three-way motorized valve is modulated between full flow to the coil and full bypass to satisfy the room thermostat.

Control of water-chilling compressors

Being large multi-cylinder machines, the chilling capacity is controlled in steps by rendering cylinder valves inoperative. Control is initiated by sensing water temperature in a storage buffer tank or by sensing return water temperature from the air-cooling coil. Small (DX) compressors are run on a start–stop basis on call from the room thermostat. Safety devices associated with compressors include:

High (gas) pressure cut-out;
Low (gas) pressure cut-out;
Oil differential cut-out (oil feed pressure to be above the crankcase pressure);
Freeze thermostat (low water temperature limit);
Single-phase protection (preventing attempts to start with loss of one phase of a three-phase electrical supply);
Time out (preventing too-frequent starting and motor burn).

Until all the switches in the interlock train are satisfied and closed the compressor will not start or will not continue to run.

16.1.7 The load on the plant

The unit of heat is the watt. However, the imperial unit should be understood as it will still be met, particularly outside Europe. The ton of refrigeration is derived from an ability to remove sufficient heat from a short ton (2000 lb) of water at 32°F to turn it to ice at the same temperature in the course of 24 h. This amounts to a heat extraction rate of 3.517 kW.

The load presented by the room is first transmitted to the room air which, in turn, passes it to the cooling coil. Other gains will occur as detailed below. The total load is best considered in two parts – sensible and latent.

Machine heat

All electrical energy fed to the room will appear as heat. This presents a load to the plant unless power is conveyed out of the room by cable or hot items are physically removed. Parts of some large computers are cooled by a direct supply of chilled water presenting a load on the refrigeration plant but reducing the load on the air side. Cold outside conditions will result in some of the internal load being met by fabric loss.

Fabric losses or gains

In winter in cold climates heat will be lost through fabric of the building. This will be advantageous at times of high internal load but will need to be considered as a heater battery duty at times of low internal load. In winter there can be considerable loads for humidifiers if the structure is not adequately vapour sealed. Weather-tightness is insufficient. For example, a computer room in the UK held at normal conditions may experience up to 100 mm water gauge vapour pressure difference, forcing moisture out. During high summer in temperate and hot climates the external water vapour pressure will be higher than in the conditioned space.

Personnel

People give off both sensible and latent heat. During light work such as in a computer suite or laboratory they emit 110 W sensible 30 W latent and while seated, 90 W sensible 20 W latent.

Air ducts

A duct carrying cool air through a warm space such as a loft will gain heat before entering the conditioned room, contributing to the load on the plant. Ducts passing through the conditioned space do not add to the load. Similarly, there will be losses in winter if they carry warm air through cold spaces and moisture gains and losses if leaky. Insulation and vapour-tight joints are necessary.

Fan heat

The power fed to the fan shaft (or the total electrical power fed to the motor if within the duct) appear as sensible heat in the system. During a heating cycle this is useful. As a rule of thumb, fan shaft power is 17 × flow m^3 × mm swg watts. This commonly lies between 2% of the cooling duty for small systems to 10% for large systems.

Compressor heat

Friction (or, if within the refrigerant path, as in hermetically sealed or semi-hermetic machines, the whole of the input to the motor) increases the amount of heat to be rejected by the final cooler or available for heat recovery. It is uneconomic to operate refrigeration plant at unnecessarily high temperatures in order to assist recovery.

Pump heat

In chilled-water systems pump shaft power adds to the heat of the circulating water. Similarly, if the chiller has a water-cooled condenser pump heat is added to that handled by the final cooler. Power is proportional to the flow and pressure:

$$\frac{102 \times \text{l/s} \times \text{metres head}}{\text{Efficiency}} \text{ watts}$$

Efficiency can vary between 35% for very small pumps and 80% for large ones, and may be found to lie between 3% for small heating circuits and 10% of the heat conveyed for large systems.

16.2 The air quantity required

16.2.1 Air change rate

Change rates (room volumes per hour) can be used to calculate the quantity to be supplied or extracted by a mechanical ventilation system. These figures also apply to parts of an air-conditioning system where stale air must not be recirculated. Recommendations and regulations are given in Chapter 15.

Volume to be supplied

For the design of close control systems or where large amounts of heat are to be removed the mass flow to be employed must be calculated. Use is made of the specific heat of air which for normal room conditions may be taken as 1.02 kJ/kg °C. 1.02 kW raises a flow of 1 kg/s by 1°C and pro rata. The volume of a given mass of air varies with change in temperature but supply volume is often more convenient to consider. Taking the specific volume to be 0.82 m^3/kg then 0.1 m^3/s will convey 1 kW with a temperature rise of 8°C and pro rata. A subsequent fall of 8°C across the cooling coil is suitable for areas controlled to 21°C 50% rh. A smaller air quantity would be too close to the limiting temperature to hold the necessary moisture when leaving the coil. Greater temperature differences may be used if the rh is to be controlled to a lower level or not controlled at all. A maximum temperature difference of 10°C supply to room is recommended where occupants are close to supply points. For comfort, larger volumes are preferable to low temperature.

In ventilation systems the temperature rise calculated on the basis of specific heat alone will be pessimistic by one or two degrees because of the effect of building mass. Unless an extract is specifically designed to remove heat from hot spots or lights, the extract and room temperature can be taken to be the same.

Outside air intake

When temperature limitation is more important than close control of conditions, a considerable economy of refrigeration plant operation results from arranging to draw in outside air when it is sufficiently cool and rejecting this back to outside after having gained heat from the room. However, where close control of temperature and humidity is required accuracy and economy ensue from minimizing the outside air intake. The quantity may be based on the number of occupants using 0.008 m^3/s minimum per person. Alternatively, if larger, use 0.002 m^3/sm^2 floor area, which is sufficient in a good quality building to keep the room pressurized to one or two mm swg. Because air loss is a function of the building surface, this is preferred to the basis of a proportion of the supply air volume, which is a function of load. If the later is used for applications such as computer rooms it is seldom necessary to use more than 2.5%.

General considerations

Outlets should not be provided in constantly recirculating systems, particularly where close control of humidity is required. The overpressure developed is far less than that exerted by the wind, and for this reason any system which does have both intake and discharge ducts should have them on the same face of the building. While care is necessary to prevent short-circuiting, this alleviates problems arising from the considerable wind pressure difference that can develop on opposite sides of a building.

Air-lock entries should be used where close control of conditions is required and for clean rooms, but conventional doors with close fit and self-closures are sufficient.

Fire authorities may stipulate pressurization of certain areas such as stairways and may require smoke-extract systems to be brought into operation automatically in the event of a fire. They should be consulted at an early stage of the design.

16.3 Heat losses and gains

16.3.1 Heat losses

Heat transfer through a partition is a function of resistance to heat flow, the temperature difference driving the heat through and the surface area. The function of heating systems is to provide the heat lost in maintaining the temperature difference. Thus

Heat flow (watts) is $U \times \text{TD} \times \text{area}$

where U is a coefficient for the partition in watts per m^2 °C. For each room served the loss through each wall, ceiling and floor should be calculated and for each part of those surfaces where differences occur, such as windows in walls or cantilevered structures of upper floors.

The U values of many partitions or composite constructions can be found in standard references but others may have to be calculated where no data are available or changes are contemplated. The resistivity (r) of each element of the partition encountered by the heat in passing through must be found and multiplied by the thickness in metres. Manufacturers are usually able to give resistivities. The sum (R total) of all the elements, including the inner and outer surface resistances and the resistance of any interleaf air gap, is found. The U value is the reciprocal of R total. For example, for a wall:

Outer surface (normal exposure)	0.055
112 mm brick skin 0.12×0.112	0.01344
50 mm uninsulated air gap	0.180
150 mm lightweight block skin 5.88×0.15	0.882
15 mm rendering 2.5×0.015	0.0375
Inner surface	0.123
R total =	1.29094
$U = 1/R =$	0.775

If this is insufficient for the purpose (e.g. the external wall of a dwelling house) the problem could be reworked using insulation in the interleaf gap.

The temperature at any intermediate point is proportional to the R total to that point. This may be used to decide whether that point is above or below the dewpoint of penetrating air.

In maritime climates such as the UK the lowest external temperatures are not sustained for long periods. The mass of the structure has a slugging effect, and it is safe to use a relatively high external design temperature (e.g. −1°C in the south of the UK and −4°C in the north). These figures do not apply to an outside air intake where the full effect of low temperature is felt immediately nor to lightweight structures. When choosing a design temperature difference, one must take into account that adjacent rooms may not always be heated to their design temperature. Heat bridges, which are weak points in the insulation, must also be considered in proportion to their areas.

Cold outside air introduced to the system by infiltration or by design of the plant will require heat to be added by the plant or directly in the room to maintain the room temperature. The heat required is

kg/s × 1.02 × °C TD kW

For infiltration it is convenient to use $m^3/h \times 0.33 \times$ °C TD watts, but where air conditioning is employed infiltration should not be allowed. Air lost from the space by pressurization carries heat away but this is not an additional load beyond that mentioned above.

The available heating capacity should exceed the

calculated figure, this being sufficient only to balance the losses under steady conditions. A 25–50% excess capacity is recommended to provide warm-up from cold and good response to controls but without excess overshoot. It is important in close control air-conditioning design to have sufficient heating capacity to raise the room temperature from cold following plant stoppage. If the room is cold the controls will see high rh and call for dehumidification upon restarting. Unless this is countered by sufficient reheat the control of conditions will not recover automatically.

16.3.2 Internal gains

Any heat liberated within the room reduces the heating effort required by the plant. At any time these gains exceed the loss more than marginally, cooling is required. Heat sources are:

1. The total electrical input to the room unless power is carried away by cable or heat by pipe. This includes lighting and it should be remembered that the input to fluorescent fittings is greater than the tube ratings. Some luminaires are designed as air-extract fittings, in which case their heat is a load on cooling plant, but does not contribute to room heating unless the plant is in a non-cooling recycling mode;
2. Heat from other processes unless carried away by pipe or items are taken out of the room;
3. Heat from personnel (see Section 16.1);
4. There should be no space heating within air-conditioned rooms but some gains may arise from hot water pipes passing through to other areas;
5. Gains from adjacent spaces held at a higher temperature;
6. Heat gains into ducts where these pass through warm areas *en route*;
7. The shaft power of fans will appear in the system as heat. Where the drive motor is in the airstream the whole of the motor input will appear as heat.

16.3.3 Heat gains

The calculation of heat gains through the building fabric is more complicated than for heat losses, taking into account the gains from both the air-temperature difference and from solar intensity. The gain varies during the day with the movement of the sun and changes in air temperature. Heavily glazed buildings are susceptible to large gains from low sun elevations at all times of year and here building orientation can have a considerable effect on plant loads.

A structure with a large mass will result in the peak gain appearing on the inside some hours after the external peak, and the gain will be attenuated since the outside condition will be reduced before that time. The time of peak gain will differ for each of the enclosing surfaces, so it is necessary to calculate each for several hours to find the peak for one room and to repeat this for each room to find the peak load on a plant serving several rooms. Computer packages of varying merit are available to undertake this laborious task but the broad method is shown below.

To avoid temperature shock and for economy, the control of comfort air conditioning may be allowed to drift with extremes of external temperature. However, where the design is to maintain specified internal conditions under virtually all external conditions data may need to be adjusted upwards.

Gains through walls

$(A \times U \times (M-T)) + (A \times U \times F \times (P-M))$ watts

where A = area (m),
U = W/m(°C,
M = mean solair temperature (°C) (solair temperature is a composite temperature taking into account outside dry bulb temperature and solar radiation),
T = fixed room temperature (°C),
F = factor depending on wall thickness (varies from 1 at zero thickness to approximately 0.05 at 0.5 m thick),
P = solair intensity at the sun time considered.

The delayed time at which this gain is manifest depends on the structure thickness and density and varies from zero for zero thickness to 15 h for 0.5 m thickness.

Gains through windows

$A \times ((C \times I) + U \times (O-T))$ watts

where A = area (m),
U = W/m °C,
C = solar gain factor,
I = direct radiation intensity,
T = room temperature (°C),
O = outside temperature (°C).

The gain is immediate.

16.3.4 Heat of outside air intake

In close control air conditioning the condition of the outside air is rarely as required for passing forward to the controlled space. Therefore it contributes to the cooling, heating, humidification or dehumidification loads on the plant. Whether computer rooms, laboratories, etc. with their sparse population are considered or auditoria where there is a larger ventilation requirement, the peak latent heat load caused by treating the outside air is greater than that from the personnel. The heat from personnel is given in Section 16.1.

The calculation of the heat of intake air is in two parts:

Mass flow kg/s × change in latent heat kJ/kg (from tables or psychrometric chart) (kW); plus
Mass flow kg/s × change in sensible heat kJ/kg (kW).

16.4 Air conditioning for computers

16.4.1 The computer

In the context of this section a computer is a large mainframe machine standing in its own room. The machine is housed in cabinets 1–2 m high with a metre or so clearance all round. Some items are purely electronic, others handle magnetic tape, magnetic discs or paper. The function may be printed output following calculations, information storage and retrieval or electronic control of a remote process. In the largest configurations output printers and communications equipment may be housed in their own adjacent rooms but with similar environmental control. Other associated rooms which require air conditioning if included in the suite are: Job Assembly, Operations Control Room, Magnetic Media Store, Computer Engineer's room, Ready-Use Stationary Store. The main stationary store and store for master tapes are not normally conditioned although are heated in cool weather. The items can be conditioned in the Job Assembly room before use. Reserve master magnetic media can be housed in a remote building or room where, in temperate and warm climates, conditions are maintained by building mass.

The computer will dissipate large amounts of heat which have to be removed to prevent temperature rise. There is usually a contractural requirement to provide conditions within fairly close limits to ensure reliability and to allow handling media at high speed. The requirements are specified by the computer manufacturer, and are typically 21°C±2°C and 50% rh±5% together with a filter performance of 80% average arrestance Euro 4/5 or 80% average efficiency Euro 4/5, at least in the outside air intake (see Section 16.7). The control should centre on the specified figures and only touch limits (if at all) with changes of load. In spite of developments in design, heat concentrations tend to rise and configurations expand.

Modern telecommunications equipment is very similar, but environmental tolerances are normally much wider, with particular emphasis on reliability.

16.4.2 Air supply

Cooling air to the cabinets is normally introduced from the room at low-level front or back with fan-assisted discharge to the rear or top. The normal practice is to introduce room-cooling air at high level. This mixes with rising hot air to give a near-uniform condition in the occupied levels. The distribution of cooling air must be carefully matched to heat load around the room. A large room (say, over 300 m^2) should have separate zones of control with their own sensors. The ceiling void should nevertheless be common. The recirculation of air will confine itself to the zones fed by individual units when all are running but will be redistributed by a ventilated ceiling in the event of partial failure of plant modules.

The easiest and most flexible method of introducing air is by a ventilated false ceiling. A great advantage of a ventilated ceiling supply is that the layout of the computer or indeed its make and heat pattern may not be known at the time of air-conditioning design, and will most certainly alter during the life of the plant. The distribution of supply air can be altered to match the load providing sufficient cooling capacity and air volume are installed.

Upward air flow is also practised. The cable void formed by the raised floor is used to supply air which enters the room via floor grilles. These can be moved to meet the pattern of heat distribution and are normally placed close to the computer cabinets, but consideration must be given to changing air conditions, intended to meet changing room load, entering the computer compartments. Care has to be taken to avoid the updraught lifting dust into the occupied space.

The extracts may be in the ceiling with the advantage of taking up the heat of lights without it passing to the room. Greater use can thus be made of a given air flow. Care has to be taken in the choice of grilles if trolley traffic is going to be present. Low supply temperatures may cause discomfort to staff.

16.4.3 Air extract

Air extract can be at one wall or, in the case of ceiling diffuser supply, extract grilles can be interleaved. The normal for all but the smallest computers is to use the underfloor cable void as a return path. This void is formed by proprietary 600 mm or 2 ft square interchangeable tiles standing on corner jacks. Some of these may be perforated or they may be fitted with grilles. A maximum velocity of 2.5 m/s over the grille face or through perforations is recommended. Depending on the detail of the tile chosen, a pressure drop of about 10 Pa can be expected, but in the case of carpet-covered tiles where the holes tend to become obstructed a downrating to 70% is recommended.

16.4.4 Chilled water

Some very large machines have a direct supply of chilled water to a heat exchanger in the central processor or its store, the final distribution within the machine being part of its design. The user's responsibility is the provision of half couplings (often Hanson, USA) and the chilled water supply rate, temperature, pressure, pH and other analyses. Welded pipe and vapour-sealed insulation are essential. End-stop valves should be installed and consideration should be give to fitting bypasses to compensate for changes in demand and to allow one computer (if there is more than one) to be fed while another is uncoupled. Consideration should also be given to the installation of moisture detectors underfloor.

The plant should be replicated for reliability and to allow maintenance (say, 3×50% capacity). Proprietary units are available specifically for this function.

The final connection to the computer is by flexible pipe, but it is still necessary to liase with others to know the position to terminate and, in due course, which floor tile is to be cut and precisely how. This can be expected to entail extra floor jacks.

16.4.5 Computer heat

Ideally, the air-conditioning designer is given the heat loads to be catered for at the outset. However, in the fast-moving field of computers this may not always be the case. Also, the life of the plant is likely to outlast more than one computer configuration but need not itself require more than re-zoning. Particularly where a computer has a number of peripherals, (input and output devices) it is unlikely to dissipate fully to the maker's data at any one time, but it is strongly recommended that further diversity on that for individual items given at the maker is not allowed in the design of the cooling capacity of the plant. It is unwise not to allow for expansion of the computer configuration. An initial design on a basis of an overall heat dissipation (say, 500 W/m^2 overall average) unevenly distributed about the room is not unreasonable for large computer halls. In this field the design engineer should not be afraid to question his brief if the client is to be well served. Flexibility to cater for changes has already been covered above.

16.4.6 Telecommunications

Much telecommunications equipment is designed for overhead cabling, which can be very extensive and dense. Any modern equipment which relies on cooling to function should have breaks in the cabling layout partly to allow cool air to fall. In the case of naturally ventilated racks the air discharge should be directed down gangways (the racks being arranged in long suites) and not directly over the racks. A ventilated ceiling supply can be confined to the racked area to allow other services to pass without breaking the ceiling. Except in smaller cases, for modern systems duct and diffuser designs may be insufficiently flexible and may not be able to meet the concentrations of heat without discomfort to attendants. However, in small installations (150 m^2) air-handling units may be employed directing air freely over the top of the cabling.

While the air distribution must match the heat distribution, the position of extracts is not important. Direct return to air-handling units mounted around the room in numbers necessary for reliability is common practice. In the event of loss of one or more cooling modules a well-designed ventilated ceiling adds to reliability by distributing the reduced amount of air in the same proportions throughout the room.

16.5 Air distribution and system resistance

16.5.1 Duct sizing

Ducts convey conditioning air from point to point at a variety of speeds. Slow speeds results in large ducts, costly in themselves and in building space. High speeds result in noise and the need for high fan powers. A good basis for air conditioning is 6–7 m/s adjacent to the plant but, as discussed below, less at distant points.

When we refer here to static pressure we mean the difference between internal and external pressure causing air to tend to flow into or out of ducts. Velocity pressure is that due to the air's forward movement. The sum of the two is total pressure.

16.5.2 Duct design

Simple runs of a few metres may be designed for a constant velocity. A supply duct is thus reduced in steps at each outlet or a return or exhaust duct similarly increased in section in the direction of flow.

A large system with branches, several inlets or outlets and some tens of metres long will be more easily controlled at the end distant from the plant if velocities are reduced as we progress down the length of the duct. If air is slowed in a controlled way with the duct sides diverging at not more than 15° included angle its velocity pressure will reduce and (ideally, without loss) its static pressure will rise to maintain a constant total head to compensate for pressure loss as the air progresses down the duct. However, at least by manual methods, design for static regain is laborious and the duct shape unconventional. It is seldom practised.

A method commonly used is that of equal surface friction per unit run. If rectangular ducts are being considered their equivalent circular diameter must be found. This may be obtained by

$$1.3 \times [(w \times d)^{\cdot 625}]/[(w+d)^{\cdot 25})]$$

Here w and d are width and depth and the units may be metres or millimetres. Alternatively, the diameter may be found from published charts. Using Figure 16.5 a pressure gradient (Pa/m) is chosen and with each change of volume flow a new diameter is found. This is converted to a convenient rectangular section equivalent to the round section. It will be seen that the velocity reduces.

Where practicable, the large surface area of wide shallow ducts should be avoided to keep pressure gradients to a minimum. There can be no hard-and-fast rule, 4:1 being a suggested limit. The system may, of course, be designed to use only round-section ducts.

16.5.3 System pressure drop

The point of interest is the path of highest pressure drop or index leg. Other parallel branches can be designed of appropriate size to pass the required amount of air, those of lower resistance than the index leg being throttled by dampers. The pressure drop is the sum of the drops caused by the following and is calculated to determine the pressure against which the fan must operate:

Surface friction of duct as discused above (Pa/m $\times$ length);
Changes in section;
Bends;
Branches;
Obstructions;
Grilles, meshes, etc.

Pressure drop is calculated by $k \times$ velocity pressure, where k is the resistance factor for the above items other than duct friction and is found from Table 16.1 or similar

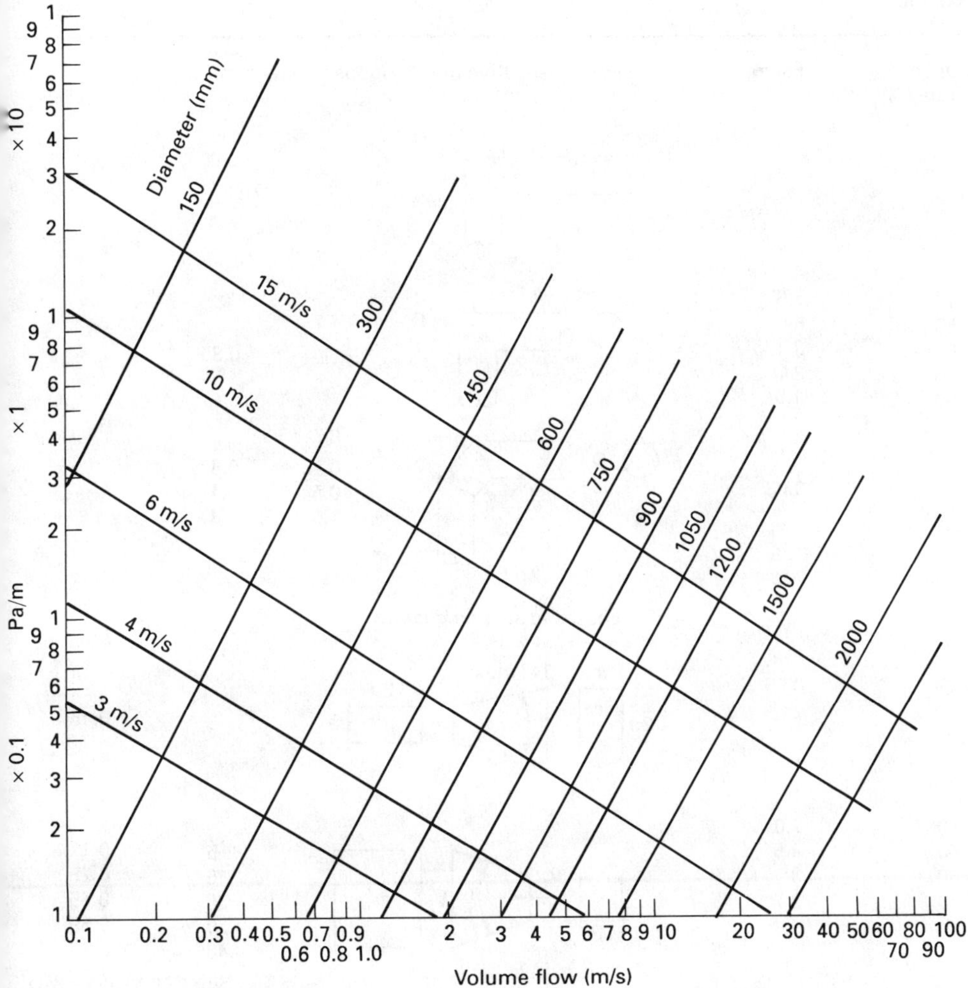

Figure 16.5 Pressure gradient in circular steel ducts

references. Velocity pressure is found by 1/2 density × velocity^2, taking standard density at 1.2 kg/m^3, or from Table 16.2.

Plant resistance

In the design of tailor-made plant it is necessary to calculate the above as they occur within the plant and add the pressure drop of all other items such as air filters at their dirty conditions, coils, etc. In the case of proprietary units it is normal for the manufacturers to quote an external pressure against which they will deliver the specified air quantity.

Duct resistance

The basis of good duct design is to arrange gradual changes to section and direction. It is sometimes necessary to construct a 90° elbow with no inside radius. The pressure drop and noise generation can be greatly reduced by incorporating turn vanes which split the air into a number of near-parallel paths. The information given in Table 16.3 assumes simple (non-aerofoil) turn vanes (see also Figure 16.6).

16.5.4 Ventilated ceilings

A ventilated ceiling is an alternative to ducts, diffusers and grilles as a means of distributing air within a room. It is suspended below the structural ceiling forming a shallow void. One type consists of perforated metal trays or tiles. Each is supplied with a bagged acoustic pad which is removed from those trays or tiles which are to ventilate. Another consists of tiles with single-row slot openings at intervals between runs of tiles, the openings being controlled by dampers integral with the slots. Any type of tile which can shed dust or is combustible should be avoided. Rising hot air and descending cold air mix a few hundred millimetres below the ceiling, giving a near-uniform condition in the occupied levels. The

Table 16.1 Resistance factors

Losses at duct entries

Entry	Open area (%)	Factor k
Plain duct end		
Grille	40	7.5
	50	5
	60	3
	70	2
	80	1.5
Bellmouth		0.03
Flanged		0.85
Louvres	40	7.0
	50	4.5
	60	3
	70	2

Losses in bends

Bend	Ratio R/D	Factor k
Round (D, R)	0.5	0.73
	0.66	0.5
	0.75	0.38
	1.0	0.26
	1.25	0.17
	2.0	0.15
Rectangular W/D = 0.5	0.5	1.3
	0.66	0.65
	0.75	0.47
	1.0	0.28
	1.5	0.18
W/D = 1 to 3	0.5	0.95
	0.66	0.45
	0.75	0.33
	1.0	0.2
	1.5	0.13

For consecutive bends opposite direction k x 4

Bend	R/D	Factor k
Square mitred bend		1.25
Round mitred bend		0.85
Round segmented bend (D, R)	0.25	0.8
	0.5	0.4
	1.0	0.3

Losses at section change

Change		Factor k
a × $2a$ to $2a$ × a, 14° max		0.15
Expansion V_1 to V_2, θ°	5	0.17
	10	0.28
	20	0.45
	30	0.59
	40	0.73
	Referred to $V_{p1} - V_{p2}$	
Contraction to V_2, θ	30	0.02
	45	0.04
	60	0.07
	Referred to V_{p2}	
Splitter expansion V_1 to V_2		0.28 ($V_{p1} - V_{p2}$)
Contraction to V, θ	60	0.06

Expansion D_1 to D_2, reference V:

D_1/D_2 = 0.5	θ	k
	5	0.2
	10	0.3
	20	0.55

V = velocity reference position, D = duct diameter

Table 16.1 *continued*

Losses at duct exits

Free area to face area (%)	Factor k
40	9
50	6
60	4
70	3
80	2.5
	1.0
100	3
150	2
	1
40	10
50	6
60	4
70	3

Fitting	Volume in branch (%)	k branch	k main
		k 0.28	
$R = D$	25	0.4	0.1
	50	0.3	0.15
	25	0.25	0.14
	50	0.3	0.17
	25	2.0	0.4
	50	1.5	0.5
	25	1.0	0.5
	50	1.2	0.6
Vol. / Vol.	25	1.1	
	50	0.8	
Vol. / Vol.	25	0.6	
	50	0.8	

Branch velocities = main velocities

Losses at obstructions

	Ratio D_1/D_2	Factor k
V, D_1, D_2 (round)	0.1	0.16
	0.25	0.5
	0.5	1.6
V, D_1, D_2 (square)	0.1	0.5
	0.25	1.1
	0.5	3.0
Idle stage of axial fan		0.1
Electric heater battery 80% free area		1.0
Finned-tube coils (referred to face velocity)		2 to 6

availability of very small areas of control can be advantageous where electronic apparatus with mild chimney effect would have its natural cooling upset by strong downdraughts, by confining air supply to gangways.

The method is useful where large quantities of air are to be introduced without draughts; where the distribution of heat load is unknown at the time of design or is likely to alter; and where, as a measure of reliability, it is required to redistribute the remaining air in the original proportions in the event of partial failure of the plant.

Ventilated ceilings can cater for 1 kW load per square metre and in excess of 100 ach. Lighting and power tracks can be incorporated and fire-detection heads fitted. In computer suites (see Section 16.4) it is normal to combine a ventilated ceiling supply with an underfloor extract, which is then not critical as to layout of grilles and assists

Table 16.2 Velocity pressure (Pa)

m/s	0	1	2	3	4	5	6	7	8	9	10
0.0	0.0	0.6	2.4	5.4	9.6	15.0	21.6	29.4	38.4	48.6	60.10
0.1	0.01	0.73	2.65	5.77	10.09	15.61	22.33	30.25	39.37	49.69	61.21
0.2	0.02	0.86	2.90	6.14	10.58	16.22	23.06	31.1	40.34	50.78	62.42
0.3	0.05	1.01	3.17	6.53	11.09	16.85	23.81	31.97	41.33	51.89	63.65
0.4	0.1	1.18	3.46	6.94	11.62	17.5	24.58	32.86	42.34	53.02	64.90
0.5	0.15	1.35	3.75	7.35	12.15	18.15	25.35	33.75	43.35	54.15	66.15
0.6	0.22	1.54	4.06	7.78	12.7	18.82	26.14	34.66	44.38	55.30	67.42
0.7	0.29	1.73	4.37	8.21	13.25	19.49	26.93	35.57	45.41	56.45	68.69
0.8	0.38	1.94	4.7	8.66	13.52	20.15	27.74	36.5	46.46	56.62	69.98
0.9	0.49	2.17	5.05	9.13	14.41	20.89	28.57	37.45	47.53	58.81	71.29

Table 16.3 Where to place splitters

Duct width (m)	*If one split r1* (mm)	*If two r1 r2* (mm)	*If three r1 r2 r3* (mm)
0.5	110	65, 190	50, 110, 230
1.0	155	85, 280	60, 150, 380
1.5	190	100, 380	65, 190, 520
2.0	226	110, 470	70, 230, 660
2.5	255	115, 540	75, 255, 790
k factors:			
If none			
1.25	0.65	0.5	0.45

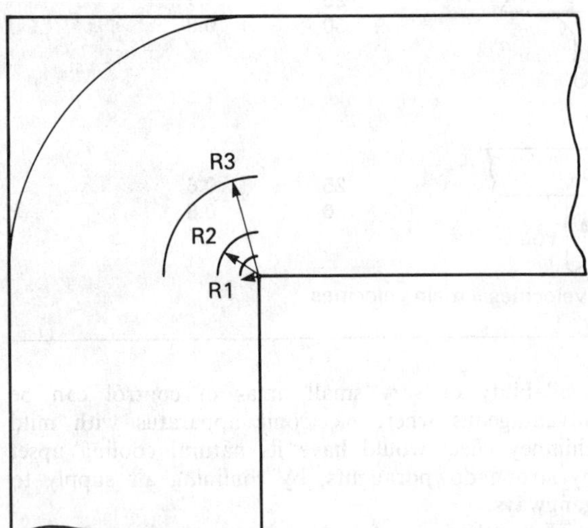

Figure 16.6 Splitters in mitre bends

any heating mode. The pressure and slot velocity are, for all practical purposes, uniform throughout the ceiling area, and the air pattern is finally set by checking the room temperature from point to point. Pressure drops are of the order of 20 Pa. Void depths are seldom critical, being a function of the throw required and the presence of obstructions such as beams or pipes. Depths of 200–1000 mm are common, with throws from the entry point of up to 20 times void depth.

The air supply is introduced by stub ducts penetrating the void by 0.5 m or less and simply cut off square. However, where there are obstructions across the void it can be advantageous to convey a proportion of the air to or past this by extended ducts. Equally, it can be advantageous to duct air part-way if it is to double back round a corner. In general, however, the system is very tolerant, including positioning air entries about the room. Stub velocities should be 3.5–5 m/s but individual manufacturers will give their own requirements.

Unless beams are present at the entry the stubs should be close to the structural ceiling. Beam fairing may be used where the void has to be shallow and this can reduce the risk of re-entrainment due to local high-velocity pressure. Zone barriers are fitted within the void where separate rooms are to be fed. The ceiling need not cover the whole room.

16.6 Fans

16.6.1 Fan selection

Fans propel the air through the system, and must be chosen to be capable of delivering the required volume flow against the calculated system static pressure, advisedly with a small margin.

Individual fans are capable of operating over a range of volumes and pressures which are interrelated, the performance being shown in manufacturers' tables and curves. Venturing outside the recommended area of operation may, depending on the type of fan, result in motor overload, motor undercooling, vibration or stall. Selection should be made for minimum power input, which is also likely to be the quietest fan for the duty. Performance is usually quoted for a standard condition of 1.2 m^3/kg. Calculations of system resistance are best carried out at the same condition. The user may find only the static pressure quoted. If total or velocity pressure are also quoted or the outlet velocity can be calculated the designer can calculate how much pressure can be recovered after the exit. Any mismatch due to difficulty in calculating system resistance will cause the volume to rise or fall, to settle on the fan characteristic curve.1

16.6.2 Types

The broad range of types used in air conditioning is:

1. Centrifugal
2. Axial
3. Propeller
4. Cross Flow

The centrifugal fan has an inlet on the axis of rotation and outlet from a scroll casing 90° from the inlet. The centrifugal fan is not reversible. The characteristics of the fan are determined by blade shape, i.e.:

1. Straight blades or paddle wheel: Low efficiency; used for conveying particle-laden air.
2. Forward-curved blades: Moves large volumes for a given size; the driving power required rises markedly with increased volume, necessitating an adequately rated drive motor.
3. Back-curved blades: Highest efficiency of the group; more stable characteristic of volume against pressure; capable of developing higher pressures; non-overloading, i.e. the power does not rise to a peak as the flow tends to zero or free flow.
4. Many developments of the centrifugal fan exist, the most commonly employed being the double-width double-inlet fan moving twice the volume for a given wheel diameter. It is possible in large tailor-made plants to arrange for the drive motor (and hence its efficiency losses) to be external to the airstream.

Axial fans have a cylindrical casing and the fan blades are mounted on an extension of the shaft on the centre line. The blades are normally of aerofoil section. Axial fans are available with adjustable blade angles which allow differing characteristics from one construction or alteration on-site by specialists. The power characteristic is usually non-overloading. In the working range the developed pressure increases with reduced volume. Stall is said to occur when the fan can no longer develop the pressure to deliver the required air volume. This is a feature of high blade-angle fans. It is best not to attempt to use such a fan to deliver too small a volume.

They are reversible with varying degrees of efficiency. Air leaves the blades with rotary motion and can be straightened by downstream vanes, leading to increased pressure development. Two-stage contra-rotating fans develop more than twice the pressure of a single stage. Further staging may be arranged.

When using conventional induction motors the available range of speeds is restricted by the fixed-supply frequency and the number of field poles in the motor. It should be noted that, outside Europe, supply frequencies other than 50 Hz are used, resulting in induction motors operating at proportionally different speeds, and that this applies to all types of fan.

Special versions are available, such as hazard-proof and for high-pressure operation, but these are not normally applicable to the field of air conditioning. Fans are available with fewer blades (part-solidity) to allow the use of smaller higher-speed motors. There are also fans with short casings having a length little more than that to accommodate the blades for use in confined spaces.

Propeller fans are of low capital cost for the volume moved but are used in applications where the resistance is very small such as unducted openings through partitions. The power required increases with resistance. Their pressure–volume characteristic changes with the relative position of the blades and mounting plate.

Cross-flow fans have a blade arrangement similar to that of a back-flow centrifugal fan but the casing arrangement allows the incoming air to enter along the width of the blades, avoiding the 90° turn. Long narrow shapes can be arranged, making them suitable for use in small air-handling units, fan heaters or air curtains.

16.6.3 Multiple arrangements

Fan may be operated in parallel but are best of similar characteristic to avoid stalling. In the event of failure of one the effect of reduced pressure drop in the system is to give a flow of about two thirds that of two. Non-return dampers should be fitted.

Fans may also be operated in series. Axial-flow fans are available having two (contra-rotating) impellers and motors in the one casing. Further staging may be arranged. If one stage fails or is switched off it will idle round with loss, but since similar losses in centrifugal fans are much higher, these are not usually operated in series. However, supply and extract fans in a recirculating system effectively in series may differ.

16.6.4 Fan laws

The laws quoted here are selected as those appropriate to the practicable applications of fans:

Volume flow varies as the speed of rotation;
Pressure varies as the square of speed of rotation;
Power varies as the cube of the speed of rotation;
At constant speed a fan delivers constant volume (m^3/s) into a fixed system in spite of change in density;
Pressure and power absorbed vary as change in density.

16.6.5 Volume regulation

This falls under two headings:

1. Where a permanent change is found to be needed during commissioning or change of duty. The volume may be changed:
   (a) By variation of system resistance by damper action. Care is needed to prevent axial fan motors overheating or reaching the stall point;
   (b) By changing the speed of the drive. In centrifugal fans the common practice is to change or adjust belt drive pulleys, taking care not to exceed the power capability of the motor. In axial fans the pitch of the blades may be adjustable by swivel mountings on the hub.
2. Where frequent change is required in the normal operation of the plant:
   (a) By switching motors in multi-stage fans;

(b) By speed control of motors electrically (e.g. by pole changing), varying the circuit resistance of wound rotor motors, or thyristor part-cycle disconnection;
(c) By damper modulation;
(d) By bypass (fan recirculation) dampers;
(e) By axial fans with pitch variable while running.

Mounting

It is common practice to support the fan on anti-vibration mountings and connect the fan casing to the duct by short lengths of flexible non-combustible material. Care should be taken during installation to see that these are aligned to prevent entry turbulence and noise generation.

16.7 Dust control and filtration

16.7.1 Dust control

Dust is continually being introduced to the conditioned space by clothing fibres, skin particles, shoe dirt, room processes and the outside air make-up. Dust particle sizes range from 1 μm to 75 μm, smaller particles being described as smoke. In an apparently clean office there may be as many as 30 million particles per cubic metre.

Control is by filtration in the plant but smoke can be removed by local recirculation through fan filter units.

For Clean Rooms (rooms of a very high standard) dust count per unit volume will be specified, but other specifications for room cleanliness are usually in terms of filtration performance against a standard test dust. Other important features are resistance to air flow and dust-holding capacity, leading to the fan energy required and filter life.

By number count the great majority of particles in the outside air are likely to be less than 1 μm. By weight these small particles will account for a very small proportion of the sample. A filter with a high efficiency measured by weight of particles trapped may be almost transparent to the small ones. Very high counts can be found in rural areas from pollen or agricultural activities.

Where a high cleaning efficiency is required it is sound practice to install a filter of lower performance upstream to trap the larger particles and prolong the life of the more expensive High Efficiency Particulate (HEPA) Air Filter. Two filters of equal merit placed in series will not be materially more efficient than one.

The action of most filters is mechanical. These are normally scrapped when fully loaded but may, for lower efficiencies, be washable. Another type uses electrostatic charges.

16.7.2 Electostatic filters

These are usually provided with a mechanical pre-filter to remove the larger particles and serve to even out the flow over the face. The air then passes between plates positively charged at 10 or more kV and the dust thus ionized is attracted to downstream plates charged at a lower voltage negatively. A mechanical after-filter may also be fitted in case electric flashover at the plates releases dirt. Cleaning is by hot water wash, necessitating drying procedures, or by mechanical agitation. Their resistance to air flow is low and their efficiency good against small particles. They are not often employed in general air-conditioning work.

16.7.3 Tests

Different test methods produce numerically different results and comparisons should only be made of results using the same test method. Tests fall into three categories: gravimetric tests, which measure ability to trap and retain dust; those which measure staining power of contaminants before and after filtration; and those measuring the concentration of a test aerosol either side of the filter by photoelectric methods. The tests which will be met most often are as follows.

BS 2831 has been withdrawn. The No. 1 test for high-grade filters has been superseded by BS 3928 for absolute filters (those having an efficiency of better than 99.9% in that or similar tests) and by ASHRAE 52-76.

ASHRAE 52-76

This forms the basis of Eurovent 4/5. There are two tests. The atmospheric dust spot efficiency test, assessing the reduction of staining power of atmospheric dust, is generally used for the higher efficiencies as well as the synthetic dust weight arrestance test, using an artificial dust. In both cases averages are quoted from a series of tests.

Filters are classified in BS 6540: Part 1 – Eurovent 4/5 but without statements of airflow rate or final pressure drop. From the lower performance upwards:

Average arrestances (Am) EU1 <65%; EU2 ⩾65 to <80%; EU3 ⩾80% to <90%; EU4 >90%
Average efficiencies (Em) EU5 ⩾40% to <60%; EU6 ⩾60% to <80%; EU7 ⩾80% to <90%; EU8 ⩾90% to <95%; EU9 >95%

Higher cleaning ability is tested to Eurovent 4/4.

BS 3928

This test uses a cloud of sodium chloride particles. The size distribution is from below 0.1 μm to approximately 2.0 μm and a figure for percentage penetration is obtained. Since the test does not materially load the filter it can be used to assess the quality of each unit before despatch. This test forms the basis of Eurovent 4/4, which differs a little in procedure.

16.7.4 Standards required

For many applications filters with an average arrestance of 80% against Eurovent 4/5 will serve, but in more critical situations and, for example, in the outside air intake of computer rooms or laboratories 80% average efficiency would be chosen.

16.7.5 Clean rooms

The dust count to be achieved may be specified in terms of US Federal Standard 209b. There are three standards: 100, 10 000 and 100 000, which are the maximum counts per cubic foot of 0.5 μm and larger particles. Maxima for 5 μm particles are also given. BS 5295 has four classes: 1 is the most stringent and calls for a marginally lower count than US Federal Standard 209b, class 100. It carries details for a range of particle sizes, room construction and air flow pattern. Laminar flow is required for the cleanest areas, which can be achieved when the ceiling or one wall of the room is composed of a louvered filter bank. Filter efficiencies ranging from 70% to 99.995% are recommended.

16.7.6 Filter life

Filter life varies with make and type, and may be limited by the ability of the fan to operate against pressure drop. It is a function of the dirtiness of the air and the amount of material packed into the filter bank. Life may be quoted in terms of dust held in g/m^2 face area. For fan selection a knowledge of pressure drop is required. Typically, a panel filter might be quoted as operating from, say, 75 Pa initially at 2.5 m/s face velocity to 250 Pa when loaded and a HEPA might operate up to 700 Pa. The pressure drop across a bank of filters is kept within bounds by changing a proportion in rotation.

16.7.7 General

Care should be taken during installation to see that there is no edge leakage. For tailor-made plants it is best if filters are kept sealed in their delivery containers until after the plant has been completed and vacuumed out. Unfinished duct ends should be covered during the progress of other work, since cleaning will be almost impossible, except by the expedient of blowing through and physically disturbing them.

Filters should meet at least the requirements of (in the UK) BS 2963 and in some cases be non-combustible to BS 476: Part 4, although construction methods render this difficult.

16.8 Humidification

The ability to add moisture to the air to raise humidity is an essential part of close control air conditioning. The need may be brought about by a change in the condition of the outside air, fabric losses or a change in the gains within the room or in plant operation.

16.8.1 Humidifier capacity

The humidifier is sized to meet the load of the outside air intake quantity. It is necessary to know the rh to be achieved, outside air intake rate (kg/s) and its lowest winter moisture content (kg/kg). This can be taken as 0.0027 kg/kg in the south UK and 0.002 for the north, corresponding to $-4°$ and $-7°$C saturated, respectively. Since the capacity will be reduced by blowdown time, allowance is necessary for control, and there will be leaks from the plant and building fabric, an addition of about 30% above the calculated value is recommended.

16.8.2 Types of humidifier

These may be Direct (i.e. installed within the conditioned room) or Indirect (i.e. installed in the conditioning plant). Water may be introduced (a) as a spray or mist or (b) as steam generated separately or by a device within the plant.

In most air-conditioning work the humidifier will be indirect. The psychrometric operation of various types is described below.

16.8.3 Pan humidifier

The steam is generated in a pan of water by electric elements. High-temperature hot water or steam coils could also be used. About 30% of the input appears as waste sensible heat, giving a a sensible heat ratio of 0.3.

Where the pan is in the airstream the condition downstream of the pan has an increased moisture content (kg/kg) found from the air flow and moisture input. On a psychrometric chart this will lie on a line of sensible-to-total heat ratio of 0.3. Thus the psychrometric plot shows a steep rise in moisture content with a small rise in dry bulb temperature. The latter is a disadvantage when cooling is required. Regular blowdown is needed, preferably controlled by a timer, to prevent furring when mains water is used, and open pans should not be utilized where the conditioned space is to be dust controlled. A small water volume is an aid to quick response.

16.8.4 Steam jet

Where it is available the source can be a separate boiler plant, but common practice is to employ purpose-made electrode boilers within or adjacent to the plant. The latter reduces sensible gains to the plant but, being essentially saturated steam, condensate return pipes are required. In addition to the rise in moisture content of the air (kg/kg) being dependent on air-flow and steam-injection rates, there is a very small increase in dry bulb temperature by the cooling of the vapour to the air temperature. The rise in total heat is: total heat of steam (kJ/kg) × quantity supplied per kg air.

Proprietary units are supplied with automatic blow-down cycles and can be matched to the broad water analysis. Cylinders have a limited life.

16.8.5 Spray humidifiers

Water is injected into the airstream in a fine mist by pumped jets or spinning disc. For practical purposes, the psychrometric plot follows a wet bulb line. The air provides the latent heat of evaporation, resulting in a fall in dry bulb temperature. If water were to be supplied at up to 100°C the humidified condition would be at a correspondingly higher total heat of 420 kJ per kg water supplied.

Where dust control is important the system should only be used with a supply of demineralized water to avoid solids being passed into the conditioned space. The temperature of the air must be sufficient to hold the quantity of moisture being supplied, any excess being deposited in the duct. Unless drained away, this can give rise to corrosion and to incorrect control by re-evaporation when the humidifier is switched off under control.

16.8.6 Air washers

Banks of sprays discharge water into the airstream with the object of achieving saturation of the whole air flow. Excess water falls into the base tank of the washer from which it is pumped back to the sprays. Downstream eliminator plates entrap any remaining free moisture, acting best within a specified velocity range. Cleanliness is essential to avoid bacterial growth. A constant bleed and make-up is normally arranged to control the accumulation of waterborne solids, but this, in turn, dilutes bacteriocides and inhibitors. The washer does not fully wash the air in the normal sense but does have the cleaning efficiency of a low-grade filter.

If the temperature of the water is not controlled it will come to the wet bulb temperature of the air passing through. Ignoring pump heat, the process is adiabatic. The psychrometric plot follows a wet bulb line.

If heat is added to the water the condition for 100% saturation takes the new wet bulb temperature of the incoming air.

If the water is chilled, cooling of the supply air takes place together with control of dewpoint and hence humidity of the room treated. Saturation efficiency is given by:

$$\frac{\text{Entering air db} - \text{leaving air db}}{\text{Entering air db} - \text{entering air wb}} \times 100\%$$

Due to bypass, a single bank of sprays might achieve 60% saturation and a capillary washer (one where the air passes through a wetted mat) might achieve 95%.

16.8.7 Humidifier run time

Humidification is an expensive process and it is useful to be able to assess energy costs when considering its inclusion in a plant. It is possible from meteorological records for any particular area to find the time in hours per year and extent to which the external moisture content is, on average, at or below a required absolute value. As an example, in the southern UK humidification would be required in varying degrees for 6700 h per year if 21°C 50% rh was to be maintained. If the characteristic of cooling coils is to dehumidify when only sensible cooling is required then the humidifier load will be greater.

16.9 Test procedure for air-conditioning systems

16.9.1 Object and application

This is an on-site test following installation. The procedure is designed to find any weaknesses in the plant and its implementation, while not cheap, is worthwhile where performance is important or in doubt. It is essential to have contractually agreed its use, who is to make the various provisions and what action is to be taken if any faults are shown and to allow sufficient time for testing and any remedial work.

It is recommended that a contractor demonstrating to a client should have tested the plant previously and should not use the test to supplant commissioning. The acceptance that individual items of plant have the necessary capacity does not necessarily mean that the whole will perform as required. Indeed, the test can find proprietary items lacking.

16.9.2 Conduct of the tests

The necessary skills should be on-site to attend to any minor faults but must not be allowed to alter settings from one test to another, except as required by the test procedure. Dummy loads should be ready for use. If sensitive equipment is to be present (as might be the case in upgrading computer room plant) then the interested parties must agree limits previously. This may also apply to parts of the plant itself such as humidity-detecting elements. The latter can be protected physically. The job specification should state which parts of the procedure apply (if not all).

The plant is tested to hold limits in the specified external conditions and is not (unless exceptionally so agreed) tested beyond them. The plant should not be overdesigned just to pass the test. For the duration of the test external conditions can never span those for which it is designed. To fully load the plant in cool weather it may be necessary to dissipate twice the design process load. The dummy load should simulate the design load, and in the case of electronic equipment rooms this may best be done by arranging fan heaters directed upwards and in the same layout as the equipment. Ideally, simulation of the heat of outside air make-up should be in the intake duct. If impractical, then it may be placed in the room close to a return grille. Room-temperature readings should not be taken near to dummy loads, and in heavily loaded rooms it may be necessary to confine readings to extract positions. Except in test houses, it is not possible to apply the full range of external conditions at the final cooler. It will therefore be appreciated that a plant should easily pass its test on a cold day. Resort could be made to certified performance curves if a doubt exists.

Misleading results can be obtained from tests of limit 'stats' and sensors if their set point is adjusted to the current conditions to bring about operation. The control points should be set and conditions adjusted until operation occurs, opportunities arising during the test procedure.

The following is an outline of the test procedure:

1. *Pre-test performance.* An assessment of whether the plant is ready for a full test.
2. *Recorders and alarms.* Continues 1 and checks the calibration of recorders and alarms.
3. *Air balance.* To check overpressure and outside air intake.
4. *Background heaters.* Checks control and capacity of back-up heating.
5. *Heating and humidification.* The plant is called upon to perform at its maximum winter design capacity.
6. *Cooling and dehumidification.* The plant is required to perform at its maximum summer capacity.
7. *Limits, interlocks and inspection.* Cut-outs, limiting controls and interlocks are tested.

Compliance with specified details and good practice are checked.

16.9.3 Pre-test performance

Allow the plant to run automatically for a period of some days either with no artificial load or with a small one of fixed value. If no means of recording are built into the plant than recorders should be introduced and placed in representative positions in the various zones for the duration of the tests. Periodically examine the traces or monitor printouts. These will give an indication as to whether the plant is controlling correctly on light load.

Observations of external wet bulb and dry bulb air temperatures or similar assessment of other forms of load may be combined with simple observations of what the plant is doing. If it were using its refrigeration capacity under cold no-load conditions then clearly something is wrong. A common fault is finding the plant calling for large amounts of humidification when the requirement as indicated by external readings is for dehumidification, this perhaps being due to incorrect SHR of the cooling coil. Periodic excursions may be found and these are often caused by external influences which might be identified by their time of occurrence.

Make no excuse for checking whether the sun shines on the thermostats or someone has washed the floor. Another common occurrence is that while the room is not being controlled as specified, the plant appears to be operating correctly. Here a series of simultaneous readings of air flows and wet and dry bulb temperatures should be taken around the whole of the air circuit – on and off coil, condition after fan, condition at inlet grilles, etc. Referring these to a psychrometric chart will usually indicate where the fault or leak lies.

16.9.4 Recorders and alarms

These are important, since they may be used in part using the remainder of the procedure and may be the only indication for future operatives as to whether conditions are correct.

1. With the plant still operating as for test 1. With hand instruments, check conditions as close as possible to the recorder's heads:

Recorder no.	Time		Date
	DB	WB	RH
Recorder reading		—	
Test sling (or other)			

Note that some recorders have a clock time difference between the dry bulb and rh to allow for pen overlap.

2. Where this can be done, moving the alarm-indicating arms to the room condition will indicate whether the alarms operate but much more satisfactory results are obtained by causing the controlled condition to swing beyond the alarm points and at the rate that would be expected in practice.

Recorder no.		Time		Date
	DB	WB	RH	
Recorder reading		—		
Test sling (or other)				
Alarm occurred		—		

16.9.5 Air balance

Room overpressure can conveniently be tested during this period. Produce a single line sketch of the conditioned space(s) including doorways. Test for air-flow direction with hand smoke tubes, both at the top and bottom of closed doors, and record the direction of flow. Crack the doors open if fully sealed.

Measure the outside air make-up quantity, taking a pattern of readings across the area of the duct or grille: Average flow velocity m/s × area $m^2 = m^3/s$

16.9.6 Background heaters

The function of these is to prevent the room becoming cold and humid in the event of the main plant being off for any reason other than power failure. For a small input they can protect the conditioned space to a large degree and greatly reduce reconditioning time upon restoration of the main plant. They would require a comparatively small input from a standby generator.

With the main plant switched off:

1. Set the humidistat (if fitted) downwards and note when the heater(s) come on. As before, more accurate results are obtained by swinging room conditions, but (a) we will have ascertained that the direction of operation is correct and (b) the method is sufficient for the relatively coarse control required. Reset.
2. Set the thermostat upwards noting when the heater(s) come on and continue to the limit of the thermostat. Take a set of readings.

Room	Date
Time	Dry bulb temperature

Assuming the heaters are electric (i.e. have a high surface temperature) and the conditioned space is uniformly surrounded by rooms or the outside then the temperature that should be achieved is the design lift in temperature from the surrounds to room temperature, above whatever the surrounding temperature happens to be on the day of the test.

If there is a mix of external walls, heated or cold surrounding rooms and rooms under similar control to the one being tested the temperature elevation of the tested room (°C) should be:

Design heat input (watts) + sum of (surface constants × TD to outside at time of test) for each surface ÷ sum of constants for all other (outside) surfaces

where the surface constant is U value × area.

3. Since any background heaters must now be allowed to operate during the cooling cycle their control must be set below the lower limit for the conditioned space. An interlock with the main plant may be included. These features should be checked and noted at this time.

16.9.7 Heating and humidification

The plant should be able to hold the room at the desired temperature when heat is not being dissipated in the room. The plant should also be able to raise the temperature of the room from cold under winter conditions (e.g. after a power failure or an outage). In considering the temperature to be achieved under test the comments given in test 2 of Section 16.9.6 also apply here. If hot water heating is used, full temperature elevation will not be achieved because high return air temperatures will reduce the coil output. For instance, where the design rise was 22°C above an external of −1°C; 19°C, 16°C and 14°C are satisfactory with ambients of 5°C, 10°C and 15°C, respectively.

If it is variable, set the outside air intake quantity to the winter value. Set the desired temperature up to a maximum, leaving the humidity setting alone. It will be appreciated that as the room temperature rises during the heating test the rh tends to fall. However, since the humidity setting remains unaltered the humidifying system will be called upon to operate until at one condition it is working at peak winter rate. Due to the faster characteristic of heaters, the rh will be found to fall but absolute moisture should be found to steadily rise. The duration of the test is normally about 3 h and final conditions should be held for half an hour to prove the moisture source.

Calculation of test condition *kg/kg*

Absolute moisture in the room at design temperature and humidity;

Absolute moisture content of winter outside air intake at design condition;

Design difference to be made up by humidifier =

Add absolute moisture content outside at time of test;

= Room moisture content to be achieved in test.

(Neglect gains from test personnel)

Room Date

DB	WB	= kg/kg

Time

Under steady conditions the plant should be able to hold the design room rh up to the temperature given in

Room Date

Supply Air	Sens. load kW	Lat. load kW	Position 1 Dry Wet RH	Position 2 Dry Wet RH	Position 3 Dry Wet RH

Time

the remainder of the test load which represents the missing summer building gains is now applied.

Design sensible heat of outside air intake kg/s × kJ/kg
Less actual on day of test kg/s × kJ/kg

Room design conducted heat gain kW

less: $\dfrac{\text{(design gain)} \times \text{actual TD}}{\text{design TD}}$ kW

Design solar gain less actual gain (under site conditions this will have to be estimated) kW

Total kW

Plus
Design outside air moisture content (kg/kg)
Less actual moisture content on day (kg/kg)
× air flow (kg/s) = kg water/s

(it will be found that kg water per hour × 2/3 closely approximates to the power in kW of test load kettles required).

psychrometric tables or a chart where the design room rh and test absolute moisture content (kg/kg) coincide. At the end of the test restore the temperature set point to the design value and any control of outside air volume to automatic. Observe that operation.

If the plant is now shut down for a period of 1 or 2 h with the conditioned space remaining closed, a slow reduction in absolute moisture content will be observed. A rapid fall will indicate a significant leak in the building or plant, would account for any difficulty in achieving the test result and should be investigated. The design vapour pressure difference may amount to several tens of millimetres water gauge acting on the vapour sealing and equally leads to moisture ingress during summer conditions.

16.9.8 Cooling and dehumidification

The test load must be calculated and is applied in two parts. The lights should be as in normal use and all plant controls normal and automatic. The background heaters (if installed) can be called as load but it is essential that thermostats and overloads do not reduce the applied load unobserved. Electrical measurements should be taken, particularly at the full-load condition, and for this purpose, trust not put in rating plates.

The test load: watts

Sensible equipment or process heat;
Sensible heat of personnel;
(Less number present)

Latent heat of process;
Latent heat of personnel;

In the case of electronic equipment rooms there will be no call for a latent test load at this stage.

Starting from zero, take a set of readings, then increase the load in increments of about 20%, repeating at 15-min intervals. Note that if the supply air passes over a structural slab its elevated temperature from test 16.9.7 will briefly add load.

To avoid unnatural conditions the latent load is applied in steps with the sensible load until full summer conditions are simulated. This should be held or 1 h or more. If the pattern of temperature readings around the space is satisfactory we have a check on the suitability of the air distribution. If any doubt exists as to the duty achieved, simultaneous readings of air flow and wet and dry bulb temperatures across the cooling coil or supply to extract can be taken and the duty calculated.

Observations should be made that the plant is not being manually coaxed, that it is not humidifying and that the compressors have not tripped but are cycling under part-load conditions. Instruments should be indicating correctly. In cases where it is impractical to fully load the plant, ascertain that an appropriate amount of refrigeration is being employed. This situation is best avoided, as much of the plant is not being demonstrated to full capacity.

16.9.9 Limits and interlocks

The items for test under this group will vary from plant to plant and with specified requirements. It is assumed that the Clerk of Works is satisfied with the following:

Interior of plant clean before run;
Interior of ducts clean before running or blown out to bags;
Ducts sealed;
Lagging sound, non-combustible, no loose material in airstream;
Vapour sealing sound;
Filters of correct grade;
Filters kept clean before installation;
Filters installed without leaks;
Water circuits leak tested;
Water circuits flushed;
Water circuits dosed;
Refrigerant circuits vacuumed and dry;
Refrigerant circuits leak tested;
Crank case oil level correct;
Plant Room ventilated;
Humidifier drainage and deep U-trap correct.

Test

For each compressor:

HP cut out;
HP cut in;
LP cut out;
LP cut in;
Oil Differential cut out;
Oil Return functioning (sight glass);
Low water temperature cut out;
Recycling timer;
Loss of phase protection;

High room humidity override;
High room temperature override;
Plant heat high temperature cut out;
Plant low air temperature cut out;
Air flow interlock;
Automatic changeovers;

Room emergency power off buttons
Emergency power off buttons reset (only when so manually operated).

17 Water and Effluents

George Solt FIChemE, FRSC

Contents

17.1 Introductory warning

Water is an essential service to any facility. The amount and quality needed vary considerably between different plants, but the essential fact important to all engineers who have to deal with water technology is that the subject of water quality and treatment is highly specialized. It deals with chemistry and microbiology, and even within those fields the technology is specialized and in the hands of experts. It is therefore a field in which plant engineers cannot be expert, and are forced to take advice from specialists. These are usually suppliers of plant or chemicals, or professional consultants.

These specialists are in a position of great responsibility, because plant engineers are so dependent on their advice. In these circumstances it is possible for 'cowboy' organizations to thrive. Past and present experience shows that these are too common. *Plant engineers must not automatically accept the lowest offer for materials or advice in this field but must first be satisfied with the specialists' competence and integrity.*

There are many sad examples of installations which have gone disastrously wrong, and these are by no means limited to small facilities or those in which water is a relatively unimportant service. For example, manufacture of microchips is wholly dependent on a supply of highly purified water. In recent years two of the largest UK manufacturers, sited at opposite ends of the country, have had to shut down and send the workforce home because their purified water facility had failed.

The golden rule, therefore, is to deal only with consultants, contractors, plant suppliers and water-conditioning experts whose experience and standing are known to be good.* If there is any doubt, references should be sought and followed up. Water and effluent installations are a relatively minor cost item in any plant, but their failure can be disastrous. It is foolish to make false economies on so essential a service.

17.2 Requirements for water

Any plant will need water for domestic purposes: in most cases this can be provided by public supply and discharged to the public sewer. Most plants have some steam-raising equipment for space heating, and steam is often required for other purposes. Steam raising always requires water conditioning and usually external treatment of some kind. Many operations require water of a specified quality, which varies over a wide range: from cooling and washing water to softened or demineralized and (in extreme conditions) ultra-pure water.

For food and drink, medical, pharmaceuticals and cosmetics production the microbiological quality of the water becomes paramount. Even in applications where biological quality is not directly important, uncontrolled growth can be a damaging nuisance. Warm-water systems and cooling circuits in particular are a potential hazard (e.g. from *Legionella*). Some water treatment or conditioning is commonly required.

Water from public supply costs money, as does that discharged to the public sewer. Except for water which is incorporated into the product of the plant or evaporated to the atmosphere, any water which enters the plant is returned to the environment as an effluent. It may be necessary to improve the quality of this before disposal to surface water or into a public sewer, and for acceptance into the public sewer the sewerage authority may levy a charge dependent on the quality. This is levied per cubic metre of water discharged, and this volume, in turn, is normally estimated on the basis of the metered incoming mains supply flow, minus some agreed factor for water retained in the product or lost by evaporation. There is therefore a double financial incentive to reduce the plant's water consumption.

17.3 Water chemistry

Raw water analyses are normally obtained from the local water supply organization. Water analysis is a specialized trade, and analysts who do not routinely carry this out can prove unreliable. It is also important to ascertain the seasonal and long-term variations to be expected.

For conventional factory boilers, most of the many items normally shown on a water supplier's analysis sheet are unimportant. The ones to look for are listed below.

17.3.1 Units of measurement

Analyses usually give concentrations in milligrams per litre (mg/l) or parts per million (ppm). For practical purposes these units are the same.

Most minerals in water exist as ions – electrically charged particles which give them an electrical conductivity. The different systems of units which measure their concentration can cause much confusion. For any calculation involving adding different ions to one another it is vital to use one of two systems of 'equivalents'.

The traditional British method is to calculate the concentration 'equivalent to calcium carbonate' and give the results as 'mg/l as $CaCO_3$'. A more modern unit, widely used throughout Europe, is the milligram equivalent or milliVal, abbreviated as meq/l or mVal. This gives the same information as 'mg/l as $CaCO_3$' but the values are one-fiftieth of those expressed 'as $CaCO_3$'. Explaining these systems is lengthy, and usually increases the non-chemists' confusion. The system 'mg/l as $CaCO_3$' is used below and the rules are as follows:

1. Concentrations of individual constituents, such as calcium, 'hardness' or 'alkalinity' should be brought to mg/l as $CaCO_3$ for any comparisons or calculations.
2. There should be a statement on the analysis sheet which makes it clear what system of units has been used. Hardness and alkalinity are often given 'as $CaCO_3$', which makes life simple.
3. Alternatively, the analysis may state that concentrations are given 'as the ion' or 'as such', or each individual constituents may say (for example) 'calcium

* Most reputable plant suppliers are members of BEWA (British Effluent and Water Association).

as Ca'. In that case one must use the following conversion table:

To convert from mg/l as such	*To mg/l as $CaCO_3$ multiply by*	*To meq/l multiply by*
Calcium (Ca)	2.5	0.050
Magnesium (Mg)	4.17	0.083
Bicarbonate (HCO_3)	0.82	0.016

4. Total Dissolved Solids (TDS) and Silica (SiO_2) are normally given 'as such' and do not require any conversion.

17.3.2 Hardness and alkalinity

Many analyses quote 'Total Hardness'. Some give 'Temporary Hardness' (or 'Carbonate Hardness') and 'Permanent Hardness (or 'Non-carbonate Hardness'), usually in consistent units so that the values can be added together to give the total hardness. The total hardness is actually the quantity of calcium (Ca)+magnesium (Mg) in the water. If the total is not given directly, the values given for these two constituents must be added, after conversion to mg/l as $CaCO_3$ if necessary.

Hardness in water varies widely, and as an arbitrary classification:

Hardness less than 50 mg/l as $CaCO_3$	Soft
50–200 mg/l as $CaCO_3$	Medium
More than 200 mg/l as $CaCO_3$	Very hard

Temporary hardness

Temporary or carbonate hardness and alkalinity frequently, but not always, mean the same thing (see below). 'Bicarbonate' or 'hydrogen carbonate' is a more scientific term since 'alkalinity' is actually the concentration of the bicarbonate (HCO_3^-) ion in the water.

If acid is dosed into a water-containing bicarbonate, the ion becomes converted to carbon dioxide (CO_2) gas, while the water becomes only slightly acidic. (Sodium bicarbonate taken against acid stomach uses this property.) The drop in pH from a given acid dose is much smaller than would result from the same amount of acid dosed into a water containing no bicarbonate. When enough acid has been added to convert all the bicarbonate to CO_2, further acid dosing leads to the sharp drop in pH, which is expected from a water containing no bicarbonate.

Another reaction of bicarbonate is that in boiling water it combines with any hardness present to produce scale, while releasing CO_2 into the steam. This hardness is called the 'Temporary Hardness'. Its concentration therefore depends on the amounts of both hardness and bicarbonate, whichever is the less. Most waters contain more hardness than bicarbonate, so that the temporary hardness is usually equal to the bicarbonate content. The 'Temporary Hardness' quoted in analyses is often the only information available on the bicarbonate content of the water.

Permanent hardness

After the temporary hardness has been removed, any calcium and magnesium which remains is still capable of forming a scum with soap, and can also react to form boiler scale. This is called the 'Permanent Hardness'.

CO_2 release

When the steam is condensed, any CO_2 released in the boiler redissolves in the condensate, making it slightly acidic and corrosive. Normally, boiler feedwater is softened and the boiler water pH is raised by addition of caustic soda. Both these measures reduce the degree to which bicarbonate breaks down and releases CO_2. Even so, very large amounts of bicarbonate entering the boiler should be avoided.

17.3.3 Total dissolved solids (TDS)

Water contains various other salts which are generally unimportant in medium-pressure boilers, except for the contribution they make to the total solids in the boiler water. They can therefore be grouped together, and since dosing water-conditioning chemicals also contributes to the boiler contents, even the total need not be known exactly. Most analyses give a figure for TDS in mg/l, but analytical methods differ and the result is not always particularly reliable. TDS is usually given as the actual weight of dissolved materials, which is 10–15% higher than the TDS measured 'as mg/l $CaCO_3$'.

Another method of estimating TDS is to measure the electrical conductivity of the water, which is usually reported as 'μho' or 'μS'. This figure is roughly double the TDS in mg/l as $CaCO_3$.

For many years water costs have been rising faster than inflation, and are set to rise even faster in the future. The belief that water is 'free' must be combated: it is usually possible to make more economical use of water. In most factories the water bills are paid centrally, and there is no system for debiting the cost to the actual water users. This makes for wasteful practices which would be improved if there were a system for monitoring where the water costs are incurred within the facility.

In some cases sensible design can lead to re-use of water, which reduces both water and effluent costs. This is best achieved by intelligent routing of the water rather than by treatment before re-use. Effluent treatment is best avoided wherever possible. For example, very slightly contaminated process wash water can be recovered for washing down floors. This reduces charges for both incoming water and effluent.

17.3.4 Silica

The silica content of a water only becomes important if it is a large proportion – say, more than 10% – of the TDS. This is unusual in normal water supplies, but can result from external water treatment (see below).

17.3.5 Types of water

Most natural waters contain more hardness than bicarbonate. Only a few sources in the UK, usually from

Table 17.1 Drinking-water quality standards (EC, 1980)

Colour	Pt/Co	20
Turbidity	JTU	4
Threshold odour no.		3
Anionic detergents	mg/l Manoxol	0.2
Pesticides	mg/l	0.5
PAH	µg/l	0.2
Phenols	µg/l	0.5
Aluminium	mg/l Al	0.2
Ammonia	mg/l NH_4	0.5
Arsenic	mg/l As	0.05
Barium	mg/l Ba	0.1 (GL)
Calcium	mg/l Ca	100 (GL)
Cadmium	mg/l Cd	0.005
Chloride	mg/l Cl	200
Chromium	mg/l Cr^{6+}	0.05
Copper	mg/l Cu	0.1 (GL)
Cyanide	mg/l CN	0.05
Fluoride	mg/l F	1.5
Hydrogen sulphide	mg/l H_2S	–
Iron	mg/l Fe	0.2
Lead	mg/l Pb	0.05
Magnesium	mg/l Mg	50
Manganese	mg/l Mn	0.05
Mercury	mg/l Hg	0.001
Nitrate	mg/l NO_3	50
Selenium	mg/l Se	0.01
Sodium	mg/l Na	150
Sulphate	mg/l SO_4	250
Zinc	mg/l Zn	5.0
Coliforms		Zero in 95% of samples

GL = Guide Level (i.e. no MAC applies)

wells in sandstone strata, contain more alkalinity than hardness. In most cases the temporary hardness greatly exceeds the permanent hardness. This is especially true of the hard alkaline waters which come from chalk and limestone measures.

Many waters from mountainous uplands such as Wales, Scotland, Yorkshire, or the moors of south-west England are surface runoff and have not percolated through mineral strata. They are low in minerals generally, and hardness and alkalinity in particular, though even there the hardness will be greater than the alkalinity. Their main characteristic is their high content of organic matter, which may give them a faint yellowish tinge, and a low pH. Many such soft and peaty supplies have lime added at the waterworks to make them less corrosive.

Some potable supplies are treated surface waters from rivers, etc. These originally derive from any of the above, but will also contain the products of human activities, which lead to increased mineral contents and possibly some undesirable materials such as detergents.

Potable water is not (and should not) be sterile. In fact, potable water mains always have a layer of living matter clinging to the walls. Any change in the flow, temperature or chemical quality of the water passing through the pipe will cause some of this to become detached. Quite large living organisms (such as freshwater shrimps, waterflies and even leeches) can occasionally be found in a potable supply. They are aesthetically disturbing but usually unimportant.

The important quality criterion is absence of pathogens, a term which covers all disease-bearing bacteria. This criterion is usually determined by test for *Eschericia coli*, a species of bacteria so common in the gut that it is a reliable indicator of any pathogenic contamination.

Some plants have to rely on a private supply, usually taken from a borehole. The water first must be tested by a competent authority. Water analysis is a specialized procedure and is preferably undertaken by an analyst who does this regularly, such as a local water supplier's laboratory. Water from a spring or a deep well which appears potable is usually found to be of reasonable quality, but it may occasionally contain some constituents (e.g. iron or manganese) in unacceptable concentrations. The nitrate content of groundwaters is generally rising throughout the UK, and if it is found to be near the legal limit, further samples should be analysed at, say, six-monthly intervals.

17.4 Building services

17.4.1 Potable water

Most plant sites in the UK have access to a public supply of water. Until recently this was legally required to be no more than 'wholesome and palatable'. Water suppliers are now responsible for meeting the EC's Directives (see Table 17.1). Wherever possible, all drinking-water taps should be served directly from the incoming main, and the plant engineer's sole responsibility is to ensure that no deterioration takes place within his system.

The water supply authority normally insist that (for uses other than drinking-water taps) their main should discharge into a break-pressure vessel, after which the water quality becomes the consumers' responsibility. The water tank should be covered against tramp dirt and access by birds, etc., and it must be shielded from sunlight to avoid the growth of algae. Nevertheless, access must be maintained for easy inspection. The distribution pipework is preferably all-plastic and lead must be avoided altogether. The use of copper is doubtful with some corrosive waters, and soldered joints in it can lead to unacceptable concentrations of lead in the water.

The removal of such impurities is relatively simple at a waterworks, but a typical plant cannot provide the chemical expertise needed to keep the process in good working order. Treatment should therefore be avoided if there is any reasonable economical alternative.

If the water is found fit for consumption, with respect to both its mineral and biological content, the problem of sanitization can still arise. Public supply invariably has a very small residual chlorine level. This suppresses biological growth and maintains water quality even when the line is stagnant. As with other forms of treatment, the scale of private supply is usually too small to allow good control of chlorination equipment.

One method of operating an unchlorinated supply safely is to ensure that the line runs constantly to waste, as do very old drinking fountains. This avoids biological growth which can accumulate in stagnant water. Dead legs in the piping system are always undesirable, especially in such cases.

Employers may provide drinking-water dispensers, supplied by commercial water purifiers with a superior grade of water and kept in the dispenser at a temperature low enough to inhibit growth. This service is at present limited to the London area, but it may become more widespread. It assuages health fears and may provide a more agreeable water quality than the local public or private supply. It would be particularly useful if there is no public supply, and the private supply is not wholly reliable. Personnel could then be warned that the piped water is unsuitable for drinking.

17.4.2 Domestic effluent

All plants produce domestic effluent, which is preferably discharged into the public sewer. If no sewer is available the plant needs a septic tank or a similar device sized for the probable demand, which is based on the number of people whom it will serve. Architects and competent building contractors are familiar with rules of thumb which apply to the design and sizing of these. Septic tanks need regular pumping out of sludge and access must be provided. Many industrial effluents interfere with the biological activity of a septic tank and must therefore be kept separate and discharged by other means.

17.4.3 Domestic water

Domestic hot and cold water, for WC flushing, hand washing, etc., will normally be supplied within a factory from the break-pressure tank fed by the incoming water main or private supply. Although these supplies do not need to conform to potable water quality standards they can provide a breeding ground for a variety of bacteria, including *Legionella*. As a precaution, many plant engineers dose sodium hypochlorite into the break tank to maintain about 0.2–0.5 mg/l of free chlorine in these supplies. Bacteria breed most prolifically at temperatures between 10° and 60°C and, wherever possible, hot water should be maintained above and cold water below this range. If the raw water is hard then consideration should be given to softening at least the hot water to prevent scale formation in calorifiers, pipework and sanitary ware.

17.4.4 Closed circuits

Closed water circuits, such as chilled water, or medium- and high-pressure hot water systems, should be initially filled with the best quality water available – de-ionized for preference but at least softened. They should then be conditioned by dosing with suitable corrosion-inhibiting chemicals and biocides. Make-up to closed systems is usually very small and raw water will often suffice, although, of course, higher-purity water is better. Routine sampling to check on inhibitor levels and bacterial growth is vital to the operation of such circuits, and most reputable conditioning chemical suppliers will undertake this work on a contract basis.

17.4.5 Open cooling-water systems

Although direct air-cooled systems are now preferred there are still many evaporative cooling towers in operation, and open systems of this type represent the most difficult of all the water-treatment situations the plant engineer is likely to meet. The systems work as follows: heat from the heat exchanger is taken up by the circulating water which is then sprayed into the top of a packed cooling tower with a forced or natural draught of air flowing through it. Some of the water evaporates, taking in latent heat from the bulk of the water which, consequently, cools down. The cooled water collects in a sump below the tower and is pumped back to the heat exchanger. The water which evaporates leaves behind any dissolved salts and other contaminants which, as a result, become more concentrated. A certain amount of water is also lost from the tower in the form of 'windage' or spray. To compensate for these losses a make-up supply of water is required. The concentration effect is most important because, if it is not properly controlled, salts – especially hardness salts – may become over-concentrated and deposit as scale in the pipework, in the heat exchanger and on the tower packing, where it causes blocking, and its added weight can lead to mechanical failure. To control the build-up of dissolved salts a bleed-off of concentrated water is necessary, and this must also be replaced by make-up water.

The circulating water comes into contact, in the cooling tower, with large volumes of atmospheric air and washes from it a variety of airborne contaminants, including pollutant gases such as oxides of sulphur and nitrogen, dust and soot particles and spores of bacteria, fungi and algae. The dissolved gases may give rise to acidic conditions which contribute to corrosion, algae can grow in the sump where warmth and sunlight provide ideal conditions while fungi may appear on the dead algae. Aerobic bacteria, including *Legionella*, grow in the circulating water and contribute, with algae and fungi, to the formation of slimes which not only cause physical blocking of plant but also set up areas of differential aeration around the system, which can promote intense corrosion. This process is aggravated by the deposition of scale and corrosion products below which corrosion takes place and anaerobic bacteria multiply.

Water losses

The rate of evaporation from a cooling tower is approximately 1% of the circulation rate for each 5°C drop in temperature across the tower, or about 7 l/h per tonne of refrigeration. Windage losses will obviously depend on the prevailing wind conditions and the design of the tower with regard to spray elimination but, typically, these are about 0.2% of the circulation rate.

The amount of bleed-off required will depend on the

nature of the make-up water and the type of conditioning chemicals used. The specialist tower manufacturer, conditioning chemical supplier or water-treatment consultant will advise the maximum concentration factor (the ratio of circulating water concentration to make-up water concentration) which can be allowed. The necessary bleed-off is then given by:

$$B = E/(C-1) - W$$

where

E = evaporation rate,
C = concentration factor (typically 1–5),
W = windage.

The bleed-off may be set by a simple manual valve running continuously to drain or by an automatic valve controlled by the conductivity of the circulating water.

Make-up water

The required make-up water, M, is given by:

$$M = S + W$$

or

$$M = (E \times C)/(C-1)$$

Many towers operate with hard water make-up, and scale-inhibiting chemicals are dosed to prevent the formation of hard, adherent scale. Hardness salts do precipitate in these systems but in the form of a mobile sludge, which is easily removed. A more satisfactory solution is to use make-up water which should, ideally, be softened or de-alkalized. However, naturally soft or artificially softened water tends to be corrosive in the conditions encountered in open cooling systems, and it is necessary to dose a corrosion-inhibiting chemical to protect metal surfaces. In either case, the accurate control of chemical dosing is absolutely critical to the reliability and integrity of the cooling system. The selection of the correct conditioning chemical(s) and operating regime will depend on many factors, including the make-up water quality, concentration, materials of construction of the system and environmental conditions. It is a task for a specialist, and chemicals should only be purchased from reputable suppliers who will provide a continuing service of monitoring and control.

Biocides

Biocides are added to cooling water to control the growth of bacteria, fungi and algae in the system. Chlorine, dosed in the form of sodium hypochlorite, is probably the best broad-spectrum biocide and, at residual levels of 0.5 mg/l, chlorine is effective against most bacteria, including *Legionella*. However, a recirculatory system means that bacteria are exposed continuously to the same chemical conditions and resistant strains, with a natural immunity to the biocide, will eventually appear and colonize the system. To prevent this, a regular 'shot dose' of an alternative biocide is advisable, and most chemical suppliers have a range of biocides, both broad spectrum and specific, for this purpose.

Filtration

Insoluble suspended matter, either picked up from the atmosphere or formed by deposition and corrosion within the system, together with slimes will, if not removed, cause blocking and abrasion problems. The build-up of such material can be controlled by 'side stream filtration', in which about 2–5% of the circulating water flow is filtered continuously. A sand filter is commonly used for this type of duty.

17.5 Boilers

17.5.1 Water for steam raising

Most plants have boilers producing steam for space heating: many need steam for other purposes as well. Boiler water requirements for boilers have changed radically over recent years. The old 'Lancashire' and 'Economic' boilers had large heating surfaces and low heat transfer rates: scale deposits would do no more harm than reduce their thermal efficiency.

Modern packaged boilers use the heat transfer surfaces much more intensively, and are endangered by scale. The boiler water acts as a coolant without which the metal of the tube overheats. Thin films of scale can obstruct heat transfer sufficiently to bring about tube failure in this way – especially if the scale deposit is siliceous, which is a particularly good insulator. Alternatively, the scale deposit may slow down heat transfer from the hot combustion gases to such an extent that the temperature at the back of the boiler rises excessively and causes tube plate cracking.

Boilermakers now recognize that their heat transfer rates had become too high, which made control of the boiler water quality unacceptably critical. They have reverted to slightly lower heat transfer rates, but poor boiler water quality remains the main single cause of boiler failure. Good water-treatment plant and boiler-water conditioning and control are still vital not only to the boiler's performance but also to its integrity.

17.5.2 Managing the steam-water circuit

Most industrial installations have a boiler of some kind: this boiler and its steam user form a circuit in which water and steam circulate. Loose use of nomenclature sometimes leads to confusion, and it is therefore useful to define the various waters in it:

Raw water: The mains supply or other external source used to prepare make-up.
Treated water: The water leaving the external treatment plant, if there is one.
Make-up: The amount of raw or treated water added to the feed.
Condensate: The water returning from the steam user(s).
Feedwater: The water entering the boiler feed heating system, which will normally be a blend of treated water and condensate.
Boiler water: The contents of the boiler.
Blowdown: The water blown down from the boiler in

order to maintain its total dissolved level below the specified limit.

In order to maintain good boiler operation, the most important rules to follow are:

1. The boiler water should be within the limits specified for that type of boiler by British Standards,* DIN and similar standards. In a conventional shell type factory boiler the most important criteria are that hardness should be present only in very small concentrations, and the TDS should be below 3000 mg/l.
2. The boiler water must at all times contain a positive residual of oxygen scavenger (usually sodium sulphite).
3. The water-containing chemicals should include phosphate or tannin to counter any residual hardness.
4. The boiler water pH must be raised to about 9 or over to avoid corrosion, to maintain silica in solution, and to reduce the release of CO_2 into the steam.
5. The bicarbonate content of feed should be moderate to avoid excessive liberation of carbon dioxide in the boiler.

Clearly, these matters are interdependent.

17.5.3 Condensate return

Condensate normally contains no hardness and is very low in dissolved solids. Unless it has been excessively exposed to the atmosphere, condensate is also very low in dissolved oxygen. Therefore it represents the ideal feedwater, and the higher the proportion of recovered condensate in the feed, the easier it will be to maintain the boiler water within the desired limits. The percentage condensate return is thus basic to all considerations of water management for the boiler circuit.

In some process applications condensate becomes contaminated: in sugar refining it may contain sugar, in paper manufacture and some other processes it may be contaminated with raw water, and where it feeds turbines or other machinery it is liable to contain oil or grease. In all these cases the condensate may still provide a better source of boiler feed than the available raw water, but it may need condensate filtration or softening plant before re-use.

17.5.4 Raw water quality

This varies widely within temperate zones, and even more so in hot countries. A wide range of possibilities therefore exists: high hardness, high alkalinity, and/or high TDS need correction. However, the degree to which this is necessary is also dictated by the percentage of condensate return – if it is high, the need for external treatment is accordingly reduced. The processes used for external treatment are described below.

17.5.5 Oxygen scavenging

Water at ambient temperature in contact with air contains about 10 mg/l of oxygen. To avoid corrosion, boiler water must contain no oxygen, and have an excess of oxygen-scavenging chemical (usually catalysed sodium sulphite) to ensure this. Ideally, sulphite is dosed into the feed heating system with the feedwater to give the hot well, etc. some protection, with a second dosing point into the boiler itself to ensure that the residual is actually maintained.

The removal of each milligram of oxygen requires about 8 mg of sodium sulphite, so that the dose should be adjusted to suit the amount of oxygen introduced, which corresponds roughly to 80 mg/l of cold make-up. A high level of condensate return therefore reduces the scavenger demand, and this is not only an economy in itself but can also mean a considerable reduction in the amount of total dissolved solids introduced into the boiler with the feedwater.

Oxygen can also be removed from feedwater by thermal de-aeration, or partially removed by skilful design of the feed heating system and blowdown recovery. These processes run without cost to the operator, but save chemicals, and, by reducing the required dose of sulphite into the system, decrease the amount of non-volatile solids added into the boiler.

17.5.6 Blowdown

All non-volatile impurities entering the boiler must build up in the boiler water. This includes the TDS in the feed, plus most of the conditioning chemicals, of which the sodium sulphite used as oxygen scavenger is usually the major contributor. To maintain the boiler water within its permitted limits some boiler water must be blown down. The rate (as a percentage of the steaming rate) is calculated by:

$$x = \frac{f \times 100\%}{b - f}$$

where f is the concentration of an impurity in the feedwater and b is its permitted concentration in the boiler water. This calculation should be made for each individual impurity specified. In practice, the TDS is usually the controlling factor in blowdown but if the make-up is treated by partial de-ionization, silica may be more important.

Blowdown costs money in terms of heat, water and chemicals, and should therefore be minimized. Control of blowdown and recovery of heat from it are important aspects of boiler operation.

Where sodium sulphite addition is a large contributor to the non-volatiles in the boiler, thermal means of reducing the oxygen can make a significant improvement in the overall operation.

17.5.7 External water treatment

In an ideal case the condensate return is high, and the raw water low in dissolved solids, hardness and alkalinity. It is then possible to operate the boiler without external water treatment, relying on conditioning of the boiler water with phosphates, tannins or other chemicals to

* BS 2486: 1978 is current at the time of writing but a more up-to-date standard is being prepared and should appear shortly.

cope with the small amount of hardness introduced with the raw water.

In practice, especially with modern boilers, it is considered essential at least to soften the raw water. Conventional softeners do not, however, remove hardness completely but allow a very small concentration to pass through.

Where the feed contains a large proportion of treated water, softening is a minimum requirement and the raw water quality dictates whether a more sophisticated form of external treatment would be preferable. If the water has a high alkalinity it calls for de-alkalization and base exchange. De-ionization is the ideal water treatment, but is usually avoided if possible because of its cost and use of corrosive chemicals. Membrane processes giving partial de-ionization are not normally installed at present, but are certain to become important in the future.

External treatment process plant should be installed only after a specialist's advice has indicated the best process, and plant should only be purchased from reputable manufacturers. The operational characteristics of the different processes are described below.

17.6 Specified purities for process use

Various processes need waters of a quality better than the public supply, or whatever source is available. Demands vary widely. Pharmaceutical and cosmetics production generally require good biological quality. So do food and drink manufacture, but brewing and soft drink manufacture often requires a specified mineral content as well. In brewing it is becoming common for water to be largely de-ionized and a wholly synthetic water to be reconstituted by chemical dosing. Membrane processes such as reverse osmosis and electrodialysis, which do not completely de-ionize the water, are increasingly used. Table 17.2 gives details of the BEWA Water Quality Classification and Table 17.3 shows typical water characteristics.

Textile products are particularly sensitive to iron, which discolours the product. Many washing operations, as in metal finishing, require softened water to avoid staining of the product. Others are very more sensitive and use de-ionized water.

De-ionized water is required for high-pressure boiler make-up and in many chemical process applications. Where a process has a large-scale steam demand, high-pressure turbine generators are often installed to generate power before providing the process with pass-out steam, thus making the most efficient use of the fuel. If the process does not return the steam as condensate, the boiler feed will be entirely treated water. This means that the external water-treatment plant has to handle an unusually large flow whose quality of make-up is critical. The world's largest de-ionization plants have been built to serve this kind of system.

De-ionized water itself has a wide range of grades. The lowest is that obtained by a simple cation–anion unit, which may contain up to 5 mg/l TDS. The highest grade is ultra-pure water, which is necessary for making microchips and has maximum total contents three orders of magnitude lower. The specifications for suspended and dissolved matter in ultra-pure water are always at the limits of detection, and are steadily becoming more stringent as chemists devise more sophisticated methods of analysis.

The plant engineer should not be expected to select the correct process for any of these: good professional advice must be taken – and followed. The account below of the processes available therefore concentrates largely on external process characteristics which affect the general operation of the facility.

17.7 Water-purification processes

17.7.1 Filtration

This deals with all equipment used for the removal of particulate matter and represents a wide range of possibilities. Several books cover this subject and only a few typical examples are quoted here.

Simple strainers remove gross materials. These should not normally occur in public supply, but strainers are sometimes fitted to protect sensitive equipment or processes against breaks in the main, etc. The commonest form contains a stainless-steel wedge wire screen and is piped with a bypass so that the screen element can be isolated and removed for cleaning when necessary. If the load on the filter makes this kind of cleaning burdensome a self-flushing filter can be used. These can incorporate strainer elements down to 50 μm.

Smaller particles can be removed by cartridge filters which can be rated for various particle sizes down to 10 μm. These are typically candle filters whose filter elements are bobbins of nylon or similar string wound onto a former. When clogged, they must be replaced.

17.7.2 Sand filters

Sand filters are widely used in water purification and remove suspended matter by a completely different mechanism. Instead of the water passing through small orifices through which particles cannot pass, it runs through a bed of filter medium, typically 0.75 mm sand 750 mm deep. The orifices between such sand particles are relatively large, but dirt is adsorbed onto the large surface area presented by the medium. The pressure loss rises as the dirt builds up and the filter must be cleaned when it reaches about 3 m WC, otherwise the dirt can be pushed right through the filter.

Filter backwashing normally needs low-pressure compressed air and a flow of filtered water about ten times the rated filter throughput. These backwashing arrangements are critical, and providing the large flow of backwash water, as well as drainage for its disposal, can often create difficulties. Given good backwash arrangements, and on a water low in suspended matter, sand filters are simple, reliable, cheap and have low operating costs.

Sand filters vary in sophistication. A simple filter will remove most particles down to 5 µm. Multi-media filters

Table 17.2 BEWA Water Quality Classification

Class	*Type*	*Typical applications*	*General notes*	*Relevant standards*
1	Natural water	Once-through cooling systems Outside wash down Irrigation Fisheries Firefighting Recreational Natural mineral waters	Level of salinity may restrict use for irrigation and fisheries Brackish water has TDS up to 10 000 Seawater has TDS up to 50 000 All characteristics highly variable	Dept of Environment (DoE) Circular 18/85 EC Directive 75/440/EEC EC Directive 78/659/EEC EC Directive 79/923/EEC EC Directive 76/160/EEC EC Directive 80/777/EEC
2	Potable water	Drinking Domestic use Food and soft drinks Cooling systems Irrigation Firefighting		WHO (ISBN 92 4 154 168 7) EC Directive 80/778/EEC DoE Circ 20/82 DoE Circ 25/84
3	Softened water	Recirculatory cooling systems Low-pressure boilers Laundries Bottle washing Closed recirculatory systems Domestic use		Industrial Water Society (IWS) BS 2486: 1978 BS 1170: 1983 BEWA: COP.01.85
4	De-alkalized water	Medium-pressure boilers Recirculatory cooling systems Brewing Food and soft drinks		IWS BS 2486: 1978
5	De-ionized water	Medium-pressure boilers Humidifiers Renal dialysis Glass washing Battery top-up Laboratories Plating industry Spirit reduction	For renal dialysis aluminium must be less than 0.01 mg/l as Al Silica removal may be required for some applications	BS 2486: 1978 BS 4974: 1975 BS 3978: 1966 American Society for Testing and Materials (ASTM) EEC Draft 85/C 150/04 Association for the Advancement of Medical Instrumentation (AAMI)
6	Purified water	Pharmaceuticals Cosmetics Laboratories Chemical manufacturing	United States Pharmaceopeia (USP) also specifies pH = 7.0	British Pharmacopoeia (BP) European Pharmaceopoeia (EP) USP US Food and Drug Administration (FDA) BS 3978: 1966
7	Apyrogenic water	Vial washing Parenteral solutions Tissue culture	BP insists on distillation for water for injection. USP allows reverse osmosis also	BP, 1980 EP USP FDA
8	High-purity water	High-pressure boilers Laboratories	De-aeration may be required	BS 2486: 1978 BS 3978: 1966 National Committee of Clinical Laboratories Standards (NCCLS) PSC-3
9	Ultra-pure water	Microelectronics Supercritical boilers Nuclear applications Analytical instrumentation	De-aeration may be required Readily picks up contamination from pipework and environment	BS 2486: 1978 Integrated Circuit Manufacturers Consortium Guidelines (ICMC) ASTM D 1193-77

which use sand and anthracite, and possibly a third medium, in discrete layers, can yield very efficient filtration down to 2 μm. Granular activated carbon can be used instead of sand to add some measure of organic removal to the filtration process. The quality produced by any filter depends largely on the efficiency of the backwash. Sand filters in some form provide a satisfactory solution for the majority of water-filtration problems.

17.7.3 Coagulation

Still smaller particles and some of the organic matter in a water can, if necessary, be removed by coagulation, in which a chemical coagulant is dosed into the water before

Table 17.3 Typical characteristics of water

Class	Conduct. μs/cm	Resist. MΩ – cm	TDS	pH	LSI	Hardness	Alkalinity	Nitrate	Sodium	Heavy metals	Silica	Suspended solids	Turbidity	SDI	Particle count	OA	TOC	Micro-organisms	Pyrogens
2	750		Max. 500	6.5 to 8.5			More than 30	Max. 50	Max. 150	Less than 0.1		Less than 1.0	Max. 5			Max. 5		Less than 100	
3					−1 to +1	Less than 20													
4					−1 to +1		Less than 30												
5	20	0.05	Max. 10	5.0 to 9.5		0.1					0.5	Less than 0.1	Less than 0.5	Less than 5					
6	5	0.2	Max. 1	6.0 to 8.5		Max. 0.1					0.1	Less than 0.1		Less than 3		Less than 0.1		Less than 10	
7	5	0.2	Max. 1	6.0 to 8.5		Max. 0.1					0.1	Less than 0.1		Less than 3	1	Less than 0.1		Less than 1	Less than 0.25
8	0.1	10	0.5	6.5 to 7.5							0.02	Less than 0.1		Less than 1.0	1			Less than 1	
9	0.06	18	0.005			0.001	0.001	0.001	0.001	0.001	0.002	ND		Less than 0.5	0.1		0.05	Less than 1	Less than 0.25

Definitions

Conductivity: the electrical conductivity of the water measured in microSiemen/cm is the traditional indicator for mineral impurities.

Resistivity: the reciprocal of conductivity, measured in Megohm-cm. It is used in some industries instead of conductivity particularly for ultra-pure water.

TDS: total dissolved solids determined by evaporating the water and weighing the residue. Units are mg/l.

pH: the acidity or alkalinity of the water expressed on a scale of 0 (acid) to 14 (alkaline). pH is considered neutral.

LSI (Langelier Saturation Index): an indication of the corrosive (negative) or scale-forming (positive) tendencies of the water.

Hardness: the total dissolved calcium and magnesium salts in water. Compounds of these two elements are responsible for most scale deposits. Units are mg/l as $CaCO_3$.

Alkalinity: the total concentration of alkaline salts (bicarbonate, carbonate and hydroxide) determined by titration with acid to pH 4.5. Units are mg/l as $CaCO_3$.

Nitrate concentration is in mg/l as NO_3^-.

Sodium concentration is in mg/l as Na^+.

Heavy metals: the total of chromium, lead, copper and other toxic metals expressed in mg/l.

Silica: soluble or 'reactive' silica concentration in mg/l as SiO_2.

Suspended solids: the concentration of insoluble contaminants in mg/l.

Turbidity: a measure of the colloidal haze present in Nephelometric Turbidity Units (NTU).

SDI: the Silt Density Index, a measure of the rate at which the water blocks a 0.45 μm filter.

Particle count: the number of particles greater than 0.5 μm in 1 ml of water (maximum particle size is 1 μm).

OA (Oxygen Absorbed): a measure of organic contaminants determined in a 4-hour test at 27°C and measured in mg/l as O_2. (Other indicative tests could be Permanganate Value (PV) or Chemical Oxygen Demand (COD).)

TOC (Total Organic Carbon): another way of expressing organics, in this case in mg/l as C.

Microorganisms: used here to mean the number of colony-forming units of total bacteria present in 1 ml of water.

Pyrogens: the endotoxins responsible for febrile reaction on injection, determined either by the rabbit test or the LAL test. Units are Endotoxin unit/ml (EU/ml).

Note: There are many different tests and different versions of the same test for water analysis. If there is any doubt as to method or interpretation consult a reputable water-treatment supplier. The letters ND in the table indicate 'not detectable'. Parts per million (ppm) is also commonly used to express concentration and is essentially identical to mg/l.

Standard tests for water analysis

For further information including test procedures the following are recommended.

BS 2486, Treatment of water for land boilers.

BS 2690, Methods of testing water used in industry.

Methods for the Examination of Waters and Associated Materials, HMSO.

ASTM Standards, Vols 1101 and 1102 (1983)

American Public Health Association, *Standard Methods for the Examination of Water and Waste Water*, 16th edn.

the filter. Unlike simple filtration, this process requires chemicals and careful control. Dosing directly before the filter will only cope with small concentrations of dirt. Larger amounts of dirt require coagulation to be followed by sedimentation and then filtration. It is a difficult process and is to be avoided if at all possible, especially on small flows. When coagulation processes go wrong, they can severely damage downstream equipment.

17.7.4 Membrane filtration

Microfiltration and ultrafiltration have recently been introduced for the removal of particles down to any desired size. Their capital cost is relatively high. Experience with them is limited, and a short trial with a small-scale pilot element is advisable. Prediction of full-scale performance from such trials is normally quite reliable.

Dead-end filtration through membrane filters is common in some industries where high purity is imperative. When clogged, the membrane has to be replaced. The water is first purified, and the filters serve as a final polisher. They are unsuitable for applications where they have to remove any significant concentration of particulate matter, as the cost of membrane replacement can become very high.

17.7.5 Water softening

Traditional water-softening processes add lime, or lime and soda ash, to the water. This produces a precipitate in the form of a sludge, which must be settled out and the clarified water filtered in a sand filter. The chemicals are cheap, but the problems of handling solid chemicals and of sludge disposal have made the processes obsolete. They are, however, simple and robust, suitable for low-technology supervision, and cope well with changing water analyses. They should not be forgotten when considering projects in underdeveloped countries.

17.7.6 Ion-exchange processes: general

Ion-exchange units physically resemble sand pressure filters, and are almost as effective in removing, and therefore accumulating, suspended matter. As they are backwashed much less vigorously they must not be fed with water containing more than about 2 mg/l of suspended matter or they will become progressively fouled and with suspended material. Public supply quality is usually (but not always) good enough for the water to be put directly onto ion exchange.

Ion exchangers are synthetic resins in the form of beads – small spheres of 0.5 to 1.0 mm in diameter. Good backwashing of the bed is important to obtain a uniform bed in which the different sizes are graded to give the minimum pressure loss. The resins have a limited life. Cation resins, used in softening and also in the first stage of de-ionization, can last for 10 or even 20 years, but anion resins used in de-ionization rarely give more than 3–5 years of use.

If the unit becomes badly fouled with suspended matter (for example, after a pipe brake has introduced excessive suspended matter into the system) it must be taken out of service and cleaned. This is done with an extended backwash, possibly at higher flow rates. If this does not remove the dirt, the manhole should be opened and the resin agitated with an air lance. Non-ionic detergents can be used, but not at the same time as the air lance or the resulting froth will be impossible to control.

Ion exchangers in general and cation resins in particular are liable to chemical attack by chlorine. The very small residual of chlorine in public supply (typically, 0.2 mg/l) has only a mineral effect, but if more chlorine has been added it must be removed (e.g. with an activated carbon filter) before ion exchange.

The anion resins used in de-ionization are prone to fouling if the water contains organic matter. The soft peaty waters mentioned above are particularly bad in this respect, and, at worst, can reduce resin life to a few weeks.

The handling, storage, measuring and dilution of regeneration chemicals requires serious consideration, especially in de-ionization, which requires strong acid and alkali. For dilution and pumping the actual regeneration solution most systems use ejectors, which avoids moving parts in corrosive solutions.

All ion-exchange processes produce waste water from backwash, regeneration and rinse. The proportion of waste depends on the concentration of hardness or to TDS being removed, and can be as high as 15% of the product flow. Any pretreatment has to take this additional flow into account.

Base-exchange softening

This is the simplest and commonest ion-exchange process. It uses only cation resins which are regenerated with strong brine to remove calcium and magnesium ions from water in exchange for sodium ions. Well-designed standard plant is available in a wide range of sizes and operates with a minimum of supervision. Conventional softeners allow 1–3% of the incoming hardness to leak through. As the hardness is exchanged for sodium the TDS remains substantially unchanged. The softened water is at the same pH as the incoming water but is rather more corrosive to steel.

There is no difficulty in handling or storage of corrosive chemicals or effluents. Salt is easily purchased and handled in bulk, and can be discharged directly into a standard commercial salt saturator.

The run and regeneration cycle on small to medium-size units is normally governed by a multiport valve, which is the only moving part required. There have been many cases where this has failed and brine was injected directly into the boiler. A conductivity meter on the make-up line would guard against this.

The operation of a base exchanger is chemically inefficient, and the spent regenerant contains large amounts of excess salt which may occasionally be difficult to dispose of. Factory softeners make a major contribution to the chloride content of the UK's industrial rivers, and in the longer term there will be heavy pressure from

Table 17.4

	Raw water	*De-alkalized*	*Base exchanged*
Total			
Hardness[a]	290	50	1
Alkalinity[a]	265	25	25
TDS[a]	334	94	94

[a] All as mg/l $CaCO_3$.

environmentalists to reduce the amount of salt being discharged.

De-alkalization/base exchange

This three-stage process is used for waters of high alkalinity and hardness. It actually removes most of the temporary hardness and so reduces the TDS of the water. However, in the process it increases the proportion of silica in the remainder. Any residual temporary hardness and the permanent hardness are softened in a conventional softener.

The effect is best illustrated by a numerical example (Table 17.4). Let us take the case of a hard and alkaline deep well water such as that found to the north of London, whose main characteristics are shown in the first column of Table 17.4. The second column shows its quality after de-alkalization has removed nine-tenths of the temporary hardness and converted it into CO_2 gas. This is removed from the water by stripping it with air in a packed 'degassing' column, and the product then softened in the third stage to yield the product shown in the third column.

The raw water silica is 22 mg/l as SiO_2, and therefore becomes a major constituent of the treated water. Silica scale must now be avoided by raising the boiler water pH and letting silica rather than the TDS control the necessary blowdown. Silica scale not only has a tenth of the heat conductivity of calcium carbonate scale but it is glassy, adherent, and extremely resistant to boiler-cleaning chemicals.

De-alkalization resins are regenerated with sulphuric or hydrochloric acid. Sulphuric acid is cheaper and easier to store and handle, but its dilution with water is potentially dangerous and must be carefully engineered. The regenerant solution must be extremely dilute to avoid the precipitation of calcium sulphate, which would clog the unit. This leads to long regeneration times and a high production of waste water. Hydrochloric acid is usually preferred on the smaller scale, where easier operation is more important than the annual cost of chemicals.

De-alkalization resins must not be over-regenerated or the product water becomes strongly acidic. The system therefore needs some measure of skilled supervision, and may depend on a pH meter – an instrument which, in turn, needs regular and skilled maintenance.

The acid regeneration is very efficient, and the amount of free acid going to drain is small. Given a reasonable amount of dilution, it may be unnecessary to take special precautions against it. Although acids are more expensive than salt, the efficiency with which they are used means that the operating cost of a de-alkalization/base exchange plant is similar to that of a simple softener. The capital cost, on the other hand, is much higher and of the same order as that of a simple de-ionizer.

The tower in which CO_2 is stripped out must run into a break-pressure tank with subsequent repumping. It will load the water with any dust, and living organisms or other particles in the atmosphere, which leads to trouble in dirty environments or in pharmaceutical works.

The de-alkalized and degassed water has a pH of 4–5 and (having just passed through an air-blown tower) is laden with oxygen and extremely corrosive. Normal practice is to dose NaOH into the degasser tower sump, at a level sufficient to approach to desired boiler water pH. If this dosing fails, severe corrosion in the degassed water pump, the softener and the feed system will result.

De-ionization

De-ionization is a two-stage process which removes all the dissolved ions from water. First, the water passes through a bed of cation-exchange resin, regenerated with sulphuric or hydrochloric acid, where all the cations are removed in exchange for hydrogen and dissolved salts are thus converted into their equivalent mineral acids. The resulting acidic water is passed through a bed of anion-exchange resin, regenerated with caustic soda, which removes all the anions in exchange for hydroxyl. The resulting water contains, except for a small leakage of dissolved salts (typically less than 1–5 mg/l), only hydrogen and hydroxyl ions which, together, form water. On large plants, above about 20 m^3/h, it is quite common to include a degassing tower between the cation and anion exchange stages. This is a purely economic measure; it removes carbon dioxide which would otherwise have to be removed on the anion resin and consequently reduces the amount of caustic soda needed for regeneration.

While it produces very pure water, de-ionization is an expensive process to operate. It uses acid and caustic for regeneration and produces an effluent which may need neutralization before it can be discharged. On the other hand, all sizes of de-ionization plants are available with fully automatic operation and control so that the plant operator's involvement can be limited to ensuring that there are adequate supplies of regenerant chemicals.

17.8 Membrane processes

Membrane processes are not yet used widely for industrial water processing, but will become more important in future. At present, they are generally more expensive than the older processes which they promise to replace, but costs are falling. Their main advantage lies in the fact that they add little or no chemicals to the aqueous environment but return to it only the material taken from the raw water.

Ion exchange, in contrast, creates an effluent which contains between two and five times the mass of inorganic material removed from the product water. Coagulation with aluminium or iron salts creates a sludge which creates a disposals problem. 'Green' pressure, especially in Switzerland and mid-west USA, which lie in the middle of large land masses, has started to force industrialists to install alternative membrane processes to avoid these discharges.

17.8.1 Reverse osmosis (RO)

In reverse osmosis water is forced by pressure through a very fine-pore membrane, which has the property of rejecting dissolved salts. The process thus removes both particulate and dissolved matter. Generally, the flux of water is extremely slow, so that large membrane areas have to be installed to achieve the desired output. Different grades of membrane show different rejections and fluxes.

A 'tight' membrane will remove over 99% of dissolved salts but requires pressures of the order of 30 bar or more to function economically. A 'loose' membrane may only take out 90% of the salts but will function at around 5 bar or even less. Both types of membrane will remove all particulate matter and large molecules. Small, undissociated molecules such as silica and dissolved gases (i.e. oxygen and CO_2) pass through unchanged.

The ability to remove particulates has made RO indispensable in the production of ultra-pure water for microchip washing. Its ability to remove large molecules enables it to produce pyrogen-free water for the pharmaceuticals industry. In the USA and elsewhere RO is permitted for producing the water used in making up injectable preparations. The European Pharmacopoeia still insists on distillation for this, but the larger amounts of water needed for ampoule washing, etc. are often purified by RO.

For conventional water purification, RO may be economical for removing the bulk of the dissolved salts in a water before ion exchange, but only if the raw water has a high dissolved solids content. In practice, substituting two processes for one is unattractive except for large throughputs.

Recently developed 'softening' membranes reject most of the hardness in water while passing sodium salts, and operate at pressures of about 5 bar. This can be used to provide substantially softened make-up water for shell boilers. About 5–10% of the raw water hardness remains in the water but, for example, if there is a sufficiently high condensate return to the system this residual hardness may be acceptable.

RO has some intrinsic disadvantages:

1. The water wastage tends to be high, especially on small plants where a complex water-saving flow sheet is inappropriate.
2. RO membranes are very sensitive to fouling by particulates, colloids and macromolecules. The Silt Density Index test for raw water is an empirical test which shows the rate at which a 0.45 μm filter disc becomes plugged, and gives a good indication of the fouling properties of the water. It should be carried out by an experienced person. Seasonal and other variations which might lead to fouling must also be considered.
   Pretreatment to avoid fouling is always complicated and risky: if it fails, rapid fouling of the whole membrane complement of a plant will result. Cleaning procedures are ineffective against severe fouling and membrane replacement is very expensive. On bad fouling waters RO should be avoided unless, for example, its use is essential to produce ultra-pure water.
3. Some anti-scaling treatment is necessary on most waters, using, for example, pH reduction, hexametaphosphate, phosphonates or polyacrylic inhibitors. The action of the various compounds available is not yet fully understood, so this can be a hit-and-miss technique.
4. As CO_2 passes through the membrane, while bicarbonate is rejected, the product water tends to be acidic and may need pH correction.

17.8.2 Ultra-filtration

As RO membranes become 'looser' their salt rejection falls (see Section 17.8.1). Eventually a point is reached at which there is no rejection of salts, but the membrane still rejects particulates, colloids and very large molecules. The membrane pore size can be tailored to a nominal molecular weight cut-off. The resulting filtering process is called ultra-filtration.

In this process good hydrodynamics on the membrane surface are required to scour away the accumulated solids and prevent the membrane being blinded. This cannot be totally effective, and in practice the nominal membrane cut-off is often masked by the tendency of particulates to form a thin layer on the membrane surface whose effective pore size may be smaller.

The process is used mostly for industrial separations, such as the removal of yeast from beer or the recovery of emulsified cutting oils. It has been proposed as a pretreatment to RO for fouling waters, but the economics do not yet look attractive.

17.8.3 Electrodialysis (ED)

This process uses membranes which remove salts from water by passing an electrical current through it. It was originally developed for the conversion of brackish water to potable (i.e. for a product of 300–500 mg/l). As the salts are removed, the conductivity of the water falls, the power used in the process increases and the cost rises. It is therefore potentially economical primarily for pretreating a highly mineralized water (say, 300 mg/l or more) before ion-exchanger de-ionization.

Unlike RO, which is essential for producing ultra-pure water, there is little experience of ED in this field. The process has some potential advantages over RO: it is less liable to fouling and it can be engineered to waste much less water. Like RO, its costs fall sharply at higher temperatures, but the prospects of improved engineering making this a reality are better than for RO. It offers some prospects particularly where the product water has to be heated in any case (e.g. boiler make-up).

A variant of ED has recently appeared on the market: the cells between the membranes are filled with ion-exchange resin to reduce electrical resistance when producing water of high purity. This makes possible the production of good-quality de-ionized water without regeneration chemicals. On the small scale (e.g. for laboratories or pharmaceutical production) the convenience of avoiding dangerous chemicals and corrosive effluent outweighs the relatively high costs of this technique.

17.8.4 Retrofit of membrane processes

As with RO, de-ionization by ED before ion exchange is rarely feasible because of the complication of two processes instead of one. A different situation arises where an existing ion exchange installation is overloaded, either because the plant needs to treat more water or because the TDS in the raw water have risen above the original design level. In such cases the simplest and most economical solution can be to install RO or ED to take up a large part of the ionic load and allow the existing ion-exchange plant to be uprated.

17.9 Effluents

Industrial effluents are a particularly difficult problem to discuss in general terms: their nature is very diverse, possible methods of treatment vary correspondingly and their acceptability depends as much on the receiving body as on their flow and contents. There are, however, some common factors which are worth mentioning.

17.9.1 Domestic sewage

Industrial plants also discharge domestic sewage. It is vital to keep this separate from any industrial effluent which may have to be treated, so that it can then be disposed of by conventional means (to the public sewer, septic tank, etc.).

17.9.2 Surface drainage

Most of the surface drainage due to rain falling on roofs, roadways, etc. will be normal, acceptable floodwater and can run off to a soakaway, storm drain, etc. in a conventional manner. Surface drainage from areas contaminated with spillage or with material deposited from a locally polluted atmosphere creates a particularly difficult effluent problem, characterized by unpredictable and violent variations in flow and concentration of pollutant. The best approach to this is to avoid conditions in which the area becomes loaded with contaminant.

Alternatively, the contaminated surface drainage must be segregated from the normal stormwater drains and may, for example, be led into a balancing tank. This tank must be large enough to even out the variations and to allow the contents to be added to the works effluent (treated or untreated) over a period of time. Provision must be made for periodically removing the inevitable accumulation of silt in the bottom of the tank.

17.9.3 Noxious effluent

Almost any impurity added to water leaving a works can, in certain circumstances, become noxious, and with some of these impurities it is not immediately obvious that they are unacceptable. The main examples are:

- Corrosive conditions – not only high or low pH, but also high levels of sulphate, which is corrosive to cement and therefore unacceptable in the public sewers. High temperatures can be unacceptable;
- Toxic constituents such as heavy metals, cyanides, etc.;
- Organic materials whose decay consumes oxygen, which includes various harmless materials (e.g. all vegetable matter, milk wastes, cellulose, sugars, etc.);
- Inert material which may settle to the bottom of a body of water and coat its beds (clay, alum sludge, etc.). Also dyestuffs which discolour the water and are aesthetically unacceptable.

Unlike potable and industrial waters, for which quality specifications are laid down, there can be no universal regulation as to the concentration or degree to which any of these is acceptable. The receiving body, who have to use their judgement in each case, will lay down allowable limits.

A sewerage authority must decide on the level of contamination which it is prepared to accept into its sewer. The amount of sulphate and of biodegradeable matter which can be allowed will depend, first, on the quantity and quality of the flow already in the sewer and available to dilute the effluent. The second decision concerns the limits of contaminants which are acceptable into the sewage works.

If the effluent is within acceptable limits, the authority is entitled to make a charge for receiving and treating it (see below). The river authority has similar decisions to make if the effluent is to be discharged directly into the environment. The acceptable limits will, of course, be much lower but no charge will be levied.

The receiving body is entitled to take samples of effluent at all times to ensure that the conditions under which they have agreed to receive it are being met. They may insist on some kind of continuous or regular monitoring by the discharger, but as analytical methods for almost every kind of effluent are difficult to automate and standardize, this is always an unsatisfactory aspect of the system.

17.9.4 Charges for effluent

Sewerage authorities normally leavy a charge on industrial effluent. The conventional system of charging uses a formula which takes into account the volume, its concentration of suspended solids and its oxygen consumption (which is a measure of the load it will put on the treatment works). Typical figures for the flow and quality of the discharge are ascertained and agreed between the sewerage authority and the plant.

The flow of effluent is often calculated on the assumption that, say, 90% of the incoming water (metered by the water authority) will re-emerge as effluent. This is, of course, not the case at a brewery, for example, where direct metering or different methods are used to ascertain the effluent flow. Direct metering of the flow is, however, always difficult and therefore uncommon.

17.9.5 Effluent management

If a works produces any effluent liable to be unacceptable, or liable to raise an effluent charge, it is important that the whole effluent system should be properly managed. This means knowing where the drains run and in which of them each effluent flows. This information is often unavailable and is difficult to ascertain. Good management often means that if treatment or tankage is required, different flows should be segregated from one another. Diluting a noxious effluent with another of a different type, or with a non-noxious one, generally aggravates the problem.

17.9.6 Effluent treatment

Where the effluent is outside the limits acceptable for discharge it has to be treated first by the works. Generally, effluent treatment is an unsatisfactory operation because it requires skills different from those available on the works, and because it is a nuisance and is not seen to be productive.

Effluent-treatment plants fail most commonly because the design is based on an inadequate or faulty definition of the effluent problem. In an existing works it may take months of painstaking research to establish the true patterns of flow and contamination. Predicting these for a new project can be extremely difficult, and serious mistakes are common.

Treatment processes vary widely, and in themselves are reasonably well understood, but even if the nature and pattern of effluent is properly defined, problems often arise from wide variations in contents and flow of the effluent. An effluent plant tends to impose limitations on the main process – for example, it may be seriously affected by rapid dumping of the contents of a tank. These limitations are often flouted through ignorance or negligence.

The best solution to any effluent problem is to avoid treatment by the works altogether. Even if it appears initially to be more costly, it will save trouble and expense in the long term.

If the process cannot be modified to avoid the effluent the sewerage authority may be persuaded to take over its treatment. Various financial arrangements have been made for this over the years, from a charge based on the volume flow to a capital contribution towards an extension to the sewage works capacity. By this means the responsibility is handed over to an organization whose purpose and technical skills are devoted to the subject.

17.9.7 Water economy and re-use

Water is historically thought to be free, and is still so cheap that its wastage is not felt to raise a serious cost. This commonly leads to pointless waste. It is unusual for a works to meter the water usage of different operations on one site, and so the cost of waste is not made clear to those who are immediately responsible for it. Occasional spot estimates of water usage (which may be possible without installing meters), with direct debiting of water cost to the operation, can do much towards water economy. In such an exercise it is important to add the cost of effluent discharge to that of water intake.

Much lip-service is paid to water re-use, which, on the whole, is more fashionable than practicable. Better opportunities arise from simple economy in water consumption at each point of use. Water re-use after effluent treatment is not often practicable. It appears, if at all, most favourable in the re-use of a slightly contaminated water for some low-grade purpose such as washing down.

Acknowledgement

The author is grateful to the BEWA for the use of their Water Quality Classification as Tables 17.2 and 17.3.

18 Pumps and Pumping

Keith Turton BSc(Eng), CEng, MIMechE
Loughborough University of Technology

Contents

18.1 Pump functions and duties

Pumps impart energy to the liquids being transferred by mechanical means using moving parts. They can be classified as rotodynamic or positive displacement. Rotodynamic pumps cause continuous flow, and the flow rate and discharge pressure are effectively constant with time. Positive displacement pumps deliver fixed quantities at a rate determined by driving speed. The main types of pump commonly used are listed in Figure 18.1.

Pumps are used to transfer liquids, moving blood and other biological fluids, delivering measured quantities of chemicals as in dosing in water treatment, in firefighting, in irrigation, moving foods and beverages, pumping pharmaceutical and toilet products, in sewage systems, in solids transport, in water supply and in petrochemical and chemical plant. They are utilized in power transfer, braking systems, servomechanisms and control, as well as for site drainage, water-jet cutting, cleaning and descaling. Pumps thus give a wide range of pressure rises and flow rates with pumping liquids which vary widely in viscosity and constituency.

18.2 Pump principles

18.2.1 Rotodynamic pumps

Taking a typical centrifugal pump (Figure 18.2) the Euler equation can be written, at best efficiency flow, in the form

$$gH = u_2^2 - \frac{Q}{A_2} u \cot \beta_1 \quad (18.1)$$

where $u_2 = \omega D_2/2$, $A_2 = \pi D_2 b_2$, Q is flow rate and H is head rise. This ignores flow losses, so that actual performance is less than the Euler (Figure 18.3). Figures 18.4–18.6 give typical pump performance curves for a constant driver speed. The inflections in the mixed and axial flow curves are due to flow instability over blades and through impeller passages.

The hydraulic efficiency

$$\eta_H = \frac{gH_{\text{Actual}}}{gH_{\text{Ideal}}} \quad (18.2)$$

and

$$\eta_0 = \frac{\text{Hydraulic power}}{\text{Input power}}$$

or

$$\eta_0 = \frac{mgH}{P_{IN}} = \frac{\rho QgH}{P_{IN}} \quad (18.3)$$

Typical pump cross-sections are shown in Figures 18.7–18.10. Figure 18.11 illustrates a multi-stage design

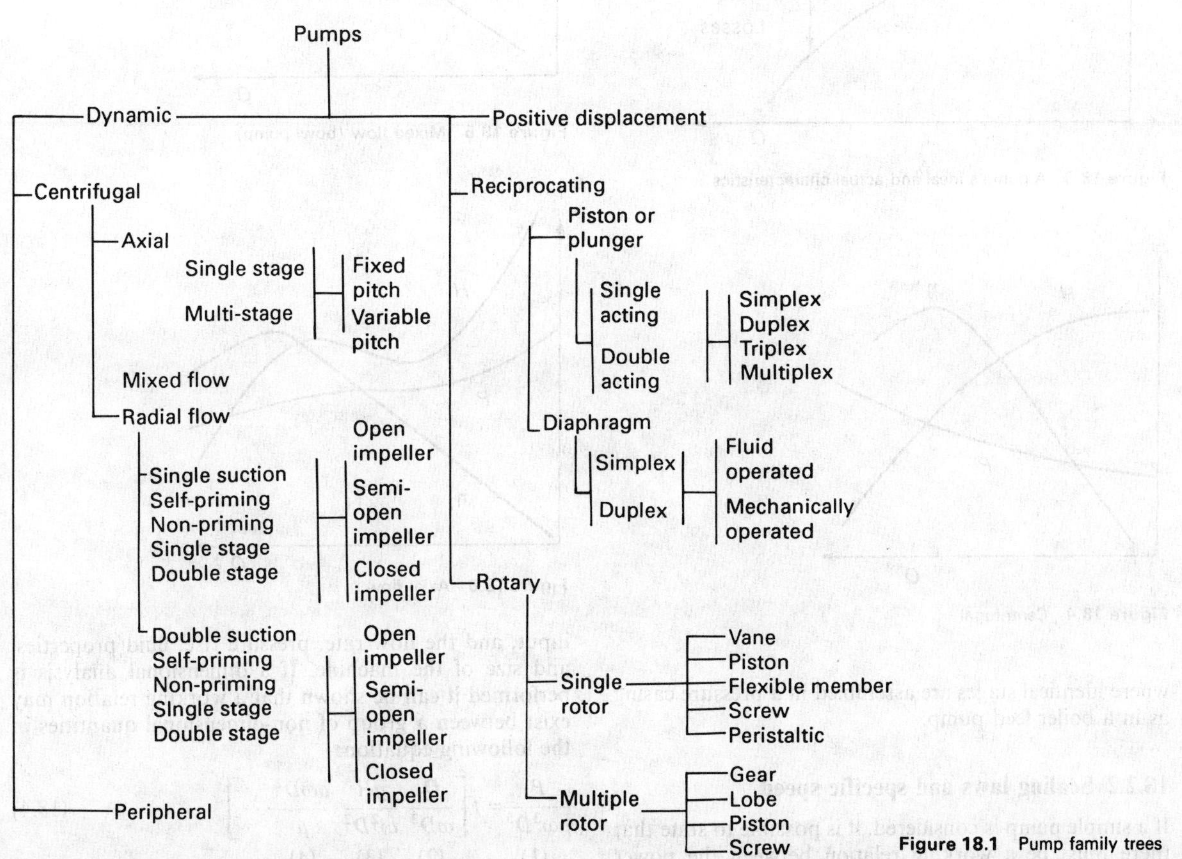

Figure 18.1 Pump family trees

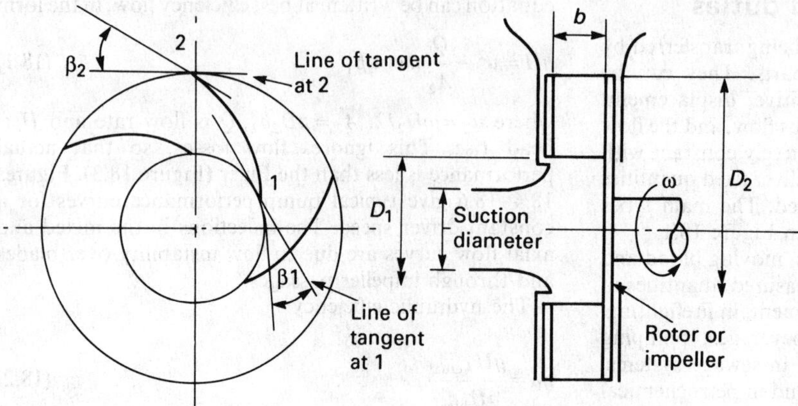

Figure 18.2 A simple centrifugal pump

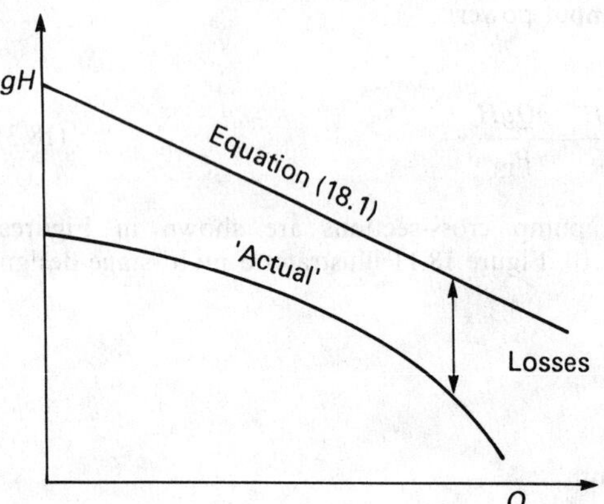

Figure 18.3 A pump's ideal and actual characteristics

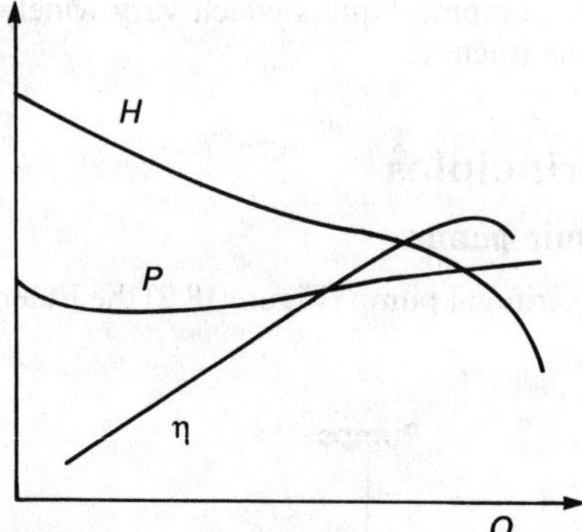

Figure 18.5 Mixed flow (bowl pump)

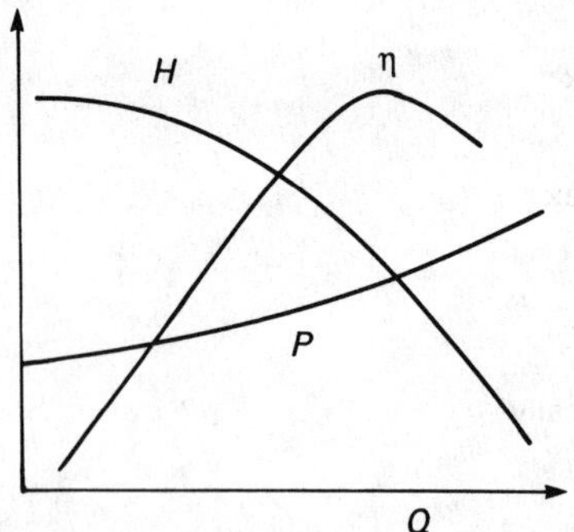

Figure 18.4 Centrifugal

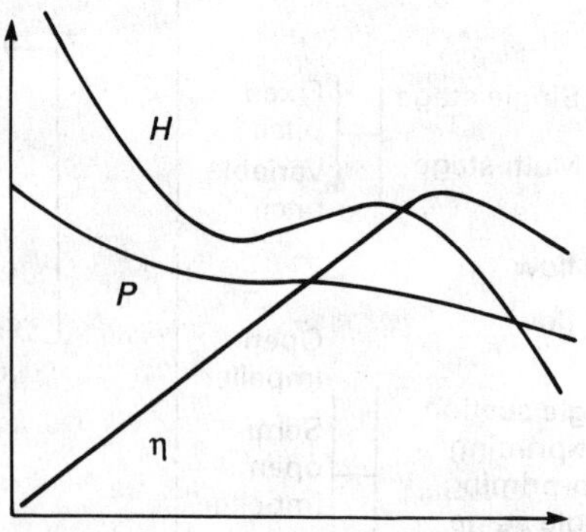

Figure 18.6 Axial flow

where identical stages are assembled in a pressure casing as in a boiler feed pump.

18.2.2 Scaling laws and specific speed

If a simple pump is considered, it is possible to state that there must be a working relation between the power input, and the flow rate, pressure rise, fluid properties, and size of the machine. If a dimensional analysis is performed it can be shown that a working relation may exist between a group of non-dimensional quantities in the following equation:

$$\underset{(1)}{\frac{P}{\rho\omega^3 D^5}} = f\left[\underset{(2)}{\frac{Q}{\omega D^3}} \cdot \underset{(3)}{\frac{gH}{\omega^2 D^2}} \cdot \underset{(4)}{\frac{\rho\omega D^2}{\mu}} \cdots\right] \tag{18.4}$$

Term (1) is a power coefficient which does not carry any conventional symbol
Term (2) can easily be shown to have the shape V/U and is called a flow coefficient (the usual symbol being ϕ).
Term (3) similarly can be shown to be gH/U^2 and is usually known as a head coefficient (or specific energy coefficient) ψ.
Term (4) is effectively a Reynolds number with the velocity the peripheral speed and the characteristic dimension being usually the maximum impeller diameter.

Since these groups in the SI system are non-dimensional they can be used to present the results of tests of pumps in a family of pumps that are geometrically and dynamically similar. This may be done as shown in Figures 18.12 and 18.13, and Figure 18.14 shows how the effect of changing speed or diameter of a pump impeller may be predicted, using the scaling laws:

$$\left.\begin{aligned} \frac{P}{\rho\omega^3 D^5} &= \text{const} \\ \frac{Q}{\omega D^3} &= \text{const} \\ \frac{gH}{\omega^2 D^2} &= \text{const} \end{aligned}\right\} \qquad (18.5)$$

The classical approach to the problem of characterizing the performance of a pump without including its dimensions was discussed by Addison,[1] who proposed that a pump of standardized size will delivery energy at

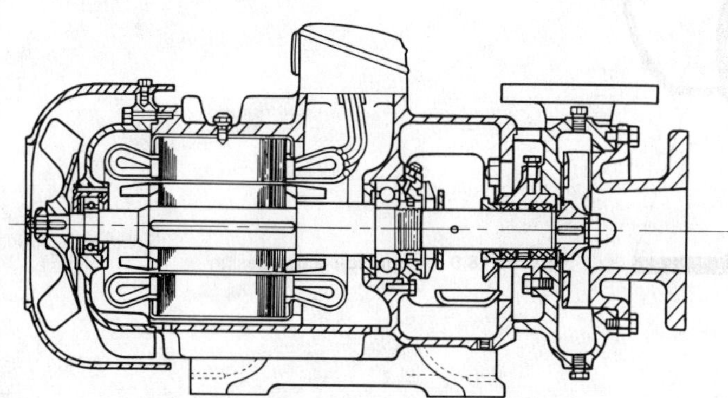

Figure 18.7 A monobloc design

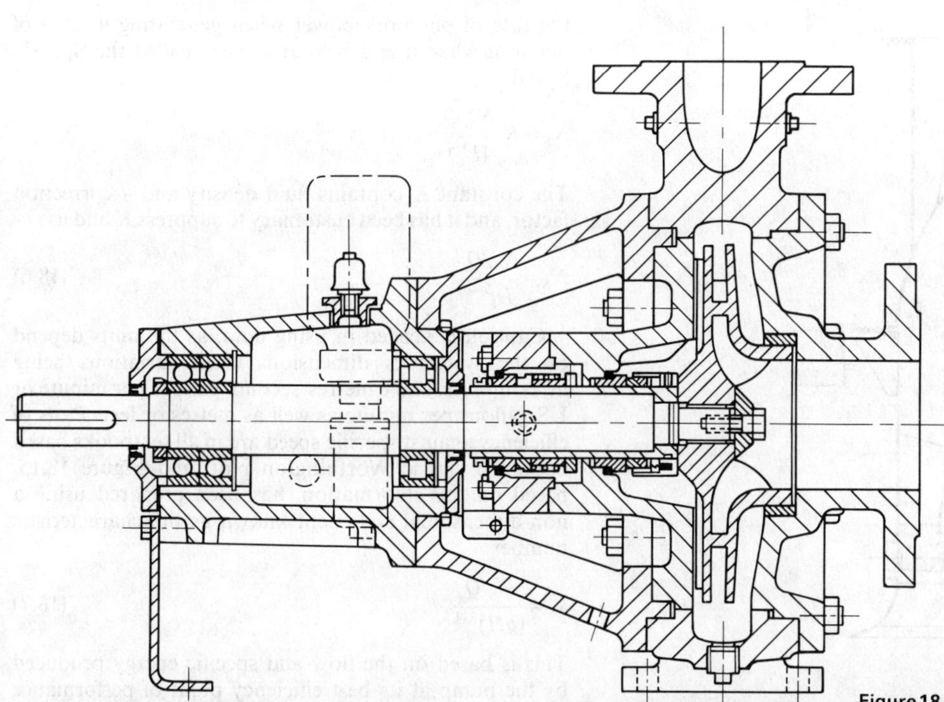

Figure 18.8 A back pull-out design

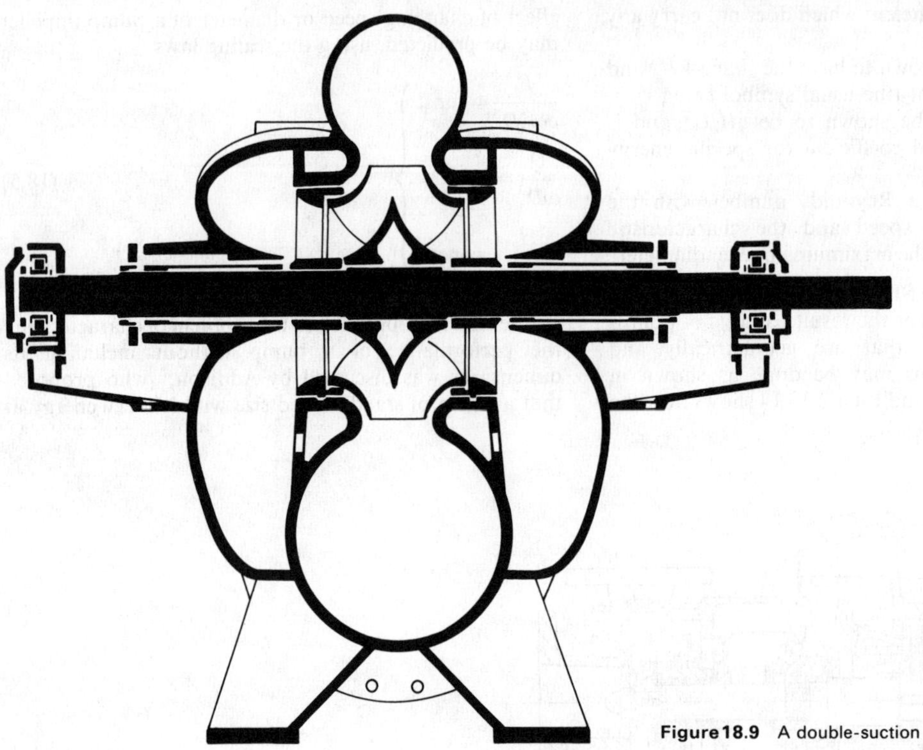

Figure 18.9 A double-suction design

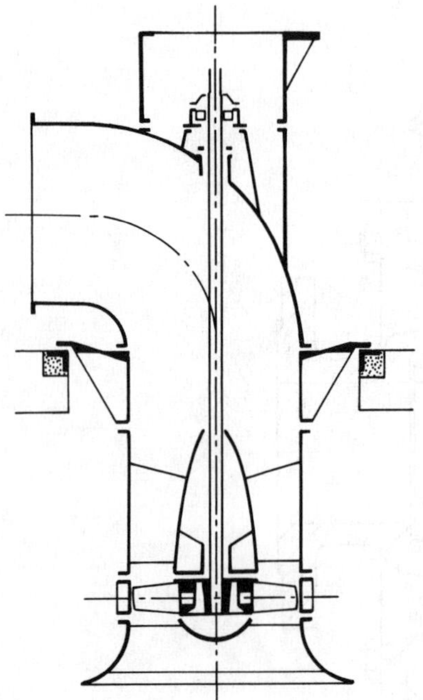

Figure 18.10

the rate of one horsepower when generating a head of one foot when it is driven at a speed called the Specific Speed:

$$N_S = K\frac{N\sqrt{Q}}{H^{3/4}}$$

The constant K contains fluid density and a correction factor, and it has been customary to suppress K and use

$$N_S = \frac{N\sqrt{Q}}{H^{3/4}} \tag{18.6}$$

Caution is needed in using data, as the units depend on the system of dimensions used, variations being litres/minute, cubic metres/second, gallons per minute or US gallons per minute as well as metres or feet. Plots of efficiency against specific speed are in all textbooks based upon the classic Worthington plot, and Figure 18.15, based on this information, has been prepared using a non-dimensional statement known as the characteristic number

$$k_S = \frac{\omega\sqrt{Q}}{(gH)^{3/4}} \tag{18.7}$$

This is based on the flow and specific energy produced by the pump at its best efficiency point of performance

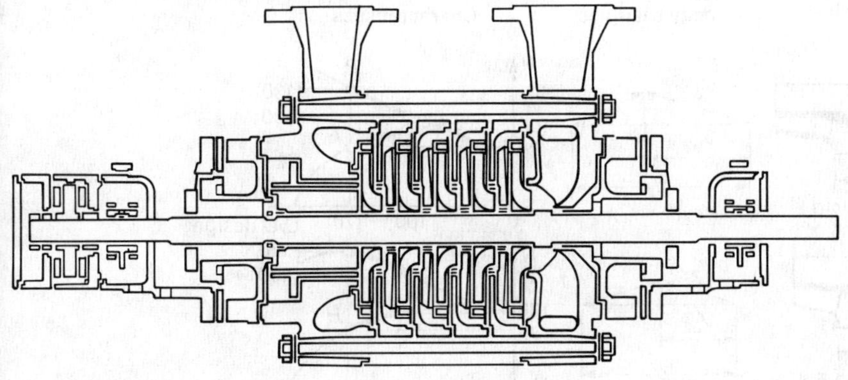

Figure 18.11 A multi-stage pump

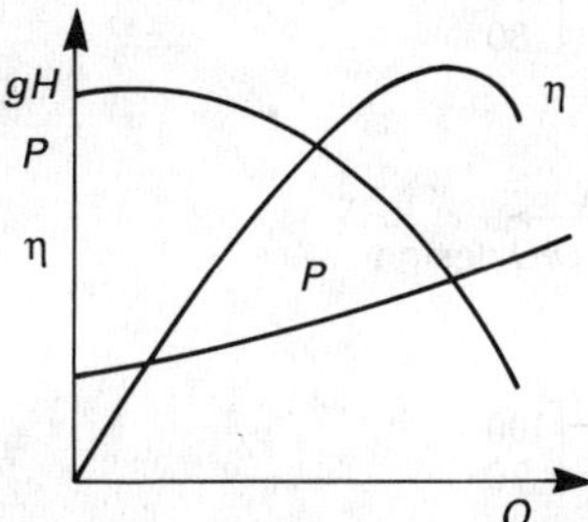

Figure 18.12 A pump's characteristic at fixed speed

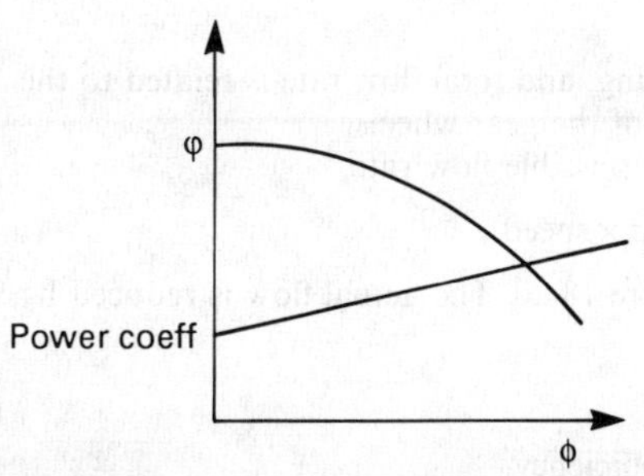

Figure 18.13 A non-dimensionalized plot

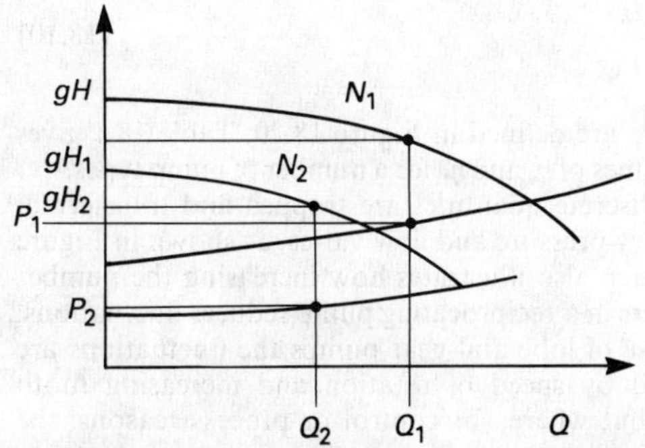

Figure 18.14 Prediction of speed change using equation (18.5)

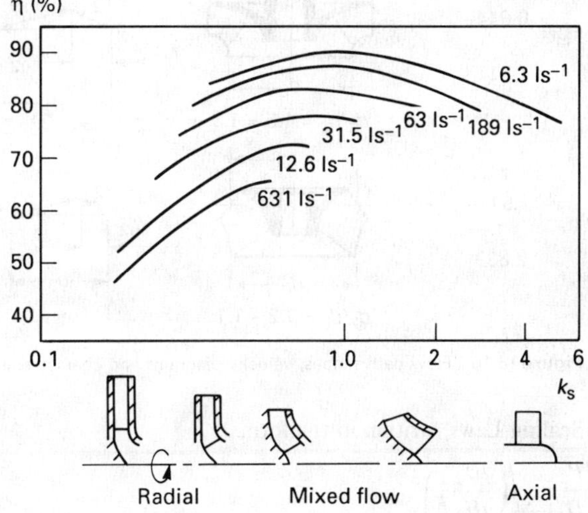

Figure 18.15 Variation of overall efficiency with non-dimensional specific speed[3]

following the approach stated by Wisclicenus: 'Any fixed value of the specific speed describes a combination of operating conditions that permits similar flow conditions in geometrically similar hydrodynamic machines.'

Figure 18.16 presents on the basis of the characteristic number the typical impeller profiles, velocity triangle shapes, and characteristic curves to be expected from the machine flow paths shown. This indicates the use of the number as a design tool for the pump engineer.

The scaling laws (equation (18.5)) may be used to predict the performance from change of speed, as indicated in Figure 18.14. In many cases the pump engineer may wish to modify the performance of the pump by a small amount. The common solution is to slightly reduce the impeller diameter. Figure 18.17 illustrates how small changes in impeller diameter can affect the performance. The figure in its original form appeared in Karrasik *et al.*[2] and has been modified to appear in metric form. The role used is often called the

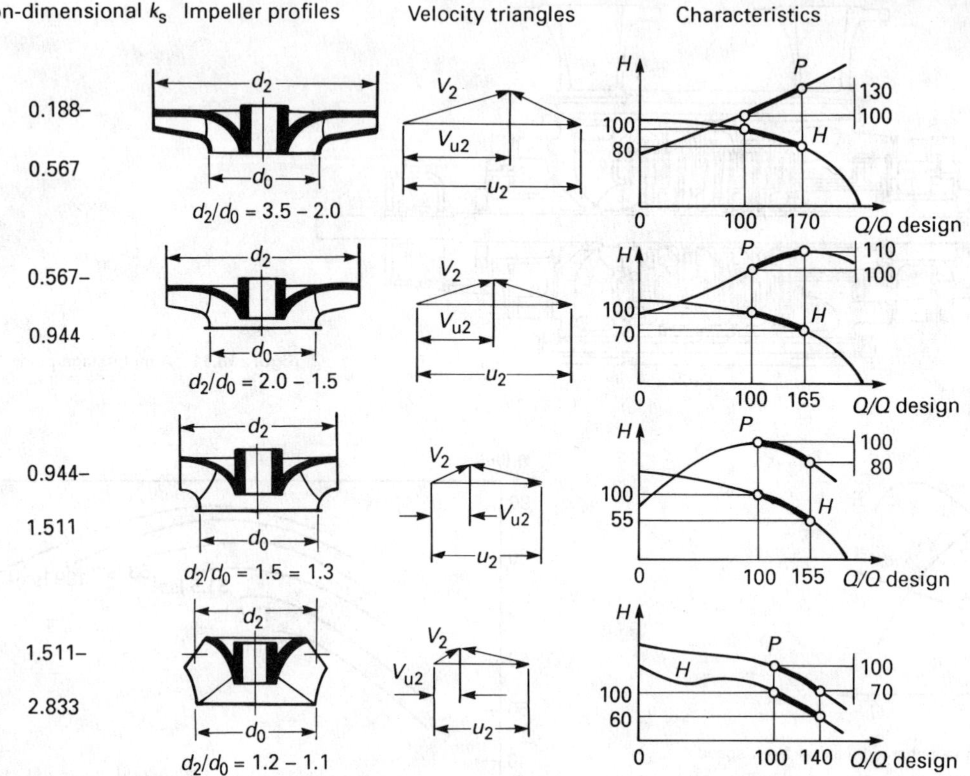

Figure 18.16 Flow path shapes, velocity diagrams and characteristics

Scaling Laws, written in the form:

$$\left.\begin{aligned}\frac{P_2}{D_1} &= \sqrt{\left(\frac{gH_2}{gH_1}\right)}\\ \frac{Q_2}{Q_1} &= \sqrt{\left(\frac{gH_2}{gH_1}\right)}\\ \frac{P_2}{P_1} &= \left(\frac{gH_2}{gH_1}\right)^{3/2}\end{aligned}\right\} \quad (18.8)$$

Other methods of adjusting the output while keeping the speed constant consist of modifying the profiles of the blades at the maximum diameter of the impeller. This technique has been used for a long time and is often used to obtain a small energy rise when the pump is down on performance when tested. The reader is referred to reference 2.

18.2.3 Positive displacement pump principles

Whether the pump is a reciprocator or a rotary design, liquid is transferred from inlet to outlet in discrete quantities defined by the geometry of the pump. For example, in a single-acting piston pump (Figure 18.18) the swept volume created by piston movement is the quantity delivered by the pump for each piston stroke, and total flow rate is related to the number of strokes per unit time. Similarly, the spur-gear pump (Figure 18.19) traps a fixed quantity in the space between adjacent teeth and the casing, and total flow rate is related to the rotational speed of the gear wheels.

The maximum possible flow rate,

Q_0 = displacement × speed

as shown in Figure 18.20. The actual flow is reduced by leakage, flow Q_L,

$Q = Q_0 - Q_L$

The volumetric efficiency

$$\eta_v = \frac{Q}{Q_0} = 1 - \frac{Q_L}{Q_0} \quad (18.9)$$

and

$$\eta_0 = \frac{\rho Q g H}{P_D + P_L} \quad (18.10)$$

P_D and P_L are defined in Figure 18.20. Table 18.1 gives typical values of η_v and η_0 for a number of pump types.

Since discrete quantities are trapped and transferred, the delivery pressure and flow varies, as shown in Figure 18.21, which also illustrates how increasing the number of cylinders in a reciprocating pump reduces fluctuations. In the case of lobe and gear pumps the fluctuations are minimized by speed of rotation and increasing tooth number, but where, for control or process reasons, the ripple in pressure is still excessive, means of damping pulsations has to be fitted. Often a damper to cope with

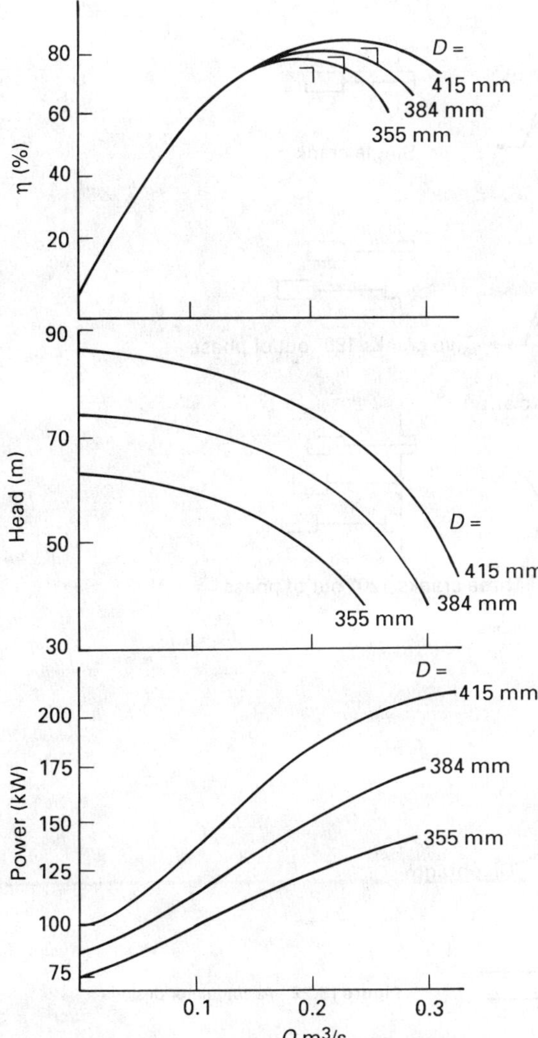

Figure 18.17 Scaling laws applied to diameter change (after Karassik *et al.*[2])

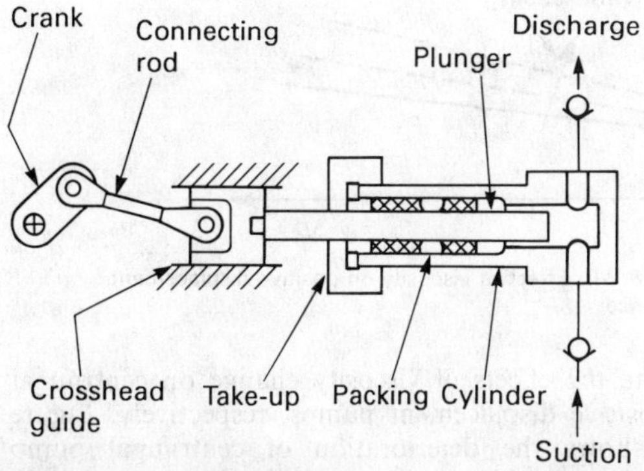

Figure 18.18 Plunger pump

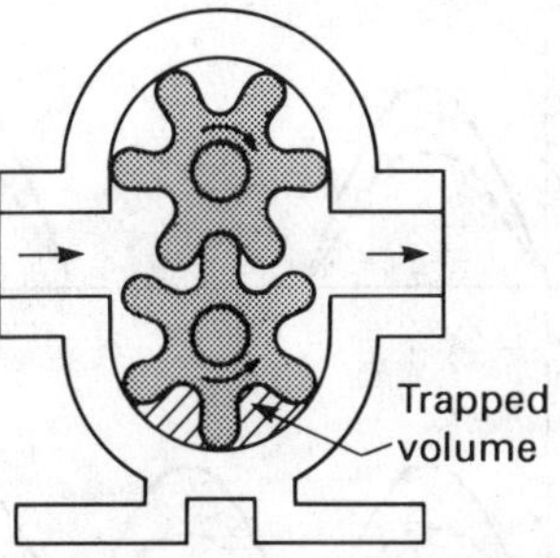

Figure 18.19 External gear pump

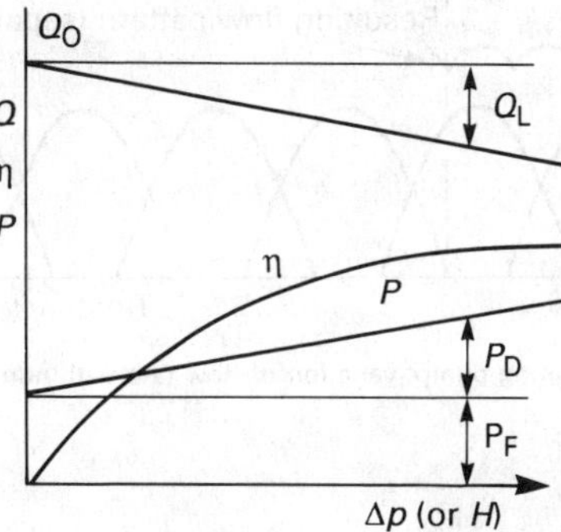

Figure 18.20 Typical characteristic at constant N

Table 18.1 Some values of η_v and η[4]

Pump	η_v (%)	η (%)
Precision gear	→98	→95
Screw	—	75–85
Vane	85–90	75–80
External gear	—	20–60
Radial – multi-piston	>95	>90
Axial – multi-piston	>98	>90

this and pressure pulses due to valve closure is fitted, and two types are shown in Figure 18.22. The capacity of the accumulator is important, and one formula based on experience for sudden valve closure is

$$Q_A = \frac{QP_2(0.016L - T)}{(P_2 - P_1)} \times 0.25 \tag{18.11}$$

Here Q_A is the accumulator volume (m^3), Q is flow rate (m^3/s), L is pipe length (m), T is valve closure time (s), P_1 is the pressure in the pipeline (N/m^2) and P_2 is the maximum pressure desired in the line (N/m^2) ($P_2 = 1.5P_1$ in many cases).

18.3 Effects of fluid properties on pump behaviour

Two effects on pump performance must be discussed: viscosity and gas content. Figures 18.23 and 18.24

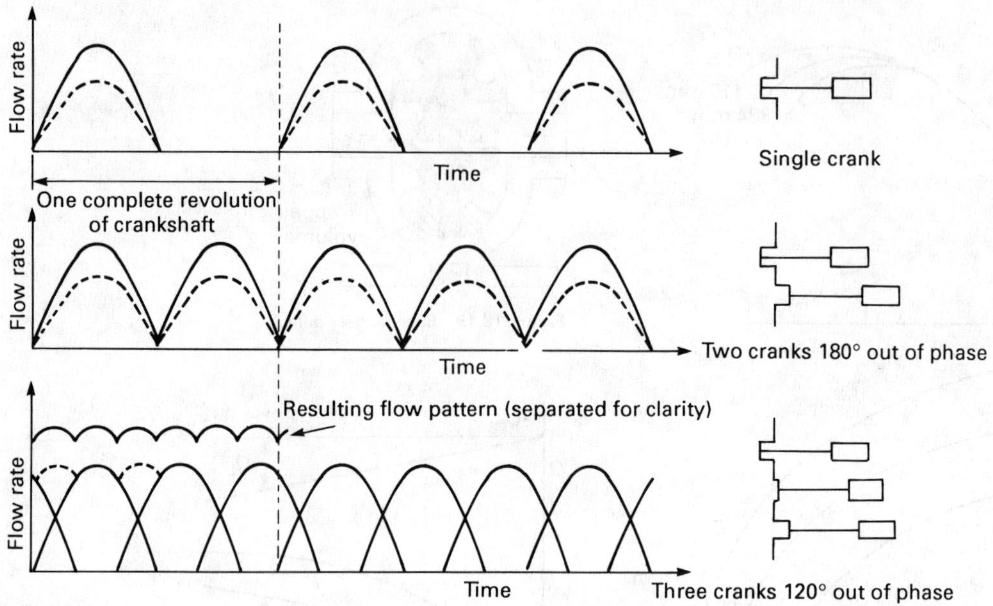

Figure 18.21 Reciprocating pump: variation of flow rate with numbers of cylinders

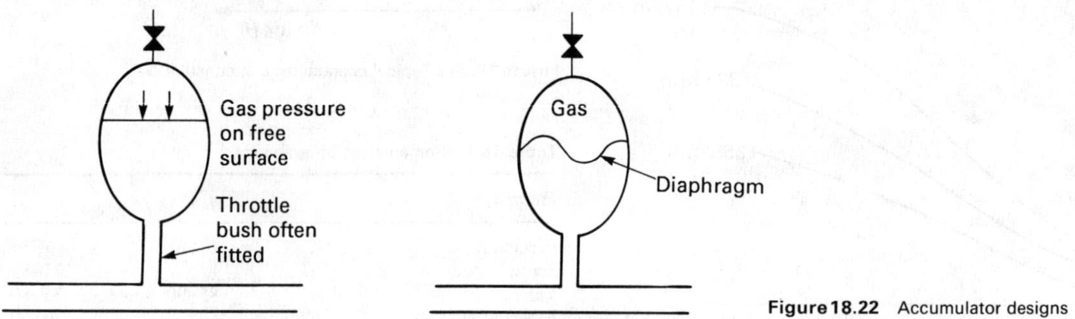

Figure 18.22 Accumulator designs

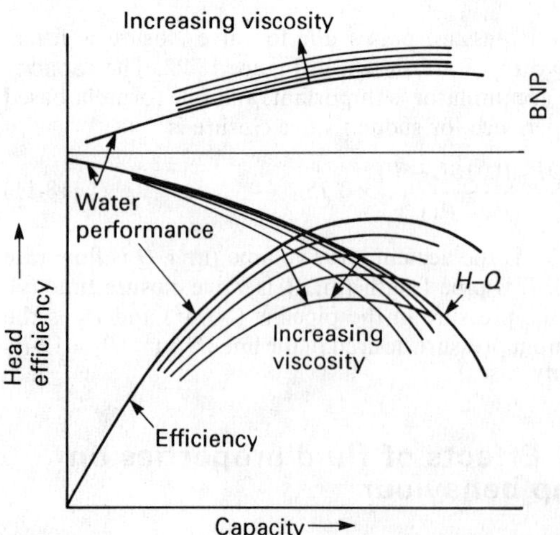

Figure 18.23 Effect of viscosity on centrifugal pump performance

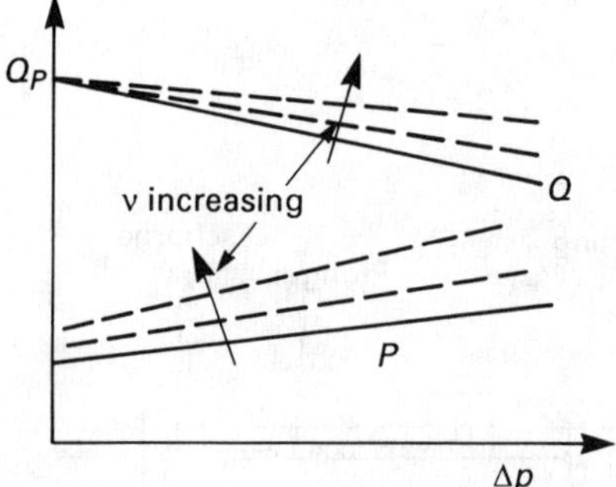

Figure 18.24 Effect of viscosity on positive displacement performance

illustrate the effects of viscosity change on centrifugal and positive displacement pumps, respectively. Figure 18.23 shows the deterioration of centrifugal pump performance, and it may be noted that if v is greater than

Table 18.2 Effect of fluid viscosity

Type of pump	*Significant[a] viscosity levels*	*Effect of viscosity level*	*Treatment and/or notes*
Centrifugal	20	—	Performance maintained similar to water performance up to this level
	20–100	Lowering of *H-Q* curve increase in input hp	General lowering of efficiency but may be acceptable
	Above 100	Marked loss of head	Considerable reduction in efficiency, but high efficiencies may still be attainable from large pumps
Regenerative	Above 100	Marked loss of performance	Pumps of this type would not normally be considered for handling fluids with a viscosity greater than 100 centistokes
Reciprocating	Up to 100	Little	Performance generally maintained. Some reduction in speed may be advisable to reduce power input required
	Above 100	Performance maintained but power input increased	Speed is generally reduced to avoid excessive power inputs and fluid heating
	Above 1000	Flow through valves may become critical factor	Larger pump size selection run at reduced speed – e.g. 3 × size at 1000 centristokes running at one-third speed. Modification of valve design may be desirable for higher viscosities
Plunger	—	—	For very high-pressure deliveries only
Sliding vane	Above 100	Sliding action impaired: slip increased	Not generally suitable for use with other than light viscosity fluids
External gear	None	Power input and heat generated increases with increasing viscosity	May be suitable for handling viscosities up to 25 000 centristokes without modification. For high viscosities: (a) Clearances may be increased (b) Speed reduced (c) Number of gear teeth reduced
Internal gear	None	Power input and heat generated increases with increasing viscosity	For higher viscosities: (a) Speed may be reduced (b) Number of gear teeth reduced (c) Lobe-shaped gears employed
Lobe rotor	250	None	(a) Speed may have to be reduced
	Above 250	Cavitation may occur	(b) Modified rotor form may be preferred
Single-screw	None	—	Nitrile rubber stator used with oil fluids
Twin- or multiple-screw	Up to 500	Little or none	—
	Above 500	Increasing power input required	Speed may be reduced to improve efficiency

[a] Viscosity in centistokes.

100 centistokes, water performance must be corrected. Figure 18.24 indicates that in a positive displacement pump the volumetric efficiency improves and power requirement increases (with increasing viscosity).

Table 18.2 summarizes the effects of liquid changes (effectively, viscosity and density changes) on pump performance and Figure 18.25 presents material presented by Sterling,[5] which illustrates how efficiency falls away with viscosity for two pumps working at the same duty point, graphically showing the rapid fall-off of efficiency as μ increases in a centrifugal pump.

Figure 18.26 demonstrates a well-known method of correcting for fluid change from water for a centrifugal pump. This allows an engineer to predict change in performance if the kinematic viscosity of the liquid to be pumped is known and the water test data are available.

Gas content is another important effect. It is well known that centrifugal pumps will not pump high gas content mixtures, as flow breaks down (the pump loses 'prime') when the gas/liquid ratio rises beyond 15%. Figure 18.27 shows how a centrifugal pump is affected, particularly at low flow rates, and the behaviour is typical

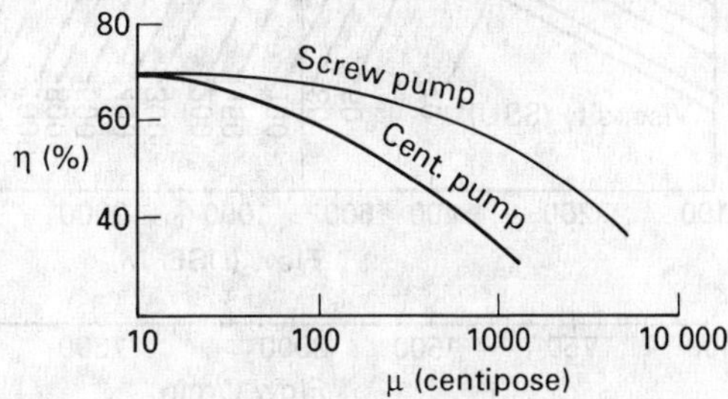

Figure 18.25 Effect of μ on screw pump and centrifugal pump compared (after Sterling[5])

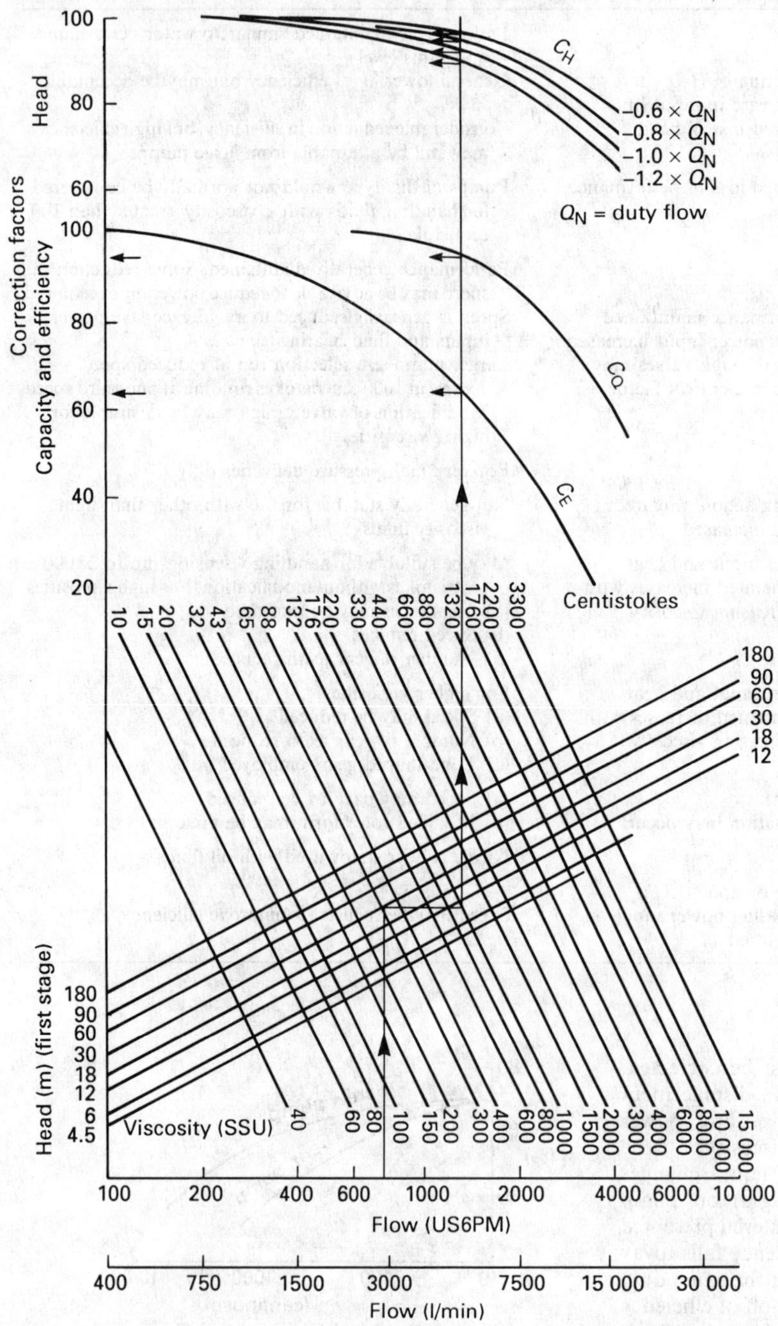

Figure 18.26 Viscosity correction curves (adapted from *Hydraulic Institute Standards*, 12th edn, Hydraulic Institute, Cleveland, Ohio, 1969). Pump to handle 750 6PM of 1000 SSU liquid against a head of 30 m. $C_E = 0.64$, $C_Q = 0.95$, $C_M = 0.92$ at duty $(1.0 \times Q_N)$. Therefore Q(water) = 789.5, Head(water) = 32.6. If η_{water} on test at 789.5 and 32.6 was 75%, $\eta_{oil} \simeq 0.75 \times 0.64 = 48\%$

of conventional centrifugal pumps. Figures 18.28 and 18.29 present well-known information on the effects of dissolved and entrained gas on the volumetric effciency of a positive displacement pump.

18.4 Flow losses in systems

18.4.1 Friction losses

The most common method of friction loss estimation is based on the D'Arcy Weisbach equation:

$$(gH)_{\text{friction}} = f\frac{L}{D}\frac{V^2}{2} \qquad (18.12)$$

Table 18.3 Some equivalent lengths for typical fittings (clean water)

		$\frac{L_E}{D}$
Gate valve (wedge:disc:plug disc)	Fully open	13
	Half open	160
	Quarter open	900
Ball and plug valves	Fully open	3
Conventional swing check valve	Fully open	135
Footvalve with strainer (hinged flap)	Fully open	75
Butterfly valve (200 mm up)	Fully open	40
90° Standard elbow		30
90° Long-radius elbow		20

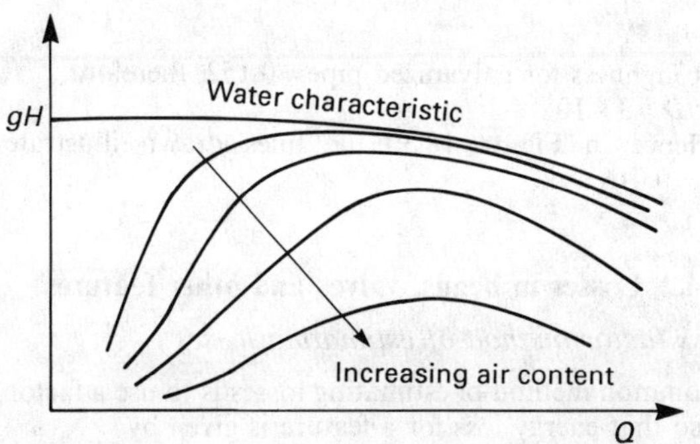

Figure 18.27 The effect of air content on pump behaviour

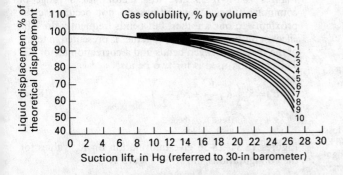

Figure 18.28 Effect of dissolved gas on liquid displacement

The friction factor f is plotted in Figure 18.30 against the Reynolds number, based on the pipe inner diameter D for circular pipes and the hydraulic diameter ($=4\times$ flow area/wetted perimeter) for non-circular pipes. Table 18.4 gives typical surface roughness. Figure 18.30 is thus used to find f as follows:

Mean velocity 7 m/s, mean diameter 50 mm, water kinematic viscosity, therefore $UD=0.35$

$R_E = 3\times10^5$

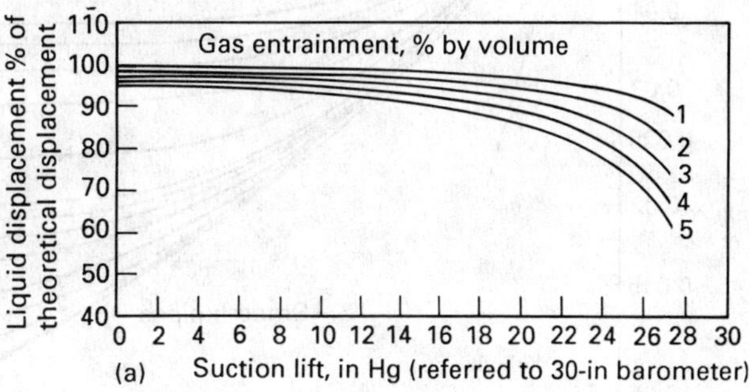

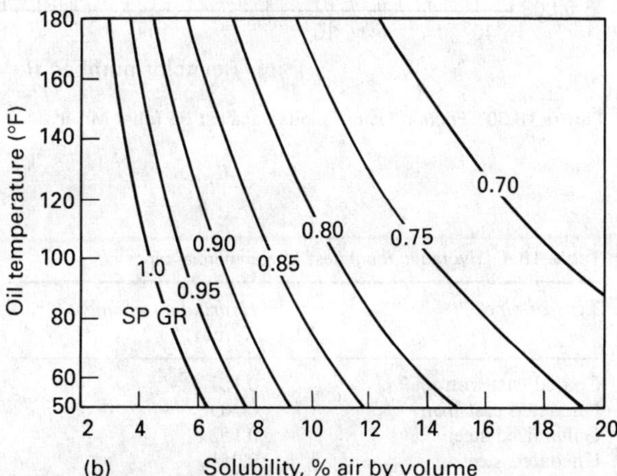

Figure 18.29 (a) Effect of entrained gas on liquid displacement; (b) solubility of air in oil. *Example*: At 5 in Hg with 3% gas entrainment by volume, pump capacity is reduced to 84% of theoretical displacement

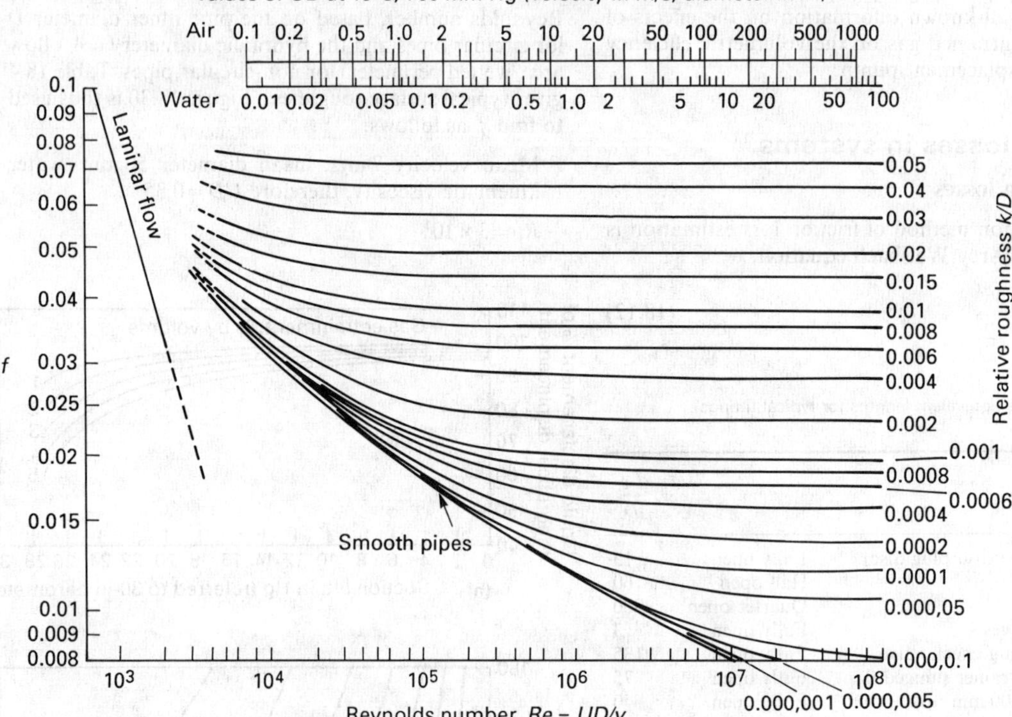

Figure 18.30 Friction factor f plotted against Re (after Miller[6])

Table 18.4 Hydraulic roughness of commercial pipes

Type of pipe	*Hydraulic roughness* K (mm)
Coated cast iron	0.127
Uncoated cast iron	0.203
Galvanized steel	0.152
Uncoated steel	0.051
Coated steel	0.076
Asbestos cement (uncoated)	0.013
Spun bitumen and concrete lined	
Drawn brass, copper, aluminium	
Glass plastic	
Concrete cast on steel forms	0.203
Spun concrete	0.076
Riveted steel – four transverse rows of rivets and six longitudinal	4.064
Two transverse rows of rivets, longitudinal seams welded	0.127
Single transverse rows of rivets	0.508
All welded	0.076
Flexible pipes	Roughness varies considerably with construction. Smooth rubber hose corresponds approximately with steel pipe. Head loss when curved can be 30% higher than when straight

Roughness for galvanized pipe = 0.152, therefore $k/D = 3 \times 10^{-3}$
Hence in Figure 18.30 the lines drawn illustrate $f = 0.016$

18.4.2 Losses in bends, valves and other features

Loss factor method of estimating losses

A common method of estimating losses is to use a factor K, so that energy loss for a feature is given by

$$\Delta(gH) = K\frac{v^2}{2} \tag{18.13}$$

Figures 18.31–18.33 give loss factors for a range of common fittings. Most fittings do not occur in close proximity to one another, but bends frequently do. The effect of putting bends close together is presented in terms of the spacing between bends and a correction factor K_p. The equation used is for two bends:

$$\Delta(gH) = K_p\left(\underset{\text{(bend 1 loss)}}{K_1\frac{V^2}{2}} + \underset{\text{(bend 2 loss)}}{K_2\frac{V^2}{2}}\right) \tag{18.14}$$

Figures 18.34–18.36 give information on K_p values for the configurations shown in Figure 18.37.

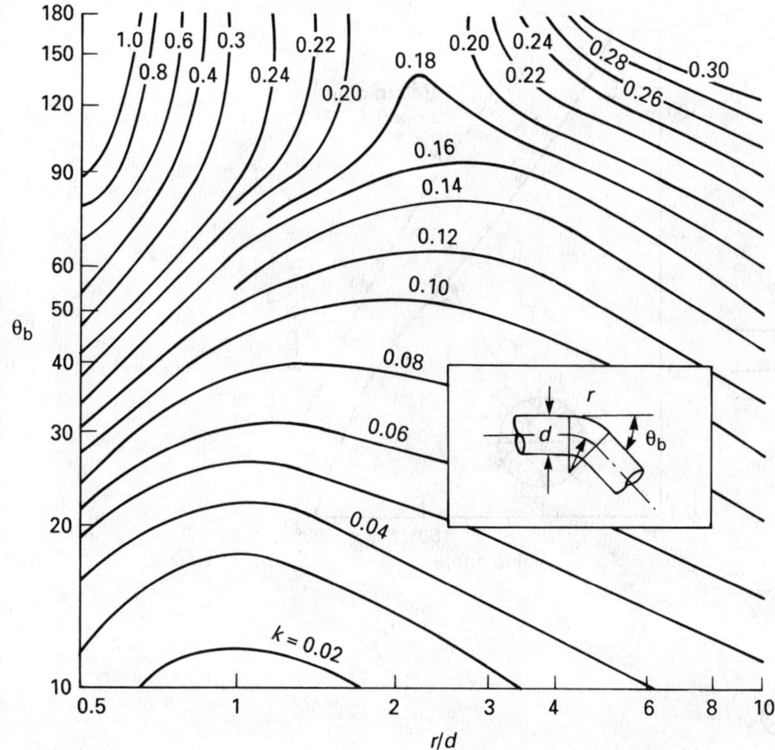

Figure 18.31 Bend loss coefficients[6]

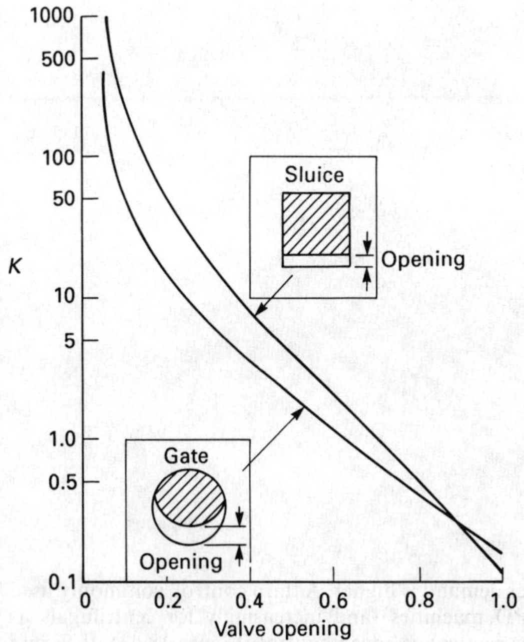

Figure 18.32 Gate valve losses[6]

Equivalent-length method

As the name suggests, the loss a feature causes is replaced by the loss due to a length of straight line which gives the same value of energy loss. Table 18.3 gives typical values of L_E, which are then added to the D'Arcy Weisbach equation to give

$$\text{Total loss in energy} = \frac{f}{D}\frac{V^2}{2}\ (\text{pipe length} + \text{sum of } L_E \text{ components}) \quad (18.15)$$

Thus for a simple line 100 m long with a bend ($L_E = 30$) a valve ($L_E = 13$) the total loss $= (fv^2/D^2)(143)$.

18.4.3 Presentation of system loss

The flow loss in a system varies as (velocity)^2 or (flow rate)^2, so the total loss imposed by a system on a pump can be shown to vary with flow rate in the way shown in Figure 18.38(a). Figures 18.38(b) and 18.38(c) illustrate how system curves vary with the proportions of static and dynamic losses.

18.5 Interaction of pump and system

18.5.1 Steady-state matching of pump and system

If the constant speed characteristic of a pump is superimposed on a system curve there is usually one intersection point, shown in Figure 18.39. If a flat system curve is being matched with a mixed or axial flow machine there can be flow instability, as illustrated in Figure 18.40, which is only corrected by changing pump speed or the static lift, or selecting a different pump.

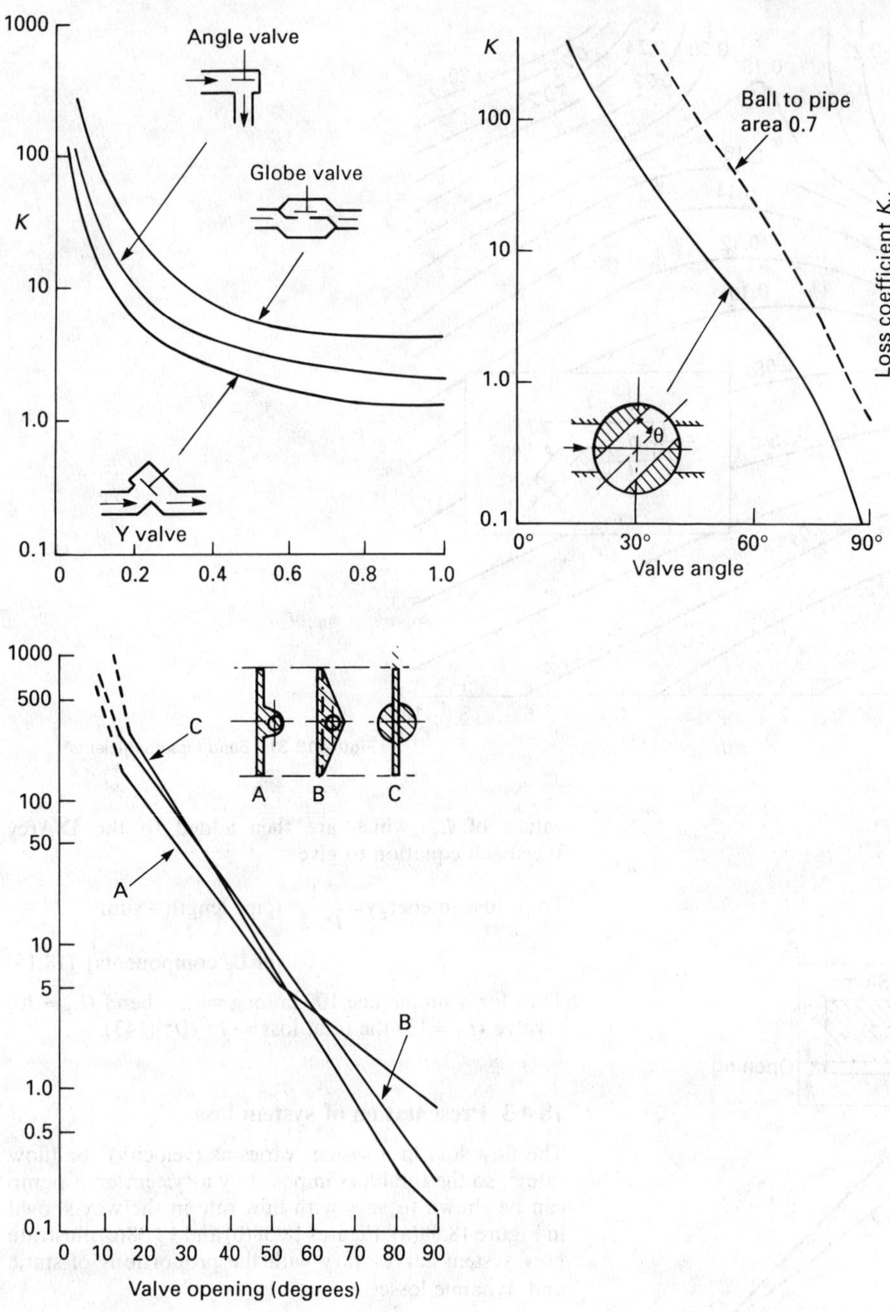

Figure 18.33 Some typical valve loss characteristics[6]

18.5.2 Flow control

The simplest flow control is by valving (Figure 18.41). The dynamic loss is changed by either opening or closing a valve in the line. The valve could be pressure controlled. A method much used in boiler feed systems because it ensures full flow through the pump and thus reduces risk of vapour locking in the bypass (Figure 18.42). In this case the pump supplies the system and bypass flow, so match point flow is higher than system demand and power demand is higher. A third control, commonly used for PD machines (and increasingly for centrifugals as prices reduce), is speed control (Figure 18.43). If Figure 18.44 is considered, achieving 90% of designed flow by speed control will happen with little sacrifice in efficiency (and probably 25% power reduction), valve control will result in a 3–4% reduction in pump efficiency (and probably 7–8% increase in pump head), bypass will probably require 15% more flow, the pump efficiency will drop 7% or 8%, and the pump may also cavitate.

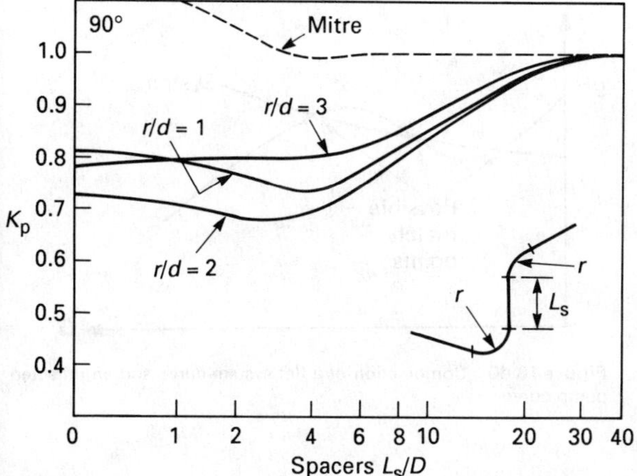

Figure 18.34 Interaction coefficient: combination bends (after Miller[6])

18.5.3 Multiple-pump layouts

If the flow range in a system is larger than can be achieved with a single machine, several similar pumps may be installed in parallel. Figure 18.45 illustrates possible layouts for centrifugal pumps and match points on a system curve for one, two or three pumps operating. Non-return valves must be installed as shown to avoid pump interaction. Note that three pumps do not give three times single-pump flow, due to the square law nature of the system curve. Figure 18.46 demonstrates how variable-speed drives will give a wide range of flows. Systems supplied by positive displacement pumps do not usually need wide flow ranges, but the same principle of parallel operation will apply.

Dissimilar pumps may be used. As Figure 18.47 shows, the resulting combined curve indicates that pump 2 will not begin to deliver until the specific energy from pump 1 drops below its shut-valve energy rise, therefore if the

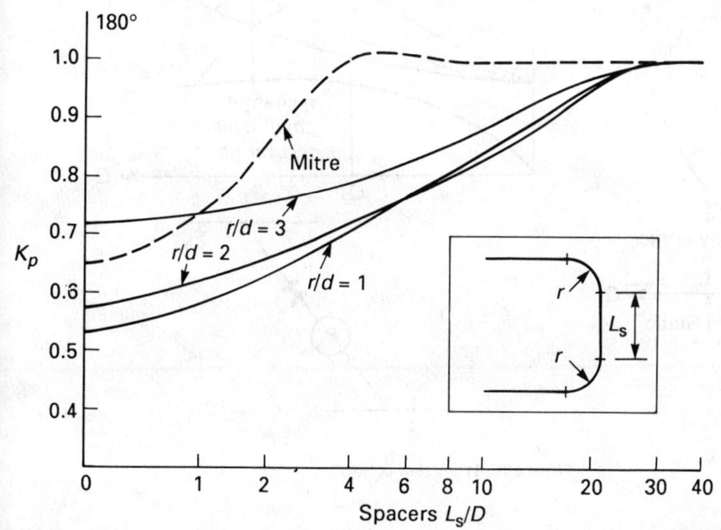

Figure 18.35 Interaction coefficient: 180° combination (after Miller[6])

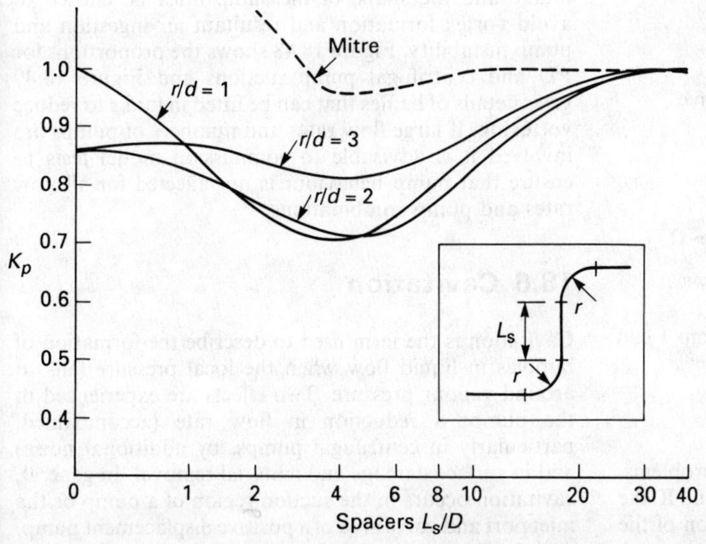

Figure 18.36 Interaction coefficient: S-bend combination (after Miller[6])

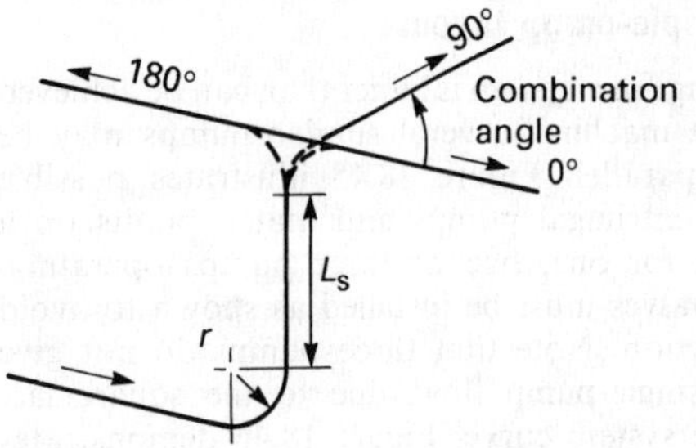

Figure 18.37 The combinations shown in Figures 18.33–18.36

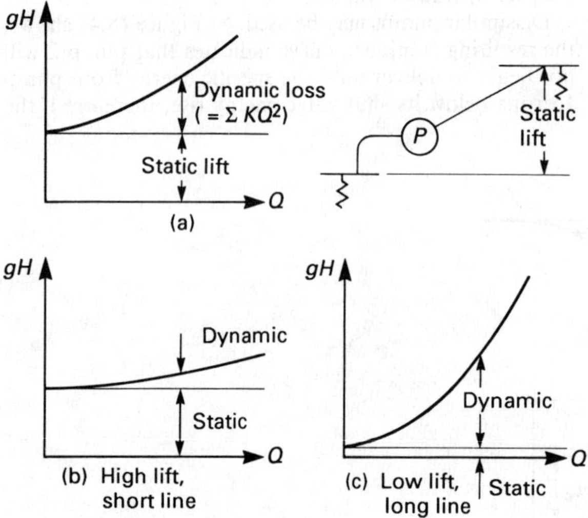

Figure 18.38 System loss characteristics

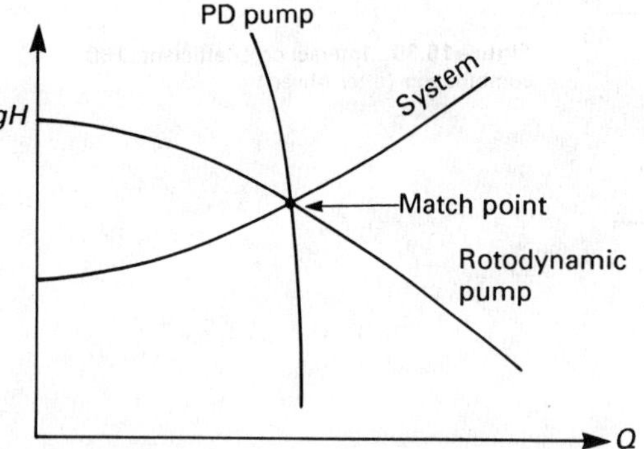

Figure 18.39 Matching constant-speed pumps to a system

non-return valve for pump 2 does not seat, pump 1 can cause it to turbine.

18.5.4 Suction systems

Poor intake layout can give rise to pump problems, unstable running, losing prime, and cavitation. If the pump is drawing water from a sump the position of the

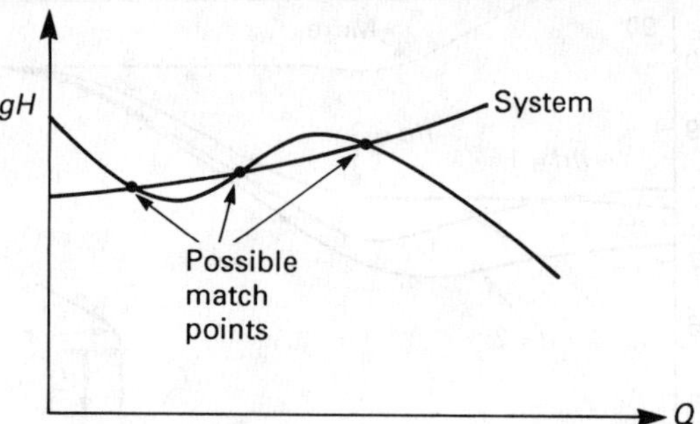

Figure 18.40 Combination of a flat system curve and an inflected pump curve

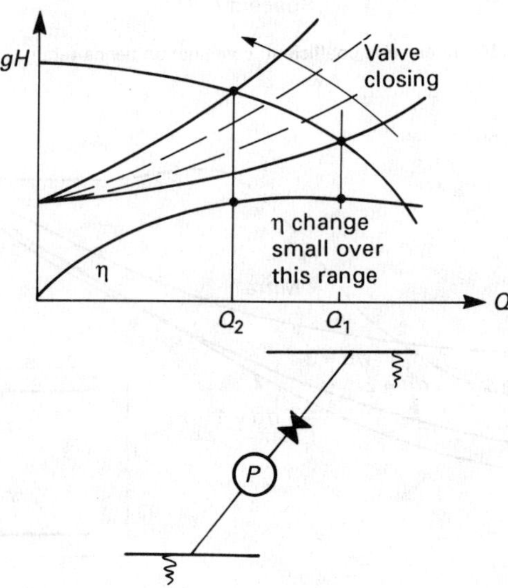

Figure 18.41 Valve control

intake and the shape of the sump must be chosen to avoid vortex formation and resultant air ingestion and pump instability. Figure 18.48 shows the proportions for PD and centrifugal pump suctions and Figure 18.49 gives details of baffles that can be fitted in tanks to reduce vortexing. If large flow rates and numbers of pumps are involved it is advisable to commission model tests to ensure that pump behaviour is not affected for all flow rates and pump combinations.

18.6 Cavitation

Cavitation is the term used to describe the formation of bubbles in liquid flow when the local pressure falls to around vapour pressure. Two effects are experienced in the pump: a reduction in flow rate (accompanied, particularly in centrifugal pumps, by additional noise) and in surface damage and material removal. In general, cavitation occurs in the suction region of a pump or the inlet port and valve area of a positive displacement pump.

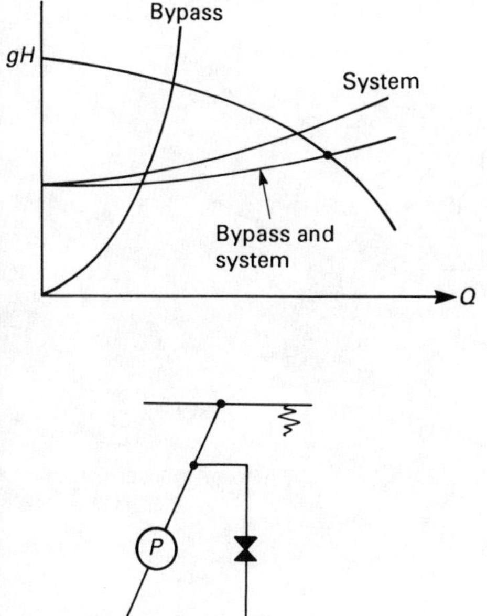

Figure 18.42 Bypass control

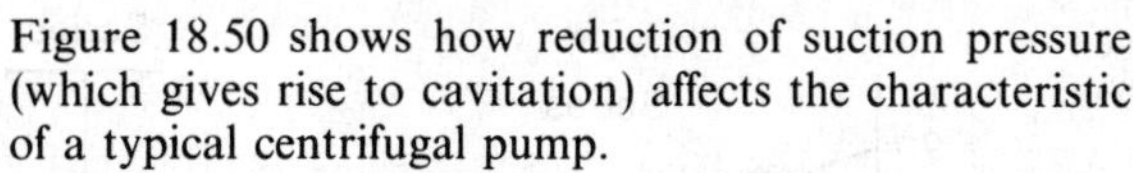

Figure 18.50 shows how reduction of suction pressure (which gives rise to cavitation) affects the characteristic of a typical centrifugal pump.

18.6.1 Net positive suction head (NPSH)

When drawing liquid through a suction a pump generates a low pressure in the suction area, which, if it reaches

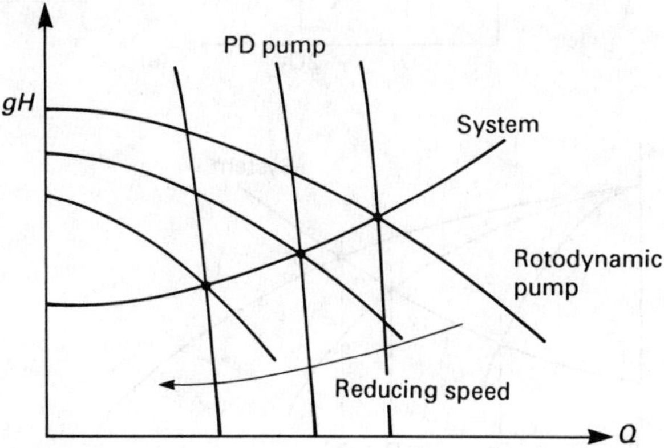

Figure 18.43 Speed control

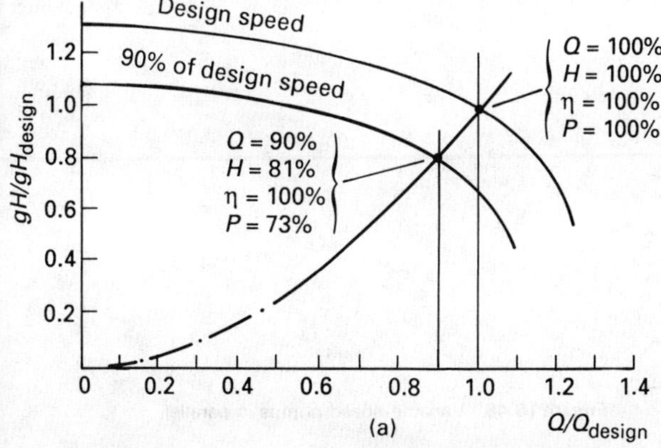

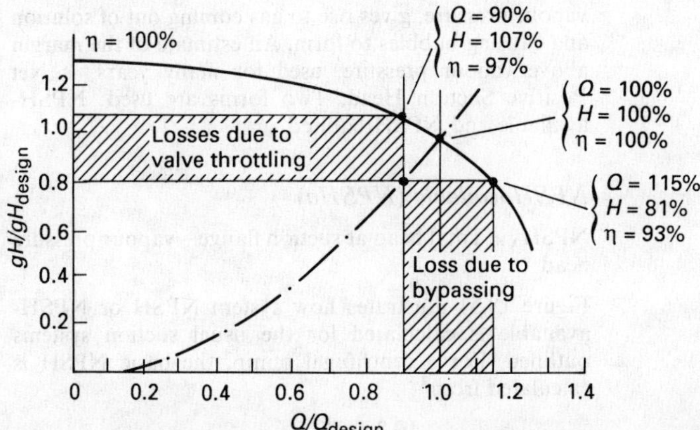

Figure 18.44 Comparison of speed, bypass and discharge regulation. (a) Speed control: (b) bypass or throttle

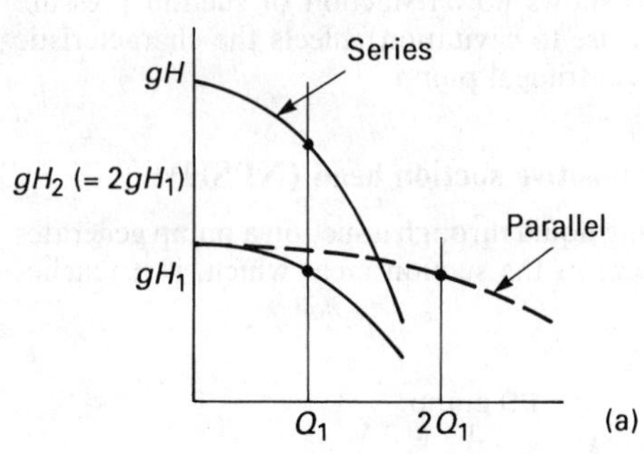

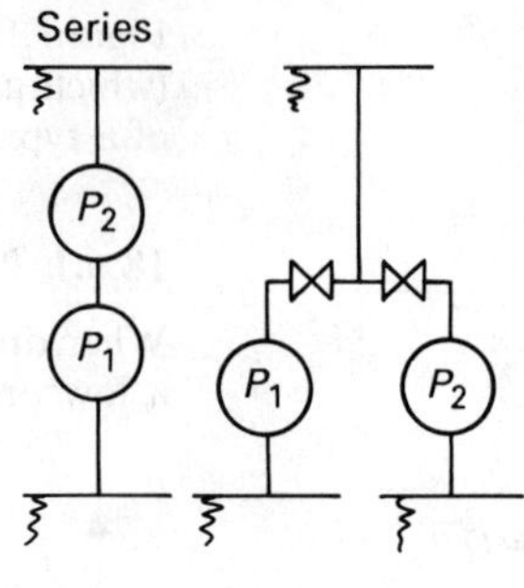

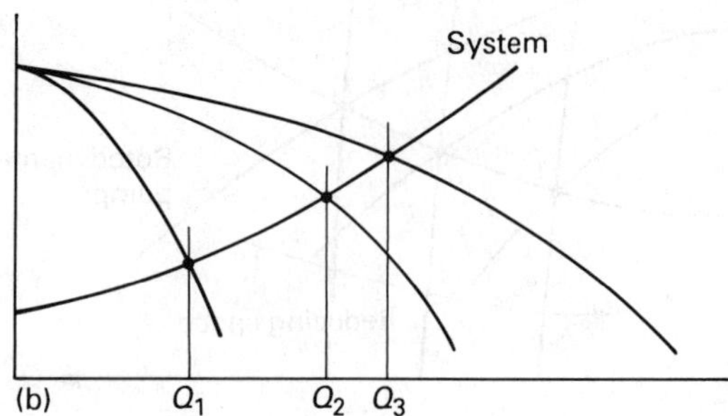

Figure 18.45 (a) Pumps in series and parallel; (b) three identical pumps running in parallel

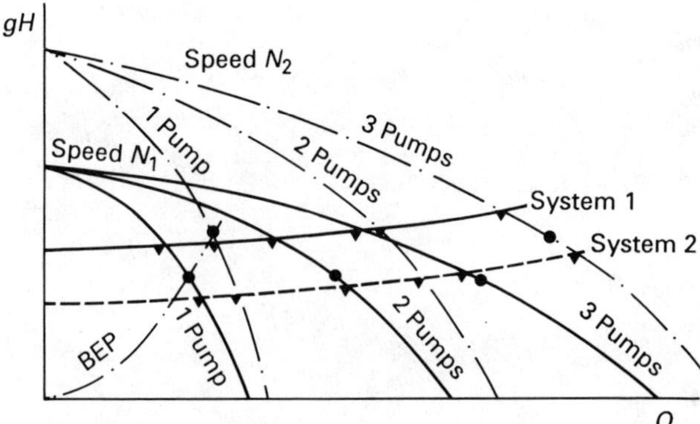

Figure 18.46 Variable-speed pumps in parallel

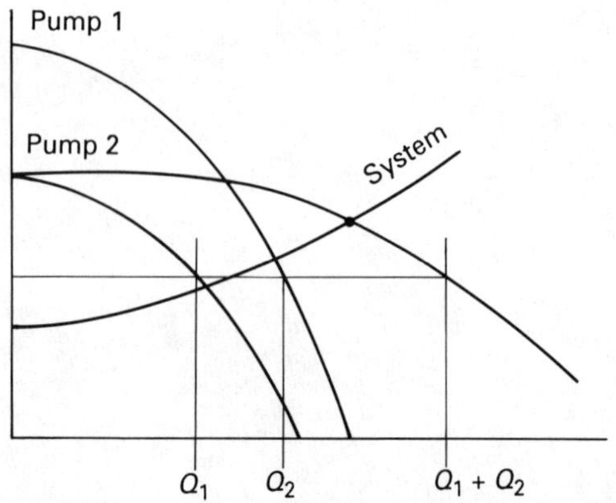

Figure 18.47 Dissimilar pumps in parallel

vapour pressure, gives rise to gas coming out of solution and causing bubbles to form. An estimate of the margin above vapour pressure, used for many years, is Net Positive Suction Head. Two forms are used: NPSH-available and NPSHrequired.

NPSHavailable (NPSHa)

$NPSH_A$ = Total head at suction flange − vapour pressure head

Figure 18.51 illustrates how system NPSH or NPSH-available is calculated for the usual suction systems outlined. For a centrifugal pump, the basic NPSH is calculated from[7]

$$\mathrm{NPSH_A} = h_s - h_f \frac{10.2}{\rho}\left(\frac{B}{1000} + P_i - P_v\right) \qquad (18.16)$$

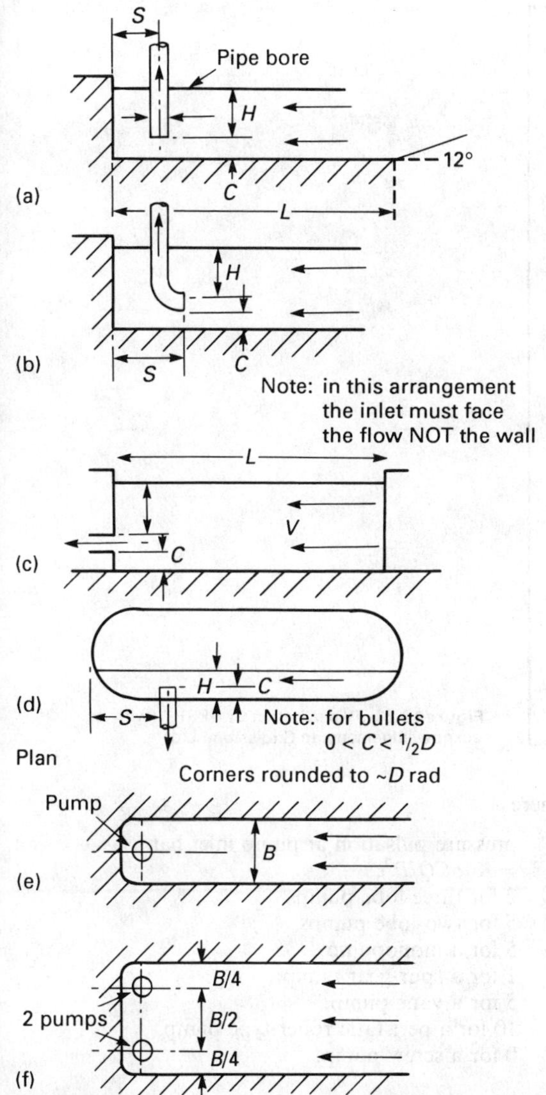

Figure 18.48 Suction bay and tank proportions (based on reference 7). Possible suction arrangements: (a)–(c) open suction layouts commonly used; (d) tank intake system (recommended limits: $3 > (S/D) > 1$; $2/3 > (C/D) > 1/2$; $(H/D) > [(V^2/3) + 1.5]$; $[(L-S)/D] > 5$); (e) single-suction cell in plan ($5 > (B/D) > 2$); (f) double-suction cell in plan ($4 < (B/D) < 10$)

where

h_f = flow losses in suction system (m),
B = minimum barometric pressure (mbar) (use 0.94 of mean barometer reading),
P_i = minimum pressure on free surface (bar gauge),
P_v = vapour pressure at maximum working temperature (bar absolute).

In the process industries h_f is calculated for the maximum flow rate and the NPSH at normal flow allowed for by using the formula

$$\text{NPSH}_A = 0.8(\text{NPSH}_{\text{basic}}) - 1) \qquad (18.17)$$

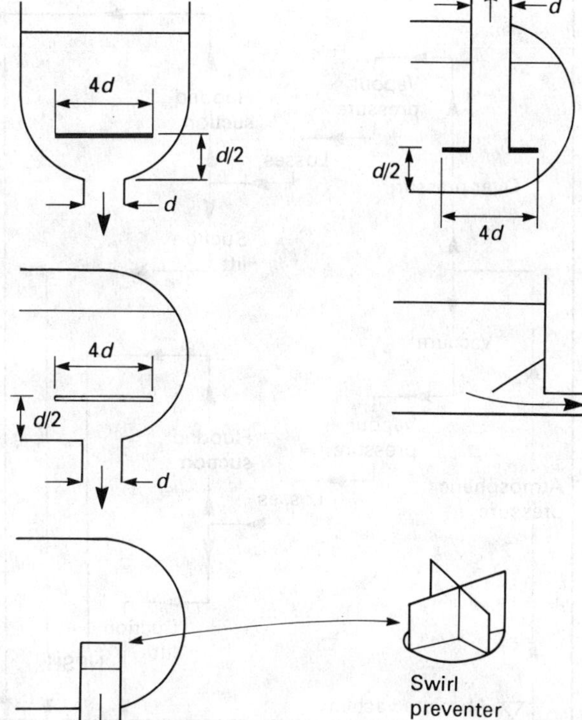

Figure 18.49 Vortex-prevention devices for tanks

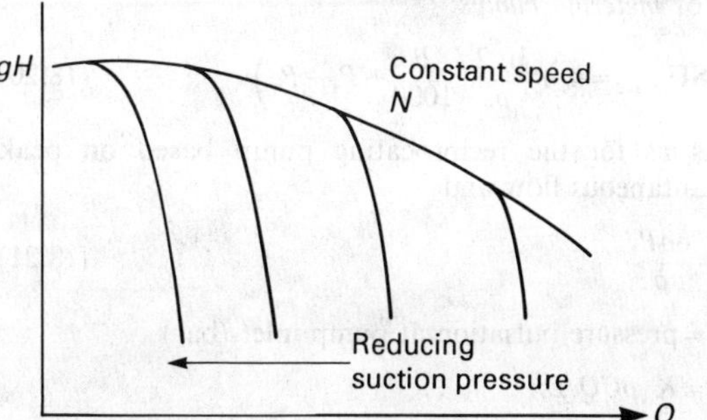

Figure 18.50 Effect of suction pressure on pump performance

This gives a 'target' value to the pump supplier that is 'worst' condition. In general, for cold water duties equation (18.16) can be used for the duty flow required. Equation (18.16) is employed for reciprocating and rotary positive displacement machines with allowance made for acceleration effects.

In reciprocators h_f is calculated at peak instantaneous flow, including maximum loss through a dirty filter, and an additional head 'loss' to allow for pulsation acceleration is used:

$$h_A = \frac{700NQ}{Z} \sum \frac{L}{d^2} \qquad (18.18)$$

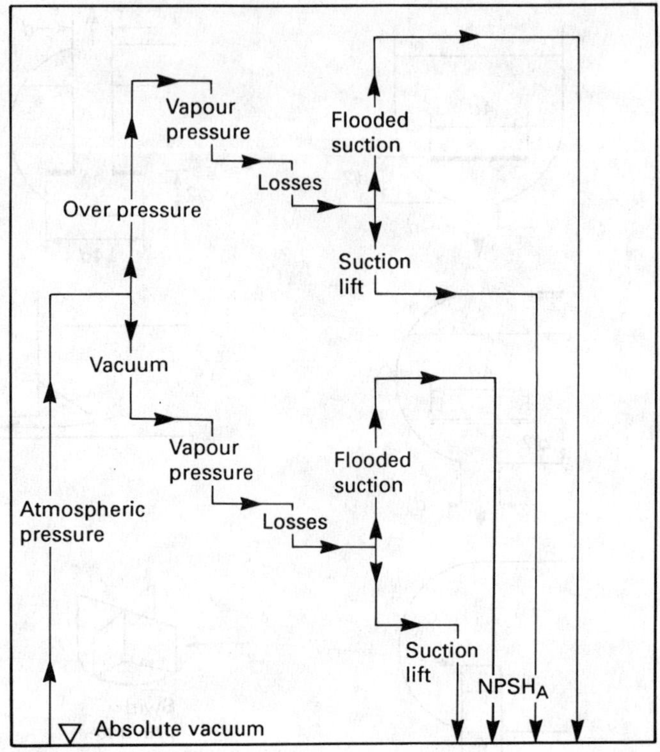

Figure 18.51 Visualization of NPSH$_A$ (with acknowledgements to Girdlestone Ltd)

and

$$\text{NPSH} = \text{NPSH}_A - h_A \tag{18.19}$$

For *metering pumps*:

$$\text{NPSH}_{\text{basic}} = h_s + \frac{10.2}{\rho}\left(\frac{B}{1000} + P_i - P_v\right) \tag{18.20}$$

h_f is as for the reciprocating pump based on peak instantaneous flow and

$$h_A = \frac{6\delta P_i}{\rho} \tag{18.21}$$

δP_i = pressure pulsation at pump inlet (bar)

$= K_1 \rho CQ/D^2$

where

$K_1 = 40$ for simplex, 20 for duplex, 3 for triplex, 1 for quintuplex,

C = velocity of sound in liquid ms^{-1} (for water at normal temperature $C = 1000$ ms^{-1})

$$\text{NPSH}_A = \text{basic NPSH} - (h_f^2 + h_A^2)^{1/2} - 1 \tag{18.22}$$

for simplex and duplex pumps and

$$\text{NPSH}_A = \text{basic NPSH} - (h_f + h_A) - 1 \tag{18.23}$$

for triplex and quintuplex pumps.

For *rotary pumps*:

$$h_A = \frac{\delta P}{\rho} \tag{18.24}$$

where

δP = pressure pulsation at pump inlet bar $= K_2 \rho CQ/D^2$,

$K_2 = 2$ for three-lobe pumps
5 for two-lobe pumps,
5 for a monopump,
1 for a spur-gear pump,
5 for a vane pump,
10 for a peristatic roller-type pump,
0 for a screw pump.

NPSHrequired (NPSHr)

This is a statement of the NPSH that the pump can sustain by its own operation, so that the operating requirement is that NPSH$_R$ > NPSH$_A$. NPSH$_R$ is more difficult than NPSH$_A$ to determine. Reference 7 suggests working relations for positive displacement machines which are outlined below.

For reciprocating metering pumps, NPSH$_R$ is related to valve loading, as shown in Figure 18.52:

$$A = \frac{24vQ\rho}{Zd_V^3} + 5 \times 10^5 \frac{\rho Q^2}{Z^2 d_V^4} \tag{18.25}$$

(where d_V = nominal valve size (mm)) for single valves, and

$$A = \frac{80vQ\rho}{Zd_V^3} + 1.5 \times 10^5 \frac{\rho Q^2}{Z^2 d_V^4} \tag{18.26}$$

for double valves. It is recommended that for hydraulically operated diaphragm pumps the extra losses imposed by

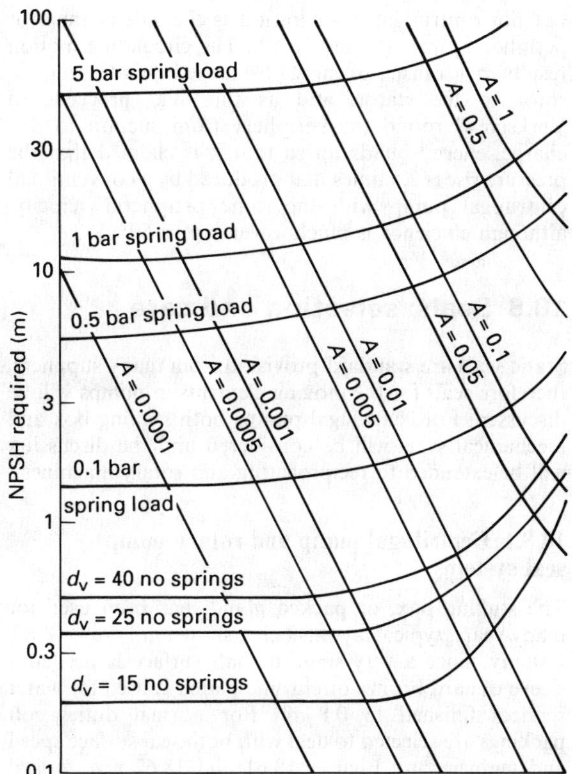

Figure 18.52 NPSHrequired for a reciprocating metering pump[7]

the diaphragm and support plate are treated as a single unloaded valve.

For other reciprocators

$$\mathrm{NPSH_R} = 5U^2 + \frac{0.12(P_d)^{0.75}}{\rho} \qquad (18.27)$$

where

U = mean plunger speed ($\mathrm{ms^{-1}}$),
P_d = discharge pressure bar (dbs).

For the centrifugal pump two terms are in common use, the Thoma cavitation number σ and the suction specific speed S_N:

$$\sigma = \frac{\mathrm{NPSH_R}}{\text{Pump head rise}} \qquad (18.28)$$

Figure 18.53 gives a typical plot of σ against k_s that may be used as a first 'design' estimate of $\mathrm{NPSH_R}$, but in many applications test data are required:

$$S_N = \frac{N\sqrt{Q}}{K(\mathrm{NPSH_R})^{3/4}} \qquad (18.29)$$

where K is a constant = 175 if $g = 9.81\ \mathrm{ms^{-2}}$, Q is in l/s, N is in revolutions/second and $\mathrm{NPSH_R}$ is metres of liquid. $\mathrm{NPSH_R}$ here is usually based on a 3% head drop, defined in Figure 18.54, for the pump tested with cold water at design flow rate and rotational speed.

18.7 Priming systems

If the suction level is above the pump there is usually no problem in 'priming' (that is, ensuring that the pump suction system is full of liquid) unless it is pumping a volatile and vapour locks when stationary. In many applications the pump is above the suction vessel or main, and the static pressure at the take-off point is too low to provide the static lift required. In these cases centrifugal and positive displacement pump suction lines are often provided with non-return valves so that, once filled they do not lose liquid on shutdown. If their operation is intermittent, a priming tank or 'jerry' is provided (Figure 18.55) to allow priming to take place. This is only possible with small machines because of the sizes needed. It is general practice that no pump should be started or run dry to avoid metal-to-metal contact and seizure. The only exceptions to this are pumps designed to provide priming – 'self-priming' machines, used in installations where the suction line empties, as in tanker-unloading bays or site-dewatering applications.

Positive displacements pumps are self-priming by their normal operating action and are designed to cope with running dry. Centrifugal pumps are not inherently

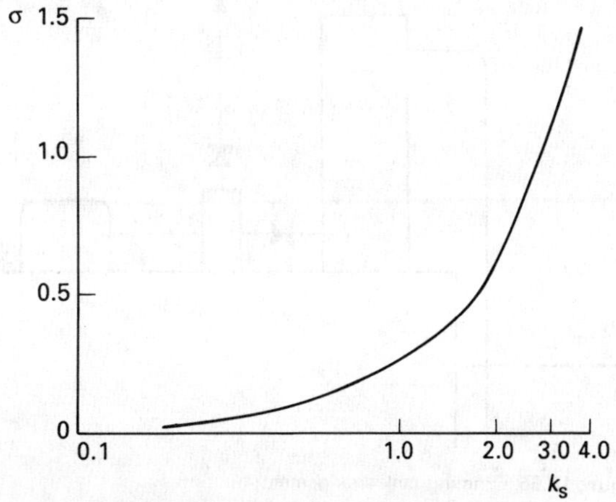

Figure 18.53 Variation of Thoma's cavitation parameter with specific speed for pumps[3]

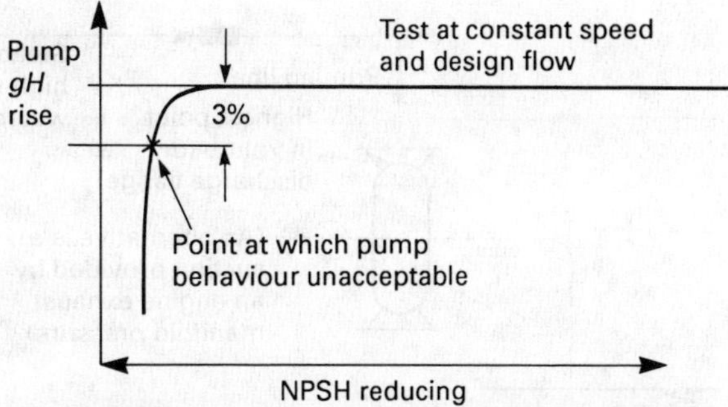

Figure 18.54

self-priming, and need to be provided with assistance in the form of ejectors, as in Figure 18.56 or dry vacuum pump systems, where a dry vacuum pump allows air to be drawn out and where an automatic valve prevents liquid entering the pump, or wet vacuum pumps, such as the liquid ring pump (Figures 18.57 and 18.58), may be used to extract the air. Such devices are common on portable fire pumps, the liquid ring pump being driven from the pump main driver in such a way as to allow disconnection when the pump is fully priming, thus avoiding extra power loss during normal operation.

Self-priming centrifugal pumps do not rely on any assisting system, but are provided with a special casing system that allows liquid retained in the pump on shutdown to recirculate to draw air from the suction line until that is full of liquid and the pump is primed. There are several designs which will pull suction lifts up to 7 m with cold water. All the designs impose a penalty of lower efficiency when in the normal pumping mode, but are effective alternative solutions to an auxiliary priming system which do not require additional equipment as bolt-on extras requiring maintenance.

One class of self-priming pumps which are rotodynamic but not centrifugal in operation is the side-channel or peripheral pump (Figure 18.59). The circulating motion results in a transfer of energy from the liquid leaving the rotor to the stator, and as the fluid proceeds to 'corkscrew' round the periphery from suction to discharge, energy builds up so that it is claimed that the pressure rise is 2.5 times that produced by a conventional centrifugal pump with the same peripheral velocity, although efficiency is much lower (up to 50%).

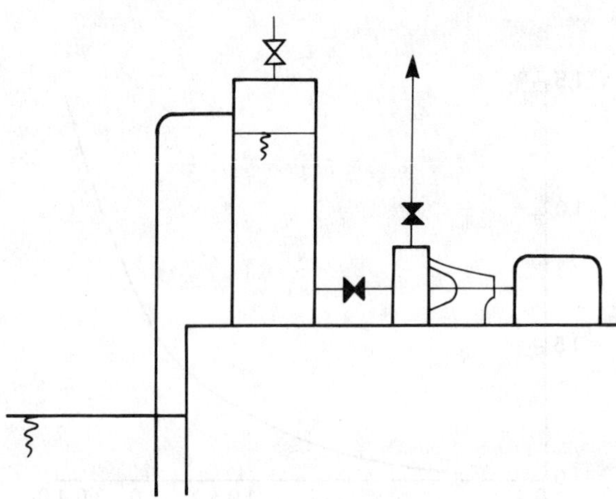

Figure 18.55 Priming tank arrangement

18.8 Seals: selection and care

Static seals are standard provision from many suppliers, therefore seals for the moving elements in pumps will be discussed. For centrifugal pumps both stuffing box and mechanical seals will be considered and the discussion will be extended to reciprocating and rotary machines.

18.8.1 Centrifugal pump and rotary pump seal systems

The stuffing box, or packed gland, has been used for many years, typical layouts being shown in Figure 18.60. Usually, since a very smooth shaft surface is needed a sleeve of hard bronze or chrome plated is used for water services finished to 0.8 μm. For normal duties soft packings are selected to deal with both seal surface speed and temperature. Figures 18.61 and 18.62 give typical materials and their capacities, and manufacturers' data sheets should be consulted when selecting the number of rings and the compression needed. When properly adjusted and 'run in', the leakage rate on a cold water duty could be about 1 l/day, which may be lost as vapour. Overtightening the packings can give rise to overheating and shaft failure or seizure. Typically, a four-ring gland will seal 3.5 to 4 bar, and absorb about 0.15 kW when run in and fitted to a 50 mm shaft.

The mechanical seal is now a standard fitting of high reliability and tends to be fitted instead of a packed gland. A typical mechanical seal is shown in Figure 18.63. The conventional design has a spring, rotating with the shaft, which holds a rotating seal ring against a stationary ring

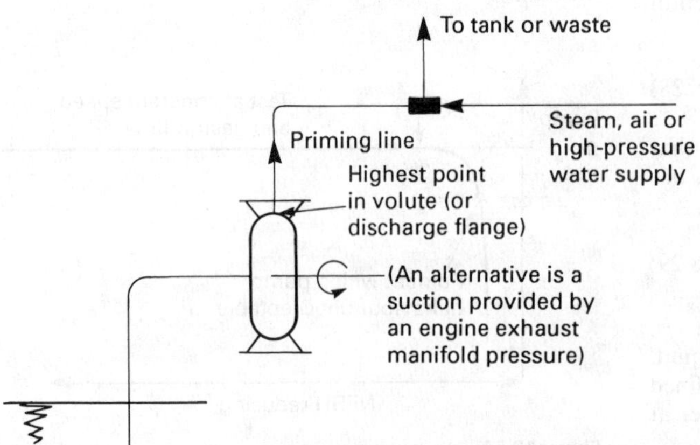

Figure 18.56 An ejector priming system

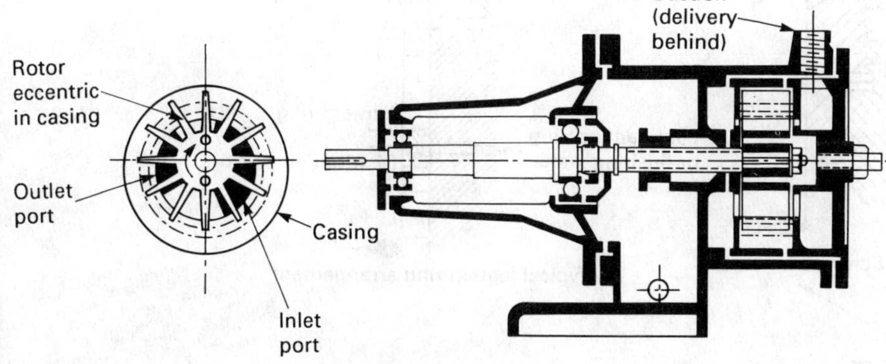

Figure 18.57 Schematic of a liquid ring pump

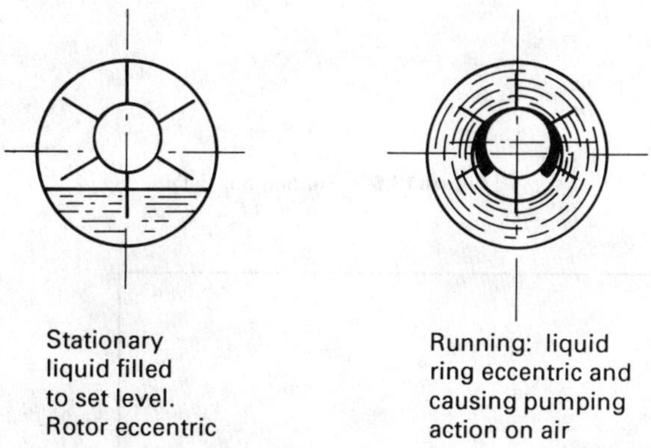

Figure 18.58 Principle of liquid ring pump

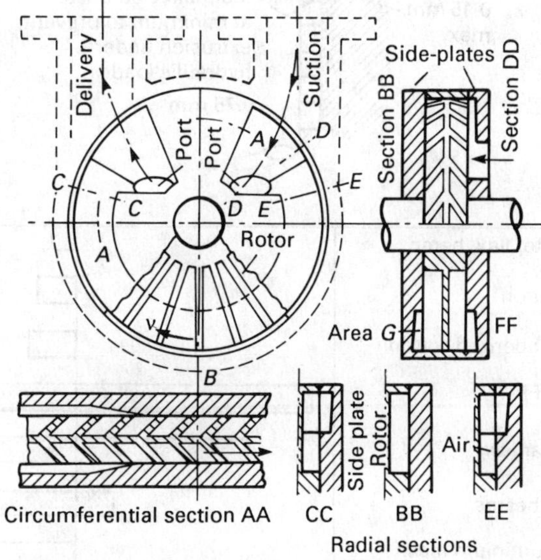

Figure 18.59 Peripheral pump layout and operating principle (after Addison[1])

to provide a sealing interface. Stationary fluid seals are provided as shown. For water and general non-corrosive duties the design is internal, so that the liquid being sealed both cools and provides a supply for the film which separates the rotating and static seal rings. Simple seals are designed to be 'balanced' or 'unbalanced'. Referring to Figure 18.64, it can be seen that a balanced seal will give a lower closing hydraulic force and thus lower contact pressure, therefore a higher sealed pressure can be sustained. Too little load, and the seal will leak, and too large will reduce life. An unbalanced seal gives a simpler shaft or sleeve design, and will seal up to 10 bar gauge, and above this, balanced seals are requested. Materials often used for standard water duties are carbon for the face and ceramic and, if condensate, carbon with a nickel–iron static seat could be fitted. The seal manufacturer will quote the appropriate materials, and will usually suggest circulation as shown in Figure 18.63. The mating sealing surfaces are lapped to within 0.5 μm, shaft run-out should not exceed 0.125 mm, and shaft finish where the seal ring has to pass should be 0.4 μm to avoid damage.

If a mechanical seal or a packed gland is fitted properly a good life can be expected if the liquid is clean. If grit is in suspension the seal manufacturer should be consulted about appropriate materials and system design.

18.8.2 Reciprocating pump seal systems

The comments on packed glands apply for shaft and rod seals. Piston-to-cylinder seals may be arranged like the packed gland, but are provided with end rings which locate and prevent the sealing rings from extrusion. Surface finishes for rod or piston will be the same as those given above with leakage rates of the same order.

18.9 Pump and drive selection

18.9.1 Pump selection

Since pumps are the essential element in a flow system, the selection of the correct machine is crucial to the success of a plant. It must be constructed from the appropriate materials, run at the right speed, be compatible hydraulically with the system at all flows, economical in first cost and in operation, and 'user-

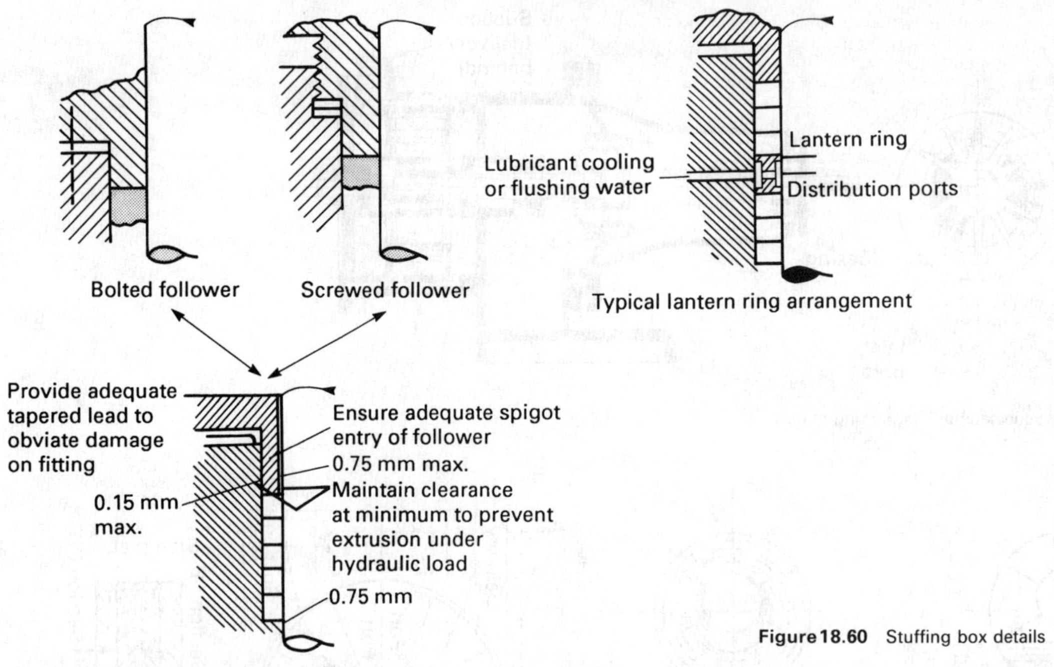

Figure 18.60 Stuffing box details

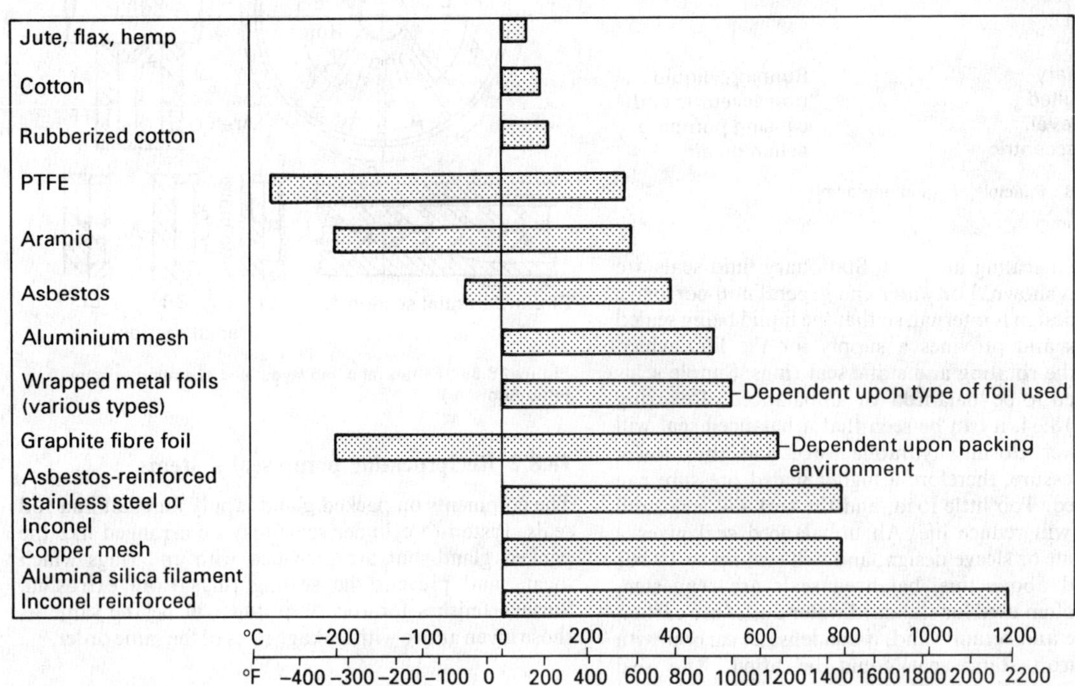

Figure 18.61 Soft packings – temperature range

friendly'. To place the process of pump selection in the plant decision tree it is necessary to follow the iterative way a complete plant evolves.

From the initial statement of need a preliminary design brief evolves containing flow rates and attendant component pressures and temperatures. From this an initial design layout is produced, the pressure losses and probable inlet and outlet pressures and attendant flows for all pumps needed estimated, and a first assessment of possible pumps made. Considerations of cavitation (and perhaps a need to rationalize on pump designs) lead to a second look at the systems. A revision of flow

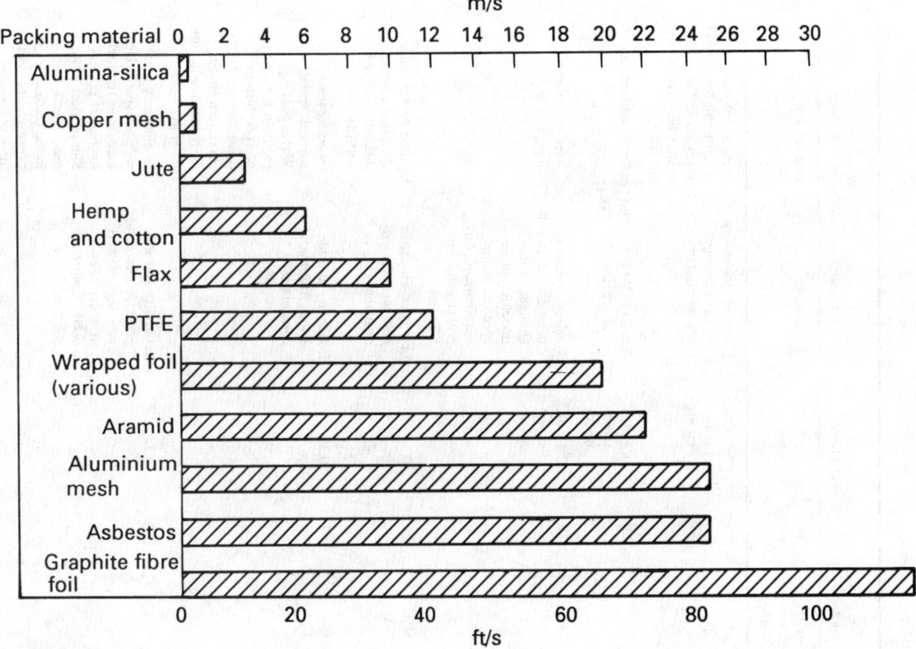

Figure 18.62 Soft packings – speed range

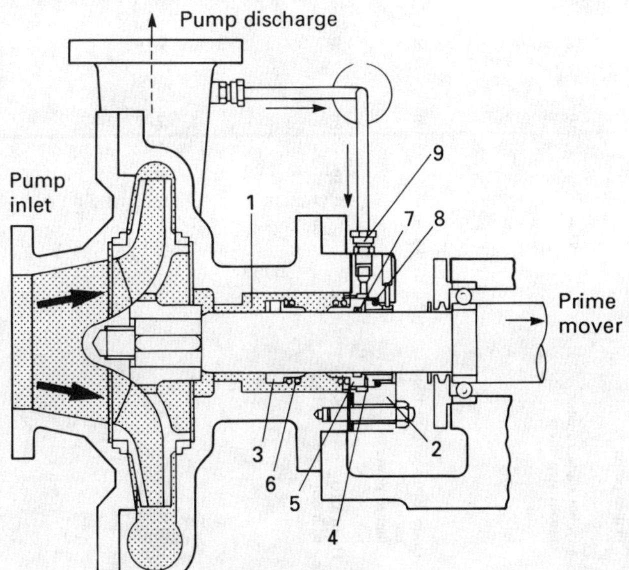

Figure 18.63 A typical mechanical seal installation. 1 Seal chamber; 2 seal plate; 3 spring sleeve; 4 seat; 5 seal ring; 6 spring; 7 dynamic secondary seal; 8 static secondary seal; 9 flush connection

schedules, pressures and temperatures results, and, by iteration, a 'final' design layout appears from which pump and equipment specifications begin to be formulated and initial enquiries are made to suppliers. Pump manufacturers will check possibilities and tender if their pumps will give the duties specified or, if invited tendering is used, will propose changes to suit their pumps.

In the consultant's design office the selection of pumps (and hence of makers) to tender must be done at the project design stage, and considerations of cost, complexity and suitability will influence the choices made. An important consideration at this stage is that the design point specified is the correction one. A common practice is the 'add a bit for ageing, add a bit for the unknowns' at the early design stage and then the next engineer in the process also 'adds a bit' since they may be on the

Table 18.5 Main characteristics of speed changing drives related to centrifugal pumps[8]

Type of drive	Power range (kW)	Speeds		Drive efficiency (%)		Overall efficiency (%) including motor				Power factor		Main characteristics related to pump drives	
		Max. rpm	Ratio	Maximum speed	Half speed	Max. speed		Half speed		Max. speed	Half speed	Advantages	Limitations
						4 kW	150 kW	4 kW	150 kW				
'V' belts or flat belts	Up to 750	5000 at limited power	8:1	'V' belts 85–90 Flat belts 90–95		70 75	80 85	40 45	65 70	0.9	0.3 with same motor	Low cost possibility if speed changes are infrequent	Increased floor space. Not suitable for outside application. Reduced bearing life if jack shaft not used
Timing belts	Up to 350	6000 at limited power	9:1	95+		80	87	50	75	0.9	0.3 with same motor	Similar to V belts but give reduced shaft loading and greater efficiency	
Gear box	Any	Any, but standard units usually step down		95+		80	87	50	75	0.9	0.3 with same motor	A robust, compact drive for infrequent changes or for matching drive speed to pump requirement. Any power available. Efficient	Correct maintenace importance
Variable-speed pulleys	Up to 125	Up to 4500 but at limited power	4:1 Std	85–90		70	N.A.	40	N.A.	0.9	0.3	Low cost. Housed-belt versions available	Limited power range. Automatic control difficult. Limited belt life
Eddy current coupling	Up to 1500	Up to 4200 with belt drive	Above 10:1	96	45–50	80	87	25	35	0.9	0.35	Reliability and life good. Automatic control easy	Low efficiency at reduced speed. Water cooling required for higher powers. Slip-in drive reduces pump maximum speed. Only speed reduction possible
Fluid couplings	15 to 12 000	3500 Std	4:1	95	45–50	80	87	24	35	0.9	0.35	Reliability and life good. Available to very high powers at which they become more cost effective	Low efficiency at reduced speed. Heat exchanger required for higher powers. Slip reduces pump max. speed. Automatic control expensive. Expensive for low powers

Positive infinitely variable chain drives	Up to 120	Up to 5500 but at limited power	10:1	85–90		70	N.A.	45	N.A.	0.9	0.3	Speed increase possible	Limited power range. Careful maintenance essential. Life limitations
Mechanical variators	Up to 75	4200 at 10 kW 2600 at 75 kW	Up to 12:1	85–95	85–85	70	N.A.	40	N.A.	0.9	0.3	Speed increase possible	Limited power range. Limited life. Careful lubrication and maintenance essential
DC motor and thyristor voltage control	Up to 2500	1500 Std 3500 avail. higher-special	Up to 30:1	See overall efficiency		80	90	45	65	0.9	0.3	Good range. Relatively good low speed efficiency. Automatic speed control easy	TEFC Expensive. High speed expensive. Harmonics generation needs consideration. More maintenance than induction motors
Voltage control of induction motor	Up to 50	3000	4:1	See overall efficiency		80	N.A.	25	N.A.	0.9	0.3	Low cost. Can use squirrel cage induction motor. Automatic control easy	Limited power range. Low efficiency at reduced speed. Only speed reduction possible. Requires at least 30% derating of motor
Schrage AC commutator motor	Up to 500	2500	4:1 Std up to 20:1	See overall efficiency		80	90	45	70	0.95	0.5	Good efficiency and power factor. Speed increase possible	Automatic control expensive. Movable brush gear increases maintenance. TEFC expensive
Stator-fed AC commutator motor	Up to 2500	2000	up to 10:1	See overall efficiency		80	90	45	70	0.9	0.4	Improved life over the Schrage motor and increased supply voltages possible. Good efficiency	Automatic control expensive. TEFC expensive
Frequency control of induction motor	Up to 75 Std 2000 special	5000	10:1	See overall efficiency		75	87	40	65	0.95	0.95/0.3	Can use squirrel cage induction motors with minimal derating. Good power factor (with most designs). Speed increase possible	High cost. Electronics more complex (and less reliable?) than other types

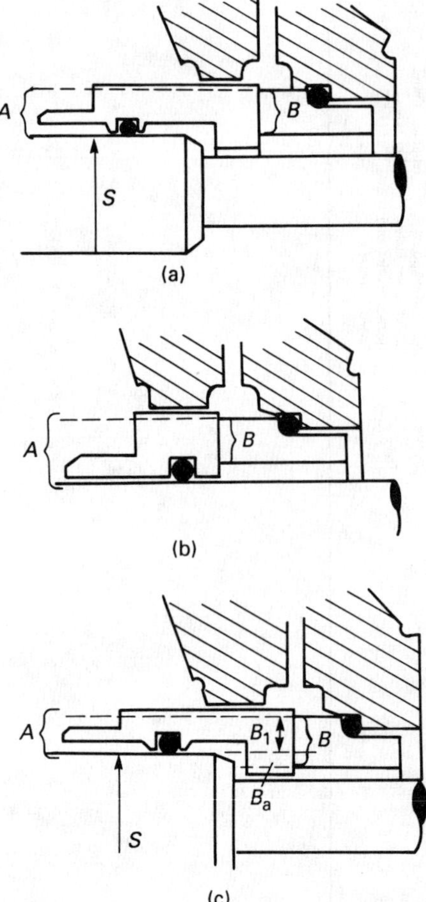

Figure 18.64 Seal balance arrangements. S = Sliding diameter or effective sealing diameter; A = hydraulic piston area of sliding ring; B = face contact area. (a) All area B is outside effective sealing diameter and areas A and B are equal. Average face contact pressure equals product pressure, i.e. seal is 100% out of balance. (b) Unbalanced seal. All area B is outside effective sealing diameter and area A is greater than area B. Average face contact pressure is greater than product pressure, i.e. seal is more than 100% out of balance, depending on the relative area ratios. This condition exists in most unbalanced seals. (c) Balanced seal. Here part of the face contact area, B_1, is outside. The sliding diameter S carries the product pressure acting on area A. The face contact area, however, equals $B_1 + B_2$, and so the unit face loading will be less than product pressure, in accordance with the area ratio, i.e.

$$\frac{B}{B_1 + B_2} = \frac{A}{B}$$

$$\text{Balance} = 1 - \frac{A}{B}$$

Generally, 0.2–0.35

low side, and thus a margin of flow is added. This process can sometimes ensure that the pump size is such that, even though the design point is that quoted, the pump will be running at low flow when the system is commissioned, and this may be in a region of instability, giving running problems on the plant. It cannot be too much emphasized that accurate estimation of the duty flow and head will assist pump application and life.

Examination of Figure 18.65 will reveal that in many cases the choice of centrifugal or positive displacement pump is dictated by the head and flow specified. Where, however, the choice may be either type, other considerations must enter the selection process. For example, first cost is an important factor. Sterling[5] has demonstrated that a screw pump and a centrifugal would give the same duty, that the screw pump would have an $NPSH_R$ less than the centrifugal, have a lower fall-off in efficiency as fluid viscosity increased, but would cost twice more than the centrifugal and be more expensive to maintain. He concluded by stating the centrifugal as being the most preferred for price and maintenance in 80% of pumping duties.

A probable selection procedure could take the following form:

> If possible, a standard centrifugal pump shall be selected, but the effect of high viscosity must be considered (see Table 18.2). If flow is low remember the efficiency fall-off, and look at alternative designs. If flow rate is important, metering pumps should be preferred, as centrifugal pumps will not give accuracy. If solids are in suspension consider working clearances, need to avoid blockage and the flushing for seals.
>
> A drive speed of 2900 rpm is preferred to give higher efficiency and smaller size and first cost (but remember noise risks). Appropriate codes (API 610, for example) should be observed.
>
> If a centrifugal is not suitable, select a rotary positive displacement pump. The design usually selects itself, but for flow rates in the centrifugal pump range, screw pumps, monopumps and wide-tooth gear pumps may be considered. Remember that if particles are present the screw pump will lose volumetric efficiency, and the mono type of machine should be selected.
>
> If the duty is metering, piston or diaphragm pumps should be selected.

8.9.2 Drive selection

For most centrifugal pumps the drive will be a synchronous motor, running at 2900 rpm for 50 Hz supply to minimize pump size and cost. Large boiler feed pumps use steam turbines which give economic output regulation since steam is available. In most plant variable-speed drives may be used to give economic flow control. A wide range of possibilities are available (Table 18.5). Bower[8] writing from the viewpoint of a pump designer, has given some comparative costs. He commented that a 10 kW drive pump running for 10 h per day will cost more than the first cost of the unit in a year. He showed that if a 75 kW pump running at 2900 rpm is to run at half flow, a variable-speed drive giving 1450 rpm will use 39 kW if supplying a low static lift system, where flow losses are important. There is thus an economic argument for a higher first-cost pump plus drive if such savings are possible.

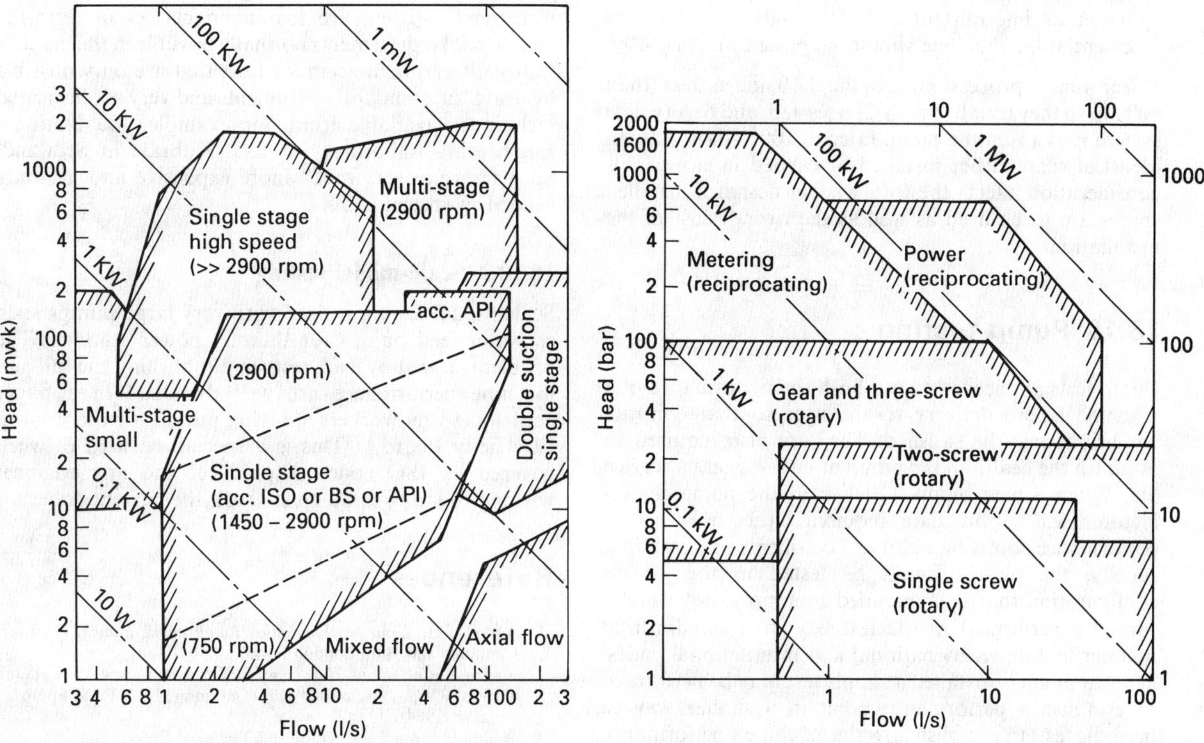

Figure 18.65 Pump range chart (courtesy Nederlandse Aardolie MIJ BV)

18.9.3 The economics of pump selection and running maintenance

With the operating economics of pump-supplied system such as the water industry networks receiving attention, it is of interest to note that recent work[9] makes two points: pump performance does fall off with time and there should be consideration of the economics of refurbishment as part of system pricing. Cullen reported a WRc study of basing figures on 50% utilization and a 3p per kilowatt hour tariff. One case involving 12 pumps, capital value £120 000, with an efficiency fall-off of 5–9%, could be refurbished for about £48 000 to give 5% efficiency uplift and show an electricity saving of £13 400 per year and a return on investment of 35% per year. A further example covering eight pumpsets, capital value £80 000 with a serious efficiency fall-off of 15–19%, could be refurbished at cost of £112 000 to give an uplift of 15%, resulting in a £31 200 per year saving and an annual investment of 25%. These examples and others emphasize that traditional oversizing is expensive in running cost, that pump mismatch should be avoided and that pump replacement or refurbishment can be economic in large installations.

18.9.4 Reliability considerations

As a further consideration of the overall system design it is necessary to examine the choice of the number of pumps installed. In the petrochemical and process industries availability is an important factor in plant operation. The Institution of Mechanical Engineers suggests several classes of pumps linked to availability:[7]

1. A Class 1 installation achieves high availability using a single pump upon which the system is completely dependent is in a single process stream where pump outage means a large loss of output. The process here demands that continuous operation is possible for at least three years without enforced halts for inspection or correction and where components must have a life expectancy of more than 100 000 h.
2. A Class 2 system is as Class 1 but, with process recovery short after shutdown, infrequent pump outage can be tolerated and an interval of about 4000 h for continuous operation can be tolerated.
3. If in a Class 1 and 2 installation a deterioration in pump operations can be tolerated or easily corrected by operators it may be categorized as Class 3.
4. A Class 4 installation is different from the first three by having standby pumps that can take over immediately a pump fails. Main pump and the standby need not be identical but should both have life expectancies exceeding 25 000 h.
5. A Class 5 installation follows the Class 4 concept, but is characterized by having pumps operating with identical spares and have one or more pumps operating in plant bays where product storage gives time to assess the problem and to take action to get the pump back on-line.

6. A further Class 6 is suggested for pumps intended for batch or intermittent duty. (If high availability is essential the machine should be placed in Class 4.)

For many process pumps the L10 life is less than 8000 h, so they usually fall in Classes 4, 5, and 6. A typical system uses a running pump rated at 100% duty with an identical spare either installed or carried in store. This consideration affects the total system design if installed spares are to be used as well as the overall cost of the installation.

18.10 Pump testing

Pump tests are needed to establish the performance of a machine before delivery to the user so that contract conditions may be satisfied. They are also required to establish the health of the pump at intervals in its service life. When a new pump is delivered the pump manufacturer will supply data required either as a set of performance points or a plot as a certificate of suitability. Ideally, the pump should be tested in the system configuration that it is intended to supply, but usually tests are performed in the test bay in a standardized manner laid down by national and international codes. When a pump is installed a simple test may be performed to establish a performance point in a similar way to periodic tests to establish how the machine's performance is being affected.

18.10.1 Factory tests

Factory tests establish the pressure head, power, efficiency and NPSH over the complete flow range the pump can deliver running at design speed. The manner of test procedure is laid down by British Standard, DIN standard or ANSI standard codes or national variations from such main codes, and a minimum requirement is quite often defined by industry codes such as API 610. This is not the place to discuss instrument accuracies, as the codes lay down the limits possible from conventional instruments. There are two main classes of test: the commercial requirements normally possible in the maker's test plant and high-accuracy tests that are only possible by using 'substandard' instruments and very sophisticated techniques available from, for example, the National Engineering Laboratory at East Kilbride in Scotland. Such tests are very much more expensive and are only needed in special cases.

18.10.2 Scale-model testing

Scale-model testing is used with very large pumps such as water feed pumps for thermal power stations. The problems posed by such tests in establishing the full-size machine performance are well discussed in a paper contributed by workers studying pumps for the Central Electricity Board.[10] This is a specialized area of work covered by ISO codes of practice, and any engineer wishing to study these must check the current codes.

References

1 Addison, H., *Centrifugal and other Rotodynamic Pumps*, Chapman and Hall, London (1955)
2 Karassik *et al.*, *Pump Handbook*, McGraw-Hill, New York (1976)
3 Turton, R. K., *Principles of Turbomachinery*, E. & F. N. Spon, London (1984)
4 *Hydraulics Handbook*, Trade and Technical Press
5 Sterling, L., 'Selection of pump type to match systems', Paper B3, BPMA 5th Tech. Conf., Bath (1977)
6 Miller, D. S., *Internal Flow Systems*, BHRA Fluid Engineering, Cranfield (1978)
7 Davidson, J. (ed.), *Process Pump Selection – A Systems Approach*, I. Mech. E., London (1986)
8 Bower, J. R., 'The economics of operating centrifugal pumps with variable speed drives', *Proc. I. Mech. E.*, Paper C108/81 (1981)
9 Cullen, N., 'Energy cost management in water supply', BPMA 11th Conference, April (1989)
10 Nixon, R. A. and Cairney, W. D., 'Scale effects in centrifugal cooling water pumps for thermal power stations', *NEL Report 505* (1972)
11 Summers-Smith, J. D. (ed.), *Mechanical Seal Practice for Improved Performance*, I. Mech. E., London (1988)

19 Cooling Towers

John Neller
Film Cooling Towers Ltd

Contents

19.1 Background

The use of water as a cooling medium has been long established, but its importance, in an industrial sense, was emphasized with the introduction of steam power. Cooling ponds are still widely used and spray ponds, which incorporate a degree or so of evaporative cooling, can be found, but the increasing requirement for control on water-cooling temperatures heralded the development of the modern cooling tower. The cost of land and the increasing expense of abstracting and of returning water, as well as its availability, ensured that the engineering and design techniques employed were sufficient to satisfy the economic factors imposed by these constraints.

The state of the art in cooling tower design is being constantly improved. Correctly designed, installed and maintained, today's cooling tower still remains the optimum selection in the great majority of cases where heat dissipation is required.

Historically, the first pack designs were random timber, to be followed rapidly by ordered timber splash bars. The concept of filming water as opposed to splash or concrete originated in England in the 1930s. The introduction of plastic packings dates from the 1950s, but this was confined to mechanical-draught towers until the 1970s, when experiments started with plastic packs in natural-draught cooling towers. Asbestos cement in flat sheets, corrugated sheets and flat bars, although widely used in the past, are now out of favour on health grounds in most developed countries. Plastic-impregnated paper is used in certain Eastern countries in air-conditioning towers, but has been unsuccessful in the West. In most cases the changes have been due to economics, but water quality and type of process can significantly affect the selection in individual cases.

19.2 Theory

Water cooling in towers operates on the evaporative principles, which are a combination of several heat/mass transfer processes. The most important of these is the transfer of liquid into a vapour/air mixture, as, for example, the surface area of a droplet of water. Convective transfer occurs as a result of the difference in temperature between the water and the surrounding air. Both these processes take place at the interface of the water surface and the air. Thus it is considered to behave as a film of saturated air at the same temperature as the bulk of the water droplet.

Finally, there is the transfer of sensible heat from the bulk of the water to the surface area. This is so slight in terms of resistance that it is normally neglected. Radiant heat transfer is also ignored for all practical design purposes.

Thus the two main processes are evaporation of water and convective cooling. The first is based on the difference in partial vapour pressure and the second upon the temperature difference.

Merkel's analysis in 1924–1925 demonstrated that for pure counterflow it is possible to combine these processes into a single term by using the enthalpy difference as a driving force. Experience over many years supports this, and cooling tower design in counterflow is universally represented by temperature–enthalpy diagrams.

For crossflow designs the additional factor of the horizontal depth of packing has to be included in the basic calculations. The accuracy of the design is directly related to the number of calculations in the selection programme. Whereas counterflow can be dealt with as a single entity, crossflow has to cope with the changes that occur at every level of pack, both vertically and horizontally.

19.3 Design techniques (see Appendix 19.1)

The performance of any cooling tower can be assessed against the following:

$$\frac{KaV}{L}=\frac{C}{(L/G)^n}$$

where KaV/L is normally given as the performance index or is quoted as KaV/L demand. L/G = water/dry air mass ratio and, taking n as average value 0.6, the constant C is proportional to the height of packing. This would cover most design requirements. Excessive temperatures and extremely high air rates will require further factors which are not necessary in the great majority of cases.

The Cooling Tower Institute (CTI) publishes sets of graphs which give demand in terms of three design temperatures and L/G. The CTI graph first published in 1967 gives KaV/L demand plotted against L/G with the approach temperature as a parameter. Each curve applies to a specified combination of wet bulb temperature and cooling range. Each cooling tower pack has its own KaV/L and responsible suppliers will supply performance graphs similar to those of the CTI.

Having selected a type and height of pack, the above equation can be plotted to intersect with the required demand curve to obtain the L/G. With the L/G and the given amount of water to be cooled, the air requirement can be calculated for:

$$\text{Air}=\frac{\text{Design water flow}}{L/G}$$

The following steps should then be:

1. Correct calculated air volume to conditions at the fan.
2. Select the air rate and calculate the plan area of the tower.
3. Select a fan and calculate the power requirements from the known air volume and the pressure drop characteristics of the selected pack.
4. Assess the cost factors applicable, in terms of fan, pump cost, price of land, maintenance and treatment costs.
5. Optimize all known factors in terms of efficiency and then economy.

19.4 Design requirements

The factors which should be applied at the design stage cover the water flow rate, the design wet bulb figure, the required return temperature at the design point, the cost of power and land, and the water analysis. Water flow is normally determined by the equipment that the cooling tower is serving (for example, heat exchangers). Process designers historically leave the cooling tower until last (it is, after all, the final heat sink). When water costs were negligible, this was acceptable, but with the increase in costs and, in certain cases, the restrictions on availability of water, this approach has had to be modified. Greater consideration must be given to the overall system. Experience in the last ten years has shown that economic optimization can lead to a more efficient cooling tower, with a corresponding drop in the cost of heat exchanger. This is particularly true in power generation and industrial processes.

Design wet bulbs can be determined from published meteorological data for the area concerned. The difficulty is deciding how to relate the annual coverage to the tower performance at any given time.

For some years it was common practice to quote three different figures, based on the tower's performance as a percentage of the year. For example. in air conditioning it could be shown that the tower would achieve its design for 95% of the year. Alternatively, a tower costing 15% less could obtain its design parameter for 85–90% of the year. Only the operator would know whether the 85–90% or less was acceptable, while the economists would welcome the saving of financial capital.

The frequent failures to achieve even the quoted reduced percentage figures led to a reappraisal, and current design is more accurate. In some respects this is also due to the improvement in pack designs, particularly in the European and American markets. However, it must be said again that in optimizing cooling tower selection the designer must be advised of all appropriate factors. Discussions with cooling tower designers at the outset can save time and money in the future.

Water quality is important, not only from an environmental point of view but also in relation to the type of packing to be specified. Analysis of the circulating water is simple to obtain, but it is very seldom offered to the cooling tower designer. The quality, or lack of it, will determine the type of pack to be used, the selection of structural materials and whether the tower should be induced or forced draught, counterflow or crossflow. Water treatment, in the shape of chemicals to control pH and to act as counter-corrosion agents or as biocides, all have a bearing on tower selection.

The '*Legionella* syndrome' has resulted in health authorities in the UK applying statutory regulations, which are directly reflected in terms of capital cost and tower material selection. To safeguard against this, responsible designers have already produced cooling tower designs which not only meet the regulations but anticipate future, more stringent, legislation.

The following list of information factors should be made available to any supplier so that discussions on the technical requirements can be carried out prior to optimization (see Appendices 19.1 and 19.2).

19.5 Materials and structure design

The great majority of towers available fall into one of two categories: counterflow or crossflow (co-current flow is available but is seldom used) (see Figure 19.1).

19.5.1 Counterflow

Counterflow designs are used throughout the entire design field, i.e.:

Natural draught – hyperbolic concrete shells
Mechanical draught – induced draught
Forced draught

19.5.2 Crossflow

Crossflow designs are also used throughout the entire field, i.e.:

Natural draught – hyperbolic concrete shells
Mechanical draught – induced, single and double sided and circular
Forced draught
Advanced fan-assisted natural draught.

Axial or centrifugal fans can be applied in most cases and are significant factors in the final selection and optimization (Figures 19.2 and 19.3).

Hybrid towers, combining wet and dry cooling, are designs to meet specific problems and require expertise from specialist suppliers. Structural materials include concrete, timber, various forms of metal (including galvanized and alloys), GRP, PVC and combinations of those along with asbestos cement, asbestos cement replacement (ACR) and, again, variations and combinations of other materials. Packing materials have an almost similar pattern but must include compressed paper and compressed asbestos cement paper, but the great majority of towers currently employ plastic, in some form or another, unless the water conditions are such that timber (or even concrete) must be used as alternatives.

19.6 Specification

The purchase of the cooling tower is, in most cases, a once in a decade operation. Where towers are bought on a regular basis, specifications are determined either by the user or by the consultant, incorporating their experience of operation and any changes required as a result of production/process alterations.

In air conditioning circles, the tower normally represents the final heat sink in a turnkey package which would include compressors/condensers, pipework, ducting, fans, pumps, control gear, etc. Where consultants and experienced contractors are concerned, the tower specification is well defined and the purchases based upon economics related to efficiency.

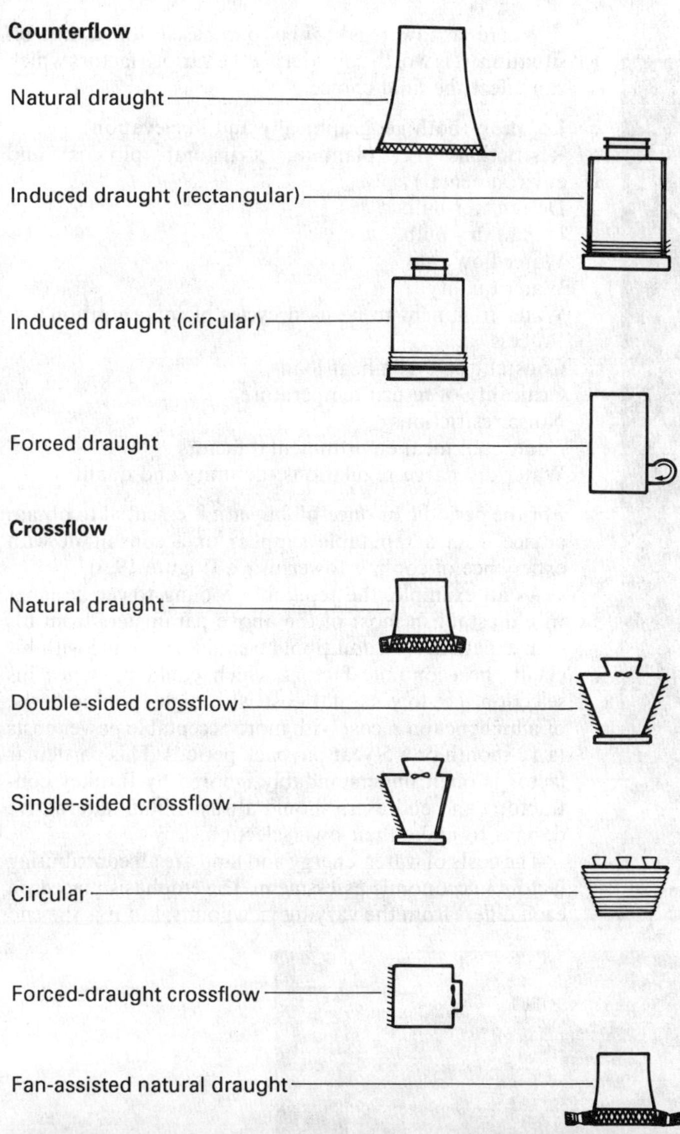

Figure 19.1 Counter- and crossflow modes

Figure 19.2 Basic crossflow units

Figure 3 Improved design of crossflow unit (note easy-removal panel)

Where a tower is to be purchased for a one-off situation, it is worth considering the various factors which can affect the final choice:

Location (both geographically and in elevation)
Restrictions (i.e. planning, structural, physical and environmental)
Design wet bulb
Design dry bulb
Water flow rate
Water quality
Water treatment to be used/cycles of concentration?
Process
Constant or cyclic heat load
Criticality of return temperature
Noise restrictions
Additional local environmental factors
Water discharge regulations (quantity and quality)

For the periodic or once-off buyer it is essential to obtain advice from a reputable supplier or a consultant with experience of cooling tower usage (Figure 19.4).

As an example, the reputable cooling tower designer would establish most of the above parameters from his own experience. In addition, he could determine with his client the economic factors which could influence his selection, i.e. low capital cost with high running costs, or a higher capital cost with more acceptable power costs (a 12-month or a 5-year payback period). This particular factor is often understandably ignored by turnkey contractors, and end users should always obtain alternative designs to make their own selection.

The costs of water, energy and land are all contributing factors in economic assessment. The emphasis placed on each differs from the varying viewpoints, but it is the end

Figure 19.4 Architect-designed tower (note water pattern in basin). The office windows are affected by chemicals in carry-over

user who has to 'foot the bill'. It is therefore in his own interests to acquire some knowledge of the cooling towers being offered.

19.7 Water quality and treatment

As, in most cases, the circulating water in any system incorporating a cooling tower is recycled (the loss being made up as a percentage of total flow), the subsequent concentration will affect the water condition. This, in turn, determines the type of treatment required to maintain pH and to control any potential biogrowth. Acidic or alkaline waters pose their own problems in terms of corrosion and material attack. The larger users probably employ their own chemical specialists, while others rely on consultants to determine the type of treatment the system requires. However, the majority of tower users have very little knowledge of the chemistry involved and depend on water-treatment organizations.

The following definitions are useful for reference to familiarize the end user with the terminology currently employed:

pH — The acidity or alkalinity of the water expressed as a scale of 0 (acid) to 14 (alkaline). pH 7 is regarded as neutral.

TDS — Total dissolved solids, expressed as ppm (parts per million) or as mg/l (milligrams per litre). Evaporate the water from a sample and the residue can be weighed.

TSS — Total suspended solids: expressed in similar terms to TDS but representing a concentration of insoluble particles.

Conductivity — Used as a measure of mineral impurities.

LSI — Langelier Saturation Index: indicates the corrosive (negative) or scale-forming (positive) characteristics.

Hardness — Expressed as $CaCO_3$, this is the total calcium and magnesium salts in the water. Hardness figures given as ppm or mg/l are important, as the compounds of these two elements are responsible for most scale deposition.

Alkalinity — Expressed as $CaCO_3$, this is the total concentration of alkaline salts (i.e. bicarbonate, carbonate and hydroxide).

BOD — Biological Oxygen Demand: expressed as ppm or mg/l, it is used as a measure of pollution.

COD — Chemical Oxygen Demand: as above, but related to chemical impurities.

Oxygen Sag — The level of oxygen in a polluted water system. Normally shown in a graph form.

Fouling Factor — This is generally applied to plastic packs in natural-draught towers but can relate to larger mechanical-draught and to the biological fouling that can occur. It also reflects on the thermal performance of the packing (CEGB published fouling factors for certain high-density packs where the supply of water was prone to seasonal biological growth and silt deposition, along with calcium hardness deposition).

While other terms are, of course, employed, the above can be useful for most end users.

19.8 Operation

Having selected and purchased a cooling tower, it needs regular maintenance, as does any other part of the plant. This is true of every cooling tower, from the largest natural-draught tower to the smallest packaged unit.

Maintenance logs should be kept. In air-conditioning installations, with the experience of *Legionella*, it is now mandatory to keep such a log, as well as a record of hygiene testing to determine the non-existence of bacteria. Chlorine dosing, as recommended by certain local authorities (including London), is essential. Tower materials have to be assessed in regard to chlorine residuals which can be damaging to galvanized metals and timber. Mechanical equipment requires regular checks, apart from the commonsense ordinary maintenance (it is surprising how often this is ignored).

The fault-finding chart (Figure 19.5) was originally produced by the Cooling Water Association (which subsequently became the Industrial Water Society), and it is practical and simple to follow. Water treatment checks (apart from the *Legionella* requirements referred to earlier) must be carried out and water samples should be analysed on a regular basis. How frequently this takes place depends on the criticality of the tower for the end user. Cooling towers are water-conservation tools as well as heat dissipators: with water costs increasing, continuous tower performance is essential. In any case, a down-time caused by lack of maintenance is costly and careless.

Remember that outside influences (for example, new building work in the vicinity of the installation) can increase air-based pollution, such as cement dust entering the tower at air inlet levels or via the forced-draught fan. Extra cleaning of the tower pack and distribution system should be undertaken under these circumstances and close checks kept on the efficiency of the overall system.

Changes of process or modification to the product can introduce new design parameters, which can affect the tower. Overloading the tower in both a thermal and hydraulic sense may be acceptable as a temporary measure, but 'temporary' is the critical word. Too long exposure to an increased temperature can affect plastic packings, unless they have been designed to withstand such an increase. (Remember to discuss future predicted problems with the cooling tower designer before installation.)

Excessive water loads can lead to malfunction of the distribution system. Higher loadings in one area can lead to pack collapse and rapid fall-off in performance.

In the majority of cases common sense, combined with

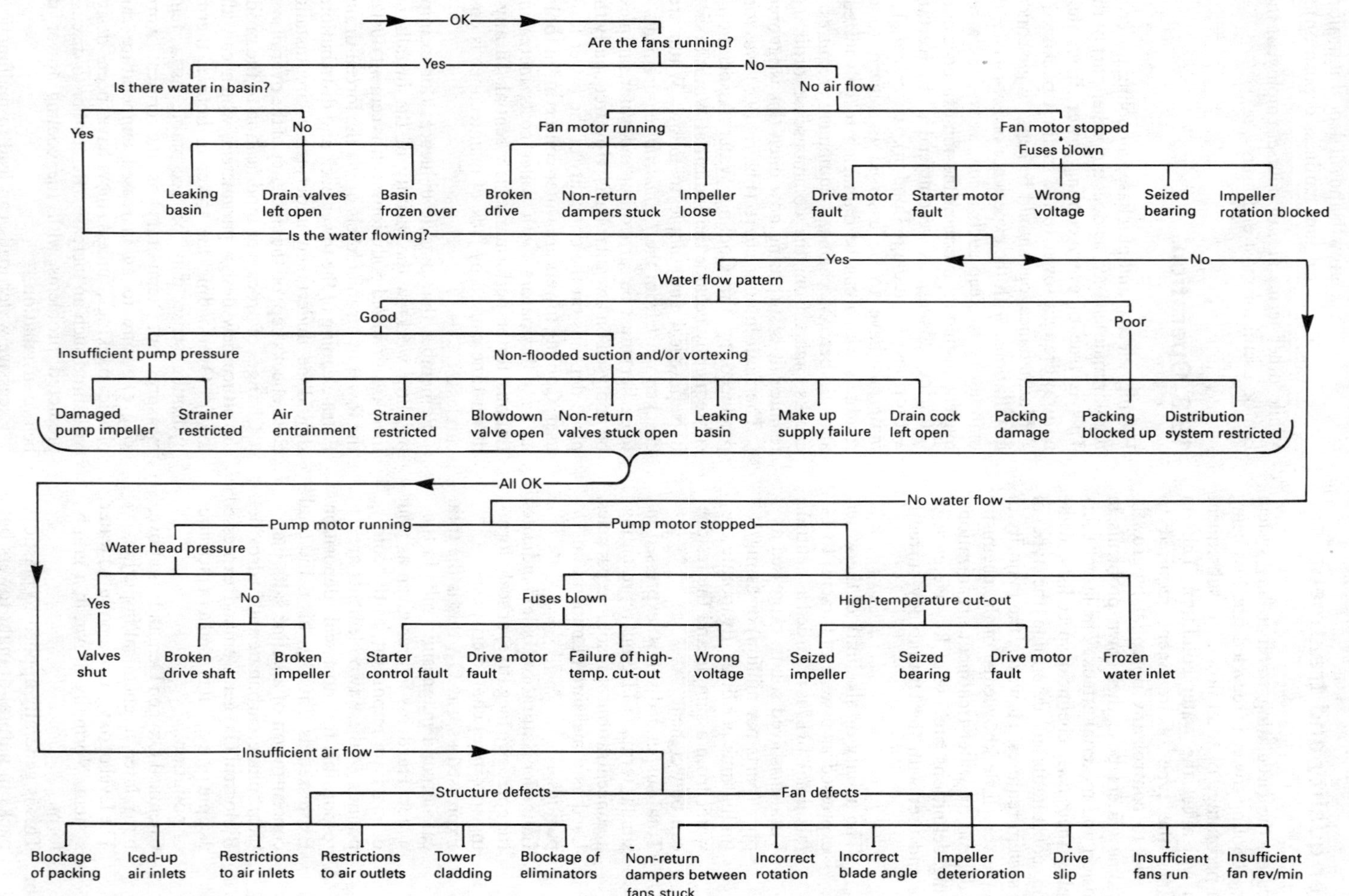

Figure 19.5 Fault-finding chart

basic engineering principles, should be sufficient to ensure good service from the tower on a continuous basis. The reputable supplier will always be ready to help and advise. If the advice is sought in time, many of the problems associated with the changes mentioned need never arise.

19.9 Modifications – retro-fits

Changes in circumstances will frequently require modifications to an existing cooling tower. Additional heat loads may be needed, and changes in the end process may cause the return temperature to increase, necessitating a new thermal load on the tower. Most frequently it is a requirement for an increase in hydraulics.

19.10 Consultation

All too often, the supplier is not consulted at the planning stage, with the inevitable result of a tower failure, not only in performance but also in pack collapse, structural failures and total shutdown. The original tower was supplied against a design water and thermal load, and changes to those parameters will affect the performance. It is possible to change existing towers, and not only is it possible, it is well-established practice. The successful amendments are those where the tower designer has been advised (in advance) of the proposed changes. He knows the limitations of his product and can advise on what can or cannot be done.

There is one area where improvements can be achieved, namely the increase in thermal performance by changing to a more advanced design of packing. Care has to be taken to ensure that the new pack configuration is compatible with the quality of water as well as its quantity. The water treatment conditions could change, and almost certainly the distribution system will need amendments. The benefits to be obtained can be listed as follows:

1. Improvement in thermal load.
2. Possible reduction in pumping head (e.g. the change from a splash pack to a high-density film-type pack can save power by installing the new pack at the bottom of the former splash area and lowering the pump inlet). If the correct design is used it may be possible to leave existing fans, thus incurring no additional power penalty. Even if fan changes are required, the economics have to be studied, but in most cases such changes will be beneficial.

Changes to a distribution system can be of assistance in minor improvements. Nozzle designs are under constant review but it is in the context of pack changes that amendments are important.

19.11 Environmental considerations

The principal areas where cooling design is affected by environmental requirements are visual and audible (i.e. aesthetic (plume) and noise).

19.11.1 Aesthetic

Planning regulations can be rigid in their attitude to the visual impact that a cooling tower may have on the surrounding area. This is not, however, confined to air-conditioning installations. Certain industrial areas, within the UK and elsewhere, limit the number of towers, their height and configuration in relation to the existing background situation. While this is understandable in environmental amenity terms, it leads to major design difficulties on the part of the supplier.

Consultations with all interested parties is essential at the planning stage. Solutions can normally be found, but the cooling tower designer must be included in these discussions. Towers can always be designed to meet the planning regulators and the demands of architects. Low towers, tall narrow towers, circular towers, multi-faceted towers, towers built into the buildings, even towers installed under ground, are all practical examples of modern installations (Figures 19.6–19.8). Obviously, the design characteristics have to be reassessed. Hence the emphasis on cooling tower designers' involvement from the outset.

As cooling towers operate on the evaporative principle, at certain temperature conditions the discharged heat vapour will appear as a plume. The amount of the pluming can be accurately assessed against the temperature conditions (both inside and outside the tower), the volume of air and the velocity of the discharge. The extent to which this can be classified as a 'nuisance' depends

Figure 19.6 Cooling plant, including towers, installed as a separate unit on a supermarket satisfies the architect, town planners and the end-user

Figure 19.7 Design requested by architect – low plume, no splash (note catchment tray), clear of office block

Figure 19.8 Cranes can be eliminated!

entirely on the location of the tower and its proximity to sensitive areas (i.e. housing, office blocks, etc.). One variation is the natural-draught tower, where geographical location may cause the plume to affect ambient conditions downwind, such as moisture deposition on roads (Figure 19.9) or, in one instance in Switzerland, where the plume from a large natural-draught tower located in a narrow valley restricted sunlight on the farming area downwind. (The predicted 7% restriction was accurate.) The drift can normally be confined within more than acceptable limits by the use of efficient drift eliminators.

The height of the discharge is important, and in air-conditioning projects it can be critical in relation to surrounding buildings. Consideration has to be given to the volume of discharge, the possibility of entrainment of water-treatment chemicals in the drift (these can cause 'etching' on glass windows), the siting of the tower in relation to the fresh air inlets to the building and the visual impact, in architectural and aesthetic terms.

Eliminator design can vary from the non-existent to an efficiency which can limit drift to 0.00005% of flow. In other words, drift can be almost undetectable, but the difference is obviously in the economics involved.

Pressure groups in the environmentally aware political parties can make excessive demands. Often these can be met by good public relations meetings, but the more reputable cooling tower suppliers can be invaluable in dealing with such matters.

Plume abatement is possible, even to the point of 'invisible plumes'. While this is not a problem in hot countries, the temperate zones (for example, Europe and the USA), will always have seasons when the plume is normally visible. To change this situation to a non-plume effect is again possible but it is expensive. As an example, one installation in the centre of a North American city, where non-pluming throughout the year was a mandatory clause in the planning permission, resulted in the tower cost being increased by a factor of 3! (A not-insignificant amount in the overall cost of the project.) Plume control can be achieved at the expense of larger installations and possible changes in the temperature levels, all of which require prior consultation with the designer.

The other source of possible drift or precipitation from cooling towers is caused by windage or blow-out from the air inlets, noticeably under strong or gusty conditions. This can occur in both mechanical- and natural-draught cooling towers, but the effect is normally localized to the immediate tower area. In the natural-draught cooling tower the extent of the problem varies according to the design of the packing.

Figure 19.9 Congested site. Restrictions on air inlet, increased pressure and air velocity results in drift deposition on road

In the case of pure counterflow packing, where the entire packing is positioned above the air inflet, the large air inlet opening can give rise to discharge of water by air entering the air inlet on the windward side, then being sucked out in the vortex depression, which generally occurs at approximately 90° to the wind direction. The resulting spray is carried as a fairly narrow band for a distance of perhaps 20 to 30 m downwind of the tower. In a mixed-flow packing, where the cooling tower packing extends down into the air inlet, there is considerable resistance to the free passage of air through the air inlet. Therefore the depressed area of the vortex is reduced and the resulting spray tendency is likely to be restricted to a fine spray, literally being blown off the peripheral pack laths. Various techniques have been adopted to minimize this effect, such as external radial baffles, internal baffles and louvres. All of these will, of course, incur additional cost and some of them may also increase the pressure drop through the air inlet and thus affect the thermal performance. Additionally, they can increase the overall dimensions of the installation.

Mechanical-draught cooling towers are normally supplied with either central baffles or inlet louvres. This depends on the tower dimensions. On these towers the wind or spray blow-out is generally confined to relatively small single-cell units where an inlet may be provided on all four faces. In this case the major remedy is to provide internal diagonal baffles to prevent crossflow of air through the air inlets.

On larger multi-celled mechanical-draught towers of both counterflow and crossflow variety, the air inlets are confined to the two opposing faces and windage or drift loss is unlikely to occur, except under exceptionally high wind conditions. Here again, remedial work, depending upon the location, can be applied but at additional cost (see Figure 19.10).

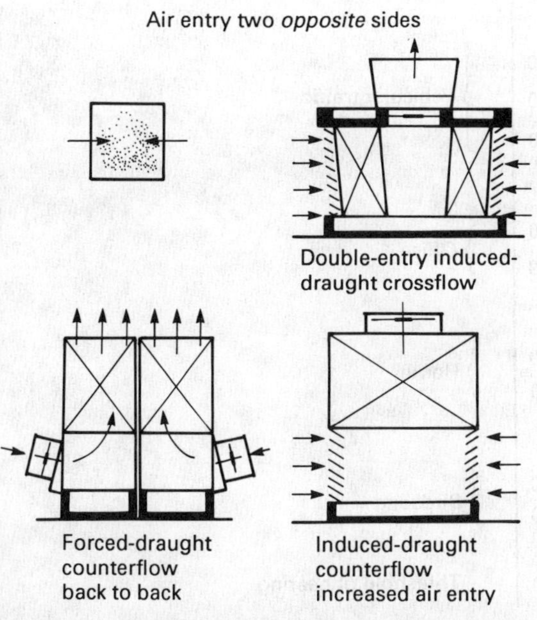

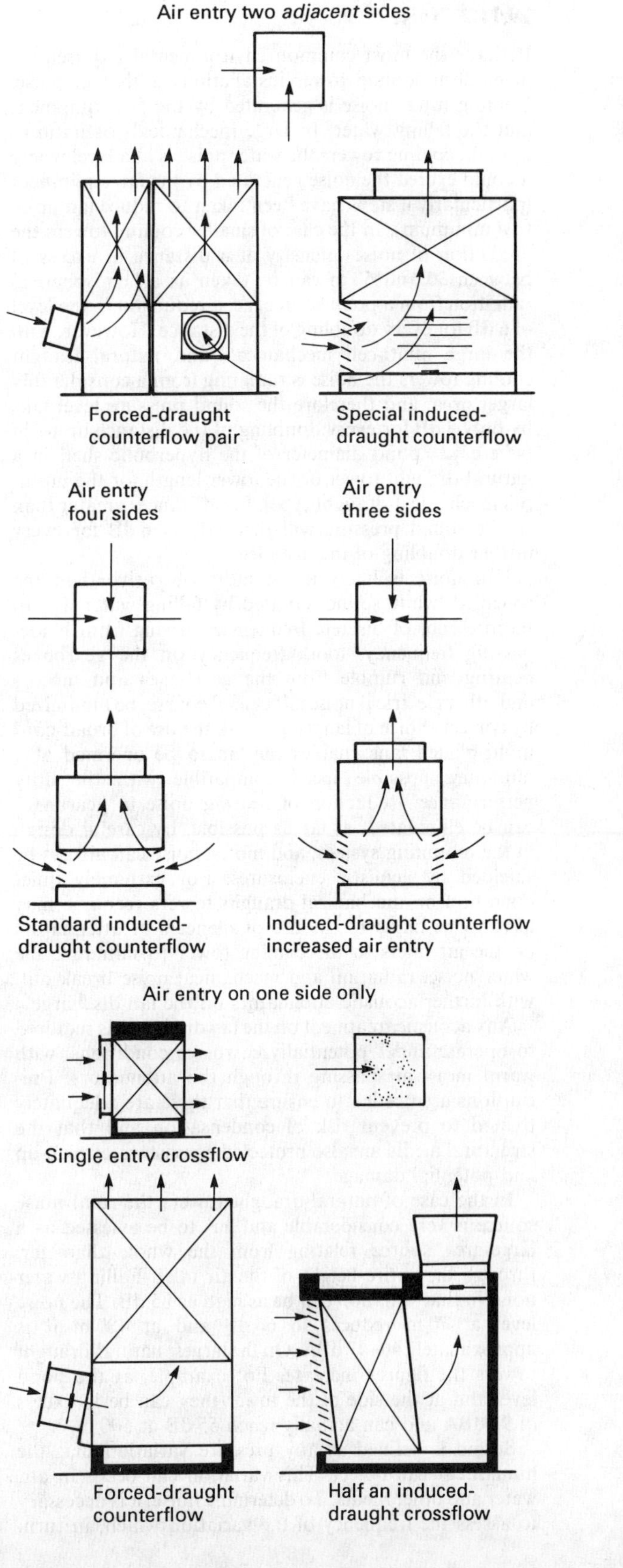

Figure 19.10 Configurations of cooling tower air flows

19.11.2 Noise

Perhaps the most common environmental requirement in modern cooling tower installations is that of noise. Cooling tower noise is generated by the fan equipment and the falling water. In large mechanical- or natural-draught cooling towers the water noise is at a level where it could exceed the noise generated by the fan equipment (particularly if steps have been taken to reduce fan noise to a minimum). In the case of smaller cooling towers the prediction of noise intensity at a distance in excess of between 30 and 50 m can be taken as a hemispherical radiation from a point source, i.e. a reduction in the level of 6 dB for every doubling of the distance. However, with the large multi-cell mechanical- and natural-draught cooling towers the noise is radiating from a considerably larger area, and therefore the sound pressure level falls by only 3 dB for every doubling of the distance up to $\frac{1}{2}d$ (where $\frac{1}{2}d$ = pond diameter of the hyperbolic shell in a natural-draught tower or the tower length for the multi-cell mechanical-draught type). For distances greater than $\frac{1}{2}d$ the sound pressure will then fall by 6 dB for every further doubling of the distance.

Fan noise is likely to be more obtrusive than the so-called 'white sound' emitted by falling water due to the presence of discrete frequencies arising from blade-passing frequency, tooth frequency on the gearboxes bearing and rumble from the gearboxes and motors and other electrical noises. It can, of course, be minimized by correct choice of fan. In general, the use of broad-cord multi-bladed fans enables the fan to be operated at a minimum possible speed compatible with the duty performance. Reduction of bearing noise in gearboxes can be eliminated, as far as possible, by careful design in the mounting system, and motor noise can always be shielded by acoustic enclosures. For extremely quiet operation on mechanical-draught towers recourse may have to be made to the use of silencers or attenuators on the air inlets to the cooling tower to minimize the water noise, radiation and mechanical noise break-out, with further acoustic attenuators on the fan discharge.

Any acoustic treatment on the fan discharge is required to operate under potentially corrosive conditions, with warm moist air passing through the attenuators. Precautions are needed to ensure that these are adequately treated to prevent risk of condensation and that the structural media are also protected against water pick-up and potential damage.

In the case of natural-draught towers the total noise source is very considerable and has to be assessed as a large-area source relating from the whole diameter, through the entire height of the air inlet. Falling water noise in that situation can be as high as 85 dB. The noise level at 70 m reduces to 66 dB and at 480 m it is approximately 46–47 dBA. On the largest natural-draught towers the figures increase. For example, at the pond level and at the side of the tower they can be in excess of 91 dBA and can actually reach 55 dB at 500 m.

Sound is defined as any pressure variation that the human ear can detect. This variation can occur in air, water and other media. To determine noise, it is necessary to assess the frequency of the variation which, in turn, can be related to the speed. For most applications the speed of sound is expressed at 340 m/s. Speed and frequency give the wavelength, i.e. the physical distance in air from one wave to the next. For example, at 20 Hz this gives 17 m while at 20 KHz one wavelength is 1.7 cm.

For convenience, the usual measurement of sound is expressed in decibels (dB), and ratings go from 'threshold of hearing' to 'threshold of pain' (135 dB). Figure 19.11 illustrates the common noise criteria, which can be expressed in sound-pressure levels (SPL). The human ear can detect 1 dB but 6 dB represents a doubling of the SPL, although it would need a 10 dB increase to make it 'sound' twice as loud.

In assessing the noise emanating from a cooling tower it is necessary to measure the main points of emission – the fan, the motor, the gearbox and the falling water. As noise bounces and can be absorbed by certain materials, it is usual, where noise restrictions apply, to map the area and calculate the SPL at a large number of points, taking into consideration interferences, bounce and absorption. The number of points measured is reflected in the accuracy of the resultant topograph.

Remember also to take background noise into your

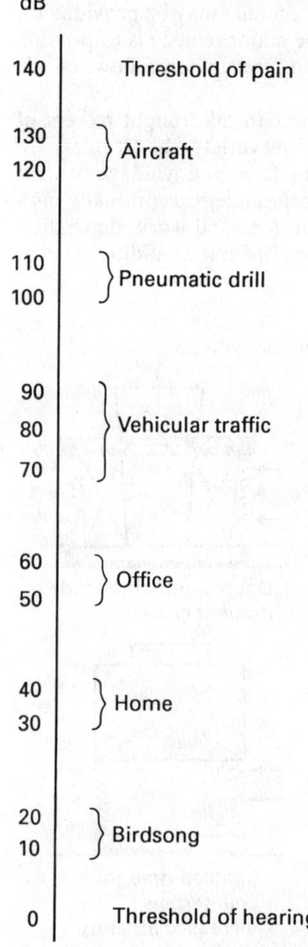

Figure 19.11 Threshold ratings for sound

calculation. Too frequently, specifications are made which ignore this, with the result that equipment is applied to a more rigid design than is absolutely necessary.

The siting, as well as the selection of type of tower, can be critical. Rotating the tower, shielding the motor, use of baffles can all help in meeting environmental noise requirements. If in doubt, consult your cooling tower designer.

19.12 Problem areas

19.12.1 Installation

Cooling towers have been called the Cinderella of the plant scene – usually unnoticed (if not even unseen), forgotten and sadly ignored. While cooling tower designers may have other ideas, they generally recognize this as being true in many cases. Designers therefore try to achieve the impossible, i.e. to build a piece of mechanical equipment that can be left alone and perform its function without fuss and attention.

While designers can claim some degree of success, there are many occasions when their products are misinstalled.

Don'ts

Mix products – placing a forced-draught tower beside an induced-draught one causes problems for both designs (Figure 19.12).
Place access panels incorrectly – the access panel is for the user's benefit. Ensure that it is accessible Figure 19.13).

Figure 19.12 Mixing forced draught with induced. The overloaded forced-draught tower with excess plume results in elevated wet bulb at air inlets on new tower. The problem is resolved by removing the forced draught and adding one more cell to the induced draught

Figure 19.13 Cyclic heat load using small tower and large water storage. This is a good idea and sound engineering, but the access panel is on the wrong side (30 ft drop to ground level)

Figure 19.14 An attempt to conceal tower and reduce noise, resulting in starvation of air and failure in performance

Starve the air inlets – insufficient air results in poor performance (Figure 19.14).
Ignore the bleed – inadequate bleed means concentration of salts, change of pH and pack fouling.
Forget about make-up – water starvation means poor performance, vortexing, pump and motor failure.
Fail to check on water treatment – inadequate treatment and haphazard slug dosing can lead to poor performance,

Figure 19.15 Ice on natural-draught tower (no de-icing ring fitted)

Figure 19.16 The effects of corrosion. Structural failure is visible (holes in fan casing)

damage to associated equipment and failure to meet discharge requirements.
Forget to install safety cut-outs, for overload and ambient changes – ice damage can be disastrous (Figure 19.15).
Allow corrosion to develop – metal failure can be costly and, at times, dangerous (Figure 19.16).

Do's

Ensure maintenance checks are carried out.
Check on power consumption, water usage, water costs.
Carry out monthly inspection – inside and out, where possible.
Check water analysis; the frequency depends on individual cases but it should be no less than quarterly.
Check mechanical equipment (i.e. fans, motors, drives). (Remember to check belt tensions where applicable.)
Check for vibration, both mechanically and structurally.
Ensure that access panels are used and replaced correctly.
Make certain that repairs, when necessary, are carried out efficiently and quickly.
If it in doubt, call the cooling tower designer.

19.13 Summary

While the majority of cooling tower installations work efficiently, the normal requirements of maintenance and good efficiency practice have still to be applied. This may not always be the case. Time for maintenance is limited, and plant engineers have other pressing problems or, as is well known, forget about the towers! With the increasing economic and environmental pressures on the use of water, this situation must change.

It may be appropriate to quote the old engineering term of 'KISS' (Keep It Simple, Stupid!) and, recognizing that the cooling tower designer does his utmost to comply, it is the responsibility of the operator to 'co-operate'!

Regular checks and an efficient logging system will ensure that the cooling tower, correctly planned, efficiently installed and adequately maintained, will give valuable service for many years. The economic returns justify a little more thought and attention than has been given to the subject in the past.

Appendix 19.1 Theoretical calculations

$$L\,dT = Ka\,dV(h_L - h_G)$$

where K = the coefficient of heat transfer for the packing in question,
a = the effective transfer surface area per unit pack volume,
V = depth of packing,
h_L = enthalpy of boundary air layer in contact with and at the same temperature as the water, and
h_G = enthalpy of bulk air passing through the packing.

Integrating this for the full depth of packing, the expression becomes:

$L \cdot \Delta T = KaV \cdot \Delta h_m$

where Δh_m is the mean enthalpy difference, otherwise known as the *Mean Driving Force* (*MDF*) (see Appendix 19.2).

This can be rearranged as:

$$\frac{KaV}{L} = \frac{\Delta T}{MDF}$$

which is the form in which it usually appears. KaV/L is commonly called the 'tower characteristic'.

Now let us refer to the right-hand side of the above expression. The mean driving force varies with the specified design temperatures and also the ratio of water/air loading (L/G). If we take a low air flow, the air soon rises in temperature and tends to reach equilibrium conditions with the boundary layer. Thus the driving force is reduced. On the other hand, excess air is unnecessary. Therefore we must adjust the air flow that supply just meets demand. A plot of L/G versus $\Delta T/MDF$ is shown in Figure 19.17. This is known as a demand curve.

The left-hand side of the above expression is a measure of the quality and quantity of the packing being used, and has been shown empirically to obey the law $(KaV/L) = c(L/G)^{-n}$ for counterflow applications only.

Cooling in the crossflow mode requires an incremental 'trial and error' technique, best suited to computer analysis. The tower characteristic KaV/L can then be plotted against varying L/G ratios, and this gives a measure of the ability of the packing to effect the transfer (Figure 19.18).

We have already equated KaV/L with $\Delta T/MDF$, therefore we can superimpose the 'supply' curve over the 'demand', the intersect being the optimum L/G ratio for the packing being considered for the duty (Figure 19.19).

It is interesting now to examine the effect of using greater or lesser depths of packing, and to consider their suitability for duties of different degrees of difficulty. In Figure 19.20 we can see the effect of using three different pack depths on a moderately easy duty. By changing from pack depth (1) to pack depth (2) we are able to use a much higher L/G ratio (which, in turn, means less air and/or a smaller tower). The increment from depth (2) to depth (3) gives a less significant increase in L/G, and therefore suggests that the optimum depth has perhaps been exceeded.

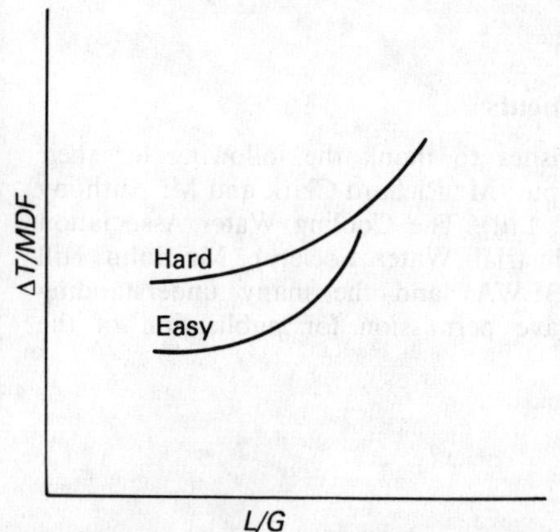

Figure 19.17

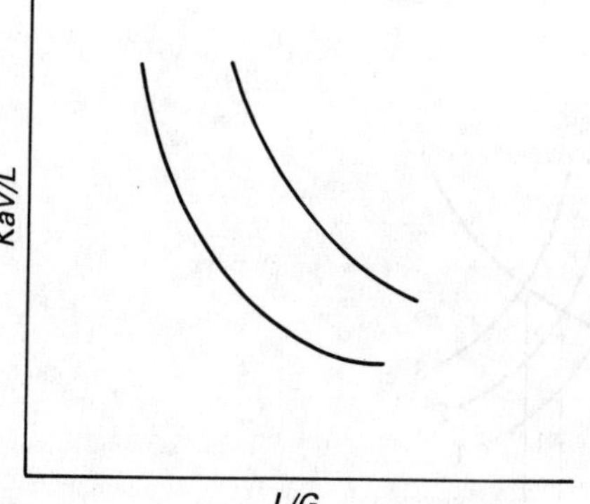

Figure 19.18

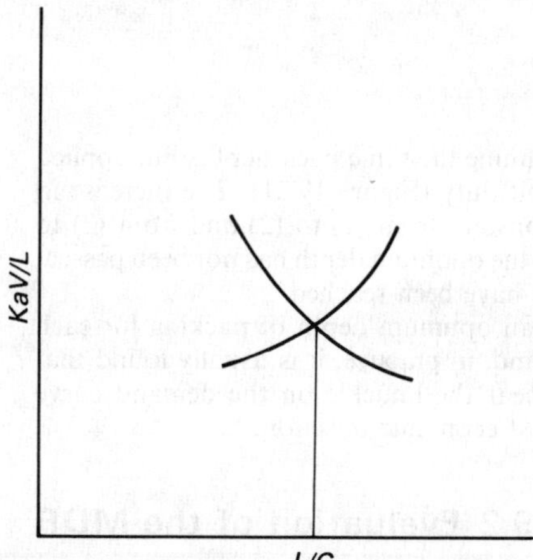

Figure 19.19

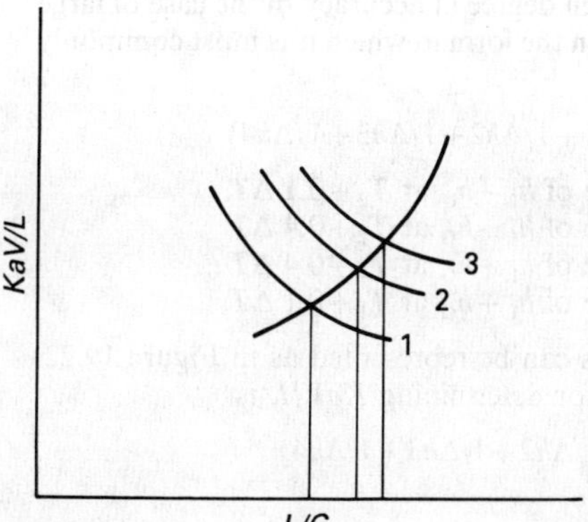

Figure 19.20

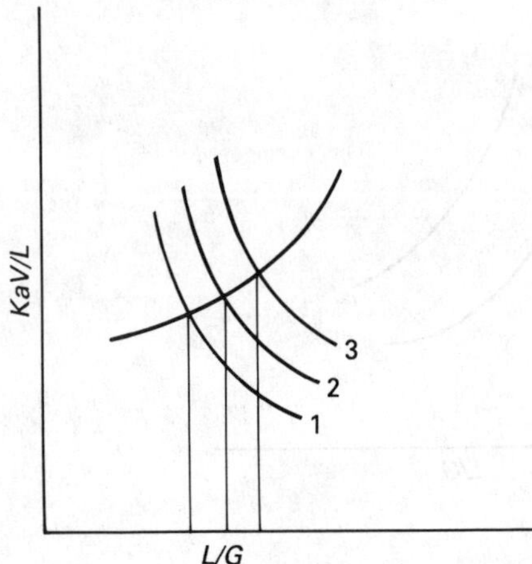

Figure 19.21

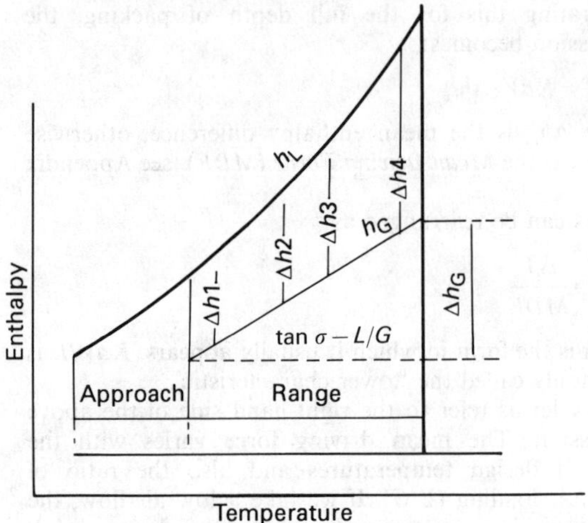

Figure 19.22

Now let us examine the same pack depths but applied to a more difficult duty (Figure 19.21). The increase in L/G is almost constant from (1) to (2) and from (2) to (3), showing that the optimum depth has not been passed, and may not yet have been reached.

Thus there is an optimum depth of packing for each individual duty and, in practice, it is usually found that an intersection near the knuckle on the demand curve produces the most economic selection.

Appendix 19.2 Evaluation of the MDF

Several methods for the evaluation of the MDF have been put forward, notably that processed by Tchebycheff, which gives a high degree of accuracy in the case of large cooling ranges. In the form in which it is most commonly used, it reads:

$$MDF = 4/(1/\Delta h1 + 1/\Delta h2 + 1/\Delta h3 + 1/\Delta h4)$$

where $h1$ = value of $h_L - h_G$ at $T_2 + 0.1\ \Delta T$,
$h2$ = value of $h_L + h_G$ at $T_2 + 0.4\ \Delta T$
$h3$ = value of $h_L + h_G$ at $T_1 + 0.4\ \Delta T$
$h4$ = value of $h_L + h_G$ at $T_1 + 0.1\ \Delta T$

Graphically, this can be represented as in Figure 19.22. The expression for determining KaV/L is

$$(\Delta I/4)(1/\Delta h1 + 1/\Delta h2 + 1/\Delta h3 + 1/\Delta h4).$$

Appendix 19.3 Technical requirements

Location
Meteorological data
Wind rose
Water flow rate
Temperature to tower (T_1)
Temperature from tower (T_2)
Cooling range
Approach
Design wet bulb
Design dry bulb
Water analysis – circulating
make-up
Cycles of concentration
Blowdown/purge rate
Power cost analysis
Drift loss requirement
Local authority requirements
Discharge qualities
Structural specifications
Pack specifications
Mechanical specifications (if applicable)
Noise specification
Impedance by adjoining structures (if applicable)

Acknowledgements

The author wishes to thank the following for their considerable input: Mr Richard Clark and Mr Anthony Kunesch (FCT Ltd), The Cooling Water Association (now The Industrial Water Society), Mr John Hill (Director of BEWA) and the many understanding people who gave permission for publication of the photographs.

20 Electricity Generation

I G Crow BEng, PhD, CEng, FIMechE, FIMarE, MemASME
Davy McKee (Stockton) Ltd
and
K Shippen BSc, CEng, MIMechE
ABB Power Ltd

Contents

20.1 Introduction

Electricity is one of the key energy sources for industry and commerce. It is normally provided by the electricity supply authorities, being generated from very large fossil-fuelled or nuclear central power stations and distributed throughout the country via high-voltage transmission systems. It can be fair to say that the electricity supply authorities play a very important part in meeting the demands of the consumer efficiently, reliably and at an economic cost. However, there will always be a role for the private generation of electrical power, either to meet the needs for security of supply or to provide the electricity more economically. For the latter this is particularly the case for the concept of combined heat and power (CHP), where the local use of a heat demand can make private generation of electricity more competitive when compared with purchase from the local grid.

This chapter aims to convey the basic technical principles involved in electricity generation for industrial and commercial applications, with supporting technical data giving examples of the performance and efficiency of various schemes. A general guide is provided on the factors which have a major bearing on choice of an electricity-generating scheme with further details of the plant, its layout and descriptions of actual installations.

20.2 Generation of electrical power

Diesels, gas turbines and steam turbines are the more commonly used prime movers for the generation of electrical power. Additionally, the steam turbine can be employed in combination with either the diesel or gas turbine for combined cycle operation. The following describes the basic operation of each of these prime movers in relation to its associated power-generating scheme and reviews the more significant factors affecting performance and efficiency. Further information on the actual plant and installation is given later in Section 20.6.

20.2.1 Diesels

Diesels are used in many industrial applications (for example, for base-load generation in mines, cement plants and in remote regions of the world). In addition, they are often utilized to provide standby power for hospitals, telecommunications, banks, computer centres and office complexes which must have full independent power capability. The diesel can be started rapidly, making it ideal for peak lopping duties to meet maximum load demands, or for emergency use in cases of power supply interruptions.

Manufacturers offer a very wide range of diesel engines, all of which fall into categories depending upon crankshaft speed. The three categories are as follows:

Low speed: below 400 rev/min
Medium speed: 400–1000 rev/min
High speed: above 1000 rev/min

The speed of a diesel engine basically dictates its cost in relation to output. The choice is an economic consideration, taking into account the application of the engine and needs for reliability and security of supply.

The diesel has a good efficiency of approximately 40% with a flat fuel consumption curve over its operating range. With a competitive capital cost this gives the diesel a distinct advantage over its competitors.

The compression ignition cycle

The diesel is an internal combustion reciprocating engine which operates in the compression ignition cycle. The familiar reciprocating mechanism consists of a number of pistons, each running in a gas-tight cylinder with connecting rod and crankshaft. The connecting rods are set at angular positions so that they contribute their power stroke in a regular sequence.

In the compression ignition cycle the air is compressed and the fuel is injected into the compressed air at a temperature sufficiently high to spontaneously ignite the fuel. The heat released is converted to mechanical work by expansion within each cylinder and, by means of the reciprocating motion of the piston, is converted to rotary motion at the crankshaft.

There are two basic cycles, the two-stroke and four-stroke, each of which is illustrated in Figures 20.1 and 20.2 with their appropriate indicator diagrams. The two-stroke engine is mechanically simplified by the elimination of the mechanically operated valves. For the same rotational speed the two-stroke engine has twice the number of working strokes. This does not, however, give twice the power. In the down-stroke of the two-stroke cycle both the inlet and exhaust ports are cleared, which allows some mixing of fuel air charge and exhaust gases, resulting in less thrust. For similar reasons, this also gives the two-stroke engine a slightly higher fuel consumption.

The overall efficiency, η_b of a reciprocating engine is the product of the thermal efficiency, η_i and its mechanical efficiency η_m. Thus:

$$\eta_b = \eta_i \times \eta_m$$

An engine of good mechanical design has a mechanical efficiency of around 80–90% at full load. The overall efficiency of a compression ignition engine depends on the type, but ranges from 30% to 40%.

One of the main reasons for the good performance of the diesel, compared with alternative machines, is due to the fact that the design is not restricted by metallurgical considerations which, for instance, limits the higher gas temperature in the gas turbine. This is because the cylinder wall is only subject to intermittent peak temperature due to combustion and its average temperature is much lower than the mean gas temperature. Therefore, the cyclic temperature can be maximized.

Turbocharging

Supercharging the inlet air by use of exhaust gas turbocharging raises the volumetric efficiency above that

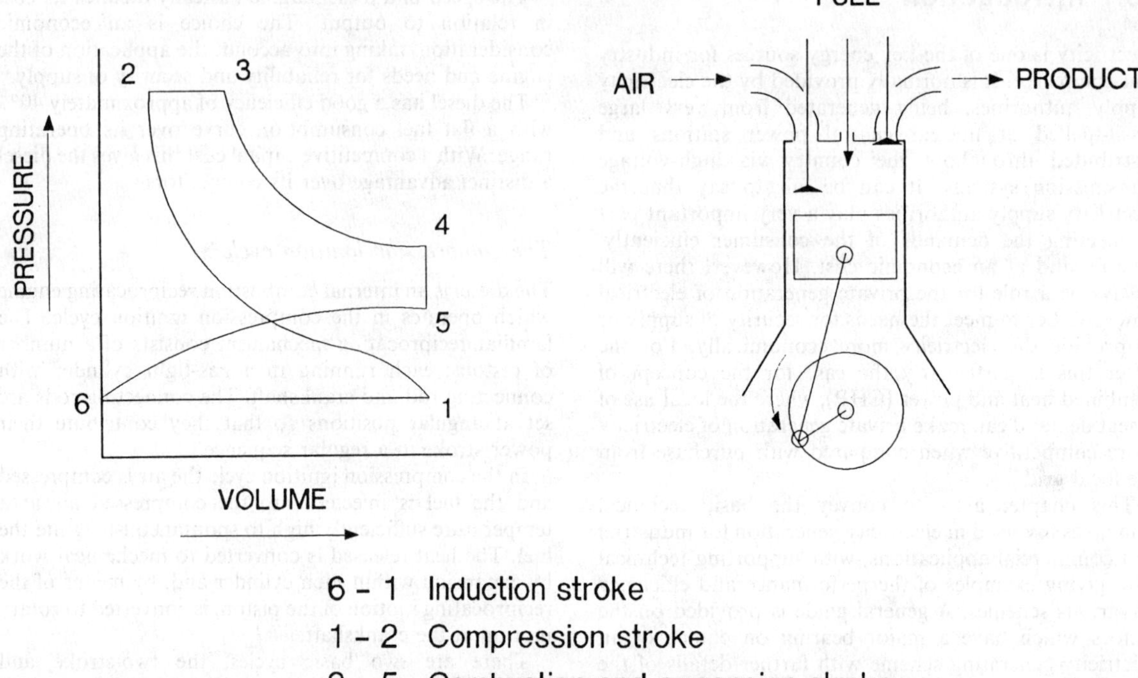

Figure 20.1 Four-stroke cycle for an internal combustion engine

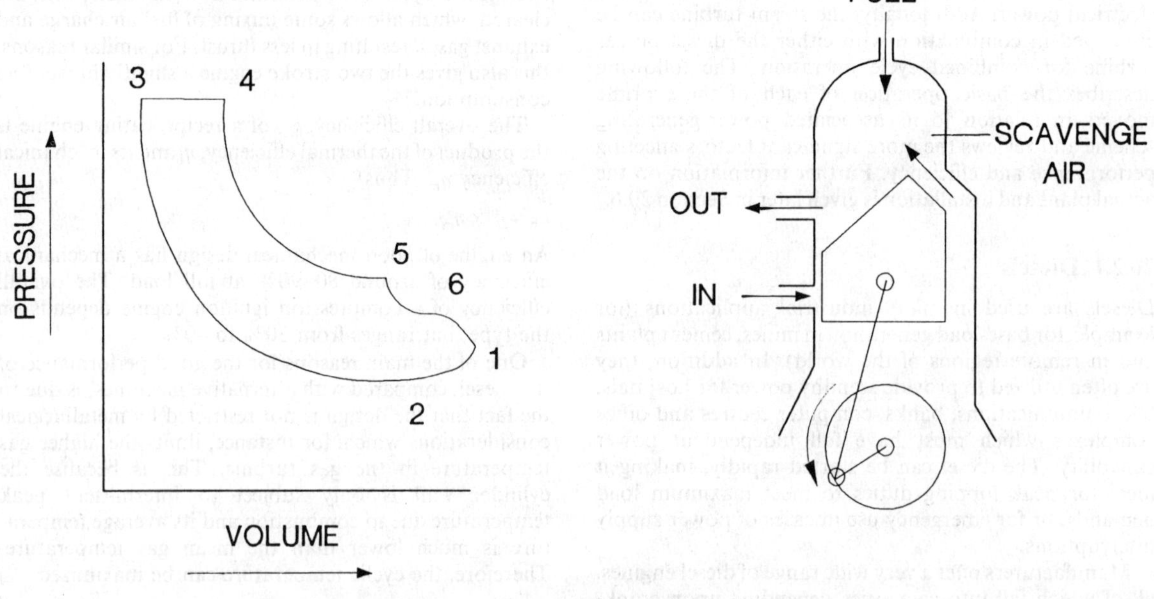

Figure 20.2 Two-stroke cycle for an internal combustion engine

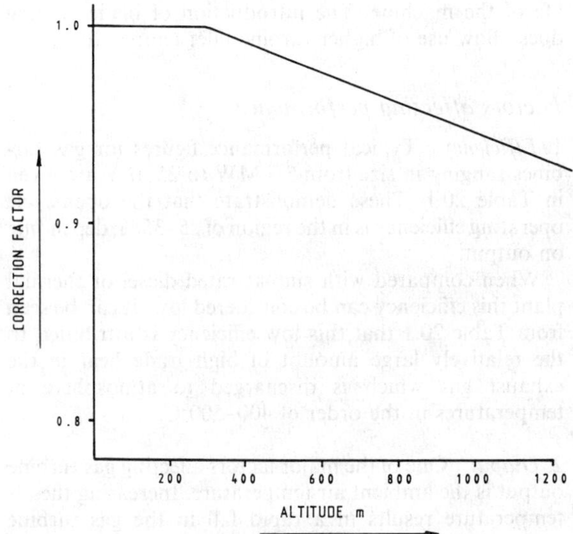

Figure 20.3 Correction to diesel power output for changes in ambient conditions

which can be obtained from normal aspiration. This gives increased output without change to the speed or capacity of the machine. The full potential of the increase in air density is affected by the use of intercooling placed downstream of the turbocharger. Naturally, the use of turbocharging also improves performance at high altitudes.

Site conditions

The site conditions can have significant effect on the output of the diesel engine. For example, British Standards 5514: 1987 Part I: sets out the de-rating method for internal combustion engines and establishes a datum level from which engine manufacturers relate their de-rating factors. Most engine manufacturers will give certain percentage reduction in power output for a certain increase in temperature and altitude above datum level.

One of the most important factors is the intercooler cooling water which has a profound effect on air manifold temperature. Use of a closed-cycle air cooling system will increase the cooling water temperature with de-rating of output. In addition, the radiators can add significantly to the parasitic load. The altitude above sea level at which the engine will be located also affects the output due to lower air density. Figure 20.3 illustrates the effect upon power output for changes in ambient conditions.

Oil consumption

One of the disadvantages of the diesel engine is its high lubricating oil consumption which, typically for a 3.6 MW engine, will be 0.035 l/kW h. Added to this quantity must be the oil changes at routine service intervals.

20.2.2 Gas turbines

The role of the gas turbine is more familiar to many of us in the aircraft field. However, since Sir Frank Whittle invented the jet engine in the early pioneering years before the Second World War there has been rapid development in both output and efficiency of these machines, and today the gas turbine is a popular choice for electricity generation.

The gas turbine is widely used by electrical supply authorities for peak lopping and standby generation. Additionally, it is often employed for base-load operation when the fuel costs are low.

The major disadvantage of the gas turbine is its low efficiency when compared with other systems. However, the gas turbine is compact, is subject to short delivery and erection period, has a competitive capital cost, and has virtually no requirement for cooling water. It can be started rapidly on fully automatic control.

Open-cycle operation

The basic gas turbine generates in the open-cycle mode and Figure 20.4 illustrates this diagrammatically. Air is drawn into a compressor and after compression passes into a combustion chamber. At this stage energy is supplied to the combustion chamber by injecting fuel, and the resultant hot gases expand through the turbine. The turbine directly drives the compressor with remaining useful work which drives a generator for electricity generation.

Figure 20.5 shows the ideal open cycle for the gas turbine which is based on the Brayton Cycle. By assuming that the chemical energy released on combustion is equivalent to a transfer of heat at constant pressure to a working fluid of constant specific heat, this simplified approach allows the actual process to be compared with the ideal, and is represented in Figure 20.5 by a broken line. The processes for compression 1–2′ and expansion 3–4′ are irreversible adiabatic and differ, as shown from the ideal isentropic processes between the same pressures P_1 and P_2:

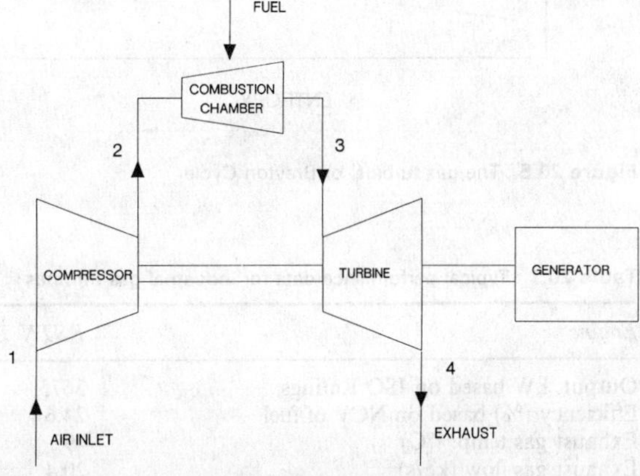

Figure 20.4 Basic gas turbine in open-cycle mode

Compressor work input $= C_p(T'_2 - T_1)$

Combustion heat input $= C_p(T_3 - T'_2)$

Turbine work output $= C_p(T_3 - T'_4)$

Therefore

Net work output $= C_p(T_3 - T'_4) - C_p(T'_2 - T_1)$

$$\text{Thermal efficiency} = \frac{\text{Net work out}}{\text{Heat supplied}}$$

$$= \frac{C_p(T_3 - T'_4) - C_p(T'_2 - T_1)}{C_p(T_3 - T'_2)}$$

By introducing isentropic efficiencies of the turbine n_t and compressor n_c the turbine work output is given as:

Net work output $= C_p[T_3(1 - 1/k)n_t - T_1(k-1)n_c]$

where

$$k = \left(\frac{P_2}{P_1}\right)^{(\gamma-1)/\gamma}$$

The efficiency of the gas turbine depends on the isentropic efficiencies of the compressor and turbine, but in practice these cannot be improved substantially, as this depends on blade design and manufacture. However, the major factor affecting both efficiency and output of a given machine is the turbine inlet (or combustor) temperature, and this is dictated by metallurgical considerations. The turbine blades are under high mechanical stress, and therefore the temperature must be kept to a minimum, taking into account the design life of the machine. The introduction of blade cooling does allow use of higher turbine inlet temperatures.

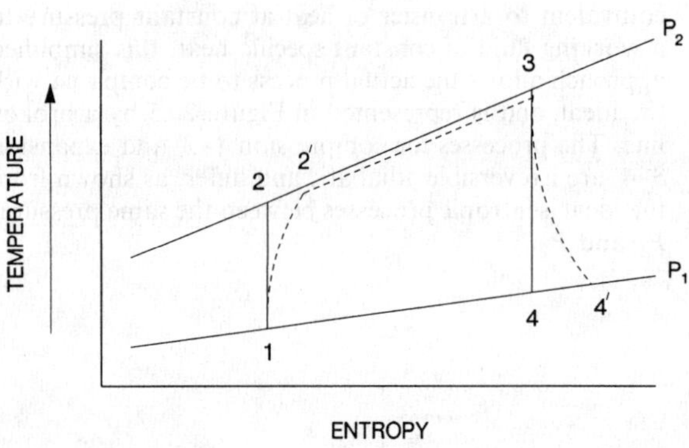

Figure 20.5 The gas turbine, or Brayton Cycle

Factors affecting performance

1. *Efficiency* Typical performance figures for gas turbines ranging in size from 3.7 MW to 25 MW are given in Table 20.1. These demonstrate that the open-cycle operating efficiency is in the region of 25–35%, depending on output.

When compared with similar rated diesel or thermal plant this efficiency can be considered low. It can be seen from Table 20.1 that this low efficiency is attributed to the relatively large amount of high-grade heat in the exhaust gas which is discharged to atmosphere at temperatures in the order of 400–500°C.

2. *Output* One of the major factors affecting gas turbine output is the ambient air temperature. Increasing the air temperature results in a rapid fall in the gas turbine output. Figure 20.6 shows a typical correction to power output curve for changes in ambient temperature.

Additionally, although to a lesser degree, output can be affected by inlet and outlet gas system pressure losses, which would be due to inlet filters and inlet and outlet

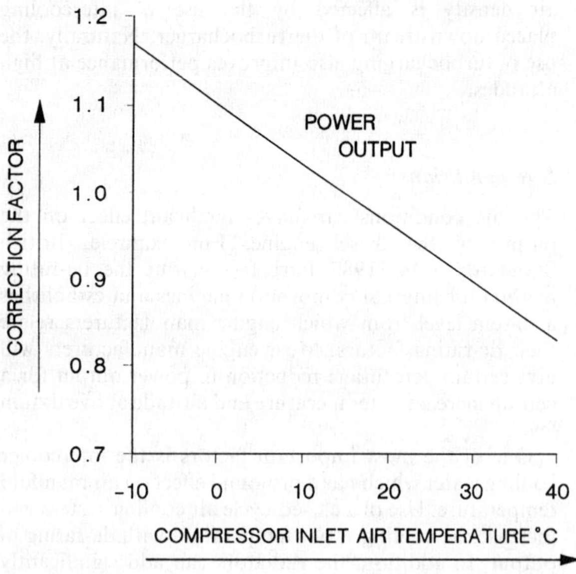

Figure 20.6 Correction to gas turbine power output for changes in ambient temperature

Table 20.1 Typical performance data for industrial gas turbines

Engine	*RSTN TB 5000*	*ABB STAL MARS*	*RB 211/COOPR*
Output, kW based on ISO Ratings	3675	10 000	24 925
Efficiency (%) based on NCV of fuel	24.6	32	34.0
Exhaust gas temp. (°C)	492	491	465
Exhaust gas flow (kg/s)	20.4	39.1	90.9

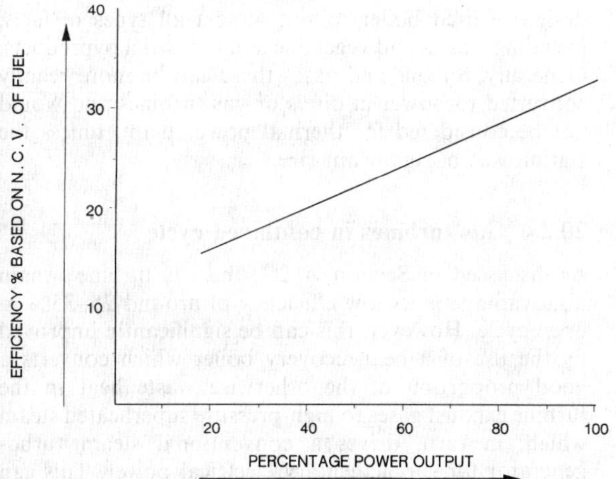

Figure 20.7 Approximate relationship showing variation of gas turbine efficiency for changes in load

ductwork and silencers. This fact must be borne in mind to ensure that in development of layout the ductwork design maintains the gas pressure losses within limits set down by the manufacturer.

3. *Part load* As the gas turbine operates with a fixed inlet air volume, its efficiency at part load deteriorates significantly. Figure 20.7 shows a typical relationship for variation in gas turbine efficiency for changes of load.

Fuels

Gas turbines are capable of burning a wide variety of fuels, both gaseous and liquid. Typical liquid fuels range from kerosene and diesel to light crudes. They are also capable of running on natural gas and industrial gases such as propane. Automatic change over facilities can be incorporated.

20.2.3 Thermal power plant

Thermal power plant is more commonly associated with very large central power stations. The capital cost for thermal power plant, in terms of cost per installed kilowatt of electrical generating capacity, rises sharply for outputs of less than some 20 MW, and for this reason the thermal power plant is not usually considered for industrial applications unless it is the combined cycle or combined heat and power modes. However, for cases where the fuel is of very low cost (for example, a waste product from a process such as wood waste), then the thermal power plant, depending on output, can offer an excellent choice, as its higher initial capital cost can be offset against lower running costs. This section introduces the thermal power cycle for electrical generation only.

Condensing steam cycle

The condensing steam cycle is shown diagrammatically in Figure 20.8. The fuel is fired in a boiler which converts the heat released from combustion to steam at high pressure and temperature. This steam is then expanded through a turbine for generation of electrical power.

Exhaust steam from the turbine is condensed in a heat exchanger and then returned via the feed pumps to the boiler. The associated Rankine Cycle is illustrated in Figure 20.9.

The efficiency of the cycle, ignoring work done by the boiler feed pumps, is given by change in enthalpy as follows:

$$\text{Efficiency} = \frac{(h_1 - h_2)}{(h_1 - h_3)}$$

and h_1 to h_2 represents the ideal isentropic expansion process in the turbine. However, in practice, because of irreversibilities, this is less than unity.

This efficiency can be improved by the use of a feed heating cycle whereby bled steam can be taken from the turbine after certain stages of expansion and then used to raise the feedwater temperature via use of feed heat exchangers. By such means the feedwater temperature, before entry to the boiler, is increased, thereby increasing the efficiency as the bled system is condensed by transfer

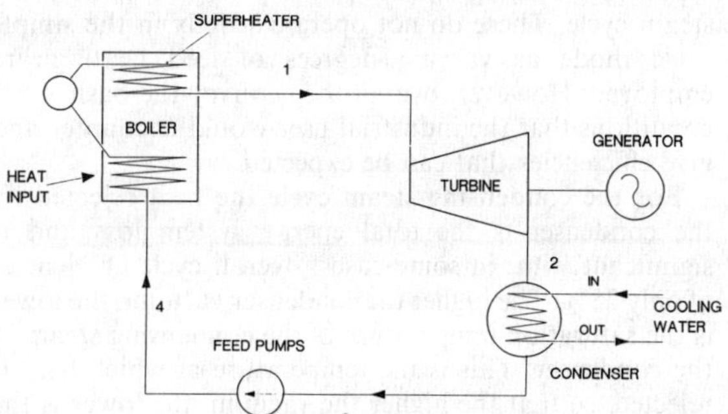

Figure 20.8 The condensing steam cycle

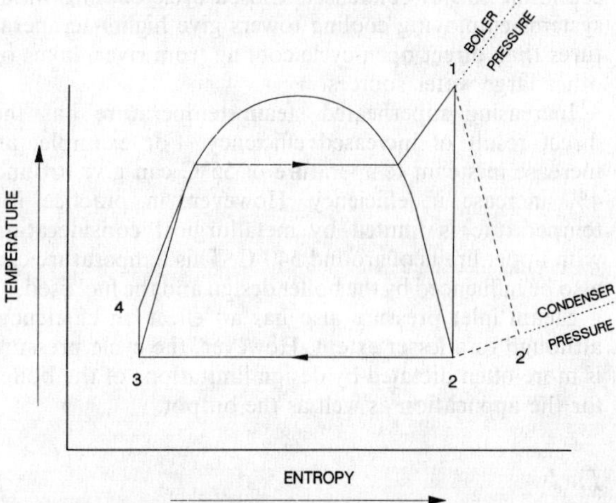

Figure 20.9 The Rankine Cycle with superheat

Table 20.2 Performance of industrial steam-generating plant

Steam conditions	17 bar/260°C	30 bar/450°C	62 bar/510°C
Generator output	2000 kW	5000 kW	10 000 kW
Overall efficiency	18.1%	21.8%	26.3%

of heat to the working fluid within the cycle, so reducing condenser heat rejection.

Large power stations use complex feed heating systems before the boiler feed pumps (LP) and after the boiler feed pumps (HP) whicn can give high overall thermal efficiencies of 39%. However, for the smaller machine it becomes uneconomic to consider multiple bleeds from the turbine, and the final choice is dictated by the extra cost for the additional complexity against lower running costs due to increased efficiency. As a minimum, a contact type de-aerator is often employed which would extract a small bleed of around 2–3 bar from the turbine.

Factors affecting performance

Table 20.2 gives performance data for typical industrial type schemes using thermal power plant in a condensing steam cycle. These do not operate strictly in the simple cycle mode as varying degrees of feed heating are employed. However, overall they convey the basic cycle conditions that the industrial user would encounter and give efficiencies that can be expected.

For the condensing steam cycle the heat rejected by the condenser is the total energy system loss and is significant with, in some cases, overall cycle efficiencies of only 25%. The higher the condenser vacuum, the lower is the saturation temperature of the condensing steam in the condenser. This is the temperature at which heat is rejected, so that the higher the vacuum, the lower is the temperature of heat rejection and consequently the greater is the cycle efficiency. The limit to the condensing temperature is set by the cooling water temperature and economic size of condenser. Closed-cycle cooling water systems employing cooling towers give higher temperatures than direct open-cycle cooling from river, lakes or other large water sources.

Increasing superheated steam temperature has the direct result of increased efficiency. For example, an increase in steam temperature of 55°C can give around 4% increase in efficiency. However, in practice the temperature is limited by metallurgical consideration with upper limits of around 540°C. This temperature can also be influenced by the boiler design and the fuel used.

Steam inlet pressure also has an effect on efficiency, although to a lesser extent. However, the cycle pressure is more often dictated by design limitations of the boiler for the application as well as the output.

Fuels

The thermal power plant uses a fired boiler for conversion of fuel to heat. It can be said that there is a design of fired boiler to suit almost all types of fuels, including wastes and vegetable and industrial byproducts. Generally, for oils and gases these can be more readily converted to power in disels or gas turbines and would not be considered for thermal power plant, unless the station was of significant size.

20.2.4 Gas turbines in combined cycle

As discussed in Section 20.2.2, the gas turbine's main disadvantage is its low efficiency of around 25–35% in open cycle. However, this can be significantly improved by the use of a heat-recovery boiler which converts a good proportion of the otherwise waste heat in the turbine exhaust gases to high-pressure superheated steam which, in turn, drives a conventional steam turbo-generator for supplementary electrical power. This can increase the overall efficiency to 50% for no further heat input as fuel.

The combination of the Brayton (gas turbine) and Rankine (steam) cycle is known as the combined cycle. A typical gas turbine in combined cycle is shown diagrammatically in Figure 20.10.

Types of combined cycle

1. *The 'fully fired' cycle* Gas turbine exhaust contains approximately 15–16% oxygen by weight, and this can be utilized to support the combustion of additional fuel, thereby enhancing the steam-raising potential and consequently the steam turbine output. The higher gas temperatures impose no limitations on either steam pressure or temperature, commensurate with current

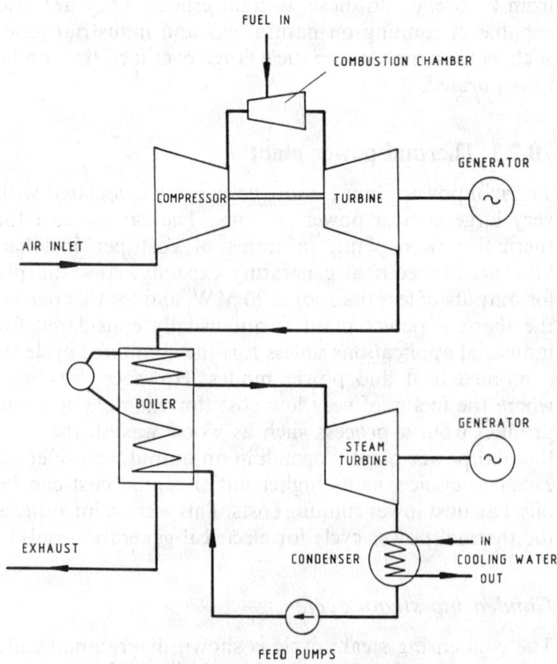

Figure 20.10 The gas turbine in combined cycle

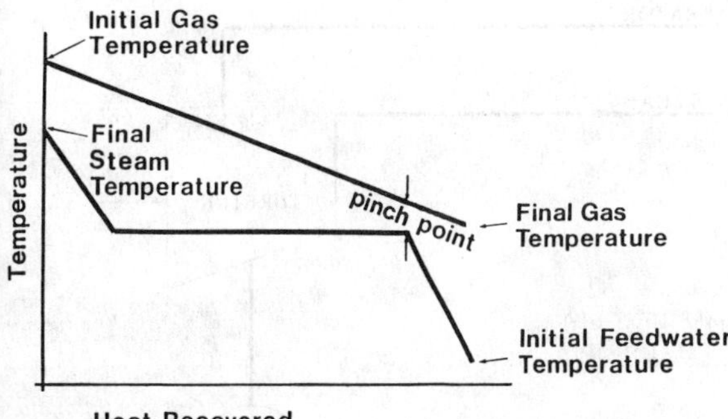

Figure 20.11 Heat-recovery diagram showing the pinch point

fired boiler/steam sturbine practice. The heat content of the exhaust gas can potentially be increased approximately fivefold. Consequently, the steam turbine output can be three to four times that of the gas turbine. Combustion efficiency is high due to the degree of preheat in the gas turbine exhaust, and the resultant cycle efficiency is much higher than that of conventional steam plant.

2. *The 'supplementary fired' cycle* The recovery of heat from gas turbine exhaust by the generation of steam has, as its basic limiting factor, the 'approach temperature' at the 'pinch point' (see Figure 20.11). The principal 'pinch point' occurs at the cold end of the evaporator and consequently the combination of the steam-generating pressure and the magnitude of the 'approach' determines the amount of steam which can actually be generated for a given gas turbine.

In single-pressure systems the only heat which can be recovered below the evaporator 'pinch point' is to the feedwater, and this, in turn, now determines the final stack temperature. If, however, a small amount of fuel is used to supplement the gas turbine exhaust heat, then while the 'pinch point' occurs at the same gas temperature level, the evaporation is increased and hence the mount of preheat required by the feedwater, resulting in a lower stack temperature. For this cycle the gas/steam power ratio is approximately 1–1.3:1.

3. *The 'unfired' cycle* This cycle is very similar to the 'supplementary fired' case except there is no added fuel heat input. The 'approach temperature' and 'pinch point' are even more critical, and tend to reduce steam pressures somewhat. Similarly, the gas turbine exhaust temperature imposes further limits on final steam temperature.

4. *General* The 'fully fired' cycle has been virtually superseded by the 'unfired' cycle, due to the former's lack of potential for increased efficiency, large cooling water requirement, larger physical size, increased capital cost and construction time. Also, because of the higher temperatures it is much less flexible in its operating capability.

The 'supplementary fired' cycle has many of the advantages of the 'unfired' cycle, and it can be argued that it is the most flexible in its operating capabilities. However, the controls are most complicated due to the afterburners and the need to maintain the steam conditions over the whole operating range. The 'unfired' cycle is the simplest in operational terms, the steam cycle simply following the gas turbine, and consequently is highly flexible in operation, having the ability to operate on sliding pressure at low loads. This cycle also has the greatest potential for increased efficiency, albeit for higher capital cost but without the need for extra fuel costs.

Single or dual pressure

There are two main pressure cycles applicable to combined cycle scheme: single or dual pressure:

1. *Single pressure* A combined cycle scheme with single pressure is shown in Figure 20.12. The waste heat recovery boiler is made up from preheater, economizer, evaporator and superheater sections giving single high-pressure steam output. The limitations imposed by the approach temperature at the pinch point, as illustrated earlier, determines the boiler heat recovery. Lowering of the boiler pressure increases the potential for heat recovery. However, the performance of the steam turbine will deteriorate at lower pressures, and therefore there is an optimum where maximum output can be achieved. The optimum pressure can be calculated, but is usually around 20–30 bara.

2. *Dual pressure* For comparison, a combined cycle scheme with dual pressure is shown in Figure 20.13. In this case the waste heat recovery boiler also incorporates a low-pressure steam generator, with evaporator and superheater. The LP steam is fed to the turbine at an intermediate stage.

As the LP steam boils at a lower temperature than the HP steam there exists two pinch points between the exhaust gas and the saturated steam temperatures. The addition of the LP circuit gives much higher combined cycle efficiencies with typically 20% more steam turbine output than the single pressure for the same gas turbine.

20.3 Combined heat and power (CHP)

The preceding section reviewed the application of popular prime movers for the generation of electrical power only. In the conversion of fuel energy to electricity it is shown that heat is rejected, either in the exhaust of a diesel or gas turbine or, alternatively, in the condenser of a thermal power plant. It can be seen that by applying these machines to provide both heat and electricity the total energy recovery can be much greater and efficiency thereby improved. Combined heat and power (CHP) schemes of this nature are well-established methods of producing both heat and power efficiently and economically.

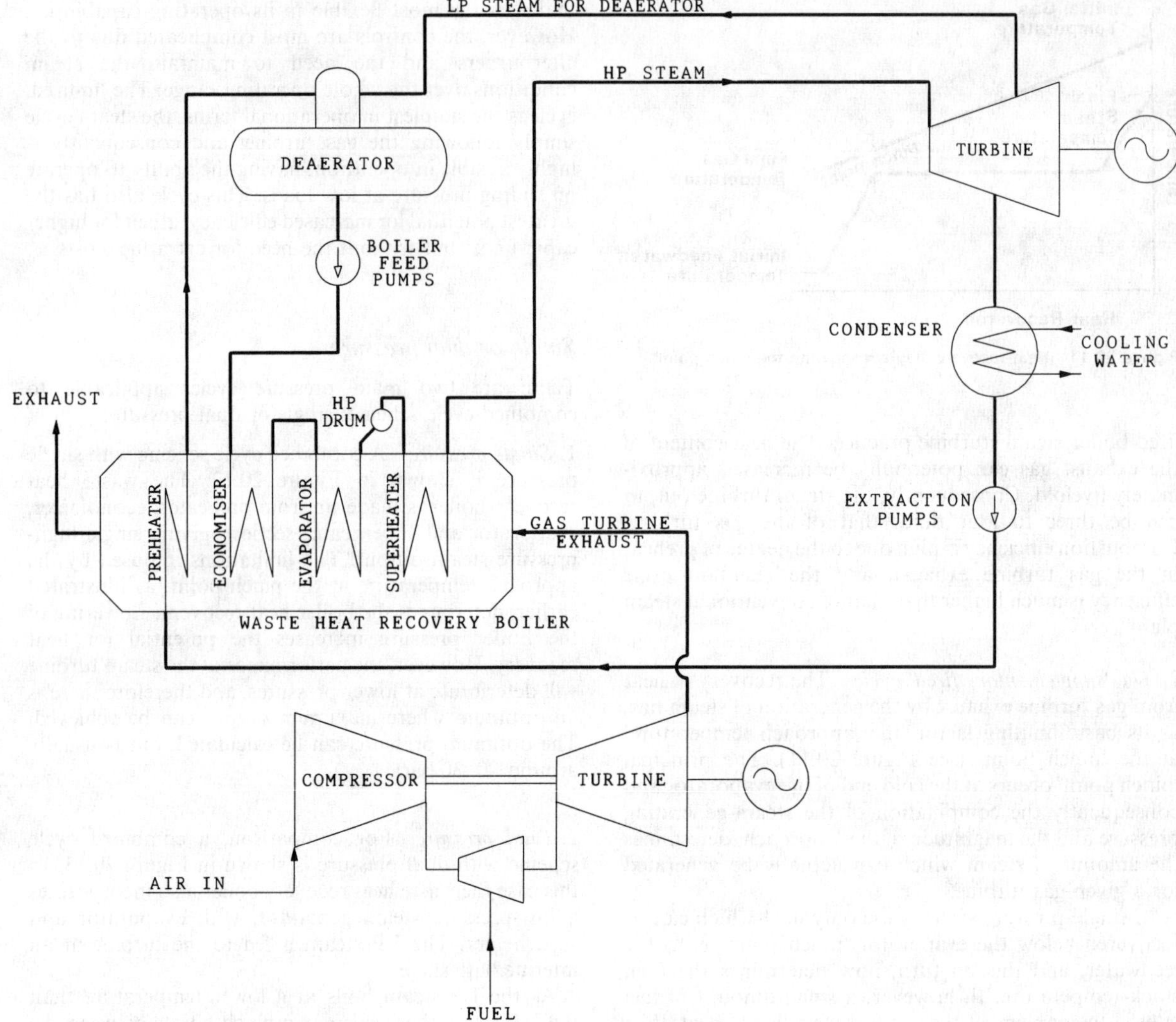

Figure 20.12 The single-pressure cycle diagram for a combined-cycle installation

20.3.1 Steam turbines for CHP

Many industrial processes require electrical power and heat. This heat is often provided from large quantities of low-pressure steam. In this section it is demonstrated that a thermal power station gives up very large quantities of heat to the cooling water in the condenser. For this purpose, the steam pressure in the condenser is usually at the lowest practical pressure (around 0.05 bara) to achieve maximum work output from the turbine.

However, if the turbine back pressure is raised to above atmospheric pressure so that the turbine exhaust steam can be transported to the process heat load then the steam will give up its latent heat usefully rather than reject this to the condenser cooling water. Although the steam turbine output is reduced, the overall efficiency is increased significantly as the generated steam is used to provide both heat and electrical power.

The alternative (but equally appropriate) logic is that a factory may use steam from low-pressure boilers. By increasing the steam pressure and then expanding this through a steam turbine to the desired process pressure additional electrical power can be provided.

An example is shown in Figure 20.14. By raising steam at high pressure (say, 60 bara and 540°C) and then expanding this through a turbine to the process steam pressure requirements of 3 bara then useful work can be done by the turbine for generation of electrical power. For this example each kg/s of steam gives 590 kW of electrical power.

There are several types of steam turbine that can be used to meet widely varying steam and power demands. They can be employed individually or in combination with each other.

Back-pressure turbine

The simple back-pressure turbine provides maximum

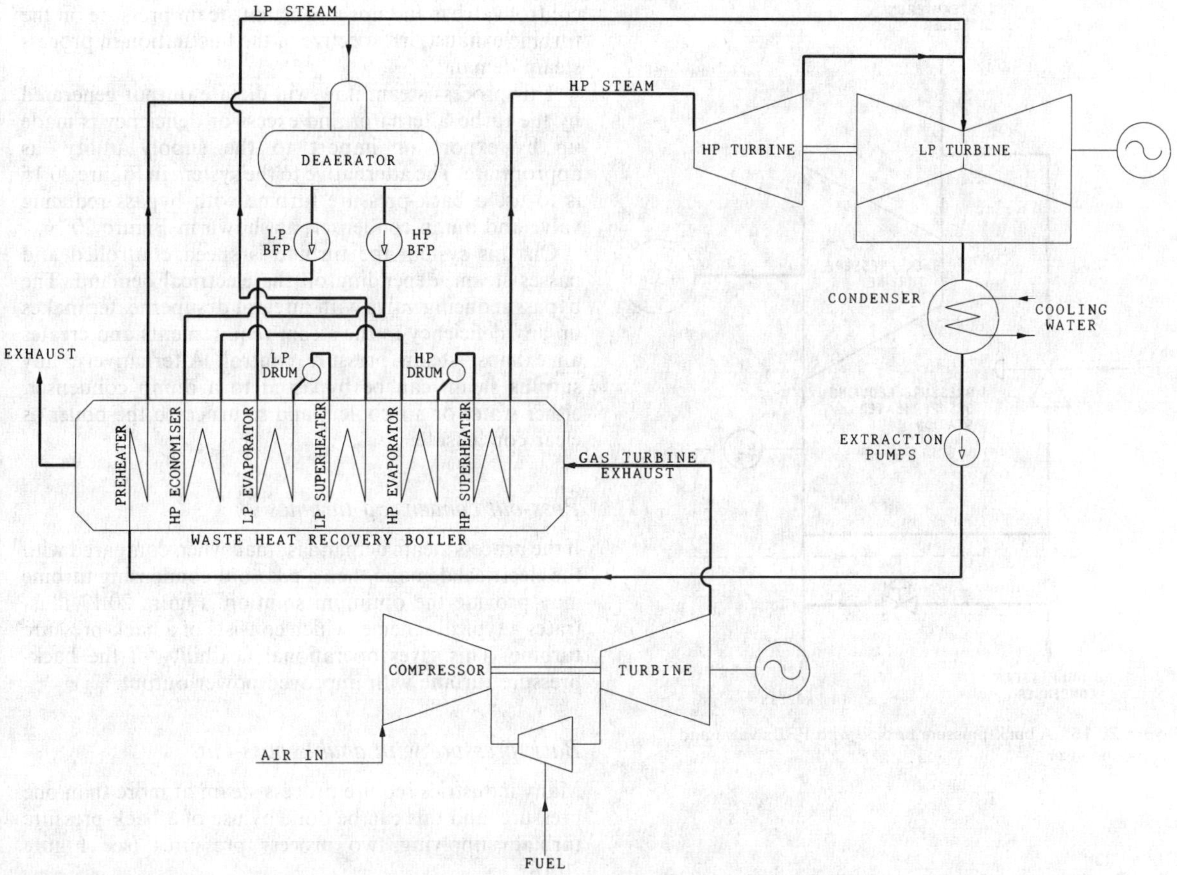

Figure 20.13 The dual-pressure cycle diagram for a combined-cycle installation

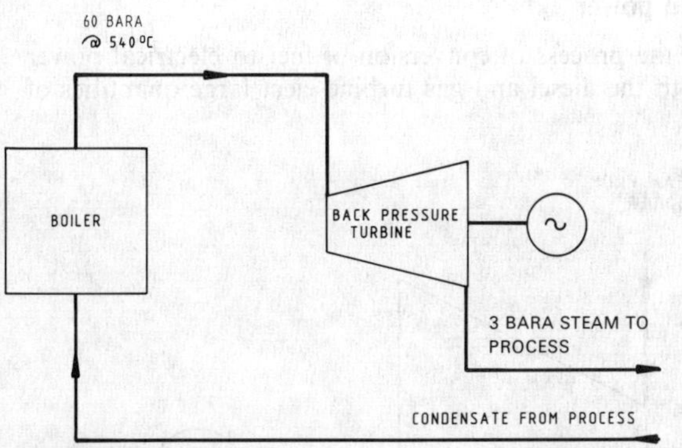

Figure 20.14 The back-pressure steam turbine in CHP application

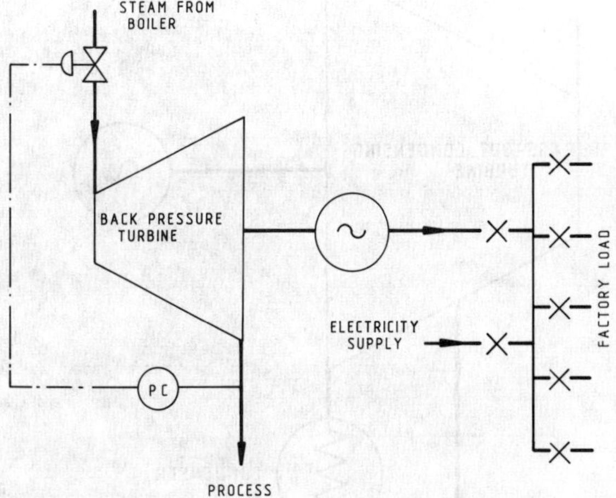

Figure 20.15 A back-pressure turbo-alternator operating in parallel with the grid supply

economy with the simplest installation. An ideal back-pressure turbogenerator set relies on the process steam requirements to match the power demand. However, this ideal is seldom realized in practice. In most installations the power and heat demands will fluctuate widely, with a fall in electrical demand when steam flow, for instance, rises.

These operating problems must be overcome by selecting the correct system. Figure 20.15 shows an arrangement which balances the process steam and electrical demands by running the turbo-alternator in parallel with the electrical supply utility. The turbine inlet

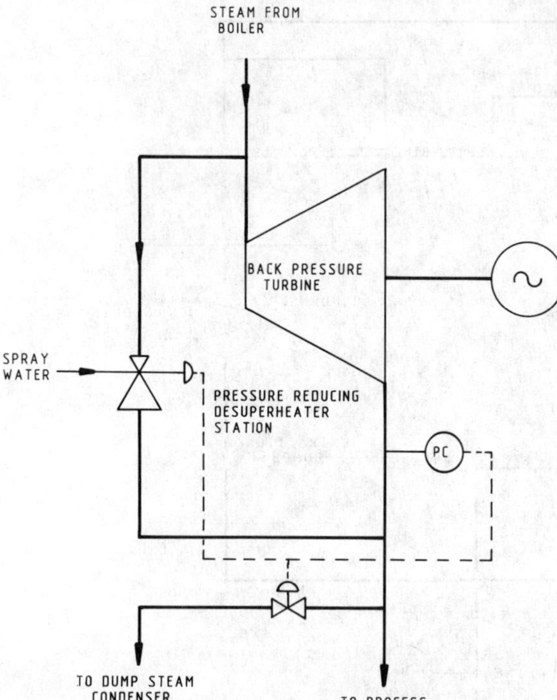

Figure 20.16 A back-pressure turbine with PRDs valve and dump condenser

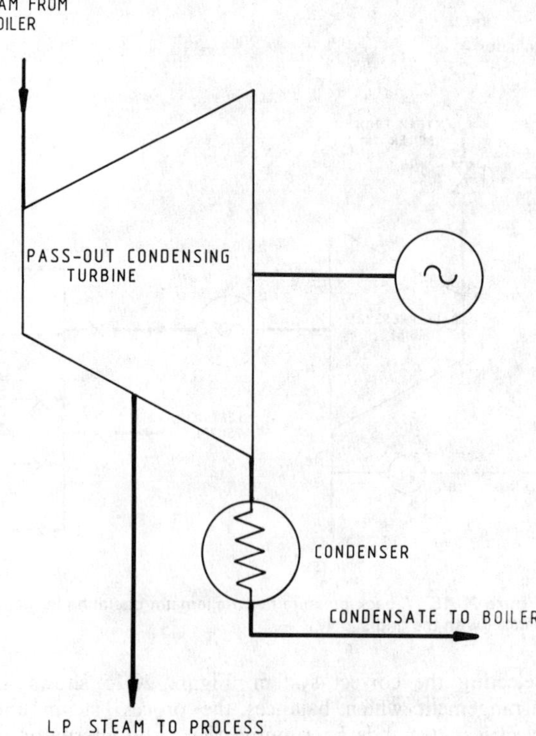

Figure 20.17 The pass-out condensing turbine

control valve maintains a constant steam pressure on the turbine exhaust, irrespective of the fluctuation in process steam demand.

This process steam flow will dictate output generated by the turbo-alternator and excess or deficiency is made up by export or import to the supply utility, as appropriate. The alternative to the system in Figure 20.15 is to use a back-pressure turbine with bypass reducing valve and dump condenser, as shown in Figure 20.16.

On this system the turbine is speed controlled and passes steam, depending on the electrical demand. The bypass reducing valve with integral desuperheater makes up any deficiency in the steam requirements and creates an exhaust steam pressure control. Alternatively, any surplus steam can be bypassed to a dump condenser, either water or air cooled, and returned to the boiler as clear condensate.

Pass-out condensing turbines

If the process steam demand is small when compared with the electrical demand then a pass-out condensing turbine may provide the optimum solution. Figure 20.17 illustrates a typical scheme, which consists of a back-pressure turbine. This gives operational flexibility of the back-pressure turbine with improved power output.

Back pressure with double pass-out

Many industries require process steam at more than one pressure, and this can be done by use of a back-pressure turbine supplying two process pressures (see Figure 20.18).

20.3.2 Diesels and gas turbines in combined heat and power

In the process of conversion of fuel to electrical power both the diesel and gas turbine eject large quantities of

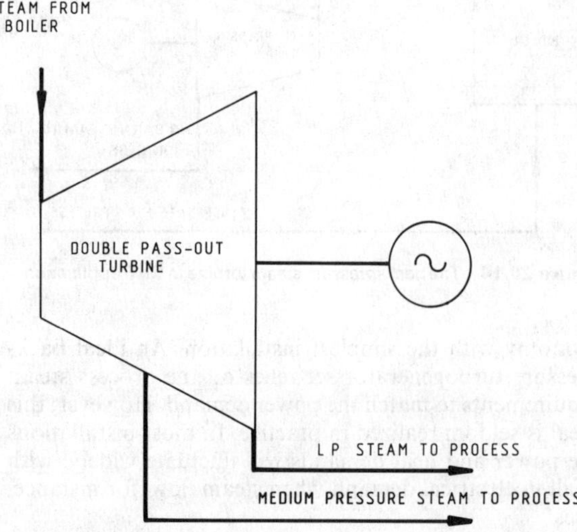

Figure 20.18 The double pass-out turbo-alternator

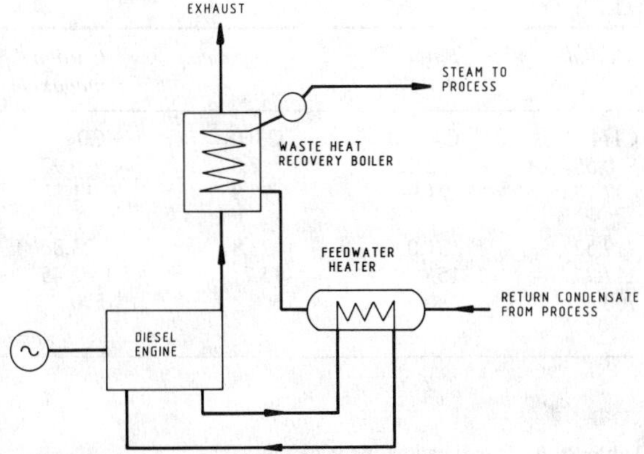

Figure 20.19 Diesel with waste heat recovery

hot exhaust gases. These gases represent a significant energy source which can be converted to useful heat by the addition of a waste heat boiler (see Figure 20.19). The boiler can be arranged to produce either steam or hot water, depending on the process needs. This heat is produced without any further fuel input and without affecting the performance of the generating machine. In addition, in the case of the diesel, additional low-grade heat can be recovered from the engine-cooling system.

The high efficiency of conversion of the diesel generator immediately restricts its potential improvement compared with gas turbines. With the simple addition of a boiler, the gas turbine can give a heat/power ratio of 2:1 compared with the diesel's 0.6:1. (The diesel heat/power ratio can be improved by 1:1 by use of the engine-cooling services, as shown in Figure 20.19.) However, both diesel and gas turbine exhaust gases contain excess oxygen (12–14% and 14–16%, respectively). This free oxygen can be utilized by supplementary firing the exhaust gases to produce additional steam in the heat-recovery boiler. By supplementary firing the ratio of heat to power can theoretically rise to 15:1 for a gas turbine and 5:1 for a diesel.

The chemical characteristics of the exhaust gases from diesel engines tend to make them less attractive for use in waste heat recovery, as they are likely to contain significant carbonaceous solid matter because of the range of residual oils burned. Fouling by soot build-up on heat transfer surfaces within the boiler can occur. In addition, the gas is likely to have significant pulsations, with a wide variation in mass flow with engine load, although the exhaust temperature fluctuation at low load is not as significant as for the gas turbine. However, there are many examples of successful CHP schemes using both the diesel and the gas turbine.

20.4 Factors influencing choice

The basic schemes for electrical power generation and combined heat and power have been identified in the earlier sections of this chapter. When assessing the alternatives to meet a specific project need there are many factors which will strongly influence the size, number, and type of generator as well as its specification and scope. While it is not intended to cover all of these aspects, the more important issues are discussed.

20.4.1 The available fuels

Oils

The petroleum industry is responsible for the production of a wide range of fuel oils for industrial use. These generally range from kerosene, gas oils and diesel fuel to residual fuel. (See Table 20.3 for typical fuel oil characteristics.)

Table 20.4 shows typical applications for the range of oil fuels for use with major prime movers. Medium/high-speed diesel engines generally use distillate fuel oils while medium/low-speed units generally burn residual fuels. The gas turbine, which normally operates on liquid distillate fuels, is capable of running on residual fuels, although examples of these are normally associated with crude production facilities.

In the case of residual fuel firing for gas turbines it is necessary to provide extensive fuel-preparation plant for the removal of vanadium, sodium, potassium, calcium and aluminium. Generally, oil fuels have good heating values and combustion characteristics, the main disadvantages being their higher sulphur content and the need for heating of the heavy fuel oils for storage and transport. Distillate grades of oil fuels may be stored and handled at ambient temperature. However, exposure to the cold for long periods is not advisable, and it is good practice to incorporate heaters to maintain the oil fuel at 5°C above freezing.

Heating facilities are required for all residual grades

Table 20.3 Typical properties of fuel oils

Oil fuel	*Kerosene*	*Diesel oil*	*Residual fuel oil*
Density (kg/)	0.8	0.85	0.99
Sulphur (%)	0.05	0.5	3.5 Max
Ash (%)	Nil	0.001	0.04
Gross calorific value (MJ/kg)	46.2	45.5	42.5
Net calorific value (MJ/kg)	43.2	42.7	39.8

Table 20.4 Application of fuel oils

System	*Spark ignition*	*Compression ignition*	*Gas turbine*
Motor gasoline	✓		✓
Middle distillates (kerosine, gas oil, diesel fuel)		✓	✓
Residual fuels		✓	✓
Crude oil			✓

Table 20.5 Typical properties of gas fuels

Gas fuel	*Town gas*	*Natural gas*	*Methane*	*Butane*	*Propane*	*Carbon monoxide*
Formulae	—	95% CH_4	CH_4	C_4H_{10}	C_3H_8	CO
Specific gravity (air = 1)	0.5	0.59	0.55	2.02	1.52	0.97
Gross CV (MJ/M^3)	18.65	37.3	37.1	121.9	93.9	11.8
Stoichiometric air						
vol/vol	4.2	9.62	9.52	31.0	23.8	23.8
kg/kg	8.6	14.8	17.2	15.4	15.7	2.45
Spontaneous ignition temperature (°C)	650	660	700	480	500	570

of oil fuels, and the following gives minimum storage and handling temperatures:

	Minimum storage temperature (°C)	*Minimum temperature at outflow from storage and for handling* (°C)
High fuel oil	10	10
Medium fuel oil	25	30
Heavy fuel oil	40	50

The tanks and pipework can be heated by use of steam, hot water, electrical heaters or a combination of these.

Gas

As a fuel, gas has many advantages in controllability and minimal effect on the environment, and it promotes reliable operation, longer life and residual maintenance compared with oil fuel. Its typical properties are shown in Table 20.5.

In many developed and developing countries there are now extensive natural gas supply networks which provide secure and reliable sources of gas fuel. The user may require some pressure reduction and metering equipment on-site, otherwise there is no requirement for storage, and the gas can be simply distributed to the generator by pipeline. A good example is in the UK, where natural gas was introduced around 1965 and since then has steadily grown to dominate the UK market.

One of the alternatives to natural gas are industrial products such as propane or butane. These fuels are transported and stored in liquid form. On-site facilities are necessary for reception, off-loading and storage.

With the trends towards conservation of energy there has been significant advances in small gasification schemes. Here the solid fuel is partially combusted to give a low calorific value gas. The gas is hot and dirty, and contains tar and particulate matter, thereby requiring some cleaning before being used as a fuel. While the scope for these schemes is limited, there have been successful examples in rural areas of developing countries where the gas produced from fuels such as wood is used to run small spark-ignition engines.

Table 20.6 Classification of coals

General description	*Volatile matter* (%)	*Specific description*
100	6–9	Anthracites
200	9–19.5	Low-volatile, steam coals
300	19.6–32	Medium-volatile coals
400	Over 32	Very strongly caking
500	Over 32	Strong caking
600	Over 32	Medium caking
700	Over 32	Weak caking
800	Over 32	Very weak caking
900	Over 32	Non-caking

Coal

In many regions of the world coal reserves represent substantial indigenous resources, and for this reason coal will always play an important role in the future of power generation. Inevitably, coal is more difficult to handle and has less appeal than other clean-burning fuels such as gas. However, substantial progress has been made in modern efficient handling techniques and combustion methods which can give clean combustion with low emissions.

For the purposes of power generation certain coal properties are significant, and a coal can be generally categorized by its rank, which embraces both volatile matter and caking properties. At the lower end of the scale are the older coals, such as anthracite with low volatile content and no caking properties. These coals are difficult to burn. As volatile content increases beyond anthracite then so does the caking quality, and prime coking coals are 12–30% volatile matter (VM). Beyond this, as VM increases further, caking properties decrease to free-burning coals of approximately 40% VM.

An example is the classification scheme used by British Coal, and is common throughout the UK. The scheme divides coals into groups, generally as shown in Table 20.6. The group, grading and modes of preparation of a coal serve to indicate the usage for which the coal is suitable for application with industrial boilers. For grading of coal the nominal sizes are shown as follows:

Table 20.7 Typical properties of some solid fuels

Fuel	*General-purpose coal*	*Wood*	*Peat*	*Lignite*
Moisture (%)	16	15	20	15
Ash (%)	8	—	4	5
Carbon (%)	61.3	42.5	43.7	56.0
Hydrogen (%)	4.0	5.1	4.2	4.0
Nitrogen (%)	1.3 }	0.9	1.5	1.6
Sulphur (%)	1.7 }			
Oxygen (%)	7.7	36.5	26.6	18.4
Gross CV (MJ/kg)	25.26	15.83	15.93	21.5
Net CV (M3/kg)	23.98	14.37	14.5	20.21

	Through (in)	*On (in)*
Doubles	2	1
Singles	1	0.5
Peas	0.5	1
Grains	0.75	0.125

Smalls constitute the fraction which passes through a screen of given size and which has no lower size limit. Smalls can be washed or untreated.

The ash content of the coal is significant, not only in its mass but also with respect to the constituents of the ash, the fusion temperature, the range of fusion temperature between initial deformation and actual fusion and the amount of coal fines present.

Use of coal requires suitable storage facilities local to the power station site. The capacity of the stockpile is a decision of strategy and depends on the frequency and size of the delivery loads. In addition, the site should be carefully chosen in view of access needs for reception and off-loading. Stored coal weighs about 4000 tonnes per 0.5 m depth per hectare, which gives a rough guide on areas for stockpile. Graded coals and smalls should always be stored separately. The large coal forms naturally, but smalls must be piled by layering and compacting to form a well-consolidated heap.

The object of compacting the smalls is to exclude air to ensure safe storage. If proper precautions are taken then coal can be stored safely. However, the user should be aware of the potential dangers of spontaneous combustion and the measures necessary to avoid this problem. For sites of limited area coal can be stored in large vertical bins.

There is a wide range of pneumatic, vibratory, belt or 'en-masse' type elevators and conveyors for transfer of coal from storage to boiler. The final scheme depends on site layout constraints and cost.

Other fuels

Solid fuels also encompass lignite, peat and wood. In its final form wood has a much greater commercial value as a product. However, some industrial processes (for example, the production of paper) create waste wood in offcuts, bark and shavings. Here the wood is ideal as an energy source. It has a high moisture content but nevertheless can be fired easily with full automation. Table 20.7 shows typical heating values for these fuels when compared with coal.

Additionally, there are food processes where the product is derived from a vegetation and a combustible fibrous waste is produced which can be used as fuel. A good example is the production of sugar from sugar cane with bagasse as waste. There are many others, and their by-products should not be overlooked in an assessment of suitable fuel sources.

20.4.2 The electrical load profile

In reviewing the size of a generating scheme it is necessary to consider the electrical load profile of the site demand, which must examine the electrical power needs over the daily and annual cycles. The capacity of the generator can be selected to cater for anticipated future growth or, alternatively, the station could be designed so that further generators could easily be added to match future load demands. At this stage it is worth noting that the power station itself will require electrical power for its auxiliaries (parasitic load), and this electrical load will be relatively constant, despite the output modulating to follow demand.

The load profiles can vary quite considerably, depending on the electrical users. For instance, for a factory the load will depend on the shift operation with a potential low demand at night and at weekends. The load can be influenced by seasonal changes. For instance, it may increase towards winter with greater lighting demands. However, the need for HVAC in computer installation may increase the load during summer. It is important that the power station be capable of meeting the peak capacity while also being able to operate efficiently at lower loads.

20.4.3 The heat load

Earlier, we reviewed the advantages of combined heat and power. It therefore follows that a local heat load, where practical, should be incorporated into the electrical scheme.

Heat loads are usually supplied by medium- to low-pressure steam and high- and low-temperature hot water. Steam is the main medium used by industry for the transport of large quantities of heat, and pressures generally range from 3 to 16 bar. Low- and high-temperature hot water systems are employed for heating, and they operate in 'closed' circulating systems with the high-temperature system (approximately 120°C) pressurized by an external source to prevent steam forming. Typical industrial and commercial applications are:

Hotels: hot water, swimming pools
Schools: hot water and steam, heating and air conditioning
Dairy: steam and hot water
Heavy manufacturing: steam, air conditioning, space heating and hot water
Hospitals: steam, hot water
Sewage plant: hot water, sludge heating
Food industries: process steam, hot water
Chemical industries: process steam, hot water

Hot water or steam can also be used in absorption-type chillers to provide chilled water with the reduction in electrical power needed for the refrigeration process.

As with the electrical load profile, it is also necessary to analyse the heat load over the daily and annual cycles. Ideally, the heat load will match the available heat from the electrical generator (however, this is rarely the case). There will be periods when supplementary output will be necessary which can be achieved by, say, supplementary firing the waste heat gases of a gas turbine, or heat output reduction is necessary by the introduction of bypass stacks. For a steam turbine installation bypass pressure-reducing valves will be necessary to supplement steam output, while a dump condenser may be needed at low-process steam demands. The nature of the electrical and heat load will obviously have significant influence in the development of the scheme and scope of equipment.

20.4.4 Power station auxiliary systems and services

A total power station project involves many aspects, covering civil works, fuel systems and storage, plant auxiliary systems and services, electrical plant and control and instrumentation. The scope will depend on the nature and size of the project envisaged. Table 20.8 gives a detailed checklist for power station auxiliary components and will provide a useful guide in developing scope. Further details of many of the components, such as ventilation systems or fuel oil handling, are given in other chapters of the book.

General information relevant to electrical system design is given in Chapter 21. Within an industrial power plant, the following key items of electrical equipment need to be addressed:

1. The power-evacuation circuit, including any generator transformer and the connection with the remaining site-supply system.
2. The supply system for the power plant auxiliaries, including the provision of supplies to any important drives and the provision of standby supply systems.
3. The protection and synchronizing schemes applicable to the generator circuit.
4. The earthing arrangements of the generator neutral.
5. The control of the excitation of the generator when operating in either isolated or parallel mode.

A typical electrical schematic for an 8 MW power station is shown in Figure 20.20.

20.4.5 Site conditions

The site conditions play an important role in the development of layout and scope. The more specific issues are as follows:

1. Naturally, the site location must be as close as practical to the electrical and heat users, thus minimizing cable and piping runs.
2. Altitude and ambient temperature, as discussed earlier, can have significant effect on the performance of the generator, and fluctuations over the annual operating cycle will similarly influence the power station output.
3. Exposure to dust or other pollutants that may exist in certain industrial processes will dictate requirements for such aspects as equipment enclosures and air filtration. For the same reasons, it may be prudent to avoid air-cooled heat transfer surfaces.
4. Adequate area will be required for water, chemical, fuel and ash storage with adequate access for reception and off-loading.
5. During the erection period access will be needed for the heavy loads, storage of equipment and for erection.
6. With the exception of the gas turbine, most generating schemes need a reliable and a good-quality water source for cooling. Alternative large air-cooled systems can be employed, but they are expensive and less efficient. Thermal power stations using steam condensers can require large cooling-water flows, although the cooling water consumption can be reduced by use of closed-cycle cooling-water systems. In these cases it is common to site the power station near a large water source such as a river or lake. The design of water-intake systems needs to recognize the changes in level and quality that can occur, depending on such aspects as time of year, rainfall and tides.
7. Boilers require very high-quality water for make-up, and for this purpose treatment facilities may be needed on-site, depending on the quality of the water source.
8. Civil engineering considerations such as ground conditions and drainage.

20.4.6 Plant availability and maintenance

It is prerequisite of most modern power stations that they achieve high availability. For normal operation it is common practice for power stations to employ standby equipment which is strategically important to overall operation. Examples are the use of 100% duty standby pumps or compressed air plant. Depending on the size of installation, 3 × 50% units can be employed with 50% standby capacity and lower initial cost. For similar reasons, security of fuel and water supplies must be addressed. Adequate fuel storage may be necessary to meet anticipated shortages in supply or, alternatively,

Table 20.8 Checklist of power station auxiliary components

Fuel system

Coal:
Rail sliding
Wagon puller
Tippler
Ground hopper
Conveyor
Crusher
Screen
Sampling
Mag. separator
Stocking-out conveyor
Conveyor (reclaim)
Transfer tower
Bunker conveyor
Dust extraction
Coal feeder
Coal mills
Pneumatic transfer system

Oil:
Bulk unloading pumps
Bulk storage tanks
Tank heater
Forwarding pumps
PH and filtering sets
Oil pipework
Steam pipework
Tracing elec./steam
Daily service tank
Fuel-treatment plant
Steam/gas turbine/diesel house
Boiler house
CW pump house

Ash handling
Clinker grinder
Clinker conveyor
Econ. dust conveyor
Air heater dust conveyor
Rotary valves
Dust collector dust conveyors
Rotary valves
Ash conditioner
Ash silo
Ash sluice pump
Make-up water pump
Pneumatic conveying system
Conditioned ash conveyor
Submerged ash conveyor

Water systems
Bar screen
Travelling-band screen
Wash-water system
Penstock
Intake pump/circulating pumps
Valves
Chlorination plant
Crane/lifting equipment
Cooling tower (ID/FD/natural)
Settling pond

Boiler water treatment
Base exchange
Dealkylization
Demineralization
Demin. water tank
Demin. water pump
Bulk acid and alkali storage
Neutralizing

Chemical dosing
LP dosing set
HP dosing set
CW dosing set

Miscellaneous
Air compressor (plant)
Air compressor (inst.)
Air compressor (coal pneumatic system)
Air compressor (ash pneumatic system)
Firefighting equipment
 Detection
 Portable
 Hydrant system
 Springer system
 Panel
 Halon/CO_2
Pressure-reducing set
Press. red. and desuperheating set
Ventilation
 Turbine hall
 Boiler house
 Switchrooms
 Battery room
 Water-treatment room
 Admin.
Air conditioning
 Control room
Hydrogen-generation plant
Electroysis
Bottle storage

Piping system
Main steam
Process steam
Feedwater
Raw water
Treated water
Potable water
Aux. cooling system
Firefighting system
Clarified water
Filtered water
Water-intake system
Circulating-water system
Chemical dosing
Station drains
Fuel oil
Fuel gas
Compressed air (plant)
Compressed air (inst.)

Electrical equipment
Generator
Unit auxiliaries
Station auxiliaries
Low-voltage auxiliaries
HV and LV switchgear
Generator circuit breaker
Batteries and charger
Main power cabling
Cabling and wiring
Lightning protection
Earthing
Telephones/intercom
Protection equipment
Electrical controls, inst. meters
Lighting and small power
Temporary supplies
Elec. trace heating
Generator
AVR
Generator VTs
Busbar equipment
Synchronizing equipment

Civil
Building/annexe for:

Water treatment
Feed pumps
Fuel handling/metering
Control room
Switch house
Offices
Admin. building
Gatehouse

Site fencing
Roads
Plant drains
Storm drains
Foundations for mech. and elec.
Plant and tanks
CW culverts
Cooling tower found.
Chimney found.
Crane rails
HVAC
Plumbing and sanitary fittings
Domestic water supply
Tanks/reservoirs
Site investigations and ground
Loading tests
Demolition of existing plant and buildings
Connection to existing plant and buildings
Temporary works

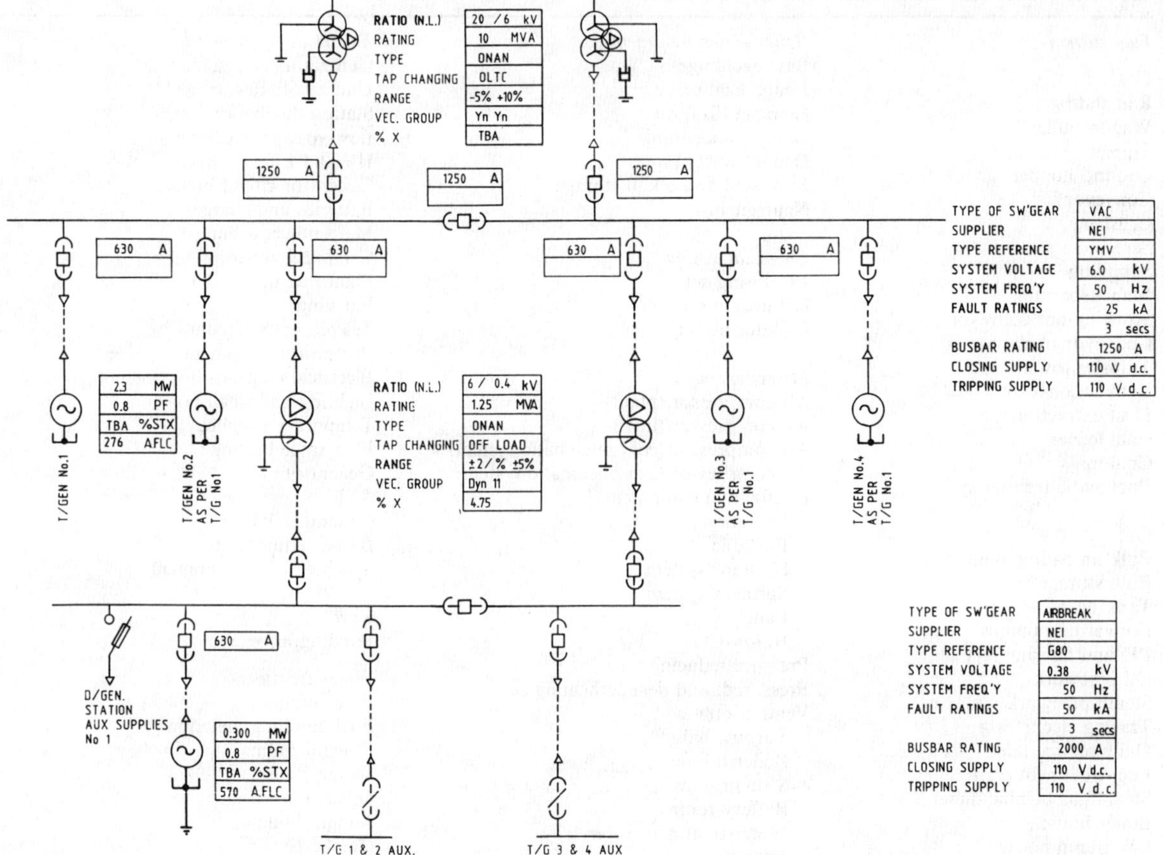

Figure 20.20 Typical electrical single-line diagram for an 8 MW power plant

standby fuel facilities incorporated. Water storage will be needed again to cater for periods when shortages occur.

Adequate consideration to the above will ensure that outages are minimized and maximum availability is achieved. However, at some stage the generator or generators will be closed for maintenance, and in these circumstances consideration must be given to making up the electrical supply. For essential electrical loads an alternative source is needed and, at a premium, this could be supplied from the electrical supply authority or from a standby generator. A standby generator will increase capital cost, but in some cases this is essential to meet the needs of total security of supply.

For locations in the Middle East subject to wide fluctuations in ambient temperature it has been common to utilize multiple gas turbines which are sized to meet the plant demands during the hot summer months and the cold winter. As the gas turbine output increases substantially there is sufficient spare capacity to allow for outage of machines without affecting the electrical power export. However, this situation is unique to the environmental conditions and type of equipment in service.

The plant components selected should be of proven design supported by strong reference installations. Modern trends are moving towards reduction in manning levels, and the use of central computer control and supervisory systems also ensures maximum efficiency of plant operation.

Specific details on maintenance of plant are given later. However, in developing the station layout adequate thought must be given to cranage, laydown areas and access for maintenance of heavy plant.

20.4.7 Environmental aspects

The effects of the power station on the local environment will be controlled by local regulations and laws. The significant considerations are:

1. Exhaust emissions of SO_x, NO_x and particulate matter.
2. Height of chimney.
3. Noise levels at the site boundary.
4. Treatment of boiler and water treatment plant effluent.
5. Architectural features and design of buildings and structures.

20.4.8 Generated voltages

The selection of generated voltage will be determined by the voltages available on-site, the amount of generation envisaged and the full load and fault current ratings of existing and proposed equipment. Following the completion of the installation, the fault levels occurring at each point within the installation (including the equipment within the supply authority system) must be within the capability of the equipment, particularly the switchgear installed.

The factors affecting the generator circuit itself are related to the relative cost and availability of equipment to be used in that circuit. Generators rated at up to 2 MVA, 2782 A at 415 V can be accommodated within standard low-voltage switchgear ratings of, typically, 3000 A full load current and 50 kA fault capacity. Generators rated larger than 2 MVA but less than 20 mVA will be usually operated at 11 kV, although at ratings of between 2 and 5 MVA, a lower-voltage 3.3 kV or 6.6 kV may be considered due to the capital cost of the generator.

20.5 The selection

It should be borne in mind that there is no single answer as to which configuration of equipment is best for any individual power generating scheme. Each alternative must be considered on its own merits, including the use of all available fuels.

20.5.1 For electrical power

For generation of electrical power only, the overriding influence in the choice of generating plant is fuel availability. In this respect the following categories can be applied:

Diesel (light and heavy fuel oils)
Gas turbine (gas and light fuel oils)
Gas turbine and combined cycle (gas and light fuel oils)
Thermal power plant (coal, lignite, wood and biomass)

For the criterion of fuel to electrical energy conversion efficiency, the diesel and gas turbine are natural choices for prime fuels such as oil and gas.

In open cycle the diesel has the higher efficiency, making it more attractive for light fuel oils. However, in combined cycle the gas turbine often has the highest overall efficiency, but there is the penalty associated with the additional cost for boiler and turbine. Nevertheless, unless the prime fuel is of low cost, the use of combined-cycle gas turbine plant will prove to be the more economic.

For industrial or commercial applications the straightforward condensing thermal power plant cannot compete with the diesel or gas turbine, as its first, or capital, cost is significantly higher and its overall efficiency is low. However, successful smaller thermal power plants have been installed where the fuel is of low cost to the user. In these cases the efficiency will be of less significance, and the much lower running costs when compared with prime fuels such as oil makes the thermal power plant an attractive alternative.

The overall efficiency of the condensing thermal cycle, as discussed, is dictated primarily by the steam conditions used. There are some small industrial stations with outputs up to 2 MW using shell-type boilers for the generation of steam. Here the steam conditions are limited to approximately 17 bara and 250°C. For larger installations these conditions will rise sharply when watertube boilers become attractive and more common steam conditions are of above 60 bara and 540°C.

Having decided the type of generator, the next questions are the capacity and unit sizes. It is usual for the total installed capacity to be capable of meeting the peak demand with spare capacity to meet outages for maintenance. However, consideration can be given to a supply from the electricity supply authority for peak demands and standby.

To meet the needs of a given output, a single generator could be provided, and this option is likely to give the lowest initial cost. However, the alternative of two or more machines, of the same total capacity, while more expensive initially gives higher availability, better flexibility of operation and improved load-following performance.

In isolated areas it may be a requirement that at least one unit is needed for standby to meet security of supply, and this can be provided by $2 \times 100\%$ or, say, $3 \times 50\%$ units. Availability is maximized as the standby unit can be brought into operation during outage of one of the normally operating machines.

This is best illustrated by the following example for a rural development project. The electrical power demand is shown in Figure 20.21 and was developed from historical data generated from similar developments in other regions. The operating statistics indicated that the maximum demand would occur at night peaking to 8 MW with a low day demand of 2 MW. Considering these power requirements, the station must be flexible and have the ability to follow the daily load pattern. In addition, because of the isolated location the power station must be sized for 'stand-alone' operation. Equally

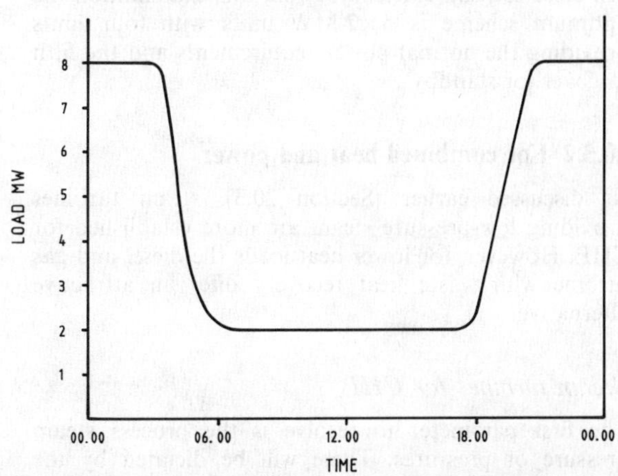

Figure 20.21 Daily load cycle for a rural development project

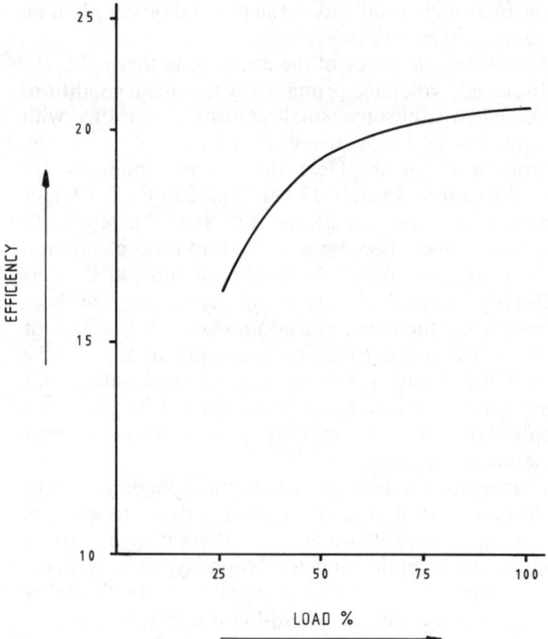

Figure 20.22 Overall efficiency of a typical coal-fired power station of 4.6 MW capacity

important factors are maximum availability and operating efficiency.

The use of a local indigenous coal source dictated the choice for thermal power plant. Considering the forecast power demand, it is feasible for an 8 MW turbine/boiler unit to be installed to provide the ultimate load requirement. However, the single power unit is not flexible in that it does not give the required security to cover for routine and statutory shutdowns for maintenance and is less efficient at part load (see Figure 20.22). Provision of multiple small units is the ideal solution, as it allows the station to be operated more flexibly, enabling the units to be put into or taken out of service as the power requirement demands so that the operating units run close to peak efficiency. From this examination the optimum scheme is 5 × 2 MW units with four units providing the normal power requirements and the fifth to cover for standby.

20.5.2 For combined heat and power

As discussed earlier (Section 20.3), steam turbines providing low-pressure steam are more established for CHP. However, for lower heat loads the diesel and gas turbine with waste heat recovery offer an attractive alternative.

Steam turbines for CHP

The first parameter to resolve is the process steam pressure or pressures. These will be dictated by the process-heating temperatures, noting that the steam will transfer its heat at constant pressure and at saturation temperature. It is in the interests of efficiency to keep the process steam pressure as low as practical. Also influencing the process steam pressure is the distribution pipeline from the turbine exhaust to process. Reductions in the pipeline pressure loss increases pipeline size and cost.

The selection is dictated by economics governing the initial plant cost versus higher turbine output. Usually, the turbine exhaust steam is designed to be slightly superheated, which is desirable, as it allows for heat loss from the steam with minimum condensate losses. At low loads from the turbine the degree of superheat can rise sharply, well in excess of the normal design conditions, and for this purpose desuperheaters are often employed to trim the steam temperature at exhaust.

The choice of boiler steam inlet conditions is usually dictated by the desire to achieve maximum output from the process steam flow. This requires high boiler steam pressure and temperature. However, there are practical considerations to observe. Above 40 bar more exacting feedwater treatment is necessary, and therefore it may be advantageous to maintain pressures below this figure. High steam and temperatures can also influence selection of boiler materials such as alloy steels. The upper limit for industrial applications is around 60 bara and 540°C.

For watertube boilers it is necessary to maintain low O_2 levels, and for this purpose a de-aerator in the feed line is required, which will also provide a degree of feed heating. The steam supply can be taken down from the low-pressure process steam main.

If a condenser is employed for a turbine with a condensing section then it is normally chosen to provide the best thermodynamic efficiency consistent with an economical capital expenditure. The condenser exhaust vacuums are usually 0.05 bara or higher.

Having ascertained the process steam flow and developed some ideas on the boiler pressure, the following step is to analyse the power available. Figure 20.23 provides a ready means of determining the approximate relationship between power available and process steam for specific steam conditions. Use of this and similar charts will allow an assessment to be made of the potential of a CHP scheme with a back-pressure turbine. The conditions can be changed to give the required balance for heat and power.

The electrical and heat analysis, as discussed in Section 20.3, will show the relationship between power and heat and how this varies over time. It may be necessary to use steam bypass and stations or dump condenser, as discussed in Section 20.2. The use of dump condensers for meeting part-load requirement is inefficient and should be avoided. It is more acceptable to reduce turbine power output accordingly and import or top-up from an alternative supply.

Diesels and gas turbines for CHP

It is generally recognized that below around 5:1 heat ratio the steam turbines in CHP become less suitable, and here the diesel and gas turbines play an important role in providing heat and power. As discussed earlier, the simple addition of waste heat boilers gives additional heat output without further input of fuel to the generator.

For flexibility supplementary firing of the oxygen rich exhaust gases can give additional heat. The following summarizes the potential of both the diesel and GT for waste heat recovery:

	Unfired	*Supplementary fired*
Diesel	1:1	5:1
	0.6:1 (excluding the cooling-water services)	
Gas turbine	2:1	15:1

This gives indicative heat/power ratios only: an actual performance will depend on the machine.

Using these guidelines it is a simple matter of matching the electrical and the heat loads. An example is a scheme which was envisaged for the supply of power and heat to a computer centre using a diesel engine with waste heat recovery. Figure 20.24 shows the proposed scheme where heat generated provides the site demand and operates with absorption-type chillers for chilled water production. The use of these chillers also reduced the electrical demand. The electrical load was a steady 2.5 MW over the year with short duration peaks of 2.9 MW. To allow for future growth, 2 × 3.6 MW diesels were selected (one operating and one standby). The electrical load calculated was 21.75×10^6 kWh/yr, and applying a useful heat recovery ratio of 0.6:1 (not including engine services), this gives 13×10^6 kWh/yr of heat energy, which is steadily available over the year. The hot water and chilled water heat loads are shown in Figures 20.25 and 20.26. As the site heat load and chilled water load peak at different times of the year, this gave

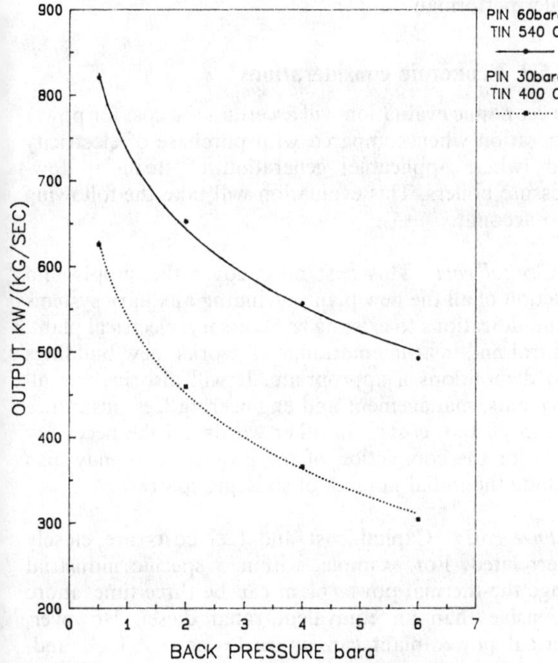

Figure 20.23 The effect of back-pressure upon steam turbine performance

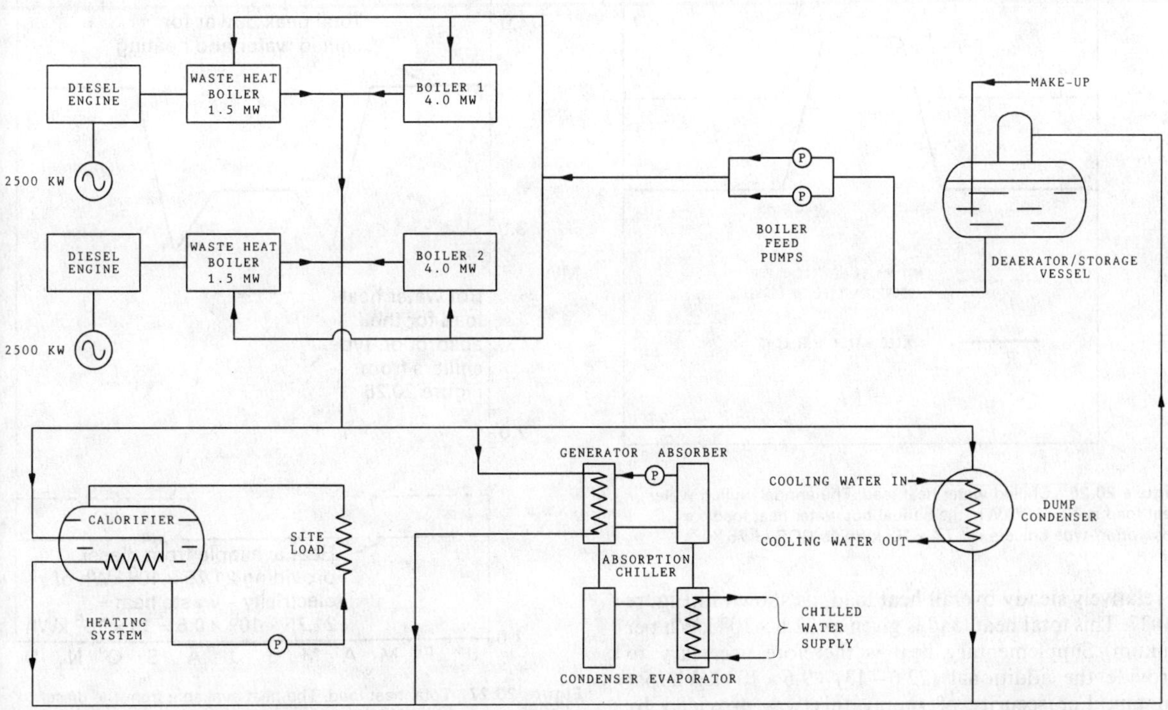

Figure 20.24 Example of a small diesel-engined combined heat and power scheme

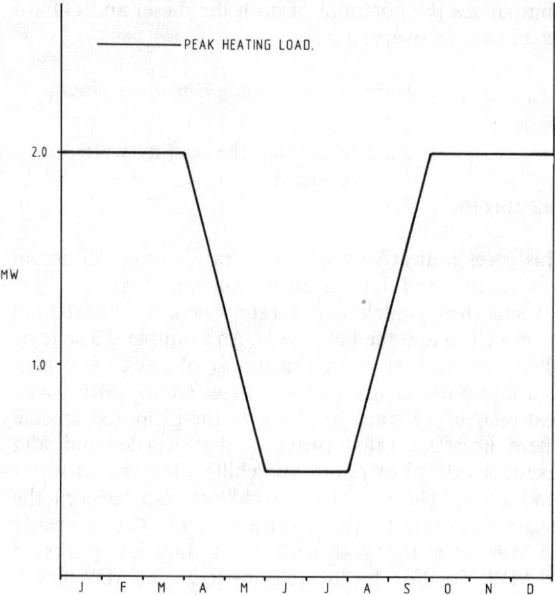

Figure 20.25 Hot water heat load. The annual heating requirement is 5.2×10^6 kWh

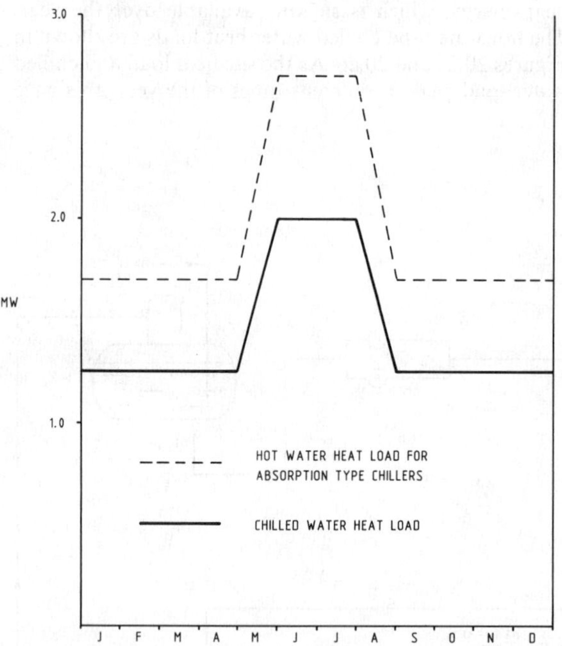

Figure 20.26 Chilled water heat load. The annual chilled water heat load is 13×10^6 kWh; the annual hot water heat load for absorption-type chillers is 17.4×10^6 kWh (a COP of 75%)

a relatively steady overall heat load, as shown in Figure 20.27. This total heat load is given as 22.6×10^6 kWh per annum. Supplementary heat is therefore necessary to provide the additional $(22.6-13)=9.6 \times 10^6$ kWh per annum. For security of supply this was provided by 4 MW boilers capable of giving full independent supply.

The diesel and gas turbine with waste heat recovery are limited in terms of fuel application being suitable for gas and oils only. Also when considering oil fuel firing careful attention must be given to the effects of sulphur in the exhaust gases, as this will place limitations on the boiler performance.

20.5.3 Economic considerations

An economic evaluation will ascertain the cost for power generation when compared with purchase of electricity and (where applicable) generation of steam in low-pressure boilers. This evaluation will take the following into account:

1. *Capital cost* This cost must cover the supply and erection of all the new plant, including auxiliary systems or modifications to existing as necessary, electrical plant, control and instrumentation, civil works, new buildings and demolitions if appropriate. It will also include all payments, management and engineering fees, insurance and inspection costs – in other words, all the necessary costs for the completion of the project. This may also include the initial holding of strategic spares.

2. *Fuel costs* Capital cost and fuel costs are closely interrelated. For example, within a specific industrial range the thermal power plant can be three times more expensive than an equivalent rated diesel. However, thermal power plant can utilize lower cost fuels and,

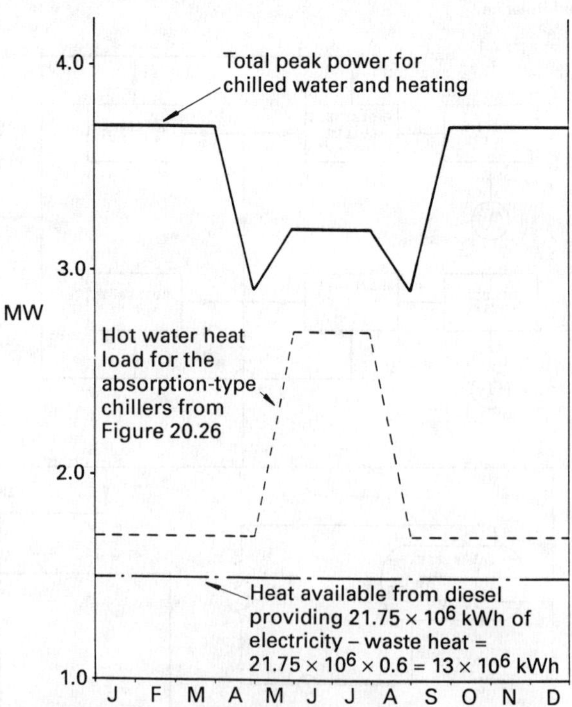

Figure 20.27 Total heat load. The heat available from the diesel waste gases of 13×10^6 kWh will provide only a proportion of the heat required for the absorption-type chiller

Table 20.9 Typical chemical requirements for a 60 MW industrial steam-generating plant

Chemical	*Location*	*Quanity*
Hydrochloric acid (HCL)	Water-treatment plant (cation)	165 kg/day at 30% solution
	Effluent neutralization	14 kg/day at 30% solution
Caustic flake (NaOH)	Water treatment plant (anion)	221 kg/day at 25% solution
	Effluent neutralization	14 kg/day at 25% solution
Oxygen scavenger (e.g. hydrazine)	LP feedwater dosing	8.5 kg/day at 25% solution
pH control (e.g. ammonia)	Ll feedwater dosing	24 l/kg/day
Phosphate	Boiler dosing	Consumption depends on blowdown rate for boiler
Corrosion inhibitor	Cooling-water system	Consumption depends on make-up
Bacterial control	Cooling-water system	Chemicals are used to provide adequate reserve

when taking into account diesel costs for consumables such as lubricating oil and maintenance spares, the thermal power plant can be attractive over the longer term. The fuel consumption must be established from an analysis of the load profile and overall efficiency of the station at various part loads. In addition, the analysis must also take into account the effects of ambient air temperature changes on air intake mass and cooling water temperatures, as these also influence efficiency. It is prudent to add a small percentage to cover for load-following and deterioration of efficiency that may occur over a prolonged operating period. The fuel consumption should also include any anticipated supplementary firing needed on CHP schemes.

3. *Chemical consumption* The chemical consumptions will be associated with boiler feed make-up water-treatment plant, dosing systems for feedwater and boiler system, treatment of cooling water circuits and effluent treatment. Typical chemical requirements for a thermal power plant are given in Table 20.9.

4. *Feedwater costs* Depending on the source of water supply, there may be charges for water consumption. This consumption will be dictated by the make-up water needs for the station.

5. *Consumables* All consumables required for normal operation should be accounted for. Typical examples here are diesel lubricating oil consumption and oil changes at service intervals.

6. *Labour costs* These will depend on the manning levels and degree of automization of the plant.

7. *Maintenance charges* Maintenance costs as an average over the plant life (based on previous experience and guidance) should be sought from the manufacturer.

8. *Purchased power costs* Depending on the schemes applicable, there may also be charges for import of electrical power. In the UK the electrical supply authority charges a sum based on a maximum demand charge, a unit charge and a fuel cost adjustment. Depending on circumstances, there may also be a fixed annual charge to cover rent of the authority's equipment.

9. *Inflation* After economic analysis it may be necessary to review how inflation may influence the costs for the items purchased.

10. *Erection periods* The period of erection can be critical, as this dictates when power will be available and income earned. Depending on the size of the station and scope, this could vary between less than 6 months to over 2 years. For systems such as the combined cycle it is practical to commission the gas turbines first in advance of the steam generators, so that useful power is generated at the earliest possible date.

20.6 Plant and installation

In order to operate the prime movers described in the previous sections it is necessary to provide auxiliary equipment for the start-up, steady operation and shut-down of the basic equipment as well as for monitoring and controlling its performance. The need also arises for the maintenance of the plant which invokes the provision of cranage and lay-down areas in the engine room. The following describes these features for the various types of prime movers. The driven machines (i.e. the electrical generators) are also reviewed in detail so that the complete picture of industrial generating stations can be obtained.

20.6.1 Diesel power plants

When driving a generator the diesel engine is usually a multi-cylinder machine similar in appearance to that shown in Figure 20.28. The figure illustrates a nominal 3 MWe engine with twelve cylinders in Vee formation, such an arrangement being attractive for its compactness. The engine will be required to operate at synchronous speed which is, in turn, dependent upon the required frequency (f) and the number of generator pole-pairs (p). We have

$$\text{Operating speed, } N = \frac{60f}{p} \text{ rev/min}$$

A typical block layout for a diesel engine room is given in Figure 20.29, showing the necessary auxiliaries and

Figure 20.28 A Vee-12 diesel engine

local control panels. Not shown but also necessary are cranes and fire-protection systems. The engine room itself must additionally provide sufficient noise attenuation as shall be deemed necessary, and there must also be adequate fuel storage, reception and handling facilities.

A key issue also concerns the foundation requirements of the generator set. Most manufacturers will provide data on their machines' requirements, but the detailed design of the block and its interface with the engine bedplate require close attention. Sometimes it is possible to provide a basement for the engine auxiliaries, while ventilation considerations (excluding combustion air requirements) should recognize that some 8% of the engine's rating will be radiated as heat, in the absence of any heat-recovery equipment.

The compression ignition engine can operate on a variety of liquid fuels and gas. Modern designs permit reliable operation on up to the heaviest residual oils, so permitting improved fuel flexibility (providing, of course, that an adequate fuel-handling capability is also present). The engine may also operate in a dual-fuel mode, using a mixture of gas and air with the option to revert to a liquid fuel in the event of any interruptions to the gas supply. Thus a more complex control and protection system must be provided in order to give modulation for both the oil and gas flows.

Air intake

The quality of the air supply directly affects the output, efficiency and life of the engine. The requirement of the induction system must therefore be to supply the engine with clean dry air close to ambient temperature conditions. Oil bath or dry (paper element) filters are adequate for low dust concentration conditions. However, as the dust burden of the air increases, centrifugal pre-cleaners become necessary. The combustion air requirements of a diesel may be taken, for estimating purposes, to be 9.5 $m^3 h^{-1}$ per kW generator rating.

Cooling

For small engines below 0.5 MW output, air cooling can be considered. However, water cooling by circulation of water through cylinder jackets is the method normally used for diesel generating plants. Fan-cooled radiators (as seen in Figure 20.29) or the system shown in Figure 20.30 are typical. As an alternative to the cooling tower, when a good source of raw water is available (such as from a river) this can be used on one side of the heat exchanger.

Engine starting

For small high-speed engines, electric start systems, incorporating lead-acid or alkaline batteries, are practical. For larger machines air-start systems are normally used. Using compressed air, starting may be effected by means of an air-driven motor engaging the flywheel or by directing air into each cylinder, in firing sequence, by means of a camshaft-driven distributor or mechanically operated valves. The air so needed can be provided from receivers charged up when the engine is running or from a small auxiliary compressor unit. The latter system is most common on the larger installations, while on critical engines an electric motor-driven compressor is backed by an IC engine-driven unit.

Control and instrumentation

Engine speed is controlled by the use of variable-speed governors which can be mechanical, mechanical-hydraulic or electronic. The last option is gaining wide acceptance for generation purposes due to its speed of response and ready integration with other control equipment used in fully automated installations (probably incorporating more than one generating unit).

Stop, start, emergency stop and speed controls are grouped adjacent to the engine, and the machine is usually equipped with instrumentation for monitoring its safe operation. Very often a centralized control room is provided, particularly when there are a number of units in operation. Typical signals monitored and recorded remotely are:

Engine and turbocharger speeds
Hours run
Lubricating oil pressures and temperatures
Starting and change air pressures

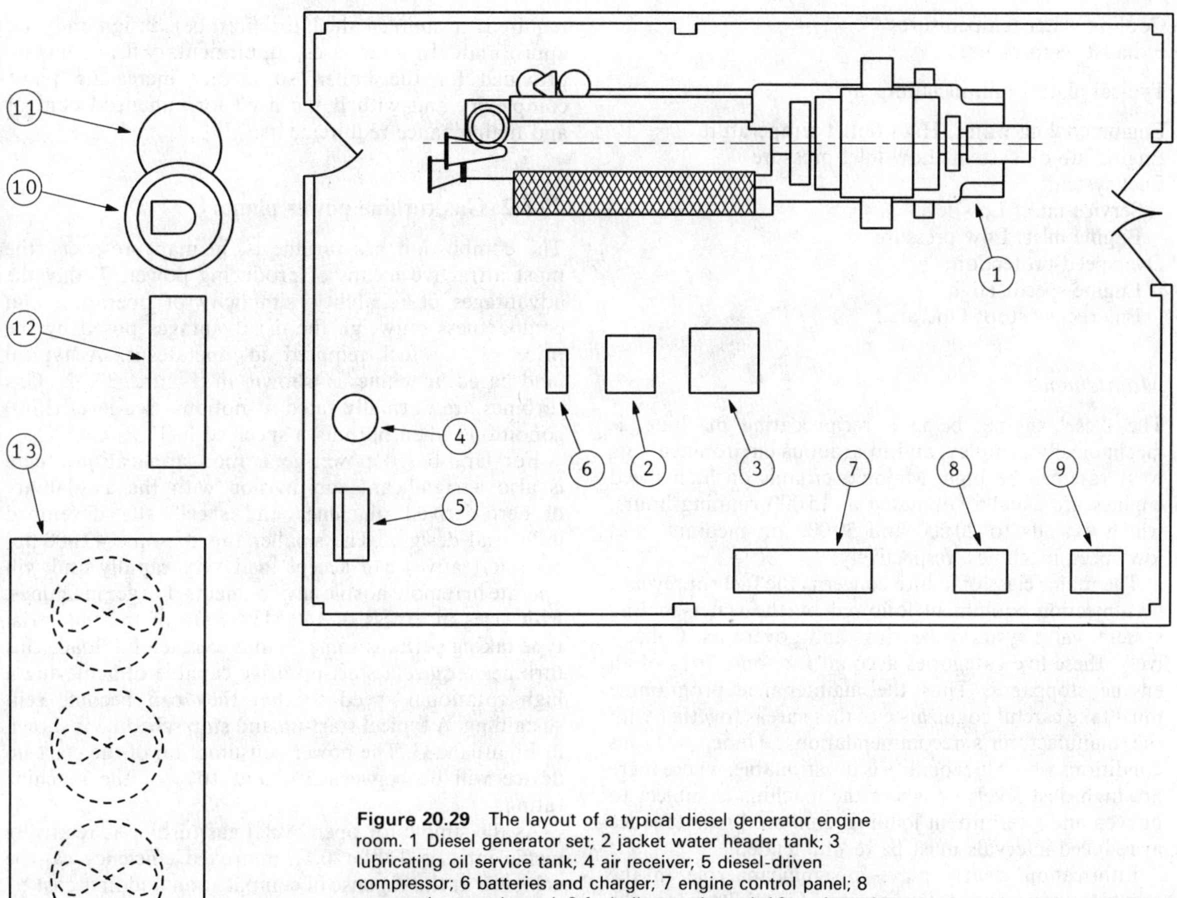

Figure 20.29 The layout of a typical diesel generator engine room. 1 Diesel generator set; 2 jacket water header tank; 3 lubricating oil service tank; 4 air receiver; 5 diesel-driven compressor; 6 batteries and charger; 7 engine control panel; 8 pneumatic control panel; 9 fuel oil control panel; 10 engine exhaust silencer; 11 charge air filter; 12 daily service fuel oil tank; 13 three-section radiator

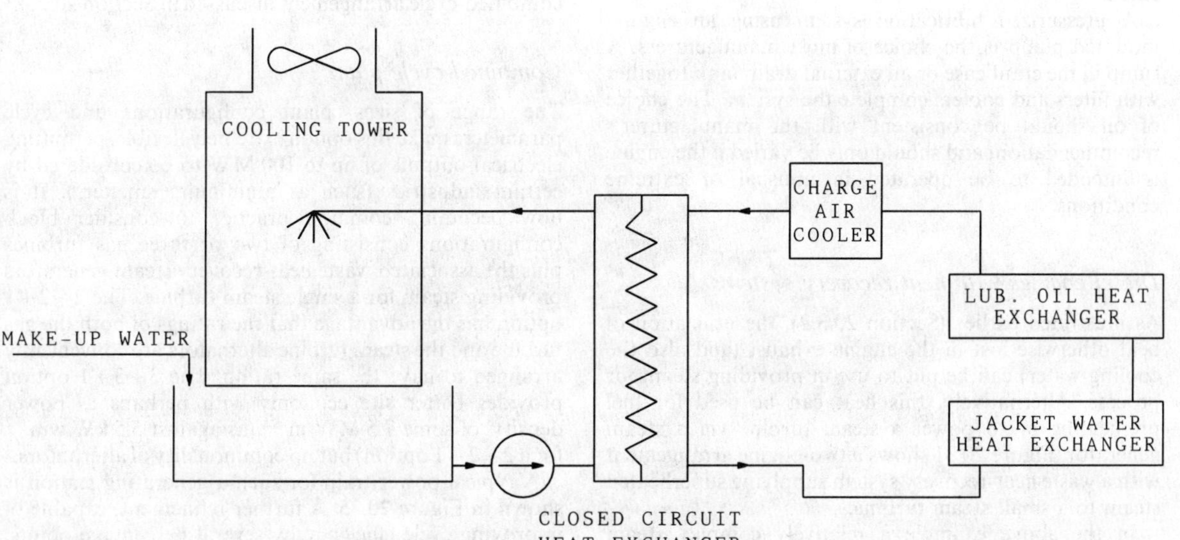

Figure 20.30 A typical closed-circuit secondary cooling system

Cooling-water temperatures
Exhaust temperature

Typical alarms annunciations are:

Engine cooling water: High outlet temperature
Engine lub oil system: Low inlet pressure
Fuel system:
  Service tank: Low level
  Engine inlet: Low pressure
Overspeed protection:
  Engine speed: High
  Emergency stop: Operated

Maintenance

The diesel engine, being a reciprocating machine, is mechanically complex, and in arduous environments its wear rate can be high. Major overhauls on high-speed engines are usually stipulated at 15 000 running hours, which extends to 20 000 and 30 000 on medium- and low-speed machines, respectively.

The major class of failure concerns the fuel supply and fuel-injection equipment followed by the water-cooling system, valve systems, bearings and governors. Collectively, these five categories account for some 70% of all engine stoppages. Thus the maintenance programme must take careful cognizance of these areas together with the manufacturer's recommendations. Under arduous conditions where fuel quality is questionable, where there are high dust levels or where the machine is subject to uneven and intermittent loading, enhanced maintenance at reduced intervals must be recommended.

Lubrication clearly plays a significant role in the reliable operation of the engine. In addition to the lubricant's primary task of reducing friction and minimizing wear it also acts as a cooling medium, a partial seal between the cylinders and piston rings, and a means of flushing combustion and other impurities out of the engine.

A pressurized lubrication system using an engine-mounted pump is the choice of most manufacturers. A sump in the crankcase or an external drain tank together with filters and coolers complete the system. The choice of oil should be consistent with the manufacturer's recommendation, and should only be varied if the engine is intended to be operated in unusual or extreme conditions.

Diesel engines with heat recovery systems

As illustrated earlier (Section 20.3.2), the utilization of heat otherwise lost in the engine exhaust (and also the cooling water) can be put to use in providing steam for process. Alternatively, this heat can be used for fuel pre-heating or to power a steam turbine via a steam generator. Figure 20.31 shows a two-engine arrangement with a waste heat-recovery system supplying superheated steam to a small steam turbine.

In the above example a relatively complex steam generator of the watertube type has been adopted. Where lower-quality steam for process or fuel heating is required, a simpler shell (or firetube) design may be appropriate. In some cases, supplementary firing may be provided for the boiler, so further increasing plant complexity and with it the need for enhanced control and maintenance requirements.

20.6.2 Gas turbine power plants

The combustion gas turbine is, in many respects, the most attractive means of producing power. Today the advantages of reliability, simplicity of operation and compactness outweigh the disadvantages posed by the price of the fuel required to operate it. A typical land-based machine is shown in Figure 20.32. Gas turbines are normally rated at notional sea-level (ISO) conditions when burning a specified fuel.

For land-based power-generation applications there is also a significant sub-division with the availability of aero-derived machines and specifically developed industrial designs. The smaller (up to some 42 000 hp) aero-derivatives can accept load very rapidly and will operate in remote hostile environments. Larger machines, with sizes in excess of 140 MWe, are of the industrial type taking perhaps some 30 min to reach full load. Gas turbines require a start-up drive capable of achieving a high rotational speed so that they can become self-sustaining. A typical start-up and stop schedule is shown in Figure 20.33. The power consumption of the start-up device will be between 5% and 10% of the machine rating.

As the simple (or open-cycle) gas turbine is relatively inefficient (see Table 20.1), improved efficiency can be achieved at the expense of complication and first cost by recovering some of the heat from the exhaust. One arrangement employing heat exchangers in a closed circuit system is shown in Figure 20.34. However, the most popular means of recovering a significant proportion of the gas turbine's utilized heat is by means of a combined-cycle arrangement discussed in Section 20.2.4.

Combined-cycle plant

The range of sizes, plant configurations and cycle parameters make this option extremely flexible, permitting electrical outputs of up to 100 MW to be considered by certain industries (such as aluminium smelters). It is now becoming common practice to consider block configurations consisting of two or three gas turbines plus the associated waste heat-recovery steam generators providing steam for a single steam turbine. The 2+2+1 option has the advantage that the ratings of both the gas turbine and the steam turbine alternators are conveniently arranged to have the same rating. The 3+3+1 option provides better site economy with perhaps a 'power density' of some 7.5 MW/m^{-2} (as against 5.5 $kW\ wm^{-2}$ for a 2+2+1 option) but no commonality of alternators.

A typical power train for such a generating station is shown in Figure 20.35. A further refinement, capable of improving cycle efficiency by several percentage points, is the introduction of dual steam pressures. If there is no requirement for the provision of process steam then it is

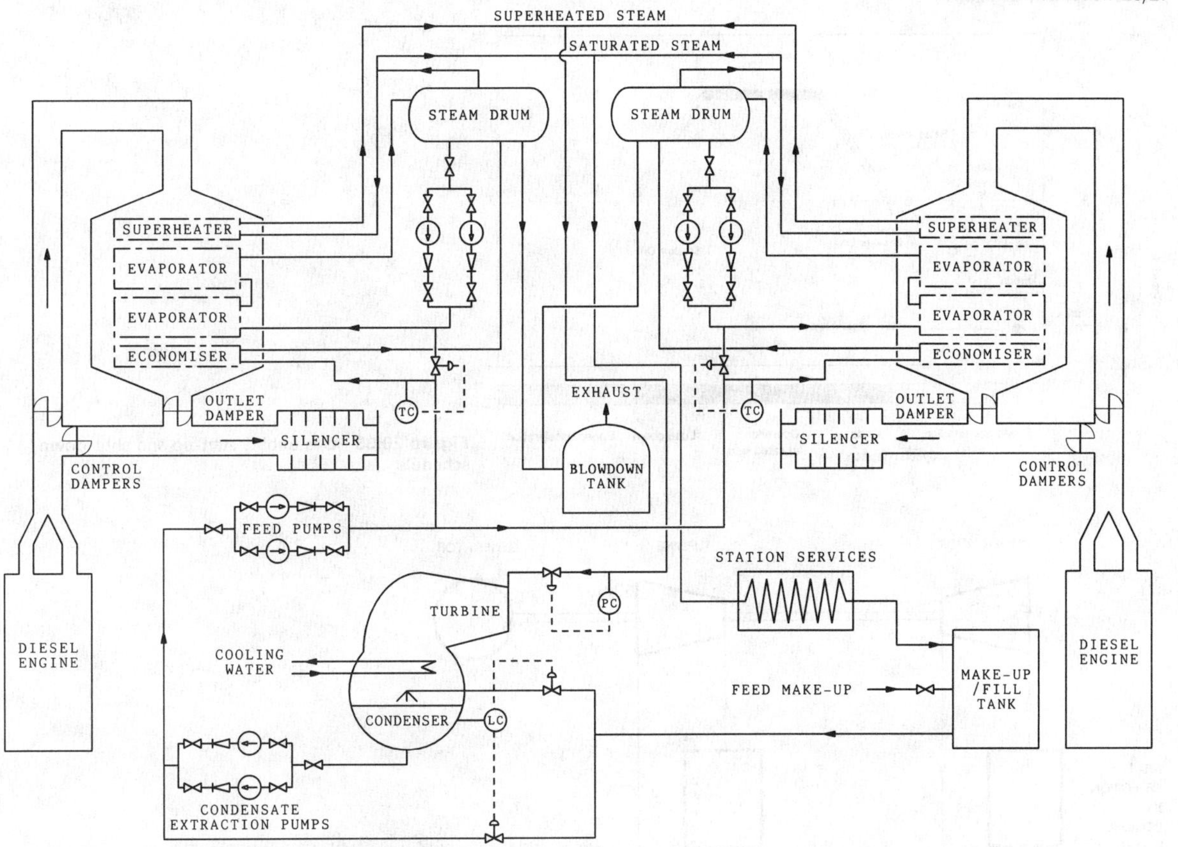

SUPERHEATED STEAM
SATURATED STEAM
STEAM DRUM
STEAM DRUM
SUPERHEATER
EVAPORATOR
EVAPORATOR
ECONOMISER
SUPERHEATER
EVAPORATOR
EVAPORATOR
ECONOMISER
EXHAUST
TC
TC
OUTLET DAMPER
OUTLET DAMPER
SILENCER
SILENCER
BLOWDOWN TANK
CONTROL DAMPERS
CONTROL DAMPERS
FEED PUMPS
STATION SERVICES
PC
TURBINE
DIESEL ENGINE
DIESEL ENGINE
COOLING WATER
FEED MAKE-UP
MAKE-UP /FILL TANK
CONDENSER
LC
CONDENSATE EXTRACTION PUMPS

Figure 20.31 A two-engine diesel power plant with heat recovery

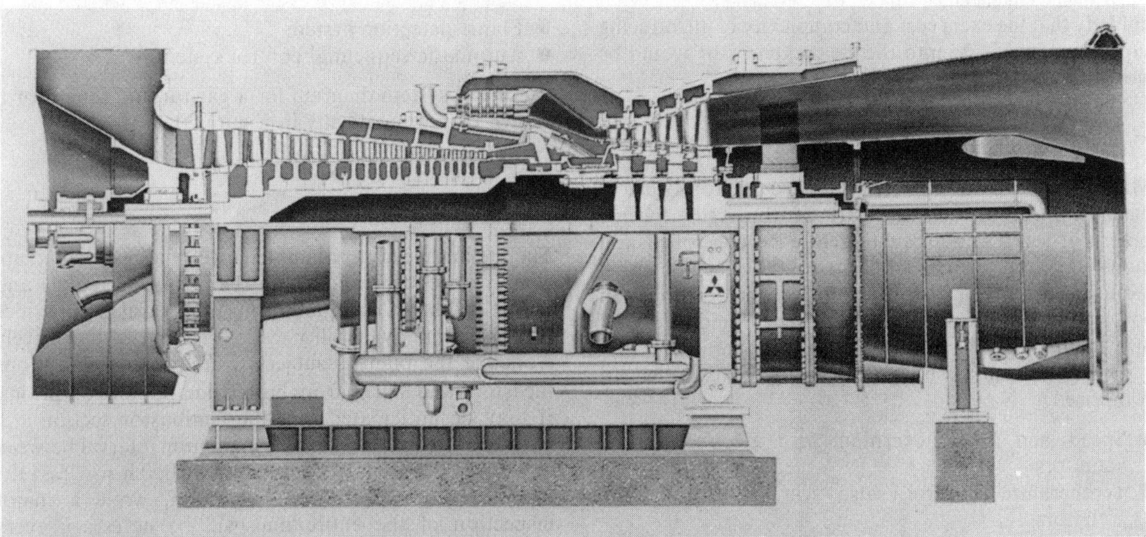

Figure 20.32 The Mitsubishi MW 701 gas turbine rated at 131 MW (ISO)

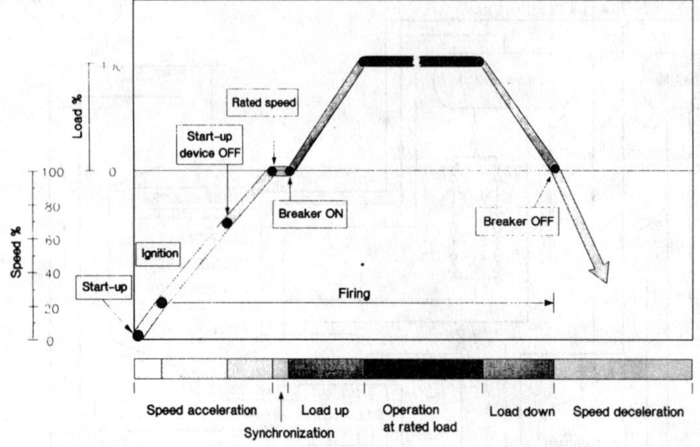

Figure 20.33 Gas turbine start-up and shut-down schedule

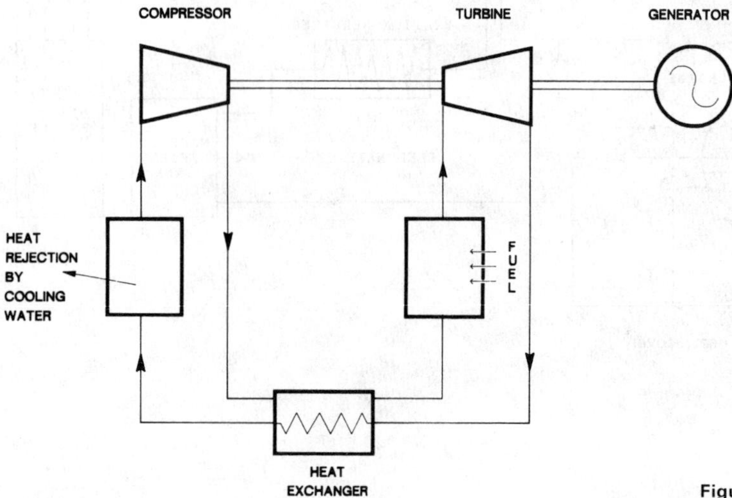

Figure 20.34 A closed-cycle gas turbine arrangement

unlikely that the extra cost and complexity of introducing supplementary firing into the steam generator would be worthwhile.

Control and instrumentation

Manufacturers now offer as standard microprocessor-based control systems as part of their gas turbine generator set. The set can therefore be controlled and monitored from panels adjacent to the machine or, if required, remotely from a central control room. In combined-cycle installations the system would be linked into the higher-level station control system.

The instrumentation may typically be expected to comprise:

- Speed and load governing with electrohydraulic actuators
- Temperature control, also via electrohydraulic actuators
- Temperature, pressure and vibration monitoring systems
- Flame-detection system
- Automatic sequential control system

A control system diagram for a gas turbine generator is shown diagrammatically in Figure 20.36.

Maintenance

The life and necessary maintenance of a gas turbine are heavily dependent upon both the operating regime and the fuel quality. Continuous firing on natural gas provides the optimum availability which will be progressively eroded if the plant is subject to frequent interruptions (i.e. stops and starts) from both cold and hot conditions. It may be anticipated that the combustor section will require most attention, with a maximum interval between inspections of some 8000 h. Every 16 000 h (or less) the turbine section will need inspection, while a major inspection of the entire unit will be necessary every 32 000 h. Under optimal conditions, the average operating availability of base-load and based gas turbine plant

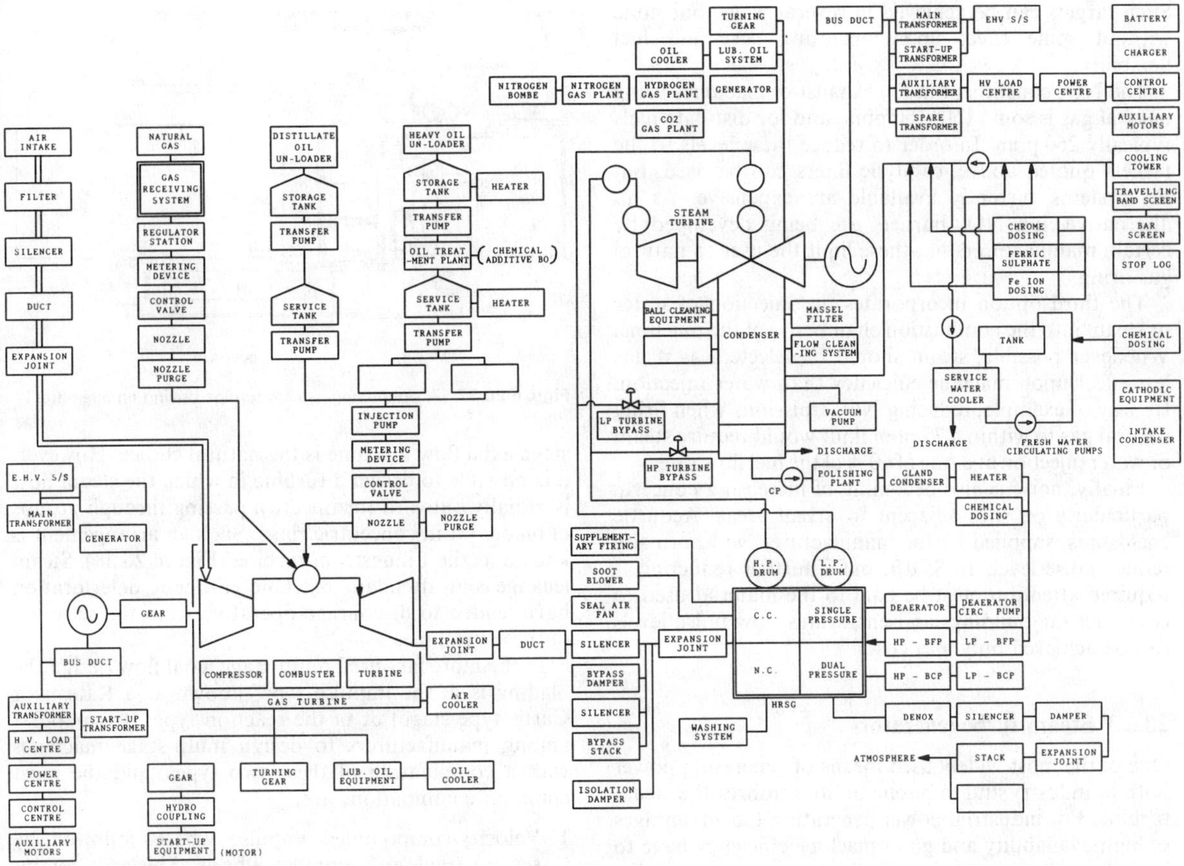

Figure 20.35 A typical combined-cycle power train

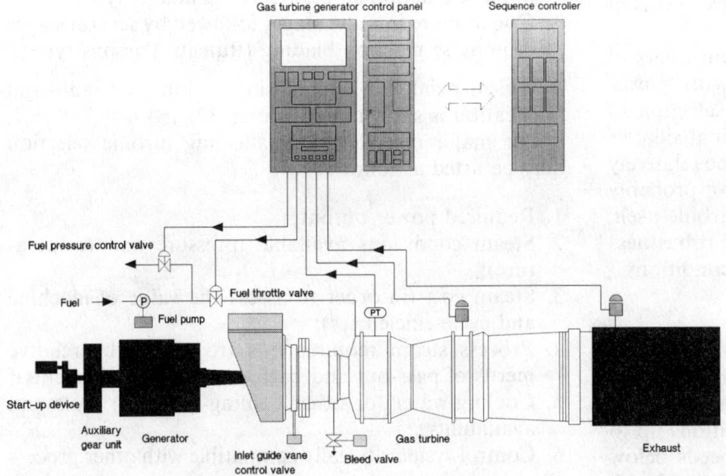

Figure 20.36 A gas turbine control system

could exceed 95% (averaged over a 4-year or 32 000-h period).

Pollution control

In the past, gas turbine installations have suffered from smoke emissions. However, today, with better combustor design and combustion control, this emission has ceased to be a nuisance. Of more concern now and resulting from higher combustor (and therefore combustion) temperatures is NO_x emissions. Current legislation is tending to demand that gas turbine installations control their nitrogen emissions to levels between 50 and 75 ppm.

Such targets can be achieved in several ways, but none without some effect upon operating cost and fuel flexibility.

The NO_x constituent in the exhaust of machines firing natural gas is some 150–160 ppm, and for distillate fuels typically 260 ppm. In order to reduce these levels to the targets quoted above, catalytic filters can be used, but the systems currently available are expensive. As an alternative, low-NO_x burners are being developed by certain manufacturers but these limit the user to natural gas firing.

The third option incorporates the injection of water or steam into the combustion chamber(s) of the machine. Whenever possible, steam should be selected, as it has less effect upon machine efficiency than water injection. By way of example, reducing NO_x emissions when firing natural gas to within a 75 ppm limit would require steam or water injection at a rate of 60% of the fuel flow rate.

Finally, noise is also becoming of increasing concern, particularly on sites adjacent to urban areas. Acoustic enclosures, supplied by the manufacturer, will normally reduce noise levels to 85 dB, but if further reduction is required attention must be paid to the material used to construct the buildings and enclosures; low noise levels can be achieved only at a cost.

20.6.3 Steam turbogenerators

One of the most widely used means of generating power both in industry and in public utilities utilizes the steam turbine. For industrial power generation the advantages of high availability and good machine efficiency have to be balanced against relatively high first costs and the need for the provision of both expensive steam-generating plant and heat-rejection equipment. However, in many instances the recovery of waste process heat by means of a steam cycle is economically attractive.

Alternatively, the option that a steam plant offers of the provision of process steam coupled with power generation may be the key element in the selection of generating plant. The turbogenerators and their auxiliaries for use in such applications tend therefore to be relatively unsophisticated, with no feed heating, except for probably the provision of a deaerator. Again, in the turbine itself, machine efficiency tends to be sacrificed for robustness and the ability to accommodate varying load conditions.

Options

The single-stage, single-valve turbine is the simplest option. Such a machine is suitable for applications requiring powers up to 300 kW, steam conditions up to a nominal 120 bar, 530°C and rotational speeds below 5000 rev/min. A typical machine is shown in Figure 20.37, characterized by an overhung rotor mounted on a stiff shaft capable of being accelerated from cold to operating speed within 10 s. The efficiency is low, but so is the cost of the installation.

For higher powers and for a wider range of steam conditions incorporating all combinations of back-pressure, condensing and pass-out systems, the multi-stage axial flow machine is the natural choice. However, it is possible to design a turbine in which the steam flow is radially outward (or inward), passing through groups of blades set in concentric rings. Such an arrangement is known as the Ljungström turbine (Figure 20.38). Steam leakage and attendant machine efficiency deterioration have tended to discourage operators from the selection of these machines.

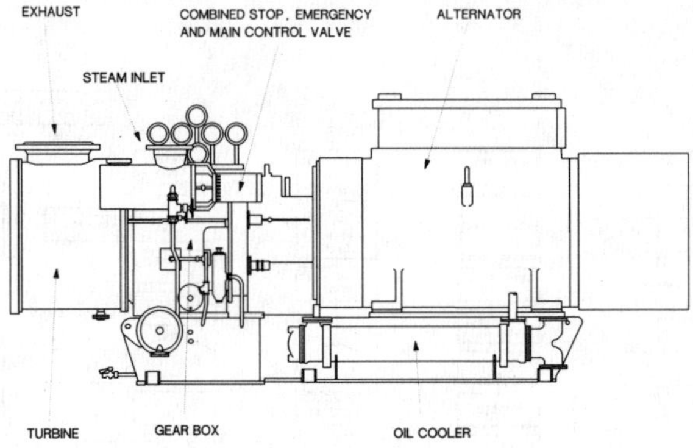

Figure 20.37 A single-stage steam turbine driving an alternator

In the more favoured multi-stage axial flow design the blading is of the impulse type (known as a Rateau or Curtis type stage) or of the reaction type. It is popular among manufacturers to design multi-stage machines with a combination of these two types, and the most common combinations are:

1. Velocity-compounded impulse wheels followed by several single-row impulse wheels. This is a combination of the Curtis and Rateau types.
2. Velocity-compounded impulse wheels followed by impulse reaction stages (Curtis–Parsons type).
3. One or more impulse stages followed by several stages of impulse reaction blading (Rateau–Parsons type).

A typical axial flow condensing turbine for industrial application is shown in Figure 20.39.

The major considerations affecting turbine selection may be listed as follows:

1. Required power output;
2. Steam conditions available (pressure and temperature);
3. Steam cost (in order to assess the value of machine and cycle efficiencies);
4. Process steam requirements (to assess the relative merits of pass-out and back-pressure arrangements);
5. Cooling-water (or other cooling medium) costs and availability;
6. Control systems (ideally compatible with other process plant instrumentation) and automation;
7. Safety features (e.g. operation in explosive environments, alarm and condition-monitoring systems).

Installation

The foundations of any turbogenerator installation play a significant part in the safe operation of the machine.

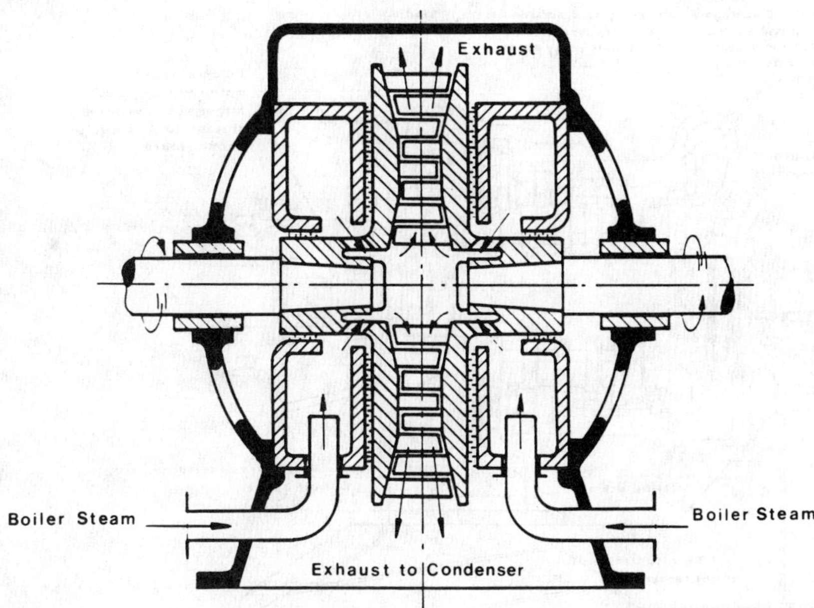

Figure 20.38 The radial-flow double-motion reaction turbine (Ljungström)

Industrial turbines will normally operate at above synchronous speed and will drive an alternator via a reduction gearbox. Any vibration or out of balance occurring under both normal or abnormal operation must be accommodated by the foundations, and their design should therefore best be undertaken by a specialist organization. The layout of the steam and water pipework, of the machine auxiliaries and, when required, of the condenser all tend to add to the complexity of the turbine island.

Where an underslung condenser has been specified, the provision of a basement to the engine room offers the attraction of compactness at the expense of enhanced civil works, while alternatively, the specification of pannier condensers can obviate the need for a basement and will simplify the foundation design, but will considerably increase the floor area requirements. The condensing plant itself consists essentially of banks of tubes through which cooling water flows and around which exhausted steam from the turbine is condensed to form a vacuum. Such tubes have traditionally been made of brass, but where severe corrosion conditions exist, cupro-nickel is sometimes used.

Control and lubrication

The turbogenerator speed must be maintained within narrow limits if it is to generate power acceptably and the control system must also be capable of preventing overspeed upon sudden loss of load. For this latter requirement fast-acting valves are necessary with full modulation within 0.5 s. The security of the turbine is also dependent upon the lubrication of its bearings, and it will be found that the control systems are closely linked with the various lubricating systems (turbine, gearbox and alternator).

A typical high-pressure, quick-response governor and lubricating system for a speed-governed turbine is shown in Figure 20.40. Basically, a speed governor controls the oil pressure supplied to the servo-operated control valves, which are arranged so that they open in sequence to ensure maximum steam economy. Other types of control systems can be incorporated with the basic speed-governed system to provide back-pressure, inlet-pressure and pass-out pressure governing. A typical fully interlinked control system for a pass-out system is shown in Figure 20.41.

Oil is supplied to these systems by a pump normally driven from the turbine rotor, with an auxiliary pump for use when starting the turbine. The main lubricating oil pump draws oil through a suction strainer and discharges to the emergency valve and hence through a cooler and filter to the lubrication system. A separate auxiliary lub-oil pump is fitted for supplying oil to the bearings when stopping and starting and to lift the emergency valve at start-up.

These systems must be inherently reliable and safe, particularly from the point of view of fire protection, and fire-resistant fluids can be used for control systems in order to reduce this risk. Fire-detection and fire-fighting systems should always be provided for any turbogenerator installation.

20.6.4 Generators

Generators for use with the prime movers previously described will almost invariably operate in conjunction with an alternating current electrical system. Such a.c. generators will operate either in parallel with other sources of supply (i.e. a grid system or other generators) or in isolation providing the whole electrical supply to

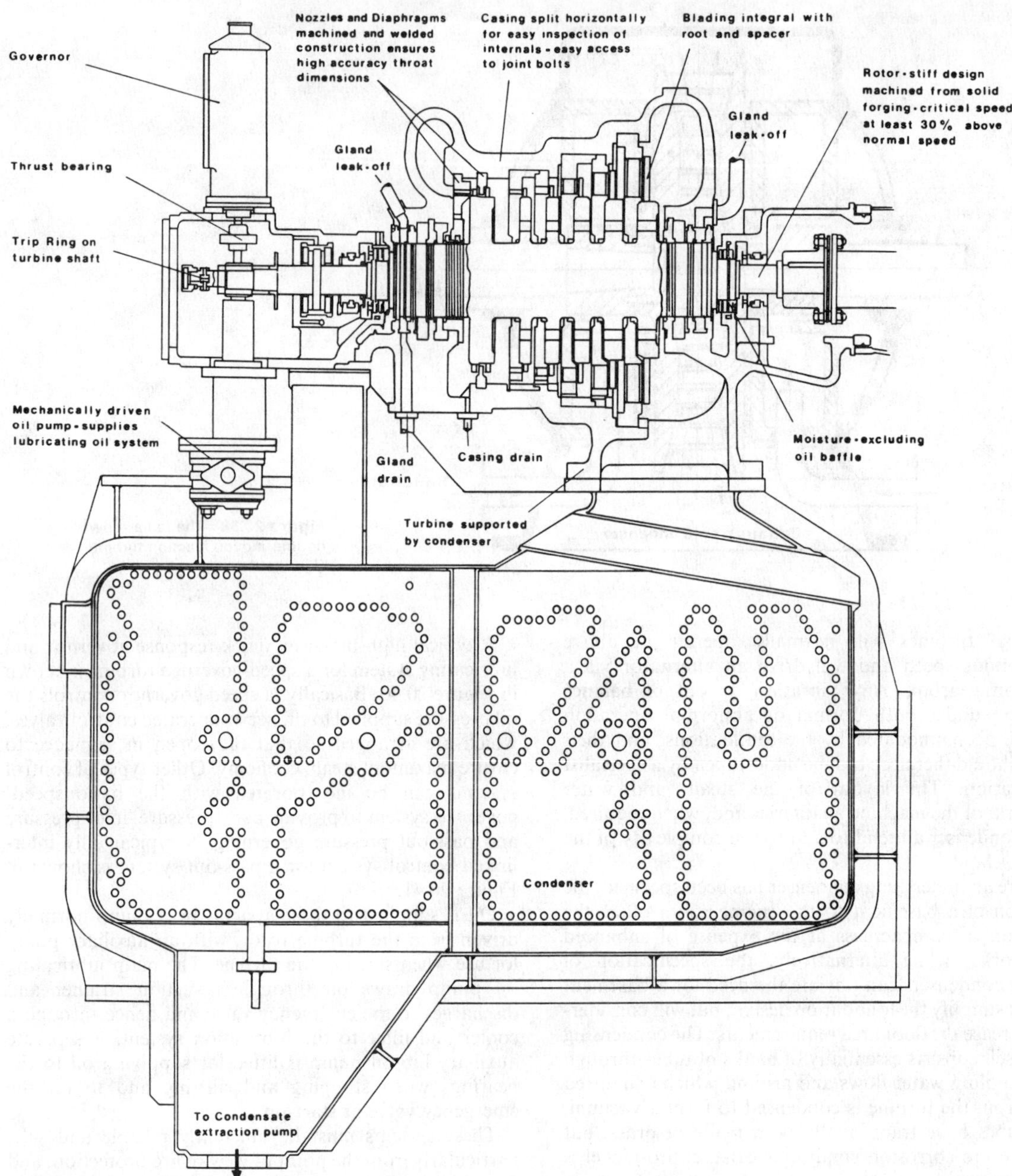

Figure 20.39 Longitudinal section of a small condensing turbine

an installation. A.C. generators can be one of the two main types:

1. *Asynchronous generators* These types are, in effect, induction machines operated at supersynchronous speed. An asynchronous generator can be either compensated type, in which excitation supplies are provided via a commutator, or uncompensated type, in which excitation supplies are provided by the supply system to which the generator is connected. Clearly, uncompensated generators are only suitable for use in parallel with other sources of supply capable of providing the excitation-magnetizing current necessary. While asynchronous generators offer a low-cost option for generation, the disadvantages of lack of control capability and the need usually for alternative sources of supply means that they are only occasionally utilized and rarely at ratings exceeding 5 MW.
2. *Synchronous generators* This type are in use for all generating appications up to the highest ratings

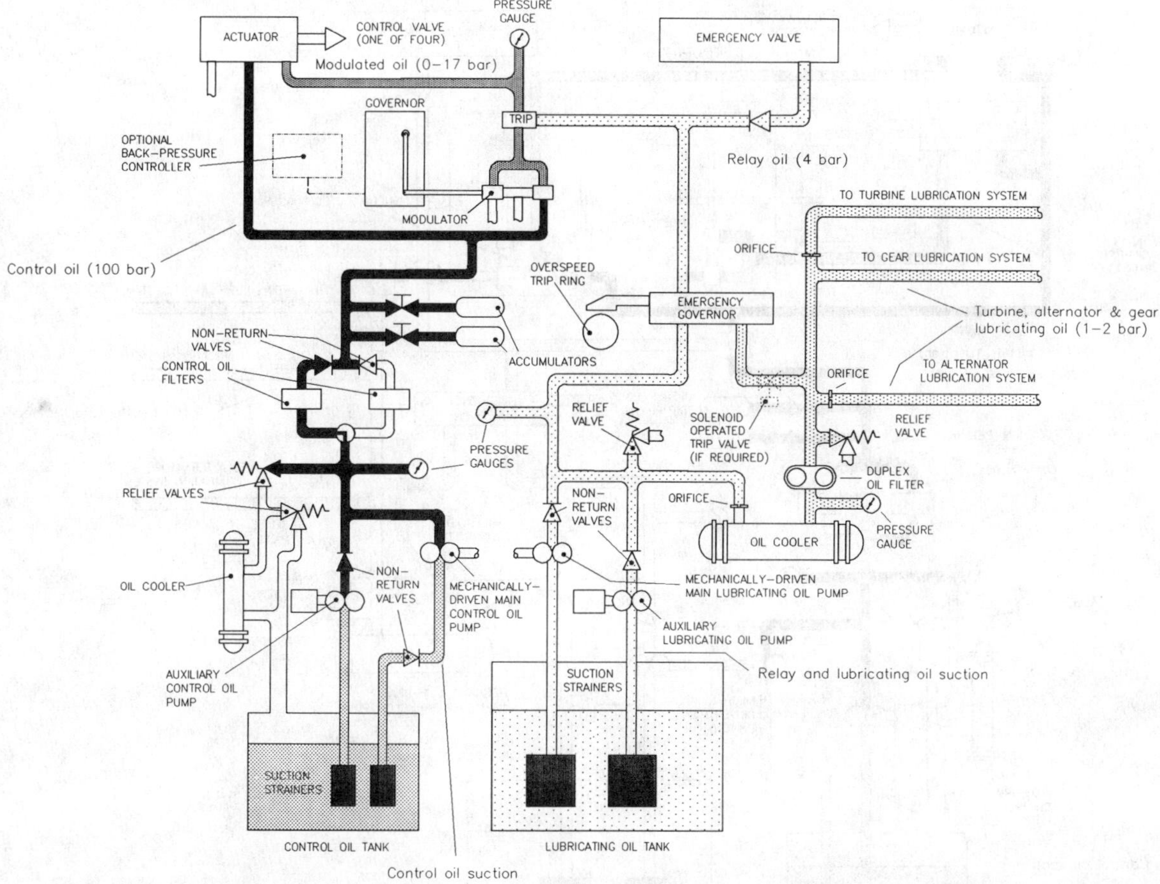

Figure 20.40 A control system for a speed-governed steam turbine

(660 MW and above). The machine is operated at synchronous speed and the terminal voltage and power factor of operation are controlled by the excitation. Principal features of synchronous generators relate to:

(a) The method of excitation;
(b) The method of cooling;
(c) Temperature rise classifications;
(d) Insulation systems.

Since synchronous machines represent the most common type of generator used in power installations, the remainder of this section is principally concerned with this type.

Excitation

The excitation system provides the magnetizing current necessary for the generator to operate at the desired voltage and, when in parallel with other generators, supplies the required amount of reactive current. In modern practice the excitation system can be either brushless or static.

In a brushless system an a.c. exciter with a rotating armature and stationary field system is provided. The voltage applied to the stationary field system is varied, thus changing the output of the rotating armature. This output is rectified via shaft-mounted diodes to produce a d.c. supply which is connected to the main generator field.

The excitation system is generally designed such that failure of a single diode within the rectifying bridge does not affect the capability of the generator to provide full load. Diode failure can be monitored by measuring the amount of a.c. ripple induced within the a.c. exciter field winding and an alarm or trip provided. In operation, however, diode failure rates are extremely low.

In static systems d.c. is supplied to the generator field winding via slip rings, the d.c. being produced from diode cubicles supplied from rectifier transformers fed from the generator output. In general, static systems are capable of producing a faster response time than brushless systems and are typically used on larger (above 70 MW) sets.

Cooling

Cooling of generators is classified according to the nature of the coolant, the arrangement of the cooling circuit and the method of supplying power to circulate the coolant.

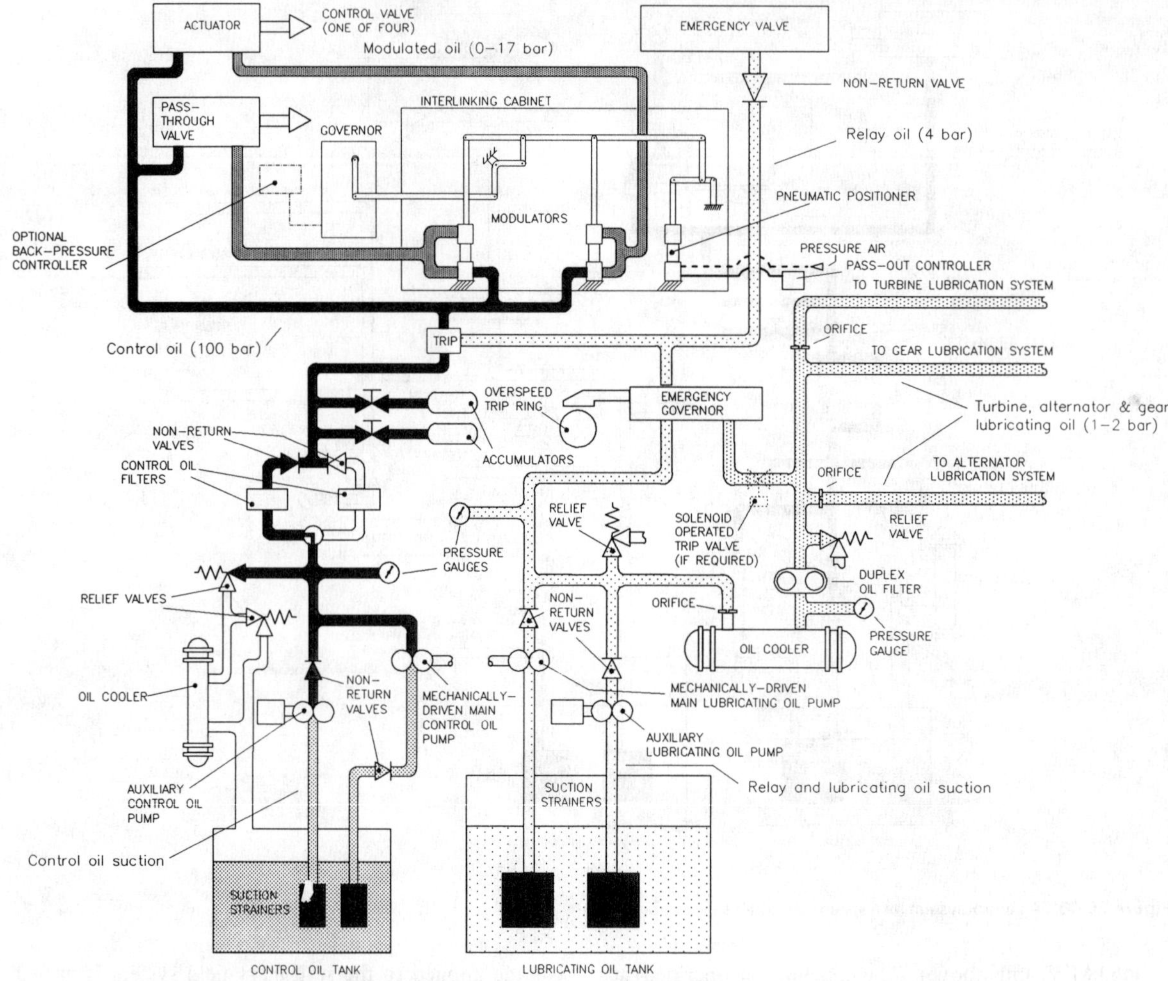

Figure 20.41 A control system for a pass-out steam turbine

For air-cooled machines a simplified coding designation is widely used to describe the cooling system. In this system the first characteristic numeral signifies the cooling circuit arrangement (e.g. inlet duct ventilated or integral heat exchanger) while the second signifies the method of supplying power to circulate the coolant (e.g. self-circulation or from an integral mounted independently powered fan). For machines with more than one cooling system (e.g. a primary and a secondary cooling circuit) a complete coding system is used. This includes the nature of the coolant within each cooling system and the means of circulating the coolant and of removing heat from the machine.

Insulation type and temperature rise

The winding insulation must remain intact with respect to both electrical insulating properties and mechanical strength for the life of the machine. Operation of insulation at temperatures higher than the design will affect the life of the insulation.

Temperature-rise limitations above an assumed ambient temperature (dependent upon location) are used to specify the requirements of the insulation. Classifications utilized for generators are typically:

Class A 105°C
Class B 130°C
Class F 155°C
Class H 180°C

Typically, insulation limited to Class B and Class F temperature rises will employ inorganic materials (e.g. mica or glass fibre) bonded with a thermosetting synthetic resin. Class H type insulation may also include silicone elastomers.

Connections

The connection of the generator to associated switchgear may be from a cable box located on the side of the generator or exposed terminals usually placed below the machine. When the terminals are the exposed type a cable connection can still be made by utilizing cable sealing ends or by phase-insulated or phase-segregated busbars. Typically, cable boxes will be provided for machines rated up to 10 MVA while for generators rated above 30 MVA, phase-isolated busbars are generally used.

21 Electrical Distribution and Installation

I G Crow, BEng, PhD, CEng, FIMechE, FIMarE, MemASME
Davy McKee (Stockton) Ltd
and
R Robinson, BSc, CEng, FIEE
The Boots Co. PLC

Contents

21.1 Introduction

This chapter introduces the basic items of design and specification for the principal systems and components of an electrical industrial installation. Electrical supply systems are discussed with regard to interface with the supply authorities and the characteristics. Salient features of switchgear, transformers, protection systems, power factor correction, motor control equipment and standby supplies are identified and discussed together with reference to the relevant Codes of Practice and standards. The equipment and systems described are appropriate to industrial plant installations operating at typically 11 kV with supply capacities of around 20 MVA.

21.2 Bulk supply

The majority of industrial installations will receive all or part of their electrical energy from an area electricity board supply authority. The factors which will affect the way in which this energy is imported to the site and subsequently controlled are:

1. The voltage and frequency
2. The load requirements
3. The fault level
4. The tariff
5. Access to equipment

The key parameters for each of these factors are discussed below.

21.2.1 Voltage and frequency

The voltage of the supply will depend upon the load requirements of the site and the relative capability of the local supply system. Voltages typically available are 33 kV, 11 kV, 6.6 kV, 4.16 kV, 415 V and 380 V. For some larger installations 132 kV (or higher) voltage supplies may be provided.

Statutory regulations imposed upon the supply authorities will normally limit variations in supply voltage to $\pm 6\%$ and in frequency to $\pm 1\%$. In order to maintain supply systems within these limits supply authorities may, in certain cases, impose restrictions upon the starting of large motors in terms of either current drawn at start-up or of frequency of start-up. Alternatively, the supply may be arranged so that the point of common coupling with other consumers is at a higher voltage system where source impedance will be lower.

The increasing usage of semiconductor equipment, particularly for variable-speed drives, has caused more attention to be given to the problem of distorting loads and the harmonics, currents and voltages they create within the supply system. The presence of harmonic voltages and currents within a supply system will affect induction and synchronous motors, transformers, power factor correction capacitors, energy metering and devices relying upon a pure sine wave for operation.

The supply authority will be concerned about the levels of each harmonic and the total percentage distortion. Within the UK, Electricity Council Engineering Recommendation G5 gives limits for harmonic voltages and currents acceptable at the point of common coupling with other consumers.

21.2.2 Load requirements

The load imposed upon the supply system will need to be analysed for load profile, i.e.:

1. Is the load constant?
2. Does the load vary significantly on a regular hourly, daily or annual cycle?
3. Is the load likely to increase as the installation expands?
4. Will the characteristics of the load affect the system supplying it?

The factors which then need to be considered are potential unbalanced loading, transient occurrences and harmonic distortion. Unbalanced loading will cause an unequal displacement of the voltages, producing a negative phase sequence component in the supply voltage. Such negative phase-sequence voltages will cause overheating, particularly in rotating plant. The starting of large motors or the operation of electric arc furnaces are two examples of loading which draw large amounts of reactive current from the supply system. These currents, flowing through the largely reactive supply system, will result in fluctuations in the supply voltage, and these fluctuations will affect other consumers (e.g. causing lighting to flicker).

21.3.3 Fault level

The fault level at the point of supply to the installation requires consideration. The minimum fault level which can occur will affect the operation of the installation, particularly with regard to voltage regulation when starting motors, while the maximum fault level will determine the ratings of equipment installed. All supply systems develop with time, and the switchgear and other equipment installed at an installation must be chosen to be suitable if the external supply system expands.

The contribution of the installation itself to fault level at the point of common coupling is another important consideration. Rotating plant, either generators or motors, will contribute to both *make* and *break* fault levels. The contribution of synchronous machines to fault current can be calculated from sub-transient and transient reactance values, and although similar calculations can be carried out for induction motors, accurate data are not generally readily available for groups of small motors operating together, as in a typical industrial installation. Generally, induction motor contribution to *make* fault levels is taken as being equivalent to motor starting current and to *break* levels as twice motor full-load currents.

21.2.4 Industrial maximum demand tariffs

The exact details of industrial maximum demand tariffs

vary between supply authorities. However, the salient features of most tariffs are as follows:

1. *Fixed charge*: This is typically a fixed cost per month plus a charge for each kVA of agreed capacity. It is important to realize that any change in the operation of a particular site may affect the supply capacity required. The supply capacity charge, although related to, is not dependent on the rating of the equipment installed at the site.
2. *Maximum demand*: These charges are related either to kVA or to kW of maximum demand. If they are of the kW type then a separate power factor charge will be made.

   The charges depend upon whether the supply is provided at high or low voltage and the time of year. They are higher if the supply is provided at low voltage since the supply authority will seek to recover the cost of the equipment and losses incurred in providing a low-voltage supply. Within the Northern Hemisphere, demand charges are highest for the months of January and December and generally reduced for February and November. They may also apply for March but are not normally incurred during the remainder of the year.
3. *Unit charge*: This is related to actual energy consumption, and two options are generally available; a rate applicable for 24 h or a split day and night rate, with, typically, a 7-hour night period duration. In per-unit terms these costs will be:

   24 h = 1 per unit

   Day/night = 17-h day, 1.04 per unit
   7-h night 0.51 per unit

   Some authorities offer reduced rates as energy usage increases.
4. *Power factor adjustment*: Although related to the average power factor of the load, the method of calculation may be based directly on measured power factor or on the measurement of reactive kVA over the period.

   Values at which power factor charges are incurred vary from 0.8 to 0.95 lag.

21.2.5 Access

The switchgear provided at the installation for the provision of the supply by the supply authority will be the property of that authority. That provided for the distribution of the electricity around the installation will be the property of the consumer, while metering for tariff purposes will be located on or supplied from the supply authority switchgear.

It is normal for the supply authority switchgear to be separate from the consumers' switchgear and frequently located in a separate room within the same building. The supply authority may insist that access to this equipment or room is limited to their own personnel.

21.3 Distribution systems

The object of any electrical power distribution scheme is to provide a power supply system which will convey power economically and reliably from the supply point to the many loads throughout the installation. The standard method of supplying reliable electrical supplies to a load centre is to provide duplicated 100% rated supplies. However, there are a number of ways in which these supplies can be provided.

The standard approach for a secure supply system is to provide duplicated transformer supplies to a switchboard with each transformer rated to carry the total switchboard load. Both transformers are operated in parallel, and the loss of a single incoming supply will not therefore affect the supply to the load feeders. With this configuration the supply switchboard must be able to accept the fault current produced when both transformers are in parallel. The system must, however, be designed for the voltage regulation to remain within acceptable limits when a single transformer supplies the load.

If the two incoming supplies are interlocked to prevent parallel operation, the power flow, fault current flow and voltage regulation are governed by a single transformer supply infeed. Although the configuration has the disadvantage that the loss of a single infeed will cause a temporary loss of supply to one set of load until re-switching occurs, the advantage is that each incoming transformer can be rated higher than the first, and thus a higher concentration of loads can be supplied.

A third alternative is to rate the transformer infeeds

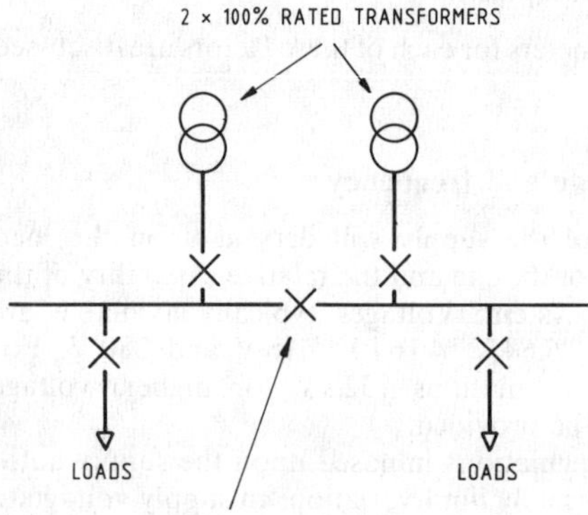

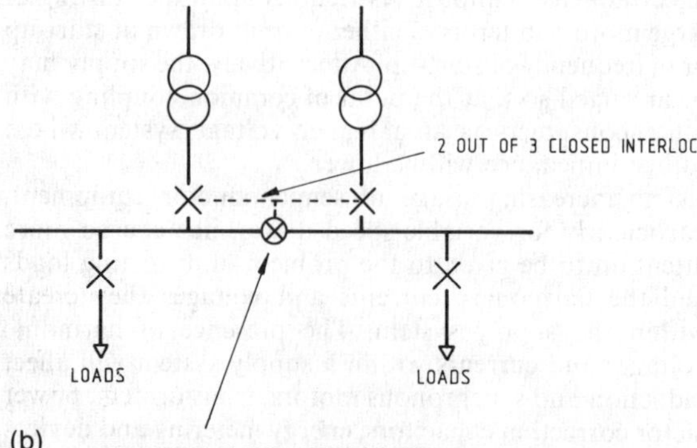

Figure 21.1 Transformer arrangement. (a) Parallel operationr; (b) 2 out of 3 interlock

Table 21.1 Load supply capabilities and single and parallel transformer combinations

Source	*Load*		*Load circuit breaker*				*Max. supply transformer rating* (MVA)			
Voltage	*Fault level*	*Voltage*	*Full load, rating A*			*Fault rating*	*Rectance (%)*			
			630	*2000*	*3000*		*6*	*10*	*6*	*10*
kV	MVA	kV		MVA		kA (MVA)	Parallel		Single	
33	1428	11	12	38.1	—	25 (476)	21	35.7	38.1	38.1
11	476	3.3	3.6	11.4	—	25 (143)	6.1	10.2	11.4	11.4
11	476	0.415	—	1.4	2.2	50 (35.9)	12	1.9	2.2	2.2

for single supplies but to arrange for automatic switching of the bus section and two incomers in the event of loss of supply to one section. A rapid transfer switching to the remaining supply in the event of loss of single supply can prevent the total loss of motor loads. In such a situation the effect of the current taken by motors to which the supply has been restored must be taken into account.

Since a group of motors re-accelerating together will draw an increased current from the supply, this current will affect voltage regulation and must be recognized when selecting protection relay settings. Figure 21.1 illustrates these arrangements of transformer operation. The effect of these combinations upon the loads which can be supplied is summarized in Table 21.1 for two typical supply voltages and fault levels, circuit breaker capabilities and transformer reactances.

In the case of parallel operation the maximum transformer rating is limited by the fault rating of the switchgear, while for a single transformer infeed the limitation is by the full-load current rating of the switchgear.

Table 21.1 takes into account only the fault contribution from the supply system. The contribution from rotating plant within an installation must also be considered when specifying switchgear and transformer ratings.

If a fully duplicated supply system is thought to be necessary, the transformer reactances can be increased in order to limit the fault level when operating in parallel mode. However, this will increase the initial capital cost, and voltage regulation with a single transformer in circuit will still need to be maintained within acceptable limits.

21.4 Switchgear

Switchgear is a general term covering circuit-interruption devices, the assemblies which enclose them and associated equipment such as current transformers, voltage transformers, disconnect switches, earthing switches and operating mechanisms. The circuit breaker is the interrupting device within the switchgear capable of closing onto, carrying and breaking those currents which can flow under healthy or fault conditions.

Switchgear is frequently described by the medium used in the interrupting device and method of mounting. The principal types of medium presently in operation are bulk oil, minimum oil, air, sulphur hexafluoride gas (SF_6) and vacuum. The most common mounts for indoor industrial switchgear are cubicle and metal clad. The main features of and the perceived advantages and disadvantages of the different types of medium are reviewed below, while international standards applicable to switchgear are listed in Table 21.2.

Table 21.2 British Standards applicable to switchgear

BS 5311, High voltage alternating current circuit breaker
BS 6867, Code of Practice for maintenance of electrical switchgear for voltages above 36 kV
BS 158, Insulating oil for transformers and switchgear
BS 5622, Insulation co-ordination
BS 5227, A.C. metal enclosed switchgear and control gear of rated voltage above 1 kV
BS 4752, Switchgear and control gear for voltages up to and including 1000 V a.c. and 1200 V d.c.
BS 3938, Current transformers

21.4.1 Bulk oil

The fact that oil was effective for the extinction of the arc formed between the opening contacts of a circuit breaker was discovered towards the end of the nineteenth century. In a bulk oil circuit breaker the oil is used both as the medium for arc extinction and as the insulating material of the contacts. The contacts are contained within a small chamber within the oil-filled breaker tank. When the circuit breaker contacts open an arc is formed which decomposes the oil, creating gases. These form a bubble which forces oil across the arc path, causing rapid extinction of the arc. Bulk oil circuit breakers have been used for many years, and it is probable that about 70% of the high-voltage circuit breakers installed in the UK are of this type.

The principal advantages of bulk oil circuit breakers are

1. Simplicity
2. Robustness in construction
3. Familiarity to users
4. Low capital cost

while the disadvantages are

1. Potential fire risk
2. Short-circuit contact life
3. Maintenance costs
4. Possible restrike problems with capacitor switching

21.4.2 Minimum oil

In a minimum oil circuit breaker only the arc control device is enclosed within an oil-filled housing. This housing is supported and insulated separately to provide the necessary phase clearances between phase connections and earth.

21.4.3 Air

Air circuit breakers can be either *air break* types (which utilize atmospheric air) or *air blast* (which use compressed air). Air-break type breakers extinguish the arc by high-resistance interruption, the arc being controlled within an arc chute. As the arc resistance increases, the current is reduced until at current zero arc extinction occurs.

Air-blast circuit breakers operate on the low-resistance principle; the arc length is minimized and a blast of air is directed across the arc to cool and remove ionized gas. Air break circuit breakers are now used extensively at medium voltages while air blast circuit breakers have been employed up to the highest voltages.

For use at normal industrial distribution voltages the principal advantages are

1. Suitability for all types of switching
2. Good fault ratings
3. No fire risk
4. Low maintenance costs

while the main disadvantages are

1. Capital cost
2. Physical size

21.4.4 Sulphur hexafluoride gas (SF_6)

Sulphur hexafluoride gas (SF_6) has a high thermal conductivity and a high electronegative attraction for ions. It has insulating properties similar to oil and arc-interruption properties comparable to compressed air at 4826 kN/m^2. Recent switchgear design utilizes SF_6 at a relatively low pressure for insulation, with compression occurring via a piston activated by the opening mechanism of the contacts.

Principal advantages of SF_6 circuit breakers are

1. Good arc-extinction properties
2. Low maintenance
3. Size

and few disadvantages are perceived.

21.4.5 Vacuum

As a replacement for bulk oil circuit breakers for distribution purposes, vacuum is a rival to SF_6. The contact interruption chamber is a sealed unit with a small contact-separation distance (typically, 16 mm for 24 kV). Choice of contact material is critical since the material affects:

1. The electrical and mechanical conductivity;
2. The production of metal vapour upon arc interruption necessary to control current chopping and permit voltage recovery; and
3. The provision of consistent contact wear and separation.

Principal advantages of vacuum circuit breakers are

1. Minimum maintenance
2. Less operating energy required
3. Size

while disadvantages are

1. Possible restriction on fault capability
2. Possible switching overvoltages

Table 21.3 Capability of distribution voltage switchgear

Medium	Full-load current (A) Typical	Full-load current (A) Max.	Fault current (kA) Typical	Fault current (kA) Max.
Bulk oil	1200	2000	26.4	43.72
Minimum oil	1600	4000	26.4	50
Air	2500	5000	26.4	52
SF_6	2000	3150	25	50
Vacuum	2000	2000	25	40

21.4.6 Range

Distribution voltage switchgear available within the UK is designed for the range of full-load current and fault-current capabilities shown in Table 21.3.

21.4.7 Construction

A.C. metal enclosed switch gear and control gear between the voltages of 1 and 72.5 kV can be manufactured to three basic designs: metal enclosed, metal clad and cubicle.

In *metal-enclosed* assemblies the equipment is contained within an earthed metal enclosure. In *metal-clad* types separate compartments are provided for the circuit interrupter, the components connected to one side of the interrupter (e.g. the cable box) and those connected to the other side of the interrupter (e.g. the busbar chamber and the control equipment).

Cubicle-type switchgear may have non-metallic components for separation between compartments, no segregation between compartments housing different components or they may be of a more open type of construction. Frequently access to cubicle-type switchgear is via doors interlocked to prevent access to energized parts. The other feature of construction of switchgear concerns the circuit breaker itself, which can be either fixed type or removable. In a fixed construction the switchgear will be fitted with isolating and earthing switches so that access to the circuit breaker is only possible when the circuit breaker is safely isolated from the supply.

In a removable design the circuit breaker is mounted

Figure 21.2 NEI Reyrolle SF_6 switchgear

on either a truck or a swing-out frame. The circuit breaker is arranged so that isolation from the busbars is only possible when the circuit breaker is open. Connections between the removable circuit breaker and the fixed assembly is either by a plug and socket arrangement or by secondary isolating contacts. Figure 21.2 illustrates a truck-mounted, vertically isolated, metal-clad, SF_6 circuit breaker design as marketed by a UK manufacturer.

21.4.8 Mechanisms

Operating mechanisms for switchgear can be either *stored-energy* or *dependent-energy closing*. Stored-energy closing is frequently achieved by means of a spring which is either manually charged (type QM) or charged by an electrical motor. Dependent-energy closing is by use of a solenoid.

The selection of operating mechanism will depend upon the control regime required for the circuit breaker. If the circuit breaker is to be operated locally (i.e. at the switchgear itself) then the manually charged spring option will be acceptable. If it is to be operated from a remote location then either motor-wound spring or solenoid operation will be required.

This choice will depend upon circuit breaker availability required, since a motor-wound spring has a finite spring-wind time. A solenoid operation will impose a significant load upon the d.c. supplies while a motor-wound spring will require a supply to the spring-wind motor.

21.5 Transformers

Transformers are classified by their method of cooling, winding connection arrangement and temperature-rise classifications. Each of these parameters is applicable, irrespective of the voltage ratio and rating required. In the UK core-type transformers are used for almost all power systems applications. In core-type transformers the primary and secondary windings are arranged concentrically around the core leg of substantially circular cross section. International standards applicable to transformers are listed in Table 21.4.

21.5.1 Cooling

Mineral oil provides a greater insulation strength than air for any given clearance. When used in a power transformer mineral oil also augments the removal of heat from the windings.

In power transformer manufacture the case of paper insulation and oil used in combination is well established. Both materials can be operated safely at the same maximum temperature (105°C) and this combination of use seems unlikely to be phased out in the near future.

For transformer windings immersed in oil, hydrocarbon oil is the most widely used, whereas in areas where fire risk is a problem, then air-cooled transformers (AN) or synthetic silicon-based liquid cooling (SN) can be specified. Silicon-based liquids do not have any of the disadvantages identified with the chlorinated biphenyls. Air-cooled transformers can be provided with Class C insulation or be cast resin insulated. The relative costs per unit of each type are:

Mineral oil	1
Silicon filled	1.25–1.5
Class C	1.8
Cast resin	2.0

21.5.2 Types

ONAN

Oil-immersed air-cooled transformers cover the majority of units installed of up to 5 MVA rating. Cooling of the oil is provided by fins or corrugations of the tank or by tube banks. For ONAN transformers of above 5 MVA rating radiator banks of elliptical tubes or banks of corrugated radiators are often provided.

ONAF

In oil-immersed air-forced transformers a direct blast of air from banks of fans is provided to the radiators which increases the rate of heat dissipation. This arrangement has no effect upon the size of the transformer itself but less space for external coolers is required.

Table 21.4 British Standards applicable to transformers and their components

BS 171, Power transformers:
- Part 1, General
- Part 2, Temperature rise
- Part 3, Insulation levels and preelectric tests
- Part 4, Tappings and connections
- Part 5, Ability to withstand short circuit

BS 158, Insulating oil for transformers and switchgear
BS 2757, Thermal classification of electrical insulation
BS 223, Bushings for alternating voltages above 1000 V
BS 5622, Insulation co-ordination
BS 2562, Cable sealing boxes for oil-immersed transformers

OFAN

Oil-immersed air natural circulation is an uncommon arrangement, and is useful only if the coolers are situated away from the transformer.

OFAF

Most larger-rated transformers are oil-forced air-forced cooled. Oil-forced circulation improves the heat dissipation around the windings, thus reducing the size of the transformer itself and air blast cooling of the radiators decreases the size of the cooling surfaces. Thermostatic control of both oil forcing and air blast is usually provided so that each is brought into service at a different oil temperature. The ONAN rating of an OFAF transformer is typically 50% of the OFAF rating, although any value can be designed when required.

AN

These are solidly insulated transformers using typically cast resin insulation. They are particularly useful when an indoor installation is necessary, since they represent a very low fire risk and oil leakage and drainage does not need to be considered. AN transformer maximum ratings are typically of the order of 1500 kVA. Forced cooling (AF) can be added in the same way as oil-immersed transformers, permitting increased output from the same size of unit.

21.5.3 Temperature rise

Temperature classes for transformers A, E, B, F, H and C are generally recognized. All classes are applicable to both oil-immersed and air-cooled transformers. Classes A, B, H and C are most commonly used but the use of Class F is increasing.

For windings immersed in either hydrocarbon oil or synthetic liquids the coil insulation is usually a Class A material. For oil-immersed transformers BS 171 specifies two maximum oil temperatures: 60°C for sealed-type transformers for those fitted with conservators and 55°C for transformers without either. In the case of air-insulated, air-cooled transformers the maximum permitted temperature rise is limited by the class of insulation used, i.e.:

Class A 60°C

Class B 90°C

Class C 150°C

These temperature rises apply above a maximum ambient temperature of 40°C and a daily average temperature of 30°C is assumed.

21.5.4 Tests

BS 171 defines both routine and type tests for transformers. Routine tests comprise the following:

1. Ratio measurement, polarity check and phase relationship;
2. Measurement of winding resistance;
3. Measurement of insulation resistance;
4. Measurement of load loss and impedance voltage;
5. Measurement of core loss and magnetizing current;
6. High-voltage withstand tests. The rated withstand voltages applicable to transformers of up to 36 kV are shown in Table 21.5.

Table 21.5 Rated withstand voltage transformers

High voltage (kV)	*Rated short-duration power frequency withstand voltage* (kV)	*Rated lighting impulse withstand voltage*[a] (kV)	(Peak)
3.6	10	20	40
7.2	20	40	60
12	28	60	75
24	50	95	125
36	70	145	170

[a] Choice of rated lighting impulse withstand voltage depends on exact site conditions and duties.

The measurement of impedance voltage is defined as the voltage required to circulate full-load current in one winding with the other windings short circuited. It is common practice to express the leakage impedance, Z, of a transformer as a percentage (or per unit) value. The per unit value is:

$$\frac{IZ}{V}$$

where I and V refer to full-load current and rated voltage and IZ is the voltage measured at rated current during a short-circuit test transformer. In the case of a transformer with tappings the impedance is conventionally expressed in terms of the rated voltage for the tapping concerned.

21.5.5 Fittings

In selecting the fittings to be provided, the specifier of a transformer has a wide choice. The most common types and their purpose are as follows.

Oil-level indicator

This is usually mounted on the transformer tank, providing a visual indication of the transformer oil level.

Thermometer pockets

These measure oil temperature from a separate temperature indicator.

Tap changing

The range and duty (i.e. off-load/on-load) can be specified to suit each particular installation. For standard distribution transformers of up to 3.15 MVA rating for use purely as load supplies, off-load tap changing for a $\pm5\%$ voltage range is normal.

Breather/conservator/gas and oil relay

This is specified where the transformer is of the breathing type. The gas- and oil-actuated relay will be fitted with contacts for remote alarm or tripping.

Pressure-relief device

This is essential on a sealed transformer and optional on a breathing unit. It can be fitted with contacts for remote alarm/tripping.

Disconnecting chambers

These are fitted between the cable boxes and the transformer windings and are accessible through a secured cover. Disconnecting chambers can be employed as a method of isolation, but their most common usage is to facilitate phase-to-phase testing of the connected cables.

Connection

Figure 21.3 illustrates the most commonly used transformer, grouping and winding arrangements and phase displacements.

21.6 Protection systems

The analysis and understanding of the operation of protection systems and the application of protective devices to power systems is wide ranging and complex. This section considers only the objectives of protection systems, discrimination and its importance to protection, and describes the most common types of relay used in industrial power systems. Many publications covering the subject of protection systems are available (e.g. *Power System Protection*, edited by the Electricity Council).

21.6.1 Objectives

All electrical systems must be provided with protection equipment, the purpose of which is to isolate faulty electrical equipment from the electrical supply system as rapidly as possible. This can be achieved by use of devices which respond directly to the current flowing (e.g. fuses) or by protective relays which respond to fault current flow and are used to initiate the tripping of other devices (e.g. circuit breakers).

So that only the faulty equipment is isolated from the supply, the protective devices provided throughout a power system must discriminate between faulty and healthy equipment. This discrimination of protective equipment is a key element in the design of electrical systems and the selection of protection devices for use within them. Each electrical system must be provided with an adequate number of suitably rated disconnecting devices located correctly throughout the power system. These must be so arranged that only appropriate devices operate to remove the faulty equipment from the supply.

21.6.2 Discrimination

The methods of achieving discrimination most commonly found in industrial power systems are as follows:

1. *By time*: Protective relays are provided with time-delay features such that the device closest to the fault will operate first.
2. *By magnitude*: In different parts of a power system fault currents of different magnitude can flow.
3. *By time/direction*: Protective devices responsive to the direction of current flow. These are necessary for parallel feeders or closed ring-main supplies.
4. *By comparison*: The currents flowing into and out of a circuit are compared. In a healthy circuit, or in a circuit in which 'through fault' current is flowing, the two currents should be equal and the protective device does not operate. (Compensation for any transformation is necessary.)

   If the two currents differ the protective device operates.

In selecting discriminating time margins between protective devices connected in series the following should be taken into account. Full discrimination is achieved between fuses of different rating when the fault is cleared by the lower rated or minor fuse without affecting the higher rated or major fuse. In order to achieve this, the total let-through energy of the minor fuse must be less than the let-through energy necessary to cause pre-arcing of the major fuse.

21.6.3 Testing

Tests carried out on protection systems comprise factory tests, on-site commissioning tests and maintenance checks. Those made on individual relays at the place of manufacture will demonstrate the compliance of the equipment with specification and the verification of its operation under simulated conditions, while tests carried out on-site prior to the equipment being put into service ensure that the full protection scheme and associated equipment operate correctly. These on-site tests must be comprehensive and should include:

1. Performance tests on current and voltage transformers to verify ratio and polarity. These may include the injection of current into the primary load circuit (primary injection).
2. Sensitivity and stability checks on each protective device. These tests may be carried out either by primary injection or by the injection of voltage into the CT secondary circuit (secondary injection).
3. Wiring checks, including insulation resistance and any pilot cables.
4. Operational checks on all tripping and alarm circuits.
5. Calibration checks on all relevant relays.

On-site commissioning tests of protection equipment

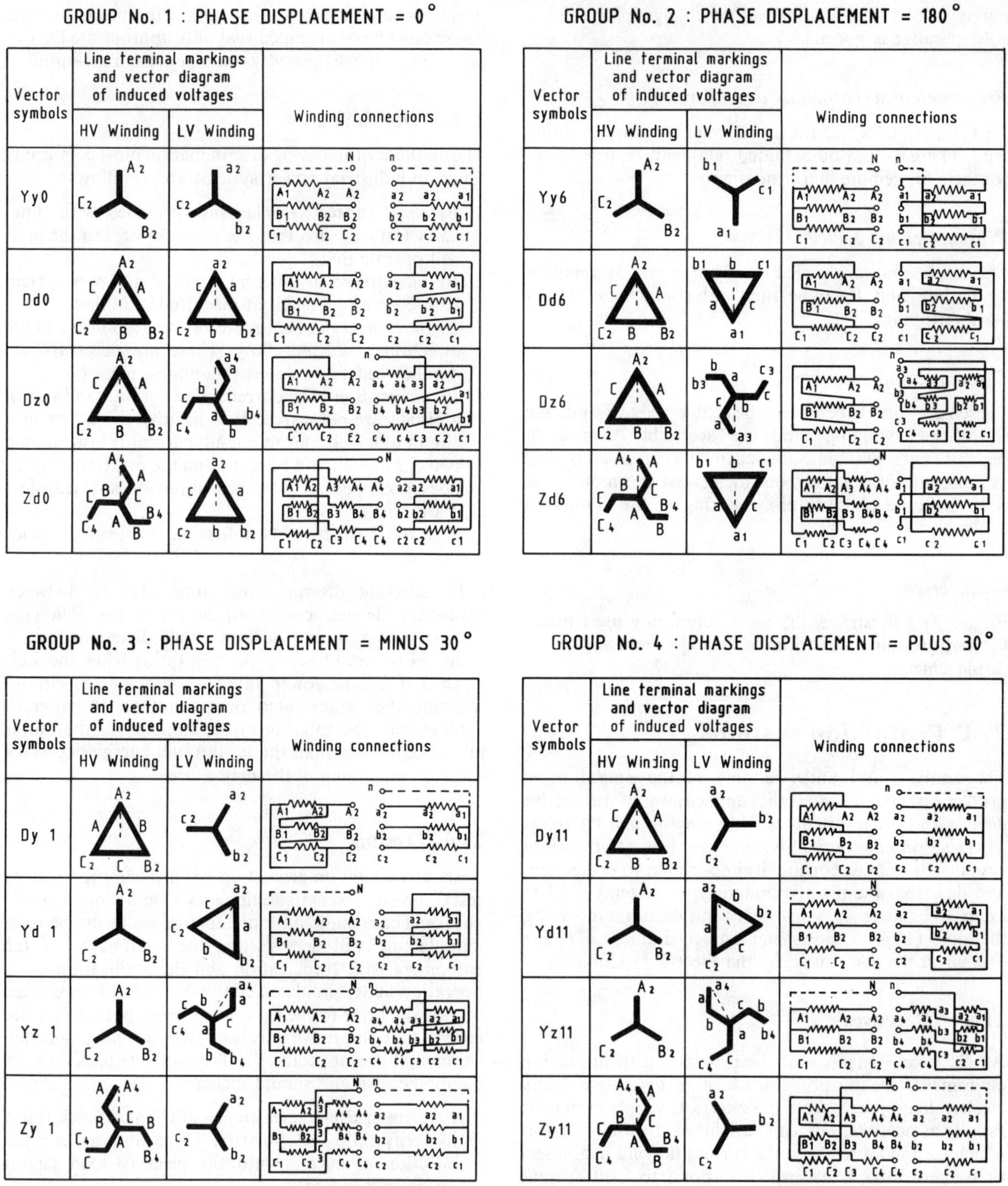

GROUP No. 1 : PHASE DISPLACEMENT = 0°

Vector symbols	Line terminal markings and vector diagram of induced voltages: HV Winding	LV Winding	Winding connections
Yy0			
Dd0			
Dz0			
Zd0			

GROUP No. 2 : PHASE DISPLACEMENT = 180°

Vector symbols	Line terminal markings and vector diagram of induced voltages: HV Winding	LV Winding	Winding connections
Yy6			
Dd6			
Dz6			
Zd6			

GROUP No. 3 : PHASE DISPLACEMENT = MINUS 30°

Vector symbols	Line terminal markings and vector diagram of induced voltages: HV Winding	LV Winding	Winding connections
Dy 1			
Yd 1			
Yz 1			
Zy 1			

GROUP No. 4 : PHASE DISPLACEMENT = PLUS 30°

Vector symbols	Line terminal markings and vector diagram of induced voltages: HV Winding	LV Winding	Winding connections
Dy11			
Yd11			
Yz11			
Zy11			

ALL TRANSFORMERS FROM ANY ONE GROUP MAY BE OPERATED IN PARALLEL WITH OTHER TRANSFORMERS FROM THE SAME GROUP.

PHASE DISPLACEMENT APPLIES FOR COUNTER CLOCKWISE PHASE ROTATION.

Figure 21.3 Typical transformer vector groupings and connection arrangements

must be carried out in a planned logical manner, and arranged so that the disturbance of tested equipment is minimized.

21.6.4 Relay types

The main relay types found within an industrial power system can be:

1. Instantaneous in operation;
2. Definite time; or
3. Inverse, in which the time of operation is dependent upon the magnitude of the current,

and such relays can be arranged to protect equipment in one of the following ways:

1. *Unrestricted*: The protection device will operate in response to any fault current flowing through the circuit.
2. *Restricted*: The protection device will operate in response to any fault current flowing within a zone restricted by the location of the current transformers. Such protection is normally applied to a single winding of a power transformer.
3. *Differential*: The protection device will operate in response to a fault current flowing within a zone bounded by current transformers located at each end of the protected circuit.

   Such protection can be used on a power transformer, a motor or generator, or a single feeder.

In addition, other types of relay are used for particular applications (e.g. transformer Buchholz gas, neutral voltage displacement and overvoltage).

21.6.5 Applications

Figures 21.4 and 21.5 show typical protection schemes for two circuits of a 60 MW generator, a generator transformer with tripping logic and a 2 MVA transformer. The figures illustrate the protection devices provided, the current and voltage transformers supplying them and the tripping scheme associated with each.

21.7 Power factor correction

21.7.1 Background

Power factor in an alternating current circuit is defined as the ratio of actual circuit power in watts (W) to the apparent power in voltage amperes (VA). The need for correction arises from fact that the majority of a.c. electrical loads take from the supply a lagging quadrature current (voltage amperes reactive, var) and thus operate at a lagging power factor due to the reactive (rather than capacitive) nature of their construction.

Most industrial installations comprise a combination of one or more of the following electrical loads:

1. A.C. motors (induction)
2. Furnaces (electric arc, induction)
3. Fluorescent or discharge lighting
4. Power transformers
5. Thyristor drive equipment (for either a.c. or d.c. drives)
6. Welding machines

All these types of load fall into the above category and operate at a lagging power factor.

Since the electrical supply system carries the full apparent power (VA) a current higher than is theoretically necessary to supply the power demand needs to be supplied. Equipment must therefore be rated to carry the full apparent power plus the losses of the supply system, which are proportional to the square of the current.

A supply system operating at a low power factor will be inefficient due to the overrating of the supply components and to the losses incurred, and supply authorities will seek to recover the costs of this inefficient operation by introducing cost penalties for consumers operating at a low power factor. In practice, these penalties are incurred on the basis of either the maximum demand of the apparent power in kVA of the load or of charges initiated if the power factor falls below a set trigger point (typically, 0.95 lag). The latter charges are monitored by measuring either the power factory directly or by the relative current drawn by the load and comparing it with the actual power drawn.

Within any particular installation, in instances where no financial penalties are incurred from the supply authority the power factor of individual circuits will influence both the losses within the installation and the system's voltage regulation. Therefore power factor correction is considered necessary in order to achieve one or more of the following objectives:

1. A direct reduction in the charges made by the supply authority;
2. A reduction in the losses within a system with consequent savings in charges;
3. An improvement in the voltage regulation of a system.

Figure 21.6 illustrates the phasor diagram applicable to power factor correction.

21.7.2 Methods of achieving correction

The power factor of an installation can be improved by the use of either a.c. synchronous machines or of static capacitor banks. An a.c. synchronous machine will either draw current from the supply or contribute current to the supply, depending on whether the machine is operating:

1. As a generator, driven by a prime mover, contributing active power;
2. As a motor-driving load, drawing active power;
3. Over-excited, contributing reactive var; or
4. Under-excited, drawing reactive var.

Figure 21.7 illustrates these four modes of operation.

When an a.c. synchronous machine operating as a motor in parallel with other loads and an external supply system is over-excited the machine will contribute reactive kvar to the supply. The net effect of this will be to reduce the amount of reactive current drawn from the supply, and this will improve the overall system power

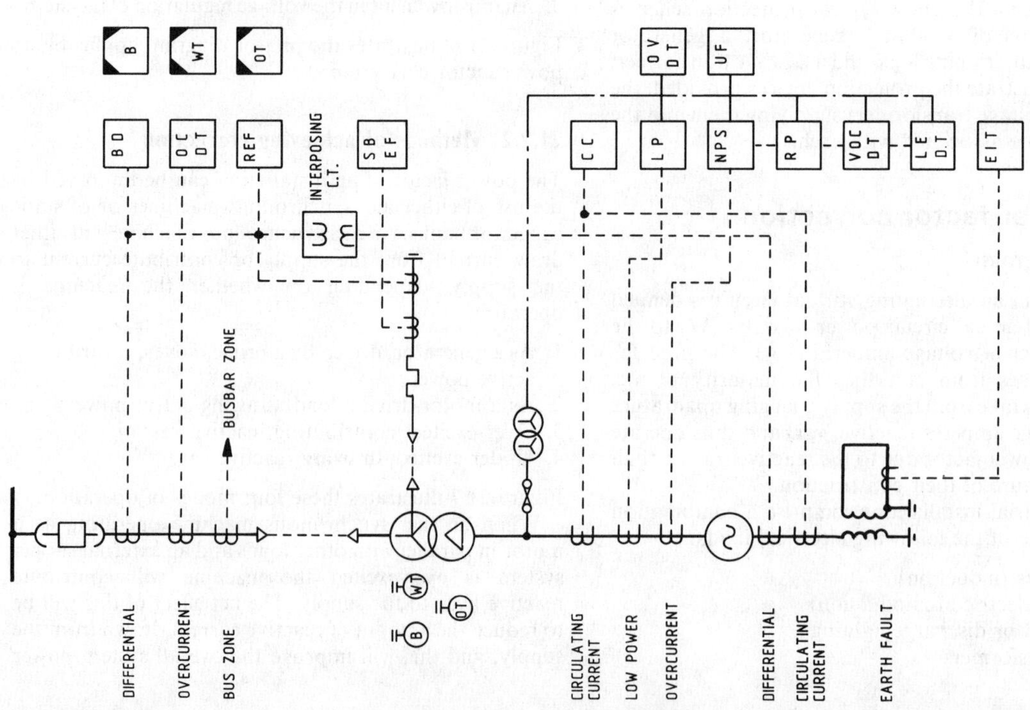

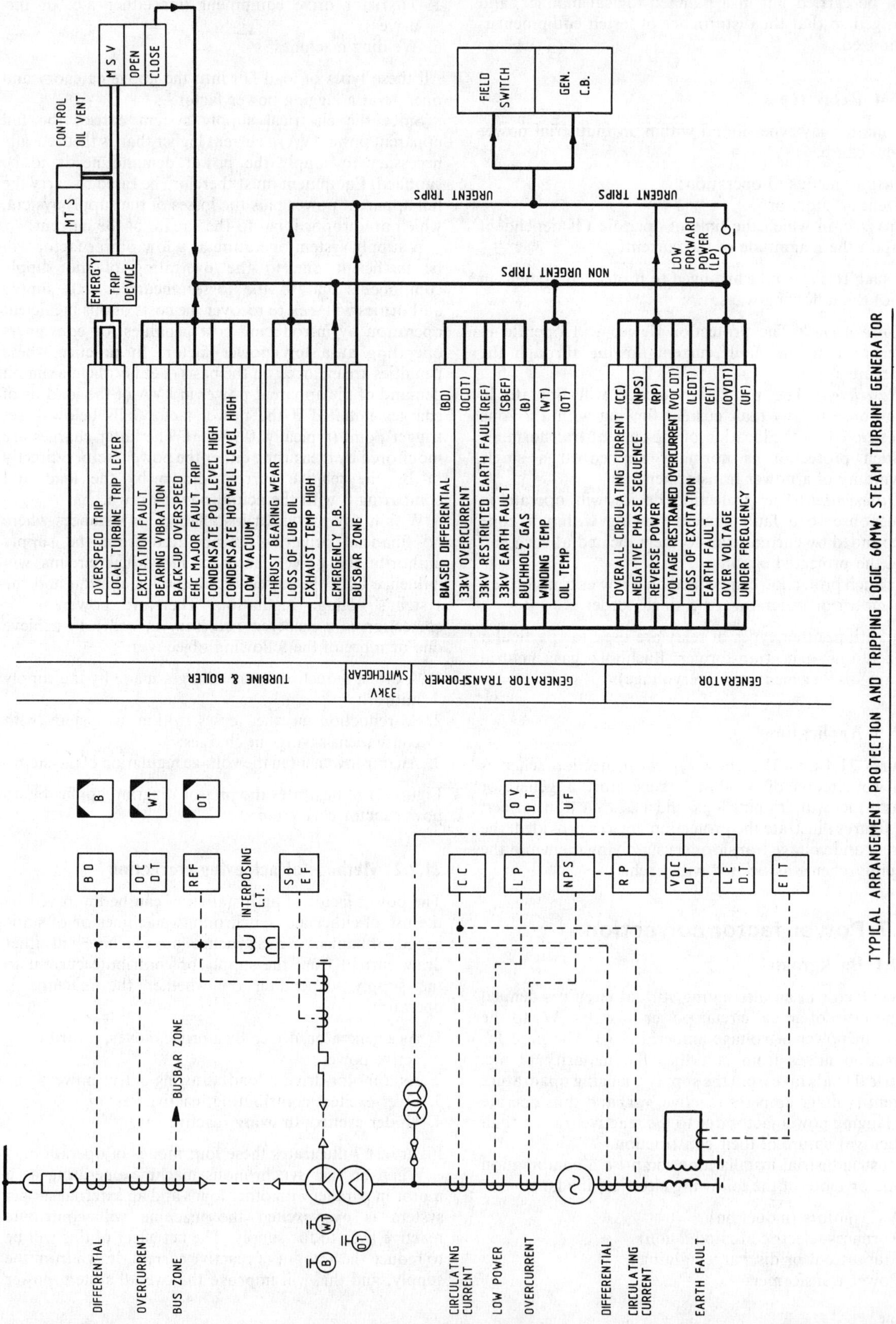

Figure 21.4 Protection and tripping logic for a 60 MW steam turbine generator

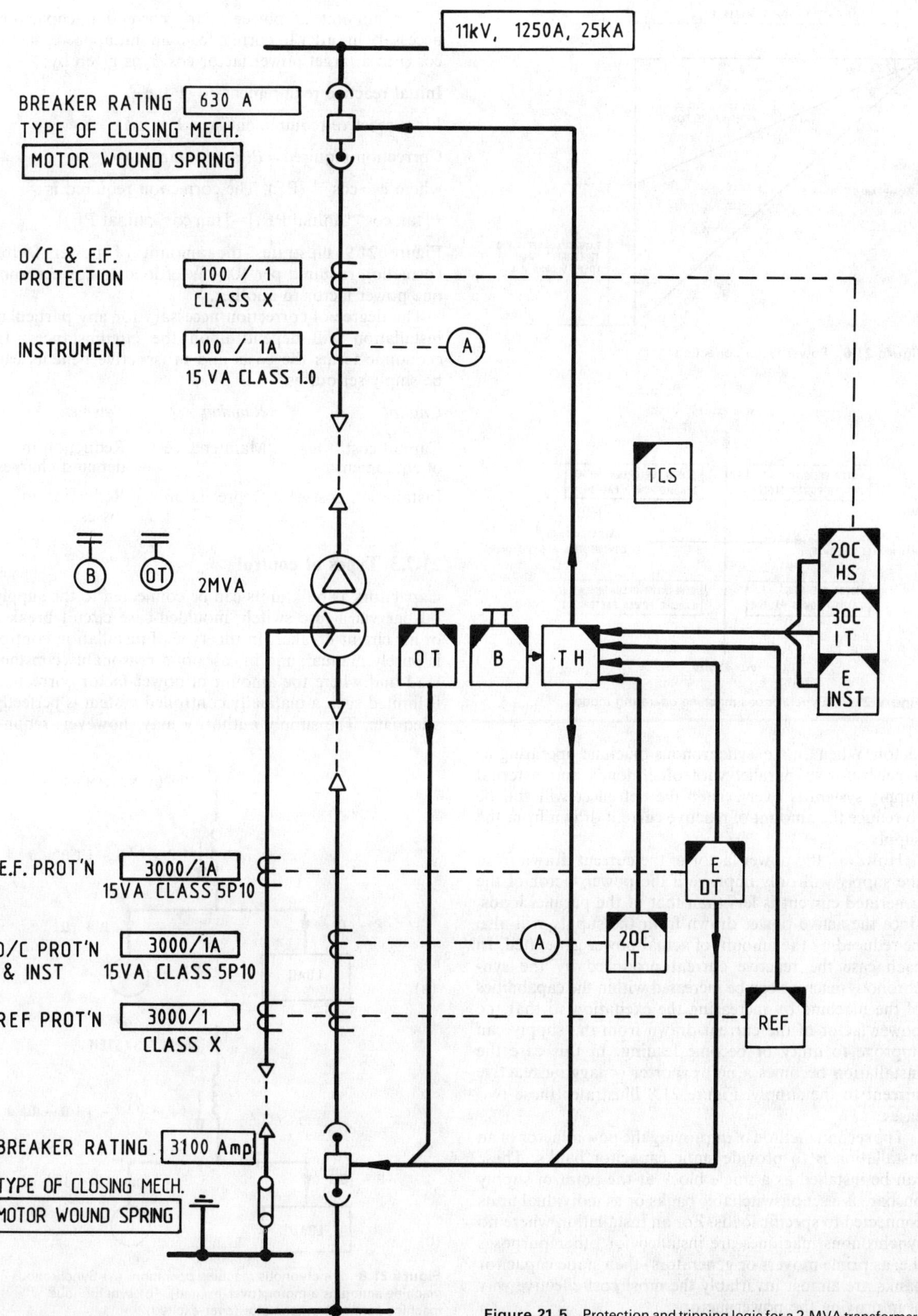

Figure 21.5 Protection and tripping logic for a 2 MVA transformer

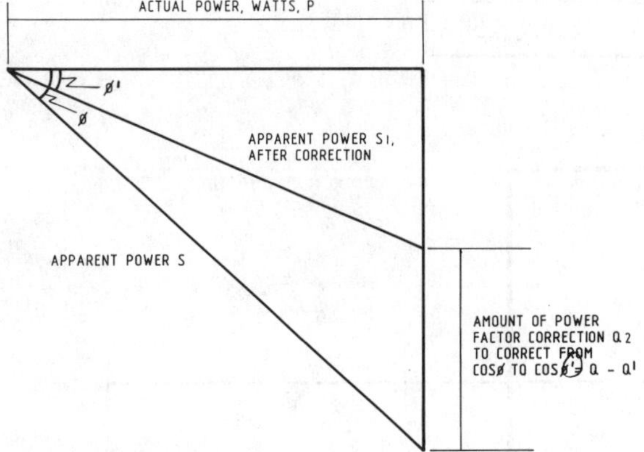

Figure 21.6 Power factor correction

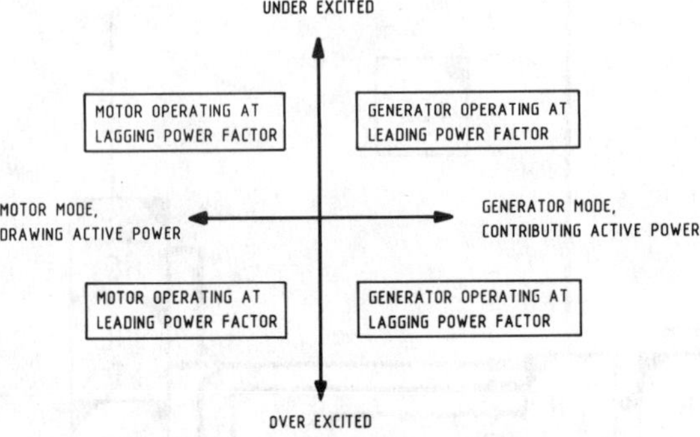

Figure 21.7 Synchronous machine operating modes

factor. When an a.c. synchronous machine operating as a generator in parallel with other loads and external supply system is over-excited the net effect will still be to reduce the amount of reactive current drawn from the supply.

However, the power factor of the current drawn from the supply will only improve if the power factor of the generated current is less than that of the parallel loads, since the active power drawn from the supply will also be reduced by the amount of actual power generated. In each case the reactive current produced by the synchronous machine can be increased within the capabilities of the machine by increasing the excitation so that the power factor of the current drawn from the supply can improve to unity or become leading. In this case the installation becomes a nett exporter of lagging reactive current to the supply. Figure 21.8 illustrates these two cases.

The second method of improving the power factor of an installation is to provide static capacitor banks. These can be installed as a single block at the point of supply busbar, as a set of switchable banks or as individual units connected to specific loads. For an installation where no synchronous machines are installed for other purposes (i.e. as prime movers or generators) then static capacitor banks are almost invariably the most cost-effective way of improving the power factor.

The amount of power factor correction capacitors necessary in order to correct from an initial power factor $\cos\phi$ to a target power factor $\cos\phi_1$ is given by:

Initial reactive requirement $\phi = P\tan\phi$

Final reactive requirement $\phi_1 = P\tan\phi_1$

Correction required $= P(\tan\phi - \tan\phi_1)$

where $\phi = \cos^{-1}$ (PF). The correction required is

$P[\tan\cos^{-1}(\text{initial PF})] - [\tan\cos^{-1}(\text{final PF})]$

Figure 21.9 illustrates the amount of power factor correction required per 100 kW of load to correct from one power factor to another.

The degree of correction necessary for any particular installation will depend upon the circumstances. In economic terms the costs and prospective benefits can be smply set out as:

Capital	*Running*	*Savings*
Capital cost of equipment	Maintenance	Reduction in demand charges
Installation costs	Depreciation	Reduction in losses

21.7.3 Types of control

Capacitors as bulk units can be connected to the supply busbar via a fuse switch, moulded-case circuit breaker or air circuit breaker. In this type of installation control is purely manual, and in cases of a reasonably constant load and where the amount of power factor correction is limited such a manually controlled system is perfectly adequate. The supply authority may, however, require

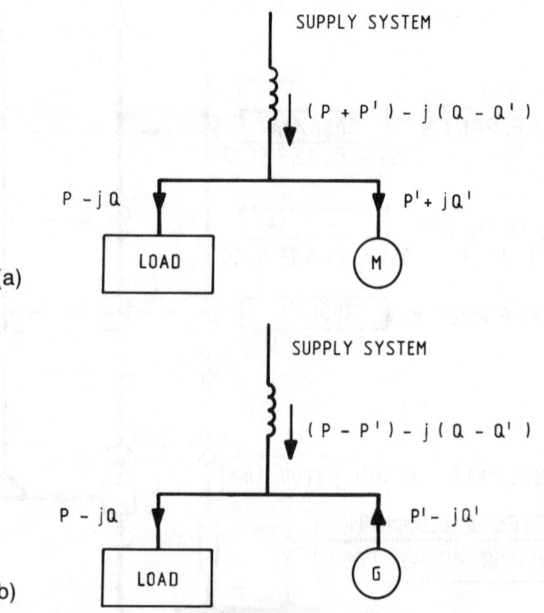

Figure 21.8 Synchronous machine operation. (a) Synchronous machine acting as a motor (over-excited); (b) synchronous machine acting as a generator (over-excited)

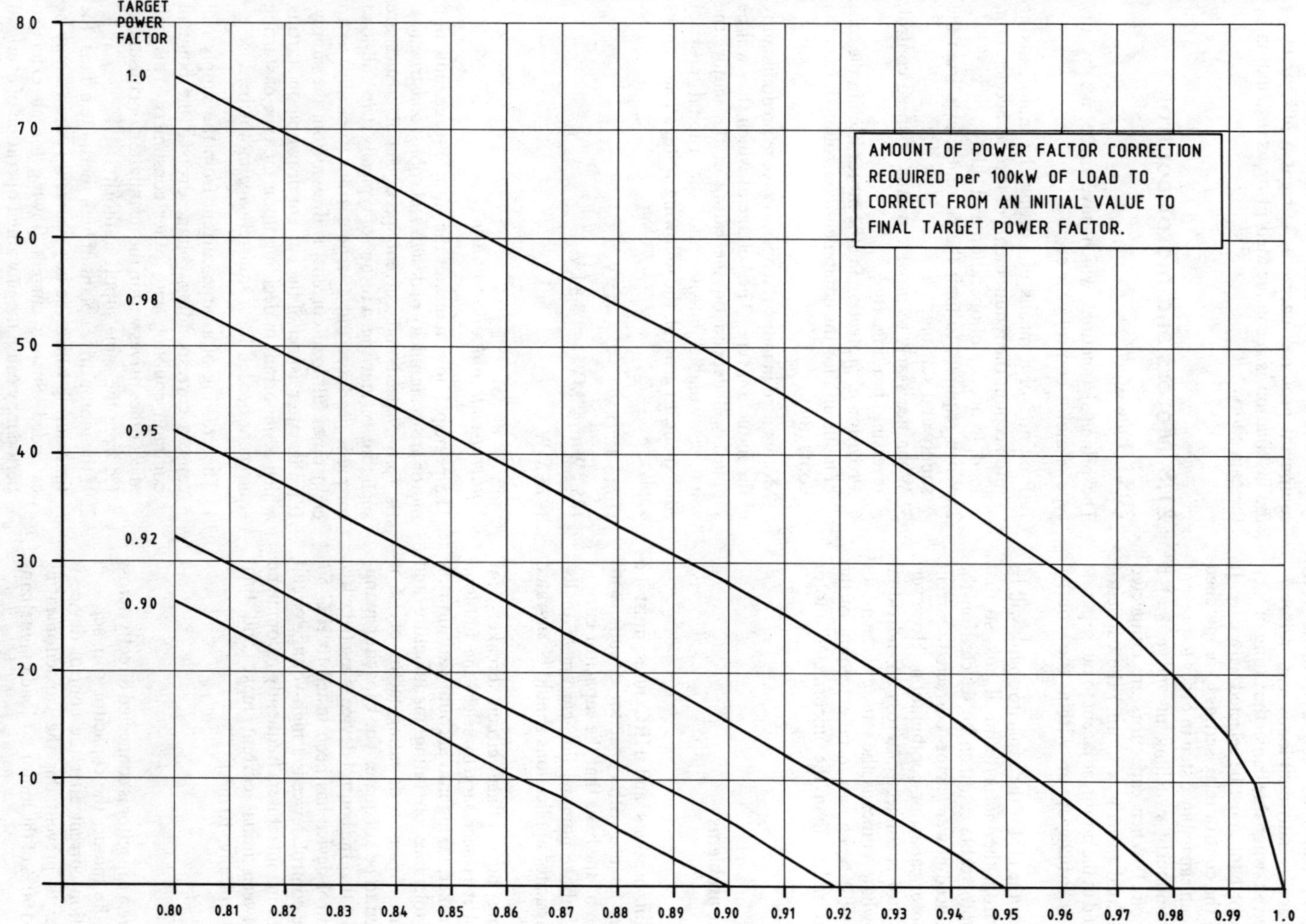

Figure 21.9 Amount of power factor correction required

to be informed that a capacitor bank is permanently connected to the supply. Capacitors are more generally connected either in banks controlled from a VAR sensitive relay or across individual loads (e.g. motors).

When connected as switchable banks the rating of each step of the capacitor bank must be selected with care. It is important that the control relay settings are matched to the ratings of each capacitance step in order to prevent hunting (i.e. continuously switching in and out at a particular load point). When capacitors are connected to one particular load (usually a motor) the capacitor bank can be located at the motor, adjacent to but separate from the control switchgear or within the control switchgear itself.

When located at the motor the capacitor bank will be normally cabled from the motor terminal box, so that the size of the motor cable can then be selected on the basis of the reduced-power factor corrected current drawn by the motor/capacitor combination. However, whenever a capacitor is connected across an individual motor circuit, by whatever means, the setting of the motor overload device must be chosen to take account of the corrected current rather than the uncorrected motor full-load current.

21.7.4 Potential problems

Control devices

Capacitors, circuit breakers and HRC fuses must be selected with care for use with capacitor circuits, with contactors chosen on the basis that the capacitor current can rise by 25% above nominal line current. Equally, HRC fuses for capacitor applications should be de-rated by a factor of 1.5.

When choosing high-voltage circuit breakers for capacitor control it is necessary to select the full-load current of the breaker, taking into account variations in supply voltage, tolerance on rating manufacture and harmonic currents. Again, a de-rating factor of 1.5 is usually considered to be adequate. The capacitor manufacturers will insist that the control device is re-strike free and that the control device has been tested to IEC 56, Part 4. When the control device is a high-voltage circuit breaker, capacitor manufacturers frequently recommend a maximum between initial current make and final contact closure (typically, 10 m/s).

Motor circuits

In instances where a capacitor is connected directly across the terminal of the motor, the capacitor can act as a source of excitation current after the control device is opened. In order to prevent this the capacitor rating should not exceed 90% of the motor no-load magnetizing current.

Harmonics

A capacitor bank will represent a reducing impedance to currents of increasing frequency. Such a reducing impedance, if matched with a similarly increasing inductance impedance of a transformer or a supply system, can cause a resonant condition. In plants where equipment produces harmonic current a full survey of the installation is recommended prior to installation of the capacitors.

21.8 Motors and motor control

21.8.1 Functions

The principal functions which have to be met by any type of motor control gear are:

1. Provision of a means of starting the motor, taking into account the requirements of torque, acceleration, load, frequency of operation and safety;
2. When required, to limit the current drawn from the supply and starting;
3. When necessary, to provide means of speed control, reversing, braking, etc.;
4. Provision of protection for the motor itself in the case of faults and for the equipment controlling it (e.g. the contactor).

Motors used in industrial operations are predominantly of the induction type. The voltage at which they will be supplied will depend on the rating of the motor, the voltages available on-site and the capital cost of equipment. In general, the following voltage/load ranges apply:

Up to 185 kW	415 V
185 kW to 2 MW	3.3 kV
1–15 MW	11 kV

Squirrel-cage induction motor

The principal advantages of the squirrel-cage induction motor are its simplicity of design and robust construction. Its torque/speed and current/speed characteristics are such that, on starting, a torque of typically *twice* full-load torque is produced but a large current (typically, six to eight times full-load current) is drawn from the supply. It is this latter aspect of the current drawn upon starting which is important in deciding upon the type of starting equipment needed for a squirrel-cage motor because

1. The drawing of a large current from the supply will cause a corresponding voltage drop at the point of common coupling with other consumers. This will affect other drives within the installation or consumers fed from the same supply authority.
2. If the motor drives plant of high inertia (e.g. fans) the time during which the large current is drawn may be extended. Such a current flowing for an extended period may cause the unwanted operation of overload or overcurrent protection relays within the supply system.

In order to overcome (1) it is often necessary to introduce a form of assisted starting, which can also help in overcoming problem (2). However, modification of the overload protection device characteristics may also be necessary.

21.8.2 Types of starter

For starting an induction motor the following types of starter are available:

1. *Direct-on-line*: Full voltage is applied to the motor;
2. *Start/delta, auto transformer, primary resistance*: Reduced voltage is applied to the motor;
3. *Electronic softstart*: The voltage and frequency applied to the motor is controlled;
4. *Stator/rotor*: Full voltage is applied with external rotor resistance.

The principal features of each type are as follows.

Direct-on-line

The equipment and connections are simple and the starter is robust. The basic equipment will comprise an isolator, high rupturing capacity fuses, a contactor, overload devices and control switches.

Star/delta starting

This is the most common form of reduced-voltage starting used in the UK. Both ends of the motor winding are cabled back to the starter. On starting, the windings are connected in star configuration, and thus the voltage applied to each winding is 57.7% of normal and the current taken from the supply is one third of that taken when started direct-on-line.

Change-over from star- to delta-connection takes place automatically by means of a timer. At change-over the motor is momentarily disconnected from the supply and the delta contactor is then closed. The closure of the delta contactor can cause a further surge of current to occur. The magnitude of the surge will depend on the speed of the motor at change-over, but it can be comparable to that taken when started direct-on-line.

Auto-transformer An auto transformer is used to provide a reduced voltage to the motor on start-up. The exact voltage applied to the motor can then be chosen, taking into account any limitation on starting current and motor torque. In an induction motor both starting current and motor torque are approximately proportional to the square of the applied voltage. Therefore for a motor which produces twice full-load torque and draws eight times full-load current on start-up, if the voltage is reduced to 75% by use of an auto transformer the starting torque $= 0.75^2 \times 2 = 1.125$ times full-load torque and starting current $= 0.75^2 \times 8 = 4.5$ times full-load current.

Primary resistance A resistance is inserted into the supply to the motor. This reduces the voltage available at the motor terminals on starting, and this voltage increases gradually as the motor current falls when the motor speeds up. Once the motor has reacted at a predetermined speed the resistance banks are short-circuited.

Depending on the value of the resistor, the motor will, in general, draw a heavier current then when started using star-delta or auto-transformer methods. The resistors must be rated to carry the limited motor starting current for the time they are in circuit.

Electronic soft-start

The most recent development in the starting of squirrel-cage induction motors is the introduction of the electronic soft-start. This principle has been derived from variable-frequency speed controllers using switched thyristor or power transistor bridges. The supply sine wave is chopped so that a reduced voltage and frequency is applied to the motor. These are gradually increased so that the motor speed rises in a controlled manner, with the starting current limited to any chosen value.

Stator/rotor

With this technique the motor has a wound rotor brought out to slip rings and an external resistance is connected into the rotor circuit. This resistance usually consists of a series of resistor banks which are switched out progressively in a number of steps as the motor accelerates. The number and rating of each step is chosen so that starting current and motor torque are within requirements.

21.8.3 Motor-starting equipment

Each component within a motor starting circuit must be selected to be suitable for the operation of the motor as it is required and for use on the electrical system from which the motor is to be operated. It must also be compatible with the other elements within the circuit. The principal components comprise:

Isolator
HRC fuses
Contactor
Overload device
Cable

The main parameters to be considered for each component are as follows.

Isolator

This must be:

1. Rated to close onto a fault at the rated fault level of the switchboard;
2. Able to carry starter rated full-load current continuously;
3. Capable of breaking a stalled motor current.

Fuses

Fuse manufacturers' catalogues give suggested fuse sizes for all 415 V motors. However, the following parameters all affect the fuse rating:

1. The starting current drawn by the motor, including all tolerances;
2. The starting time of the drive under worst-case conditions;
3. Full-load current, including efficiency and power factor;
4. Number of consecutive starts;
5. Short-time capability of contactor, overload, cable and isolator.

Contactor

A contactor must have full-load rating and be coordinated to proven capability with a selected fuse.

Overload device

Overload devices in current use are typically thermal overload relays to BS 4941, motor starters for voltages up to and including 1000 V a.c. and 1200 V d.c. or BS 142 electrical protection relays. Relays to BS 4941 generally provide overload and single-phasing protection. Those complying with BS 142 are also frequently fitted with instantaneous earth fault and overcurrent trips.

While applications vary from one industry to another, relays complying with BS 142 are typically specified for all motors larger than 45 kW.

Cable

The cable must be selected on the basis of full-load current rating, voltage drop on both running and start-up, and short-time current rating.

21.9 Standby supplies

21.9.1 Definitions

Standby supply systems within an electrical installation provide supplies to critical loads on loss of normal power supplies. In an installation one or more of the following types of standby supply systems are likely to be found:

1. Battery systems (either d.c. or inverted);
2. Uninterruptible a.c. power systems;
3. Standby diesel generating sets.

Each system type is utilized to satisfy different needs, and these and the salient points relevant to each type are as follows.

21.9.2 Battery systems

D.C. systems

These provide supplies to equipment which requires a d.c. supply both during normal operating conditions and when a.c. supplies have been lost. The loads can comprise:

1. Essential instruments
2. Control schemes
3. Switchgear closing and tripping
4. Telecommunications
5. Protection schemes
6. Interlocking
7. Alarms
8. Emergency lighting
9. Emergency drives (e.g. rundown lube oil pumps)
10. Standby diesel starting systems

Voltages in use range from 24 to 250 V. A d.c. installation will comprise a charger (capable of both float and boost charging), the battery itself and a distribution outlet.

The charger is normally arranged so that boost charging is carried out off-load, and should be designed so as to recharge the battery, after an emergency discharge has occurred, within a quarter to a half of the normal charting time. When the charger is operating in *float charge* mode it must be rated to supply the d.c. load requirements as well as float charging the battery.

Chargers can be fitted with an integral voltage-sensing device measuring circuit voltage at the charger output terminals. In the event that the terminal voltage rises significantly above the normal float voltage of the battery, the circuit will be arranged to trip the charger and provide an alarm. A separate alarm will be provided during boost charging. Alarms normally fitted comprise:

Charger fail
Rectifier fail
Battery earth fault
Battery boost charge
Battery low volts

Battery types are either lead–acid or nickel–cadmium cells. Lead–acid types have been used for a long time and, when correctly maintained, have a working life of 25 years. Nickel–cadmium batteries offer the same working life as lead–acid but are smaller in weight and volume, generally with a higher initial capital cost.

Loads applied to d.c. systems can be categorized into three types:

1. *Standing levels*: Loads which impose a consistent and continued load upon the battery (e.g. alarm facias).
2. *Emergency loads*: Loads which are only applied on loss of normal supplies. These are usually supplied only for a fixed period of time (e.g. turbine emergency rundown lube oil pump).
3. *Switching loads*: Loads imposed by the operation of both opening and closing of switchgear.

These loads, the expected duration of emergency loads and the required switching regime, and number of operations must be specified at the time of order placement.

21.9.3 UPS-a.c. systems

A.C. uninterruptible supply systems are used for the provision of supplies to those loads which:

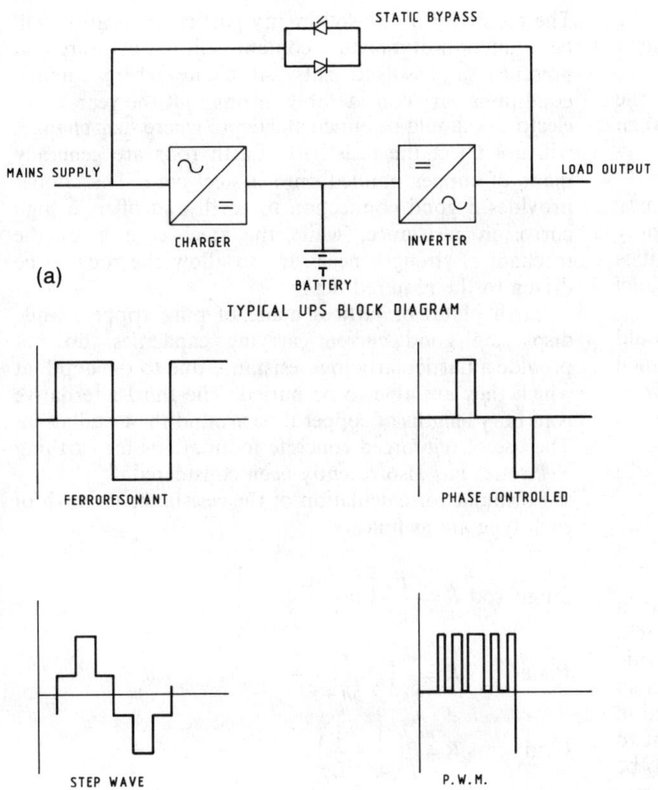

Figure 21.10 (a) UPS block diagram; (b) typical output waveforms

1. Are required for post-incident monitoring and recording following a loss of normal supplies;
2. Require a high-quality supply in terms of voltage, frequency and waveform. These comprise loads which would not give an adequate standard of reliability if operated directly from the normal a.c. supply system;
3. Are necessary to monitor plant, particularly generating plant, during a shutdown operation or loads necessary to assist in a rapid restart.

A UPS installation will typically comprise a rectifier/battery, charger, storage battery, static inverter and static bypass switch. With a UPS installation the output, which is derived from the battery via the static inverter, is completely unaffected by variations of either a steady state or transient nature in the a.c. supply (Figure 21.10).

UPS units are generally categorized by the design of the static inverter used to produce the a.c. output. The principal types used are:

1. *Ferroresonant*: The d.c. is switched via thyristors to produce a square wave output.
2. *Phase controlled*: A single pulse of variable time is produced per half-cycle.
3. *Step wave*: The output wave form is switched more frequently, producing a step wave form.
4. *Pulsewidth modulation*: Each half-cycle of output is made up of pulses of varying duration but equal magnitude.

The main problem in specifying UPS equipment is that of inadequately identifying the characteristics of the loads which the UPS is to supply, and the following must be specified:

Steady-state voltage regulation
Acceptable transient voltage variation
Non-linearity characteristics of load current
Percentage distortion in voltage
Frequency tolerance
Power factor of load
Inrush upon switch-on

21.9.4 Standby diesel generators

The selection of a diesel generator is governed basically by the electrical load and the standard engine frame sizes available. In specifying a diesel generator for use as a standby supply the electrical load must be analysed in terms of final running load, load profile, starting characteristics of individual motor loads and required operating time. Chapter 20 gives a full description of the types of diesel-driven time movers available. Electrical characteristics to be considered include:

1. Type of generator;
2. Type of excitation (brushless or static);
3. Transient performance (the capability of the generator to maintain the output voltage when a load is applied).

It must be remembered that the impedance of the generator will be greater than that of the supply system which it has replaced;

4. Method of alternator cooling (the heat gain to the building housing the diesel generator must be taken into account);
5. Method of neutral earthing;
6. Method of testing. If the diesel is to be tested it may be necessary for it to operate in parallel with the supply system in order to apply load. Synchronizing facilities will be needed and the fault levels during parallel operation must be assessed.
7. Electrical protection schemes. The installation should be carried out in accordance with Electricity Council Engineering Recommendation G26, *The installation and operational aspects of private generating plant.*

21.10 Earthing

21.10.1 General

Electrical supply systems and equipment are earthed in order to maintain the voltage at any part of the system at a known potential relative to true earth and to provide a path for current flow under earth fault conditions so that protective devices operate correctly. The connection to earth should be such that the flow of fault current to earth does not cause voltages or voltage gradients to be of sufficient magnitude or duration likely to cause danger.

An electrical earth system comprises the provision on the supply system of an earth connection to facilitate earth fault current flow, the connecting of all exposed metalwork within the installation to a common earthing terminal and the connection of this terminal to earth. For a complete installation the principal factors which need to be considered are:

The earth electrode
Neutral earthing
Substation earthing
Supply system earthing
Bonding

21.10.2 Earth electrode

The connection to earth can be made utilizing earth rods, earth plates, earth lattices or grids, or buried strip conductors. Selection depends on the locality and type of ground conditions at the site, and since the effectiveness of the earth connection is, in turn, dependent on the resistivity of the soil at the site, the type of soil and climatic conditions have a direct effect upon the resistance of the connection made. Typical values of resistivity (Ω-m) for different types of soil are:

Garden loam	4–150
Chalk	30–200
Clay	5–100
Sand	50–800
Rock	Above 1000

The resistivity of the soil in any particular location will be a function of moisture content, soil temperature and presence of dissolved salts. At a site where climatic conditions vary considerably throughout the year, earth electrodes should be buried at a depth where such changes will not affect the resistivity. Earth rods are generally made of copper bonded onto a steel core. The copper provides a good connection to earth and offers a high corrosion resistance, while the steel core gives the mechanical strength necessary to allow the rods to be driven to the required depth.

Earth plates or lattices made of pure copper, while displaying good current-carrying capacities, do not provide a particularly low resistance due to the depth at which they are able to be buried. The third alternative is to bury lengths of copper tape around the installation. The use of reinforced concrete foundations for earthing electrodes has also recently been considered.

Formulae for calculation of the resistance to earth of each type are as follows:

Single rod $R=\frac{P}{2\pi L}\left[\ln\left(\frac{8L}{\delta}\right)-1\right]$

Plate $R=\frac{e}{8r}\left(\frac{r}{2.5h+r}+1\right)$

Grid $R=P\left(\frac{1}{4r}+\frac{1}{L}\right)$

where

R = resistance,
e = resistivity,
L = rod or conductor length,
δ = rod diameter,
r = radius of plate; for a rectangular plate or grid

$$r=\frac{\sqrt{a}}{\pi}$$

h = depth of plate,
P = soil resistivity.

The earth electrode system must be designed to be capable of carrying without damage to the full earth fault current of the supply system.

21.10.3 Neutral earthing

An earth can be established on an electrical supply system by the connection to earth of the neutral point of the supply transformers, or generators or the use of inter-connected star-wound earthing transformers. The earthing must be arranged to ensure that an earth is provided on the supply system at all times and that the resistance of the connection to earth must be such that earth fault protection operates correctly. In instances where generators are to be run in parallel with each other or in parallel with a grid supply the neutral earthing arrangements will require special consideration.

21.10.4 Earthing in substations

The earth systems associated with high-voltage substations must satisfy the following conditions:

1. The resistance of the system to earth must be low enough to ensure that earth fault protection equipment operates correctly.
2. In the event of an earth fault the difference in potential between a person's feet (step potential) should not reach dangerous levels.
3. In the event of an earth fault the difference in potential between any point which may be touched and the ground (touch potential) should not reach dangerous levels.
4. The substation and its environment should not be connected to metal objects such that the voltages arising within the substation under fault conditions can be transferred to a point remote from the substation.
5. The earthing conductors must be capable of carrying maximum earth fault current without overheating or causing mechanical damage.

21.10.5 Types of system earthing

The IEE Regulations for Electrical Installations recognize the following designations of earthing systems using a two-, three- or four-letter code as follows:

First letter (denoting the earthing arrangement at the source)
T – Direct connection of one or more points to earth;
I – All live parts isolated from earth or one point connected to earth through an impedance.
Second letter (denoting the relationship of the exposed conductive parts of the installation to earth)
T – Direct connection of the exposed conductive parts to earth, independently of the earthing of any point of the source of energy;
N – Direct electrical connection of the exposed conductive parts to the earthed point of the source of energy (usually the neutral).

The designation TN is subdivided, depending on the arrangement of neutral and protective conductors. S indicates that neutral and protective functions are provided by separate conductors and C that neutral and protective functions are provided by a single conductor.

Examples of these designations and their usage are as follows:

1. *TN-C system*: Neutral and protective functions combined in a single conductor throughout the system.
2. *TN-S system*: Separate neutral and protective conductors throughout the system;
3. *TN-C-S system*: Neutral and protective functions combined in a single conductor in a part of the system;
4. *TT system*: All exposed conductive parts of an installation are connected to an earth electrode which is electrically independent of the source earth.

The following publications contain further information on and expansion of the factors relating to earthing.

Electricity Council Engineering Recommendation G26, *The installation and operational aspects of private generating plant*
BS Code of Practice CP1013, Earthing
IEE Regulations for Electrical Installations

21.11 Cables

Electrical power distribution within an industrial installation is most often at a voltage up to and including 33 kV. This section describes the types of cable suitable for power circuits for use up to 33 kV and considers the factors which will influence the current-carrying capacity of such cables.

The principal components of a power distribution cable are the conductors and the insulators. The cable is completed by the provision of armouring, overloads and other features designed to protect the cable within its installed environment.

21.11.1 Conductors

Material used for conductors comprise copper or aluminium in either stranded or solid form. Copper is the most common type of conductor due to its good conductivity and ease of working. Despite having a conductivity of only 61% of that copper, aluminium can be used as the conductor material. The lower density of aluminium results in the weight of an aluminium cable offsetting, to a certain extent, that of the additional material necessary to achieve the required current-carrying capacity.

Although aluminium cable has a proven record of satisfactory operation, the problems of terminating such cables must be recognized at the time of specifying. Since an aluminium cable will be larger than an equivalent copper cable of the same current-carrying terminal, enclosures and cable support systems must be designed to suit. Table 21.6 illustrates current-carrying capacity, overall diameter and weight for a typical range of 3.3 kV cables.

21.11.2 Insulation

Insulation materials used at voltages up to 33 kV comprise paper, PVC and polythene (XLPE). Paper-insulated cables have been used for the complete voltage range up to 33 kV throughout this century, although today they are limited in general to 6.6 kV and above. Paper cable is usually designed to operate at a conductor temperature not exceeding 80°C and PVC cable has largely replaced paper cables at voltages up to 3.3 kV. The relative imperviousness of PVC to moisture has contributed greatly to the design of the cables, making them easier to handle and install. The excellent dielectrics of properties of cross-linked polythene, together with the fact that XLPE does not soften at elevated temperatures, means that XLPE cables can operate with high continuous operating temperatures and short-circuit temperatures. The maximum operating temperature of XLPE is 90°C,

Table 21.6 Comparison of parameters for 1900/3300 V XLPE insulated, armoured, three-core cable

	Copper			*Aluminium*		
Nominal area of conductor (mm^2)	*Current-carrying capacity* (A)	*Weight* (kg/1000 m)	*Overall diameter* (mm)	*Current-carrying capacity* (A)	*Weight* (kg/1000 m)	*Overall diameter* (mm)
50	210	2850	34.7	160	2075	33.0
70	265	3625	38.0	200	2450	36.0
95	325	4500	41.4	245	2875	39.1
185	495	8075	51.9	370	4650	48.7
240	580	10375	56.9	440	5625	53.2

Table 21.7 Comparison of full-load capabilities for paper, PVC and XLPE insulated copper cable

Cable size (mm^2)	*Paper*[a]	*PVC*[b]	*XLPE*[c]
	Current-carrying capacity in air (A)		
16	91	87	105
35	150	142	170
50	180	172	205
95	280	268	320
185	430	407	490
240	510	480	580

[a] 600/1000 V: Three-core, belted, paper-insulated, lead-sheathed, stranded-copper conductor.
[b] 600/1000 V: Three-core, PVC-insulated, SWA, PVC, stranded-copper conductor.
[c] 600/1000 V: Three-core, XLPE-insulated, PVC-sheathed, SWA-stranded copper conductor.

with a maximum under short-circuit conditions of 250°C. Table 21.7 gives a comparison of the full-load capabilities of a range of cable sizes.

21.11.3 Selection

In selecting a cable for a particular installation the following factors need to be considered:

1. The steady-state voltage drop of the circuit;
2. The voltage drop in starting if the cable is for a motor circuit;
3. The current-carrying capacity under full-load conditions;
4. The maximum fault current which can flow and its duration.

For (3) and, to a certain extent, (4) the exact details of the cable installation will have a direct bearing on the cable chosen.

The cable rating will be affected by the ambient temperature of the installation, and account should be taken of any local high temperatures caused by proximity of process plant, the presence of and location of other cables and any thermal insulation which may prevent or inhibit heat dissipation. Cable ratings applicable to a variety of cable-installation techniques, together with the correction factors applicable, are given in a number of sources, including manufacturers' catalogues, the IEE Wiring Regulations and reports published by ERA Technology and IEC.

22 Electrical Instrumentation

H Barber BSc
Loughborough University of Technology

Contents

22.1 Introduction

The plant engineer needs to know the circumstances under which the plant for which he is responsible is operating. Hence he needs to be able to measure the inputs to the plant and the power inputs to the individual sectors. It will also be necessary to measure the plant outputs. In most cases the instrumentation uses electrical or electronic techniques, i.e. it measures either electrical quantities *per se* or quantities such as temperature or pressure which have been converted to electrical signals by means of transducers. The degree of accuracy to which these measurements are required will vary, depending on their purpose. In the case where the measurement is used as a direct basis for charging for energy then the accuracy must be very high. This is also the case where precise process control is involved. In other instances it is sufficient for the plant engineer to have a less accurate measurement of what is going on. As well as being technically suitable for the purpose, the instrument chosen will reflect these considerations. In the case of process control the instrumentation must be reliable and it must yield information, often over very long periods of time, which represents the state of the plant or the process and its past history. It is on the basis of this information that the plant engineer will make decisions, many of which will affect the economic viability of the process and some of which will have a direct impact on the safe operation of the plant.

Most plants will be supplied either directly or through the factory distribution system from the electricity supply authority. In the UK this supply is usually a.c. at a frequency of 50 Hz and at a voltage depending on circumstances, but which, in the case of large power demands, can be 11 000 V or even higher. Thus meters are required to measure these and related quantities. A variety of transducers is available for the measurement of physical quantities. These either give a direct readout to the operator or initiate some form of control action. In addition, there is a requirement for test and diagnostic equipment to determine if the measurement and control equipment is operating correctly and, if failures occur, to aid in tracing the source of the problem.

The plant engineer will need to familiarize himself with what is available in all these categories so that he may make the most appropriate equipment selection. He will also need to be familiar with what to expect from his instrumentation both in terms of accuracy and in order to know how to interpret the information correctly.

22.2 Electricity supply metering

22.2.1 Metering requirements

The electricity supply to the factory is usually obtained from the public supply authority, who will need to install metering in order to be able to assess the charges due in accordance with the agreed tariff. This metering will be the responsibility of the energy supplier but, in order to ensure that the rights of the customer are protected, the design of this instrumentation, and the accuracy limits within which it must operate, are controlled by statute. The customer will usually either have visual access to this instrumentation in order to confirm the readings or will install his own, duplicate, instrumentation as a check.

In any event, in most plants the plant engineer will have a need to monitor the electricity consumed by the various sectors on the site and he will need similar instrumentation to do this. The normal method of charging for electrical energy is based on two components, since the true cost is a function of (1) the total amount of primary (e.g. coal, nuclear) energy and the costs of converting it to electricity (these are the 'unit costs') and (2) the cost of providing the necessary plant to carry out the conversion and to transport the electrical energy to the customer (the 'maximum demand charge'). The first of these is primarily a function of the energy (kWh) supplied while the second is a function of the maximum value of the electric current drawn at any time from the supply and the voltage at which it is supplied. Thus two forms of instrumentation are needed, an energy (kWh) meter and a maximum demand (kVA) meter.

22.2.2 Energy meters

Energy is usually measured using an induction integrating meter (BS 37). In the case of a single-phase load the arrangement is as shown in Figure 22.1, this consists essentially of two electromagnets, mounted one above the other, with a light-weight aluminium disc free to rotate between them, as shown in the figure. The upper electromagnetic is energized by the system voltage (V), or a voltage proportional to it, and a current proportional to the system current (I) flows through the coil on the lower magnet. The electromagnetic flux created by the two magnets interact in the disc in which a torque is produced proportional to the product of the two multiplied by the cosine of the phase angle between them. Thus this torque is proportional to (VI cos), the electrical power supplied to the system, and the speed of rotation of the disc is proportional to the torque. Hence the

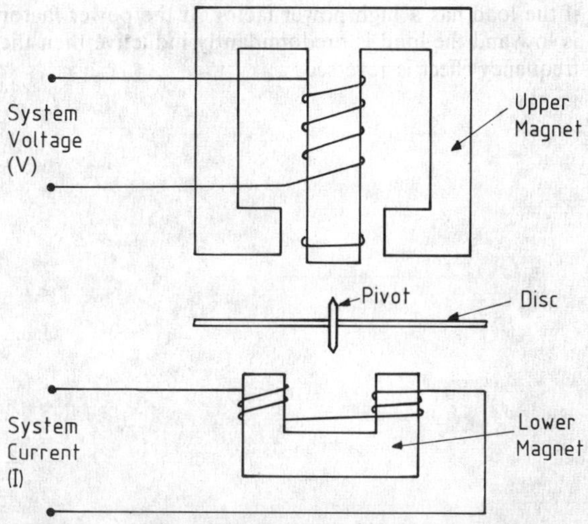

Figure 22.1 Schematic diagram of an induction meter

number of revolutions made by the disc in an hour is a measure of the energy used, i.e. the watt-hours (Wh) or, more usually, the kilowatt-hours (kWh).

The number of rotations of the disc is registered and can be recorded on a dial. As noted above, the voltages and currents energizing the coils are the normal system voltages unless these are too large for convenience, in which case voltage and current transformers are used as appropriate.

A high degree of accuracy is required over a very wide range of input currents. At very low loads (i.e. very low input currents) this accuracy can be affected by pivot friction, but this can be allowed for in the instrument design. The disc rotation is converted to a dial reading using a gear train (the 'register') and this reading is then used as a basis for charging under the tariff arrangements in use. In some cases two-part tariffs are involved, the simplest form being that in which kWh (units) above a certain number are charged at a different rate. This adjustment can be made after the reading is taken. In other cases units used during specific periods of the day are charged at a rate different from that in force during the remainder of the time. This can be catered for by incorporating a time switch into a meter with two sets of registers. At the appropriate time the switch operates and an electromagnet changes the drive to the second register. Alternatively, an entirely separate instrument can be used, with the change-over again being controlled by the time switch.

22.2.3 Error limits

The permissible error limits are defined in BS 37. In general, the error must not exceed $\pm 2\%$ over the working range, although in practice the accuracy is usually much higher than this. Temperature changes are compensated for in the design and the instrument is likely to be unaffected by variations in voltage and frequency provided that these are within the limits laid down in the Electricity Supply Acts ($\pm 4\%$ in voltage and $\pm 0.2\%$ for frequency). Very low voltages and reduced supply frequencies will both cause the instrument to read high if the load has a high power factor. If the power factor is low and the load is predominantly inductive then the frequency effect is reversed.

2.2.4 Three-phase metering

For large loads supplied with three phases then a three-phase meter is necessary. In these cases three single-phase units with their rotor torques combined mechanically are used, the whole being contained in one case. If the load current is above 150 A per phase then it is common to use instruments which combine the measuring elements, each with their own current transformer (see below) so that the combination can be calibrated as one unit (Figure 22.2).

22.2.5 Maximum-demand metering

As noted above, the tariff may also contain a component based on the maximum demand taken from the supply in a given period. In this case a separate maximum-demand meter is required. This is a modified energy meter which has a special register such that after a given period (usually half an hour) the dials are disconnected. During the next half-hour the register mechanism still operates but the clutch driving the dials is only energized if the total exceeds that already on the meter. This continues for the whole of the recording period, and thus, at the end, the reading on the meter is the maximum obtained during any one of the half-hour periods. This value is then doubled to yield the maximum kW taken in any half-hour.

It should be noted that this is not necessarily the actual instantaneous maximum demand. The quantity measured on this meter is, of course, the maximum kW demand, but some forms of tariff are based on the maximum kVA demand in order to encourage power factor improvement (see below).

22.2.6 Summation metering

In some instances it is required to measure the sum of the energy supplied along several feeders. This is the case, for example, when a factory site is fed by more than one cable and the tariff is based on the overall energy and maximum demand used. The usual method of doing this is to summate the currents in the individual feeders by installing separate current transformers in each and connecting their secondaries in parallel (Figure 22.3).

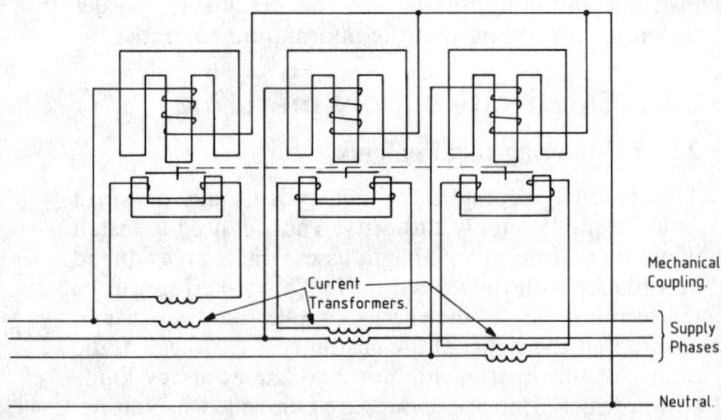

Figure 22.2 Three-phase meters

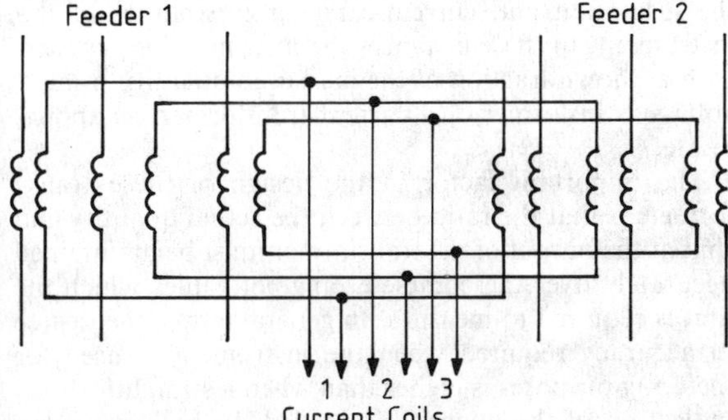

Figure 22.3 Summation current transformers

The current transformers must have identical ratios. The assumption is that all the feeders have voltages which are equal in magnitude and phase. If the feeders are physically close together it may be possible to use one current transformer as a summation unit. The current transformer has a number of primary windings, each carrying the current of one feeder. The magnetic fluxes due to the individual currents are summed in the transformer core and the output is obtained from a single secondary winding. An alternative is to use impulsing techniques whereby each meter incorporates a device which generates pulses at a rate depending on the energy being used. The pulses are transmitted to a single instrument in which they are recorded to give the total energy used over the whole group.

The pulses can be generated mechanically by a cam driven from an auxiliary spindle by using photo-transistors illuminated when a slot on a disc mounted on the main spindle passes a small lamp or by employing inductive techniques in which a shield on the disc passes a pick-up coil each time it rotates. This action modifies the e.m.f. output from the coil, so generating a pulse.

Electronic techniques can generate a larger number of pulses in a specified time and are therefore more accurate than mechanical devices. At the receiving end the pulses are used to determine the state of a series of bistable networks. These are scanned and reset sequentially and the total number of pulses recorded.

22.3 Power factor correction

Electricity is normally charged for on the basis of power (kilowatts) and the supply authority must install plant whose rating (and therefore cost) is a function of the voltage of the system and the current which the consumer takes (i.e. kilo-volt-amps). The relationship between the two is:

$$\text{kW} = \text{kVA} \times \cos\phi$$

where $\cos\phi$ is the power factor and is less than 1.0. In the case of loads which have a low power factor the supply authority is involved in costs for the provision of plant which are not necessarily reflected in the kWh used. A penalty tariff may then be imposed which makes it economically worthwhile for the consumer to take steps to improve his power factor. Low power factors occur when the load is predominantly either inductive or capacitive in nature (as opposed to resistive). In most industrial circumstances where the load includes a preponderance of motors the load is inductive (and the power factor is therefore 'lagging'). Consequently if the power factor is to be brought nearer to unity the most obvious method is to add a significant capacitive component to the load.

It is important to ensure that, as far as possible, this additional equipment is purely capacitive, i.e. that it has no resisitive component, otherwise the customer will face an increase in his energy bill. The situation is illustrated by the circuit and phasor diagrams of Figure 22.44. In the figure V represents the system voltage, I is the current drawn from the supply before the addition of power factor correction equipment and $\cos\phi$ is the associated power factor. If a capacitance, C, is connected in parallel with the load then this will introduce an additional current component I_c, the resultant total current drawn from the supply is I' and it will be seen from the figure that the new power factor is $\cos\phi'$, where $\cos\phi' > \cos\phi$.

Even if the power factor correction capacitance consumes no energy it will need capital investment, and therefore the consumer must balance the capital charges of this equipment against the savings which it produces in the energy bill. It is not normally economic to correct the power factor to its theoretically maximum value of unity, and a value of 0.9–0.95 is more usual.

22.3.1 Equipment

For relatively small loads the power factor correction equipment usually takes the form of static capacitors. In

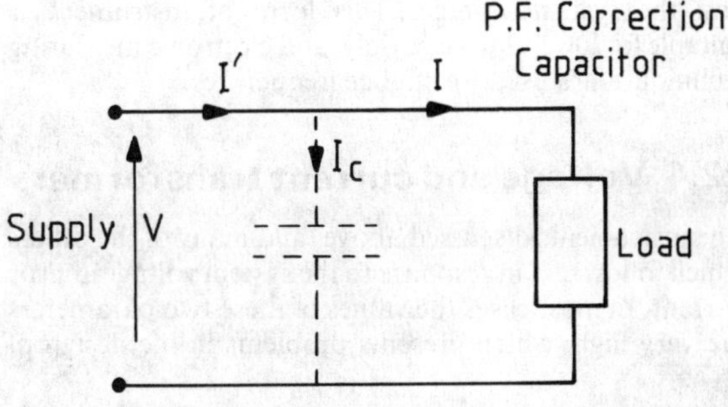

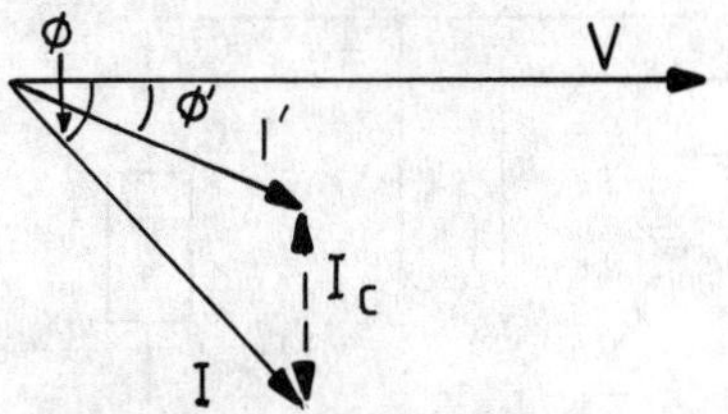

Figure 22.4 Power factor correction

larger installations it may be more economic to install an a.c. synchronous motor which, if its excitation is adjusted correctly, can be made to draw a 'leading' current from the supply. In most industrial plants the load is variable, and to gain the maximum benefit from the power factor correction plant this must also be varied to suit the load conditions.

If static capacitors are employed this can be achieved by using several capacitors arranged in units (banks) which can be switched in or out as required. This variation can only be carried out in discrete steps. In the case of the a.c. machine (the synchronous condenser) it is possible to obtain a continuous variation. The switching of the equipment can be carried out by an operator or automatically in response to the output from a power factor sensing instrument.

22.3.2 Power factor measurement

Instruments are available to measure power factor. One form is an electrodynamic instrument employing the same basic principle referred to above but with two coils mounted at right angles to each other on the same shaft (Figure 22.5). These coils are situated in a field produced by a third coil carrying the line current. One of the coils on the moving element carries a current derived from the supply voltage and is connected in series with a pure resistance. The second carries a similar current, but in this case a pure inductance is connected in series. The torque produced by each coil is a function of the current which it carries and therefore depends on the impedance of the particular coil circuit. There is also a third torque component due to the mutual inductive effects between the fixed and moving coils. The nett result is that the moving element is displaced by an amount which is a function of the phase angle between the system current and the system voltage. This form of instrument is suitable for low frequencies only, and electronic measuring techniques are used for higher frequencies.

22.4 Voltage and current transformers

The instruments discussed above (and many of the others which follow) act in response to the system voltage and/or current. In most cases the values of these two parameters are very high, which presents problems in the design of the insulation and current-carrying capabilities of the instrument. In these instances the instrument is supplied with a known fraction of the measured quantity using a voltage transformer or a current transformer, as appropriate.

The important factor in the design of these transformers is that the ratio between the actual quantity and that at the output of the transformer must be maintained accurately over the complete range of values which the unit is required to measure. In general terms, the degree of accuracy required when the instrument is used for metering purposes is higher than when a straightforward indication of the quantity is needed. In both cases the accuracy must conform to the relevant British Standard.

22.4.1 Voltage transformers

There are fundamental differences between the behaviour of the two types of transformer. The voltage transformer is shown diagrammatically in Figure 22.6. The system voltage V_p is applied across the primary winding which has N_p turns. The secondary voltage V_s is:

$$V_s = V_p(N_s/N_p)$$

neglecting errors. These errors are small and constant provided that the transformer is used over a small voltage range as specified by the manufacturer and the loading imposed by the measuring instrument connected to the secondary winding is between 10 and 200 VA.

Voltage transformers are classified into types AL, A, B, C and D, in descending order of accuracy. The ratio errors for small voltage changes (within ±10% of the rating) vary between 0.25% and 5% and the phase errors between ±10 min and ±60 min.

For all but the highest voltages the construction of the voltage transformer is very similar to that of a normal power transformer at voltages of magnitudes normally met with in plant engineering. However, at primary voltages of 100 kV and above it may be necessary to use alternative means. The type of voltage transformer described above is suitable for relatively low alternating frequency supplies. It cannot be used at d.c. and there is a possibility of large errors occurring at high

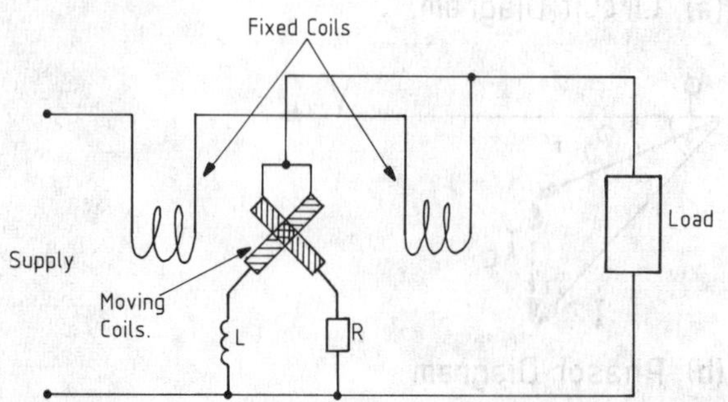

Figure 22.5 Schematic diagram of power factor meter

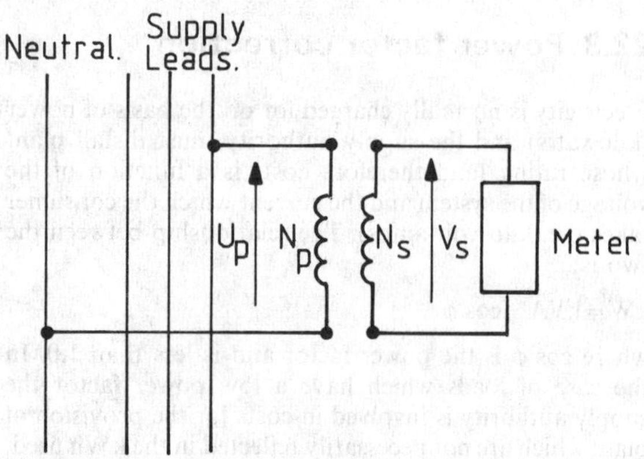

Figure 22.6 Voltage transformer

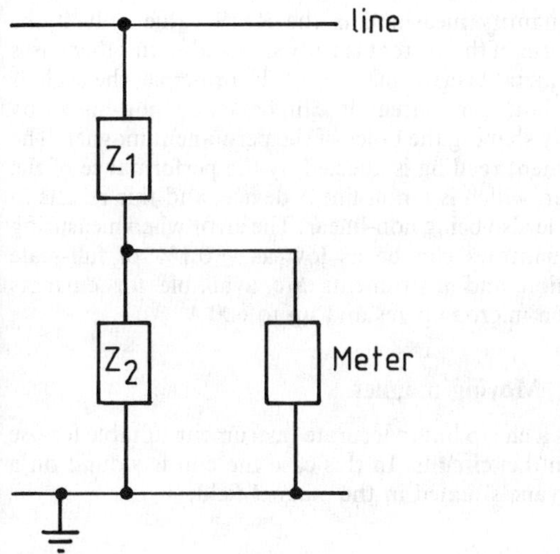

Figure 22.7 Voltage divider

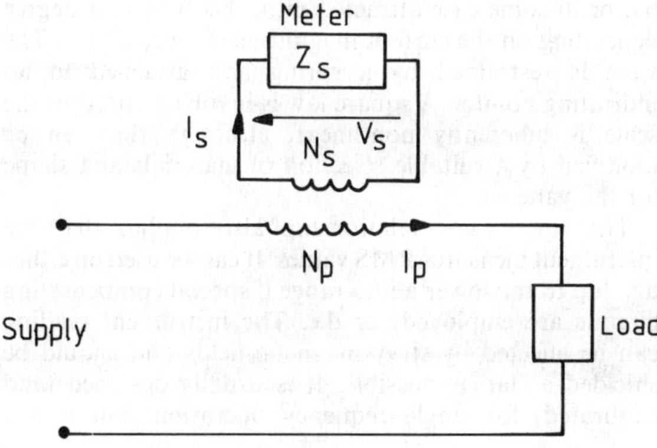

Figure 22.8 Current transformer

frequencies. In these instances some form of voltage divider is necessary. The arrangement is as shown in Figure 22.7. The two impedances, Z_1 and Z_2, are connected across the supply with the indicating instrument connected across Z_2. Then (referring to the figure):

$V_s = V_p[Z_2/(Z_1+Z_2)]$

The load (or instrument) impedance is connected in parallel with Z_2 and should be very much larger than Z_2 if accuracy is to be maintained. If d.c. voltages are involved then the impedances can be purely resistive and, in the steady state, any leakage inductance or capacitance has no effect. At high a.c. frequencies, however, it is important that they are non-reactive and that stray inductances and capacitances to earth, etc. are taken into account.

22.4.2 Current transformers

The current transformer is arranged with its primary winding in series with the supply (Figure 22.8). It thus carries the load current. The measuring instrument is connected across the secondary as shown. The ideal theoretical relationship between the currents and the number of turns on the primary and secondary is:

$I_s = I_p(N_p/N_s)$

The voltage appearing across the secondary terminals is:

$V_s = I_s Z_s$

where Z_s is the impedance of the measuring instrument connected to the secondary terminals (the 'burden').

If this impedance is high then so also is this voltage. The primary of the device should therefore never be energized when the secondary is open circuit, since in that case Z_s, and therefore V_s, is infinite. Standard secondary current ratings are 1 A and 5 A, and the preferred range of values for the burden is between 1.5 VA and 30 VA.

Ratio and phase angle errors also occur due to the need for a portion of the primary current to magnetize the core and the requirement for a finite voltage to drive the current through the burden. These errors must be small and current transformers are classified (in descending order of accuracy) into types AL, AM, BM, C and D. Ratio errors in class AL must be within the limits of $\pm 0.1\%$ for AL and $\pm 5\%$ for D, while the phase error limit on class Al is ± 5 min to ± 2 min in types CM and C.

Again, there are problems both at d.c. and at high frequencies. In the former case a current sample is obtained by connecting a very low resistance, R_s, in series with the load and using a voltmeter to measure the voltage across it. This voltage is equal to $I_s R_s$ and, provided R_s is known, then I_s is specified.

At very high frequencies the current is measured by assessing one of the effects which it produces. Several techniques are possible, e.g. (1) measuring the temperature rise when the current flows through a known resistance or (2) using a Hall-effect probe to measure the electromagnetic field created by the current.

22.5 Voltmeters and ammeters

These are manufactured according to several different principles, each being suitable for a particular application. The more common types are discussed briefly below.

22.5.1 Moving coil

This consists of a coil mounted between the poles of a permanent magnet. When a current flows through the coil a magnetic field is created and the interaction of this field with that of the magnet produces a torque, proportional to the current in the coil, which causes the coil to rotate against the action of a spring. A pointer is attached to the coil which gives an indication of the magnitude of the coil current (Figure 22.9).

Unlike most other electrical measuring instruments, the scale is inherently linear. The moving parts of the instrument must be as light in weight and as free from

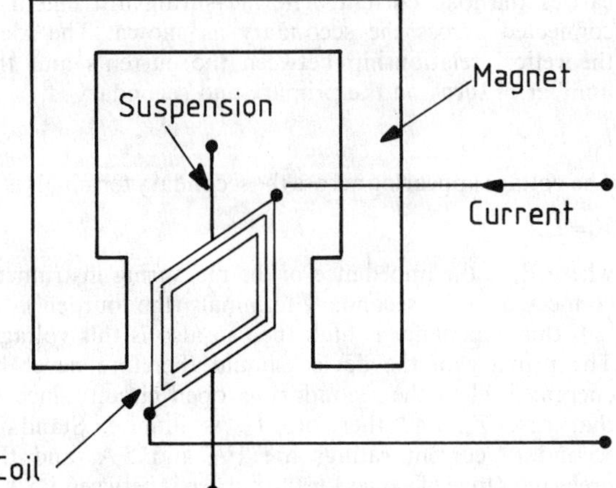

Figure 22.9 Moving-coil instrument

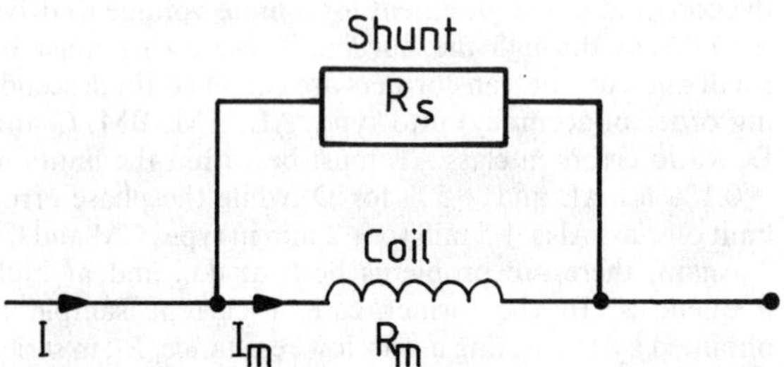

Figure 22.10 Use of shunt

friction as possible in order to preserve accuracy. This implies that the coil is capable of carrying only small currents. If higher currents are required to be measured it is necessary to employ a shunt to divert some of the main current away from the instrument. The arrangement is as shown in Figure 22.10, where R_m is the resistance of the meter and R_s that of the shunt. The relationship between the total system current (I) and that in the instrument (I_m) is:

$$I_m = IR_s/(R_m + R_s)$$

Many commercial instruments have the shunt incorporated into the unit, and the user need only concern himself with the full-scale deflection as shown on the scale. General-purpose instruments with a range of built-in shunts are available and the user then selects the appropriate range. The instrument is delicate and easily damaged by excess currents. It can also be damaged by connecting it in the circuit with reversed polarity. The instrument may also be used as a voltmeter by connecting it, in series with a large resistor, across the supply, and in most commercial instruments the resistor is incorporated into the case.

The direction of rotation depends on the direction of the current in the coil, and thus the instrument is only suitable for d.c. It is, however, possible to incorporate a full-wave rectifier arranged as shown in Figure 22.11 in order to allow the instrument to measure a.c. quantities. The quantity measured is the RMS value only if the waveform of the current is truly sinusoidal. In other cases a considerable error may result. In principle, the scale is linear but, if required, it can be made non-linear by suitably shaping the poles of the permanent magnet. The instrument reading is affected by the performance of the rectifier, which is a non-linear device, and this results in the scale also being non-linear. The error when measuring d.c. quantities can be as low as $\pm 0.1\%$ of full-scale deflection and instruments are available for currents between microamperes and up to 600 A.

22.5.2 Moving magnet

This is a cheap but inaccurate instrument suitable for use only in d.c. circuits. In this case the coil is wound on a small vane situated in the magnet field.

22.5.3 Moving iron

The current to be measured is used to energize a coil. This creates a magnetic field and a moving vane is repelled by, or in some case attracted into, this field to a degree depending on the current magnitude (Figure 22.12). The vane is restrained by a spring and attached to an indicating pointer. A square law is involved and thus the scale is inherently non-linear, although this can be modified by a suitable selection of materials and shape for the vane.

The square law relationship also implies that the instrument measures RMS values. It can be used on either a.c. (up to the lower audio range if special compensating circuits are employed) or d.c. The instrument reading can be affected by stray magnetic fields and should be shielded as far as possible. It is usually designed (and calibrated) for single-frequency operation, but it will

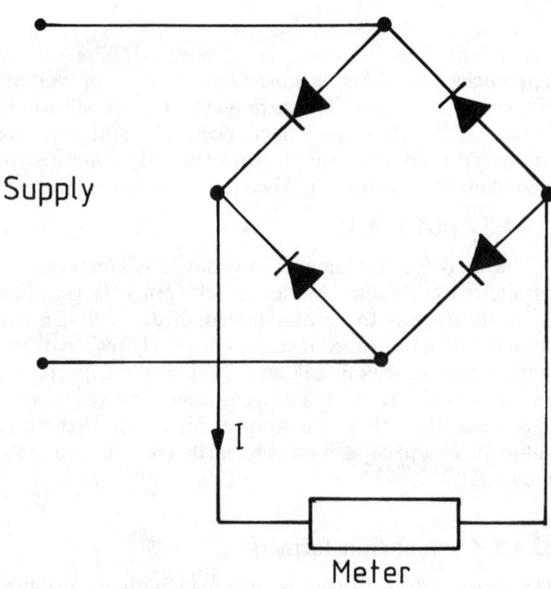

Figure 22.11 Full-wave recitification

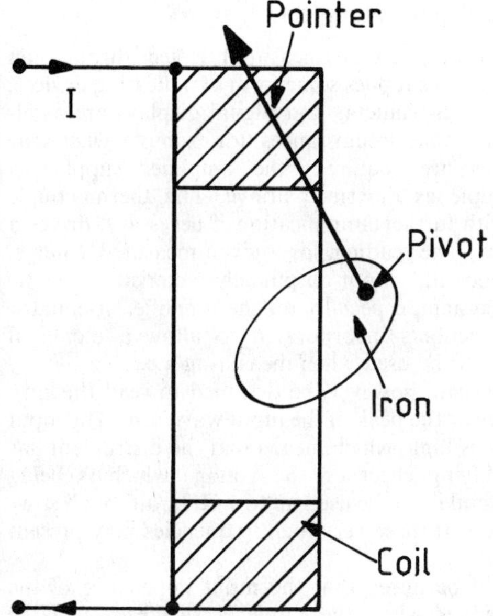

Figure 22.12 Moving-iron principle

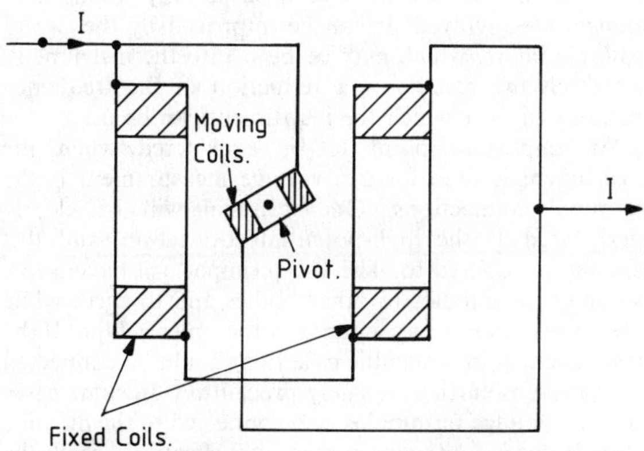

Figure 22.13 Schematic diagram of electrodynamic instrument

continue to give accurate results (in terms of r.m.s. values) even if the waveform contains high-frequency harmonics. The smallest units are capable of measuring a few milliamps and the upper limit is fixed only by the current-carrying capacity of the coil. Shunts can, of course, be incorporated for different current ranges. This type of instrument is cheap and robust.

22.5.4 Electrodynamic instruments

These incorporate an electromagnet in place of the permanent magnet of the moving-coil instrument (Figure 22.13). The current flows through both the moving and fixed coils and a torque is produced on the former which carries the pointer and which is restricted by a spring. Again, the deflection obeys a square law and the scale is cramped at the lower end. Since with a.c. the current in both coils changes direction, this instrument can be used on either d.c. or a.c. Although with a.c. the meter actually measures the mean of the square of the current, the scale is usually calibrated in terms of the rms value. With the appropriate modifications this type of instrument can be used as a wattmeter, a VAR meter, a power factor meter or a frequency meter.

22.5.5 Thermocouple

This instrument was developed from the 'hot-wire' ammeter, some examples of which can still be found. In the modern equivalent the current to be measured (or a known proportion of it) flows through a small element which heats a thermocouple, so producing an rms voltage at its terminals, which is a function of the current. This voltage then supplies a current to a permanent-magnet, moving-coil movement.

The instrument measures the true rms values and units are available to cover the frequency range from d.c. to 100 MHz. The major drawback is the susceptibility to damage by overloads (even those of short duration) unless appropriate protection is provided. The accuracy can also be affected by ambient temperature changes. Since the instrument responds to an heating effect it is slow to respond to changes in the measured quantity.

22.5.6 Electrostatic voltmeter

This consists of a pair of plates, one fixed and the other moving, mounted parallel to each other. When a voltage is applied across the pair the fixed plate moves against the action of a retaining spring. Again, the characteristic is basically square law. It has an inherently high level of insulation, due to the air gap between the plates, and can be used to measure voltages up to 15 kV. It consumes no power and therefore has an extremely high impedance.

22.6 Frequency measurement

Low frequencies can be measured using an electromechanical moving-coil instrument. Current from the supply flows through two parallel fixed coils and then through a moving coil mounted between them (Figure 22.14). The two fixed coils are tuned to slightly different frequencies and the resulting fields set up a torque which is proportional to frequency.

The nominal frequency at which the instrument can be used depends on the tuning of the coils and the instrument is only accurate to within plus or minus a few percent of this frequency. A ratiometer (Figure 22.15) can be used at higher frequencies up to 5 kHz or so. This consists of two moving coils arranged at right angles and mounted between the poles of a permanent magnet. The system current is fed into the two coils through separate phase-shifting networks and the result is to produce a torque proportional to frequency.

At frequencies higher than this a solid-state counter must be used. This is based on a stable oscillator and, in effect, counts the pulses generated during one cycle of

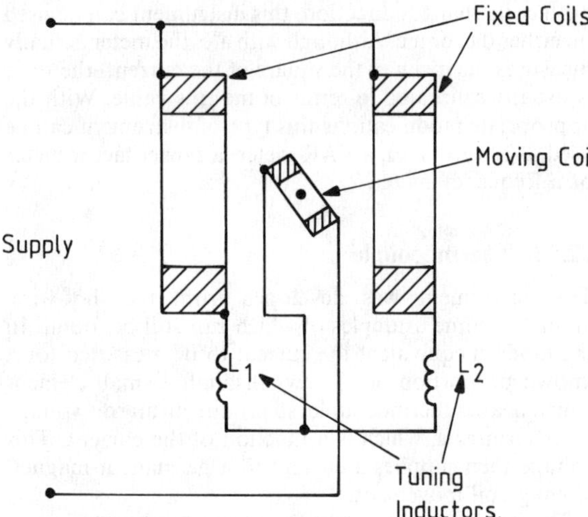

Figure 22.14 Frequency meter

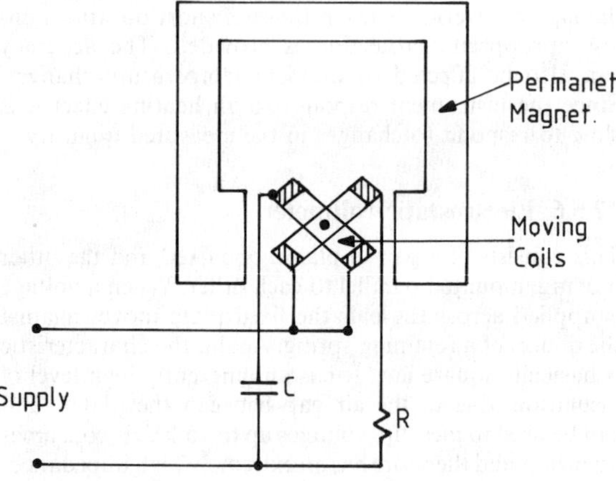

Figure 22.15 Frequency ratiometer

the supply frequency. The range and accuracy of the instrument depends on the master oscillator frequency, but units capable of use over the whole range up to 600 MHz are available, although some discrimination is likely to be lost at the lower end.

22.7 Electronic instrumentation

Electronic instrumentation is available for the measurement of d.c. and a.c. voltage, current and power as well as impedance. Such instruments usually have higher sensitivities, operating frequencies and input impedance than is normally found in the electromechanical instrumentation described above. However, they may need to incorporate amplifiers and they invariably need power to operate the final display. Hence an independent power source is needed. Both mains and battery units are available. The accuracy of measurement is very dependent on the amplifier, and bandwidth and adequate gain are important qualities.

22.7.1 Analogue voltmeters

The a.c. voltage to be measured is fed through an amplifier which supplies some form of indicating devices. Moving-coil instruments and digital displays are available, and in some circumstances (for example, when true rms values are required) the amplifier supplies a thermocouple as described above. The thermocouple output, with further amplification if necessary, drives a d.c. display. The reading, for a given measured voltage, depends on the input/output characteristics of the complete system, especially on the amplifier/attenuator and any rectifiers incorporated to allow a d.c. final instrument to be used when measuring a.c.

The unit can, however, be designed to read the rms, the average or the peak of the input waveform. The input impedance is high, which means that the instrument has little modifying effect on the voltage which is being measured and it can be used up to 1 GHz, although stray capacitances at these very high frequencies may present a problem.

It should be noted that the input impedance of the meter changes when the range is switched. A large amount of spurious electronic signals ('noise') may be present in some systems. This is a potential source of error in the instrument reading, especially when small signals are involved. It can be improved by the use of suitable filters, which may be built into the instrument, although the penalty is a reduction in the frequency bandwidth over which the instrument can be used.

An important point to be considered when the instrument is used for a.c. voltage measurement is the terminal connections. One terminal will be clearly designated as the high-potential connection, and this should be adhered to. The HT terminal will have a low value of capacitance to other bodies and to earth while the corresponding capacitance of the other is high. If the instrument is in a metallic case this should be connected to the mains earth as a safety precaution. In some cases the low-voltage terminal is also connected to the metallic case. If this is so, the instrument will effectively earth the circuit under test, which may give rise to problems.

Instruments with a balanced input circuit are available for measurements where both input terminals are normally at a potential other than earth. Further problems arise due to common-mode interference arising from the presence of multiple earth loops in the circuits. In these cases the instrument may need to be isolated from the mains earth. Finally, high-frequency instruments, unless properly screened, may be subject to radiated electromagnetic interference arising from strong external fields.

22.7.2 Digital multimeters

A number of different types are available but they all share the common characteristic of a very high input impedance, which means that they are suitable for measuring voltages with minimal interference to the actual circuit. They are used whenever accuracy, good resolution and high levels of stability are important. As their name implies, these instruments are capable of measuring both a.c. and d.c. voltages and currents as well as resistance, and incorporate a wide range of scales. When used with a.c. the input signal is divided down to a level suitable for the input to an analogue to digital converter. This needs to be preceded by other function converters and range-switching units. In some instruments of this type the range is selected automatically by sensing the level of the input signal and selecting the appropriate range electronically.

The analogue-to-digital converter has a fixed range, usually either 0–200 mV or 0–2 V, and the input divider network is used to obtain a signal within this range. This implies that there may be considerable differences in input impedance between different ranges. When used to measure alternating voltage the appropriate function converter may be a transformer followed by rectification and smoothing. This produces a direct voltage equivalent to the average value of the a.c. voltage and, provided the input signal is purely sinusoidal, the indicator is then calibrated in rms values.

More sophisticated instruments use special circuitry to obtain the rms values of non-sinusoidal signals. The current convertor usually consists of low-value resistors in the input, and one of these is chosen as a shunt resistor (depending on the scale chosen) and the voltage developed across this is measured by the instrument. Resistance is measured by using the resistive chain at the input as a reference. When a voltage is applied across this and the unknown resistor the two voltage drops are proportional to the respective resistance values. The voltage across the unknown resistor is applied to the A–D converter followed by that across the known resistor with which it is compared and the ratio is used to determine the value of the former. It should be noted that the instrument is essentially a sampling device. The input signal is periodically sampled and the result processed using one of a number of techniques, examples of which are given below:

1. The *linear ramp* instrument uses a linear time-base to determine the time taken for an internally generated voltage to reach the unknown voltage, V. The limitations are due to small non-linearities in the ramp, instability of the electronic components and a lack of 'noise' rejection.
2. The *charge balancing* voltmeter employs a pair of differential input transistors used to charge a capacitor which is then discharged in small increments. The individual discharges are then sensed and the total number, which is a function of the measured voltage, stored. It has improved linearity and high sensitivity as well as a reduced susceptibility to 'noise'. In addition to its use as a straightforward voltmeter, it can be interfaced directly with other measurement devices such as thermocouples and strain gauges.
3. The potentiometer principle is used in the *successive approximation* instrument, where the unknown voltage is compared with a succession of known voltages until balance is obtained. The total time for measurement and display is about 5 ms.
4. In the *voltage-frequency* unit the unknown voltage is used to derive a signal at a frequency proportional to it. This frequency is then applied to a counter, the output of which is thus a measure of the voltage input.
5. The *dual-slope instrument* incorporates an integrator operated in conjunction with a time difference unit, and this is one of a range of units employing a combination of the above techniques designed to take advantage of the favourable characteristics of each. It is subject to the same limitations as the linear ramp instrument described above but it does have the advantage of inherent noise rejection.

22.8 Instrument selection

The selection of the instrument for a specific purpose should take account of the following points:

1. Use no more precision (i.e. no more digits) than is absolutely necessary in order to keep the cost to a minimum.
2. Make sure that the input impedance is adequate in the circuit in which the instrument is used. This applies not only to the normal (or static) input impedance but also to the dynamic impedance which occurs when the instrument is subjected to transient surges with significant components at frequencies other than that at which the measurement is desired.
3. Take adequate steps to achieve noise rejection both by the choice of a suitable instrument and by using external networks if necessary.
4. Ensure that the instrument has appropriate output ports to drive other devices such as storage and computational facilities. Wherever possible, these ports should conform to the IEEE interface standards.

Most instruments of this type incorporate built-in self-checking facilities but the check can only be as accurate as the internal reference which, itself, will only be accurate over a specific range. The instrument should be calibrated against a reference source at periodic intervals.

Modern electronic instruments are much more than simple but accurate methods of measuring and can include many additional facilities such as:

1. Variable reading rate from one reading per hour up to several hundred per second;
2. Storage capability allowing many hundreds of separate readings to be stored internally and recalled subsequently, if necessary, at a different rate;
3. The ability to perform calculations on the readings and to display the derived result;
4. The ability to indicate if specific read values are outside pre-set limits and to store (and subsequently recall) maximum and minimum readings.
5. Self-calibration and/or testing according to predetermined programming.

22.9 Cathode ray oscilloscopes

The cathode ray oscilloscope (CRO) is an extremely versatile instrument for monitoring electrical signals, for

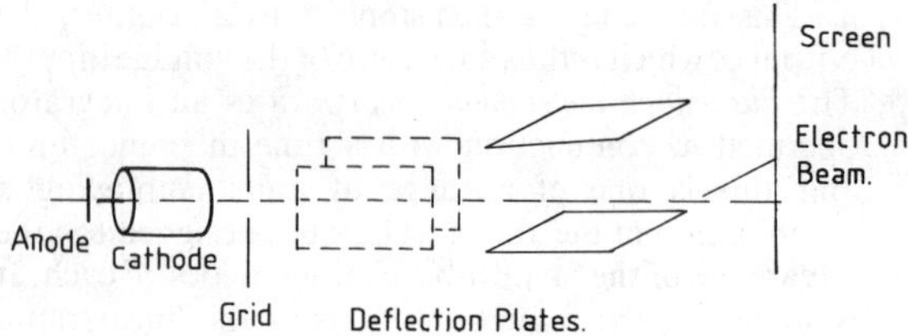

Figure 22.16 Principle of CRO

diagnostic testing and for studying time-varying phenomena. A stream of electrons is produced from a heated cathode and accelerated through an electron 'gun' towards a fluorescent screen (Figure 22.16). The stream passes through a coil which focuses it into a beam which then passes between pairs of vertical and horizontal plates. Electric fields are created between these plates which cause the beam direction to change in response to them.

In the simplest form the vertical plates are supplied with a voltage which is a linear function of time and, in the absence of a signal on the horizontal plates, the beam will move horizontally from left to right across the screen. The speed of movement is controlled by the rate at which the voltage applied to the vertical plates is increased, and at the end of its travel the beam flies back to its starting point.

If now a second, time-varying, voltage is applied to the horizontal plates this will cause a vertical deflection of the beam, and the result is a trace on the screen which corresponds to the variation of this second voltage with time. The action of the beam on the screen causes a fluorescent trace to appear on the screen as the beam is deflected. The time for which this trace persists is a function of the electron density in the beam and the material with which the screen is coated.

The accuracy of the device is limited by the linearity of the circuit which moves the beam horizontally in response to time (i.e. the 'time base') and the response of the input circuit. The 'time base' is generated by a sweep circuit which is triggered to instigate the start of the movement. The triggering signal may be derived from the voltage which is being measured or it may be obtained from an entirely separate source. This can cause difficulties in some applications. If it is required to examine a voltage pulse then the simplistic approach is to allow the pulse itself to trigger the signal. This is unsatisfactory if it is the leading edge of the pulse which is of interest, and therefore a delay is built into the instrument which allows the trace to be triggered immediately the pulse is applied but which delays the application of the signal to the vertical deflection plates for a short time in order to enable the pulse to be displayed in the middle of the screen.

The velocity with which the spot moves from left to right across the screen is adjustable, enabling single or multiple cycles of an a.c. signal to be observed. Typical instruments can be used to measure a wide range of voltages on a series of different scales. The input signal may be either a.c. or d.c. and in the latter case the operation of the appropriate switch on the front panel short-circuits the input capacitor. In each case the voltage to be measured needs to be amplified before it is applied to the plates. A different amplifier is used for each scale range and more sophisticated instruments may include a range of plug-in amplifiers in addition to those built into the instrument. The important amplifier characteristics are the gain, the bandwidth, the amplitude of signal which can be handled without distortion and the input impedance. Since the inherent sensitivity of a typical oscilloscope is likely to be of the order of 50 mV/cm the amplifier gain must be large enough to give an acceptable deflection when the input signal amplitude is very small. On the other hand, if high input signals are involved then an attenuator, rather than an amplifier will be needed (see below).

Reproducibility of the input signal is all-important, and therefore the amplifier must be capable of handling the input signal magnitude involved. Frequency response is also important and the amplifier in use must have a bandwidth sufficient to accommodate the frequency of the input signal. The input impedance should be high in order to avoid loading the circuit under test. As with the amplifier, the attenuator must produce a known degree of change (in this case a reduction) in the input signal within the specified frequency band. It must have a high input impedance which should not alter when the input sensitivity switch setting is changed. The oscilloscope is normally connected directly to the circuit under test. However, this may not be possible in all cases and a probe may have to be used. The probe may simply serve as an isolator, so avoiding problems caused by, for example, shunt capacitances in the measuring leads or it may contain built-in amplifiers (or attenuators) to extend the range of the instrument. The probe should be matched to the instrument and the use of a probe produced by one manufacturer in conjunction with an oscilloscope from another is not recommended.

Many variants of the basic type are possible and some of the more usual are listed below:

1. *Low frequency* (up to around 30 MHz): typical deflection sensitivities are between 0.1 mV/cm and 10 V/cm. This is a general-purpose instrument and can incorporate a second trace energized by a voltage different from the first and using the same or different deflection speeds for the time base. An alternative is to use a single beam but to apply a 'chopped' signal to the deflection plates. The applied signal is derived by sampling the two separate signals for brief intervals of time and applying each in turn to the plates. The luminescent persistence of the screen then results in two 'continuous' traces (one corresponding to each of the two applied voltages) to be observed.
2. *Medium frequency* (up to 100 MHz): The features described above (with the exception usually of the chopped-signal variant) are available but the timebase components and amplifiers are designed for much wider frequency bands.
3. *High frequency* (up to 400 MHz): Again, the difference between this and the above is the use of higher-quality components and circuits to give linearity over a much wider frequency range.

22.9.1 Storage

Most of the types described above have the facility for single-shot operation if it is necessary to measure single 'events' (i.e. transients). In these cases the timebase is triggered by the start of the transient. The limitation is the persistence of the screen luminescence since the event only occurs once, rather than a repetitive series of events, as happens with a periodic waveform where the trace is, in effect, overwritten during each operation of the timebase.

It is possible to sample and store the input transient in digital form by frequent sampling of the input waveform. These data can be retrieved (repetitively if necessary) from the store and displayed on the screen at the operator's leisure. The frequency range over which this type of instrument can be used depends on the sampling network, and is usually limited to a few MHz. Waveforms and transients corresponding to much higher frequencies can be stored and displayed in this way, with the storage being carried out by optical methods.

22.9.2 Sampling oscilloscopes

These are used for very high-frequency work. The waveform of the applied signal, which must be periodic in nature, is sampled at intervals which bear a definite relationship to some reference point on the wave. These samples are stored and subsequently used to reconstruct a display of the actual signal on the oscilloscope screen.

The non-storage oscilloscope can be found in most electronic test situations, from sophisticated research laboratories to production engineering plants. The storage unit is most widely used in medical work and in electromechanical applications, particularly where very high-speed transients need to be recorded, while, as noted above, the sampling type finds its main use in the evaluation of ultra-high-frequency equipment.

Appropriate connections must be used to pick up the required signal from the circuit under investigation. This can be done directly for low- and medium-voltage applications and, if a measurement of current is required, by measuring the voltage across a resistor of known value. However, for accurate work it is important that neither the impedance across which the leads are connected nor the impedance of the leads themselves affects the result. It is also important to avoid errors being introduced, particularly in high-frequency work, from spurious signals induced in the leads if they are situated in electromagnetic fields. The latter can be overcome in most cases by using co-axial leads with the outer screening connected to the case of the instrument and thence to earth. The former problem can be more difficult to deal with, since it implies that the load across the oscilloscope terminals, including the effects of the leads, must be matched to the input impedance of the instrument.

These problems can be minimized in the great majority of cases by using specially designed probes. The user will need to know the frequency range (i.e. the bandwidth) over which the probe can be used satisfactorily and also the scaling factor introduced by the probe.

22.9.3 Use of oscilloscope

The cathode ray oscilloscope is a multi-purpose instrument with several ranges, switching between which is carried out by controls on the unit. The required values are obtained by measuring the height of the trace of the screen or the time between different events in the horizontal deflection of the beam. These measurements are then converted to the required values using the scale values obtained from the control switch settings. It is thus of paramount importance to make sure that the scale settings are recorded. The vertical scale will normally be calibrated in volts (or fractions of a volt) per centimetre and the horizontal scale in seconds (or fractions of a second) per centimetre. The accuracy of the measurement will then depend on (1) the accuracy with which the beam deflection can be measured on the screen and (2) the accuracy of the built-in attenuators and amplifiers applicable to the range in use.

A graticule is normally provided on the screen to assist in the measurement. Parallax errors can cause problems unless this graticule is actually engraved on the screen itself, and the final accuracy will depend on the degree of beam focusing which can be achieved. Accuracies of better than 1.5% should be possible in most cases.

As noted above, the operator will select a convenient range which will allow for the whole of the signal he wishes to observe to be displayed on the screen. It is important that the internal circuits so selected are linear over this range and again, with most instruments, values of better than 1% should be achievable.

22.10 Transducers

Many of the quantities required to be measured in process plant operation are not, in themselves, electrical, and if electrical or electronic instrumentation is to be used then these need to be converted to electrical signals using a transducer. The transducer is a physical object and its presence will have an effect on the quantity being sensed. Whether or not this effect is significant will depend on the particular application, but in all cases it is advisable to consider carefully the balance between the requirement that the transducer should, on the one hand, cause the minimum interference with the quantity being measured and, on the other, that it should be intimately associated with the effect being measured.

Temperature measurement is a case in point. A large transducer in close contact with the body whose temperature is being measured will act as a heat sink and consequently produce a localized reduction at the point where the temperature is being measured. On the other hand, if an air gap exists between the transducer and the hot surface then the air (rather than the surface) temperature will be measured.

The important characteristics of a transducer used in conjunction with an electronic measurement system are accuracy, susceptibility, frequency, impedance and, if appropriate, the method of excitation. The transducer is likely to be the least accurate component in the system, and it should be calibrated (and recalibrated) at frequent intervals. It is likely to be subject to a range of different physical conditions, some of which it is there to detect and others by which it should remain unaffected (for example, a pressure transducer should be unaffected by any changes in temperature which it might be called upon to experience). Some types of transducer are not suitable for use under d.c. conditions and all will have an upper limit of frequency at which accuracy is acceptable. Many types of transducer are also affected by stray electromagnetic fields.

The impedance of the transducer is important if it provides an output signal to an electronic device (an amplifier, for example) and the impedance of the two must be matched for accurate measurement. Some transducers (thermocouples, for example) generate their output by internal mechanisms (i.e. they are self-excited). Others such as resistance thermometers need an external source and an appropriate type must be available. Transducers used in the measurement of the more common physical quantities are discussed below.

22.10.1 Temperature

There are several instruments for measuring temperature as follows.

Resistance thermometers

These utilize the fact that the resistance of most materials changes with temperature. In order to be useful for this purpose this change must be linear over the range required and the thermal capacity must be low. Although the above implies a high resistivity and temperature coefficient, linearity and stability are the paramount considerations, and suitable materials are platinum, nickel and tungsten.

The sensor usually consists of a coil of wire made from the material which is wound on a former and the whole sealed to prevent oxidization, although a film of the metal deposited on a ceramic substrate can also be used. The resistor is connected in a Wheatstone bridge network (Figure 22.17), using fixed resistors in the other three arms. The instrument connected across the bridge is calibrated directly in terms of temperature. The range is limited by the linearity of the device and the upper temperature, which can be measured, must be well below the melting point of the material.

Thermocouples

These are active transducers in the sense that they act as a generator of emf whose magnitude is a function of the temperature of the junction of two dissimilar metals. Rare-metal combinations are used for high accuracy and at high temperatures, but for most engineering applications one of the following is suitable, depending on the temperature:

Copper/constantan	(up to 670 K)
Iron/constantan	(up to 1030 K)
Chromel/constantan	(up to 1270 K)
Chromel/alumel	(up to 1640 K)

If the instrument is to be direct reading the second (or 'cold') junction must be kept at a constant reference temperature. If high temperatures are to be measured then the terminals of the detector can be used as the cold junction without an unacceptable loss of accuracy.

The voltage output of the more common types of thermocouple is of the order of 50 V/C and the output is either read on a sensitive moving-coil meter or on a digital voltmeter. The reading is converted to temperature using a calibration chart supplied with the thermocouple. Some commercial units are available in which the thermocouple and instrument is supplied as an integral unit with the scale directly calibrated in temperature. If a separate instrument is to be used then it should be noted that the thermocouple resistance is only of the order of 10 Ω and for maximum sensitivity the meter resistance should be matched to this.

The commercial units have a very low thermal capacity and very high response speeds. Some are available with several independent channels and a common 'cold' junction. Each channel is scanned in turn by the instrument, and the readings either displayed or stored for future recovery. Accuracies of better than 0.2% are possible. Thermocouples are available to cover a very wide range of temperatures, their cost is low and they have a small mass, so minimizing the intrusive effect on the surface at the point where the temperature is being measured. The output characteristics (output voltage versus temperature) are reasonably linear but the measurement accuracy is not particularly high.

Thermistors

These are semiconductor devices which have a high resistivity and a much larger temperature/resistance coefficient than the materials available for resistance thermometers. However, there is a temperature limitation of around 550 K and the characteristics are excessively non-linear, although it is possible to obtain matched pairs to measure differential temperatures. Thus, for example, they can be used to measure air flow in ducts. The change in resistance is again detected using a Wheatstone bridge network. The materials used are manufactured from sintered compounds of copper, magnanese, nickel and cobalt. Conventional semiconductor diodes in the form of p–n junctions can also be used in this way.

Although thermistors with a positive temperature characteristic are available, the negative characteristic type are more common. Devices which are physically small have low power-handling capabilities and some form of output amplification is usually needed. They do,

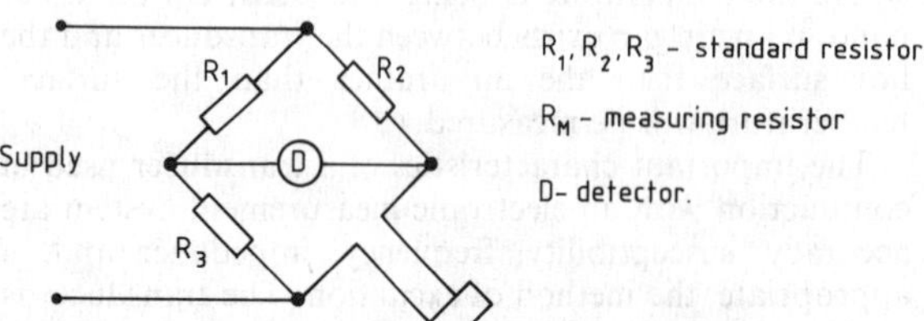

Figure 22.17 Resistance temperature measurement

however, have a fast response. Larger units are available in which amplification can be dispensed with but the response is much slower. These are used for indication (rather than control) purposes.

Pyrometers

These are non-contact instruments for measuring temperature, and as such their thermal mass is relatively unimportant. They can be used for very high temperatures. Radiation pyrometers focus the infrared radiation emitted by the body onto a thermocouple contained within the instrument. The voltage produced by the thermocouple is then read on a dial calibrated directly in terms of temperature. If required, this voltage can be amplified and used as a control signal. The actual temperature measured is that integrated over the whole of the surface 'seen' by the pyrometer and the reading may be affected by infrared absorption in the medium between the hot surface and the instrument. This also applies to optical pyrometers, which are usually of the disappearing-filament type. The instrument contains a lamp the filament of which is viewed against the background of the hot body. The current through the lamp is adjusted, so varying the brightness of the filament until it merges into the background. The current required to achieve this is measured and converted into a temperature reading on the dial.

22.10.2 Force

For the present purposes this includes the measurement of pressure, acceleration and strain.

Piezoelectric crystals

These use crystalline materials in which the electrical properties of the material are changed when it undergoes slight deformation by, for example, the application of mechanical pressure. The principal effect is to cause a change in the frequency at which the material resonates. This change in resonant frequency can be detected and measured, so giving an indication of the change in pressure.

Quartz is a natural crystalline material which exhibits this form of behaviour, although its relevant properties are highly temperature dependent and synthetic materials have been developed which, although fundamentally less accurate, are more stable under varying temperature conditions.

One application is the accelerometer, in which the acceleration force of a mass is made to increase (or decrease) the pressure produced on the crystal by a spring. This, in turn, produces the required electrical change, the effect of which is amplified. It is important to select units appropriate for the expected changes, which should be within the frequency range from almost zero to the natural frequency of the crystal.

Strain gauges

These units are rigidly attached to a surface. Any small deformation in that surface due to the application of an external force changes the dimension of the strain gauge, and this change is then detected to give a value of the applied force or the strain in the material. Resistance strain gauges consist of a grid resistance wire cemented between two films. This resistance element is connected in one arm of a measurement bridge with a similar, compensating, gauge in a second arm. The bridge is rebalanced when the strain is applied, and the changes required to do this are measured and converted, using the calibration information, to the quantities required.

Modern techniques use thin-film resistors deposited directly on the area and semiconductor units are available which are considerably more sensitive than the resistive type. Dynamic measurements can also be made. The change in resistance unbalances the bridge, causing a voltage to appear across the detector terminals. This voltage is then amplified and applied to a CRO or the information can be stored digitally for future retrieval.

Strain gauges can also be used to measure pressure by bonding the gauge to a diaphragm in the wall of a liquid or gas container, acceleration, by sensing the relative change along one or more sensing axes, and torque (by sensing angular strain).

Temperature-sensing devices may also be used for the measurement of flow in pipes or ducts and for an indication of the level of liquid in a tank. The device is indirectly heated, using a separate heating filament, and the flow of liquid cools the sensor at a rate which is a function of the velocity of the flow. Thus by incorporating the device into a bridge circuit it is possible, after appropriate calibration, to obtain a direct measurement of the flow velocity. Similarly, several such devices can be placed, one above the other, in a tank, containing liquid. The liquid acts as a heat-transfer medium, those devices below the surface are cooled as the liquid level falls and the temperature and therefore the resistance of the devices increase. This can be detected, and the whole chain is thus used as a level detector.

Impedance transducers

Strictly, the strain gauges referred to above come into this category, since in such cases the change in the measured quantity causes a corresponding change in the resistance of the element. However, the principle has a much wider application, using changes in either the inductive or capacitive reactance of electrical circuit elements.

The inductance of an iron-cored inductor varies with any air gap included in the iron core. Thus physical movement can be detected by allowing this movement to displace part of the core, so changing the width of the gap. The detection of very small movements is possible in ths way, and instruments based on the principle are used to measure acceleration, pressure, strain, thickness and a variety of other changes.

The effects usually need to be amplified, and in some cases signal processing is necessary in order to obtain the derived information. Two electrical conductors at different potentials exhibit the property of capacitance (i.e. energy storage) between them. The value of this capacitance changes when the relative position of the plates alters or when the medium between them is changed. Such changes can be detected by employing the capacitor in an oscillatory circuit. The resonant frequency of this circuit is proportional to the capacitance which it contains and therefore the detection of such a frequency change allows the original quantity to be determined.

This technique can be used to measure displacement where, in effect, the two electrodes are connected to the two bodies. It has also other applications (for example, in moisture meters where the presence of water vapour between the electrodes causes the capacitance change).

Photo-sensors

Photovoltaic sensors are semiconductor devices and have the property that when light falls on them an electrical voltage is produced across them high enough to drive a current through a resistive load. Banks of these units (solar cells) can be used to produce significant amounts of electrical power, but the technique can also be used for measurement purposes, optical tracking and reading bar codes and punched tape. Photoresistive sensors, are essentially inactive devices, but they have the property of exhibiting a change in resistance when light falls on them. This change is detected by an appropriate circuit and is used to measure the light falling on the unit (e.g. in photography exposure meters) or to trigger a control action (for instance, in intruder alarm systems). Photoemission is the property of some materials to emit electrons when light falls on them. These materials are used as a cathode and the electrons are collected by an anode. One application is in the photomultiplier unit which is used in counters.

Photovoltaic cells are independent of an electrical supply but, in general, they lack sensitivity as compared with photoemissive units.

22.11 Spectrum analysers

Much present electronic equipment deals with signals which are not sinusoidal and which may not even be periodic in nature. However, these signals can be divided into a series of components, each of a single specific frequency, and each can then be studied in turn in order to determine its characteristics. Where plant instrumentation is concerned, this technique can be particularly useful for diagnostic purposes.

The instrument required is a spectrum analyser which must be capable of operating over a wide range of frequencies to cope with all those present in the input signal. Some of these signals may be close together in frequency and they may have widely varying magnitudes. The real-time technique utilizes a series of fixed filters designed to separate contiguous frequency bands. Each supplies its own detector, and the outputs of these are then scanned to produce individual traces on an oscilloscope. The resolution is governed by the bandwidths of the individual filters and the instrument can be used from d.c. to a few kHz.

The Fourier transform analyser uses digital techniques in order to carry out the frequency separation. This can cope with signals up to several hundred kilohertz and swept-tuned analysers employ a tuned filter technique. From the user's point of view the desirable characteristics are high stability (much higher than the frequency being analysed) and good resolution (the ability to distinguish between different signals which are close together in frequency). It must also be able to detect small signals (i.e. its sensitivity must be adequate) while at the same time have an adequate dynamic range. The latter is the ratio of the largest to the smallest signal which can be detected without distortion in either. Good modern instruments of this type incorporate a considerable degree of automation whereby inherent accuracies can be corrected, the control settings are recorded and the data manipulated to yield additional information as required, all with the aid of a desktop computer.

22.12 Bridge measurements

Reference has been made above to the simple Wheatstone bridge. This and the developments arising from it are an essential component in many of the instruments already referred to. These bridges (whose operation basically depends on comparing an unknown quantity with a known series of quantities in order to measure the former) are also used as stand-alone units for such diverse purposes as fault detection and the determination of the parameters of electronic components. The accuracy of the measurement is governed by the accuracy to which the value of the components against which the comparison is made is known, and the care which is taken in the measurement process. Brief descriptions of some of the more common methods are given below together with their applications. The theory of operation may be found in the many textbooks on the subject.

In its simplest form the *Wheatstone bridge* is used on d.c. for the measurement of an unknown resistance in terms of three known resistors. Its accuracy depends on that of the known units and the sensitivity of the detector. It is also used for sensing the changes which occur in the output from resistance strain-gauge detectors. The latter instruments can be made portable and can detect variations of less than 0.05%.

The *Kelvin double bridge* is a more sophisticated variant used for the measurement of very low resistance such as ammeter shunts or short lengths of cable. This is also operated on d.c. In industrial terms the *digital d.c. low-resistance instruments* are more convenient although somewhat less accurate.

A.C. bridges suffer from the complication that they must allow for stray inductances and capacitances in the circuit and that the measurements are, in general, frequency dependent. Inductance is measured by the *Maxwell bridge* and its development, the *Wien bridge*, the commercial variants of which operate at frequencies up to 10 kHz with accuracies of better than 1%. The corresponding unit for capacitance is the *Schering bridge*. The major problems in the latter case are usually concerned with stray capacitances to earth. In order to overcome these, the Wagner earth system is used.

The *Wayne–Kerr bridge* employs a modified principle involving the balancing of the output between from two windings of a transformer. It can be used up to high frequencies (100 MHz) for the measurement of impedance. However, at these frequencies it is now more usual to make use of the resonance principle, whereby a variable-frequency source is used to supply the unknown impedance. The frequency is adjusted until resonance is obtained and the value of this frequency together with the amplitude of the signal is used to determine the impedance. This is the principle used in the *Q-meter*, and is the basis of many of the automatic impedance-measuring instruments now commercially available.

22.13 Data recording

There are many industrial applications in which permanent records (extending over long periods of time) of the instrument readings are required. Chart recorders of various forms are available for this purpose. The most common general-purpose unit is the digital strip chart recorder, in which the input signal is used to drive the movement of a recording arm which passes over a paper chart in the *y*-direction. At the same time the chart is being driven forward at a known speed in the *x*-direction. A stylus on the end of the arm marks a series of dots on the paper, using either electrostatic or thermal means, and thus a continuous record of the signal is obtained. The recorder, as its name implies, operates digitally but analogue signals can be dealt with by the internal analogue-to-digital converters in the instrument. It is possible to cope with several independent inputs by using a number of recording arms and these instruments can be used up to about 25 kHz, although the quality of the trace suffers at these high frequencies.

The digital chart recorder is now gradually replacing the *x–y* chart recorder which, although the basic principle is similar, is analogue in operation and uses a pen on the end of the recorder arm to mark the paper. The response of these instruments is relatively slow, which limits their application to low-frequency work. The underlying limitations on accuracy and frequency of both this and the digital recorder is imposed by the mechanical movement of the recording arm. Higher speeds can be obtained with ultraviolet recorders, in which the input signals are fed to coils mounted in a magnet field and which therefore deflect in response to the signal. The coil suspension also carries a lightweight mirror which deflects a beam of ultraviolet light onto photo-sensitive paper. Thus the inertia of the moving parts is replaced by a light beam.

22.14 Acoustic measurements

The plant engineer is often concerned with the measurement of noise arising from the operation of machinery, especially since permissable noise levels are closely specified in current Health and Safety at Work legislation. The measurement of noise involves the use of a pressure transducer. This is a microphone in which the change of pressure on the diaphragm causes a corresponding change in its electrical impedance which can be detected by using appropriate circuitry. The output signal is amplified, rectified and applied to a d.c. meter to give an indication of the rms value and therefore of the power independently of the waveform.

22.15 Centralized control

In many applications, especially where process plant is concerned, the measurement information is required to be conveyed to a central point, which may be a control room manned by operators or a computer which carries out the control functions automatically. The connecting link between the various measurement points and the central control is usually a telemetry system, although in some cases the distances involved are so large that radio links need to be used.

The measurement data are derived from some form of detector or transducer, the output of which is an electrical signal that must then be conditioned to a value suitable for the input to the telemetry system (typically, of the order of 10 V). In some cases amplification is required and in others attenuation is needed.

The signals transmitted over the network will have a value corresponding to the output from the individual source, and it must be possible to identify the origin of the signals at the receiving end. On the other hand, a discrete connection between each source and the control centre may be prohibitively expensive, and there is a requirement to use a single channel for many separate inputs in such a way that the channel can be monitored and the individual inputs separated from each other. This involves the use of some form of multiplexing. Earlier systems used frequency modulation (FM) techniques in which each measurement signal is used to modulate the output from an oscillator operating over a dedicated part of the frequency spectrum. All signals are transmitted simultaneously over the network, and at the receiving end the signals at the different frequencies are separated out, using a demodulator, to give the individual measurement information.

The requirement for a different frequency band for each measurement channel is restrictive, and later systems make use of some form of time-division multiplexing, where all channels use the same part of the frequency spectrum but not at the same time. The channels are sampled in sequence and an instantaneous value of the signal is obtained from each. Obviously, the sampling rate must be extremely high (many times per second) and the sampling at the various channel inputs must be synchronized exactly with the corresponding sampling of the output at the control centre. Pulse amplitude modulation systems (PAM) use analogue signals while in pulse code modulation (PCM) systems the signals are converted to digital form. At present the choice usually lies between the FM and the PCM systems. Generally, the former are the cheaper but the latter give a much better accuracy.

Further reading

Baldwin, C. T., *Methods of Electrical Measurement*, Blackie, Glasgow (1953)

Bibby, G. L., 'Electrical methodology and instrumentation', in *Electrical Engineers Reference Book*, Butterworths, London (1985)

Coombs, C. (ed.), *Basic Electronic Instrumentation Handbook*, McGraw-Hill, New York (1972)

Cooper, W. D., *Basic Electronic Instrumentation Handbook*, Prentice-Hall, Englewood Cliffs, NJ (1972)

Dalglish, R. L., *An Introduction to Control and Measurement with Microcomputers*, Cambridge University Press, Cambridge (1987)

Drysdale, A. C., Jolley, A. C. and Tagg, G. V., *Electrical Measuring Instruments*, Chapman & Hall, London (1952)

Golding, E. W. and Widdis, F. C., *Electrical Measurements and Measuring Instruments*, Pitman, London (1963)

Hnatek, E. R., *A User's Handbook of D/A and A/D Converters*, John Wiley (Interscience), New York (1976)

Lion, K. S., *Elements of Electrical and Electronic Instrumentation*, McGraw-Hill, New York (1975)

Neubert, H. K., *Instrument Transducers*, Oxford University Press, Oxford (1975)

Norton, H. N., *Handbook of Transducers for Electronic Measuring Systems*, Prentice-Hall, Englewood Cliffs, NJ (1969)

Ott, H. W., *Noise Reduction Techniques in Electrical Systems*, John Wiley, New York (1976)

Rudkin, A. M. (ed.), *Electronic Test Equipment*, Granada, St Albans (1981)

de Sa, A., *Principles of Electronic Instrumentation*, Edward Arnold, London (1990)

Stout, M. B., *Basic Electrical Measurements*, Prentice-Hall, Englewood Cliffs, NJ (1960)

Strock, O. J., *Introduction to Telemetry*, Instrument Society of America, New York (1987)

Tagg, G. F., *Electrical Indicating Instruments*, Butterworths, London (1974)

Usher, M. J., *Sensors and Transducers*, Macmillan, London (1985)

Van Erk, R., *Oscilloscopes*, McGraw-Hill, New York (1978)

Wobschal, D., *Circuit Design for Electronic Instrumentation*, McGraw-Hill, New York (1987)

Wolf, S., *Guide to Electronic Measurements and Laboratory Practice*, Prentice-Hall, Englewood Cliffs, NJ (1983)

23 Lighting

Hugh King
Thorn Lighting

Contents

23.1 Lighting theory

23.1.1 The nature of light

Light is a form of electromagnetic radiation similar in nature and behaviour to radio waves at one end of the frequency spectrum and X-rays at the other. The range to which the human eye is sensitive covers a waveband of approximately 380–760 nanometers (nm), a nanometer being a wavelength of one millionth (10^{-6}) of a millimetre. Within these limits there are differences in colour, white light being that in which an approximately equal amount of power occurs at all frequencies within the waveband.

23.1.2 Frequency and colour

The human eye is more sensitive to frequencies in the middle of the spectrum (green and yellow light) than those at the ends (red and violet). Consequently, more power must be expended to obtain the same effect on the eye from red or violet light than from green or yellow.

The regions of radiation just outside the visible spectrum are known as the ultraviolet (UV) and infrared (IR). Both are present in natural daylight and they are emitted in varying amounts by artificial light sources. IR gives the effect of heat, while UV radiation is used to excite fluorescence in the powders in fluorescent lamps. Short-wave UV causes skin tanning and can damage the eyes, but the long-wave UV, used for theatrical or display purposes and in discotheques, is virtually harmless.

23.1.3 Basic lighting units and terms

The average illumination on a surface will diminish as the square of its distance from the source. This is known as the *inverse square law* and is common to all forms of radiation.

Light is *reflected* from a polished (specular) surface at the same angle as it strikes it. A matt surface will reflect it in a number of directions and a semi-matt surface will behave in a manner between the two.

Refraction occurs when light passes through a surface between two media (e.g. air and water, air and glass). The 'ray' of light is bent and this property is used in the manufacture of lenses and prisms.

Diffusion is the scattering of light when it passes through an obscured medium (such as opal glass or plastic). The medium contains a number of particles of matter which reflect and scatter the light passing through it. A uniform diffuser scatters the light equally in all directions, and partial diffusers allow some directional flow of light. All these properties are used to control the direction and distribution of light from artificial light sources.

Excessively bright areas in the visual field can, separately or simultaneously, impair visual performance (disability glare) or cause visual) discomfort (discomfort glare). Lighting systems in most working interiors are unlikely to cause significant direct disability glare but a degree of discomfort glare is probable.

The unit of luminous intensity is the *candela*. This is the illuminating power of a light source in a given direction. The candela is the standard international (SI) unit.

The unit of light flux is the *lumen*. This is the amount of light contained in one stearadian from a source with an intensity of one candela in all directions. Alternatively, it is the amount of light falling on a unit area of the surface of a sphere of unit diameter from a unit source.

The unit of illuminance is the *lux*. This is the illumination produced by one lumen over an area of one square metre. The non-metric equivalent, the lumen per square foot (lm/ft^2), is still used in the UK and is called a 'foot-candle' in the USA. It is approximately equal to 10 lux.

The units of measured brightness (luminance) are the *candela per square metre* and the *apostilb* (lumens emitted per square metre). The imperial units are the *candela per square foot* and the *foot-lambert* (one lumen emitted per square foot of surface). Note that *luminance* should not be confused with *illuminance*. Illuminance is the measure of light falling on a surface and luminance that of the light reflected from it (or, in some case, emitted by it).

Two other terms which can easily be confused are *luminance* and *luminosity*. Luminance is used to describe the measured brightness of a surface and luminosity is apparent brightness.

Efficacy (or luminous efficiency) of lamps is measured in lumens per watt (lm/W).

23.1.4 Colour appearance and colour rendering

We see objects by the light reflected from them. If they have the property of absorbing some wavelengths of light and reflecting others they will appear to be coloured. Some colours are 'pure', that is, they consist of only one waveband of light. Others are obtained by mixing frequencies (for example, mixing red and green light gives the effect of yellow).

Obviously, only those colours which fall on a surface can be reflected from it. Consequently if it is lit by light of one colour only (e.g. yellow light from a sodium lamp) it will appear to be that colour. If the light falling upon it is deficient in some colours (e.g. red and blue light) those colours will appear weak or may not be seen at all. The eye is unable to distinguish between pure colours and those obtained from mixing, so that the *colour appearance* of a light source (the apparent colour of the light emitted from it) does not necessarily indicate its *colour rendering* (the colour of objects seen in its light).

One way of defining colour appearance is by *colour temperature*, measured in degrees Kelvin (K). Colour rendering can be quantified by a general colour rendering index (RA), of which 100 is the best number to achieve.

23.2 Electric lamps

23.2.1 Light production

Light can be produced from electricity in three main ways:

1. *Incandescence* or thermoluminescence is the production of light from heat. Light from a filament lamp

is produced in this way. Electricity is used to raise the temperature of the filament until it is incandescent.

2. *Electrical discharge* is the production of light from the passage of electricity through a gas or vapour. In lamps using this principle the atoms of the gas are agitated or excited by the passage of the electric current and this atomic excitation produces light and sometimes UV and IR energy.
3. *Phosphorescence* and *fluorescence* are the processes of converting invisible UV energy emitted normally from an electrical discharge into visible light. Material called phosphors cause UV energy to make this transition into visible light.

23.2.2 Incandescent lamps

The following abbreviations are used in the UK to define these lamps:

GLS General lighting service.
TH Tungsten-halogen.
PAR Followed by lamp nominal diameter in eighths of an inch. (Pressed glass filament lamps with an internal reflector coating.)
R Followed by lamp nominal diameter in millimetres (previously in eighths of an inch). (Blown-glass filament lamps.)

Comparisons of luminous efficiency and light output are given in Figures 23.1 and 23.2.

GLS lamp

A tungsten-coiled filament is held by molybdenum filament supports and connected to lead wires. An electric current passing through the filament raises the temperature of the filament so it becomes incandescent, emitting light and heat. The outer glass bulb is usually filled with amounts of argon and nitrogen to inhibit the evaporation of tungsten and arcing across the filament. The bulb can be clear or internally etched to give a pearl finish, resulting in diffused light. The lamp has a suitable lamp cap for its size and wattage (see Figure 23.3).

Application

Domestic, commercial and industrial uses. It is the most familiar type of light source, with many advantages (for example, low initial cost, immediate light when switched on, good colour rendering, no control gear, easily dimmed). For these reasons there are numerous applications. The very low efficacy is often the factor that precludes the selection of GLS lamps.

Average life

1000 h.

Similar lamps

Double-life lamp with average life of 2000 h
Coloured lamps
Netabulb (mushroom lamp)
Nightlight

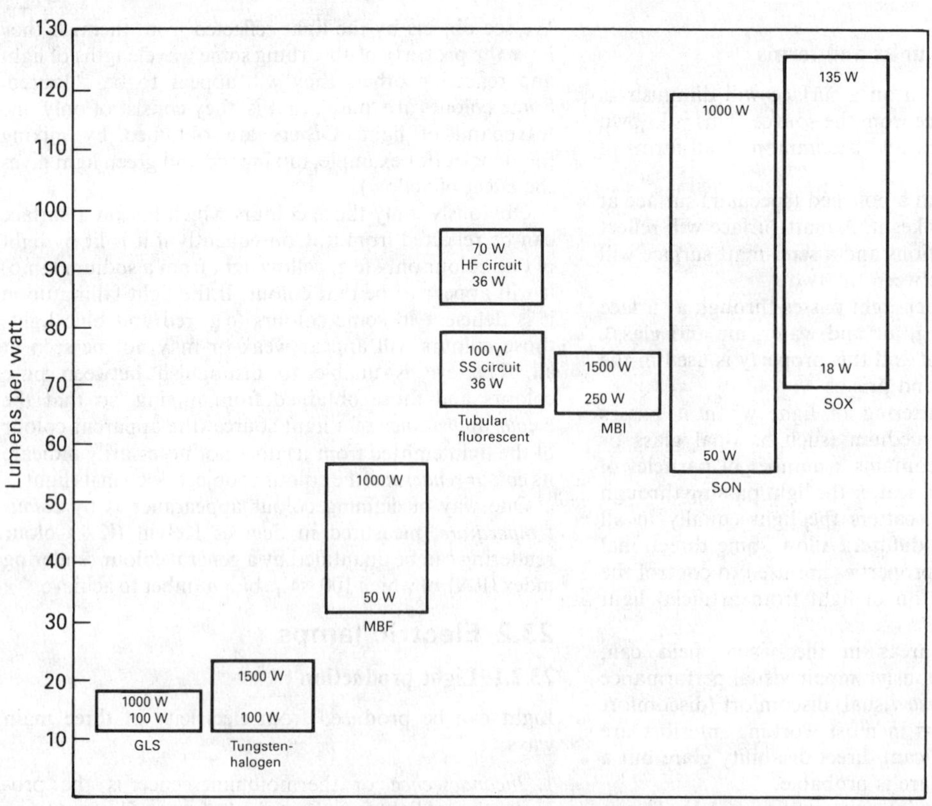

Figure 23.1 Luminous efficacy comparison based on initial lumens and total circuit watts

Lumens (thousands): 10, 20, 30, 40, 50, 60, 70, 80, 90, 100, 110, 120, 130

- Fluorescent: 36 W, 58 W, 100 W
- GLS: 150 W, 300 W, 500 W, 750 W, 1000 W
- Tungsten-halogen: 200 W, 300 W, 500 W, 750 W, 1000 W, 1500 W
- SOX: 18 W, 35 W, 55 W, 90 W, 135 W
- MBF: 50 W, 80 W, 125 W, 250 W, 400 W, 700 W, 1000 W
- MBI: 150 W, 250 W, 400 W, 750 W, 1000 W, 1500 W
- SON: 50 W, 70 W, 100 W, 150 W, 250 W, 400 W, 1000 W

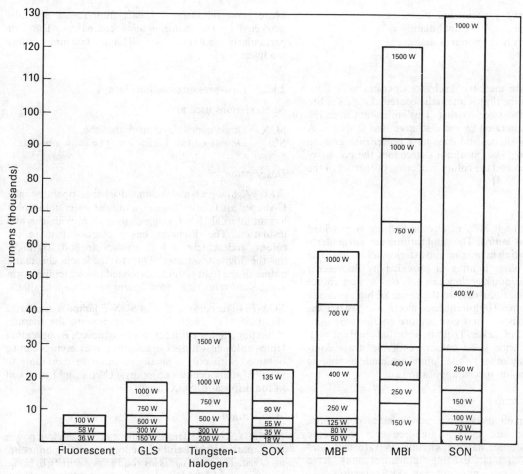

Figure 23.2 Light output comparison

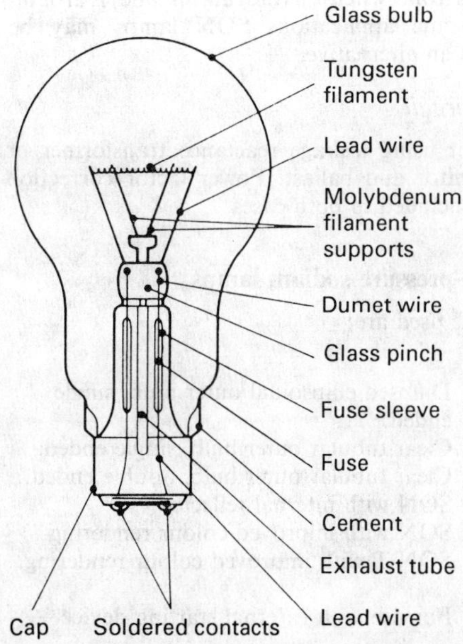

Figure 23.3 Typical construction of a GLS lamp

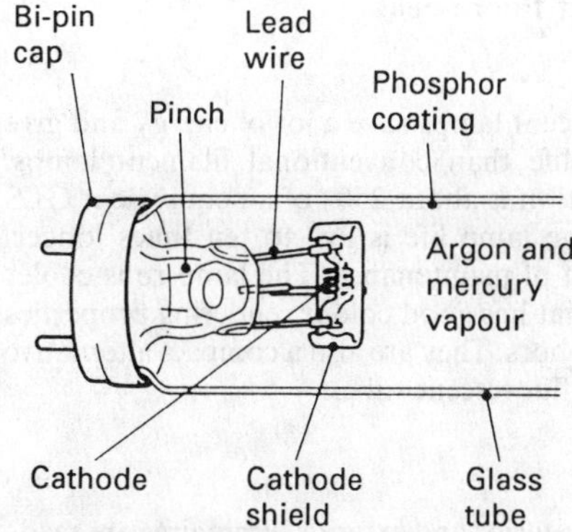

Figure 23.4 Typical construction of a fluorescent tube

23.2.3 Fluorescent lamp MCF

Abbreviations used are:

MCF Tubular fluorescent lamps.
MCFA Tubular fluorescent lamp with earth strip cemented to outside of tube. Restricted range, mainly for cold environments.

T12 38 mm ($1\frac{1}{2}$ in) nominal diameter.
T8 26 mm (1 in) nominal diameter.
T5 15 mm ($\frac{5}{8}$ in) nominal diameter.

Description

A low-pressure mercury discharge operates in a thin-walled glass tube that is internally coated along its entire length by a phosphor coating. This emits light from the UV energy generated by the discharge that strikes it. A wide range of shapes and sizes have power ratings from 4 W to 125 W. The phosphor determines the colour of light produced and the colour rendering properties of the lamp (Figure 23.4).

Application

Worldwide, about 80% of electric lighting is provided by fluorescent lamps. The applications are particularly diverse, operating in interior and exterior luminaires. The majority of office lighting is provided by fluorescent lamps and in industry they can still be the best choice for new installations despite the range of high-pressure discharge lamps. High-frequency electronic ballasts increase their efficacy and give a more comfortable light by the absence of flicker. Triphosphors that combine high light output with good colour rendering are recommended. Many other 'white' lamps are made as well as coloured ones for special applications.

Operating circuit

The basic circuit usually has power factor correction. Various types are used – high-frequency electronic ballast, electronic starting circuits (Vivatronic) and switch-start circuits. Existing luminaires may have quick-start and semi-resonant start (SRS) circuits.

23.2.4 Compact fluorescent

Description

Compact fluorescent lamps save a lot of energy and give a much longer life than conventional filament lamps. Power consumption is about 25% of an equivalent GLS lamp and average lamp life is five to ten times longer, lowering the cost of maintenance. The lamp runs cooler and gives light that has good colour rendering properties by using triphosphors. They are also a compact alternative to conventional fluorescent tubes.

Application

A wide range of interior and exterior luminaires are made for commercial, leisure, civic, domestic and other uses. In terms of light output the Thorn 2D 16 W is comparable to a 75 W GLS lamp. As the lamps are cool running small luminaires can be designed in a great variety of materials.

Average life

8000 h for 2D 16 W;
10 000 h for larger wattage versions.

More powerful linear compact fluorescents are now marketed by the major lighting companies. They are particularly suitable for 600 mm × 600 mm ceiling modules.

23.2.5 Low-pressure sodium lamps

Abbreviations used are:

SOX Single-ended, U-shaped arc tube.
SLI Double-ended. Linear are tube (now obsolete).

Description

SOX A low-pressure sodium discharge operates in a U-shaped arc tube enclosed in a tubular outer bulb which has an internal IR reflecting coating to provide thermal insulation. The discharge has a characteristic yellow colour, and achieves a high efficacy around 160 lm/W for the highest wattage. Due to the monochromatic nature of the light output, correlated colour temperature and colour rendering index figures are not applicable.

SOX-E The construction of SOX-E lamps is similar to standard SOX except that, by improving the thermal insulation, higher efficacies can be achieved. By operating lamps at lower wattages significant power savings can be obtained with only a small reduction in light output. (The 'E' suffix denotes economy.) Care should be taken in the disposal of SOX lamps.

Application

SOX and SOX-E The high efficacy results in low power costs, particularly useful where long operating hours are involved. The main use is for roadways, security lighting, subways and footpaths where the poor colour rendering is not so important. The time it takes the lamp to reach full brightness from switch-on (the run-up time) is around 10 min. In some applications SON lamps may be considered as an alternative.

Operating circuit

Ballast circuit using leakage reactance transformer or electronic ignitor and ballast. Power factor correction capacitor is included in both cases.

23.2.6 High-pressure sodium lamps

Abbreviations used are:

SON
SON-E Diffused ellipsoidal outer bulb, single ended.
SON-T Clear tubular outer bulb, single ended.
SON-TD Clear tubular outer bulb, double ended.
SON-R SON with internal reflector.
SON DL-E SON with improved colour rendering.
SON DL-T SON-T with improved colour rendering.
⚠E For use with external starting device.
⚠I Contains internal starting device.

Description

SON-T A high-pressure sodium discharge operates in a sintered alumina arc tube contained in a clear tubular-shaped bulb. This is a high-efficiency light source that gives a long life while maintaining the lumen output well. The warm coloured light allows some colour discrimination but not accurate colour rendering. As the arc tube is small, precise optical control can be achieved by suitable luminaire design.

SON-E This is similar but has an elliptical outer bulb with a diffuse internal coating.

SON-TD A high-pressure sodium discharge operates in a sintered alumina arc tube contained in a clear quartz sleeve with lamp caps at each end.

SON-R A 70 W high-pressure sodium arc tube is mounted axially inside a 96 mm diameter soft glass reflector bulb which has a diffuse front to remove striations from the beam.

Application

SON-T/SON-E Suitable for many interior and exterior industrial applications, floodlighting, street and precinct lighting. It is now the most common form of streetlighting for city centres. The warm colour of SON frequently improves the appearance of stone and brickwork. The long life combined with good lumen maintenance throughout its life makes the lamp suitable where access is difficult or expensive, as may be the case in lofty industrial interiors such as steelworks, power stations and aircraft hangars. A wide range of lamp wattages is available from 50 W to 1000 W. Energy savings and reduced maintenance can often result in the refurbishment of industrial installations. Another aspect of SON is the ability to operate at low temperatures (−40°C).

SONDL-T This is a high-pressure sodium lamp with a discharge arc that operates at a higher pressure and temperature than is the case with the standard SON-T lamps, resulting in a great improvement in the colour-rendering properties. It is mainly used in commercial interiors, leisure centres and institutional and social areas. Swimming pools and sports halls are common applications, which are often the same scale as the industrial interiors for which these lamps are also used. Another frequent application is with uplighters in offices.

SON-TD The lamp is specfically for use in floodlights where precise optical control can be attained. Floodlights can be compared because the lamp is slim and if the control gear is mounted remotely the floodlight can be relatively low weight and easy to handle.

SON-R Typical applications include lobbies, foyers, pubs and clubs where the warm colour can create a warm and informal atmosphere with a reduced heat load when compared against conventional downlights.

SON lamps

Operating circuit

Ballast circuit with power factor correction capacitor and electronic ignitor.
Lamp uses an external ignitor.

23.2.7 Linear tungsten-halogen lamps

Description

A tungsten filament is supported between molybdenum foils connected to contacts in ceramic caps at each end of a slim quartz-enclosed tube. The gas filling inside the tube includes halogen, which regenerates the loss of tungsten from the filament. It operates at a higher temperature and in a gas at a higher pressure than GLS lamps. The lamp provides a whiter light and improved efficacy. K-type lamps have double the average life of a standard GLS lamp.

Application

Display, exhibition, commercial, photographic and car lighting, and security and floodlighting. It is suitable where the requirement is for immediate light or good colour rendering with a warm colour appearance. The initial capital cost is low. The lamp is also used in display lighting because the source is compact and strong modelling and 'sparkle' can be produced. Since there is no control gear required, small luminaires can be made for tungsten halogen lamps.

Average life

4000 h for 100–3000 W (K type);
2000 h for 500–2000 W (K type).

23.2.8 Low-voltage tungsten-halogen lamps

Description

A compact tungsten-halogen lamp operating at low voltage (12 V) is combined with a precision-faceted glass reflector. Due to its dichroic coatings, the reflector allows about 50% of the heat to be transmitted backwards while reflecting light forwards in a controlled beam. A choice of beam angles is available. The lamp provides a white light, improved efficacy and beam control, and longer average life than conventional display lamps. Cooler interiors, reduced maintenance and running costs can result when compared against conventional PAR 38 display spots. Clip-on colour filters extend the scope for display purposes.

Application

These lamps are suitable for display, exhibition, museum, commercial and domestic lighting. Their compactness has enabled diverse spotlight designs, some incorporating the necessary transformer to step down the voltage to 12 V.

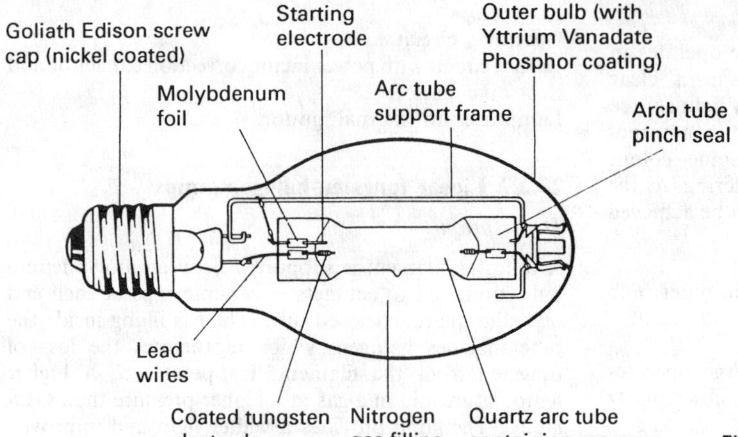

Figure 23.5 Typical construction of a mercury lamp

Average life

3000 h.

Operating circuit

Wirewound transformer or electronic transformer for connection to 240 V supply.

23.2.9 High-pressure mercury lamps (Figure 23.5)

Abbreviations used are:

MBF High-pressure mercury with phosphor coating.
MBFR MBF with internal reflector.
MBTF Combination of MBF lamp and filament lamp.

Description

MBF A high-pressure mercury vapour discharge operates in a quartz arc tube. The internal surface of the outer elliptical bulb is coated with a phosphor which converts UV radiation from the discharge into light.

MBFR This is an identical lamp except the outer bulb is shaped to form a reflector. Titanium dioxide is deposited inside the conical portion of the outer bulb and this deposit is coated with a phosphor. About 90% of the light is emitted through the front face of the lamp in a directional manner.

Application

MBF Used in industry for low initial cost where colour rendering is not a major factor. Also employed in streetlighting, mainly for replacement or matching existing installations.

MBFR Suitable for industrial applications where minimum maintenance is important (for example, in a foundry or processing plant with a dusty atmosphere). The internal reflector is not affected by dirt or a corrosive atmosphere. The outer bulb, which is made of hard glass, allows exterior use.

Both types of high-pressure mercury lamp have a life-survival characteristic similar to SON lamps but MBF lumen maintenance reduces more throughout its life.

Operating circuit

Ballast circuit with power factor correction capacitor. No ignitor is needed.

MBFSD A high-pressure mercury super de-luxe lamp that is identical to the MBF lamp except that it has a different phosphor coating to create more red light. The colour rendering is improved and the colour temperature reduced.

MBTF This is similar to the MBF high-pressure mercury lamp but with a coiled tungsten filament connected to the arc tube that acts as a ballast within the outer glass bulb. Therefore no external control gear is needed.

Application

MBFSD Suitable for commercial interiors such as shops, supermarkets, public concourses and reception foyers. It is capable of blending well with 'white' fluorescent tubes in adjoining areas.

MBTF A direct replacement for tungsten filament lamps giving a higher light output and longer life with a cooler colour appearance. The tungsten filament emits light immediately when switched on. The efficacy is higher than an equivalent tungsten filament GLS lamp.

Operating circuit

MBFSD Ballast circuit with power factor correction capacitor. No ignitor is needed.

MBTF No control gear is needed.

23.2.10 Metal halide lamps

Abbreviations used are:

MBI High-pressure discharge with metallic halides.
MBIL Double-ended linear arc tube.
MBIF MBI with phosphor coating.
CSI Compact source discharge with metal iodides.

Description

MBIF A high-pressure discharge in mercury vapour with metallic additives operates in a quartz tube. The metallic additives are introduced as halide compounds which control the dosing and ensure that the metallic elements mix well with the mercury vapour. An elliptical outer bulb is internally coated with phosphor.

MBI This lamp is similar but has a clear outer bulb.

Application

Commercial and industrial interiors and exterior applications where good colour quality white light is required. The efficacy is better than the same wattage of MBF lamps. These lamps are suitable for televising events when the appropriate illuminances are provided. Uplighting is a typical commercial use when the white quality of light is important.

Operating circuit

Ballast circuit with power factor correction capacitor and ignitor. The lamp uses an external ignitor.

Lamp orientation

BUH Base up to horizontal. Operates in any position between cap up and cap 15° below horizontal.

H horizontal Primarily designed with cap ±15° of horizontal, but can be used up to ±60° of horizontal BU Base up. Cap up ±30°.

Description

MBI-T A high-pressure mercury vapour discharge with metal halides operates in a quartz bulb. These halides are chosen to give high efficiency and good colour-rendering properties. The electrode spacing is only 6 mm, so the arc is extremely compact, making it suitable for precise optical control in small luminaires. An outer quartz bulb gives thermal stability and protection. This bulb is mounted in a ceramic bi-pin cap.

Application

Suitable for display lighting, uplighting and floodlighting. A choice of colour temperatures (3000 K and 4000 K) can blend in with tungsten-halogen and selected fluorescent lamps. Efficacy is around four times higher and the average life three times longer (at 6000 h) than comparable tungsten-halogen lamps. This makes it an attractive alternative to some former tungsten-halogen applications.

Operating circuit

Ballast circuit with power factor correction capacitor and electronic ignitor.

Lamp orientation

BDH Base down or horizontal.

23.2.11 Lamp selection

1. First decide what minimum standard of colour rendering is needed. Except in the case of specialist applications, the CIBSE/CIE colour-rendering classes (Table 23.1) are ideal for this purpose.
2. Do not select lamps from the appropriate class in the colour-rendering guide shown in Table 23.2. Lamps in a better class may be more efficient or preferable for some other reason. Instead, make a note to reject any lamps which are in inferior classes.
3. Now decide which colour appearance groups will be unacceptable, e.g.:
4. Note that some applications which call for precise colour judgement must use lamps with a specific colour appearance (often cool or cold). Make a note of what lamp appearances to reject from the colour-appearance guide in Table 23.3.
5. Decide how the lighting will be controlled and how rapidly light will be required. This can be compared to the run-up and re-strike times of different lamps to determine which are unacceptable (auxiliary lamps can be used for rapid illumination). Make a note of the lamps that are unacceptable from the run-up and re-strike times shown in Table 23.4.
6. By deleting lamps which have unsuitable colour appearance and colour-rendering properties, and those with unacceptable run-up and re-strike characteristics, a list of suitable lamps can be compiled. It is possible to rank this list in terms of efficacy, life and other factors which the designer considers to be important so that a decision can be made.
7. One of the other factors which may be a limitation to the use of certain lamp types is the ambient temperature.

Table 23.1 CIBSE/CIE colour-rendering classes

Minimum class needed	*Description*
1A	Excellent colour quality. Where accurate colour matching is required (e.g. colour printing inspection)
1B	Very good colour quality. Where accurate colour judgement or good colour rendering is required for reasons of appearance (e.g. merchandising)
2	Good colour quality. Where moderate colour rendering is required, good enough for merchandising
3	Poor colour quality. Where colour rendering is of little importance. Colour can be distorted but marked distortion is not acceptable
4	Very poor colour quality. Colour rendering is of no importance and severe distortion of colours is acceptable

Table 23.2 Colour-rendering guide

Class	Type	Thorn description
1A	MCF	Northlight
		Artificial daylight
		De-luxe natural
	TH	Tungsten-halogen
	GLS	General lighting service
1B	MCF	Kolorite
		Polylux 4000
		Polylux 3500
		Polylux 3000
		Polylux 2700
		Rosetta
	CSI	Compact source iodide
	MBI-T	Arcstream 4000
		Arcstream 3000
2	MCF	Natural
		De-luxe warm white
	MBI/MBIF	Metal halide
	MBIL	Linear metal halide
	SONDL	SON de-luxe
3	MCF	Pluslux 4000
		Pluslux 3500
		Pluslux 3000
		Cool white
		White
		Warm white
	MBFSD	High-pressure mercury – super de-luxe
	MBF/MBFR	High-pressure mercury
4	SON	High-pressure sodium
	SON-R	SON
	SON-TD	Reflector/double ended
None	SOX	Low-pressure sodium

Table 23.3 Colour-appearance guide

Colour temperature (K)	Type	Description
		Cold
6500	MCF	Northlight
		Artificial daylight
		Intermediate–cool
5200	MBIL	Linear metal halide
4000	MCF	Kolorite
		Natural
		Polylux 4000
		Pluslux 4000
	MBI/MBIF	Metal halide
	MBI-T	Arcstream 4000
	CSI	Compact source iodide
		Intermediate
3800	MBF/MBFR	High-pressure mercury
3500	MCF	Polylux 3500
		Pluslux 3500
		White
		Warm
3300	MBFSD	High-pressure mercury super de-luxe
3200	MCF	Rosetta
3300	MCF	Polylux 3000
		Pluslux 300
		De-luxe warm white
		Warm white
2900	TH	Tungsten halogen
	MBI-T	Arcstream 3000
2700	MCF	Polylux 2700
	GLS	General lighting service
2200	SONDL	SON de-luxe
2200	SON	High-pressure sodium
None	SOX	Low-pressure sodium

23.3 Luminaires

23.3.1 Basic design requirements

The majority of light sources emit light in all directions, but for most purposes this is both wasteful and visually uncomfortable. The function of most luminaires is therefore:

1. To redirect the light of a bare electric lamp in a preferred direction with the minimum of loss;
2. To reduce glare from the light source;
3. To be acceptable in appearance and, in some cases, make a contribution to the decor of the surroundings.

In addition to optical design there are other aspects which must be considered:

1. *Lamp protection*: Apart from requiring electrical connections, the electric map must be mechanically supported and protected. The extent of this protection will depend on the application for which the luminaire is intended.
2. *Electrical safety*: The lampholder and associated wiring must be supported and protected, as well as the control gear if used to operate the lamp.
3. *Heat dissipation*: The heat generated by lamps and control gear must be conducted away from heat-sensitive parts and surfaces adjacent to the luminaire.
4. *Finishing*: Any exposed metalwork must be painted or plated to protect it against corrosion. The degree of protection necessary will depend on the luminaire's application. Some *proof luminaires* designed for use in particularly hostile, corrosive atmospheres may be made in plastic materials rather than metals.

In addition to the functional aspects of mechanical, electrical thermal and chemical protection, the designer will also select materials and processes to achieve the desired external form and appearance of the design, bearing in mind production costs for either small or large batch quantities. Consideration must also be given to 'built-in' design features aimed at making the installation and maintenance of the luminaires easier (Figure 23.6).

23.3.2 Luminaires and lighting terms

There is an abundance of lighting terms in common use and the following list attempts to cover the main ones.

Accent lighting See *display*.

Air terminal devices The means by which air is supplied or extracted from a room. This term applies to all such equipment whether called air diffusers or grilles.

Table 23.4 Run-up and re-strike times

Type	Rating (W)	Run-up time (min)	Re-strike time (min)
TH	All	Instant	Instant
GLS	All	Instant	Instant
MCF	All HF	Instant[a]	Instant[a]
	All others less than 1[a]		Less than 1[a]
CSI	1000	1	10
MBI-T	150	1	4
MBI	250	2	7
	400	2	7
	1000	2	7
MBIL	750	2	8–12[b]
	1500	2	15–20[b]
SON, SON-R	50	3	Less than 1
and SON-TD	70	3	Less than 1
	100	4	Less than 1
	150	6	Less than 1
	250	6	Less than 1
	1000	6	3
MBIF, MBFSD	50	5	4
and MBFR	80	3	4
	125	3	4
	250	4	4
	400	4	4
	700	3	6
	1000	2	7
SONDL	150	8	Less than 1
	250	8	Less than 1
	400	7	Less than 1
SOX and SOX-E	18	12	Instant
	35	9	Instant
	55	9	Instant
	90	9	10
	135	8	10

[a] Striking time.
[b] In floodlight.
HF: High-frequency circuit.

Airfield lighting This is covered by a range of specialized luminaires intended primarily as an aid to navigation and include approach lighting (line and bar, VASI and PAPI), runway and taxiway lighting.

Amenity lighting This is most commonly applied to outdoor public areas and is primarily intended to provide safe movement. It can incorporate decorative elements, both in terms of the appearance of the equipment and the lighting effect produced. In its most utilitarian form it is also applied to exterior private industrial areas intended mainly for pedestrian use.

Barndoors Four independently highed matt-black flaps mounted on a rotate frame solely intended for reducing spill light or shaping the soft-edged beam of a spotlight.

Batten This refers to a fluorescent luminaire with no attachments; just the control gear in a channel and a bare tube. It is often designed to take various attachments as the basis of a complete range.

Bulkhead fittings These luminaires are often compact, designed for horizontal or vertical mounting outdoors or indoors. A wide selection of lamps are used, including miniature fluorescent tubes and high-pressure sodium lamps. The front is sometimes prismatic of either plastic or glass and the fittings are usually intended for exterior use over doors and stairways.

Figure 23.6 A selection of modern luminaires. Included are two SON floodlights; a linear tungsten-halogen floodlight; two compact fluorescent bulkheads and two low-voltage spotlights

Ceiling systems A grid and panel system providing a false ceiling to incorporate lighting and other building services.

Columns Poles for mounting roadlighting lanterns or floodlights usually from 3 to 12 m constructed in steel, aluminium or concrete.

Commercial These luminaires are designed to have a clean shape and efficient design but are devoid of decorative treatment. The term is applied to interior fluorescent diffuser luminaires as well as tungsten and discharge units for similar applications. There is frequently a grey area between this term and decorative or display luminaires.

Controller A clear plastic or glass with prism shapes in the surface. The overall shape may be flat or formed. This is distinct from opal plastic or glass diffusers.

Dark light See *Downlighters*.

Decorative (1) Decorative 'fittings' designed primarily to have a pleasing appearance, efficiency not always being of major importance. (2) Decorative 'lighting', however, refers to the effect produced and may require optically efficient fittings which may or may not have an external decorative appearance.

Diffuser An attachment intended to reduce the lamp brightness by spreading the brightness over the surface of the diffuser. Many types of fluorescent, tungsten and discharge lamp luminaires use diffusers.

Display Lighting intended to attract attention and show objects to good effect.

Domestic fittings Strictly, equipment intended to light the private home, and as such it includes decorative fittings.

Downlighters This term is frequently misused to refer to any luminaire which produces no upward light, whereas the object of a downlighter is to provide illumination without any apparent source of light. They are therefore extremely 'low-bright' luminaires, referring particularly to small (usually circular) recessed, semi-recessed or surfaced 'can' luminaires with mains or low-voltage tungsten reflector lamps. Some high-pressure discharge versions are also available.

Egg-crate See *Louvres.*

Emergency lighting Lighting provided for use when the mains lighting fails for whatever reason. There are two types: (1) Escape lighting; (2) Standby lighting.

Emphasis lighting See *Display*

Escape lighting See *Emergency lighting.*

Eyeball fitting A recessed adjustable display spotlight.

Flameproof An enclosure capable of withstanding the pressure of an internal explosion and preventing transmission of the explosion to gases and vapours outside the luminaire.

Floodlight A very general term usually applied to exterior fittings housing all types of lamps and producing beams from very narrow to very wide. They usually employ a specular parabolic reflectors, and their size and shape is determined by the type of lamp employed and the distribution of light required.

Fresnel lantern A soft-edge spotlight with a stepped front lens which gives a wide range of beam-angle adjustments by moving the lamp and reflector relative to the front lens.

General lighting An interior installation designed to provide a reasonably uniform value of horizontal illumination over a complete area.

High bay As the name implies, these are for use when mounting heights of around 8–10 m or above are encountered. They have a controlled light distribution to ensure that as much light as possible reaches the working plate. Cut-off angle is 60–65°. Although primarily intended for industrial areas, they also find application in lofty shopping areas and exhibition halls.

High mast lighting This was originally developed as a system for lighting exterior parking areas and complex road intersections using high-pressure discharge lamps in luminaires very similar to industrial high bay reflectors at heights of 20–50 m. Since then, many installations using directional floodlights have been installed and the same term is used for this.

Hood A baffle projecting forward from the top of the floodlight or display spotlight to improve the cut-off of spill light above the horizontal to reduce glare and light pollution.

Indirect lighting System of illumination where the light from lamps and luminaires is first reflected from a ceiling or wall. Traditionally, tungsten or fluorescent lamps were mounted in a cove and cornice to light a ceiling or behind a pelmet or in 'wall washer' units to light walls. For modern high-efficiency applications of this technique, see *Uplighting.*

Industrial These luminaires are designed for high optical efficiency and durability. They range from simple fluorescent reflector luminaires to complex proof luminaires housing all modern light sources for both interior and exterior use.

Inverter Electronic device for operating discharge lamps (usually fluorescent) on a d.c. supply obtained from either batteries or generators.

Lantern Apart from reproduction period lighting units, this is an alternative terms for 'luminaire' or 'fitting', usually restricted to roadlighting and stage lighting equipment.

Local lighting Provides illumination only over the small area occupied by the task and its immediate surroundings.

Localized lighting An arrangement of luminaires designed to provide the required service illuminance over the working area together with a lower illuminance over surrounding areas. This is an alternative system to installing 'general' lighting. See also *Uplighting.*

Louvres Vertical or angled fins of metal, plastic or wood arranged at right angles to linear lamps or in two directions. The object is to increase the cut-off of the luminaire and so reduce glare from critical angles. Louvres can also form ceiling panels with lighting equipment placed in the void above.

Low bay Luminaires housing high-pressure discharge lamps (usually mounted horizontally) to provide a wide distribution with good cut-off at mounting heights around 4–8 m. Apart from industrial applications, these are used in many sports halls and public concourses. They are often fitted with louvres or clear visors.

Low brightness This term is usually applied to commercial fluorescent luminaires where, by use of specular reflectors or low-reflectance louvres, the brightness of the luminaire is strictly limited above 50° from the vertical. They are particularly appropriate for offices with VDUs or if it is wished to make the luminaires in an interior particularly unobstrusive (see *Downlighters*).

Maintained emergency lighting An emergency lighting circuit which remains energized at all times both when the main supply is available and when it has failed for any reason.

Mast Mounting columns higher than 12–15 m are generally referred to as masts and are used to support floodlighting for lighting large areas. For heights above 30 m access ladders are usually mounted inside the shaft and the base compartment is large enough to house control gear.

Optic The reflector and/or refractor system providing the light control for the luminaire.

Perimeter lighting See *Wall washer* and *Security lighting.*

Post top lantern Road or amenity lighting luminaire which mounts directly onto the top of a column without a bracket or out-reach arm. These are available for all lamp types with systematical and asymmetric distributions for mounting heights from 3 m to 20 m. The external design can vary from purely decorative to highly functional.

Precinct lighting See *Amenity.* This is generally associated with public rather than private areas.

Presence detector A device activated either by sound or heat energy to detect the occupation of an area or an intruder and thereby control the lighting approximately.

Projector A term often applied to any type of floodlight. More correctly it is an alternative to 'profile spotlight'.

Proof Applied to all types of luminaires which have a higher degree of protection from the ingress of solids and liquids than standard interior lighting luminaires.

Rack lighting Warehouse lighting where goods are stored on racking systems. The lighting must provide adequate vertical illumination on the faces of the racking which form narrow gangways.

Recessed The luminaire mounting arrangement where the whole or part (semi-recessed) of the luminaire body is set into a ceiling or wall or floor surface. With modular ceilings the luminaire fills the space provided by the removal of one or more tiles, but the term applies equally to where any type of luminaire is set into a false ceiling, wall or floor or solid structural slab.

Refractor Clear glass or plastic, panel or bowl where an array of prisms is designed to redirect the light of the lamp into the required distributions.

Residential lighting An installation for subsidiary roads and areas with little or no vehicular traffic except that generated by the residents themselves. Requirements are covered by BS 5489: Part 3 (see *Amenity*).

Road lighting An installation intended primarily to serve the needs of drivers. Although this is usually associated with public motorways, trunk and feeder roads and tunnels, the same principles apply to private, industrial and commercial roadways with vehicular traffic. It is covered by BS 5489: Parts 1–9.

Safety/security lighting An installation designed to protect property, plant and people from criminal activity and to assist guards in detecting the presence of intruders. In the case of prisons the objective is reversed, in that the aim is to prevent escape. In many cases, other detection devices, such as closed-circuit television and vibration detection, are used in conjunction with the lighting. All security lighting tends to be concentrated around the perimeter of the site and it is therefore sometimes referred to as 'perimeter' or 'perimeter fence' lighting. In certain high-security areas invisible IR systems are used and linked with closed-circuit television facilities.

Semi-recessed See *Recessed.*

Spine See *Batten.*

Spotlight An adjustable interior luminaire with controlled beam using a reflector lamp or optical system (see *Display*, *Floodlight*, *Fresnel*, *Profile*).

Stage lighting An installation designed to enable an audience to view a 'performance', ranging from legitimate theatre in all its forms to a lecture.

Standby lighting See *Roadlighting.*

Studio lighting Lighting for film and television has similar requirements for *Stage lighting* but involves a live audience less often. The higher lighting levels dictate the use of 5–10 kW lamps.

Tower Lattice, steel or concrete structure from 15 m to 50 m to support area floodlighting equipment (see also *Mast*).

Track system A linear busbar system providing one to three main circuits or a low-voltage supply to which display lighting can be connected and disconnected at will along the length of the system. The luminaires must be fitted with an adaptor to suit the particular track system in use.

Troffer Fluorescent or discharge luminaire designed to recess into suspended ceilings and fit the module size of the ceiling (see *Recessed*).

Trunking Apart from standard wireway systems in ceilings and floor, trunking, associated specifically with lighting, usually provides mechanical fixings for the luminaires as well as electric connection. These are usually favoured where (1) there is a lack of fixing points for separates luminaires and the trunking can span between trusses (to carry the wiring and support the luminaires); (2) a simple means must be provided for repositioning lighting equipment with changes in use in the area and where track systems would not be appropriate; (3) additional support for suspended ceilings is required.

Uplighting Indirect lighting system where task illuminance is provided by lighting reflected from the ceiling. When used as a functional rather than a decorative system it usually employs high-pressure discharge lamps and high-efficiency reflector systems. The luminaires can be floor standing, wall mounted or suspended, and can be arranged to provide either general or localized lighting. They are particularly suitable for areas with VDUs although they have many other applications.

VDU lighting See *Low brightness* and *Uplighting.*

Vizor Clear or diffused glass or plastic, bowl or flat panel closing the mouth of a luminaire. It is fitted either to protect the lamp or to prevent falling debris from the luminaire. In the case of diffusing vizors the object is to reduce lamp brightness and, to a certain extent, reduce glare.

Wall washer Interior floodlight or spotlight with asymmetric distribution intended to provide uniform lighting of walls from a close offset distance. Units may be recessed or surface mounted and house tungsten or sometimes high-pressure discharge lamps, mounted in a pelmet unit. This is commonly used to light shelf units in supermarkets and other large stores.

Wellglass These consist of a lamp surrounded by an enclosure of transparent or translucent glass or plastic. They are usually proof luminaires and often used outdoors fixed to a bracket.

Zones 0, 1 and 2 Classification of various hazardous areas.

23.3.3 The International Protection (IP) code

This code is now widely used and classifies luminaires under two headings:

1. Protection against electric shock and the ingress of solid foreign bodies, including fingers or tools at one end of the spectrum and fine dust at other;
2. Protection against the ingress of liquids. The system is referred to in BS 4533: Part 1, section 2.3, and is fully explained in BS 775: Part 1, Appendix B.

An IP number consists of two numerals, the first in the left-hand column and the second in the right-hand column of Table 23.5. For example, a dust-proof luminaire which can be hosed down would carry the number 55 and the symbols shown in Figure 23.7. A rainproof fitting with protection from finger contact within the enclosure would bear the number 23 and only the symbol in the right of the figure. The higher the numerals of the first and second characteristic, the greater the degree of protection the enclosure offers. For example, Jetproof IP55 meets all the less onerous degrees such as IP22, IP23, IP34 and IP54.

Table 23.5 Defining IP numbers for luminaires (from BS 775)

Protection of persons against contact with live or moving parts inside the enclosure and protection of equipment against ingress of solid foreign bodies. Protection against contact with moving parts inside the enclosure is limited to contact with moving parts inside the enclosure that might cause danger to persons

First characteristic numeral	*Degree of protection*
0	No protection of persons against contact with live or moving parts inside the enclosure No protection of equipment against ingress of solid foreign bodies.
1	Protection against accidental or inadvertent contact with live or moving parts inside the enclosure by a large surface of the human body, e.g. a hand, but not protection against deliberate access to such parts. Protection against ingress of large solid foreign bodies.
2	Protection against contact with live or moving parts inside the enclosure by fingers. Protection against ingress of medium-size solid foreign bodies.
3	Protection against contact with live or moving parts inside enclosure by tools, wires or such objects of thickness greater than 2.5 mm. Protection against ingress of small solid foreign bodies.
4	Protection against contact with live or moving parts inside the enclosure by tools, wires or such objects of thickness greater than 1 mm. Protection against ingress of small solid foreign bodies.
5	Complete protection against contact with liver or moving parts inside the enclosure. Protection against harmful deposits of dust. The ingress of dust is not totally prevented, but dust cannot enter in an amount sufficient to interfere with satisfactory operation of the equipment enclosed.
6	Complete protection against contact with live or moving parts inside the enclosure. Protection against ingress of dust.

Protection of equipment against ingress of liquid

Second characteristic numeral	*Degree of protection*
0	No protection.
1	Protection against drops of condensed water. Drops of condensed water falling on the enclosure shall have no harmful effect.
2	Protection against drops of liquid. Drops of falling liquid shall have no harmful effect when the enclosure is tilted at any angle up to 15° from the vertical.
3	Protection against rain. Water falling in rain at an angle equal to or smaller than 60° with respect to the vertical shall have no harmful effect.
4	Protection against splashing. Liquid splashed from any direction shall have no harmful effect.
5	Protection against water jets. Water projected by a nozzle from any direction under stated conditions shall have no harmful effect.
6	Protection against conditions on ship's decks (deck watertight equipment). Water from heavy seas shall not enter the enclosure under prescribed conditions.
7	Protection against immersion in water. It shall not be possible for water to enter the enclosure under stated conditions of pressure and time.
8	Protection against indefinite immersion in water under specified pressure. It shall not be possible for water to enter the enclosure.

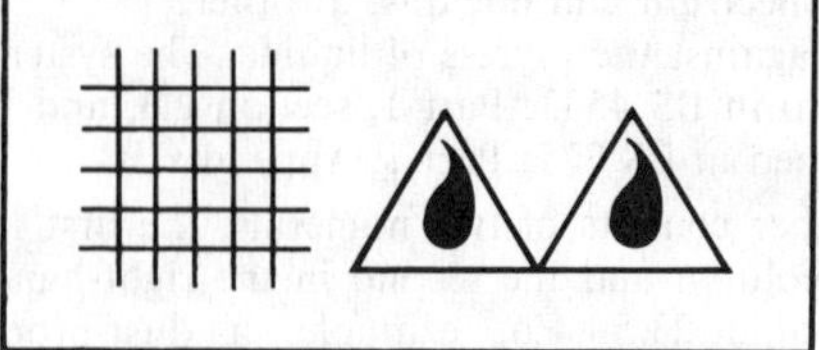

Figure 23.7 IP symbols

23.3.4 Explosion hazards

If a luminaire is to be used in an explosive atmosphere protection over and above that described in the IP code is needed. Explosive atmospheres are a potential problem in various industries where flammable materials are manufactured, processed or stored.

Electrical regulations over the years and, more recently, the Health and Safety at Work, etc. Act 1974, stipulate that any possible danger to safety be recognized and the necessary action taken to ensure adequate protection of the workforce. This places considerable responsibility on the user and installer, and it is most important that the correct type of luminaire be selected and that the installation is in accordance with the recognized safety standard.

Unlike the IP code, the equipment is classified according to the nature of the area in which it is to be known as the *zonal classification*. The latest UK Code of Practice (BS 5345) covers the selection, installation and maintenance of electrical apparatus for use in potentially explosive atmospheres (other than mining applications or explosives processing and manufacture). Therefore the user or engineer handling plant design should refer to this before equipment is specified.

23.3.5 Safety and quality

On 19 February 1973 the European Commission issued Directive No. 73/23/EEC, which has subsequently become known as the 'The Low Volt Directive'. In effect, Article 2 of this directive calls upon member states to ensure that all electrical appliances placed on the market are safe. Other articles call for the establishment of common safety standards throughout EC member states so that free movement of goods within the Community shall not be impeded for reasons of safety.

The government has implemented this directive in the UK via the Factories Inspectorate and the Department of Prices and Consumer Protection (DPCP). The Health and Safety at Work, etc. Act 1974 covers the industrial and commercial sectors. In the domestic sector, the Secretary of State has issued Regulations 1975, to be enforced by local weights and measures inspectors. Under this legislation it is now unlawful to make, to hold in stock or to offer for sale any electrical appliance which is unsafe.

As it affects lighting equipment, BS 4533 (Luminaires) is accepted by the government as a 'safety' specification and the BSI Safety Mark gives an independent guarantee to all concerned that a luminaire has been designed and made in accordance with good engineering practice, that it has been tested and complies with BS 4533 and that its manufacturing quality is monitored regularly by inspectors of the Quality Assurance Department of the British Standards Institution.

23.4 Control gear

23.4.1 General

The electrical characteristics of many lamp types dictate that they cannot be operated directly from the normal a.c. mains supply. Most discharge lamps need a current-limiting device in the circuit to control lamp power. To provide the necessary control, electrical components, collectively known as control gear, are needed. There may also be components in the circuit which have no direct effect on lamp operation. They are, strictly speaking, control gear, although they are not always thought of as such. Power factor correction capacitors and fuses are examples.

23.4.2 Ballast

This is the general term for control gear inserted between the mains supply and one or more discharge lamps, or fluorescent lamps, which, by means of inductance, capacitance or resistance, singly or in combination, serves mainly to limit the current of the lamp(s) to the required value. A ballast may also incorporate means of:

Transforming the supply voltage
Providing a starting voltage
Providing a preheating current
Improving cold starting
Reducing stroboscopic effects
Correcting power factor
Suppressing radio interference

Other components are as follows.

Choke

A simple low power factor inductive ballast.

LPF switchchart ballast

A one-piece unit with a choke and a starter switch socket.

HPF switchstart ballast

A one-piece unit containing choke, starter switch socket and power factor correction capacitor.

Leakage reactance autotransformer

A one-piece unit with a special magnetic circuit which provides a high starting voltage and subsequent ballast control.

Transformer

A wire-wound unit which steps up or steps down its supply voltage.

Lagging (inductive) circuit

A circuit in which the current waveform leads the voltage waveform.

Lead/lag circuit

A twin-lamp circuit comprising one leading circuit and one lagging circuit. Such an arrangement overcomes stroboscopic problems and also results in a circuit with near-unity power factor.

Ignitor

A starting device, intended to generate voltage pulses to start discharge lamps, which does not provide for the preheating of electrodes.

Electronic starter

A starting device, usually for fluorescent tubes (such as Vivatronic starters), which provides the necessary pre-heating of the electrodes and, in combination with the series impedance of the ballast, causes a surge in the voltage applied to the lamp.

Glow-starter switch

A starter, which depends for its operation on a glow

discharge in a gaseous atmosphere, closes or opens the preheating circuit of a fluorescent tube and, in combination with the series impedance of the ballast, causes a surge in the voltage applied to the lamp.

An energy limiter

An energy limiter is a device which may be incorporated into the electrical distribution board in a building to reduce the supply voltage and hence the power consumed by fluorescent lighting installations. Such voltage reductions also cause a drop in light output and changes to other lamp operation characteristics, depending on lamp type, ballast circuit and luminaire.

Temperature

Rated maximum ambient temperature (t_a) – the temperature assigned to a luminaire to indicate the highest sustained temperature in which the luminaire may be operated under normal conditions.

Rated maximum operation temperature (t_c) – the highest permissible temperature which may occur at any place on the outer surface of a device under normal operation conditions.

Rated maximum operating temperature of a winding (t_w) – the operating temperature of a ballast winding which gives an expectancy of 10 years' continuous service at that temperature.

Rated temperature rise of a ballast winding (Δt) – the temperature rise of the winding under the conditions specified in the relevant BS or IEC publication.

23.4.3 Power factor correction

Inductive devices are the major source of current control in discharge lamp circuits but the inductance causes the supply current to lag the supply voltage – they become out of phase. This effect is quantified by the power factor (PF) of a circuit which is the relationship between the power and volt/amps:

$$PF = \frac{P}{V \times I}$$

where

P is the circuit power input (W), V is the circuit supply voltage (V), I is the total circuit current (A).

The circuit power factor can be no greater than one. When there is no phase difference $P = V \times I$ and the power factor is unity.

With a low power factor the circuit's apparent power ($V \times I$) and current are large compared with its power demand. Unnecessarily high currents require large cables and this is wasteful. As a result, the supply authority may impose extra tariffs where the overall power factor is low. There are therefore good reasons for operating at a high power factor.

In an inductive circuit the current lags the voltage; in a capacitive circuit the reverse is true. A combination of inductance and capacitance can be arranged so that these effects at least partially cancel each other out to produce a small phase difference and consequently a high power factor. This is a form of power factor correction.

More commonly, a capacitor is included connected across the circuit supply between 'live' and 'neutral' in what was a low power factor inductive circuit to reduce the supply current and bring the power factor closer to unity. The capacitor is referred to as a power factor correction (PFC) capacitor. Capacitors are usually included in discharge lamps circuits for power factor correction.

Large installations

Where the lighting installations will be a high proportion of the load on a distribution transformer, liaison with the transformer manufacturer at the design stage is advisable to ensure that natural resonance between transformer leakage reactant and the total ballast parallel PFC capacitor load does not occur at the lower harmonic frequencies of the main supply. Increased line current will be avoided if natural resonance occurs at frequencies above 400 Hz, where the voltage level of the harmonics is normally very low.

23.5 Emergency lighting

23.5.1 General

In the UK the Fire Precautions Act 1971 and the Health and Safety at Work, etc. Act 1974 make it obligatory to provide adequate means of escape in all places of work and public resort. Emergency lighting is an essential part of this requirement. BS 5266 (BSI 1975b) lays down minimum standards for design, implementation and certification of such installations.

The UK has a long history of fire and safety legislation (a legacy of the Great Fire of London in 1666). In other countries the legislation ranges from none to different regulations with similar weight and intent. The nature of the legislation and the variation from one country to another make it impossible to deal with this topic in a general way. The reader must refer to the regulations which apply to the locality and comply with them.

The comments in this section refer to UK practice, but even within the UK there is considerable local variations in the regulations (by means of bye-laws) and approval should be obtained from the appropriate authority.

Escape lighting

This type of emergency lighting is provided to ensure the safe and effective evacuation of the building. It must:

1. Indicate the escape routes clearly and unambiguously.
2. Illuminate the escape routes to allow safe movement towards and out of the exits.
3. Ensure that fire-alarm call points and fire equipment provided along the escape route can be readily located.

Standby lighting

Some building areas cannot be evacuated immediately in the event of an emergency or power failure because life would be put at risk (for example, a hospital's operating theatre or chemical plants where shutdown procedures must be used). In these circumstances the activities must be allowed to continue and standby lighting is required. The level of standby lighting will depend on the nature of the activities, their duration and the associated risk, but the provision of 5–25% of the standard service illuminance is common.

Standby lighting can be regarded as a special form of conventional lighting and can be dealt with accordingly, but escape lighting requires different treatment, and is the major concern of BS 5266. Like most British Standards, this is not a legal document, but it can acquire legal status by being adopted as part of the local bye-laws. Although most local authorities quote BS 5266, many modify the conditions (for example, by insisting on higher illuminances). It is left to the enforcing authority to decide for what duration emergency lighting shall be required to operate. This is normally longer than the time needed for an orderly evacuation, for the sake of rescue services (typically, 1, 2 or 3 hours).

23.5.2 Escape lighting

Marking the escape route

All exists and emergency exists must have exit or emergency exit signs. Where no direct sight of an exit is not possible or there could be doubt as to the direction, then direction signs with an appropriate arrow and the words 'Exit' or 'Emergency Exit' are required. The aim is to direct someone who is unfamiliar with the building to the exit. All signs must be illuminated at all reasonable times so that they are legible.

Illuminating the route

The emergency lighting must reach its required level 5 seconds after failure of the main lighting system. If the occupants are familiar with the building this time can be increased to 15 seconds at the discretion of the local authority. The minimum illuminance along the centreline of the escape route must be 0.2 lux or more. In an open area such as a hall or open-plan office, this applies to all areas where people may walk on their way to the exits. The ratio of the maximum illuminance to the minimum illuminance along the centreline of the escape route should not exceed 40:1. Once again, in open areas, this applies to all parts of the floor where people may walk on their way to the exits.

Glare

Luminaires should not cause disability glare and should therefore be mounted at least 2 m above floor level. They should not be too high or they may become obscured by smoke.

Exits and changes of direction

Luminaires should be located near each exit and emergency exit door, and at points where it is necessary to emphasize the position of potential hazards, such as changes of direction and floor level, and staircases.

Fire equipment

Firefighting equipment and fire-alarm call points along the escape route must be adequately illuminated at all reasonable times.

Lifts and escalators

Although these may not be used in the event of a fire, they should be illuminated. Emergency lighting is required in each lift car in which people can travel. Escalators must be illuminated to the same standard as the escape route to prevent accidents.

Toilets and control rooms

In toilets of over 8 m^2 gross area, emergency lighting should be installed to provide a minimum of 0.2 lux. Emergency lighting luminaires are required in all control rooms and plant rooms.

23.5.3 Types of systems

There are two main types of supply for emergency lighting: generators and batteries. Few generators will run up to provide the required illuminance within 5 or 15 seconds, in which case they must either be running continuously or be backed up with an auxiliary battery system. Generators require considerable capital investment and are difficult to justify except for standby use.

A battery system can one of two types; either a central system, where the batteries are in banks at one or more locations, or a self-contained system, where each individual luminaire has its own battery.

Central systems

These have battery rooms or cubicles in which the charger, batteries and switching devices are located. Modern systems tend to have several cubicles to improve system integrity in the event of fire damage to one cubicle.

Self-contained luminaires

These are self-powered and operate independently in an emergency. Thus although an individual luminaire may be destroyed in a fire, the other luminaires will be unaffected. The fact that each luminaire is an independent unit means that maintenance must be thorough. For most applications self-contained luminaires must operate for a period of 1–3 hours. Most designers base their designs on the safer 3-hour standard, irrespective of the requirements of the enforcing authority.

Maintained

In this mode the lamp is on all the time. Under normal conditions it is powered directly or indirectly by the mains. Under emergency conditions it uses its own battery supply.

Non-maintained

Here the lamp is off when mains power is available to charge the batteries. Upon failure the lamp is energized from the battery pack.

Sustained

This is a hybrid of the previous two modes. A lamp is provided which operates from the mains supply under normal conditions. Under emergency conditions a second lamp, powered from the battery pack, takes over. Sustained luminaires are often used for exit signs. Although self-contained luminaires are the easiest and most flexible to install, maintenance and testing must be thorough if operation in the event of an emergency is to be guaranteed. Maintained systems continuously demonstrate the integrity of the lamp, but the lamp is continually ageing and an operating lamp gives no indication of the state of the battery or charger circuits. In non-maintained and sustained systems the lamps do not age, but only regular testing will reveal faults.

A variation of the self-contained luminaire is the standard luminaire with an inverter unit and battery pack. These luminaires are normally conventional types used for general lighting (e.g. batten fittings, recessed reflector fittings, etc.) but have a change-over device, inverter and battery pack which will run one of the fluorescent lamps at reduced output if the mains fails.

23.5.4 Planning

When planning an emergency lighting system the following sequence is recommended:

1. Define the exits and emergency exists.
2. Mark the escape routes.
3. Identify any problem areas (for example, areas that will contain people unfamiliar with the building, toilets over 8 m^2, plant rooms, escalators, etc.).
4. Mark location of exit signs. These can be self-illuminated or illuminated by emergency lighting units nearby. Indicate these on the plan.
5. Where direction signs are required, mark these and provide necessary lighting.
6. Identify the area of the escape route which is already illuminated by the lighting needed for the signs.
7. Add extra luminaires to complete the lighting of the escape route, paying particular attention to stairs and other hazards. Remember to allow for shadows caused by obstructions or bends in the route.
8. Add extra luminaires to satisfy the problem areas identified in (3) above. Make sure that the lighting outside the building is also adequate for safe evacuation.
9. Check that all fire-alarm points and fire equipment have been adequately dealt with.

Testing

The local inspector will normally require a written guarantee that the installation conforms to the appropriate standards. Drawings of the system must be provided and retained on the premises (this also applies to modifications).

All self-contained luminaires and internally illuminated signs must be tested for a brief period once a month. The battery system must be used twice a year in a simulated mains failure for a period of at least one hour. The charging system must also be checked. The system must be operated to its full duration (normally 3 hours) at least once every three years.

23.6 Lighting design

23.6.1 General

Lighting is an art as well as a science. This implies that there are no hard-and-fast rules of design nor will there be one ideal or optimum solution to a particular lighting problem. More often than not, the lighting designer is presented with a list of conflicting requirements to which priorities have to be allocated before a satisfactory compromise can be found. Here we can therefore do little more than give general guidance.

This section should be read in conjunction with the current CIBSE Code, which includes appropriate illuminances and glare indices for various activities, together with useful notes and flow charts for easy planning (see Figure 23.8).

23.6.2 Choice of lighting system

One should not confuse choosing a lighting system with selecting lamps, luminaires or other equipment. This is a strategic decision which determines the relationship between daylight and artificial light and the degree to which the lighting system is tailored to match the local needs within the space. One basic system decision is to choose between general, localized or local lighting.

Uniform or general lighting

Lighting systems which provide an approximately uniform illuminance on the horizontal working plane over the entire area are called general or uniform lighting systems. The luminaires are arranged in a regular layout, giving a tidy appearance to the installation. General lighting is simple to plan and install, and requires no coordination with task locations which may not be known or which may change.

The greatest advantage of such a system is that it permits complete flexibility of task location. Its major disadvantage is that energy is wasted in illuminating the whole area to the level needed only for the most critical tasks.

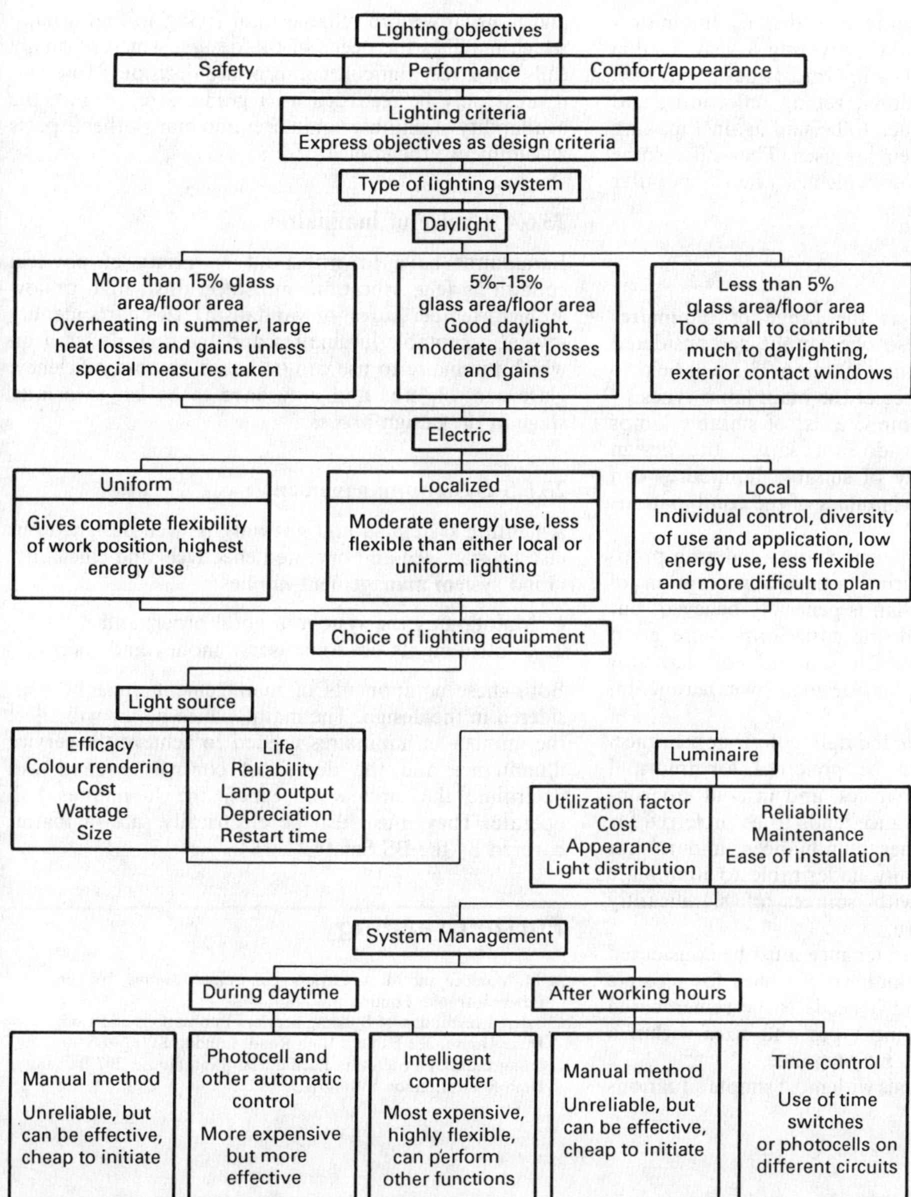

Figure 23.8 An interior lighting design flowchart

Localized lighting

Localized lighting systems employ an arrangement of luminaires related to the position of tasks and workstations. They provide the required service illuminance on work areas together with a lower level of general illumination for the space. By careful luminaire positioning, good light utilization is achieved with few problems from shadows, veiling reflections and discomfort glare. Localized systems normally consume less energy than do general systems, unless a high proportion of the area is occupied by workstations.

Local lighting

Local lighting provides illumination only over the small area occupied by the task and its immediate surroundings. A general lighting system is installed to provide sufficient ambient illumination for circulating and non-critical tasks. Therefore the local lighting system simply supplements the general lighting to achieve the necessary service illuminance on tasks. The system is sometimes referred to as 'task-ambient lighting'. It is very efficient, particularly when high standards of task illuminance are required.

Local lighting is commonly provided by luminaires mounted on the workstation, providing a very flexible room layout. Such local units must be positioned carefully to minimize shadows, veiling reflections, and glare. There is therefore much to be said against making the units adjustable by their lay uses. They offer some personal control of lighting, which is often a popular feature in a large open office.

23.6.3 Choice of lamp

The choice of lamp affects the range of luminaires available, and vice versa, so one cannot be considered without reference to the other. (Section 23.2 will help to differentiate the performance of the main lamp types.)

The designer should compile a list of suitable lamps by rejecting those which do not satisfy the design objectives. The availability of suitable luminaires can then be checked and the economics of the combinations assessed.

Lamps must have satisfactory colour-rendering properties. Visual tasks requiring accurate perception of colour are less common than is generally believed, but there are many merchandising situations where good colour rendering is desirable. The suitability of a lamp for a particular application is best decided by experimental tests.

Lamps must also provide the right colour appearance. A warm colour tends to be preferred for informal situations, at lower illuminances, and in cold environments, while a cool appearance tends to be preferred for formal situations, at higher illuminances, and in hot environments. It is normally undesirable to illuminate visibly adjacent areas with sources of significantly different colour appearance.

Lamp life and lumen maintenance must be considered in conjunction with the maintenance policy for sensible economic calculations, and it needs to be remembered that standardization of lamp types and sizes within a particular site can simplify maintenance.

This selection process deals with most simple situations and is usually more reliable than trying to find a lamp which matches the needs of the design, which normally calls for a very difficult compromise decision. However, it must only be regarded as a guide. Size, output, the availability of suitable luminaires and many other aspects will influence the choice.

23.6.4 Choice of luminaire

Luminaires have to withstand a variety of physical conditions (e.g. vibration, moisture, dust, high or low ambient temperatures, or vandalism). IR charts identify several acceptable luminaires and the final decision on which luminaire to use can be based on cost, efficiency, glare control, etc., and may have to be left to a later stage in the design process.

23.6.5 System management

A lighting system should not only be well designed but also be managed and operated effectively and efficiently. Good system management implies:

1. Maintaining the system in good order; and
2. Controlling its use to conserve money and energy.

Both these components of management must be considered in the design. The maintenance policy will affect the number of luminaires needed to achieve the service illuminance and the degree of control required will determine the circuits in which the luminaires will operate. They must also be electrically safe (a matter covered by the BS Safety Mark).

Further reading

A. M. Marsden and M. A. Cayless, *Lamps and Lighting*, 3rd edn, Edward Arnold, London

Chartered Institution of Building Services Publications and Codes, Delta House, 222 Balham High Road, London SW12 9BS

Lighting Industry Federation Factfinders, Swan House, 207 Balham High Road, London SW17 7BQ

24 Compressed Air Systems

British Compressed Air Society

Contents

24.1 Assessment of a plant's air consumption

The main consideration in the selection of a compressor plant is the production of an adequate supply of compressed air at the lowest cost consistent with reliable service. The installation of a compressed air system, as with all forms of power transmission, calls for capital investment with consequent operating and maintenance costs. The information on which the selection of plant is based should be as accurate as possible. Important factors to be considered are the following.

24.1.1 Operating pressure

Most compressed air equipment operates at about 6 bar (gauge) and it is usual for the compressor to deliver air into the mains at 7 bar (gauge) in order to allow for transmission losses (see Tables 24.1–24.4). If some of the air is to be used at a lower pressure (for example, for instrument control) the pressure is reduced by means of a pressure regulator to the required line pressure.

All equipment connected to the system shall either have a design pressure greater than the maximum output pressure of the compressor or special precautions shall be taken to ensure that, if its design pressure is lower than the output pressure of the compressor, it cannot be subject to excessive pressure (see Section 24.3.1). If there is a requirement for a large volume of air at a higher or lower pressure it may be more economical to install a separate compressor to deal solely with that.

Table 24.1 Maximum recommended flow through main lines[a]

Nominal bore (mm)	*Actual bore* (mm)	*Rate of air flow at 7 bar* (l/s)
6	6	1
8	9	3
10	12	5
15	16	10
20	22	17
25	27	25
32	36	50
40	42	65
50	53	100
65	69	180
80	81	240
100	100	410
125	130	610
150	155	900

Velocity of air flow must be restricted to less than 6 m/s as shown if carry over of moisture past drain legs is to be avoided.
[a] Medium-weight steel tube to BS 1387, table 2, or ISO 65.

24.1.2 Maximum and average load

Ideally, the total capacity would be based on exact knowledge of the equipment or process requirements. If this is underestimated the compressor plant will be too small, and will be unable to maintain the required pressure in the system. Conversely, if the total air consumption is greatly overestimated, there may be excessive capital investment. Furthermore, any arrangement which results in significant off-load running wastes energy. However, it is safer to err on the high side with a slight overestimates, as in most installations the use of compressed air will increase and soon take up any surplus capacity.

24.1.3 Use factor

Before deciding the capacity of the compressor required it is necessary to calculate the air consumption expected. It is recommended that reference is made to Table 24.5 and Figure 24.1, which show typical 'use factors' for various types of pneumatic equipment. In some cases, where there is experience of a similar installation, a fairly accurate analysis can be made by plotting data obtained from past activity.

Table 24.2 Maximum recommended flow through branch lines of steel pipe[a]

Applied gauge pressure[b] *bar*	*Nominal standard pipe size (nominal bore) (mm)* 6	8	10	15	20	25	32	40	50	65	80
0.4	0.3	0.6	1.4	2.6	4	7	15	25	45	69	120
0.63	0.4	0.9	1.9	3.5	5	10	20	30	60	90	160
1.0	0.5	1.2	2.8	4.9	7	14	28	45	80	130	230
1.6	0.8	1.7	3.8	7.1	11	20	40	60	120	185	330
2.5	1.1	2.5	5.5	10.2	15	28	57	85	170	265	470
4.0	1.7	3.7	8.3	15.4	23	44	89	135	260	410	725
6.3	2.5	5.7	12.6	23.4	35	65	133	200	390	620	1085
8.0	3.1	7.1	15.8	29.3	44	83	168	255	490	780	1375
10.0	3.9	8.8	19.5	36.2	54	102	208	315	605	965	1695
12.5	4.8	10.9	24.1	44.8	67	127	258	390	755	1195	2110
16.0	6.1	13.8	30.6	56.8	85	160	327	495	955	1515	2665
20.0	7.6	17.1	38.0	70.6	105	199	406	615	1185	1880	3315

[a] Maximum recommended air flow (l/s free air) through medium series steel pipe for branch mains not exceeding 15 m length (see BS 1387). The flow values are based on maximum recommended peak flows. Normal steady state air consumption should not exceed 80% of these figures in pipe sizes 6–15 mm nominal bore and 60% of these figures in pipe sizes 20 mm nominal bore and above.
[b] Applied pressures selected from ISO 2944: Fluid Power Systems – nominal pressures.

Table 24.3 Pressure loss through steel fittings – equivalent pipe lengths

Item	*Equivalent pipe length* (m)									
	Inner pipe diameter (mm)									
	15	20	25	40	50	80	100	125	150	200
Gate valve fully open	01.	0.2	0.3	0.5	0.6	1.0	1.3	1.6	1.9	2.6
half closed		3.2	5	8	10	16	20	25	30	40
Diaphragm valve fully open	0.6	1.0	1.5	2.5	3.0	4.5	6	8	10	
Angle valve fully open	1.5	2.6	4	6	7	12	13	18	22	30
Globe valve fully open	2.7	4.8	7.5	12	15	24	30	38	45	60
Ball valve (full bore) fully open	0.5	0.2	0.2	0.4	0.3	0.4	0.3	0.5	0.6	0.6
Ball valve (reduced bore) fully open	3.4	4.9	2.4	2.2	5.0	2.6	4.1	3.3	12.1	22.3
Swing check valve fully open		1.3	2.0	3.2	4.0	6.4	8.0	10	12	16
Bend R = 2*d*	0.1	0.2	0.3	0.5	0.6	1.0	1.2	1.5	1.8	2.4
Bend R = *d*	0.2	0.3	0.4	0.6	0.8	1.3	1.6	2.0	2.4	3.2
Mitre bend 90°	0.6	1.0	1.5	2.4	3.0	4.8	6.0	7.5	9	12
Run of tree	0.2	0.3	0.5	0.8	1.0	1.6	2.0	2.5	3	4
Side outlet tee	0.6	1.0	1.5	2.4	3.0	4.8	6.0	7.5	9	12
Reducer		0.3	0.5	0.7	1.0	2.0	2.5	3.1	3.6	4.8

The table shows the length of pipe with equivalent pressure loss in a given size and type of fitting.

Table 24.4 Pressure loss through ABS fitting – equivalent pipe lengths

	Equivalent pipe length (m)							
Pipe outside diameter	16	25	32	50	63	75	90	110
90° elbow	0.34	0.5	0.65	1.0	1.26	1.5	1.88	2.58
45° elbow	0.16	0.24	0.32	0.52	0.63	0.75	0.95	1.33
90° bend	0.1	0.16	0.22	0.34	0.44	0.56	0.75	1.00
180° bend	0.28	0.41	–	–	–	–	–	–
Adaptor union	0.14	0.21	0.25	0.4	0.5	–	–	–
Tee in line flow	0.123	0.19	0.23	0.36	0.45	0.56	0.69	0.95
Tee in line to branch flow	0.77	1.17	1.47	2.21	2.98	3.68	4.57	6.00
Reducer	0.22	0.31	0.37	0.51	0.80	1.11	1.34	1.58

The table shows the length of pipe with equivalent pressure loss in a given size and type of fitting. For example, a 50 mm 90° elbow has a pressure loss equal to 1.0 m of 50 mm ABS pipe.

24.1.4 Future expansion

Future expansion should always be taken into account when installing new plant. Increasing compressor capacity presents no problem provided that the rest of the plant installation has been planned accordingly.

24.1.5 Allowance for air leakage

Experience has shown that the initial estimate of the total compressor capacity should include an allowance for leakage. Leakage in the pipelines can be overcome by proper installation practice. A large proportion of the total leakage occurs at hoses, couplings and valves.

For installations with regular inspection and maintenance a factor of 5% minimum should be adequate. The importance of this is obvious when one remembers that while a tool or appliance may use a considerable amount of air, it is only working intermittently, whereas any leakage, even from a small hole, is both continuous and significant.

24.2 Compressor installation

24.2.1 Type of installation

When planning a compressor installation one of the first matters to be decided is whether there should be a central compressor plant or a number of separate compressors near the main points of use. The following comments can be no more than general. In order to select the type and size of installation which will be adequate for both immediate and future requirements it is advisable to consult the supplier. Points to be considered are as follows.

Centralized installation

1. Lower total installed compressor capacity and, perhaps, lower initial cost;
2. Possibly a higher efficiency, and thus lower power cost, due to larger units;
3. Lower supervision cost.

Table 24.5 Air consumption of pneumatic equipment

Example of calculation

The following calculation is typical of a medium-sized engineering workshop including a foundry, where a high degree of mechanization is to be carried out by means of compressed air-driven machines and tools. Listed in the table are the tools and other pneumatic devices which are expected to be included in the installation at full production capacity. The use factor of the different tools is calculated in connection with production planning and thus it is possible to establish the average total air consumption.

Machine or tool	*Air consumption per unit* (l/s)	*Quantity*	*Maximum air consumption* (l/s)	*Use factor*	*Average air consumption* (l/s)
Foundry					
Core-shop (I)					
Core blowers	11	3	33	0.50	16.5
Bench rammers	4	2	10	0.20	2.0
			43		18.5
Machine moulding (II)					
Moulding machines	12	5	60	0.30	18.0
Blow guns	8	5	40	0.10	4.0
Air hoist – 500 kg	33	2	66	0.10	6.6
			166		28.6
Hand moulding (III)					
Rammers					
– medium	6	1	6	0.20	1.2
– heavy	9	2	18	0.20	3.6
Blow guns	8	3	24	0.10	2.4
Air hoist – 500 kg	33	1	33	0.10	3.3
			81		10.5
Cleaning shop (IV)					
Chipping hammers					
– light	6	2	12	0.35	4.2
– medium	8	3	24	0.35	8.4
– heavy	13	2	26	0.20	5.2
Grinders					
– 75 mm	9	2	18	0.30	5.4
– 150 mm	25	3	75	0.45	33.8
– 200 mm	40	1	40	0.20	8.0
– medium	23	2	46	0.10	4.6
– heavy	42	2	84	0.10	8.4
Sandblast units					
– light	32	1	32	0.50	16.0
– heavy	53	1	53	0.50	26.5
			410		120.5
Total for foundry			700		178.0
Workshop					
Machine shop (V)					
Blow guns	8	10	80	0.05	4.0
Operating cylinders for jibs, fixtures and chucks			12	0.10	1.2
			92		5.2
Sheet metal shop (VI)					
Drills					
– light	6	1	6		1.2
– medium	8	1	8		1.6
– 12 mm	15	2	30		9.0
– angle	8	1	8		1.6
– screwfeed	52	1	52		2.6
Tapper	8	1	8	0.20	1.6
Screwdrivers	8	2	16	0.10	1.6
Impact wrench					
– 20 mm	15	1	15	0.20	3.0
– 22 mm	23	1	23	0.10	2.3
Grinders					
– 150 mm	25	2	50	0.30	15.0
– 200 mm	40	1	40	0.20	8.0
– medium	23	2	46	0.30	13.8
– heavy	42	1	42	0.20	8.4

Table 24.5 Continued

Machine or tool	*Air consumption per unit* (l/s)	*Quantity*	*Maximum air consumption* (l/s)	*Use factor*	*Average air consumption* (l/s)
Riveting hammers					
– medium	18	1	18	0.10	1.8
– heavy	22	1	22	0.05	1.1
Chipping hammers					
– light	6	2	12	0.20	2.4
– medium	8	2	16	0.20	3.2
– heavy	13	1	13	0.10	1.3
Air hoist – 5 tonne	97	1	97	0.05	16.2
Blow guns	8	1	16	0.10	1.6
			538		97.3
Assembly shop (*VII*)					
Drills					
– light	6	3	18	0.20	3.6
– medium	8	5	40	0.30	12.0
– 12 mm	15	6	90	0.35	31.5
– angle	8	2	16	0.10	1.6
– heavy	22	1	22	0.10	2.2
– heavy	33	1	33	0.10	3.3
Tappers	8	2	16	0.10	1.6
Screwdrivers	8	2	16	0.20	3.2
Impact wrenches					
– light	6	1	6	0.20	1.2
– 20 mm	15	2	30	0.20	6.0
– 22 mm	23	1	23	0.10	2.3
Grinders					
– 75 mm	9	2	18	0.20	3.6
– 150 mm	25	1	25	0.10	2.5
– medium	23	2	46	0.20	9.2
Air hoists					
– 500 kg	33	1	33	0.10	3.3
– 1 tonne	33	1	33	0.10	3.3
Blowguns	8	5	40	0.05	2.0
			505		92.4
Painting shop (*VIII*)					
Grinders and polishers					
– angle	8	1	8	0.20	1.6
– medium	23	1	23	0.30	6.9
Sandblast unit	38	1	38	0.50	19.0
Blow guns	8	1	8	0.10	0.8
Air hoist – 5 tonne	97	1	97	0.05	4.9
Spray painting guns	5	2	10	0.50	5.0
			184		38.2
Total for the workshop			1319		233.1

Decentralized installation

1. Output and/or pressure can be varied to suit each particular plant section.
2. Pipe sizes can be reduced, thus minimizing leakage and cost.
3. Compressors and/or associated equipment can be shut down during periods of low demand or for preventative maintenance with only a localized effect.

24.2.2 Compressor siting

The requirements for a compressor site will be affected by location and climate as well as by the equipment to be installed. The following aspects should be considered.

Foundations

The compressor plant should be located in a place with good ground conditions. In some cases the compressor foundation may have to be isolated from that of the main building so that vibration is not transmitted from compressor to the building structure, and from heavy plant to the compressor. Where the vibrations are slight, resilient pads may be used to advantage. Receiver-mounted units should always be either free-standing or mounted on resilient pads. If in doubt, the suppliers should be consulted.

Servicing facilities

For small and medium compressors lifting gear is

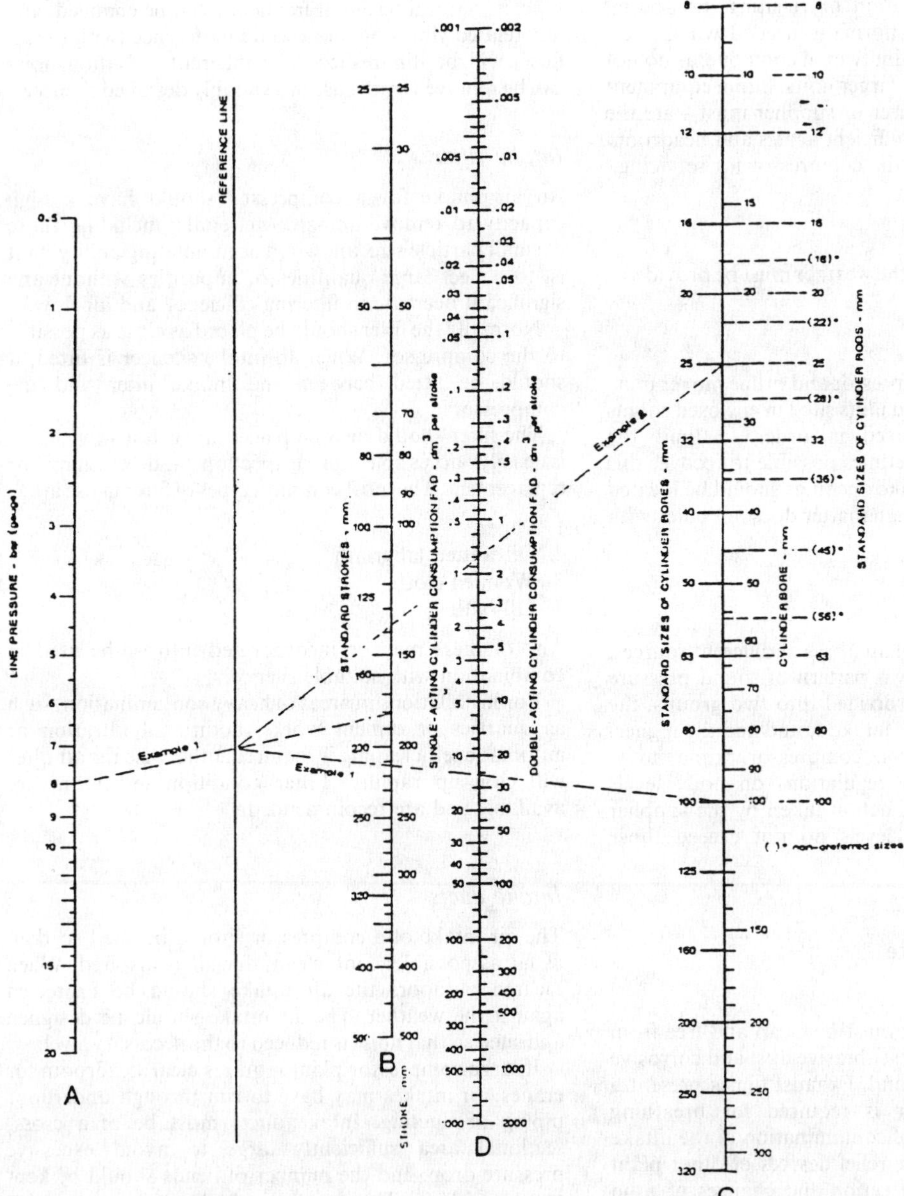

Figure 24.1 Air consumption of cylinders (metric)

How to use the nomogram

First connect line pressure (A) to stroke (B); then from the point where this cuts the reference line, connect across to cylinder bore size (C). Read off consumption where this line cuts the consumption scale (D). Figures on the left of this scale are for single-acting cylinders. Figures on the ring are for double-acting cylinders, neglecting the effect of the rod. This is accurate enough for most purposes. However, if the correct (theoretical) consumption is required for double-acting cylinders, go back to the point on the reference line and connect across to the rod *diameter* size, entered on the 'cylinder bore' scale. This figure should then be deduced from the consumption arrived at with the first solution.

Example 1: Find the (nominal) consumption of a double-acting 100 mm bore cylinder with a stroke of 180 mm operating at 8 bar line pressure.

Answer: 25.4 l per stroke.

Example 2: The cylinder above has a 25 mm diameter rod. Find the true (theoretical) consumption.

Connection from the same point on the reference scale to the rod diameter size on the bore scale gives a single-acting cylinder consumption figure of 0.8 l per stroke. Deduce this from the solution found in Example 1 to give the true consumption figure, i.e. $25.4 - 0.8 = 24.6$ l per stroke

necessary only for installation or re-siting, no special hoisting equipment being normally needed when overhauling the units provided individual components do not exceed 16 kg in weight. On larger units, lifting equipment is essential; the manufacturer or supplier must state the maximum hoisting load. Sufficient access and headroom must be provided around the compressor for servicing.

Weather protection

Adequate protection from the weather must be provided.

Ventilation

Heat generated by the compressor and prime mover must be dispersed. For air-cooled units sited in enclosed rooms this heat must be removed in order to limit the temperature rise. It is sometimes possible to recover this heat for use elsewhere. Intake openings should be located so that dust and other foreign matter does not enter with the air.

Noise

Noise from a compressor plant arises at different sources, and each source has its own pattern of sound pressure levels. Noise levels can be divided into two groups, the low-frequency pulsating air intake sound and the higher-frequency machine noise from compressor, prime mover and fans. Local statutory regulations on noise levels should be determined, and action taken by the supplier to ensure that the noise levels do not exceed those stipulated.

24.2.3 Compressor intake

General

The compressor intake air must be clean and free from solid and gaseous impurities; abrasive dust and corrosive gases are particularly harmful. Exhaust fumes present a hazard if compressed air is required for breathing purposes. The possibility of contamination of the intake by discharge from pressure-relief devices of other plant must be taken into consideration and changes of wind direction must not be overlooked.

For maximum efficiency the intake air should be as cold as possible. A temperature decrease of 3°C will increase the volume of the delivered air by 1%. The air intake system should be sized to give a minimum pressure drop. Each compressor should have its own intake filter.

Intake silencing

The reciprocating compressor inspires air in a series of pulsations which causes an equivalent variation in pressure in the intake system. Dependent on the length of the intake pipe, resonance may occur; this can decrease the compressor output and produce disturbing noise levels and stresses sufficient to cause damage. By fitting a pulsation dampener or changing the length of the intake pipe, its natural resonant frequency can be changed, and any related vibration, noise and interference (with the air flow) will be diminished. The inherent pulsation noise can be removed by the use of a suitably designed silencer.

Intake filter

An air intake for a compressor should have a high capacity to remove abrasive materials, including those of small particle size, and good accumulating ability, that is, to collect large quantities of impurities without any significant decrease in filtering efficiency and air flow.

Normally the filter should be placed as close as possible to the compressor. When an intake silencer is fitted, it should be fitted between the intake filter and the compressor.

The filter should also be placed in such a way that it is easily accessible for inspection and cleaning or replacement. The most common types of filter in use are:

1. Paper
2. Oil-wetted labyrinth
3. Woollen cloth
4. Oil-bath

Any of these may be incorporated into or be used in combination with suitable silencers.

For installations in areas of heavy contamination, such as quarries or cement works, additional filtration or automatic self-cleaning is required, otherwise the air filter will clog up rapidly. Filter condition indicators are available and are recommended.

Intake ducts

The air intake of a compressor should be sited so that, as far as possible, cool, clean, dry air is inspired. When located outdoors the air intake should be protected against the weather. The air intake should be designed and sited so that noise is reduced to the necessary level.

If large compressor plant requires clear headroom for cranes, air intakes may have to run through underfloor piping or ducting. Intake ducts must be of a cross-sectional area sufficiently large to avoid excessive pressure drop, and the number of bends should be kept to a minimum. The ducts should be of non-corrosive material and care should be taken that extraneous material cannot enter the duct. The duct should be cleaned thoroughly before connection to the compressor.

Intake pipes may be subject to pulsations and should be too rigidly attached to walls or ceilings, since vibration may be transmitted to the building structure.

Corrosive intake gases

In certain plants, especially in the chemical industry or in the neighbourhood of such plants, the air is often polluted with acidic and corrosive gases which can cause corrosion in the compressor and the compressed air system. Special filtration methods and/or materials may have to be used and the supplier should be consulted.

24.2.4 Compressor discharge

Discharge pipe specification and siting

The diameter of the compressor discharge pipe should not normally be smaller than the compressor outlet connection and should be arranged with flanged fittings or unions to permit easy access to the compressor and components at any time. The possibility of vibration should be taken into account. The compressor discharge pipe will attain a high temperature and precautions must be taken to prevent this being a source of danger.

The interior of the pipelines through which the discharge air passes to the aftercooler or air receiver should be cleaned regularly so that a build-up of combustible oily carbon deposits is avoided. All the piping should slope downwards in the direction of air flow to a suitable drain point at the lowest point of the pipeline.

Discharge pipes can be located in trenches covered by floor plants, and there is no technical reason against laying the pipes directly on the, but provision must be made for drainage.

Any pocket unavoidably formed after the compressor discharge shall be provided with a drain valve or trap at the lowest point so that any oil and condensate can be removed.

Under certain conditions of installation and operation, pulsations may be set up in the compressor discharge lines. It is essential to consult the suppliers for their recommendations.

Thermoplastics shall not be used for a compressor discharge pipe and inflammable materials shall be kept away from it.

Isolating valve

Where an isolating valve is installed in the discharge pipework the pipeline on the compressor side of the valve shall be protected by a suitable safety valve. This safety valve must be of sufficient size to pass the full output of the compressor without the pressure rising more than 10% above the maximum allowable working pressure (BS 6244, Section 21).

Multiple compressors

Where two or more compressors feed into a single air line, the discharge line from each compressor shall be fitted with a non-return valve and isolation valve at the furthest point from the compressor or outlet, just prior to where the discharge pipe enters the common manifold feed pipe. A safety valve is fitted on the compressor side of the isolation valve, upstream of the aftercooler.

Non-return valves

Non-return valves used in compressor delivery lines must be designed to withstand the pressure, temperature and pulsations of compressed air.

24.2.5 Cooling-water system

General

Where water is used as a cooling medium for compressor and ancillary equipment it should be within the temperature and pressure levels prescribed by the compressor supplier and should be free from harmful impurities. The cooling water should have a low inlet temperature in order to assist in achieving a high volumetric efficiency in the compressor and to cool the air passing through the aftercooler to a temperature adequate for effective condensation of water vapour.

Overcooling

The compressor should not be overcooled so as to cause condensation in and on the compressor.

Water quality

Good-quality cooling water is essential.

Re-cooling the cooling water

In order to achieve economy in the use of water it will have to be re-cooled. This is achieved by transferring heat to the ambient air by means of cooling ponds, towers, tanks or mechanical coolers. Temperature regulators may assist control and conserve energy.

Mechanical coolers

The cost of cooling water is an important factor and mechanical coolers are in most cases more economical than allowing the water to run to waste. A forced-draught type of cooler consists of a casing with a water header at the top and a sump at the bottom. A series of cooling elements is provided which offers a large area for the transfer of heat between the water to be cooled and the cooling air. The hot water enters the top header and runs through the elements to a sump from which it is pumped through the compressor plant. A fan forces the colder ambient air through the elements to absorb the heat from the water as it passes through the elements.

Where this type of cooler is installed inside a building it is essential to duct away the warm air discharged by the fan. Consideration must be given to protection against frost.

Cooling towers

These operate by setting the cooling air in motion over a surface of water. This can be done by either natural convection or a fan. In order to provide a good transfer between water and air, towers usually have internal arrangements for spreading the water as a thin film. With cooling towers a final water temperature of about 5°C above the ambient air temperature can be expected. In general, good cooling can be obtained even at relatively high ambient temperatures, since the relative humidity is, in such a case, usually low. However, extreme tropical climates are an exception. The amount of water which is lost as vapour during re-cooling must be replaced by the addition of 'make-up' water. This quantity is considerably smaller than that consumed in open-flow cooling. Cooling towers should not be used in heavily contaminated atmospheres.

Cooling ponds

A cooling-water pond is the simplest form of cooling arrangement. The pond should be located so that an unrestricted air circulation is obtained. Vaporizing ability is improved if the hot water is returned to the pond by some kind of sprinkler device. Most of the cooling effect is caused by vaporization and the water thus lost must be replaced. Cooling ponds should not be used in heavily contaminated atmospheres.

Cooling tanks

A cooling tank is, in effect, a small cooling pond. However, because of the difficulty in keeping the water clean, this method is not recommended.

Keeping the cooling system clean

Cooling water should be free from solid impurities which could damage pumps and cause blockages and filtered, with filters cleaned regularly. The whole cooling system should be inspected and cleaned regularly. Sand, sludge, rust, etc. can be removed by flushing against the normal direction of flow. Lime deposits are more difficult to remove. Such deposits can usually be avoided by keeping the water outlet temperature at a low level. If excessive deposits do occur, a specialist should be called in to clean the system by chemical methods.

24.2.6 Ventilation

In compressor operation, part of the heat given off by the compressor and motor is transmitted to the surrounding air. For plants located in closed rooms this heat must be removed to limit the rise in temperature of the ambient air. Some of the heat dissipates through walls, windows, floor and roof, but this heat removal is seldom sufficient. The compressor room should be ventilated and the heat removed with the ventilating air. Sometimes the heat can be recovered and used for heating purposes. In an entirely water-cooled compressor installation the heat to be removed by ventilation is relatively small, since the major part is taken away by the cooling water.

Insufficient ventilation shortens the life of the electric motor. In installations where the intake air is drawn from the compressor room poor ventilation may also damage the compressor, as the temperature of the discharge air increases in proportion to that of the intake air. The compressor room should always be placed so that ventilation air is available without the need for long ducts. The intake should be sited low down on the coldest wall, whereas the ventilation air outlet should be situated high up on the opposite wall in order to avoid temperature stratification.

Modern completely air-cooled compressor plants have aftercoolers with fans. The aftercooler should be arranged so that it assists in the ventilation of the room. For the major part of the year the aftercooler fan will handle room ventilation. Extra fans may be needed only during hot months in the summer.

24.3 Overpressure protection

1. If any equipment having a design pressure lower than the maximum output pressure of the compressor is used, or if an increase of pressure above normal operating pressure will cause a malfunction, it shall be protected against overpressure by suitable means.
2. Any relief valve or safety valve shall have a design flow capacity such that when subjected to the maximum output pressure and flow of the compressor, and taking into account the flow restriction caused by the upstream pipework and fittings, it will prevent the pressure in that part of the system exceeding the design pressure of the equipment. This requirement may, in certain circumstances and depending upon compressor pressure and design pressure of the equipment, imply the need for a relief valve having a port size at least twice the nominal diameter of the pipework and pressure regulator feeding the equipment.
3. An alternative method of protection is the use of a smaller relief valve in conjunction with an automatic isolating valve which shuts off the air supply to the equipment if the pressure rises more than 20% above the blow-off pressure of the relief valve. If this method is used it is essential to ensure that sudden cessation of air supply to the equipment cannot cause a hazard.

24.4 Selection of compressor plant

24.4.1 Air compressors

The principal types of compressors and their basic characteristics are outlined below (see also Figure 24.2).

Positive displacement compressors

Positive displacement units are those in which successive volumes of air are confined within a closed space and elevated to a higher pressure. The capacity of a positive displacement compressor varies marginally with the working pressure.

Reciprocating compressors The compressing and displacing element (piston or diaphragm) has a reciprocating motion. The piston compressor is available in lubricated and non-lubricated construction.

Helical and spiral-lobe compressors (*screw*) Rotary, positive displacement machines in which two intermeshing rotors, each in helical configuration displace and compress the air; available in lubricated and non-

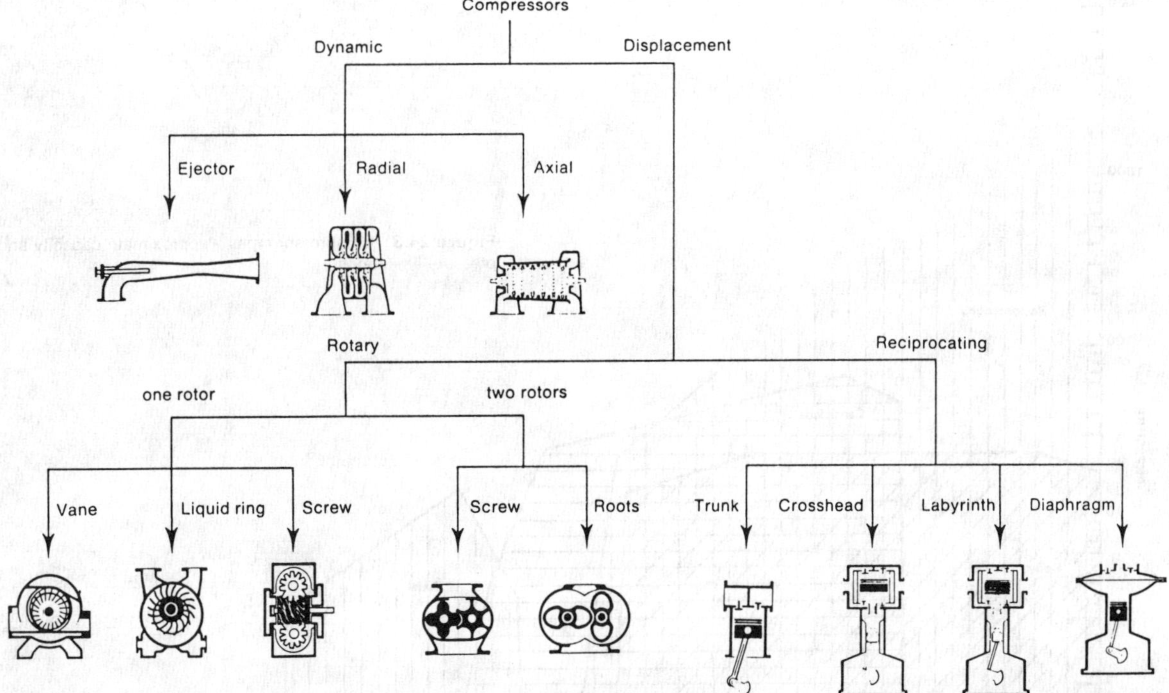

Figure 24.2 Basic compressor types

lubricated construction; the discharge air is normally free from pulsation; high rotation speed.

Sliding-vane compressors Rotary, positive displacement machines in which axial vanes slide radially in a rotor mounted eccentrically within a cylindrical casing. Available in lubricated and non-lubricated construction; the discharge air is normally free from pulsation.

Two impeller straight-lobe compressors and blowers Rotary, positive displacement machines in which two straight, mating but non-touching lobed impellers trap the air and carry it from intake to discharge. Non-lubricated; the discharge is normally free from pulsation; low pressure; high rotation speed.

Dynamic compressors

Dynamic compressors are rotary continuous-flow machines in which the rapidly rotating element accelerates the air as it passes through the element, converting the velocity head into pressure, partially in the rotating element and partially in stationary diffusers or blades. The capacity of a dynamic compressor varies considerably with the working pressure.

Centrifugal compressors Acceleration of the air is obtained through the action of one or more rotating impellers; non-lubricated; the discharge air is free from pulsation; very high rotation speed.

Axial compressors Acceleration of the air is obtained through the action of a bladed rotor, shrouded at the blade ends; non-lubricated; very high rotation speed; high-volume output.

Specific power consumption

This varies with the size and type of compressor; consultation with the supplier is advised.

24.4.2 Capacity and pressure limitations

Figure 24.3 shows the approximate capacity and pressure limitations of each type of compressor. There are areas where more than one type of compressor will provide the required capacity and pressure. In such cases other characteristics such as those given above and the type and pattern of use will govern the selection. Consultation with the supplier is advised.

Compressor standby capacity

On many installations it is normal to plan the number of compressor units and their output so that there is a standby capacity to permit one unit to be shut down for servicing. Where a constant supply of air is essential to operations, standby compressors are a necessity.

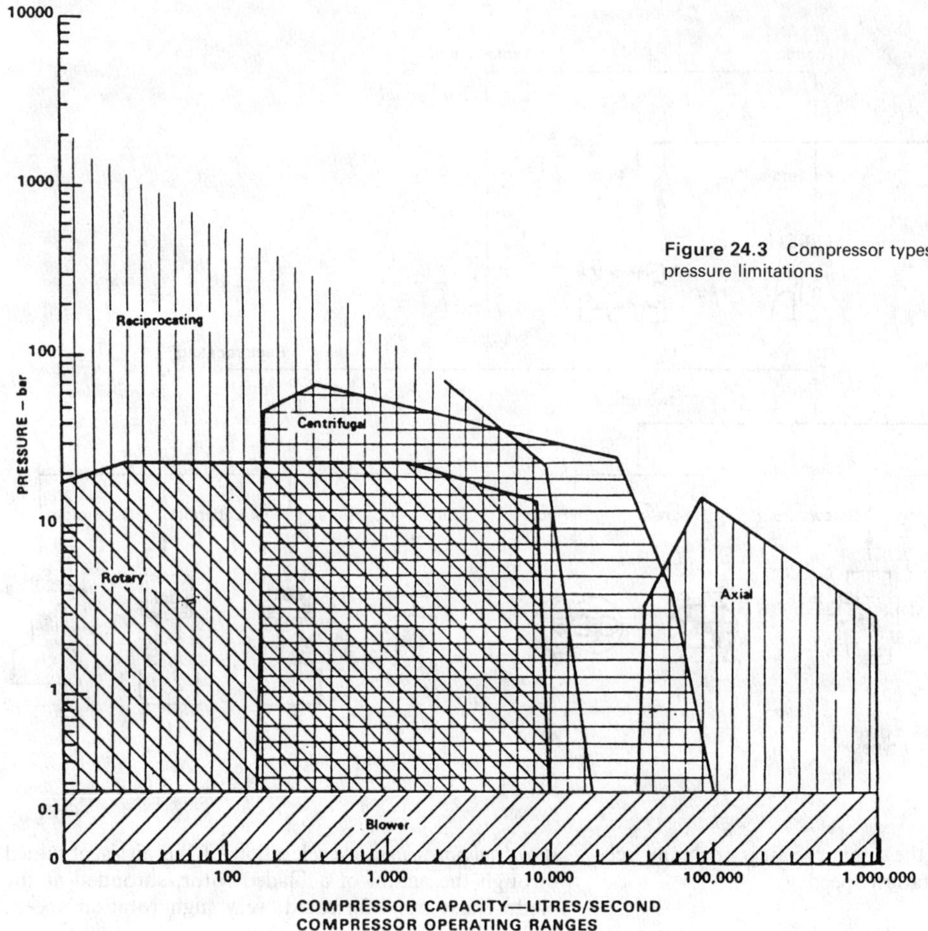

Figure 24.3 Compressor types – approximate capacity and pressure limitations

Load splitting

In all installations consideration should be given to having at least two compressors to allow for conditions of light load and for maintenance.

Closed-loop systems

The same general considerations listed above apply to two level closed-loop systems but the required input power to the compressor will be considerably reduced or a smaller compressor can be used.

24.4.3 Output control

A wide range of controls is available to match compressor output to demand. Consultation with the equipment supplier is essential. The output of a compressor can be controlled by several methods as outlined below. The following functions can be performed by pneumatic, hydro-pneumatic and electronic devices:

1. *Reciprocating compressors*:
   (a) Intermittent operation using automatic stop/start mechanism;
   (b) Constant speed running with inlet valve blocking or intake throttling or external bypass or inlet valve unloading or clearance pocket;
   (c) Variable speed;
   (d) Combinations of (a)–(c) above.
2. *Rotary sliding-vane compressors*:
   (a) Intermittent operation using automatic stop/start mechanism;
   (b) Constant-speed running with inlet valve blocking or intake throttling or external bypass;
   (c) Variable speed; minimum rotational speed must be high enough to ensure that the blades remain in full contact with the stator.
3. *Rotary screw compressors*:
   (a) Constant speed running with external bypass or intake throttling coupled with blow-off to atmosphere;
   (b) Variable speed.
4. *Dynamics*:
   (a) Constant-speed running with intake throttling coupled with blow-off to atmosphere;
   (b) Variable speed.

Advice should be sought from the supplier as to the best type of control to suit a particular application.

24.4.4 Selection of compressor prime movers

An important factor in obtaining an economical plant is the selection of an appropriate compressor drive. The most common power units are:

Electric motor
Engine (diesel, petrol, gas, etc.)
Turbine (gas, steam, etc.)

Among the advantages of electric motor drive are compactness and ease of control. The internal combustion engine is preferred for mobile units, emergency standby units, or where electric power is not available.

A turbine drive is preferred where it helps balance the energy system of a plant or where the steam or gas can be further used. This type of drive permits easy speed control and conserves energy.

Regardless of the type of prime mover, professional advice should be taken in matching prime mover to compressor.

Application requirements

To avoid delays in the preparation of estimates and unnecessary expense for both buyer and supplier it is important that all necessary data should be available and recorded. The parameters which must be established are outlined below.

Compressor output conditions

1. Volume of free air required (l/s); including an allowance for future expansion;
2. Minimum discharge pressure required to maintain an acceptable working pressure at the point of use;
3. Quality of air; degree of cleanliness required (see PNEUROP Publication 66110);
4. The purpose for which the air is to be used;
5. The pattern of demand for air; continuous or intermittent consumption;
6. Estimated operating hours per day/week;
7. Type of control;
8. The need for an air receiver;
9. Any special conditions which the compressor must satisfy;
10. Requirement for ancillary equipment (for example, water pumps, valves, piping, anti-vibration mountings, aftercoolers, dryers, intake filters and silencers, etc.).

25 Noise and Vibration

Roger C Webster BSc, MIEH
Environmental Consultant

Contents

25.1 Introduction: basic acoustics

Sound can be defined as the sensation caused by pressure variations in the air. For a pressure variation to be known as sound it must occur much more rapidly than those of barometric pressure and the degree of variation is much less than atmospheric pressure.

Audible sound has a frequency range of approximately 20 Hz to 20 kHz and the pressure ranges from 20×10^{-6} N/M to 200 N/M. A pure tone produces the simplest type of wave form, that of a sine wave (Figure 25.1). The average pressure fluctuation is zero, and measurements are thus made in terms of the root mean square (rms) of the pressure variation. For the sine wave the rms is 0.707 times the peak value.

Since rms pressure variations have to be measured in the range 20×10^{-6} N/M to 200 N/M (a range of 10^7) it can be seen that an inconveniently large scale would have to be used if linear measurements were adopted. Additionally, it has been found that the ear responds to the intensity of a sound (αP^2) in a logarithmic way. The unit that has been adopted takes these factors into account and relates the measured sound to a reference level. For convenience, this is taken as the minimum audible sound (i.e. 20×10^{-6} N/M) at 1 K.

The logarithm (to the base 10) of the ratio of the perceived pressure (squared) to the reference pressure (squared) is known as the Bell, i.e.

$$B = \log \frac{P^2}{P_{ref}^2}$$

Since this would give an inconveniently small scale (it would range from approximately 0 to 14 for a human response) the bell is divided numerically by 10 to give the decibel. The equation therefore becomes:

$$\text{dB} = 10 \log \frac{P^2}{P_{ref}^2}$$

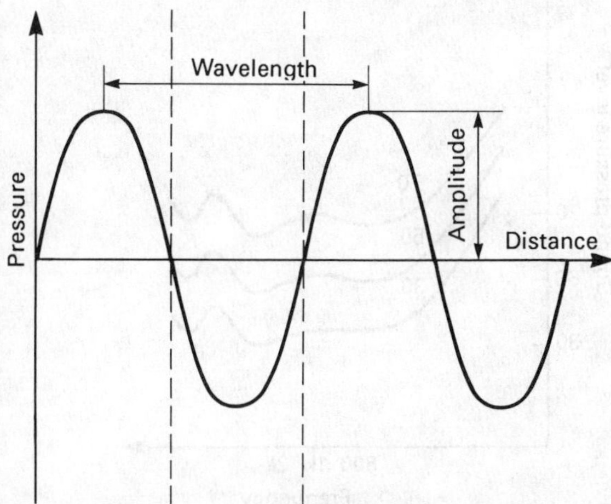

Figure 25.1 Sine wave

25.1.1 Sound intensity

Sound intensity is a measure of energy, and its units are watts per metre. Intensity is proportional to the square of pressure:

$$\frac{I}{I_0} = \left(\frac{P}{P}\right)^2$$

Sound intensity level is defined in a similar manner to sound pressure level. In this case the equation is

$$\text{(dB) (Sound intensity level)} = 10 \log \frac{I}{I_{ref}}$$

I_{ref}, the reference level, is taken as 10^{-12} W/M^2.

25.1.2 Sound power

The power of a source (measured in watts) can be similarly expressed in terms of decibels (in this case, called the sound power level):

$$\text{(dB) (Sound power level)} = 10 \log \frac{W}{W_{ref}}$$

where W_{ref} is taken as 10^{-12} W.

It can thus be seen that it is important not only to express the unit but also to state sound pressure level, sound intensity level or sound power level.

25.1.3 Addition and subtraction of decibels

For coherent sound waves addition of values is possible. It will be apparent that as the scale is logarithmic, values cannot merely be added to one another. Intensities can, however, be added and thus the equation becomes:

$$\text{SIL (total)} = 10 \log \frac{I_1 + I_2}{I_{ref}}$$

i.e. 70 dB + 73 dB

$$70 = 10 \log \frac{I_1}{I_{ref}}$$

$$I = \frac{\text{Antilog } 7}{I_{ref}}$$

$$73 = 10 \log \frac{I_2}{I_{ref}}$$

$$I = \frac{\text{Antilog } 7.3}{I_{ref}}$$

$$\begin{aligned}\text{SIL (total)} &= 10 \log (\text{Antilog } 7 + \text{Antilog } 7.3)\\ &= 10 \log (10^7 + 1.99526 \times 10^7)\\ & 10 \log (2.99526 \times 10^7)\\ &= 74.76 \text{ dB}\end{aligned}$$

The square of individual pressures must be added, and thus the equation in this case must utilize:

$$P \text{ (total)} = \sqrt{(P_1^2 + P_2^2)}$$

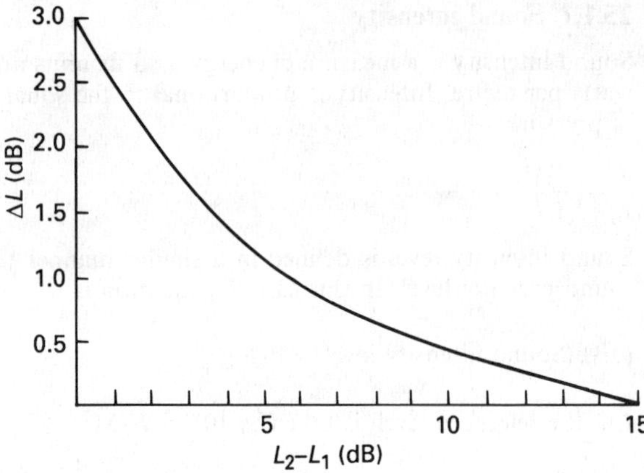

Figure 25.2 Noise-level addition graph

25.1.4 Addition of decibels: graph method

It is possible to use a graph to calculate the addition of decibels, even in the case of multiple additions (Figure 25.2). The graph is used in the following way:

1. *In the case of the addition of two levels.* The difference between the higher and the lower levels is plotted on the lower scale of the graph. The correction is then read from the vertical scale by projecting a horizontal line across to this scale from the point on the graph. The correction is added to the highest original level to give the total level.
2. *In the case of the subtraction of levels.* The difference between the total sound level and the one to be subtracted is plotted onto the graph and the correction obtained as above. In this case the correction is subtracted from the total level to give the remaining sound level.
3. *In the case of multiple additions.* If there are more levels to be added, the first two levels are added using the graph and then the third is added to the resultant using the same method.

25.1.5 The relationship between SPL, SIL and SWL

The total acoustic power of a source can be related to the sound pressure level at distance r by the following equation (assuming spherical propagation):

$W = P^2/\rho_c \cdot 4\pi r^2$

where ρ = density of the medium and C = velocity of sound in the medium. By substituting this back into the SPL equation we obtain:

$\text{SPL} = \text{SWL-}20 \log r - 11$ (spherical propagation)

It is also possible to derive equations for the other common situations, i.e.

Point source on a hard reflecting plane
Line source radiating into space
Line source on a hard reflecting plane

These equations are:

$\text{SPL} = \text{SWL-}20 \log r - 8$ (hemispherical propagation)

$\text{SPL} = \text{SWL-}10 \log r - 8$ (line source in space)

$\text{SPL} = \text{SWL-}10 \log r - 5$ (line source radiating on a plane)

The above equations are useful for calculating distance-attenuation effects.

If the sound pressure level at a distance r_0 is known it is possible to calculate the sound pressure level at positions r_1 quite easily:

$$\text{SPL}_0 - \text{SPL}_1 = 20 \log r_1 - 20 \log r_0$$

$$\text{SPL}_0 - \text{SPL}_1 = 20 \log \frac{r_1}{r_0} \text{ dB}$$

If r_1 is double r_0 it will be seen that $\text{SPL}_0 - \text{SPL}_1$ will be approximately equal to 6 dB (20 log 2). This gives us the principle of a decrease in level of 6 dB per doubling of distance (inverse square law). For the line source the same calculation produces a difference of only 3 dB per doubling of distance.

25.1.6 Frequency weighting and human response to sound

In practice, noises are not composed of one single pure tone but are usually very complex in nature. It is essential that more than the overall noise level (in dB) is known in order to appreciate the 'loudness' of a noise, as the ear does not respond uniformly to all frequencies.

As previously stated, the ear can respond from 20 Hz to 200 kHz, and this response can be demonstrated by equal-loudness contours (Figure 25.3). It can be seen from the figure that there is a loss in sensitivity (compared to 2 kHz) of approximately 60 dB at the low-frequency end of the chart. It will also be noted that all the curves are approximately parallel, but there is a tendency towards linearity at the higher noise levels. In order to produce meaningful readings it is therefore important to state the sound pressure level in decibels and the frequency of the noise.

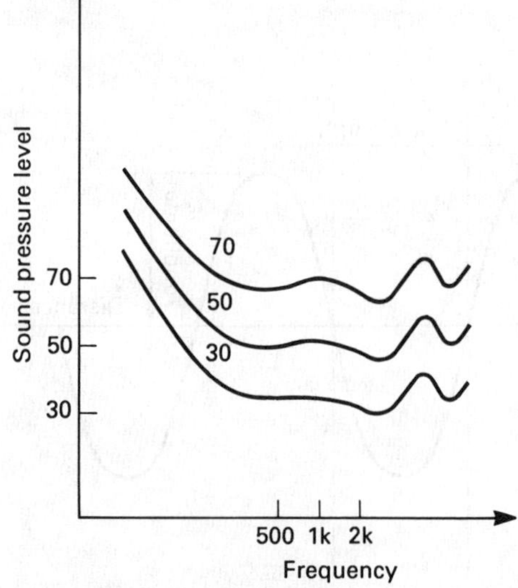

Figure 25.3 Equal-loudness contours

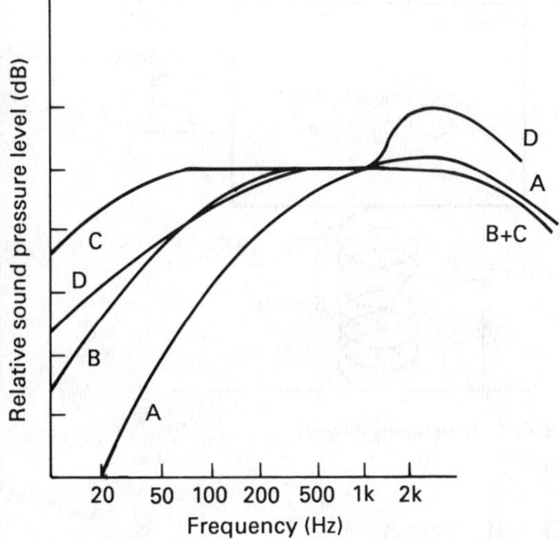

Figure 25.4 Weighting networks

A weighting can be imposed on noise readings which corresponds to the inverse of the equal-loudness contours. If this weighting is used, all readings which are numerically equal will sound equally loud, regardless of frequency.

Originally, three networks were proposed (A, B, and C) and it was suggested that these be used for low, medium and high noise levels, respectively. It was shown in practice that this introduced numerous difficulties, particularly with a rapidly changing noise when a change of filter network was necessary. It was also found that at all noise levels the A weighting network corresponded well to annoyance levels. It was therefore decided that the A weighting would be used as the norm for noise readings concerning human response. Another weighting network (the D network) is used for aircraft noise measurement (Figure 25.4).

If it is necessary for engineering purposes to know the tonal make-up of a noise, several approaches are possible. The noise can be processed by a bandpass filter. The most common filters are octave band filters, and the agreed centre frequencies are as follows:

31 63 125 250 500 1 K 2 K 4 K 8 K 16 K (Hz)

If further resolution is necessary one-third octave filters can be used but the number of measurements required to be taken is most unwieldy. It may be necessary to record the noise onto tape loops for the repeated re-analysis that is necessary. One-third octave filters are commonly used for building acoustics, and narrow-band real-time analysis can be employed. This is the fastest of the methods and is the most suitable for transient noises. Narrow-band analysis uses a VDU to show the graphical results of the fast Fourier transform and can also display octave or one-third octave bar graphs.

25.1.7 Noise indices

All the previous discussions have concerned steady-state noise. It will, however, be apparent that most noises change in level with time. It may therefore be necessary to derive indices which describe how this happens. The most common of these are percentiles and equivalent continuous noise levels.

Percentiles are expressed as the percentage of time (for the stated period) during which the stated noise level was exceeded, i.e. 5 min L_{90} of 80 dB(A) means that for the 5-min period of measurement for 90% of the time the noise exceeded 80 dB(A). Therefore L_0 is the maximum noise level during any period and L_{100} is the minimum. Leq (the equivalent continuous noise level) is the level which, if it were constant for the stated period, would have the same amount of acoustic energy as the actual varying noise level.

25.1.8 Noise-rating curves

These are a set of graphs that are used as a specification for machinery noise. They are similar to noise criteria curves (used in the USA to specify noise from ventilation systems). The rating of a noise under investigation is the value of the highest noise-rating curve intersected by the readings when plotted on the graphs (Figure 25.5).

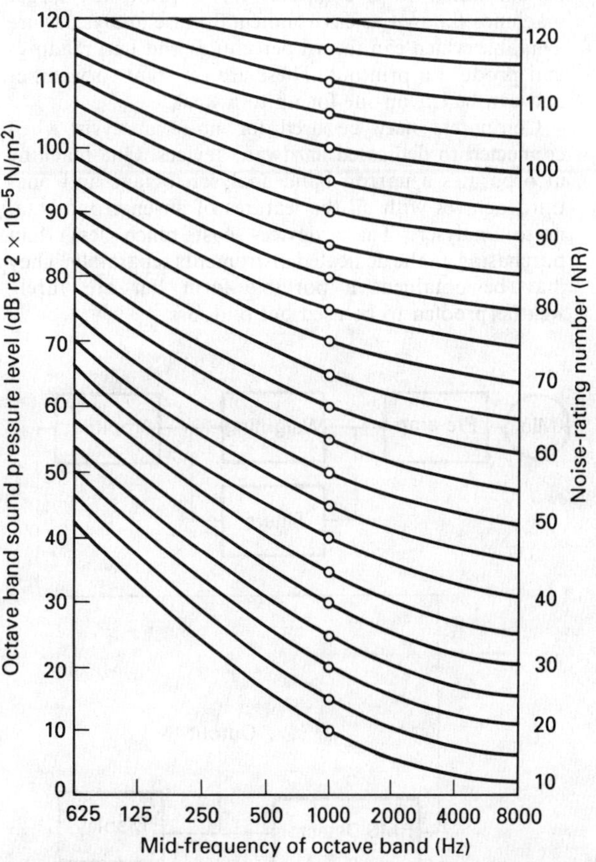

Figure 25.5 Noise-rating curves

25.2 Measurement of noise

The simplest sound level meter consists of a microphone, an amplifier and a meter of some type. Sound level meters are graded according to British and international Standards, and the most common type used for accurate measurement purposes it known as the Precision Grade or Type 1 meter. In practice, a basic sound level meter will incorporate weighting networks with either in-built octave filters or provision for connecting an external filter set (Figure 25.6).

The meter will also have a control for the time constant for the display (i.e. the 'speed' of the meter response), and the two common time constants are 'Fast' and 'Slow'. Others may have an impulse and peak hold facility. More complex meters incorporate Leq-measuring devices, and these are also available as hand-held Type 1 meters with filters as in the basic meter.

Outputs are available in either analogue (d.c. or a.c.) or digital form. Digital output may connect the meter to computers (either portable or office based) for more complex calculations or to produce larger graphical displays.

Portable sound level meters are also available which can measure percentiles. These either hold the results in a memory which can be separately interrogated or may be connected to a computer for a printout. Larger machines (known as environmental noise analysers) are available which can record percentiles and Leq readings and produce a printout. These are resistant to weather and can be left on-site for up to a week.

Computers may be used for noise analysis when connected to dedicated hardware devices. One machine incorporates a narrow-band analyser, octave and one-third octaves with all the features of an environmental noise analyser. These devices cost much less than purchasing all the dedicated instruments separately. They may be obtained in portable form but are rarely weatherproofed to be used out of doors.

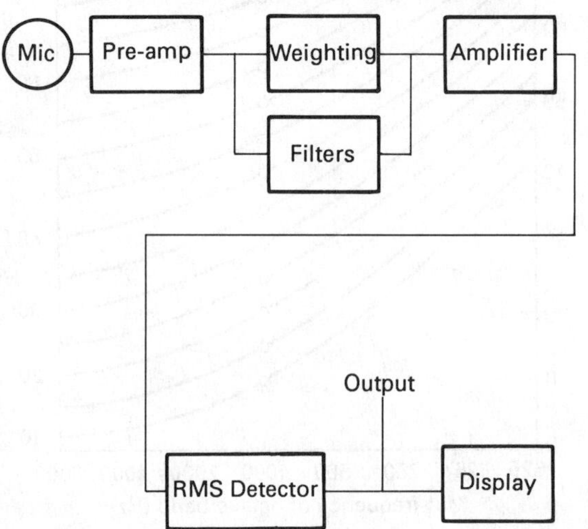

Figure 25.6 Schematic diagram of a sound level meter

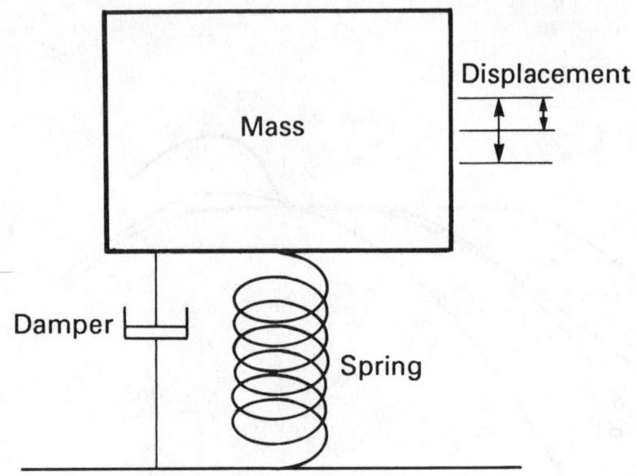

Figure 25.7 Mass/spring diagram

25.3 Vibration

This may be defined as the oscillatory movement of a mechanical system, and it may be sinusoidal or non-sinusoidal (also known as complex). Vibration can occur in many modes, and the simplest is the single freedom-of-movement system. The vibration of a system can be explained by a mass/spring diagram (Figure 25.7).

The displacement of the object from its rest position can be derived from

$X = X_{\text{peak}} \text{ sine } (\omega t)$

where X = displacement at time t, X_{peak} = peak displacement and ω = angular velocity ($2\pi f$). The velocity of the object is proportional to displacement and frequency and its acceleration is proportional to the displacement and the square of frequency. There is a phase angle of 90° between displacement and velocity and a further 90° between velocity and acceleration. The units of measurement may be made directly, i.e.

Displacement in metres
Velocity in metres per second
Acceleration in square metres per second

Vibration may also be expressed in decibels, and the standard reference levels used are:

$10\ \text{m}^{-8}/\text{s}$ for velocity
$10\ \text{m}^{-5}/\text{s}^2$ for acceleration

In practice, measurements are made with the use of an accelerometer. This device is connected to a sound level meter and may make measurements of acceleration in terms of decibels (or by changing scales or use of a device similar to a slide rule, in direct terms). An integrater can be connected between the accelerometer and the meter to express the results in terms of velocity or displacement.

25.3.1 Effects of vibration on people

If a body is vibrating at a frequency within the audible range, sound will be radiated. Vibration may be transmitted considerable distances through buildings, ground structures or rock strata, and then reradiated as noise at the receiver position. For example, low-frequency noise nuisance has been caused at a residence on the banks of the River Thames, caused by vibration from a pump on the opposite bank. In this case the vibration passed under the river along a curved layer of harder rock.

Vibration can be perceived from a frequency of approximately 3 Hz. The lower frequencies appear to cause the most discomfort. At high levels (above 120 dB) vibration can be physically damaging to people, and resonances of the human body occur at the following frequencies:

3.6 Hz: Thorax, abdomen
20–30 Hz: Head, neck
60–90 Hz: Skull

High vibration levels at these frequencies can therefore be the most disturbing.

Various sources publish data of permissible vibration levels for employees (usually as a graph of acceleration against frequency). These are designed to avoid injury and cannot be used as a guide to the degree of disturbance caused by vibration. Vibration caused by neighbouring industrial premises when received at a residence as vibration (i.e. no noise implications) has been considered a nuisance when it is just perceptible.

25.4 Noise and vibration control

Noise and vibration must be controlled in order to avoid

1. Nuisance;
2. Physical injury to workers;
3. Damage to plant/machinery/building structures.

25.4.1 Noise nuisance

Nuisance is not defined as such. Common-law nuisance is used as a guide and is divided into two types, public and private. Private nuisance relates to premises, and is a tort that can be defined as 'the unlawful interference or annoyance which causes damage to an occupier or owner in respect to his use and enjoyment of his land'. Public nuisance is an unlawful act or omission to discharge a legal duty which can endanger the life, safety, health or comfort of the public or some section of it, or by which act the public are denied some common right. It should be noted that more than one person must be affected. Public nuisance is a criminal offence and, as such, may be tried on indictment in the Crown Court. The Attorney General may initiate an action in the High Court.

25.4.2 Legislation

Noise nuisance is controlled primarily by the Environmental Protection Act 1990. Section 79 of the Act places a duty on a local authority to inspect their area for nuisances. Section 80 places a duty on a local authority to serve a legal notice on persons responsible for a situation when a nuisance has occurred and is likely to recur, or where, in the opinion of the local authority, the nuisance is likely to occur. Section 82 enables an individual to complain to a magistrate's court about a noise nuisance. If convinced, the magistrate may issue an Order telling the person causing the nuisance to cease the activity. This section may be used where the local authority cannot detect the nuisance. Under Section 60/61 of the Control of Pollution Act 1974 the local authority may issue a notice to a person carrying out construction work. This may require the person to adopt the best practical means of minimizing the noise from the site. (There is no reference to the noise having to be a nuisance in this section.) Section 62 relates to noise in the street.

25.4.3 Environmental Protection Act 1990, Section 80

A notice may be served where a nuisance has occurred or the local authority think a nuisance may occur. Noise nuisance is not defined as such, but includes vibration. The notice may not be specified and may merely require the abatement of the nuisance. A notice may, however, require the carrying out of works or specify permissible noise levels. The time period for compliance is not specified in the Act, but must be reasonable.

Appeals against a Section 80 notice must be made to the magistrate's court within 21 days of the serving of the notice. The grounds of appeal are given in the Statutory Nuisance (Appeals) Regulations 1990 and are as follows:

1. That the notice is not justified by the terms of Section 80. The most common reason for this defence is that the nuisance had not already occurred, and that the local authority did not have reasonable grounds to believe that the nuisance was likely to occur.
2. That there had been some informality, defect or error in, or in connection with, the notice. It may be that the notice was addressed to the wrong person or contained other faulty wording.
3. That the authority have refused unreasonably to accept compliance with alternative requirements, or that the requirements of the notice are otherwise unreasonable in character or extent, or are unnecessary. This defence is self-explanatory.

   The local authority are only permitted to ask for works that will abate the noise nuisance. Other works (perhaps to comply with legislation) should not be specified in the notice. They may, however, be contained in a letter separate from the notice. An example of this would be where food hygiene requirements were breached by the fitting of acoustic enclosures to food-manufacturing machines. Readily cleanable enclosures may be a requirement of the Food Hygiene Regulations, but it should not be contained in a Section 58 Control of Pollution Act notice.
4. That the time (or, where more than one time is specified, any of the times) within which the requirement of the notice are to be complied with is not reasonably sufficient for the purpose.
5. Where the noise to which the notice relates is that

caused by carrying out a trade or business, that the best practicable means have been used for preventing or for counteracting the effects of the noise. 'Best practicable means' incorporates both technical and financial possibility. The latter may be related to the turnover of a company. Therefore a solution that may be the best practicable means for one company may not be so for another.

6. That the requirements imposed by the notice are more onerous than those for the time being in force in relation to the noise to which the notice relates of
   (a) Any notice under Section 60 or 66; or
   (b) Any consent given under Section 61 or 65; or
   (c) Any determination made under Section 67.
   Section 60 relates to a construction site notice. Section 61 is a consent for construction works. Sections 65–67 relate to noise-abatement zones (see below).
7. That the notice might lawfully have been served on some person instead of the appellant, being the person responsible for the noise.
8. That the notice might lawfully have been served on some person instead of, or in addition to, the appellant, being the owner or occupier of the premises from which the noise is emitted or would be emitted, and that it would have been equitable for it to have been so served.
9. That the notice might lawfully have been served on some person in addition to the appellant, being a person also responsible for the noise, and that it would have been equitable for it to have been so served.

25.4.4 The assessment of nuisance

In assessing whether a certain noise is a nuisance the local authority may make use of BS 4142, method of rating industrial noise affecting mixed residential and industrial areas. This standard forms the basis of local authorities' assessment of nuisance, and relates the background level (i.e. the noise that would pertain if not for the offending noise) to the measured noise. If the background cannot be found a method is given for deriving a notional background. However, the use of this notional background has fallen into disuse due to its unreliability.

The standard rates the offending noise according to its nature: 5 dB(A) is added where the noise has a definite continuous note and a further 5 dB(A) added for noise of an intermittent nature. The number of occasions that happen in an 8-hour period is then plotted on a graph and the correction for intermittency is derived. When these calculations have been performed, the noise level is compared to the background level. The standard states that where the noise exceeds the background by 5 dB or more, the nuisance is to be classed as marginal, and where the background is exceeded by 10 dB(A) or more, complaints are to be expected.

It is stressed in this standard that it is not intended to be a criterion but most local authorities tend to use it as a guide. Other methods are available or may have value for individual circumstances. Particular problems occur when a noise is of a very tonal nature or contains discernible information (i.e. music or voice). It is quite possible for music to become a nuisance at no more than 3 dB(A) above ambient background level.

By inference, if BS 4142 is used as a guide to nuisance, 5 dB(A) above background would not be cause for serving a notice. If this were permitted in areas where there were several sources of noise a 'creeping' ambient problem can occur. If this is the case, the local authority may have to control noise levels such that no increase in ambient is allowed. This may be particularly severe for a large developer in an already noisy area, but is necessary. If no addition is permitted the design noise level for the new development must be at least 10 dB(A) *below* background.

If there are tonal noise problems the local authority may use more complex measurements to specify the required reduction. Noise rating may be used or octave or one-third octave band levels specified.

25.4.5 Offences and higher court action

It is an offence to cause a noise nuisance while in breach of a notice. Proceedings in the magistrate's court can result in a fine of up to £2000 for each offence. It is also possible that the court may impose a daily penalty for continuing nuisances.

If the authority are of the opinion that magistrate's court action will not give an adequate remedy a complaint may be made to the High Court. This court will issue an injunction prohibiting the repeat of the nuisance. Non-compliance with an injunction constitutes contempt of court and penalties include imprisonment.

25.4.6 Noise-abatement zones

Local authorities are empowered by the Control Of Pollution Act 1974 to designate areas as noise-abatement zones. Within these areas noise levels are measured and entered into a register. It is an offence to increase noise levels beyond register levels unless a consent is obtained. If the local authority are of the opinion that existing noise levels are too high, noise-reduction notices can be served.

In the case of new premises the local authority will determine noise levels which it considers acceptable, and these will be entered into the noise level register. Appeals against notices or decisions can be made to the Secretary of State.

25.4.7 Planning application conditions

Local authorities are empowered to impose conditions on planning applications to protect environmental amenities of neighbours. Noise is commonly controlled by planning conditions, which are designed to avoid reduction in amenity of neighbours. This may mean that a process has to be almost inaudible (particularly in the case of light industrial consents). Appeals against planning conditions may be made to the Secretary of

State. Local authorities may ask for more onerous controls on planning conditions than the mere avoidance of nuisance.

25.5 Avoiding physical injury to workers

Noise levels between 85 and 120 dB(A) affect the hearing of exposed workers on a dose-related basis. The Leq of the noise is calculated and compared to the criterion currently employed. Under Section 2 of the Health and Safety at Work, etc. Act 1974 an employer has to take all steps, as far as is reasonably practicable, to ensure the health, safety and welfare of his workers. Permissible noise levels are defined by codes of practice, and the figure permitted for employee exposure is 90 dB(A) 8-hour Leq.

Low-frequency noise (in the range 3–50 Hz) may have other injurious effects on the body. Research has also indicated that a type of fatigue caused by low-frequency noise has a similar effect to that caused by alcohol. Infrasound (low-frequency sound) also has a synergistic effect with alcohol. Low-frequency noise is particularly important in the case of workers operating machinery (e.g. vehicles, cranes, etc.). It must also be remembered that very high power levels may be generated at low frequency and may not be readily detected by the ear. Attenuation of low-frequency noise is very difficult (see Section 25.7).

25.6 Avoidance of damage to plant/machinery/building structures

Plant and machinery may be directly affected by excessive vibration but may also be damaged due to operator fatigue (caused by working in high noise-level areas) or to operators not being able to hear unusual noises from machines before it is too late. Vibration may be caused either by direct transmission or by excitation. Damage to machines may occur in many ways, including the accidental impact of surfaces on one another and the fatigue of shafts, etc. caused by bending moments.

In extreme cases building structures can be damaged by vibration. Damage can be caused either directly to the structure or by settlement of foundations due to vibration. It is possible to calculate the effects of the former (graphs are obtainable from various sources) but not the settlement effect. However, this will occur only on certain strata and in extreme cases.

25.7 Noise-control engineering

Before attempting noise control it is important to consider the nature of the problem:

Source of sound energy → Transmission pathway → Receiver

There are many proprietary systems available for controlling noise and it is easy to become blinkered in one's approach to noise control.

The first consideration is the source of the noise itself. In the case of a new project the first option must be to select the quietest machine available that will perform the required task. With an existing problem this may still be the cheapest option, particularly in the case of a machine which is coming to the end of its useful life. Spin-offs may include more efficient operation (less fuel costs), greater reliability or quieter operating conditions.

If the decision is made to retain the original machine the design of the noise-control devices to be employed should be considered. These may affect:

Access to machine	(may hinder cleaning, tool changing, etc.)
Heat build-up	
Safety, etc.	

In nearly all cases noise control at source is the best option. Typical examples are:

1. *Presses*: Is the degree of impact necessary? Can it be adjusted? Can the press operate by pressure alone?
2. *Air discharge – use of air tools and nozzles*: The turbulence in the boundary layer of air between the rapidly moving airstream and the atmosphere is heard as noise. Can the air stream be diffused (silencers fitted to the exhaust)? Nozzles used for cleaning can have devices fitted which give a gradual transition from the rapidly moving air to atmosphere by the use of an annular ring of small nozzles round the central nozzle. These silence with very little loss in efficiency.
3. *Reciprocating compressors*: These cause very high noise levels at low frequency (typically, below 250 Hz). The low-frequency noises are very difficult to attenuate, and the most popular solution is to use rotary (vane type) compressors. These are inherently quieter and have the further advantage that the noise they generate is at high frequency (typically, above 1 kHz) and is therefore easy to attenuate.
4. *Cutting machines*: Modifications can be made to the method of restraining material being cut to reduce 'ringing' (i.e. reducing free lengths of material).

25.7.1 Noise-reduction principles

There are many noise-control devices. All, however, rely on one or more of the three basic noise-control principles: insulation, absorption and isolation.

25.7.2 Insulation

The simplest insulator is a sheet of material placed in the sound-transmission pathways. Sound energy reaches the surface in the form of a pressure wave. Some energy passes into the partition and the rest is reflected.

Energy that passes into a partition may be partially absorbed and transformed into heat. This is likely to be very small in a plain partition. The remainder of the energy will then pass through the partition by displacement of molecules and pass as sound in the same way that sound travels in air. This can then pass to the edge of the partition and be reradiated as sound from other elements of the structure. This is known as flanking transmission. In a thin partition by far the greatest amount of energy will pass through the partition by actually causing the partition to vibrate in sympathy with the incident sound and hence reradiating the sound onto the opposite side. The amount of sound transmission through a partition is represented by the ratio of the incident energy to the transmitted energy. This factor, when expressed as decibels, is known as the sound-reduction index (SRI):

$$\text{SRI} = 10 \log \frac{1}{\text{Transmission coefficient}} \text{ dB}$$

The moment of the panel (and hence its resistance to the passage of sound) is controlled by a number of factors:

1. The surface mass affects the inertia of panel. Greater mass causes a corresponding greater inertia and hence more resistance to movement. At high frequencies this becomes even more significant. The mass law can be expressed as

   $$\text{SRI} = 20 \log mf - 43 \text{ dB}$$

   where m is the superficial weight (kg/m^2) and f the frequency (Hz).
2. At very low frequencies the movement of the panel will be controlled by the stiffness, as inertia is a dynamic force and cannot come into effect until the panel has measurable velocity. Stiffness controls the performance of the panel at low frequencies until resonance occurs. As the driving frequency increases, the resonance zone is passed and we enter the mass-controlled area. The increase in the sound-reduction index with frequency is approximately linear at this point, and can be represented by Figure 25.8.
3. A panel will have a bending mode when a wave travels along the length of the sheet of material. The frequency of this bending mode is known as the critical frequency. This mode of bending will be introduced by sound incident at angles greater than 0°. At the critical frequency coincidence will only occur for a sound wave with a grazing incidence (90°). At greater frequencies the partition will still be driven, but in this case by progressively lower angles of incidence. The coincidence dip is therefore not a single dip but will result in a loss of SRI at progressively higher frequencies. The desirable insulaton panel will therefore be very large but not stiff.

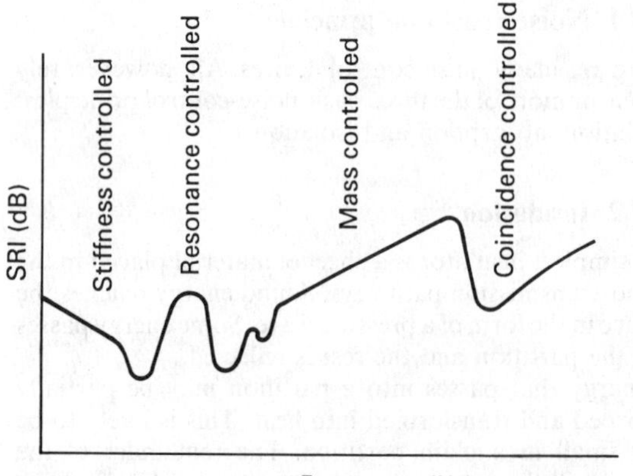

Figure 25.8 Typical insulation characteristics of a partition

25.7.3 Absorbers

Porous

As sound passes through a porous material, energy is lost by friction within the material. The material is usually attached to various surfaces in a room. The absorber will have the highest efficiency when positioned where the air molecules are moving the fastest (and hence more energy is absorbed). At the wall surface the molecules are stationary. If we plot a single-frequency graph we find that the maximum particulate velocity occurs at $\lambda/4$ (one quarter wavelength) from the surface. In practice, incident sound is rarely of single frequency, but the principle can be observed that the absorber must be one quarter of the wavelength away from the wall (for the frequency of the sound to be absorbed). This can be arranged either by having a thickness greater than $\lambda/4$ for the lowest frequency to be absorbed or by mounting the absorber on a frame some distance away from the wall so that the centre of the absorber is at $\lambda/4$ for the frequency to be absorbed.

Resonant

The simplest resonant absorber is known as the Helmholtz resonator. This consists of a chamber connected to the duct (or whatever area is to be controlled) by a narrow neck. The volume of air in the chamber will resonate at a frequency determined by the volume of the chamber, the length of the neck and the cross-sectional area of the neck:

$$F_{\text{res}} = 55 \sqrt{\frac{S}{IV}} \text{ (Hz)}$$

where S = cross-sectional area (m), I = length of neck (m) and V = volume of enclosure (m). As the chamber resonates, air is forced through the narrow neck and hence energy is absorbed in overcoming the resistance.

The degree of attenuation at the critical frequency can be very large, but this type of silencer has a very narrow bandwidth. This device may be suitable when the machine being dealt with emits sound predominantly of a single wavelength. The absorber bandwidth of a Helmholtz resonator can be expanded by lining the chamber with absorbers, but this has the effect of reducing the efficiency. The perforated absorber which forms the basis of many acoustic enclosures and silencers is a development of the resonator principle.

As stated previously, the bandwidth may be broadened by packing the chamber with an absorber, but this lowers efficiency. It may be overcome by using multiple absorbers in the sound path, and this can be done by placing a perforated sheet some distance away from the rigid outer wall of the enclosure and filling the cavity with absorber. It is not necessary to use cross walls between the 'chambers' so formed. In this case the equation becomes:

$$F_{\text{res}} = 5000 \frac{P}{I(t+0.8d)} \text{ Hz}$$

where I = depth of airspace, t = thickness of panel, d = diameter of holes and p = percentage open area of panel.

Panel

This absorber is basically a panel attached to a structural wall which is designed to absorb energy. The absorber is therefore frequency dependent and has an absorption peak at its resonant frequency. This type of absorber is not commonly used.

25.7.4 Vibration isolation

Vibration in machinery or plant can be induced in a number of ways, e.g.:

Out-of-balance forces on shafts;
Magnetic forces in electrical apparatus;
Frictional forces in sliding objects.

The first course of action in vibration isolation is reduction at source. This may be achieved by balancer shafts in engines, stiffer coils in electrical apparatus or better lubrication between adjacent sliding surfaces.

When all possible vibration reduction has been obtained, the machine must be isolated from the structure. This is achieved by some form of spring mounting. Spring mounts have a resonant frequency dependent on the stiffness of the spring and the weight of the object placed on it. It will be apparent that the

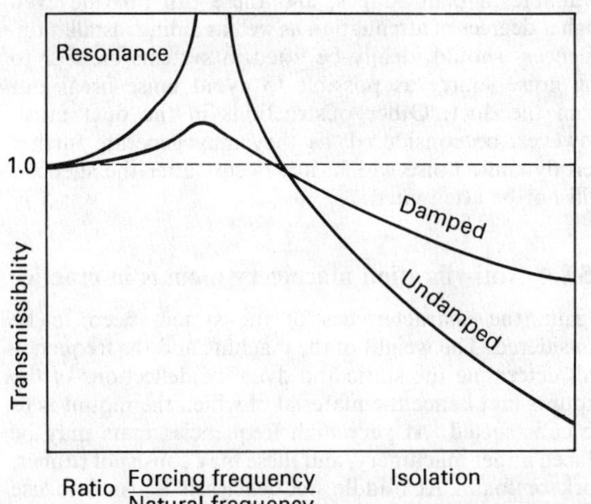

Figure 25.9 Performance of anti-vibration mounting

static deflection of the spring will also be proportional to the resonant frequency.

As the driving force of the mass/spring increases from zero up to the resonant frequency, the amount of transmission of the vibration increases until resonance is reached and the transmission becomes infinitely large. As the resonant point is passed, the transmission begins to reduce until at some point the transmissibility falls below one (see Figure 25.9), i.e. isolation occurs.

In practice, however, spring systems have some in-built damping, and this will have the effect of reducing the amplitude of the resonance below infinity. This is very necessary in real systems to avoid excessive excursions of mounted machinery. A damped mounting will follow the second curve on the graph in Figure 25.9, and it will be noted that the vibration isolation at high-frequency ratios is less than that for undamped systems. It is important therefore to use the lowest degree of damping necessary.

25.8 Practical applications

25.8.1 Acoustic enclosures

Panels of multi-resonator material are made from perforated plate sandwiched with solid plate and an intermediate absorber layer between them. These panels can be built up in enclosures, taking care to seal all junctions adequately. Typically, these enclosures are made to surround small machines (e.g. compressors). They may be fitted together with spring catches to allow for dismantling for maintenance purposes.

Ventilation may be a problem but can be dealt with in several ways:

1. Acoustic louvres constructed of the absorbent panel material (suitable for a small degree of noise reduction only);
2. A silencer fitted to the ventilation duct (see below);
3. Baffled enclosures to the ventilation duct.

25.8.2 Building insulation

Single-panel insulators have been described above, but in buildings it is usual to provide double insulation. In theory, if the insulation panels had no interconnection it should be possible arithmetically to add the sound reduction of the two elements of the structure. In practice, it will be found that there is bridging, by the structure, wall ties or flanking transmission or by the air between the two elements acting as a spring. If the two elements of the wall were in rigid connection the insulation would be 3 dB more than the single element alone (mass law), and if totally separated it would be the sum of the figures. In practice, a cavity wall with ties and a 50 mm cavity gives approximately 10 dB more reduction compared to a single-skin wall of half the surface mass.

Double-glazed windows work on the same principle. It is important to avoid the coincidence of the resonant frequencies of the two elements, and hence it is usual to arrange for the glazing panels to have different thicknesses (and hence a different resonant frequency). This is not necessary if one element is sub-divided by glazing bars to give different-size panes to its opposing element.

The reveals of a double-glazed window should be lined with acoustically absorbent material to damp the sound within the cavity. The width of the cavity should not be less than 150 mm.

If insulation panels are not of uniform construction (as in the case of a wall containing a window) the average sound-insulation value must be derived for use in calculations. The total transmission coefficient for the composite panel will equal the sum of the individual coefficient times their respective areas divided by the total area. Thus:

$$t_{\text{average}} = \frac{t_1 s_1 + t_2 s_2 + t_3 s_3}{s_{\text{tot}}} \text{ etc.}$$

and the SRI of the total panel is derived from

$$\text{SRI} = 10 \log \left(\frac{1}{t_{\text{av}}}\right) \text{dB}$$

25.8.3 Control of noise in ducts

Fans produce the least noise when operating at their maximum efficiency. It is therefore important to select the correct fan for the airflow and pressure characteristics required. It is also important to remember that the noise generated within the system (as opposed to at the fan) is a factor of air velocity, and hence, for a required airflow rate, a larger cross-section duct (with a correspondingly lower velocity) will give quieter results. It will also have other advantages when providing extra noise attenuation and fitting silencers.

In the design of systems it is most important to eliminate as much turbulence as possible, and to achieve this the fans should be mounted some distance away from bends (at least one and a half duct diameters). Junctions between pipes and connectors should present a smooth internal profile and inlets to systems must be tapered and not plain. Outlet grills should be of larger diameter than the duct and have aerodynamically smooth profiles where possible.

If it is necessary to add extra attenuation to a duct it is essential to decide on the required amount. If only a relatively small degree of absorption is required, first a part of the duct must be lined with absorber. The length of duct to be lined will be determined by the degree of attenuation required and the thickness by the noise frequency. Data for these factors are available from many sources and are usually published as tables.

For further attenuation it is necessary to provide a centre-pod-type attenuator (Figure 25.10). This increases the area of the absorber and also aids low-frequency attenuation. For further low-frequency attenuation an in-line splitter silencer is employed (Figure 25.11). These are capable of providing a high degree of attenuation, dependent on the width between the elements. The smaller the gap, the higher the attenuation. Again, performance tables are published by the major manufacturers. In order to decide on the design of the silencer to be installed it is necessary to know the required attenuation (and the frequency/noise level profile) and the permitted pressure loss in the system. Manufacturers' data can then be consulted. Splitter silencers are also available in bent shapes, and these can provide even higher degrees of attenuation as well as aiding installation. Silencers should ideally be fitted in systems as near to the noise source as possible to avoid noise break-out from the duct. Other obstructions in the duct must, however, be considered, as they may generate further aerodynamic noise which, if it occurs after the silencer, will not be attenuated.

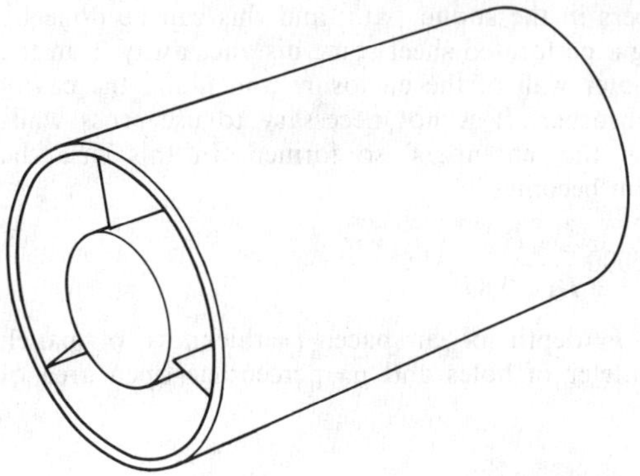

Figure 25.10 Centre-pod silencer

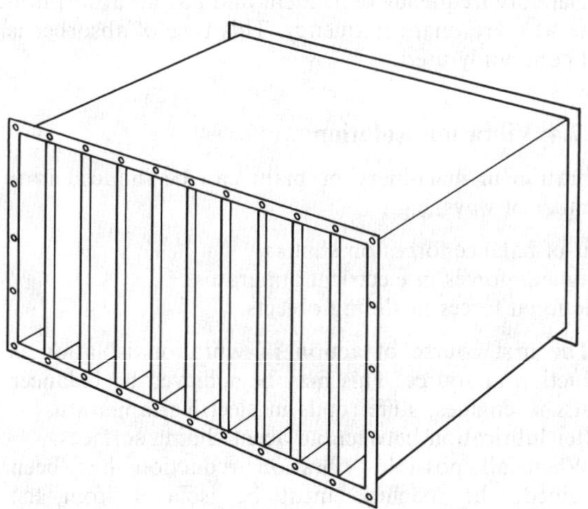

Figure 25.11 Splitter silencer

25.8.4 Anti-vibration machinery mounts in practice

Again, the characteristics of the system need to be considered. The weight of the machine and the frequency will determine the static and dynamic deflections of the mounts and hence the material of which the mount is to be constructed. At very high frequencies mats may be placed under machinery, and these may consist of rubber, cork or foam. At middle frequencies it is usual to use rubber in-shear mounts. At low frequencies metal spring mounts are employed.

25.8.5 Mounts

Anti-vibration mats are very useful for frequencies above 25 Hz. They have the disadvantage of being liable to attack by oils and, if they become saturated or deteriorated, they will compress and lose their efficiency.

25.8.6 Rubber mounts

Although these are loosely termed rubber mounts, they are often composed of synthetic rubbers which are not readily attacked by oils and can operate over a much wider temperature range. Typical maximum static deflections are 12.5 mm.

25.8.7 Steel spring mounts

Steel springs have the disadvantage of transmitting the high frequencies along the length of the spring. It is usual to mount the spring with a rubber or neoprene washer under its base. Steel spring mounts are also most vulnerable to resonance problems, and the solution is to build in a damper device. This has the disadvantage of reducing the isolator's efficiency.

25.8.8 Positioning of anti-vibration mounts

Machinery must be positioned so that all mounts are equally loaded, and failure to do so will result in the possibility of a rocking motion developing. This may require mounting the machine on a sub-frame. If this is not possible, the load should be assessed at each mounting point and mounts of different stiffness used.

25.8.9 Installation

Mounts should be installed so that the whole machine is isolated from the structure. Services (e.g. power, hydraulics, etc.) should also be mounted flexibly. Bridging is the most common fault when providing vibration isolation to machines and building structures, and should be carefully avoided. Services should be designed to withstand the degree of movement permitted by the anti-vibration mounts without suffering damage.

26 Air Pollution

Roger C Webster BSc, MIEH
Environmental Consultant

Contents

26.1 Introduction

A pollutant may be defined as any substance or condition which, when present in a sufficient concentration, may have a detrimental effect on the Earth's biosphere. Thus pollution may take place in the form of gases, vapours, particulates or temperature variations (these may be directly or indirectly caused).

The Health and Safety Executive publish exposure levels for many pollutants. It should be noted that these are in terms of short- or long-term exposure levels and are based on an average working week's exposure and an average working life. They do not directly relate to exposure outside the workplace.

It is an accepted practice when assessing the environmental effects of pollution on man and his place of abode to use a divisor of 40 (some agencies may divide by 30) against the long-term exposure level in the Health and Safety Executive's published Occupational Exposure list (known as EH40/90). Much lower exposure limits are necessary due to the much longer term of exposure in the domestic situation. The section of the population most likely to spend long periods of time in the home are those most susceptible to the detrimental effects of pollutants, i.e. the young, the elderly or the infirm. For short-term exposure the known data can be used directly from the list or from animal-exposure data.

26.2 Effects on plants, vegetation, materials and buildings

26.2.1 Vegetation

Plants have evolved and adapted to suit the atmospheric conditions in which they find themselves. This atmosphere may contain traces of gases which we would classify as pollutants. However, in many cases these may be necessary for the plants' existence. Near to industrial centres, the relative concentrations of the various gases change, and this can have an adverse effect on the plants' development.

The effects of 'acid rain' on vegetation has perhaps received the greatest publicity in recent times and forms an interesting case study. Numerous agencies have reported damage to forests caused by acid rainfall, but experiments have revealed that the direct impingement of acidic material on the plant cannot produce the effects as noted. Rather, it may be the result of the combination of other environmental factors, along with acid rainfall, which is responsible. It is known that calcium and magnesium are more readily leached out of the soil under acidic conditions and that this can affect aluminium levels in the plants (primarily pine forests), causing them to die. Unfortunately, these effects are very long term, and as the chemistry is most complex the problem is only just beginning to be understood, and it may already be too late to save many forests both in the UK and elsewhere. Acid rain also may cause heavy metals to enter man's food chain via fish or water supplies.

26.2.2 Materials and buildings

Acid rain erodes buildings, particularly those constructed from limestone. It has been reported that the Acropolis in Athens has suffered more deterioration in the last 20 years than in the previous 2000! Acidic gases are produced directly by the combination of oxides of sulphur and oxides of nitrogen with water and also by more complex processes involving unburnt hydrocarbons and ozone in the atmosphere.

It has been found that particulate pollution, while causing soiling of materials, may also be responsible for increasing corrosion levels (compared to the corrosion that would be caused by the same level of acid impingement alone) by a process of adsorption. Also, particulates can react synergistically with the acid deposition to cause much greater damage.

26.3 Effects on weather/environment

By far the most serious effect under consideration today is the 'greenhouse' effect. The combustion of fossil fuels has resulted in an increase in the carbon dioxide content of the atmosphere. Carbon dioxide has the effect of insulating the Earth against the loss of heat, and hence the mean temperature is in part dependent on carbon dioxide levels. It has been predicted that the carbon dioxide level may double in the next 50 years and produce a global warming of up to 3°C. Scientists currently disagree as to the rate of this warming, but all agree that the situation is very serious.

It may be expected to that, with this degree of warming, flooding will result as the polar ice-caps melt, and some coastal towns in the UK will be lost due to this process. Ozone in the atmosphere is also affected by pollution and has an effect that interreacts with that of carbon dioxide.

Ozone forms a layer around the Earth that insulates against thermal radiation, and this layer is being destroyed by pollutants (principally fluorocarbons). The effect of the depletion of the ozone layer is to warm the Earth (and hence exacerbate the greenhouse effect) and may also lead to an increase in the incidence of skin cancers.

26.4 Legislation on air pollution of concern to the plant engineer

26.4.1 Alkali and Works Regulation Act 1906

This Act has been much amended and now forms a 'relevant statutory provision' for the Health and Safety at Work, etc. Act 1974. Originally it provided for the registration and control of certain classes of chemical works. The Act has now been amended and the classes of premises, the 'scheduled works', are now included in the Health and Safety (Emission into Atmosphere) Regulations 1983 (amended 1989). These works are controlled by Her Majesty's Inspectorate of Pollution (HMIP). There are 62 main categories of works, and in some cases only certain processes within a category are covered:

Acetylene works	Large combustion works
Acrylates works	Large glass works
Aldehyde works	Large paper pulp works
Aluminium works	Lead works
Amines works	Lime works
Ammonia works	Magnesium works
Anhydride works	Manganese works
Arsenic works	Metal recovery works
Asbestos works	Mineral works
Benzene works	Nitrate and chloride or iron works
Beryllium works	Nitric acid works
Bisulphite works	Paraffin oil works
Bromine works	Petrochemical works
Cadmium works	Petroleum works
Carbon disulphide works	Phosphorus works
Carbonyl works	Picric acid works
Caustic soda works	Producer gas works
Cement works	Pyridine works
Ceramic works	Selenium works
Chemical fertilizer works	Smelting works
Chlorine works	Sulphate of ammonia works and chloride of ammonia works
Chromium works	Sulphide works
Copper works	Sulphuric acid (Class 1) works
Di-isocyanate works	Sulphuric acid (Class 11) works
Electricity works	Tar works and bitumen works
Fibre works	Uranium works
Fluorine works	Vinyl chloride works
Gas liquor works	Zinc works
Gas and coke works	
Hydrochloric acid works	
Hydrofluoric acid works	
Hydrogen cyanide works	
Incineration works	
Iron works and steel works	

26.4.2 The Clean Air Acts 1956 and 1968

The Clean Air Act 1956

These Acts control smoke from chimneys and open sites. 'Smoke' is not defined but includes soot, ash, etc. The emission of dark smoke (as dark or darker than shade 2 on the Ringleman chart) from chimneys and open sites is made an offence (exceptions are provided for short periods of time).

New furnaces have to be constructed (as far as is practicable) so as to operate smokelessly. Chimney heights are controlled (see below). Smoke Control Orders can be introduced (to control domestic smoke) and grants are available to convert fireplaces to burn authorized fuels. Smoke (other than dark smoke, which is already controlled) is dealt with by Section 16 of the 1956 Act and is, for the purposes of Part III of the 1936 Public Health Act, to be considered as a statutory nuisance.

Railway engines are dealt with in the same way as premises for the purposes of Section 1 of this Act (which controls dark smoke) except that the owner of the engine is to be held responsible. Other Sections of the Act do not apply.

Exemptions from the terms of the Clean Air Act are possible, on application to the local authority, for appliances used for the purposes of investigation and research.

New furnaces shall be fitted with plant to arrest grit and dust if they burn

1. Pulverized fuel; or
2. Solid fuel (or solid waste)

at a rate of 1 ton per hour or more. This plant should be approved by the local authority prior to installation (this Section is now replaced by the Clean Air Act 1968, Section 3).

In the case of large installations (burning pulverized fuel, or solid fuel at a rate of 100 lb or more per hour or liquid or gaseous fuel at a rate of 1.25 million BTUs or more) local authorities may serve notice to direct that Regulations shall apply concerning the measurement of grit and dust. These Regulations (the Clean Air (Measurement of Grit and Dust from Furnaces) Regulations 1971) require the owner of a furnace to make measurements and keep records of the grit and dust emissions from the furnace. Local authorities have the power to require information concerning furnaces and the fuel consumed to enable them to perform functions under this Act.

Chimney heights on new buildings are controlled by this Act. However, Section 10 was repealed by the 1968 Act (see below) as far as chimneys serving furnaces are concerned. This Section of the Act does not apply to domestic chimneys or shops or offices.

The Clean Air Act 1968

Section 1 of the Act prohibits dark smoke from trade premises (the 1956 Act only controlled smoke from chimneys). Bonfires are thus now included. Section 2 controls the rate of grit and dust emission from furnaces and the Minister may make Regulations. These are known as the Clean Air (Emission of Grit and Dust from Furnaces) Regulations 1971 and the Clean Air (Emission of Grit and Dust from Furnaces) (Scotland) Regulations 1971. A breach of these emission levels is an offence under Section 2(2) of this Act. Best practicable means may be used as a fence against this action, and the Regulations prescribe to which classes of appliance they apply. Schedule 1 furnaces are rated by heat output and are boilers or indirect heating appliances where the material heated is a gas or a liquid. The maximum continuous rating concerned are from 825 000 to 475 million BTUs.

Schedule 2 furnaces are rated by heat input and are indirect heating appliances or appliances in which the combustion gases are in contact with the material being heated (but the material does not contribute to the grit and dust). These are a heat input in the range 1.25 to 575 million BTUs. Tables of permissable grit and dust emission rates are given.

Section 3 Subject to Section 4 (which allows for exemptions to Section 3) any furnace to which Section 2 applies shall be fitted with grit and dust arrestment plant if burning

1. Pulverized fuel;

2. Solid fuel at a rate greater than 100 lb or more per hour;
3. Any liquid or gaseous matter at a rate of 1.25 million BTUs or more.

Section 6 of the 1956 Act shall not apply to furnaces if covered by this Section.

Approval is given by the local authority (in writing) to plants installed under this Section. It is an offence to use plant which does not have approved grit and dust arresters. If a local authority refuses approval, an appeal is to be made within 28 days of refusal of consent.

Section 4 – Exemptions The Minister may make Regulations to exempt furnaces of certain classes. These Regulations are known as the Clean Air (Arrestment Plant) (Exemption) Regulations 1969(and similarly for Scotland). The exemptions are:

1. *Mobile furnaces* while providing a temporary source of heat or power during building operations, or for the purpose of investigation or research, or for the purposes of agriculture;
2. *Furnaces* other than those burning solid fuel at a rate of greater than one ton per hour – of the following classes, used for any purpose other than the incineration of refuse:
   (a) Burning liquid or gases;
   (b) Hand-fired sectional burning solid matter at not more than 25 lb/h/ft^2 of grate surface;
   (c) Magazine-type grate furnaces burning less than 25 lb/h/ft^2 of grate surface;
   (d) Underfed stokers less than 25 lb/h/ft^2 plan area of stoker;
   (e) Chain grate stokers less than 25 lb/h/ft^2 of grate surface;
   (f) Coking stokers less than 25 lb/h/ft^2 of fire bars excluding area of the solid coking plate.

Under sub-Section 2 of the Act the local authority may grant exemptions on application provided that the furnace will not give rise to a nuisance or be prejudicial to health if permitted to operate without arrestment plant. The exemption must specify the purpose for which the furnace is permitted to be used.

Section 5 empowers the Minister to make Regulations to substitute new ratings for the requirement to measure grit and dust and fumes from furnaces under Section 7 of the 1956 Act. The Minister is not empowered to reduce the rate without approval of both Houses of Parliament.

Sub-Section 3 enables an owner to serve a notice on the local authority requiring them to take the measurements that he would be required to obtain under Section 7 of the 1956 Act. The ratings of such furnaces shall be less than one ton per hour of solid fuel other than pulverized fuel or liquid or gaseous fuelled furnaces with a rating of less than 28 million BTUs.

Section 6 This Section expands on Section 10 of the 1956 Act. Under this Section, the local authorities must approve chimney heights for all furnaces with ratings as per Section 3(1).

Section 10 of the principal Act shall cease to have effect as respect chimneys serving furnaces. This Section applies to all new furnaces and chimneys to furnaces which have had the combustion space increased and to installations which have replacement furnaces of greater ratings than that previously installed. The method specified in the Chimney Heights, Third Edition of the 1956 Clean Air Act Memorandum is used to determine chimney heights (see below).

Section 7 enables the Minister (Secretary of State for the Environment) to apply certain Sections of the Act to fumes by the making of Regulations. None have yet been made.

26.4.3 Relation of the Clean Air Acts to the Alkali Acts

Section 11 of the 1968 Act states that Sections 1–10 of the 1956 Act and Sections 1–16 of the 1968 Act shall not apply to works subjected to the Alkali Act (those premises now listed in the Health and Safety (Emission into Atmosphere Regulations)). These premises are therefore subject to enforcement by HMIP. However, sub-Section 3 of Section 11 does contain a proviso for the local authority, upon application to the Minister, to ask for an Order applying the Acts to the whole or part of the schedule works. If an Order is made, best practical means is applied to all (alkali) works whether or not provided for in the two Clean Air Acts.

HMIP publish sets of Guidance Notes which relate to activities they control. These notes are known as Best Practicable Means (BPM) Notes. BPM is explained in Note BPM 1/88, Best Practicable Means. General Principles and Practices. Copies of these notes may be obtained from the Department of the Environment Publications Sales Unit, Building 1, Victoria Road, South Ruislip, Middlesex HA4 0NZ, telephone: 081-841 3425.

It has been proposed that a set of scheduled works may be assigned to local authorities. This class of works has been discussed in documents published by the Department of the Environment Air Quality Division and legislation came into force on 1 April 1991. It is proposed that these schedule (B) works will be licensed in much the same way as the existing scheduled works and that prior consent will be needed before operations of this type commence. This will give local authorities a much stronger hand in pollution abatement, and they will be able to avoid the establishment of premises in unsuitable areas or without adequate pollution-abatement equipment. At present, local authorities rely on planning conditions or nuisance provisions.

26.4.4 Health and Safety at Work, etc. Act 1974

Section 5 of this Act places a duty on persons having control of premises to take the 'best practicable means' to prevent the emission into the atmosphere of noxious or offensive substances. This Section is used by HMIP in the enforcement of best practical means for the scheduled processes.

26.4.5 Control of Pollution Act 1974

Under this Act the local authorities may undertake research relevant to the problem of air pollution. The authority may obtain information concerning this subject by serving a notice requiring this information. If they so do, they should specify the information they require and the regularity of the measurements. The authority are not entitled to ask for returns of less than three-monthly intervals and they may not ask for information covering more than one year on any one notice.

If such a notice is received, an occupier may serve a counter-notice on the local authority requesting them to carry out the measurements themselves. The converse may also happen, in that the local authority may serve a notice which empowers them to enter premises to take measurements (after 21 days' notice). If this notice is received an owner can (by notice) require the local authority to allow him to carry out the measurements himself.

26.4.6 Chimney height calculations: Third Edition of the 1956 Clean Air Act Memorandum

The Memorandum is published to help local authorities and industry to derive an approximate height for chimneys. It is stated that the Memorandum should not be used as an absolute criterion but merely as a guide. Valleys, hills and other geographical features will modify the height as derived from the Memorandum.

The Memorandum covers furnaces with a gross heat input in the range of 0.15–150 MW. (The 1968 Clean Air Act, Section 6, requires chimney height approvals for furnaces for burning fuel at a rate greater than 1.25 million BTUs per hour. This is equivalent to 0.375 MW.)

The method is designed to ensure the adequate dispersal of sulphur dioxide (and hence a different method is used for very low-sulphur fuels). It assumes certain efflux velocities, i.e.

For boilers up to 2.2 MW – 6 m/s (full load)
For boilers up to 9 MW – 7.5 m/s (full load)
For boilers where the rating is greater than 135 MW – 15 m/s
Between 9 and 135 MW – pro rata.

The Memorandum does accept that the figure of 6 m/s may be difficult to achieve with small installations.

Very low-sulphur fuels

The first stage of the calculation is the uncorrected chimney height. This is obtained from

$$U=1.36Q^{0.6}[1-(4.7\times10^{-5}Q^{1.69})]$$

(For heat inputs less than 30 MW the part of the equation in brackets may be omitted.) Q may be obtained from

$$Q=\frac{WB}{3600}$$

where Q is the heat input (MW), W is the maximum rate of combustion of fuel (kg/h for mass or m^3/h for volume) and B is the gross calorific value (MJ/kg or MJ/m^3).

Other fuels

The rate of emission of sulphur dioxide is first calculated from

Oil firing: $R=0.020WS$
Coal firing: $R=0.018WS$

where R is the rate of sulphur dioxide (kg/h), W is the maximum rate at which fuel is burned (kg/h) and S is the sulphur content of the fuel (%). The equations are different, as it is assumed that in the case of coal firing a certain amount of sulphur will be retained in the ash.

The area in which the chimney is situated is then considered. The classes are:

A: underdeveloped area where development is unlikely
B: partially developed area with scattered houses
C: a built-up residential area
D: an urban area of mixed industrial and residential development
E: a large city or an urban area of mixed heavy industrial and dense residential development.

The uncorrected chimney height is then calculated by plotting the emission rate and the classification on a graph supplied in the Memorandum or by the multiplication of factors supplied:

Type of district	*Factor*
A	0.55
B	0.78
C	1.00
D	1.30
E	1.60

The uncorrected chimney height (U) is calculated from the sulphur dioxide emission rate (R_a) as previously:

If R_a is less than 10 kg/h: $U=6R_a^{0.5}$
If R_a is from 10 to 100 kg: $U=12R_a^{0.2}$
If R_a is from 100 to 800 kg/h: $U=5R_a^{0.5}-0.9R_a^{0.67}$

Corrected chimney heights

A procedure is given to take account of tall buildings near to the stack or if the building to which the stack is attached is less than $2.5U$. The Memorandum suggests that $5U$ should be used as a radius around the chimney as the definition of 'near', but in practice, interpretation and local knowledge may expand this circle.

In the simplest case with one tall building near to the stack (wider than it is tall) the simple equation is

$$C=H+0.6U$$

where C=corrected chimney height and H=building height. In other cases (with more than one building, for instance) a more complicated procedure is necessary.

All buildings within $5U$ of the stack should be measured and the height and width recorded. The following factors are then determined:

H_m – the largest value of the building height
K – for each building, the lesser of building height or building width

T – for each building ($T = H + 1.5$ K)
T_m – the largest value of T

If U is greater than T_m then U becomes the corrected chimney height. If not, calculate C from

$$C = H_m + U(1 - H_m/T_m)$$

When C is derived certain overriding factors are applied:

1. A chimney should not be less than 3 m from any area where there is access (i.e. ground level, roof areas or openable windows).
2. C should never be less than U.
3. A chimney should never be lower than the attached building within $5U$.

Fan dilution

This is also known as air dilution, and is a common procedure for gas-fired boilers. It is allowed for any very low-sulphur fuels (less than 0.04% sulphur) or for fuels in the range 0.04–0.2%S where the VLS fuels chimney heights calculation gives a lower value of U than the 'other fuels' calculations. In these cases, the flue may terminate at the uncorrected height U so long as:

1. The flue gas is diluted with air at a ratio that is known as F, i.e. V/V_0, where V = actual flue gas volume and V_0 = the stoichiometric combustion volume.
2. The actual value of F is a compromise and may be in the order of 20-1. F is used to determine other factors.
3. The emission velocity must be at least $75/F$ m/s.
4. The outlet must not be within $50U/F$ of a fan-assisted air intake (except for intakes of combustion and/or dilution air for the boiler in question).
5. The outlet must not be within $20U/F$ of an openable window on the emitting building.
6. The distance to the nearest building must be greater than $60U/F$.
7. The lower edge of the outlet should be at least 2 m high for boilers related below 1 MW or 3 m for others.
8. The outlet should be directed at an angle above the horizon (preferably about 30°) and must not be under a canopy, or emit into an enclosed wall or courtyard.

The measurement of 'smoke': the Ringleman chart

This chart was devised in the nineteenth century and consists of a white card with black cross hatching. The percentage of hatching increases by divisions of 20% for each Ringleman number. Therefore 0 = white, 1 = 20% obscuration, 2 = 40%, 3 = 60%, 4 = 80% and 5 = black.

When placed at a distance from the observer's eye the cross hatch lines merge and appear as a uniform shade of grey. There is a standard size for this chart (BS 2742C: 1957), and this has the five shades from 0 to 4. The chart is usually viewed from a distance of 15 m and hence has to be mounted on a tripod which should not shadow the card. The chart should be set up in such a position that the smoke being measured has the same sky background as the chart. It should not be placed so that the sun is either directly behind the chart or directly in front of it. Comparisons can then be made between the shade of the smoke and that on the card.

It will be noted that the chart may be impractical to use on a regular basis and thus two other Ringleman charts are used:

1. *The miniature smoke chart*: This does not have a cross-hatched lines but rather shades of grey which correspond to the numbers. The chart can be viewed from approximately 1.5 m.
2. *The micro Ringleman*: This is a photographically reduced Ringleman (i.e. with cross hatching). It has a slot in the middle through which the smoke can be viewed at arm's length.

Instruments have been developed through which the smoke can be viewed and a shaded filter superposed on part of the image. By far the most common of these is the telesmoke. This device can be carried in the Inspector's pocket and most local authorities possess at least one of these.

While the use of a Ringleman chart is to be encouraged, it is not necessary to show, in any legal action, that a chart was used as a reference. Section 34(2) of the 1956 Act states 'for the avoidance of doubt, it is thereby declared that, in proceedings brought under or by virtue of Section 1 or Section 16 of this Act, the Court may be satisfied that smoke is or is not dark smoke as herein before defined [i.e. as dark or darker than Shade 2] notwithstanding that there has been no actual comparison thereof with a chart of the said type'. This would also apply to the 1968 Act as the expression 'dark smoke' is referred back to the 1956 Act. Interestingly, the expression 'black smoke' is not defined in either Act; it is only described in the Dark Smoke (Permitted Periods) Regulations 1958, where it is defined as 'as dark or darker than Shade 4'. Section 34(2) of the 1956 Act would not therefore apply and, strictly, reference to a Ringleman chart should be necessary for any court action to succeed.

Definition of the terms fume, smoke, dust and grit

The most common air-pollution descripter is the expression 'smoke'. For the purposes of the 1956 and 1968 Acts, smoke includes soot, ash, grit and gritty particles emitted in smoke. Smoke is intended to mean the visible products of combustion and not the invisible ones (CO_2, SO_2, NO_x, etc.) and is used to indicate the degree of completeness of combustion (if combustion is 100% no 'smoke' is produced).

Smoke is taken as having a particle size of less than one μm. Dust consists of particles 1–76 μm in diameter. Grit can be interpreted as particles larger than dust. These definitions were taken from the Beaver Report of November 1954 which formed the basis for the 1956 Act.

Fumes are defined in the 1956 Act as being smaller than dust and thus are a constituent of smoke if visible and smaller than 76 μm. However, fumes are traditionally classified as being smaller than 1 μm (see Table 26.1).

Table 26.1 Definition of terms

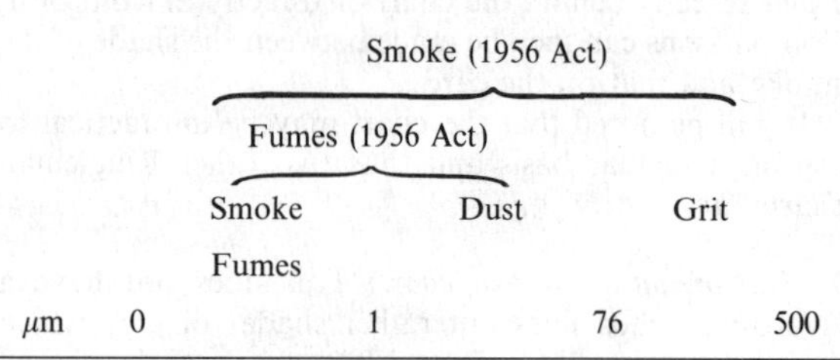

Definition of the term nuisance

Statutory nuisance While certain occurrences are declared to constitute statutory nuisances, there has never been a formal statutory definition of the term 'nuisance'. The term has been in use since the thirteenth century, and no doubt the legislators felt that there was no need to explain a well-understood concept.

'Nuisance' covers a great variety of areas but it must involve activities which interfere with the activities of one's neighbours or denies the public their rights. It can involve the blocking of highways or allowing one's animals to escape onto neighbouring premises. The two main types of nuisance are private and public.

Private nuisance Private nuisance is related to the ownership of land and has been defined as 'where a person is unlawfully annoyed, prejudiced or disturbed in the enjoyment of their land, whether by physical damage to the land or by other interference with the enjoyment of land or with his health, comfort or convenience as occupier'. The degree of disturbance must be examined. If physical damage to structures or to health can be shown, nuisance is likely to be proven. To interfere with comfort, substantial interference must occur. The amount of occurrences is important, and case law has stated that a single occurrence does not constitute a nuisance.

Also, the area under question must be considered. Air-pollution levels on a general industrial estate may be considered acceptable, but those same levels may not be so in a residential area. In one case (*Sturgess* v. *Bridgam* (1879)) the judge stated that 'what would be a nuisance in Belgrave Square would not necessarily be so in Bermondsey'.

Sensitivity on the part of the recipient is not considered, even though actual damage may result. Another legal quote is often used: 'Ought this inconvenience to be considered, not merely according to elegant or dainty modes of habit or living, but according to plain and simple and sober notions amongst English people?' (*Walter* v. *Selfe* (1851)).

Prescriptive right is also important in private nuisance The law of prescription states that if things are done to the knowledge of the occupier of land (likely to be affected) for 20 years a right may exist. It is important to note that the person 'suffering' the nuisance must be aware of the nuisance. In *Sturgess* v. *Bridgam* (1879) this was contested. A doctor had built a consulting room at the end of his garden and was affected by the noise of machinery from his neighbours' house. However, while his neighbour claimed prescriptive right as the noise had been made for more than 20 years, this was refused. As the doctor had not been aware of the nuisance, the right could not exist.

Thus if new houses are built near to a factory the owners of the houses will still have a right to complain about any nuisance caused.

Public nuisance Public nuisance is the interference with the lawful activities of Her Majesty's subjects or a substantial section of them. One person cannot suffer a public nuisance. Public nuisance is a crime and is actionable by the Attorney General or (under Section 2.2.2 of the Local Government Act 1972) by the local authority. There is no prescriptive right to commit a public nuisance.

Air pollution other than smoke

EC Directive levels for SO_2, NO_x and lead have recently been incorporated into UK legislation in the form of the Air Quality Standards Regulations 1989. SO_2 is generated when high-sulphur fossile fuels are burned. It is thus a factor of the amount of coal and heavy oil (predominantly) being burned.

The recent tendency towards burning more gas and light oils has tended to reduce the industrial emission of SO_2, although power stations remain the major source. European pressure is leading to the reduction of SO_2 from large plant. Existing power stations are to be fitted with flue-gas desulphurization plant and other new plants are likely to use low-SO_2 technology (e.g. pressurized fluidized bed, etc.). Flue-gas desulphurization is unlikely to be a viable proposition for medium-sized plant. It is possible that the new regulation limits may be approached in certain parts of the UK and if this happens, plant design will have to take account of permitted sulphur-emission figures. Designers of any medium-sized plant should consult the local authority at an early stage (before chimney heights application) to discuss this possibility.

NO_x levels are particularly disturbing in some large towns and cities in the UK and are increasing. Whereas SO_2 is generated from sulphur contained within the fuel, NO_x is primarily a combination of atmospheric nitrogen and oxygen, and is generated at a greater level in high-temperature combustion conditions. As the efficiency of furnaces/engines increases, so does the rate of NO_x emission. Levels (other than at roadsides) are less likely to approach regulation levels currently, but are on the increase. A major source of NO_x generation is the internal combustion engine.

NO_x measurement is much more difficult than that of SO_2 and levels seem to vary considerably, depending on the position of the measuring apparatus. SO_2 measurements are much more consistent.

Gas turbines and power stations are particularly prone to generate NO_x and the search for the 'low-NO_x' burner that will operate at high efficiency (i.e. with low hydrocarbon emissions) continues. The principle of the low-NO_x burner is to slow the rate of combustion by dividing it into several stages by the gradual mixing of the combustion gases with the stoichimetric air volume.

Steam or water injection may also be used, and under some circumstances this can reduce NO_x emissions without lowering plant efficiency. Such injection seems to operate in two ways. First, it cools the combustion and hence slows its rate (therefore acting in the same manner as a low-NO_x burner) and, second, it converts NO_x into nitric acid, which is emitted from the chimney as acidic vapour.

Atmospheric dispersion theory

Fumes and vapours discharged to the environment via a chimney form a plume which is approximately cone shaped. Mathematical modelling of dispersal rates is possible. The Gaussian dispersion model is commonly used to calculate the concentration of pollutants at coordinate positions *X*, *Y* and *Z*. (The coordinates are measured from the plume centre line.) The equation used is:

$$C_{(X,Y,Z)}=\frac{Q}{2\pi\sigma_y\sigma_z u}\exp\left[-\frac{1}{2}\left(\frac{y}{y_0}\right)\right]^2 \times\left\{\exp\left[-\frac{1}{2}\left(\frac{2-H_e}{\sigma_z}\right)^2+\exp\left[-\frac{1}{2}\left(\frac{2+H_e}{z}\right)^2\right]\right]\right\}$$

where

C = concentration at points x, y, z (μg/m),
Q = pollutant emission rate,
u = mean wind speed affecting plume (m),
z = standard deviation of plume concentration in the vertical at distance X (m),
y = standard deviation of plume concentration in the horizontal at distance X (m),
X, Y and Z are the coordinates with the base of the stack as the origin,
H_e is the effective stack height (see below).

The equation simplifies considerably when we consider only ground-level concentrations directly in line with the plume, i.e. X and Z are equal to zero:

$$C_{(X)}=\frac{Q}{\pi\sigma_y\sigma_z u}\exp\left[-\frac{1}{2}\left(\frac{H_e}{\sigma_z}\right)^2\right]$$

Effective height (H_e)

The amount by which a plume rises above the top of a chimney can be derived mathematically.

Hot buoyant plume

$\Delta H=20.5Q_h^{0.6}H_s^{0.4/u}$

where

Q_h = rate of sensible heat emission from the chimney (MW),
u = wind speed at the top of the stack,
H_s = actual chimney height,
$H_e=H_s+H$

Table 26.2 Pasquill's stability categories

	Day			Night	
	Incoming solar radiation				
	Strong	*Moderate*	*Slight*	*<3/8 Cloud*	*>4/8 Low cloud*
<2	A	A–B	B	G	–
2–3	A–B	B	C	F	E
3–5	B	B–C	C	E	D
5–6	C	C–D	D	D	D
>6	C	D	D	D	D

A and B are the most unstable, G is the most stable and D is neutral.

Cold plume

$\Delta H=3W/ud$ if $W/u>$ or $=4$

where

W = efflux velocity of plume at chimney top (m/s)
d = internal diameter of top of chimney (m).

If $W/u<4$, plume rise above top of chimney should be ignored. Again, $H_e=H_s+\Delta H$.

Atmospheric stability

Atmospheric stability and mechanical turbulence (important near to the ground) are used to derive the vertical and horizontal dispersion coefficients. Table 26.2 shows Pasquill's stability categories used to derive the coefficients by reference to standard graphs.

The dispersal of the plume at X, Y and Z is determined by the values of σ_y and σ_z. Small-scale eddies can affect dispersal near to the source and larger-scale ones are needed before effects are noted at greater distances from the source. Y and Z thus have large orders of magnitude furthest from the source and increase if a larger time period is used for sampling. It is thus important to state the sampling period used. The trend towards changes in σ_y and σ_z are thus

Near to stack – ground turbulence dominates (coefficients much greater in urban areas);
Far distance – coefficients much smaller with stable conditions.

Other models (or combinations of them) are often employed when computers are used to analyse dispersal. These can give an acceptable degree of accuracy when combined with detailed weather data. Short-exposure modelling is the most difficult and is liable to the greatest degree of error. It is for this reason that such models are not accurate when dealing with odour nuisances. The problem of modelling odour dispersal is dealt with below.

Other special atmospheric conditions can interfere with the modelling process and the most common of these is temperature inversion. This condition is so called because the air temperature increases with height above the ground, the converse of the situation that pertains for most of the time.

Temperature inversions

There are two common types of inversion as follows.

Ground base inversions On clear nights when there is strong radioactive cooling of the ground, the inversion starts at ground level and extends upwards to 100 m or more.

Elevated inversions Elevated inversions begin at some height above the ground. Inversions affect flue gas dispersal in the following ways:

1. *Ground based*: Upward dispersal is very slow. Horizontal dispersal proceeds normally. For low chimneys, ground-based concentrations can be high, especially when the ground is heated in the mornings and eddies are caused which bring down plume gases.
2. *Elevated*: Plumes can be trapped either above or below the base of the inversion and held in a horizontal plane. Again, these can be brought down to ground level by eddies. This process is known as 'fumigation' and can result in short-term high-level concentrations.

Odour dispersal

Odours may be detected for a very short exposure period, perhaps less than one second. It is thus necessary to determine the likely one-second peaks knowing the concentrations derived from the Gaussian equation. This is based on 10-minute average period. The equation to convert the time-averaging period is:

$$C_t = C_{10}(t_0/t)^p$$

where $p = 0.17$–0.2. Therefore in this case

$$C_t = C_{10}(600/1)^{0.17} = C_{10} \times 2.96$$

We must thus multiply the Gaussian concentrations by approximately three to obtain the short-term peaks. Other sources have suggested that the multiplication factor may be higher.

If the chimney was designed to avoid ground-level concentrations exceeding the nuisance level based on the Gaussian equation we would find short-term peaks that would produce nuisance. To avoid nuisance for a 5:1 ratio of 10 minutes to peak exposure we would need to double the chimney height from that derived for the Gaussian equation.

Nuisance levels for odours are not absolute, but are related to the minimum detectable level for 50% of the population. These levels have been explored by Warren Spring Laboratories, who have concluded that five times the minimum detectable level is likely to give rise to complaint.

The maximum ground-level concentration calculated on a 10-minute basis for no nuisance should therefore be equal to the detection threshold, i.e.

Peak $= 5 \times$ 10-minute level
Nuisance $= 5 \times$ detection level

Therefore for no nuisance, 10-minute level = detection level.

Note that the ground-level concentrations do not depend on the flue gas concentrations but rather on the amount of pollutant emitted. It is therefore not worth diluting the flue gases with fresh air (other than to raise the efflux velocity).

27 Dust and Fume Control

Brian Auger IEng, FIPlantE, MBES
J. B. Auger (Midlands) Ltd

Contents

27.1 Introduction

Dusts and fumes have been a part of industrial life for many years and the hazards associated with them are well known. The diseases and respiratory disorders found in foundries, potteries and cotton works are examples which are familiar to many. These need not occur with a better understanding of control measures and more efficient equipment.

Attitudes to dusts and fumes have changed over the years. It is no longer either permissible or tolerated for workers to be subjected to unhealthy or unsafe conditions at work. Society as a whole is demanding a cleaner environment, and this has led to new legislation being enacted. The effect of these changes on the plant engineer have resulted in a great increase in the burden of work and in the content and level of knowledge of the subject. He now has to take account of the workings and activities within the plant and its effect on people and surroundings outside it. To ensure that dust and fume control is effective in meeting its objectives it is imperative that those responsible for the specification and running of the plant are well versed in the nature of dusts and fumes, methods of entrainment, available equipment and test and inspection techniques.

27.2 The nature of dusts and fumes

Dust can be said to be a solid broken down into powder, and the form that it takes will have different effects on the body. Fibrous dusts can attack tissue directly while others may be composed of poisons which are absorbed into the bloodstream. For the purpose of this chapter, fumes can be regarded as very small particles resulting from the chemical reaction or condensation of vapour, which can have the same two effects. It is not necessary for plant engineers to have an in-depth knowledge of pathology, only that they must be aware of the possible results of exposure to dusts and fumes.

The standard unit normally used for measuring dust particles is the micron (μm: one-thousandth of a millimetre). The smallest particle visible to the unaided eye is between 50 and 100 μm and the most dangerous sizes are between 0.2 and 5 μm. Particles larger than this are usually unable to penetrate the lung defences and smaller ones settle out too slowly. Some dusts can be both toxic and fibrous (e.g. asbestos) and are therefore harmful even outside these parameters. It may therefore be assumed that dusts which are visible (i.e. between 50 and 100 μm) are quite safe. However, this is not the case, as dust clouds never consist solely of particles of one size. Analysis would show percentages of all sizes, and it is for this reason that special care is needed in measuring dust clouds and concentrations.

The next most important factor to consider when assessing dust clouds is the actual amount of dust present. This is known as the concentration, and is defined as follows. A substantial concentration of dust should be taken as concentration of 10 mg/m^3 8-hour time-weighted average of total inhalable dust or 5 mg/m^3 8-hour time-weighted average of respirable dust where there is no indication of the need of a lower value. There are now many lists available for consultation which set out the safety limits as they are known at present. If a substance is not listed this is not always an indication that it is safe, and the general rule should be applied that dust in any substantial concentration is hazardous. Even with dusts and fumes that are listed and have set limits it must be borne in mind that our knowledge is always growing, and that the standards of today may be obsolete tomorrow.

Clearly, the assessment of dust clouds and concentrations cannot be left to the casual practitioner, and this has now become the specialist field of the industrial hygienist.

27.3 Control of dusts and fumes

The purpose of the control plant is to maintain a working environment that is acceptable in terms of any statutory regulations and the custom and practice within an industry. The effectiveness of a control system is measured by the amount of dust or fume it controls. Efficiency, on the other hand, is measured by the amount of power it takes to do the work. It is the job of the dust-control engineer to produce the most effective plant in the most efficient way, and the techniques of control will vary from one industry to another. All control plants will have either four or five elements, as shown in Figure 27.1, i.e.

Hoods
Ducting
Fan
Collector
Disposal

The collector is not always used, as many systems still discharge untreated air to atmosphere. The growing awareness of environmental matters will, in time, see all such systems having collectors or treatment plants.

27.3.1 Hoods

This is the inlet into the system, and will be the single most important element in determining the effectiveness of the control plant. A study of the dust- or fume-producing process is necessary to ensure that the twin aims of effectiveness and efficiency are met. Hoods that totally enclose the process for maximum effectiveness may, however, prevent the operator from carrying out the process for which the control was needed in the first place.

There are four rules for the design of effective hoods, irrespective of the process:

1. The hood should be as close to the source of generation as possible.
2. The location and shape of the hood is such that the contaminant is thrown into the hood.

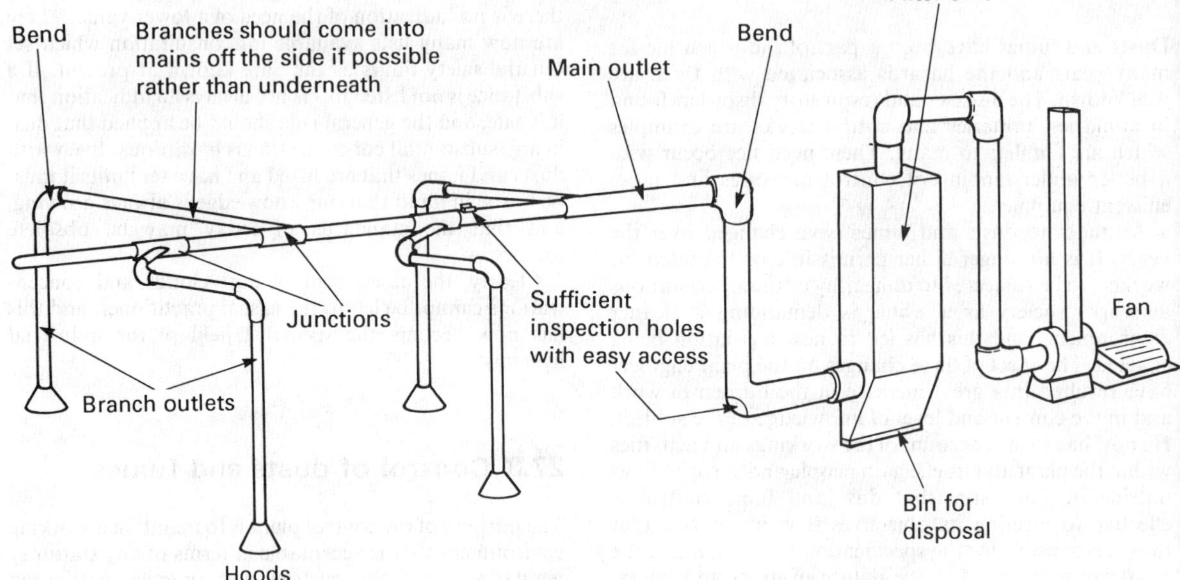

Figure 27.1 Horizontal duct with adequate carrying velocity. Ducting of increasing size to accommodate increasing air flow

3. The air flowing past the source and into the hood has a velocity at the origin greater than the velocity of escape of the contaminant.
4. The hood is located so that the operator is never between the source and the hood.

With these rules four types of hoods have evolved.

Total enclosures

These ensure that the process is totally enclosed, which prevents any leakage to the workplace. It is rarely used except for very dangerous materials such as radioactive particles in the nuclear industry or biological particles in the drugs industry. The enclosure is subject to negative pressure, and because the source is inside, any work done on the source must be carried out by manipulators. A modification of the total enclosure is one where part of the enclosure is removed for manipulation of loading. In this case the velocities across the open faces of the hood must be sufficiently high to prevent emissions or escape. This technique is used on bucket elevators, conveyors and holding bins.

Booths

These are really enclosures with one whole side removed where the source is deep within the booth. They are particularly suitable where the particles are not moving at high speed (e.g. filling and weighing operations). When the booth is used where high-speed particle generation occurs (e.g. grinding and fettling) careful thought must be given to the depth. As booths tend to be large, the restraining velocities across the open face must only be sufficient for control, otherwise the efficiency of the plant would be low with very high velocities and hence volumes.

A development of the simple booth is the laminar flow system, which uses a nominal velocity at the face of 0.5 m/s. As the air flow is laminar, all parts of the booth are subjected to its effect.

Captor hoods

In many processes it is not possible to use enclosures or booths without imposing unacceptable operational restraints (see Figure 27.2). The hood must be placed at some distance from the source. The natural projection of the particles will not necessarily be into the hood, and

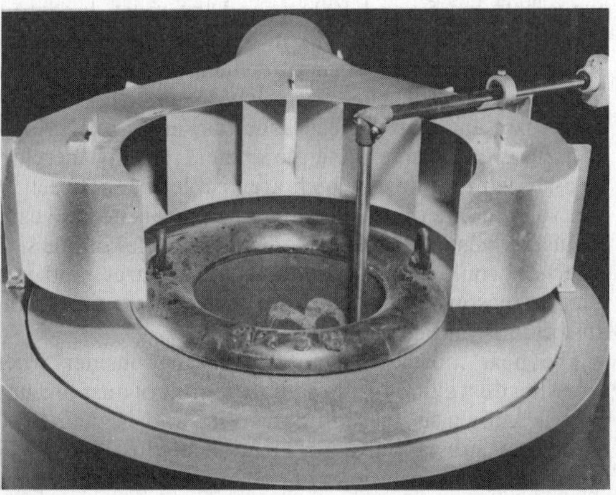

Figure 27.2 Captor-type claw hood for a furnace

may indeed be in the opposite direction. To be effective, these hoods require high face velocities to give control (e.g. up to 10 m/s). Hoods for controlling fumes fall into this category, especially on tanks and vats.

Receptor hoods

These derive their name from the method of entrainment. The hood is placed in the path of the particle and uses the momentum of the particle to assist in control. It is important to have a clear understanding of the process, the direction particles take and the movement of air created by the process. Grinding wheels release dust downwards and over the top of the wheels, and considerable air movement is generated by the process. These are ideally suited to control by receptor hoods (see Figure 27.3).

There are many machines which arrive on the factory floor with correctly designed and tested hoods supplied by the machine manufacturers. This is a trend that has grown over the years, and is one to be encouraged. Hoods supplied in this way can have the volume specified, which, in turn, will ensure that the control meets with the appropriate regulations.

27.3.2 Ducting

The purpose of the ducting is to convey the entrained contaminants away from the sources to a collection or disposal point. It is very often a neglected part of the system. Different processes require different specifications, and the following will act as a guide in selecting which type of ducting is necessary:

1. The gauge of ducting should take account of the nature of the particles and the operating pressures of the system.
2. The methods of construction usually encountered are (a) lockformed, (b) welded, (c) slip jointed and (d) flanged.
3. The construction materials should be selected to withstand the operating conditions and the condition of the pollutants. Galvanized sheet steel and black mild steel are the most common for general work. Corrosion or heat applications will have ducting constructed in stainless steel or plastic.
4. The standard gauges (mm) for varous work are:

General dust work		*Light-duty work*
0–450	1.0	0.8
Bends	1.5	0.8
450 and over	1.5	1.0
Bends	2.0	1.0
All ducts over 450 flanged		
Foundry work		*Oily and wet fumes*
0–450	1.5	1.0
Bends	2.0	1.0
450 and over	2.0	1.5
Bends	3.0	1.5
All ducts over 450 flanged		

5. The design of the ductwork must ensure that the plant is both effective and efficient. Sharp bends and abrupt entries of branches into mains cause unnecessary pressure losses. Incorrectly sized ducts result in high pressure losses or blockages due to fallout from velocities being too low.

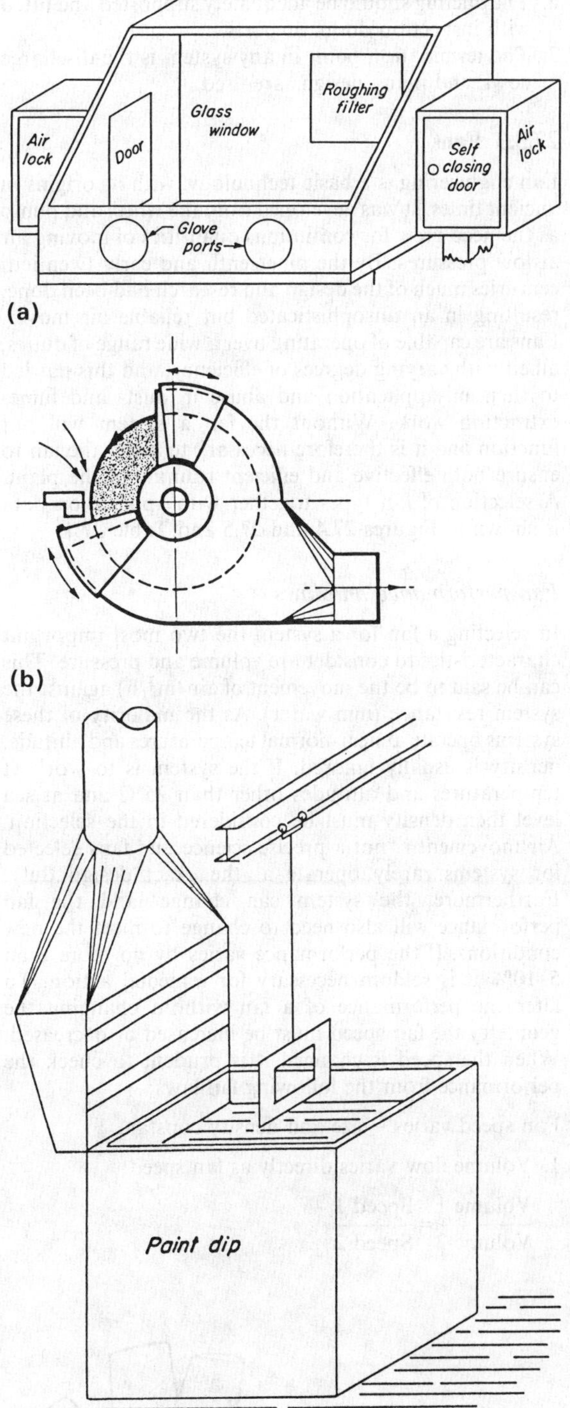

Figure 27.3 Types of hoods. (a) Enclosure; (b) receptor; (c) captor

6. The ducting should be adequately supported and fitted with inspection doors or ports.
7. The termination point in any system is the discharge cowl, and many designs are used.

27.3.3 Fans

Fan engineering is a basic technology, with its origins in ancient times. It was developed from the wheel and pump as the need grew for continuous quantities of moving air at low pressures. By the nineteenth and early twentieth centuries much of the design and research had been done, resulting in an unsophisticated but reliable air mover. Fans are capable of operating over a wide range of duties, albeit with varying degrees of efficiency, and this has led to their misapplication and abuse in dust- and fume-extraction work. Without the fan a system will not function and it is therefore necessary to select the fan to ensure both effective and efficient running of the plant. A selection of fan types together with application data is shown in Figures 27.4 and 27.5 and Table 27.1.

Fan performance and laws

In selecting a fan for a system the two most important characteristics to consider are volume and pressure. This can be said to be the movement of air (m^3/h) against the system resistance (mm water). As the majority of these systems operate at near-normal temperatures and altitude, density is usually ignored. If the system is to work at temperatures and altitudes other than 25°C and at sea level then density must be considered in the selection. Air movement is not a precise science and fans selected for systems rarely operate at the exact design duty. Furthermore, the system can change, and the fan performance will also need to change to meet the new conditions. If the performance varies by no more than 5–10% it is seldom necessary for remedial action. To alter the performance of a fan without changing the geometry the fan speed must be increased or decreased. When the speed is changed it is prudent to check the performance from the following fan laws:

Fan speed varies – size and density constant

1. Volume flow varies directly as fan speed:

$$\frac{\text{Volume 1}}{\text{Volume 2}} = \frac{\text{Speed 1}}{\text{Speed 2}}$$

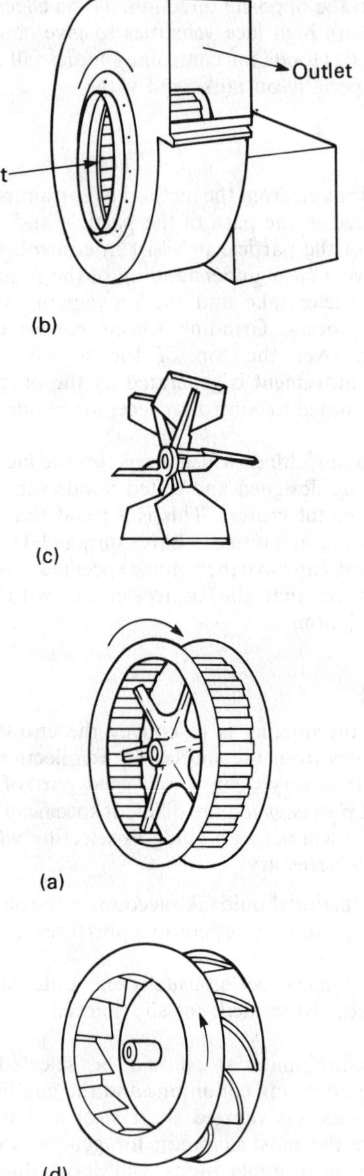

Figure 27.5 Centrifugal types of fan. (a) Centrifugal; (b) paddle or radial bladed; (c) forward curve; (d) backward curve

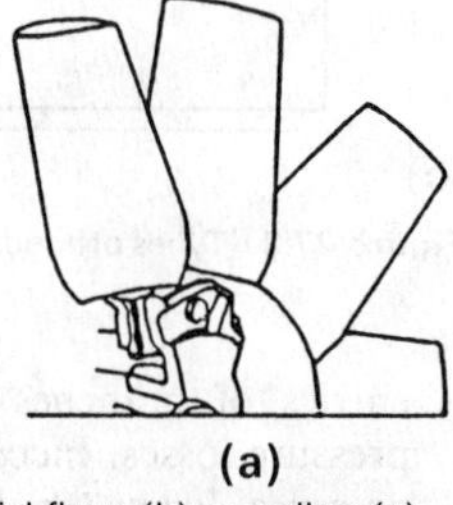

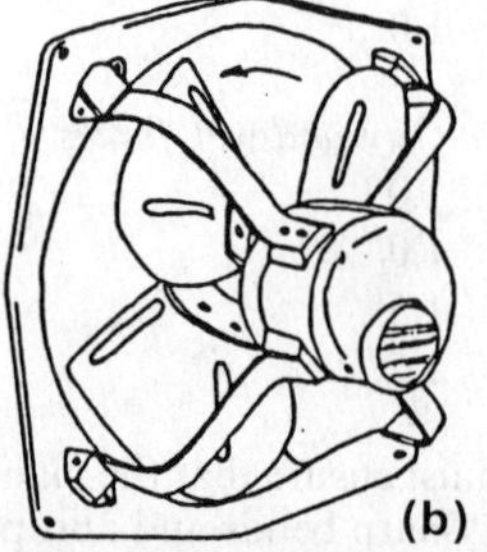

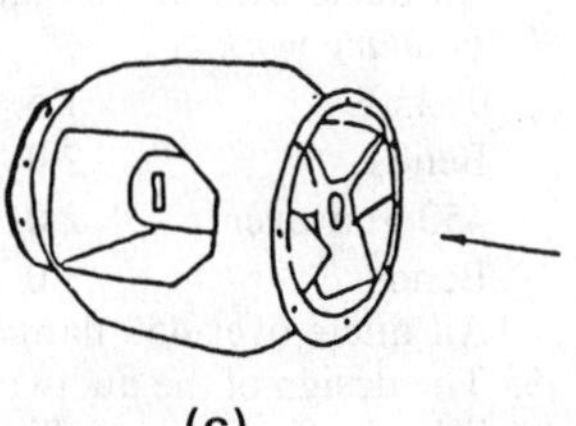

Figure 27.4 Axial-flow fans. (a) Axial flow; (b) propellor; (c) bifurcated

Table 27.1 Application data for fans

Type	*Pressure volume*	*Efficiency*	*Industry*	*Normal drive*	*Application*
Axial	2.5 in w.g. high volume	Very high	H and V	Direct	General use for ventilation, heating and minor fume work on low-pressure systems
Propeller	0.4 in w.g. high volume	Low	H and V	Direct	Usually applied on free air work, such as input and output units for buildings due to pressure limitations
Bifurcated	1.0 in w.g. up to approx. 10 000 CFM	Medium	Fume	Direct	Motor not in air-stream. Used on explosive fume, wet fume, high-temperature work and severe applications
Paddle	12 in w.g. up to approx. 30 000 CFM	Medium	Dust and fume	Vee and direct	General dust and fume. Will handle air containing dust and chippings. Wide application in wood-waste extraction plants
Forward	6.0 in w.g. very high volume	High	H and V	Vee and direct	Will only handle clean air. Compact and quiet running. Used on heating, ventilation and air-conditioning work
Backward	20 in w.g. high volume	High	Dust and fume	Vee and direct	General dust and fume. High-pressure systems and on dust-collector plants. Will handle some dusty air
Blowers	42 in w.g. usually low volume	Medium	General	Direct	Furnace blowing, cooling, conveying and where there is a need for high pressures

2. Pressure varies as the square of the fan speed:

$$\frac{\text{Pressure 2}}{\text{Pressure 1}} = \frac{(\text{Speed 2})^2}{(\text{Speed 1})^2}$$

3. Power varies as the cube of the fan speed:

$$\frac{\text{Power 2}}{\text{Power 1}} = \frac{(\text{Speed 2})^3}{(\text{Speed 1})^3}$$

A typical example will illustrate their use. A 600 mm diameter paddle-bladed centrifugal fan delivers 5100 m^3/h against a resistance of 200 mm water and absorbs 5.16 kW when running at 1665 rev/min. The motor for driving the fan is 7.5 kW at 2-pole speed through vee belts and pulleys. What would be the new speed, pressure and power required for the volume to be increased to 7000 m^3/h?

$$\frac{7000}{5100} = \frac{\text{New speed}}{1665}$$

$$\text{New speed} = \frac{7000 \times 1665}{5100} = 2285.29 \text{ rev/min}$$

$$\frac{\text{New pressure}}{200} = \frac{(2285.29)^2}{(1665)^2}$$

New pressure = 376.77 mm

$$\frac{\text{New power}}{5.16} = \frac{(2285.29)^3}{(1665)^e}$$

New power required = 13.32 kW

For the fan to operate within the same duct system at the increased duty to give the new volume an uprated drive motor of at least 11 kW is required. Care must always be exercised in using the above laws so as not to exceed the drive capacity of the motor or the critical speed of the fan. No absolute guidelines can be given to cover all fans at all speeds due to the wide range of designs and materials used. If, on recalculation, the new fan speed at the blade tip exceeds 76 m/s no action should be taken without consulting the manufacturer. In the above example the tip speed is 71.8 m/s and would, in most circumstances (depending on the condition of the fan), be safe to run.

27.3.4 Collectors

The function of the collector is to separate the entrained dust or fume from the airstream and deposit it in a convenient form for ultimate disposal. The four most common collectors are:

Cylones
Fabric filters
Wet collectors
Electrostatic filters

Each of the above types has been developed to meet the needs of industry and a brief description here will enable the plant engineer to make a start in the selection of a particular application.

Cyclones

There are two basic designs in use. The first is the general-purpose large-diameter cyclone, traditionally used in metal polishing and wood-waste extraction. The second is the so-called high-efficiency small-diameter cyclone, used in groups for pre-separation or in metal grinding exhausts. If the cyclone is to remain in use it will be confined to those operations where further collection of fines is part of the overall system. This is because they are relatively inefficient on the smaller particle sizes and release an unacceptable amount of discharge into the atmosphere.

Wet collectors

There are two types: self-induced spray and pressure spray. Many designs have been developed to suit specific dust or fume problems, and it is now possible to use wet collectors on the treatment of solids and fumes down to very small particle sizes.

Self-induced spray wet collectors This is the most common type, and relies on its separating action by the induced air from the fan pulling the contaminated air through a curtain of water, It is simple in operation with no pumps or moving parts except for the fan, which is set on the clean side of the collector. The scrubbing action is dependent on the pressure drop across the collector. When set, this is constant and is determined by the water level within the collector. The removal of sludge is either by automatic ejection or manual drag-out.

This type of collector has found wide application in general engineering and very high collection efficiencies are possible, but at the expense of considerable power requirements. General-purpose collectors at pressure drops across the collector of 150 mm will have collection efficiencies of 98% at 10 μm and above. Units with pressure drops of 800 mm and efficiencies of more than 99% on sub-micron particles are available.

Spray-type collectors In this system water is sprayed or cascaded onto the contaminated air directly or through packed towers, and the fumes or dust are washed away by absorption. These collectors are used extensively on the treatment of fumes of all types and have low-pressure drops and hence low power requirements compared to induced spray. A development of this collector is the venturi scrubber, which injects high-pressure water into a venturi through which the fume-laden air is passing. The intimate contact of the two ensures absorption and removal from the airstream. These collectors are used in fume removal and have efficiencies of more than 99% on sub-micron particles.

The general rule with wet collectors is that the higher the collection efficiency, the greater the pressure drop and hence the power absorbed.

Fabric filters

Fabric filters are capable of separation efficiencies approaching 100% if correctly applied. The ideal filter is permeable to the airstream but not to the dust requiring separation. Separation occurs by impaction of the dust particles upon fabric fibres, resulting in a dust cake forming on the fabric. Although aiding filtration, this cake does increase the resistance to air flow. If allowed to go unchecked it would result in a reduction of the total air being exhausted in the system.

Two methods are used to remove the dust cake, both of which require interruption of the air flow. The difference in dust-cake removal conveniently divides filters into intermittent and continuous rating. In the intermittent type the pressure increases (with time) up to a pre-arranged level. The air flow is then stopped and the fabric is mechanically shaken. In the continuously rated filter the pressure drop rises to a low set point, after which it remains constant across the filter as a whole. The cleaning is done by isolating a part of the filter from the airstream and that section is cleaned.

Intermittent filters are best suited to small applications which will allow the process to be stopped at intervals. The interval used is 4 h (i.e. a morning or afternoon shift). Mechanical shaking is done by either hand or electric motor. The application of these filters is limited to the incoming dust burden of the order of 5 g/m^3 and is known as nuisance dust.

Continuously rated filters have whole sections of the filter shut off from the air flow and then those sections are shaken or cleaned. Shaking is carried out in sequence, usually by electric motor. Where the filter is cleaned it is done by a jet of compressed air being blown in reverse to the air flow through the fabric. This system does not require whole sections to be shut down, as the reverse blow is carried out when the filter is on-stream. The time of blow is very small and are measured in parts of a second rather than in minutes, as in the case of shaking filters. The application of these filters is in continuous processes and where the dust burdens are high (in excess of 100 g/m^3).

The shaking and continuous filters are regenerative, but there is a third group usually associated with ventilation work rather than dust and fume. These are throw-away filters which, as the name implies, means that when they become too caked with dust to operate correctly the filters are removed and replaced with new ones. They will only handle low incoming dust burdens, but their efficiencies are the highest of any filter. Typical applications are fresh air input plants, clean-room filtration and nuclear processes.

Electrostatic precipitators

These remove particles by means of applied electrical forces, and are used extensively on cement and fly ash removal from airstreams. The particles are first given an electric charge and are then passed through an electric field to apply a precipitation to them. They are then captured on the collecting surface (normally an electrode). The collection surface is the cleaned by rapping or water wash and the dust collected in a hopper below. The efficiencies achieved are 99% at particle sizes above 1 μm. They are cheap to run but the capital costs are high, especially for large collectors. In recent years small units have been developed to control oil and welding fumes.

Disposal

Many control plants fail in their objective when the collected waste is removed from the inlet. The practice of dropping dust into sealed bins at the base of the collector is both sensible and practical. It is when the bins are to be emptied that a secondary dust problem arises. If the waste is simply put into a larger container with no control, dust will be released back into the workplace. If dangerous dusts are being collected, sealed inner liners to the dustbins can be used, thereby preventing this release. On the larger installations the collected dust is retained under sealed conditions at all times and the discharge from collectors is by rotary valve and screw conveyor. These will feed into bulk containers for further processing. In the case of fumes these are collected by absorption into liquors, and these liquors are treated in an effluent plant separate from the fume plant.

Table 27.2

Type of machine	*Size* (mm)	*Air volume* (m^3/h)
Grinders (double end)	200–250	680
Grinders (double end)	280–406	1189
Grinders (double end)	430–455	1495
Grinders (double end)	480–560	1870
Grinders (double end)	585–762	2720
Grinders (double end)	787–915	3738
These figures are for bottom connections only and average duties		
Toolroom grinders (double ends)	Up to 200	510
Cutter grinders (single end)	Up to 150	510
Banding machines		
Horizontal	Up to 100	595
Horizontal	125–200	1019
Backstands	38 belt	595
Backstands	50 belt	866
Backstands	75 belt	1053
Backstands	100 belt	1257
Backstands	125 belt	1699
Backstands	150 belt	2243
Double-end hand-polishing machines	Up to 150	1019
	178–225	1359
	250–355	2209
Wire mops	Up to 200	1699
Cut-off machines		
Abrasive discs	Up to 405	934
Abrasive discs	430–610	1699

27.4 System design and application

In selecting the best system to control the hazard the following should be noted:

1. The problem must be surveyed under actual working conditions and data collected and recorded in a logical manner. The survey must establish:
   (a) The origin of the dust or fume and its nature. Is it toxic, explosive or hazardous in any way?
   (b) The process which produces the contaminant. Is it wet or dry?
   (c) Whether the problem can be solved by elimination of the process;
   (d) The source or sources of the problem. Does it occur at more than one point?
   (e) Do any special regulations apply to the hazard and are the materials being handled listed in any published form as having control limits?
   (f) Is a control system in use at the time of the survey?

Having completed the survey, the next stage is to draw the system showing the sources of dust and the duct runs. An assessment must then be made on the air volumes and velocities required to give control. This is largely a matter of experience, as air-entrainment rates are derived empirically. It is possible to calculate the rates but is unusual in general engineering. There are published lists for air rates and many companies have their own standards. Table 27.2 shows rates for metal-working machines. When the total air volume has been established the collector and fan can be sized. The total air volume and the type of dust will determine the size and type of collector to be used. Table 27.3 shows the types of collector for various applications.

To size the fan it is necessary to know the total air volume and the pressures in the system. These are calculated from the losses in the system on the longest or index leg, and begin with the hood. The hood entry loss can be expressed as 0.6 of the velocity head and is accurate enough for first estimates. The losses are then calculated on the velocities in the ducts. Each change of direction means a small loss in each length of duct. Added to the pressure drop loss across the collector and the outlet losses, these give the total static pressure required in the system.

The formula used in the calculation of system losses for first estimates favoured here is:

Loss in water (mm)/length of duct (m)

$$= \frac{(\text{Velocity (m/s)})^2 \times 1.243}{\text{Diameter of duct (mm)}}$$

Velocity head or velocity pressure is the pressure required to accelerate the flowing mass from rest to its flowing

Table 27.3

Application	*Type of collector*
Sand foundry dusts	Self-induced wet collectors
Magnesium and aluminium working	Self-induced wet collectors
Hot applications	
Sparks from grinding	Self-induced wet collectors
Heat treatment	
Small grinding machines, polishing machines	Self-contained intermittent filters
Batch operations (e.g. filling, tipping, mixing)	Self-contained intermittent filters
Process plant	Continuously rated filters
Heavy dust burdens	Continuously rated filters
Diffiuclt dusts (e.g. carbon black sugar dust)	Continuously rated filters
Cement and fly ash production	Electrostatic precipitators
Woodworking and initial separation	Cyclones

velocity:

$$\text{Velocity pressure in water (mm)} = \frac{(\text{Velocity (m/s)})^2}{(4.04)^2}$$

Static pressure is the pressure to overcome resistance and is expressed in millimetres of water. Total pressure is the sum of velocity and static pressures.

The basic formula for air flow under all conditions is:

$$Q = A \times V$$

where Q = volume (m^3/s), A = area (m^2) and V = velocity (m/s). The nominal velocities used in dust and fume control depend on the materials being handled, but the following will suffice for most work:

Metalworking	Ducts 23 m/s	Branches 20 m/s
Woodworking and light dusts	Ducts 20 m/s	Branches 18 m/s
Gases and vapours	Ducts 10 m/s	Branches 9 m/s

It is now possible to calculate the total system resistance and select a suitable fan and collector. The following will illustrate the use of the formula.

A simple duct system of length 50 m serving a double-end grinding machine requires an exhaust volume of 6000 m^3/h. The duct velocity would be 23 m/s, giving a diameter of 300 mm and a velocity pressure of 32.41 mm.

$$\text{The hood loss} = 0.6 \times 32.41 = 19.45 \text{ mm}$$

$$\text{Duct losses in 50 m} = \frac{(23)^2 \times 1.243}{300 \text{ mm}} = 109.59 \text{ mm}$$

$$\text{The total pressure} = VP + SP = 32.41 + 109.59 + 19.45$$

$$= 161.45 \text{ mm}$$

To this figure must be added the pressure drop across the collector, which in this case would be a wet unit having a drop of 150 mm.

The fan would then be selected to handle 6000 m^3/h at a total pressure of 311.45 mm water. As this is a high-pressure fan, it would be a centrifugal backward laminar type.

27.5 Testing and inspections

After the plant has been installed it will require a test to determine whether the design meets the original objectives. Prior to 1 October 1989 except in a limited number of cases, the test was only to establish whether the commercial contract had been fulfilled. It is now part of the duty under the Control of Substances Hazardous to Health Regulations 1988 (COSHH) that an assessment is made of the control system. In this assessment the effectiveness of the plant must be established and at the same time a record made showing the engineering parameters. A typical assessment sheet is shown in Figure 27.6.

The precise nature of the test will depend on the particles being controlled. The simple observation tests carried out by Tyndall lights and smoketubes will suffice for the majority of dust and fume systems. It is only where the contaminants are listed and known to be dangerous that special testing needs to be done. This work requires on-site monitoring of the workplace using air samplers and the expert services of an industrial hygienist.

27.6 Legislation

As mentioned earlier, new legislation has been put on the statute book in response to public opinion and other pressures to protect people at work. The three most significant are:

1. The Factories Acts 1961
2. The Health and Safety at Work, etc. Act 1974
3. The Control of Substances Hazardous to Health Regulations 1988

The provisions contained in the Acts must be obeyed because they are part of the criminal law. It is common practice for Parliament to lay down only broad general duties, and in such cases the statute is referred to as a framework Act or an enabling legislation.

Parliament will grant power to government agencies to make detailed Regulations under the Acts, and this system has been successful. The COSHH Regulations are made under Section 16(1) of the Health and Safety at Work, etc. Act 1974. Codes of practice are published by the Health and Safety Executive to give practical guidance in respect of regulations.

The COSHH Regulations extend the responsibilities of the employer and the self-employed beyond the factory. Any duties placed by these Regulations on employers in respect of his employees also apply, as far as is reasonably practical. These Regulations are important with respect

SHEET 1	
COSHH LEV RECORD	
NAME OF EMPLOYER	DATE
ADDRESS	
DEPARTMENT OR SITE OF PROCESS	REFERENCE NUMBER
LOCATION OF LEV PLANT	TYPE OF LEV PLANT
IS THE EXHAUST AIR RETURNED TO THE WORKPLACE AND IF SO IN WHAT CONCENTRATION	IS SECONDARY FILTRATION PROVIDED
SUBSTANCE/PROCESS CONTROLLED	HAZARD ASSESSMENT
ASSESSMENT OF LEVEL OF CONTROL AND TYPE OF TEST	CONDITION AT TIME OF TEST
FREQUENCY OF TEST	REGULATION
COMMENTS	
NAME OF PERSON RESPONSIBLE	DATE FOR TEST
SIGNATURE	EMPLOYER

Figure 27.6 A typical assessment sheet

SHEET 2

COSHH LEV RECORD

LAYOUT SHOWING POINTS OF MEASUREMENT

Date Installed

PRIMARY COLLECTOR/FILTER		SECONDARY FILTER	
Make Size/Area Media	S.P.Outlet S.P.Inlet P.D.Across	Type Size/Area Media	S.P.Outlet S.P.Inlet P.D.Across
MOTOR		**FAN SET**	
Supply Amps Power Starting	Make Type Speed Encl.	Size Type Handling Drive	Volume Speed S.P.Inlet S.P.Outlet

HOOD AND TRANSPORT SIZES, VELOCITIES AND VOLUMES

Point	Hood Size Area	Duct Size Area	V.P. in.W.G.	S.P. in.W.G.	Velocity	Volume	Assessment of Control Remarks

Notes

SHEET 3

LEV NUMBER

COSHH LEV REPORT

NAME AND ADDRESS OF EMPLOYER	DEPARTMENT OR SITE OF PROCESS

REGULATION	PERIOD BETWEEN TESTS	LAST TEST

HOODS, ENCLOSURES AND DUCTING

Point	S.P. in.W.G.	Are Elements in Working Order Answer Yes, No or N/A.				Assessment of Control. State Tests Used.	Remarks
		Hood	Enclos.	Duct	Fan		

COLLECTORS/FILTERS	SECONDARY FILTER	
S.P.Outlet S.P.Inlet P.D.Across	Quantity of Contaminant in Returned Air	Monitoring Equipment if Fitted
Condition of Collector/ Filter	Test Method Used	Condition of Filter

Particulars of Repairs or Modifications Required to Ensure that the LEV Plant Effectively Controls the Dust or Fumes

SIGNATURE

DATE

to dust and fumes, for they state that dust in any substantial concentration is hazardous, and the concentration is specified. Furthermore, these control measures must be maintained and tested at regular intervals. In practice, this will mean that all dust and fume plants will now require at least an annual test and inspection. The Regulations revoke many previous Regulations, and it is current practice, where inspection intervals existed, that these intervals should also be used under COSHH.

All the Regulations will have a beneficial effect on the long-term health of workers. Equipment and techniques for control will improve and the pressures of economics will ensure that more efficient systems are developed.

28 Insulation

F T Gallyer
Pilkington Insulation Ltd

Contents

28.1 Introduction

Insulation is one of those ubiquitous techniques that is always around, always impinging on our work, social and domestic activities and yet for most of the time is hardly noticed. Insulation is a passive product; once installed, it works efficiently, quietly and continually, usually out of sight, enclosed within a structure or a casing or under cladding.

It comes to the fore when new design of buildings, plant, equipment or production processes is being considered. It is at this stage that the right specification must be made. Any shortfall in the thickness or error in the type and application details will prove costly to rectify at a later date.

28.1.1 Why insulate?

There are many reasons why professional engineers, architects and laymen use insulation, e.g.:

1. To comply with mandatory legislation (i.e. Building Regulations);
2. To reduce heat loss/heat gain;
3. To reduce running costs;
4. To control process temperatures;
5. To control surface temperatures;
6. To reduce the risk of freezing;
7. To provide condensation control;
8. To reduce heating plant capacity;

Other reasons why insulation is used are to provide:

9. Acoustic/correction and noise control;
10. Fire protection.

28.1.2 Scope

This chapter will deal primarily with thermal insulation. Acoustic and fire-protection properties and applications will be treated as subsidiary to the thermal insulation aspects.

28.1.3 Thermal insulation

A thermal insulation material is one which frustrates the flow of heat. It will slow down the rate of heat loss from a hot surface and similarly reduce the rate of heat gain into a cold body. It will not stop the loss or gain of heat completely.

No matter how well insulated, buildings will need a continual input of heat to maintain desired temperature levels. The input required will be much smaller in a well-insulated building than in an uninsulated ones – but it will still be needed. The same applies to items of plant – pipes, vessels and tanks containing hot (or cold) fluids. If there is no heat input to compensate for the loss through the insulation the temperature of the fluid will fall. A well-insulated vessel will maintain the heat of its contents for a longer period of time but it will never, on its own, keep the temperature stable.

Thermal insulation does not generate heat. It is a common misconception that such insulation automatically warms the building in which it is installed. If no heat is supplied to that building it will remain cold. Any temperature rise that may occur will be the result of better utilization of internal fortuitous or incidental heat gains.

28.2 Principles of insulation

28.2.1 Heat transfer

Before dealing with the principles of insulation it is necessary to understand the mechanism of heat transfer. When a hot surface is surrounded by an area that is colder, heat will be transferred and the process will continue until both are at the same temperature. Heat transfer takes place by one or more of three methods:

Conduction
Convection
Radiation

28.2.2 Conduction

Conduction is the process by which heat flows by molecular transportation along or through a material or from one material to another, the material receiving the heat being in contact with that from which it receives it. Conduction takes place in solids, liquids and gases and from one to another. The rate at which conduction occurs varies considerably according to the substance and its state.

In solids, metals are good conductors – gold, silver and copper being among the best. The range continues downwards through minerals such as concrete and masonry, to wood, and then to the lowest conductors such as thermal insulating materials.

Liquids are generally bad conductors, but this is sometimes obscured by heat transfer taking place by convection. Gases (e.g. air) are even worse conductors than liquids but again, they suffer from being prone to convection.

28.2.3 Convection

Convection occurs in liquids and gases. For any solid to lose or gain heat by convection it must be in contact with the fluid. Convection cannot occur in a vacuum. Convection results from a change in density in parts of the fluid, the density change being brought about by an alteration in temperature.

Convection in gases

If a hot body is surrounded by cooler air, heat is conducted to the air in immediate contact with the body. This air then becomes less dense than the colder air further away. The warmer, light air is thus displaced upwards and is replaced by colder, heavier air which, in turn, receives heat and is similarly displaced. There is thus developed a continuous flow of air or convection around the hot body removing heat from it. This process is similar but reversed if warm air surrounds a colder body, the air becoming colder on transfer of the heat to the body and displaced downwards.

Convection in liquids

Similar convection processes occur in liquids, though at a slower rate according to the viscosity of the liquid. However, it cannot be assumed that convection in a liquid results in the colder component sinking and the warmer one rising. It depends on the liquid and the temperatures concerned. Water achieves its greatest density at approximately 4°C. Hence in a column of water, initially at 4°C, any part to which heat is applied will rise to the top. Alternatively, if any part is cooled below 4°C it, too, will rise to the top and the relatively warmer water will sink to the bottom. It is always the top of a pond or water in a storage vessel which freezes first.

Natural convection

The process of convection that takes place solely through density change is known as 'natural convection'.

Forced convection

Where the fluid displaced is accelerated by wind or artificial means the process is called 'forced convection'. With forced convection the rate of heat transfer is increased – substantially so in many cases.

28.2.4 Radiation

The process by which heat is emitted from a body and transmitted across space as energy is called radiation. Heat radiation is a form of wave energy in space similar to radio and light waves. Radiation does not require any intermediate medium such as air for its transfer. It can readily take place across a vacuum.

All bodies emit radiant energy, the rate of emission being governed by:

1. The temperature difference between radiating and receiving surfaces;
2. The distance between the surfaces;
3. The emissivity of the surfaces. Dull matt surfaces are good emitters/receivers, bright reflective surfaces are poor ones.

28.2.5 Requirements of an insulant

In order to perform effectively as an insulant a material must restrict heat flow by any (and preferably) all three methods of heat transfer. Most insultants adequately reduce conduction and convection elements by the cellular structure of the material. The radiation component is decreased by absorption into the body of the insulant and is further reduced by the application of a bright foil outer facing to the product.

Convection inhibition

To reduce heat transfer by convection an insulant should have a structure of a cellular nature or with a high void content. Small cells or voids inhibit convection within them and are thus less prone to excite or agitate neighbouring cells.

Conduction inhibition

To reduce heat transfer by conduction, an insulant should have a small ratio of solid volume to void. Additionally, a thin-wall matrix, a discontinuous matrix or a matrix of elements with minimum point contacts are all beneficial in reducing conducted heat flow. A reduction in the conduction across the voids can be achieved by the use of inert gases rather than still air.

Radiation inhibition

Radiation transfer is largely eliminated when an insulant is placed in close contact with a hot surface. Radiation may penetrate an open-cell material but is rapidly absorbed within the immediate matrix and the energy changed to conductive or convective heat flow. It is also inhibited by the use of bright aluminium foil, either in the form of multi-corrugated sheets or as an outer facing on conventional insulants.

28.3 Calculation of heat loss

28.3.1 Glossary of terms

Terms and symbols used in computing heat loss are as follows.

Heat (J/s or W)

The unit of quantity of heat is the joule (J). Heat flow may be expressed as joules per second (J/s), but as a heat flow of one joule per second equals one watt the unit watt (W) is usually adopted for practical purposes.

Temperature (°C or K)

For ready identification, actual temperature levels are expressed in degrees Celsius (°C) while temperature difference, interval or gradient is expressed in kelvins (K).

Thermal conductivity (λ)

Thermal conductivity, now denoted by the Greek letter lambda (previously known as the *k*-value), defines a material's ability to transmit heat, being measured in watts per square metre of surface area for a temperature gradient of one kelvin per unit thickness of one metre. For convenience in practice, its dimensions Wm/m^2K be reduced to W/mK, since thickness over area m/m^2 cancels to 1/m.

Thermal resistivity (r)

This is the reciprocal of thermal conductivity. It is expressed as mK/W.

Thermal conductance (C)

Thermal conductance defines a material's ability to transmit heat measured in watts per square metre of surface area for a temperature gradient of one kelvin in terms of a *specific thickness* expressed in metres. Its dimensions are therefore W/m^{2}K.

Thermal resistance (R)

Thermal resistance is the reciprocal of thermal conductance. It is expressed as m^{2}K/W. Since the purpose of thermal insulation is to resist heat flow it is convenient to measure a material's performance in terms of its thermal resistance, which is calculated by dividing the thickness expressed in metres by the thermal conductivity. Being additive, thermal resistances facilitate the computation of overall thermal transmittance values (U-values).

Surface coefficient (f)

This is the rate of heat transfer from a surface to the surrounding air (or fluid) due to conduction convection and radiation. It is generally used only in still-air conditions and when the temperature difference between surface and ambient is of the order of 30 K. It is obtained by dividing the thermal transmission per unit area in watts per square metre by the temperature difference between the surface and the surrounding air. It is expressed as W/m^{2}K.

Surface resistance (R_s)

Surface resistance is the reciprocal of surface coefficient. It is expressed as m^{2}K/W.

Thermal transmittance (U)

Thermal transmittance (U-value) defines the ability to an element of structure to transmit heat under steady-state conditions. It is a measure of the quantity of heat that will flow through unit area in unit time per unit difference in temperature of the individual environments between which the structure intervenes. It is calculated as the reciprocal of the sum of the resistance of each component part of the structure, including the resistance of any air space or cavity and of the inner and outer surfaces. It is expressed as W/m^{2}K.

28.3.2 Fundamental formulae

In calculating heat loss from surfaces freely exposed to air it is necessary to deal separately with both radiant and convective losses.

Radiation

The following Stefan–Boltzmann formula applies to both plane and cylindrical surfaces:

$$Q_r = 5.673 \times 10^{-8} E((T_s + 273.1)^4 - (T_{rm} + 273.1)^4) \quad (28.1)$$

Natural convection – plane surfaces

$$Q_c = C(T_s - T_a)^n \quad (28.2)$$

The factor C and index n can be assumed to have the following values for plane surfaces of differing orientations:

Horizontal – downward heat flow	$C=1.3$	$n=1.25$
Vertical	$C=1.9$	$n=1.25$
Horizontal – upward heat flow	$C=2.5$	$n=1.25$

Natural convection – horizontal cylinders and pipes

$$Q_c = 0.53 \frac{\lambda_a}{d_o} (Gr \cdot Pr)^{0.25} (T_s - T_a) \quad (28.3)$$

However, as the Grashof and Prandtl numbers can be different to determine, the following formula, which gives a close approximation, can be used for cylinders freely exposed to air:

$$Q_c = (T_s - T_a) 1.32 \left(\frac{T_s - T_a}{d_o} \right)^{0.25} \quad (28.4)$$

In the above equations:

Q_r = heat loss by radiation (W/m^2),
Q_c = heat loss by convection (W/m^2),
E = surface emissivity (see Section 28.4) (dimensionless),
T_s = surface temperature (°C),
T_a = ambient air temperature (°C),
T_{rm} = Mean radiant temperature of enclosure (°C),
Gr = Grashof number (dimensionless),
Pr = Prandtl number (dimensionless),
λa = thermal conductivity of air (W/mK),
d_o = outside diameter of cylinder/pipe (m).

Forced convection

It is not proposed to deal with forced convection here. Experimental work has yielded considerably differing results for ostensibly similar conditions. It is sufficient to note that forced convection affects small-bore pipes to a greater extent than large-bore and is dependent on temperature differences. While the heat loss from uninsulated surfaces may increase by a factor of up to 200–300%, the increase in heat loss from the insulated surface would be considerably less (of the order of 10%).

28.3.3 Insulated heat loss

Heat loss through insulation can be calculated from a knowledge of the thickness and thermal conductivity of the insulation and the emissivity of the outer surface of the insulation/cladding system.

Single-layer insulation

$$q=\frac{T_1-T_2}{R}=\frac{T_1-T_m}{R+R_s} \quad (28.5)$$

$$q_1=10^{-3}\pi d_o q \quad (28.6)$$

For plane surfaces:

$$R=10^{-3}\frac{L}{\lambda} \quad (28.7)$$

$$R_s=\frac{1}{f} \quad (28.8)$$

For cylindrical surfaces:

$$R=10^{-3}\frac{d_o}{2\lambda}\ln\frac{d_1}{d_o} \quad (28.9)$$

$$R_s=\frac{d_o}{fd_1} \quad (28.10)$$

Multi-layer insulation

$$q=\frac{T_1-T_2}{R_1+R_2+\cdots R_n}=\frac{T_1-T_m}{R_1+R_2+\cdots R_n+R_s} \quad (28.11)$$

$$q_1=10^{-3}\pi d_o q$$

For plane surfaces:

$$R_1=10^{-3}\frac{L_1}{\lambda_1} \quad (28.12)$$

$$R_2=10^{-3}\frac{L_2}{\lambda_2} \quad (28.13)$$

$$R_n=10^{-3}\frac{L_n}{\lambda_n} \quad (28.14)$$

$$R_s=\frac{1}{f} \quad (28.15)$$

For cylindrical surfaces:

$$R_1=10^{-3}\frac{d_o}{2\lambda_1}\ln\frac{d_1}{d_o} \quad (28.16)$$

$$R_2=10^{-3}\frac{d_o}{2\lambda_2}\ln\frac{d_2}{d_1} \quad (28.17)$$

$$R_n=10^{-3}\frac{d_o}{2\lambda_n}\ln\frac{d_n}{d_{(n-1)}} \quad (28.18)$$

$$R_s=\frac{d_o}{fd_n} \quad (28.19)$$

Symbols

In the above formulae the following symbols apply:

q = heat loss per square metre of hot surface (W/m^2)
q_1 = heat loss per linear metre of pipe (W/m)
R = thermal resistance of insulation per m^2 (m^2K/W)
R_1 = thermal resistance of inner layer of insulation (m^2K/W)
R_2 = thermal resistance of second layer of insulation (m^2K/W)
R_n = thermal resistance of nth layer of insulation (m^2K/W)
R_s = thermal resistance of outer surface of insulation system (m^2K/W)
T_1 = temperature of hot surface (°C)
T_2 = temperature of outer surface of insulation (°C)
T_m = temperature of ambient still air (°C)
d_o = outside diameter of pipe (mm)
d_1 = outside diameter of inner layer of insulation (mm)
d_2 = outside diameter of next layer of insulation (mm)
d_n = outside diameter of nth layer of insulation (mm)
λ = thermal conductivity of insulating material (W/mK)
λ_1 = thermal conductivity of inner layer of insulation (W/mK)
λ_2 = thermal conductivity of second layer of insulation (W/mK)
λ_n = thermal conductivity of nth layer of insulation (W/mK)
f = surface coefficient of outer surface (W/m^2K)
ln = natural logarithm

Note: the values taken for $\lambda, \lambda_1, \lambda_2, \ldots, \lambda_n$ should be those applicable to the mean temperature of the hot and cold surfaces of the appropriate layer.

28.3.4 Surface emissivity

Emissivity values are needed to calculate the radiation component of heat loss from high-temperature surfaces. Accurate values which reflect actual conditions are difficult to obtain, as the emissivity value varies with temperature and with contamination or oxidation of the surface.

For most non-critical calculations the following values may be used:

Steel, paint (matt surface)	0.90–0.95
Galvanized steel (new)	0.40
Galvanized steel (weathered)	0.85
Aluminium (polished)	0.05

28.3.5 Surface coefficients

These present similar problems to emissivities, and BS 5422 has standardized on three values:

1. $f=5.7$ for surfaces of low emissivity (e.g. polished aluminium);
2. $f=8.0$ for surfaces of medium emissivity (e.g. planished or galvanised steel, aluminium paint, stainless steel and aluminium/zinc amalgam);
3. $f=10.0$ for surfaces of high emissivity (e.g. matt-black surfaces, steel, brick and plain insulation surfaces).

28.3.6 *U*-value calculation

The insulation effectiveness of elements of building structures is represented by the *U*-value or thermal transmittance. As defined in Section 28.3.1, the *U*-value is the reciprocal of the sum of the thermal resistances and can be expressed as:

$$U = \frac{1}{R_{so} + R_1 + R_2 \cdots R_n + R_{as} + R_{si}} \qquad (28.20)$$

where

R_{so} = thermal resistance of outer surface,
R_1 = thermal resistance of first material,
R_2 = thermal resistance of second material,
R_n = thermal resistance of nth material,
R_{as} = thermal resistance of any air space in the construction,
R_{si} = thermal resistance of inner surface.

In calculating R-values of the material elements of construction their λ values should be taken at 10°C mean temperature (which is assumed to be normal building temperature).

28.3.7 Standardized resistances

Actual thermal resistances of surface and air spaces within a construction vary according to size, exposure and nature of the material. Standardized resistance values were adopted by the Building Research Establishment in their BRE Digest 108 (August 1969) to ensure a constant base for the calculation of U-values. These values are now collected in the CIBSE Guide A3, from which Tables 28.1–28.4 were prepared

Table 28.1 Internal surface resistance, R_{si} (m²K/W)

Building element	*Direction of heat flow*	*Surface emissivity*	*Surface resistance*
Walls	Horizontal	High	0.12
		Low	0.30
Ceilings, roofs and floors	Upward	High	0.10
		Low	0.22
Ceilings and floors	Downward	High	0.14
		Low	0.55

Source: Chartered Institution of Building Services Engineers Guide, Section A3.

Table 28.2 External surface resistance, R_{so} (m²K/W)

Building element	*Surface emissivity*	*Surface resistance for stated exposure*		
		Sheltered	*Normal*	*Severe*
Walls	High	0.08	0.06	0.03
	Low	0.11	0.07	0.03
Roofs	High	0.07	0.04	0.02
	Low	0.09	0.05	0.02
Floors	High	0.07	0.04	0.02

Source: Chartered Institution of Building Services Engineers Guide, Section A3.

Table 28.3 Standard thermal resistances for unventilated air spaces, R_{as} (m²K/W)

Width of airspace (mm)	*Surface emissivity*	*Thermal resistance for heat flow in stated direction*		
		Horizontal	*Upward*	*Downward*
5	High	0.10	0.10	0.10
	Low	0.18	0.18	0.18
25 or more	High	0.18	0.17	0.22
	Low	0.35	0.35	1.06
High-emissivity and corrugated sheets in contact		0.09	0.09	0.11
Low-emissivity multiple-foil insulation with air space on side		0.62	0.62	1.76

Source: Chartered Institution of Building Services Engineers Guide, Section A3.

Table 28.4 Standard thermal resistances for ventilated air spaces, R_{as} (m²K/W)

Air space thickness 25 mm minimum	
Air space in cavity-wall construction	0.18
Air space between tiles and roofing felt on pitched roof	0.12
Air space behind tiles on tile-hung wall	0.12
Loft space between flat ceiling and pitched roof lined with felt	0.18
Loft space between flat ceiling and pitched roof of aluminium cladding, or low-emissivity upper surface on ceiling	0.25
Loft space between flat ceiling and unsealed fibre cement or black metal cladding to pitched roof	0.14
Air space between fibre cement or black metal cladding with unsealed joints and low-emissivity surface facing air space	0.30
Air space between fibre cement or black metal cladding with unsealed joints and high-emissivity lining	0.16

Source: Chartered Institution of Building Services Engineers Guide, Section A3.

28.4 Standards of insulation

In the UK the Building Regulations contain the mandatory requirement for thermal insulation of building structures, and heating and hot water services. Thermal insulation of cold water supply pipes is dealt with under the water supply bye-laws, but there are no national requirements to insulate pipes, vessels or equipment used in commercial or industrial processes.

28.4.1 Building Regulations 1990

These Regulations came into force on 1 April 1990 and apply to England and Wales only. It is the government's intention that these levels be incorporated into the regulations for Scotland and Northern Ireland.

The requirement

The mandatory requirement is that 'reasonable provision

shall be made for the conservation of fuel and power in buildings'. This requirement applies to dwellings and all other buildings whose floor area exceeds 30 m^2.

28.4.2 Approved Document L

The Department of the Environment's interpretation of reasonableness is given in the 1990 edition of Building Regulations 1985, Approved Document L, Conservation of Fuel and Power, available from HMSO. This deals with three areas of energy conservation:

The building fabric
Controls of heating/hot water systems
Insulation of hot water systems

Limitation on heat loss through the building fabric

This section of Approved Document L allows three methods of compliance:

Elemental approach
Calculation procedure 1
Calculation procedure 2

Elemental approach The insulation standards are set by the elemental approach, which establishes maximum U-values for the various elements of structure and also maximum glazing area. These are detailed in Tables 28.5 and 28.6. The Regulations also introduce two new categories of structure which have previously not needed to be insulated:

1. Semi-exposed walls and floors, which can be considered to be structures between a heated and an unheated part of a building;
2. Ground floors, which because of their inherently good U-values are unlikely to need insulation unless they are of a small domestic size or are of long narrow buildings (see Figure 28.1).
Calculation procedure 1 This is an alternative to the elemental approach and allows variation in the levels of insulation and of the glazing area. The calculation should show that the rate of heat loss through the envelope of the proposed building is not greater than that through a notional building of the same size and shape which is designed to comply with the elemental approach.

Calculation procedure 2 This procedure is the calculation of an energy target and allows for a completely free design using any valid energy-conservation measure. The procedure is intended to allow for useful solar heat gains and fortuitous internal gains. The requirement of Building Regulations will be met 'if the calculated annual energy use of the proposed building is less than the calculated energy use of a similar building designed to comply with the elemental approach'.

Limitations on calculation procedures Limitations are placed on the use of high (e.g. poor U-values) and of glazing areas when using the calculation of methods of complying. As a general rule:

1. *In dwellings*: The U-value of roofs and exposed walls should not be greater than 0.35 W/m^{2}K and 0.6 W/m^{2}K, respectively.
2. *In all other buildings*: The U-value of roofs and exposed walls should not be greater than 0.6 W/m^{2}K.
3. *Glazing areas*: If areas of glazing smaller than those allowed in Table 28.6 are used in the 'proposed building' then this smaller area should also be used in the 'notional building' calculation.

Insulation thickness While the actual thickness of insulation needed to comply with the Building Regulations will vary depending on insulation type, thermal conductivity and the structure into which it is fitted, some general guidelines can be given:

1. *Domestic roofs*: $U = 0.25$ W/m^{2}K
150 mm mineral wool loft insulation
2. *Other roofs*: $U = 0.45$ W/m^{2}K
80 mm in steel-frame metal-clad roofs
40–60 mm in flat roots
3. *External walls*: $U = 0.45$ W/m^{2}K
80 mm in steel-frame metal-clad wall
50 mm in aerated concrete inner-leaf cavity wall
65 mm in high-density masonry cavity walls
4. *Semi-exposed elements*: $U = 0.6$ W/m^{2}K
0–50 mm, depending on construction
5. *Ground floors*

Table 28.5 Building Regulations: maximum U-values (W/m^{2}K)

Structure	*Dwelling*	*All other buildings*
Exposed walls	0.45	0.45
Exposed floors	0.45	0.45
Ground floors	0.45	0.45
Roofs	0.25	0.45
Semi-exposed walls and floors	0.6	0.6

Table 28.6 Building Regulations: maximum single-glazed areas of windows and rooflights

Building type	*Windows*	*Rooflights*
Dwellings	Windows and rooflights together 15% of total floor area	
Other residential, including hotels and institutional buildings	25% of exposed wall area	20% of roof area
Places of assembly, offices and shops	35% of exposed wall area	20% of roof area
Industrial and storage	15% of exposed wall area	20% of roof area

Note: In any building the above glazing areas may be doubled when double glazing is used and trebled when triple glazing, or double glazing with low-emissivity coating, is used.

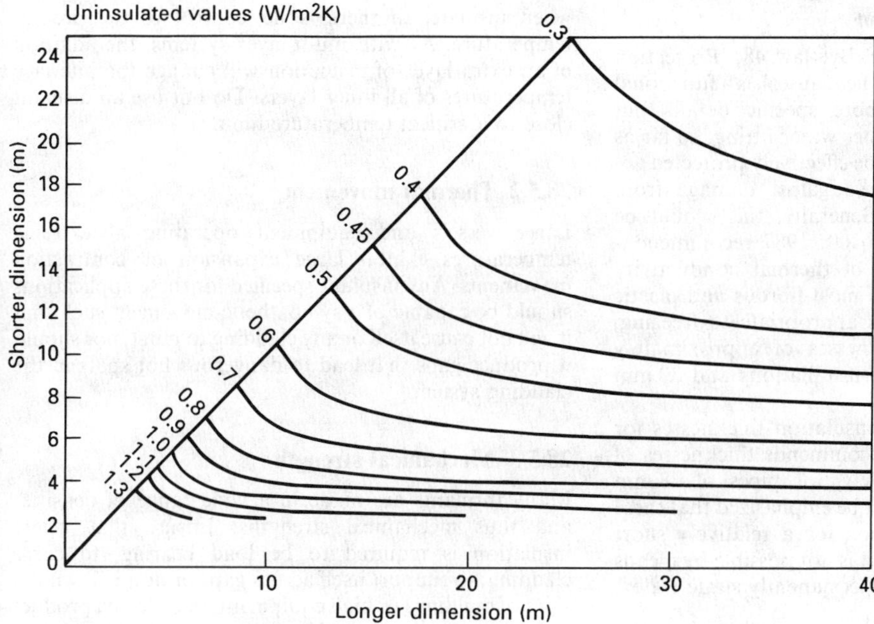

Figure 28.1 *U*-values of uninsulated solid concrete ground floors with four exposed edges. (Source: Eurisol UK, *U*-value Guide)

None needed in commercial or industrial-sized floors exceeding about 15 m × 15 m. Actual *U*-values can be obtained from Eurisol's graph shown as Figure 28.1.

Controls for space heating and hot water systems

This section of Approved Document L does not have any relevance to insulation.

Insulation of hot water storage vessels, pipes and ducts

This third and last section of Approved Document L is not intended to apply to storage and piping systems for commercial and industrial processes. It concerns only the central heating and the domestic hot water supply of all buildings. The standards are presented in three sections for:

Storage vessels
Pipes
Ducts

Insulation of hot water storage vessels The requirements for these tanks and cylinders will be met if the heat loss is not greater than 90 W/m^2. The thickness of insulation needed will therefore vary not only according to its thermal conductivity but also to the temperature of the water being stored. In practice, as long as the water is not greater than 100°C the insulation thickness needed is likely to be of the order of 20–35 mm.

As an alternative to the heat loss specification, domestic cylinders complying with BS 699: 1984, BS 1566: 1984: Parts 1 and 2, BS 3198: 1981 or having a cylinder jacket complying with BS 5615: 1985 will all meet the requirement.

Insulation of pipes The requirement for pipes is that they should be insulated to a thickness at least equal to the outside diameter, and with an insulant of thermal conductivity not greater than 0.045 W/mK. Two limitations are imposed: the insulation thickness need not be greater than 40 mm whatever the pipe outside diameter and insulation is not needed on pipes whose heat loss contributes to the useful heat requirements of the room or space through which it runs.

As an alternative option, the insulation should meet the recommendations of BS 5422: 1977. This Standard tabulates thicknesses of insulation too numerous to mention here, according to whether (1) the pipes carry central heating or domestic hot water, (2) the system is heated by gas and oil or solid fuel, (3) the water temperature is 75°C, 100°C or 150°C and (4) the thermal conductivity of the insulant is 0.04, 0.55 or 0.70 W/mK at the appropriate mean temperature.

Insulation of ducts The requirement for warm air heating ducts is that they also should meet the recommendations of BS 5422: 1977. As with pipes, if the ducts offer useful heat to the areas through which they run then they need not be insulated.

The lowest thicknesses of insulation recommended in Table 10 of BS 5422 are for insulants having a thermal conductivity of 0.04 W/mK. It recommends thicknesses of 38 mm, 50 mm and 63 mm when the warmed air to ambient air temperature differences are 10 K, 25 K and 50 K, respectively.

28.4.3 Water supply bye-laws

Like the Building Regulations, Bye-law 48, 'Protection from damage by freezing and other causes', is a functional requirement. It does not quote specific details but requires that 'every pipe or other water fitting, so far as is reasonably practicable shall be effectively protected . . . by insulation or other means against damage from freezing and other causes'. Generally, this would be satisfied by insulating to BS 6700: 1987 recommendations which, for an insulant of thermal conductivity 0.035 W/mK (which applies to most fibrous and plastic insulants) at the temperatures appropriate to freezing, would mean insulation thicknesses of approximately 25 mm for all pipes in indoor installations and 30 mm for those outdoors.

BS 5422 (which also gives insulation thicknesses for protection against freezing) recommends thicknesses of 32 mm and 38 mm, respectively, for pipes of 48 mm outside diameter or less. It must be emphasized that these thicknesses only give protection for a relatively short period of time (i.e. overnight). It is not possible by means of insulation alone to protect permanently static water.

28.4.4 BS 5422, Specification for the use of thermal insulating materials

Much mention has been made of this Standard with regard to insulation. It contains definitions, physical characteristics and recommended thicknesses of insulation for a wide range of industrial applications, including:

1. Refrigeration;
2. Chilled and cold water supplies for industrial applications;
3. Central heating and domestic hot and cold water supplies;
4. Process pipework and equipment.

It also contains heat loss calculation methods. This Standard should be the basis for any specification drawn up by a company that does not have its own in-house insulation standards manual.

28.5 Product selection

In selecting an insulation product for a particular application consideration should be given not only to its primary function but also to the many secondary functions, and often unappreciated requirements which the insulation of pipe or vessel may place on the insulation product. Some of these product requirements are discussed below.

28.5.1 Limiting temperatures

Ensure that the insulation selected can operate effectively and without degradation at temperatures beyond the design temperature called for. Temperature control of the process system can fail and systems can overheat. Hot surfaces exposed to ambient air will become hotter when insulated if there is no control of the process temperature. As with multi-layer systems, the addition of an extra layer of insulation will change the interface temperatures of all inner layers. Do not use an insulant close to a critical temperature limit.

28.5.2 Thermal movement

Large vessels and equipment operating at extreme temperatures exhibit large expansion or contraction movements. Any insulant specified for these applications should be capable of a sympathetic movement such that it will not cause itself or any cladding to burst, nor should it produce gaps that lead to dangerous hot spots in the cladding system.

28.5.3 Mechanical strength

Many insulants are made in a wide range of densities and thus mechanical strengths. Ensure that where insulation is required to be load bearing, to carry cladding, to support itself across gaps or drape down the sides of buildings or high equipment, the selected product has the necessary strength to accommodate these mechanical requirements as well as the primary thermal one.

28.5.4 Robustness

Insulation is subjected to abuse on-site, in storage and often in transit. To ensure minimum wastage through breakage, contamination or deformation, select products which are resilient or robust enough to tolerate site conditions and malpractice.

28.5.5 Chemical resistance

No matter how well installed, insulation is always at risk of contamination from outside sources. Overfilling of vessels, leaking valves and flanges, oil thrown off from rotating transmission shafts and motors can all penetrate protective cladding or lining systems. Consideration of any insulant's compatibility with possible contaminants should be considered in these situations.

28.5.6 Weather resistance

As mentioned above, insulation applied to externally located equipment can be subjected to rain and weather contamination if the outer cladding fails. Insulants with water-repellant, water-tolerant or free-draining properties offer an additional benefit in this type of application. In the structural field insulants used as cavity-wall fills must be of those types specially treated and designed for this application.

28.5.7 Surface emissivity

The effects of surface emissivity are exaggerated in high-temperature applications, and particular attention should be paid to the selection of the type of surface of the insulation system. Low-emissivity surfaces such as bright polished aluminium reduce heat loss by inhibiting the radiation of heat from the surface to the surrounding ambient space. However, by holding back the heat being transmitted through the insulation a 'dam' effect is created and the surface temperature rises. This temperature rise can be considerable, and if insulation is being used to achieve a specified temperature the use of a low-emissivity system could necessitate an increased thickness of insulation. For example, a hot surface at 550°C insulated with a 50 mm product of thermal conductivity 0.055 and ambient temperature of 20°C would give a surface temperature of approximately 98°C, 78°C and 68°C when the outer surface is of low (polished aluminium), medium (galvanized steel) or high (plain or matt) emissivity, respectively. Conversely, to achieve a surface temperature of 55°C with the same conditions the insulation would need to be approximately 120 mm, 87 mm and 70 mm, respectively.

28.5.8 Acoustics

Because of their cellular or open-matrix construction, most insulants have an inherent ability to absorb sound, act as panel dampers and reduce noise breakout from plant by their ability to be a flexible or discontinuous link between an acoustically active surface and the outer cladding. This secondary aspect of thermal insulation specification will gain more prominence when the UK adopts the EC Directive 86/1888, 'Protection of workers from the risks related to exposure to noise at work'.

28.5.9 Fire safety

Large volumes of insulation are used in industry, and most is hidden under cladding behind linings and between sheeting. However, much is located in voids and open to view in workplaces.

Fires do occur and accidents happen. In selecting the appropriate material, consideration should be given to its fire-safety properties and also to methods of maintaining the integrity of the protective cladding in fire situations.

Fire properties of insulation materials range from the highest to the lowest, from non-combustible to flammable with toxic fume emission. Generally, inorganic materials tend to be non-combustible while organic (or oil-based) materials are combustible, but many have surface treatments to improve their fire-safety rating.

28.6 Thermal conductivity

Thermal conductivity is not a static property of a material. It can vary according to the density, operating temperature and type of gas entrapped within the voids.

28.6.1 Density effects

Most materials achieve their insulating properties by virtue of the high void content of their structure. The voids inhibit convective heat transfer because of their small size. A reduction in void size reduces convection but does increase the volume of the material needed to form the closer matrix, thus resulting in an increase in product density. Further increases in density continue to inhibit convective heat transfer, but ultimately the additional benefit is offset by the increasing conductive transfer through the matrix material and any further increase in density causes a deterioration in thermal conductivity (see Figure 28.2).

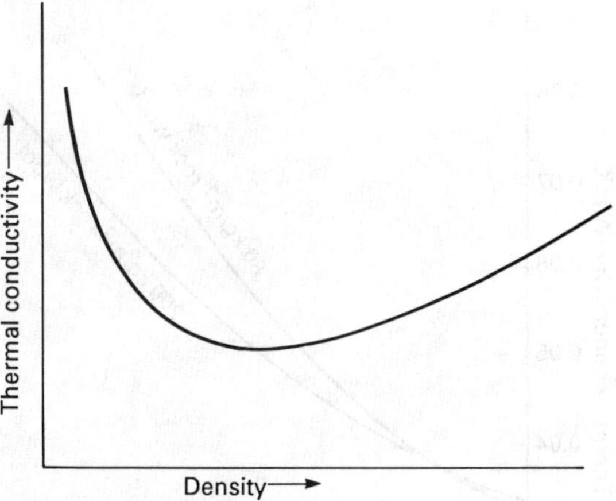

Figure 28.2 Typical relationship between thermal conductivity and density

Most traditional insulants are manufactured in the low- to medium-density range and each particular product family will have its own specific relationship between conductivity and density. One particular group of products (insulating masonry) is manufactured in the medium- to high-density range. Thermal conductivity is improved by reducing density.

28.6.2 Temperature effects

Thermal conductivity increases with temperature. The insulating medium (the air or gas within the voids) becomes more excited as its temperature is raised, and this enhances convection within or between the voids, thus increasing heat flow. This increase in thermal conductivity is generally continuous for air-filled products and can be mathematically modelled (see Figure 28.3). Those insulants which employ 'inert gases' as their insulating medium may show sharp changes in thermal conductivity, which may occur because of gas condensation. However, this tends to take place at sub-zero temperatures.

28.6.3 Chlorofluorocarbons

Much has been written lately about CFCs. This gas is used in the manufacture of a range of plastic insulants, including polyurethanes, phenolics, polyisocyanurates and extruded polystyrenes.

The manufacturers and their suppliers are actively seeking alternatives, and some CFC-free polyurethanes

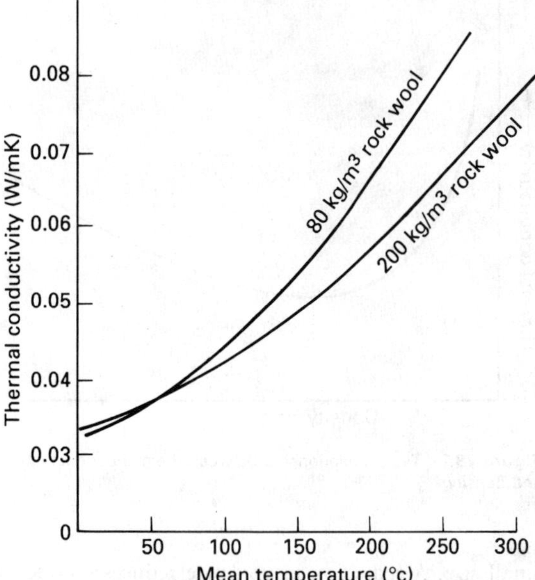

Figure 28.3 Typical relationship between thermal conductivity and mean temperature

are already appearing on the market. However, these new products do not have such good insulating properties (thermal conductivities of the order of 0.03 W/mK being quoted, as against the 0.02/0.025 W/mK of the originals).

Where CFC-bearing products are described in Section 28.9 the thermal conductivities quoted are those of current (CFC) manufacture.

28.7 Physical forms

The extensive range of insulation types and the numerous forms in which they are manufactured ensures that any listing can never be comprehensive. It becomes even more so when one acknowledges the fact that within the insulation industry there is a secondary network of fabricators and laminators who will take manufacturers' basic products and cut, shape, mould, laminate and enclose them to almost any requirement. The following gives the basic physical forms and some of the uses to which insulation materials are put.

28.7.1 Slab or board

Material of a rectangular shape and of uniform thickness. The basic product supplied by almost all manufacturers in forms ranging from 'flexible' to 'rigid'. Dimensions determined by manufacturing facility and handling ability, but usually in 600 mm and 1200 mm modules. May be used in most applications on plane surfaces. As lining panels, under cladding, within casings, in partitions, in buildings or plant and equipment.

28.7.2 Pipe sections

Rolled, moulded or cut to shape as annular cylinders for the insulation of pipes. Made by most manufacturers usually in one or two segments but occasionally in three or four. Sizes available for most steel and copper pipes of up to approximately 1120 mm outside diameter.

28.7.3 Bevelled lags

Slabs of narrow width bevelled on edges (and sometimes radiused on faces). Fitted longitudinally on large-sized pipes with face conforming closely to the circumference. Another product available from most manufacturers.

28.7.4 Loose fill

Granulate, particle-size material specifically manufactured or produced from basic product off-cuts. Supplied by most manufacturers and used for packing or pouring into irregular-shaped enclosures.

28.7.5 Mat or blanket

Flexible material of low density and long length generally supplied in rolls available with or without facings. Mineral wool and flexible foams are the most usual type of product. Widely used in buildings as roof insulation and in steel- and timber-frame walls. Sound insulation in partitions, on low-temperature heating ducts and in equipment casings.

28.7.6 Rolls

Mat or blanket supplied in the form of helically wound cylindrical packs.

28.7.7 Mattress

Medium- to heavy-density mineral wool mat with wire netting or expanded metal mechanically secured on one or both sides. Used for irregular-shaped surfaces and on large pipes and vessels. Also used as large-cavity fire barriers.

28.7.8 Quilt

Medium-density mat with flexible facings of paper or scrim stitched through. Used for acoustic absorption behind perforated or slatted panels and ceilings.

28.7.9 Lamella

Paper- or foil-backed mineral wool product fabricated from low-density slabs in which the slats stand upright with the fibres predominantly perpendicular to the major faces. Used on circular and elliptical ducting, pipes and vessels to give a compression-resistant insulant. Supplied in roll form.

28.7.10 Blowing wool

Nodulated mineral wool produced for application through pneumatic hoses into areas of restricted access.

28.7.11 Sprayed insulation

Insulation dry mixed in factory with inorganic fillers and binders for application by wet spraying. Usually mineral wool or vemiculite based, used for thermal and acoustic treatments and for fire protection of steelwork.

28.7.12 Sprayed foam

A cellular plastic – usually of the urethane or isocyanurate families – where chemical components are mixed at the spray head.

28.7.13 Moulded products

Insulation formed by slurry casting or heat curing under pressure in moulds in a number of insulation types. Most common mouldings are preformed bends, valve boxes and flange covers.

28.8 Facings

While all insulation products are supplied in their natural 'as-produced' finish, they are also available with a variety of facings which are either applied 'on-line' during manufacture or as a secondary 'off-line' process. The facings are applied for functional, technical or aesthetic reasons, generally as follows.

28.8.1 Paper

Brown Kraft paper is generally used on mineral wool products to give added tensile strength, ease handling, aid positive location between studs or rafters, or prevent contamination when used under concrete screeds. Paper is often laminated with polythene to give vapour control layer properties.

28.8.2 Aluminium foil

Glass-reinforced aluminium foil with either a bright polished or white lacquer surface is utilized with most types of insulant. Primarily it is used as a vapour control layer or as a means of upgrading the fire properties of plastic foams, but it does give a semi-decorative finish to the insulation. It is therefore often use where the insulation is open to view but located away from direct risk of mechanical damage.

28.8.3 PVC

Generally used with mineral wool products where, in its decorative forms, it gives attractive facings to ceilings and wall tiles and enhances their sound-absorption characteristics. PVC is also used as a vapour control layer facing.

28.8.4 Tissue

Glass-fibre tissue or non-woven fabrics are used for decorative purposes on many insultants. They also give improved strength to foam plastics and enhanced sound-absorption characteristics to mineral wools.

28.8.5 Glass cloth

As with tissue, woven glass cloth is used for decorative or acoustic purposes. Additionally, close-woven fabrics give improved fire-safety properties and are resistant to mechanical abuse. Glass cloth or scrim of an extremely open weave is used on insulants as a key for mastic or hard-setting finishes.

28.8.6 Wire netting

Used to give mechanical strength to mineral wool mattresses and is also an aid to application. Also a key for mastic or hard-setting finishes applied on-site.

28.8.7 Laminates

Insulation is widely used in laminate form with plaster-board, chipboard, cement board, and metal sheeting, etc. Applications are generally as lining board systems in buildings but prefabricated metal panels and ceiling tiles are also produced.

28.9 Insulation types

It is not possible to detail all the many different types of insulation used in production, service and building industries. The following attempts to give a summary of the composition, properties and major areas of use of a representative range of insulation types.

28.9.1 Mineral wool

Mineral wool is perhaps the best known of the whole range of insulation types. It is widely used in all sectors of industry, transport and building for thermal, acoustic and fire-protection purposes. Despite this, there is still a common misconception that mineral wool is a specific product type – it is not. Mineral wool is a generic name for a range of man-made non-metallic inorganic fibres. The following definitions should help to clarify the situation:

1. *Mineral fibre*: A generic term for all non-metallic inorganic fibres.
2. *Mineral wool*: A generic term for mineral fibres of a woolly consistency, normally made from molten glass, rock or slag.
3. *Glass wool*: A mineral wool produced from molten glass.
4. *Rock wool*: A mineral wool produced from naturally occurring igneous rock.
5. *Slag wool*: A mineral wool produced from molten furnace slag.

From these it can be seen that rock wool, slag wool and glass wool are all mineral wools.

28.9.2 Glass wool

Glass wool is made from borosilicate glass whose principal constituents are sand, soda ash dolomite, limestone, ulexite and anhydrite. They are melted in a furnace at about 1400°C and then fed along a channel to a forehearth, where the glass flows through bushings into spinners. These rapidly rotating spinners have several thousand small holes around the perimeter through which the glass is forced by centrifugal force to become fiberized. Immediately after formation, the fibres are sprayed with resinous binders, water repellents and mineral oils as appropriate and fall under suction onto a moving conveyor which takes the wool to one of three production lines (Figure 28.4).

Main line

The wool passes through an oven which cures the resinous binder and determines the thickness of the product. On leaving the oven, the insulation is trimmed, slit and chopped into the appropriate product length prior to reaching the packing station, where it is either packaged as rolls or slabs.

Pipelines

The uncured wool from the forming conveyor is separated into 'pelts' which are converted into pipe sections by being wrapped around a heated mandrel and the wall thickness set by counter rollers. The sections are then passed through a curing oven before being trimmed, slit, covered and packaged.

Blowing wool line

The water-repellent wool is shredded by a flail and pneumatically transferred to a rotating drum nodulator to be further processed before bagging.

Product range

A wide product range, from lightweight mats through flexible and semi-rigid, to rigid slab. Pipe sections, loose wool, blowing wool, moulded products and mattresses.

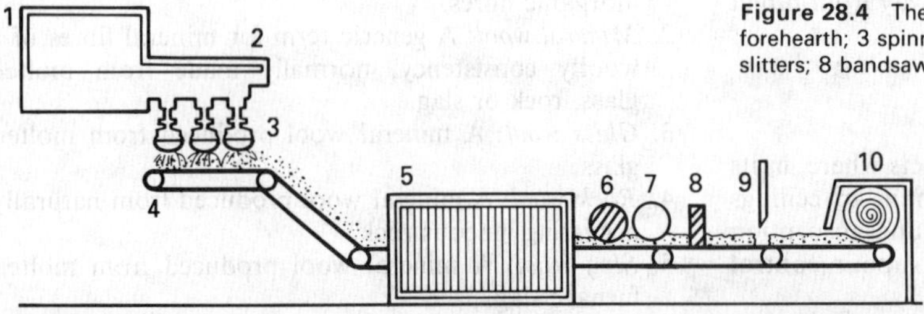

Typical properties

Glass wool products have a limiting temperature of 540°C but are mostly used in buildings and H & V applications where a limiting working temperature of 230°C is recommended.

Fire safety Basic wool is non-combustible to BS 476: Part 4.

Density range 10–75 kg/m^3.

Thermal conductivity 0.04–0.03 W/mK 10°C mean: 0.07–0.044 W/mK at 100°C mean.

Special applications

Acoustic absorption in duct lining and splitters. Decorative facings for ceiling and lining panels. Tensile strength for good draping qualitites. Loft insulation up to 200 mm thick, cavity wall insulation. Fire stopping and small-cavity fire barriers.

28.9.3 Rock wool

Rock wool is made from basalt or dolomite rocks. The crushed rock, together with limestone and coke, is loaded into a cupola furnace where, with the aid of oxygen, it is melted at about 1500°C. The molten rock then runs down channels and cascades onto a train of rotating discs which sling the melt off as fibres. Resin binders, water repellents and mineral oil are sprayed onto the fibres as they leave the discs and fall under suction onto the forming conveyor. From this stage the wool follows a similar processing sequence to glass wool.

Product range

Lightweight mats, flexible, semi-rigid and rigid slabs, pipe sections, loose wool, blowing wool mattresses, sprayed fibre, lamella mats.

Typical properties

Rock wool special products can be used up to 1100°C, but generally a maximum operating temperature of 850°C is recommended. Like glass wool, the lower-density products used in buildings and H & V applications have recommended limits of 230°C.

Figure 28.4 The glass wool manufacturing process. 1 Tank; 2 forehearth; 3 spinners; 4 conveyor; 5 curing oven; 6 trimmers; 7 slitters; 8 bandsaw; 9 guillotine; 10 rolling machine.

Fire safety Basic wool is non-combustible to BS 476: Part 4.

Density range 23–200 kg/m^3.

Thermal conductivity 0.037–0.033 W/mK at 10°C mean: 0.052–0.042 W/mK at 100°C mean.

Special applications

High-temperature work, fire protection, acoustics, sprayed fibre, moulded products, cavity-wall insulation, loft insulation.

28.9.4 Ceramic fibres

As with mineral wools, there are different types of ceramic fibre, but basically they are all made from a combination of alumina, silica and china clay and may be made by blowing or extruding the liquid melt.

Product range

A wide basic range of blankets, slabs, block and pipe sections, especially moulded products, felts and gaskets.

Typical properties

Ceramic fibres are used at temperatures up to 1600°C but their melting point can be in excess of 2000°C.

Fire safety Basic wool is non-combustible to BS 476 Part 4.

Density range 50–300 kg/m^3.

Thermal conductivity 1.8–0.8 W/mK at 500°C mean.

Special applications

Furnace linings, 'refractory blocks', gaskets, expansion joints, very high-temperature work.

28.9.5 Magnesia

Magnesium carbonate is extracted from dolomite rock, and after mixing with fibre reinforcement is slurry cast into appropriate moulds. After drying, the products are machined to size.

Product range

Rigid slabs, pipe sections, bevelled lags and moulded products. Also as a powder mix for use in a wet cement form.

Typical properties

Temperature range Up to 315°C.

Fire safety Non-combustible to BS 476: Part 4.

Density 190 kg/m^3.

Thermal conductivity 0.058 W/mK at 100°C mean.

Special applications

Where load-bearing capacity is required and in food, pharmaceuticals and cosmetics processing industries.

28.9.6 Calcium silicate

A chemical compound of lime and silica with fibre reinforcement for added strength. It is cast as a wet slurry into moulds and charged into autoclaves, finally being machined to accurate size.

Product range

Rigid slabs, pipe sections, bevelled lags, moulded products and as a dry mix for wet plastic application.

Typical properties

Standard products have limiting temperatures of 800°C but special formulations enable applications up to 1050°C.

Fire safety Non-combustible to BS 476: Part 4.

Density range 240–400 kg/m^3.

Thermal conductivity 0.054–0.068 W/mK at 100°C.

Special applications

Industrial process applications where compressive strength is needed. Also used in underground pipework district heating mains.

28.9.7 Cellular glass

Powdered glass and crushed carbon are placed in moulds and heated to 1000°C, at which temperature the carbon is oxidized, forming gas bubbles which causes expansion of the glass mix. The cellular material is then annealed and, after cooling, cut to size.

Product range

Rigid slabs, pipe sections, bevelled lags, moulded products.

Typical properties

Temperature range −260°C to 430°C.

Fire safety Non-combustible to BS 476: Part 4.

Density range 125–135 kg/m^3.

Thermal conductivity 0.043–0.45 W/mK.

Special applications

Its closed-cell structure makes cellular glass particularly suitable for refrigeration applications on pipes and cold stores. High load-bearing capacity enables it to be used under rooftop car parks.

28.9.8 Exfoliated vermiculite

Vermiculite is a naturally occurring group of hydrated aluminium–iron–magnesium silicates having a laminate structure. When subjected to direct heat in a furnace, the pulverized material 'exfoliates' or expands in size, and then consists of a series of parallel plates with air spaces between.

Product range

Base product produced as a granular loose fill which can be bonded to form boards or dry mixed with fillers and binders for spray application.

Typical properties

Limiting temperatures 1100°C.

Fire safety Non-combustible to BS 476: Part 4.

Density 60–80 kg/m^3.

Thermal conductivity 0.062 W/mK at 10°C mean.

Special applications

Primarily used in fire-protection applications as dense boards or wet sprays.

28.9.9 Expanded polystyrene

Expandable polystyrene grains are usually heated by stream, which causes them to expand. They are then conditioned and, as they cool, the steam in the voids within the beads condenses, thus permitting air to diffuse into them. After conditioning, the granules are placed in moulds through which more steam is blown, resulting in further expansion. As they are enclosed in the mould the granules tend to fuse together, forming a rigid block. The blocks are later cut to shape by the hot-wire technique.

Product range

Rigid slabs, pipe sections, tiles, and loose granules.

Typical properties

Limiting temperatures 80°C.

Fire safety Flame-retardent grade. Class P not easily ignitable to BS 476: Part 5.

Density 12–32 kg/m^3.

Thermal conductivity 0.037–0.034 W/mK at 10°C mean.

Special application

Used mainly for building insulation and as a laminate. Also employed in cold water tanks and pipes.

28.9.10 Extruded polystyrene

This is produced by a continuous extrusion process which gives the product a smooth-surface skin and enhances the mechanical properties.

Product range

Rigid slabs.

Typical properties

Limiting temperature 75°C.

Fire safety Classified as Type A to BS 3837.

Density 28–45 kg/m^3.

Thermal conductivity 0.025–0.033 W/mK at 10°C mean.

Special applications

Used mainly in structural applications. Is suitable for refrigeration and cold stores.

28.9.11 Rigid polyurethane foam

Polyurethanes are manufactured by the mixing of various resins, isocyanates and catalysts to produce an exothermic reaction which liberates the foaming agent and causes the mix to expand. They are made in large block moulds as a batch process or are continuously foamed onto a paper or polythene substrate on a conveyor system.

Product range

Rigid slabs, pipe sections, bevelled lags.

Typical properties

Limiting temperature 110°C.

Fire safety Class P not easily ignitable to BS 476: Part 5. Can achieve Class 1 surface spread of flame to BS 476: Part 7 when faced with aluminium foil.

Density 30–160 kg/m^3.

Thermal conductivity 0.023 W/mK at 10°C mean.

Special applications

Used mainly for refrigeration and cold water services,

also low-temperature tankage. When laminated to suitable facings is employed in building applications.

28.9.12 Polyisocyanurate foam

Polyisocyanurates are manufactured in a similar way to polyurethanes, the chemical components being selected to enhance their fire-safety properties.

Product range

Slabs, pipe sections, bevelled lags.

Typical properties

Limiting temperature 140°C.

Fire safety Class 1 surface spread of flame to BS 476: Part 7. Can achieve Class O to Building Regulations when faced with aluminium foil.

Density 32–50 kg/m^3.

Thermal conductivity 0.023 W/mK at 10°C mean.

Special applications

For cryogenic and medium-temperature pipework and equipment, especially in oil petrochemical industries and on refrigerated road and rail vehicles.

28.9.13 Phenolic foam

Similar production process to polyurethane. Has the best fire-safety properties of all the rigid foams.

Product range

Slabs, pipe sections, pipe bends, bevelled lags.

Typical properties

Limiting temperature 150°C.

Fire safety Class O to Building Regulations.

Density 35–200 kg/m^3.

Thermal conductivity 0.02 W/mK to 10°C mean.

Special applications

Commercial and Industrial H & V applications and where Class O fire rating and low smoke-emission characteristics are required.

29A Economizers

Colin French CEng, FInstE, FBIM
Saacke Ltd

Contents

29A.1 Introduction

Economizers for boilers have been available for nearly 150 years, almost as long as boilers themselves. For shell boilers, increasing effiencies have made it increasingly difficult to justify the use of an economizer, the final decision being based in terms of payback period, which is also heavily dependent on fuel prices. Watertube boilers, on the other hand, need an economizer section in the gas passes in order to obtain satisfactory efficiency. For this reason, the economizer is integrated into the overall design, normally between the convective super-heater and the air heater if fitted.

In shell boilers with a working pressure of between 7 and 17 bar the temperature of the mass of water in the boiler is typically in the range of 170–210°C. Allowing for, say a temperature difference of 30–50°C between the exhaust gases and the water temperature, the boiler exit gas temperature cannot be practically reduced beneath about 200–260°C, dependent on the operating pressure. It becomes necessary therefore to modify the process on principles to achieve further heat utilization and recovery. In the case of economizers this is done by conducting the feedwater supply via an economizer wherein the exhaust gas passes over tubes carrying the feedwater. The feedwater, normally at temperatures between 30° and 100°C, represents a further cooling medium for the exhaust gases and provides the potential for the extra heat utilization. This is shown in Figure 29A.1.

29A.2 Oil and coal applications

Although it would be possible to design for an economizer gas exit temperature of 30–50°C above the feedwater temperature, this would result in a temperature too close to the acid dewpoint of the gases. The acid dewpoint is the temperature at which acidic gases begin to condense out of the exhaust gas mixture. This is principally sulphuric acid due to the sulphur contained in the oil or coal. Although the bulk gas temperature may be satisfactory, in practice, the surfaces close to the tube wall will be nearly at water temperature due to the high conductivity of the metal tube. This limits the minimum practical gas exit temperature from the economizer to, say, 170–180°C, remembering, of course, that at low fire this will have fallen closer still to the acid dewpoint (which is typically in the range of 125–140°C, dependent on the excess air and fuel sulphur content). The cold temperature of the heat transfer surface gives rise to heavy corrosion which would reach a peak at about 95°C.

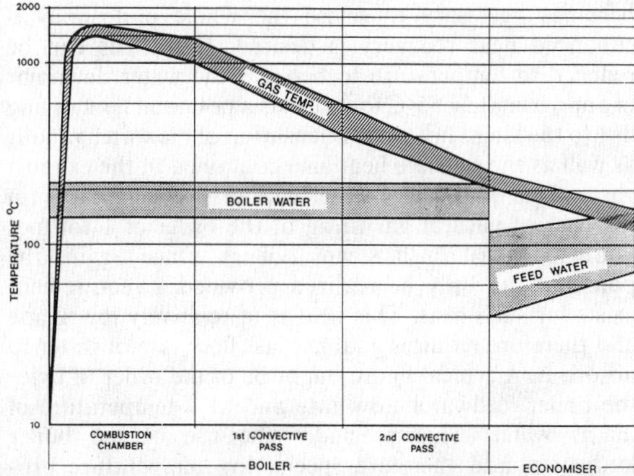

Figure 29A.1 Temperatures in three-pass boilers and economizers

The potential to recover heat from the gases of an oil- or coal-fired boiler is therefore limited to a tempera-ture drop from 240°C to 170°C. This results in a 3% saving. The average saving would be somewhat lower than this since fouling of the economizer surface is inevitable from the carbonaceous emissions of the firing equipment.

The design of the economizer must be robust enough to survive occasional excursions beneath the acid dewpoint and the effects of the methods used to clean the economizer periodically. This may take the form of rapping equipment to shake off deposits, sootblowing by steam or air and water washing with lances.

Average savings of 2–3% combined with the cost necessary for a robust design have therefore limited the use to times when fuel prices are high or special applications where the boiler exit temperature is higher than usual. Developments using an additive to suppress the dewpoint are worth consideration, since this extends the heat-saving potential but, of course, there is an additional burden with the cost of the additive. Mag-nesium oxide is the most commonly used suppressant and this is injected into the gas stream to combine with the sulphuric acid to form magnesium sulphate (which also has to be removed regularly in addition to the soot).

29A.3 Gas-fired economizers

Far greater potential exists for gas-fired economizers, since the gas is virtually free of sulphur. The limitation on gas temperature is the ability of the water to extract the heat, although the water vapour in the gas caused by the combustion of hydrogen does give rise to a water dewpoint at 55°C. This should be avoided, since general corrosion can take place in the latter rows of the economizer and in the exhaust gas ducts and chimney. Normally, this only occurs for short periods during starting from cold, but it should be minimized.

Since no serious problems exist with corrosion, the materials of construction can be cheaper and the thicknesses reduced. Fouling coefficients do not need to be incorporated into the design calculations as the surfaces remain clean indefinitely. This in itself provides a secondary benefit in that the pitching of fins may be reduced without the risk of reducing flow.

A saving of (typically) 5% can be made with gas-fired economizers, and as this is related to the boiler output it represents a saving of 6.25% in fuel consumed. Gross heat-saving potential related to feedwater temperature and excess air level are shown in Figure 29A.2.

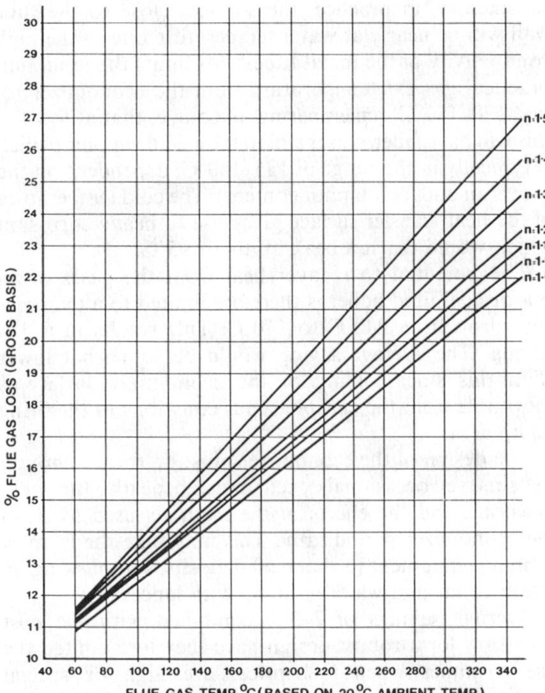

Figure 29A.2 Total heat in gas-fired boiler exhaust

29A.4 Design

The driving force for heat transfer is temperature difference, and this is maximized through the economizer by arranging for concurrent flow through the passages. In other words, the colder feedwater is made available to the coolest gases. This gives the highest overall average temperature difference throughout the length of the economizer, and is characterized by the logarithmic temperature difference ΔTlm:

$$\Delta Tlm = \frac{(Tg_1 - Tw_2) - (Tg_2 - Tw_1)}{\log n[(Tg_1 - Tw_2)/(Tg_2 - Tw_1)]}$$

The amount of heat (Q_t) that may be transferred is controlled by:

$$Q_t = \Delta Tlm \cdot U \cdot A$$

where U = overall heat transfer coefficient and A = heat transfer area. To achieve a cost-effective design of economizer it is necessary to maximize the overall heat transfer coefficient and the surface area within the economizer.

The heat transfer coefficient, U, can be maximized by the highest practical velocities that can be achieved for both the water and the gas. Water-side pressure loss is limited by the available spare pressure rise from the feed pump. This is often limited to 0.2–1.0 bar, as it is not generally economic to replace the feed pump. Gas-side velocities are generally limited by the available spare fan head from the burner. Once again, it is not normally economic to replace this in the case of retrofitted economizers or, for that matter, to install an induced-draught fan downstream of the economizer. In addition to maximizing fluid velocities, careful design to promote turbulence by staggered pitching enhances the heat transfer coefficient.

The heat transfer area, A, can be greatly increased by using finned tubes, but care must be taken to ensure good conduction of heat away from the fin into the tube and subsequently into the water. Some common fin types are shown in Figure 29A.3 while Figure 29A.4 shows some of the attachment methods employed to ensure satisfactory fin to tube heat transfer.

Mechanical design in the UK is normally to BS 1113: 1989, which is the standard for watertube boilers. Although it is theoretically possible to design to a less stringent standard than this for shell boilers, in reality the requirements of the insurance companies are that this standard is applied for all boiler types. The result of this is that design standards are very high and that problems are rarely found. Typical constructional arrangements (including tube layout) are shown in Figure 29A.5.

29A.5 Installation

Due to interruptible gas tariffs, it is often necessary to adopt gas as the primary fuel and burn oil in periods of peak loads. This means that the economizer has to be arranged so that when oil firing, the flue gases are bypassed around the economizer. The bypass duct must also contain a damper to simulate the economizer gas resistance so that the burner back pressure remains the same for both fuels. Figure 29A.6 shows a typical installation layout.

29A.6 Condensing economizers

The restrictions on feedwater temperature dictated by condensate rates and the need to minimize oxygen corrosion limit further development of the performance of conventional economizers. As mentioned earlier, it becomes necessary to revise the whole philosophy if enhanced heat recovery is desired. If the gas can be reduced in temperature to beneath the water dewpoint of approximately 55°C there exists a potential heat saving due to the latent heat of condensation of the water vapour as well as the sensible heat also contained in the exhaust gases. This quantity of water vapour is considerable (in the case of natural gas firing of the order of 1 ton per hour for a 10 ton/h steam boiler). Once again, the potential can only be realized provided a requirement exists for such heat. This heat is of relatively low grade and therefore requires a large mass flow rate of water to absorb it. A typical figure might be of the order of twice the boiler feedwater flow rate and at a temperature of mains water. This precludes the use of the boiler feedwater, and therefore special low-temperature processes must be integrated into condensing economizer applications. The temperature rise of this water would

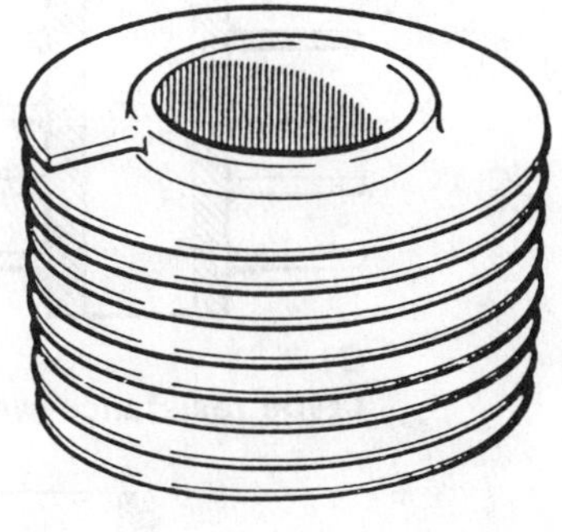

a)
Wound on - Plain fin

b)
Wound on - With turbulence inducing segments

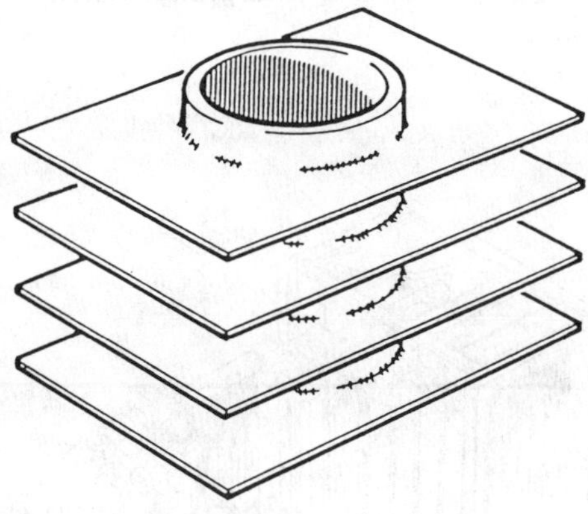

c)
Lamel fin - Square or rectangular continuous

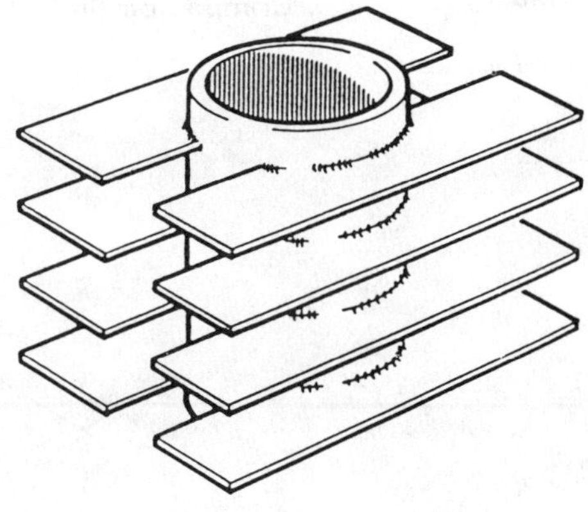

d)
Lamel fin – non continuous

Figure 29A.3 Common fin types

probably be about 45°C. Assuming the exhaust gases are cooled from, say, 230°C to 40°C, the gross heat saving would be approximately 13%, yielding a fuel saving of over 15%.

The term 'fuel saving' needs to be qualified, since the use of the heat in the water will be given to a process which may be unrelated to boiler demand. Examples of suitable applications occur in laundries, agricultural soil heating, food industries, abattoirs and swimming-pool heating.

The success of such schemes is highly dependent on matching supply and demand together with a basic low-grade heat requirement. The total heat remaining in the exhaust gases of the boiler are shown in Figure 29A.7 illustrating the potential savings.

Condensing economizers are constructed from corrosion-resistant materials (notably aluminium or stainless steel), since the condensed water vapour in the gas is slightly acidic (typically, with a pH of 3–5). This is because some carbonic and nitric acid is formed in the condensing of the products of combustion and also as a result of chlorofluorocarbon propellants in the atmosphere. Provided the correct choice of materials is made, corrosion life should not become problematic.

A typical arrangement of a condensing economizer is shown in Figure 29A.8. Note that an induced-draught fan is almost mandatory, since a high pressure loss is inevitable with such low-grade heat recovery. The design of such economizers often takes the form of a large shell and tube heat exchanger, but the conventional economizer construction of watertubes is quite feasible. Condensate from the exhaust gases are normally discharged to drains.

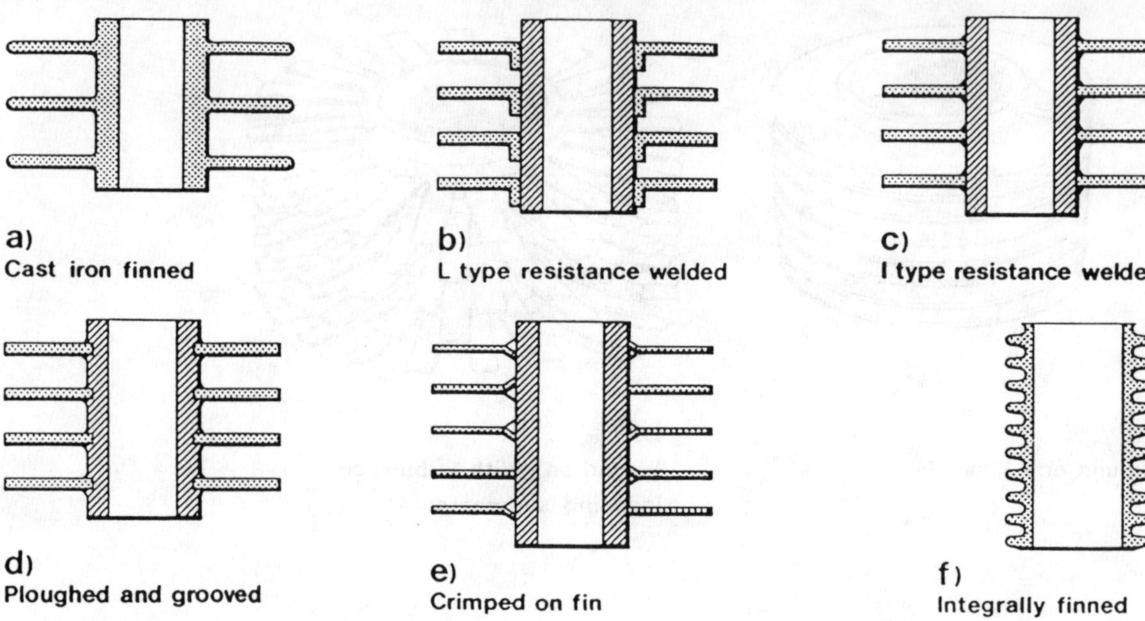

Figure 29A.4 Fin attachment methods

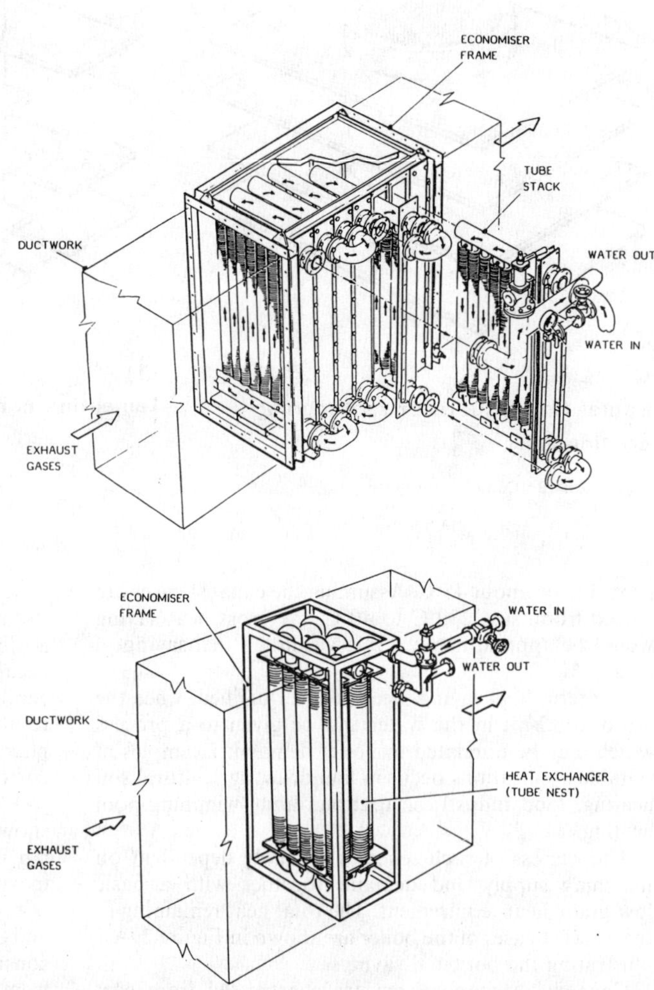

Figure 29A.5 Economizer – typical designs

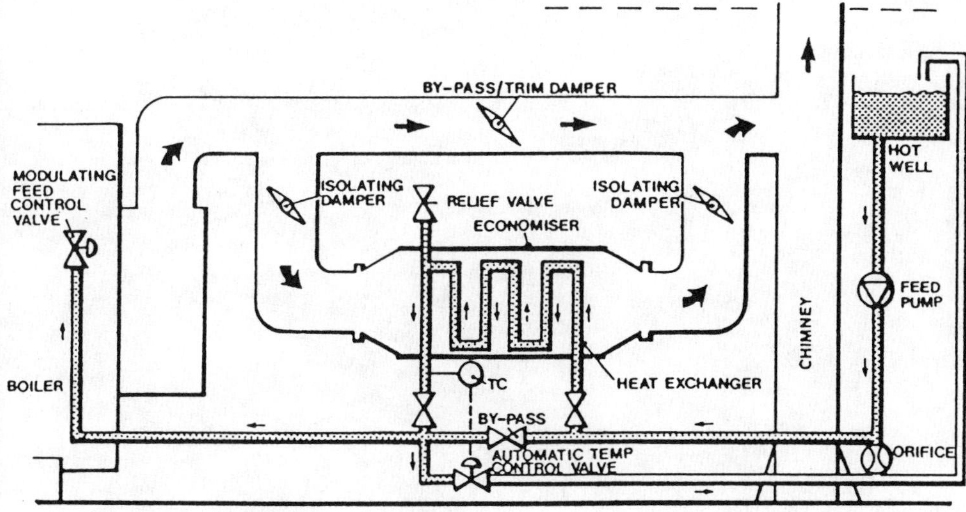

Figure 29A.6 Typical installation layout

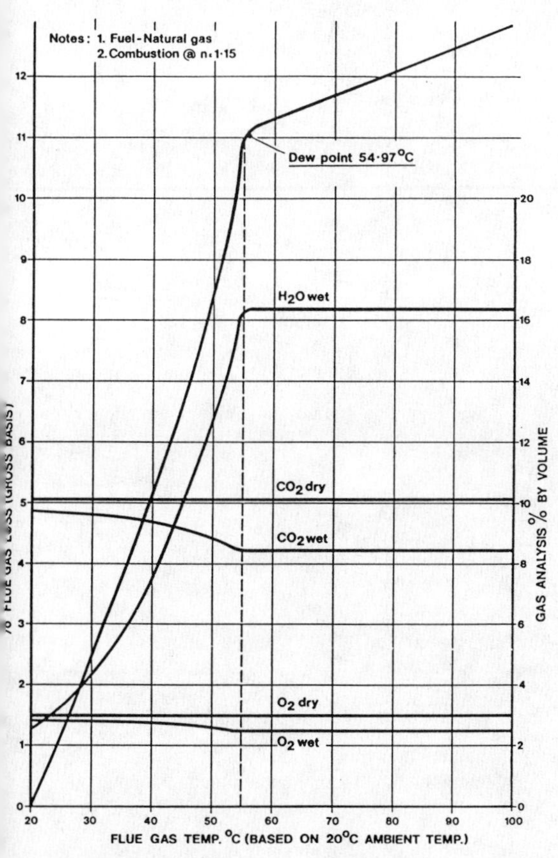

Figure 29A.7 Sensible and latent heat savings potential and gas analysis with condensing economizers

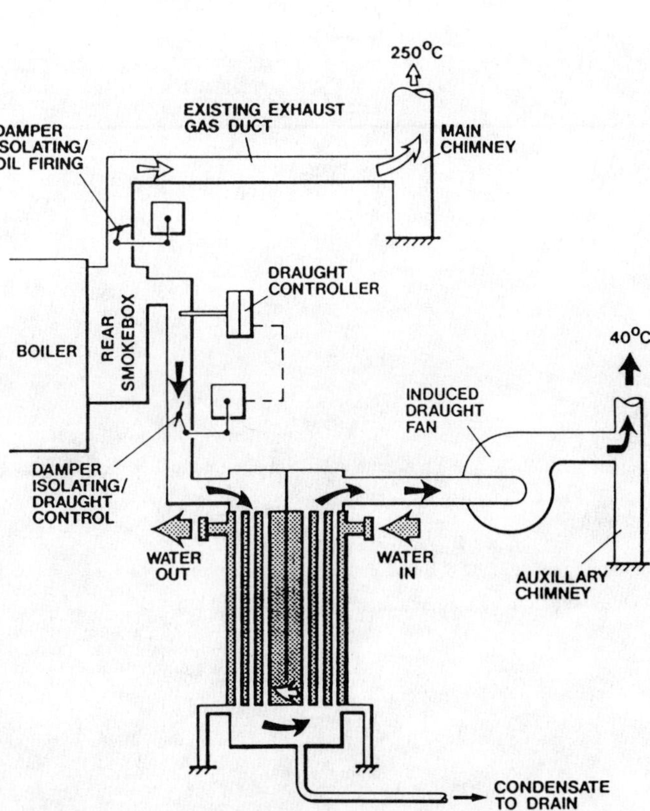

Figure 29A.8 Condensing economizer layout

29B Heat Exchangers

Brian R Lamb CEng, MIChemE
APV Baker Ltd

Contents

29B.1 The APV Paraflow

While the original idea for the plate heat exchanger was patented in the latter half of the nineteenth century, the first commercially successful design was introduced in 1923 by Dr Richard Seligman, founder of APV. Initially, a number of cast gunmetal plates were enclosed within a frame in a manner quite similar to a filter press. The early 1930s, however, saw the introduction of plates pressed in thin-gauge stainless steel. While the basic design remains unchanged, continual refinements have boosted operating pressures from about 1 to 20 kgf/cm^2 in current machines.

The APV Paraflow plate heat exchanger consists of a frame in which closely spaced metal plates are clamped between a head and follower. The plates have corner ports and are sealed by gaskets around the ports and along the plate edges. A double seal forms pockets open to atmosphere to prevent mixing of product and service liquids in the rare event of leakage past a gasket.

The plates are grouped into passes with each fluid being directed evenly between the paralleled passages in each pass. Whenever the thermal duty permits, it is desirable to use single-pass, counterflow (Figure 29B.1) for an extremely efficient performance. Since the flow is pure counterflow, correction factors required on the LMTD approach unity. With all connections on the head of the unit, the follower is free for very quick access to cleaning and maintenance. The effect of multipass operation will be discussed later in this chapter.

29B.1.1 Comparative plate arrangements

Clarification of Paraflow plate arrangements with those for a tubular exchanger is detailed in Figure 29B.2. Essentially, the number of passes on the tube size of a tubular unit can be compared with the number of passes on a plate heat exchanger. The number of tubes per pass can also be equated with the number of passages per pass for the Paraflow. However, the comparison with the shell side is usually more difficult, since with a Paraflow the total number of passages available for the flow of one fluid must equal those available for the other fluid to within ± 1. The number of cross passes on a shell, however, can be related to the number of plate passes, and since the number of passages/pass for a plate is an indication of the flow area, this can be equated to the shell diameter. This is not a perfect comparison but it does show the relative parameters for each exchanger.

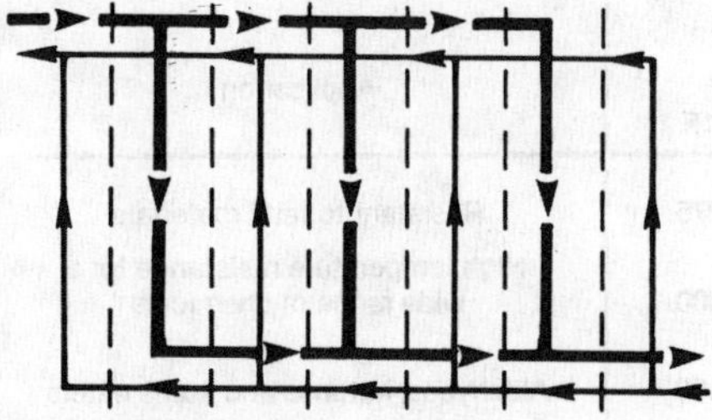

Figure 29B.1 Single-pass operation

	Shell and tube	Plate equivalent	
Tube side	Number of passes	Number of passes	Side one
	Number of tubes/pass	Number of passages/pass	
Shell side	Number of cross-passes (Number of baffles + 1)	Number of passes	Side two
	Shell diameter	Number of passages/pass	

Figure 29B.2 Comparison of pass arrangement: plate versus tubular

An important, exclusive feature of the plate heat exchanger is that by the use of special connector plates it is possible to provide connections for alternate fluids so that a number of duties can be done in the same frame.

29B.1.2 Plate construction

Paraflow plates are pressed from stainless steel, titanium, Hastelloy, Avesta 254 SMO, Avesta 254 SLX or any material ductile enough to be formed into a pressing. The specious design of the trough pattern strengthens the plates, increases the effective heat transfer area and produces turbulence in the liquid flow between plates. Plates are pressed in materials between 0.5 and 1.2 mm and the degree of mechanical loading is important. The more severe case occurs when one process liquid is operating at the highest working pressure and the other at zero pressure. The maximum pressure differential is applied across the plate and results in a considerable unbalanced load. There are two alternative trough forms, one using deep corrugations to provide contact points for about every 650–1950 mm^2 of heat transfer surface, the other criss-cross shallow troughs with support maintained by corrugation/corrugation contact. Alternate plates are arranged so that corrugations cross to provide a contact point for every 100–600 mm^2 of area. The plate then can handle large differential pressures and the cross pattern forms a tortuous path that promotes substantial liquid turbulence and a very high heat transfer coefficient. The net result is high rates with moderate pressure drop.

Plates are available with effective heat transfer area from 0.03 to 3.5 m^2 and up to 700 can be contained within the frame of the largest Paraflow, providing over 2400 m^2 of surface area. Flow ports and associated pipework are sized in proportion to the plate area and control the maximum liquid throughput.

Figure 29B.3 shows the relationship between port diameter and fluid velocity at 4 and 7 m/s and highlights the nominal maximum velocities for various plates. As the flow through the machine increases, the entry and exit pressure losses also increase. The nominal maximum flow rate for a plate heat exchanger limits these losses to

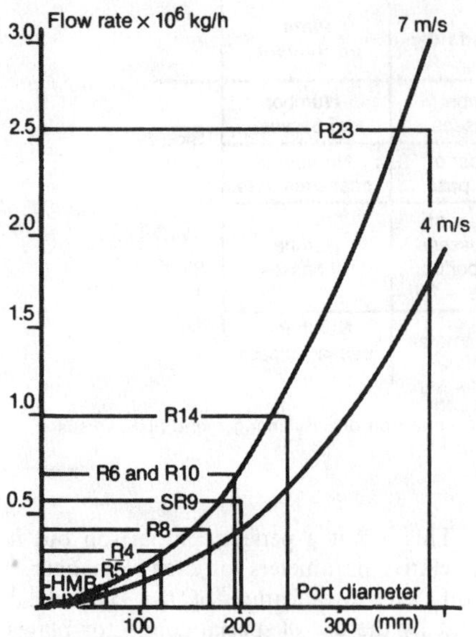

Figure 29B.3 Throughput versus port diameter

an acceptable proportion of the total pressure losses, and is therefore a function not only of the port diameter but also of the nature of the plate which has been empirically determined.

These velocities seem, at first, rather high compared to conventional pipework practice, but they are very localized in the exchanger and are progressively reduced as distribution into the flow passages occurs from the port manifold.

29B.1.3 Gasket materials

As detailed in Figure 29B.4, various gasket elastomers are available which have chemical and temperature resistance coupled with good sealing properties. The temperatures shown are maximum, therefore possible simultaneous chemical action must be taken into account.

29B.1.4 Thermal performance

Data on thermal performance are not readily available on all heat exchangers because of the proprietary nature of the machines. To exemplify typical thermal data, heat transfer can best be described by a Dittus–Boelter type equation:

$$Nu = A(Re)^n(Pr)^m\left(\frac{\mu}{\mu_w}\right)^x$$

Reported values of the constant and exponents are

$A = 0.15–0.40 \qquad m = 0.30–0.45$

$n = 0.65–0.85 \qquad x = 0.05–0.20$

where

$$Nu = \frac{hd}{d} \qquad Re = \frac{Vdp}{u} \qquad Pr = \frac{Cp\mu}{k}$$

d is the equivalent diameter defined in the case of the plate heat exchanger as 2× the mean gap.

Typical velocities in plate heat exchangers for waterlike fluids in turbulent flow are 0.3–0.9 m/s but true velocities in certain regions will be higher by a factor of up to 4 due to the effect of the corrugations. All heat transfer and pressure drop relationships are, however, based on either a velocity calculated from the average plate gap or on the flow rate per passage.

Figure 29B.5 illustrates the effect of velocity for water at 16°C on heat transfer coefficients. This graph also plots pressure drop against velocity under the same conditions. The film coefficients are very high and can be obtained for a moderate pressure drop.

One particularly important feature of the plate heat exchanger is that the turbulence induced by the troughs reduces the Reynolds number at which the flow becomes laminar. If the characteristic length dimension in the Reynolds number is taken as twice the average gap between plates, the *Re* number at which the flow becomes laminar varies from about 100 to 400, according to the type of plate.

To achieve these high coefficients it is necessary to expend energy. With the plate unit, the friction factors normally encountered are in the range of 10 to four hundred times those inside a tube for the same Reynolds number. However, nominal velocities are low and plate lengths do not exceed 2.3 m, so that the term $V^2L/2g$ in the pressure drop equation is very much smaller than one would normally encounter in tubulars. In addition,

Gasket material	Approx. maximum operating temp. °C	°F	Application
Paracril (medium nitrile)	135	275	Resistant to fatty materials
E.P.D.M.	150	300	High temperature resistance for a wide range of chemicals
Paratherm (resin cured butyl)	150	300	Aldehydes, ketones and some esters
Paradur (fluorocarbon rubber base)	177	350	Mineral oils, fuels, vegetable and animal oils

Figure 29B.4 Gasket materials, operating temperatures and applications

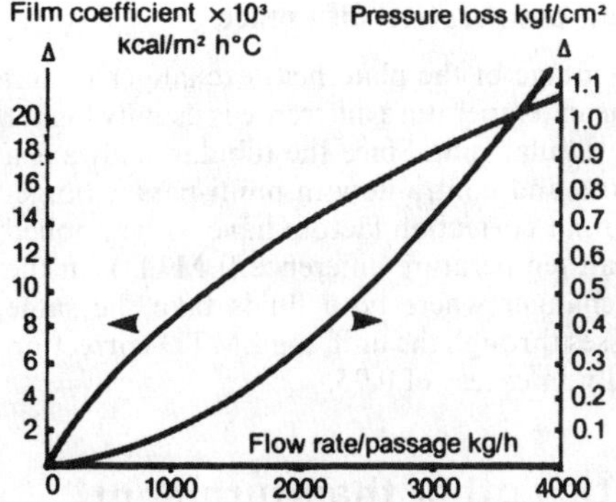

Figure 29B.5 Performance details: Paraflow

single-pass operation will achieve many duties, so that the pressure drop is efficiently used and not wasted on losses due to flow direction changes.

The friction factor is correlated with:

$$\Delta p = \frac{f \cdot LpV^2}{2g \cdot d} \qquad f = \frac{B}{Re^y}$$

where y varies from 0.1 to 0.4 according to the plate and B is a constant characteristic of the plate. If the overall heat transfer equation $H = US\,\Delta T$ is used to calculate the heat duty it is necessary to know the overall coefficient U (sometimes known as the K factor), the surface area S and the mean temperature difference ΔT.

The overall coefficient U can be calculated from

$$\frac{1}{U} = r_{fh} + r_{fc} + r_w + r_{dh} + r_{dc}$$

The values of r_{fh} and r_{fc} (the film resistances for the hot and cold fluids, respectively) can be calculated from the Dittus–Boelter equations previously described and the wall metal resistance r_w from the average metal thickness and thermal conductivity. The fouling resistances of the hot and cold fluids r_{dh} *and* r_{dc} are often based on experience, but a more detailed discussion of this will be presented later in this chapter.

The value taken for S is the developed area after pressing. That is, the total area available for heat transfer and due to the corrugations will be greater than the projected area of the plate, i.e. 0.17 m² versus 0.14 m² for an APV HX plate.

The value of T is calculated from the logarithmic mean temperature difference multiplied by a correction factor. With single-pass operation, this factor is about 1 except for plate packs of less than 20, when the end effect has a significant bearing on the calculation. This is due to the fact that the passage at either end of the plate pack only transfers heat from one side and therefore the heat load is reduced.

When the plate unit is arranged for multiple-pass use a further correction factor must be applied. Even when two passes are counter-current to two other passes, at least one passage must experience co-current flow. This correction factor is shown in Figure 29B.6 against a number of heat transfer units (HTU = temperature rise of the process fluid divided by the mean temperature difference). As indicated, whenever unequal passes are used, the correction factor calls for a considerable increase in area. This is particularly important when unequal flow conditions are handled. If high and low flow rates are to be used, the necessary velocities must be maintained with the low fluid flow rate by using an increased number of passes. Although the plate unit is most efficient when the flow ratio between two fluids is in the range of 0.7–1.4, other ratios can be handled with unequal passes. This is done, however, at the expense of the LMTD factor.

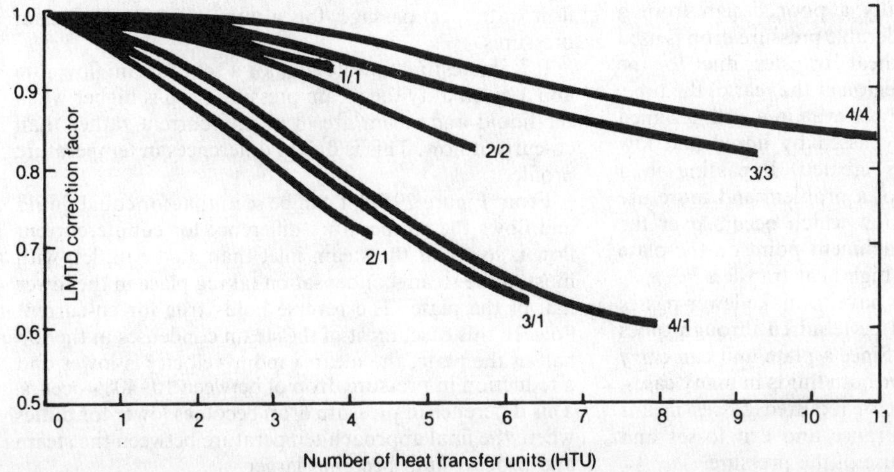

Figure 20B.6 LMTD correction factor

29B.2 Comparing plate and tubular exchangers

29B.2.1 Ten points of comparison

In forming a comparison between plate and tubular heat exchangers there are a number of guidelines which will generally assist in the selection of the optimum exchanger for any application. In summary, these are:

1. For liquid/liquid duties, the plate heat exchanger will usually give a higher overall heat transfer coefficient and in many cases the required pressure loss will be no higher.
2. The effective mean temperature difference will usually be higher with the plate heat exchanger.
3. Although the tube is the best shape of flow conduit for withstanding pressure it is entirely the wrong shape for optimum heat transfer performance since it has the smallest surface area per unit of cross-sectional flow area.
4. Because of the restrictions in the flow area of the ports on plate units it is usually difficult (unless a moderate pressure loss is available) to produce economic designs when it is necessary to handle large quantities of low-density fluids such as vapours and gases.
5. A plate heat exchanger will usually occupy far less floor space than a tubular for the same duty.
6. From a mechanical viewpoint, the plate passage is not the optimum and gasketed plate units are not made to withstand operating pressures much in excess of 20 kgf/cm^2.
7. For most materials of construction, sheet metal for plates is less expensive per unit area than tube of the same thickness.
8. When materials other than mild steel are required, the plate will usually be more economical than the tube for the application.
9. When mild steel construction is acceptable and when a closer temperature approach is not required, the tubular heat exchanger will often be the most economic solution since the plate heat exchanger is rarely made in mild steel.
10. Plate heat exchangers are limited by the necessity that the gasket be elastomeric.

29B.2.2 Heat transfer coefficients

Higher overall heat transfer coefficients are obtained with the plate heat exchanger compared with a tubular for a similar loss of pressure because the shell side of a tubular exchanger is basically a poor design from a thermal point of view. Considerable pressure drop is used without much benefit in heat transfer due to the turbulence in the separated region at the rear of the tube. Additionally, large areas of tubes even in a well-designed tubular unit are partially bypassed by liquid and low heat transfer area are thus created. Bypassing in a plate-type exchanger is less of a problem and more use is made of the flow separation which occurs over the plate troughs since the reattachment point on the plate gives rise to an area of very high heat transfer.

For most duties, the fluids have to make fewer passes across the plates than would be required through tubes or in passes across the shell. Since a plate unit can carry out the duty with one pass for both fluids in many cases, the reduction in the number of required passes means less pressure lost due to entrance and exit losses and consequently more effective use of the pressure.

29B.2.3 Mean temperature difference

A further advantage of the plate heat exchanger is that the effective mean temperature difference is usually higher than with the tubular unit. Since the tubular is always a mixture of cross and contra-flow in multi-pass arrangements, substantial correction factors have to be applied to the log mean temperature difference (LMTD). In the plate heat exchanger where both fluids take the same number of passes through the unit, the LMTD correction factor is usually in excess of 0.95.

29B.3 Duties other than turbulent liquid flow

20B.3.1 Beyond liquid/liquid

The plate heat exchanger, for example, can be used in laminar flow duties, for the evaporation of fluids with relatively high viscosities, for cooling various gases, and for condensing applications where pressure-drop parameters are not excessively restrictive.

29B.3.2 Condensing

For those condensing duties where permissible pressure loss is less than 0.07 kpf/cm^2 there is no doubt but that the tubular unit is most efficient. Under such pressure-drop conditions only a portion of the length of a plate heat exchanger plate would be used and a substantial surface area would be wasted. However, when less restrictive pressure drops are available the plate heat exchanger becomes an excellent condenser, since very high heat transfer coefficients are obtained and the condensation can be carried out in a single pass across the plate.

29B.3.3 Pressure drop of condensing vapour

The pressure drop of condensing steam in the passages of plate heat exchangers has been investigated experimentally for a series of different Paraflow plates. As indicated in Figure 29B.7, which provides data for a typical unit, the drop obtained is plotted against steam flow rate per passage for a number of inlet steam pressures.

It is interesting to note that for a set of steam flow rate and a given duty the steam pressure drop is higher when the liquid and steam are in countercurrent rather than co-current flow. This is due to differences in temperature profile.

From Figure 29B.8 it can be seen that for equal duties and flows the temperature difference for countercurrent flow is lower at the steam inlet than at the outlet, with most of the steam condensation taking place in the lower half of the plate. The reverse holds true for co-current flow. In this case, most of the steam condenses in the top half of the plate, the mean vapour velocity is lower and a reduction in pressure drop of between 10–40% occurs. This difference in pressure drop becomes lower for duties where the final approach temperature between the steam and process fluid becomes larger.

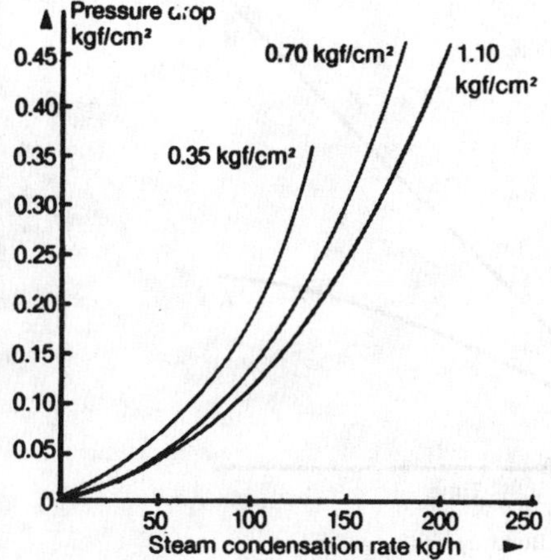

Figure 29B.7 Steam-side pressure drop

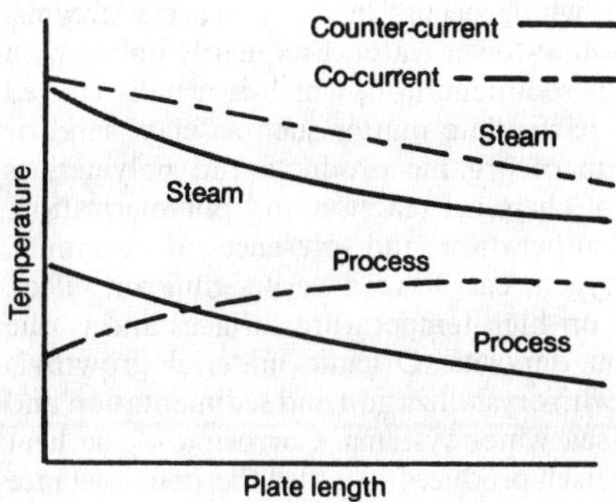

Figure 29B.8 Temperature profile during condensation of steam

The pressure drop of condensing steam is therefore a function of steam flow rate, pressure and temperature difference. Since the steam pressure drop affects the saturation temperature of the steam, the mean temperature difference, in turn, becomes a function of steam pressure drop. This is particularly important when vacuum steam is being used, since small changes in steam pressure can give significant alterations in the temperature at which the steam condenses.

29B.3.4 Gas cooling

Plate heat exchangers also are used for gas cooling. The problems are similar to those of steam heating since the gas velocity changes along the length of the plate due either to condensation or to pressure fluctuations. Designs usually are restricted by pressure drop, therefore machines with low-pressure drop plates are recommended. A typical allowable pressure loss would be 0.035 kgf/cm^2 with low gas velocities giving overall heat transfer coefficients in the region of 244 kcal/m^{2}h°C.

29B.3.5 Evaporating

The plate heat exchanger can also be used for evaporation of highly viscous fluids when the evaporation occurs in plate or the liquid flases after leaving the plate. Applications generally have been restricted to the soap and food industries. The advantage of these units is their ability to concentrate viscous fluids of up to 50 poise.

29B.3.6 Laminar flow

One other field suitable for the plate heat exchanger is that of laminar flow heat transfer. It has been previously pointed out that the Paraflow can save surface by handling fairly viscous fluids in turbulent flow because the critical Reynolds number is low. Once the viscosity exceeds 20–50 cP, however, most plate heat exchanger designs fall into the viscous flow range. Considering only Newtonian fluids since most chemical duties fall into this category, in laminar ducted flow the flow can be said to be one of three types:

1. Fully developed velocity and temperature profiles (i.e. the limiting Nusselt case);
2. Fully developed velocity profile with developing temperature profile (i.e. the thermal entrance region); or
3. The simultaneous development of the velocity and temperature profiles.

The first type is of interest only when considering fluids of low Prandtl number, and this does not usually exist with normal plate heat exchanger applications. The third is relevant only for fluids such as gases which have a Prandtl number of about one. Therefore, let us consider type two.

As a rough guide for plate heat exchangers, the rate of the hydrodynamic entrance length to the corresponding thermal entrance length is given by

$$\frac{l_{\mathrm{TH}}}{l_{\mathrm{HYD}}}=1.7Pr$$

Correlations for heat transfer and pressure drop in laminar flow are:

$$Nu=\mathrm{e}\left(\frac{Re\cdot Pr\cdot d}{L}\right)^{1/3}\left(\frac{\mu}{\mu_{\mathrm{w}}}\right)^{n}$$

where

Nu = Nusselt number (Ld/k),
Re = Reynolds number (vdf/μ),
Pr = Prandtl number ($cp\mu/k$),
L = nominal plate length,
d = equivalent diameter (2 × average gap),
μ/μ_{w} = Sieder Tate correction factor,
c = construct for each plate (usually in the range 1.86–4.50),
n = index varying from 0.1–0.2, depending on plate type.

Pressure drop

For pressure loss in a plate, the friction factor can be taken as

$$f=\frac{a}{Re}$$

where a is a characteristic of the plate.

It can be seen that for heat transfer, the plate heat exchanger is ideal because the value of d is small and the film coefficients are proportional to $d^{-2/3}$. Unfortunately, the pressure loss is proportional to d^{-4}, and pressure drop is sacrificed to achieve the heat transfer.

From these correlations it is possible to calculate the film heat transfer coefficient and the pressure loss for laminar flow. This coefficient, combined with that of the metal and the calculated coefficient for the service fluid together with the fouling resistance, are then used to produce the overall coefficient. As with turbulent flow, an allowance has to be made to the LMTD to allow for either end-effect correction for small plate packs and/or concurrency caused by having concurrent flow in some passes. This is particularly important for laminar flow since these exchangers usually have more than one pass.

29B.4 The problem of fouling

29B.4.1 The fouling factor

In view of its complexity, variability and the need to carry out experimental work on a long-term basis under actual operating conditions, fouling remains a neglected issue among the technical aspects of heat transfer. However, the importance of carefully predicting fouling resistance in both turbular tubular and plate heat exchanger calculations cannot be overstressed. This is well illustrated by the following examples.

Note that for a typical water/water duty in a plate heat exchanger it would be necessary to double the size of the unit if a fouling factor of 0.0001 was used on each side of the plate (i.e. a total fouling of 0.0002). Although fouling is of great importance, there are relatively few accurate data available, and the rather conservative figures quoted in Kern (*Process Heat Transfer*) are used all too frequently. It also may be said that many of the high fouling resistances quoted have been obtained from poorly operated plants. If a clean exchanger, for example, is started and run at the designed inlet water temperature, it will exceed its duty. To overcome this, plant personnel tend to turn down the cooling water flow rate and thereby reduce turbulence in the exchanger. This encourages fouling, and even though the water flow rate is eventually turned up to design, the damage will have been done. It is probable that if the design flow rate had been maintained from the onset, the ultimate fouling resistance would have been lower. A similar effect can happen if the cooling-water inlet temperature falls below the design figure and the flow rate is turned down again.

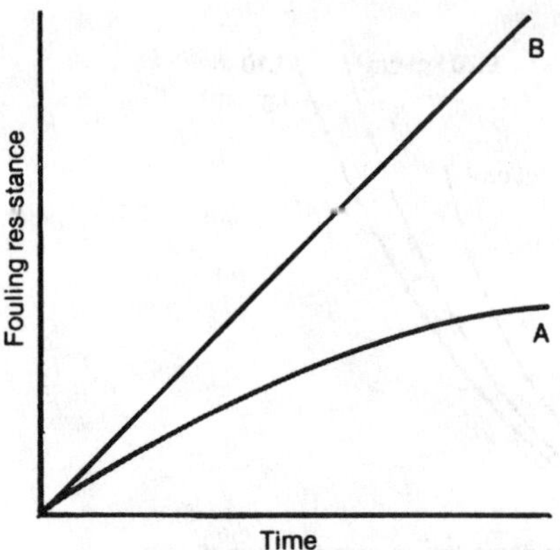

Figure 29B.9 Build-up of fouling resistance

29B.4.2 Six types of fouling

Types of fouling can be divided into six distinct categories. The first is crystallization – the most common type of fouling, which occurs in many process streams, particularly cooling-tower water. Frequently linked with crystallization is sedimentation, which is usually caused by deposits of particulate matter such as clay, sand or rust. A build-up of organic products and polymers is often a result of chemical reaction and polymerization. The surface temperature and presence of reactants, particularly oxygen, can have a very significant effect. Coking occurs on high-temperature surfaces and is due to hydrocarbon deposits. Organic material growth is usually linked with crystallization and sedimentation and is common to sea water systems. Corrosion of the heat transfer surface itself produces an added thermal resistance as well as a surface roughness.

In the design of the plate heat exchanger, fouling due to coking is of no significance, since the unit cannot be used at such high temperatures. Corrosion is also irrelevant, since the metals used in these units are non-corrosive. The other four types of fouling, however, are most important. With certain fluids such as cooling-tower water, fouling can result from a combination of crystallization, sedimentation and organic material growth.

29B.4.3 A function of time

From Figure 29B.9 it is apparent that the fouling process is time dependent, with zero fouling initially. Fouling then builds up quite rapidly and in most cases, levels off at a certain time to an asymptotic value as represented by curve A in the figure. At this point the rate of deposition is equal to that of removal. Not all fouling levels off, however, and curve B shows that at a certain time the exchanger would have to be taken off-line for cleaning. It should be noted that a Paraflow is a particularly useful exchanger for this type of duty because of the ease of access to the plates and the simplicity of cleaning.

In the case of crystallization and suspended solid fouling, the process is usually of type A. However, when

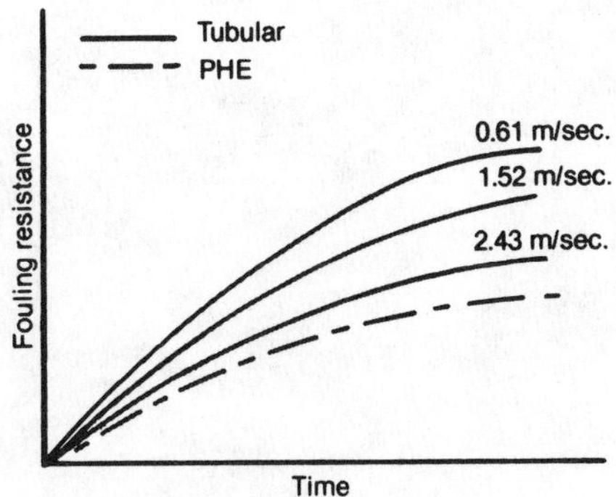

Figure 29B.10 Effect of velocity and turbulence

the fouling is of the crystallization type with a pure compound crystallizing out, the fouling approaches type B and the equipment must be cleaned at frequent intervals. In one particularly severe fouling application three Series HMB Paraflows are on a $4\frac{1}{2}$-hour cycle and the units are cleaned in place for $1\frac{1}{2}$ hours in each cycle.

Biological growth can present a potentially hazardous fouling, since it can provide a more sticky surface with which to bond other foulants. In many cases, however, treatment of the fluid can reduce the amount of biological growth. The use of germicides or poisons to kill bacteria can help.

29B.4.4 Lower resistance

It generally is considered that resistance due to fouling is lower with Paraflow plate heat exchangers than with tubular units. This is the result of four advantages of Paraflow:

1. There is a high degree of turbulence, which increases the rate of foulant removal and results in a lower asymptotic value of fouling resistance.
2. The velocity profile across a plate is good. There are no zones of low velocity compared with certain areas on the shell side of turbular exchangers.
3. Corrosion is maintained at an absolute minimum.
4. A smooth heat transfer surface can be obtained. If necessary, the plate can be electropolished.

The most important of these is turbulence. HTRI (Heat Transfer Research Incorporated) has shown that for tubular heat exchangers, fouling is a function of low velocities and friction factor. Although flow velocities are low with the plate heat exchanger, friction factors are very high, and this results in lower fouling resistance. The effect of velocity and turbulence is plotted in Figure 29B.10. The lower fouling characteristic of the plate heat exchanger compared to the tubular has been conclusively proved by HTRI's work.

Tests have been carried out which tend to confirm that fouling varies for different plates, with the more turbulent type of plate providing the lower fouling resistances.

30 Corrosion

Michael J Schofield BSc, MSc, PhD, MICorrT
Cortest Laboratories Ltd

Contents

30.1 Corrosion basics

30.1.1 Definitions of corrosion

Corrosion is generally taken to be the waste of a metal by the action of corrosive agents. However, a wider definition is the degradation of a material through contact with its environment. Thus corrosion can include non-metallic materials such as concrete and plastics and mechanisms such as cracking in addition to wastage (i.e. loss of material). This chapter is primarily concerned with metallic corrosion, through a variety of mechanisms.

In essence, the corrosion of metals is an electron transfer reaction. An uncharged metal atom loses one or more electrons and becomes a charged metal ion:

$$M \rightleftharpoons M^{+} + \text{electron}$$

In an ionizing solvent the metal ion initially goes into solution but may then undergo a secondary reaction, combining with other ions present in the environment to form an insoluble molecular species such as rust or aluminium oxide. In high-temperature oxidation the metal ion becomes part of the lattice of the oxide formed.

30.1.2 Electrochemical corrosion

The most important mechanism involved in the corrosion of metal is electrochemical dissolution. This is the basis of general metal loss, pitting corrosion, microbiologically induced corrosion and some aspects of stress corrosion cracking. Corrosion in aqueous systems and other circumstances where an electrolyte is present is generally electrochemical in nature. Other mechanisms operate in the absence of electrolyte, and some are discussed in Section 30.1.4.

Figure 30.1 depicts a metal such as iron, steel or zinc immersed in electrolyte such as sodium chloride solution. The fundamental driving force of the corrosion reaction is the difference in the potential energies of the metal atom in the solid state and the product which is formed during corrosion. Thus corrosion may be considered to be the reverse of extractive metallurgy. Metals are obtained by the expenditure of energy on their ores. The greater the energy that is required, the more thermodynamically unstable is the metal and the greater its tendency to revert to the ore or any other oxidized form. Thus gold is found native (i.e. as the metal) and it resists corrosion. Iron, aluminium and most metals exist naturally combined as oxides or sulphides and, once reduced to the metallic state, they attempt to revert to the combined state by corrosion.

In the systems illustrated in Figure 30.1 the anodic reaction has to be electrically balanced by the cathodic reaction, since electrical charge cannot build up at any location. A continuous electrical circuit is required through the metal (for electron conduction) and the environment (for ionic conduction).

There are four main aspects to the corrosion chemistry of this solution:

1. The presence of dissolved ions facilitates the passage of an ionic current through the solution.
2. Certain anions, especially chloride, penetrate the protective films which are naturally present on some metals (e.g. aluminium and its alloys and stainless steels). This process is an initiator for corrosion, especially for localized corrosion.
3. The pH of the solution is a measure of the availability of one of the two most important cathodic agents, H^{+} (the other being dissolved oxygen, see below). It is also largely responsible for determining whether the metal surface will be active or passive (see below).
4. Dissolved oxygen (and even water) can passivate certain metals such as titanium, aluminium and

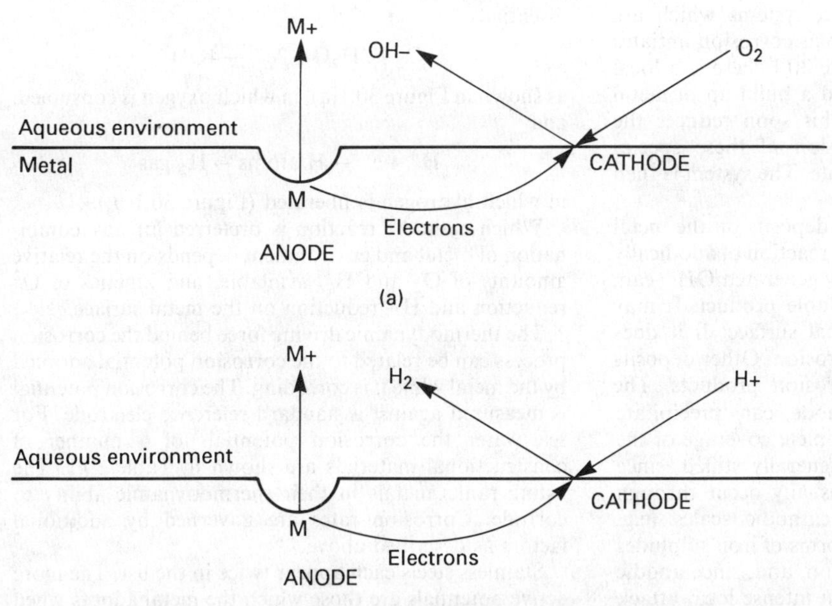

Figure 30.1

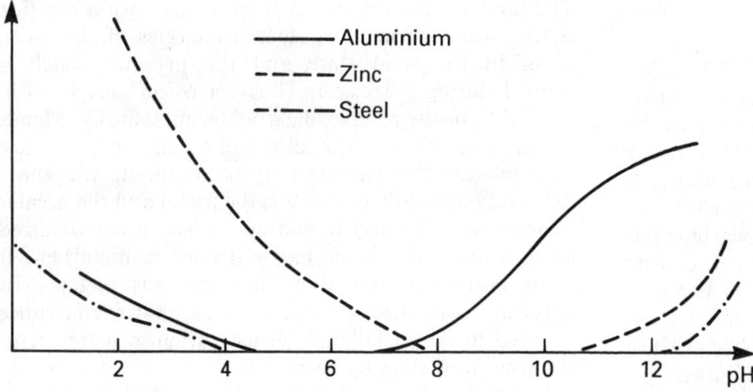

Figure 30.2 Variation in corrosion rate with pH

stainless steels, thereby rendering them, to varying degrees, corrosion resistant. Because oxygen is a strong cathodic agent it stimulates corrosion of many metals, especially those such as iron and zinc, which tend not to form stable, adherent oxides.

Thus in the system depicted in Figure 30.1, for any metal which is thermodynamically unstable with respect to its dissolved ions in the solution (and this includes most metals of industrial importance) corrosion will occur. The rate of corrosion is determined by several factors:

1. The conductivity of the environment; low conductivity hinders the ionic current flow hence distilled water is less corrosive than a solution of sodium chloride with the same pH and dissolved oxygen content.
2. The presence of cathodic agents; dissolved oxygen and H^+ promote corrosion of most metals in an electrolyte solution (see Figure 30.2).
3. The supply of cathodic agents to the metal surface: stirring a solution or increasing the flow rate in a pipe will increase corrosion rates in systems which are under diffusion control. As soon as corrosion initiates in the system depicted in Figure 30.1 there is a local depletion of cathodic agent and a build-up of metal ions. In stagnant conditions this soon reduces the corrosion rate, and the diffusion of these species governs the overall corrosion rate. The system is then under diffusion control.
4. The build-up of scales or other deposits on the metal surface can stifle corrosion. The reaction of anodically generated M^+ with cathodically generated OH^- can, depending on M^+, be an insoluble product. It may or may not adhere to the metal surface. If it does adhere strongly it can stifle corrosion. Other deposits can form, in addition to corrosion products. The locally high pH at the cathode can precipitate insoluble carbonates. With complete coverage of the cathodic surface corrosion is generally stifled, since the cathodic reaction cannot usually occur through the deposit. However, some cathodic scales (e.g. magnetite (Fe_3O_4)) and some forms of iron sulphides can support the cathodic reaction, and, since anodic area is limited, this can result in intense local attack leading to pitting corrosion at bare areas.
5. The effect of temperature is generally to increase corrosion rate as temperature rises. This is a chemical–kinetic effect and can sometimes be overridden by physical effects such as deposition of scales. The solubility of oxygen in a saline solution decreases from approximately 8–10 ppm at 5°C to virtually zero near the boiling point. Thus in a system in which dissolved oxygen is the only cathodic agent corrosion rates would pass through a maximum and decrease to very low levels as temperature is increased and the availability of cathodic agent is diminished.

In Figure 30.1 the region from which metal is lost is the anode. The atoms of the metal are ionized and release electrons:

$$M \rightarrow M^{n+} + ne^-$$

where M denotes metal, M^{n+} its ion and e^- is an electron.

This reaction can proceed to a significant extent only if the electrons which are released are consumed by a cathodic reaction. The commonest cathodic reactions are essentially:

$$O_2 + 2H_2O + 2e^- \rightarrow 4OH^-$$

as shown in Figure 30.1(a), in which oxygen is consumed, and

$$H^+ + e^- \rightarrow H \text{ atoms} \rightarrow H_2 \text{ gas}$$

in which hydrogen is liberated (Figure 30.1(b)).

Which cathodic reaction is preferred for any combination of metal and environment depends on the relative amounts of O_2 and H^+ available, and kinetics of O^2 reduction and H^+ reduction on the metal surface.

The thermodynamic driving force behind the corrosion process can be related to the corrosion potential adopted by the metal while it is corroding. The corrosion potential is measured against a standard reference electrode. For sea water the corrosion potentials of a number of constructional materials are shown in Table 30.1. The listing ranks metals in their thermodynamic ability to corrode. Corrosion rates are governed by additional factors as described above.

Stainless steels each appear twice in the list. The more active potentials are those which the metal adopts when corroding as in a pit. The more cathodic potential is that

Table 30.1 Galvanic series in sea water[a]

Metal	*Corrosion potential*[b] (V)
Magnesium and alloys	−1.6
Zinc	−1.0
Alluminium alloys	−0.8/−1.0
Steel (incl. low alloys)	−0.55/−0.75
Stainless steels (304, 321, 410, 430)	−0.5[c]
Stainless steels (316, 317)	−0.4[c]
Copper	−0.35/−0.2
Bronzes (Si, Sn)	−0.35/−0.25
Stainless steels (410, 430)	−0.3/−0.2[d]
Lead	−0.25[e]
Copper nickel alloys	−0.25
Bronze (Ni–Al)	−0.15
Nickel	−0.15
Stainless steels (304, 321, 316, 317)	−0.1−0[d]
Titanium	0/−0.1
Platinum, graphite	+0.2/0.3

[a] These values are roughly constant across a range of electrolyte environments except where noted but the variations between alloys, heat treatment conditions, etc. creates a range for each metal. For some metals such as iron and steel the range is low (±100 mV), but for lead, nickel, stainless steels a range is given.
[b] The corrosion potential is reported with respect to the saturated calomel reference electrode.
[c] When active, as in a pit or a crevice or when depassivated by mechanical damage of oxide film or chemical removal in non-oxidizing acid.
[d] When passive, on a bare surface around a pit or crevice in normal conditions.
[e] −0.5 V in environments low in chloride ion content.

adopted by the bare surface around the pit. The potential difference constitutes a significant driving force, analogous to the situation where the coupling of dissimilar metals such as copper and iron promotes the corrosion of the more anodic of the two (see below).

Pitting corrosion

Pitting is a form of localized corrosion in which part of a metal surface (perhaps 1% of the exposed area) is attacked. Rates of pitting penetration can be very high; type 316 stainless steel in warm sea water can suffer pit penetration rates of 10 mm per year. This is a natural consequence of the low ratio of anodic area to cathodic area, coupled with the self-acidification processes that occur in some metals.

Figure 30.3 illustrates a pit in a stainless steel such as type 304 or 316 austenitic alloy. Pitting starts at a heterogeneity in the steel surface, such as an outcropping sulphide inclusion, the shielded region beneath a deposit or even a discontinuity in the naturally present oxide film caused by a scratch or embedded particle of abrasive grit. This initiation phase of pitting corrosion may take seconds or weeks to develop into a feature large enough to be recognized as a pit. Initiation is facilitated by oxidizing conditions. Thus hypochlorite ions are particularly aggressive towards stainless steels since they raise the potential of the steel to such a high level that the protective oxide layer breaks down, allowing localized attack.

The second phase of pitting corrosion is propagation. Because of the geometry of the small pit that has been created, oxygen diffusion to the corroding region is hindered. This creates a potential difference between the pit and the surrounding region. The solution in the pit acidifies through hydrolysis of the Fe^{2+} and Cr^{2+} ions present. The pH in the pit falls to very low values (0.6 in the case of 316 stainless steel), irrespective of the bulk solution conditions, which may be neutral or even alkaline. This enhances corrosion by dissolving the pit walls and ensures that the pit walls cannot be repassivated. The surrounding metal surface is passive (i.e. is covered with an adherent oxide film of mixed $Fe_2O_3 \cdot Cr_2O_3$ oxide). This is dissolved in strongly acidic conditions such as those prevailing in the pit. Because the conditions in the pit promote self-acidification, pit penetration rates often increase with time.

A third phase is sometimes identified in pitting corrosion, i.e. termination. Pits can become stifled by the build-up of insoluble corrosion products at their mouths. Removal of these mounds of corrosion products, either mechanically or through some change in the environmental chemistry, can allow the pits to restart growth.

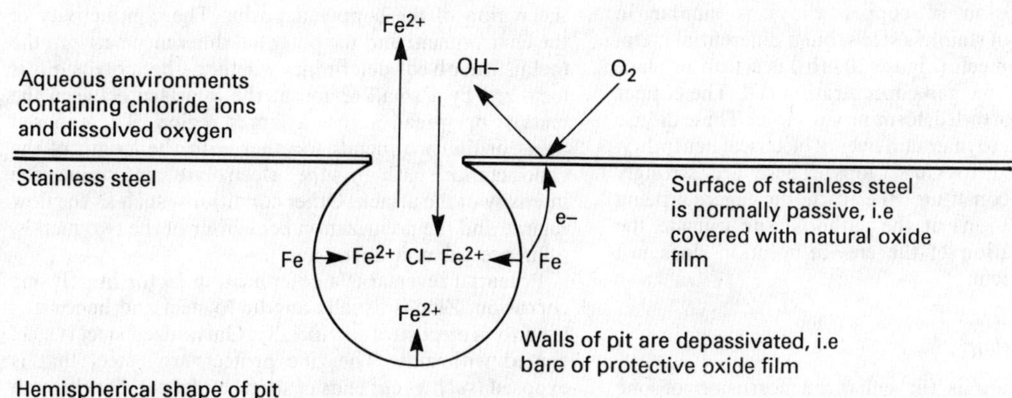

Figure 30.3 Pitting corrosion of stainless steel

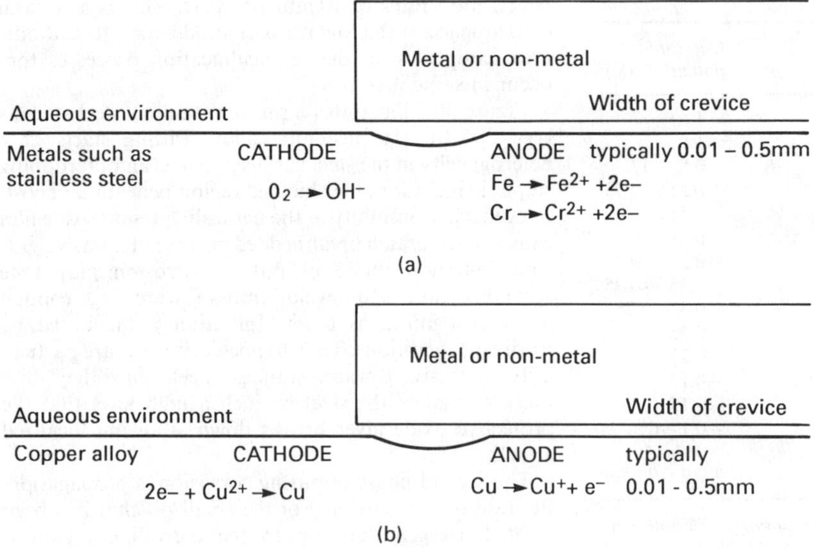

Figure 30.4 Crevice corrosion driven by (a) a differential aeration cell and (b) a differential metal ion concentration cell

Crevice corrosion

Crevice corrosion of stainless steels (Figure 30.4(a)) has a similar mechanism to pitting corrosion. The initiation phase is assisted by the creation of a crevice-suitable geometry. Crevices are formed by certain fabricational processes including riveted seams, incompletely fused welds, interference fits. O-rings, gasketed joints and even paint markings of components. The solution in the crevice becomes acidified and locally the metal surface anodic compared with the surrounding bare metal, as described above. Crevice initiation is largely dependent on the geometry of the crevice that is present: shorter initiation times are brought about by tighter crevices, since the primary effect is the limitation on inward diffusion of oxygen to maintain passivity. The lower limit of crevice width that will permit crevice corrosion is the point at which the crevice is watertight.

Crevice corrosion of copper alloys is similar in principle to that of stainless steels, but a differential metal ion concentration cell (Figure 30.4(b)) is set up in place of the differential oxygen concentration cell. The copper in the crevice is corroded, forming Cu^+ ions. These diffuse out of the crevice, to maintain overall electrical neutrality, and are oxidized to Cu^{2+} ions. These are strongly oxidizing and constitute the cathodic agent, being reduced to Cu^+ ions at the cathodic site outside the crevice. Acidification of the crevice solution does not occur in this system.

Galvanic corrosion

Galvanic corrosion is the enhanced corrosion of one metal by contact with a more noble metal. The two metals require only to be in electrical contact with each other and exposed to the same electrolyte environment. By virtue of the potential difference that exists between the two metals, a current flows between them, as in the case of copper and zinc in a Daniell cell. This current dissolves the more reactive metal (zinc in this case), simultaneously reducing the corrosion rate of the less reactive metal. This principle is exploited in the cathodic protection (Section 30.7.2) of steel structures by the sacrificial loss of aluminium or zinc anodes.

The metal that corrodes in any couple is that which has the most negative corrosion potential in the galvanic series in that environment. As a guidance, Table 30.1 is of general but not universal applicability. Thus in the case of copper and aluminium the aluminium corrodes unless corrosion inhibitors are employed. Copper and aluminium are therefore not generally used in close contact with each other, but copper ions in solution from the corrosion of copper pipework can plate out onto aluminium items (including saucepans) downstream in the system. Rapid corrosion of the aluminium occurs in the region of the copper deposits. The conductivity of the environment and the potential difference between the metals involved determine whether the corrosion is localized by a small region at the interface between the metals, or spread across a larger region. The exposed areas of the two metals together with the length of the conductance path in the electrolyte determine the intensity of the attack. Other conditions, such as the flow regime and the polarization behaviour of the two metals, influence galvanic corrosion.

Potential reversal is a complicating factor in galvanic corrosion. Zinc is usually anodic to steel and hence it is used to protect steel sacrificially. Galvanized steel is steel coated with zinc. The zinc protects any steel that is exposed (such as cut ends of sheet) and scratches through the zinc layer. At temperatures above 80°C, in low-conductivity water (e.g. condensation) the metals undergo a reversal of polarity. The steel corrodes in preference to the zinc, and this causes rapid loss of the zinc in sheets as corrosion proceeds along the steel surface undermining the zinc. Tin is usually more corrosion resistant than steel, hence its use in tinplate (tin-plated steel). In some organic acids, including some of those encountered in fruit juices, the relative potentials of tin and steel are reversed and the tin corrodes, contaminating foodstuffs with tin corrosion products.

The corrosion potentials of metals and alloys are temperature dependent. The potential of a steel distillation column, heat exchanger or other item of plant varies in the temperature zones in which it is operating. Under some circumstances it is possible to create galvanic corrosion cells between different areas of one item fabricated from one material. This causes localized corrosion problems.

To avoid galvanic problems, different materials of construction may have to be electrically isolated or at least the electrical resistance between them increased to a level sufficient to reduce the corrosion current to an acceptable value. In some instances it is more practical to paint or otherwise coat the more cathodic of the two parts of the couple. The anodic material should not be coated, since even more rapid penetration would occur at any breaks in the coating. Guidance on the galvanic hazards posed by the contact of different materials of construction is given in the British Standard publication PD 6484.

Microbially induced corrosion

Various bacteria are involved in corrosion processes. Of these the most important are some *Thiobacillus* types (these generate sulphuric acid which dissolves concrete) and sulphate-reducing bacteria (SRB). SRB are ubiquitous but only create significant corrosion problems in conditions suitable for rapid growth. They thrive best in anaerobic conditions, with a supply of sulphate ions and a carbon source. Anaerobic conditions are created in stagnant areas of chemical plant, under slime in industrial cooling systems and soils and seabed muds. Sulphate ions are widely available, especially in the sea and some chemical processes. Suitable carbon sources include organic materials such as acetic acid. SRB produce sulphide ions from sulphate ions. In conditions such as beneath a colony of these bacteria corrosion processes as illustrated in Figure 30.5 occur. Many metals are susceptible to this form of attack, including steel, stainless steels, copper alloys and aluminium alloys. Localized pitting rates in excess of 10 mm/yr are possible on steel in some circumstances. Some *Cladosporium* fungi cause pitting attack on aluminium alloys. *Gallionella* bacteria thrive in solutions containing Fe^{2+} and produce deposits under which anaerobic processes can occur.

30.1.3 Cracking mechanisms

Cracking mechanisms in which corrosion is implicated include stress corrosion cracking, corrosion fatigue, hydrogen-induced cracking and liquid metal embrittlement. Purely mechanical forms of cracking such as brittle failure are not considered here.

Stress corrosion cracking

The conjoint action of a tensile stress and a specific corrodent on a material results in stress corrosion cracking (SCC) if the conditions are sufficiently severe. The tensile stress can be the residual stress in a fabricated structure, the hoop stress in a pipe containing fluid at pressures above ambient or in a vessel by virtue of the internal hydraulic pressure created by the weight of its contents. Stresses result from thermal expansion effects, the torsional stresses on a pump or agitator shaft and many more causes.

The corrosive environments which cause SCC in any material are fairly specific, and the more common combinations are listed in Table 30.2. In the case of chloride stress corrosion cracking of the 300 series austenitic stainless steels it is generally considered that the risk is minimal below approximately 60°C. Above this threshold, if the chloride ion content of the environment is low a higher tensile stress can be tolerated without SCC developing, and vice versa. It is difficult to design for the control of SCC since it is the local stress and corrodent concentration rather than average or bulk values that are relevant.

The essential feature of most SCC mechanisms is the rupture of the protective film on the metal (passive oxide layer in the case of stainless steels and aluminium alloys) which allows the corrodent to reach the bare surface of the metal. Cracking is the result of intense corrosion, concentrated by the tensile stress. In austenitic stainless steels, chloride SCC is transgranular in nature and the cracks are typically high branched. In ferritic materials cracking is largely due to a hydrogen embrittlement

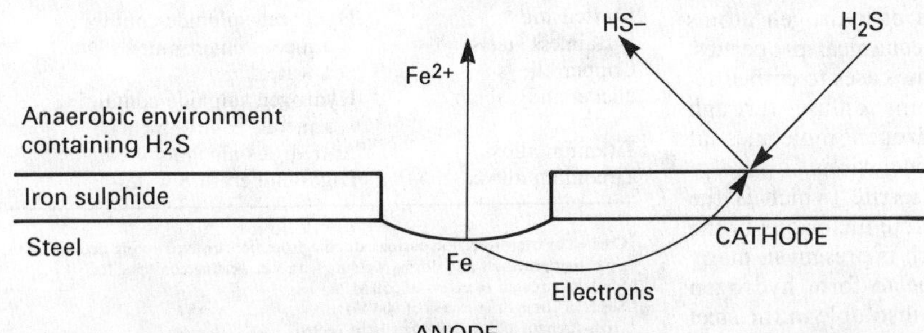

Figure 30.5

Table 30.2 Stress corrosion cracking

Metal	*Agent*
Steels	Nitrate ion
	Strong alkali
	Carbonate/bicarbonate
	Liquid ammonia[a]
	Hydrogen sulphide (aqueous)[b]
	Cyanide ion
Austenitic stainless steels	Chloride (and other halide) ion
	Polythionic acids
	High-temperature water
	Hydrogen sulphide (aqueous)
Copper alloys	Ammonia/amines
	Nitric acid/nitrate fumes
Nickel alloys	Strong alkali
	Polythionic acids
	Chloride ion
Titanium alloys	Alcohols[a,b]
	Nitric acid plus nitrogen oxides
Magnesium alloys	Chloride ion[c]
Aluminium alloys	Chloride ion[c]
	Organics
	High-purity hot water

Environments are aqueous unless specified otherwise.
[a] Inhibited by 1000 ppm of water.
[b] Sometimes considered to be a form of hydrogen-induced cracking.
[c] In oxidizing conditions.

mechanism brought about by hydrogen created during corrosion.

Corrosion fatigue

Many metals suffer fatigue, which is the result of the application of an alternating stress. Corrosion fatigue results from the exposure of a metal, which is subjected to alternating stress, in a corrosive environment. Unlike stress corrosion cracking, corrosion fatigue does not require a specific corrosive environment for any metal. Moist air is adequate in many cases, thus the majority of fatigue failures are actually corrosion fatigue in nature. The alternating stress pattern requires only a small tensile component. This tensile stress may be small relative to the metal's ultimate tensile strength and millions of cycles of the stress pattern may be required to cause failure.

Hydrogen-induced cracking

Hydrogen-induced cracking (HIC) is most commonly encountered in steels but other metals are susceptible, as shown in Table 30.3. The presence of hydrogen atoms in a metal degrades some of its mechanical properties, especially its ductility, leading in some cases to embrittlement. Additionally, hydrogen atoms diffuse through metals and coalesce to form hydrogen molecules at certain preferred locations such as inclusions.

Hydrogen atoms are soluble in ferrite (which is the major phase of most steels). At discrete inclusions in the steel (e.g. manganese sulphide, which is present in many steels) the hydrogen atoms combine to form hydrogen molecules. Hydrogen molecules are insoluble in the steel lattice and these molecules associate, producing high tensile stresses which can initiate hydrogen cracking. When laminations are present in the steel hydrogen molecules form at the interface between them, creating blisters. At high temperatures hydrogen atoms can react with carbides in the steel producing methane gas. Methane is insoluble and causes blistering or cracking. In clean steels containing none of these features the hydrogen atoms pass straight through the steel, causing no damage.

Hydrogen can originate from several sources:

1. Electroplating (e.g. cadmium or hard chromium plating);
2. Acid pickling (e.g. prior to galvanizing);
3. Cathodic protection, especially if overprotected, through inadequate potential control;
4. Corrosion, where H^+ reduction is a cathodic reaction (Section 30.1.2);
5. The process inside the plant.

The amount of the hydrogen that is liberated on or near a metal surface which then enters the metal varies according to the environment and condition of the metal. The main factor that promotes the entry of hydrogen into a metal is the presence on the metal of a surface poison such as sulphide or other species which inhibit the hydrogen recombination reaction.

Liquid metal embrittlement

This is a form of stress corrosion cracking in which the corrodent is a liquid metal. Mercury at ambient temperature and metals including zinc (from galvanized steelwork) and copper (from electric cables) when melted during welding or in a fire cause rapid failure of certain metals.

Steels and austenitic stainless steels are susceptible to molten zinc, copper, lead and other metals. Aluminium and copper alloys are attacked by molten mercury, zinc

Table 30.3 Hydrogen-induced cracking

Metal	*Agent*
Steel	Hydrogen gas
	Atomic hydrogen[a]
	Hydrogen sulphide-containing aqueous environments[b]
	Water[c]
Martensitic stainless steels	Hydrogen sulphide-containing aqueous environments
Copper alloys	Sea water[d]
Nickel alloys	Hydrogen sulphide-containing aqueous environments
Titanium alloys	Anhydrous alcohols
Zirconium alloys	High-temperature water

[a] Created by overprotection during cathodic protection, evolved during electroplating or pickling processes and during welding with wet welding consumables.
[b] Steels of strength in excess of 550 MPa.
[c] Steels of strength in excess of 900 MPa.
[d] High-strength alloys, cathodically protected.

Table 30.4 Oxidation in air

Steels	480°C
Low-alloy steels	560°C
Ferritic stainless steel (type 410)	650°C
Austenitic stainless steel (type 316)	900°C
Nickel-based superlloys	1100°C
Superaustenitic stainless steel (type 310)	1150°C
Cobalt-based superalloys	>1150°C

1. Temperatures are those above which oxidation is generally too rapid to permit the metal's safe or economic use.
2. Atmospheres contaminated with acid or halogen gases reduce these values.
3. The mechanical properties including strength, ductility and creep resistance can be affected below these temperatures.

and lead. Nickel alloys are attacked by mercury, zinc, silver and others. Other low-melting point metals which can attack common constructional materials include tin, cadmium, lithium, indium, sodium and gallium.

30.1.4 Non-electrochemical corrosion

Although all the types of corrosion discussed in this section result in the oxidation of metal and some involve direct electron transfer, they can be understood without reference to electrochemistry.

Oxidation

In oxidizing atmospheres (i.e. in the presence of oxygen or a source of oxygen from which oxygen can be derived, such as water vapour) most metals are unstable with respect to their oxides. Oxidation to create the metal oxide depends on several factors, the most important being temperature (and hence the thermodynamics of the process), availability of oxygen and the protective action of any oxide that is created. Oxygen can migrate through the oxide, where it reacts with the metal. In other metal-oxide systems, metal ions migrate outwards through the oxide. The structure and the stability of the oxide, together with its semiconducting properties, control the mechanism of oxide. Oxidation rates can be linear, parabolic (tending to a steady state oxide thickness) or exponential (termed catastrophic, because the rate increases with time). Table 30.4 gives an indication of the temperature limits for some metals through oxidation in a cleaned atmosphere.

Sulphidation is analogous, but catastrophic sulphidation is common because of the generally lower melting points of sulphides than corresponding oxide. This is especially true in the case of nickel alloys, when a nickel/nickel sulphide eutectic is formed.

Fretting corrosion

Fretting corrosion occurs as a consequence of oscillating relative motion between touching surfaces. As little as 3×10^{-9} m lateral movement is required. The amount of damage increases as the normal force between the surfaces is increased. In dry conditions the corrosion product is usually the oxide. In the case of steel this is wustite. This is otherwise the high-temperature form of the oxide, which infers that locally high temperatures are created on the fretting surfaces. The surfaces weld together in the high stress conditions at points of contact and are torn apart by the relative motion of fretting surfaces.

Materials with hard oxides, including stainless steels and aluminium and titanium alloys, are particularly susceptible to this form of attack. In steel it is also known as 'false Brinelling' because of the high surface hardness that can be created in work-hardening grades.

Molten salt corrosion

Any salt that is present on a hot metal surface can cause corrosion of that surface. This mechanism is often rapid and is due to straightforward dissolution of the metal and any oxide which may be present on the surface. The mechanism is similar to that of aqueous corrosion. The high rate of attack is a consequence of the high activities of the ions present in the molten salt. The problem is greatest for salts of low melting point, since these are present over a wide temperature range and can be very fluid. Low melting-point salts include many chlorides, sulphides and sulphates.

30.2 The implications of corrosion

30.2.1 Economics

The principal economic implications of corrosion of a plant are the initial cost of construction, the cost of maintenance and replacement, and the loss of production through unplanned shutdowns. The initial cost of the plant is influenced by material selection, and a choice of material which is more corrosion resistant than is necessary for the safe operation of the plant over its design life is a very expensive error. This cost involves initial outlay of money, and plants have been built which could never be profitable because of the inappropriate materials selection.

Maintenance costs arising from corrosion can also prove to be unacceptably high. For plant which requires regular maintenance for other reasons (e.g. desludging or batch operation) it is often accepted that certain items are expendable and annual or even monthly repair or replacement with low-cost materials is preferable to the use of expensive ones. Plant handling hot, strong hydrochloric acid is often regarded in this way, since the only possible construction materials which have acceptably low corrosion rates are too expensive to use. Plant which requires to be run continuously or maintenance-free for long periods necessitates high-integrity design and hence the use of more reliable corrosion-resistant materials.

Significant savings can be achieved by optimum material selection (guidance is given in Section 30.3), by considering corrosion at the design stage (Section 30.4), by employing corrosion-control techniques (Section 30.7) and corrosion monitoring (Section 30.8). In the industrialized countries of the Western world, corrosion is estimated to cost some 4% of a country's GNP. In the

UK this represents £12 billion per annum, of which a significant proportion could be saved by the use of existing knowledge. It cannot all be saved, since the gradual deterioration of a plant over its operational life is one of the costs incurred in the process.

30.2.2 Safety

The failure of plant by corrosion can be gradual or catastrophic. Gradual failure has few implications for safety providing it is monitored. Direct corrosion-monitoring techniques are described in Section 30.8. Indirectly, the correct interpretation of records relating to metal contamination of products or the loss of efficiency of heat exchangers, etc. can provide useful information.

Sudden or catastrophic failure of plant through corrosion can result in the loss of product at high velocity from a failed reactor vessel, high-energy steam from a steam line, toxic or inflammable materials from storage vessels. Incidents of this nature from plants around the world have led to the death and injury of plant operatives and nearby householders. The majority of serious incidents in the chemical process industry, however, have been attributed to mechanical failure of plant.

30.2.3 Contamination of product

The metal lost from the inside of pumps, reaction vessels, pipework, etc. usually contaminates the product. The implications of this depend upon the product. White plastics can be discoloured by ppb levels of iron, though at this level the effect is purely cosmetic. The taste of beer is affected by ppm levels of iron and other metals. Products sold to compositional requirements (such as reagent-grade acids) can be spoiled by metal pick-up. Pharmaceutical products for human use are often white tablets or powders and are easily discoloured by slight contamination by corrosion products.

30.3 Materials selection

The materials selection procedure for new or replacement plant is crucial to the safe and economic operation of that plant. There is no one correct way to select the appropriate construction material since for small plant handling toxic or inflammable products the integrity of the plant is high in priority, whereas large plant producing bulk material in a competitive market is more likely to be made of cheaper constructional materials, and the occasional leak or failure may have fewer safety implications.

Wherever possible, items should be made from materials which have proved themselves in similar service. Essential items of plant for which long delivery times are expected should be considered most carefully. The possible interactions between items fabricated from different materials should be considered. Section 30.1.2 discusses galvanic corrosion.

30.3.1 Sources of information

The first source of information for the behaviour of a material in the proposed service environment is the potential supplier of the item of plant. Except for new (or significantly modified) processes specialist suppliers or fabricators have relevant information and service experience. The supplier should be provided with all process or environmental details that are of possible relevant to corrosion. The most important are listed below:

1. *Environmental chemistry*: including complete typical and worst-case compositional analyses, pH, redox potential, dissolved gas content;
2. *Physical conditions*: temperature ranges in bulk and at surface, heat transfer, mass flow, fluid velocities, the presence of entrained particles and gases;
3. *Operating cycles*: commissioning procedures, heating, cooling and pressure cycling, stagnant periods, cleaning procedures;
4. *External environment*: corrosive gases, humidity temperature;
5. *Mechanical aspects*: the presence of tensile or cyclic stresses.

If the potential suppliers cannot provide documented service performance for similar plant on similar duties, materials suppliers should be consulted.

For all materials other than basic constructional steels and cast irons, reputable suppliers have information bases and applications laboratories from which information can be obtained. Trade organizations representing categories of materials suppliers are excellent sources of information; some are listed at the end of this chapter. The materials suppliers should be consulted in conjunction with equipment suppliers in order to ensure that the information generated is fully applicable to the end use to which the material is to be put. Fabricational techniques should be agreed between the two types of suppliers, since some materials cannot be cast or welded and some items cannot be made by forging.

For some new or modified processes it will not be possible to obtain sufficient good information to permit material selection to be made. Advice is available from professional corrosion consultants and corrosion-testing facilities are available at a number of laboratories in the UK.

The following guidance is intended as a checklist for general use, since there is no universal, definitive guide to the selection of materials. It is based on corrosion and not on mechanical property requirements such as strength, toughness, hardness or fatigue resistance.

30.3.2 Aqueous systems

Water supply

For the purpose of corrosion, water is scaling or non-scaling. Scaling water tends to deposit generally protective hardness scales. Soft water does not scale and hence is potentially more corrosive, especially when it contains dissolved gases such as oxygen, carbon dioxide, ammonia and sulphur oxides.

For drinking water, copper has now replaced the use of lead, except in areas of cuprosolvency. The local water authority provides information including analytical data. Stainless steels are being introduced especially for domestic systems. Experience on the Continent is mixed, with some corrosion problems such as pitting and stress corrosion cracking reported.

For process water, steel pipes are used unless iron pick-up is to be minimized. Plastic pipes (polyethylene and polyvinylchloride) are used but they sometimes need external protection from solvents present in industrial atmospheres, ultraviolet radiation (including sunlight), freezing and mechanical damage.

For fire water, steel pipes are used but corrosion products can block sprinklers. Cement asbestos pipes are utilized but pressure limitations restrict their use. For critical applications, including offshore oil installations, cupronickel alloys and even duplex stainless steels are used. Fire-retardant grades of fibre-reinforced plastics are now available.

Borehole waters are generally very hard and cast iron pipes are still used because of the low internal corrosion rates permitted by the scaling which occurs naturally. Acidic waters cause graphitic attack on cast irons.

Boiler water, steam and condensate

The water supply for boilers is usually treated. Treatment depends on the quality of the water supply, the pressure of the boiler, the heat flux through the tube walls and the steam quality required. Most waters require dealkalization. The water produced in this process is non-scaling and potentially corrosive (see above).

The quality of water thus produced is generally adequate for low-pressure boilers, but deoxygenation is usually achieved through the addition of sodium sulphite in a feedwater tank. This should be designed to prevent further ingress of oxygen, employing a nitrogen blanket if necessary. Pipework from the feedwater tank (or 'hot well' if condensate is returned to it) to the boiler is steel, since the water is low in corrosivity.

For higher-pressure boilers demineralization is necessary to minimize total dissolved solids in the boiler. This water is normally carried in steel pipework, but if condensate is returned and the condensate has become contaminated (for example, with carbon dioxide or copper ions) more corrosion-resistant materials such as copper are required. Downstream of the boiler, steam pipework is usually steel with steel or stainless steel expansion bellows.

The presence of corrosive species in steam, however, creates the following corrosion problems:

1. Oxygen renders the steam corrosive towards steel. If the feed is not deoxygenated a corrosion allowance should be added to the steam piping wall thickness, an oxygen scavenger such as hydrazine can be used or a volatile corrosion inhibitor could be employed. Candidate inhibitors include filming amines such as ethoxylated soya amines. Some amine-based inhibitors can break down and cause corrosion of copper in the steam or condensate system. Some products such as hydrazine are toxic or carcinogenic and cannot be used in food-processing plant, breweries and steam sterilizing equipment for hospitals.
2. Carbon dioxide, from the decomposition in the boiler of temporary hardness salts present in some waters, causes corrosion of steel steam pipework and cast iron valves and traps. Corrosion inhibitors may be used, but the choice of inhibitor must take into account the other materials in the system. Neutralizing amines such as morpholine or cyclohexylamine are commonly used.
3. Condensate return lines are often copper. Copper has good corrosion resistance to oxygen and carbon dioxide individually. When both gases are present in the condensate, copper is susceptible to corrosion. Copper picked up in the condensate system and returned to the boiler causes serious corrosion problems in the boiler and any steel feedwater and steam pipework. Boiler tubes should last for 25 years but can fail within one year in a mismanaged or ill-designed boiler system suffering from these faults.
4. Boiler salts can contain chloride ions. When carried over into the steam (e.g. during priming) this can result in chloride stress corrosion cracking of austenitic stainless steel expansion bellows. In steam systems where freedom from chloride cannot be guaranteed, bellows can be supplied in 9% Cr steel or austenitic alloys containing 40% nickel.

Aqueous processes

The principal corrosive species as described in Section 30.1.2, combined with the reducing or oxidizing nature of the environment, can be used to select candidate materials for any aqueous system. The process variables discussed in Section 30.3.5 must also be considered.

Steel is the most common constructional material, and is used wherever corrosion rates are acceptable and product contamination by iron pick-up is not important. For processes at low or high pH, where iron pick-up must be avoided or where corrosive species such as dissolved gases are present, stainless steels are often employed. Stainless steels suffer various forms of corrosion, as described in Section 30.5.2. As the corrosivity of the environment increases, the more alloyed grades of stainless steel can be selected. At temperatures in excess of 60°C, in the presence of chloride ion, stress corrosion cracking presents the most serious threat to austenitic stainless steels. Duplex stainless steels, ferritic stainless steels and nickel alloys are very resistant to this form of attack. For more corrosive environments, titanium and ultimately nickel–molybdenum alloys are used.

The following brief summary of materials for acid duty is very simplified, and further guidance should be obtained from the sources of information outlined in Section 30.3.1:

1. *Sulphuric acid*: Plain carbon steel is used at strengths in the range 70–100% at temperatures up to 80°C. In flowing conditions the corrosion rate increases, thereby

rendering steels unsuitable for pumps. Storage tanks for sulphuric acid require care in design to prevent the possible ingress of water vapour. This creates a layer of dilute acid on top of the bulk acid, creating rapid attack at the fill line. At strengths below 70°C chemical lead is the preferred material for tanks but at temperatures above 120°C corrosion becomes significant. For castings such as pumps, valves, fittings and some heat exchangers cast iron containing 15% silicon is used at strengths up to 100%, at temperatures up to the boiling point. Where iron pick-up must be avoided, conventional stainless steels can be used in certain ranges of strength and temperature. Anodically protected titanium is used at 70% (60°C) and 40% (up to 90°C). For higher temperatures and strengths above 100%, nickel-based alloys were molybdenum are used. Glass, gold and platinum are required for specific combinations of acid strength and temperature.

2. *Hydrochloric acid*: The selection of materials for handling hydrochloric is exceptionally difficult on account of the sensitivity of corrosion to minor impurities. Oxidizing hydrochloric acid is highly corrosive; dissolved oxygen or trace levels of metal ions in high oxidation states (especially ferric or cupric) ions are the common oxidizing species. Dilute hydrochloric acid (2% at up to 70°C or 10% at 20°C) is handled by copper, nickel and some of their alloys. Moderate-strength acid (up to 10% at up to 70°C, up to 30% and 30°C and up to 40% at 15°C) requires the use of more resistant materials such as cast iron alloyed with silicon and molybdenum or, in the absence of oxidizing agents, silicon bronze. Strong hydrochloric acid at moderate temperatures (below 70°C) can be handled by some of the nickel–molybdenum alloys or non-metallic linings such as rubber. At higher temperatures silver, platinum, zirconium and tantalum are suitable across the whole range of acid strength.
3. *Nitric acid*: Stainless steels are the most commonly used materials for nitric acid duty. Because molybdenum is widely considered to be detrimental to corrosion behaviour in nitric acid, type 304 is generally preferred to type 316L. It is used at temperatures up to 100°C and acid strengths up to 70%. At 100% strength the temperature limitation is 30°C. Higher alloys, based on 20% Cr, 25% Ni, extend the temperature range to nitric acid's boiling point at concentrations up to 50%. An austenitic stainless steel containing 18% chromium, 10% nickel and 2–3% silicon was developed specifically for this service. Cast iron with 15% silicon is used for valves, pumps, etc. in acid strengths of 0–40% (up to 70°C) and 40–80% (up to 110°C). Some aluminium alloys may be used for nitric acid above 80% strength at temperatures up to 35%. Titanium gold, platinum and tantalum are used to extend the temperatures range where required.

For alkaline duty, steel can sometimes be used up to approximately pH 11. Zinc, aluminium and similar metals and their alloys have limited use in alkaline conditions because they dissolve, giving complex anions. Iron and steel react in this way above about pH 12. Approximate limits of use for zinc are pH 6–12 and aluminium alloy pH 4–8. Stainless steels, including the lower grades, can be used even in the presence of chloride ions at pH levels of approximately 8. In the absence of halides they can be used up to about pH 13.

Materials selection cannot be based on any simple combination of common corrosive species. There are many complicating factors, including the harmful or beneficial effects of contaminants at the p.p.m. level, the relative proportion in which certain combinations of species are present (H^+ and Cl^- are often synergistic in their effect, whereas SO_4^{2-} and Cl^- often counter each other) and the presence of naturally occurring corrosion inhibitors.

30.3.3 Non-aqueous processes

Solvent systems

Most organic solvents, except for alcohols, have reasonably low ionic conductivities and hence do not support electrochemically corrosion to any significant extent. Steel is commonly used except in systems in which water can separate and where the conductivity is sufficient to permit the flow of ionic current.

Dry gases

Standard materials for handling dry gases include: for chlorine, UNS N10276 (type stainless steel is used for liquid chlorine); for bromine, UNS N10276 (below 60°C); for fluorine, copper; for hydrogen chloride, UNS N10276; for hydrogen sulphide (sulphur dioxide and trioxide), type 316 stainless steel. The presence of water in even trace quantities changes this significantly.

30.3.4 High-temperature environments

Chimneys, flues and ducts

The main corrosion process that occur in these items arise from condensing liquids on the internal surface. Although often lagged, heat loss frequently causes internal skin temperatures to fall below the dewpoint of one or more components of the gas stream, albeit locally, such as at support points. Even at temperatures above its dewpoint a gas can dissolve in condensed water. Rapid corrosion can then occur in this thin film of corrosive liquid.

Condensing species of relevance to corrosion include water and all acid gases. The dewpoint of water is obtained from standard tables, requiring only the water content (i.e. relative humidity) of the gas stream. Above the water dewpoint, corrosion problems include condensing acids (Section 30.3.2), dry acid gases (Section 30.3.3) and erosion. Below the water dewpoint acid gases dissolve in the water film to create an acidic solution, including:

1. Carbon dioxide produces a solution of carbonic acid (as in boiler condensate, see Section 30.3.2). Carbon steel is often employed but corrosion rates of up to 1 mm/yr can be encountered. Coatings and non-metallic materials may be employed up to their

temperature limits (Section 30.5.6). Basic austenitic stainless steels (type 304) are suitable up to their scaling temperatures.
2. Sulphur oxides produce sulphurous/sulphuric acid. Because oxygen and halides are often also present, high-nickel alloys are required. Current practice for some flue gas desulphurizer ducts operating in the temperature range 60–120°C involves lining the steel structure with 2 mm thick UNS N10276 or UNS N06022. Non-metallic linings, including glass flake epoxy materials, have limited application in the lower part of the temperature range.

Other acid gases such as hydrogen chloride and oxides of nitrogen produce similar corrosion problems. The corrosion effects produced by acid condensates are amplified by the motion of the gas stream (typically 20–30 m/s) and erosion effects due to entrained solids and impingement at bends, damper plates, reheaters, etc.

In the complete absence of acid gases, steel or galvanized ducting can be suitable but 80°C is often the operational temperature limit for galvanized steel).

Furnaces

Table 30.4 gives scaling data for classes of heat-resisting alloys in air. Mechanical properties, particularly creep strength, may be reduced significantly at temperatures well below scaling temperatures.

In addition to scaling and loss of mechanical properties the third major factor in material performance in furnaces is the detailed chemical environment. Specific problems are encountered with the following:

1. Oxygen or steam in reformer tubes (typically up to 1000°C), steam superheater tubes (700–900°C), regenerators, etc. promote oxidation. Catastrophic oxidation occurs in some materials including molybdenum-bearing stainless steels because of the volatility of MoO_3. Alloys containing 25% Cr and 20% Ni are used up to 1100°C but it is sometimes necessary to employ 35% Cr, 25% Ni or even 45% Cr, 25% Ni types. Type 316 austenitic stainless steel can suffer catastrophic oxidation at temperatures as low as 760°C, although it is often used up to nearly 900°C provided that a free flow of gas is achieved over the entire surface.
2. Sulphur compounds, whether organic or inorganic in nature, cause sulphidation in susceptible materials. The sulphide film which forms on the surface of many construction materials at low temperatures becomes friable and melts at higher temperatures. The presence of molten sulphides (especially nickel sulphide) on a metal surface promotes the rapid conversion to metal sulphides at temperatures where these sulphides are thermodynamically stable. High-alloy materials such as 25% Cr, 20% Ni alloys are widely used, but these represent a compromise between sulphidation resistance and mechanical properties. Aluminium and similar diffusion coatings can be of use.
3. Carbon sources permit carburization. In cracker furnaces, ethylene pyrolysis tubes, etc. carburization results in rapid metal loss in the temperature range 850–1100°C. A lower-temperature form of this mechanism (sometimes termed 'metal dusting') occurs in the range 400–800°C in the carbon monoxide/carbon dioxide/hydrogen/hydrocarbon gas streams. Preferred materials include HK40 (25% Cr, 20% Ni, 0.4% C) or higher alloys (35% Ni, 25% Cr or 30% Cr, 30% Ni), depending on the severity of the environment.
4. Nitrogen-containing compounds, including ammonia and amines, cause nitriding in susceptible materials. Ammonia synthesis (500–600°C, 250–400 bar pressure) requires the use of austenitic stainless steels or nickel alloys. The rate of penetration of nitriding, which causes severe embrittlement, can be 0.2 mm/yr in basic austenitic grades (e.g. type 304), falling by more than an order of magnitude for 40% Ni base alloys.
5. Molten salts promote rapid corrosion of many constructional materials at relatively low temperatures. Low melting-point salts include sodium salts from saline atmospheres, fireside ash, silicate insulation, contaminants in the feed, etc. Corrosion rates of several mm/yr can be observed at temperatures as low as 520°C. High chromium- and nickel-containing alloys up to 50% Cr/50% Ni are employed.

Other detrimental factors which require to be taken into account in the materials selection process include temperature cycling and the presence of halide gases. Specialist alloys containing rare earth element additions such as cerium, lanthanum and yttrium have been developed for use in certain environments up to 1300°C.

30.3.5 Influence of process variables on material selection

Flow

The main effects of flow on corrosion of most materials is deleterious. Increasing flow through the laminar region serves to create a thinner hydrodynamic boundary layer on the metal surface. This increases the rates of diffusion of cathodic species and anodic products, generally increasing corrosion rates. Protective deposits and scales are removed or prevented from forming. Damage is caused at locations such as those shown in Figure 30.6.

The most important exception to this is the behaviour of stainless steels in sea water. Grades which are susceptible to pitting or crevice corrosion, including types 304 and 316, are liable to greatest pitting attack in stagnant conditions where debris settles onto upward-facing horizontal surfaces. In conditions of higher flow (greater than 1 m/s) the debris is less likely to settle and thus pitting (Section 30.5.2) is less likely.

If flow becomes turbulent the corrosion rate increases even more rapidly. In practice, most engineering materials have a 'critical velocity' above which the corrosion rate is unacceptably high. This does not correspond with the laminar-to-turbulence transition. Surface roughness is an important consideration.

Impingement attack (sometimes termed erosion corrosion) is a result of the combined effect of flow and corrosion on a metal surface and it occurs when metal

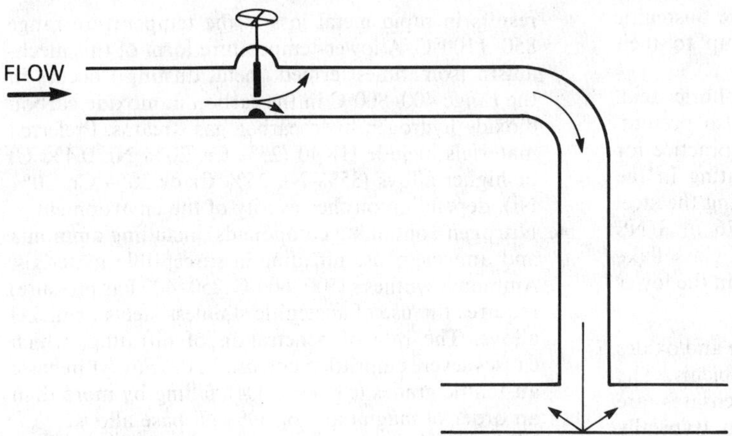

Figure 30.6 Areas affected by erosion corrosion

is removed from the surface under conditions where passivation is insufficiently rapid. It is a function of flow, corrosion and passivation.

At higher flow rates cavitation is a serious degradation mechanism, where vapour bubbles created by pressure fluctuations brought about by the flow of liquid past the surface collapse on the metal surface with tremendous force. This damages any protective oxide which may be present, leading to pitting corrosion. It also causes mechanical damage to the metal.

The corrosive and mechanical effects of flow are observed in pipes, especially at bends and downstream of flow disturbances, tube and shell heat exchangers, valves and pumps. More corrosion and/or harder materials are used in such areas. Austenitic stainless steels work harden and hence are superior in flowing conditions to ferritic stainless steels of otherwise similar corrosion resistance. Hard facings, such as hard chromium electroplate or cobalt based alloys, are used for local improvement of flow-assisted corrosion resistance where it is not required, practical or economic to fabricate the complete item of high resistant material. Hard facing by cobalt-based alloys can be achieved by weld overlay and plasma spray techniques. Correct choice of alloy is important to optimize corrosion resistance, though the resistance of most of the alloys in use is superior to the more common grades of stainless steel. Hard chromium plating contains micro cracks; this allows corrodents to gain access to the substrate. Layers of copper and/or nickel are often required to prevent the corrosion of underlying steel.

The alternative option for counteracting cavitation damage is the use of a resilient material such as rubber. The mechanical forces attendant on collapse of the bubbles are absorbed by elastic deformation of the resilient material.

Intermittent operation

This is generally harmful to plant, since deposits can form in periods of stagnation. The common grades of stainless steel, such as types 304 and 316 and other materials (including aluminium alloys and some copper alloys), are susceptible to under-deposit attack and are at risk in circumstances where the process liquid is left to stagnate in between operation cycles. Evaporation can also take place, creating more aggressive conditions than occur during operation. Condensation can form on vessel walls, roofs and support points as heat is lost. This condensation can absorb corrosive gases, creating localized corrosion effects of greater severity than the bulk environment that is normally present.

In rare circumstances, however, intermittent operation is beneficial. Plant which would pit if operated continuously can be operated successfully if the operational cycle is shorter than the initiation period for pitting attack and it is cleaned between batches. This is mainly the case for stainless steel plant operating with high enough chloride levels (or a low pH or high temperature) to initiate pitting if operated continuously. The cleaning regime between batches promotes repassivation; hot water, steam, chloride-free caustic or sodium orthophosphate solution may be required to achieve the required rate of passivation. Plant is rarely designed on this principle, but a significant number of cooking vessels, especially in the food and drink industries, operate this way.

Commissioning

The first use of new plant, or start-up after a shutdown, poses corrosion hazards additional to those encountered in normal operation. New plant such as boilers requires special water treatment, involving boil-out, passivation and possible chemical cleaning. Actual requirements depend on the boiler type, the proposed service, the quality of water available during commissioning and the internal condition of the boiler. The condition of the boiler depends on for how long and in what conditions it has been stored. The presence of any salts, dirt or rust is harmful. Boiler internals are covered by an adherent, protective layer of magnetite in normal operation. To create an intact layer of magnetite of the correct thickness, crystal type and morphology all debris must be removed from the boiler, manually or by means of a vacuum cleaner. If rusting is more than superficial, or if millscale is present on the steel of the boiler tubes or shell, chemical cleaning is necessary. BS 2486 contains details of approved chemical cleaning techniques. These are based on inhibited acids (e.g. citric acids). The most appropriate technique is determined by a number of factors, and the most important include the nature of deposits to be removed (i.e. rust, hardness scales, millscale and metallic copper).

Degreasing may be necessary, especially on new plant, and alkaline cleaning techniques are described in BS 2486. For optimum boiler life, passivation treatment is recommended. This is described in BS 2486.

All boiler treatment and cleaning should be carried out by reliable contractors, preferably within the guidelines contained in BS 2486. Boilers are fired pressure vessels and are subject to mandatory insurance inspections. Significant benefits in safe and economic operation, particularly by reducing unnecessary chemical treatment, can be achieved by monitoring the condition of boilers.

Items of plant fabricated from stainless steels should be inspected before first use and after any maintenance work or unplanned shutdown. All materials which rely for their corrosion resistance on the presence of an oxide or similar passive layer are susceptible to localized attack where that layer is absent or damaged. Damage is most commonly caused by scratching, metallic contamination (nearby grinding or touching with ferrous tools), embedding of grit and weld spatter.

Serious damage of this nature should be rectified before the plant is used. Scratches, embedded grit and weld splatter can be removed by careful grinding. All grinding and abrading on stainless steel must be carried out using alumina or a similar abrasive, rather than silicon carbide consumables. Wire brushes must be stainless steel, rather than carbon steel. Local dressing should be fine-finish, since coarse scratches provide sites for the initiation of pitting attack. In critical applications, if stainless steel is to be used near its limit (in terms of corrosion), and for cases such as welds, where a good finish cannot be otherwise achieved, additional passivation is required. Nitric acid (10–15% by volume) is the best passivator. It also dissolves iron contamination. In circumstances where the use of nitric acid is not possible for safety or physical reasons (such as the underside of vessel roofs) passivation paste is appropriate. Both materials are used at ambient temperature and require a contact time of approximately 30 min. They must be removed by thorough rinsing with low chloride-content water.

New plant can be contaminated with corrosive species such as sea salt if it has been transported by sea as deck cargo. Corrosive species present on building sites can gain ingress into items of plant. Many metallic materials, particularly aluminium, zinc alloys and galvanized steel, are susceptible to attack by alkaline cement dust and acid floor cleaners. Significant corrosion can occur by any of these agents before or during commissioning. Titanium and, to a lesser extent, tantalum are at risk of hydriding attack when contaminated by iron smears (see Section 30.5.5).

The materials selection procedure should take into account the commissioning requirements of all parts of the plant in order not to include any materials that cannot safely be degreased or chemically cleaned.

Heat transfer

During the materials selection procedure isothermal corrosion testing may indicate the suitability of a material for handling a corrosive process fluid. In many cases where heat transfer is involved the metal wall temperature experienced in service is higher than the bulk process fluid temperature. This, and the actual heat transfer through the material, must be taken into account since both factors can increase corrosion rates significantly.

Liquid line

The corrosion conditions can be different at the fluid line from the bulk condition. Aqueous liquids have a concave meniscus which creates a thin film of liquid on the vessel wall immediately above the liquid line. Some corrosion processes, particularly the diffusion of dissolved gases, are more rapid in these conditions. Additionally, the concentration of dissolved gases is highest near the liquid surface, especially when agitation is poor. Locally high corrosion rates can therefore occur at the liquid line, leading to thinning in a line around the vessel. This effect is reduced if the liquid level in the vessel varies with time. Any corrosion tests undertaken as part of the materials selection procedure should take this effect into account.

30.3.6 Influence of external environment

The external environment experienced by plant can be more corrosive than the internal process stream. Any construction material must be chosen to withstand or be able to be protected from external corrosive agents.

Corrosive atmospheres

Corrosive species in the atmospheres include water, salts and gases. Clean atmospheres contain little other than oxygen, nitrogen, water vapour and a small quantity of carbon dioxide. These species are virtually non-corrosive to any of the common constructional materials for plant at normal temperatures. Steel is susceptible to corrosion in even fairly clean air where water can exist as liquid. For plant operating at temperatures up to approximately 100°C coatings are employed to protect steel if required. In clean air corrosion rates are low, and corrosion is primarily a cosmetic problem, although it may be necessary to prevent rust staining of nearby materials. Organic coatings (i.e. anti-corrosion paints) are used. Above 100°C the steel will be dry provided it is sheltered from rain. It is also likely to be lagged and therefore not required to be further protected from corrosion. Few organic coatings are suitable for use in this temperature range, but epoxy-based coatings are used up to 120°C and vinylesters up to 160°C. Silicones and inorganic and metallic coatings can be used at higher temperatures. At temperatures above 480°C, oxide scaling is a serious problem for steel, and alternative materials such as stainless steels or nickel-base alloys are required.

Contaminated atmospheres create additional corrosion problems. In marine environments (e.g. coastal power stations, offshore oil platforms and above the splash zone on ships) the chief corrosive agent is sea salt. The chloride content of sea salt precludes the use of stainless steel and bare steel or aluminium in these areas. Steel requires to be coated as described in the previous section. Any coating scheme used for the protection of steel must be properly selected and applied. BS 5493 contains guidelines for coating systems for a variety of environments. The most common reasons for coating failure are incorrect application and inadequate surface preparation. The use of an independent coating inspector, as discussed in Section 30.7, is recommended. Aluminium alloys can be coated with plastics, such as U-PVC, epoxy finishes or conventional paints, with the correct etch primer where applicable to ensure adhesion. Some alloys are suitable for anodizing, but the corrosion protection afforded by anodizing is variable and requires tight specification and control to ensure correct anodized film thickness and sealing efficiency. The lower grades of stainless steels are not suitable for external exposure in marine atmospheric conditions unless they are cleaned regularly with fresh water. Higher grades of stainless steel, nickel alloys and most copper alloys do not normally benefit from additional protection. The development of the coloured patina on copper alloys can, however, be avoided by coating with inhibited lacquer.

Industrial environments can be contaminated with sulphur oxides, other acid gases, ammonia, hydrogen sulphide, nitrogen oxides, chlorine, etc. Most of these are corrosive towards steel, which is normally protected as outlined above. Copper alloys are corroded by hydrogen sulphide at levels above 60 p.p.b. although generally the corrosion product is generally stable and adherent and further protection is not usually required. Corroded in this way, copper is unsightly and the functioning of electronic equipment is affected. Ammonia corrodes and causes stress corrosion cracking of some copper alloys and if it is present in significant quantities, most alloys require protection by coating. Chlorine and most acid gases are sufficiently corrosive towards aluminium alloys and the lower grades of stainless steel that protection is required. Coatings are the only practical solution if these gases are present at high enough levels.

Corrosion under lagging

Thermal insulation of vessels and pipework usually employs glass fibre or foamed polyurethane products. In their pure forms these pose little corrosion risk. Generally, they are contaminated with leachable acids and/or chloride ions. Chloride-free lagging can be specified and this should be used for contact with metals which are susceptible to chloride pitting or chloride-induced stress corrosion cracking. The lower-grade austenitic stainless steels, including types 304, 321, 347, 316, 317 and their low-carbon variants, are at greatest risk. If the lagging remains dry there is no corrosion problem. Lagging can become wet through rain ingress at damage sites, spillage around entry ports or leakage from steam coils or the lagged vessel wall. Chloride leached from the lagging or introduced with the water tends to concentrate on the pipe or the vessel, and even trace quantities create serious corrosion problems. If chloride-free lagging is not used, or the risk of water ingress cannot be removed, the vessel can be coated prior to lagging. Any barrier coating can be used but if it is damaged it will not prevent corrosion. Aluminium, in the form of foil, is used to prevent chloride-induced stress corrosion cracking of austenitic stainless steels. Applied to the vessel or pipe prior to lagging, it acts as a barrier to water and possibly by slight cathodic protection.

Corrosion in concrete

Intact, good-quality, void-free concrete creates an alkaline environment and protects embedded steelwork such as reinforcing steel. Penetration by chloride ions, carbon dioxide, oxygen and other corrosive species reduces this protection. Voids in the concrete created by inadequate compaction at the time of application accelerate this penetration. Carbon dioxide reacts chemically with the concrete, reducing its alkalinity. Chloride and oxygen reach the steel and initiate corrosion. The corrosion products occupy a greater volume than the steel from which they were formed. This expansion of some three to fourteen times creates a pressure greater than the tensile strength of the concrete. Spalling of the concrete allows further corrosion to take place. Footings for structural steelwork and vessel supports disintegrate and blocks of detached concrete fall off the roofs and walls of buildings. For reinforced concrete bases for vessels, especially where acidic or chloride- or sulphate-containing spillages are anticipated, the following options should be considered at the materials selection process: acid- and sulphate-resisting concrete are available; specialist coatings can be applied to concrete to prevent the ingress of species that would attack the concrete or corrode the reinforcing steel or the process vessel that is encased in the concrete; reinforcing material can be galvanized steel, stainless steel (BS 6743) or plastic coated; reinforced steel can be cathodically protected.

Corrosion in soil

Pipework is the item of plant that is most usually buried in soil. In addition to the internal environment, the soil can present a significant corrosion hazard.

Light, sandy, well-drained soil of high electrical resistivity is low in corrosivity and coated steel or bare stainless steels can be employed. It is unlikely that the whole pipe run would be in the same type of soil. In heavier or damp soils, or where the quality of backfilling cannot be guaranteed, there are two major corrosion risks. Steel, copper alloys and most stainless steels are susceptible to sulphide attack brought about by the action of sulphate-reducing bacteria in the soil. SRB are ubiquitous but thrive particularly well in the anaerobic conditions which persist in compacted soil, especially clay. The mechanism of corrosion where SRB are involved is described in Section 30.1.2. Cathodic protection alone, described in Section 30.7.2, is not completely effective against this type of corrosion. Where SRB activity is likely or the resistivity of the soil is less than 4000 Ω cm the pipeline should be coated. Anti-corrosion paint systems at a minimum thickness of 0.2 mm and rubber, plastic or bituminous coatings are preferred protection systems. Close control of backfill quality is necessary to ensure that sharp stones do not damage the coating, and full on-site inspection is required to ensure defect-free application of site-applied coatings and field joints. It is usually the case that a combination of coating and cathodic protection is the most economic alternative. For critical application or the upgrading of old pipes, cathodic protection is generally suitable. Whether or not cathodic protection is used, care is required to avoid stray current corrosion problems through interaction with nearby cathodic protection or other electrical systems such as electric railways.

Corrosion in timber

Metals in contact with or in the proximity of timber can suffer enhanced corrosion attack. Some species of timber, especially oak and Douglas fir, contain high levels of acetic acid. These are volatile and cause corrosion of nearby metals, especially iron, steel and lead alloys.

Metals in contact with timber can be corroded by the acetic acid of the timber and by treatment chemicals present in it. Treatment chemicals include ammonium sulphate and ammonium phosphate flame retardants. These are particularly corrosive towards steel, aluminium and copper alloys. Preservative treatments include copper salts which, at high timber moisture contents, are corrosive towards steel, aluminium alloys and zinc-coated items.

30.4 Design and corrosion

The design of a plant has significant implications for its subsequent corrosion behaviour. Good design minimizes corrosion risks whereas bad design promotes or exacerbates corrosion.

30.4.1 Shape

The shape of a vessel determines how well it drains (Figure 30.7). If the outlet is not at the very lowest point process liquid may be left inside. This will concentrate by evaporation unless cleaned out, and it will probably become more corrosive. This also applies to horizontal pipe runs and steam or cooling coils attached to vessels. Steam heating coils that do not drain adequately collect condensate. This is very often contaminated by chloride ions, which are soon concentrated to high enough levels (10–100 ppm) to pose serious pitting and stress corrosion cracking risks for 300-series austenitic stainless steel vessels and steam coils.

Flat-bottomed storage tanks tend to suffer pitting corrosion beneath deposits or sediments which settle out. Storage tanks may be emptied infrequently and may not experience sufficient agitation or flow to remove such deposits.

Flange face areas experience stagnant conditions. Additionally, some gasket materials, such as asbestos fibre, contain leachable chloride ions. This creates crevice and stress corrosion cracking problems on sealing surfaces. Where necessary, flange faces which are at risk can be overlaid with nickel-based alloys. Alternatively, compressed asbestos fibre gaskets shrouded in PTFE may be used.

Bends and tee-pieces in pipework often create locally turbulent flow. This enhances the corrosivity of the process liquid. These effects should be minimized by the use of flow straighteners, swept tees and gentle bends. Flow-induced corrosion downstream of control valves, orifice plates, etc. is sometimes so serious that pipework requires lining with resistant material for some twelve pipe diameters beyond the valve.

30.4.2 Stress

The presence of tensile stress in a metal surface renders that surface more susceptible to many kinds of corrosion

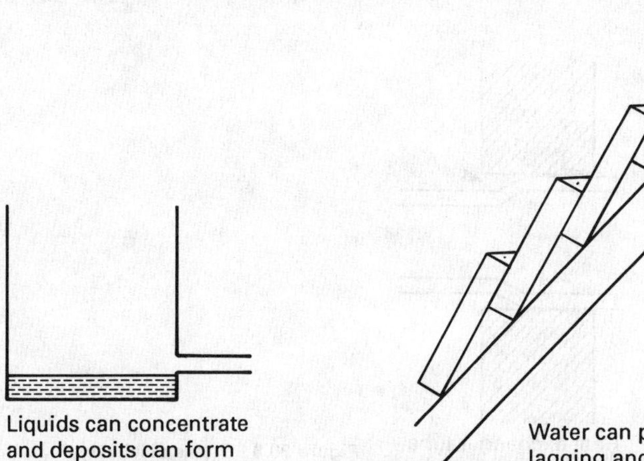

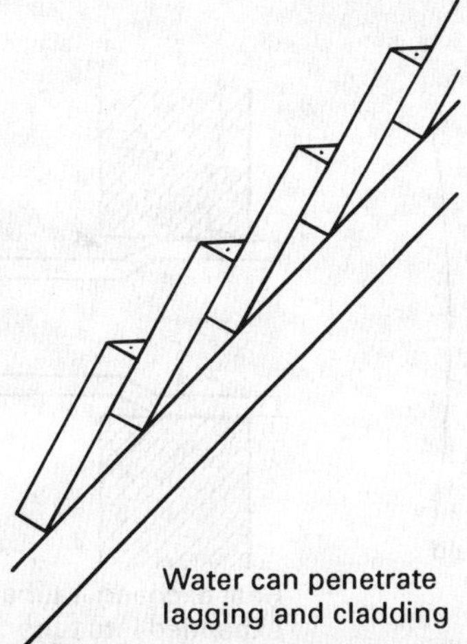

Figure 30.7 Details of design creating corrosion problems

than the same material in a non-stressed condition. Similarly, the presence of compressive stress in the surface layer can be beneficial for corrosion behaviour.

Tensile stresses can be residual, from a forming or welding operation, or operational from heating–cooling, filling–emptying or pressurizing–depressurizing cycles. The presence of a tensile stress from whatever origin places some materials at risk from stress corrosion cracking, as described in Section 30.1.3. Some items of plant can be stress-relieved by suitable heat treatment, but this cannot prevent operational stress arising.

Cyclic stresses can also give rise to fatigue or corrosion fatigue problems. Information relating to the fatigue life of the material in the service environment is required, together with the anticipated number of stress cycles to be experienced by the item over its operational life. The fatigue life (the number of cycles to failure) or the fatigue strength (the design strength below which it does not exhibit fatigue problems) is then used in the design.

The presence of stress raisers, including sharp corners and imperfect welds, produces locally high stress levels. These should be avoided where possible or taken into account when designing the materials for use in environments in which they are susceptible to stress corrosion cracking or corrosion fatigue.

30.4.3 Fabrication techniques

Most fabricational techniques have implications for corrosion performance. Riveted and folded seam construction creates crevices as shown in Figure 30.8. Those materials which are susceptible to crevice corrosion should be fabricated using alternative techniques (e.g. welding). Care should be taken to avoid lack of penetration or lack of fusion, since these are sites for crevice corrosion to initiate.

Welding should be continuous, employing fillets where possible, since tack welds create locally high stresses and leave crevice sites. Welding consumables should be chosen to create weld metals of similar corrosion resistance to the parent material. This often requires the use of a slightly over-alloyed consumable, to allow for loss of volatile alloying elements during the welding process and to compensate for the inherently poorer corrosion resistance of the weld metal structure. Strongly over-alloyed weld consumables can create galvanic corrosion problems if the weld metal is significantly more noble than the parent material. In all welds the heat-affected zone is at risk. The new structure which forms as a consequence of the thermal cycle can be of lower corrosion resistance, in addition to the often poorer mechanical properties, than parent material. Austenitic steels such as type 304 and 316 are also susceptible to sensitization effects in the heat-affected zone. In these materials carbide precipitation during the welding thermal cycle denudes the parent material of chromium. This creates areas of significantly diminished corrosion resistance, resulting in knife-line attack in many corrosive environments. This is avoided by the use of the low-carbon equivalents (304L, 316L, etc.) or grades such as type 321 or 347 which are stabilized against sensitization. With correct welding techniques, however, this should be necessary only with thick sections (5 mm for 304 and 8 mm for 316).

Some materials, particularly certain aluminium alloys, duplex stainless steels in certain reducing environments and most steel plate, are susceptible to end-grain attack. Penetration along the end grain can be very rapid, with corrosion exploiting the potential differences that exist between inclusions and ferrite crystals in steel and between austenitic and ferrite grains in duplex stainless steel. Where end-grain attack is significant this should not be exposed to the corrosive environment. It can be covered by a fillet 'buttering' weld if necessary.

30.4.4 Design for inspection

Unseen corrosion can be the most damaging type of attack. Items should be designed to permit periodic inspection. This involves the provision of sufficiently large manways, the installation of inspection pits, the placing of flat-bottomed vessels on beams instead of directly onto concrete bases and the facility for removal of thermal insulation from vessel walls.

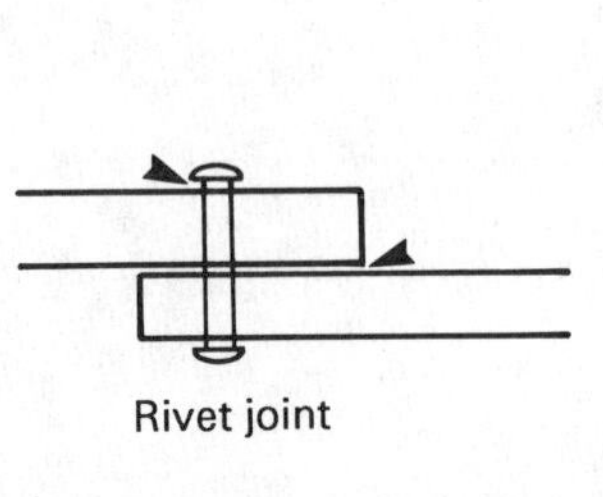

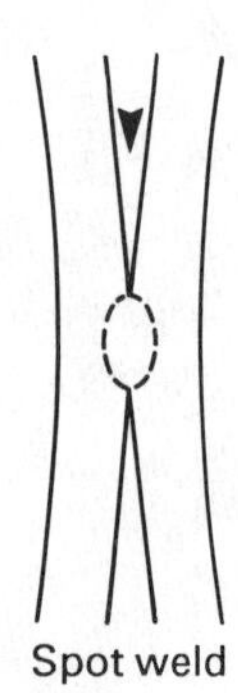

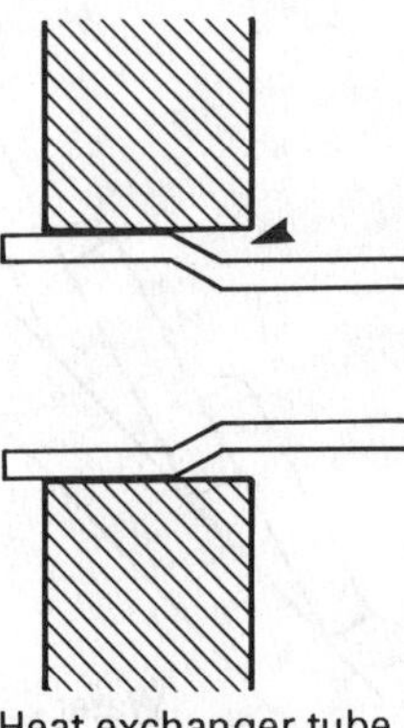

Figure 30.8 Details of jointing processes creating additional corrosion risks (crevices and stress concentrations)

30.5 Uses and limitations of constructional materials

30.5.1 Steels and cast irons

Steel is essentially iron with a small amount of carbon. Additional elements are present in small quantities. Contaminants such as sulphur and phosphorus are tolerated at varying levels, depending on the use to which the steel is to be put. Since they are present in the raw material from which the steel is made it is not economic to remove them. Alloying elements such as manganese, silicon, nickel, chromium, molybdenum and vanadium are present at specified levels to improve physical properties such as toughness or corrosion resistance.

Steels are used primarily for their strength and other mechanical properties rather than for corrosion resistance. The corrosion rates of steels is very low only in a few environments, i.e. clean or hard water, certain non-electrolytes, clean atmospheres and properly constituted and compacted ordinary Portland cement. The economics of any high-volume chemical processes, including oil and gas production and transport systems, permits the widespread use of steels. By incorporating a corrosion allowance into the design of a plant, corrosion rates of up to 1 mm/yr can be accommodated, particularly for plant designed for an operational life of only a few years. Steels generally corrode in a uniform manner. This is beneficial, since localized attack, as experienced by stainless steels, is very difficult to allow for in plant design. The use of corrosion inhibitors (Section 30.7.4) extends the economic range of use of steels. Steel is also available coated with zinc (galvanized, see BS 729) or aluminium-zinc (BS 6830); see also BS 2659, BS 1706 and BS 4921.

Limitations on the use of steels include:

1. Acidic solutions below about pH 5 due to general corrosion;
2. Alkaline solutions above about pH 11 due to general dissolution and caustic cracking;
3. Marine and industrial atmospheres (without the use of protective coatings), because of high corrosion rates;
4. Atmospheric exposure at temperatures in excess of 480°C, because of oxide scaling;
5. Certain environments containing nitrate, cyanide, carbonate, amines, ammonia or strong caustic, due to the risk of stress corrosion cracking. Temperature is an important factor in assessment of each cracking environment;
6. Solutions of hydrogen sulphide, because of hydrogen-induced cracking. Grades of steel are available for certain ranges of pH and hydrogen sulphide partial pressure.

Cast irons are iron with high levels of carbon. Heat treatments and alloying element additions produce grey cast iron, malleable iron, ductile iron, spheroidal cast iron and other grades. The mechanical properties vary significantly. Nickel-containing cast irons have improved hardness and corrosion resistance. Copper or molybdenum additions improve strength.

Chromium, silicon and other alloying elements are used to create cast irons for corrosion resistance in specific environments. Silicon-containing cast irons are used for sulphuric acid duty.

Limitations on the use of cast irons are similar to those for steel, since in many environments most cast irons have poor corrosion resistance. Most grades are also susceptible to graphitization (the loss of iron, leaving a weak structure of graphite) in acidic environments below a pH of approximately 5.5. This attack occurs in soils.

30.5.2 Stainless steels

Stainless steels are iron-based alloys which contain at least 11% chromium. Hundreds of grades of stainless steel are available with alloying elements including nickel, molybdenum, manganese and copper. Stainless steels are not electrochemically inert, but are protected by a thin layer of oxide. This passive layer is unstable in oxidizing and reducing environments. If lost completely, the underlying steel corrodes rapidly in corrosive (particularly acidic) environments. Where the oxide is lost locally, pitting, crevice corrosion or stress corrosion cracking can proceed at rates of several millimetres per year. Chemical species including the halide ions assist in the breakdown of the passive film in even mild conditions such as neutral sodium chloride solution. For satisfactory service, stainless steels should be passivated continuously (or intermittently, see Section 30.3.5) with a suitable oxidizing species. The redox potential of an environment is of prime importance in the selection of a stainless steel. In reducing conditions the protective oxide cannot form, and general corrosion takes place; in strongly oxidizing conditions pitting corrosion is possible.

Stainless steels are used in a wide variety of applications and are most often selected because steel or cast iron would corrode at an unacceptably high rate or produce high levels of iron contamination in the proposed service environment. The main limitations on their uses are:

1. The lower grades of austenitic stainless steels suffer chloride-induced stress corrosion cracking at temperatures in excess of 60°C. Grades containing high levels of nickel are more resistant. Sensitization (Section 30.4.3) occurs near welds in certain non-stabilized grades. The maximum hardness limitations described in NACE MR-01-75 are necessary to prevent sulphide stress corrosion cracking in hydrogen sulphide-containing environments.
2. Ferritic stainless steels have inferior corrosion resistance compared with austenitic grades of equivalent chromium content, because of the absence of nickel. Stress corrosion cracking can occur in strong alkali.
3. Martensitic stainless steels are of limited use in chemical environments because of their inferior corrosion resistance compared with ferritic and austenitic grades. Their mechanical properties allow them to be used where high hardness is required, but they are susceptible to hydrogen embrittlement.
4. Duplex stainless steels are mostly composed of alternate austenite and ferrite grains. Their structure improves resistance to chloride-induced stress corrosion cracking. In certain reducing acids, such as

acetic and formic, preferential attack of the ferrite is a serious problem.

In all categories the lower grades suffer pitting and crevice corrosion in even low-chloride environments. As the chloride content or temperature increases, or the pH falls, higher grades are required. Reducing acids, including acetic and formic, attack most grades, particularly at elevated temperatures. Free-cutting grades, containing deliberate additions of sulphur or selenium, have significantly poorer corrosion resistance than the corresponding standard grade.

30.5.3 Nickel alloys

Commercially pure nickel has good corrosion resistance to a variety of aggressive environments and is specially used for hot caustic service. It has moderate resistance to acid attack but cannot be used in oxidizing acidic conditions, i.e. in oxidizing acids or acidic media containing oxidizing agents. Nickel is widely used for electroplating steel. Electroless nickel plating is used, especially for items of complex shape that cannot be successfully electroplated. As with all coatings, the presence of defects such as pores allow the corrosive environment to reach the substrate. Since nickel is more noble than steel, rapid, galvanically assisted pitting of the substrate steel occurs at defect sites. Nickel is not resistant in the combined presence of ammonium compounds and oxidizing agents.

Nickel is usually alloyed with elements including copper, chromium, molybdenum and *then* for strengthening and to improve corrosion resistance for specific applications. Nickel–copper alloys (and copper–nickel alloys; see Section 30.5.4) are widely used for handling water. Pumps and valve bodies for fresh water, sea water and mildly acidic alkaline conditions are made from cast Ni–30%Cu type alloys. The wrought material is used for shafts and stems. In sea water contaminated with sulphide, these alloys are subject to pitting and corrosion fatigue. Ammonia contamination creates corrosion problems as for commercially pure nickel.

Nickel–chromium alloys can be used in place of austenitic stainless steels where additional corrosion resistance is required. These alloys are still austenitic but are highly resistant to chloride-induced stress corrosion cracking when their nickel content exceeds 40%.

Molybdenum-containing nickel alloys (the Hastelloys™) are a family of alloys which include grades for handling hot hydrochloric, sulphuric, oxidizing, reducing and organic acids. Some grades are susceptible to attack in ammonia- or sulphide-containing environments.

30.5.4 Copper alloys

Copper has excellent resistance to some corrosive environments, including fresh waters and fluoride-containing atmospheres. Alloying is necessary to achieve good strength, but copper limiting with steel for strength is an alternative (BS 5624). Copper and some of its alloys are susceptible to crevice corrosion, but the mechanism is different from that which affects stainless steels.

Alloying with zinc produces brasses. Brasses are generally suitable for fresh and potable waters but dezincification is a problem for grades containing more than 12% zinc, especially in acidic/alkaline conditions. The addition of small amounts of tin, with arsenic or phosphorus, prevents dezincification and is particularly necessary for sea water applications. For handling flowing sea water or other corrosives, copper–nickel alloys are favoured. Brasses are susceptible to stress corrosion cracking in the presence of ammonia or ammonium compounds. This includes lavatory fittings.

Copper–aluminium alloys (aluminium bronze) have good general corrosion resistance but de-aluminification is possible in some grades in certain environments. Copper–nickel alloys are used where flow-related corrosion is a problem, particularly in sea water heat exchanger tubes. The grades containing a small iron addition have the highest critical velocity. These alloys are all susceptible to pitting by sulphides (created in stagnant conditions in sea water systems), especially in the presence of chlorine and stress corrosion cracking in the presence of ammonia. To minimize their susceptibility to sulphide attack in estuarine or other polluted waters it can be advisable to pre-passivate and, if necessary, inhibit with ferrous sulphate.

30.5.5 Miscellaneous metallic materials

Aluminium alloys

Commercially pure aluminium has good resistance to atmospheric corrosion, except where chloride is present in significant quantities, i.e. within 1–2 km of the coast and in the vicinity of chemical plants. To achieve reasonable strength, alloying additions of silicon, manganese, zinc and copper are made. Aluminium–copper alloys have poor corrosion resistance and should not be used in corrosive environments. The other aluminium alloys have good corrosion resistance to most near-neutral media but are attacked in acidic and alkaline conditions. They are susceptible to pitting corrosion, especially where deposits form. Chloride ions also induce pitting of these alloys; chloride is often generated by the hydrolysis of chloride containing organic chemicals. Other organic materials can be handled safely when dry, except alcohols. Metal ions such as copper, which may be introduced by the corrosion of other plant, storage vessels or pipework, can plate out onto the aluminium and create a galvanic cell which produces intense pitting.

The use of anodic films on aluminium alloys is only applicable to some mildly corrosive environments, including architectural purposes (BS 3987) and where abrasion resistance is required (BS 5599).

The presence of liquid mercury poses a liquid metal embrittlement problem for many alloys. This occurs by the spillage of mercury in aircraft and the condensation of mercury vapour from mercury pumps in cryogenic applications.

Titanium

Titanium is available unalloyed in several grades of

purity, and alloyed. Each grade or alloy has specific uses. Generally, titanium has excellent resistance to sea water and oxidizing conditions, including acids and concentrated hypochlorite solution. It is susceptible to corrosion in reducing acids, rapid attack in dry chlorine gas, cracking in alcohols and embrittlement by hydriding. Hydriding is possible when titanium or a metal with which it is in contact corrodes, liberating hydrogen. For this reason, titanium plant should be inspected carefully for traces of metallic contamination (e.g. scratch marks from steel-nailed boots). Titanium can be used in the anodized form; a thick anodic coating greatly reduces hydrogen absorption. It is used with anodic protection for handling 40–70% strength sulphuric acid and 36% hydrochloric acid.

Other metals

Other metals are available which have very specific uses. Their cost is very high but their use is often justified in certain processes such as fine chemical manufacture. They are generally used as thin cladding or loose lining supported by steel or stainless steel.

1. *Gold* is resistant to many strong acids, but not cyanides or high levels of fluoride, hydrogen fluoride and chlorine.
2. *Platinum* is resistant to strong acids and most halogen gases.
3. *Silver* has reasonable resistance to inorganic and organic acids (not strongly oxidizing acids), hot alkaline conditions and some fluoride-containing environments. It tarnishes in the presence of sulphides.
4. *Tantalum* is inert to many environments including hydrochloric acid at strengths up to 25% at 90°C. It is attacked by fluorine, hydrofluoric acid, fuming sulphuric acid containing free sulphur trioxide, and many alkalines. Above 300°C it reacts with many gases including air. It is subject to hydrogen embrittlement in a similar manner to titanium.

30.5.6 Linings and coatings

Many materials with good corrosion resistance have inadequate strength or other mechanical properties to enable their fabrication or economic use. Organic linings and coatings are used in a range of plant. Organic materials are generally susceptible to organic solvents and have use only within limited ranges of temperature. All coatings and linings must be applied in a manner to control defects within them, since the substrate on which they are applied will be corroded by any contact with the process fluid.

Inspection techniques are available for specific coating and lining systems. Proper bonding to substrate is necessary to avoid blistering and disbonding in case of a vacuum being created in the vessel. BS 6374 provides guidance on materials selection, design application and repair.

Resins

Epoxy, polyester, phenolic and other resins are used as coatings and linings with or without reinforcement (see BS 6464 and BS 4994). Glass fibre, silica, carbon and many other materials can be used as filters or reinforcement to produce materials with specific properties of strength, flexibility, wear resistance and electrical conductivity.

Plastics

A large range of man-made polymeric materials is available, from polyethylene, which is attacked by most organic chemicals, to fluorinated products such as polytetrafluoroethylene and polyethyletherketones, which have outstanding resistance to virtually all chemicals. All polymers have their own adhesive, welding and fabricational limitations which must be taken into account in the design of the coated item. These materials can also be used in solid form.

Elastomers

Elastomers are used for their flexibility in seals, gaskets and hoses and to resist abrasion (through absorption of the kinetic energy of the impinging particles). The range of materials includes natural and synthetic rubbers and modern elastomers with chemical resistance.

Glass

Glass-lined reactor vessels are widely used in pharmaceutical and fine-drug manufacturing processes in which metallic contamination of the products has to be minimal. Most glasses are susceptible to attack in strong acid and alkali media. Manufacture and design of glass-lined vessels is a very specialized area. Such vessels require periodic inspection to ensure the integrity of the lining, which is susceptible to mechanical damage and repair is difficult.

30.6 Specifying materials

Having selected the appropriate material for an item of plant, it is imperative that this material is used in the required state of heat treatment, surface finishes, etc. for its construction.

30.6.1 Compositional aspects

1. *National standards*: British Standards, DIN (Deutsch Industries Normen, from Germany) and ASTM/AISI (American Society for Testing of Materials/American Iron and Steel Institute) are those in most common use for metallic materials. It is always preferable to select a material for which national or international specifications exist. Many materials, including steels and the 300-series austenitic stainless steels, have wide ranges of chemical composition. If an alloy or a grade

of stainless steel is to be used in a marginal duty (i.e. where relatively small variations in alloying element content, heat treatment or surface condition could markedly affect its corrosion behaviour) this should be taken into account when the material is specified. For stainless steels and some other materials of construction it is possible to purchase, at a premium rate, the material with specified minimums of alloying elements. Type 316 stainless steel, when supplied to AISI specification, can have 10–14% nickel, 16–18% chromium and 2–3% molybdenum. This wide specification produces material across a range of corrosion resistance. It also creates a latitude for mechanical and physical properties, including strength and magnetic permeability. For demanding applications, but where the more resistant type 317 material is not justified, type 316 with a minimum molybdenum content may be specified.
2. *Proprietary alloy designation*: Many alloys are the subject of patents and are the registered trade mark of the producer. When produced by the holder of the patent, these alloys are generally of the highest standard, with tight control of composition, and exhibit consistent corrosion behaviour. They are mostly originally developed for a specific corrosive service, but are subsequently marketed more widely. These materials can also be manufactured under licence or, when the patent has expired, by other producers using a generic name. Produced in this way, these materials can be less reliable, with the introduction of unwanted tramp elements. Where proving tests have indicated the suitability of a proprietary alloy for specific application, the same source of material should be used for construction. Published compositional data are typical values; cast analyses should be requested for critical applications.
3. *Government standards*: These are developed by government departments, especially the military, and are primarily for aerospace and marine service materials.
4. *In-house specifications*: End users requiring larger amounts of material and fabricators of specialty plant benefit from preparing their own in-house specifications. These are often based on existing national standards modified to include additional requirements. They include tigher compositional ranges of alloying elements, mechanical properties such as strength and surface finish, inspection and certification. These specifications must be written by experts in order to satisfy the users' real requirements without compromising any aspect of the materials' performance.

30.6.2 Mechanical properties

Many available standards, particularly national standards, either do not cover the material physical and mechanical properties or leave the user to specify from a range of options. The aspects of mechanical and physical properties which have specific implications for a material's corrosion behaviour include:

1. Strength, particularly where the material will be subjected to fatigue loading or is susceptible to degrading mechanisms such as hydrogen-induced cracking;
2. Hardness, including surface hardness, especially for materials for sour service or environments in which stress corrosion cracking is expected. It is also important where erosion corrosion is likely;
3. Surface finish, since good surface finish retards the onset of certain types of corrosion attack, including pitting and stress corrosion cracking.

All relevant physical and mechanical property requirements should be included in the material specification.

30.6.3 Certification

All aspects of the material's chemical, mechanical and physical properties which are included in the specification should be capable of measurement and certification. For critical duties all material supplied should be fully tested and certified by competent approved, independent test laboratories. The nationally recognized laboratory approval scheme in the UK is NAMAS (National Measurement Accreditation Service), operated by the Department of Trade and Industry through the National Physical Laboratory. For less demanding duties material should be accompanied by test certificates from the supplier. All items of plant should be purchased with material certification. Additional certification is required in cases where the fabricator, in manufacturing an item of plant, used techniques such as welding or heat treatment which may affect the corrosion behaviour of the construction materials.

All bought-in items of plant, especially those supplied without adequate materials certification, should be subject to random inspection. Portable instruments are available for many types of non-destructive examination, chemical analysis and mechanical testing of fabricated items of plant.

30.7 Corrosion-control techniques

Economic reasons may dictate the use of an inexpensive constructional material (i.e. steel protected by one of the methods discussed below) in place of a more resistant but more expensive material. Although all these corrosion control techniques could be used on a variety of materials, with appropriate design and safeguards, they are only rarely employed to protect any material other than steel.

30.7.1 Painting

Paints are complex formulations of polymeric binders with additives including anti-corrosion pigments, colours, plasticizers, ultraviolet absorbers, flame-retardant chemicals, etc. Almost all binders are organic materials such as resins based on epoxy, polyurethanes, alkyds, esters, chlorinated rubber and acrylics. The common inorganic binder is the silicate used in inorganic zinc silicate primer for steel. Specific formulations are available for application to aluminium and for galvalized steel substrates.

Because of their generally poor resistance to solvents,

acids, alkalis and other corrosive agents, paints are not normally used to protect plant internals handling anything other than waters. Even clean (i.e. potable or better) water is damaging to most epoxy-based coatings. Limitations on the use of paint systems include:

1. *Erosion resistance*: The mechanical properties of paints prevent their use in conditions where impingement or erosion by entrained solids is expected.
2. *Solvent resistance*: This varies from very poor (for chlorinated rubbers) to good (for polyurethanes). All paint binders have specific susceptibilities and the presence of small quantities of the appropriate solvent in the atmosphere in the region of an item of plant can cause rapid failure.
3. *Saponification*: Paints are most commonly used to protect steel from corrosion by sea water in marine applications and soil in the case of buried structures. Additional protection is often supplied by the application of cathodic protection to the steel. Any paint coating used in conjunction with cathodic protection must be resistant to the alkali which is produced on the steel at defect sites in the coating. The amount of alkali generated depends on the potential to which the steel is polarized. Some paint binders such as alkyds and vinyl ester are very susceptible to saponification, and should not be used on cathodically protected structures. Cathodic disbondment testing, to BS 3900: Parts F10 and F11, should be undertaken if the relevant information is not available.
4. *Temperature*: Thermal breakdown of the binder limits the service temperature for painted items to 70°C (e.g. chlorinated rubbers and polyvinylacetates), 120°C (many epoxies) and 160°C (for vinylesters). The commonest reason for coating failure is incorrect application. This is a skilled task, involving surface preparation (usually by grit blasting) and a tight control over environmental factors (temperature and relative humidity of the local atmosphere), the cleanliness of the steel (national standards, including Swedish Standard SIS 05 59 00, BS 4232 and ISO 8502, cover the testing of residual contamination and freedom from millscale, rust, etc.), the paint and the application procedure. All coating should be supervised by an independent, qualified inspector certified to a recognized standard such as the National Association of Corrosion Engineers International Coating Inspector Training and Certification Scheme.

The selection of an anti-corrosion paint coating scheme should be undertaken by a qualified expert. BS 3900 covers most aspects of paint coating testing.

30.7.2 Cathodic protection

Cathodic protection (CP) is an electrochemical technique of corrosion control in which the potential of a metal surface is moved in a cathodic direction to reduce the thermodynamic tendency for corrosion. CP requires that the item to be protected is in contact with an electrolyte. Only those parts of the item that are electrically coupled to the anode and to which the CP current can flow are protected. Thus the inside of a buried pipe is not capable of cathodic protection unless a suitable anode is placed inside the pipe. The electrolyte through which the CP current flows is usually sea water or soil. Fresh waters generally have inadequate conductivity (but the interiors of galvanized hot water tanks are sometimes protected by a sacrificial magnesium anode) and the conductivity of air is far too low to permit the cathodic protection of buildings or vehicles.

The operating principle of CP is shown in Figure 30.9. Good CP practice is embodied in BS Code of Practice CP 1021. For buried pipelines the power source is most frequently d.c. voltage derived from a rectified mains power supply, or in suitable climates such as Australia and the Middle East, solar cells. Such systems operate by 'impressed current' and suitable anodes are platinized titanium or niobium wire, cast iron, and graphite. They are not consumed, but can be damaged by the chlorine generated in sea water CP systems. Reinforcing steel in concrete jetties, bridge and carpark decks and some offshore structures are generally protected by means of impressed current systems.

Sacrificial anode systems operate without external power source. The anodes are reactive metals such as magnesium and zinc or aluminium alloys. The energy for the process is derived from the anode material. Careful design is required to match the output and lifetime of

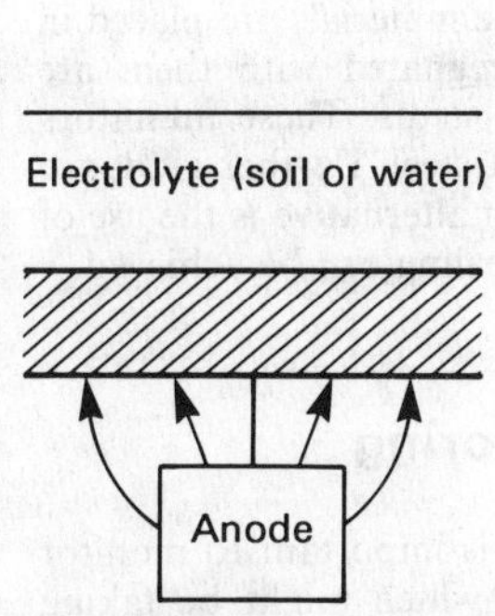

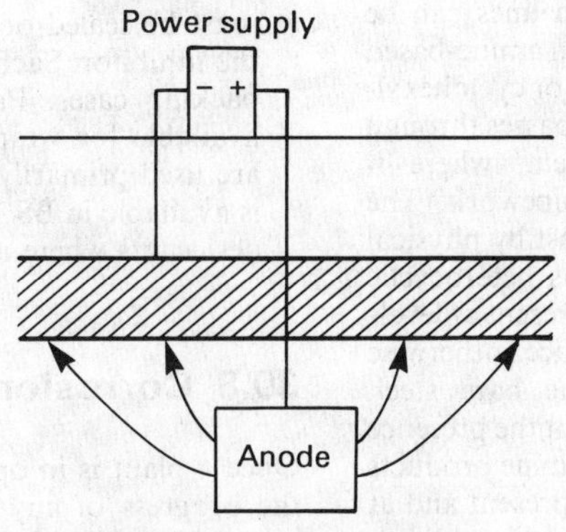

Figure 30.9 Cathodic protection

the anodes with the polarization and life-expectancy requirements of the plant. Sacrificial anode CP is used for offshore platforms, subsea pipelines and the inside of ballast tanks on tanker ships.

Correct design is necessary to achieve full protection without overprotection and to minimize wastage of power or anode materials. Overprotection is undesirable because alkali is generated at the cathode (this degrades many plant systems) and hydrogen can be evolved. Hydrogen is deleterious to some mechanical properties of steels and pipelines have failed through this mechanism. Steel is also susceptible to cracking in carbonate-containing environments in certain potential ranges. A reduction in the power requirements of a CP system can usually be achieved by coating the protected structure.

Stainless steel pipes (buried in the ground) and the interiors of stainless steel heat exchangers have been successfully cathodically protected, but CP is rarely used for materials other than steel. The protection potential usually adopted for steel is -850 mV to the saturated calomel reference electrode. This varies with temperature and the presence of other aggressive species in the environment.

30.7.3 Anodic protection

Anodic protection is possible only for material-environment combinations that exhibit fairly wide passive regions. Examples include type 304 stainless steel in phosphoric acid and titanium in sulphuric acid, steel and stainless steels in 98% sulphuric acid. The effect of the anodic polarization is to shift the steel into a region of passivity. This promotes the formation of a protective film on the steel surface, preventing general corrosion. Inadequate or over-polarization creates corrosion and hydrogen evolution problems. As with CP, only wetted surfaces can be protected; this excludes condensed films or droplets on vessel roofs and walls above the liquid line. Such areas are subject to rapid corrosion attack.

30.7.4 Corrosion inhibitors

The use of cheap constructional materials such as steel can be tolerated in certain instances where corrosion inhibition is possible:

1. *Boilers and steam systems*: Steel steam lines can be inhibited by the use of a volatile amine-based inhibitor such as ammonia, morpholine or cyclohexylamine introduced with the feedwater. It passes through the boiler and into the steam system, where it neutralizes the acidic conditions in pipework. The inhibitor is chemically consumed and lost by physical means. Film-forming inhibitors such as heterocyclic amines and alkyl sulphonates must be present at levels sufficient to cover the entire steel surface, otherwise localized corrosion will occur on the bare steel. Inhibitor selection must take into account the presence of other materials in the system. Some amine products cause corrosion of copper. If copper is present and at risk of corrosion it can be inhibited by the addition of benzotriazole or tolutriazole at a level appropriate to the system. All aspects of boiler water treatment are covered in BS 2486 (see also Section 30.3.2).
2. *Cooling waters*: Once-through cooling systems cannot usually be inhibited because of the expense of the chemicals required and the problems of the disposal of the treated water. Open recirculating systems lose water by evaporation from the tower and pick up atmospheric dust and other detritus. Inhibition of such a water is part of the overall water treatment in which scaling tendency and other factors are also controlled. Phosphates and phosphonates together with zinc ions are commonly used. Biocide additions are also required and algal growth control in the case of systems into which sunlight penetrates. Closed recirculating systems, including those in motor vehicles, can be treated with the most efficient inhibitors such as chromates. Disposal problems have restricted the use of these chemicals, but borates, benzoates and nitrites are used. If there is copper in the system, this must be inhibited also. If aluminium is in the system, nitrite poses a pitting problem in certain pH ranges. Nitrite is also aggressive towards lead–tin soldered joints and is ineffective as an inhibitor for zinc and its alloys. Aluminium is only properly protected by benzoate in acidic solutions; in alkaline conditions it is not fully effective (borates are preferred). The compatibility of all materials, including stainless steels and galvanized steel, must be ascertained before any corrosion-inhibition system is employed.
3. *Heating systems*: Where they are virtually sealed except for a make-up tank (as in the case of domestic central heating) systems can be treated as for closed recirculating cooling systems, and benzoates, nitrites and borates are used. In a correctly designed system such that oxygen ingress is minimal and little make-up is required the oxygen content diminishes as corrosion proceeds and the system can require no inhibitor treatment. The use of certain toxic corrosion inhibitors is not permitted in hospitals, food-manufacturing plant and breweries, where a leak could result in contamination of food or exposure of people to the chemical.
4. *Vapour phase inhibitors*: These are used for the temporary protection of new plant in transit or prior to commissioning. Volatile corrosion inhibitors such as cyclohexylamine derivatives are used. The plant must be sealed or contained to prevent rapid loss of the inhibitor. Sachets of these materials are placed in packing cases. Papers impregnated with them are available for wrapping steel items. These inhibitors are used primarily to protect steel. Further guidance is available in BS 1133(6). An alternative is the use of desiccants where adequate sealing can be achieved.

30.8 Corrosion monitoring

Once a plant is in operation it is important to monitor the progress of any corrosion which might be taking place. The four approaches described below vary in sophistication and cost. The most appropriate for any plant is determined by a number of factors, including the mechanisms of corrosion which are anticipated and the implications of catastrophic or unexpected failure. Key areas of the plant require closer monitoring than ready replaceable items. The measures described below do not replace the mandatory inspections of pressure vessels, etc. for insurance purposes. The overall philosophy of corrosion monitoring is to improve the economics of the plant's operation by allowing the use of cheaper materials and generally reducing the over-design that goes into plant to combat corrosion.

30.8.1 Physical examination

Full records of all constructional materials that are used in the plant should be maintained and updated when repairs are undertaken. The exteriors of all parts of the plant should be subjected to frequent visual examination and the results reported and stored for future reference. This maximizes the warning time before corrosion failures occur, since the majority of failure mechanisms cause leaks before bursting. Key items of plant, those in which some degree of corrosion is anticipated and those which might suffer catastrophic failure should be examined in greater detail. Internal visual inspection during shutdowns is sufficient to identify most corrosion effects. Cracking can usually be seen with the naked eye but where cracking is considered to be a possible mechanism an appropriate non-destructive test method should be employed. In items of plant which are shut down only infrequently (relative to the timescale of possible corrosion or cracking failure) external non-destructive testing is often possible.

Candidate non-destructive test methods include:

1. *Ultrasonic techniques*: Wall thickness can be measured to monitor the progress of general corrosion, cracks can be detected and hydrogen blisters identified. Certain construction materials such as cast iron cannot be examined by ultrasound. Skilled operators and specialist equipment is required. Plant can be examined *in situ* except when it is above 80°C.
2. *Magnetic particle inspection*: Surface emergent and some sub-surface cracking can be detected in ferromagnetic materials. The technique must be used on the side of the material in contact with the corrodent.
3. *Dye penetration inspection*: This is a simple technique, requiring a minimum of operator training. In the hands of a skilled operator it is capable of detecting fine cracks such as chloride stress corrosion cracks in austenitic stainless steels and fatigue cracks.

30.8.2 Exposure coupons and electrical resistance probes

If changes have been made to the process (e.g. if incoming water quality cannot be maintained or other uncertainties arise concerning the corrosion behaviour of the construction materials it is possible to incorporate coupons or probes of the material into the plant and monitor their corrosion behaviour. This approach may be used to assist in the materials selection process for a replacement plant. Small coupons (typically, 25 × 50 mm) of any material may be suspended in the process stream and removed at intervals for weight loss determination and visual inspection for localized corrosion. Electrical resistance probes comprise short strands for the appropriate material electrically isolated from the item of plant. An electrical connection from each end of the probe is fed out of the plant to a control box. The box senses the electrical resistance of the probe. The probe's resistance rises as its cross-sectional area is lost through corrosion.

The materials should be in the appropriate form (i.e. cast/wrought/welded, heat treatment and surface condition). Metal coupons should be electrically isolated from any other metallic material in the system. They should be securely attached to prevent their being dislodged and causing damage downstream. Simple coupons and probes cannot replicate the corrosion effects due to heat transfer but otherwise provide very useful information. It should be noted that any corrosion they have suffered represents the integrated corrosion rate over the exposure time. Corrosion rates often diminish with time as scaling or filming takes place, thus short-term exposures can give values higher than the true corrosion rate.

30.8.3 Electrochemical corrosion monitoring

A number of corrosion-monitoring techniques, based on electrochemical principles, are available. These give an indication of the instantaneous corrosion rate, which is of use when changing process conditions create a variety of corrosion effects at different times in a plant. Some techniques monitor continuously, others take a finite time to make a measurement.

1. *Polarization resistance*: The current-potential behaviour of a metal, externally polarized around its corrosion potential, provides a good indication of its corrosion rate. The technique has the advantage of being well established and hence reliable when used within certain limitations. This technique can only be used for certain metals, to give general corrosion rate data in electrolytes. It cannot be employed to monitor localized corrosion such as pitting, crevice corrosion or stress corrosion cracking, nor used in low-conductivity environments such as concrete, timber, soil and poor electrolytes (e.g. clean water and non-ionic solvents). Equipment is available commercially but professional advice should be sought for system design and location of probes.
2. *Impedance spectroscopy*: This technique is essentially the extension of polarization resistance measurements into low-conductivity environments, including those listed above. The technique can also be used to monitor atmospheric corrosion, corrosion under thin films of condensed liquid and the breakdown of protective paint coatings. Additionally, the method provides mechanistic data concerning the corrosion processes which are taking place.
3. *Electrochemical noise*: A variety of related techniques are now available to monitor localized corrosion. No external polarization of the corroding metal is required,

but the electrical noise on the corrosion potential of the metal is monitored and analysed. Signatures characteristic of pit initiation, crevice corrosion and some forms of stress corrosion cracking are obtained.

30.8.4 Thin-layer activation

This technique is based upon the detection of corrosion products, in the form of dissolved metal ions, in the process stream. A thin layer of radioactive material is created on the process side of an item of plant. As corrosion occurs, radioactive isotopes of the elements in the construction material of the plant pass into the process stream and are detected. The rate of metal loss is quantified and local rates of corrosion are inferred. This monitoring technique is not yet in widespread use but it has been proven in several industries.

Further reading

Dillon, C. P., *Corrosion Control in the Chemical Process Industries*, McGraw-Hill, New York (1986)

Fontana, M. G. and Greene, N. D., *Corrosion Engineering*, 2nd edn, McGraw-Hill, New York (1978)

The Forms of Corrosion Recognition and Prevention, Corrosion Handbook No. 1, NACE, Houston, Texas (1982)

Lees, F. P., *Loss Prevention in the Process Industries*, Butterworths, London (1983)

Shreir, L. L. (ed.), *Corrosion*, 2nd edn, London (1976)

Journals

British Corrosion Journal

Industrial Corrosion is the main journal of the Institution of Corrosion Science and Technology, PO Box 253, Leighton Buzzard, LU7 7WB.

Corrosion is the scientific journal of the National Association of Corrosion Engineers, PO Box 21840, Houston, Texas, USA

Materials Performance is the engineering journal of the National Association of Corrosion Engineers

Metal Producers' Trade Associations

Nickel Development Institute
The Holloway
Alvechurch
Birmingham
B48 7QB

Copper Development Association
Orchard House
Mutton Lane
Potters Bar
EN6 3AP

Aluminium Federation Ltd
Broadway House
Calthorpe Road
Birmingham

Zinc Development Association
34 Berkeley Square
London W1

31

Paint Coatings for the Plant Engineer

D A Bayliss FICorrT, FTSC
ITI Anti-Corrosion Ltd

Contents

31.1 Definition and function of coatings

The term 'paint' is normally used to describe the liquid material before application and 'coating' after it has been applied, dried and cured. Organic-based coatings form the largest use for protection and decoration, and these can vary widely in properties and characteristics, ranging in thickness from a few microns to several millimetres. The term 'coating' is not confined to thick films.

Metal coatings (e.g. zinc on steel) can also provide a useful and economic form of protection for steelwork and plant in the appropriate circumstances. Some metallic coatings (e.g. hot-dip galvanizing) act as a complete barrier to the environment, but most corrosion-protection metal coatings and all organic coatings are permeable to moisture and gases to some extent. The rate of permeability varies with different types of coating, but in all cases protection of the substrate is mainly afforded by adhesion. Good adhesion, which is largely a product of adequate surface preparation, prevents the lateral spread of moisture and contaminants which would undermine the coating and lead to disbondment or corrosion and eventual breakdown.

The thickness of the coating also affects permeability, and in general, there is a correlation between thickness and life. However, with relatively thick coatings as obtained with some modern materials such as epoxies the relationship is not so well established. There may be a limiting thickness above which either little additional protection is obtained or the increase in cohesive strength reduces its adhesive strength. Coatings should always be applied as closely as possible to the manufacturer's recommended thickness.

Coatings will also react with the environment over a period of time. Metal coatings corrode, albeit at a slower rate than steel. They have a finite life in a specific environment, depending on the thickness of the coating. Organic coatings react in a different way, and generally a type can be chosen which will have superior resistance to chemical attack. However, they also deteriorate in time, the main natural destructive influences being sunlight and moisture.

The normal visible deterioration of a paint coating is by the appearance of 'chalking' on its surface. Chalking is the term used to denote the powdery material that appears as the binder slowly disintegrates and exposes the pigments.

31.2 The constituents of paint

All paints consist of a binder (sometimes called a medium) and pigment. Materials consisting of binder only are called varnishes. Most paints and varnishes contain solvent in order to make the binder sufficiently liquid to be applied. The combination of binder and solvent is called the vehicle. Some paints are available without solvent (e.g. solventless epoxies) but these generally require special methods of application (e.g. application of heat) to reduce the viscosity.

31.2.1 Paint binders

Binders are the film formers. After application, they turn from liquids to solids. Different types of paint have, in some cases, radically different mechanisms for this process. This gives important differences in properties between the paints made from different binders. In all cases, however, binders are a vital constituent of paint and provide its main mechanical and physical properties. Binders may vary from 20% to 50% by weight of the paint but will be a higher percentage in the coating film.

31.2.2 Paint pigments

Pigments are the solid constituents added during paint manufacture. Their function is not solely to provide colour but also opacity. Since pigments are opaque they also help to protect the binder from the harmful ultraviolet content of sunlight. In general, varnishes are less durable than paints. Pigments also reduce permeability, particularly when they are in the lamellar form (e.g. micaceous iron oxide, aluminium and graphite), since they increase the moisture path through the film.

Some pigments exert an inhibitive effect on the corrosion of metal. The mechanisms are complex and not always fully understood. Inhibitive pigments include red lead, zinc phosphate and zinc chromate.

Most pigments can be used in any type of binder, therefore paints cannot be identified by pigment type alone. For example, micaceous iron oxide pigment is traditionally in an oil-based binder but is being increasingly used in epoxies, etc. In the paint coating film the pigment content may vary from 15% to 60%. In the special case of zinc rich primers it is over 90%.

The proportions of pigment to binder are a critical factor in paint formulation not only in providing the optimum strength and impermeability but also in the finished appearance (e.g. high pigmentation can give a matt surface, low pigmentation a high gloss). All other things being equal, a high gloss is generally more durable than a matt finish.

31.2.3 Other constituents

Extenders

These are similar to colour pigments but are generally inorganic. Although often considered as cheap fillers, they can form a useful function as a reinforcement to the coating film.

Diluent

This is a volatile liquid which, while not a solvent for the binder, can be used in conjunction with the true solvent to reduce cost without precipitation of the binder.

Anti-skinning agent

This is added to reduce or prevent skin formulation in partially filled containers.

Water-based corrosion inhibitor

This is added to water-based paints to reduce corrosion of the container.

31.2.4 Solvents

Solvents are used to reduce the viscosity of a paint so that it is suitable for application. They play no direct part in the protective value of the coating and are lost to the atmosphere. Amounts are kept as low as possible (usually 5–40% by weight).

The type of solvent used depends on the binder. There is normally more than one type of solvent in a paint, particularly for spray application. Highly volatile solvents are needed to reduce the viscosity during atomization and then disperse as quickly as possible, but lower volatile solvents are necessary to remain momentarily to ensure that there is sufficient flow to form a continuous film.

Organic solvents are a toxic, fire and explosive hazard to varying degrees, depending on type. Solvent entrapment is one of the most common forms of premature failure of modern, high-build, fast-drying materials. Paint manufacturers are urgently developing materials which can be substantially water based. Currently, this can only be achieved with a comparative loss of durability and some application problems, but development of suitable materials is inevitable for the future.

31.3 Types of coating and their uses

A convenient method of classifying paint coatings is by their curing mechanism. The process by which the wet film turns into a solid one has a significant effect on the properties of both application and durability. However, within a particular generic type the formulation can vary, and one manufacturer's product may be superior to another's. The general lack of standards and changes in the use of raw materials by manufacturers means that different products cannot be directly compared except by testing.

Given below are the main types of paint, their properties and some general guidance on their use. For any specific requirements it is always advisable to check with paint manufacturers.

31.3.1 Air-oxidizing coatings

Examples

Alkyds
Silicone alkyds
Urethane alkyds
Epoxy esters
Tung oil phenolic

General advantages

1. Relatively cheap;
2. Easy to apply by brush or spray;
3. Generally formulated with mild solvents, such as white spirit;
4. Readily available in a wide range of colours;
5. Reasonable gloss retention on exterior exposure;
6. Resistant to mineral and vegetable oils;
7. Available in high gloss.

General disadvantages

1. Poor alkaline resistance (e.g. not suitable for application to concrete);
2. Not suitable for very wet, immersed or condensation conditions;
3. Limited resistance to solvents;
4. Relatively slow drying (e.g. touch-dry 4–6 h, dry to recoat 12 h (min.);
5. Possible problems of intercoat adhesion with unweathered high-gloss coats;
6. Can only be applied at limited film thickness.

Comparison of paints within this classification

Alkyds These are the most widely used of all air-oxidizing coatings and have the broadest use both industrially and domestically. They are usually classified according to the proportion of drying oil to synthetic resin (known as 'oil length'). The oil length influences all properties (e.g. chemical resistance, viscosity, flexibility and hardness).

Long oil-length alkyds (e.g. 60% and over of oil) are the most suitable materials for site application on a wide range of surfaces under normal exterior or interior environments. They are not suitable for very wet conditions, immersion or condensation, or for application to alkaline surfaces such as concrete.

The shorter oil-length alkyds are quicker-drying, harder finishes which are most suitable for works application. The very short oil-length alkyds are used for stoving finishes on industrial and domestic equipment, etc.

Silicone alkyds These are alkyds modified with silicone resin. They have superior resistance to weathering (particularly gloss retention) than pure alkyds but are generally significantly more expensive. They are useful for exterior use where appearance is important.

Urethane alkyds These are formed by the reaction of an isocyanate with alkyd, although the curing remains substantially through the oil-oxidation reaction. They have properties similar to alkyds although superior resistance to abrasion is claimed.

Epoxy ester Epoxy esters are a type of alkyd where a high molecular weight resin is reacted with alkyd resin. The curing mechanism remains primarily through the oil-oxidation reaction and their properties are in no way similar to the chemically reacted epoxies. They have similar properties to alkyds although with improved chemical resistance but inferior appearance. They form a reasonably hard, oil-resistant coating which can sometimes be suitable for machinery enamels, but are primarily for interior use, since they tend to chalk rapidly on exteriors. Their best use is for chemical or water resistance where circumstances dictate that more superior finishes cannot be used.

Tung oil phenolic Tung oil is a natural, fast-drying material which, when combined with phenolic resins, forms an air-drying film of good film hardness, water resistance, flexibility and toughness. Prior to the development of more modern materials it was widely employed as a corrosion-resistant coating or varnish, particularly for marine use. It tends to yellow on ageing, so is not suitable for decorative use but when pigmented with micaceous iron oxide is still widely used for exterior structural steelwork painting.

Its best use is for bridges, gantries, conveyor-belt steelwork, etc. where the relatively drab, matt appearance can be tolerated. The Department of Transport uses this material on inland motorway bridges, and has found that, if applied to a correctly prepared and primed surface, it can be substantially maintenance-free for 20 years.

31.3.2 Solvent-dry coatings

Examples

Chlorinated rubber
Vinyls
Acrylated rubber
Modified chlorinated rubber
Bituminous
Vinyl tar

General advantages

1. Quick drying (typically touch-dry 2 h, dry to recoat 8 h);
2. Low water permeability, suitable for wet conditions and condensation;
3. Good acid resistance;
4. Good alkaline resistance, therefore suitable for application to concrete, etc.;
5. Single-pack material;
6. Very easy to recoat in both new and old condition;
7. No danger of loss of adhesion between coats;
8. Can be used for water immersion (e.g. swimming pools) but its thinner film generally makes it less durable than two-pack, chemically reacted materials (e.g. epoxies);
9. With the exception of bitumen, available in a limited range of colours, but light shades are generally more durable externally;
10. Forms a solid film entirely by solvent evaporation and therefore does not rely on temperature, oxidation, etc. to cure. For internal, closed conditions adequate ventilation is the main requirement.

General disadvantages

1. Limited temperature resistance (generally, maximum 60°C dry heat);
2. Very limited resistance to solvents, therefore it tends to be 'lifted' or 'pickled' if overcoated with paints containing stronger solvents (e.g. two-pack epoxies or urethanes);
3. Thermoplastic material, so it moves with temperature and should not be overcoated with harder, thermoset materials such as alkyds, since these will eventually crack and craze;
4. They contain a high percentage of strong solvents (typically 60–75%), therefore a limit to film thickness and always a risk of solvent entrapment;
5. Relatively more difficult to apply than other types, particularly by brush. Since the paint always remains soluble in its solvent the brush tends to 'pick up' the previous coat.

Comparison of paints within this classification

Chlorinated rubber These are formed by the reaction of chlorine on natural rubber. They become a very hard resin, lacking the elastic and resilient characteristics of rubber. They have a specific gravity of 1.64, which is almost twice that of pure rubber. Chlorinated rubber resin is odourless, tasteless and generally non-toxic (although it is sometimes alleged that they can contain residual carbon tetrachloride from the manufacturing process). They are generally less critical in application than vinyls but are less resistant to oils and fats. They will not support combustion or burn and need careful spray technique to avoid 'cobwebbing'. The best use is for surfaces liable to condensation and for coating concrete.

Vinyls Vinyl chloride co-polymer resins were developed in the USA in the late 1930s. They have better weather and slightly more chemical resistance than chlorinated rubber paints. They are generally resistant to crude oil but application is more critical. For example, they are particularly sensitive to moisture present on a surface during painting and this can lead to adhesion failure. They are also more prone to solvent entrapment than chlorinated rubber paints.

Acrylated rubber These are based on styrene butadiene and have become commercially available only relatively recently. They are manufactured in several grades but most have the advantage over other materials in this class of being based on white spirit solvent rather than the stronger and more obnoxious xylol. In other respects they are similar to chlorinated rubber and cost approximately the same, although they are easier to airless spray and the dried film contains less pores. They are considered to have superior weather resistance to chlorinated rubber and vinyls but, as yet, long-term experience is lacking.

Modified chlorinated rubber In these, part of the resin is replaced with an alkyd or oleo-resinous material. The chlorinated rubber content can still remain high but the modification gives better adhesion to steel and easier application properties. However, there is some sacrifice in corrosion resistance. In other versions the alkyd resin becomes the major ingredient and these are claimed to have good weatherability, adhesion, gloss and brushability, plus some improved chemical and water resistance over the alkyd alone. The best use is as a primer for steel under chlorinated rubber top coats. They are widely used for this purpose by the Department of Transport for painting bridges in adverse conditions.

Bituminous This term is used for products obtained from both petroleum and coal-tar sources but the petroleum products are the more widely used. These materials are very resistant to moisture and tolerant to poor surface preparation. They are only available as black, dark brown or aluminium pigmented. The last has reasonable outdoor durability but, without the aluminium, the film will crack and craze under the influence of sunlight. Normally they cannot be overcoated with any other type of paint, because not only will harder materials used for overcoating tend to crack or craze but there is also a possibility that the bitumen will 'bleed through' subsequent coats. The best use is as a cheap waterproofing for items buried or out of direct sunlight.

Vinyl tars

These are a combination of vinyl resins and selected coal tars. They are claimed to be similar to the two-pack coal tar epoxies but with the advantages that they are a single pack, not dependent on temperature for curing, and are easy to recoat at any stage.

Their disadvantages, particularly in comparison with coal tar epoxies, include the fact that they are limited in film thickness per coat and therefore require multi-coated application. They have a higher solvent content and therefore there is an increasing risk of solvent entrapment, and the slower cure may limit their use in a tidal zone. The best use is for immersed conditions.

31.3.3 Chemically cured coatings

Examples

Two-pack epoxies
Coal tar epoxies
Two-pack urethanes
Moisture-cured urethanes

General advantages

1. Very resistant to water, chemicals, solvents, oils and abrasion;
2. Tough, hard films that can, if correctly formulated, be applied to almost any thickness.

General disadvantages

1. The majority are two-pack materials that must be mixed correctly before use;
2. Their cure, overcoating properties, pot life (e.g. the time the mixed components are usable) are all temperature dependent. This means, for example, that an increase of 10°C can halve the reaction times;
3. They are more difficult to overcoat then most types, and this can lead to problems of intercoat adhesion. Cured films require abrading before subsequent coats will adhere;
4. As a rule, they cannot be applied over other types of paint;
5. The high cohesive strength developed during the curing of these materials tends to place stress on their adhesive properties. Therefore unless they are specifically modified they require a higher standard of surface preparation than other types of paint.

Comparison of paints within this classification

Two-pack epoxies These were first patented in 1938 but were not in general production until 1947. They have been very widely used over the last decade. Produced from the by-products of the petroleum industry, the basic epoxy resins may be in the form of relatively low-viscosity liquid resins or they may be solid resins of increasing hardness. Both solid and liquid resins can then be reacted with a number of different curing agents. This means that almost any type of film and with any required properties can be made.

These materials are now widely used for coating both steel and concrete surfaces that are subject to a particularly aggressive environment (e.g. North Sea oil platforms). There is less validity for their use under normal atmospheric conditions since they are relatively expensive and tend to chalk on exposure to sunlight. However, their use as zinc phosphate, pretreatment or blast primers for blast-cleaned steel which is subsequently overcoated by any other paint system is an extremely valuable contribution to the painting of new steel work.

Coal tar epoxies These are a combination of epoxy resins and selected coal tars. Properties can vary, depending on the coal tar-to-epoxy ratio. The ideal compromise appears to be approximately 50/50. Coal tar epoxies are only available in black or dark brown. They cost less than straight epoxies and generally have better wetting properties, so they can be used on slightly less than perfect surface preparation. Similar recoating problems as for the two-pack epoxies.

Their best use is for water immersion, particularly sea-water tidal situations. Performance in sunlight varies and should be checked with the paint supplier, as should use for lining potable water tanks.

As with the straight epoxies, there are a number of curing agents that can be used. The advantages and disadvantages of the main types are listed below:

1. *Polyamine curing*: This provides very good chemical and corrosion resistance, good solvent and water resistance and excellent resistance to alkali. The resistance to weathering is poor since there is a tendency to chalk heavily from exposure to sunlight. These materials can form an amine bloom on their surface if applied under conditions of high humidity. Unless this bloom is removed there is likely to be a

loss of intercoat adhesion. The amines in these materials also present some handling hazards, since they are moderately toxic skin irritants which can cause allergic reactions.
2. *Polyamide curing*: This is the most widely used of epoxy-curing agents and gives a good compromise of properties. The resultant films are generally softer, more resiliant and flexible than with amine curing. They have excellent alkali resistance but poorer acid resistance than the amine cured. Generally, the two components that have to be mixed are of similar volume, which also makes for easier mixing. There is less tendency to bloom and better wetting characteristics than the polyamine type. Both the polyamine and polyamide types will cure slowly at low temperature and will not cure below 5°C.
3. *Isocyanate curing*: These are used when the ambient temperatures are low because they cure rapidly and at temperatures below 5°C. There is a tendency for such films to embrittle with age, and it is preferable not to use such materials at high ambient temperatures, since the pot life and other properties may become unacceptably short.

Two-pack polyurethanes Also called urethanes, these materials are similar to two-pack epoxies in that they can be formulated to provide different properties. They can be made into foams or soft, rubbery materials, as well as very hard, tough, abrasion-resistant coatings.

In comparison with epoxies, the correctly formulated urethane can possess considerably superior gloss retention and flexibility. They also form a film so hard and solvent resistant that they can be used as an anti-graffiti coating. The graffiti can be removed with solvent without damaging the urethane. Since these materials give off toxic fumes in a fire it is preferable that they are not used in confined spaces.

The paint is sensitive to moisture during storage and application but then becomes exceptionally water resistant. Its best use is for exterior exposure in an aggressive environment but where the maximum gloss and colour retention is required. It is preferable to apply the urethane as a finishing coat over epoxy undercoating and priming.

Moisture-cured urethanes These are the only paints in this class of materials that are single pack. The curing is provided by moisture from the atmosphere. This has the advantage that the material can tolerate a degree of dampness in the atmosphere and on the surface to be coated. Its disadvantage is that the film thickness per coat is limited and that, once opened, all the container must be used immediately. Conversely, if the humidity is very low the cure may be lengthened or stopped.

It is claimed that the cured film has similar properties and durability to the two-pack urethane. It also is used as an anti-graffiti finish.

During application of this material the spray mist is harmful and fresh-air masks to BS 4667: 1974: Part 3 must be worn during application. This material should also not be used under confined conditions.

31.3.4 Heat-condensation coatings

Examples

Phenolic coatings
Epoxy phenolic coatings

General advantages

1. These materials have exceptional chemical and solvent resistance;
2. In the fully cured form they are odourless, tasteless and non-toxic;
3. They have excellent resistance to boiling water, acids, salt solutions, hydrogen sulphide and various petroleum products.

General disadvantages

1. These materials can only be cured at relatively high temperatures (typically, 130–300°C). Therefore the process is normally carried out at works. With the use of special reactors it can be accomplished on-site but the operation is difficult. In all cases the application is critical, and variations in temperature and cure time will result in different film-forming properties. The application should always be left to specialists.
2. The resins need careful storage (possibly under refrigeration) before application.

Comparison of coatings within this classification

Phenolic coatings Phenolic is reacted with formaldehyde under heat to form a completely insoluble material. They are usually applied in an alcohol solution by spray, dip or roller. During their curing they release water which must be removed. They have maximum chemical and solvent resistance but poor alkaline resistance.

Epoxy phenolic coatings These materials are also cured at relatively high temperatures and are made by the reaction of the epoxy resin with the phenolic resin. They are slightly less critical in application requirements, are less sensitive to curing conditions and can be applied in thicker coats. The best use for both these materials is as tank linings used for the storage or food products, drinks, etc. or for process plant, evaporators, etc. that contain boiling water.

31.3.5 100% Solids coatings

Examples

Coal tar enamel
Asphalt
Unsaturated polyesters, glass flake and fibre reinforced
Powder epoxies
Powder polyesters
Hot-applied solventless epoxies
Hot-applied solventless urethanes

General advantages

1. In general, exceptionally thick films, very fast curing;
2. Can provide exceptional water and chemical resistance;
3. They contain no solvents to pollute the atmosphere.

General disadvantages

1. Generally require specialist, skilled application techniques;
2. Materials and application costs can be high, but for large surfaces the quick, thick single-coat application can make them cost effective;
3. Cannot be repaired or recoated easily on-site. In some instances solvent-based versions are needed for repair and maintenance.

Comparison of coatings within this classification

Coal tar enamel This is derived from the coking of coal and is further distilled to produce coal tar pitches. It is used for hot application on-site. It will crack and craze if exposed to sunlight but has been employed successfully for over 50 years for the protection of underground or immersed structures. The main use is now for the exteriors of buried or immersed pipelines. Different types of enamel are available to give various degrees of heat resistance. It is now generally used for pipelines below 155 mm diameter.

Asphalt Asphalt is a natural occurring mineral or as the residue from the distillation of asphaltic petroleum. It is less brittle and has better resistance to sunlight and temperature changes than coal tar enamel. Its water resistance is good but less than for coal tar enamel. It is not resistant to solvents or oils. It may crack at low temperatures and age at elevated ones. Like coal tar enamels, it is primarily black in colour and difficult to overcoat with other materials. Its main use is for the *in-situ* coating of roofs or above-ground steel structures.

Unsaturated polyesters The formation of the coating occurs *in situ* by the reaction between polyester resin and styrene, activated by a catalyst such as organic peroxide. The main use as coatings is in the formation of glass fibre or glass flake, reinforced plastics.

It has good resistance to acid, oils and oxidizing materials but poor resistance to alkalis. Application is difficult and hazardous and is generally a specialist operation.

Shrinkage on curing means that a high standard of visual cleanliness and an exceptionally coarse surface profile is required on blast-cleaned steel. Its main use is for tank linings.

Powder epoxies Applied as powder either by electrostatic gun or fluidized bed, these materials must be cured at 200–250°C. The coating cures within a few seconds and is normally water quenched to control cure time. They have good resistance to water but will tend to chalk on outdoor exposure.

Normally a specialist works application, these materials are useful protective coatings for such items as cable trays and switch boxes. They are also used for the externals of pipelines.

Powder polyesters These are similar to powder epoxies but with superior resistance to weathering and less chemical and corrosion resistance. They are used mainly for outdoor decorative use on equipment, etc.

Solventless epoxies These use heat rather than solvent to reduce viscosity, and are generally applied by airless spray, which requires specialist skills. The films cure very rapidly and can be built up in one coat to at least 1000 μm. They need a very high standard of surface preparation. The films have exceptional water and abrasion resistance, and are usually only economic for use with large structures (e.g. the external and internal surfaces of the Thames Barrier gates).

Solventless urethanes These are similar to solventless epoxies but can be applied in even thicker films without adhesion problems. They are considered to have better water and abrasion resistance than the epoxy but cost more and application is even more critical. Their main use would be for large areas requiring abrasion resistance.

31.3.6 Non-oxidizing coatings

Examples

Grease paint
Wax anti-corrosive compounds

General advantages

1. Cheap;
2. Easy to apply, even with areas of difficult access;
3. Surface preparation not critical;
4. Can be applied in relatively thick films.

General disadvantage

The coating remains relatively soft and flexible, and is not suitable for areas where people may walk or touch the surface. The best use is for steelwork in enclosed areas or with difficult access where surface preparation and application are not easy.

31.3.7 Heat-resistant coatings

Examples

Modified oleo-resinous binder with aluminium pigment
Oleo-resinous binder with zinc dust and graphite pigments
Modified or pure silicone resin with aluminium
Zinc silicate

General requirements

In general, the requirements of heat resistance limit film thickness and therefore corrosion resistance. This is a particular problem when surfaces fluctuate between hot and cold. Coatings should be selected carefully, depending on the exact maximum temperature that will be experienced. Wherever possible, conventional materials should be used. The majority of air-oxidation coatings will be satisfactory up to 95°C and epoxies up to 175°C continuous dry heat.

Comparison of coatings within the classification

Modified oleo-resinous binder with aluminium pigment These materials are sometimes called general-purpose aluminium paint. They are suitable for use up to 200°C but for corrosion resistance need to be applied over a corrosion-resistant primer such as zinc silicate.

Oleo-resinous binder with zinc dust and graphite pigments Generally called zinc dust graphite paint, this is a relatively cheap material that can resist temperatures in the range of 200–300°C, but regular maintenance repainting will be required.

Modified or pure silicone resin with aluminium These are silicone-based aluminium paints that can be used for temperatures between 260° and 540°C. They require a minimum temperature for curing (usually about 260°C) and in general have poor corrosion and weather resistance.

Zinc silicate This material has good corrosion resistance and can withstand temperatures up to 540°C, particularly when overcoated with silicone-based aluminium. The zinc silicate requires a high standard of surface preparation before application.

31.3.8 Coatings for concrete

Examples

Emulsion paints
Chlorinated rubber
Vinyls
Two-pack epoxy
Water-based epoxy

General requirements

Concrete may be coated for:

1. Decoration;
2. Waterproofing: correctly formed concrete will not transmit liquid water but will always remain permeable to moisture vapour;
3. Protection from freeze–thaw cycles: because concrete absorbs moisture it is very susceptible to damage by water freezing. The physical forces of the ice are greater than the concrete's strength and spalling and shattering occurs. Coatings can help to keep the concrete as dry as possible but they also must withstand the freeze–thaw conditions.
4. Protection for the reinforcing steel: high-performance coatings can help to prevent chloride ions, etc. permeating to the reinforcing steel. However, care must be taken that the use of relatively impermeable coatings does not trap water and make the situation worse. There is a growing tendency to coat the reinforcing bars themselves with powder epoxies;
5. Decontamination: since concrete is a porous material it can absorb oil, rusty water, etc. and coatings can reduce the absorption.

Comparison of coatings within the classification

Emulsion paints These materials are water based and cure by coalescence as the water evaporates. Films formed in this manner have sufficient porosity to allow water vapour to escape from the concrete but at the same time to retard the entry of water. The composition is typically vinyl acrylic (e.g. a combination of vinyl and acrylic resin). The acrylic portion has superior weathering properties. Emulsions are mainly used for decoration.

Chlorinated rubber See Section 31.3.2. This is used for decoration but also provides a relatively impermeable coating. Its best use is for concrete surfaces that have to be kept clean by regular washing.

Vinyls Section 31.3.2. These have been used on concrete for a variety of situations. Thinned down, they have good penetration of the surface and provide a good base for subsequent applications of heavily bodied, thick viscous vinyl coatings. However, since they dry quickly but contain a high proportion of solvent, care must be taken to avoid solvent entrapment in the pores of the concrete.

Two-pack epoxy Section 31.3.3. These are widely used to give the maximum protection to concrete surfaces, floors and walls. They can be applied as relatively thin coatings by spray or as thick epoxy surfaces applied by trowel. In all cases the application must be preceded by adequate surface preparation (see Section 31.6.1). To allow maximum penetration into the concrete the first coat must have a low viscosity. Coal tar epoxies are used where protection is the main requirement.

Water-based epoxy These are sometimes referred to as acrylic epoxies. They are generally inferior to solvent-based epoxies in durability and resistance but have less toxic hazards and low solvent odour.

31.3.9 Coatings for concrete floors

Examples

Seals
Thin coating
Trowelled coatings
Self-levelling screeds

General requirements

An important requirement for all floor coatings is adequate surface preparation (see Section 31.4.3). In general, abrasion resistance is proportional to thickness and the thick single-colour epoxy screeds will give the maximum life before wear is obvious. Thin coatings will need regular maintenance but will be considerably cheaper, and this must be weighed against ease of access for any particular situation.

Anti-slip properties are generally required. Nearly all surfaces have poor slip resistance when wet. Self-levelling finishes tend to be glossier than other types, and this also has a psychological effect. Some surfaces with good initial slip resistance can become dangerous after heavy wear over a long period and should be inspected regularly. Surfaces with the highest slip resistance, particularly if this involves the incorporation of coarse aggregate, will also retain the most dirt and can be costly and difficult to clean.

Comparison of coatings within the classification

Seals Seals are normally transparent, quick-drying, highly penetrating materials that are used to stop cement dusting and contamination by oils, etc. The moisture-cured urethanes are particularly suitable, since they penetrate to produce a hard-wearing surface. However, if safety factors preclude their use there are also epoxy and even oleo-resinous products available.

Thin coatings These pigmented materials, commonly known as floor paints, are often based on chlorinated rubber (see Section 31.3.2) or epoxy ester (see Section 31.3.1). They have limited life in heavy traffic but are easy to apply.

Trowelled coatings There are a number of proprietary products available but the majority are based on epoxy resins and applied at thicknesses ranging from 0.5 to 5 mm, depending on exposure requirements.

Self-levelling These are generally based on solvent-free epoxy and there are a number of proprietary products available. Generally, they are 3–8 mm thick, but if the base floor is too rough it is advisable to first level this with a cement/resin screed.

31.3.10 Coatings for wood

Examples

Oleo-resinous primers plus alkyd finish
Aluminium-based primer plus alkyd finish
Emulsion-based primer plus alkyd finish
Wood stains

General requirements

Wood is painted in order to preserve from decay, to minimize changes in moisture content that lead to distortion, and for decoration. It is a waste of time and effort to apply expensive paints or stains to exterior wood not previously treated with preservative. Preservatives are designed to penetrate the timber to protect against microorganisms and insect attack. Modern preservatives can be overcoated with paint without problems.

Oleo-resinous primers plus alkyd finish These are normally used for soft wood and should be designed to be flexible and smooth for subsequent paint.

Aluminium-based primer plus alkyd finsh The binder for these primers is normally oleo-resinous but the pigmentation is aluminium flake. This type is preferred for hardwoods and softwoods where knots or resinous areas predominate. They do not give such a smooth surface as primers without aluminium.

Emulsion-based primer plus alkyd finish These are based on acrylic resin dispersions and have the advantage of a rapid rate of drying. They generally have excellent adhesion and flexibility but lack the sealing properties of aluminium primers.

Wood stains The problem with normal paint systems on wood is that if moisture penetration occurs (e.g. in the untreated end-grain) then the impermeability of the paint causes blistering, cracking and flaking. New developments with translucent finishes for wood incorporate pigments which deflect ultraviolet but do not obscure the wood grain. They are also designed to permit moisture movement but contain wax additives which shed surface water. They also degrade by gradual erosion rather than cracking and flaking. If the appearance is acceptable these form a useful, economic answer to wood protection.

31.3.11 Road and floor markings

Examples

Road paints
Thermoplastic road-marking materials

General requirements

Road markers require ease of application and very quick drying.

Road paints These are normally chlorinated rubber alkyds and are easy to apply but are inherently less durable than the thermoplastic materials.

Thermoplastic road-marking materials These require special application apparatus. Although they are thicker and last longer than the road paints their tendency to discolour may cancel this advantage.

31.3.12 Anti-condensation paints

General requirements

These materials are designed to reduce water from condensation dripping on equipment, etc. They often incorporate particles of cork so that water is absorbed. They are generally thick films to provide some insulation and have a rough textured surface finish to increase the surface area and encourage water evaporation. In general, physical methods of prevention such as adequate ventilation, etc. are more effective.

31.4 Surface preparation and priming

31.4.1 Structural steel (thickness greater than 5 mm)

Abrasive blasting

Abrasive blasting is generally the most suitable and reliable method of obtaining a visually clean surface and a satisfactory surface profile. The visual cleanliness can be specified by reference to the photographs in Swedish Standard SIS 05 59 00. This Standard is incorporated into a new International Standard (ISO 8501, Part 1) and also a new British Standard (BS 7079: Part A1). The surface profile can be determined by the comparator method, as specified in ISO 8503, Part 2 and BS 7079, Part C.

Dry abrasive blast cleaning should be used on new steelwork where the main contaminant is millscale. For heavily rusted and pitted steelwork, increased durability can be obtained by the use of wet abrasive blasting where this is practicable. The water will be more effective in removing the potentially destructive and corrosive soluble iron-corrosion products that form at the bottom of corrosion pits.

Hand or mechanical tool cleaning

This method is generally not capable of achieving a uniform standard of cleanliness on structural steel. It is not effective in removing intact millscale or corrosion products from pitted surfaces. The durability of subsequent coats is therefore variable and unpredictable, and depends on the thoroughness of the operation and the exact nature of the contaminants left on the surface. The method should be confined to non-aggressive environments or where short-term durability is economically acceptable.

Acid pickling

This is only suitable for works application and is only economically viable for repetitive cleaning of relatively small simple-shaped items, such as tank plate. The lower profile achieved by this method may not be suitable for high-performance coatings.

Priming

Priming can be carried out before or after fabrication.

Before fabrication This has economic advantages and cleaning can be of a higher standard and before any destructive corrosion has occurred. Pre-fabrication primers need to be extremely quick drying (e.g. within minutes) to allow handling and ensure coverage of the blast profile peaks without attenuation. Normally, the specifier will have to accept the type and make of primer used at a particular fabricator because, unless it is a major job, it is difficult and expensive to change primers on automatic plant. Most fabricators now use a zinc phosphate two-pack epoxy primer, which is a sound base for any subsequent paint system. At 25 μm these primers are generally capable of being welded through, and although they are based on two-pack epoxy they are so formulated that the cure is delayed, and they remain overcoatable for at least one year. Zinc-rich epoxies, which were formerly popular for this application, are now out of favour because the formation of zinc salts on the surface after outdoor exposure can result in adhesion failure of subsequent coats.

After fabrication If a structure consists of a large area of welds it is inadvisable to carry out pre-fabrication blasting and then prepare the weld areas to a much lower standard. In those cases the structure should be blast cleaned after fabrication. This has the advantages that a thicker primer can be used and also that the entire paint system can be completed without delay and possible contamination between primer and top coats.

31.4.2 Zinc metal coatings

Hot-dip galvanizing (new)

Etch primer pretreatment This is suitable for overcoating with most coatings but is sensitive to moisture during application. It must not be applied as a thin, transparent coating (typically, 10 μm) or intercoat adhesion loss can take place.

T-wash pretreatment This is an acid-mordant solution which turns the surface black when correctly treated. Problems can occur with use of such an acid solution *in situ* and also from its pungent odour. It must be applied to zinc in a bright condition without corrosion products on its surface. The paint manufacturer's advice must be sought before using under thick coats of two-pack epoxy or urethane.

Sweep blast or abrading This is a suitable method before application of thick two-pack materials but care is required not to remove too much zinc coating.

Hot-dip galvanizing (weathered)

Weathered zinc surfaces that have lost their initial bright appearance and which are to be painted with thin paint systems need only to have dirt and zinc-corrosion products removed by brushing. For high-performance coatings it would be advisable to remove zinc-corrosion products by abrading or sweep blasting and treat as for new galvanizing.

Sherardizing

No special treatment is required other than the removal of any zinc-corrosion products, dirt or oil, etc.

Zinc or aluminium metal spray

Sprayed metal coatings are porous and should be sealed after application by applying a sealer coat (i.e. a thin coat such as an etch primer) or a thinned version of the final coating system. Oil-based systems should not be used. Metal spray coatings can have excellent durability without overcoating with paint and, particularly for aggressive conditions, it is preferable to leave them with sealer only.

31.4.3 Concrete

General requirements

Surface preparation of concrete consists mainly of removing laitence, form oils and air pockets. Laitence is the fine cement powder that floats to the surface of concrete when it is placed. Coatings applied over such a powdery, weak layer will lose adhesion. Form oils are used for the easy stripping of forms or shuttering. Their presence will also cause loss of adhesion of subsequent coatings. Forms should be coated with non-migratory hard coatings and the use of oils or waxes prohibited.

Air pockets or bubbles are left on the surface of all concrete. Good vibration and placing techniques will reduce their number but not eliminate them entirely. Many air pockets have a small opening on the surface in relation to their size. Paints will not penetrate into such holes, with the result that air or solvent is trapped and subsequent expansion will cause the coating to blister. In addition, some air pockets are covered with a thin layer of cement which also has no strength and will cause loss of adhesion.

The following alternative methods of surface preparation can be used for new concrete.

Dry abrasive cleaning

This should be carried out using a finer abrasive, lower pressure and a faster movement across the surface with the blast nozzle further from the work than for blast cleaning steel. It provides a roughened, irregular surface with the laitence removed and all holes and voids opened up so they can be more easily sealed.

Wet abrasive cleaning

As for dry abrasive cleaning, but with the advantage that there is reduced dust nuisance.

High-pressure water jetting

This is an effective treatment on eroded and weak surfaces but will not open up the sub-surface voids or pockets or provide a surface profile on dense concrete.

Acid etching

Hydrochloric acid (10–15%) is generally used. If applied correctly it will remove laitence and provide an adequate surface for adhesion. It will not open up air pockets and voids and is difficult and hazardous to handle and apply. It is most suitable for use on floors.

Hand or power tools

These are generally more time consuming and costly than abrasive cleaning and can give variable results.

31.5 Specifications

The main functions of a painting specification, whether for new or maintenance work, are as follows:

1. To state the means by which the required life of the coating is to be achieved. This includes surface-preparation standards, paints and systems, application, storage, handling and transport, quality control;
2. To service as a basis for accurate costing;
3. To provide a basis for resolving disputes.

31.6 Economics

The greatest problem in making an economic assessment regarding coatings is the prediction of the coating life. With paint coatings this is particularly difficult, because there are many stages to completion of the coating process and many opportunities in practice for problems to arise. Furthermore, if there is poor workmanship at the outset and this is not detected, the potential life may fall short of expectations by a very wide margin.

However, some companies consider that they have sufficient experience of their requirements to use computer programs to provide economic assessments of candidate protective systems. The following summarizes typical steps to be taken.

1. Select suitable systems based on technical requirements.
2. Estimate life to first maintenance of each system.
3. Prepare sound specifications for each system.
4. Estimate maintenance costs at present values for each system (e.g. by using Net Present Value).
5. Select the most economic system if the difference in costs is greater than 10%. Otherwise, select the system for which the most experience is available.

If at the end of the exercise the cheapest system is beyond the budget available then either the budget must be increased or changes in the initial protection will be required. Whenever possible, the level of surface preparation should not be lowered, since if at a later date more money is available the protective coating can be built up but the surface cannot be re-prepared.

31.7 Painting inspection

The potential durability of a coating system can be realized only if it is applied to a suitably prepared surface, in the correct manner under correct conditions. Paintings differs from any other industrial processes in that it is not susceptible to operator abuse or adverse environmental influences throughout all stages of the work.

Furthermore, it is generally difficult to deduce from examination of the completed work what has occurred previously.

The above is quoted from BS 5493. Many engineers appreciate that painting, which should be a minor part of an engineering project, can assume major proportions if there are problems or premature failures. For any painting work for which premature failure is economically or practically unacceptable it is advisable to use full-time, qualified paint inspection. Note that part-time inspection, or visit inspection, can in some ways be worse than no inspection at all.

The National Association of Corrosion Engineers (NACE) in the USA has a comprehensive process for the training and certification of painting inspectors. The following are their typical duties:

1. Measure, at regular intervals, the air temperature, steel temperature, relative humidity and dewpoint in the area where blast cleaning or painting is to take place.
2. Decide whether, in the light of these measurements, the ambient conditions are within the specification and therefore whether blast cleaning or painting can proceed.
3. In outdoor conditions assess the way in which the weather is likely to change and decide how this may affect the progress of the job.
4. Examine the abrasive to be used in the blasting process and record the name of the manufacturer and the type and grade of the abrasive.
5. Check that the abrasive is not contaminated with moisture, dirt, spent abrasive, etc., that the blasting equipment can deliver the abrasive at an adequate pressure and that the airlines are fitted with a water trap.
6. Check that the freshly blast-cleaned surface is of the specified standard (e.g. Sa2$\frac{1}{2}$, Sa3, etc.) at all points.
7. Measure the surface profile of the freshly blast-cleaned surface and ensure that it is within specification.
8. Check freshly blasted surface for steel imperfections, laminations, weld spatter, etc. and ensure that any necessary metal dressing is carried out.
9. Ensure that the newly blast-cleaned surface is primed within the specified overcoating time.
10. Record the name of the paint manufacturer, the manufacturer's description of each paint used in the system, the reference number and the batch number.
11. Ensure that the paints are applied in the correct sequence, by the correct method and that they are of the correct colour.
12. Ensure that, where possible, paint with the same batch number is used on any particularly identifiable unit of the total job.
13. Ensure that the correct overcoating times are observed.
14. Ensure that all paints, particularly two-pack, are thoroughly mixed in accordance with the manufacturer's instructions.
15. Ensure that the pot life and shelf life of paints are not exceeded.
16. Take wet-film thicknesses for each coat applied and confirm that they are such as to yield the specified dry-film thicknesses.
17. Measure the dry-film thickness of each coat over a representative area and ensure that the specified film thickness has been attained.
18. Ensure that each coat has been evenly applied and is free from runs, sags, drips and other surface defects.
19. In the event of discovering any paint defects, mark them up and ensure that the necessary remedial steps are taken.
20. When required, perform adhesion tests, curing tests or tests for surface contamination.
21. Ensure that all blasting and painting operations are carried out without contravention of safety regulations.
22. Report on manning levels and time lost due to weather, mechanical breakdowns, labour disputes, etc.
23. Submit full daily inspection reports.

31.8 Factors influencing the selection of coating systems

31.8.1 Environmental conditions

For aggressive conditions only highly resistant coatings can be considered; for milder environments, virtually the whole range of coatings can be used.

31.8.2 Access for maintenance

The cost of maintenance may be greatly influenced by the disruption caused by the access requirements. This involves not only direct costs such as scaffolding but indirect ones such as disruption of a process plant, closure of roads, protection of plant and equipment in the vicinity, etc.

31.8.3 Maintenance requirements

In some situations it may be impossible for physical or legal reasons to use blast cleaning or to spray paint when maintenance is required. In such cases, systems that can be easily maintained or very long-life systems should be considered.

31.8.4 Facilities for coating

Suitable skilled labour may not be available, at a particular site, to use complicated equipment (e.g. hot airless spray). On a foreign site or in remote areas even conventional equipment may be unfamiliar or unobtainable.

31.8.5 Handling, storage and transport

The choice of coating will be influenced by the need to

avoid damage during handling. Where considerable handling is involved this may be a major factor.

31.8.6 Application properties

Many of the modern, high-performance coatings require a high degree of skill in application and are considerably less tolerant than older, conventional materials. Caution is required in choosing coatings when there is no record of sound work by contractors.

31.8.7 Experience of coating performance

New types of coating are regularly developed and marketed. Before selecting such coatings the specifier should be satisfied that the test results and practical experience is sound.

31.8.8 Special requirements

Sometimes the type of coating is determined by special requirements such as abrasion or heat resistance. The coating may have to withstand specific chemicals or solvents. All coatings have to be a compromise of properties. A gain in one may be a loss in another.

31.8.9 Importance of the structure

The importance of the structure in both technical and aesthetic aspects is an obvious factor to be taken into account.

31.8.10 Costs

Cost is always a primary factor. However, the actual cost of paint is generally a small percentage of the total.

31.8.11 Other factors

Other factors may include: colour, toxicity and safety, temperature restrictions, ability to cure at service temperature, resistance to bacterial effects, shelf life and pot life of the paints.

31.8.12 Limited choice of coating systems

In practice, the selection of coatings is often the opposite to what might be expected. There are a multitude of proprietary materials but the choice of generic types is limited. Section 31.3 of this chapter is designed to illustrate the basic properties of these generic types as a preliminary guide to selection by the engineer. The specifier tends to examine the available materials to see if they will fit the requirements, rather than vice versa.

31.9 Sources of advice

Since the subject of paints and coatings is complicated, covers a large number of uses and conditions but, in general, is not a major factor in any of them, there are no publications that describe all aspects that are particularly suitable for an engineer. There are limited publications on such specialized subjects as the protection of iron and steel.

The technical service departments of paint manufacturers are a valuable source of free information on their own products. For independent advice on new materials, comparison between materials, specifications not tied to proprietary types and failure investigations, there are consultants and test laboratories that will generally assist on a fee-paying basis. Suitable names can be obtained from the Institute of Corrosion, PO Box 253, Leighton Buzzard, Beds LU7 7WB.

Further reading

Bayliss, D. A. and Chandler, K. A., *Steelwork Corrosion Protection*, Elsevier Applied Science, Barking (1991)

British Standards Institution, BS 5493: 1977, Code of Practice for protective coatings of iron and steel structures against corrosion

British Standards Institution, BS 6150: 1982, Code of Practice for painting of buildings

Chandler, K. A. and Bayliss, D. A. *Corrosion Protection of Steel Structures*, Elsevier Applied Science, Barking (1985)

Department of Trade and Industry, Committee on Corrosion Guides
No. 3: *Economics*
No. 6: *Temporary protection*
No. 12: *Paint for the protection of structural steelwork*
No. 13: *Surface preparation for painting*
No. 16: *Engineering coatings – their applications and properties*

Durability of steel structures, CONSTRADO

Painting steelwork, CIRIA Report 93

32 Maintenance

George Pitblado IEng, MIPlantE, DipSM
Support Services

Contents

32.1 Introduction

The implementation of a programme for the management of a company's assets has never been so important now that we are more involved with the European market and, indeed, will be an integral part from 1992. Chief engineers, group engineers, works engineers, building services managers, premises managers – the list is never ending with respect to job title and designation. While the individual's title may change, the responsibilities rarely alter, in that, it is from the person in those positions upon whom the employer depends so much. Building premises, factories, production equipment, facilities, service utilities and local services must be maintained in the most sound, safe and economical way to ensure that the client or user is provided with the appropriate goods, materials and working environment, as all these fall within the remit of the engineering maintenance department.

In a constantly changing work environment, with the introduction of new techniques, work practices and legislation the engineering department must ensure that they maintain an adequate level of continuous educational training. Major changes include the wide implementation in construction, installation and service fields of BS 5750, Quality assurance, the introduction of more specific legislation, codes of practice, guidance notes for health and safety, and the standards being introduced by such bodies as the British Standards Institute. All are to improve the end product, service, etc., but they place an additional demand on existing resources that cannot be ignored. Therefore in an endeavour to minimize these demands, and in so doing to maximize the benefits from available resources, a 'planned' maintenance programme should be implemented.

32.2 A 'planned' maintenance programme

The development of any programme must take into consideration the maintenance tasks that need be carried out and the resources available, thereby ensuring that both product and safety standards are met. It follows that operational demands, whether they be from a service utility installation, a production line, a mainframe computer or an office environment, will play a major part in reaching the decision as to what type of maintenance programme requires implementation.

Plant and equipment that provide a service or is required to operate for 24 hours a day, 7 days a week, presents a different proposition to the maintenance programming requirements of, say a heating pump with a standby, which is required for a heating installation in an office with a set number of working hours from Monday to Friday. Alternative methods of maintenance programming may require the planner to allow for total replacement on plant failure or have available replacement units, so that when an item of plant or equipment fails, it is removed and a replacement installed.

The ideal method would be that all plant or equipment units have a duplicate standby, which on failure would automatically be brought into service. This is satisfactory when considering small, less expensive units, which are installed to ensure that the services they are providing are not disrupted (this could also include auxiliary items on major plant and equipment). However, due to the capital costs of the major unit itself, it would be difficult to establish adequate economic grounds to duplicate these items to the same degree (although this may be done in a computer environment due to the high costs involved in downtime).

'Planned' maintenance programmes are an essential weapon in a department's armoury to ensure that the services it is called on in meeting its responsibilities are fully met. The traditional method of working from pieces of paper or individuals' 'own' notebooks as to when maintenance is to be carried out or when the insurance representative is due to visit to carry out an inspection are no longer satisfactory. This is especially the case when the skilled resources necessary to carry out the work are more difficult to obtain.

It is therefore essential that a planned maintenance programme be established, which can encompass all (or elements of) the different maintenance methods of establishing the frequency and/or work to be carried out. This programming requires skills that, in most instances, can have only been gained by experience in the field of maintenance and operation. Operation must play an important part in the programming. If the planned maintenance programme is prepared without due consideration to the demands placed on the operation element the planned maintenance programme would probably collapse when the plant's equipment could not be released (i.e. switched off) when the maintenance technician arrived at the plant to carry out his duties indicated on his work docket.

Maintenance procedures that should be considered when preparing the planned maintenance programme include:

1. Carrying out repairs needed when plant or equipment breaks down;
2. Predicting, from a history of breakdowns, the life expectancy of parts, bearings, etc., the tasks to be carried out and the frequency to be established;
3. Checking the condition throughout the plant of equipment, its running hours, readings of different responses (e.g. vibration, temperatures, current, etc.);
4. Monitoring the operating cycle and, where appropriate, seasonal shutdowns of plant, equipment (e.g. production process, 24-hour duty, etc.).

32.2.1 Maintenance systems

The benefits to be accrued from the implementation of a programme of planned maintenance can be found in the efficient and economical operation of the plant and equipment and the utilization of resources (i.e. plant and equipment and manpower) while also maintaining a sound standard of safe working and environmental conditions for operators, other occupants and employees within the workplace. Maintenance systems vary, depending on the location of the plant and equipment and/or company policy. Systems can range from the complete maintenance of plant and equipment using all available methods to their replacement on failure. To meet the company's requirements it is then necessary to decide on the maintenance system that provides the most satisfactory benefits overall.

The most commonly maintenance systems in use are: planned, preventive, scheduled, corrective and emergency.

Planned maintenance is work having benefited from information issued by manufacturers and suppliers, the experience and knowledge of the service department staff, and reports and records from previous service visits.

Preventive maintenance is work to be carried out at a specific frequency as indicated by potential failures or known reduction in efficiency of the plant and equipment, thereby avoiding failures or a decrease in performance.

Scheduled maintenance is work based on known information, such as number of operations, hours run, mileage, etc., and can therefore be carried out at a predetermined time interval.

Corrective maintenance is work carried out following the failure of the plant and equipment, and is so designed to return the component to its normal operating condition.

Emergency maintenance is that work which is required to be actioned without delay due to a failure of a component which, if not implemented, would lead to further failures or even permanent damage, resulting in the total loss of the plant and equipment. Plant and equipment in such a condition may also be dangerous to personnel.

As planned maintenance encompasses all types covered within the preventive or scheduled systems, this can be examined in more detail. In preparing a planned maintenance system all the available sources of information should be used. These include manufacturers' and suppliers' literature, trade associations, professional institutions, knowledge and experience from within the department and history and feedback from previous work for the specific type of equipment. Condition monitoring, life-cycle costing and predictive maintenance procedures should all be considered during the preparation of the planned maintenance system.

Planned maintenance systems should not be complicated. The simpler the system is to meet the requirements of the department and company, the more likelihood of it being used with satisfactory results. This aspect is of the greatest importance when, due to the size of the organization, engineers may be transferred from one department to another to gain a greater knowledge of the total company. This may cause the implemented planned maintenance system to fail or not be used to its full effect, due to the incoming engineer not understanding it fully.

In preparing a planned maintenance system the opportunity should be taken to involve the whole department. This can be achieved by using the operatives who will subsequently action the work to carry out the initial survey of plant and equipment.

32.3 A manual planned maintenance system

The planned maintenance programme forms the basis of a system whereby an in-house department may prepare and implement its own maintenance programme or introduce a trial system along similar lines. This would be prior to seeking the assistance of a consultant to provide guidance on the system that would satisfy the demands placed on the department by others and yet remain under the department's control. Irrespective of whether it is a manual or a computer system that is introduced, the elements of its composition vary little. Depending on the reports required from the implementation planned maintenance programme, the more satisfactory method of obtaining this result, both in quantity and quality of information, is by running the programme on a computer.

To assist engineers to implement a manual planned maintenance system that can be of benefit to the department it is important that a programme be set with respect to the system's implementation. Items that should be considered are:

1. Departments to be covered;
2. Plant and equipment to be included;
3. Technician and craftsmen trades available;
4. Person responsible for preparation and implementation of the system;
5. Time scale for preparation and implementation;
6. Administrative support.

It should be noted that initial interest in the preparation and implementation of planned maintenance systems can gradually decrease if only one person (preferably an experienced engineer) is given the responsibility to ensure that the proposal is carried out to its satisfactory conclusion. It is essential that the nominated person is given adequate support when necessary, to ensure that the planned maintenance system's introduction into the working operation of the department meets with success.

A suitable planned maintenance system, irrespective of the location or type of business, is compiled from a number of standard elements (see the appendix at the end of this chapter):

1. Asset register;
2. Maintenance and repair record;
3. Technician and craftsmen guidance notes;
4. Planning schedule;
5. Week tasks;
6. Work dockets;
7. Year visual aid plan.

32.3.1 Asset register

Each item of plant and equipment is allocated a specific asset number. This number can be either for a complete boiler (with associated equipment) or a specific asset number for the boiler and individual asset numbers for the associated equipment. It is advisable to restrict this numbering sequence to a minimum while ensuring that it meets the specific needs of the company and location. Care must be exercised in determining the asset numbering during this manual phase if it is envisaged that, on completing a satisfactory trial period, the planned maintenance system will be transferred onto a computer.

To assist in the numbering of the assets, each type of plant and equipment can be given a predetermined plant code reference number. In this case, the boiler and all its associated equipment are given the same asset number. The asset number is built up from certain elements, e.g.:

Location: Plant Code: Plant/equipment number

e.g.

Plant room 1: Boiler: No. 1 Boiler = PR1-01-01

Pump units, including valves, gauges, etc., are classified as individual assets. To enable the asset number to signify the different process that the pump is serving, the Plant Code 05 is suffixed, e.g.:

Roof Plant Room: Heating Circulating Pump: No. 1 = RPR-05/02-01

(02 indicates that the pump is installed in the heating system).

Items of equipment such as emergency lights may be grouped as one asset, supported by a checklist, which details the numbers, types and locations of the individual units in a predetermined area, e.g.:

First Floor: Emergency Lights = F1-34-01

(01 indicates it is No. 1 in the series of emergency lights assets). General area services such as lighting, heating, hot and cold water and air conditioning can be registered in a way similar to that of emergency lights.

In designing the asset register format for the specific location items such as manufacturer/supplier, purchase price and date, order no., cost code, function, parts/spares, guidance note reference and insurance inspections should be catered for. While carrying out the survey for the asset register, all the information found on plant and equipment nameplates should be recorded as, during their life, these tend to be lost or painted over.

To simplify this task it is an advantage if there is a set format prepared on the survey forms for specific types of plant and equipment. This enables the surveyor to enter the relevant information against the appropriate elements (e.g. manufacturer; volts; amps; bearings; etc.).

32.3.2 Maintenance and repair record

This is designed to record *all* work carried out and parts fitted on each asset item of plant and equipment. Service visits by contractors and insurance inspections are also entered. This record, which provides the history of the asset, may either be placed on the reverse of the asset record (thereby ensuring that all relevant information on a specific asset can be found in one place) or it may be an individual assets record form inserted within the asset register, next to the specific asset record.

To reduce the amount of information entered in the record, predetermined work of a planned maintenance nature can be entered in code form, while additional work or breakdowns may be more fully detailed.

As the maintenance and repair record indicates the cost of maintenance for the specific item of plant and equipment, a simple system may be used to provide recognition of the different methods by which the work has been carried out, e.g.

Planned maintenance work is entered in *black* pen.
Planned or corrective work is entered in *blue*.
Emergencies (i.e. breakdowns) are entered in *red*.

From this method of entry the engineer can observe at a glance if the planned maintenance programme is effective with respect to corrective work or emergencies. Further examination of the operation of the asset and its records are necessary in determining if the frequency of planned maintenance is correct or whether the plant and equipment should be replaced.

32.3.3 Guidance notes

The guidance notes can be produced either as a composite handbook containing task instructions for all types of plant and equipment for each trade group, or as specific task/advice notes for each asset of the service requirements of the plant equipment. When used, the handbook method provides the necessary work instructions for all similar types of plant and equipment throughout the location. This method of operation reduces the number of task/advice notes issued and therefore the system's work load, as well as the demands placed on the administration of the system. Either method eliminates the requirement of entering work instructions, etc. on the work docket/advice notes before they are issued.

The handbook can be prepared either for a specific trade or for all the trades involved in the maintenance of the plant and equipment. Its contents are as follows:

1. *Health and Safety*: Stresses the importance of carrying out the work in a safe and responsible manner.
2. *Introduction*: Details how the handbook contents and the individual's responsibility in the application of his skills are to be implemented.
3. *Plant Code, Frequency of Services and Work Tasks*: Details the work to be carried out, and at what frequency, as indicated by the instruction on the work docket.
4. *Plant Code*: Lists the plant codes for the plant and equipment covered by the handbook, prefixed by the appropriate trade reference (e.g. Mechanical Fitter = M).

The handbook is designed in an A5 loose-leaf format so that additional entries and amendments can be made as and when relevant.

To enable the maximum input and experience of the craftsmen to be introduced into the system the pages of the handbook should not be encapsulated thereby enabling the craftsman to enter additional information and comments that could increase the performance of the plant and equipment as well as amended instructions in the guidance notes. These amendments can then be issued to all staff in a similar trade.

32.3.4 Planning schedule

It is at this stage in the preparation of the planned maintenance that the engineer's knowledge and experience of maintenance is essential. In preparing the planned maintenance schedule it may be found that information on maintenance received from the manufacturer and supplier is no longer available. Technicians' and craftsmens' knowledge can play a major role in this planning stage.

Experience indicates that when preparing the planning schedule this should be carried out for each trade group. Prior to entering any asset detail on the planning schedule, items such as holiday periods and seasonal or shutdown programmes should be indicated. This can be done by the use of highlighter pens on the calender weeks that require specific attention to loading of the relevant work tasks.

In scheduling the assets and work tasks to be carried out it is recommended that the plant room and department that require the greatest resources (i.e. man-hours) should be entered on the planning schedule first. The scheduling for each item of plant and equipment follows the same pattern in that the location (e.g. boiler-house) is entered as a heading, then the assets and their asset numbers are entered for each specified location.

Planning the frequency, work tasks and man-hours for each asset then follows. Choose the week in which the least-frequency service is to be undertaken (e.g. yearly; enter a 'Y/'). Other frequencies can then be entered (e.g. quarterly, 'Q/'; monthly, 'M/'). To complete the planning scheduled for the specific asset the hours required to carry out the work at the nominated frequencies are then entered (e.g. 'Y/12'; 'Q/4'; 'M/1', etc.). (*Note*: It is recommended that this scheduling be done in pencil so that amendments can be easily made.)

Plant and equipment that require a service on completion of a certain specified period of 'hours' may, through experience, be catered for on a fixed frequency basis. If employed, this method avoids the need to record running hours (or, in the case of transport, mileage) on a daily or weekly basis to schedule the relevant planned maintenance. If it is essential that the maintenance of the plant and equipment be carried out on the completion of a certain number of operating hours or mileage, then this must be allowed for in the allocated work hours of the relevant trade group.

Peaks and troughs in the man-hours allocated weekly for planned maintenance can be avoided if hours entered are added up for each week after a number of assets have been scheduled. Treating each plant room or department in this way provides the number of man-hours required for the respective plant room and/or department. All assets to be covered by the planned maintenance system are scheduled in the same format, thereby providing the engineer responsible for allocating the work tasks with total man-hours for each trade group.

There are two methods of entering man-hours:

1. The actual hours necessary to carry out the planned maintenance work task;
2. The 'total' hours to complete the planned maintenance work task (including non-productive hours). Non-productive hours would include such items as collecting spares, tea breaks, discussions, etc.

The planned maintenance system most commonly used is that indicated in (2), whereas the method of calculating the hours as in (1) is preferred if there is a productivity scheme in operation.

Service contractors' visits can also be indicated in the planning schedule (with 'C' for contractors or another symbol indicating a different contractor). To highlight different grades of service visits (e.g. yearly) 'Y/C' can be entered.

A major benefit of carrying out this planning schedule phase for all the plant and equipment is that the numbers of each trade required are known. Also, having carried out the exercise for all plant and equipment, if the decision is to be made as to which is to be covered by the scheme or contracted out, a complete picture of the total planned maintenance requirements for the department/company is available.

32.3.5 Week tasks

The information required to compile this form is obtained from the planning schedule. To minimize 'administrative' tasks in prearing the work dockets the information contained on the week task forms is presented in a format that enables that information to be easily transferred onto the work dockets, e.g.:

Location/Area: Plant/Equipment: Plant/Equipment No.:

Job Code: Check List

Week task forms are completed for each week. They may be designed as a separate form for each trade or as one covering all trades. The latter will, of course, depend on the number of tasks per week for the individual trades.

Tasks that are of a weekly nature (e.g. visual inspection) are entered onto a separate week task form. To avoid issuing work dockets for such inspections, the week task form may then be used as a work docket, a tick being placed against the tasks when they are completed.

Week task forms should only require amending when either additional plant and equipment is introduced or the frequency of planned maintenance tasks is being adjusted to meet revised operational or maintenance demands. Departments in which work is to be carried out may be issued a copy of their department's week tasks four weeks in advance. This then enables them to programme their operation, where necessary, so that the maintenance work can be carried out without disruption to the department's output. Alternatively, on receipt of this prior notice the departmental head can contact the engineer controlling the work to discuss departmental matters that may affect the proposed planned maintenance. This may, in some instances, require the engineer to reschedule the planned maintenance programmed for that department.

32.3.6 Work dockets

On examining the week task form for the forthcoming week the engineer will decide on the work dockets that are to be prepared for issue. To enable the engineer to carry out this function satisfactorily, a knowledge of manpower resources and operational demands on the plant and equipment is essential.

It is at this stage that, having the information above, a decision can be made on which planned maintenance tasks are to be carried out and which may be postponed or cancelled. This may be due either to insufficient resources (e.g. man-hours or a forthcoming planned shutdown) or to planned maintenance that need not be carried out for another reason.

Recourse to the maintenance and repair records or a visual aid enables the engineer to establish if the planned maintenance tasks that are not to be carried out had been maintained either during the previous scheduled visit or, when next due, a planned maintenance visit. If not carried out as programmed, the engineer must ensure that the period over which no planned maintenance is proposed does not exceed any known maintenance/operational requirement.

The engineer or supervisor responsible for planned maintenance may issue the work dockets on an individual, daily or weekly basis. When issued on a weekly basis, the technician/craftsman responsible for a specific plant room or departmental area can then plan how the work should be programmed throughout the week, having gained experience in identifying which plant and equipment and also in which area planned maintenance can be actioned on or at specific periods of the week.

The work dockets would be normally issued on Fridays for the following week. This enables the technician/craftsman to plan his work for the forthcoming week. All completed work dockets should be returned to the engineer's office daily.

Work dockets partially completed (i.e. with the work task incomplete) should be discussed with the engineer. This then enables the engineer to make a decision on how or when the outstanding items on the work docket may be actioned.

Work dockets that have not been actioned should also be returned for the engineer's attention, with comments on why they have not been actioned. The engineer may then decide on whether the work need be rescheduled.

The engineer controlling the planned maintenance function within the department ensures that all work dockets are returned. Any comments entered on the work dockets are noted and, where further action is required, plans are prepared accordingly. (*Note*: A copy of the week tasks will suffice in maintaining a record of work dockets issued. If necessary, the technician/craftsman's initials can be entered on the week task form as a reminder to whom the work docket was issued.) Technicians/craftsmen should be advised of any matter that delays implementation of action required, as reported in the returned work docket, so that, if no corrective action has taken place on their next scheduled visit they understand why.

Further to the engineer's examination of the returned work dockets, purchase requisitions can be placed for parts, specialist contractors' attendance, etc. On completion of all the technical aspects of the process, the information regarding the service carried out on the plant and equipment can then be entered in the maintenance and repair record by the administration.

32.3.7 Year visual aid

Visual aids are developed from the planning schedule, in that assets and asset numbers are repeated and service visits indicated by a symbol. Hence the estimated man-hours for the specific task per frequency are not displayed for general review.

The benefit to be gained by the engineer and technician/craftsmen from the visual aid is that it provides them with a visual picture of the full year's programme, covering all the recorded assets and their associated work tasks. The format of this visual aid depends on the number of assets within the system, and each visual aid may cover approximately 100 assets.

If there is not a large number of assets a visual aid can be prepared for each trade, whereas when there is a greater number (e.g. 400+) a visual aid that covers all trades suffices. Indication markers on the aid in this instance would signify the least frequent service of the collective trades (e.g. mechanical: monthly visit and electrical: quarterly visit). The last would be indicated.

Visual aids can be purchased on which the information regarding assets and service visits would be affixed or, as in the examples shown in the Appendix, a negative (A1 size) can be drawn covering all the assets. Draft copies can then be printed for each trade's planning schedule.

The year visual aid visits indicated by colour spots relating to frequency is covered by a firm plastic sheet, so that the issue and return of work dockets can be indicated on the plastic by a china-graph pencil (e.g. '/': docket issued; 'X': work completed).

32.3.8 Comments

The above planned maintenance manual system provides a firm basis on which a computer-planned maintenance system may be developed. Having such knowledge and experience from working the manual system, the benefits that can be accrued from planned maintenance systems compared to the corrective or other non-planned methods previously used are numerous. It can be found that, due to the large amount of assets to be managed by the administration, it would be beneficial for the planned maintenance system to be transferred onto a computer system. A major non-technical weekly task is one of producing work dockets for issue and the subsequent entry into the maintenance records on completion of the work task.

Technical benefits are also obtained more readily from a computer system in that if sound identification of assets is established through their asset number, codes etc., similar assets can be examined in the event of a failure of any similar asset, and spares can be held in stock to cover the range of assets for which they are required, rather than for each individual asset. In preparing the asset register and respective work tasks these can also be more easily duplicated with respect to similar assets than is possible with a manual system.

Scheduling of work tasks and therefore labour utilization can be carried out in a number of different methods on a computer, depending on the software system purchased. These methods can be similar to the manual system or they may take the form of automatic scheduling. That is, the work tasks are entered into the computer system with the appropriate hours and trades and the software program automatically schedules the hours evenly over the weeks of the year. In implementing such scheduling care has to be taken to ensure that such items as weekends, statutory holidays, annual leave, annual shutdown and seasonal requirements are all considered prior to loading the asset information onto the computer system.

Many engineering maintenance departments allocate more hours to new or extra works than they do to maintenance. Therefore the introduction of either type of planned maintenance system highlights this matter to both the department and the company generally when appropriate.

When considering a computer system, aspects that require attention are hardware and software (planned maintenance system). During the initial enquiry as to which hardware and software would best serve the company or department both must be considered.

If there is no restriction placed on the proposal by having to use the current in-house computer system, the choices of hardware/software packages are numerous. The software must be user friendly, complementary to existing systems to avoid duplication of asset registering, etc., and have the facility to expand (i.e. have other programs (modules) that enable greater use be made of the information held within the system). The program should also be compatible with any in-house computer system where possible.

The hardware should be dedicated to the maintenance department, as they then have access to the maintenance programmes as and when necessary. Systems are implemented whereby the software is loaded onto the company in-house hardware, which can lead to periods when access is limited by other demands on computer time (e.g. accounts departments). The ideal situation would be for the maintenance department to have their own computer, with the back-up facility provided by the in-house computer, thereby ensuring that copy is held.

It follows that if the planned maintenance system is to be shown as effective, the number of breakdowns reported must reduce, irrespective as to whether it be a manual or computer system. A satisfactorily installed system allows the engineer to plan his department's work and, when necessary, action emergency items or additional work, without their being detrimental to the department's overall performance.

The benefits from a planned maintenance system cannot be achieved without the commitment of the person responsible for its implementation. To ensure that the system operates to the satisfaction of operators and users, discussions should take place during the initial decision making and with the relevant departments when the planning schedules are being prepared.

32.4 Computer systems

Before committing the company or department to a system of planned maintenance it is essential that there is a complete overview of the total demand on plant, equipment and resources and the benefits from the implementation of such a system, manual or computerized. Varying systems and/or part systems exist in almost every engineering maintenance department, which are due to company policy, changes in departmental staff or workforce, commitment of persons responsible, or, as is the situation in a number of instances, a previous breakdown, which the planned system did not prevent. In this case the system is then often abandoned as unworkable.

On reaching the decision that there are benefits to be obtained from a planned system and that the system that will be most beneficial to the company generally is to be computer based, then the following aspects must be examined more fully:

1. Has the company or department the necessary knowledge of the computer system to prepare and implement a planned maintenance programme by this method? The benefits to be gained from the use of computers requires sufficient understanding of both computers and maintenance to foresee the advantages over those obtained from the manual system.
2. Has a budget cost been established on hardware and software?

Questions to be raised include:

1. Is the program to be run on the company's main computer or is the department to be given its own hardware? While assessing this, consideration must be given to which computer program is to be given priority (e.g. finance, purchasing, deliveries or maintenance). It is generally found that maintenance is given the lowest priority. The decision to 'go it alone' (i.e. purchase hardware specifically for the maintenance program) requires that the following details of the system's implementation be examined:
   (a) Will there be only one terminal and printer?
   (b) Due to the complexity of the company premises (e.g. dispersed locations, a number of large departments and operation and maintenance sections) it may be that although there is only one main computer source for information and records there are a number of terminals and printers distributed around the other demand areas. To ensure that records and information are not corrupted or destroyed it is normal practice for the remote terminals to have access only to infor-

mation. This enables the information held within the program to be interrogated by authorized persons with respect to history, spare parts, inventory, etc. It could be considered good policy to implement the maintenance program with only one workstation (i.e. a central computer and printer) initially to gain experience of its operation before extending it into other departments. This method enables bugs to be ironed out before the system becomes too large for any necessary modifications to be introduced.

2. Are there skills within the company or organization that can develop a cost-effective programme specifically for the respective department, or would it be more economical in terms of costs and the proposed timetable for implementation of the programme to purchase an 'off-the-shelf package'? Caution on which route should be taken must be exercised in that in-house staff may be available in the initial programming stages but as the system becomes more developed, they may not be free due to other priorities discussed above.
3. Who will prepare and then implement/operate the programme? What resources will be made available (e.g. facilities, resources, accommodation and administration assistance)? To implement the proposal the staff forming the team will need to include a computer-literate operator (i.e. someone who can correct the faults that develop during the planning and programming stages), a person knowledgeable in the programming of the engineering tasks (e.g. an engineer) and also staff to inspect the plant and equipment and enter all the information into the computer's maintenance program. If there are not the skills available as mentioned above, it is imperative that the engineer responsible for the implementation of the computer program be given adequate training in the specific system to be purchased. If this is not done, substantial costs could be increased in continual recourse to the suppliers for assistance.
4. What assets, departments, etc. are to be included within the proposed program? This plays a major part in the decision to be reached on to what hardware and software is to be purchased. To support a system covering a number of areas and departments it will be necessary to consider network systems. Therefore the computer maintenance system to be purchased must accommodate such arrangements.
5. What details must be sought from the suppliers of both hardware and software packages? How can the computer best meet the department's current requirements and how can it be added to if and when necessary (e.g. increased demand from planned maintenance)? How can it be extended to cover other departments (network)? Energy consumption, input from condition monitoring, stores inventory, project work and possible interaction with space allocation must also be considered.

32.4.1 Computer system checklist

A checklist of items that may assist the intended purchaser of a computer system would include:

1. *Costs*
   Computer and printer;
   Maintenance package (i.e. number of modules required to provide a satisfactory system);
   Modifications of maintenance package to meet user demands;
   Modifications of computer so that it is compatible with the company's main computer;
   Other possible add-on packages (list);
   Network system, and installation costs (ensure compatibility with computer and maintenance package);
   Stationery (include cost for modifying supplied material to meet user demands);
   Maintenance agreement for computer;
   Support from suppliers included in initial package;
   Support from suppliers during operation;
   Training;
   Input of information into system;
   Cover for staff involved in implementation of system;
   Additional administration staff required for the system's operation.
2. *Software package*
   Is it compatible with the company's main computer?
   Is the total package user friendly or will it require a large amount of support?
   Does it require modifications to meet the department's requirements?
   Does it have the facility to copy information/details of similar assets, thereby reducing initial input of these?
   Does it have the facility to copy information/details of common work tasks, thereby reducing initial input of these? (This would include labour requirement.)
   Does it have the facility to search out similar assets (by manufacturer, duty, task etc.)?
   To what extent is asset information held? Is there a limitation on the amount of information that can be stored in a readily useable format (e.g. history, costs, spare parts, etc.)?
   Is there adequate help when entering information into the program (i.e. simple keyboard entries)?
   When entering information, does it require to be entered in a number of different fields?
   Is labour scheduling by a team or individual trades?
   When loaded automatically, are work tasks scheduled to ensure an even weekly output?
   How does the program cater for seasonal work (e.g. heating boilers, air condition)?
   How does the program cater for seasonal work (e.g. plant shutdowns, holidays)?
   How is scheduling of work tasks handled (e.g. different frequencies)?
   Does the system only produce work dockets of the least frequent service (i.e. when yearly and quarterly service visits are due, will the quarterly visits also be printed)?
   Is it easy to carry out simple corrections when incorrect information has been entered by mistake?
   How is breakdown work handled? Is there a simple entry format so that a work docket can be issued

without going through a lengthy keyboard process?
Is there a suitable format within the process to view pending work (e.g. job file) for a specific forthcoming period? Is it available in print?
Is there a facility for viewing the job file on the screen indicating docket status (e.g. work overdue, work due, work in progress, etc.)?
Is there a facility for viewing the job file on the screen indicating docket status (e.g. work overdue, work due, work in progress, etc.)?
Is the keyboard instructions simple with respect to the printing of work dockets and the entering of history relevant to these dockets on completion/part completion or cancellation of work?
Does the system require 'tidying' (e.g. transferring information from 'working file' to 'history file')?
Is there a facility to enter 'free format' information into e.g. history file, breakdown work dockets?
If purchasing a stores/spare parts module, does it have the facility to link similar parts for all assets to reduce stock holding? Does it automatically enter into the issue of these?
What reports can be produced (e.g. labour in trade groups and/or for specific assets, departmental, premises)?
Can a bonus system be incorporated into the labour element?
What facility is there for evaluation (e.g. life-cycle costing, capital write-off/replacement, design, reliability)?
Is the cost module compatible with the company's cost procedures?
Can the reports be produced in chart or table form?

If, after assessing the various computer and software packages, the decision is made to implement the proposed maintenance system on computer rather than by manual methods caution must be exercised. Irrespective of the above-mentioned benefits, the use of a computer does not increase work output but it does provide the department with a very sound method of improving the management of the company's assets and resources.

In establishing a sound maintenance programme, budget costs for both operation of the department and associated costs of the user department's assets will be readily obtained.

The proposed maintenance programme could be likened to a car. Both manual and automatic models can take one from A to B, and although the automatic may be easier to drive, both still require a driver (in the case of the maintenance programme, the engineer) to manage them.

32.4.3 Comment

Irrespective of the method of planned maintenance to be implemented, benefits to either will be accured if all possible information/methods/systems, supported by the appropriate training, are also considered. Much will depend on the demands placed on the plant and equipment and what method is to be adopted in establishing the most economical means of maintaining their life. It is only when the specific plant and equipment is known that the appropriate maintenance programme can be implemented.

32.5 Life-cycle costing

This is the method of establishing the cost of the plant and equipment over its recognized life cycle. In maintenance situations the design and specification elements may be ignored, but those that must be considered in establishing the life-cycle cost of plant and equipment will include:

Purchase
Installation
Commissioning
Trials and tests
Operation
Maintenance
Replacement

For production plant and equipment life-cycle costs are important in that the revenue from a product must cover all aspects of expenditure. For plant and equipment such as boilers and air-conditioning units, establishing life-cycle costs can play a major part in assessing when these items should be replaced. For plant and equipment such as self-contained units (e.g. pumps with sealed bearings, etc.) life-cycle costing may be used to determine their replacement date.

The information on life-cycle costs gained through the use of certain items of plant and equipment need not be true for a similar item in another location, as specific details on operating hours and maintenance attendance can vary markedly. Also, it may be an auxiliary part (e.g. an electric motor) of a major component. Therefore it may have an entirely different set of parameters. An important element that must be examined in detail when establishing life-cycle costs of a specific item of plant and equipment is the demand placed on energy resources by them.

To establish costs with regards to a specific item of plant, sound feedback must be an element of the company's programme and subsequent records must be maintained. To be worthwhile these must include *all costs*: if not, the assumed life-cycle cost over a number of years could be misleading.

It follows that any maintenance programme introduced for specific plant and equipment must be designed so that all maintenance costs are recorded. This would then be taken into consideration when the life-cycle cost of the item is assessed, as it should be added to the loss of production if this has occurred due to any maintenance malfunction. Care must be taken when presenting the life-cycle cost to ensure that all elements within the cost cycle have been effective.

Life-cycle costing must be considered when purchasing or taking over new plant or equipment and include all aspects as listed above.

32.6 Condition monitoring

This is the method of establishing the condition of the respective plant and equipment, usually while it is in operation. Monitoring can be carried out continuously or at periodic intervals, depending on the plant or equipment being monitored. Plant and equipment that is required to run for a predetermined period over many weeks would require continuous monitoring, whereas periodic monitoring would suffice for, say, a heating pump/motor unit, fitted with sealed bearings and a mechanical seal on the pump element.

With continuing developments in the electronics field, care should be taken when considering which type of monitoring equipment is to be purchased. Another aspect to be examined is the equipment's compatibility with any installed building management or energy management systems.

To establish the condition of a certain item of plant or equipment, there are a variety of monitoring methods. The most common equipment in use is for temperature, vibration and oil condition, i.e.:

1. *Temperature meters:* Temperature meters are now available in the form of infrared cameras which enable the operative to take readings from a distance. This eliminates the need to gain access to the item of plant or equipment which may present a hazard due to its being in an operating mode, behind safety guards or at high/low level where access equipment would be necessary.
2. *Vibration monitors*: In most situations whereby the plant is a major source of energy or production, vibration monitors are built into the monitoring system controlling the item of plant. This provides an early warning of any deterioration in bearings, gear damage and wear, valve leaks, imbalance of rotating parts, misalignment or effects caused by incorrect displacement of fluids. Hand-held vibration monitors are also available. These are ideal for periodic checks on a wide variety of plant and equipment.
3. *Fluid condition monitors*: These units are designed to monitor the build-up of ferromagnetic wear debris in samples of lubrication and hydraulic oils. This method identifies whether there is any trace of wear from moving parts and, if found, enables the appropriate maintenance to be carried out.

With the high degree of accuracy available from such monitoring equipment the demand on skilled resources can be reduced. With suitable training and established control parameters for each monitor and item of plant or equipment an unskilled person can carry out this task. The information gathered (in some instances directly into the computer program of the more developed monitors) can then be transferred into the computer-planned maintenance programme so that the appropriate maintenance may be implemented. The readings could result in either the plant and equipment being brought out of service to avoid undue damage or the frequency of the service being increased.

Other readings that assist in obtaining an overall view of the condition of an item of plant or equipment would include voltage, starting and running amperage and wattage. An individual's knowledge gained from working on such plant and equipment should also not be ignored.

32.7 Training

A major concern within maintenance departments and service contractors is the lack of experienced maintenance technicians and craftsmen. This has been allowed to occur through cutbacks in staffing levels accompanied by a reduction in training. Maintenance training is provided by a few companies, and it could be said that these are progressive in that they acknowledge that, by providing training for their workforce, they will not find themselves in the situation of searching for the limited skilled labour now available.

Planned maintenance can reduce the demand for highly experienced craftsmen in that, if the instructions issued in work docket and advice/guidance notes are adequate, a less-skilled person should be able to perform the work task correctly. It follows that adequate training must be provided to ensure that, irrespective of the degree of the operative's skills and experience, the work task given will be carried out to the required standard.

There are a number of suitable training establishments available, but for the specific plant and equipment within the organization, in-house training would provide a better result. Carrying out in-house training reduces costs and ensures that the subject matter is relevant. Support in this training can be complemented by the use of manufacturers, suppliers or specialists (e.g. control and/or service engineers). It will be found that such companies encourage this method, as it helps to reduce the demand on their own highly skilled staff. Trained operatives can carry out basic fault finding either alone or by telephone.

With the changing face of technology it is imperative that service departments keep up to date with the latest practices. Training should not be restricted to any specific individual but should include engineers, technicians, craftsmen and, where appropriate, semi-skilled staff.

32.8 Health and safety

No maintenance task can be issued without the person responsible for controlling workforce establishing that the work is to be carried out safely. Areas where this is essential would include boilers, pressure systems, electrical systems, confined spaces, hot work, deep sumps/shafts, tanks, and lone working.

To reduce the hazard from any of the above the risk to the individual must be assessed and, where necessary, a safe working system/procedure prepared and/or a permit to work issued. It follows that, whatever procedure is adopted, it can only be effective if closely monitored.

32.9 Information

Information on maintenance is available from a large number of sources. The initial one is that of the manufacturer/supplier for new plant and equipment. Where this plant and equipment forms a part of a process, the party responsible for the overall design must prepare the appropriate operation and maintenance and instructions. These should be supported by technical manuals, which should include spare parts lists, control measures and, where possible, fault-finding charts.

Problems arise in older plant and equipment, as there may have been no information supplied when the plant was first installed. Alternatively, the information may have been lost through the passage of time or by changes in department location and/or staff.

For an enquiring mind (which is an essential tool in the field of maintenance), sources of information outside the service department are available and include manufacturers, suppliers, trade associations, engineering institutions (e.g. plant, mechanical, electrical, etc.), consultants, suppliers of maintenance systems, libraries, technical journals. Standards are available from bodies such as the British Standards Institute, The Institution of Electrical Engineers. Health and Safety information is available from HMSO and the HSE.

32.10 Conclusion

This chapter should be sufficient to guide the engineer or manager through the initial stages of introducing a sound and effective planned maintenance system. However, as in any establishment – production, administration or (as in this case) engineering – without the commitment and endeavours of all those involved, the system will not produce the required results.

On setting out on the planned maintenance route, if the percentage of the work to be carried out under the plan is not sufficient (say, 80%), the department will continue to operate as before (i.e. firefighting) with the control of that department being in the hands of others. Any planned maintenance system is a *tool*. It cannot do the work itself but, employed correctly, the sound management results will prove that the efforts involved in setting up the system were worthwhile.

Appendix: Elements of a planned maintenance system

Plant codes

Plant code	*Plant/equipment*
01	Boilers
02	Heating systems
03	Domestic hot and cold water
05	Pumps
06	Water treatment
10	Ventilation
12	Chiller/refrigeration
13	Chilled water systems
15	Cooling water systems
31	Electrical distribution
33	Electrical lighting and power
34	Batteries/emergency lighting
41	Kitchen equipment
44	Fixed equipment
45	Portable equipment
46	Laundry equipment
70	Lifts

Checklist

Emergency lights (E34) Floor/Zone..

This service covers all emergency lights.
Indicate thus M: N: S/1: 2: 3, etc. for type of unit and duration of emergency light.
e.g. Maintained/3 hours: M/3.

Luminaire No.:	Type/Hour:	Comments/Report/Recommended action

Asset register

Asset: Description	No.
Location	G. N.
Description of plant and auxiliary equipment	Drg. No's
Manuf./supplier	Ord. No. Date
Schedule of plant and components/additional technical data/remarks	

Survey form

Asset: Description (e.g. Pump)	No.
Location	G. N.
Description of plant and auxiliary equipment	Drg. No's
Manuf./supplier	Ord. No. Date
Schedule of plant and components/additional technical data/remarks	

Unit model
Size/Cap
Head
Brg. NDE
DE
Lubr't
Motor
Manuf.
Mod/Type
Frame
V/A
HP/W
RPM
Frame
Rating
Class Ins.
Brg. NDE
DE
Lubr't.
Isolator
Starter
Dist. Brd.

Maintenance and repair record

Date	Report	Parts	Costs Labour	Total

Guidance notes

Health and safety

Health and safety is everyone's responsibility.
Plant and equipment must be made safe before any work is carried out.
A permit to work may be required for certain work tasks.
Protective clothing/equipment should be used when necessary.
Defective tools or access equipment must not be used.
All safety rules and regulations must be adhered to.
On completion of a work task, the plant/equipment must be left in a safe and operable condition.
Safety guards, where fitted, must be replaced and secured.

Technicians'/craftsman's responsibility

The instructions given in these guidance notes do not cover every component part, but the technician/craftsmen is expected to carry out any necessary maintenance in accordance with their normal trade practice.
Manufacturers' maintenance handbooks, when available, should be followed with respect to recommendations for work to be carried out and at which frequency.

Work tasks

Heating Systems Plant Code E02.
Half-Yearly (H) Service.
5. Fan Assisted Heaters.
Blow out motor windings, lubricate as necessary.
Check insulation resistance and test earth continuity.
Examine apparatus for general mechanical condition, including guards and other safety devices.
Test mechanical action and polarity of equipment switch.
Clean and test switches, pilot lamps, thermostats and timing devices.
Ensure that writing and terminal connections are secure.
Test under normal running conditions.

Planning schedule

Appendix 9
Planning Schedule

MAINTENANCE PROGRAM

ASSET-LOCATION	ASSET NO.	JANUARY				FEBRUARY				MARCH				APRIL				MAY				JUNE	
		1	2	3	4	5	6	7	8	9	10	11	12	14	15	16	17	18	19	20	21	22	23
PLANTROOM 1.																							
BOILER NO.1	PR1–01–01										Q/8												Y/20
AIR HANDLING UNIT 1.	PR1–10–01				M/1				M/1				Y/8				M/1				M/1		
																						gmp	

Week tasks

Week One

Mechanical

Location/Area	Plant/Equipment	Plant/Equipment No.	Job Code	Checklist
Plant Room 1	Boiler 1	PRI-01-01	M91/3,4	

Electrical

Location/Area	Plant/Equipment	Plant/Equipment No.	Job Code	Checklist
Plant Room 1	Boiler 1	PRI-01-01	E01/3,4	

Week dockets

Location/Area	Plant/Equipment	Plant/Equipment No.	Job Code	Checklist	Week

Work carried out:

Recommendation/Action:

Date: Time taken: Signature:

Visual aid

Appendix 12
Year "Visual Aid" Plan

MAINTENANCE PROGRAM

ASSET-LOCATION	ASSET NO.	JANUARY				FEBRUARY				MARCH				APRIL				MAY				JUNE	
		1	2	3	4	5	6	7	8	9	10	11	12	14	15	16	17	18	19	20	21	22	23
PLANTROOM 1.																							
BOILER NO.1	PR1–01–01										B												R
AIR HANDLING UNIT 1.	PR1–10–01				G				G				G				G				G		

FREQUENCY INDICATION
GREEN FOR MONTHLY
BLUE FOR QUARTERLY
RED FOR YEARLY

33 Energy Conservation

Eur Ing M G Burbage-Atter, BSc, CEng, FInstE, HonFIPlantE, FCIBSE
Heaton Energy Services

Contents

33.1 The need for energy conservation

Energy conservation has often been referred to as the 'Fifth Fuel', the other four being the so-called primary or 'fossil' fuels of coal (solid), oil (liquid), gas and nuclear/hydro-electricity. This emphasizes the importance of reducing the amount of energy used, not only nationally but also internationally.

The simple fact is that the world's reserves of fossil fuels will eventually run out, depending on the rate of use, and therefore, if the consumption of these forms of energy are reduced, the existing reserves will last longer. Research and experimentation could lead to those reserves currently available but uneconomic to recover and use being rendered economic, thus extending further the number of years before these non-renewable sources of energy do eventually run out.

The amount of worldwide energy reserves and life is variable according to the source of information used, but the position is of the order shown in Table 33.1. Thus at some time in the not too distant future (less than 100 years) oil, gas and uranium will no longer be available. Every effort is required to reduce the world's energy demand to cater for this event. In the case of the UK the situation is very similar.[1]

The economically recoverable coal reserves are of the order of 4.2×10^9 tce, which, with an annual consumption of around 111×10^6 tce, gives a life of 38 years. However, the recoverable coal reserves are much greater, at 30×10^9 tce or 270 years' supply.

Recoverable oil reserves are given as 1.23×10^9 toe. The UK consumption is only 70×10^6 toe, some 60×10^6 toe being exported from the total annual production of 130×10^6 toe. At this total production level there is only 10 years' supply (but the cessation of exports would virtually double the life). Gas reserves are similarly limited, and are said to be equivalent to about 40 years. Hydro-electric sources of power are being exploited to the full in the UK, and the contribution of nuclear power is subject both to the supply of uranium ore and to environmental problems. At present, nuclear generation only contributes around 7% of the total UK energy consumption.

The major problem in energy conservation is, however, not the concept but the economics. Saving money by energy conservation is preferred to saving energy unless

Table 33.1 World energy reserves

		Oil $(10^9$t$)$	*Coal* $(10^9$t$)$	*Natural gas* $(10^{12}$t$)$	*Uranium* $(10^6$t$)$	*Shale oil and tar sands* $(10^9$t$)$
Present reserves:						
Lowest		89	480[b]	77	2.0	80
Highest		96	630	96	2.2	90
Life at:						
1971 consumption[a,e] rate:	Lowest	32	60	33	—	39
	Highest	36	1000	45	—	48
1984 consumption rate:	Lowest	31	150	49	64	28
	Highest	34	190	62	70	32
1971 future view:	Lowest	16	30	15	16	Extend oil by 9
	Highest	18	190	19	50–100[d]	Extend oil by 11
New future rate:	Lowest	30	120	40	37	Extend oil by 25
	Highest	32	160	50	38	Extend oil by 30
Future reserves:						
Lowest		200	7000[b]	185	1.2	370
Highest		350	10 000[c]	295	1.4	1300
Life at:						
1984 consumption	Lowest	70	>2000	120	38	130
	Highest	125	>3000	190	45	>450
1971 future view:	Lowest	30	150	25	20 (50–100)[d]	Extend oil by 10
	Highest	40	250	40	37 (50–100)[d]	Extend oil by 17
New future rate:	Lowest	67	>1700	95	21	Extend oil by 120
	Highest	116	>2500	150	24	Extend oil by 430

Notes:
Units: Oil, shale oil and tar sands are expressed in billion (10^9) tonnes of oil.
Coal is expressed in billion tonnes of coal (multiply by 0.66 to give billion tonnes of oil equivalent).
Natural gas is expressed in trillion (10^{12}) cubic metres (multiply by 0.86 to give billion tonnes of oil equivalent).
Uranium is expressed in million (10^6) tonnes of uranium (multiply by 0.245 to give billion tonnes of oil equivalent).
Life is in years.

[a] Data taken from *Energy from the Future*.
[b] Excludes lignite.
[c] Includes 2400 billion tonnes lignite.
[d] Figure takes into account fast reactors. It is stated in original document that predictions are 'particularly speculative'.
[e] 1971 consumption rates were taken as oil 2500 Mtoe; gas 900 Mtoe; coal 1500 Mtoe.

the energy saving occurs with the main effect required of cost saving. Hopefully, the energy manager will be able to persuade colleagues that both financial economy and energy reduction go together, hand in hand. He must take positive action within his organization at all levels, from the board of directors to the shopfloor to achieve the stated objectives.

33.2 Energy purchasing

The major energy cost-reduction exercise which can be undertaken prior to an energy audit is that of ensuring that energy is purchased at the most economical price. This is a relatively simple matter, and can be undertaken by the energy manager and his staff, if they have the expertise. An alternative is to use the fuel suppliers, such as the local electricity companies who undertake such work free of charge in the case of their own tariff investigations, or by independent energy consultants. In the case of the latter, fees will be payable, generally based on a percentage of the cost saving achieved over a period of time. These fees can vary from 50% of the annual savings for a period of five years down to 50% of the actual savings for a twelve-month period only. When using consultants the energy manager should inquire around in order to find the practice which bests suits his needs.

The price of energy varies throughout the UK, but as a guide, the following figures (1989) prices are typical:

Industrial coal	£50.00 per tonne	16p per therm
Gas oil	12.0p per litre	33p per therm
Heavy fuel oil	5.0p per litre	13p per therm
Gas – interruptible	—	16p per therm
Gas – firm	—	30p per therm
Electricity – normal	4.5p per unit	132p per therm
Electricity – night time	2.1p per unit	61p per therm
Liquefied petroleum gas	30.0p or kg	63p per therm

In some cases there is a wide variation in the price paid for energy due to a number of factors.

33.2.1 Industrial coal

The basic price of coal at the pit is based on the coal gross calorific value, with allowances then made for the ash, sulphur and chlorine contents. The haulage charges depend on the distance from the pit to the site and on the method of delivery. Tipper-vehicle deliveries are cheaper than conveyor vehicles which, in turn, are cheaper than the pneumatic (blower) vehicles. The method of delivery will obviously be decided by a combination of space and cleanliness factors. In the case of certain customers, special agreements may be available at special rates where the annual coal consumption is large and the supplier wishes to retain the market.

33.2.2 Oil

The prices of the various grades of oil are highly competitive due to the relatively low cost of crude oil as compared to the situation a few years ago. Oil is available in two grades, that of gas oil (35 secs) and heavy fuel oil (3500 secs) as compared to the four previous grades.

The prices charged may vary between supplying companies within the same area by up to 2.0p per litre. Thus it is desirable to enquire around for the best contract.

The basic prices of the two grades of oil are interrelated to the prices of general tariff gas (non-interruptible) and contract gas (interruptible). The price also varies according to location, generally being lower around the coast or close to oil refineries.

33.2.2 Gas

Variations in contract gas prices are due to the availability of the fuel. Thus a non-interruptible supply costs more than an interruptible one by a quite significant figure. The energy manager considering a change of fuel source to gas should maintain his options by installing dual-fuel plant rather than a single-fuel-fired plant in order to gain advantage from the fluctuating fuel price market.

33.2.4 Electricity

Each separate electricity company purchases its electricity from the Generating Companies and each area company has a large variety of tariffs available for the customer. These tariffs include:

Monthly Maximum Demand with both day and night-time variations;
Quarterly Non-domestic Tariffs with variations for day and night time, evening and weekend usage;
Seasonal Time of Day Tariffs – a recent addition to those available;
Off-peak Tariffs – these are still in use but generally apply to previous supplies only and are not available to new customers.

Finally, there are tariffs available for those organizations which generate their own electricity and export/import to/from the Distribution Grid system.

A typical example of the number of choices can be gained from the following example of a small company in the Yorkshire area using 100 000 units of electricity per annum (1989 tariffs):

1. *Normal General Quarterly Tariff*

Quarterly standing charge 4 × £7.90	£31.60
Primary units 1000 × 4 × 8.13p	£325.20
Excess units 96 000 × 6.09p	£5846.40
Total	£6203.20

2. *Economy Seven Quarterly Tariff*

Night-time electricity usage = 12 500 units

Quarterly standing charge 4 × £10.20	£40.80
Night units 12 500 × 2.07p	£258.75
Primary units 1000 × 4 × 8.13p	£325.20
Excess units 83 500 × 6.41p	£5352.35
Total	£5977.00

3. *Weekend/Evening Quarterly Tariff*
   Weekend and evening usage = 50 000 units
   Quarterly standing charge 4 × £10.20 £40.80
   Weekend/evening units 50 000 × 3.85p £1925.00
   Normal day primary units £392.80
   1000 × 4 × 9.82p
   Normal day excess units 46 000 × 7.92p £3643.20
   Total £6001.80
4. *Economy Seven Weekend/Evening Quarterly Tariff*
   Quarterly standing charge 4 × £12.50 £50.00
   Night units 12 500 × 2.07p £258.75
   Weekend/evening units 37 500 × 3.85p £1443.75
   Normal day primary units £392.80
   1000 × 4 × 9.82
   Normal day excess units 46 000 × 7.92p £4036.00
   Total £5788.50
5. *Monthly Maximum Demand Tariff with Night Usage*
   Maximum demand = 35 kVA
   Availability charge £403.20
   12 × 35 kVA × 96p/kVA
   Maximum demand charges: £147.00
   Nov./Feb. 2 × 35 kVA × £2.10/kVA
   Dec./Jan. 2 × 35 kVA × £6.50/kVA £455.00
   Night units 12 500 × 2.06p £257.50
   Day units 87 500 × 4.62p £4042.50
   Total £5305.20
6. *Monthly Maximum Demand Tariff without Night Usage*
   Availability charge (as above) £403.20
   Maximum demand charges (as above) £602.50
   Unit charges 100 000 × 4.58p £4590.00
   Total £5595.70
7. *Seasonal Time of Day Tariff*
   High-rate usage = 2400 units.
   Medium-rate usage = 6000 units
   Fixed charge 12 × £24.00 £288.00
   Availability charge £378.00
   12 × 35 kVA × 90p/kVA
   Unit charges: £257.50
   Night 12 600 × 2.06p
   High rate – Nov./Feb. 2400 × 15.25p £366.00
   Dec./Jan. 2400 × 31.50p £756.00
   Medium rate – Nov./Feb. 6000 × 6.75p £405.00
   Dec./Jan. 6000 × 7.15p £429.00
   Other 70 700 × 4.05p £2863.35
   Total £5742.85

A comparison of the costs of these seven tariffs clearly indicates that in this case the company should purchase its electricity on the Monthly Maximum Demand Tariff with Night Usage. In order to benefit from the much lower price for night-time units (as compared to the slightly higher day-time ones) the night-time consumption need only be some 1% of the total consumption.

Regular meter readings are required to ascertain the breakdown of the total load and these will form a useful basis for the energy audit. Regular readings throughout the working week (say, at hourly or even half-hourly intervals for a week or fortnight around periods of peak energy consumption) along with details of the plant in use give an indication of the possibilities for peak lopping in order to reduce the maximum demand. Alternatively, suitable monitoring equipment with printout facilities would be of great benefit in many cases, especially where fluctuating loads are concerned. It should be remembered that the highest maximum demand incurred applies not only for the relevant monthly account but also for the availability charge for the next twelve months unless exceeded, when the new, higher, figure applies.

Maximum demand alarms are available for individual organizational use and these can be made to shut down non-essential plant in order to maintain a lower actual maximum demand. Where automatic shutdown is not in use alarms can be used to trigger a manual shutdown procedure.

33.3 The energy audit

The basic energy facts must be established in an organization's energy conservation campaign. Any organization will purchase energy, even if this is limited solely to electricity. Thus, as a start, the actual energy accounts are available for use. These will have been employed already in determining the most economical purchase price (see Section 33.2). The accounts, however, only indicate the total site consumption, generally during the last three months, or monthly, depending on the plant size.

In order to gain an accurate picture of the site energy usage it is necessary to:

1. Provide metering facilities for each energy source for each major cost area (e.g. the boiler house or a particular production process);
2. Read these metering facilities regularly. This can best be undertaken weekly and, at worst, monthly, but the frequency will depend on the load pattern and level of consumption.

On the basis of these regular readings, energy consumptions and costs can be allocated to particular cost centres, items of plant or process. The energy audit can be undertaken by using the organization's staff, such as the energy manager and his department. As an alternative, an energy consultant can be called in. It is suggested that costs and levels of consumptions should, where possible, be in some way related to production output, operating hours or process requirements.

The energy consultant may be better employed in investigating a specific cost centre, as detailed investigation could be required such as energy readings every hour, or half an hour or even every few minutes (e.g. a boiler test). As a result of an energy audit it should be possible to build up the pattern of energy usage of the whole site, or a particular cost centre. This can be expressed in energy terms in a Sankey diagram (Figure 33.1). The figure shows the uses made of the various energy sources purchased, and where this energy is eventually lost from the site to waste. It is essential to maintain a regular monitoring system, recording energy consumptions and costs. The use of computers provides a ready-made facility for this and software packages are available (see Section 33.5).

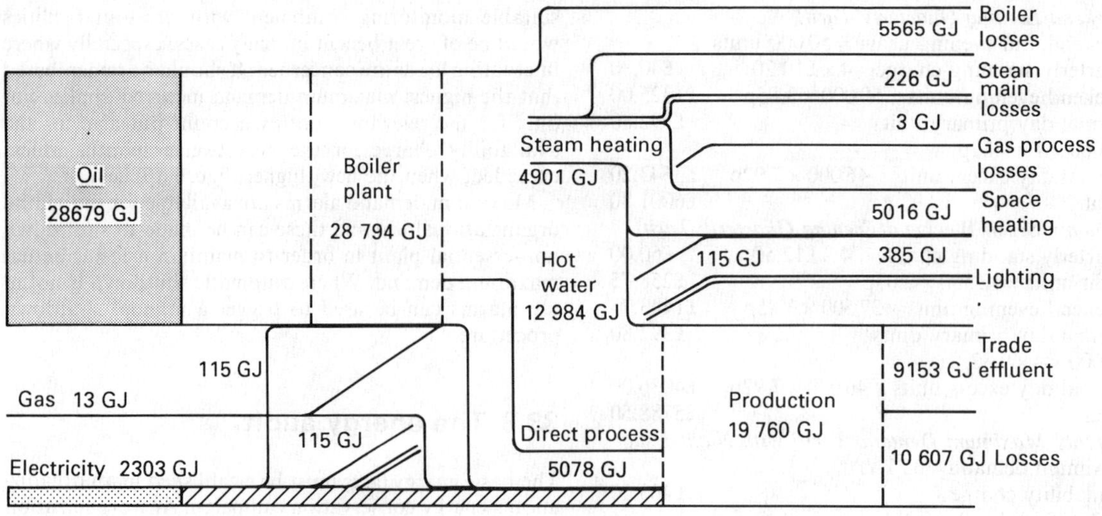

1 GJ = 1×10^9 J

Figure 33.1 A typical Sankey diagram for a dyehouse

33.4 Energy management

Energy management will cost money whether the staff are employed by the organization or outside consultants are used. Large organizations, who already employ their own engineering staff, probably add energy mangement duties to an existing member of staff, or may appoint a specialist in such matters. Extra staff may be appointed, depending on the nature of the business.

In this case, there is an on-going cost, and this must be borne in mind when embarking on an energy-conservation campaign. Within a large organization with a large number of buildings, 80% of the energy consumption takes place in 20% of the buildings, and initially there are major savings to be made. However, when the initial and obvious energy-conservation schemes have been implemented then there only remains the effort to maintain the results achieved and to investigate any possible new schemes.

Within a large organization there may well be sufficient continuous work to keep a small team of energy-conservation engineers busy. However, in many cases, once the initial work has been completed then perhaps only one or, at most, two persons are required. If this is likely to be the case then the use of energy consultants may prove attractive for the investigation of individual areas of energy conservation, as there is no problem of long-term employment costs. In the case of small organizations, unless there is a very high energy consumption then full-time company employment cannot be justified.

It may be possible to employ a suitable engineer on a part-time basis, but perhaps the best alternative is to use either an energy consultant, as and when required, or a suitable student from a technical college, polytechnic or university during the vacations or during their year out of college on industrial placement. The advantage of the latter is that an eminently suitable student may then be offered full-time employment upon successful completion of his course, which could include further energy work.

The financial arrangements could be in the form of:

1. Energy consultant:
   An agreed hourly rate or day rate within a specified period;
   An annual retainer with a lower hourly or day rate;
   A percentage of the capital cost of the scheme involved;
   A percentage of any savings obtained.
2. A student:
   An agreed hourly or weekly rate for the specified period;
   An agreed salary.

Continuous energy management is required and the person so employed must have access to the organization's highest management or be a member of that team.

Many excellent energy-conservation schemes have failed or been altered by senior management due to inadequate representation from the energy-conservation staff concerned. The approach to energy conservation and energy management must be whole-hearted and enthusiastic. Only then can the best results be achieved.

33.5 Energy monitoring

Reference has been made earlier to the need for energy monitoring as part of the energy audit (Section 33.3). The monitoring system used can be as simple or as sophisticated as required. In its simplest form, energy monitoring consists of recording the billing details of the various energy suppliers and comparing these bills by reference to the previous bill or to the previous bill for the period twelve months earlier. Such comparisons are not the best basis for commenting on the performance of the plant concerned. Changes in plant and equipment, and the use of that plant and equipment, may give rise to large variations in energy consumptions, hence costs.

Thus for the best comparisons not only must the energy usage be monitored more frequently, but at weekly intervals (monthly at worst), with other parameters being recorded. These could include the following:

Plant/equipment
Period of time and times in use, each day of the week;
Load factor on that plant – goods produced;
Conditions of the plant – pressure, temperature, etc.;
The identity of the plant operator.

Buildings
Period of occupancy each day;
Internal environmental temperature and humidity;
Number of occupants;
Nature of the building construction;
Internal heat gains from lighting, machinery, etc.;
External heat gains from the sun;
Climatic conditions such as cloud cover, wind speed, external air temperature and humidity.

From this information it should be possible to ascertain the basic standing loads and the variable production loads which make up the total site load.

It is necessary to record many data, which is time consuming when undertaken by individual members of staff. Meters are available for the direct measurement of such energy sources as oil, gas and electricity and also for the measurement of water, steam and air flows. Steam can also be measured in the absence of any metering by diverting the condensate into a drum of water and weighing the drum prior to and after the test, noting the actual time for the test. Any steam system will have steam traps, and in order to check that these are working correctly, sight glasses should be provided after the trap. Solid fuel such as coal, coke and wood can best be measured by weighing the fuel on an appropriate system such as a belt weighing machine for large quantities of fuel down to a simple spring balance and small drum for small amounts.

Many of these parameters can be easily measured and recorded by modern devices, utilizing a minimum of manpower. A large variety of modern computer-based energy management systems will monitor, record and store the required data. Some will also control the required parameters within closely defined limits.

Energy management systems can vary in cost from low to high (£50 000–100 000) for the system itself. In addition, energy controls may have to be provided either to replace existing controls which are not compatible or which are not operating correctly or have never been provided in the past.

This additional cost to facilitate and implement the energy management system can vary from £100 000 to £300 000. In the long term, the centralization of energy monitoring can yield large financial savings, coupled with a considerable reduction in energy consumption.

With such systems it is possible to obtain individual room space heating and humidity control without the need to despatch an engineer to check every room in the building. Manpower costs can be reduced, with the staff being required for checking and maintenance purposes only. The cost of an energy management system can be recovered in a relatively short period of time (of the order of 2–3 years).

The cost of an energy management system at one site in the North of England was of the order of £500 000 and the payback period was just over 2 years. In view of the high cost of such systems it may be possible to obtain a system on a loan basis, the system being rented from the installer or by lease purchase. All options should be considered before entering into a contract.

A further advantage of an energy-management system is that the quantity of production can be included for process plant and a correlation made between the product and the energy required to produce it. Such a figure is useful in comparing similar processes within the same organization and in comparison with similar ones. It also allows the energy cost per unit of production to be calculated.

33.6 Energy targeting

Having initially ascertained the basic energy data, energy targeting is the next logical step. It is desirable to know if the energy consumption of a particular site or piece of equipment can be reduced without detriment to the product or personnel involved.

The easiest way is to alter the parameters downwards and note the effect, while monitoring the plant conditions and energy consumptions. The result of this is that eventually the product will become useless and/or the personnel aggrieved at the change in environmental conditions. This method is frequently used by making any downward adjustment small and monitoring the effect, adjusting downwards again after a short period of time.

It is possible to undertake theoretical calculations to ascertain the amount of energy required for the operation and to compare this with the actual consumptions. To do this requires the knowledge of a large number of parameters but it can be undertaken, especially if an energy management system is installed.

The amount of energy required for space heating in a particular space can be estimated from the various parameters and compared with the actual energy consumption. The difference indicates the scope for further improvement. The following is a typical case of a warehouse used basically for the storage of goods only but with access by staff for goods movement:

Warehouse size – 91.5 m long by 61.0 m wide by an average height of 8.13 m, giving a total floor area of 5582 m^2 and a volume of 45 400 m^3.
Modern construction with a wall and roof U-value of 0.50 W/m^2C.
Internal environmental temperature required = 15°C.
External air temperature = −3°C.
Natural air infiltration rate = $\frac{1}{2}$ air change per hour = 22 700 m^3/h

Calculated heat losses – Building fabric loss = 146 000 W
Air infiltration loss = 152 000 W
Total loss = 298 000 W

Type of heating in use – Thirteen downdraught-type steam unit heaters, controlled by an optimizer and internal air thermostats.
Hours of occupancy – Five days per week, 16 hours per day.
Calculated annual steam requirement = 692 000 kg (1 525 000 lb)
Calculated steam cost at £9.68 per 1000 kg = £6700.00.

In this case the actual annual steam consumption which was metered was 1 906 000 kg (4 202 000 lb).

In the case of a process, such a basic estimate is harder to achieve, but this can be undertaken. The heat requirement for most chemical reactions is known and this method can be used in most manufacturing processes to determine the energy requirement.

In a series of chemical process vessels where heating of each vessel takes place it may be possible to reduce this amount of heating by ensuring that the product from the first vessel is hot enough for the next. This can be achieved by alteration to the liquid boiling point by variation in the vessel pressure. A reduction in pressure reduces the boiling point. The rescheduling of production may also lead to economies.

The dyeing of cloth is a good example. The dye master of old always insisted that the dyeing process in all the required vessels be started at the same time. This process began with the boiling of large quantities of cold water in the dye vessels by the direct injection of steam through 50-mm diameter pipes, the steam load being extremely high. To meet this load it was customary to install a battery of boilers. After 30 minutes the steam load reduced to a minimum, as all the vessels were boiling. Fours hours later this same process would be repeated. The additional cost of the boiler capacity and of the fuel to keep the boilers alight between successive loads was excessive. A simple alteration to the production schedules by staggering this boiling of water and the introduction of steam accumulators reduced costs significantly.

Many computer software packages are available for energy-targeting purposes, but care should be taken in specifying the requirements and insisting on performance satisfaction and testing before final payment. Such software packages may be expensive and could require modifications for actual use.

There are many building space heating/humidity energy usage programmes available. A large number assume the building ventilation or wind speed rates. In most buildings, the major factor in any discrepancy between the theoretical and actual energy consumptions is due to the variation in wind speed. This is highlighted in the example of the warehouse above. The large discrepancy between the theoretical and actual steam consumptions was investigated in detail, and it was found that one of the large doors giving access into the warehouse was open for most of the time the warehouse was in use by the staff, and that this door was subject to the prevailing wind. In order to reduce the steam consumption a rapid-opening door was installed, with the result that the steam consumption was halved. The energy target set must be realistic and capable of achievement.

33.7 Major areas for energy conservation

For most organizations the major energy use is for building space heating, ventilation, air conditioning, domestic hot water purposes and lighting. Where energy is used for process, this may be secondary energy generated from a primary fuel (e.g. steam generated in a fossil fuel-fired boiler plant).

33.7.1 Boiler plant

Boiler plants are a major user of energy. The combustion efficiency of a boiler plant can easily be set at the optimum, and Table 33.2 suggests the parameters for this for various fossil fuels:

Exit flue gas temperature = 200°C
Ambient air temperature = 15°C

Close control of the amount of excess air is possible by the use of oxygen trim control equipment. Such equipment will control the flue gas oxygen content within the range of 2.0–3.0% as compared to the normal 3.0–5.0%. The improvement in boiler plant efficiency is of the order of 1.0–2.0%.

The combustion conditions suggested above should be achieved bearing in mind the fact that the lower the flue gas oxygen content, the greater the risk of incomplete combustion, whil the higher the oxygen content, the greater the flue gas losses (see Figures 33.2–33.6). Poor overall boiler performance outside these parameters is due to the radiation and other loss factors.

The surface loss from a boiler is fixed once the physical size, insulation and boiler working conditions are known. The usual radiation loss figure provided is that for a boiler operating at least at 80% maximum continuous rating (MCR) or more. In this case, the surface loss is relatively low in percentage terms (around 3–5%). Experience of boiler loadings have indicated that very few boilers work at such high loads during their periods of use. Typical annual boiler load factors are only around 40%. As the boiler load factor decreases, the radiation loss becomes a much higher factor, and at 20% boiler load amounts to 25% and at 10% load factor to a loss of 50% (see Figure 33.7).

Many boilers do operate at low load factors and consequent poor annual efficiencies. This can be avoided

Table 33.2

	Coal	*Oil*	*Natural gas*
Excess air (%)	30	20	10
Dry flue gas CO_2 content (%)	12.0	13.3	10.5
Dry flue gas O_2 content (%)	7.5	3.5	2.3
Dry flue gas loss (%)	13.2	7.4	6.1
Moisture in the flue gas loss (%)	4.9	6.5	11.3
Unburnt/ash loss (%)	1.3	Nil	Nil
Radiation and other losses (%)	3.1	3.1	3.1
Total losses (%)	22.5	17.0	20.5
Inferred boiler plant efficiency (%)	77.5	83.0	79.5

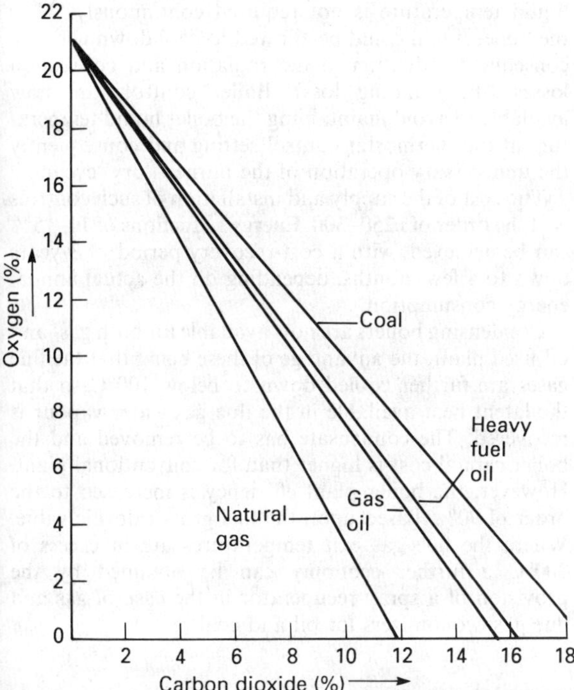

Figure 33.2 The relationship between oxygen and carbon dioxide in the products of combustion for various fuels

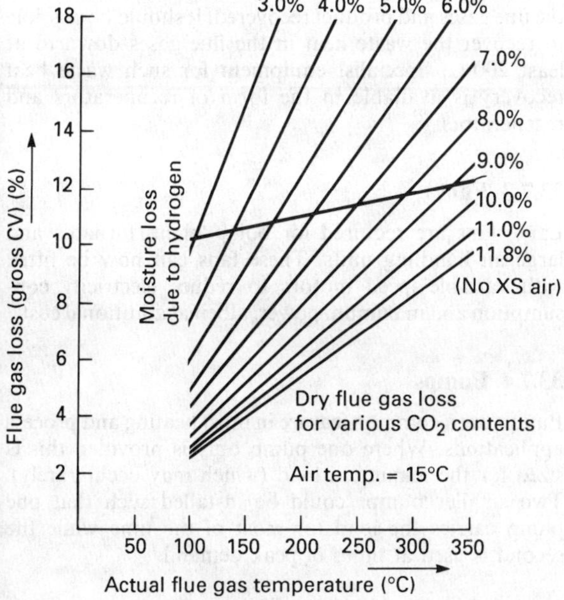

Figure 33.3 Flue gas losses – natural gas

by providing boiler plant with little or no margin over the actual required capacity and by installing multi-boilers or two or three smaller boilers. Boilers of this modular type are available for low-pressure hot water (LPHW) purposes, but obviously cost more than a single boiler. This will also increase the maintenance, number

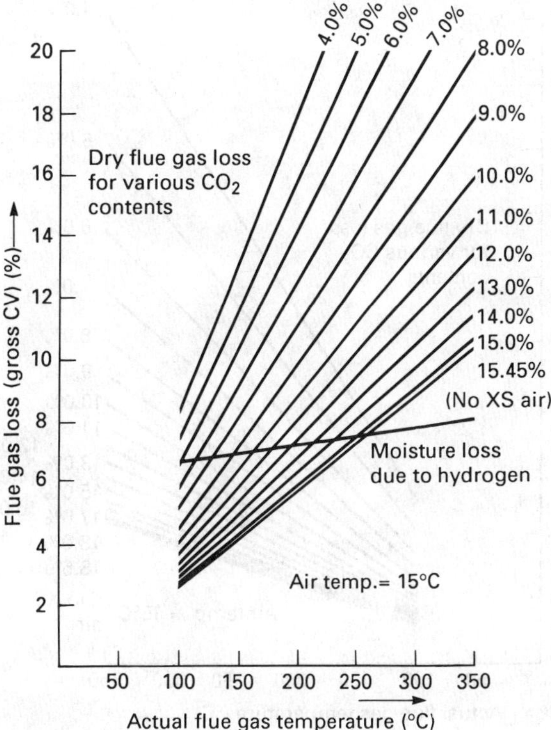

Figure 33.4 Flue gas losses – gas oil

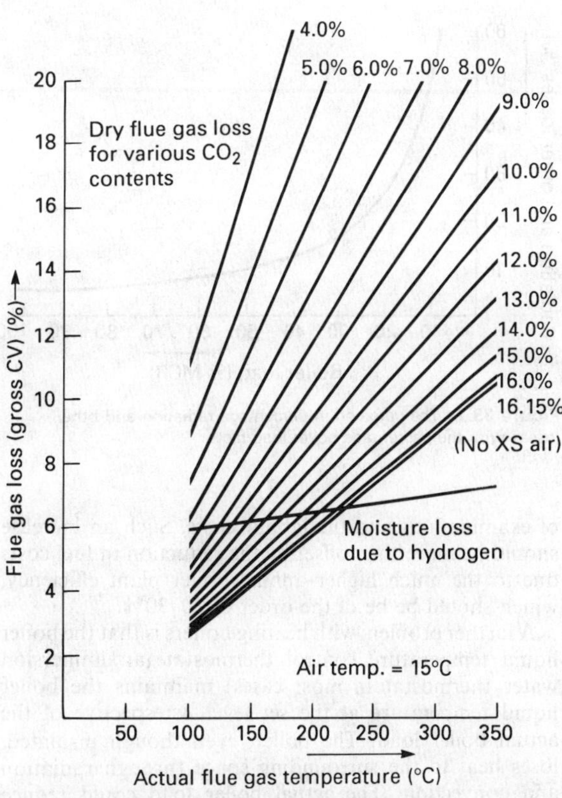

Figure 33.5 Flue gas losses – heavy fuel oil

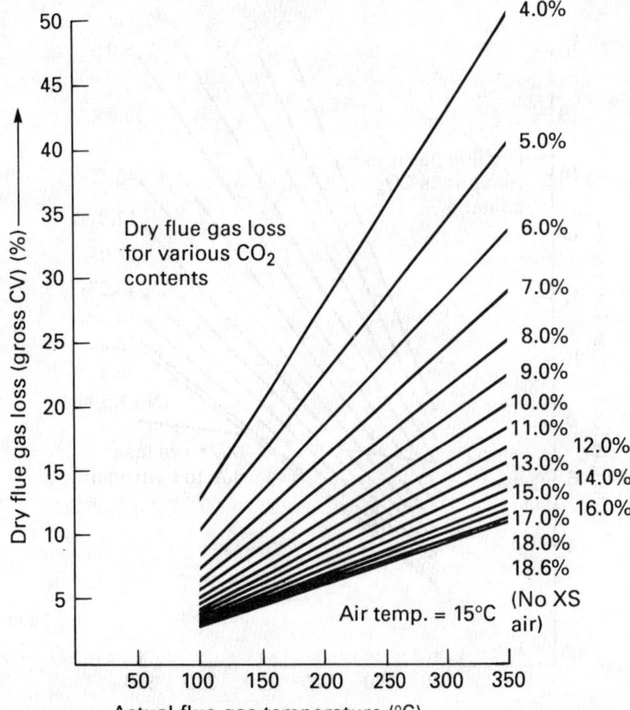

Figure 33.6 Flue gas losses – coal (Glasshoughton washed singles)

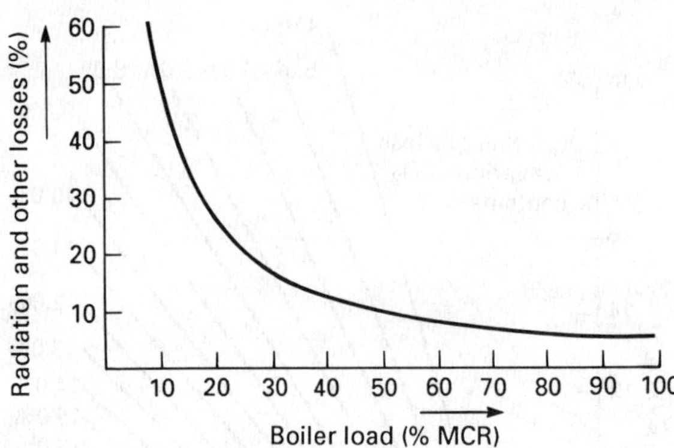

Figure 33.7 The variation in percentage radiation and other losses with the percentage boiler load factor

of examinations and the labour costs. Such an increase should be more than offset by the reduction in fuel costs due to the much higher annual boiler plant efficiency, which should be be of the order of 20–30%.

A further problem with heating boilers is that the boiler liquid temperature control thermostat (an immersion water thermostat in most cases) maintains the boiler liquid temperature at the set level, irrespective of the actual boiler load. The boiler, even though insulated, loses heat to the surrounding space through radiation and convection. The actual boiler load could reduce during the day such that the maintenance of the set boiler liquid temperature is not required continuously. Thus the boiler liquid could be allowed to cool down with the consequent reduction in the radiation and convection losses (the standing loss). Boiler controls are now available to avoid maintaining the boiler liquid temperature at the thermostat control setting and consequently the unnecessary operation of the burner ('dry' cycling).

The cost of the supply and installation of such controls is of the order of £250–300. Energy reductions of 10–15% can be achieved, with a cost-recovery period of 3 years down to a few months, depending on the actual annual energy consumption.

Condensing boilers are now available for both gas- and oil-fired plant, the advantage of these being that the flue gases are further cooled down to below 100°C so that the latent heat available in the flue gas water vapour is recovered. The condensate has to be removed and the boiler capital cost is higher than for conventional plant. However, the boiler plant efficiency is increased to the order of 90%, based upon the fuel gross calorific value. Where the flue gas exit temperatures are in excess of 200°C a further economy can be obtained by the provision of a spray recuperator in the case of gas and flue gas economizers for oil and coal.

33.7.2 Furnaces

Furnaces are large users of energy, and in order to reduce costs, such equipment should be well insulated, used to maximum capacity and most of the waste heat in both the flue gases and product recovered. It should be possible to recover the waste heat in the flue gases down to at least 200°C. Specialist equipment for such waste heat recovery is available in the form of recuperators and regenerators.

33.7.3 Fans

Large fans are required for boiler plant, furnaces and large air-handling units. These fans can now be fitted with variable-speed motors to reduce electricity consumption and maximum power, albeit at additional cost.

33.7.4 Pumps

Pumps are a common feature in most heating and process applications. Where one pump only is provided this is sized for the maximum load (which may occur rarely). Two smaller pumps could be installed such that one pump carries the load for most of the time while the second is used at times of peak demand.

33.7.5 Air compressors

Compressed air systems should be checked frequently for air leakage, as the loss of air is frequently unnoticed due to the noise of escaping air being masked by other production noises. Modern air compressor systems composed of multi-compressors produce much heat, and it is now possible to recover this for space heating purposes.

33.7.6 Space heating

The provision of space heating, especially in large non-compartmented buildings, can be reduced by:

1. The installation of direct gas-fired units, where the products of combustion are discharged into the space but are so diluted as to be harmless. The efficiency of these units is thus 95–100%, depending on the siting of the actual unit.
2. The provision of overhead radiant heating instead of convective heating. In this case the building heat requirements are some 15% lower for radiant heating than convective, with the consequent saving in both capital and running costs. It is also believed that the comfort conditions are improved with less draughts. Radiant heating is also beneficial in buildings which are sparsely occupied, such heating being provided only in the area of the occupants.
3. The provision of ceiling-mounted fans to recirculate the warm air at ceiling level to the ground. These fans can vary from the small three-bladed slow speed fan up to much larger units, recirculating large quantities of air. The operation of these fans can be thermostatically controlled and fuel savings of 10–15% are quoted.

33.7.7 Insulation

Buildings should be insulated to as high a standard as possible, as should any part of energy-using equipment. As a guide to buildings, the following overall heat transfer coefficients (or *U*-values) are suggested, which are currently of lower value than the requirements of the Building Regulations:

Roofs	0.20 W/m^2°C
Walls	0.30 W/m^2°C
Windows	2.90 W/m^2°C
Exposed floors	0.30 W/m^2°C

The current Building Regulations (1990) call for:

Roofs	0.25 W/m^{2}K (Dwellings)
	0.45 W/m^{2}K (Other buildings)
Walls	0.45 W/m^{2}K (All buildings)
Ground floors	0.45 W/m^{2}K (All buildings)

In order to improve the standards of existing building insulation the following methods are available.

Pitched roofs

Provide a blanket of 200 mm thickness of fibre glass or mineral wool insulation on top of the ceiling plaster.

Flat roofs

Provide a similar blanket of insulation on top of any suspended ceiling which is 300 mm below the flat roof or an external hot liquid spray-on type of insulation to the roof finish.

Cavity walls

Fill the existing wall cavity with a suitable insulant such as fibre glass, rock wool, polystyrene beads or foam. The greater the thickness of the cavity, the lower the *U*-value.

Solid walls

Solid walls can only be improved either by removing the internal plaster layer and providing suitable blanket insulation fixed to battens and then covering with the required new internal finish or by applying an external insulant to the wall and a new external finish.

Windows

Double or triple glazing may be provided in this case, the main criterion being that the cavity thickness between the layers of glass must be at least 20 mm.

Exposed floors

If access is available to the area below the floor, provide 50 mm of rigid insulation between the floor joists, leaving an air gap between the insulation and the underside of the floorboards.

Air infiltration

Provide good draughtproofing strip around the door and openable window frames, leaving sufficient fresh air ingress for any combustion appliances.

On completion of such measures, the use of thermography should be considered as a survey, undertaken when the building is in use and heated, will highlight hot spots. Any hot spots may be due to poor workmanship or unnoticed building faults and can then be rectified.

33.7.8 Controls

In order to achieve optimum energy usage (hence maximum energy conservation) it is essential to provide accurate controls designed to suit the application and for those controls to be correctly installed. It is no use providing insulation and highly efficient energy or heat generators if that energy or heat generated is simply lost in the system by providing better standards of comfort than those required. In the absence of such controls on space heating, the provision of building insulation simply increases the internal temperature conditions and results in windows and doors being opened to lose this heat.

Controls are expensive and quickly become outdated by the increase in technology. They are also the items which are omitted or their quality is reduced if the total project cost becomes excessive. The difference in the quality of the controls against cost can be determined and should be included as part of the project budget or application.

A typical example is the provision of a single room-air thermostat to provide whole building control of space heating. It will work, but far greater economy can be achieved by providing building zonal controls or, better still, individual room heating control. Zonal control is available in the form of either two- or three-port motorized valves which open or close in conjunction with an internal air thermostat. Individual room heating control is in the form of individual thermostatic radiator valves and in internal air thermostats controlling on/off or three-port motorized valves.

Individual room temperature control is difficult with warm air systems, as any temperature control damper is likely to pass air at all times. Air thermostats are not accurate detectors of temperatures and temperature overshoot occurs, with consequent increases in energy consumptions and costs. The typical limits of these thermostats is $\pm 1^{\circ}$C, and as the air in the thermostat is not at the temperature of the air in the room because the thermostat is remotely sited (usually on an internal wall) the actual air temperature in the room can be 1.5–2.0°C above the desired temperature. Modern temperature detectors are accurate to $\pm 0.5^{\circ}$C and give much closer and better control.

Steam flow to vessels can be controlled by motorized valves, allowing timed control of the valve. Such valves can also be made to modulate rather than on/off operation.

33.7.9 Heat recovery

All the heat used to supply heating in buildings (and much of that provided in any process) is ultimately lost from the building. Recovery of as much waste heat as possible reduces the need to purchase fresh energy from the suppliers. Much waste heat is produced in the form of contaminated hot moist air and in process liquids which require cooling.

There are many ways of recovering the waste heat from these sources and the following should be considered:

The heat wheel
The heat pipe
The run-round coil
The heat pump

The great benefit of the heat pump is the possibility of upgrading the heat as compared to the other methods, and is popular in swimming-pool applications, where the heat recovered from the exhaust air can be used to heat the incoming fresh air and the pool water and to provide hot water for showers and washing purposes.

33.7.10 Lighting

Artificial lighting is used throughout the UK and this can now be provided by energy-efficient lamps which give high lumens per watt. A new range of fluorescent lamps is available and similarly high-efficiency lamps have been developed to replace the conventional tungsten lamp.

The problem of lamps being left on when not required can be solved by provision of automatic switching, of which there are a number of systems available. These can be time-switch controlled with manual override at the point of use.

33.8 The justification for energy-conservation measures

In many cases, areas of obvious energy conservation are not considered simply because a process or some space heating has always been undertaken that way. In order to persuade the organization's management to embark on such energy-conservation measures overwhelming evidence is required that the theoretical results suggested can be achieved. However, most energy-conservation measures do offer large financial savings with very short payback periods and lead to increased organizational profitability. The cost of such schemes has to compete with others for new capital equipment and their success could depend on the effect of energy costs on the total organization costs.

It is obvious that a scheme to reduce energy costs by, say, 10% may fail if the total energy bill is only 3% or 4% of the organization's costs when compared to, say, a new production machine costing ten times as much but which reduces the organization's costs by 10%.

If an energy-conservation scheme is to succeed it must be well prepared and presented to senior management. In the project application and budgets the objectives of the scheme must be clearly defined and the capital costs fully detailed. This may involve preparing several budget costs from various equipment manufacturers and contractors. In such cases it is necessary to read and note the 'small print', as any comparison has to be on a similar basis. This is extremely difficult to achieve, as both manufacturers and contractors omit certain items which may be required to be added later. The estimated energy and cost savings have to be identified as accurately as possible, and this is very difficult in view of the large number of parameters involved.

In formulating these savings the parameters used should be clearly defined so that the estimates can be revised for any alterations. Indeed, in some cases a computer program can be utilized which allows such changes to be evaluated very quickly (e.g. the TAS package for buildings). In the end the submission will stand or fall on the overall impact on the organization, so that it is essential to get it right first time. That saving on the bottom line is the major item.

A further factor to be considered is that of the national energy situation. Fortunately, in the short term there are available sufficient reserves of gas, oil and coal to meet the national need, so supplies should not be a problem. However, the prices of these fuels are subject to international as well as national policies. Currently, there is a relationship between heavy fuel oil and coal and also between gas and gas oil (see Section 33.2). In the long term, as supplies of oil and gas are reduced it is likely that increases in the price of gas will outstrip those of oil.

33.9 The mathematics of the presentation

Any presentation to the organization's senior management, besides giving the technical details, should include a financial appraisal which could be on a simple payback basis. However, other methods are available, such as return on capital employed and discounted cash flow (DCF).

33.9.1 Simple payback

This method of comparing the initial capital cost against the annual energy/cost savings is relatively easy. Unfortunately, it does not take into account certain factors such as:

1. The life of the project;
2. Any benefits after the payback period.

Normally, as the required payback period is short (i.e. 2 or 3 years maximum) and the length of life of the equipment is much longer than this, on the simple payback basis there are usually large financial benefits in the long term.

Let us consider two projects both costing £20 000 with the following savings:

Project 1: Annual saving £10 000, life of equipment 10 years
Project 2: Annual saving £8000, life of equipment 15 years

In terms of simple payback, project 1 has an excellent payback of only 2.0 years, while project 2 has a payback of 2.5 years. Thus it is likely that project 1 will be selected. However, project 1 has a life of 8 years after the initial payback period, the financial saving being £80 000, the equipment then requiring replacement. Project 2 has a life of 12.5 years after the initial payback period, giving a total financial saving of £100 000. Thus project 2 is the better case in the long term.

33.9.2 Return on capital employed

This method compares the initial capital cost with the cash flow over the life of the project. Thus in this case an accurate estimate of the equipment life is required. When a project has a long life (considerably longer than the simple payback period) the results can be different.

There are four ways in which return on capital employed can be expressed:

1. Gross return on capital employed;
2. Net return on capital employed;
3. Average gross annual rate of return;
4. Average net annual rate of return.

In this method each year's savings need to take account of the fact that, as the equipment gets older, the cost of maintenance and repairs is likely to increase. The effect of inflation can also be allowed for in assuming the annual cash flows through the life of the equipment.

Using the two projects in Section 33.9.1 as an example, the returns shown in Table 33.3 are likely. Of the four methods, the net return is generally favoured to the gross return as the loss of capital is a major factor.

33.9.3 Discounted cash flow (DCF)

This method, which is favoured by many accountants today, takes into account the concept that money has a time value. This is based on the fact that £1000 in 10 years' time is not the same as £1000 now. Similarly, if a project earns £1000 in 10 years' time this is not the same as £1000 spent now to help finance the project. If, instead of spending this £1000, it had been invested at compound interest, then in 10 years' time it would have become a much larger amount than £1000, depending on the interest rates in the prevailing period. The way that this is taken into account is called 'discounting'. Trading income produced for each future year by the project is discounted after allowing for maintenance and repairs, giving present values for that income.

There are various ways of discounting, i.e. 'net present value' (NPV) or 'internal rate of return' (IRR). Discounting should ideally be carried out over the whole life of the project but generally, after 10 years, the discount factors are small and make little difference.

The selection of the discount factor depends on the financial policy of the business, but is usually 2–3% above the current interest rates. Use of discounting methods will determine whether the project cost will produce a better return than by simply investing the capital involved at the highest compound interest rate or, if the capital

Table 33.3

Method	*Project 1*	*Project 1*	*Ratio*
Gross return on capital employed	$\frac{1000\,000 \times 10 \times 100}{20\,000}$ = 500%	$\frac{8000 \times 15 \times 100}{20\,000}$ = 600%	1:1.2
Net return on capital	$\frac{[(10\,000 \times 10)20\,000] \times 100}{20\,000}$ = 400%	$\frac{[(8000 \times 15) - 20\,000] \times 100}{20\,000}$ = 500%	1:1.25
Average gross annual return on capital employed	$\frac{(10\,000 \times 10) \times 100}{(10 \times 20\,000)}$ = 50%	$\frac{(8000 \times 15) \times 100}{(15 \times 20\,000)}$ = 40%	1:0.30
Average net annual return on capital employed	$\frac{[(10\,000 \times 10) - 20\,000] \times 100}{(10 \times 20\,000)}$ = 40%	$\frac{[(8000 \times 15) - 20\,000] \times 100}{(15 \times 20\,000)}$ = 33.3%	1:0.83

cost has to be borrowed, whether the rate of return is much higher than the cost of borrowing.

33.10 Third-party energy management and finance

As indicated in Section 33.7, energy-conservation measures cost money, and in spite of the likely results indicated by any of the methods available (simple payback, discounted cash flow, etc.) it is possible that the organization's funds are not available for such schemes. Alternatives that should be considered are:

1. Leasing;
2. Hire purchase.

Many organizations now have their energy requirements managed by a heat service company. In this case the company will provide and operate all the energy-producing equipment on the organization's site. This could include space heating, air conditioning, combined heat and power (CHP), steam boiler plant and other services.

The contractor will purchase all the fuel required and provide all labour, repairs and maintenance, thus relieving the organization of all responsibility. In return, the contractor will require a contract (generally of up to 10 years) with the fees charged comprising a monthly fixed standing charge and a monthly variable charge, depending on the amount of energy used. Such an arrangement is useful for obtaining new equipment when finance is not available within the organization itself.

A variation of the heat management service is to have contract energy management (CEM). In this case the energy contractor will generally survey the existing plant and equipment and take it over, running it for the organization. As equipment becomes due to replacement the contractor will undertake this part of the contract.

The contract is usually for a period ranging from 3 to 9 years, but in this case the amount payable to the contractor is the same or similar to the costs prior to contractor take-over. A further clause is added such that a proportion of the savings made by the contractor are returned to the client (perhaps 2–10% of the previous annual costs). It may also be possible for any further large saving to be shared in agreed proportions between the two parties involved.

It is essential to have any agreement properly drawn up in conjunction with the organization's legal section and to use reputable and busy heat service companies. It is important to lay down strict details of the period of the contract and of the services to be provided so that future legal difficulties can be avoided.

33.11 Motivation

Energy conservation will succeed if all the particles involved wish it to succeed. It only requires ill will on one person's behalf to negate all the positive efforts made. Typical problems which can occur are:

1. The unauthorized alteration of internal air thermostats and time switches;
2. The use of supplementary heating (e.g. electric fires without permission).

The unauthorized alteration of set parameters is the most common problem and it is possible to install tamperproof equipment, albeit at additional cost. There is also the problem of the relationship with the existing management (e.g. the senior manager who views the growth and success of the energy management team as a threat to the supervisor or department manager).

There are a variety of methods to avoid this, and the more senior the management structure that takes an interest in energy matters, the greater the success. Ideally, this should start with the chairman, managing director or chief executive. Once they believe in it, then their staff will follow suit!

These people are extremely busy with the overall organization's objectives and time for energy matters may be very limited. They should therefore delegate this to a senior officer. Energy matters may be left to the works engineer as an additional duty. Consideration should be given to the appointment of an energy manager who is qualified for the post and provided with a suitable job description and objectives. Such a job description could be based on the following:

1. The energy manager will be responsible for the total energy function of the organization.
2. He will be responsible to the chief executive and will achieve the objectives laid down by senior management.
3. The duties consist of (a) the most economic purchase of all forms of energy – in conjunction with the purchasing section; (b) the optimum use of all forms of energy for both process and building environment use – in conjunction with the various departmental and engineering managers.
4. The energy manager needs to form a team – an energy committee – to help him and this could comprise the managers of each department, whether production or otherwise, and other suitable people. The team must be small to be effective (less than 10 persons). Each department could then have its own energy committee.

Hopefully, the enthusiasm for energy conservation will spread downwards to all employees. Their help can be obtained by the use of competitions and suggestions.

33.12 Training

Few senior staff, works engineers or even junior staff have had training in energy matters. When energy was cheap, this was a small company cost (say, 1%) so that any cost-reduction measures were directed at other areas. Universities and colleges now include energy management as part of the curriculum in engineering and related courses. Newly appointed staff will be energy orientated.

For those staff who have never had the benefit of education and training in energy matters there are courses both by part-time attendance at college and by correspondence. Details of the courses can be obtained from the Secretary, the Institution of Plant Engineers. Throughout the UK there are a large number of energy management groups where the members (who are nearly all energy managers) meet regularly to discuss energy matters. Thus it is possible to obtain training at all levels required.

Reference

1 *Pattern of Energy Usage – Energy for the Future*, Institute of Energy, London (1986)

34 Insurance: Plant and Equipment

A P Hyde
National Vulcan Engineering Insurance Group Ltd

Contents

34.1 History

It is important to understand how the relationship between inspection and insurance, which is available from the major engineering insurers, came about. From the outset the objectives have been towards the safe and efficient operation of all types of machinery and plant used in industry and commerce.

The early nineteenth century saw the beginning of factory production systems, particularly in the cotton mills of Lancashire and woollen mills in Yorkshire. Accidents arising from the use or misuse of steam plant, particularly boilers, became common and led to not only damage and destruction to property but also to death and bodily injury to persons in or about the scene of the explosions.

In 1854 the Manchester Steam Users Association was formed to help with the prevention of explosions in steam boilers and also to find efficient methods in their use. To achieve this, the Association employed the first boiler inspectors, whose services were then made available to the Association's members. Within a short space of time the members became convinced that insurance to cover the high cost of repair or replacement of damaged boilers was desirable, and this resulted in the first boiler insurance company (The Steam Boiler Assurance Company) being formed in 1858. The scope of the services for inspection and insurance later extended to include pressure vessels, steam engines, cranes, lifts and electrical plant, the insurance protection in each case being supported by an inspection service carried out by qualified engineer surveyors.

The development of engineering insurance has been closely linked with legislation. The Boiler Explosions Act was passed in 1882 and this empowered the Board of Trade to hold enquiries into the causes of all boiler explosions except where the boilers were used in the service of the Crown or for domestic purposes, and to charge the cost of the enquiry against any person held to have been responsible for causing that explosion. This provided a strong incentive to insure boilers, and in 1901 the passing of the Factory and Workshops Act made the regular inspection of steam boilers in factories compulsory. This was later extended by legislation to include steam boilers in mines and quarries.

From those early days a great deal of legislation has been passed, including the Factories Act 1961 (which repealed earlier acts), the Power Presses Regulations 1965, the Offices Shops and Railway Premises (Hoists and Lifts) Regulations 1968 and the Greater London Council (General Powers) Act 1973, which was specifically directed at self-operated laundries. These have now all been embodied within the Health and Safety at Work, etc. Act 1974 which, although it does not change existing regulations regarding frequency of inspections, does provide that duties under earlier regulations are now enforced within this Act.

Engineer surveyors employed by engineering insurance companies have always been regarded as being 'competent persons' as required by these various Acts. The knowledge and experience which has been accumulated by the companies over the years enables them to offer to industry and commerce a service of exceptional quality.

34.2 Legislation

It can be seen from Section 34.1 that as legislation has grown and changed so the role of the Independent Engineering Inspection Authority has had to change to meet the demands and needs of industry and commerce. Legislation, particularly that related to health and safety at work, is something which never remains static. As new areas of potential hazard are identified and their implications are discussed to the point that new legislation is needed, the engineering insurers are making their contribution to discussions. Currently such matters as the Pressurized Systems Regulations and those related to the Control of Substances Hazardous to Health are implemented, and the engineer surveyors are being trained and retrained to enable them to cope with the requirements of this new legislation.

The following sections include the relevant sections of the complete Act and indicate those parts of the legislation within the different plant categories to which the inspections provided by the independent engineering inspection companies will conform. It should, however, be appreciated that while the inspections provided will fulfil the statutory requirements for inspection, the actual responsibilities under the various Acts to conform remains the responsibility of the plant owner/user. It must be appreciated that the interpretation of any Act is a matter for the local Health and Safety Inspectorate, who should be consulted at all times if doubt exists as to whether any item of machinery and plant requires inspection to comply with a statutory provision.

A general guide is set out in matrix form in Figure 34.1, indicating by business and trade the types of machinery and plant that are likely to be found in general usage and the normally accepted position of inspections which may be required to conform with the legislation.

34.3 The role of the inspection authority

Inspection by independent persons or bodies for safety purposes goes back to the middle of the nineteenth century. At that time the focus of concern was the explosion of steam boilers, and this hazard was most prevalent in the textile industry. Consequently a group of public-spirited individuals formed the Manchester Steam Users Association for the Prevention of Boiler Explosion. This body carried out boiler examinations and later added insurance as an inducement to the plant owners. By the turn of the century steam and gas engines and electrical machines had been added, followed by lifts, cranes and hoisting machines.

The inspection companies shared their technical knowledge, and 1917 saw the formation of the Associated

Business	Electrical plant wiring installation	Refrigeration, air conditioning	Process machinery	Power press	Dust extraction plant	Lifts	Motor vehicle lifting table	Manual block & sling	Electric block	Fork lift truck	Builders hoist	Excavator	Lorry loader crane	Mobile crane	Power crane	Other steam plant & ovens	Air receiver	Hot water boiler	Steam boiler
Bakers	●	●	●			○		○	○	○			○			○	○		○
Brewers	●	●	●			○		○	○	○			○			○	○	●	○
Building contractors	●			○		○		○	○	○	○	○	○	○	○	○	○	●	○
Churches, schools and halls	●	●				●												●	
Clothing factories	●					○		○	○	○			○			○	○	●	○
Dept stores	●	●				○		●	●	●						●	○	●	○
Docks	●	●						○	○	○			○	○	○	○	○	●	○
Dry cleaners	●		●			○		○								○	○	●	○
Engineering works	●	●	●	○	○	○		○	○	○			○	○	○	○	○	●	○
Farms	●	●	●					●		●	●		●			●	●		●
Flats	●	●				●												●	
Food manufacturers, canners	●	●	●	○		○		○	○	○			○	○		○	○	●	○
Garages	○	●		○	○	○	◑	○	○	○			○	○		○	○	●	○
Hotels, etc.	●	●				○										○	○	●	○
Launderers	●		●			○		○	○	○						○	○	●	○
Millers	●		●		●	○		○	○	○			○			○	○	●	○
Nurserymen	●	●														●		●	●
Nursing homes	●	●				●										●	●	●	●
Office buildings	●	●				○										●	●	●	●
Printing works	●		●			○		○	○	○			○	○		○	○	●	○
Provision shops	●	●				○				●							●	●	
Quarries	○		●		●			●	●	●		●	○	○		○	○		○
Scrap yards	●		●	○	○			○	○			○	○	○	○	○	○		○
Ship yards	●		●	○	○	○		○	○	○	○		○	○	○	○	○		○
Stonemasons	●		●		○			○	○	○			○	○		○	○	●	○
Supermarkets	●	●				○			○	○						●	●	●	●
Theatres, cinemas	○	●				○		●	●							●	●	●	●
Timber merchants	●		●		●			○	○	○			○	○		●	○	●	○
Warehouses, wholesalers	●	●			●	○		○	○	○			○				○	●	○

Key ○ Statutory need for inspection.
◑ Statutory need for inspection in some cases in other cases inspection recommended.
● Inspection recommended.

Figure 34.1 Statutory requirements

Offices Technical Committee, the founding members being:

The British Engine Boiler and Electrical Insurance Co. Ltd
The National Boiler and General Insurance Co. Ltd
The Ocean Accident and Guarantee Corporation Ltd
The Scottish Boiler and General Insurance Co. Ltd
The Vulcan Boiler and General Insurance Co. Ltd

This Committee published its own technical rules, many of which were later incorporated into British Standards. Its continues to make an active contribution to standards, guidance notes, legislation and international policy on inspection. Some independent inspection bodies have no connection with insurance but many do, and these companies have the advantage of feedback from an analysis of insurance claims.

Legislation introduced the requirements for statutory examinations by 'a competent person', the responsibility for ensuring the competence of the examiner resting on the plant owner. There is no requirement for the competent person to be independent of the owner's organization, but independence does have clear advantages.

It will be seen that owners and users of plant which can be hazardous to employees or to the general public come at the top of a list of inspecting authorities' clients. Safety is the first objective, but avoidance of unscheduled stoppages and objective assessments of plant condition aimed at timely replacement are powerful commercial inducements. The client may be an individual owner of a vintage traction engine or a large public utility, but the advantages are essentially the same. National and local government bodies, including the Health and Safety Executive, also rely on inspecting companies to warn the enforcing authority of potential dangers and to provide 'fitness for use' certificates for hazardous plant items or installations.

Inspection is also an important activity in the regulation of international trade through the certification of vehicles and containers used for transporting hazardous products and for providing foreign purchasers with evidence that manufactured goods comply with specification before they leave the country of origin. The essential characteristics of and requirements for an inspection authority are:

Technical competence
Maturity and integrity of judgement
Confidentiality
Access to a database of relevant plant histories
Ability to communicate

The person who carries out the examination is the 'competent person', although this term may also be used in a corporate sense. One very concise description of the competent person which originated in a law case is 'he must know what to look for and how to recognize it when he sees it'. This has been expanded both officially and unofficially, and a current guidance note on this subject is paraphrased as follows:

The Competent Person should have sufficient practical and theoretical knowledge and actual experience of the types of plant he is charged with examining and testing as will enable him to detect faults and weaknesses. He should have the maturity to seek such specialist advice and assistance as will be required to enable him to make the necessary judgements. He should also be able to assess the importance of any faults or weaknesses in relation to the strength and function of the plant before he is required to certify it as suitable to carry out its specified duty. The Competent Person should have access to the equipment, specialist's support and laboratory services necessary to enable him to carry out the relevant examination and testing. Where he is unable to carry out all parts of the examination and testing himself, he must be a proper judge of the extent to which he can accept the supporting opinion of other specialists. An inspecting authority may be appointed as the 'Competent Person'. In this case where more than one engineer employed by the inspecting body carries out the examination and testing, it is appropriate for a nominated person within the Company to sign the certificate on behalf of the inspecting body.

The larger inspecting companies carry their own specialist support. Typically, this will cover:

Non-destructive testing
Failure analysis using modern techniques such as finite element stress analysis and fracture mechanics
Metallurgical and weld analysis
Chemical engineering and process capability
Quality assurance

Clearly, there are 'horses for courses'. The practising plant engineer may have access to a local specialist in certain types of non-destructive testing who provides an excellent service and value for money. At the same time, the plant engineer must be vigilant regarding the limitations of such support.

The role of the inspecting authority must clearly be sensitive to change in the light of both technical advances and alterations in the political climate. The response time will sometimes depend on whether the authority concerned has addressed itself to a rather narrow sector of the inspection field or operates over a wide spectrum.

The 1982 White Paper on Standards, Quality and International Competitiveness was concurrent with increasing interest in the techniques of quality assurance and the need for international harmonization of standards and the reciprocal recognition of certification. It is worth noting that the ISO 9000 series of standards on Quality Systems: 1987 followed the layout of BS 5750: 1979 almost clause by clause. ISO Guide 39 covers the general requirements for inspection bodies and, at the time of writing, a National Certification Scheme for In-Service Inspection Bodies (NCSIIB) is in preparation under the auspices of the Institution of Mechanical Engineers and the Health and Safety Executive. Auditing on behalf of certification bodies is part of the inspecting authorities' role.

It is to be hoped that the variety of certification schemes will not multiply, or the client is likely to be confused rather than informed. When making his choice he may usefully remember that the criteria of good quality systems are as applicable to service organizations as they are to manufacturers, and that clarity regarding the objectives of the service offered together with practical common sense in providing this service are the real indicators.

34.4 Types of plant inspected

34.4.1 Boiler and pressure plant

The type of plant which would be included in the above generic description will include many items which, because of difference in size, function and appearance, will appear to be unrelated. They do, however, have one common factor in that all operate to some degree under pressure.

To illustrate this the use of the term 'Boilers' can vary considerably not only in its design and construction but also in the class for which they are intended, for instance:

1. Steam-generating boilers
2. Low-pressure hot water heating boilers
3. Low-pressure steam heating boilers
4. Domestic hot water supply boilers
5. High-pressure hot water heating boilers
6. Process vessels

In the first category the type most often found in use today will be what is referred to as the steam package boiler of the multi-tubular type, otherwise referred to as the shell boiler. This can equally be used to describe vertical multi-tubular shell boilers, locomotive-type boilers or even watertube boilers more closely associated with power generation or, alternatively, where there is a high and regular steam demand from within a factory. Shell-type steam generating boilers are also used for high-pressure hot water heating where there is the need to supply hot water under a pressure and temperature equal to that of the steam in the boiler for circulation throughout heating or process plants within a building. The circulation arrangement used with these boilers is often similar to those found in connection with low-pressure hot water heating systems, but where the water is at a much higher temperature and the pressure is, equally, at a much higher level. This system can incorporate the use of its own separate pressurization unit.

By far the most numerous class of boiler within the UK must be the low-pressure hot water heating or hot water domestic supply boiler, and these are found installed in various types of situations (e.g. offices, factories, shops, hotels) and now widely used in connection with private dwelling houses.

34.4.2 Steam and air pressure vessels

The diverse types of plant in this category would be too numerous to list, but include steam-jacketed pans, steam calorifiers, steam heating and drying units and all types of air-pressure vessels. These are manufactured in a wide range of materials (e.g. stainless steel) and may then be lined with materials such as rubber or glass, depending on the needs of the process in which the vessel will be used.

34.4.3 Lifts, cranes and other mechanical handling plant

Items in this category can range from small manual items of lifting tackle (e.g. rings, hooks, slings, chains and blocks) to the very large fixed overhead or mobile powered cranes, fork lift trucks, straddle carriers, side loaders, loading shovels and excavators as well as the general type of passenger lift to be found in most modern offices and hotels. Cranes can range from the most basic manual fixed-pillar jib upon which an electric or manual block is hung to the modern specialist container crane or order picker.

34.4.4 Elecrtrical and mechanical plant

Plant falling within this description includes everything from small fractional horse-power motors and their driven components to the major steam or gas turbine and their generators; from small printing machines to complete continuous process lines such as may be found in steel production. In addition, this category would include any item which incorporates either static or moving electrical or mechanical parts.

34.5 Insurance covers on inspected plant

34.5.1 Boiler and pressure plant

Boilers and pressure vessels can be insured against explosion and collapse, which provides for damage to the insured plant, surrounding property and, in addition, liability for damage to third-party property and injury to third-party persons. This cover can be extended to include full sudden and unforeseen damage which, as far as boilers and pressure vessels are concerned, includes overheating, frost, cracking, water hammer action, leakage between the sections of sectional heating boilers and damage by external impact. Often insurances are written so that the sudden and unforeseen damage extension provides for resultant water damage to surrounding property and will also provide for costs up to defined limits to expedite repair. There is also provision to insure on a reinstatement basis.

While boiler explosions fortunately do not occur too often today, because of the existence of extensive safety devices as well as the regular programme of inspection, their effects can be catastrophic. Similarly, sudden and unforeseen damage caused by the overheating of multi-tubular steam boilers due to lack of water can lead to eventual furnace collapse, with very extensive repair costs as well as lost production.

34.5.2 Storage tanks and their contents

Items of this type are linked to the boiler and pressure plant generic type but are given their own specific covers which are for sudden and unforeseen damage, including collapse, impact and frost as well as damage to the surrounding property, cleaning-up costs and the costs of expediting repair. The actual contents of the tanks, which can be very varied and expensive, can be insured against accidental leakage, discharge, overflowing and even contamination.

34.5.3 Cranes and lifting machinery

The types of plant falling under this heading are many and varied, and insurance covers which are offered are equally varied, falling into the categories of:

1. Breakdown;
2. Accidental extraneous damage only; or
3. Sudden and unforeseen damage which will include, in addition to the electrical and mechanical breakdown, accidental causes such as toppling, overloading and collision.

Depending on the type of machine in question and the operations or use to which the machine is put, there are numerous extensions to the basic covers available. These include third-party liability, cover while the plant is hired out, indemnity to the first hirer, damage during erection and dismantling, damage to goods being handled, damage to surrounding property and the cost of hiring a replacement machine following an indemnifiable insured loss or accident.

34.5.4 Lifts and hoists

Lifts and hoists can be insured against breakdown only or for full sudden and unforeseen damage, which includes, in addition to electrical and mechanical breakdown, accidents arising from impact, fatigue, malicious damage and entry of foreign bodies. The cover can be extended to include liability to third parties and damage to goods being carried arising out of the use of an insured item.

34.5.5 Electrical and mechanical plant

Electrical and mechanical plant will include all types of process machinery, engines, generators, pumps, fans, furnaces, transformers and refrigeration plant, to list but a few. Plant of this type can be insured against breakdown only or full sudden and unforeseen damage, which will include, in addition to electrical and mechanical breakdown, accidents arising from faulty insulation, failure of wiring, short circuiting, excessive or insufficient voltage, non-operation of safety or protective devices, renewal of insulating oil or refrigerant and the consequences of impact from an external source.

34.5.6 All machinery and plant insurances

While the normal engineering insurance covers on inspected plant are designed for a high degree of selection of specific items which will be identified in a policy schedule, there are now covers available designed to meet the needs of protection for 'All Machinery and Plant' without the requirement of scheduling of the items separately. The machinery and plant can be insured against breakdown only or for full sudden and unforeseen damage risks, which will include electrical and mechanical breakdown and can be extended to take in the explosion and collapse cover normally found under the boiler and pressure plant section.

In instances where cover is written on this basis it is normal that higher than usual excess is applied, reflecting the type of plant involved. While it is normal on the more traditional basis for the insurance cover to be complementary by inspection of the insured items in the case of covers written in this manner, it is usual for the inspection of all statutory plant to be incorporated together with those items where the insurer considers inspection to be necessary as part of the risk management of the cover being provided.

34.6 Engineer surveyors

34.6.1 Selection

The AOTC member companies use common criteria for selection of candidates to train as engineer surveyors, and these may be considered typical. Traditionally, surveyors have been recruited from among sea-going engineers who had a first-class Board of Trade Certificate of Competency or a similar certificate from the Engineering Branch of the Royal Navy. Sea-going engineers have suitable practical experience, and are also likely to have had sufficient practice in working relationships to enable them to deal with clients' personnel, including plant engineers, on an equal basis.

As this source has progressively dried up, engineers with an industrial background have been recruited from those with a Technician Engineer qualification (now known as Incorporated Engineer). The minimum age for entry is 25 years, and since maturity of judgement is also required (see Section 34.3), considerable weight is attached to the selection interview. A significant number of engineer surveyors now have BSc/BEng degrees or equivalent, some having obtained these during their employment in this field.

In addition, at age 26 engineer surveyors can apply for Associate Membership of the Bureau of Engineer Surveyors, who are a division of the Institution of Plant Engineers. Transfer to membership is possible after a further two years' employment as an engineer surveyor.

34.6.2 Training

The period of training following engagement varies, depending on previous experience, and is normally of between 3 and 6 months' duration. Traineers are usually streamed into specialities (i.e. boiler surveyor, lift and crane surveyor or electrical surveyor). Alternatively, a 'composite' surveyor's training embraces all these technologies. Further specialization sometimes occurs following training and some experience in the field. For instance, surveyors may become specialists in air conditioning and dust extraction, power presses or new construction of pressure containing plant or of rotating machines.

The scope of initial training always covers an induction

period of perhaps four weeks, during which the trainee learns about the structure of the company which he has joined and the administration which he will be expected to carry out in the field. It is important that he learns about the legislation which applies to his job and the statutory forms which he must use. Surveyors are often required to keep a notebook, the purpose of which is similar to that of a police officer. He will later be required to cover the interests of all the clients who are his 'district', and he must therefore set up in his home the necessary administrative and record-keeping facilities to enable him to do this effectively and economically.

After initial induction training, technical training is provided which is tailored to meet the individual's needs, based upon background, experience and ability to absorb the information. Quite early in this period of training the trainee will probably spend time with an experienced surveyor in the field in order to become familiar with as wide a cross section of plant and type of examination as is possible in the time available. Each type of inspection is discussed to explore the various ways in which it may be carried out, types of defect found and specific preparation or tests which may be used. The trainee makes out a report for each examination together with a dimension sheet for each class of plant.

The trainee then returns to head office and continues training under a head office engineer, who may be specially skilled in training techniques or may be the supervising or section engineer to whom the trainee will later be responsible.

The trainee will visit other departments in the head office to learn about the supporting facilities which are available such as non-destructive testing, metallurgical and failure analysis, hazard analysis and quality assurance. Emphasis is laid on safety, i.e. the integrity and safety of the plant in operation and also on the surveyor's own personal safety. It is important that the trainee understands the techniques involved, particularly the limitations of their application.

During the latter part of training, considerable periods are likely to be spent in the field operating alone but returning to head office to discuss the examinations and reports with the training engineer. During this period the trainee's ability to operate alone is assessed. At various points throughout the training, written examinations may be undertaken, and an essential feature of the training process is that the head office engineer responsible prepares a report on the trainee's progress and ultimately recommends to his manager that the trainee should be assigned to a district.

A surveyor's district is a geographical area of such a size as will keep him fully employed carrying out examinations of plant and supporting his company's clients in the district. He is required to live in or close to his district. He may report to a regional office or direct to his company's head office. Because he operates away from base, his training is recognized as being of particular importance, and it will later be supplemented with regular in-service refresher courses. It is assumed that he will be self-motivating and he is required to provide his clients with a fully professional service, mobilizing the back-up services of his company when necessary on their behalf.

34.6.3 Recognition of defects in operational plant

During this training, an engineer surveyor can expect to be shown examples of the defects which he is likely to encounter. Many inspecting authorities have a 'black museum' of such defects or they are described and illustrated by photographs in the training literature. He will also be instructed, both in the classroom and on-site, about locations where defects are likely to occur. Most surveyors develop an 'instinct' which helps them to find defects, and this can only be acquired through practical experience of different types of plant.

Having found a defect, the surveyor must be able to assess the possible consequences. Construction standards are not of much use, since they are written around what it is possible for a good workman to produce under reasonably good working conditions. For example, BS 4153 and BS 5500 allow intermittent undercut at the edge of welds up to 0.5 mm deep. The new surveyor will have been taught that fatigue cracks often originate from the toe of a weld, so he must be cautious in distinguishing between harmless undercut and an incipient fatigue crack. This is a simple example which will almost certainly have been thoroughly covered during training, but it illustrates the importance both of experience and of having the backing of NDT specialists who will settle the doubtful cases.

There is no doubt that the judgement as to whether a defect is harmless or may lead to a failure is the most difficult type of decision that any engineer, be he a engineer surveyor or not, is asked to make. His first judgement will be whether the defect was created during original manufacture. If it has developed during operation he must then assess what has caused it and whether this causative factor is still active. Unless he is certain that defect growth has stopped, he can only assume that it will continue to grow at approximately the same rate. If the surveyor has examined the plant previously and knows its history he will be better placed to estimate the growth rate and to decide whether he can allow the plant to operate until the next examination is due or until some future date which he agrees with the plant owner. Some theoretical knowledge of stresses and of the mechanisms of defect growth are valuable, but experience of similar defects is likely to be the best basis for a judgement.

An engineer surveyor employed by an insurance company will often be used to investigate claims, and each claim investigation adds something to his experience of failures. At the same time, he must observe strict confidentiality. To quote from a quality assurance manual:

> An essential part of the Engineer Surveyor's role is the detection of defects in plant or of operating practices which could lead to a dangerous occurrence. This knowledge is part of his experience.
> Nevertheless he will never disclose to another client or to a third party the sources of his experience in such a way that breach of confidentiality could result.

All surveyors need a database of technical information which should include reports of accidents or of defects which could lead to accidents derived from other sources.

He also needs a regular supply of technical documentation from his head office which keeps him abreast of technical developments. If his company operates an effective quality assurance system they will periodically check that he is keeping these data properly and they may control the indexing of them.

The reader may remember the brief definition of a competent person, i.e. 'he must know what to look for and how to recognize it when he sees it'. Perhaps what has been described in this section gives further insight as to how appropriate this definition is.

34.6.4 Reporting

An inspecting authority should always make a written report to its client when the inspection is completed or progressively as it proceeds. In the case of statutory examinations, the format of the report is prescribed in the legislation. The authority must clearly identify both itself and the individual who is responsible for the contents of the report. This will usually be the engineer surveyor who has made the inspection in his capacity as 'competent person'. Forms 55 and 55A which are prescribed for examination of steam boilers (required at intervals not exceeding 14 months) are interesting because they cover a two-part examination. These are the cold or 'thorough' examination with the boiler shut down and prepared for examination and also the supplementary or 'working' examination, when the boiler is under operating conditions. As a matter of interest, the 14-month interval starts from the date of the supplementary examination, provided this is done within two months of the cold examination, so a total of 16 months is possible.

Report forms act as checklists to ensure that the essential aspects are covered. However, these aspects will change from one type of plant to another. In general, the report must state:

Identity and address of owner/operator
Location and identity of plant
Age of plant and date of last examination
Scope of examination (itemized on the form)
Parts which could not be examined
Quantitative values where possible
Description of any defects
Statement of any limitations of use
Repairs required and a limiting period for their execution
Observations not covered by the above
Identity of examiner
Countersignature of inspecting authority if required

It is the surveyor's task to report facts, and every report should be made 'without fear or favour'. However, every report also implies an opinion, which inevitably contains a subjective element. This opinion is that the plant will be safe to operate until the next examination is due. The plant owner should remember that engineer surveyors are not infallible, nor do they possess 'second sight'. However, their implied opinion is likely to be the most reliable view that the owner can obtain without spending a great deal more money. The opinion will be additionally reliable if the surveyor has personal knowledge of the plant history. Consequently, if the plant owner thinks of changing his inspecting authority, he might remember that in doing so he will jettison accumulated knowledge which the new authority will take time to acquire.

It has to be recognized that some examiners are better than others in their use of language, and many inspection authorities have standard phrases which their surveyors are advised to use. Some use a shorthand system to facilitate the production of reports whereby the surveyor sends back a string of numbers to his head office which are then converted into standard phrases in the report. Inevitably, these reports will look stilted. It should also be remembered that when a surveyor reports regularly on the same item – perhaps at three-monthly intervals on some types of lifting equipment – a good deal of duplication from one report to the next is inevitable. A word processor in the surveyor's home, connected by electronic mail to his head office plus an electronic signature produced by a method which is acceptable to the Health and Safety Executive, is therefore the engineer surveyor's dream.

Having set out the factors which may tend to produce a rather stilted report, we should also consider the positive side. The last thing an inspecting authority wants is for his client to look at the report and say 'Why am I paying so much money for this?'.

Most inspection contracts – whether applied to insurance or not – are aimed at obtaining assurance that plant is safe to operate. If the client needs more than this he should therefore define his additional requirements with care and make sure that these have been effectively communicated. This will probably result in the inspection contract being treated as a 'one-off' basis and a more personalized service should result.

It is extremely important that the obligations on both sides should be recognized. The owner/operator of statutory plant cannot pass on his legal obligations to have his plant properly examined on time. He can, of course, make his inspecting authority contractually responsible for carrying out the examination and providing the report on time, but that is not the same thing. The client also has an obligation to prepare the plant for examination, and this usually involves shutting it down and perhaps dismantling some of its parts. The engineer surveyor has normally learnt that outage time is costly to his client, and will do his utmost to be on time on the appointed date. He will, however, have a natural aversion to remaining inside unpleasantly hot or exceedingly dirty equipment, and a less effective examination may result. It is also essential that both the client and the inspecting authority keep in mind the requirement that 'the engineer surveyor must be allowed the time he considers necessary and sufficient to complete his examination in a satisfactory manner'.

When inspecting certain statutory items of plant, the inspecting authority has a legal obligation to inform the enforcing authority of any plant which is considered to be dangerous. When this occurs, the surveyor makes out a 'site defect notice' at the time of the examination which is signed both by him and by the client's representative. The written report is then sent both to the client and to the Factory Inspectorate branch of the Health and Safety Executive. This is the only occasion when an inspectorate is obliged to break confidentiality with the client, and it should be the only occasion when it does so.

Although written reports are the inspecting authority's end product they by no means comprise the whole of the professional service which is supplied or is available on request. The authority's quality assurance objective is likely to be on the lines of 'client satisfaction allied to compliance with contractual obligations'. Although reports are almost certainly monitored, they are a poor indicator of the quality of the examination, and the authority's quality control will rely less on report inspection and more on surveillance of the whole process which it operates. This will be done through careful attention to:

Surveyor training
Surveyor aids (software and hardware)
Surveyor audits
An administration which removes obstacles which might prevent the surveyor performing at this best.

The authority should also take pains to promote a good working relationship between the engineer surveyor and the client's representative on-site.

34.7 Technical services

The majority of the independent engineering inspection authorities now provide a wide range of engineering inspection services quite apart from the in-service inspections of insured plant. These services are available to industry at large and, although widely used by presently insured clients, they are frequently employed by organizations who have no specific insurance or inspection involvement with the chosen engineering inspection authority.

The inspections/services provided under the heading of 'Technical Services' form either independent or combined operations of the various technical department disciplines that the chosen inspection authority has to offer. Inspections will range from lifts, cranes and other specialized machinery items, and include electrical inspections of wiring, switchgear, motors, etc. Special inspections of used plant or in-service plant are carried out for a wide variety of clients, and the inspection of pressure plant during construction, including design assessment prior to manufacture, is undertaken by a department specializing in this type of service.

Most inspection authorities have metallurgical and chemical laboratories and a separate non-destructive testing department within their head office complex, undertaking separate inspection/service operations as well as providing a comprehensive support to all of the other technical service functions. It is a requirement under the Health and Safety at Work, etc. Act 1974 that designers, manufacturers, importers or suppliers of articles or substances for use at work must ensure that, so far as reasonably practicable, they are safe and properly used. They must test articles for safety in use, or arrange for this to be done by a competent authority. They must also supply information about the use for which an article was designed, and include any conditions of use regarding its safety. Anyone who installs or erects any article for use at work must ensure that, so far as reasonably practicable, it does not constitute a risk to health and is safe for use.

The major engineering insurance companies are recognized as competent independent inspection authorities having a range of services which they have developed during many years of service to the engineering industry, and therefore the following inspection procedures indicate many of the standard inspection services in use. However, special inspection procedures are frequently drawn together to meet particular circumstances, and the inspection authorities offering these services are always ready to discuss any special inspection requirements.

34.7.1 Pressure plant

Pre-commissioning inspection services are designed to examine key stages of production and witness final tests on completion at the manufacturer's works and/or at site as appropriate. It is frequently arranged for the inspection authorities personnel to attend pre-design meetings to discuss the best and most effective inspection service suitable for the item or projecting being considered. The main starting point, however, is normally the design drawing or drawings, together with calculations, which are reviewed to check the integrity of the design in accordance with the specified construction code, which is often to a British Standard.

Once the drawing is approved, the field staff make inspections at key stages during manufacture and follow up on-site when necessary to inspect during prior to commissioning work and to witness tests. The normal stages of inspection for pressure plant would include identification of materials of construction, weld procedures, operator performance qualifications being checked to establish that they are relevant and up-to-date and random stage inspection during welding continues, at intervals, during the fabrication. These latter tests will include heat treatment recording and non-destructive results being subject to assessment.

On completion of the item it is examined externally while subject to hydraulic test, followed by internal examination so far as construction permits. All the inspection stages are reported at regular intervals during construction, thus providing a prompt written account of inspections during progress of manufacture. A similar design assessment and stage inspection procedure is available for reinforced plastic vessels and tanks.

34.7.2 Electric passenger and goods lifts

Pre-commissioning inspection service for lifts is normally undertaken in one of three ways:

1. At works, at site and during test
2. At site and during test
3. During test only.

It is recommended that the lift specification should call

for compliance with British Standards, and the service usually commences with perusal of drawings and specifications. The inspection service can include machine parts prior to assembly and a similar examination of electrical components. Guide rails, brackets and supporting structures are examined and the lift manufacturer's shop tests on motor-control equipment and high-voltage tests of electrical equipment are witnessed. A car sling, safety gear and counterweight assembly are checked prior to dispatch and shop tests on the governor unit are witnessed. The on-site inspection would include supporting steelwork, the alignment of the car and counterweight guides and the door or gate mountings. The mechanical and electrical parts of the lift would be examined on completion of erection. The manufacturers' tests would be witnessed by the engineer surveyor, including balance tests, performance tests, round trip with 10% overload, safety gear test, limit switch tests, static testing of brake, gate/door lock as well as earth continuity tests and insulation tests. All these inspections are reported with the reports dispatched at intervals during manufacture and a final report is submitted on completion.

34.7.3 Cranes

The inspection of cranes has the three major categories (1)–(3) as applicable to the lift inspection outlined above. The service would commence with perusal of drawings and specifications which should call for compliance with British Standards. Inspection during manufacture would include mechanical parts prior to assembly and selected structural members as well as the examination of motors and electrical equipment and the checking of test certificates for electrical equipment. On completion, the crane would be examined by the engineer surveyor, who would visit the site during erection. The site-erected crane would be examined and proof-load and performance tests would be witnessed. Test certificates for ropes and hooks would be perused and all the inspection stages would be reported progressively. Although the description applies to electric overhead travelling cranes and dockside portal cranes, a similar service is available for power-driven mobile cranes, manual overhead travelling cranes, box trolley runway tracks and lifting tackle.

34.7.4 Electrical inspection

Inspection services relevant to electrical plant during production includes inspection at several stages of manufacture and witnessing of tests on completion. Depending on the item concerned, examinations are carried out of mechanical details prior to assembly and of insulating materials, conductors, coils, supports, wedges and windings. The witnessing of tests of complete component parts would include commutators, slip rings and brush gear. This is followed by inspections during the winding of motors and transformers and, on completion, witnessing of manufacturer's tests to determine the performance and reliability of the items under normal working conditions. Examination and witnessing manufacturer's tests of auxiliary equipment, including control gear, is also undertaken.

34.7.5 Turbogenerators

The schedule of inspection for turbogenerators covers key stages during manufacture and normally includes witnessing of mechanical tests on sections from the motor forgings and on completion of machining of mechanical components. Insulating materials, conductors, controls, supports, wedges, slip rings, commutators and windings are examined at appropriate stages during progress of the work on the stator, rotor and exciter. The tests carried out by the manufacturer during progress and on completion are all witnessed at the appropriate stages.

34.7.6 Centrifugal pumps

The services provided could include the witnessing of mechanical tests on specimens representing the main castings and forgings and inspection of the pump casing, bed plates and other principal components after machining. This would include the pump shaft, impeller, guide-vanes and division plates. The engineer surveyor would witness hydraulic tests on pressure-retaining parts and also the test carried out by the manufacturer under operating conditions.

34.7.7 Non-destructive testing

The principal engineering insurers all now operate NDT services with fully experienced, qualified non-destructive testing engineers available in all the main centres of industry throughout the UK. These engineers are able to undertake most forms of non-destructive testing at short notice, being equipped with modern ultrasonic, radiographic (X-ray and gamma-ray) testing equipment, digital sound-velocity measurement instruments, electromagnetic, eddy-current and spark-testing instruments as well as a range of equipment for all forms of magnetic particle and dye penetrant testing. In addition, many types and forms of equipment are used as aids to visual inspection, including closed-circuit TV using miniaturized cameras with a facility to record results.

34.7.8 Metallurgical testing

The prime function of the metallurgical laboratories is to investigate failures of all descriptions and prepare illustrated technical reports based on their findings. From these, the causes of any incident can be assessed and recommendations made to minimize the possibilityof a recurrence. Equipment will include several light microscopes, both laboratory based and portable, for site work and replication testing on-site. A scanning electron microscope with magnification up to 200 000 and a feature enabling X-ray micro-analysis for any particular element to be made should also feature in the equipment available to the engineering insurers providing these services.

34.7.9 Laboratory chemical section

This is primarily engaged in analysis of boiler water treatment matters and involves on-site studies of various problems and the chemical examination of corrosion products, boiler scales, etc. It can also carry out certain types of metallurgical, fuel and inorganic analysis. Normal wet methods of analysis coupled with a visible ultraviolet and atomic absorption spectrophotometer are used for a wide range of analytical applications. Equipment in use by the engineering insurers providing these services can include an ion chromatograph, spectrometer equipment, atomic absorption spectrophotometer, flue gas analysis equipment and testing equipment for transformer and switchgear oil.

34.7.10 Other services

Other inspection services available include the examination of steel structures (new and existing), electrical wiring installations, containers (to meet Statutory Instrument No. 1890), dangerous substances (carriage by road in road tankers or tank containers) to meet Statutory Instrument No. 1059, examination of second-hand plant prior to purchase, plant undergoing repair or modification, the Control of Industrial Major Accident Hazard Regulations (CIMAH) Statutory Instrument No. 1902 and Control of Substances Hazardous to Health (COSHH) and Pressure Systems Regulations.

34.8 Claims

The vigilant plant user is well informed about the terms of the policies which cover the plant in his charge, including any excesses which may apply. As soon as an incident occurs which may give rise to a claim on one of the policies in force he will ensure that the insurer concerned is notified immediately, as failure to do so many invalidate a claim.

If the incident constitutes a 'reportable accident' as defined in legislation the plant engineer will also ensure that the incident is immediately reported to the Factory Inspectorate. The Inspectorate will decide whether or not they wish to carry out an enquiry and, particularly if there has been loss of life, the accident site may be compulsorily isolated. Unless this is so, investigation by the insurer or a loss adjuster acting on his behalf may proceed.

Smaller claims may be dealt with by completion of a claim form, perhaps a report from the repairer if appropriate and sight of the repair or replacement invoice. For larger claims a visit to the site by the insurer's representative may be necessary. If the plant is inspected by the insurer the engineer who normally carried out the inspections will usually be asked to report on the incident. The insurer's representative will seek to establish the facts, the cause, nature and extent of the physical damage and perhaps the consequences in terms of interruption to production. An engineer surveyor is not usually concerned with the commercial outcome and may be expressly excluded from any commercial negotiations. A loss adjuster, as well as seeking to establish the facts, will also negotiate settlement of the claim once it has been established that a valid claim exists.

Both the engineer surveyor and the loss adjuster may advise on how best to effect a repair or replacement of the damaged item and will bear in mind the client's need to minimize their loss of production. If, for any reason, there is a delay in having damaged plant examined by the insurer's representative the insurer will normally be agreeable to the insured giving his own instructions for repairs to proceed, provided that any damaged parts replaced are kept for examination by the insurer with suitable protection against further damage. The agreement of the insurer to the insured giving his own instructions for repairs to proceed is not confirmation that the incident constitutes a valid claim. This will only be decided when the insurer's investigations are complete.

In any case, the plant engineer has a responsibility to keep careful account of all the costs incurred, which may be recoverable in whole or part under the policy, and submit appropriate invoices to the insurer. He also has an obligation to minimize these costs insofar as this is reasonably compatible with achieving his objective, which will be to restore normal production. The insurer may have access to sources of replacement plant items or to specialist repairers of whom the plant engineer is not aware, and advice on these matters is part of the service provided by the insurer.

When the cause of the incident has been established and the costs of rectification finalized, these will be compared with the insurance cover provided by the policy and the extent of the insurer's liability, if any, determined. The policy will normally be one of indemnity, i.e. returning the insured to the same position after an accident as he was before. This may be achieved by repairing or replacing what is damaged or by paying the amount of the damage. It may be necessary to carry out modifications to prevent a recurrence of the accident or desirable to uprate the specification for better performance or the life of the machine may have been extended by the repairs carried out. In this case a degree of 'betterment' is involved which will be reflected in the settlement by a contribution by the insured to the cost of repairs.

On occasions, the amount of a claim will be found to exceed the sum insured, and as the sum insured is the maximum amount the insurer will pay, the insured will not be fully compensated for his loss. It is the responsibility of the insured, with the assistance of his broker or other intermediary, to ensure that sums insured are adequate and are not eroded by inflation. The plant engineer also has a responsibility to inform his broker if he increases the value of plant items, adds substantially to the total value of items at any one location or, where items are insured individually, takes additional items into service or removes them from service.

34.9 Sources of information

The plant engineer may progressively become a mine of information in his field, but it is more important that he knows where to find the information when he needs it and, equally important, that this information should be up to date.

34.9.1 British Standards

British Standards tend to be concerned with new construction rather than with plant already in service. In general, although there are exceptions, they are also concerned more about minimum standards than with standards of excellence. There are notable exceptions: for instance, BS 5500 is arguably the best guide to design and construction of pressure vessels which has ever been produced in the English language. In other technical fields the standards of other nations take pride of place. For instance, the Americans tend to rate well in the oil and petrochemical sector and ANSI B.31.3 is particularly useful for pipework.

The plant engineer should strive to make sure that, whenever possible, his purchase orders specify compliance with the correct standard. Many manufacturers can truthfully claim that their specifications exceed the minimum requirements of a material standard, in which case they will have no objection to its inclusion in the order. Those who protest too much that its requirements are superfluous may deserve further scrutiny. Choosing the correct specification is not always easy, and requires expertise in the relevant field. This expertise is available from consulting engineers or from an inspection authority, who will also appraise the vendor's designs and, if required, will witness critical stages of construction and test.

The *BSI Catalogue* (previously known as the *BSI Yearbook*) is not an expensive investment for a company which spends appreciable sums on purchased plant and materials.

34.9.2 The Health and Safety Executive

The Factory Inspectorate branch of the HSE issue a series of booklets with the prefix HS/E. For example, HS/G 34, entitled *The storage of LPE at fixed installations*, may be considered essential reading for plant engineers who use LPG in more than small quantities. The HS(R) Series is another collection of booklets on safety, ranging from HS(R) 1, *Packaging and labelling of dangerous substances*, to HS(R) 19, *A guide to the Asbestos (Licensing) Regulations 1983*. The booklets in this series are guides which often contain explanations of and quite extensive extracts from legislation. They are highly recommended reading.

The TDN series are Technical Data Notes, which cover a particularly wide variety of subjects: for instance, TDN 53/3 is entitled *Creep of metals at elevated temperatures*. These notes are being progressively replaced by the Plant and Machinery (PM) series, of which PM5, *Automatically controlled boilers*, is perhaps the best known.

All the booklets issued by the Health and Safety Executive are available from HMSO, and as the series is continually expanding the reader is advised to obtain an up-to-date list from the nearest Stationery Office.

34.9.3 Publications by other organizations

Other organizations which represent groups of companies with a common interest also issues rates, regulations and guidance notes on subjects which come within their respective orbits. Occasionally, the HSE will advise users of a particular type of plant, commodity or material to observe the appropriate document(s), and the user will then ignore this advice at his peril. Apart from this aspect, they do not have any official force. At the same time, most of them contain excellent information and guidance. A non-exhaustive list of the organizations in alphabetical order, together with the address last known to the author, is given below:

The Associated Offices Technical Committee (AOTC). St Mary's Parsonage, Manchester M60 9AP
The Chemical Industries Association

Appendix 1: Glossary

AHEM	The Association of Hydraulic Equipment Manufacturers Limited, 192–198 Vauxhall Bridge Road, London SW1V 1DX
API	American Petroleum Institute, 1271 Avenue of the Americas, New York, USA
ASB	Association of Shell Boiler Makers, c/o David L. Chaplin, The Meadows, Ryleys Lane, Alderley Edge, Cheshire SK9 7UV
ANSI	American National Standards Institute, 1430 Broadway, New York, NY 10018, USA
ASME	The American Society of Mechanical Engineers, 345 East 47th Street, New York, NY 10017, USA
BASEEFA	British Approvals Service for Electrical Equipment in Flammable Atmospheres, Harper Hill, Buxton, Derbyshire
BCAS	British Compressed Air Society, Leicester House, 8 Leicester Street, London WC2H 7BN
BMEC	British Mechanical Engineering Confederation, 112 Jermyn Street, London SW1Y 4UR
BSC	British Safety Council, 62–64 Chancellors Road, London W6 9RS
BSI	British Standards Institution, Linford Wood, Milton Keynes, MK14 6LE
BCGA	Federation of British Electro Technical & Allied Manufacturers Association, Leicester Street, London WC2H 7BN
BI/NDT	The British Institute of Non-Destructive Testing, 53–55 London Road, Southend-on-Sea, Essex SS1 1PF
BRE	Building Research Establishment, Building Research Station, Garston, Watford WD2 7JR
CBMPE	Council of British Manufacturers of Petroleum Equipment, 118 Southwark Street, London SE1 0SU

CIA	Chemical Industries Association, Alembic House, 93 Albert Embankment, London SE1 7TU
CIIA	Council of Independent Inspecting Authorities, c/o Parklands, 825a Wilmslow Road, Didsbury, Manchester M20 8RE
CWSIP	Certification Scheme for Weldment Inspection Personnel, Abington Hall, Abington, Cambridge CB1 6AL
SCI	Society of Chemical Industry, 14–15 Belgrave Square, London SW1X 8PS
DTI	Department of Trade and Industry, Room 320, Kingsgate House, 66/72 Victoria Street, London SW1E 6SW
DOT	Department of Trade, 1 Victoria Street, London SW1H 0ET (International Trade, Policy Division Room 450 or 455)
DOE	Department of Energy, Thames House South, Millbank, London SW1P 4QJ
EEMUA	Engineering Equipment & Materials Users Association, 14/15 Belgrave Square, London SW1X 8PS
EFTA	The European Free Trade Association, 9–11 Rue de Varembe, CH-1211 Geneva 20, Switzerland
FPA	Fire Protection Association, 140 Aldersgate Street, London EC1A 4HX
HSE	Health and Safety Executive, Hugh's House, Trinity Road, Bootle, Merseyside L20 3QY
ICE	The Institution of Chemical Engineers, George E. Davies Buildings, 165, 171 Railway Terrace, Rugby CU21 3HQ
ICE	The Institution of Civil Engineers, Great George Street, London SW1
IEE	The Institution of Electrical Engineers, Savoy Place, London WC2R 0BL
IMechE	Institution of Mechanical Engineers, 1 Birdcage Walk, Westminster, London SW1H 9JJ
IP	Institute of Petroleum, 61 New Cavendish Street, London W1H 8AP
LPGITA	Liquified Petroleum Gas Industry Technical Association (UK), 17 Grosvenor Crescent, London SW1X 7ES
NRPB	National Radiological Protection Board, Chilton, Dicot, Oxfordshore OX11 0RQ
PPA	Process Plant Association, Leicester House, 8 Leicester Street, London WC2H 7BN
RoSPA	Royal Society for the Prevention of Accidents, Cannon House, The Priory, Queensway, Birmingham B4 6BS
THE	Technical Help to Exporters, BS1, Linford Wood, Milton Keynes MK14 6LE

Appendix 2: Statutory report forms

Statutory form no.	*Title of regulation*	*Title of form*
F54	Factory Act 1961 and the Offices, Shops and Railway Premises (Hoists & Lifts) Regulations 1968	Report of Examination of Hoist or Lift
F55	Factory Act 1961 and the Examination of Steam Boiler Regulations 1964	Report of Examination of Steam Boilers other than Economisers, Superheaters, Steam Tube Ovens and Steam Tube Hotplates
F55A	As F55 above	Report of Examination of Steam Boiler under normal Steam Pressure
A2197	Factory Act 1961 and the Power Presses	Report of thorough Examinations and Test and Record of Repairs
F260 and F262 (also available in book form)	Mines and Quarries Act 1954 and The Quarries (Electricity) Regulations 1956	Tests of Insulation Resistance and Conductivity of Earthing Conductors and Earth Plates of Electrical Apparatus
	Conveyance by Road in Road Tankers and Tank Containers 1981 (Statutory Instrument No. 1059)	Report of Periodic Examination and Test Certificate of Suitability of a carrying tank/tank container and its fittings for the conveyance of dangerous substances by road

In any correspondence relating to this report please quote

also identification number

THE OFFICES, SHOPS AND RAILWAY PREMISES (HOISTS AND LIFTS)
Regulations 1968 Regulation 6
FACTORIES ACT 1961 Sections 22, 23 and 25
FORM PRESCRIBED FOR THE
REPORT OF EXAMINATION OF HOIST OR LIFT

F54

Occupier (or Owner) of premises

Address

1 (a) *Type of hoist or lift and identification number and description*
(b) *Date of construction or re-construction (if ascertainable)*

2 *Design and Construction*
Are all parts of the hoist or lift of good mechanical construction, sound material and adequate strength (so far as ascertainable)?
Note: Details of any renewals or alterations required should be given in (5) and (6) below)

3 *Maintenance*
Are the following parts of the hoist or lift properly maintained and in good working order? If not, state what defects have been found
(a) *Enclosure of hoistway or liftway*
(b) *Landing gates and cage gate(s)*
(c) *Interlocks on the landing gates and cage gate(s)*
(d) *Other gate fastenings*
(e) *Cage or platform and fittings, cage guides, buffers, interior of the hoistway or liftway*
(f) *Over-running devices*
(g) *Suspension ropes or chains, and their attachments*
(h) *Safety gear, i.e. arrangements for preventing fall of platform or cage*
(j) *Brakes*
(k) *Worm or spur gearing*
(l) *Other electrical equipment*
(m) *Other parts*

4 *Whart parts (if any) were inaccessible*

5 *Repairs, renewals or alterations required to enable the hoist or lift to continue to be used with safety-*
(a) *immediately*
(b) *within a specified time, the said time to be stated*
If no such repairs, renewals or alterations are required, enter "NONE"

6 *Defects (other than those specified at 5 above) which require attention*

7 *Maximum safe working load subject to repairs, renewals or alterations (if any) specified at 5*

8 *Other observations*

I certify that on I thoroughly examined this hoist or lift and that the foregoing is a correct report of the result.

Signature Date

Qualification – Engineering Surveyor to Insurance Company

L54AJ (12/88) TO BE ATTACHED TO THE GENERAL REGISTER

In any correspondence relating to this report please quote

HEALTH & SAFETY EXECUTIVE
FACTORIES ACT 1961, Sections 32–34 and the Examination of Steam Boilers Regulations 1964
FORM PRESCRIBED FOR

REPORT OF EXAMINATION WHEN COLD OF STEAM BOILERS OTHER THAN ECONOMISERS, SUPERHEATERS, STEAM TUBE OVENS, AND STEAM TUBE HOTPLATES†

F55

1 Name of Occupier

2 Address of
(a) Factory
(b) Head Office of Occupier

NOTE: Adress (b) is required only in the case of a boiler used on a temporary location, e.g., on a building operation, work of engineering construction.

3 Description and distinctive number of boiler and type

4 If the boiler is one of those described in regulation 4(2), this should be stated and the appropriate sub-paragraph (a), (b) or (c)) should be given.

5 Date of construction
The history should be briefly given, and the examiner should state whether he has seen the last previous report.

6 Date of last hydraulic test (if any), and pressure applied.

7 Quality and source of feed water

8 Is the boiler in the open or otherwise exposed to the weather or to damp?

9 Boiler:
(a) What parts of seams, drums or headers are covered by brickwork?
(b) Date of last exposure of such parts for the purpose of examination
(c) What parts (if any) other than parts covered by brickwork and mentioned above were inaccessible?
(d) What examination and tests were made? (see Note 2 overleaf). If there was any removal of brickwork, particulars should be given here
(e) Condition of boiler — *External:* / *Internal:*
(State any defects materially affecting the maximum permissible working pressure)

10 Fittings and attachments *(a) Are there proper fittings and attachments?* *(b) Are all fittings and attachments in satisfactory condition (so far as ascertainable when not under pressure)?*		Subject to further report after examination under normal steam pressure

11 Repairs (if any) required, and period within which they should be executed, and any other conditions which the person making the examination thinks it necessary to specify for securing safe working

12 Maximum permissible working pressure calculated from dimensions and from the thickness and other data ascertained by the present examinations due allowance being made for conditions of working if unusual or exceptionally severe.

13 Where repairs affecting the working pressures are required, state the maximum permissible working pressure. *(a) Before the expiration of the period specified in (11)* *(b) After the expiration of such period if the required repairs have not been completed* *(c) After the completion of the required repairs*	*(a)* *(b)* *(c)*

14 Other observations	

Subject to the reservation (noted above) of certain points for examination under steam pressure,

I certify that on the boiler above described was sufficiently scaled, prepared, and (so far as its construction permits) made accessible for thorough examination and for such tests as were necessary for thorough examination and that on the said date I thoroughly examined this boiler including its fittings and attachments and that the above is a true report of the result.

Signature — Counter-Signature

Date — Chief Boiler Engineer

Qualification – Engineer Surveyor to Insurance Company
Date

- Delete if not required. †See overleaf.

B55J (8/88)

TO BE INSERTED IN THE GENERAL REGISTER

In any correspondence relating to this report please quote

also distinctive number

HEALTH AND SAFETY EXECUTIVE
FACTORIES ACT 1961, Sections 32–34 and the
Examination of Steam Boilers Regulations 1964
PRESCRIBED FORM FOR

REPORT OF EXAMINATION OF STEAM BOILER UNDER NORMAL STEAM PRESSURE

F.55A This form may also be used (as far as possible) for supplementary reports on Economisers and Superheaters

Name of Occupier
Address of
(a) Factory
(b) Head Office of Occupier

NOTE. – Adress (b) is required only in the case of a boiler used in a temporary location.

3 Description and distinctive number of boiler and type

The next thorough examination to be completed on or before:

4 If the boiler is one of those described in regulation 4(2), this should be stated and the appropriate sub-paragraph (a), (b) or (c) should be given

5 Conditions (External)

6 Fittings and Attachments
(a) (i) Is the safety valve so adjusted as to prevent the boiler being worked at a pressure greater than the maximum permissible working pressure specified in the last report (F.55) on examination when cold?
(ii) (If a lever safety valve). Is the weight secured on the lever in the correct position?
(b) Is the pressure gauge working correctly?
(c) Is the water gauge in proper working order?

7 Repairs (if any) required, and period within which they should be executed and any other conditions which the person making the examination thinks it necessary to specify for securing safe working

8 Other observations

I certify that on I examined the above-mentioned boiler under normal steam pressure and that the above is a true report of the result.

Signature

Date

Counter-Signature

Chief Boiler Engineer

Date

Qualification – Engineer Surveyor to Insurance Company

B55SJ (2/89)

TO BE INSERTED IN THE GENERAL REGISTER

In any correspondence relating to this report please quote

also identification number

HEALTH AND SAFETY AT WORK etc. ACT 1974
FACTORIES ACT 1961
The Power Presses Regulations
1965 (SI 1965 No. 1441) and 1972 (SI 1972 No. 1512)
Power Presses and Safety Devices Thereon

REPORT OF THOROUGH EXAMINATION AND TEST AND RECORD OF REPAIRS

(FORM APPROVED BY H.M. CHIEF INSPECTOR OF FACTORIES UNDER REGULATIONS 5 AND 6)

F2197

See Notes and space for continuation of entries overleaf.

Name of Occupier

Address of Factory

1 Make, type and date of manufacture (if known) — *(a) Power press*; *(b) Safety device(a)*[1]

2 Identification mark or number — *(a) Power press* — *Maker's* — *Occupier's*; *(b) Safety device(s)* — *Maker's* — *Occupier's*

3 Are the following parts of the press and safety device(s) in good working order?[2] *If note, state what defects have been found*

(a) Power Press
- *(i) Clutch mechanism*
- *(ii) Clutch-operating controls*
- *(iii) Brake*
- *(iv) Flywheel bearing(s)*
- *(v) Other parts affecting safety at the tools*

(b) Safety device(s)
- *(i) Interlocking guard*
- *(ii) Automatic guard*
- *(iii) Fixed fencing (inc. that associated with (i), (ii) & (iv)).*
- *(iv) Other type of safety device (e.g. photo-electric)*

4 What parts (if any) were inaccessible?

5 Repairs, renewals or alterations to the power press and safety device(s) to remedy defects which are or may become a cause of danger to employed persons (Regulation 6(1)) (see note 2)

(a) before press is used again
(b) within a specified time, the said time to be stated
If no such repair, renewals or alterations are required enter "NONE".

6 Defects (other than those specified at 5 above) which require attention.

7 Repairs, renewals or alterations required in item 5 of this report, completed at the time of the thorough examination and test.

I hereby certify that on I thoroughly examined and tested the power press and the safety device(s) thereon specified above and the result of my examination and test are as shown.

Signature

Date

Qualification – Engineer Surveyor to Insurance Company

If any defects require action as noted at 5 above please give date of notification of such defects (under Regulation 6(1)) to the occupier..

L30J (10/88)

(Continued overleaf)

Quarries and Miscellaneous Mines (M. & Q. Forms 260 and 262). **Electrical Installation Tests.** (See also separate report of same date).	Sheet number Affix this sheet to page of record book

Number and address of quarry or mine	

Tested by the undersigned Engineer Surveyor	*Type of earth plate*	*Location*	*Resistance to earth (ohms)*	*Date of test*

Circuit	*Insulation Resistance (Megohms)*	*Resistance in Ohms*		*Percentage Conductivity L/E × 100*
		Line Conductor	*Earth or Metallic Covering*	

Signature Date

Qualification – Engineer Surveyor to Insurance Company

Countersigned Manager Owner Date

3J (10/85)

In any correspondence relating to this report please quote

Report of Periodic Examination and Test Certificate of Suitability of a carrying tank/tank container* and its fittings used for the conveyance of dangerous substances by road. (SI 1059)

*Delete as appropriate

1 Name of Operator

2 Address of Operator

3 Description and distinctive number of tank	*4 Max. permissible working pressure*
5 Give: (a) interval between periodic exam. as specified in written scheme. *(b) Was this period shortened at the last exam?*	*(c) If so, give shortened period*

6 Nature of Examination
Condition:
(a) External

(b) Internal

(c) Parts inaccessible

7 Brief description and results of any tests

8 State: (a) Repairs/modifications which should be carried out*

(b) Whether tank may continue to be used for conveyance of dangerous substances listed in initial Certificate

9 Further exam. (a) If interval between exams. is to be shortened state new intervals	*(b) State latest date before next examination must be carried out.*

10 Other observations:

I examined (and tested*) tank and fittings No. on in accordance with the requirements of the written scheme relating to the periodic examination and test of the tank as required in Regulation 7(2)(a) of the Dangerous Substances (conveyance by Road in Road Tankers and Tank Containers) Regulations 1981.

*The tank and its fittings must be repaired and/or modified in accordance with Paragraph 8 above and then re-examined and tested before being used for the conveyance by road of dangerous substances.

*I am satisfied that the tank and its fittings are suitable for the purposes and under the conditions specified in the current Certificate of Initial Examination and Test dated:

Signature Date

Qualification – Engineer Surveyor to Insurance Company

B99J (11/86)

Appendix 3: Report forms – non-statutory

Report form titles

1. Report of Examination and Test of Electrical Installation
2. Inspection Certificate – Based on requirements to comply with the IEE Regulations for the Electrical Equipment of Buildings
3. Hazardous Substance Pressure Vessel Inspection Report

In any correspondence relating to this report please quote

REPORT OF EXAMINATION AND TESTS OF ELECTRICAL INSTALLATION

Name

Address

Description of installation and nature of supply

DETAILS OF TESTS

Earth plate resistance to general mass of earth (ohms)

Circuit number	*Description of circuit*	*Fuse rating or circuit breaker setting (amps)*	*Insulation resistance (megohms-min.)*		*Earth continuity conductor resistance (ohms-max.)*	*Earth fault loop impedance (ohms-max.)*
			To earth	*Between conductors*		

GENERAL REMARKS. Including defects requiring attention and alterations since last examination or any departure from the relevant regulations.

Continued on sheet number

Date of Examination

Signature

Date

Qualification – Engineer Surveyor to Insurance Company

D 5J (8/86)

In any correspondence relating to this report please quote

INSPECTION CERTIFICATE
Based on requirements to comply with I.E.E. WIRING REGULATIONS

The accessible parts of the Electrical Installation at

Name

Address

Have been visually inspected and tested in accordance with requirements of The I.E.E. Wiring Regulations and that the results are as indicated below.

I RECOMMEND that (due to age and condition) this installation be further tested after an interval of not more than .... years

ITEMS INSPECTED OR TESTED

METHOD OF EARTHING

Cable sheath. Additional overhead line conductor Protective multiple earthing (P.M.E.) Buried Strip/rod/plate Earth-leakage circuit breaker, Voltage-operated Earth-leakage circuit breaker, Current-operated

State which:

	TESTS	RESULTS
a)	*Resistance of each earth Continuity Conductor*	
b)	*The total earth loop impedance for ready operation of the largest rated excess current protective device relied upon for earth leakage protection.*	
c)	*Earth leakage protection Current Operated/Voltage Operated*	
d)	*Polarity throughout installation*	
e)	*All Single Pole Control Devices*	
f)	*Insulation Resistance of Fixed Wiring Installation. (minimum required 1 Megohm)*	
g)	*Insulation Resistance to Earth of each separate item. (Minimum Requirement 0.5 Megohm)*	

Each item of apparatus tested, all flexible cords, switches, fuses, plugs and socket outlets are in good serviceable condition except as stated. There is no sign of overloading of conductors or accessories except as stated. Apparatus tested includes/does not include Portable Appliances.

COMMENTS

Continued on Sheet number

Date of Examination

Signature

Date

Qualification – Engineer Surveyor to Insurance Company

D13J (4/90)

HPV1

Associated Offices Technical Committee

British Engine Insurance Ltd.
Eagle Star Insurance Company Ltd.
National Vulcan Engineering Insurance Group Ltd
Plant Safety Ltd.

HAZARDOUS SUBSTANCE PRESSURE VESSEL INSPECTION REPORT

1. Owner's name and address

1a. Plant address or location

2. Description (including substance contained)

3. Maker, date of construction, code of construction and serial no.

4. Date of last hydraulic test and pressure applied.

5. Parts inaccessible

6. Nature of examination.

7. External condition

8. Internal condition

9. Condition of fittings and safety valve setting

10. Maximum internal pressure and temperature

11. Minimum internal pressure and temperature

12. Maximum permissible vacuum

13. Filling ratio from BS 5355 kg/litre of Capacity

14. The stresses due to the supporting arrangement have been checked under the following conditions and are considered satisfactory. This assesment does not include foundation:–

15. Other observations:–

This assessment does not include foundations.

Date of examination:

....
Inspecting Authority

....
Engineer Surveyor

Date of next thorough examination on or before:

....
for Member Company

35 Insurance: Buildings and Risks

Risk Control Unit
Royal Insurance (UK) Ltd
Liverpool

Contents

35.1 Insurance

It is very important to understand that an insurance policy is a legal document, and insurance negotiations should be approached on this basis. Insurance contracts have a detailed legal background of both case law and common law in the interpretation of their provisions. Professional insurance advice from a reputable source, whether an insurance company or an insurance broker, is essential if problems are to be avoided.

It is important to understand that insurers are separate commercial organizations in competition and each trying to make a profit. This affects the insurers' approach to loss prevention and it does mean that some insurers will have different priorities in individual cases.

Insurers trade in a competitive market – there are some 200 separate insurers in the market in the UK. Because insurance is such a competitive business, insurers do not have unlimited freedom to act. They occasionally have to accept less than the optimum in terms of safety. However, an individual insurer can always refuse the business if it is judged that the conditions are bad – and this is done from time to time.

There are occasions where somewhat less serious problems are encountered and the client refuses to take any action. In these circumstances statutory bodies like the Health and Safety Executive or the fire brigade can enforce their views through legislation. Insurers are not in that position. They can and do increase premiums for bad risks, but they also try to persuade managements by commonsense argument that changes should be made, even if this is not necessarily an economic proposition from the point of view of insurance cost.

Behind the insurance companies is the influence of the Association of British Insurers and the Loss Prevention Council, which includes the Fire Protection Association. Both these trade bodies carry out a range of functions, but the most relevant in this context is the literature that they produce. This is for guidance of insurers and their clients in combating fire problems and encouraging good standards of general fire protection. The various codes and standards are issued using the experience of insurers combined with industry, fire-protection equipment manufacturers, users and other interested parties with a contribution to make. A comprehensive list of these publications appears in Section 35.11.

35.2 Fire insurance

Fire is the major insurable risk to property on land. Each industry, trade or manufacturing process has its own fire problems. There are, in addition, causes of fire which are found in all occupancies (e.g. arson, smoking, misuse of electricity). With such a wide field it is only possible to touch lightly on the many important aspects of the subject.

The normal fire policy covers the fixed assets (buildings, machinery and stock) of the business against loss or damage by fire, lightning and explosion. Assets consumed by the fire damage caused as a direct result of fire (e.g. by smoke or water used to extinguish the fire) are included in the cover. Damage caused by equipment overheating or by something which is being fired but is burnt incorrectly is not covered by this type of policy. Lightning damage is self-explanatory.

Explosion damage cover means damage resulting from a wide variety of trade-related explosions (e.g. dust, vapour or uncontrolled chemical reactions). The explosion or collapse of boilers and pressure plant, in which internal pressure is due to steam or other fluids, is covered by engineering (not fire) policies.

It is the responsibility of the owner of the property to fix the figure for which the assets are insured. This must be a realistic assessment of the value, as insurers settle claims on the actual loss incurred and do not automatically pay the figure insured in the policy.

The policy cover may be extended to include damage to assets from extra perils if the necessary additional premium is paid. These include storm damage, floodwater, burst water pipes or tanks, aircraft, riot, malicious damage or impact by mechanical vehicles. It is also possible to include an item to cover architects', surveyors' and consultants' fees and legal fees all incurred in the reinstatement of the property insured, as well as a sum to cover the costs of removing debris from the site before rebuilding can start.

35.3 Business interruption insurance

Business interruption policies (also known as loss of profits or consequential loss) are designed to cover the trading loss due to the occurrence of the fire or other insured peril. This loss is normally identified either by a reduction in turnover as a result of the disruption caused to the business or by increased costs incurred to minimize the loss of turnover, or indeed a combination of the two. The cover under the policy does not last for an indefinite period after the loss but is restricted to a timescale expressed in the policy as the maximum indemnity period. This time limit is chosen by the management of the business, and is the time they think that they would need to recover the trading position of the company following the incident. With fire insurance there is a limit to the amount payable, which is the sum insured chosen by the management of the business.

35.4 Insurance surveys

It is normal practice in insurance for surveyors employed by the insurance company or the insurance broker to inspect premises that are to be insured and prepare reports for the underwriters. A major part of the survey report is an assessment of the quality of the fire protection relative to the level of hazard in the premises. Obviously, the business being carried on has a considerable influence on the risk of fire or explosion.

35.5 Fire legislation

It is not the intention in this chapter to cover fire-protection legislation. However, legislation plays such an important part in trying to ensure that adequate fire precautions are established to protect people that a brief mention is appropriate.

The basic legislation on fire precautions is covered by the Fire Precautions Act 1971. The intention of this Act is to require a fire certificate to be issued by the local fire authority for certain categories of buildings according to occupation, i.e. hotels, boarding houses, offices, shops, railway premises and factories.

The 1971 Act controls the following major areas:

1. Means of escape;
2. Securing the safe use of the means of escape, i.e. fire-resisting doors, protected corridors and stairways;
3. Fire-warning systems;
4. Firefighting equipment.

The Act was restrictive in its scope but this was changed by the Health and Safety at Work, etc. Act 1974, which enabled the Secretary of State to designate 'use as a place of work' as an occupancy. There is therefore a general duty of employers under the Health and Safety at Work Act, etc. 1974 to ensure the health and safety of their employees and those not in their employment, including members of the general public 'so far as is reasonably practicable'. This does, of course, include fire hazards.

The Home Office reviewed the operation of the Fire Precautions Act in 1985 because of the widespread criticism of its inflexibility in that it applies equally to all designated premises, whatever the actual degree of risk and expense of implementation. The Fire Safety and Safety of Places of Sport Act 1987 is the result. Basically, the 1971 Act still stands but with some important changes.

Fire authorities may exempt 'low-risk premises' from requiring a fire certificate. These premises are required to have adequate means of escape and suitable means for fighting a fire. If these conditions prove not to be met the occupier or owner is guilty of an offence.

Codes of practice are being issued and guidance will be given on the definition of 'low risk'. Occupiers of such 'low-risk' premises are required to notify the authority of changes which may turn the premises into 'normal risk' or 'high risk'.

Another change is the introduction of Improvement Notices and Prohibition Notices. When it considers that the duty to provide means of escape or means for firefighting has been breached the fire authority can issue an Improvement Notice specifying the measures necessary to remedy the breach. When the fire authority considers that a premises constitutes an excessive risk to persons in case of fire, the Act provides for the fire authority to issue a Prohibition Notice, which in serious circumstances could have the effect of closing down a building or premises.

The definition of premises is being revised for the purposes of fire precautions so that it includes open spaces adjacent to buildings. Consideration is also being given to the provision of a power to enable adequate fire safety standards to be achieved in open-air workplaces. Further, the Act enables fire authorities to take into account the provision of automatic means for firefighting before issuing a fire certificate. This constitutes a major change in approach to life safety.

The modified system of control of fire precautions is more flexible and discriminating than the previous system. It shows a better balance between enforcement and compliance costs and puts a less demanding administrative burden on fire service resources. It should also identify where fire risks are greatest and raise the standard of 'high-risk' premises. The burden on small businesses should also be lightened in many cases.

In premises of 'normal' and 'high fire risk' there should be a greater availability of experienced fire-prevention staff to advise on practical problems of fire safety. The 1987 Act changes the law governing fire precautions 'across the board', not just sports grounds, as could be inferred from the title. Special prominence has been given to sports grounds as a result of the fire at Bradford City Football Club in 1985, but the changes are not a direct result of this fire. The appropriate authorities, whether fire brigade or local government, must be consulted for advice.

35.6 Fire protection

Every year there are a number of fires in the UK costing £500 000 or more, but the greater part of the total annual loss is made up from a large number of smaller fires. Most fires are initially small, and if tackled manually or automatically at this early stage can be put out before any really serious damage is done. Rapid action can be taken either with portable firefighting equipment with hose reels or by fixed systems which are brought into operation automatically by the heat or smoke of a fire. The type and quantity of firefighting equipment necessary will vary greatly. Advice can be obtained from the fire brigade or the Health and Safety Executive.

In many cases insurers allow a discount from fire insurance premiums where equipment is installed providing the equipment complies with insurance standards. These may differ from those recommended for life safety under the Factories Act.

35.7 Extinguishers

The effectiveness of portable extinguishers is limited by the need to keep their weight, and therefore the amount of extinguishing agent they contain, within the limits necessary for quick and easy handling. The maximum discharge time which can be expected from any portable extinguisher is approximately 60–120 seconds.

For firefighting in occupied rooms with no exceptional fire risk, portable extinguishing equipment will normally suffice, provided that efficient appliances of the right type are properly positioned and will be handled by people trained to use them effectively. A fixed system will usually be necessary, on the other hand, to ensure adequate protection in factories where there is a real risk of fire breaking out after working hours, to ensure effective firefighting in workrooms where a fire is likely to spread too fast to be controlled by portable extinguishers, and in parts of buildings which have an appreciable fire risk but which are not normally occupied. Such a system should always be in addition to and not a substitution for portable extinguishers.

It is important to provide the right type of extinguisher for protection against a given risk. Some extinguishing agents which are outstandingly effective against fires involving certain substances may be useless or positively dangerous when applied to fires of another kind. Where more than one type of extinguisher is provided, the type of fire for which each is designed should be clearly indicated and staff instructed in the correct methods of use.

35.7.1 Water

Due to its cooling power, water is the most effective extinguishing agent for many types of fire. It is particularly suitable for fires in carbonaceous materials. Portable extinguishers provide a limited quantity of water using gas pressure. Extinguishers should have a nozzle fitted so that the direction of the jet can be properly controlled.

Hose reels connected to the public water supply have the advantage that their supply of water is unlimited, although the pressure may be less than that generated in a portable extinguisher. Very large hoses are difficult to handle and lengths should not normally exceed 35 m. Water should be immediately available when the control valve is opened.

35.7.2 Dry powder

A range of powders are available in portable extinguishers for fighting fires in flammable liquids and are also suitable for certain solid materials, including special metals. Discharge is by gas pressure and a hose is essential.

35.7.3 Gas

Carbon dioxide and certain halon compounds have a specialized application for fires in electrical equipment where a non-conducting medium is important. All are toxic to a degree, and operate either by smothering the fire or by a chemical reaction which inhibits combustion. Gas extinguishers must not be used in a confined space because of the toxic risk or the risk of asphyxiation.

35.7.4 Installation and maintenance

Equipment should be purchased from reputable manufacturers, preferable those registered to BS 5750 (Quality systems). They should be located at clearly defined fire points. All equipment should be recharged after it has been used. Manufacturers' instructions for maintenance and recharging must be followed closely. Regular inspection and maintenance of all firefighting equipment is essential, otherwise it is liable to deteriorate and prove unserviceable when needed.

Staff who may require to use the extinguishers must be trained both in selecting the appropriate extinguisher and in handling it properly. There is no substitute for actual 'hands on' use of extinguishers, and fire brigades are usually very pleased to help.

35.8 Auto-sprinkler installations

Automatic sprinkler systems have the great advantage that they are comparatively simple in concept and operate automatically, whether or not there are people present on the premises. Water is supplied from the public mains or tanks and pumps into a network of distribution pipes at ceiling level, which covers the whole premises. Water is discharged through nozzles or heads sited at regular intervals in the pipework, which are normally sealed with a heat-sensitive device. These devices are actuated by the heat of the fire, therefore the system only discharges in the area of the fire. The sprinkler heads can be arranged to operate at suitable temperatures, taking into account ambient conditions. Where heating is not always available, the pipework can be charged with air during the winter months. An audible alarm is given when the installation operates, and this can be relayed automatically to the fire brigade. As with all fire equipment, maintenance is essential, as it is not in regular use.

35.9 Automatic fire alarms

Fire alarms are intended to give an early warning of the outbreak of fire. This may take the form of a local alarm or it may be signalled to the fire brigade. The disadvantage over, for example a sprinkler installation is that the system does not attack the fire, depending upon the human element for any firefighting to take place. Its advantages over a sprinkler installation is that it is more sensitive and detection is quicker. This is a particular advantage if quick action can be taken.

There are various types of detectors which recognize heat and/or smoke utilizing fused bimetallic strips, ionization chambers and the interruption of a light beam by smoke or other combustion products. It is important to select the most appropriate form of detector for the environment. Insurers give a modest discount from premiums if the alarm installation complies with the insurance rules.

35.10 Trade hazards

The insurance survey referred to in Section 35.4 aims to identify plant, processes and storage of materials that are of significance in terms of fire hazard. Some examples include:

Drying
Dust hazards

Paint spraying and coating
Use of highly flammable liquids
Waste collection and disposal

Insurance companies are particularly interested in the standards to which the plant is operated and how the hazardous features are controlled.

35.10.1 Flammable liquids

Flammable liquids are widely used in many types of factories, and their misuse is responsible for many outbreaks of fire. The fire risks from the flammable liquids in common use such as petrol, paraffin, white spirit, cellulose solutions and thinners are well known, but these are only a few of the liquids which present hazards in industry. The variety of flammable liquids used in processes as solvents or carriers and for other purposes is constantly extending.

Processes involving coating, spreading and printing usually have a considerable area of exposure. If materials which include flammable liquids as solvents are sprayed, large quantities of vapour and fume are produced.

Many flammable liquids are used for a variety of purposes in bench work, either in semi-closed containers or in open trays. If spillage occurs due to breakage of apparatus or plant, or carelessness in handling, the liquid is distributed over a large area and considerable vapour is produced. Fires due to the ignition of vapours spread with extreme rapidity over the exposed surface of the liquid and the amount of flame and heat given off quickly increases.

35.10.2 Flammable gases

All the gases which are used for heating or lighting are easily ignitable, and there is grave risk of fire if the gas escapes because of leakage from a container or from piping. The only effective action against a gas fire is to stop the flow of the gas. If the fire is extinguished and the gas still allowed to flow, the escaping gas will form an explosive mixture with the surrounding air and produce conditions which are potentially far more dangerous than the fire. In addition, flammable gases are almost inevitably toxic, and hence a health hazard is produced.

Oxygen is not flammable, but leakages from oxygen supply pipes enrich the surrounding air and increase the ignitability of various materials within the enriched atmosphere. Some materials which would normally require a source of heat for ignition become pyrophoric. In the event of a fire the supplies of all flammable gases and oxygen should be cut off as quickly as possible.

35.10.3 Electrical equipment

In electrical equipment fires are caused by arcing or overheating, but continued combustion is usually due to insulation, oil or other combustible material associated with the installation. If the electrical apparatus remains alive, only extinguishing agents which are non-conductors of electricity should be used.

35.10.4 Management

Insurers are aware of the importance of a management that, at all levels, is fully conscious of the fire and other hazards within its premises and consequently acts in a highly responsible way. An opinion of the standard of management is an important part of the insurance surveyor's report referred to earlier, and is obtained by observation of conditions in the premises and a scrutiny of the organization and systems that are in operation.

35.11 Recommended references

35.11.1 Legislation

Fire Precautions Act 1971
Health and Safety at Work Act 1974
Fire Safety and Safety of Places of Sport Act 1987

35.11.2 Loss Prevention Council

List of approved products and services covering: building products, fire detection and alarms, fire break doors and shutters, portable fire extinguishers, sprinkler and water spray systems.

35.11.3 Loss Prevention Council rules and recommendations

The rules give fundamental information related to loss prevention necessary for satisfactory property underwriting as follows:
Automatic fire alarm installations for the protection of property
Automatic sprinkler installations
Construction and installation of fire break doors and shutters
Provision of fire extinguishing appliances

Recommendations give practices which insurers expect to be adopted:

Emergency power, heating and lighting
Builders' or contractors' operations
Protection of computer installations against fire
Liquefied petroleum gas
Welding and other hot work processes
Oil-fired installations
Installation and maintenance of grain dryers
Mobile power-driven appliances
Spraying and other painting processes involving flammable liquids
Portable and transportable space heaters
Deep fat frying equipment
Shrink wrapping
Warehouses and other storage places

35.12 Fire Protection Association Compendium of fire safety data

The Compendium on fire safety data brings together advisory information on all aspects of fire, its prevention and control. Each item in the compendium is produced as a separate data sheet and these build up into a comprehensive work of reference.

35.12.1 Volume 1 – Organization of fire safety

ORO 1	Fire safety structure in the United Kingdom
OR 1	Fire protection law
OR 2	Fire brigades
OR 3	The Health and Safety Commission and the Health and Safety Executive
OR 4	Government departments
OR 5	Fire research
OR 6	Test facilities
OR 7	Security organizations
OR 8	Industrial safety organization

Management of fire risks

MR 2	Fire safety and security planning in industry and commerce
MR 3	Fire facts and figures
MR 4	Fire safety patrols
MR 5	Security against fire-raisers
MR 6	Fire precautions during stoppages of work
MR 7	Procedure in event of fire
MR 8	Fire damage control: emergency planning
MR 9	First day induction training
MR 10	Fire safety training
MR 11	Occupational fire brigades and fire teams: organization, role and functions
MR 12	Fire safety and training of people at work

35.12.2 Volume 2 – Industrial and process fire safety

FS 6011	Flammable liquids and gases: explosion hazards
FS 6012	Flammable liquids and gases: explosion control
FS 6013	Flammable liquids and gases: ventilation
FS 6014	Flammable liquids and gases: electrical equipment
FS 6015	Explosible dusts, flammable liquids and gases: explosion suppression
FS 6016	Hydraulic oil systems: fire safety
FS 6017	Piped services
FS 6018	Corrosive smoke: planning to minimize the acid damage hazard
S 14	Electro-plating
FS 6021	Explosible dusts: the hazards
FS 6022	Explosible dusts: control of explosions
FS 6023	Explosible dusts: elimination of ignition sources
FS 6024	Explosible dusts: extraction
FS 6025	Electric soldering irons
FS 6027	Bund walls for flammable liquid storage tanks
FS 6028	Safety containers for flammable liquids
S 11	Non-sparking tools
FS 6030	Cleaning contaminated electronic equipment
FS 6031	Shrink wrapping
FS 6032	Vapour degreasing
FS 6033	Crushing and grinding
FS 6934	Reciprocating compressors
FS 6035	Electro-discharge machining
FS 6036	Electrostatic wet paint spraying
FS 6037	Heat treatment: controlled atmosphere furnaces
FS 6038	Heat treatment: molten salt baths
FS 6039	Heat treatment: quench tanks
FS 6041	Flammable liquids and gases: application of HAZCHEM code to these and other substances

Occupancy fire safety

OCC 1	Fire hazards in the cake and biscuit industry
OCC 2	Cake and biscuit ovens
OCC 4	Fire dangers in the construction industry
OCC 5	Fire dangers in shops and stores
OCC 6	Fire protection in warehouses and storage

35.12.3 Volume 3 – Housekeeping and general fire precautions

GP 1	Guide to fire precautions with outside contractors
GP 2	Cutting and welding
GP 3	Hot work permits
GP 4	Blowlamps
GP 5	Safe practice in production areas
GP 6	Safe practice in storage areas
GP 7	Outdoor storage
GP 9	Fire dangers of smoking
GP 10	Waste incinerators
GP 11	Oil soaked floors
GP 12	Guide to fire doors and shutters
GP 13	Foundry pattern stores
GP 14	Flame retardant treatment for textiles

Nature and behaviour of fire

NB 1	The physics and chemistry of fire
NB 2	Ignition, growth and development of fire
NB 3	The physiological effects of fire
NB 4	The combustion process
NB 5	Auto-ignition temperature (AIT)
NB 6	Ignition of flammable gases and vapours by sparks from electrical equipment

35.12.4 Volume 4 – Hazardous materials

Information sheets on a wide range of hazardous materials

Information on uses, hazards, precautions, firefighting, regulations and characteristics

35.12.5 Volume 5 – Fire-protection equipment and systems

PE 1	Automatic fire detection and alarm systems
PE 2	Fire points
PE 3	Fire blankets

PE 4	Portable fire extinguishers
PE 5	Fire-aid firefighting: training
PE 6	Fixed fire-extinguishing equipment: the choice of a system
PE 7	Fire firefighting equipment: hose reels
PE 8	Hydrant systems
PE 9	Automatic sprinklers: an introduction
PE 10	Automatic sprinklers: components of a system
PE 11	Automatic sprinklers: design and installation
PE 12	Automatic sprinklers: care and maintenance
PE 13	Automatic sprinklers: safety during sprinkler shutdown
PE 14	Security equipment and systems
PE 17	Automatic fire ventilation systems

Arson

AR 1	Management guide to fire investigation
AR 2	Prevention and control of arson in warehouses and storage buildings
AR 3	Prevention and control of arson in school buildings

35.12.6 Volume 6 – Buildings and fire

FPDG 1	Fire and the law
FPDG 2	Site requirements: space separation
FPDG 3	Control of fire and smoke within a building
FPDG 4	Planning means of escape
FPDG 5	Firefighting facilities for the fire brigade
FPDG 6	Equipment for detection and warning of fire and fighting fire
FPDG 7	Fire hazard assessment
FPDG 8	Fire insurance
FPDG 9	Building services
FPDG 10	Cavities and voids in building construction
FPDG 11	Suspended ceilings
FPDG 12	Construction of roofs
FPDG 13	Flame retardant treatment for wood and its derivatives
FPDG 14	Fire doors and shutters
FPDG 15	Fire safety bibliography
FPDG 16	Air conditioning and ventilation systems
J 99	FPA guide to how factory buildings can be adapted to control fire spread

Information sheets on building products

B 1	Wood wool building products
B 2	Asbestos insulation boards and asbestos wallboards
B 3	Rock fibre building insulation products
B 4	Gypsum plasterboard
B 5	Vermiculite-silicate
B 6	Sprayed vermiculite cement
B 7	Gypsum plasters
B 8	Sprayed mineral fibre
B 9	Glass reinforced cement
B 11	Fibre reinforced calcium silicate insulating boards
B 12	Fibre reinforced Portland cement
B 14	Steel and mineral fibre reinforced cement composite panels
B 15	Fire-resisting glazing – wired glass
B 16	Fire-resisting glazing – copper-light glazing
B 17	Fire-resisting glazing – clear borosilicate
B 18	Fire-resisting glazing – laminated wired glass

All are available from the Loss Prevention Council, 140 Aldersgate Street, London EC1A 4HX.

35.13 Security insurance

While a number of different classes of insurance cover relate to crime risks, the principal one which is sought by most businesses is known as 'theft' or 'burglary' insurance. The terms 'theft' and 'burglary' are defined in the Theft Act 1968, but while insurers use 'theft', the cover provided by this insurance is much narrower than the legal definition of that term in that it applies only in specified circumstances.

The basic theft policy covers the property insured against loss, destruction or damage by theft or attempted theft involving:

1. Entry to or exit from the premises by forcible and violent means; or
2. Actual or threatened assault or violence or use of force at the premises (this relates to the legal definition of 'robbery').

The theft policy extends to cover damage to premises caused in furtherance of theft or attempted theft, but malicious damage itself without theft is normally covered as an extension to the fire policy.

Theft by employees is another major exclusion from the theft policy, although the insurance market is prepared to cover this risk (subject to an excess to take care of petty pilfering) through a fidelity policy.

Theft insurance premiums vary from risk to risk. The principal factors which insurers take into account are:

1. The geographical location of the insured property (experience shows that metropolitan and urban areas have a crime incidence which is above average for the UK); and
2. The type and valuable of commodity insured (some items are particularly attractive and are always a major target for thieves).

In all cases, however, the theft insurance premium assumes that the security standards applied and put into operation are commensurate with the theft risk.

35.14 Theft insurance policy terms and conditions

Under most theft insurance policies the liability of the insurance company is conditional on the client's compliance with any policy terms and conditions. There is a Reasonable Precautions Condition under which the client is required to take reasonable precautions to safeguard the insured property and secure the premises, including the installation, use and maintenance of any security precautions stipulated or agreed with the insurance company.

In some instances special terms may be applied. For example, where the insurance company requires the installation of an intruder alarm, the protected premises must not be left unattended unless the alarm is put into full operation. In other cases, cover under the policy may apply only when the property insured is left within agreed and designated areas.

From these examples it will be seen that the client's obligations do not end with the mere provision of security hardware, etc. but extend to include procedures and routines which ensure that it is used effectively.

35.15 Risk assessment

Insurance companies employ staff who have been trained in crime-prevention techniques, and when theft insurance is provided/requested for commercial premises it is normal practice for the insurer to carry out a security survey of the premises to prepare a report for the underwriters. The report is based on the surveyor's assessment of the risk, which will consider the following main factors:

1. Type of property and values at risk;
2. Construction of premises;
3. Location/situation of the premises;
4. Extent and periods of unoccupancy;
5. Nature and degree of accessibility to thieves;
6. Existing protections:
   (a) Locks, bars, bolts, etc.;
   (b) Intruder alarms;
   (c) Safes and strongrooms;
   (d) Surveillance (e.g. security patrols, CCTV);
   (e) Site security (perimeter fencing, security lighting, etc.);
7. Previous theft history of the premises.

The assessment will establish whether or not the existing security arrangements equate to the standards normally looked for by the insurance company for the specific theft insurance exposure. If the security falls short of these standards, the insurance company will submit a list of security items requiring improvement. Pending satisfactory completion of these items, insurers may:

1. Withhold theft cover entirely; or
2. Provide provisional cover but usually subject to some restriction such as a significant excess or a percentage of self-insurance being carried by the client.

35.16 Planning for security

Like all forms of asset protection, security needs to be properly designed, planned and co-ordinated if it is to be both effective and cost beneficial. From previous comments in this chapter it will be noted that the insurance company's attention is largely focused on the security of movable assets (e.g. stock (both raw materials and finished goods), plant (including vehicles) equipment and tools) against the risk of theft. However, every industrial and commercial operation may have its own security vulnerability (e.g. arson bomb threats, fraud, information theft, industrial espionage, malicious contamination of products, etc.).

While the insurance company's security requirements against the risk of theft may go some way towards countering these other vulnerabilities, some of these issues present significantly different security problems and they may require separate assessment and control. This emphasizes the need for a total plan to cover all aspects of security. If planning is piecemeal, some aspect may be overlooked and the benefits (financial and operational) of a totally integrated approach will be lost.

Similarly, security should be a function of senior management, and it should rank alongside other management responsibilities. This will help to demonstrate that the company is committed to a security programme, that appropriate resources are allocated, and that protective systems and procedures are properly used and maintained.

35.17 Security objectives

The principal aims of security should be to forestall both organized and opportunist crime by cost-effective measures. While there is considerable variety in the type of criminal attack and skills involved, fortunately there is a tendency for them to act in a similar way, and this enables a common philosophy to be applied when determining countermeasures.

Based on the initial risk assessment, it will be obvious that as the exposure (commodity, value, location) increases so must the standard of security which is necessary for the risk. It is also the case, however, that items of comparatively low value will be stolen if they can be easily reached and if their removal does not represent any undue risk to the thief.

Experience has shown that thieves are strongly influenced by a number of factors:

1. Thieves are mainly opportunists, so there is a greater likelihood that they will be deterred by more apparent security.
2. They wish to act quickly, so any delay will complicate matters for them and discourage or even frustrate them.
3. They wish to remain undetected while working.

An important principle is that security must be built in depth – otherwise known as 'defence in depth'. In this context, it may be helpful to think of security as a set of concentric rings, where the target is located at the centre. Each ring represents a level of physical protection (perimeter fence, building shell, security case) but the number of rings and security resistance will vary relative to the risk. The spaces between the rings may represent other defensive measures such as closed-circuit television (CCTV), security lighting, intruder alarm systems, etc.

It will be apparent that delay (by physical features) and detection (by alarm, etc.) are entirely separate concepts. They should not be treated as being one and the same thing, nor should they be regarded as alternatives when arranging security. Delay, on its own, will have little security value unless it keeps the thief in a conspicuous place where there is a distinct likelihood of being seen (detection). Similarly, detection (by alarm) of the thief, where there are no obstacles to ensure subsequent delay, is unlikely to deter when it is known that the nature of property at risk may enable the removal of items of a high value in a short time. The message is that delay and detection, however achieved, should be used together and then with intelligence.

By using risk assessment techniques to determine the exposure of the property it is possible to determine the appropriate standard of security for a risk. The following simple example will clarify this point.

(A) Property at Risk: cigarettes

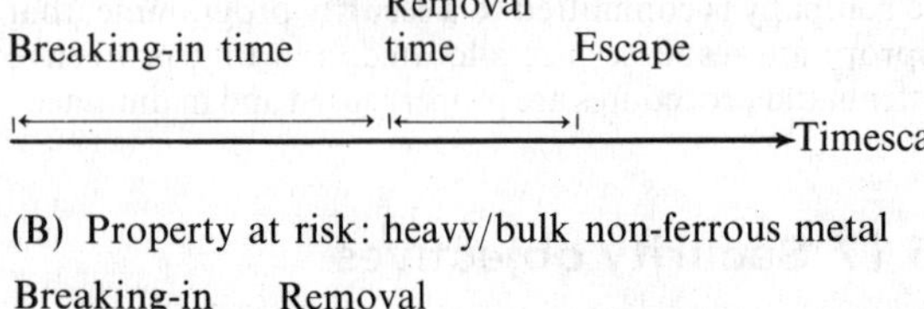

(B) Property at risk: heavy/bulk non-ferrous metal

Breaking-in time | Removal time | Escape | Timescale

In example (A) the easy transportability of cigarettes requires a very high level of physical protection symbolized by the lengthy 'breaking-in time'. When considering the necessary complementary alarm protection it is obvious that this must occur at the earliest possible time to enable assistance to be summoned and to arrive while the intruders are still on the premises but before they have reached their target. To arrange for detection to coincide with the beginning of removal time is much too late.

In example (B) the exposure of the property to risk is rather less both by way of the value of the goods and the problems posed by their transportability. While, therefore, it is still desirable that detection should occur as early as possible, it will be seen that its concurrence with the beginning of the removal time would still prevent the intruders from being successful.

Security is intended to frustrate thinking people, and it will be obvious that its effectiveness will be determined by the strength of its weakest part. Care must always be taken to ensure that there is no known vulnerable point, since this will be exploited to the full, thereby nullifying the security equipment otherwise provided. Once the standard of security has been determined it must be uniformly and intelligently applied.

35.18 Location

The location of any premises has a considerable influence on the overall standard of security. With new building, and other things being equal, the site with a low level of local crime should be chosen. In most instances, however, location is a *fait accompli*, and the protection of the site should recognize the pattern of crime in the district. This information is normally available from the local crime prevention officer. It must be remembered that crime patterns are never constant, and it is important to keep abreast of local trends and, when necessary, be prepared to modify (usually improve) security standards.

35.19 Site perimeter security

35.19.1 Fences

The security of any premises may be enhanced by perimeter fencing. Not only does the fence present an obstacle for intruders to overcome but it establishes the principle of defensible space and constitutes a psychological barrier to access. A perimeter fence, when supplemented by gates, traffic barriers and gatehouses (manned by security personnel or some other system of access control), allows the site operator to have control and supervision over all vehicles and pedestrians entering and leaving the site.

There are three types of fencing suitable for security applications. In increasing order of security, these are:

1. Chain link: BS 1722: Part 10
2. Welded mesh: BS 1722; Part 14 (in course of preparation)
3. Steel palisade: BS 1722: Part 12

The following features are necessary for any security application:

1. Total vertical height must be not less than 3 m;
2. Mesh size should not exceed 50 mm² to make footholds difficult;
3. Top edge of chain link and welded mesh should be barred and the fence topped by three strands of barbed wire on cranked arms extending at 45° out from the external face of the face, where possible. Care should be taken to ensure that any barbed wire is not a danger to pedestrians. If barbed razor type is used, notices should be posted warning of the danger of serious injury if climbing the fence is attempted;
4. The bottom edge of the fencing material should be 'rooted' into a continuous-cast concrete sill set into the ground to prevent the material being bent up and burrowing;
5. Gates should be designed to provide the same barrier to intruders as the fence;
6. If a gross attack (using a vehicle to ram the fence) is envisaged, protection can be provided either by erecting a purpose-designed vehicle barrier or, where ground permits, digging a trench on the exposed side.

35.19.2 Security lighting

The purpose of security lighting is to deny the criminal the cover of darkness which he uses to conceal his activities. In addition, strategically designed lighting can considerably increase the security value of other measures, and it can also assist the detection capabilities of surveillance (manned or CCTV) systems.

Lighting need not be devoted to security, and indeed most lighting installations have a dual role. Where security is a required feature, the installation needs to be very carefully specified and designed to achieve optimum protection. Advice on lighting and the choice of equipment is discussed in the Electricity Council publication *The essentials of security lighting*.

35.20 Building fabric

35.20.1 Outer walls

The outer walls of a building should be commensurate with the security risk. The use of robust construction and materials (e.g. those resistant to manual attack and damage) is essential to the initial provision of security. Weaknesses in walls will be exploited by criminals. Where lightweight claddings are employed, security can be enhanced by a reinforcing lining such as welded steel mesh. Where the security risk is high it may be necessary to reinforce even brick/blockwork walls in a similar manner.

35.20.2 Roofs

Roofs are a common means of access to criminals. Where necessary, access can be minimized or deterred by:

1. Restricting access to the rooftop by physical barriers;
2. Applying anti-climb paint or spiked collars to access routes such as downpipes;
3. Use of strategic lighting of the roof area;
4. Fitting expanded metal or welded mesh below the roof covering, sandwiched between the skins of double-skinned roof coverings or within the roof space;
5. Roof openings, including rooflights and ventilation, should be treated, from a security point of view, in the same way as openings, located elsewhere in the building fabric.

35.21 Doors and shutters

External doors and shutters are often the focus of criminal attention. The range and construction of doors, shutters and their securing devices is too numerous to cover but the points which need to be considered are:

1. *Siting*: Wherever practicable, doors should be located where there is some element of surveillance;
2. *Numbers*: These should be kept to the minimum, consistent with operational requirements and any other requirements imposed under Building Regulations and fire/health and safety legislation;
3. *Construction*: Preference should be given to solid wood, steel or steel-faced doors. Glazing should be avoided but, where necessary, should be as small as practicable;
4. *Frames*: The frame should be of sound solid construction, with a substantial rebate and securely fixed into the building fabric;
5. *Condition*: All doors should be a good fit in the frame with no excessive gaps, and maintained in good condition;
6. *Hinges*: Outward-opening doors normally have exposed hinges which are vulnerable to attack. These should be supplemented by fitting hinge or dog bolts on the back edge of the door;
7. *Locks*: There is a wide variety of locks and security fittings, many with special applications. Advice from locksmiths or other security experts should be taken before deciding upon a type of lock, but, wherever practicable, only those locks which require a key to lock and unlock them should be chosen. BS 3621 establishes a useful minimum standard but is not applicable to all lock types;
8. *Emergency exits*: These doors should be identified and the local fire prevention officer consulted before any key-operated locks are fitted.

35.22 Windows

Accessible windows are a high security risk. Wherever practicable, storage areas in particular should have no windows. For a variety of reasons, this is not always feasible, and in these cases the following features should be considered:

1. Windows not required to be openable should be permanently fixed closed;
2. Where windows are required to open, fit limitations (e.g. stops or cradle bars) to restrict the extent to which the window may be opened;
3. Where there is a security risk, fit internal bars or grilles properly secured to the building fabric. Specifications for bars and grilles are available from the local crime prevention officer, the insurance company or from proprietary grille manufacturers.

35.23 Intruder alarms

In many cases it is necessary to complement physical security by the installation of an intruder alarm system in order to achieve the standard of security commensurate to the risk exposure. The scope of protection to be afforded by the alarm system depends on the security risk, but it may embrace fences, windows, doors, roofs, walls, internal areas, yards and external open areas, and vehicles inside and outside buildings. There is a comprehensive range of detection devices, but the choice of detector is critical to ensure that it provides the desired level of protection and is stable in the particular environment.

Intruder alarms are designed to give a warning of the presence of an intruder within, or attempting to enter, the protected area. Alarm systems may act as a deterrent to the casual or opportunist thief but they will do little or nothing to prevent a determined intrusion, and to be effective they must provoke an early response from the appropriate authority (in most cases the police). The warning may be a local audible device, but normally the alarm signal is transmitted by the telephone network to a central station operated by a security company on a 24-hour basis.

The alarm and telephone companies provide a comprehensive range of facilities for the transmittion and receipt of alarm signals. The choice of signalling system will be determined by the security risk.

A number of alarm systems suffer from repeated false alarms. This may be due to the poor design and selection of detectors, inadequate standards of installation or preventative maintenance, and faulty operation of the system by key holders. Recurrent false calls will quickly discredit any system and may result in the withdrawal of police response to any alarm activation.

While alarm systems may be complex, false alarms can be minimized by the careful choice of installer, who should:

1. Be on the current List of Approved Installers maintained by the National Approval Council of Security Systems (NACOSS);
2. Be acceptable to the local police and appear on their list of recognized companies;
3. Install and maintain the system in full conformity with BS 4747 or, where the security risk is high, BS 7042.

35.24 Closed-circuit television (CCTV)

CCTV can make an important contribution to the security of a site, but it should be regarded as complementing and not replacing other security measures. Ideally, it should form part of an integrated system. In particular, CCTV allows greater flexibility in the deployment of security personnel by providing the facility whereby a single observer can remotely monitor several areas from one location. CCTV systems are not cheap, and need to be carefully designed by specialist CCTV engineers who will ensure that the necessary standard of reliability is obtained by proper planning and specification of the type, quality and installation of the equipment.

35.25 Access control

Although many access-control arrangements are designed to control access and egress during business hours they are an essential part of a total security programme. Leaving aside any arrangements previously discussed, access control is usually achieved by use of:

1. Security staff; or
2. Electronically operated entry/exit systems

or a combination of both.

35.25.1 Security staff

The employment of full-time security staff either in-house personnel or contracted out to a professional guarding company (preferably one that is a member of the British Security Industry Association (BSIA) Manned Services Inspectorate) can make a significant contribution to overall security. Consideration should be given to the range of duties to be performed by security staff, their location, and how they will interface with other security measures and external agencies.

35.25.2 Electronically operated access/egress systems

These systems are used in conjunction with some form of physical barrier (e.g. door, gate, turnstile) and comprise:

1. Recognition equipment such as a token or card and appropriate reader;
2. Electrically activated release hardware; and
3. In some systems, central computerized control and monitoring equipment.

The choice of access-control technology depends upon the level of security to be achieved, number of users and traffic at peak periods, etc. Proper design of the system is critical to its performance (and user acceptance), and reliability and expert advice should be obtained from equipment manufacturers and system installers.

35.26 Recommended references

BS 1722: Fences
Part 10: Anti-intruder chain link fences
Part 12: Steel palisade fences
Part 14: Welded mesh fences (at draft stage)
BS 3621: Specification for thief resistant locks
BS 4737: Intruder alarm systems
Part 1:1986, Specification for installed systems with local audible and/or remote signalling
Part 4: Codes of Practice
Section 4.1: 1987, Planning and installation
Section 4.2: 1986, Maintenance and records
BS 5051: Security glazing
BS 5357: Code of Practice for installation of security glazing
BS 5544: Specification for anti-bandit glazing (glazing resistant to manual attack)
BS 5979: Specification for direct line signalling systems and remote centres for intruder alarm systems
BS 7042: Specification for high security intruder alarm systems in building
BS 8220: Guide for security of buildings against crime
Part 2: Offices and shops
Part 3: Warehouses and distribution units (at draft stage)
Building Regulations: 1985 (HMSO)
Code of Practice on noise for audible intruder alarms: Department of the Environment, HMSO (1982)
Essentials of security lighting (periodically updated) published by and available from the Electricity Council, 30 Millbank, London SW1P 4RD
Fire Precautions Act 1971 (HMSO)
Health and Safety at Work, etc. Act 1974 (HMSO)
List of Approved Installers (periodically updated): published by and available from The National Approval

Council of Security Systems, Queensgate House, 14 Cookham Road, Maidenhead, Berks SL6 8AJ
Security Precautions: Security Data Sheets
SEC 3 – Fences, Gates and Barriers
SEC 4 – External Security Lighting
SEC 5 – Closed Circuit Television (CCTV)
SEC 6 – Doors
SEC 7 – Windows and Rooflights
SEC 9 – Locks
all published by and available from the Loss Prevention Council, 140 Aldersgate Street, London EC1A 4HX

35.27 Liability and liability insurance

One of the basic requirements of society is to conduct ourselves so that we do not injure other people or damage their property. A business or any other undertaking is equally subject to this requirement.

For many years the social and legal climates have had a growing involvement in requiring and encouraging business undertakings to protect the welfare of persons and their property. As a result, a body of common law has developed and Acts of Parliament have been passed which impose a duty of care upon any business undertaking to conduct its activities responsibly in order to avoid causing injury to persons, or damage to their property.

A breach of duty under either common law or statute which causes injury or damage may result in those responsible having a legal liability to pay compensation (termed 'damages') to the aggrieved party.

35.28 Employer's liability

The liability of an employer to his or her employees in respect of injury arising out of or in the course of their employment usually falls for consideration under either common law (the law of tort) or statute.

35.28.1 Tort

The common law liability of an employer to an employee can arise, *inter alia*, as follows:

1. Acts of personal negligence;
2. Failure to exercise reasonable care in the selection of competent servants (including contractors);
3. Failure to take reasonable care that the place of work is safe;
4. Failure to take reasonable care to provide safe machinery, plant and appliances and to maintain them in a proper condition;
5. Failure to provide a safe system of working.

35.28.2 Breach of statutory duty

There is a wide range of legislation laying down standards of safety and health, the most notable and wide ranging being the Factories Act 1961, the Health and Safety at Work Act, etc. 1974 and the Control of Substances Hazardous to Health Act 1988, breach of any of which can result in liability to civil and/or criminal action against the offenders.

35.29 Third-party liability

Liabilities to third parties may arise from breach of contract or tort.

35.29.1 Breach of contract

A breach of contract is a violation of a right created by an agreement or promise, and an action for damages or other recompense can normally only be brought by the aggrieved party against the other party or parties to the agreement.

35.29.2 Tort

Tort (which is part of common law but may be modified by statute) always implies breach of duty. This may take various forms, the most common applicable to the engineering and production fields being as follows.

Negligence

This has been defined as 'the failure to do something a reasonable person would do, or to do something a reasonable person would not do'. For an action in liability to succeed in negligence the following conditions must be satisfied:

1. The defendant must have owed a duty of care to the plaintiff;
2. There was a breach of that duty;
3. The plaintiff sustained injury or damage as a result of that breach.

Nuisance

This has been defined as 'a wrong done to a man by unlawfully disturbing him in the enjoyment of his property or in the exercise of a common right'. Examples of this are the emission of toxic fumes from a factory affecting occupiers of surrounding property or of toxic effluents affecting adjacent watercourses.

35.30 Liability insurance

As previously stated, a business can be held legally liable to pay compensation (damages) for injury or damage caused by its activities, and a successful action against it may result in large financial demands upon the business. Liability insurance ensures that, subject to satisfactory compliance with specified conditions and procedures by the insured, funds are available for a business if it is held legally liable to pay damages (and associated legal costs) which are awarded following injury to persons or damage to property.

Under the Employers Liability (Compulsory Insurance) Act 1969 there is a legal requirement upon most employers to have in force employer's liability insurance covering injury to their employees. Public liability insurance, while not compulsory, is strongly recommended, both in terms of third-party liability (arising from the effects of corporate activities) and product liability (arising from the effects of corporate products.

35.31 The cover provided by liability insurance

The following provides a general summary of the cover provided by standard liability insurance policies.

35.31.1 Employer's liability

Indemnity for an employers' legal liability to his employees against damages awarded for injury or disease happening during the period of insurance and arising out of and during the course of their employment, excluding radioactive contamination. An employee is defined as any person under a contract of service or apprenticeship with the insured, or any person supplied to, or hired to or borrowed by the insured (includes labour-only sub-contractors).

Cover normally applies to employees in the UK, Northern Ireland, the Isle of Man and the Channel Islands or while persons normally resident in those territories are temporarily engaged in the business outside of these territories. Indemnity is unlimited in amount.

35.31.2 Public liability (third party)

Indemnity against legal liability to third parties against damages awarded for:

1. Injury or disease;
2. Loss of or damage to property or personal effects of visitors or employees

happening in connection with the business and during the period of the insurance. Limited indemnity is awarded for financial losses not flowing from injury or damage but which result from escape or discharge of substances from the insured's premises or from stoppage of or interference with pedestrian, vehicular, rail, air or waterborne traffic. Wide contractual liability cover excludes only

1. Liquidated damages, fines or penalties;
2. Injury or damage caused by products;
3. Financial loss

arising solely because of a contract.

Indemnity is awarded in respect of injury to third parties and damage to property arising out of or in connection with the exercise of professional engineering skill. The following are excluded:

1. Radioactive contamination;
2. War and kindred risks;
3. Liability arising solely because of a contract for:
   (a) Liquidated damages, fines or penalties;
   (b) Financial loss;
4. Cost of rectifying defective work;
5. Liability arising out of the use of or caused by craft or vehicles;
6. Deliberate acts or omissions;
7. Principal's professional risk.

Worldwide indemnity is normally provided for injury or damage resulting from:

1. Negligent acts in the UK, Northern Ireland, the Isle of Man or the Channel Islands;
2. Temporary work undertaken abroad.

The limit of indemnity is selected at the time of insurance or renewal thereof by the management of the insured. It applies to any claim or number of claims arising out of any one cause, and is unlimited in total in the period of insurance.

35.31.3 Public liability (products)

Products means goods (including containers and packaging) not in the custody or control of the insured, sold or supplied by the insured in connection with the business from any premises in the UK, Northern Ireland, the Isle of Man or the Channel Islands, including errors in the sale, supply or presentation of such goods. The cover provides indemnity against legal liability for injury or damage happening during the period of insurance and caused by products. This cover also applies in respect of injury or design arising from the design of or formula for products.

Exceptions to the cover are as for public liability (third party) policies plus:

1. Financial loss caused by products;
2. Aircraft products.

It should be noted that very few standard product liability policies provide cover for liability for the costs of repairing or replacing defective products or those which fail to perform as intended, nor for the costs of any necessary product recall. The insurer's liability in any one period of insurance for injury or damage caused by products during that period shall not exceed the selected limit of indemnity.

It cannot be sufficiently emphasized that individual liability policies do differ in detail from the general standard policy, and reference to the policy in force in any individual set of circumstances is most strongly recommended.

35.32 Points to be considered

As stated in previous sections of this chapter, there exists at all times a statutory and common law duty on all employers (which includes all engineers) to maintain a safe working environment for all their employees and the public at large. Additionally, it is a contractual obligation on the insured under liability insurance policies to maintain the best reasonable standards of working procedures, equipment and the environment at all times. Consequently, there is a duty on all engineers to conduct their professional duties at all times in a manner which does not downgrade the safe environment of their sphere of influence.

While not exhaustive, the following points for consideration will go far towards optimizing the control of safety and liability standards within the engineer's sphere of influence.

35.33 Employee safety and employer's liability

35.33.1 Premises/buildings

Design

Suitability for purpose/occupation?
Best 'state of the art' design?

Construction

Suitability for identified operating conditions?
Adequate strength of roofs and control of access thereto?
No built-in accident traps?

Maintenance programme

Schedule of maintenance relevant to the needs for the operation?

Floors, aisles, stairs and floor openings

Evenness of floor surfaces?
Adequate provision and standards for handrails, toeboards and gates?
Adequate stairway capacity for number of persons likely to use them at one time?

Lighting, ventilation and heating

Adequate for identified conditions?
Compliance with Factories Act requirements?

Roadways, pavements, yards, car parks and open areas

Proper maintenance of surfaces?
Separation of pedestrian traffic from vehicular traffic (as far as possible)?
Speed limits (speed ramps, etc.)?
Adequacy of access?
Crane headroom (overhead wires), etc.?

35.33.2 Machinery and plant

Type and design

Adequate engineering design for the duty intended?
Hazard studies ('What if?') carried out?

Power sources and transmissions

Electrically safe (if applicable)?
Guarded (especially on rotating parts)?

Guarding/fencing

Compliance in all reports with Factories Act requirements?
No unguarded access to moving parts?
Safe provisions for maintenance, etc.?
Interlocking of guards with power supply?

Maintenance/adjustment procedure

Scheduled maintenance programme?
Appointment of machinery attendants?
Isolation procedures laid down?
Adequate oil containment?

'Permit to work' systems

Adequate specification for identified needs?
Responsibilities allocated for issue, receipt, etc. of 'permits to work'?

Access, scaffolds, ladders

Correct maintenance schedules and procedures?
Designated certification and test/inspection procedures?

35.33.3 Tools

Power tools

Correct specification/design for duty envisaged?
Adequate inspection/maintenance procedures?
Regular insulation testing?
Safe power supplies?
Power safety cut-out provisions?
Safe storage when not in use?

Hand tools

Correct specification of duty?
Adequate maintenance?
Safe storage when not in use?

35.33.4 Electrical installations

Type and design

Correct specification for duty envisaged?
Suitability for hazardous environment (if required)?

Maintenance

Schedule of maintenance and maintenance procedures specified?
Earth testing schedules?

Operating procedures

Adequacy of 'lock-off'/isolation systems?
Conformity with the Electrical Regulations?
'Permit to work' systems for maintenance?

Overhead power lines

Maintenance of adequate clearance with satisfactory signing of hazard?

Control of access

Adequate access control to areas of concentration of electrical equipment?

35.33.5 Pressure equipment

Type and design

Adequately engineered for duty envisaged?
Safe working practices clearly deployed and adhered to?
Possible abnormal fluctuations in pressure identified and allowed for in design?
Adequate design for safe relief of overpressure?

Inspection/maintenance

Inspection schedule in force in conformity with Pressure Vessel Regulations?

Gas cylinders

Correct storage facilities for full/empty cylinders?
Segregation of combustible from oxidizing gases?
Compliance with HFL (highly flammable liquids) Regulations where applicable?

35.33.6 Materials handling and storage

Cables, ropes, chains and slings

SWL (safe working load) marked thereon strictly adhered to?
Certification and retest procedures specified and implemented?
Adequate and suitable storage facilities?

Lifts and hoists

Statutory inspection schedule in force?
Safe barriers/chains, etc. on teagle openings?
Runway beams tested and SWL displayed?

Cranes

SWL schedule (including operating radius variations) displayed and adhered to?
Overload devices fitted?

Conveyors

Adequate guarding (if powered)?
Adequate tripwires/buttons?
'Hand-traps' eliminated

Power/hand trucks

Correct specification/design for duty?
Laid-down code of operation?
Full and correct training/retraining/licencing of operators?

Manual lifting/handling

Training in safe techniques of lifting/handling?

Storage/piling of materials and goods

Safe condition of pallets?
Observance of specified maximum stacking heights?
Safe conditions of stacks?

35.33.7 Hazardous materials

Flammable/explosive substances

Properties known and documented?
Compliance with Storage Regulations?
Quantities within licence/regulation limits?
Tank vents safe and adequately sized?
Compatibility of materials in storage acceptable?

Toxic substances

Properties known and documented?
Compliance with storage regulations?
Quantities within licence/regulation limits?
Compatibility with other materials in storage acceptable?
Adequate personal protective equipment available for handling and use of material specified?
Antidotes/emergency facilities available?

Corrosive materials

Properties known and documented?
Adequate emergency washing facilities available?

Disposal of hazardous waste

Regular removal?
Use of experienced/licensed removal contractors?
Conformity with Hazardous Waste Regulations?
Safe operation of site incinerators, settling lagoons, etc.?
Adequate containment of hazardous spillage?

35.33.8 Hazardous processes

General

Hazard studies carried out to identify normal and abnormal hazards and circumstances leading to such hazards?

35.33.9 Health hazards and their control

Dust and fume

Optimum suppression of dust/fume at point of origin?
Enclosure of dust/fume producing processes?
Superior ventilation at point of dust/fume origin?
Priority to cleaning methods (vacuum, etc.)?
Provision of adequate respiratory equipment and instructions on its correct use?
Correct (statutory) procedures for entry into enclosed areas which may contain fumes?

Noise

Minimization at source?
Regular monitoring of noise levels?
Enclosure of excessive noise sources?
Provision of personnel ear protection and instructions on its correct use?
Regular audiometry of employees at risk?

Skin disease (dermatitis)

Identification of potential causes?
Adequate hygiene in use of dermatitic substances?
Provision of protective equipment, barrier creams, etc. and instruction on their use?
Medical screening of employees at risk?

Carcinogenic materials

Identification of potential causes of cancer in the workforce?
Serious considerations of alternative materials?
Adequate containment of identified carcinogens?
Medical screening of employees at risk?

Radioactive materials and X-ray/ionizing radiation equipment

Minimization of exposure time?
Identification and maintenance of safe working distances between sources and employees?
Correct shielding of sources?
Adequate monitoring of exposure (film badges, dosimeters, etc.)?
Warning notices?

35.33.10 Personal protective equipment

Goggles/face screens, gloves/protective clothing, ear protection, barrier creams

Risks identified and the appropriate level of protective equipment determined?
Best available equipment specified for identified risks with instructions on correct use?
Effects of use of safety equipment assessed apposite possible creation of additional hazard (e.g. loss of ability to hear emergency signals while wearing ear protection)?

35.34 Safety of the public and public liability (third party)

In general, all points already listed under employer's liability apply equally in public liability (third party) situations. Additionally, the following should be considered.

35.34.1 Pollution

Solid wastes

Safe means of disposal, both on-site and at the disposal site, dump, etc. (long- and short-term effects)?
Safe and full incineration (if employed) with safe and full treatment of off-gases?
Correct pre-treatment of waste before disposal (if need identified)?

Liquid effluents

Correct design to optimize discharges?
Containment of spillages?
Discharges within consent levels?
Adequate effluent treatment plant (if need identified)?
Adequate monitoring of effluent streams with relevant recording?

Gaseous emissions

Correct design to optimize discharge within requirements of consent, Regulations, etc.?
Adequate monitoring of emission quality with relevant recording?

Noise

Adequate provision for limitation of noise levels (by distance or by reduction at source) at site boundary and beyond?

35.34.2 Visitors

Contractors

Adequate advice/training on-site, hazard site rules, etc.?
Competence assessed?
Liaison personnel appointed?

Invitees

Potential risks to visitors assessed?
Adequate advice on alarm systems, etc.?
System of escorts on-site implemented?

Trespassers

Optimization of site perimeter security and its maintenance in satisfactory condition?
Security patrols off-site?
Specified policy and means of dealing with trespassers found on premises?

35.34.3 Of-site work ('work away')

General

Adequate and competent supervision?
Adequate control of all operations involving the application of heat?
Adequate control of any operations (e.g. excavation) which may cause damage to or weaken third-party property?

Further reading

Abrasive Wheels Regulations 1970
Asbestos Regulations 1969
Carcinogenic Substances Regulations 1967
Compressed Air Regulations 1960
Construction (Health and Welfare) Regulations 1966
Control of Substances Hazardous to Health Act 1988
Factories Act 1961
Health and Safety At Work, etc. Act 1974
Highly Flammable Liquids and Liquefied Petroleum Gas Regulations 1972
HSE Health and Safety At Work Booklets 1–50 *et seq.*
Ionising Radiations (Sealed Sources) Regulations 1969
Offices, Shops and Railway Premises Act 1963
Power Press Regulations 1965
Protection of Eyes Regulations 1974
Woodworking Machines Regulations 1974
Unfenced Machinery (Prescribed Leaflet) Order 1967

36

Health and Safety

George Pitblado IEng, MIPlantE, DipSM
Support Services

Contents

36.1 Introduction

The initial introduction of the Health and Safety at Work, etc. Act 1974 enabled provisions in the Act to supersede legislation contained in 31 relevant Acts and 500 subsidiary regulations in use at that time. The 1974 Act is the framework which, over the years, will enable the gradual improvement of existing health and safety requirements. This will be achieved by revising and updating current provisions in the form of regulations and approved codes of practice prepared in consultation with industry. These provisions will maintain or improve the health and safety standards established by existing legislation. The objective of the Act is not only that it be introduced to rationalize existing provisions but also to ensure that standards of health and safety are improved, thereby providing greater protection to persons at work and to the general public.

36.2 Legislation

Health and safety legislation is in four separate categories as follows.

36.2.1 Acts of Parliament

The Factories Act 1961 and the Health and Safety at Work, etc. Act 1974 are the result of Bills which, after being debated in Parliament, have received the Royal Assent and now form part of criminal law.

36.2.2 Regulations

The 1974 Act is all encompassing. It lays down only general duties and grants the appropriate minister the power to make detailed Regulations to cover relevant aspects referred to in general legislation. That is, the Act, being an up-to-date form of legislative control, differs from that found in previous formats of the Factories Act 1961.

The Control of Substances Hazardous to Health Regulations 1988 (COSHH), the Noise at Work Regulations 1989, the Pressure Systems and Transportable Gas Containers Regulations 1989 and the Electricity at Work Regulations 1989 are examples of the Regulations being introduced. For example, the Electricity at Work Regulations which came into effect from 1 April 1990 revoke the Electricity Regulations 1908 and the Electricity (Factories Act) Special Regulations 1944. Within the structure of Regulations provisions may be granted for exemptions from their requirements or application in some form, with respect to specific premises, processes or industries.

36.2.3 Codes of Practice

Codes of Practice, where appropriate, will be issued to supplement Regulations. These Codes, although having a special legal status, are not statutory requirements but may be used in criminal proceedings as evidence that statutory requirements have been contravened. Prior to replacing or amending existing Codes of Practice discussions are normally held with interested parties before the Health and Safety Commission approves them for use under the Health and Safety at Work, etc. Act 1974. In some cases these discussions could be supported by comments from interested bodies on a previously issued draft Code of Practice.

36.2.4 Guidance notes

Guidance notes relevant to sound health and safety practices are published from time to time by the Health and Safety Executive. They have no legal status nor a significance such as found in Codes of Practice. However, as their content is normally based on a wealth of practical experience, it would be expected by the enforcing authorities that employers would follow the advice contained in them.

36.3 Administration of the Health and Safety at Work Act

The bodies set up to ensure the satisfactory implementation and operation of the Health and Safety at Work, etc. Act 1974 (and the Employment Medical Advisory Service Act 1972) under Section 10 of the Act are the Health and Safety Commission and the Health and Safety Executive.

36.3.1 The Health and Safety Commission

This consists of a chairman appointed by the Secretary of State for Employment and a committee of not less than six and not more than nine members. The composition of the Commission is drawn from different groups: three members may be from employers' organizations and three from employees' organizations. The three remaining vacancies may be filled from such organizations as local authorities and professional bodies, as the Secretary of State considers appropriate.

36.3.2 The Health and Safety Executive

Whereas the Health and Safety Commission is responsible for the development of policies in the health and safety field, the Health and Safety Executive, being a separate statutory body appointed by the Commission, will work in accordance with directions and guidance given by the Commission. The Executive will provide advice on health and safety to both sides of industry and will also enforce legal requirements. The major inspectorates which normally worked within the structure of each relevant government department now report to the Executive.

36.3.3 Local authorities

Under the Commission's guidance these will enforce legislation in certain areas of employment. In general, theirs will not be industrial activities.

36.3.4 Fire authorities

The Health and Safety at Work, etc. Act 1974 gives the relevant fire authority the power to designate a place of work as premises requiring a fire certificate. This certificate will indicate the measures to be taken with respect to means of escape in the event of a fire when the structural design of the premises is taken into consideration. The fire authority in most areas has the responsibility for enforcing the provisions of the Act, but the Health and Safety Executive are responsible for issuing fire certificates in certain classified premises (e.g. those used for storage, manufacture and use of highly flammable liquid).

36.4 General duties

Employers and others who now have duties under the 1974 Act will have had some form of responsibilities under previous legislation (e.g. the Factories Act 1961) to ensure the health and safety and welfare of people at work. However, a number of employers will be subject to such legislation for the first time.

Much of the earlier legislation has been retained, although as the respective regulations are amended they are presented within the remit of the 1974 Act, such as the Electricity at Work Regulations 1989, which revoked the Electricity Regulations 1908 and the Electricity (Factories Act) Special Regulations 1944. The earlier legislation, where it remains, has been brought under the same method of enforcement as those under the Act, and is applied universally.

36.4.1 Duties of employers

It is the duty of employers, as far as is reasonably practicable, to safeguard the health, safety and welfare of the people who work for them and also others who may be affected. This applies in particular to the provision and maintenance of safe plant and equipment and methods of work, and covers all plant, equipment and substances used. Specific areas that require attention include:

1. Is the plant and equipment designed, installed and operated to the accepted standards?
2. Are there safe systems of work implemented during the operation and maintenance of the plant and equipment? (This may necessitate a permit to work to be issued by an authorized competent person.)
3. Is the work environment monitored, as required under the COSHH Regulations, to ensure that there is no hazard from, for example, toxic contaminants?
4. Has the monitoring of any control measure been implemented?
5. Is there a programme in place whereby all equipment and appliances found necessary to ensure the safety and health of those likely to be affected are regularly inspected?
6. Have the risks to health from the use, storage, handling or transport of articles or substances been kept to the minimum? (Expert advice should be sought in this area if there is no sufficient in-house knowledge).

36.4.2 Safety policies: organization and arrangements

It is a statutory requirement that an employer with five or more employees prepares a written statement of the company's general policy, organization and arrangements for health and safety at work. This policy statement should be revised at regular intervals and, where appropriate, amended.

36.4.3 Safety training and information

Employers have a duty under the 1974 Act to provide, as necessary, training and information to ensure that there is no risk to the health and safety of their employees. It may be found necessary to provide operators and maintenance employees with specific training to carry out certain processes or work tasks.

36.4.4 Duties to others

Employers must, as far as is reasonably practicable, have regard for the health and safety of contractors' employees or the self-employed who may be affected by the company's operations and for the health and safety of the general public. This covers, for example, the emission of noxious or offensive gases and dust into the atmosphere, or danger from plant and equipment to which the public or those not directly employed by the company have access.

36.4.5 Responsibilities of the self-employed

The self-employed have a similar duty to that of employers to ensure that there is no risk or danger to the health and safety of themselves or any other persons.

36.4.6 Duties of manufacturers and suppliers

Section 6 of the Act has been amended by the Consumer Protection Act 1987 and now imposes specific duties on manufacturers, importers, designers and suppliers to ensure that articles and substance supplied for use at work are safe and without risk to health.

To assist manufactures and suppliers attain the required level of health and safety, the Health and Safety Commission have commenced approving standards laid down by such bodies as the British Standards Institution.

36.4.7 Duties of employees

The Act also places a responsibility on the employees in that they must take reasonable care to avoid injury to themselves or to others by their work activities, and also cooperate with their employer and others in meeting all statutory requirements. The Act also requires that employees do not interfere with or misuse anything designed to protect their health, safety or welfare.

36.4.8 Enforcement

On finding that there has been a contravention of either an existing Act or Regulation, or of a provision of the Health and Safety at Work, etc. Act 1974, an inspector can:

1. Issue a prohibition notice, if there is a risk of serious personal injury, to stop the work or process giving rise to this risk until remedial action specified by the inspector in the notice has been undertaken. This notice can be served on the person carrying out the activity or in control of it at the time the notice is served.
2. Issue an improvement notice if there is a legal contravention of any statutory legislation relevant to the activity in that the activity cannot be continued until the remedial action specified in the notice is implemented. This notice can be served on the person who is deemed to be contravening the legal provision or who is responsible for the activity, whether they are an employer, employee or a supplier of equipment or materials.
3. Prosecute any person who may be contravening a relevant statutory provision instead of (or in addition to) serving a notice.

36.5 Safety policy

The safety policy for the company, although common in most parts with the safety policy for every company, may require to be prepared in greater detail. This would naturally cover any process, hazardous product or procedure specific to the organization. In preparing the policy it is essential to ensure that parts referred to in Section 2(3) of the 1974 Act are covered, i.e.:

1. The general policy statement;
2. The organization and arrangements for carrying out the policy.

In large companies these two aspects may be dealt with separately. In larger organizations it can be found that it is of greater benefit if the policy document is produced in two separate sections. These would include:

1. A concise statement of the company's general policy, organization and arrangements in a single document. It follows that this document can then be issued individually to all members of staff and contractors who work on the premises.
2. A more detailed document in the form of a health and safety manual. This manual would include the company's policy statement, company rules, safe working procedures, etc., and would normally be located in a nominated office. It is necessary in using this method that all staff are made aware of its contents and its location.

36.5.1 Policy statement

The company's general policy statement should be a declaration of the employer's intent to provide a safe and healthy workplace for all employees and should also include the request that the employees provide the necessary support towards achieving the company's aims.

36.5.2 Levels of responsibility

The policy statement should give the name, designation and office location of the nominated senior member of the company designated as the responsible person within the organization for ensuring that the company's policy statement is complied with. This nominated person should have high position within the company (e.g. director, senior manager, company secretary).

While the overall policy responsibility for health and safety rests with senior managers, all employees within the organization, irrespective of their duties, have some degree of responsibility for carrying out the policy. Where appropriate, nominated persons with specific responsibilities for health and safety should be named with a summary of their responsibilities defined. It follows that there should be a procedure established whereby deputies are available during the nominated person's absences.

Where specialist knowledge is required, the relevant aspects should be clearly established and the respective persons made aware of such. This expertise can be obtained, for example, from the company's safety officer, chemist, etc. Finally, the policy statement should make clear the level of responsibility of every employee.

36.5.3 Safety representatives and joint safety committees

Where health and safety consultation is in place (e.g. joint safety committees) the structure with regards to members' representation should be described.

36.5.4 Training and supervision

The policy statement should indicate how the company proposes to carry out training with respect to health and safety, but equally important is the responsibility placed on managers, supervisors, etc. to ensure that the individual who has been given the work task has the knowledge to carry it out without risk to themselves or others. This task may be operating a piece of equipment on which all safety measures must be in order or carrying out a maintenance task where there is an acknowledged hazard. It is therefore equally important that managers and supervisors are suitably trained in both the technical and safety aspects of the work.

36.5.5 Hazard details

Many accidents occur because the operator or maintenance person does not understand the hazards involved or has not been instructed on the precautions to be taken. The policy document should identify the main hazards within the company with advice on which rules must be obeyed while carrying out a hazardous task. General rules should also be included to cover items such as untidy work areas, replacement of guards, the use of protective clothing/equipment where appropriate, safe working practices in handling goods and materials, etc. It is essential that on the introduction of new products, processes, operations or plant and equipment that any hazards associated with these are brought to the attention of all concerned.

36.5.6 Reporting accidents

Procedures must be prepared to ensure that the reporting of accidents meets the requirements as set out in the Reporting of Injuries. Diseases and Dangerous Occurrences Regulations 1985. Accidents should also be recorded in the Accident Book (Form BI 510). The procedure adopted should include a measure of recording and reporting of accidents, which should then form the basis of discussion by management and/or joint safety committees, with a view to identifying any hazard that gives rise for concern and introducing the appropriate corrective action where necessary. The number of accidents could be taken as a measure of how the organization is performing overall with respect to the health and safety of its workforce and others.

36.5.7 Policy review

The policy document should be reviewed at regular intervals to ensure that any changes from the introduction of new legislation (i.e. regulations, codes of practice, information concerning safe working procedures) are implemented.

36.6 Information

Health and safety information can be obtained from the following Health and Safety Executive offices:

Public enquiry points: HSE Library and Information Services
St Hugh's House
Stanley Precinct
Trinity Road
Bootle
Merseyside L20 3QY
Tel. 051 951 4381

Broad Lane
Sheffield S3 7HQ
Tel. 0742 752539

Baynards House
1 Chepstow Place
Westbourne Grove
London W2 4TF
Tel. 071 221 0870

Enquiries with respect to health and safety at the workplace should be directed to the local area office of the HSE or the local authority's Environmental Health Department. The name of the authority responsible for the health and safety in the organization should be displayed prominently within the premises on the poster 'Health and Safety Law, What You Should Know'. This poster gives a brief guide to health and safety law and specific details on key points must be obtained from the relevant legislative documents.

36.7 HSE Inspectorates

HSE inspectors systematically visit and review a wide range of work activities, giving expert advice and guidance where necessary. The Inspectorates, within the HSE, which monitor or enforce standards and provide advice in specific sectors of industry or employment are as follows:

1. *HM Factory Inspectorate*: Manufacturing and heavy industrial premises and processes, as well as construction activities, local authority undertakings, hospitals, schools, universities and fairgrounds;
2. *HM Agriculture Inspectorate*: Farming, horticulture and forestry;
3. *HM Explosives Inspectorate*: The manufacture, transport, handling and security of explosives;
4. *HM Mines and Quarries Inspectorate*: All mines, quarries and landfill sites;
5. *HM Nuclear Installations Inspectorate*: On HSE's behalf licences nuclear installations ranging from nuclear power stations and chemical works to research reactors.

36.8 The Employment Medical Advisory Service (EMAS)

This is an organization of doctors and nurses based in area offices and forms the medical arm of the HSE. The EMAS provides advice, at the request of the employer, employee, self-employed, trade union representative or medical practitioner, on the effects of work on health.

36.9 HSE area offices

The Health and Safety Executive operate from area offices as follows.

South-West
Inter City House
Mitchell Lane
Victoria Street
Bristol BS1 6AN
Tel. 0272 290681
(Avon, Cornwall, Devon, Gloucestershire, Somerset, Isles of Scilly)

South
Priestly House
Priestly Road
Basingstoke RG24 9NW
Tel. 0256 473181
(Berkshire, Dorset, Hampshire, Isle of Wight, Wiltshire)

South-East
3 East Grinstead House
London Road
East Grindstead
West Sussex RH19 1RR
Tel. 0342 26922
(Kent, Surrey, East Sussex, West Sussex)
London North
Maritime House
1 Linton Road
Barking
Essex IG11 8HF
Tel. 081-594 5522
(Barking and Dagenham, Barnet, Brent, Camden, Ealing, Hackney, Haringey, Harrow, Havering, Islington, Newham, Redbridge, Tower Hamlets, Waltham Forest)
London South
1 Long Lane
London SE1 4PG
Tel. 071-407 8911
(Bexley, Bromley, City of London, Croydon, Greenwich, Hammersmith and Fulham, Hillingdon, Hounslow, Kensington and Chelsea, Kingston, Lambeth, Lewisham, Merton, Richmond, Southwark, Sutton, Wandsworth, Westminster)
East Anglia
39 Baddow Road
Chelmsford
Essex CM2 OH1
Tel. 0245 284661
(Essex)
Northern Home Counties
14 Cardiff Road
Luton
Beds LU1 1PP
Tel. 0582 34121
(Bedfordshire, Buckinghamshire, Cambridgeshire, Hertfordshire)
East Midlands
Belgrave House
1 Greyfriars
Northampton NN1 2BS
Tel. 0604 21233
(Leicestershire, Northamptonshire, Oxfordshire, Warwickshire)
West Midlands
McLaren Building
2 Masshouse Circus
Queensway
Birmingham B4 7NP
Tel. 021 200 2299
(West Midlands)
Wales
Brunel House
2 Fitzalan Road
Cardiff CF2 1SH
Tel. 0222 497777
(Clywd, Dyfed, Gwent, Gwynedd, Mid-Glamorgan, Powys, South Glamorgan, West Glamorgan)
Marches
The Marches House
Midway
Newcastle-under-Lyme
Staffs ST5 1DT
Tel. 0782 717181
(Hereford and Worcester, Shropshire, Newcastle-under-Lyme, Staffordshire)
North Midlands
Birbeck House
Trinity Square
Nottingham HB1 4AU
Tel. 0602 470712
(Derbyshire, Lincolnshire, Nottinghamshire)
South Yorkshire
Sovereign House
40 Silver Street
Sheffield S1 2ES
Tel. 0742 739081
(Humberside, South Yorkshire)
West and North Yorkshire
8 St Pauls Street
Leeds LS1 2LE
Tel. 0532 446191
(North Yorkshire, West Yorkshire)
Greater Manchester
Quay House
Quay Street
Manchester M3 3JB
Tel. 061 831 7111
(Greater Manchester)
Merseyside
The Triad
Stanley Road
Bootle L20 3PG
Tel. 051 922 7211
(Cheshire, Merseyside)
North-West
Victoria House
Ormskirk Road
Preston PR1 1HH
Tel. 0772 59321
(Cumbria, Lancashire)
North-East
Arden House
Regent Centre
Regent Farm Road
Gosforth
Newcastle-upon-Tyne NE3 3JN
Tel. 091 284 8448
(Cleveland, Durham, Northumberland, Tyne and Wear)
Scotland East
Belford House
59 Belford Road
Edinburgh EH4 3UE
Tel. 031 225 1313
(Borders, Central, Fife, Grampian, Highland, Lothian, Tayside, and the island area of Orkney and Shetland)
Scotland West
314 St Vincent Street
Glasgow G3 8XG
Tel. 041 204 2646
(Dumfries and Galloway, Strathclyde, and the Western Isles)

36.10 Health and Safety procedures

Note that the following procedures only provide a framework in which specific health and safety procedures for the individual works or premises may be prepared. The foreword to a company's procedure should be written by a senior member of the company or the person nominated as responsible for health and safety in the organization. To emphasize its importance it should also include a part of the company's health and safety statement followed by an instruction such as 'You are required to read the following procedures carefully and comply with the sections relevant to your workplace'.

36.11 Fire and first-aid instructions

36.11.1 Fire

Fire bells are installed in works or premises:

1. To alert all staff in the event of a fire alarm;
2. To order complete evacuation of the works or premises (in a number of workplaces such evacuation may not be necessary).

On hearing the alarm all personnel should leave the building by the nearest safe fire exit. They should not use the lifts or re-enter the building until given permission by the person controlling the incident. Staff should be aware of all fire-precaution notices and procedures (including the fire brigade's telephone number), fire call points, extinguishers, fire exits, escape routes and assembly points.

36.11.2 First aid/medical services

Staff should make themselves familiar with the first-aid procedures in the event of an accident and know the location of the nearest first aid officers and the first-aid room (if there is one on the premises). They should also be aware of all first aid/medical notices and procedures and the telephone number to call for assistance during and outside normal hours. (These are likely to differ unless the works or premises operates 24 hours a day.)

36.12. Good housekeeping

Good housekeeping can play a major part in maintaining a safe and environmentally sound place of work. Many accidents are caused by tripping over material not tidied away. Another source of potential injury is in the lack of secure storage of cleaning equipment, tools, etc.

Liquids allowed to overflow and those not adequately cleaned off the floor also present potential hazards by persons slipping.

General hints on keeping the workplace clean are:

1. Use waste bins. Special containers may be required for specific hazardous waste material.
2. Return everything to its correct place.
3. Always tidy up after completing the task.
4. Keep passageways and footpaths clear at all times.
5. Store materials and equipment in a safe manner.
6. Store hazardous material in its correct location.
7. Encourage everyone to develop 'clean' habits.

36.13 Protective clothing and equipment

1. Eye protection should comply with the Protection of Eyes Regulations 1974 and relevant British Standard (e.g. BS 2092, BS 1542, BS 679).
2. Safety helmets should comply with the relevant British Standards (e.g. BS 5240, BS 4033). If engaged in construction, the Construction (Head Protection) Regulations 1989 must be applied.
3. The Noise at Work Regulations 1989 has set action levels which must be complied with:
   (a) Where employees are exposed to between 85 dB(A) 90 dB(A) they are entitled to ask for ear protectors and the employer must provide them. There is no obligation on the employer to ensure that protectors are used or on the employee to use them.
   (b) At or above 90 dB(A) and the peak action level of 200 Pa or 140 dB re 20 μPa the employer must provide suitable ear protectors and enforce their use, and employees are under an obligation to use them. Hearing protectors should comply with BS 6344, with sound-attenuation measurement to BS 5108.

   Where exposure is at or above 90 dB(A) the employer is required to reduce exposure to noise as far as is reasonably practicable by means other than ear protectors. As with other regulations, it is now a duty of the employer to assess the noise exposure and advise the employee as appropriate. In a noisy environment the employer may have to seek expertise from a competent person to carry out the necessary assessments and monitoring of the premises.
4. Safety footwear *must* be worn where applicable (moving heavy objects, working with hazardous materials (e.g. chemicals), working in wet conditions). Footwear should comply with the following Standards as appropriate: BS 1870, BS 4972, BS 5131, BS 5145, BS 5451, BS 5462, BS 6159.
5. Safety belts should comply with BS 1397 and BS 6868.
6. Protective clothing (including gloves) *must* be used where applicable (e.g. normal duties; welding/cutting hot work; water treatment; application/use of chemicals; during servicing of batteries; descaling/chemical cleaning). Clothing should comply with BS 1542, BS 5426, BS 2653, BS 4171, BS 6408 and BS 4724 and gloves with BS 1651, BS 697 and BS 6526.
7. Respirators should comply with BS 2091, BS 4555, BS 4558, BS 4771 and BS 4275. (Refer to Form F2486, published by HSE for suppliers of respirators suitable for use against aerosols containing *Legionella*.)

When in doubt as to which standard of protection is necessary, consult the manufactures, suppliers, HSE, EMAS, the local authority or the Environmental Health Officer.

36.14 Safe working areas

It is essential when carrying out any operation or task that cannot be undertaken standing on a floor that an appropriate working platform is used. This can take many forms, but the methods that should be used fall within the following types: stepladders; trestles with suitable boards; towers (mobile); proprietary scaffolding or general-access scaffolding. Such access equipment should be to the relevant British Standard, e.g.

Stepladders: BS 1129, BS 2037;
Trestles: BS 1129, BS 2037, BS 1139;
Scaffolds; BS 1139, DD72, CP 5973, BS 2482, BS 6037, BS 5974, BS 6289.

The use of access equipment should meet the acceptable standards as laid down in the Construction (Working Places) Regulations 1966.

Items that should be checked on access equipment before use include:

1. General condition – is there any damaged section or parts?
2. Is it the correct access equipment to carry out the task safely? For example, is a ladder adequate or should a tower be used?
3. Is it positioned correctly (e.g. ladders at the correct ratio of 4:1; scaffold on a sound footing)?
4. Is the ladder or scaffold secured safely?
5. What is the condition of access equipment after inclement weather?
6. Is there any risk to its use in a specific area (e.g. close to overhead power lines)?
7. Does it have handrails and toeboards fitted? (Regulations insists on these when working height is at 2 m or more. This should not preclude taking the same measures at lower heights.)
8. Does it provide a safe working area? If in doubt seek advice, as there are many injuries each year caused by falls or by material dropping from heights. The wearing of safety helmets is a secondary protection and should not be allowed to encourage carelessness.

36.15 Materials handling

Whenever possible, lifting of excessive weights should be by mechanical means and should only be done manually if there is no practicable method of obtaining access for mechanical equipment. It follows that material must only be moved manually if it is within the capability of the persons involved and that there is no risk of injury.

36.15.1 Manual lifting

Correct manual lifting and handling of material prevents strains and injury and also reduces the effort required. Persistently incorrect lifting and handling may lead to the person suffering from permanent back strain or other health problems. Points to consider when preparing to lift material are:

1. Can you lift it yourself?
2. Does it have sharp corners or edges?
3. Is there any obstruction where you have to walk or place the material?

When lifting the material:

1. Stand with your feet apart, with one foot in the direction you intend to move.
2. Grip the object with the palms of your hands.
3. Keep a straight back and, bending slightly at the knees, use your thigh muscles to lift.
4. With elbows in and arms close to the body, slowly rise.
5. Keeping your chin in and raising your head, lift the object to the height required, adjusting the position of your feet to ensure that you maintain your balance.
6. Take care when a hazardous substance (e.g. acid) is lifted.
7. Always wear the correct protective clothing.
8. Get help with heavy loads.

36.15.2 Mechanical lifting

In addition to the general rules of the Health and Safety at Work, etc. Act 1974 there are specific requirements under the following legislation with respect to chains, ropes and lifting tackle:

Factories Act 1961
The Construction (Lifting Operations) Regulations 1961
The Docks Regulations 1934
The Shipbuilding and Ship-Repairing Regulations 1960

Depending on the location of the workplace, the regulations will indicate the frequency of tests and examination of the equipment.

36.16 Portable tools and equipment

1. Portable electrical handtools and equipment shall be properly earthed and wound to operate on 110 V a.c. centre tapped to earth supply, and shall only be connected to the system by permanent joints or proper connections.
2. Portable lighting, when used in wet conditions or confined spaces, should operate at no more than 25 V a.c. single phase, and must be fitted with the correct guard.
3. Persons operating abrasive wheels (e.g. bench grinders, machine grinders, disc grinders) should be trained in their use. To dress or mount (change) a wheel or a disc, the operative must have attended a training course in compliance with the Abrasive Wheels Regulations 1970. Attendees at such course must be recorded in the company register (F2346) for the purpose of the Abrasive Wheels Regulations 1970.
4. It follows that with any tool or equipment that is potentially dangerous or operators who may suffer injury from its use must be protected. To meet this requirement, training must be provided.

36.17 Confined spaces

Some plant and equipment are immediately classified as confined spaces, but extreme caution is necessary in the assessment of other areas. Section 2 of the Health and Safety at Work, etc. Act 1974 requires employers to ensure the health and safety at work of their employees; this duty is 'so far as is reasonably practicable'. Therefore as work in confined spaces is potentially dangerous this Section of the Act clearly requires employers to ensure that there is no risk to their employees when working in such an area.

Where work is to be carried out in a confined space within a factory Section 30 of the Factories Act 1961 lays down specific requirements when work is performed on or in a number of items of plant and equipment. No person must enter or be in a confined space unless he or she has been authorized to do so, and all the appropriate safety precautions have been strictly adhered to prior to entry and while within it.

Points that must be checked prior to the issue of the appropriate permit to work should include the following:

1. Are there fumes present?
2. Is there an adequate supply of maintainable oxygen?
3. Will breathing apparatus be required? Has the operative had training in its use and is there adequate support?
4. Is there any hazardous residue in the 'space'?
5. Has the space been tested for being clean and gas free?
6. Is there a supporting person standing by in the event of an emergency?
7. Is there a safety harness and rope available for use when the work is to be carried out? The operative entering the space will wear a harness while the operative standing by the entrance will retain the free end of the rope attached to the safety harness.

36.18 Electricity

The Electricity at Work Regulations 1989 revoke, among others, the Electricity Regulations 1908 and the Electricity (Factories Act) Special Regulations 1944, which have been the most important electricity regulations for many years. These regulations, effective from 1 April 1990, place a responsibility on both employer and employee to comply as far as it is reasonable and practicable within their control. Major changes from the previous regulations are as follows.

Regulation 13: Precautions for work on equipment made dead

In order to prevent danger while work is carried out on or near electrical equipment which has been made dead, adequate precautions shall be taken to prevent it from becoming electrically charged during that work if danger may arise thereby.

Regulation 14: Work on or near live conductors

No person shall be engaged in any work activity on or near any live conductor (other than one suitably covered with insulating material) so that danger may arise unless:

1. It is unreasonable in all circumstances for it to be dead;
2. It is reasonable in all circumstances for the person to be at work on or near it while it is live; and
3. Suitable precautions (including, where necessary, the provision of suitable protective equipment) are taken to prevent injury.

Regulation 16: Persons to be competent to prevent danger and injury

No person shall be engaged in any work activity where technical knowledge or experience is necessary to prevent danger or, where appropriate, injury, unless he or she possesses such knowledge or experience, or is under such degree of supervision as may be appropriate having regard to the nature of the work.

Additional safety precautions

1. You must not work on or interfere with any electrical equipment unless you have been authorized to do so.
2. Do not use electrical equipment or switchrooms as a storage area.
3. Assume that all electrical distribution circuits are live.
4. Obtain the necessary authority/permit to work before commencing any work on electrical systems.
5. Do not improvise: electricity can kill.

36.19 Plant and equipment

36.19.1 Compressed air

Compressed air can be dangerous if used incorrectly:

1. Never use compressed air for cleaning. The pressure from nozzles may blow particles of dirt and dust into the eyes, ears or skin of the person using the nozzle for cleaning or of others in the vicinity.
2. Do not dust yourself down with the compressed air nozzle.
3. Indulging in horseplay with compressed air can have disastrous results.

Persons employed to use compressed air in carrying out a specific work task must be correctly equipped to prevent injury. The minimum requirement is to wear eye protectors which conform to BS 2092.

36.19.2 Steam boilers

Section 38 of the Factories Act 1961 defines a steam boiler as a 'any closed vessel in which for any purpose steam is generated under pressure greater than atmospheric pressure'. Economizers used to heat water being fed to such a vessel and superheater for heating steam are also included. Every boiler must be fitted with the recommended safety measures (e.g. safety valve, stop valve, water gauge, low-water alarms, pressure gauges, etc.).

Prior to being brought into use in a factory, boilers must have been thoroughly examined (when cold) by a competent person in accordance with the Examination of Steam Boilers Regulations 1964. Following subsequent examination at the recommended frequency, the results on the prescribed forms must be attached to the General Register (F31) within 28 days.

Entry into boilers are controlled by Section 34 of the Factories Act 1961, which states that no person shall enter or be in any steam boiler which is one of a range of two or more boilers unless:

1. All inlets through which steam or hot water might otherwise enter the boiler from any part of the range are disconnected from that part;
2. All valves or taps controlling the entry of steam or hot water are closed and securely locked, and where the boiler has a blow-off pipe in common with one or more other boilers or delivering into a common blow-off vessel or sump, the blow-off valve or tap on each boiler is so constructed that it can only be opened by a key which cannot be removed until the valve or tap is closed and is the only key in use for that set of blow-off valves or taps.

36.19.3 Pressure systems

The Pressure Systems and Transportable Gas Regulations 1989, effective from 1 July 1990 in parts, will be fully effective from 1 July 1994. Aspects that became effective from 1 July 1990 are:

1. The supply of information for new plant;
2. The proper installation of plant; and
3. The establishment of safe operating limits on existing plants.

The Regulations relating to gas cylinders apply to suppliers, importers, fillers and owners of cylinders from 1 January 1991 if:

1. The plant contains a compressed gas (such as compressed air or liquified gas) at a pressure greater than 0.5 bar (about 7 psi) above atmospheric;
2. If steam systems with a pressure vessel are used at work, irrespective of the pressure;
3. The plant is used at work by employees and/or the self-employed.

For full details of the requirements under these Regulations reference should be made to the following:

The Pressure Systems and Transportable Gas Containers Regulations 1981
Approved Code of Practice, Safety of pressure systems
Approved Code of Practice, Safety of transportable gas containers
HS(R) 30, A Guide to the Pressure Systems and Transportable Gas Containers Regulations 1989

The Approved Codes of Practice list relevant HSE Guidance Notes.

36.19.4 Machinery

The requirement for safety precautions when operating or working on all types of machinery cannot be overstressed, in that the likelihood of an accident occurring on moving machinery is invariably high. It is essential that all these who work on such machines are adequately trained as to the operation and the safeguards fitted. Regulations that must be complied with include,

Health and Safety at Work, etc. Act 1974, Sections 2 and 33
Factories Act 1961, Section 14
Offices, Shops and Railway Premises Act 1963, Section 17

Other information can be found in BS 5304, Code of Practice for safety of machinery.

In premises in which the Factories Act 1961 apply, Section 15 of the Act acknowledges that certain dangerous parts of machinery can only be adjusted or lubricated while in motion. Employees carrying out this work must be properly trained and their tasks must be specified in writing by their employers.

Details of the appointed persons must be entered in Register F31. Also, such persons must be instructed in the requirements of the Operations at Unfenced Machinery Regulations 1938 and issued with the precautionary leaflet F2487. General precautions to observe include:

1. Machines must not be used unless properly guarded (or operatives are trained). All damaged guards/security measures must be reported.
2. Never reach over operating machinery or wear loose clothing when working at or on machinery.
3. Never start up a machine until you are certain that no one can be injured by doing so.
4. Keep the floor area around the machine clean.
5. Do not clean the machine when it is in operation.
6. Wear appropriate protective clothing.

36.20 Safety signs and pipeline identification

36.20.1 Safety signs

The Safety Signs Regulations 1980 apply to signs which give a health and safety message to people at work by using certain shapes, colours and pictorial symbols. There are four basic types of safety signs which the Regulations require to conform to BS 5378: 1980, Safety signs and colours:

1. Solid blue circle: mandatory ('You must do');
2. Red circle with a red band across the diameter: prohibition ('Do not do');
3. Yellow triangle with black border: caution ('Warning of danger');
4. Solid green square: information ('Safe conditions').

Graphic symbols may be placed in the above shapes to give them more meaning and there may also be text affixed to them.

36.20.2 Pipeline identification

Pipelines should be marked clearly as to the contents. This can be achieved by marking them in accordance with BS 1710, System (contents) and BS 4800, Colours (indicating contents).

36.21 Asbestos

In the 1987 Regulations, asbestos is defined as any of the following minerals; crocidolite, amosite, chrysotile, fibrous anthophyllite, fibrous actinolite, fibrous tremolite and any mixture containing any of these. Before carrying out work on any substance suspected of being asbestos a competent person must be called to advise on its possible composition. Such a person will advise on the measures that must be taken to avoid any hazard to the occupants or others likely to be affected. Companies with asbestos on their premises should hold the appropriate documents. Current Regulations include

Asbestos (Licensing) Regulations 1983
Asbestos (Prohibitions) Regulations (amended) 1985
Control of Asbestos at Work Regulations 1987

36.22 COSHH

The Control of Substances Hazardous to Health (COSHH) Regulations 1989 cover virtually all substances hazardous to health. Only asbestos, lead, materials producing ionizing radiations and substances below ground in mines (which all have their own legislation) are excluded. The Regulations set out measures that employers must implement. Failure to comply with COSHH, in addition to exposing employees and others to risk, constitutes an offence and is subject to penalties under the Health and Safety at Work Act, etc. 1974.

Substances that are hazardous to health include:

1. Those labelled as dangerous (i.e. very toxic, toxic, harmful, irritant or corrosive);
2. Agricultural pesticides and other chemicals used in farming;
3. Those with occupational exposure limits;
4. Harmful microorganisms;
5. Substantial quantities of dust;
6. Any material, mixture or compound used at work, or arising from work activities which can harm people's health.

In works, premises, factories, etc. there will be substances in use that come within the control of COSHH. Seek advice!

The following publications give detailed information on COSHH and its requirements:

1. Control of Substances Hazardous to Health Regulations 1988, Approved Code of Practice Control of substances hazardous to health and Approved Code of Practice Control of carcinogenic substances;
2. Occupational Exposure Limits: Guidance Note EH40 (revised and issued yearly);
3. Control of Substances Hazardous to Health Regulations 1988 and Approved Code of Practice Control of substances hazardous to health in fumigation operations;
4. Approved Code of Practice Control of vinyl chloride at work;
5. COSHH Assessments (a step-by-step guide to assessment and the skills needed for it);
6. Also available are free leaflets from the HSE.

36.23 Lead

The Control of Lead at Work Regulations 1980 apply to work which exposes persons (both employees and others) to lead as defined in Regulation 2(1) (i.e. to the metal and its alloys to compounds of lead, which will include both organic and inorganic compounds, and to lead as a constituent of any substance or material) when lead is in a form in which it is liable to be inhaled, ingested or otherwise absorbed by persons. Relevant publications are:

1. Approved Code of Practice Control of lead at work;
2. Guidance Notes;
   EH 28 Control of lead: air sampling techniques and strategies
   EH 29 Control of lead: outside workers
   EH 30 Control of lead: pottery and related industries;
3. Also available are free leaflets from the HSE.

36.24 Other information

There are numerous leaflets that can be obtained from the local offices of the Health and Safety Executive or HSE Information Points. Remember, if you have any doubts about the health and safety risks involved with with respect to any substance you may consider using, or any work you are planning to undertake, seek advice. This information may be obtained from your company safety officer, consultants, manufacturer/supplier or equipment/material/substance, HSE or local Environmental Health Officer. Use their expertise and knowledge. Do not be a statistic.

36.25 Assessment of potential hazards

36.25.1 Purpose

A assessment of potential hazards (APH) shall be carried out prior to work tasks being issued to determine:

1. Physical hazards;
2. Access/exit;
3. Potentially unsafe practices likely to be implemented in carrying out the work;
4. Safe working procedures;
5. Whether a permit to work is required.

36.25.2 Working practices

An assessment of each element of the programme of work

is carried out using the Assessment of Potential Hazards Form APH1 (see below), which is to be completed by the person responsible for the task to be carried out. This may include 'hot' work; cutting/grinding; work on electrical systems; working alone; use of hazardous materials (e.g. chemicals, solvents, etc.); confined spaces or any other work which may be of a hazardous nature.

In establishing potential hazards in carrying out specific tasks, other sources may have to be consulted (e.g. manufacturer, supplier, HSE, environmental health officer, consultants, etc.) and protective clothing and equipment must be considered.

Prior to implementing any APH within the Method Statement for the task the APH and Method Statement must first be approved by the responsible person's senior and/or the manager of the department responsible for the plant and equipment. The approved APH and Method Statements will be brought to the attention of all persons called upon to carry out the specific task. Copies of the APHs will be issued to:

The department responsible for the person carrying out work;
The department in which work is being done;
Health and safety/security departments.

APHs will be reviewed annually to ensure compliance with current plant and equipment, procedures and changes in health and safety recommendations.

In the event of accidents, an investigation should include examination of all aspects of the APH.

Assessment of Potential Hazards – Analysis Form APH1
Job Description/Classification Department Date

Task Elements	Hazards	Safe Working Practices

Completed by:............ Department..........
Approved by:............ Department..........

36.26 Alternative method of assessing hazards

An alternative method of assessing the hazards of a specific work task may be as follows. The work task envisaged, to be carried out by a contractor, is that of connecting an additional outlet to fuel oil tanks situated within an oil tank room, which is located below ground level.

36.26.1 Work on fuel oil tanks

Relevant information:
Guidance Note CS1: Industrial use of flammable gas detectors (HMSO)
Guidance Note CS2: Storage of highly flammable liquids (HMSO)
Guidance Note CS15: The gas-freeing and cleaning of tanks containing flammable residues (HMSO)
Guidance Note GS5: Entry into confined spaces (HMSO)
Hot Work on Tanks and Drums: IND(G)35(L) M20 (HSE)

36.26.2 Procedures: 1

1. Identify the work task.
2. Examine alternative methods for carrying out the task.
3. Select the appropriate method.
4. Plan how the work will be carried out.
5. Who is to carry out the work task and when?
6. Carry out the work task.
7. Continually monitor the work task and the safety measures taken.
8. Review the procedures.

36.26.3 Procedures: 2

1. Identify the work task. Can an alternative method be suggested to the client, hence eliminating potential hazards from the work task?
   (a) Task 1: Ensure that there is no hazard from tanks, etc. that have to be worked on or from the locality of work (i.e. tanks have been emptied, degreased and gas-freed and checked and tested);
   (b) Task 2: As per client's instructions (client provides Method of Work Statement).
2. Examine alternative methods for carrying out the task.
   (a) Can the work be carried out without the use of heat (i.e. welding, burning, electrical equipment (arcing), cutting causing heat or sparks, assembled externally and flanged rather than welded?
   (b) Will the work be carried out in stages, i.e. will some of the adjacent tanks remain charged?
3. Select the appropriate method.
4. Plan how the work will be carried out.
5. Discuss with the client:
   (a) The period during which work can be carried out;
   (b) Their involvement and responsibility with respect to fire precautions, extinguishers, permit to work, liaison with fire brigade, fire officer attendance;
   (c) Who is responsible for emptying the fuel tanks;
   (d) Cleaning, degreasing and gas freeing the tanks;
   (e) Testing for explosions or oxygen deficiency.
6. Nominate a person to be in charge.
7. Prepare a programme of work.

8. Is training required on use of respirators, entry into confined spaces, fire prevention and use of extinguishers?
9. Who is to carry out the work task and when?
   Task 1:
   (a) Contractor;
   (b) Nominated specialist contractor;
   (c) Client;
   Task 2:
   (a) Contractor;
   (b) Nominated specialist contractor.
10. Carry out the work task *only* when a permit to work has been issued which states that the work area is safe for working and details the safety checks that have been made and precautions to be taken.
11. Continually monitor the work task and the safety measures taken.
    (a) Ensure that Task 1 has been carried out satisfactorily and check report/record;
    (b) Continually monitor working environment for explosion or fire risk and/or oxygen deficiency.
12. Review the procedures.
    (a) Ensure that the company's safety officer checks the procedures;
    (b) Have the procedures been effective or should they be amended? If so, why?

If specialist contractors are to be engaged a Method Statement should first be requested from them on how they plan to carry out the work.

36.27 Permits to work

Where there is a risk to the employee carrying out work, in addition to a Method Statement a permit to work should be used. A permit to work is a formal written means of making sure that potentially dangerous jobs are examined first, before authorizing the work to commence. Its task is twofold: it ensures that the person making the system safe and the person who is to carry out the work have both checked that it is safe to carry out the work task. Permits to work cannot be transferred to other parties. If any circumstances change from the issue of the original permit to work, it must be cancelled and a new one prepared.

Permits to work can be of differing formats. Therefore it is essential that when a permit to work has been issued it should be read carefully and understood and all possible measures taken to reduce or eliminate the known danger. Permits can be issued for a variety of work tasks (e.g. work on high-voltage electrical systems, steam boilers, hot work, confined spaces, etc.). The permit below is an example of differing types of work tasks.

36.27.1 Permission to Work on Electrical Systems Certificate (not exceeding 1000 V)

This certificate must always be raised by the authorized person and sections 1 to 3 inclusive completed before any work is allowed to be carried out. A consecutive serial number must be entered on the certificate.

Where the authorized person is to carry out the work, he must complete section as the person responsible for work being carried out and retain the top copy on him throughout the period the work is being carried out. The bottom control copy must remain in the register.

Where a competent person is to carry out the work he must complete section 3 as the certificate holder. The authorized person then issues him the top copy which he must retain on him throughout the period the work is being carried out. The bottom copy is retained in the register.

On completion or suspension of the work the responsible person must return the top copy to the register, so that the authorized person ensures that sections 7 and 8 are completed on both top and bottom copies of the certificate. The bottom copy must always be retained in the register for record purposes. Before any suspended work is allowed to be restarted, a new certificate must be raised and processed by the authorized person as above.

Examples of tasks that will require the issuing of the Permission to Work on Electrical Systems Certificate are indicated in Method Statements for Work on Electrical Systems.

Note

1. Authorized and competent persons are nominated by the company after they have been given the appropriate training in both the tasks to be carried out and their responsibility as nominated persons.
2. The electrical engineer would normally stipulate the tasks that require the use of Method Statements for Work on Electrical Systems and would also prepare the Method Statements.
3. A similar format of certificate would suffice as a permit to work for high-voltage systems (i.e. exceeding 1000 V).

An example of this certificate as well as a Hot Work Permit are given below.

36.28 Working alone

36.28.1 Company and contractor's responsibilities for the lone worker

Manager's responsible for services within works, offices or premises have an additional task in that when engaging service contractors they then have joint responsibility under the Health and Safety at Work, etc. Act 1974 for the health and safety of the contractor's employees while on their premises. When engaging contractors to carry out work within the premises, systems must be implemented by which the contractor's employee works in a safe manner and does not create a hazard to the premise's occupants or staff while carrying out this work. This responsibility is greater when there is an employee or service contractor working alone, as in most instances the premises' communications do not allow for such circumstances (e.g. the lone employee may be working in remote areas such as plant rooms).

Company **Serial No . . . .**

PERMISSION TO WORK ON ELECTRICAL SYSTEMS CERTIFICATE
(not exceeding 1000 V)

1. *This Certificate is issued for the following Authorized Work. No other work may be carried out. Entrance to work area is limited to those authorized. Tools/access equipment etc. must be approved by authorized person prior to them being taken into the work area.*
2. **Work Details/Job No./Method Statement Ref. No**
. . . .
. . . .
. . . .
3. **Work Location:**
. . . .
. . . .
4. **Method of Isolation:**
The equipment on which the work is to be carried out has been proven dead after isolation at: . . . .
. . . .
Care must be exercised as the equipment is still live at: . . . .
. . . .
Safety screening has been applied at: . . . .
. . . . . Lock-Off Key Nos . . . . Danger Notices Posted: Y/N
Warning Signs Displayed: Y/N Safety Barriers Erected: Y/N
5. **Authorized Person issuing Certificate:**
. . . . Designation . . . .
Date . / . / . Time . . . . Signature . . . .
6. **Person Responsible for Work being carried out:**
Company/Dept . . . . Name . . . .
Designation . . . . Signature . . . .
7. **Work Completed/Suspended:** Date . / . / . Time . . . .
Person Responsible for Work being carried out:
Company/Dept . . . . Name . . . .
Signature . . . .
8. **Cancellation of Certificate by Authorized Person:**
Receipt of Certificate indicating work is
Complete/Suspended:
Name . . . . Signature . . . .
Designation . . . . Date . / . / . Time . . . .

On cancellation of ths certificate, no other work must be carried out on the plant/equipment without a further certificate being issued.

To enable both company and contractor to comply with their specific responsibilities under the Health and Safety at Work, etc. Act 1974, a Lone Workers' policy, supported by a method of assessment of potential hazards relating to the work to be carried out, should be prepared. Such a system would prove beneficial when, for any reason, the regular lone worker cannot be found.

36.29 Safety policy for lone workers

In the Health and Safety at Work, etc. Act 1974 the following Sections apply:

Section 2: General duties of employers to their employees
Section 4: General duties of persons concerned with premises to persons other than their employees

Working alone presents a specific health and safety problem, and companies, department managers and contractors must ensure that there is a safe system whereby:

1. The person working alone is adequately trained for the work that is to be carried out;
2. Having carried out an assessment of potential hazards (APH), a Method Statement (including advice on whether a permit to work is required) is prepared;
3. The correct tools, equipment and manufacturers instructions are available;
4. Contact is maintained by:
   (a) Means of telephone at regular intervals;
   (b) Visits by others at regular intervals;
   (c) Persons passing through a 'Lone Worker' area are made aware of the lone worker's presence;
   (d) The intervals between contact being made should not exceed one hour.

Company **Serial No . . . .**

AUTHORIZATION TO WORK ON ELECTRICAL/MECHANICAL APPARATUS CERTIFICATE

1. This Certificate is issued for the following Authorized Work. *No other work other than that detailed must be carried out.*

Work Details:

. . . .

. . . .

Work Location:

. . . .

. . . .

Method of Isolation/Making Safe:

. . . .

. . . .

Lock-Off Key Nos . . . .

Danger Noticed Posted: Yes/No . . . .

Warning Signs Displayed: Yes/No . . . .

Authorized Person issuing Certificate . . . .

Designation . . . . Date . / . / . Time . . . .

Person Responsible for Work being carried out:

Company . . . . Name . . . .

Designation . . . . Signature . . . .

2. Work Completed: Date . / . / . Time . . . .

Person Responsible for Work carried out:

Company . . . . Name . . . .

Designation . . . . Signature . . . .

3. Authorized Person:

Receipt of Certificate indicating work is complete:

Designation . . . . Date . / . / . Time . . . .

Name . . . . Signature . . . .

N.B. Copy of this Certificate to be issued to Person responsible for carrying out Work.
Signature of Receipt . . . .

A lone worker must:

1. Comply with any Method Statement or safe system introduced for his or her health and safety;
2. Work safely at *all* times, and not take risks;
3. Understand the hazards/risks associated with the work to be carried out;
4. Understand the hazards/risks when *working alone*;
5. Ensure, where necessary, that a permit to work has been requested and issued.

36.30 Contractor's conditions and safe working practices

The use of contractors to carry out work that may in the past have been done by direct labour or during a normal sub-contract does not relieve the company from the responsibility of ensuring that all those working on their premises do carry out their work in a safe manner. The following contractor's conditions and safe working practices should form a part of the company's health and safety policy manual.

36.30.1 Introduction

These conditions and safe working practices form part of the contract and are to be read in conjunction with the general conditions of the contract form. The company reserves the right to make alterations or additions to those conditions and practices as and when necessary. Contractors, sub-contractors or their employees must not communicate to persons other than the company staff any information relating to the work being carried out by the contractor or sub-contractors. Information, drawings, etc. must not be removed from company premises without prior permission from the company contact, who is the responsible person for arranging the appropriate clearance relating to that information.

36.30.2 Contents

Local conditions

1. On arrival at company premises
2. Prior to commencement of work
3. Attendance at company premises for more than one day.

HOT WORK PERMIT

Required when any of the following tasks are to be carried out:
Cutting; Welding; Soldering, Brazing and also during the use of any equipment which can produce heat, naked flames or dust (i.e. dust likely to affect Detector Heads)

Period of Work:
Commence: . . . . hrs . . . . day . . . . month . . . . .year
Completion / . . . . hrs . . . . day . . . . month . . . . .year
Expiry

Location of Work:
. . . .
. . . .
. . . .
. . . .

Description of Work:
. . . .
. . . .
. . . .
. . . .

Precautions:
Department in which work is to be carried out has been advised
Cutting/Welding/Brazing/Hot Work Equipment is in good order

Fire Alarm: ON/OFF **Halon System**: ON/OFF **Sprinkler System**: ON/OFF
Fire Notices Posted: YES/NO Insurance Informed YES/NO
Fire Detectors are protected

Precautions within 35 ft (10 m) of work area:
Floors and areas clear of combustibles
Walls and floor openings covered with e.g. Fire Blanket
Combustibles if not moved protected by Fire Blanket
Fire Watch will be provided for at least 30 minutes after work (incl. breaks)
Fire Watch is supplied with suitable extinguishers
Fire Watch is trained in use of extinguishers and method of raising alarm

---

I have examined the location of the work, checked that all precautions and preventive measures have been implemented.
Authorizing Signature . . . . **Name** . . . .
Designation . . . . hrs 90

I hereby declare that no work other than that stated above will be carried out.
Signed . . . . **Name** . . . . **Design'n/Company** . . . .

I hereby declare that the work has been completed and that the work and adjacent areas have been checked for fire 30 minutes after the completion of the work task.

Signed . . . . **Name** . . . . **Design'n/Company** . . . .
Authorizing Signature . . . . **Name** . . . .
Designation . . . . hrs 90

Safe working practices and procedures

1. Health and safety
2. Emergency procedures

36.31 Safe working practices and procedures

36.31.1 Health and safety

1. Health and safety is each employer's responsibility with respect to their employees under the Health and Safety at Work, etc. Act 1974.
2. Contractors must comply with all statutory and site regulations.
3. It is the contractor's responsibility to ensure that all work is carried out safely and by competent persons.
4. Method statements as to how the work is to be carried out, clearly indicating matters relating to health and safety, must be passed for approval to the nominated company officer prior to the commencement of work.
5. The successful contractor is responsible for ensuring that all items above are brought to the attention of any contractors to whom work is sub-contracted.

36.31.2 Protective clothing and equipment

1. Eye protection must comply with the Protection of Eyes Regulations 1974 and BS 2092, BS 1542 and BS 769.
2. Helmets must comply with BS 5240 and BS 4033.
3. Noise must be comply with the Noise at Work Regulations and ear protectors with BS 6344, Industrial hearing protectors.
4. Safety footwear *must* be worn where applicable (e.g. moving heavy objects; working with hazardous materials and in wet conditions). Footwear should comply with the following as appropriate: BS 1870, BS 4792, BS 5131, BS 5145, BS 5451, BS 5462 and BS 6159.
5. Safety belts must comply with BS 1397 and BS 6858.
6. Protective clothing (including gloves) *must* be used where applicable (e.g. normal duties; welding/cutting hot work; water treatment; application/use of chemicals; during servicing of batteries; descaling/chemical cleaning).
   Clothing: BS 1542, BS 5426, BS 2653, BS 4171, BS 6408, BS 4724;
   Gloves: BS 1651, BS 697, BS 6526.
7. Respirators must comply with BS 2091, BS 4555, BS 4558, BS 4771 and BS 4275. Also refer to Form F2486, published by HSE for suppliers of respirators suitable for use against aerosols containing *Legionella*.

When in doubt, consult with manufactures, suppliers, HSE, EMAS, the local authority or the environmental health officer.

36.31.3 Access equipment

1. Stepladders: BS 1129, BS 2037
2. Trestles: BS 1129, BS 2037, BS 1139
3. Scaffolds: BS 1139, DD72, CP 5973, BS 2482, BS 6037, BS 5974, BS 6289

Scaffold and other access equipment should comply with a standard as for the Construction (Working Places) Regulations 1966.

36.31.4 Warning signs and good housekeeping

All appropriate warning signs must comply with the Safety Signs Regulations 1980 and must be displayed prior to commencement and for the duration of the work to be carried out (see BS 5378). Workplaces must be kept safe and tidy and all access/exit areas clear of obstruction. Additional information is available from HSE publications.

36.31.5 Tools and equipment

The contractor will be responsible for the supply of all tools and equipment, unless otherwise agreed by company representatives:

1. Portable electrical handtools and equipment shall be properly earthed and wound to operate on 110 V a.c. centre tapped to earth supply, and shall only be connected to the system by permanent joints or proper connections.
2. Portable lighting must operate at no more than 25 V a.c. single phase, and must be fitted with the correct guard.
3. Equipment issued by the company is so provided on the understanding that it will be properly used by competent workers under adequate supervision, and that the contractor shall be solely responsible for and idemnify the company against loss or damage to such equipment and all or any loss, injury, death of any person or damage to any property whatsoever arising out of the use of such equipment.

36.31.6 Materials/parts/spares

1. Materials used must be of the highest standard, complying as necessary with the appropriate British Standard.
2. Materials supplied by the company must only be used for the work task for which they have been issued.

36.32 Emergency procedures

36.32.1 Fire

When arriving at the Company premises and being shown their work area the contractor's employees must:

1. Check the location of the nearest fire exit.
2. Make themselves familiar with fire alarm procedures.
3. Ensure that means of escape are kept clear at all times.

36.32.2 First aid

The contractor's employees must:

1. Make themselves familiar with the first aid procedures in the event of an accident.
2. Check the location of the nearest first aid persons and the first aid room if there is one on the premises.

36.33 *Contractor's Guide*

This is to be issued to an individual on arrival at a company's premises. All contractors' employees must adhere to the following procedures.

36.33.1 On arrival at company premises

1. Report to reception
2. Advise reception/office of:
   (a) Company contact who requested attendance;
   (b) Work to be carried out, stating department/location;
3. On being signed in/issued identity badge be advised of:
   (a) Whereabouts of work location; or
   (b) Be escorted to work location by member of staff; or
   (c) Be escorted by company contact;
4. Be issued with a copy of the *Contractor's Guide*.

36.33.2 Prior to commencement of work

The company contact/coordinator will:

1. Discuss items in the *Contractor's Guide* relevant to the work to be carried out;
2. Ensure that adequate notice has been issued to all other parties, e.g.:
   (a) Members of department, including deputy;
   (b) Others who may be carrying out work in that area;
   (c) Contact security/loss prevention coordinator for issue of permits to work, advice/instructions on health and safety, specific hazards relevant to work and work area and issue of fire extinguishers/fire blankets;
3. Give permission to commence work.

36.33.3 Security

If attendance at company premises is required on more than one day to complete the work the contractor's employees must:

1. Sign in/out at reception/security;
2. Collect/return;
   (a) Identity badge;
   (b) Permits to work (not to be removed from premises);
   (c) Keys (not to be removed from premises).
3. Report proposed absence from/return to premises to:
   (a) Contact;
   (b) Security/loss prevention coordinator if work requires a permit to work;
   (c) Facilities coordinator/department head if work involves general premises/site facilities/services.

36.33.4 Welfare/canteen

1. These facilities will be used by contractors subject to discussion with the company contact.

36.33.5 Useful contacts

These include the telephone numbers of contact, security, first aid, etc.

36.34 Addresses for health and safety information

BRE Building Research Station
Bucknalls Lane
Garston
Watford
WD2 7JR
Tel. 0923 894040

BRE Fire Research Station
Melrose Avenue
Borehamwood
Herts
WD6 2BL
Tel. 081-953 6177

British Compressed Air Society
Leicester House
8 Leicester Street
London
WC2H 7BN
Tel. 071-437 0678

British Non-Ferrous Metals Federation
10 Greenfield Crescent
Edgbaston
Birmingham
B15 3AU
Tel. 021 456 3322

British Occupational Hygiene Society
1 St Andrews Place
Regent's Park
London
NW1 4LB
Tel. 071-486 4860

British Safety Council
National Safety Centre
62/64 Chancellor's Road
London
W6 9RS
Tel. 081-741 1231/2371

British Standards Institution
Linford Wood
Milton Keynes MK14 6LE
Tel. 0908 220022

Building Employers Confederation
82 New Cavendish Street
London
W1M 8AD
Tel. 071-580 5588

Building Services Research and Information Association
Old Bracknell Lane West
Bracknell
Berks RG12 4AH
Tel. 0344 426511

Chain Testers Association of Great Britain
21–33 Woodgrange Road
London
E7 8BA
Tel. 081-519 3702

Chartered Institute of Building
Englemere
King's Ride
Ascot
Berkshire
SL5 8BJ
Tel. 0990 23355

Chartered Institution of Building Services Engineers
Delta House
222 Balham High Road
London
SW2 9BS
Tel. 081-675 5211

Chemical Industries Association
King's Buildings
Smith Square
London
SW1P 3JJ
Tel. 071-834 3399

Construction Health and Safety Group
c/o John Ryder Training Centre
St Anne's Road
Chertsey
Surrey KT16 9AT
Tel. 0932 561871

Construction Industry Training Board
Bircham Lewton
King's Lynn
Norfolk
PE31 6RH
Tel. 0553 776677

Department of Education and Science
Elizabeth House
39 York Road
London
SE1 7PH
Tel. 071-934 9000

Department of the Environment
2 Marsham Street
London
SW1P 3EB
Tel. 071-276 3000

Department of the Environment for Northern Ireland
Stormont
Belfast
BT4 3SS
Tel. 0232 63210

Ergonomics Society
c/o Dept of Human Sciences
University of Technology
Loughborough
Leicestershire
LE11 3TU
Tel. 0509 234904

Fire Extinguishing Trades Association
48a Eden Street
Kingston-upon-Thames
Surrey
KT1 1EE
Tel. 081-549 8839

Health and Safety Agency for Northern Ireland
Canada House
22 North Street
Belfast
BT1 1NW
Tel. 0232 243249

Health and Safety Commission
Baynard's House
1 Chepstow Place
London
W2 4TF
Tel. 071-229 3456

Health and Safety Executive
Baynard's House
1 Chepstow Place
London W2 4TF
Tel. 071-229 3456

Health Education Authority
Hamilton House
Mabledon Place
London
WC1H 9TX
Tel. 071-631 0930

Heating and Ventilating Contractors' Association
ESCA House
34 Palace Court
London
W2 4JG
Tel. 071-229 2488

HMSO
PO Box 276
London
SW8 5DT
Tel. 071-873 9090/0011

Industrial Safety (Protective Equipment)
Manufacturer's Association
69 Cannon Street
London
EC4N 5AB
Tel. 071-248 4444

Institution of Electrical Engineers
Savoy Place
London
WC2R 0BL
Tel. 071-240 1871

Institution of Chemical Engineers
165–171 Railway Terrace
Rugby
Warwickshire
CV21 3HQ
Tel. 0788 78214

Institution of Occupational Safety and Health
222 Uppingham Road
Leicester
LE5 0QG
Tel. 0533 768424

Institution of Plant Engineers
77 Great Peter Street
Westminster
London SW1P 2EZ
Tel. 071-233 2604

International Labour Office
Vincent House
Vincent Square
London SW1P 2NB
Tel. 071-828 6401

Loss Prevention Certification Board
Melrose Avenue
Borehamwood
Hertfordshire
WD6 2BJ
Tel. 081-207 2345

National Association of Scaffolding Contractors
82 New Cavendish Street
London
W1M 8AD
Tel. 071-580 5588

National Radiological Protection Board
Chilton
Didcot
Oxon
OX11 0RQ
Tel. 0235 831600

National Society for Clean Air
136 North Street
Brighton
East Sussex
BN1 1RG
Tel. 0273 26313

Paintmakers Association of Great Britain
6th Floor
Alembic House
93 Albert Embankment
London
SE1 7TY
Tel. 071-582 1185

Public Health Laboratory Service Board
61 Colindale Avenue
London
NW9 5DF
Tel. 081-200 1295

Royal Society for the Prevention of Accidents (ROSPA)
Cannon House
The Priority Queensway
Birmingham
B4 6BS
Tel. 021 200 2461

Society for the Prevention of Asbestosis and Industrial Diseases
38 Draper's Road
Enfield
Middlesex EN2 8LU
Tel. 081-366 1640

Sports Council
16 Upper Woburn Place
London
WC1H 0QP
Tel. 071-388 1277

Support Services Consultants
13 Abbots View
Kings Langley
Herts WD4 8AW
Tel. 0923 264758

Suspended Access Equipment Manufacturers' Association
82 New Cavendish Street
London
W1M 8AD
Tel. 071-580 5588

Timber Research and Development Association
Stocking Lane
Hughenden Valley
High Wycombe
Bucks
HP14 4ND
Tel. 024024 3091/2771/3956

Trades Union Congress
Congress House
23–28 Great Russell Street
London WC1B 3LS
Tel. 071-636 4030

Water Services Association
1 Queen Anne's Gate
London
SW1H 9BT
Tel. 971-222 8111

Welding Institute
Abington Hall
Abington
Cambridge
CB1 6AL
Tel. 0223 891 1162

Welding Manufacturers' Association
Leicester House
8 Leicester Street
London
WC2H 7BN
Tel. 071-437 0678

World Health Organization
H1211
Geneva 27
Switzerland
Tel. 022 91 21 11

37 Education and Training

Roger S Pratt, ALU, CEng, MIMfgE, MBIM, MSAE
Secretary-General, The Institution of Plant Engineers

Contents

37.1 The professional plant engineer

The profession of engineering, in contrast to many others, is extremely wide ranging in the spread of topics, technologies and specializations included under the overall heading. The early engineers, the creative geniuses of their day, encompassed all these latter-day specializations, famous examples being Brunel, Stephenson and Telford. Engineers have been at the heart of all technological and scientific progress. Without them the world as we know it today would not exist.

This has been despite the fact that the UK has developed with a culture which is indifferent to engineering, the 'respectable' professions being those such as law or medicine, offering more money and prestige. This deeply rooted British attitude was supported by an education system in which on the whole applied science – engineering – was not studied in schools or universities. This contrasts with the rest of Europe, where such studies were an important part of the curricula of many schools and universities as early as the eighteenth century. Engineering was not considered suitable for those with the ability to enter a British university, where arts and sciences were studied.

The need for education in engineering in the UK was met by the development of Mechanics Institutes. By the middle of the nineteenth century around 120 000 students per annum attended some 700 institutes on a part-time basis, thus laying the foundations for the pattern of engineering education in the UK. In 1840 the first chair in an engineering discipline (civil engineering) was established, at Glasgow University, soon to be followed by one at University College, London. Oxford and Cambridge were late on the scene, establishing chairs in engineering in 1875 and 1910, respectively.

Also peculiar to the UK is a somewhat confusing array of professional engineering institutions. These were originally learned societies where like-minded people met to exchange views and information. They developed into qualifying bodies by setting levels of experience and academic attainment for different grades of membership. The oldest professional engineering institution in the UK is the Institution of Civil Engineers, established in 1818. The Institution of Mechanical Engineers was established in 1847 and the Institution of Electrical Engineers in 1871. Three quarters of the approximately 50 institutions which are the Nominated Bodies of the Engineering Council were founded in the twentieth century, some quite recently, reflecting the growth of certain engineering disciplines such as nuclear engineering, computing and electronics.

37.2 The Institution of Plant Engineers

The Institution of Plant Engineers (IPlantE) had its origins in the Second World War, when engineers who found themselves responsible for the operation and maintenance of the large excavators and other mobile plant brought from the USA to work open-cast coal met together for the exchange of information and to discuss their problems. These meetings were so successful that the engineers concerned decided to continue them in a more formal manner through the medium of a properly incorporated body. The Memorandum of Association of 'Incorporated Plant Engineers' was subsequently signed on 3 September 1946.

The concept of an engineering institution which covered a wide field attracted engineers from many different areas of activity, including industrial, municipal and service establishments, civil engineering projects, transport undertakings, design, research and education. By 1947 branches of the Institution were holding monthly meetings in London, Birmingham, Manchester, Leeds, Newcastle, Glasgow and Bristol, and in the following year six more branches were established. There are now 20 branches in the UK and also a large number of members in other countries.

In January 1959 the Board of Trade gave permission for a change of title from 'Incorporated Plant Engineers' to 'The Institution of Plant Engineers'. This marked an important stage in the Institution's development, enabling it to take its place alongside other established engineering institutions. The Bureau of Engineer Surveyors, whose members had particular interests and expertise in relation to the safety and insurance aspects of plant operation and maintenance, merged with the Institution in 1987, forming the basis of a new specialist division.

The Institution of Plant Engineers is therefore in many ways a small-scale reflection of the engineering profession as a whole, embracing a wide range of disciplines and activities. The Institution's members work in the fields of, and have responsibility for, designing, specifying, building, installing, overseeing, commissioning, operating and monitoring the efficiency of plant of all kinds. This can include most types of building, plant and equipment used in the manufacturing, chemical and process industries, educational establishments, warehouses, hospitals, office and residential accommodation, hotels, banks, theatres, concert halls and all types of transportation systems. In the broadest sense of the term, these are the assets of the organization in question, without which it could not function. The plant engineer thus carries out a key role as the practical manager of these assets.

37.3 Aims of the Institution

The aims of the Institution of Plant Engineers are:

1. To bring together those already qualified by the attainment of such standards of knowledge, training, conduct and experience as are desirable in the profession of plant engineering;
2. To promote the education and provide for the examination of students in the profession of plant engineering;
3. To encourage, advise on and take part in the education, training and retraining of those engaged in plant engineering activities at all levels;
4. To diffuse knowledge of plant engineering by every means, including lectures, papers, conferences and research;

5. To increase the operational efficiency of plant for the greater benefit and welfare of the community, bearing in mind the importance of the conservation of the environment and the preservation of amenity.

37.4 Organization

Overall direction of the Institution is vested in its Council, but much of the Institution's detailed work is carried out by committees and panels of members. Branches and divisions of the Institution are run by their own committeess, which arrange programmes of visits, lectures and other appropriate activities, spread throughout the year. Non-members are very welcome to attend most Institution events. The Institution publishes its journal, *The Plant Engineer*, and other technical information, and also organizes national conferences and exhibitions.

The Institution's permanent staff are always available to give help and advice on matters relating to membership, education and training, and Engineering Council registration.

37.5 Membership

Membership of the Institution of Plant Engineers is the hallmark of the professional plant engineer and is often a prerequisite for successful career progression. This will become increasingly so in post-1992 Europe, when evidence of appropriate professional qualifications may be a legal requirement for employment in many engineering appointments.

37.5.1 Membership requirements

A summary of the grades of membership and the personal requirements for each of these grades is shown in Table 37.1.

37.5.2 Courses leading to a career in plant engineering

The main courses leading to a career in plant engineering are the Business and Technician Education Council's (BTEC) Technician Certificate or Diploma in Plant Engineering and Higher National Certificate or Diploma in Plant Engineering. In Scotland the equivalents are the Scottish Technician and Vocational Education Council's (SCOTVEC) Technician Certificate in Mechanical Engineering (Plant Engineering Options) and Higher Certificate in Mechanical Engineering (Plant Engineering Options). Additionally, certain other BTEC Certificates and Diplomas and Higher National Certificates and Diplomas in subjects other than 'Plant Engineering' have been assessed by the Institution and approved for membership purposes. Degrees, degree course options and diplomas and higher degree course options in plant engineering are available at certain universities and polytechnics in the UK. Further guidance on courses and their entry requirements may be obtained from technical colleges, polytechnics or universities or from the Institution's membership department.

Table 37.1 Summary of IPlantE membership requirements

Class of membership	*Minimum age (years)*	*Minimum academic qualifications*	*Evidence of competence*	*Minimum responsibility*
Student Member	16	Engaged in engineering studies and training		
Graduate Member	18	BTEC NC/ND or HNC/HND or degrees in EC approved subjects	Engaged on an EC approved system of training and experience	
Associate	—	—	Employed in an allied industry or profession	
Associate Member (AMIPlantE)	21	BTEC NC/ND or ONC/D or CGLI Part II in EC approved subjects	4 years combined training and experience	
Member (MIPlantE) (i)	23	BTEC HNC/HND or HNC/D or CGLI FTC in EC approved subjects	4 years combined training and experience	2 years of responsible experience
Member MIPlantE) (ii)	35	Technical Paper and Interview	15 years combined training and experience	2 years of responsible experience
Member (MIPlantE) (iii)	26	At Membership Panel's discretion	8 years combined training and experience	2 years of responsible experience
Fellow (FIPlantE) (i)	25	EC approved degree and interview	4 years combined training and experience	2 years in responsible appointments
Fellow (FIPlantE) (ii)	35	Technical Paper and interview	15 years combined training and experience	2 years in responsible appointments
Fellow (FIPlantE) (iii)	35	At Membership Panel's discretion	15 years combined training and experience	2 years in responsible appointments

37.6 Registration with The Engineering Council

An individual engineer's registration with The Engineering Council is a further valuable indication of professional attainment and standing. The Engineering Council was established by Royal Charter in 1981 to:

1. Promote and develop the science and best practice of engineering in the UK;
2. Ensure the supply and best use of engineers;
3. Coordinate the activities of the engineering profession.

The Charter empowers The Engineering Council to establish and maintain a Register of qualified engineers. The registrants may, where appropriate, use one of the following titles and designatory letters:

Chartered Engineer (CEng)
Incorporated Engineer (IEng)
Engineering Technician (EngTech)

Each of these three qualifications is obtained in three stages. Stage 1 indicates attainment of the required academic standard, Stage 2 that approved training has been received and Stage 3 that responsible experience has been gained. The titles may only be used at Stage 3.

The Institution of Plant Engineers is a Nominated Body of The Engineering Council (EC) and is thus able to nominate members in appropriate membership grades for EC registration.

37.7 Registration as a European Engineer

Registration with the European Federation of National Engineering Associations (FEANI) is now open to UK engineers, and may be helpful to careers in post-1992 Europe. Such registration is available at two levels, Group 1 and Group 2. Group 1 is normally appropriate for engineers having the education, training and experience to qualify them for the title Chartered Engineer. Group 2 is approximately appropriate for those qualified to Incorporated Engineer level, but at the time of writing (late 1989) the matter has not been finalized. Further information and FEANI application forms are available from the IPlantE's membership department.

As mentioned above, FEANI Group 1 registration is for those registered as Chartered Engineers. Registration with FEANI will allow the engineer concerned to use the title European Engineer. This title has the designatory letters Eur Ing, which should be used as a prefix (for example, Eur Ing John B. Smith, CEng MIPlantE).

37.8 Professional engineering development

Throughout the professional life of most engineers there is a need to acquire new knowledge to enable them to tackle the technical and managerial problems that they face from day to day. Recent advances in technology, materials and processes emphasize this need, but with ever-increasing demands on time, opportunities to attend full-time courses are few. The plant engineer must therefore rely upon a Continuing Education and Training (CET) programme to enable successful updating to take place, thus enhancing his or her professional development.

The Engineering Council places considerable emphasis on CET as an essential part of a professional engineer's development, anticipating that in due course CET will form a normal part of an engineer's career and that such CET activity will be noted in his or her personal career record.

To enable those engineers engaged in plant engineering to look to the future, the Institution of Plant Engineers has formulated a simple procedure for recording an engineer's attendance at activities which contribute to CET and have been approved by the Insitution for that purpose. Further information may be obtained from the Institution.

37.9 Addresses for further information

The Institution of Plant Engineers
77 Great Peter Street
London SW1P 2EZ
Telephone 071-233 2855

Business and Technician Education Council
Central House
Upper Woburn Place
London WC1 0HH
Telephone 071-388 3288

Scottish Vocational Education Council
Hannover House
24 Douglas Street
Glasgow G2 7NQ
Telephone 041-248 7900

City and Guilds of London Institute
76 Portland Place
London W1
Telephone 071-580 3050

The Engineering Council
Canberra House
Maltravers Street
London WC2R 3ER
Telephone 071-249 7891

38 Lubrication

Stuart McGrory
BP Oil UK Ltd

Contents

38.1 Introduction

This chapter examines the need for lubrication and the types of lubricant available. Various applications are considered, including engines gears, hydraulic equipment, machine tools, metal cutting and working fluids, compressors, turbines and electrical oils. The care of lubricants on-site, application of planned lubrication and inclusion within overall maintenance management are also examined.

The plant engineer's objective must be to ensure that plant operates at a profit. If overall efficiency of operation is to be achieved, and the working costs of plant kept within acceptable bounds, time must be set aside for the control and application of lubrication. The evolution of lubricants and their application, has continued ever since the beginning of the Industrial Revolution, and as the pattern of industry becomes increasingly more complex, the standard of performance of lubricants becomes progressively more important.

38.2 Lubrication – the added value

All machines depend for their accuracy on the strength of their component parts, their bearings and on the type and efficiency of their lubrication systems. Many machine bearings are subjected to extremely heavy shock loads, or intermittent loads, or exposure to unfavourable environmental conditions, yet in spite of this they must always maintain their setting accuracy.

Accuracy and reproducibility are of vital importance to industry. Quite apart from the effect of these factors on the final product, several plant items are frequently links in a continuous chain of production processes. A sizing error in one machine, for example, could overstress and damage the succeeding machinery. Similarly, an error in a press may increase stress on the tool and could necessitate an additional operation to remove excessive 'flash'. Wear in a material preparation unit could allow oversize material to be passed to a moulding machine, creating an overload situation with consequent damage.

The reduction of friction is only one of the functions of a lubricant. It must remove heat (often in large amounts), protect bearings from damage and preserve the working accuracy and alignment of the structure. It must also protect bearings, gears and other parts against corrosion, and must itself be non-corrosive. Sometimes it may be required to seal shafts and bearings against moisture and the ingress of contaminating particles. The lubricant must be of the correct viscosity for its application and may need additives to meet specification requirements. It must also be non-toxic, and both chemically and physically stable.

Lubrication plays a vital role in the operation of industrial plant. For example, in a heavy rolling mill the lubrication system, though mostly out of sight in the oil cellar, may have a capacity of many thousands of litres and exceed in bulk the mill itself. Lubrication systems of this size and complexity are usually fully automatic, with many interlocks and other safety features. Even with the smallest machines, automatic lubrication is becoming more popular. Where an automatic system would be impracticable or uneconomical, as with many older or less complex machines, it is nevertheless important that lubrication be carried out in accordance with a planned schedule.

It can be seen therefore that heavy demands are made on the plant concerned, and hence on the lubricants required for its efficient operation. The importance of the correct selection and application of lubricants, in the correct amount and at the right time, will be readily appreciated. The cost of providing high-quality lubrication is negligible compared with the material return it will bring in terms of longer working life, higher output of work and reduced maintenance costs.

38.3 Why a lubricant?

When the surfaces of two solid bodies are in contact a certain amount of force must be applied to one of them if relative motion is to occur. Taking a simple example, if a dry steel block is resting on a dry steel surface, relative sliding motion will not start until a force approximately equal to one fifth the weight of the steel block is applied. In general, the static friction between any two surfaces of similar materials is of this magnitude, and is expressed as a coefficient of friction of 0.2. As soon as the initial resistance is overcome, a very much smaller force will keep the slider moving at uniform velocity. This second frictional condition is called dynamic friction. In every bearing or sliding surface, in every type of machine, these two coefficients are of vital importance. Static friction sets the force required to start the machine and dynamic friction absorbs power which must be paid for in terms of fuel consumed. Also, friction resistance of unlubricated surfaces causes heating, rapid wear and even, under severe conditions, actual welding together of the two surfaces.

Lubrication, in the generally accepted sense of the word, means keeping moving surfaces completely separated by means of a layer of some liquid. When this is satisfactorily achieved the frictional resistance no longer depends on the solid surfaces but solely on the internal friction of the liquid which, in turn, is directly related to its viscosity. The more viscous the fluid, the greater the resistance, but this is never comparable with that existing between unlubricated surfaces.

38.3.1 Types of lubrication

Lubrication exists in one of three conditions:

1. Boundary lubrication
2. Elastohydrodynamic lubrication
3. Full fluid-film lubrication

Boundary lubrication is perhaps best defined as the lubrication of surfaces by fluid films so thin that the friction coefficient is affected by both the type of lubricant and the nature of the surface, and is largely independent of viscosity. A fluid lubricant introduced between two surfaces may spread to a microscopically thin film that reduces the sliding friction between the surfaces. The peaks of the high spots may touch, but interlocking occurs only to a limited extent and frictional resistance will be relatively low.

A variety of chemical additives can be incorporated in lubricating oils to improve their properties under boundary lubrication conditions. Some of these additives react with the surfaces to product an extremely thin layer of solid lubricant, which helps to separate the surfaces and prevent seizure. Others improve the resistance of the oil film to the effect of pressure.

Elastohydrodynamic lubrication provides the answer to why many mechanisms operate under conditions which are beyond the limits forecast by theory. It was previously thought that increasing pressure reduced oil film thickness until the asperities broke through, causing metal-to-metal contact. Research has shown, however, that the effect on mineral oil of high contact pressure is a large increase in the viscosity of the lubricant. This viscosity increase combined with the elasticity of the metal causes the oil film to act like a thin solid film, thus preventing metal-to-metal contact.

Full fluid-film lubrication can be illustrated by reference to the conditions existing in a properly designed plain bearing. If the two bearing surfaces can be separated completely by a fluid film, frictional wear of the surface is virtually eliminated. Resistance to motion will be reduced to a level governed largely by the viscosity of the lubricating fluid.

To generate a lubricating film within a bearing, the opposed surfaces must be forced apart by pressure generated within the fluid film. One way is to introduce the fluid under sufficient pressure at the point of maximum loading, but this hydrostatic method, although equally effective at all speeds, needs considerable power and is consequently to be avoided whenever a satisfactory alternative exists.

Above a certain critical speed, which depends mainly on the size and loading of the bearing and the viscosity of the lubricant, hydrodynamic forces are set up that part the surfaces and permit full fluid-film lubrication. At rest, the fluid film has been squeezed from beneath the shaft, leaving only an absorbed film on the contacting surfaces. As the shaft starts to revolve, friction between the journal and the bearing bore causes the shaft to climb up the inside of the bearing until torque, together with the increased thickness of lubricant film, overcomes frictional resistance and the shaft starts to slip at the point of contact. The rotating shaft then takes up its equilibrium position, where it is supported on a fluid film drawn beneath it by viscous friction (see Figure 38.1).

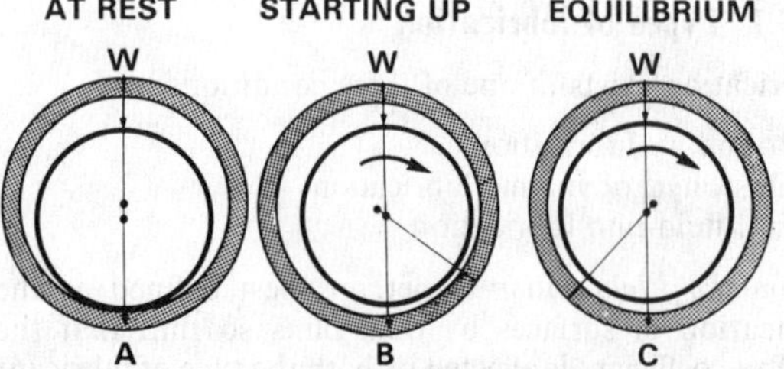

Figure 38.1 Journal positions during start-up while a hydrodynamic oil 'wedge' is being established

38.4 Physical characteristics of oils and greases

Reference will be made to the physical characteristics of lubricants as they affect their selection for various applications. These terms are well known to the lubricant supplier but are not always fully understood by the user. Brief descriptions of these characteristics are therefore given so that their significance may be appreciated.

38.4.1 Viscosity

This is the most important physical property of a lubricating oil; it is a measure of its internal friction or resistance to flow. In simple terms, it provides a measure of the thickness of a lubricating oil at a given temperature; the higher the viscosity, the thicker the oil. Accurate determination of viscosity involves measuring the rate of flow in capillary tubes, the unit of measurement being the centistoke (cSt). As oils become thinner on heating and thicker on cooling a viscosity figure must always be accompanied by the temperature at which it was determined.

The number of commercial viscosity systems can be confusing, and as kinematic viscometers are much more sensitive and consistent, there is a growing tendency to quote kinematic viscosities. The International Standards Organization (ISO) uses kinematic viscosity in its viscosity grade classification (Table 38.1). These ISO grade numbers are used by most oil companies in their

Table 38.1 ISO viscosity grade chart

ISO viscosity grade	*Mid-point kinematic viscosity*	*Kinematic viscosity limits cSt at 40°C (104°F)*	
		min.	*max.*
2	2.2	1.98	2.42
3	3.2	2.88	3.52
5	4.6	4.14	5.06
7	6.8	6.12	7.48
10	10	9.00	11.0
15	15	13.5	16.5
22	22	19.8	24.2
32	32	28.8	35.2
46	46	41.4	50.6
68	68	61.2	74.8
100	100	90.0	110
150	150	135	165
220	220	198	242
320	320	288	352
460	460	414	506
680	680	612	748
1000	1000	900	1100
1500	1500	1350	1650

industrial lubricant nomenclature. This provides the user with a simple verification of conformity regarding viscosity between plant manufacturer and oil supplier recommendations and also in the monitoring of correct oil usage on his plant.

38.4.2 Viscosity Index (VI)

This is a way of expressing the rate of change of viscosity with temperature. All oils become less viscous as the temperature increases. The rate of change of viscosity varies with different oils and is mainly dependent on the type of crude from which the oil is derived and the refining method. The higher the VI figure, the lower is the variation in viscosity relative to temperature. The VI of an oil is an important property in applications where the operating temperature is subject to considerable change.

38.4.3 Pour point

This is a rough measure of a limiting viscosity. It is the temperature 2.5°C above that at which the oil ceases to flow when the vessel in which it has been cooled is held horizontally for 5 s. The pour point is a guide to behaviour and care should always be taken that operating temperatures are above the figure specified by the oil manufacturer as the pour point of a given oil.

38.4.4 Flash point

The flash point is an oil is the temperature at which it gives off, under specified conditions, sufficient vapour to form a flammable mixture with air. This is very different from the temperature of spontaneous combustion. The test is an empirical one and the result depends upon the instrument used and the prescribed conditions. For example, the flash point may be 'closed' or 'open', depending on whether the test apparatus has a lid or not. As far as lubricating oils are concerned, the test is of limited significance, although it can be indicative of contamination (for example, the dilution of crankcase oil by fuel).

38.4.5 Penetration of grease

The most important physical property of a lubricating grease is its consistency, which is analogous to the viscosity of a liquid. This is determined by an indentation test in which a weighted metal cone is allowed to sink into the grease for a specified time. The depth to which the cone penetrates, in tenths of a millimetre, is a measure of the consistency. There is a widely accepted scale, that of the American National Lubricating Grease Institute (NLGI), that relates penetration to a consistency number (Figure 38.2).

The penetration test is used mainly to control manufacture and to classify greases and is, within limits, a guide to selection. Penetrations are often qualified by the terms 'worked' and 'unworked'. As greases are thixotropic, that is, they soften as a result of shear but harden again after shearing has stopped, the worked

Figure 38.2 Grease penetrometer

Table 38.2 NLGI consistency classification for greases

NLGI number	*ASTM worked penetration at 77°F*
000	455–475
00	400–435
0	355–385
1	310–340
2	265–295
3	220–250
4	175–205
5	130–160
6	85–115

penetration for a particular grease may be appreciably greater than the unworked penetration. The difference between these two figures may be a useful guide to the selection of greases for operating conditions that involve much churning – as small a difference as possible being desirable (see Table 38.2).

38.4.6 Drop point of grease

The drop point of a grease is an indication of change from a soft solid to a viscous fluid; its value depends completely on the conditions of test, particularly the rate of heating. The grease sample, which is held in a small metal cup with an orifice, is heated at a predetermined rate. The drop point is the temperature at which a drop of the sample falls from the cup.

The drop point is of limited significant as far as the user is concerned, for it gives no indication of the condition of the grease at lower temperatures, or of change in consistency or structure with heat. It is a very rough indication of a grease's resistance to heat and a guide to manufacture. The difference between the highest temperature at which a grease can be used and the drop point varies very much between types. It is at its maximum with some soda greases and much smaller with multi-purpose lithium products and modern complex greases.

38.5 Additives

Much highly stressed modern machinery runs under conditions in which a straight mineral oil is not adequate. Even the highest quality mineral oil can be unsatisfactory in response of its resistance to oxidation and its behaviour under pure boundary conditions, but it is possible to improve these characteristics by the addition of relatively small amounts of complex chemicals. This use of additives resembles in many ways the modification of the properties of steel by the addition of small amounts of other chemicals. It will be of value to have some knowledge of the effect of each type of additive.

38.5.1 Anti-oxidants

When mixed with oxygen, lubricating oil undergoes chemical degradation resulting in the formation of acidic products and sludge. This reaction, which is affected by temperature, the presence of catalysts such as copper and the composition of the oil, can be delayed by the inclusion of suitable additives.

Anti-oxidants are the most extensively used additives and will be found in oils and greases which are expected to operate for considerable periods or under conditions which would promote oxidation. Typical examples are crankcase oils and bearing greases.

38.5.2 Anti-foam

The entrainment of air in lubricating oil can be brought about by operating conditions (for example, churning) and by bad design such as a return pipe which is not submerged. The air bubbles naturally rise to the surface, and if they do not burst quickly, a blanket of foam will form on the oil surface. Further air escape is thus prevented and the oil becomes aerated. Oil in this condition can have an adverse affect on the system which, in extreme cases, could lead to an adverse affect on the system which, in extreme cases, could lead to machine failure. The function of an anti-foam additive is to assist in the burst of air bubbles when they reach the surface of the oil.

38.5.3 Anti-corrosion

The products of oil oxidation will attack metals, and this can be prevented by keeping the system free from pro-oxidative impurities and by the use of anti-oxidants. These additives will not, however, prevent rusting of ferrous surfaces when air and water are present in the mineral oil. The presence of absorbed air and moisture is inevitable in lubricating systems and therefore the oil must be inhibited against rusting. These additives, which are homogeneously mixed with the oil, have an affinity for metal, and a strongly absorbed oil film is formed on the metal surface which prevents the access of air and moisture.

38.5.4 Anti-wear

The increasing demands being made on equipment by the requirement for increased output from smaller units create problems of lubrication, even in systems where full-fluid film conditions generally exist. For instance, at start-up, after a period of rest, boundary lubrication conditions can exist and the mechanical wear that takes place could lead to equipment failure. Anti-wear additives, by their polar nature, help the oil to form a strongly absorbed layer on the metal surface which resists displacement under pressure, thereby reducing friction under boundary conditions.

38.5.5 Extreme pressure

Where high loading and severe sliding speeds exist between two metal surfaces, any oil film present is likely to be squeezed out. Under these conditions very high instantaneous pressures and temperatures are generated. Without the presence of extreme pressure additives the asperities would be welded together and then torn apart. Extreme pressure additives react at these high temperatures with the metal or another oil component to form compounds which are more easily deformed and sheared than the metal itself, and so prevent welding. Oils containing extreme pressure additives are generally used in heavily loaded gearboxes which may also be subjected to shock loading.

38.5.6 Detergent/dispersant

The products of combustion formed in internal combustion engines, combined with water and unburnt fuel, will form undesirable sludge which can be deposited in the engine and so reduce its operation life and efficiency. Detergent/dispersant additives prevent the agglomeration of these products and their deposition in oilways by keeping the finely divided particles in suspension in the oil. They are used in engine-lubricating oils where, when combined with anti-oxidants, they prevent piston-ring sticking. They are essential for high-speed diesels, and also desirable for petrol engines.

38.5.7 Viscosity Index improvers

When mineral oils are used over an extended temperature range it is frequently found that the natural viscosity/temperature relationship results in excessive thinning out in the higher-temperature region if the desired fluidity is to be maintained at the lower region. The addition of certain polymers will, within limits, correct this situation. They are of particular value in the preparation of lubricating oils for systems sensitive to changes in viscosity such as hydraulic controls. They are also used in multigrade engine oils.

38.6 Lubricating-oil applications

There is a constant effort by both the supplier and consumer of lubricants to reduce the number of grades in use. The various lubricant requirements of plant not only limit the extent of this rationalization but also create the continuing need for a large number of grades with different characteristics.

It is not possible to make lubricants directly from crude oil that will meet all these demands. Instead, the refinery produces a few basic oils and these are then blended in varying proportions, together with additives when necessary, to produce an oil with the particular characteristics required. In some instances the continued increase in plant performance is creating demands on the lubricant which are at the limit of the inherent physical characteristics of mineral oil. Where the operational benefit justifies the cost, the use of synthetic base stocks is being developed.

Where these are considered for existing plant, seal and paint compatibility needs to be reviewed before such products are introduced. The problems which face the lubricant supplier can best be illustrated by looking at the requirements of certain important applications.

38.7 General machinery oils

These are lubricants for the bearings of most plant, where circulating systems are not involved. These are hand, ring, bottle or bath lubricated bearings of a very wide range of equipment; line shafting, electric motors, many gear sets and general oil-can duties. The viscosity of these oils will vary to suit the variations in speed, load and temperature.

While extreme or arduous usage conditions are not met within this category, the straight mineral oils which are prescribed must possess certain properties. The viscosity level should be chosen to provide an adequate lubricant film without undue fluid friction, though this may also be influenced by the method of application. For instance, a slightly higher viscosity might be advisable if intermittent hand oiling has to be relied upon. Although anti-oxidants are not generally required, such oils must have a reasonable degree of chemical stability (Figure 38.3).

38.8 Engine lubricants

The type of power or fuel supply available will influence the decision on prime mover to be used. This is often electric power, but many items of plant such as compressors, generators or works locomotives, will be powered by diesel engines, as will most of the heavy goods vehicles used in and outside the works.

The oils for these engines have several functions to perform while in use. They must provide a lubricant film between moving parts to reduce friction and wear, hold products of combustion in suspension, prevent the formation of sludges and assist in cooling the engine. Unless the lubricant chosen fulfils these conditions successfully, deposits and sludge will form with a consequent undesirable increase in wear rate and decrease in engine life.

38.8.1 Frictional wear

If the effects of friction are to be minimized, a lubricant film must be maintained continuously between the moving surfaces. Two types of motion are encountered in engines, rotary and linear. A full fluid-film between moving parts is the ideal form of lubrication, but in practice, even with rotary motion, this is not always achievable. At low engine speeds, for instance, bearing lubrication can be under boundary conditions.

The linear sliding motion between pistons, piston rings and cylinder walls creates lubrication problems which are some of the most difficult to overcome in an engine. The ring is exerting a force against the cylinder wall while at the same time the ring and piston are moving in the cylinder with a sliding action. Also, the direction of piston movement is reversed on each stroke. To maintain a full fluid oil film on the cylinder walls under these conditions is difficult and boundary lubrication can exist. Frictional wear will occur if a lubricant film is either absent or unable to withstand the pressures being exerted. The lubricant will then be contaminated with metal wear

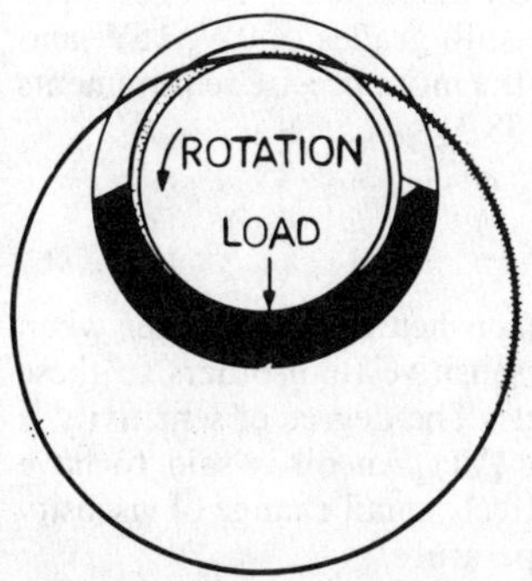

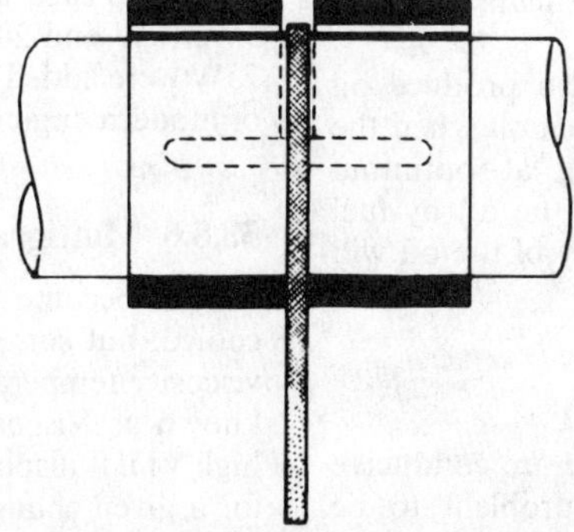

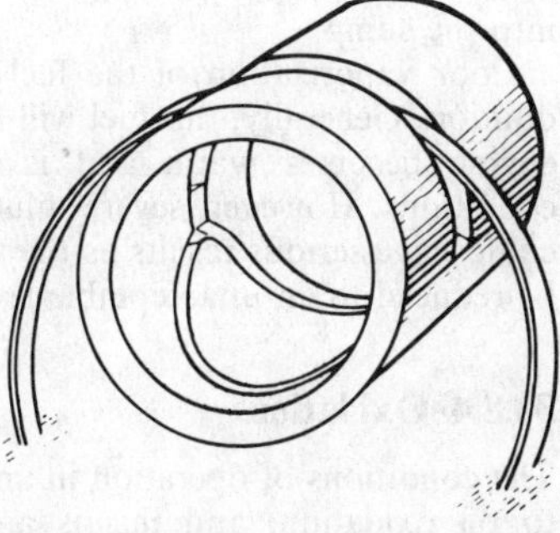

Figure 38.3 Ring oiled bearings

particles which will cause wear in other engine parts as they are carried round by the lubricant.

38.8.2 Chemical wear

Another major cause of wear is the chemical action associated with the inevitable acidic products of fuel combustion. This chemical wear of cylinder bores can be prevented by having an oil film which is strongly adherent to the metal surfaces involved, and which will rapidly heal when a tiny rupture occurs. This is achieved by the use of a chemical additive known as a corrosion inhibitor.

38.8.3 Products of combustion and fuel dilution

As it is not possible to maintain perfect combustion conditions at all times, contaminations of the oil by the products of combustion is inevitable. These contaminants can be either solid or liquid.

When an engine idles or runs with an over-rich mixture the combustion process is imperfect and soot will be formed. A quantity of this soot will pass harmlessly out with the exhaust but some will contaminate the oil film on the pistons and cylinders and drain down into the crankcase. If there is any water present these solids will emulsify to form sludges which could then block the oilways. Filters are incorporated into the oil-circulation system to remove the solid contaminants together with any atmospheric dust which bypasses the air filters.

One of the liquid contaminants is water, the presence of which is brought about by the fact that when fuel is burnt it produces approximately its own weight in water. When the engine is warm this water is converted into steam, which passes harmlessly out of the exhaust. However, with cold running or start-up conditions this water is not converted and drains into the sump. Having dissolved some of the combustion gases, it will be acidic in nature and will form sludges.

Another liquid contaminant is unburnt fuel. A poor-quality fuel, for example, may contain high boiling point constituents which will not all burn off in the combustion process and will drain into the sump. The practice of adding kerosene to fuel to facilitate easy starting in very cold weather will eventually cause severe dilution of the lubricating oil. Excessive use of over-rich mixture in cold weather will mean that all the fuel is not burnt because of the lack of oxygen and again, some remains to drain into the sump.

Poor vaporization of the fuel will also produce oil dilution. Generally, this fuel will be driven off when the engine becomes warm and is running at optimum conditions. However, severe dilution of the oil by fuel could have serious results as the viscosity of the oil will be reduced to an unacceptable level.

38.8.4 Oxidation

The conditions of operation in an engine are conducive to oil oxidation, and this is another problem to be overcome by the lubricant. In the crankcase, the oil is sprayed from various components in the form of an oil mist which is in contact with a large quantity of air and at a fairly high temperature. Oxidation produces complex carbonaceous products and acidic material and these, combined with fuel contaminants, will form stable sludges. In the combustion chamber, where the temperatures are very much higher, the oil is scraped up the cylinder walls by the piston ascending at very high speeds and is again present in the form of an oil mist. A form of carbon deposit is produced by a combination of heat decomposition and oxidation. Some of this deposit will remain, but some will pass into the sump. The effect of oxidation adds to the problem of oil contamination by the products of combustion, resulting in the formation of a resin-like material on the pistons and hot metal parts known as 'laquer' and acidic material which will attack bearing metals such as copper-lead.

These problems of engine lubrication can be overcome by using a highly refined oil. The resistance to oxidation is further enhanced by the use of anti-oxidants. The addition of corrosion-inhibitors counters acidic materials produced by combustion at low engine temperatures.

Detergent–dispersant additives are incorporated so that the carbonaceous matter produced by imperfect combustion is retained in suspension in the oil, preventing it from being deposited on the engine surfaces. Such an oil is known as a fully detergent-type lubricant. All these additives are gradually consumed during operation and the rate of decline in their usefulness will determine the oil-change period. This rate is, in turn, influenced by the conditions of operation.

38.8.5 The SAE viscosity system

This classification was devised by the Society of Automotive Engineers (SAE) in America by dividing the viscosity span into four and giving each of the divisions a number – SAE 20, 30, 40 and 50. The thinnest (SAE 20), for example, covered the range 5.7–9.6 cSt specified at 210°F, which was considered to be a temperature typical of a hot engine. (The SAE originally specified temperatures in °F, because they was the convention. Today, temperatures are quoted in °C.)

Later, the SAE series was extended to include much thinner oils because of the growing demand for easier winter starting. The viscosities of the three new grades were specified at 0°F (typical of cold morning temperatures) and each was given the suffix W for Winter – SAE 5W, 10W and 20W. Later still, grades of 0W, 15W and 25W were added to satisfy the more precise requirements of modern engines (Table 38.3).

38.8.6 Multigrades

All oils become thinner when heated and thicker when cooled, but some are less sensitive than others to these viscosity/temperature effects. The degree of sensitivity is known as Viscosity Index (VI). An oil is said to have high VI if it displays a relatively small change of viscosity for a given change of temperature.

In the 1950s, development in additive technology led to the production of engine oils with unusually high VIs,

Table 38.3 Viscosity chart

	Maximum viscosity cP at °C[a]	*Maximum borderline pumping temperature (°C)*[b]	*Viscosity (cSt) at 100°C min.*	*max.*
0W	3250 at −30	−35	3.8	—
5W	3500 at −25	−30	3.8	—
10W	3500 at −20	−25	4.1	—
15W	3500 at −15	−20	5.6	—
20W	4500 at −10	−15	5.6	—
25W	6000 at −5	−10	9.3	—
20	—	—	5.6	9.3
30	—	—	9.3	12.5
40	—	—	12.5	16.3
50	—	—	16.3	21.9

[a] As measured in the Cold Cranking Simulator (CCS).
[b] As measured in the Mini Rotary Viscometer (MRV).

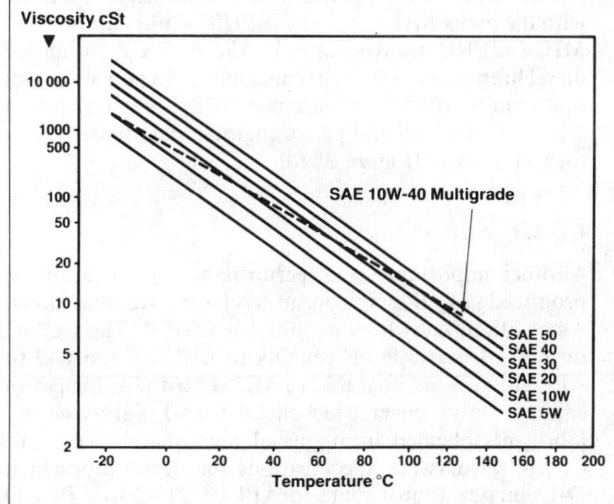

Figure 38.4 Multigrade chart

known as multigrade oils. A multigrade oil's high resistance to temperature change is sufficient to give it the combined virtues of a thin grade at low (starting) temperatures and a thick one at running temperatures. An SAE 20W-40 multigrade, for example, is as thin at −20°C as a 20W oil, but as thick at 100°C as an SAE 40 oil. Thus the multigrade combines full lubrication protection at working temperatures with satisfactorily easy starting on frosty mornings. Figure 38.4 is a viscosity–temperature graph for six monograde oils and a 10W-40 multigrade, showing how the multigrade has the high-temperature properties of an SAE 40 oil and the low-temperature properties of an SAE 10W. Thus the multigrade is suitable for all-year-round use.

38.8.7 Performance ratings

The SAE numbering system refers purely to the viscosity of the oil, and is not intended to reflect lubricating performance (there is no such thing as an 'SAE quality' oil, for example). Engine oils are marketed in a range of performance levels, and need to be classified according to the severity of service conditions in which they are designed to operate. Accordingly, the American Petroleum Institute (API) has drawn up a coding system in which oils are subjected to a series of classifying bench-tests known as the 'Sequence' tests.

38.8.8 The API service classifications

In the API system the least demanding classification for a petrol engine was originally designated SA. The most demanding is, at present, SG. (The S stands for Service Station.) Constant development of both engines and oils means that from time to time the highest ratings are superseded by even higher ratings. The API system also classifies diesel engine oils by their severity of service. Here the categories have the prefix C, which stands for Commercial.

Petrol engines

SA Service typical of engines operated under mild conditions. This classification has no performance requirements.

SB Service typical of engines operating in conditions such that only minimum protection of the type afforded by additives is desired. Oils designed for this service have been used since the 1930s; they provide only anti-scuff capability and resistance to oil oxidation and bearing corrosion.

SC Service typical of petrol engines in 1964–1967 cars and trucks. Oils designed for this service provide control of high- and low-temperature deposits, wear, rust and corrosion.

SD Service typical of 1967–1970 petrol engines in cars and some trucks; but it may apply to later models. Oils designed for this service provide more protection than SC against high- and low-temperature deposits, wear, rust and corrosion; and may be used where SC is recommended.

SE Service typical of petrol engines in cars and some trucks in 1972–1979. Oils designed for this service provide more protection against oxidation, high-temperature deposits, rust and corrosion than SD or SC, and may be used where those classifications are recommended.

SF Service typical of petrol engines in cars and some trucks from 1980. Oils developed for this service provide better oxidation stability and anti-wear performance than SE oils. They also provide protection against engine deposits, rust and corrosion. Oils meeting SF may be used wherever SE, SD or SC is recommended.

SG Service typical of petrol engines in present cars, vans and light trucks. Oils developed for this service provide improved control of engine deposits, oil oxidation and engine wear relative to oils developed for previous categories. Oils meeting SG may be used wherever SF, SE, SF/CC or SE/CC are recommended.

Diesel engines

CA Service typical of diesel engines operated in mild to moderate duty with high-quality fuels. Occasionally this category has included petrol engines in mild service. Oils designed for this service were widely used in the late 1940s and 1950s; they provided protection from bearing corrosion and light-temperature deposits.

CB This category is basically the same as CA, but improved to cope with low-quality fuels. Oils designed for this service were introduced in 1949.

CC Service typical of lightly supercharged diesel engines operated in moderate to severe duty. Has included certain heavy-duty petrol engines. Oils designed for this service are used in many trucks and in industrial and construction equipment and farm tractors. These oils provide protection from high-temperature deposits in lightly supercharged diesels and also from rust, corrosion and low-temperature deposits in petrol engines.

CD Service typical of supercharged diesel engines in high-speed high-output duty requiring highly effective control of wear and deposits. Oils designed for this service provide protection from bearing corrosion and high-temperature deposits in supercharged diesel engines running on fuels of a wide quality range.

CDII Service typical of two-stroke cycle diesel engines requiring highly effective control over wear and deposits. Oils designed for this service also meet all the requirements of CD.

CE Service typical of certain turbocharged or supercharged heavy-duty diesel engines operating under both low speed–high load and high speed–low load conditions. Oils designed for this service must also meet the requirements specified for CC and CD classifications.

Before an oil can be allocated any given API performance level it must satisfy requirements laid down for various engine tests. In the SG category, for example, the engine tests are as follows:

Service IID measures the tendency of the oil to rust or corrode the valve train and to influence the value lifter operation.
Sequence IIIE measures high-temperature oil oxidation, sludge and varnish deposits, cam-and-tappet wear, cam and lifter scuffing and valve lifter sticking.
Sequence VE evaluates sludge deposits, varnish deposits, oil-ring clogging and sticking, oil-screen plugging and cam wear.
Caterpillar IH2 determines the lubricant effect on ring sticking, ring and cylinder wear, and accumulation of piston deposits.
CRC L-38: the characteristics assessed are resistance to oxidation, bearing corrosion, tendency to formation of sludge and varnish, and change of viscosity.

In the CE category the tests are:

Caterpillar IG2: the lubricant characteristics determined are ring sticking, ring and cylinder wear, and accumulation deposits under more severe test conditions than those for Caterpillar IH2.
Cummins NTC-400 measures crownland and piston deposits, camshaft roller follower pin wear and oil consumption.
Mack T6 assesses oil oxidation, piston deposits, oil consumption and ring wear.
Mack T7 evaluates oil thickening.
CRC L-38 (as above).

Other specifications

Various authorities and military bodies issue specifications relating to the service performance of engine oils. In some instances the ratings are almost identical with those of the API, but most of them are not precisely parallel because they cover performance factors encountered in particular engines and particular categories of service.

The most common of the other specifications are those with the prefix MIL, issued by the US military authorities. MIL-L-2104E approximates to the API CE rating for diesel lubricants, although it also relates to petrol engines that require API SE performance. MIL-L-46152D covers oils for both diesel and petrol engines, and approximates to API SG/CC (Figure 38.5).

CCMC ratings

Another important set of performance specifications is produced by the European Vehicle Manufacturers' Association, known by its initials CCMC*. The CCMC rating G-1 corresponds roughly to API SE, and G-2 to API SF. G-3 (comparable to MIL-L-46152B, for petrol engines only) covers fuel-efficient and light-viscosity lubricants blended from special high-quality base oils. CCMC also issues specifications for diesel lubricants: D-1 equates approximate to API CC, D-2 to API CD and MIL-L-2104D, and D-3 to API CE and MIL-L-2104E.

To qualify for the CCMC categories G and D, an oil must meet the requirements of the following tests in addition to the relevant API classification tests.

For the G category

Ford Kent, which evaluates cold ring sticking, piston skirt varnish, oil thickening and consumption.
Fiat 132, to evaluate the tendency of the oil to cause pre-ignition.
Daimler Benz OM 616 to evaluate wear of cylinders and cams.
Bosch Injector Rig, measuring the mechanical stability of the oil to assess its shear stability.
Noack Test, to measure the weight loss due to evaporation of the oil.

* Comité des Constructeurs d'Automobiles du Marché Commun represents joint industry opinion on factors such as lubricant specifications, emissions, vehicle design and safety standards. With regard to crankcase lubricants, CCMC defines sequences of engine tests, and the tests themselves are defined by CEC (Coordinating European Committee for the Development of Performance Tests for Lubricants and Engine Fuels: a joint body of the oil and motor industries).

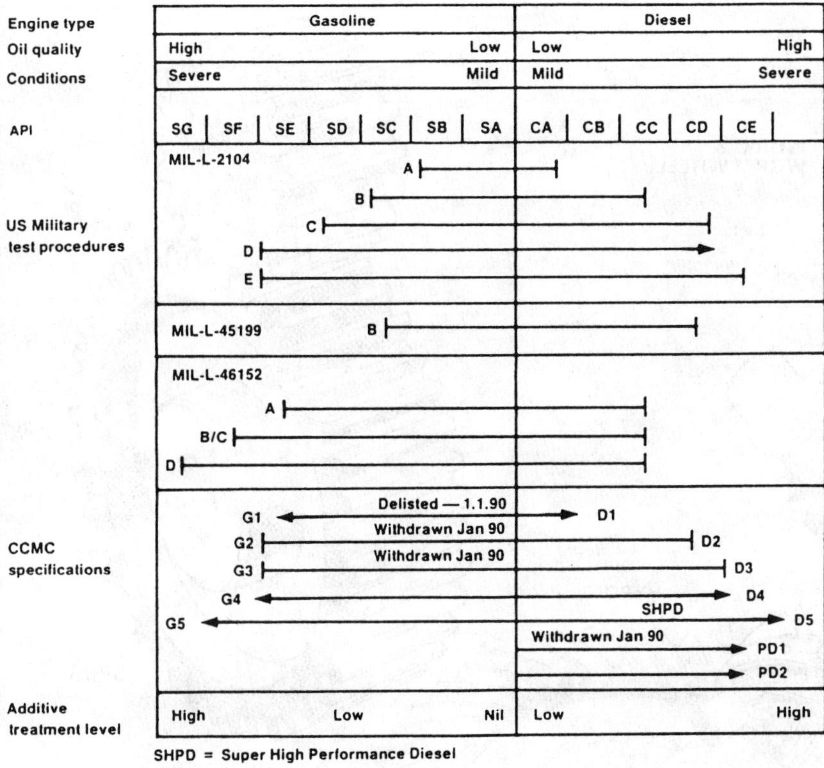

Figure 38.5 Approximate relationship between classifications and test procedures

High shear/high temperature viscosity test, to assess the oil's capability for resisting shear, and so retaining its viscosity, at high temperatures.
Tests for oil/seal compatability and oil consumption are still to be established.

For the D category

Bosch Injector Rig, *Noack* and *D-B OM* 616 tests as above together with:

For D1 and D2 only, *MWM-B* evaluating varnish, carbon deposits, and ring-sticking;
For D3 only, *D-B OM 352A* bore polishing and piston cleanliness;
For PD1 only, *VW 1.6L* to evaluate ring sticking and piston cleanliness.

38.9 Gear lubricants

There are few engineering applications in which gears do not play an essential part. They can be used to reduce or increase speed, transmit power and change the direction or position of a rotating axis. There are several types of gears to suit these varying operational conditions, such as spur, helical, plain and spiral bevels, hypoid, worm and wheel (Figure 38.6).

Extremely high pressures are developed between meshing teeth as, in theory, they only have point or line contact. Together with the sliding between mating surfaces which is always present it is clear that, if there is metal-to-metal contact, rapid wear will occur. The function of the lubricant is to provide and maintain a separating film under all the variations in speed, load and temperature. It must also act as a coolant and protect the gears against corrosion.

The lubrication of gears is not a simple matter, because of their shape and variability of motion. Fundamental factors which affect their lubrication are gear characteristics, materials, temperature, speed, loading, method of applying the lubricant and environment.

38.9.1 Gear characteristics

Spur, bevel, helical and spiral bevel gears

Providing the speed is sufficient and the load does not squeeze out the lubricant, the effect of rotating these gears is to produce a hydrodynamic wedge of a relatively thick lubricating film between the meshing surfaces. If the load is increased, the pressure within the contact zone increases, causing a reduction in film thickness until the high spots on meshing teeth begin to touch; wear will then occur.

At low speeds, when the high spots make contact due

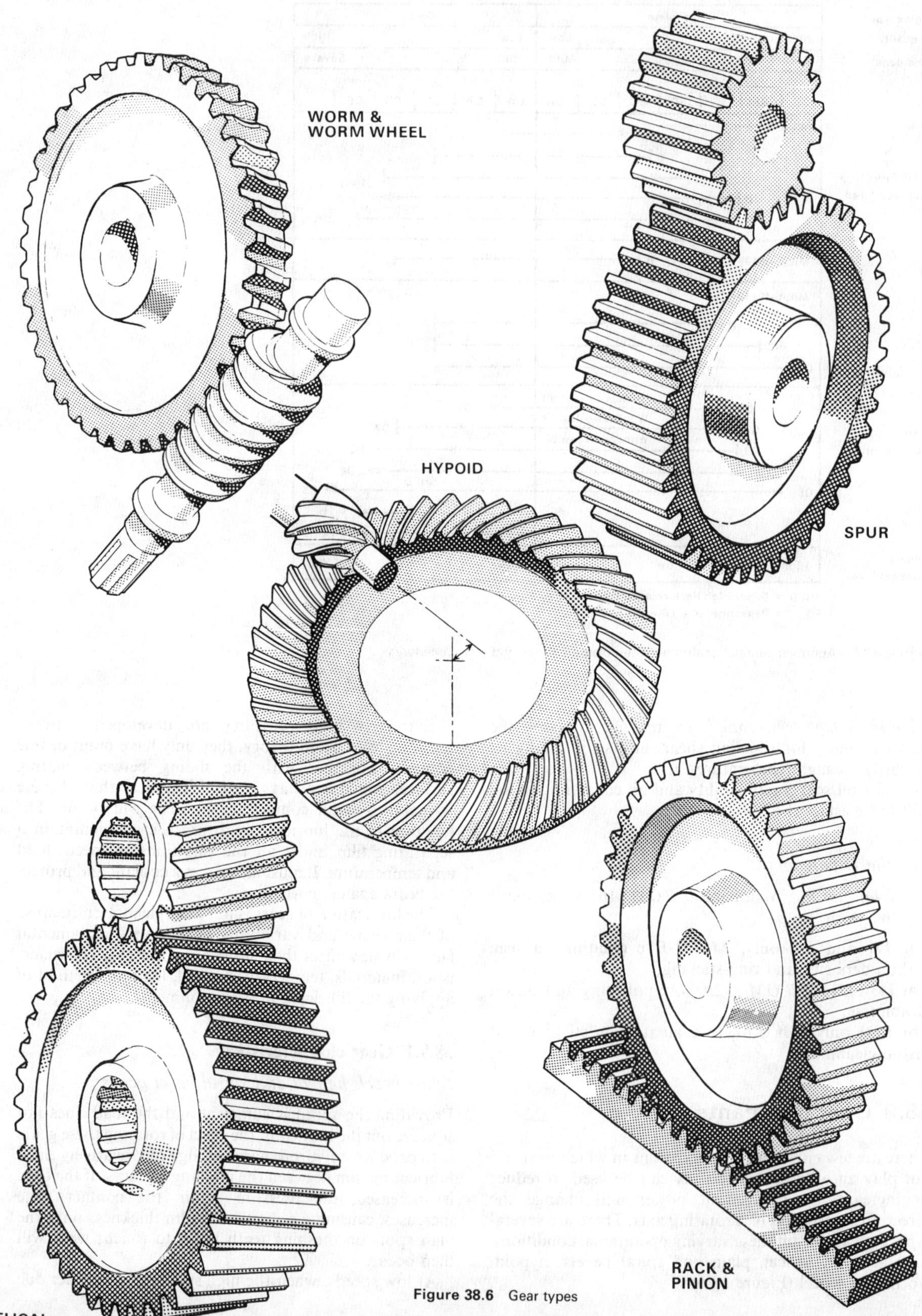

Figure 38.6 Gear types

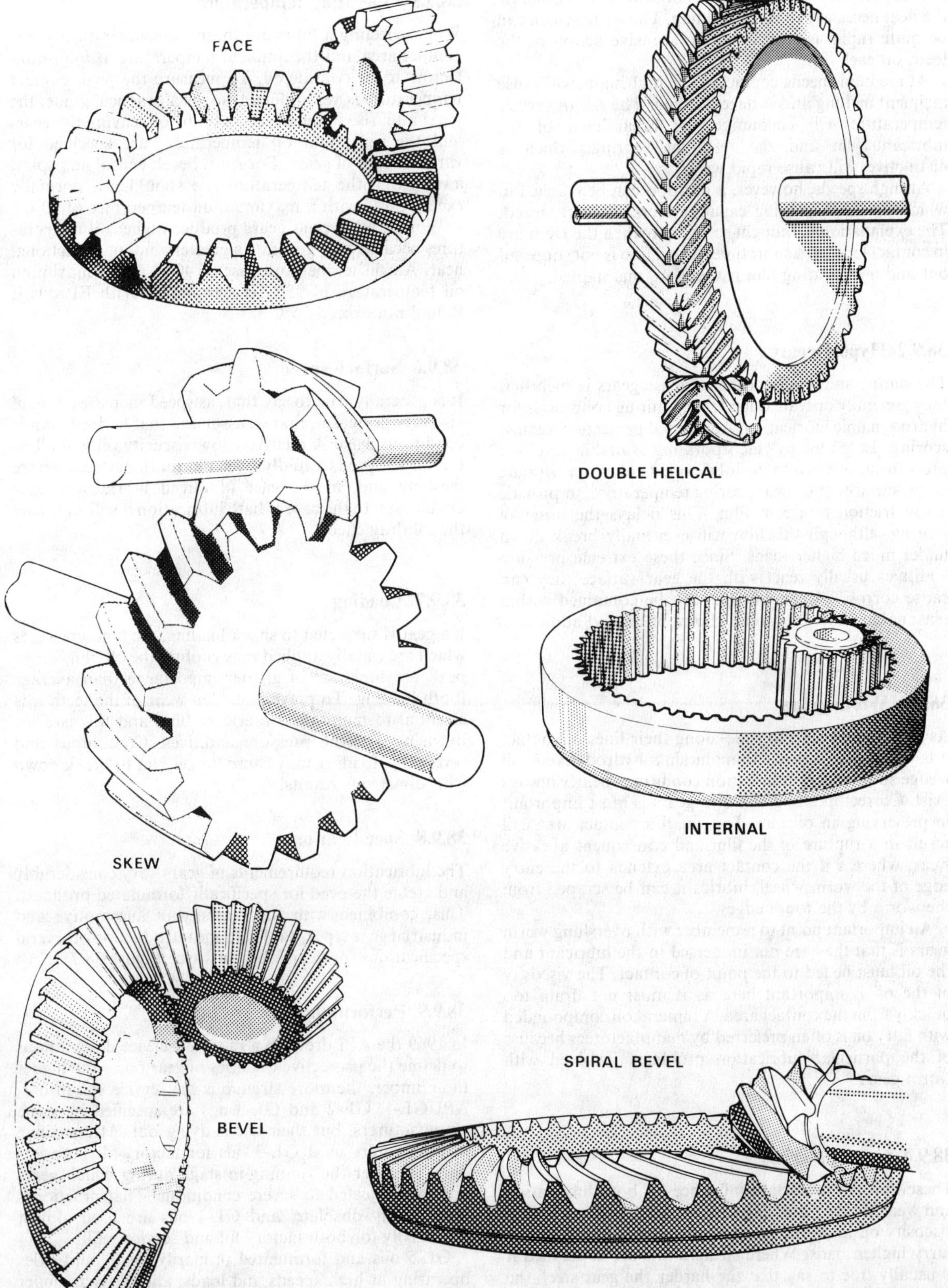
FACE
DOUBLE HELICAL
SKEW
INTERNAL
SPIRAL BEVEL
BEVEL

to load, the sliding velocities are not sufficient to generate the heat necessary to cause welding. The wear, which can be quite rapid, is caused by the abrasive action of the teeth on each other.

At medium speeds contact of the high spots will cause incipient melting and surface welding. The rise in surface temperature will encourage the breakdown of the lubricant film and the resultant scoring, which is distinctive, will cause rapid wear.

At high speeds, however, a position can be reached in which the load-carrying capacity increases with speed. The explanation is thought to be that when the teeth are in contact for such a short time, the oil film is not squeezed out and a separating film is therefore maintained.

38.9.2 Hypoid gears

The sliding and loading effect of these gears is such that they generally operate outside the limiting conditions for hydrodynamic lubrication. Additional protection against scoring is given by incorporating suitable extreme pressure additives into the lubricant. These react with the metal surfaces at or near scoring temperature, to provide a low-friction lubricant film. This delays the onset of scoring, although this film will eventually break down under much higher loads. Since these extreme pressure additives usually react with the gear surface they can cause corrosive wear, but this can be contained within reasonable limits by careful choice of oil and additives.

38.9.3 Worm gears

As worm gears tend to slide along their lines of contact it is virtually impossible to maintain a hydrodynamic oil wedge so boundary lubrication conditions nearly always exist. Correct meshing of worm gears is most important in preserving an oil film. Too small a contact area can result in a rupture of the film and consequent abrasive wear, whereas if the contact area extends to the entry edge of the worm wheel, lubricant can be scraped from the worm by the tooth edges.

An important point to remember with overslung worm gears is that they are not immersed in the lubricant and the oil must be fed to the point of contact. The viscosity of the oil is important here as it must not drain too quickly from the contact area. A mineral oil compounded with fatty oil is often preferred by manufacturers because of the particular lubrication problems involved with worm gears.

38.9.4 Gear materials

These have an important influence on both lubrication and wear. At the same speed, steel gears require a higher viscosity oil than cast iron or bronze, because they can carry higher loads. Where straight oils are concerned it is usually true to say that the harder the gear steel, the higher the viscosity needed.

38.9.5 Operating temperature

When selecting a lubricant, both the temperature at the contact area and the ambient temperature at important factors to be considered. To measure the peak contact temperature is very difficult, so, as a practical guide, the maximum rise in temperature of the oil leaving the gears and the maximum oil temperature, are specified for various types of gears. For spur, bevel, helical and spiral level gears, the temperature rise should not normally exceed 30°C with a maximum oil temperature of 70°C.

Worm and hypoid gears produce higher oil temperatures because they generate a greater amount of frictional heat. An oil temperature rise of 40°C and a maximum oil temperature of 95°C is acceptable. With EP oils it should not exceed 75°C.

38.9.6 Surface speed

It is generally true to say that, as speed increases, the oil viscosity decreases, that is, assuming that hydrodynamic conditions exist. A relatively low viscosity oil will allow the oil to spread rapidly over the tooth surfaces before meshing and, in the case of forced lubrication, ease circulation. In the case of bath lubrication it will eliminate the oil drag effect.

38.9.7 Loading

If a gear is subjected to shock loading, the high pressures which are rapidly applied may rupture the oil film. These peak pressures are of greater importance than average tooth loading. To prevent sudden wear of the teeth it is essential to maintain a lubricant film, and this is done by using extreme pressure additives. Continuous and severe overloading may cause the oil film to break down with disastrous results.

38.9.8 Specifications

The lubrication requirements of gears vary considerably and create the need for specifically formulated products. This, combined with the diversity of automotive and industrial gear types, has led to the introduction of several specifications for gear lubricants (see Figure 38.7).

38.9.9 Performance

In 1969 the API drew up a range of service designations to define the protective qualities of gear oils – the higher the number, the more strenuous the service conditions. API GL-1, GL-2 and GL-3 oils are specified by some manufacturers, but their use is dying out. At one time, manufacturers used GL-3 oils for factory-fill purposes only, to cover the running-in stage during which gears are not subjected to severe conditions. That practice is now largely obsolete, and GL-4 oils are used almost universally for both factory fill and service refill.

GL-5 oils and formulated primarily for hypoid axles operating at high speeds and loads, and likely to suffer shock loads. The GL-6 classification was introduced to

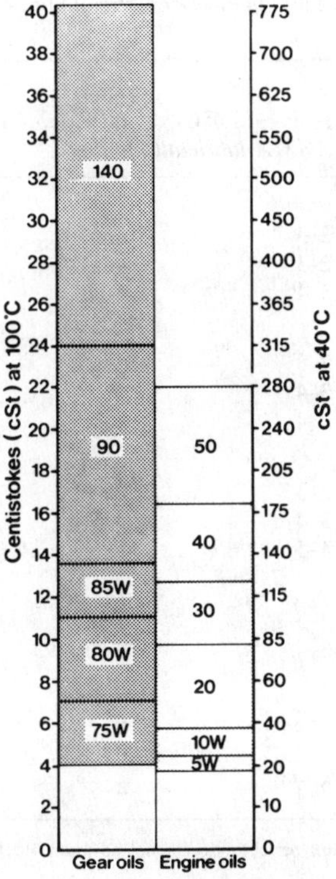

Figure 38.7 The SAE numbering systems for engine oils and gear oils are not related and must not be confused. This figure illustrates the differences

to cover a Ford requirement for special hypoid gears operating in high-performance conditions at high speeds. In the event, however, the Lotus Elan was virtually the only European car to require this category. GL-6 is now regarded as obsolete, but oils meeting its requirements may still be used to overcome isolated service difficulties.

Manufacturers and other bodies have issued their own designations, which may be encountered occasionally. Most important is the US Military specification MIL-L-2105D (which replaced MIL-L-2105C) corresponding to API GL-5, and introduced approval for multigrade gear oils.

38.9.10 AGMA specifications

For industrial gears and circulating oils there is a specification published by the American Gear Manufacturers Association which expresses the viscosities corresponding to the AGMA Lubricant Numbers in Saybolt Universal Seconds only. The approximate viscosities in centistokes are provided in Table 38.4 for convenience. The recommendations in Tables 38.5 and 38.6 are intended as a guide for lubricant selection.

38.9.11 Other specifications

Among other specifications that the plant user will encounter are those, for example, issued by David Brown Gears, the German DIN 51517, and individual suppliers' recommendations on the grades of oil to be used in their units.

Table 38.4 AGMA lubricant numbering system

Rust and oxidation inhibited oils	*EP lubricants*[a]	*Viscosity (40°C)*[b]
AGMA lubricant number	AGMA lubricant number	Kinematic cSt
1		41.4–50.6
2	2 EP	61.2–74.8
3	3 EP	90–110
4	4 EP	135–165
5	5 EP	198–242
6	6 EP	228–352
7 comp[c]	7 EP	414–506
8 comp	8 EP	612–748
8A comp		900–1100

[a] EP lubricants should be used only when recommended by the gear manufacturer.
[b] As per ASTM 2422 (also BS 4231).
[c] Oils marked 'comp' are compounded with 3–10% fatty or synthetic fatty oils.

38.9.12 Method of lubrication

The method of application, whether splash, spray, jet or mist, will have an effect on the lubricant. If an oil is atomized its surface area is increased and it is much more easily oxidized. The rate of oxidation will be increased by the frictional heat generated by the meshing teeth. The dissiption of the heat absorbed by the oil is important and this will be influenced by the lubrication system. The length of pipe run, the temperature and size of bearings, pumps, other surfaces with which the oil comes into contact, whether filters and oil coolers are fitted, all affect the heat content of the oil.

38.9.13 Splash lubrication

Some enclosed gears are lubricated by oil splashing at random within the casing. If the pitch-line speeds are low (say, less than 100 m/min), paddles and scrapers attached to the gear wheels may be needed to pick up the required amount of oil. If the speed is excessively high, however, the lubricant may be flung off the gears before it reaches the meshing zone, while that in the reservoir portion may be overheated by churning. The upper speed limit is usually about 800 m/min, but this can sometimes be exceeded where shields are close to gear peripheries.

Although a gear train may comprise wheels of markedly different diameters, as in multiple-reduction sets, the degrees of immersion in the trough must be approximately constant. If not, the smallest or highest wheel may have adequate pick-up while the largest is over-immersed. The trough should therefore be stepped or the bearing-lubrication system extended to lubricate the smaller or higher gears.

Optimum depth of immersion depends on pitch-line

Table 38.5 Lubrication of industrial enclosed gear drives. AGMA Lubricant Number recommendations for helical, herringbone, straight bevel, spiral bevel and spur[a]

Centre distances: main gear, low speed	*Ambient temperatures*[b] *(−40–17.8°C)* *Other lubricants*	*(−28.9 to −3.9°C)*	*(−9.4–15.6°C)* *AGMA lubricant number*[c,d]	*(10–51.7°C)*
Parallel shaft (single reduction):	Automatic transmission fluid or similar product[e]	Engine oil SAE 10W-30, or similar product[e]		
Up to 8 in (200 mm)			2–3	3–4
Over 8 in (200 mm) and up to 20 in (510 mm)			2–3	4–5
Over 20 in (510 mm)			3–4	4–5
Parallel shaft (double reduction):				
Up to 8 in (200 mm)			2–3	3–4
Over 8 in (200 mm) and up to 20 in (510 mm)			3–4	4–5
Over 20 in (510 mm)			3–4	4–5
Parallel shaft (triple reduction):				
Up to 8 in (200 mm)			2–3	3–4
Over 8 in (200 mm) and up to 20 in (510 mm)			3–4	4–5
Over 20 in (510 mm)			4–5	5–6
Planetary gear units:				
Outside dia. of housing up to 16 in (410 mm)			2–3	3–4
Outside dia. of housing over 16 in(410 mm)			3–4	4–5
Spiral or straight bevel gear units:				
Cone distance up to 12 in (300 mm)			2–3	4–5
Cone distance over 12 in (300 mm)			3–4	5–6
Geared motors and shaft-mounted units			2–3	4–5

[a] Drives incorporating over-running clutches as backstopping devices should be referred to the clutch manufacturers, because certain types of lubricants may adversely affect clutch performance.
[b] The pour point of the lubricant should be at least 10°C lower than the expected minimum ambient starting temperature. If the ambient starting temperature approaches the pour point, sump heaters may be required to ease starting and ensure adequate lubrication.
[c] Viscosity ranges are provided to allow for variations such as surface finish, temperature rise, loading and speed.
[d] AGMA viscosity number recommendations listed above refer to R & O (rust and oxidation) inhibited gear oils, which are widely used for general bearing, turbine and gear lubrication systems. EP gear lubricants in the corresponding viscosity grades may be substituted, where they are deemed necessary by the gear drive manufacturer.
[e] Where they are available, good-quality industrial oils having similar properties are preferred to the automotive oils. The recommendation of automotive oils for use at ambient temperatures below −10°C is intended only as a guide pending widespread development of satisfactory low-temperature industrial oils. It is advisable to consult the gear manufacturer.
High-speed units are those operating at speeds above 3600 rev/min or pitch line velocities above 1520 m/min. (Refer to Standard AGMA for High Speed Helical and Herringbone Gear Units for detailed lubrication recommendations.)

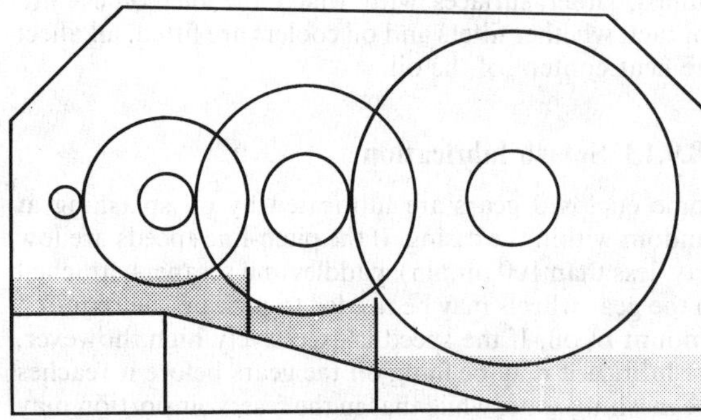

Figure 38.8 In a splash-fed gear the depth of immersion must be reasonably constant for all the gears, therefore the casing should incorporate stepped troughs for the oil

speed. Deep immersion is permissible at low speeds because power losses from churning are low, but at moderate speeds the immersion depth should not be more than three times the tooth height. At the highest speeds, only the addenda of the teeth need be submerged (Figure 38.8).

When gears are rotating inside a casing some oil is being carried around by them and some is adhering to the walls, so the oil in the trough is lower than when at rest. The sight-glass or dipstick should therefore be calibrated when the gears are in motion. With bevel, hypoid and worm sets, the relative positions of the various elements can vary so much that the 'correct' oil level cannot always be accurately defined.

In splash lubrication the oil absorbs heat, which is transmitted to the casing. Gearbox designers sometimes augment the cooling effect by finning the casing to increase the area of metal exposed to atmosphere. In some designs a fan is mounted on the driving shaft, to force cooling air over the casing. At the installation stage the cooling of gear units can sometimes be facilitated by intelligent siting.

38.9.14 Forced circulation

Circulation by pumping is, of course, more expensive than simple splash lubrication, but the expense is usually justified. Some of the many advantages are as follows:

- Oil supplied to each meshing area can be metered to provide the optimum flow for lubrication and cooling.
- Heat exchangers can be installed to minimize changes of temperature. This means that the user can select

Table 38.6 Enclosed cylindrical and double enveloping worm-gear drives

Worm centre distance	*Worm speeds* (rev/min) *up to*	*Ambient temperatures* 15–60°F (−9.4–15.6°C)	*Ambient temperatures* 50–125°F (10–51.7°C)	*Worm speeds* (rev/min) *up to*[b]	*Ambient temperatures* 15–60°F (−9.4–15.6°C)	*Ambient temperatures* 50–125°F (10–51.7°C)
		AGMA lubricant no.			*AGMA lubricant no.*	
Up to and including 6 in (150 mm):						
Cylindrical worm	700	7 comp	8 comp	700	7 comp	8 comp
Double-enveloping worm	700	8 comp	8A comp	700	8 comp	8 comp
Over 6 in (150 mm) and up to 12 in (300 mm):						
Cylindrical worm	450	7 comp	8 comp	450	7 comp	7 comp
Double-enveloping worm	450	8 comp	8A comp	450	8 comp	8 comp
Over 12 in (300 mm) and up to 18 in (460 mm):						
Cylindrical worm	300	7 comp	8 comp	300	7 comp	7 comp
Double-enveloping worm	300	8 comp	8A comp	300	8 comp	8 comp
Over 18 in (460 mm) and up to 24 in (610 mm):						
Cylindrical worm	250	7 comp	8 comp	250	7 comp	7 comp
Double-enveloping worm	250	8 comp	8A comp	250	8 comp	8 comp
Over 24 in (610 mm):						
Cylindrical worm	200	7 comp	8 comp	200	7 comp	7 comp
Double-developing	200	8 comp	8A comp	200	8 comp	8 comp

[a] The pour point of the oil used should be less than the minimum ambient temperature expected. Consult the gear manufacturer for recommendations regarding ambient temperatures below −10°C.

[b] Worm gears of either type operating at speeds above 2400 rev/min or 610 m/min rubbing speed may require force-feed lubrication. In general, a lubricant of lower viscosity than recommended in the above table may be used with a force-feed system.

Warning: Worm gear drives operate satisfactorily on R & O gear oils, sulphur phorphorus EP oils or lead naphthenate EP oils. These oils, however, should be used only with the approval of the gear manufacturer.

viscosity solely on the grounds of conditions at the tooth face, and can virtually ignore other requirements. A thin oil, for example, can be cooled to prevent it becoming too thin during continuous circulation through meshing zones, or to prevent overheating; a thick oil can be heated to encourage rapid initial flow through feeder lines.

- Filtration units can be incorporated into circulation lines, thus ensuring that clean oil is fed to gears and bearings.
- Gear casings can be small, since their oil reservoirs can be sited remotely. In most rolling mills, for instance, an oil 'cellar' houses storage tanks, pumps, heat exchangers and filters.

Oil is fed to meshing zones by wide jets or sprayers, and uniform distribution across the tooth faces is necessary to avoid local distortion. Clearly, the rate of flow is important; the effects of too little oil are easily imagined. A sound practice is to position an extra jet on the 'exit' or disengagement side of a meshing zone, where the teeth are at their highest temperature. If gears run in either direction, sprayers must be installed on both sides.

Formulae exist for calculating the optimum rates of flow, based on considerations of the heat to be removed – which calls for more oil flow than does lubrication alone – but they are complex and require data that may not be available at the design stage. A reasonable estimate is 0.05–0.10 l/min for each millimetre of tooth width.

Heat from bearings is an important fraction of the total heat to be dissipated from a gear system, and must be taken into account in calculations of cooling requirements. Excessive churning of oil causes unnecessary loss of power, and the resultant combination of high temperature and atomization of the oil raises the rate of oxidation. Careful attention to the design and siting of

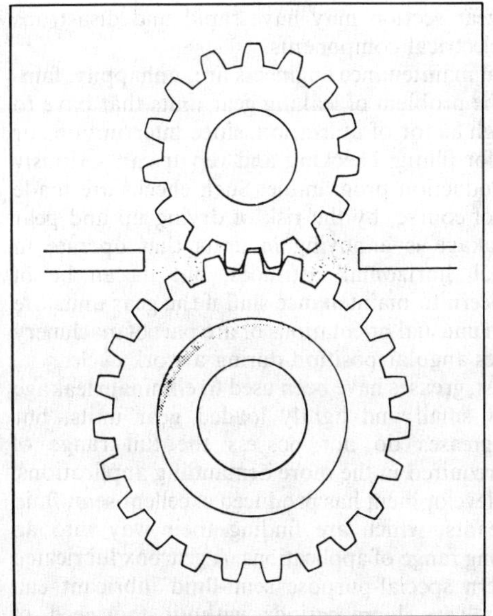

Figure 38.9 Oil is fed to some gears by jet or spray, so that the full benefits of lubrication and cooling are obtained precisely where they are needed. An extra jet at the 'exit' side of a meshing zone ensures that the teeth are cooled where the temperature is highest

sprayers minimizes these effects, keeps the oil cooler and prolongs its service life (Figure 38.9).

38.9.15 Oil-mist lubrication

Although not widely used for gears, oil-mist lubrication is nevertheless worth mentioning here. It is a 'total loss' technique in which the oil is supplied in the form of fine droplets carried by compressed air. Two virtues are that the lubricant can be carried long distances through pipes without severe frictional losses, and that no oil pumps are needed since the motive power is provided by factory compressed-air lines. However, unless such systems are totally enclosed, the exhaust can create a build-up of oil-mist in the atmosphere. In order to maintain good standards of industrial hygiene, it is recommended that the concentration of oil-mist in the working environment should not exceed 5 mg/m^3.

38.9.16 Semi-fluid lubrication

Gear oils possess most of the attributes required in the lubricating function, but the need for an alternative lubricant is encountered in some industrial applications. One such need rises from the leakage caused by slight wear of oil-seals – which is inevitable in any gear unit. Oil escaping past seals seeps along input and output shafts, from where it is probably thrown onto adjacent walls, floors or machinery. Loss of lubricant is particularly undesirable in a geared motor, where leakage from the reduction-gear section may have rapid and disastrous effects on electrical components.

Industrial maintenance engineers are, unhappily, familiar with the problem of leaking gear units that have to run at a high factor of utilization, since interruptions or downtime for filling, checking and repair can seriously retard a production programme. Such checks are made necessary, of course, by the risk of drying up and gear failure. Leakage is annoying in units that operate in conventional 'horizontal' attitudes, but it can be of serious concern to maintenance staff if the gear units are operating in unusual orientations or are part of machinery that changes angular position during a work cycle.

In the past, greases have been used to eliminate leakage from a few small and lightly loaded gear units, but orthodox greases do not possess the full range of properties required in the more demanding applications. However, development has produced excellent semi-fluid gear lubricants, which are finding their way into an ever-widening range of applications. A gearbox lubricated by a modern special-purpose semi-fluid lubricant can operate for very long periods without the need of maintenance. The semi-fluid lubricants formulated for gear lubrication include a mineral based for spur, helical and bevel gears, and synthetic based for worm gears.

38.9.17 Spur, helical and bevel gears

The latest mineral-base grease is thickened, not by a metallic soap, but by a polymer, and its additives include an anti-oxidant and a modern EP agent of the sulphur/phosphorus type. Such a grease has flow characteristics that vary with shearing conditions. It is like a gear oil in the tooth-contact regions, where shearing is severe, but is more resistant to flow in regions of low shear – as in oil seals. Consequently, the grease resists leakage at the seals and joints, and the lubrication function is fulfilled, especially in heavily loaded steel gears. Another important virtue is that the product does not readily drain from metallic surfaces: dry starts are thus avoided (Figure 38.10).

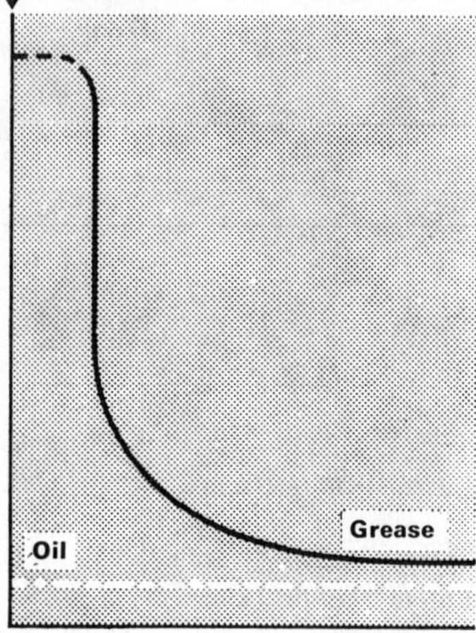

Figure 38.10 Comparison of the apparent viscosities of a gear oil and a gear grease. The grease flows readily in the tooth-contact region, where shear stress is high, but resists leakage at seals and joints

38.9.18 Worm gears

As already mentioned, the severe loading and sliding conditions at the tooth faces of worm gears generate a great deal of heat. EP or anti-weld additives are ineffective in these circumstances, and can even be harmful. Consequently the synthetic oils that have been developed specifically for worm gears are incorporated into semi-fluid grease formulations for applications in which leakage must be avoided. One of their primary requirements, of course, is high shear stability.

Several of the current synthetic semi-fluid lubricants for worm gears are based on a polyalkylene glycol – which possesses 'slipperiness'. Oxidation and rust-inhibitors are usually incorporated, and they are carefully selected to be compatible with both the polyalkylene and the grease-thickener system.

38.9.19 Oil or grease?

In summary, it can be said that the best choice of lubricant for a gearbox depends on the service conditions. If leakage is a problem, or if, during shutdown, the maintenance of a thick film of lubricant is necessary to inhibit corrosion and prevent 'dry-state' operation, greases show to advantage. If generation of heat and consequent excessive temperatures are a problem, users will find it better to use liquid lubricants, which have the advantage of high rates of heat transfer.

38.9.20 Automatic transmissions

Frictional behaviour

Satisfactory gear-changing in an automatic transmission unit is a compromise between smooth slippage during take-up and firm grip during drive. The frictional properties of an ATF can be adjusted with friction-modifying additives during the formulation stages to ensure careful matching with the requirements of transmission manufacturers. Two kinds of frictional behaviour have to be considered in combination:

1. Static friction, which exists when sliding motion starts from rest;
2. Dynamic friction, which comes into effect when the mating surfaces are in motion.

The transition stage from one to the other is critical, and is the point at which any problem of noise or chattering occurs. Some transmission designs have a take-up characteristic best suited by low static friction and some by a high static friction. On no account should the wrong ATF be employed when a transmission unit is being completely filled, but a limited degree of topping up with the wrong fluid is permissible in most cases. (The need for fluids of high static friction is receding now that related designs are being phased out.)

Automatic transmission fluids (ATFs)

An ATF has to satisfy several requirements. It must:

1. Lubricate the gears in the gearbox;
2. Fill the torque converter (or 'automatic clutch') and thus convey power from the engine to the gearbox;
3. Act as hydraulic medium by conveying signals from the valves in the control unit to the internal clutches/brakes that engage the gears;
4. Ensure smooth but rapid take-up of power between the friction faces of the oil-immersed clutches/brakes;
5. Remove frictional heat and cool the whole transmission unit.

An ATF requires high lubrication performance, high levels of oxidation stability to withstand the locally high operating temperatures, a degree of fluidity that ensures easy operation in cold weather, stable viscosity properties, closely controlled frictional behaviour to ensure quiet, consistent and chatter-free take-up of power in the internal brakes and clutches. It must preserve all these properties throughout its service life, so as to maintain consistent lubrication performance and gear-change characteristics.

Over the years, GM and Ford have progressively raised their levels of demand for oxidative/thermal stability, to cater for cars with higher power and thus greater likelihood of high transmission temperatures. GM's latest designation is Dexron IID, and the latest Ford specification is M2C 33-G, commonly referred to as the Ford G fluid. The Daimler-Benz requirement is that a fluid should retain its frictional characteristics throughout its life (in other words, throughout the life of the car). Additive systems have been developed with properties acceptable to Daimler-Benz, however, and a 'universal' ATF can now be blended.

38.9.21 Open and shielded gears

In general, the lubricant for open or partly enclosed gears is used on a 'total loss' basis – it is neither recirculated nor recovered. Rates of application must be minimal, both for economy and because the accumulation of used lubricant creates problems. A distinguishing feature of this category is that separate provision must be made for lubricating the bearings – by grease cups, oil bottles or mechanical lubricators.

Open gears usually run at low pitch-line speeds, below, say, 150 m/min, and are not machined to high standards of accuracy. If their lubricant is to be applied intermittently by hand, the most satisfactory results are obtained with special high-viscosity 'residual' oils that have strong properties of adhesion. Very viscous and heavy lubricants may be heated and spread over the gear teeth, which must be clean and dry, with a brush or paddle. In this case, the oil must not be heated so that it becomes too fluid, for although it will then be easier to apply there will be considerable wastage. It is equally important not to use too much oil, for the excess will immediately be flung off when the gears are in motion. In addition, over-oiling can lead to the formation of hard pads of lubricant at the bottom of the teeth, which could stress or distort the shaft or cause wear in the shaft bushes. To make heating unnecessary, the lubricant may be supplied in a suitable solvent that evaporates after cold application to the gears.

Under better conditions, mechanical lubrication may be used. Force-feed lubricators can be installed that will provide a continuous and measured supply of oil to the meshing teeth or by spraying atomized lubricant onto the gears.

Some slow-running shielded gears, particularly those operating in relatively clean conditions, can be lubricated by thinner oils. In that case, the teeth of the largest wheel can dip into a 'slush pan' and carry oil to the rest of the gear train.

38.9.22 Environment

While a gear oil will protect the metal surfaces against corrosion appreciable amounts of corrosive contaminants will affect the properties of the lubricant. Water is particularly troublesome in this respect. Different oils, while quite suitable for gear lubrication under normal conditions, may react very differently when contaminated with water. With active extreme pressure type oils, water leads to severe corrosion problems.

All these factors have to be borne in mind when blending gear oils. The oil must have a viscosity capable of maintaining a hydrodynamic lubricating film under, as far as possible, all working conditions. Its viscosity index should be high enough to maintain the viscosity within permitted limits at any operating temperature. It must be adhesive enough when on the gear teeth to resist removal by wiping or centrifugal force and not rupture under heavy or shock loads. The chemical stability of the oil must be such that it will not break down under the action of temperature, oxidizing agents and contaminants. It must provide protection against rusting and corrosion. It should not foam in service and should have the ability to separate from any water that may enter the system.

38.9.23 Gear wear and failure

A gearbox can be regarded as having four main components – gears, casing, bearings and lubricant – and failure can be caused by shortcomings in any one of them. Gear teeth that are inaccurate through design or manufacture can cause poor meshing, noisy running, overheating or surface failure due to overloading. Excessive flexing of the casing would allow misalignment of the gear shafts, which would also lead to surface failure due to overheating. Unsuitable or badly fitted bearings may fail because of their own surface damage, and this creates shaft misalignments and produces debris that can quickly damage the gear teeth. Lubrication failure can arise from unsuitable lubricant or from the inability (through bad design) of the lubricant to get where it is needed.

38.9.24 Manufacture

The makers of gear units have quite a choice of gear-cutting and gear-forming techniques at their disposal, including hobbing, shaping, grinding, broaching, milling and rolling. Every production process is inevitably a compromise between ease of machining, good surface finish, profile accuracy, speed of output and economic production. All metal-removal operations leave minute ridges and scratches, caused by tool shapes or by the motion of rotary cutters; 'perfect' profiles and finishes are not obtainable with normal industrial processes. Tooling marks and deep scratches act as stress raisers, which occasionally lead to fatigue failure and tooth breakage. Such a failure almost invariably starts in the fillet radius at the root of a tooth, and may be caused by the use of a cutting tool with too small a radius.

Most gears benefit from careful running-in under light load, since in this way the teeth acquire the surface finish needed for good lubrication and smooth running. In effect, they are given a final production operation, without which the process of surface failure might be initiated as soon as service operation begins.

The life of a gear can also be limited by poor assembly. If a pinion is clumsily pressed onto a shaft, for example, it is likely to operate in an overstressed condition and fail at an early stage, even through running well within its designated rating.

38.9.25 Materials

Gear failure is rarely attributable to the type of material selected. More often it is the result of defects such as cracks, casting inclusions and poor heat-treatment. A badly case-hardened steel might fail by the flaking of its hard outer layers and a through-hardened gear might have hard or soft patches. Distortions due to substandard hardening can modify tooth profiles to such an extent that load concentrations will break the ends of teeth.

The term 'failure' is relative, because severe operating conditions affect different materials in different ways. An example of this might be seen in a high-reduction gear unit whose pinion is very small and is therefore made of harder material, to avoid premature wear relative to the larger gear. In arduous running the two gears might not display the same kind of wear pattern.

38.9.26 Operation

The three most likely types of operational service misuse are overloading, incorrect lubrication and the presence of contaminants. Overloading is primarily due to the use of too small or too weak a gear unit, and this may be the result of false economy (installing an available unit for an application beyond its capacity) or failure to cater for the effects of shock loads in calculations of power rating.

Incorrect lubrication can take many forms. One example is the use of oil that is too thick or too thin, or is incompatible with the metal of the gears. Others include unsuitable methods of application, bad filtration, inadequate maintenance, filling to the wrong level, and poor standards of storage and handling.

Ideally, the lubricant should remain free from contaminants that might cause rust, abrasion or chemical attack; contamination may be caused by the presence of machining swarf, casting sand or even small items of rubbish left in the casing during manufacture. During service, dust and dirt can enter through the breathers, through neglected inspection covers or in additions of poorly maintained lubricant. Again, other contaminants have their origin inside the unit – rust particles, fine metal dust produced during the wearing process, and substances that result from the lubricant's gradual deterioration. Thick layers of sludge inside the casing can hinder the dissipation of heat, and thereby cause overheating; thick layers of dirt outside the casing can have exactly the same effect.

38.9.27 Surface damage

The types of wear and failure that occur in metallic gears are distinctive and have been subjected to close analysis over many years. The more common types of failure are abrasion, scuffing, pitting, corrosion, plastic yielding and fracture. Often the original cause, which may be misalignment, shock loading, faulty lubrication or one of numerous other possibilities, can be traced from the appearance of the gear teeth. It is important, also, that various kinds of wear may be combined in a single failure, making interpretation more difficult.

38.10 Hydraulic fluids

The wide application of hydraulic systems has undoubtedly been stimulated by the increasing use of fully automatic controls for sequences of operations where the response to signals must be rapid and the controls themselves light and easily operated. These needs are met by hydraulic circuits which, in addition, provide infinitely variable speed control, reversal of high-speed parts without shock, full protection against damage from overhead and automatic lubrication.

Over the years the performance standards of hydraulic equipment have risen. Whereas a pressure of about 7000 kPa used to be adequate for industrial hydraulic systems, nowadays systems operating with pressures of 15 000–25 000 kPa are common. Pressures above 35 000 kPa are to be found in applications such as large presses for which suitable high-pressure pumps have been developed. Additionally, systems have to provide increased power densities, more accurate response, better reliability and increased safety. Their use in numerically controlled machine tools and other advanced control systems creates the need for enhanced filtration. Full flow filters as fine as 1–10 μm retention capability are now to be found in many hydraulic systems.

With the trend toward higher pressures in hydraulic systems the loads on unbalanced pump and motor components become greater and this, coupled with the need for closer fits to contain the higher pressures, can introduce acute lubrication problems. Pumps, one of the main centres of wear, can be made smaller if they can run at higher speeds or higher pressures, but this is only possible with adequate lubrication. For this reason, a fluid with good lubrication properties is used so that 'hydraulics' is now almost synonymous with 'oil hydraulics' in general industrial applications. Mineral oils are inexpensive and readily obtainable while their viscosity can be matched to a particular job.

The hydraulic oil must provide adequate lubrication in the diverse operating conditions associated with the components of the various systems. It must function over an extended temperature range and sometimes under boundary conditions. It will be expected to provide a long, trouble-free service life; its chemical stability must therefore be high. Its wear-resisting properties must be capable of handling the high loads in hydraulic pumps. Additionally, the oil must protect metal surfaces from corrosion and it must both resist emulsification and rapidly release entrained air that, on circulation, would produce foam.

Mineral oil alone, no matter how high its quality, cannot adequately carry out all the duties outlined above and hence the majority of hydraulic oils have their natural properties enhanced by the incorporation of four different types of additives. These are: an anti-oxidant, an anti-wear agent, a foam-inhibitor and an anti-corrosion additive. For machines in which accurate control is paramount, or where the range of operating temperatures is wide – or both – oils will be formulated to include a VI improving additive as well.

38.10.1 Viscosity

Probably the most important single property of a hydraulic oil is its viscosity. The most suitable viscosity for a hydraulic system is determined by the needs of the pump and the circuit; too low a viscosity induces back-leakage and lowers the pumping efficiency, while too high a viscosity can cause overheating, pump starvation and possibly cavitation.

38.10.2 Viscosity Index

It is desirable that a fluid's viscosity stays within the pump manufacturer's stipulated viscosity limits, in order to accommodate the normal variations of operating temperature. An oil's viscosity falls as temperature rises; certain oils, however, are less sensitive than others to changes of temperatures, and these are said to have a higher VI. Hydraulic oils are formulated from base oils of inherently high VI, to minimize changes of viscosity in the period from start-up to steady running and while circulating between the cold and hot parts of a system.

38.10.3 Effects of pressure

Pressure has the effect of increasing an oil's viscosity. While in many industrial systems the working pressures are not high enough to cause problems in this respect, the trend towards higher pressures in equipment is requiring the effect to be accommodated at the design stage. Reactions to pressure are much the same as reactions to temperature, in that an oil of high VI is less effected than one of low VI. A typical hydraulic oil's viscosity is doubled when its pressure is raised from atmospheric to 35 000 kPa (Figure 38.11).

38.10.4 Air in the oil

In a system that is poorly designed or badly operated, air may become entrained in the oil and thus cause spongy and noisy operation. The reservoir provides an opportunity for air to be released from the oil instead of accumulating within the hydraulic system. Air comes to

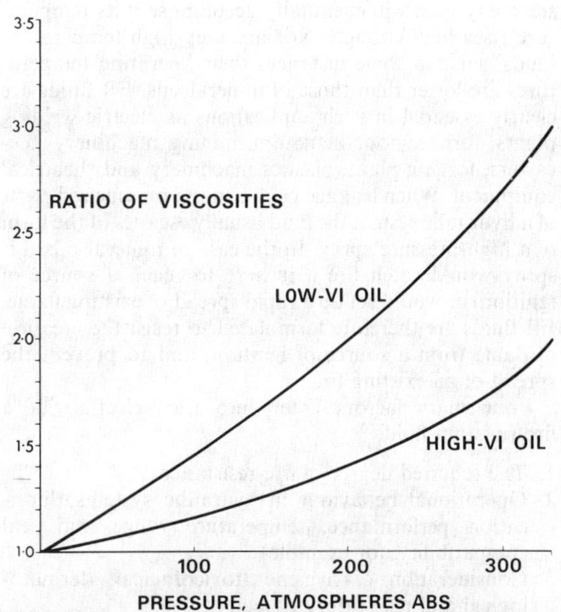

Figure 38.11

the surface as bubbles, and if the resultant foam were to become excessive it could escape through vents and cause loss of oil. In hydraulic oils, foaming is minimized by the incorporation of foam-breaking additives. The type and dosage of such agents must be carefully selected, because although they promote the collapse of surface foam they may tend to retard the rate of air release from the body of the oil.

38.10.5 Oxidation stability

Hydraulic oils need to be of the highest oxidation stability, particularly for high-temperature operations, because oxidation causes sludges and lacquer formation. In hydraulic oils, a high level of oxidation stability is ensured by the use of base oils of excellent quality, augmented by a very effective combination of oxidation inhibitors.

A very approximate guide to an oil's compatibility with rubbers commonly used for seals and hoses is given by the Aniline Point, which indicates the degree of swelling likely to arise; a high figure indicates a high level of compatibility. This system has been superseded by the more accurate Seal Compatibility Index (SCI), in which the percentage volume swell of a 'standard' nitrile rubber is determined after an immersion test in hot oil.

38.10.6 Fire-resistant fluids

Where fire is a hazard, or could be extremely damaging, fire-resistant hydraulic fluids are needed. They are referred to as 'fire resistant' (FR) so that users should be under no illusions about their properties. FR fluids do not extinguish fires: they resist combustion or prevent the spread of flame. They are not necessarily fireproof, since any fluid will eventually decompose if its temperature rises high enough. Nor are they high-temperature fluids, since in some instances their operating temperatures are lower than those of mineral oils. FR fluids are clearly essential in such applications as electric welding plants, furnace-door actuators, mining machinery, die-casters, forging plant, plastics machinery and theatrical equipment. When leakage occurs in the pressurized parts of a hydraulic system the fluid usually escapes in the form of a high-pressure spray. In the case of mineral oils this spray would catch fire if it were to reach a source of ignition, or would set up a rapid spread of existing flame. FR fluids are therefore formulated to resist the creation of flame from a source of ignition, and to prevent the spread of an existing fire.

Four main factors enter into the selection of a fire-resistant fluid:

1. The required degree of fire-resistance
2. Operational behaviour in hydraulic systems (lubrication performance, temperature range and seal compatibility, for example)
3. Consideration of hygiene (toxicological, dermatological and respiratory effects)
4. Cost

Table 38.7 CETOP classifications of fire-resistant hydraulic fluids

Class	*Description*
HF-A	Oil-in-water emulsions containing a maximum of 20% combustible material. These usually contain 95% water
HF-B	Water-in-oil emulsions containing a maximum of 60% combustible material. These usually contain 40–45% water
HF-C	Water–glycol solutions. These usually contain at least 35% water
HF-D	Water-free fluids. These usually refer to fluids containing phosphate esters, other organic esters or synthesized hydrocarbon fluids

CETOP: Comité European des Transmissions Oleohydrauliques et Pneumatiques.

Table 38.8 Comparison of oil and FR fluids

	Mineral oil	*Water-in-oil emulsion*	*Water–glycol*	*Phosphate ester*
Fire resistance	Poor	Fair	Excellent	Good
Relative density	0.87	0.94	1.08	1.14
Viscosity Index	High	High	High	Low
Vapour pressure	Low	High	High	Low
Special seals	No	Partly	Partly	Yes
Special paints	No	No	Yes	Yes
Rust protection	Very good	Good	Fair	Fair

37.10.7 Types of fluid

The fluids available, cover a range of chemical constituents, physical characteristics and costs, so the user is able to choose the medium that offers the best compromise for operational satisfaction, fire-resistance and cost effectiveness. Four basic types of fluid are available and are shown in Table 38.7.

In a fully synthetic FR fluid the fire resistance is due to the chemical nature of the fluid; in the others it is afforded by the presence of water. The other main distinction between the two groups is that the fully synthetic fluids are generally better lubricants and are available for use at operating temperatures up to 150°C, but are less likely to be compatible with the conventional sealing materials and paints than are water-based products.

When a water-based fluid makes contact with a flame or a hot surface its water component evaporates and forms a steam blanket which displaces oxygen from around the hot area, and this obviates the risk of fire. Water-based products all contain at least 35% water. Because water can be lost by evaporation, they should not be subjected to operating temperatures above about 60°C. Table 38.8 shows a comparison of oil and FR fluids.

38.10.8 High water-based hydraulic fluids

For a number of years HF-A oil-in-water emulsions have been used as a fire-resistant hydraulic medium for pit props. Concern over maintenance costs and operational life has created interest in a better anti-wear type fluid. Micro-emulsions are known to give better wear protection than the normal oil-in-water emulsions. At the same time the car industry, in attempts to reduce costs especially from leakages on production machinery, has evaluated the potential for using HWBHF in hydraulic systems. As a result, in many parts of industry, not only those where fire-resistant hydraulic fluids are needed, there is a increasing interest in the use of HWBHF.

Such fluids, often referred to as 5/95 fluid (that being the ratio of oil to water), have essentially the same properties as water with the exception of the corrosion characteristics and the boundary lubrication properties which are improved by the oil and other additives. The advantages of this type of fluid are fire resistance, lower fluid cost, no warm-up time, lower power consumption and operating temperatures, reduced spoilage of coolant, less dependence on oil together with reduced transport, storage, handling and disposal costs, and environmental benefits.

In considering these benefits the user should not overlook the constraints in using such fluids. They can be summarized as limited wear and corrosion protection (especially with certain metals), increased leakage due to its low viscosity, limited operating temperature range and the need for additional mixing and in-service monitoring facilities.

Because systems are normally *not* designed for use with this type of fluid, certain aspects should be reviewed with the equipment and fluid suppliers before a decision to use such fluids can be taken. These are compatibility with filters, seals, gaskets, hoses, paints and any non-ferrous metals used in the equipment. Condensation corrosion effect on ferrous metals, fluid-mixing equipment needed, control of microbial infection together with overall maintaining and control of fluid dilution and the disposal of waste fluid must also be considered. Provided such attention is paid to these design and operating features, the cost reductions have proved very beneficial to the overall plant cost effectiveness.

38.10.9 Care of hydraulic oils and systems

Modern additive-treated oils are so stable that deposits and sludge formation in normal conditions have been almost eliminated. Consequently, the service life of the oils which is affected by oxidation thermal degradation and moisture is extended.

Solid impurities must be continuously removed because hydraulic systems are self-contaminating due to wear of hoses, seals and metal parts. Efforts should be made to exclude all solid contaminants from the system altogether. Dirt is introduced with air, the amount of airborne impurities varying with the environment. The air breather must filter to at least the same degree as the oil filters.

It is impossible to generalize about types of filter to be used. Selection depends on the system, the rate of contamination build-up and the space available. However, a common arrangement is to have a full-flow filter unit before the pump with a bypass filter at some other convenient part of the system. Many industrial systems working below 13 500 kPa can tolerate particles in the order of 25–50 μm with no serious effects on either valves or pumps.

Provided that the system is initially clean and fitted with efficient air filters, metal edge-strainers of 0.127 mm spacing appear to be adequate, although clearances of vane pumps may be below 0.025 mm. It should be remembered that an excessive pressure drop, due to a clogged full-flow fine filter, can do more harm to pumps by cavitation than dirty oil.

If flushing is used to clean a new system or after overhaul it should be done with the hydraulic oil itself or one of lighter viscosity and the same quality. As the flushing charge circulates it should pass through an edge-type paper filter of large capacity. It is generally preferable to use a special pump rather than the hydraulic pump system, and the temperature of the oil should be maintained at about 40°C without local overheating.

38.11 Machine tools

Lubricants are the lifeblood of a machine tool. Without adequate lubrication, spindles would seize, slides could not slide and gears would rapidly distintegrate. However, the reduction of bearing friction, vital though it is, is by no means the only purpose of machine-tool lubrication. Many machines are operated by hydraulic power, and one oil may be required to serve as both lubricant and hydraulic fluid. The lubricant must be of correct viscosity for its application, must protect bearings, gears and other moving parts against corrosion, and, where appropriate, must remove heat to preserve working accuracies and alignments. It may additionally serve to seal the bearings against moisture and contaminating particles. In some machine tools the lubricant also serves the function of a cutting oil, or perhaps needs to be compatible with the cutting oil. In other tools an important property of the lubricant is its ability to separate rapidly and completely from the cutting fluid. Compatibility with the metals, plastics, sealing elements and tube connections used in the machine construction is an important consideration.

In machine-tool operations, as in all others, the wisest course for the user is to employ reputable lubricants in the manner recommended by the machine-tool manufacturer and the oil company supplying the product. This policy simplifies the selection and application of machine-tool lubricants. The user can rest assured that all the considerations outlined above have been taken into account by both authorities.

The important factors from the point of view of lubrication are the type of component and the conditions under which it operates, rather than the type of machine into which it is incorporated. This explains the essential similarity of lubricating systems in widely differing machines.

38.11.1 Bearings

As in almost every type of machine, bearings play an important role in the efficient functioning of machine tools.

38.11.2 Roller bearings

There is friction even in the most highly finished ball or roller bearing. This is due to the slight deformation under load of both the raceway and the rolling components, the presence of the restraining cage, and the 'slip' caused by trying to make parts of different diameter rotate at the same speed. In machine tools the majority of rolling bearings are grease-packed for life, or for very long periods, but other means of lubrication are also used (the bearings may be connected to a centralized pressure-oil-feed system, for instance). In other cases, oil-mist lubrication may be employed both for spindle bearings and for quill movement. In headstocks and gearboxes, ball and roller bearings may be lubricated by splash or oil jets.

38.11.3 Plain journal bearings

Plain bearings are often preferred for relatively low-speed spindles operating under fairly constant loads, and for the spindles of high-speed grinding wheels. These bearings ride on a dynamic 'wedge' of lubricating oil. Precision plain bearings are generally operated with very low clearances and therefore require low-viscosity oil to control the rise of temperature. Efficient lubrication is vital if the oil temperature is to be kept within reasonable limits, and some form of automatic circulation system is almost always employed.

38.11.4 Multi-wedge bearings

The main drawback of the traditional plain bearing is its reliance on a single hydrodynamic wedge of oil, which under certain conditions tends to be unstable. Multi-wedge bearings make use of a number of fixed or rocking pads, spaced at intervals around the journal to create a series of opposed oil wedges. These produce strong radial, stabilizing forces that hold the spindle centrally within the bearing. With the best of these, developed especially for machine tools, deviation of the spindle under maximum load can be held within a few millionths of a centrimetre.

38.11.5 Hydrostatic bearings

To avoid the instabilities of wedge-shaped oils films, a lubricating film can be maintained by the application of pressurized oil (or, occasionally, air) to the bearing. The hydrostatic bearing maintains a continuous film of oil even at zero speed, and induces a strong stabilizing force towards the centre which counteracts any displacement of the shaft or spindle. Disadvantages include the power required to pressurize the oil and the necessary increase in the size of the filter and circulatory system.

38.11.6 Slideways

Spindles may be the most difficult machine-tool components to design, but slideways are frequently the most troublesome to lubricate. In a slideway the wedge-type of film lubrication cannot form since, to achieve this, the slideway would need to be tilted.

38.11.7 Plain slideways

Plain slideways are preferred in the majority of applications. Only a thin film of lubricant is present, so its properties – especially its viscosity, adhesion and extreme-pressure characteristics – are of vital importance. If lubrication breaks down intermittently, a condition is created known as 'stick-slip' which affects surface finish, causes vibration and chatter and makes close limits difficult to hold. Special adhesive additives are incorporated into the lubricant to provide good bonding of the oil film to the sliding surfaces which helps to overcome the problems of table and slideway lubrication. On long traverses, oil may be fed through grooves in the underside of the slideway.

38.11.8 Hydrostatic slideways

The use of hydrostatic slideways – in which pressurized oil or air is employed – completely eliminates stick-slip and reduces friction to very low values; but there are disadvantages in the form of higher costs and greater complication.

38.11.9 Ball and roller slideways

These are expensive but, in precision applications, they offer the low friction and lack of play that are characteristic of the more usual rolling journal bearings. Lubrication is usually effected by grease or an adhesive oil.

38.11.10 Leadscrews and nuts

The lubrication of leadscrews is similar in essence to that of slideways, but in some instances may be more critical. This is especially so when pre-load is applied to eliminate play and improve machinining accuracy, since it also tends to squeeze out the lubricant. Leadscrews and slideways often utilize the same lubricants. If the screw is to operate under high unit stresses – due to pre-load or actual working loads – an extreme-pressure oil should be used.

38.11.11 Recirculating-ball leadscrews

This type was developed to avoid stick-slip is heavily loaded leadscrews. It employs a screw and nut of special form, with bearing balls running between them. When the balls run off one end of the nut they return through an external channel to the other end. Such bearings are usually grease-packed for life.

38.11.12 Gears

The meshing teeth of spur, bevel, helical and similar involute gears are separated by a relatively thick hydrodynamic wedge of lubricating oil, provided that the rotational speed is high enough and the load light enough so as not to squeeze out the lubricant. With high loads or at low speeds, wear takes place if the oil is not able to maintain a lubricating film under extreme conditions.

Machine-tool gears can be lubricated by oil-spray, mist, splash or cascade. Sealed oil baths are commonly used, or the gears may be lubricated by part of a larger circulatory system.

38.11.13 Hydraulics

The use of hydraulic systems for the setting, operation and control of machine tools has increased significantly. Hydraulic mechanisms being interlinked with electronic controls and/or feedbacks control systems. In machine tools, hydraulic systems have the advantage of providing stepless and vibrationless transfer of power. They are particularly suitable for the linear movement of tables and slideways, to which a hydraulic piston may be directly coupled.

One of the most important features for hydraulic oil is a viscosity/temperature relationship that gives the best compromise of low viscosity (for easy cold starting) and minimum loss of viscosity at high temperatures (to avoid back-leakage and pumping losses). A high degree of oxidation stability is required to withstand high temperatures and aeration in hydraulic systems. An oil needs excellent anti-wear characteristics to combat the effects of high rubbing speeds and loads that occur in hydraulic pumps, especially in those of the vane type. In the reservoir, the oil must release entrained air readily without causing excessive foaming, which can lead to oil starvation.

38.11.14 Tramp oil

'Tramp oil' is caused when heat slideway, gear, hydraulic and spindle lubricants leak into water-based cutting fluids and can cause problems such as:

- Machine deposits
- Reduced bacterial resistance of cutting fluids and subsequent reduction in the fluid life
- Reduced surface finish quality of work pieces
- Corrosion of machine surfaces

All these problems directly affect production efficiency. Recent developments have led to the introduction of synthetic lubricants that are fully compatible with all types of water-based cutting fluids, so helping the user to achieve maximum machine output.

38.11.15 Lubrication and lubricants

The components of a hydraulic system are continuously lubricated by the hydraulic fluid, which must, of course, be suitable for this purpose. Many ball and roller bearings are grease-packed for life, or need attention at lengthy intervals. Most lubrication points, however, need regular replenishment if the machine is to function satisfactorily. This is particularly true of parts subjected to high temperatures.

With the large machines, the number of lubricating points or the quantities of lubricants involved make any manual lubrication system impracticable or completely uneconomic. Consequently, automatic lubrication systems are often employed.

Automatic lubrication systems may be divided broadly into two types: circulatory and 'one-shot' total-loss. These cover, respectively, those components using relatively large amounts of oil, which can be cooled, purified and recirculating, and those in which oil or grease is used once only and then lost. Both arrangements may be used for different parts of the same machine or installation.

38.11.16 Circulatory lubrication systems

The circulatory systems used in association with machine tools are generally conventional in nature, although occasionally their exceptional size creates special problems. The normal installation comprises a storage tank or reservoir, a pump and filter, suitable sprays, jets or other distribution devices, and return piping. The most recent designs tend to eliminate wick feeds and siphon lubrication.

Although filtration is sometimes omitted with non-critical ball and roller bearings, it is essential for most gears and for precision bearings of every kind. Magnetic and gauze filters are often used together. To prevent wear of highly finished bearings surfaces the lubricant must contain no particle as large as the bearing clearance.

Circulatory systems are generally interlocked electrically or mechanically with the machine drive, so that the machine cannot be started until oil is flowing to the gears and main bearings. Interlocks also ensure that lubrication is maintained as long as the machine is running. Oil sight-glasses at key points in the system permit visual observations of oil flow.

38.11.17 Loss-lubrication systems

There are many kinds of loss-lubrication systems. Most types of linear bearings are necessarily lubricated by this means. An increasingly popular method of lubrication is by automatic or manually operated one-shot lubricators. With these devices a metered quantity of oil or grease is delivered to any number of points from a single reservoir. The operation may be carried out manually, using a hand-pump, or automatically, by means of an electric or hydraulic pump. Mechanical pumps are usually controlled by an electric timer, feeding lubricant at preset intervals, or are linked to a constantly moving part of the machine.

On some machines both hand-operated and electrically timed one-shot systems may be in use, the manual system being reserved for those components needing infrequent attention (once a day, for example) while the automatic systems feeds those parts that require lubrication at relatively brief intervals.

38.11.18 Manual lubrication

Many thousands of smaller or older machines are lubricated by hand, and even the largest need regular refills or topping up to lubricant reservoirs. In some shops the operator may be fully responsible for the lubrication of his own machine, but it is nearly always safer and more economical to make one individual responsible for all lubrication.

38.11.19 Rationalizing lubricants

To meet the requirements of each of the various components of a machine the manufacturer may need to recommend a number of lubricating oils and greases. It follows that, where there are many machines of varying origins, a large number of lubricants may seem to be needed. However, the needs of different machines are rarely so different that slight modification cannot be made to the specified lubricant schedule. It is this approach which forms the basis for BS 5063, from which the data in Table 38.9 have been extracted. This classification implies no quality evaluation of lubricants, but merely gives information as to the categories of lubricants likely to be suitable for particular applications.

A survey of the lubrication requirements, usually carried out by the lubricant supplier, can often be the means of significantly reducing the number of oils and greases in a workshop or factory. The efficiency of lubrication may well be increased, and the economies effected are likely to be substantial.

38.12 Cutting fluids

New machining techniques are constantly being introduced. Conventional workpiece materials have improved progressively through close control of manufacturer and heat treatment, and new materials have been fostered by the aeronautic and space industries. The results have been ever-improving output, dimensional control and surface finish. The continuous development of cutting fluids has enabled these increasingly severe conditions to be accommodated.

38.12.1 Functions of a cutting fluid

The two main functions of a cutting fluid are to cool and to lubricate. During a machining operation the cutting tool induces a continuous wave of dislocations, which travel ahead of the cutting edge, deforming and shearing the workpiece metal into continuous swarf or broken chips. The energy required for this deformation is converted into heat, and simultaneously frictional heat is created at the points where the tool rubs against the newly exposed surface of the workpiece and the chip rubs against the top face of the advancing tool.

These combined effects raise the temperature of the tool, workpiece and chip to a level that would be unacceptable in most operations if it were not for the presence of cutting fluid. Since the cooling effect is the greatest benefit, cutting fluids are widely known as 'coolants', In practice, the coolant cannot penetrate to the extreme tip of the tool, of course, because the tip is so well shrouded by the metal being cut from the workpiece. Nevertheless, the coolant performs its function successfully because it enters a zone some way back from the tip, where the chip is passing over the tool at high rubbing velocity and where most of the frictional heat arises (Figure 38.12).

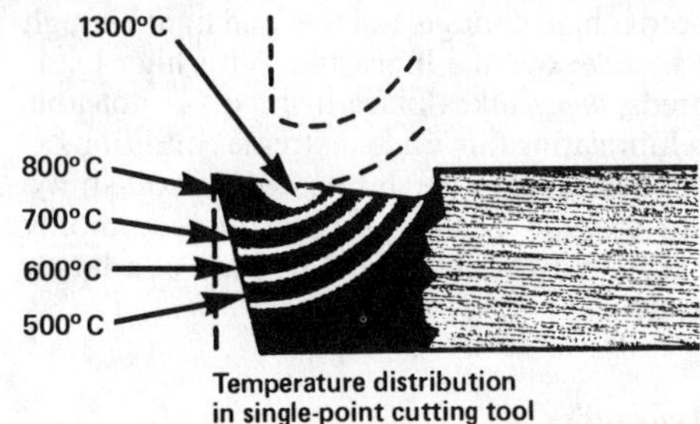

Figure 38.12 Heat is generated both by deformation of workpiece material and by the friction of the chip across the tool tip. A cutting fluid acts as a lubricant to reduce frictional heat and simultaneously cools the whole cutting zone

38.12.2 Benefits of a coolant

Cutting fluids provide benefits such as extended tool life, dimensional accuracy and good surface finish, all of which contribute to high rates of production. Almost invariably, the coolant that adheres to workpieces is relied on to protect ferrous components against corrosion while they are awaiting further machining or assembly operations.

The benefits to be obtained by the correct selection and application of coolant are considerable. The cost is hardly a major factor in production operations, however, because, on average, the price of coolant amounts to about 1% of the finished cost of a machined component.

38.12.3 Types of coolant

Water-based emulsions

Water, of course, is the cheapest cooling medium generally available. Its high specific heat makes it particularly effective as a coolant and its low viscosity allows high rates of flow through the coolant system and so enables it to penetrate the cutting zone and make contact with hot metal surfaces. Unfortunately, it corrodes most of the ferrous and some of the non-ferrous metals (a process hastened by high temperatures), and therefore is unsuitable in most machining.

However, its cooling properties can be utilized when other materials are added to improve machining performance. One of the most common water-based cutting fluids is the so-called 'soluble' oil which, in fact, is not a true solution but an oil-in-water emulsion in which very fine droplets of oil are suspended in water. Such a fluid has very effective cooling power and its lubricating and protective properties are provided by the petroleum oil and its additives. The proportion of oil may be as low as 2%, but is normally between 5% and 10%.

Water-based coolants, as diluted and used in service, are much cheaper than neat oils, and they offer remarkable economic benefits in the applications to which they are best suited. (Neat oils, however, are preferred for some machining processes – usually involving severe operations – where they give overall production economies and reduced machine-tool maintenance.)

Soluble oils are delivered, 'concentrated', to the user and contain an emulsifying agent to ensure that a stable emulsion forms when added to water. This additive does not mix readily with mineral oil, however, so to overcome this a 'coupling' agent is included in the formulation.

The performance of soluble oils is made possible not only by their high specific heat and thermal conductivity but also by their low viscosity, which permits good penetration into the very fine clearances around the cutting zone. Consequently, these fluids are used mainly where cooling is the primary requirement. Lubricating properties can be improved by polar additives, which are agents that enhance the 'oiliness' or anti-friction characteristics. Further improvements can be effected by EP (extreme-pressure) additives, which are usually compounds of sulphur or chlorine.

Some of the recently developed high-performance EP soluble oils have a cutting performance that almost matches that of additive-type neat oils, and they are particularly suitable for demanding operations in machine tools whose design allows the use of water-based fluids.

Water-based solutions

In recent years another kind of water-based coolant, the 'synthetic' or non-petroleum cutting fluid, has been used increasingly. Coolants of this kind contains agents that provide the necessary lubricating and anti-corrosion properties. They were originally introduced to satisfy the special demands of grinding operations, but now they have been developed to a state where they can cope with general machining requirements. Their use in CNC machines and flexible manufacturing systems is a significant result of this development.

Because the synthetic fluid is transparent, operators can see the cutting area at all times – a feature particularly useful for intricate machining and grinding operations. Workpieces, gauges and tools in the vicinity of the cutting zone do not become coated with an oily film, which is a pleasing feature for tool setters who have to handle spanners and precision tools in confined spaces. Occasionally a transparent dye is included in the formulation of a synthetic cutting fluid, both for identification and to give a pleasant appearance.

Synthetic cutting fluids are suitable for a wider range of uses than any other single cutting fluid, and can be employed in any machine designed for operation with water-based fluids. Applications include turning, milling, drilling and screw-cutting on all but the toughest materials; and also cylindrical, surface and centreless grinding. Most metals can be machined effectively; cast iron, which is prone to rusting when machined with conventional soluble oils, shows negligible signs of corrosion when the synthetic fluid is used. Copper-base alloys may display a slight degree of staining if workpieces are not washed immediately after the machining operation.

Bacteria

Machine operators working with emulsions can become susceptible to skin infections because of the combination of the de-fatting effect of soluble-oil emulsifiers and the abrasive action of metallic swarf, but bacteria in cutting fluids are seldom the source of such infections. High standards of personal hygiene and the use of barrier creams should prevent such problems. A more difficult situation arises when a soluble-oil emulsion becomes infected with a bacteria capable of utilizing the emulsifier and mineral-oil components in the system. Even in clean conditions, untreated soluble-oil emulsions and solutions cannot remain completely sterile for any length of time.

The more common bacteria found in infected soluble oil systems can degrade the inhibitors, emulsifiers and mineral oil components. They they cause a loss of anti-corrosion properties, increase of acidity and deterioration of the emulsion. These bacteria thrive in well-aerated systems, and are termed aerobic.

When a system becomes infected with sulphate-reducing bacteria it develops the obnoxious smells associated with putrefaction. Bacteria of the species mainly responsible for such conditions thrive in the absence of oxygen, and are described as anaerobic. The foul smells of hydrogen sulphide are produced mainly in stagnant soluble-oil systems, particularly where the emulsion has a surface layer of oil that tends to exclude oxygen. The hydrogen sulphide arises from chemical reduction of either the inorganic sulphates that can give water its hardness or the petroleum sulphonate component of the soluble oil.

Bacterial examination could be used to confirm the presence of (and perhaps identify) the species. Where a soluble oil is being degraded, or where a noticeable loss of anti-corrosion properties has been detected, a check on the pH (alkalinity/acidity) of the system would be advantageous.

Cleanliness and good housekeeping in machine shops do much to avoid bacterial infection, and their importance cannot be overstressed. Various techniques such as heat treatment, centifuging and filtration can be used to advantage, although economic considerations may restrict their use to systems containing large volumes of soluble oil. Chemical sterilization with bactericides can be more convenient.

Bactericides are substances that destroy bacteria, and they can be used in various ways. They may be incorporated into the soluble-oil concentrate, either at concentrations suitable to protect the oil in storage, or at levels sufficient to provide a persistent bactericidal effect on the emulsion in service. The cost of providing sufficient bactericide to cover the use of the soluble oil at a high dilution might prove prohibitive. Continued use of the same bactericide may produce resistant strains of bacteria.

Bactericides can be added to the soluble-oil system as a shock treatment when infection occurs, but the user must bear in mind that a badly infected and degraded emulsion cannot later be reclaimed. The bactericidal treatment may have to be repeated periodically thereafter, and the effect of the bactericide must be monitored.

Table 38.9 Classification of lubricants

Class	*Type of lubricant*	*Viscosity grade no. (BS 4231)*	*Typical application*	*Detailed application*	*Remarks*
AN	Refined mineral oils	68	General lubrication	Total-loss lubrication	May be replaced by CB 68
CB	Highly refined mineral oils (straight or inhibited) with good anti-oxidation performance	32 68	Enclosed gears – general lubrication	Pressure and bath lubrication of enclosed gears and allied bearings of headstocks, feed boxes, carriages, etc. when loads are moderate; gears can be of any type, other than worm and hypoid	CB 32 and CB 68 may be used for flood-lubricated mechanically controlled clutches; CB 32 and CB 68 may be replaced by HM 32 and HM 68
CC	Highly refined mineral oils with improved loading-carrying ability	150 320	Heavily loaded gears and worms gears	Pressure and bath lubrication of enclosed gears of any type, other than hypod gears, and allied bearings when loads are high, provided that operating temperature is not above 70°C	May also be used for manual or centralized lubrication of lead and feed screws
FX	Heavily refined mineral oils with superior anti-corrosion anti-oxidation performance	10 22	Spindles	Pressure and bath lubrication of plain or rolling bearings rotating at high speed	May also be used for applications requiring particularly low-viscosity oils, such as fine mechanisms, hydraulic or hydro-pneumatic mechanisms electro-magnetic clutches, air line lubricators and hydrostatic bearings
G	Mineral oils with improved lubricity and tackiness performance, and which prevent stick-slip	68 220	Slideways	Lubrication of all types of machine tool plain-bearing slideways; particularly required at low traverse speeds to prevent a discontinuous or intermittent sliding of the table (stick-slip)	May also be used for the lubrication of all sliding parts – lead and feed screws, cams, ratchets and lightly loaded worm gears with intermittent service; if a lower viscosity is required HG 32 may be used.

HM	Highly refined mineral oils with superior anti-corrosion, anti-oxidation, and anti-wear performance	32 68	Hydraulic systems	Operation of general hydraulic systems	May also be used for the lubrication of plain or rolling bearings and all types of gears, normally loaded worm and hypoid gears excepted, HM 3X and HM 68 may replace CB 32 and CB 68, respectively
HG	Refined mineral oils of HM type with anti-stick-slip properties	32	Combined hydraulic and slideways systems	Specific application for machines with combined hydraulic and plain bearings, and lubrication systems where discontinuous or intermittent sliding (stick-slip) at low speed is to be prevented	May also be used for the lubrication of slideways, when an oil of this viscosity is required

Class	*Type of lubricant*	*Consistency number*	*Typical application*	*Detailed application*
XM	Premium quality multi-purpose greases with superior anti-oxidation and anti-corrosion properties	1 2 3	Plain and rolling bearings and general greasing of miscellaneous parts	XM 1: Centralized systems XM 2: Dispensed by cup or hand gun or in centralized systems XM 3: Normally used in prepacked applications such as electric motor bearings

Note: It is essential that lubricants are compatible with the materials used in the construction of machine tools, and particularly with sealing devices.
The grease X is sub-divided into consistency numbers, in accordance with the system proposed by the National Lubricating Grease Institute (NLGI) of the USA. These consistency numbers are related to the worked penetration ranges of the greases as follows:

Consistency number	*Worked penetration range*
1	310–340
2	265–295
3	220–250

Worked penetration is determined by the cone-penetration method described in BS 5296.

Studies of systems infected with bacteria show that a process of continued re-infection occurs from the residual oil left in the system whenever the fluid is changed. As far as possible, the system should be designed to avoid traps that might retain coolant. An effective treatment however, is the use of a bactericide in the form of a detergent sterilizing solution to cleanse the system. All traces of it must be removed from the system before a new charge of coolant is introduced, otherwise emulsion instability may develop in the fresh coolant. Thorough cleansing, combined with the use of a suitable bactericide in the fresh charge of soluble oil, should overcome most problems of bacterial infection.

Neat oils

Two factors militate against the universal use of water-based fluids. Very severe machining operations call for a lubrication performance that is beyond the capacity of such fluids, and the design of some machine tools means that water cannot be used because of the risk of cross-contamination with machine lubricants. In these instances, 'neat' cutting oil is the only fluid that can provide the required performance.

Neat oil is the name given to an orthodox petroleum cutting fluid, whether or not it contains additives, to enhance cutting properties. Oils of this sort are available in a very wide variety, and many combinations of workpiece material, machining characteristics and tooling requirement justify special formulations. The neat oils have lower specific heat than water, so they have to be fed to the cutting zone in copious amounts to provide the optimum cutting effect.

The viscosity of a neat cutting oil is chosen to suit the machining application. Some operations need a very light oil, others a heavier viscosity because of its better load-carrying capacity. Oils are therefore available in a wide selection of viscosities. Most general-purpose cutting oils have viscosities of about 20 cSt at 40°C, but a lighter viscosity is used in operations such as broaching and deep-hole boring/drilling, when chips have to be flushed away from the confined areas around the tool. Whereas for machining – such as some milling, turning and gear cutting – heavier viscosities, up to 64 cSt at 40°C, are preferred.

Additives in neat oils

The increasing diversity of operations, new materials and processes and the constant demand for improved production efficiency can only be met by various additives and compounding agents being blended into the oil to enhance its performance. Additives tend to be expensive and the selection of enhanced cutting fluid is only justified by overall production economies.

Fatty oils

Small additions of fatty oils improve the 'oiliness' or anti-friction characteristics, and cutting oils reinforced in this way are known as compounded oils. Such oils are particularly useful in the machining of difficult yellow metals and aluminium alloys, giving excellent tool life and good finish without staining yellow metals.

EP oils

For the more difficult operations, neat oils containing EP (extreme-pressure) additives have to be used. The EP cutting oils usually contain additives based on sulphur or chlorine, or combinations of them.

The sulphur in an EP oil can be present in two forms. In the inactive fluid it is chemically combined with a fatty-oil additive, which is blended with mineral oil to produce a sulphurized fatty oil. The active version, on the other hand, contains sulphur in elemental form, dissolved in mineral oil; the fluid is known as sulphurized mineral oil. Chlorine is usually present only as a chlorinated paraffin, which is blended sometimes singly with mineral oils and sometimes in combination with fatty oils and sulphurized additives.

Combinations of additives

Combinations of sulphur and chlorine additives are employed because the chlorine-based agents react at a lower temperature than do the sulphur-based ones, and in some applications this progressive action is advantageous. Occasionally two kinds of additives, when used in combination, have a greater effect than the sum of their individual contributions. This phenomenon is known as a 'synergistic' effect, and it gives the lubricant technologist more scope for matching the properties of the cutting fluid to the requirements of the machining operation.

Electro discharge machining (EDM) oils

Electro discharge (spark erosion) techniques rely heavily on the ability of an EDM oil to act as an electrical insulant, to dissipate heat from the electrode, and to flush away erosion debris from the workpiece. EDM oils also are suitable for all die-sinking spark erosion operations. They should have low aromatic levels, good filterability, low fuming, high dielectric strength, excellent oxidation resistance and low colour level.

A low-viscosity grade helps provide maximum flushing and cooling for delicate work and close tolerances. When rough and finishing machining are combined a medium viscosity is recommended for the good flushing needed. In roughing operations a high-viscosity oil enables fast metal removal rates to be achieved.

38.12.4 Selection of cutting fluid

Selection of the correct fluid is essential if the maximum benefit is to be obtained for the user. In some machining operations and with some workpiece metals, of course, the choice is fairly straightforward, but in the majority of cases selection is inevitably a compromise in which several factors have to be weighed. The two most important considerations are the workpiece material and the type of machining operation (which, in turn, involves considerations of the tool material).

38.12.5 Workpiece material

A general guide to metals and their difficulty of machining, beginning with the most difficult, could be listed as:

1. Titanium
2. Nimonic alloys
3. Inconel
4. Nickel
5. Stainless steel
6. Monel metal
7. High-tensile steel
8. Wrought iron and cast alloys
9. High-tensile bronze
10. Copper
11. Medium- and low-carbon steels
12. Free-cutting mild steel
13. Aluminium alloys
14. Brass
15. Bronze
16. Zinc-based alloys
17. Free-cutting brass
18. Magnesium alloys

38.12.6 Machining operations

Operations vary in severity, but the following list gives a general indication of how they compare, starting with the most severe:

Broaching (internal and surface)
Gear-cutting and shaving
Deep-holeboring (Sandvick ejector drill or gun drilling)
Tapping and threading
Multi- and single-spindle automatics
Capstan lathes, central lathes
Milling and drilling
Sawing

38.12.7 Tool material

A third important factor in the economies of machining is the material of the cutting tool. This largely determines the rates of metal removal, the standards of surface finish and the frequency at which the tool needs to be reground – all of which are interrelated. These can be broadly grouped in three categories, each separated by a factor of 10 in terms of performance.

1. Ceramic-coated disposable inserts, including silicon nitride, boron nitride, titanium nitride (TIN), titanium carbide (TIC) and sintered synthetic diamond;
2. Tungsten carbide, usually in the form of tipped tooling;
3. High-speed steel.

38.12.8 Ancillary factors

Although the three sets of factors just given can be considered singly or in combination, and although a coolant-selection chart can be used as a guide, the most obvious choice of coolant is not always satisfactory from every aspect. Among other (and probably overriding) considerations are the economics of production, scope for reclaiming the coolant, health and safety requirements, and finally the suitability of the machine's design – because of possible incompatibility of the coolant and the machine's lubricant. In time, moreover, the performance of any chosen coolant can be seriously affected by variations of cutting speed, feed rate, tool-setting accuracy and even the condition of the machine. Incorrect tool geometry and inaccurate grinding may have their own adverse effects.

The chosen fluid, to be most effective, must be kept in good condition, supplied in the correct quantity and directed accurately towards the cutting area. If its pressure and rate of flow are too high, the resultant splattering causes wastage of coolant and untidy splashes around the machine and may well lead to overheating in the cutting zone because insufficient fluid is reaching it. The fluid must be applied where it is most effective – pointed either directly at the tip of the tool or flooding the complete cutting zone. Too often the flow is directed vaguely at the workpiece, only to be deflected by the workpiece itself or by ancillary parts of the machine.

Poor maintenance practices lead to dirty and degraded coolant, which affects tool life, surface finish and perhaps the welfare of the operator. Proper filtration is vital, and care must be taken that contamination from other coolants or machine lubricants is kept to a minimum.

38.12.9 Grinding

Although primarily regarded as a finishing process, grinding has also been developed for heavy stock removal. In effect, the grinding wheel is a multi-point machining tool because the metal is cut by many fragments of very hard material (the grit) embedded in a softer matrix (the bond). Each cutting edge requires a lubricant for clean cutting and a coolant for the dissipation of heat. Clearly, the cutting fluid must satisfy the requirements of both the workpiece material and the grinding wheel, and before a grinding fluid is selected a check must be made that the correct grade of wheel is being used for the operation.

The cutting fluid has to cool and lubricate without allowing grinding debris to clog (or load) the surface of the wheel. This could produce a burnished surface rather than a ground one – which might appear satisfactory but would be metallurgically unsound.

For many grinding operations, translucent soluble oils or synthetic solutions give the best results. Synthetic solutions are especially favoured for their very high clarity on intricate internal grinding. On the more complex and heavy-duty operations involving form-grinding, more importance is attached to lubricating properties, so EP neat oils are favoured. The level of EP activity required is determined by the severity of the form being ground.

38.12.10 Storage

Extremes of temperature should be avoided, particularly in the storage of coolants that contain water, because the balance of constituents can be upset. High temperatures lead to evaporation of water. Low temperatures can cause the separation of some of the components in additive oils; in particular, the natural fatty oils are susceptible to coagulation. Soluble oil, as supplied in concentrated form, usually contains a small amount of water for ensuring rapid dispersion during subsequent dilution. If this water freezes, it does not re-blend when the temperature rises again, and subsequently the oil proves very difficult to mix.

38.12.11 Water quality

Water from towns' main supplies is usually suitable for the preparation of water-based cutting fluids. That from factory boreholes is also generally suitable, although occasionally it contains excessive amounts of corrosive salts. Water from rivers, canals and ponds almost always contains undesirable contaminants, and should be tested before use. A good first test is to mix a small quantity of emulsion and allow it to stand for 24 h: in this time, no more than a trace of the oil should separate. If serious separation occurs, the water should be analysed to indicate the sort of remedial treatment required.

A laboratory check is normally desirable to assess the amounts of organic and mineral acids present, and a check on hardness is usually necessary in any case. Hardness in water is due to dissolved salts – mainly of calcium, magnesium and iron, and occasionally of aluminium. Softening may be required if the water is extremely hard, because the salts react with the emulsifier in the soluble oil to form an insoluble scum that floats on the surface of the emulsion. The scum may not in itself be harmful, but its formation uses up some of the emsulsifier and causes the emulsion to be unstable.

38.12.12 Cutting fluids in service

Neat oils

During service, neat cutting oils suffer negligible deterioration of quality; their service life is almost indefinitely long, provided they are kept clean and free from contamination. They suffer very little depletion of their additives through cutting action, and quality is adequately maintained by additions of new oil to make up for 'drag-out' losses (those caused by oil adhering to both the swarf and the work). Serious leakage of hydraulic fluid or lubricating oil into the coolant system can have a diluting effect, which reduces additive concentration and hence the performance of the cutting oil.

Water-based fluids

As long as they are kept clean and maintained at the correct concentration, water-based fluids have a long service life. The concentration of an emulsion can be affected in two ways: by evaporation of water and by drag-out of the oil content. In many cases the effects counteract each other, so the top-up fluid is given the same concentration as the working fluid. Solutions, on the other hand, tend to become more concentrated during service because they suffer from evaporation of water but not from drag-out losses of the other constituents, so top-up solutions are generally weaker than working solutions.

It is advisable to monitor water-based fluids at frequent intervals so that their water content can be brought to within tolerance before the machining operation is affected. Maintenance of the correct concentration ensures that cutting performance and rust-preventive properties stay at the required level. A simple technique makes use of the fact that a fluid's density influences its refractive index. In this method a very small amount of the fluid is viewed in a pocket-size instrument called a refractometer, and the density is indicated by the edge of a dark shadow falling on a graduated line. The line gives a figure which, in some cases, is a direct reading of dilution, but in others it may have to be correlated with dilution on a graph drawn for the particular fluid.

The strength of a solution or emulsion is, of course, corrected by additions of stronger or weaker fluid. The percentage concentration of make-up fluid is calculated by the formula:

$$C=\frac{C_e}{Q_r}(C_f-C_e)+C_f$$

where C_f = required final concentration,
C_e = existing concentration,
Q_e = existing quantity,
Q_r = quantity required of make-up fluid.

Cleaning procedures

Although some neat cutting fluids can be used almost indefinitely if well maintained, most fluids eventually reach the end of their useful life, so the system has to be drained and refilled with new fluid. On such occasions the system should be thoroughly cleaned out before the new charge is poured in. The nature of cleaning depends on the type of cutting fluid.

Normally a system containing net oil requires only manual cleaning because the debris contained is likely to be mainly metallic, but water-based fluids call for more searching treatment. Caustic-soda and proprietary solutions are available for the purpose of flushing away slimy deposits. If there is evidence of bacterial degradation, however, bactericides must be introduced. These tend to persist in remote pockets and in the more stagnant parts of a coolant system, and manual cleaning by itself is inadequate. The risk of persistence can be significantly minimized at the planning stage if the system is designed to be free of regions to trap stagnant fluid.

38.13 Compressors

Compressors fall into two basic categories: positive-displacement types, in which air is compressed by the

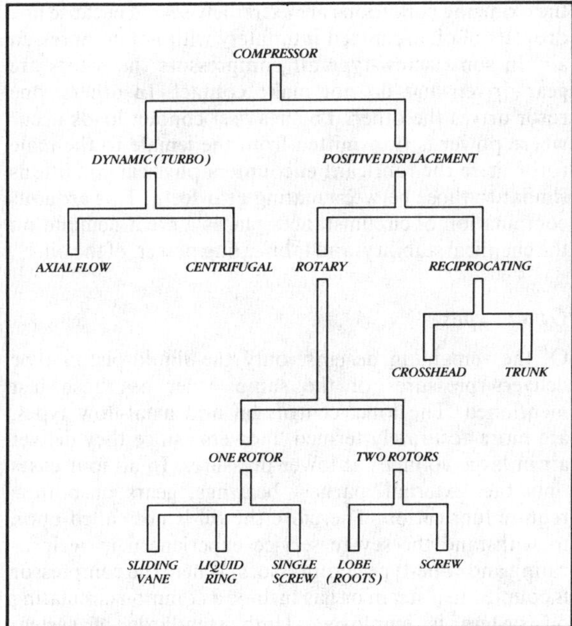

Figure 38.13 Compressor types

'squashing' effect of moving components; and dynamic (turbo)-compressors, in which the high velocity of the moving air is converted into pressure. In some compressors the oil lubricates only the bearings, and does not come into contact with the air; in some it serves an important cooling function; in some it is in intimate contact with the oxidizing influence of hot air and with moisture condensed from the air. Clearly, there is no such thing as a typical all-purpose compressor oil: each type subjects the lubricant to a particular set of conditions. In some cases a good engine oil or a turbine-quality oil is suitable, but in others the lubricant must be special compressor oil (Figure 38.13).

38.13.1 Quality and safety

Over the years the progressive improvements in compressor lubricants have kept pace with developments in compressor technology, and modern oils make an impressive contribution to the performance and longevity of industrial compressors. More recently a high proportion of research has been directed towards greater safety, most notably in respect of fires and explosions within compressors. For a long time the causes of such accidents were a matter of surmise, but it was noticed that the trouble was almost invariably associated with high delivery temperatures and heavy carbon deposits in delivery pipes. Ignition is now thought to be caused by an exothermic (heat-releasing) oxidation reaction with the carbon deposit, which creates temperatures higher than the spontaneous ignition temperature of the absorbed oil.

Experience indicates that such deposits are considerably reduced by careful selection of base oils and anti-oxidation additives. Nevertheless, the use of a top-class oil is no guarantee against trouble if maintenance is neglected. For complete safety, both the oil and the compressor system must enjoy high standards of care.

38.13.2 Specifications

The recommendations of the International Standards Organization (ISO) covering mineral-oil lubricants for reciprocating compressors are set out in ISO DP 6521, under the ISO-L-DAA and ISO-L-DAB classifications. These cover applications wherever air-discharge temperature are, respectively, below and above 160°C. For mineral-oil lubricants used in oil-flooded rotary-screw compressors the classifications ISO-L-DAG and DAH cover applications where temperatures are, respectively, below 100°C and in the 100–110°C range. For more severe applications, where synthetic lubricants might be used, the ISO-L-DAC and DAJ specifications cover both reciprocating and oil-flooded rotary-screw requirements.

For the general performance of compressor oils there is DIN 51506. This specification defines several levels of performance, of which the most severe – carrying the code letters VD-L – relates to oils for use at air-discharge temperatures of up to 220°C.

The stringent requirements covering oxidation stability are defined by the test method DIN 51352, Part 2, known as the Pneurop Oxidation Test (POT). This test simulates the oxidizing effects of high temperature, intimate exposure to air, and the presence of iron oxide which acts as catalyst – all factors highly conducive to the chemical breakdown of oil, and the consequent formation of deposits that can lead to fire and explosion.

Rotary-screw compressor mineral oils oxidation resistance is assessed in a modified Pneurop oxidation test using iron naphthenate catalyst at 120°C for 1000 h. This is known as the rotary-compressor oxidation test (ROCOT).

38.13.2 Oil characteristics

Reciprocating compressors

In piston-type compressors the oil serves three functions in addition to the main one of lubricating the bearings and cylinders. It helps to seal the fine clearances around piston rings, piston rods and valves, and thus minimizes blow-by of air (which reduces efficiency and can cause overheating). It contributes to cooling by dissipating heat to the walls of the crankcase and it prevents corrosion that would otherwise be caused by moisture condensing from the compressed air.

In small single-acting compressors the oil to bearings and cylinders is splash-fed by flingers, dippers or rings, but the larger and more complex machines have force-feed lubrication systems, some of them augmented by splash-feed. The cylinders of a double-acting compressor cannot be splash-lubricated, of course, because they are not open to the crankcase. Two lubricating systems are therefore necessary – one for the bearings and cross-head slides and one feeding oil directly into the cylinders. In some cases the same oil is used for both purposes, but the feed to the cylinders has to be carefully controlled, because under-lubrication leads to rapid wear and over-lubrication leads to a build-up of carbon deposits in cylinders and on valves. The number and position of cylinder-lubrication points varies according to the size and type of the compressor. Small cylinders may have a single point in the cylinder head, near the inlet valve; larger ones may have two or more. In each case the oil is spread by the sliding of the piston and the turbulence of the air.

In the piston-type compressor the very thin oil thin has to lubricate the cylinder while it is exposed to the heat of the compressed air. Such conditions are highly conducive to oxidation in poor-quality oils, and may result in the formation of gummy deposits that settle in and around the piston-ring grooves and cause the rings to stick, thereby allowing blow-by to develop.

Rotary compressors – vane type

The lubrication system of vane-type compressors varies according to the size and output of the unit. Compressors in the small and 'portable' group have neither external cooling nor intercooling, because to effect all the necessary cooling the oil is injected copiously into the incoming air stream or directly into the compressor chamber. This method is known as flood lubrication, and the oil is usually cooled before being recirculated. The oil is carried out of the compression chamber by the air, so it has to be separated from the air; the receiver contains baffles that 'knock out' the droplets of oil, and they fall to the bottom of the receiver. Condensed water is subsequently separated from the oil in a strainer before the oil goes back into circulation.

Vane-type pumps of higher-output are water-jacketed and intercooled: the lubricant has virtually no cooling function so it is employed in far smaller quantities. In some units the oil is fed only to the bearings, and the normal leakage lubricates the vanes and the casing. In others, it is fed through drillings in the rotor and perhaps directly into the casing. This, of course, is a total-loss lubrication technique, because the oil passes out with the discharged air.

As in reciprocating units, the oil has to lubricate while being subjected to the adverse influence of high temperature. The vanes impose severe demands on the oil's lubricating powers. At their tips, for example, high rubbing speeds are combined with heavy end-pressure against the casing.

Each time a vane is in the extended position (once per revolution) a severe bending load is being applied between it and the side of its slot. The oil must continue to lubricate between them, to allow the vane to slide freely. It must also resist formation of sticky deposits and varnish, which lead to restricted movement of the vanes and hence to blow-by and, in severe cases, to broken vanes.

Rotary compressors – screw type

The lubrication requirements for single-screw type compressors are not severe, but in oil-flooded rotary units the oxidizing conditions are extremely severe because fine droplets of oil are mixed intimately with hot compressed air. In some screw-type air compressors the rotors are gear driven and do not make contact. In others, one rotor drives the other. The heaviest contact loads occur where power is transmitted from the female to the male rotor: here the lubricant encounters physical conditions similar to those between mating gear teeth. This arduous combination of circumstances places a great demand on the chemical stability, and lubricating power, of the oil.

Other types

Of the remaining designs, only the liquid-piston type delivers pressures of the same order as those just mentioned. The lobe, centrifugal and axial-flow types, are more accurately termed 'blowers', since they deliver air in large volumes at lower pressures. In all four cases only the 'external' parts – bearings, gears or both – require lubrication. Therefore the oil is not called upon to withstand the severe service experienced in reciprocating and vane-type compressors. Where the compressor is coupled to a steam or gas turbine a common circulating oil system is employed. High standards of system cleanliness are necessary to avoid deposit formation in the compressor bearings.

Refrigeration compressors

The functions of a refrigerator compressor lubricant are the same as those of compressor lubricants in general. However, the close association between refrigerant and lubricant does impose certain additional demands on the oil. Oil is unavoidably carried into the circuit with refrigerant discharging from the compressor. In many installations provision is made for removal of this oil. However, several refrigerants, including most of the halogen refrigerants, are miscible with oil and it is difficult to separate the oil which enters the system which therefore circulates with the refrigerant. In either case the behaviour of the oil in cold parts of the systems is important, and suitable lubricants have to have low pour point and low wax-forming characteristics.

Effects of contamination

The conditions imposed on oils by compressors – particularly by the piston type – are remarkably similar to those imposed by internal combustion engines. One major difference is, of course, that in a compressor no fuel or products of combustion are present to find their way into the oil. Other contaminants are broadly similar. Among these are moisture, airborne dirt, carbon and the products of the oil's oxidation. Unless steps are taken to combat them, all these pollutants have the effect of shortening the life of both the oil and the compressor, and may even lead to fires and explosions.

Oxidation

High temperature and exposure to hot air are two influences that favour the oxidation and carbonization of mineral oil. In a compressor, the oil presents a large surface area to hot air because it is churned and sprayed in a fine mist, so the oxidizing influences are very strong – especially in the high temperatures of the compressor chamber. The degree of oxidation is dependent mainly on temperature and the ability of the oil to resist, so the problem can be minimized by the correct selection of lubricant and by controlling operating factors.

In oxidizing, an oil becomes thicker and it deposits carbon and gummy, resinous substances. These accumulate in the piston-ring grooves of reciprocating compressors and in the slots of vane-type units, and as a result they restrict free movement of components and allow air leakages to develop. The deposits also settle in and around the valves of piston-type compressors, and prevent proper sealing.

When leakage develops, the output of compressed air is reduced, and overheating occurs due to the recompression of hot air and the inefficient operation of the compressor. This leads to abnormally high discharge temperatures. Higher temperature leads to increased oxidation and hence increased formation of deposits, so adequate cooling of compressors is very important.

Airborne dirt

In the context of industrial compressors, dust is a major consideration. Such compressors have a very high throughput of air, and even in apparently 'clean' atmospheres, the quantity of airborne dirt is sufficient to cause trouble if the compressor is not fitted with an air-intake filter. Many of the airborne particles in an industrial atmosphere are abrasive, and they cause accelerated rates of wear in any compressor with sliding components in the compressor chamber. The dirt passes into the oil, where it may accumulate and contribute very seriously to the carbon deposits in valves and outlet pipes. Another consideration is that dirt in an oil is likely to act as a catalyst, thus encouraging oxidation.

Moisture

Condensation occurs in all compressors, and the effects are most prominent where cooling takes place – in intercoolers and air-receivers, which therefore have to be drained at frequent intervals. Normally the amount of moisture present in a compression chamber is not sufficient to affect lubrication, but relatively large quantities can have a serious effect on the lubrication of a compressor. Very wet conditions are likely to occur when the atmosphere is excessively humid, or compression pressures are high, or the compressor is being overcooled.

During periods when the compressor is standing idle the moisture condenses on cylinders walls and casings, and if the oil does not provide adequate protection this leads to rusting. Rust may not be serious at first sight, and it is quickly removed by wiping action when the compressor is started, but the rust particles act as abrasives, and if they enter the crankcase oil they may have a catalytic effect and promote oxidation. In single-acting piston-type compressors, the crankcase oil is contaminated by the moisture.

38.14 Turbines

38.14.1 Steam

Although the properties required of a steam-turbine lubricant are not extreme it is the very long periods of continuous operation that creates the need for high-grade oils to be used. The lubricating oil has to provide adequate and reliable lubrication, act as a coolant, protect against corrosion, as a hydraulic medium when used in governor and control systems, and if used in a geared turbine provide satisfactory lubrication of the gearing. The lubricant will therefore need the following characteristics.

Viscosity

For a directly coupled turbine for power generation a typical viscosity would be in the range of 32–46 cSt at 40°C. Geared units require a higher viscosity to withstand tooth loadings typically within the range of 68–100 cSt at 40°C.

Oxidation resistance

The careful blending of turbine oils, using components which, by selective refining, have a reduced tendency to oxidize, produces the required long-term stability. The high temperatures and pressures of modern designs add to these demands, which are combatted by the incorporation of suitable anti-oxidant additives.

Demulsibility

The ability of the lubricant to separate readily and completely from water, in either a centrifuge or a settling tank, is important in a turbine lubricant. Otherwise the retained water will react with products of oxidation and particle contaminants to form stable emulsions. These will increase the viscosity of the oil and form sludges which can result in a failure. Careful and selective refining ensures a good demulsibility characteristic. Inadequate storage and handling can seriously reduce this property.

Corrosion resistance

Although the equipment is designed to keep the water content at a minimum level, it is virtually impossible to eliminate it entirely. The problem of rusting is therefore overcome by using corrosion inhibitors in the lubricant formulation.

Foaming resistance

Turbine oils must be resistant to foaming, since oil-foam reduces the rate of heat transfer from the bearings, promotes oxidation by greatly extending the area of contact between air and oil. It is also an unsatisfactory medium for the hydraulic governor controls. Careful refining is the primary means of achieving good resistance to foaming. Use of an anti-foam additive may seem desirable but this should be approached with caution. If it is used in quantities higher than the optimum it can in fact assist air entrainment in the oil by retarding the release of air bubbles.

38.14.2 Gas

The lubricants generally specified for conventional gas turbines invariably fall within the same classification as those used for steam turbines and are often categorized as 'turbine oils'. In those cases where an aircraft type gas turbine has been adapted for industrial use the lubricant is vitally important to their correct operation. Specifications have been rigidly laid down after the most exhaustive tests, and it would be unwise, even foolhardy, to depart from the manufacturers' recommendations. No economic gain would result from the use of cheaper, but less efficient, lubricants.

38.14.3 Performance standards

In the UK there is BS 489: 1983. In Europe there is DIN 51515 together with manufacturers' standards such as those set by Brown Boverie and Alsthom Atlantique. In the USA there are the ASTM standards and the well-known General Electric requirements.

The total useful life of a turbine oil is its most important characteristic. ASTM method D943 (IP 157) measures the life indirectly by assessing the useful life of the oxidation inhibitor contained in the formulation and is often referred to as the TOST 'life' of the oil. Rust prevention is generally assessed by the ASTM D665 (IP 135) method.

There are many other specifications designed by equipment builders, military and professional societies, as well as users. Care always needs to be taken when purchasing turbine oil to specification. The cheapest oil, albeit conforming to the specification, may not necessarily be the best within that specification for the particular purpose. For instance, the additive package is rarely (if ever) defined, so that unexpected reactions can occur between oils which could affect overall performance.

38.15 Transformers and switchgear

The main requirement for a power-transmission equipment oil is that it should have good dielectric properties. Oil used in transformers acts as a coolant for the windings; as an insulant to prevent arcing between parts of the transformer circuits; and prevents the ionization of minute bubbles of air and gas in the wire insulation by absorbing them and filling the voids between cable and wrapping. In switchgear and circuit breakers it has the added function of quenching sparks from any arc formed during equipment operation. Oils for use in power transmission equipment should have the following properties; high electric strength, low viscosity, high chemical stability and low carbon-forming characteristics under the conditions of electric arc.

38.15.1 Performance standards

The efficiency of transformer oils as dielectrics is measured by 'electric strength' tests. These give an indication of the voltage at which, under the test conditions, the oil will break down. Various national standards exist that all measure the same basic property of the oil. In the UK it is BS 148: 1984. There is an international specification, IEC 296/1982, which may be quoted by equipment manufacturers in their oil recommendations.

38.15.2 Testing

How frequently the oil condition should be tested depends on operating and atmospheric conditions; after the commissioning sample, further samples should be taken at three months and one year after the unit is first energized. After this, under normal conditions, testing should be carried out annually. In unfavourable operating conditions (damp or dust-laden atmospheres, or where space limitations reduce air circulation and heat transfer) testing should be carried out every six months.

Testing should include a dielectric strength test to confirm the oil's insulation capability and an acidity test, which indicates oil oxidation. While acid formation does not usually develop until the oil has been in service for some time when it does occur the process can be rapid. If acidity is below 0.5 mg KOH/g no action would seem necessary. Between 0.5 and 1 mg KOH/g, increased care and testing is essential. Above 1 the oil should be removed and either reconditioned or discarded. Before the unit is filled with a fresh charge of oil it should be flushed. These suggestions are contained in a British Standards Code of Practice.

Sludge observations will show if arcing is causing carbon deposits which, if allowed to build up will affect heat transfer and could influence the oil insulation. There is also a flash point test, in which any lowering of flash point is an indication that the oil has been subjected to excessive local heating or submerged arcing (due to overload or an internal electrical fault). A fall in flash point exceeding 16°C implies a fault, and the unit should be shut down for investigation of the cause. Lesser drops may be observed in the later stages of oil life, due to oxidation effects, but are not usually serious. A 'crackle' test is a simple way of detecting moisture in the oil. Where water is present the oil should be centrifuged.

38.16 Greases

Grease is a very important and useful lubricant when used correctly, its main advantage being that it tends to remain where it is applied. It is more likely to stay in contact with rubbing surfaces than oil, and is less affected by the forces of gravity, pressure and centrifugal action. Economical and effective lubrication is the natural result of this property and a reduction in the overall cost of lubrication, particularly in all-loss systems, is made possible.

Apart from this, grease has other advantages. It acts

both as a lubricant and as a seal and is thus able, at the same time as it lubricates, to prevent the entry of contaminants such as water and abrasive dirt. Grease lubrication by eliminating the need for elaborate oil seals can simplify plant design.

Because a film of grease remains where it is applied for much longer than a film of oil it provides better protection to bearing and other surfaces that are exposed to shock loads or sudden changes of direction. A film of grease also helps to prevent the corrosion of machine parts that are idle for lengthy periods.

Bearings pre-packed with grease will function for extended periods without attention. Another advantage is the almost complete elimination of drip or splash, which can be a problem in certain applications. Grease is also able to operate effectively over a wider range of temperatures than any single oil.

There are certain disadvantages as well as advantages in using grease as a lubricant. Greases do not dissipate heat as well as fluid lubricants, and for low-torque operation tend to offer more resistance than oil.

38.16.1 Types of grease

The general method of classifying greases is by reference to the type of soap that is mixed with mineral oil to produce the grease, although this has rather less practical significance nowadays than it had in the past. One example of this is the multi-purpose grease that may replace two or three different types previously thought necessary to cover a particular field of application. Nevertheless, there are unique differences in behaviour between greases made with different metal soaps, and these differences are still important in many industrial uses, for technical and economic reasons.

Calcium-soap greases

The line-soap (calcium) greases have been known for many years but are still probably the most widely used. They have a characteristic smooth texture, thermal stability, good water resistance and are relatively inexpensive. The softer grades are easily applied, pump well and give low starting torque. Their application is limited by their relatively low drop points, which are around 100°C. This means that, in practice, the highest operating temperature is about 50°C.

Nevertheless, they are used widely for the lubrication of medium-duty rolling and plain bearings, centralized greasing systems, wheel bearings and general duties. The stiffer varieties are used in the form of blocks on the older-type brasses. Modifications of lime-base grease include the graphited varieties and those containing an extreme pressure additive. The latter are suitable for heavily loaded roller bearings such as in steel-mill applications.

Sodium-soap greases

The soda-soap (sodium) greases were, for some considerable time, the only high-melting-point greases available to industry. They have drop points in the region of 150°C and their operating maximum is about 80°C. These greases can be 'buttery', fibrous or spongy, are not particularly resistant to moisture and are not suitable for use in wet conditions. Plain bearings are very frequently lubricated with soda-based greases.

For rolling-contact bearings, a much smoother texture is required, and this is obtained by suitable manufacturing techniques. Modified grades may be used over the same temperature range as that of the unmodified grade and, when they are correctly formulated, have a good shear resistance and a slightly better resistance to water than the unmodified grades.

Lithium-soap greases

These products, unknown before the Second World War, were developed first as aircraft lubricants. Since then the field in which they have been used has been greatly extended and they are now used in industry as multi-purpose greases. They combine the smooth texture of the calcium-based greases with higher melting points than soda-soap greases, and are almost wholly manufactured in the medium and soft ranges. Combined with suitable additives, they are the first choice for all rolling-contact bearings, as they operate satisfactorily up to a temperature of 120°C and at even higher for intermittent use. Their water resistance is satisfactory and they may be applied by all conventional means, including centralized pressure systems.

Other metal-soap greases

Greases are also made from soaps of strontium, barium and aluminium. Of these, aluminium-based grease is the most widely used. It is insoluble in water and very adhesive to metal. Its widest application is in the lubrication of vehicle chassis. In industry it is used for rolling-mill applications and for the lubrication of cams and other equipment subject to violent oscillation and vibration, where its adhesiveness is an asset.

Non-soap thickened greases

These are generally reserved for specialist applications, and are in the main more costly than conventional soap-based greases. The most common substances used as non-soap thickeners are silicas and clays prepared in such a way that they form gels with mineral and synthetic oils. Other materials that have been used are carbon black, metal oxides and various organic compounds.

The characteristic of these non-soap greases which distinguishes them from conventional greases is that many of them have very high melting points; they will remain as greases up to temperatures in the region of 260°C. For this reason, the limiting upper usage temperature is determined by the thermal stability of the mineral oil or synthetic fluid of which they are composed. Applications such as those found in cement manufacturing, where high-temperature conditions have to be met, require a grease suitable for continuous use at, say, 204°C.

Although it is difficult to generalize, the non-soap products have, on the whole, been found to be somewhat less effective than the soap-thickened greases as regards lubricating properties and protection against corrosion, particularly rusting. Additive treatment can improve non-soap grades in both these respects, but their unique structures renders them more susceptible to secondary and unwanted effects than is the case with the more conventional greases.

Filled greases

The crude types of axle and mill grease made in the early days frequently contained large amounts of chemically inert, inorganic powders. These additions gave 'body' to the grease and, possibly, helped to improve the adherence of the lubricating film. Greases are still 'filled' but in a selective manner with much-improved materials and under controlled conditions. Two materials often used for this purpose are graphite and molybdenum disulphide.

Small amounts (approximately 5%) of filler have little or no effect on grease structure, but large amounts increase the consistency. However, the materials mentioned are lubricants in themselves and are sometimes used as such. Consequently it is often claimed that when they are incorporated into the structure of the grease the lubricating properties of the grease are automatically improved. A difference of opinion exists as to the validity of this assumption, but it is true that both molybdenum disulphide and graphite are effective where shock loading or boundary conditions exist, or when the presence of chemicals would tend to remove conventional greases.

Mixing greases

The above comments on the properties of the various types of grease has shown that very real differences exist. Each one has its own particular type of structure, calls for individual manufacturing processes and has its own advantages and disadvantages. It is because of these distinct differences that the mixing of greases should never be encouraged. If greases of different types are mixed indiscriminately there is a risk that one or other of them will suffer, the resulting blend being less stable than either of the original components and the blend may even liquefy.

38.16.2 Selecting a grease

A few brief notes on the fundamental factors that influence a choice of grease may be found helpful. The first essential is to be absolutely clear about the limitations of the different types, and to compare them with the conditions they are to meet. Table 38.10 gives the characteristics of high-quality greases.

Greases with a mixed base are not shown in the table because, in general, they are characterized by the predominant base; for example, a soda-lime grease behaves like a soda grease. Temperature limits may be modified by the required length of service. Thus, if a soda grease requires to have only a short life, it could be used at temperatures up to 120°C.

Table 38.10 Characteristics of high-quality greases

Grease (type of soap)	*Recommended maximum operating temperature (°C)*	*Water resistance*	*Mechanical stability*
Lime	50	Good	Good
Soda	80	Poor	Good
Lithium	120	Good	Good
Aluminium	50	Fair	Moderate

When the type most suitable for a particular application has been chosen, the question of consistency must be considered. The general tendency over the last two decades has been towards a softer grease than formerly used. Two factors have probably contributed to this trend; the growth of automatic grease dispensing and the use of more viscous oils in grease making.

In practice, the range of grease consistency is quite limited. For most general industrial applications, a No. 2 consistency is satisfactory. Where suitability for pumping is concerned, a No. 1; for low temperatures, a No. 0; and for water pumps and similar equipment, a No. 3.

38.16.3 Grease application

In applying lubricating grease the most important aspect is how much to use. Naturally, the amount varies with the component being serviced, but some general rules can be laid down. All manufacturers agree that anti-friction bearings should never be over-greased. This is particularly true of high-speed bearings, in which the churning of excess lubricant leads to overheating. The rise in temperature of a bearing as the amount of grease increases has been recorded. With the bearing housing one-third full, the temperature was 39°C; at two-thirds full the temperature rose to 42°C; and with a full charge of grease it went up to 58°C.

The general recommendations for grease packing are:

1. Fully charge the bearing itself with grease ensuring that it is worked around and between the rolling elements.
2. Charge the bearing housing one-half to two-thirds full of grease.

Churning, and its attendant high temperature, may change the structure of the grease permanently, in which event softening may result in leakage and stiffening in lubricant starvation. There is no fixed rule for the period between re-greasings, since this depends on the operating conditions. Most recommendations suggest inspection and possible replenishment every six or twelve months, though the general tendency as grease quality improves has been to extend this period. The higher the temperature of a machine, the more frequently it must be greased because of possible losses of softened lubricant or changes in its structure.

It is not always incorrect to over-grease. With a sleeve bearing, for instance, gun pressure may be maintained until old grease exudes from the ends of the bearing, and the same is true of spring shackles. For the sake of economy and clealiness, however, this should never be overdone.

38.17 Corrosion prevention

Most plant has to work under adverse conditions, in all sorts of weather, and subject to contamination by various agents. However, as long as it is in use it can be reasonably sure of receiving at least a minimum amount of regular maintenance and attention, and this will reduce the likelihood of working parts being attacked by corrosion when plant is in service. However, when plant has to be laid up until required, no matter how carefully matters have been planned, corrosion is always a serious possibility. Modern machinery, with highly finished surfaces, is especially susceptible to atmospheric attack. The surfaces of components also require protection during transport and storage.

Even today, rusting of industrial plant and material is accepted by some as an inevitable operating expense. There is no necessity for this attitude, however, as the petroleum industry has evolved effective, easily applied temporary protectives against corrosion, where are well suited to the conditions met in practice.

38.17.1 Categories of temporary corrosion preventives

Temporary corrosion preventives are products designed for the short-term protection of metal surfaces. They are easily removable, if necessary, by petroleum solvents or by other means such as wiping or alkaline stripping. Some products for use in internal machine parts are miscible and compatible with the eventual service lubricant, and do not, therefore, need to be removed.

The major categories of temporary corrosion preventives are:

Soft-film protectives
Dewatering fluids giving soft/medium films
Non-dewatering fluids giving soft films
Hot-dip compounds
Greases

Hard-film protectives
Oil-type protectives
General-purpose
Engine protectives

The development of products in these categories has been guided by known market demands and many manufacturers have made use of established specifications for temporary protectives. In the UK, for example, British Standard 1133, Section 6 (covering all categories) and British Government Specifications CS 2060C (PX10 dewatering fluid) are frequently followed.

38.17.2 Selection of a corrosion preventive

Temporary corrosion preventives are in some cases required to give protection against rusting for periods of only a few days for inter-process waiting in factories. Where the protected components are not exposed to the weather, protection can be given for up to a year or more for stored components in internal storage conditions. On the other hand, components may require protection for a few days or even weeks under the most adverse weather conditions. Some components may have to be handled frequently during transit or storage. In general, therefore, the more adverse the conditions of storage, the longer the protective periods, and the more frequent the handling, the thicker or more durable the protective film must be.

Because of the wide variation in conditions of exposure it is not possible to define the length of protection period except in general terms. Solvent-deposited soft films will give protection from a few days to months indoors and some weeks outdoors; a solvent-deposited medium film will give long-term protection indoors and medium-term protection outdoors. Hot-dip compounds and cold-applied greases give films that can withstand considerable handling and will give medium to long protection. Solvent-deposited hard-film protectives will give long-term protection but are fairly difficult to remove. Oil protectives give short- to medium-term protection of parts not subjected to handling and are also much used for the preservation of internal working parts; they need not be removed and can in some instances serve as lubricating oils.

'Short term', 'medium term' and 'long term' are expressions that are not rigorously defined but are generally accepted as meaning of the order of up to 6 months, 12 months and 18 months, respectively, in temperate climates. Where local conditions are more severe (in hot, humid climates, for example) the protection periods are less. These protection periods are related to the preventive film alone, but where transit or storage conditions call for wrapping or packaging then longer protection periods can be obtained.

The distinction between a simple part and a complex assembly is an important factor in selecting a temporary protective. The solvent-containing protectives may not be suited to treating assemblies, because:

1. Assemblies may contain non-metallic parts (rubber, for example) that could be attacked by the solvent;
2. The solvent cannot evaporate from enclosed or shielded spaces and the intended film thickness will not be obtained;
3. Evaporated solvent could be trapped and could then leach away the protective film.

Hence the hot-dip compounds, or greases smeared cold, are better for assemblies with non-metallic parts masked if necessary. Solvent-containing protectives therefore find greater application in the protection of simple parts or components. The available means of application, the nature of any additional packaging and the economics and scale of the protective treatment are further factors that influence the choice of type of temporary corrosion preventive.

38.18 Spray lubricants

There are several applications where the lubrication requirement is specialized and very small, needing precise applications where access is limited because of equipment design or location. In these instances lubricant application by aerosol is the most suitable method. Extreme-pressure cutting flud for reaming and tapping, etc., conveyor and chain lubricant, anti-seize and weld anti-spatter agents, release agents, electrical component cleaner and degreasants are examples of the ever-widening range of products available in aerosol packs.

38.19 Degreasants

Often, before any maintenance work starts it is necessary (and desirable) to remove any oil, grease and dirt from the equipment concerned. It may also be necessary to clean replacement components before their installation. Solvents, emulsions and chemical solutions are three broad types of degreasants. The method of degreasing (direct onto the surface, by submersion, through degreasing equipment or by steam cleaners), component complexity and the degree of contamination will all have to be taken into account when selecting the type of product to be used.

38.20 Filtration

Some 70–85% of failures and wear problems in lubricated machines are caused by oil contamination. Clean oil extends machine and oil life and gives greater reliability, higher productivity and lower maintenance cost. Hence some type of filter is an essential part of virtually all lubrication systems.

Cleaning of oil in service may be accomplished quite simply or with relatively complex units, depending on the application and the design of the system. Thus for some operations it is enough to remove particles of ferrous metal from the oil with a magnetic system. In a closed circulatory system, such as that of a steam turbine, the nature of the solids and other contaminants is far more complex, and the treatment has therefore to be more elaborate. In an internal-combustion engine both air and fuel are filtered as well as crankcase oil.

The efficiency of filtration must be matched to the needs of the particular application, and, this is true both quantitatively (in relation the anticipated build-up of solids in the filters) and qualitatively (in relation to the composition of the contaminants and their size). Dirt build-up varies considerably, but it is probably at its maximum with civil engineering equipment. In this field, diesel engines in trucks will steadily accumulate something like 0.3 kg of solids in the crankcase oil within a month.

Particle size is naturally important. It is generally assumed that particles of less than 3 μm in diameter are relatively harmless. However, this is on the assumption that the oil film is itself of this, or greater, thickness; in other words, that full fluid-film hydrodynamic lubrication persists during the whole working cycle of the machine. This is seldom the case, for there are either critical areas or critical phases at or during which mixed or even wholly boundary conditions prevail – when, in fact, the oil film is less than 3 μm thick. The tendency of modern industrial equipment to operate at higher speeds and under greater pressures leads to higher wear rates. Increased pump capacity, as in hydraulic circuits, coupled with a decreased oil volume means a relatively greater amount of contamination. All in all, much more is demanded of the filter today, whatever the application, than at any time in the past.

38.20.1 Types of filter

The terms 'filter' and 'strainer' are in common use and may lubricant systems contain both. The word 'strainer' is often associated with the removal of large particles, and though it is true that in the majority of cases a strainer is in fact employed to remove coarse particles, the fundamental difference between it and a filter is not one of porosity but purely one of geometry. In a strainer the liquid passes through in a straight line, but in a filter a far more devious route is followed.

Strainers are usually made from woven wire gauze, like a sieve, and though today the pre-size can be made very small indeed (BSI 300 mesh gauze separates particles of roughly 50 μm) they are mainly included for the exclusion of large particles. Filters deal with the removal of very much smaller particles.

Naturally from the above definition there is some unavoidable overlapping, and a really fine strainer of, say, stainless steel 'cloth' is regarded as a filter. There are five main types of filtering units as follows.

Surface films

These are usually constructed of woven metal gauze, paper or cloth. The paper filter may have the working surface enlarged by pleating and the paper impregnated and strengthened. As an example, one proprietary pleated model gives, from an element 11.5 cm long and 8.5 cm in external diameter, a filtering surface of some 3250 cm^2. This type, sometimes described as a radial-fin unit, has a good throughput and is easy to clean or replace. Filters in this class generally have porosities from 100 μm down to 10 or, in extreme cases, even down to 2 μm.

Edge filters

A typical unit comprises a pack of metal or paper discs with a washer between each, the gauge of the latter governing the degree of filtration. The oil flows from the outside and is discharged through a central channel. Some designs can be cleaned without dismantling or interrupting the flow.

An alternative method of manufacturing is to employ a coil of flat metal ribbon as the element, each turn spaced) from the next by small lateral protuberances. The principle of filtration is the same. Porosities of both types are identical and cover a wide range, usually from 100 μm down to 0.5 μm.

Depth filters (absorption-type filters)

1. *Chemically inactive*: These are made from a variety of

materials that include wound yarn, felt, flannel, cotton waste, wood pump, mineral wool, asbestos and diatomaceous earths. The solid particles are trapped and retained within the medium. Certain types will remove water, as well as large and small particles of solids in a range down to 10 μm. Ceramics are sometimes employed for depth filtration, as also are special sintered metals.

2. *Chemically active*: These filters are similar in design to the non-active depth units but the filtering media used are so chosen that contaminants adhere by chemical attraction. Thus there is a dual action, mechanical and chemical. The materials used include various activated clays, Fuller's earth, charcoal and chemically treated paper. Their cleansing action is much more thorough than that of the purely mechanical devices, for they are capable of removing matter actually in solution in the oil.

Magnetic and combined magnetic filters

In its simplest form the magnetic filter comprises a non-magnetic outer casing with an inner permanent magnetic core round which the liquid flows. Because of the magnetic anisotropy of the field the ferrous particles are continuously diverted to the area of strongest attraction coinciding with the direction of flow. A more elaborate design of magnetic clarifier has its elements mounted in a rotating disc. The dirty fluid flows through the chamber in which the disc dips, and ferrous particles adhering to the magnetized areas are removed by the action of scrapers and collected in containers. The capacity of one such disc has been given as 2250 l/h with a range of sludge removal as high as 30 kg/h. Combined units may have the magnet located within a coil of wire that forms the permeable, mechanical filter.

For its specialized application (cleaning the coolants used for metal-machining operations such as grinding and honing) the magnetic filter is easily maintained and cleaned. It has a high throughput and will remove ferrous particles as small as 1 μm. Some of the non-magnetic material is associated with the ferrous particles suspended in the fluids and this is also removed with them.

The centrifugal filter

This is a specialized design and is, in effect, a true centrifuge of small size that operates on the reaction turbine principle, an oil-circulating pump providing the necessary power. One advantage claimed for this type is that it operates at a steady flow rate, whereas the flow rate through a felt or paper element diminishes as the bed of dirt is built up. The centrifugal filter has been successfully applied to diesel engines where the greater part of the dirt particles are under 2 μm in diameter.

38.21 Centrifuging

The centrifugal separation of solid impurities is adopted either as an alternative to filtration or combined with it. For example, a lubricant circulating system can be cleaned by having fixed-element filters that arrest larger particles, and a centrifuge system that removes the finer solids in suspension together with any water contained in the oil.

The centrifuge is a powerful tool. The magnitude of the available centrifugal force – the product of the mass of the particle and its acceleration – is easily appreciated when the speeds and dimensions of a commercial unit are considered. A vessel with a diameter of 25.4 cm spinning at 1700 rev/min gives an acceleration at the centrifuge wall of some 400 g. In terms of settling this means that centrifuging a crude oil for 30 s is at least equivalent to simple gravitational settling over a 24 h period.

The advantage of the modern continuous centrifuge is the rapidity with which it will separate both solids and immiscible liquids. Another stems from the larger volume of oil it can handle in a given time.

38.22 Shaft seals

Both physical and process-material considerations influence the choice and use of shaft seals in the industry. It is essential that the seal material is compatible with any process material and the operating variables.

38.22.1 Seal materials

Nitrile rubbers, including fibre-reinforced varieties, are used both as radial shaft-seal materials and as moulded packings for reciprocating shafts. They have excellent resistance to a considerable range of chemicals, with the exception of strong acids and alkalis, and are at the same time compatible with petroleum-based lubricants. Their working temperature range is from −1°C to 107°C continuously and up to 150°C intermittently. When used on hard shafts with a surface finish of, at most, 0.00038 mm root mean square (RMS), they have an excellent resistance to abrasion.

Polyacrylic resins, which have similar chemical resistance to the nitriles, can be used for slightly higher temperature conditions, but because their abrasion resistance is not as good as that of the nitriles they need continuous lubrication.

Leather is a useful material in association with a whole range of organic liquids and oils. It retains its sealing properties very well over a temperature range of −45°C to 93°C. Its abrasion resistance is good and it will tolerate low speeds on 0.00076 mm RMS shafts with only intermittent lubrication.

Silicone rubber as a shaft seal and backing material has a number of special applications. It can be used over a temperature range of −60°C to 260°C in air or suitable fluids. Its abrasion resistance is good with hard shafts having a 0.000254 mm RMS surface finish. Commercial grades of silicone rubber are compatible with most industrial chemicals up to 260°C. In lubricating oils the limiting temperature is 120°C, but special types have been developed for use up to 200°C.

Face-seal materials can be chosen from filled, moulded or reinforced resins with which water, hydraulic fluids, mineral oils or synthetic oils are all compatible. Their maximum temperature in service depends on the brittle point of the resin but, generally, the range is from −50°C to 100°C. Abrasion resistance is generally good but, as far as possible, resins are not used in the presence of foreign solids.

Carbon is an excellent material, being unaffected by most industrial chemicals including corrosives, oils, solvents, water and steam. It has a temperature range of up to 260°C, coupled with a high resistance to abrasion in clean fluids.

Ceramics are highly resistant to wear, chemical attack and high temperatures and, consequently, have many applications. They are frequently bonded to metal in making seal components or applied as a coating for shafts where seal elements introduce wear.

Metal can cover many requirements in seal design; it may be used for faces and other parts such as rings and bellows. The specification of the metal depends on the degree of corrosion resistance necessary, the thermal expansion and the material of the mating face.

Rubber bellows casings and sealing rings are vulnerable to temperature and are thus limited to a range of between 0°C to 100°C. Rubber is compatible with most industrial chemicals and is particularly useful in the presence of solids.

Sealing rings that are inert to most chemical corrosives and solvents are usually manufactured from PTFE or one of the synthetic elastomers.

38.22.2 Packing materials

Nitrile and polyacrylic materials have temperature ranges of −34°C to 107°C and −34°C to 177°C, respectively. Both require continuous lubrication.

Silicone rubber is not recommended for use with hot organic materials. If, for specific reasons, its use is contemplated, the manufacturer should be consulted regarding its possible use for a particular application.

Neoprene, a well-proved packing material, has a temperature range of −60°C to 107°C but because it is unsuitable for use with many organic liquids it is normally used only with aqueous solutions. It possesses good resistance to abrasion, which makes it particularly useful with slurries or where there is abrasive contamination.

Metallic packings of aluminium, copper and lead cover a very wide temperature range up to 540°C. They should be matched to the shaft material, since shaft finish and tolerance are critical factors in overall efficiency. The great limiting factor in the use of metallic packings is the difficulty of ensuring corrosion resistance with all the likely process materials.

The textile fibres have a limited range of application. They may be used with hot or cold water, steam, oils and ammonia up to a maximum temperature of 100°C. Asbestos fibre is more versatile, can be be used as compression packing material with hot water, superheated steam, hot oils and gases up to 310°C.

38.23 Centralized lubrication

Manual application of lubricants has the inherent risk of failure due to omission. With the increasing complexity of plant, the costs of lost production and of manpower to try to prevent such omissions are becoming unacceptable.

Mechanized methods of pumping oil and grease to bearings and other components are becoming increasingly utilized. Some of these systems are fundamentally suited to either oil or grease, but others, including all those where continuous circulation is involved, are suitable only for oil.

Built-in mechanized grease lubrication is nearly always of the centralized 'one-shot' variety, in which a single pump stroke supplies grease simultaneously to a number of bearings. The amount supplied to each station is regulated by suitable valves or adjustable metering orifices. The pump may be manually operated or connected to a suitable machine component, whereby grease is fed only when the machine is actually running and at controlled temperatures. Pneumatic or electric pumps are also used, set in operation at regular intervals by an automatic timing device.

One-shot metered lubrication is eminently suited to oiling systems and can be employed either in an 'all-loss' arrangement or as part of a circulatory system. Sight-glasses or other indicators should be incorporated, since such lubricating mechanisms are nowadays so reliable that a blockage or other failure might not be suspected until too late.

Circulatory systems often use an intermediate header tank, from which the bearings are supplied by gravity. The complete system may comprise, in addition and according to the size of the installation, heat exchangers or coolers, filters, strainers, settling tanks, centrifuges and other purifying equipment.

Oil mist feeds are used less for plain bearings than for lubricating some other types of machine parts, but applications are increasing in number. A stream of dry compressed air is used both to generate the mist and to carry it to the bearing. The atomized oil droplets are released from air suspension at points of turbulence around bearings, gears and other moving components or in a special re-classifying fitting at the end of the supply line. Reclassifiers are generally employed when plain bearings are to be lubricated by oil mist, but the method is fundamentally unsuited for bearings requiring hydrodynamic thick-film lubrication.

Special precautions must be taken with oil-mist feeds to ensure that the compressed air, which greatly enhances the rate of heat dissipation, can escape from the housing. If vents or other outlets become blocked, the back pressure may stop the flow of lubricant.

38.24 Storage of lubricants

It cannot be emphasized too strongly that dirt and correct lubrication are incompatible. The lubricant manufacturer has a comprehensive system of classification, filtration and inspection of packages which ensures that all oils and greases leaving his plant are free from liquid and solid contaminants. It is in his own interests that the user should take the same care to ensure that the lubricant enters his machinery in as clean a condition as that in the bulk tank or barrel. The entry of abrasive dust, water and other undesirable matter into bearings and oilways may result if lubricants are handled carelessly.

The conditions in a plant are often far from ideal and usually storage facilities are limited. This, however, should serve as a constant reminder of the need for continual care, the adoption of suitable dispensing equipment, organized storekeeping and efficient distribution methods. Furthermore, the arrangements on any particular site will be governed by local organization and facilities. Technical personnel from lubricant suppliers are available to assist and advise plant management on the best methods for a particular site. The general recommendations given about the care of lubricants consist of elementary precautions which are mainly self-evident and yet, unfortunately, are often ignored.

The modern steel barrel is reasonably weatherproof in its original condition, but if stored out of doors and water is allowed to collect in the head, there may, in time, be seepage past the bung due to the breathing of the package. Exposure may also completely obliterate the grade name and identification numbers, as is evidenced by the frequent requests made to sample and test lubricants from full packages that have been neglected on-site because no other method of identification is possible. Unless it is absolutely unavoidable, packages should never be stored in the open and exposed to all weather. Even an elementary cover such as a sheet of corrugated iron or a tarpaulin may provide valuable protection.

However, rudimentary the oil stores, the first essential is cleanliness; the second is orderliness. These two essentials will be easily achieved if maximum possible use is made of bulk storage tanks. In the case of bulk storage of soluble oils the need for moderate temperatures is vital, and the tanks should be housed indoors to protect their contents against frost. There are several other benefits to be derived from the use of tanks, i.e. reduction in storage area, handling of packages and, possibly, bulk-buying economics. All barrels should be mounted on a stillage frame of suitable height, fitted with taps and the grade name clearly visible. The exterior surfaces of both tanks and barrels should be kept scrupulously clean and each container provided with its own drip tray or can.

The storage and handling of grease presents more problems than are encountered with fluid lubricants, as the nature of the material and design of the conventional packages make contamination easier. Lids of grease kegs must be kept completely free from dust and dirt, and should be replaced immediately after use. The most common way in which solids enter a grease package is by the user carelessly placing lid either on the ground or on some other unsuitable surface. Fortunately, there are available today a number of simple dispensing units which can entirely obviate this danger and which can be adapted to all types of packages.

Wherever manual distribution has to be adopted, containers should be reserved for the exclusive use of specific units and their operators and, as far as possible, for a particular grade. When not in use they must be stored away from all possible sources of contamination. To promote economy and reduce waste due to spillage, their shape and proportions must be suited to the application.

While it is impossible to describe a system of storekeeping and distribution suitable for every site there are certain essential principles which should be adhered to if cleanliness, order and economy are to be maintained. How these principles should be applied is for individual managements to decide. The keynote, however, should be simplicity. Distribution should be controlled by a storekeeper familiar with both grades and needs. While the lubrication schedule for any particular unit is generally the concern of the operator, the storekeeper must equally be aware of it and have a comprehensive list of the different grades, their applications, quantities, daily and other periodic needs. On such a basis he will be able to requisition and store the necessary lubricants in the most convenient and economic quantities and packages, and ensure that supplies are used on a 'first in, first out' basis.

Care and good housekeeping at every stage from handling, stacking and storage, right through to dispensing and application will:

- Ensure that the correct product reaches the point of application and is free from contamination;
- Help towards maximum efficiency in the use of lubricants and the equipment in which they are employed;
- Avert accidents and fire hazards arising from mishandling;
- Prevent any adverse effects on people, equipment and the environment.

38.25 Reconditioning of oil

Reconditioning is the removal of contaminants and oxidation products (at least in part) but not previously incorporated additives. It may also involve the addition of new oil and/or additives to adjust the viscosity and/or performance level. This process is sometimes referred to as 'laundering' or 'reclamation'. The method treats used lubricating oil to render it suitable for further service, either in the original or a downgraded application. Two types of treatment are generally employed.

1. Filtration to remove contaminants, followed by the

addition of new oil and/or additives to correct performance level;
2. A simple filtration process to remove contaminants.

In practice, treatment (1) usually involves a contractor collecting a segregated batch of oil, reconditioning and returning it for re-use. The simple filtration process can be carried out by a contractor, but is more usually done on-site. Re-refining is the removal of contaminants and oxidation products and previously incorporated additives to recover the lube base stock for new lubricant or other applications.

38.26 Planned lubrication and maintenance management

Having the correct lubricant in each application will only give the maximum benefit if and when it is applied at the correct frequency and quantity. With the increasing complexity of plant this is becoming more vital and, at the same time, more difficult to achieve. The solution to this problem is planned lubrication maintenance, which, in essence, is having the right lubricant in the right place at the right time in the right amount.

Most oil companies offer a planned lubrication maintenance (PLM) service that will meet these requirements with the minimum of effort on the part of the customer. These schemes provide logical routing for the lubrication operative, balanced work loads and clear instructions to those responsible for specific tasks associated with lubrication and fault-reporting facilities. Many schemes are now designed for computer operation which also accommodate plant and grade changes, operation costings and manpower planning. It is essential that any such scheme should be adaptable to individual requirements.

There are a few computerized PLM schemes which are dynamic systems and can be integrated into an overall maintenance management information system. These contain maintenance, inventory and purchase order modules and go far beyond 'just another work order system'. They provide the necessary information to control complex maintenance environments, thereby improving productivity and reducing operational costs.

38.27 Condition monitoring

Condition monitoring is an established technique which has been used by capital-intensive or high-risk industries to protect their investment. The concept has developed radically in recent years largely due to advances in computerizations which offer greater scope for sophisticated techniques. These fall into three types of monitoring: vibration, performance and wear debris. The last monitors particulate debris in a fluid such as lubricating oil, caused by the deterioration of a component.

Oil-related analysis encompasses a variety of physical and chemical tests such as viscosity, total acid number and particulate contamination. This is often extended to include the identification of wear debris, as an early warning of component failure, by either spectrographic analysis or ferrography or both. The former is commonly used in automotive and industrial application for debris up to 10 μm and the latter mainly for industry users covering wear particles over 10 μm. Ferrography is relatively expensive compared with many other techniques, but is justified in capital-intensive areas where the cost is readily offset by quantifiable benefits such as longer machinery life, reduced loss of production, less downtime, etc.

38.28 Health, safety and the environment

There are a wide variety of petroleum products for a large number of applications. The potential hazards and the recommended methods of handling differ from product to product. Consequently, advice on such hazards and on the appropriate precautions, use of protective clothing, first aid and other relevant information must be provided by the supplier.

Where there is risk of repeated contact with petroleum products (as with cutting fluids and some process oils) special working precautions are obviously necessary. The aim is to minimize skin contact, not only because most petroleum products are natural skin-degreasing agents but also because with some of them prolonged and repeated contact in poor conditions of personal hygiene may result in various skin disorders.

38.28.1 Health

It is important that health factors are kept in proper perspective. What hazards there may be in the case of oil products are avoided or minimized by simple precautions. For work involving lubricants (including cutting fluids and process oils) the following general precautions are recommended:

- Employ working methods and equipment that minimize skin contact with oil;
- Fit effective and properly positioned splash guards;
- Avoid unnecessary handling of oily components;
- Use only *disposable* 'wipes';
- Use soluble oils or synthetic fluids at their recommended dilutions only, and avoid skin contact with their 'concentrates'.

In addition to overalls, adequate protective clothing should be provided. For example, a PVC apron may be appropriate for some machining operations. A cleaning service for overalls should be provided and overalls should be cleaned regularly and frequently. Normal laundering may not always be sufficient to remove all traces of oil residues from contaminated clothing. In some instances dry cleaning may be necessary. Where this applies to cotton overalls they should first be dry cleaned and then laundered and preferably starched, in order to restore the fabric's oil repellancy and comfort. As a general rule, dry cleaning followed by laundering is always preferable to minimize the risk of residual contamination wherever heavy and frequent contamination occurs and when the type of fabric permits such cleaning.

Overalls or personal clothing that become contaminated with lubricants should be removed as soon as possible – immediately if oil soaked or at the end of the shift if contaminated to a lesser degree. They should then be washed thoroughly or dry cleaned before re-use.

Good washing facilities should be provided, together with hot and cold running water, soap, medically approved skin-cleansers, clean towels and, ideally, showers. In addition, reconditioning creams should be available. The provision of changing rooms, with lockers for working clothes, is recommended.

Workers in contact with lubricants should be kept fully informed by their management of the health aspects and the preventive measures outlined above. Any available government leaflets and/or posters should be prominently displayed and distributed to appropriate workers.

It should be made clear to people exposed to lubricants that good standards of personal hygiene are a most effective protection against potential health hazards. However, those individuals with a history of (or thought to be particularly predisposed to) eczema or industrial dermatitis should be excluded from work where, as in machine-tool operation, contact with lubricants is virtually unavoidable.

Some industrial machining operations generate a fine spray or mist of oil, which forms an aerosol – a suspension of colloidal (ultra-microscopic) particles of oil in air. Oil mist may accumulate in the workshop atmosphere, and discomfort may result if ventilation is inadequate. Inhalation of high concentrations of oil mist over prolonged periods may give rise to irritation of the respiratory tract, and in extreme cases to a conditions resembling pneumonia. It is recommended that the concentration of oil mist in the working environment (as averaged over an 8-h shift) be kept below the generally accepted hygiene standard of 5 mg/m^3. This standard does, however, vary in some countries.

38.28.2 Safety

In the event of accident or gross misuse of products various health hazards could arise. The data provided by the supplier should outline these potential hazards and the simple precautions that can be taken to minimize them. Guidance should be included on the remedial action that should be taken to deal with medical conditions that might arise. Advice should be obtained from the supplier before petroleum products are used in any way other than as directed.

38.28.3 Environment

Neat oils and water-based coolants eventually reach the end of their working lives, and then the user is faced with the problem of their correct disposal. Under *no* circumstances should neat oils and emulsions be discharged into streams or sewers. Some solutions can, however, be fed into the sewage system after further dilution – but only where permitted.

There are many companies offering a collection service for the disposal of waste lubricating oil. The three main methods employed are:

1. Collection in segregated batches of suitable quality for use by non-refiners
2. Blending into fuel oil
3. Dumping or incineration

If method (3) is used due regard must be paid to the statutory requirements that must be met when disposing of waste material. These are covered in two main items of legislation; namely, the Deposit of Poisonous Waste Act 1972 and the Control of Pollution Act 1974. It is the responsibility of the producer of waste oil to ensure that the waste is disposed of in the correct manner, to ensure that no offence is committed and that the contractor is properly qualified to execute the service.

Acknowledgements

The author is grateful to BP Oil UK Ltd for their help in writing this chapter and for their permission to reproduce the figures and tables.

Further reading

BP publications

Industrial lubrication
Machine tools and metal cutting
Lubricants for heavy industry
Gear lubrication
Hydraulic fluids
Machine shop lubricants
Cutting fluds
Compressor lubrication
Greases
Temporary corrosion preventives
Degreasants
Storage and handling of lubricants
Aerosols
Health, safety and environmental data sheets

Further information on lubrication can be obtained from:

Booklets and leaflets published by most oil suppliers
Libraries of Institute of Plant Engineers
Institution of Production Engineers
Institute of Petroleum
Institution of Electrical Engineers
Institution of Mechanical Engineers
Literature published by additive companies
Literature published by the American Society of Lubrication Engineers
Libraries of universities
National Centre of Tribology

With such a wide and important subject it is not possible to provide a full list but the above will indicate some initial contact points. The author hopes that no offence is caused by any omission.

Index